Helmholtz

Helmholtz

A Life in Science

David Cahan

The University of Chicago Press   Chicago and London

The University of Chicago Press, Chicago 60637
The University of Chicago Press, Ltd., London
© 2018 by The University of Chicago
Published 2018

Printed in the United States of America

27 26 25 24 23 22 21 20 19   2 3 4 5

ISBN-13: 978-0-226-48114-2 (cloth)
ISBN-13: 978-0-226-54916-3 (e-book)
DOI: 10.7208/chicago/9780226549163.001.0001

Library of Congress Cataloging-in-Publication Data
Names: Cahan, David, author.
Title: Helmholtz : a life in science / David Cahan.
Description: Chicago : The University of Chicago Press, 2018. | Includes bibliographical
references and index.
Identifiers: LCCN 2018003417 | ISBN 9780226481142 (cloth : alk. paper) |
ISBN 9780226549163 (e-book)
Subjects: LCSH: Helmholtz, Hermann von, 1821–1894. | Scientists—Germany—Biography.
Classification: LCC Q143.H5 C34 2018 | DDC 509.2 [B]—dc23
LC record available at https://lccn.loc.gov/2018003417

♾ This paper meets the requirements of ANSI/NISO Z39.48-1992 (Permanence of Paper).

To the memory of my father and mother,
Haskell and Sylvia Cahan

The wise man

Seeks the trusted law in chance's horrifying wonders,

Seeks the resting pole in phenomena's flight.

Schiller, "The Walk," 1795

Contents

Introduction

It is more than time for a new, fresh biography of the German scientist Hermann von Helmholtz (1821–94). The standard and nearly oldest biography of him, Leo Koenigsberger's *Hermann von Helmholtz* (1902–3), was published more than a century ago.[1] Though it has long been essential for scholarly use, it is an uncritical assessment, embellishing if not heroizing its subject; indeed, it helped to create a mythological figure of the man, making him an icon and an idol.

The present biography of Helmholtz, by contrast, seeks to be a comprehensive, balanced, and thematic study of Helmholtz's life and science, setting both deeply in their time and place and attempting to understand the influences upon Helmholtz and his upon his age. It endeavors to take the full measure of the man, doing so critically in the multiple contexts in which he lived and worked. It utilizes all of Helmholtz's published and known unpublished writings: his scientific, philosophical, and popular articles and books; his extant correspondence (both published and unpublished); all known pertinent official documents concerning his academic career; third-party correspondence; and the excellent, extensive secondary literature on or about Helmholtz. It utilizes many previously unknown or scarcely known sources, as well as older, better-known sources, to illuminate Helmholtz's life, work, and career in a fresh way. It seeks neither to lionize nor to demonize him but aims to provide an authoritative, critical account of his life and work as these developed in their historical context, that of an emerging and evolving scientific community.

::::

The intellectual leitmotifs and the driving forces of Helmholtz's creative scientific, philosophical, and aesthetic life were threefold: a passion to unify the sciences, both as individual disciplines and collectively; vigilant epistemological attention to the sources and methods of knowledge; and acute appreciation of the complementary and mutually stimulating roles of the arts for the sciences and vice versa. These leitmotifs and passions—which constitute the theme of this biography—bobbed in and out of his life and work, and they took different forms on different occasions. They expressed themselves as his drive to unify the different branches of physics (at first by means of the law of conservation of energy, then later in his career by means of the principle of least action); as his drive to find a common foundation for physics and physiology; and as his desire to bring some unity to all the natural sciences. He sought laws in science.

Helmholtz also deeply respected and analyzed the methods of the *Geisteswissenschaften*: the human sciences, that is, the humanities and, to speak somewhat anachronistically, the "social sciences." They, too, found a place—subordinate though it may have been—in his vision of the wholeness and unity of knowledge. To be sure, since the human sciences inherently involved human psychology, he did not believe they could produce laws—an all-important point for him, as we shall see. To the intellectual and psychological driving force of his career must be added his passion for and scientific analysis of music and painting. In dealing with their respective forms of tone and color, he showed how a combination of acoustics and optics, and the closely associated physiological acoustics and optics, could illuminate understanding of the fine arts. For Helmholtz, the arts—not only music and painting, but also literature and theater—were at once a source of relaxation from his demanding scientific work and a source of inspiration for that work. He thought that artists, in their own and different ways, were, like scientists, seeking to express nature's laws. He was a Renaissance man.

Helmholtz was driven in no small part, this biography argues, by a psychological need to find or create laws in science. It was a need that originated in childhood as a means of compensating for what he later described as the shortcomings of his deficient memory. By adulthood it had evolved into the centerpiece of his philosophy of science. In his view, law brought understanding of *and* mastery over nature. Along with Schiller, he believed that "the wise man/seeks the trusted law." These lines concerning "laws" encapsulated in a striking, poetic way the crux of what Helmholtz sought to establish in prose, that is, in science and in his epistemological reflection on it.[2]

Helmholtz was an intellectual risk-taker in his search for laws; he sought after far more from scientific investigation than merely new factual data. As a young scientist, he hoped to find the "causes"—sometimes he called them the "forces"—that supposedly lay behind laws, and thus he hoped ultimately to provide an "objective" picture of the natural world, one that would somehow link together the different fields of natural science. Ultimately, he modified these aims. In 1891 he wrote: "To discover the causal connections of phenomena has guided me throughout my life."[3] Indeed, as this biography shows, there were strong connections between Helmholtz's physiology and parts of his physics, between his physiology and his development of non-Euclidean geometry, between his physics and his geometry, between his physics and his chemical thermodynamics, and between his physics and his meteorology, not to mention, more broadly, the implications of scientific laws and other findings for several other fields (medicine, experimental psychology, philosophy, music, and painting). Human sight and sound, and bodily heat, could be understood in no small part by understanding the physical and physiological laws that they obeyed. In this quest he demonstrated a perhaps unmatched, and certainly uncanny, ability to synthesize ideas, concepts, theories, and results from different fields of science.

By the 1860s or so, Helmholtz had come to favor a less demanding, and perhaps less philosophical goal for scientific laws or theories, a goal recognizing that perhaps all a scientist could ever realistically hope for was to articulate limited laws (and theories) on the basis of known phenomena. A "world picture" (*Weltbild*) ultimately eluded him, and with this, as with everything else in life, he learned to come to terms. He was a scientific "genius," but he never forgot that theories must always be tied to empirical reality. Still, to the end of his life, he believed that the unity of the sciences implied the unity of all knowledge, and hence of the producers of knowledge—the scientific community as a whole. This in turn, he further believed, meant increased civilization for all of humanity. Science, in his vision, meant a civilizing power for everyone.[4]

: : :

This biography presents Helmholtz's life also as a private individual, both because that life is intrinsically interesting and because it helps make coherent the broad arc of his scientific and social impact. I thus seek to elucidate not only his personality and ambitions, but also the familial, educational, social, and political worlds in which he developed and lived. His families, friendships, loves and marriages; his career hopes, tensions, and moves; his love of music, painting, theater, literature, and the other arts; his travels, and even the state of his health—all these are presented not merely as isolated biographical details but rather as constituting his support system for his complex scientific thought and practice, as manifestations of the psychological, emotional, and intellectual aspects of his personality, of his moral and intellectual character, and of what helped stimulate and sustain him throughout his life. They serve as a means of further appreciating the motivations behind his scientific thought and the larger ends he hoped it might serve. The narrative of this biography delineates Helmholtz's passions and ambitions: his decision to pursue science in the first place, his drive for human excellence in general, and his competitiveness and cooperation with others. All of these helped shaped the contour of his half-century-long scientific life. The biography seeks, in short, to illuminate the dynamic relationship between Helmholtz's passionate self and the world of reason that became the hallmark of his life and legacy.

The overall nature, structure, and development of Helmholtz's scientific work; his principal scientific achievements; and his role as a public figure of science are presented. However, aiming for breadth, I do not offer an in-depth analysis of every one of Helmholtz's scientific theories, observations, and experiments, or of his philosophical essays. This is a biography in the broadest sense of the term—a cultural and not a narrowly scientific biography. In relating the breadth and interconnectedness of Helmholtz's work, I have drawn upon many excellent analyses by historians of science and of philosophy concerning specific aspects of Helmholtz's scientific work and philosophical views.

Helmholtz's scientific achievements and his philosophical reflections on science ultimately became an oeuvre of seven thick volumes: three of collected scientific papers (containing about 175 original papers plus five or six dozen reprinted versions or translations), a three-part tome on physiological optics, a volume on physiological acoustics and music, and two volumes of essays on popular science and philosophy, as well as a six-volume (in seven) set of lectures on theoretical physics that was assembled posthumously (and apparently considerably recomposed) by several of his last students. As this publishing record suggests, Helmholtz was a workhorse and, at times, a workaholic as well.

Helmholtz was, by any measure, an intellectually off-scale individual, a scientific genius, to use the term that was once reserved solely for artists but which in the nineteenth century became increasingly used to describe scientists as well. Such individuals were the effective leaders of their scientific communities, and they helped determine who was "merely" an ordinary scientist. (The same can be said of our age, of course.) At the same time, Helmholtz's intellectual gifts differed from those of Darwin or Einstein, for example, who concentrated their efforts almost exclusively on elucidating foundational theories in one field, such as biology or physics. Helmholtz's achievements, by contrast, ranged across the physical and life sciences (including medicine), and he did transformative work in others. He also helped construct and single-handedly directed three scientific institutes (one for physiology in Heidelberg and two for physics in Berlin); and he helped to popularize science.

The extremely broad range of Helmholtz's scientific activities perhaps explains why historians of science, not to mention the general public, do not remember his name in the way or to the extent that they remember Darwin's or Einstein's. Though early in his career he had hoped to develop an overarching set of foundational principles that would integrate all of science, his own scientific work and institute-building in fact contributed mightily to the specialization of the sciences that emerged so distinctly in the nineteenth century. This biography shows that despite his own desires and efforts, and despite the interdisciplinary character of much of his work, Helmholtz never achieved the intellectual unity of the sciences that he had hoped for—though it also shows that he never abandoned that hope, even if it later assumed other forms. Though he was (and is) often referred to as a natural philosopher—here meaning a figure of broad intellectual outlook concerned with the natural world as a whole and with providing a causal, lawlike account of it—in fact he embodied rather well the era's novel conception of the scientist as a professional who undertakes specialized and manageable, if not indeed definitely solvable, scientific problems. Indeed, he turned at least one philosophical issue, that of space, into a scientific or mathematical matter. At the same time, however, he helped set the research frameworks and agendas (including unresolved issues) of several disciplines, he synthesized previous results of several fields of science and medicine, and he developed key new instruments. Even though the totality of his accomplishments did not add up to a unified natural philosophy, Helmholtz helped set the direction in a range of scientific fields during the second half of the nineteenth century. He remained an inspirational and instructive figure for many twentieth- and twenty-first-century scientists, from physicists like Einstein to cognitive scientists like V. S. Ramachandran.

Helmholtz was shaped by and also helped to shape the German academic system and the scientific community of his day. Studying his rise through it—from Gymnasium student to medical student and army physician, from extraordinary (associate) to ordinary (full) professor of physiology and, eventually, of physics—illuminates in a concrete way the changing nature and operation of that system and community, and with it the institutional development of science in nineteenth-century Germany and beyond. To follow his career is to follow something of the construction of the infrastructure of nineteenth-century science: its institutes, laboratories, journals, disciplinary organizations, national and international meetings, and the like. As already suggested, Helmholtz's work affected numerous scientific disciplines, and the course of his professional work and advancement brought him into contact with a wide range of competitive and cooperative colleagues—not only natural scientists, mathematicians, and medical researchers and clinicians, but also humanists and social scientists—and with many students, both the ordinary and the extraordinary, both Germans and non-Germans. To follow his life and career—starting in Potsdam and Berlin, then at Königsberg, Bonn, Heidelberg, and finally back to Berlin; as an assistant, then as an extraordinary and an ordinary professor of physiology and physics; and as a university dean and rector—is thus to learn something about nineteenth-century science as a whole: to see scientists communicating with and visiting one another, to meet instrument makers and audiences, to see his intellectual and institutional leadership within the scientific community, and to learn something about his few peers and many colleagues in science.

Competitiveness was a distinct part of Helmholtz's personality. Without this psychological drive, how can one explain why his research performance was both steady and long-lasting, from his first publication at age twenty-two in 1843 to his last at nearly age seventy-three in 1894? He continued to conduct direction-setting research and to publish long after he had reached (between the 1850s and the 1870s) the pinnacle of outward professional success. His deeply felt need to discover the laws of science, his equally deep sense of obligation as a paid German university professor, and the general leadership positions that he held as the director of several scientific institutes pushed him to ceaselessly conduct scientific research and report on it to others (both professionals and lay audiences). From about 1847 onward, he exhibited what in other creative individuals—for example, gifted writers or top-performing athletes—is sometimes called "flow." His passion for science never declined. It may be that the combined work process, from thinking about and selecting a problem, to observing or experimenting on specific phenomena, to developing theoretical (and mathematical) structures and their consequences, to writing up results, to seeing manuscripts through to publication, gave him a sense of psychological well-being, of high self-worth, and of importance and belonging to a community. Continued professional success both satisfied him and became a habit and a need.

Finally, this biography also seeks to show how Helmholtz established himself as a popularizer and a statesman of science, to display the persona that he gradually developed as a public scientist and public figure. As a spokesman for science, and as a general promoter and man of influence within the scientific community, he was second to none during his time. As a result, he came to have considerable influence within academic, governmental, and high-society circles. His only shortcoming lay in his lecture-hall ability as a teacher at the introductory and intermediate levels. Here he was apparently perceived as disorganized, uninspired, even dull. He gave minimal effort to such teaching, probably sacrificing preparation time and energy for the sake of his research. At the advanced level, by contrast, particularly after he moved to Berlin in 1871 to teach physics, he managed to inspire many young physicists.

Through many popular scientific and other public addresses, Helmholtz helped explain and justify science to the general public as well as to educated elites—governmental, cultural, commercial, and industrial. He communicated its principal empirical findings, laws, goals, methods, and values; he related the different branches of science to one another; and he encouraged and helped justify further spending on science. He became, in sum, a public scientist. In carrying out these activities, he demonstrated a sense of social responsibility, even as he took self-interested pleasure in these efforts. Helmholtz, in effect, inherited from Alexander von Humboldt the mantle of Germany's premier public figure of science, a mantle that in subsequent generations was worn by Max Planck and Werner Heisenberg. In Germany and beyond, Helmholtz's persona and reputation helped shape the educated public's image of the scientist, and analysis of his popularization of science illuminates the relationship of science and society during the second half of the nineteenth century.

Helmholtz became deeply involved in German culture in a broad sense that transcended science proper. His work was widely read within and beyond the German scientific community. Readers included those interested in the relations of the sciences to one another, the meaning of the freedom of science, the scientific foundations of music and painting, philosophy, the new laws of thermodynamics and their implications for the "heat death" of the universe, and the new non-Euclidean geometry. A wide range of leading nineteenth-century intellectuals and artists read him: George Bancroft, George Eliot, George Henry Lewes, Herbert Spencer, Karl Marx, Friedrich Engels, Friedrich Max Müller, Friedrich Nietzsche, Charles Sanders Peirce, Wilhelm Dilthey, William James, Sigmund Freud, Franz Boas, and Max Weber, not to mention musicians and painters. Helmholtz's popular and semipopular writings continued to find a readership among intellectuals and academics well into the twentieth century, and indeed they continue to do so into our own day.

Though Helmholtz's fundamental research interests were those of pure science, he also demonstrated positive, practical results that might issue from it. With his invention of the ophthalmoscope in 1850–51, his name spread rapidly throughout the medical world, the German civil service, and the public at large. Helmholtz himself argued that science was the fundamental source of technology, and he contributed to both medicine and acoustics, influencing piano making and tuning and deepening scientific understanding of the new telephone. He also advanced the understanding of meteorology and the microscope and generally supported the instrument-making trade. Through his decisive leadership role at the International Electrical Congress (especially the first one, of 1881) and his cofounding and directing of the Physikalisch-Technische

Reichsanstalt (Imperial institute of physics and technology, or PTR) and associated work in metrology, he participated in the establishment of national and international metrological units and standards for the burgeoning electrical industry as well as for the older heating, optical, and mechanical industries.

Helmholtz belonged to the German *Bildungsbürgertum*, that part of the German middle class that owed its status largely to its educational attainments. Like many others in that social group, he hoped for German national unity. At least until German unification came in 1871, German nationalism was closely tied to German liberalism and expressed itself in good measure through the notion of the cultural state (*Kulturstaat*). Science broadly conceived (*Wissenschaft*) was seen by the educated classes (especially) as critical to this political-cultural movement, and the institutionalization of science was considered increasingly pertinent to industry. In the eyes of many, science meant rationality and progress, and the pursuit of it helped legitimate the modern nation-state.

Helmholtz's general political outlook might, within the context of his time and place, be broadly characterized as conservative liberalism. Only on the rarest of occasions was he a chauvinist, and he was never a political professor. He took great pride in German scientific and other achievements, yet he unhesitatingly recognized the contributions of scientists from other nations. Like many of his colleagues at home and abroad, he spoke simultaneously of science as being national and being international. He served proudly as a professor (a state civil servant) and a director of a national institute (the Reichsanstalt); he proudly accepted numerous state awards and medals, foreign as well as domestic; and, especially during the final phase of his career, in Berlin, he often hobnobbed with monarchs, aristocrats, and senior civil servants. But he almost never thought of himself as being political, and he did not exploit his connections for political purposes or professional advancement. He saw himself as belonging to Germany's intellectual, cultural, and social elite, not to its political establishment. He saw himself as a *Kulturträger*, a bearer of culture. As much as anyone else of his day, he embodied the German ideal of the man of knowledge, standing as a symbol of a broad and critical intellectual spirit that ranged beyond any individual science and beyond science itself, holding a vision of the unification of the sciences and their utility for human welfare. It was a liberal, Enlightenment spirit that, however inadequately and insufficiently, coexisted with the politically conservative, reactionary, and militaristic tendencies of modern German history. Nonetheless, as the epilogue argues, the memory of Helmholtz in modern Germany, from his death in 1894 down to the present, has often been selective and was often used and occasionally abused to create a mythological figure who somehow lived above history. This biography seeks to ground him in the real world of his day and to show that no simplistic representation of his life and work will do him justice and give us accurate historical understanding. It seeks to show, finally, what the creative life of a scientist can be like, and how his individual life needed and in turn enhanced understanding of the arts and of culture broadly conceived. It seeks to show how the sciences and the humanities were interwoven in the life of one highly motivated, energetic, and gifted person.

PART I

The Making of a Scientist

1

The Boy from Potsdam

Ancestry

Family stood at the heart of Hermann Helmholtz's being, so to know something of his family is to begin to know something of him. Throughout nearly all of his life, he was surrounded and supported by family.

Helmholtz's paternal grandfather, August Wilhelm Helmholtz, was a Prussian government tobacco-warehouse inspector who had married into a family of Huguenot *réfugiées* (the Sauvage family); to that extent Helmholtz's ancestry was French. Helmholtz's father, August Ferdinand Julius Helmholtz, was born in Berlin in 1792 and christened into the Evangelical (i.e., the Lutheran) faith. In the winter of 1807–8, while attending a Berlin Gymnasium and while Prussia was occupied by Napoleon's forces, Ferdinand heard the philosopher Johann Gottlieb Fichte deliver his highly nationalistic addresses to the German nation. He was deeply moved by these, and he came to hate the French occupier. Shortly thereafter he met Fichte's son, Immanuel Herrmann Fichte, and the two became soul mates. The elder Fichte's wife, Marie Johanne Fichte, also became his confidante. In 1811 Ferdinand enrolled in the Theological Faculty of the new, reform-minded University of Berlin and devoted himself to attending Johann Gottlieb Fichte's classes and learning his philosophical system of transcendental idealism. Fichte's philosophy remained the intellectual center and inspiration of Ferdinand's life. Fichte had brought religion back into philosophy, and Ferdinand believed that to understand Fichte's philosophy required "absolute inspiration and religiosity." For Ferdinand, Fichte was a man "full of endless sources of deep, Godly wisdom," one in whom the life of God himself spoke.[1] The great attraction of Fichte's philosophy for Ferdinand thus lay in its religious basis, and therein lay one source of future tension between father and son. In becoming an adept of Fichte, moreover, Ferdinand effectively linked his son to one of Germany's premier philosophical traditions and one of its principal universities.

Following Napoleon's setbacks in Russia and his army's retreat westward, Prussia's monarch, Frederick William III (1797–1840), called his subjects to fight for the fatherland against Napoleon, and in March 1813 Ferdinand, like most of his fellow university students, withdrew from the university and joined the Prussian militia. He was probably equally moved to do so by Fichte's lectures of 1812–13; his emphasis on freedom and the role of the state inspired many of his students to rally to the Prussian colors. Ferdinand became a soldier in Prussia's so-called Wars of Liberation (1813–15) against the French invader, a participant in the nationalist and anti-French movement that swept across Prussia. Stationed in the Silesian mountains (the Prussian monarch had retreated to Breslau), Ferdinand told his "brother" Immanuel Herrmann Fichte that his military duties were dreadfully boring—he worked with cannons and "with the most impoverished people"—but that as long as there was no fighting, army life was something of a vacation. Even so, he hated the constant cleaning and marching, and above all not getting paid; "I certainly want to refresh my weakened body," he said. Like many other ultrapatriotic Prussians, he also despised the Russians, who, along with the British and the Austrians, were now allied with Prussia against France, and who had destroyed the region: "The Russians are incredibly uncivilized; they are almost like wild nomads (their entire way of fighting still has this character)." He thought little more of the French forces and could not believe that Prussians had run away from the French troops and "surrendered [this] beautiful country to their plundering." He envied Immanuel Herrmann, who had remained at home, where he could continue his intellectual life and live with his "dear, glorious parents." He considered himself a "poor devil," one who "thirsts for salvation" even though he had already "been reborn again." His fate, he said, lay fully in God's hands. Two months later, Napoleon defeated the Prussian forces at the Battle of Dresden. Ferdinand retained a lifelong hatred of the French emperor if not of the French in general. Sixty-five years later, Hermann Helmholtz wrote of the men of the Wars of Liberation: "The older ones among us still knew the men of that period: men who had entered into the army as the first volunteers; who were always ready to immerse themselves in the discussion of metaphysical problems; who were well read in the works of Germany's great poets; and who still burned with anger when Napoleon I, inspiration and pride, and the acts of the War of Liberation were discussed."[2] Hermann inherited Ferdinand's wartime memories and his attitudes toward the French.

By September 1813 Ferdinand had managed to earn a second-lieutenancy. He planned to return to his university studies, and he asked Johann Gottlieb Fichte what professional course he should pursue. Fichte curtly replied, however, that he should make such a decision himself. After the Treaty of Paris, Ferdinand was released from the army on grounds that he suffered from a chronic, low-level, and unspecified nervous disease. He reenrolled at the university in October, this time to study philology, but he did so only to earn a living: philosophy remained his true passion.[3] Yet, after Fichte's death in 1814, and with his philosophical system now becoming unfashionable—Hegel succeeded him in the chair of philosophy at Berlin in 1818—Ferdinand was left intellectually isolated.

But for moral support and love, he could still turn to "Mother Fichte," as Fichte's widow, who was a niece of the poet Friedrich Gottlieb Klopstock and a most pious woman, called herself. While Ferdinand underwent a bathing cure for his bad nerves, "Mother Fichte" worried about his delicate health and instructed him on how to improve it. He hoped her love would aid his healing and "would make him worthy of the great, moral Christian community." He hoped, too, that he would not become a burden to her, though he knew he would always need her "love and help." His own family, he explained to "Mother Fichte," sought to pull him back to its way of life—presumably meaning that of a lower-middle-class, minor government official—"while we individuals [who] develop ourselves further and apart from them have raised ourselves above them." He felt ashamed and guilt-ridden in telling her this and feared that he had "revealed the picture" of his "confusion, frivolousness, and folly . . . before [her] holy view."[4]

After graduating from the university, Ferdinand taught for several years at Berlin's Cauer Institute, a private secondary school oriented toward the natural sciences. (Among its pupils was Heinrich Gustav Magnus, a future Berlin chemist and physicist, who played a supportive role in Hermann's professional career.) While the position provided for Ferdinand's material needs, his emotional state remained fragile. He unburdened himself to Immanuel Herrmann Fichte as someone who "is scared of life," is afraid to try anything new, who "sees the world timidly and darkly," and who fears its love and inspiration. Ferdinand felt filled with gloom and condemned a world that would not allow him to fulfill his ideals but, rather, forced upon him "the old concern for existence and pleasure." His attempts to rise above material existence to higher things had greatly weakened his power, will, and body. In these moments of depression, self-hate, Weltschmerz, and need for redemption—moments that came upon him all too often—he became diffident and filled with self-doubt, self-misery, and self-reproach. He told Fichte, as he had told "Mother Fichte," of his need to escape from his family, whose "existence and nature is so alien to me." He wanted Fichte to travel with him to Italy, where he hoped to cure his body and soul. He felt weak, dispirited, and disheartened. He wrote: "You know, it would be the highest goal of my scientific striving if I could once illuminate . . . the history of nature from the higher light of the *Wissenschaftslehre* [i.e., from Johann Gottlieb Fichte's theory of scientific knowledge]. It is a matter of action, not scholarship and books." He hoped God and Italy's lush environment would inspire his spirit and restore his body. But instead of visiting Italy, he and Immanuel Herrmann had to settle for hiking around Germany and Switzerland until their funds ran out. Ferdinand never saw Italy; and after 1822, when Fichte moved to Saarbrücken, where he became a Gymnasium teacher, he never saw his closest friend again.[5]

In the autumn of 1820, two events relieved Ferdinand's depression, at least temporarily. On 1 October he became a teacher at the Potsdam Gymnasium, where, after a probationary year, he remained for the rest of his career. The new job meant more security. The other event was marrying Caroline Auguste Penne.

Precious little is known about Helmholtz's mother, who was called Lina by Ferdinand and their friends. She was born in Breslau in 1797, the daughter of a Hanoverian artillery officer. Helmholtz himself later declared she was "of an emigrated English family," and it is widely believed that she descended on her father's side from William Penn, the Anglo-American founder of the Quaker movement and of the colony of Pennsylvania.[6] Yet Helmholtz's claim, apparently made only once, in 1876, is the only known (written) claim that he ever made to descending from an English family. Moreover, the vast genealogical literature on Penn and his family gives no indication of any connection to Helmholtz's mother or her possible ancestors. In any case, Lina, like Ferdinand, descended (at least on her mother's side) from a Huguenot *réfugiée* family, and, indeed, one whose family name was also Sauvage.

His mother had only a limited formal education, and it appears that her literacy was limited. Her family was without wealth or social position. She was, however, a "sensitive" woman. Immanuel Herrmann Fichte told a mutual friend that his "beloved H." was in love with "a most worthy and *completely* admirable girl." (Lina had also "known and loved" Fichte's mother.) Though she was not worldly, her soul was of "rare depth and fervor." Fichte thought the couple very much in love and rightly believed that they would soon marry. On 5 October 1820, four days after his appointment to the Gymnasium, they did so.[7]

A week later Ferdinand told Fichte that, for once, he was "so happy, so completely ineffably well! For a lovely, holy angel smiles at me always with heavenly joy." He believed that nothing was greater than "the love of a woman"; "Truly, the angels in heaven cannot be purer, holier, more innocent than the loving wife.

. . . With her I want to arise upwards from the Earth and soar up into the eternal spring of another life." His new job and Lina's love had saved him; he had (momentarily) found heavenly bliss on earth, which included Lina's excellent terrestrial (read: domestic) skills.[8] These she now exercised in their new home in Potsdam.

Potsdam

The city of Potsdam lies immediately southwest of Berlin, encircled by a chain of picturesque, forested hills surrounding the sandy, damp Havel river and valley. Its magnificent lakes, which were unspoiled territory, and its proximity to Berlin made it an ideal location for the Prussian monarch's country residence as well as for many nobles. For the Hohenzollerns and their entourages, Potsdam was to Berlin what Versailles was to Paris for the Bourbons: a royal palace and park (Sanssouci), and a second residence for the king. But it was also a garrison city: it housed Prussia's military headquarters and was filled with soldiers. Potsdamers saw the Prussian state and military up close. They knew, too, the city's splendid gardens, its botanical school, and its Nikolaikirche and Garnisonkirche, two of its principal landmarks. It was home to numerous Huguenot descendants, many of whom had settled near the city's center, the Wilhelmsplatz, giving Potsdam a *touche française*.[9] See figure 1.1.

Yet the Napoleonic Wars had greatly damaged Potsdam's economy, and it took nearly two generations—including nearly all of Helmholtz's youth—for the city to fully recover. Napoleon's troops forced the Prussian high command and its army to abandon the city, and for two years (1806–8) he quartered his troops there and generally ravaged the city. Local industries were shut down, and depression ensued. Though the Prussian army returned, industry crept back only slowly, and the recovery remained limited in scope. The city did grow from about twenty-one thousand (in 1821) to about thirty-one thousand (in 1849), but the poor were as present as the king's soldiers. Although the construction of several villas during Frederick William III's reign brought some economic benefits, the city's poverty increased and its industry and trade remained at low levels until 1850.

On the other hand, Potsdam was also home to numerous civil servants. The monarchy and its retinue did much to shape the city. Frederick William III induced Karl Friedrich Schinkel, Prussia's premier architect, and Peter Joseph Lenné, its premier landscape architect, to help renovate Potsdam, and together they transformed the city's appearance. After 1826 Schinkel rebuilt the Nikolaikirche, and it became the city's architectural focal point. When Frederick William IV (1840–61) assumed power, he lavishly enhanced Potsdam's parks, gardens, and landscape and sought to give them an Italian flavor. Moreover, in

Fig. 1.1 Friedrich August Schmidt, *Potsdam vom Babelsberg, 1830/40*. Horst Drescher and Renate Kroll, *Potsdam aus drei Jahrhunderten* (Weimar: Hermann Böhlaus Nachfolger, 1981), Nr. X.

1832 an optical telegraph was built on Potsdam's Telegrafenberg, linking various parts of Prussia with one another and with Potsdam and making the latter into something of a Prussian communications center. The telegraph line and the Berlin-Potsdam railway line (opened in 1838) became symbols of modernity. While Potsdamers continued to experience economic hard times, the monarchical and governmental presence, and the spending on public goods and services, helped the city to recover.[10]

Potsdam's liberal, middle-class citizenry also helped the city to develop culturally. Many of them joined voluntary associations (*Vereine*) devoted to various public and private causes. Among them was the Potsdam Art Association (*Kunstverein*), founded in 1834 by Wilhelm Puhlmann, a military doctor and friend of the Helmholtzes. Puhlmann became a leading art collector and an important figure in promoting the arts in the Berlin region; he was a very close, lifelong friend of the painter Adolph von Menzel. Moreover, Puhlmann and Ferdinand Helmholtz were both active in Potsdam's Peace Society (*Friedensgesellschaft*), a civic organization that provided stipends to help support impoverished but talented youths in the Berlin-Potsdam region who wanted to study art.[11]

In short, the Potsdam in which Helmholtz was born and raised was at once a royal, military, civilian, and artistic center seeking to recover from the devastation of the Napoleonic Wars, a city whose surrounding natural beauty, small population, struggling economy, and art and architecture all contributed to his sense of *Heimat*.

Family and Early Childhood

Home for the newlywed Helmholtzes was a house at Wilhelmsplatz 14. It was here, on 31 August 1821, that Lina bore the first of their six children: Herrmann Ludwig Ferdinand Helmholtz. The first of Helmholtz's three Christian names was the most Germanic one, symbolizing both German military strength and unity; he owed that name to Ferdinand's love of his "brother," Immanuel Herrmann Fichte. (Until Helmholtz's medical-school years, he usually spelled it with two r's; only then did he drop the archaic for the modern form.) His second name he owed to Christian Ludwig Mursinna, an elderly great-uncle and the surgeon general of the Prussian army. His third name, given at Lina's insistence, he owed to Ferdinand himself. With these three Christian names his parents sought to give him a sense of tradition, family, and honor and so to shape his identity. As for his surname, it derived from an ancient Germanic name, perhaps Helmbold or Helmhold. Like his parents, he was christened into the Evangelical (Lutheran) faith; on 7 October, he was baptized at the Holy Ghost Church (Heilig-Geist-Kirche), with no fewer than twenty-three godparents present.[12] From the outset of his life, and indeed throughout its entirety, family surrounded and protected him.

In the weeks following his birth, he gave his mother "unspeakable pains" during breastfeeding. (Much later in life, he turned this intimate physical act into an epistemological point in favor of his empiricist theory of perception, declaring that breastfeeding was a learned act on the newborn's part.) He gave his father, who as usual felt overworked and filled with "care and anxiety," much pride. Ferdinand wrote Fichte that the latter's godson was growing "healthier and stronger daily." He was his father's "Man-Son, a true little giant, as everyone says, and beyond the level of intelligence for his age."[13]

In October 1822, the family moved into a two-story, three-bedroom house at Hoditzstrasse 10 (modern Wilhelm-Staab-Strasse 8), near the Wilhelmsplatz. It was here that Helmholtz grew up, and the home's culture gradually became his. The influence of the Fichtes could be literally seen, since Ferdinand furnished the parlor with a cherished reading desk and couch that had once belonged to the master himself. The room also contained a large bookcase that had once belonged to "brother" Fichte and that Ferdinand used to shelve his "better philological and scientific works." A small mahogany bookcase had works by Shakespeare, Pedro Calderón de la Barca, and Johann Diederich Gries, among others. Icons of high culture adorned bookshelves and other places in the Helmholtz home: busts of Venus, Socrates, Aristides, and Goethe, as well as one of Johann Gottlieb Fichte. Before "brother" Fichte moved away from Berlin, Ferdinand had borrowed numerous books from him, including novels by Sir Walter Scott, whom he read with great relish, and the works of one of the era's most prominent romantics, Lord Byron.[14] The Helmholtz home, thanks to Ferdinand, was enveloped in high culture, and the family belonged to the broad cultural elite of German society, to its *Bildungsbürgertum*, the educated middle classes. This home and objects like these represented Ferdinand's cultural reality and aspirations, and they set the home's tone and served as model and inspiration for the young Hermann.

One-year-old Hermann, according to Ferdinand, was healthy and good-looking, behaved well, and brought him joy. He devoted much time to his son's education. Both parents were anxious about money, feeling that Ferdinand's salary was insufficient; it forced them to take in boarders.[15]

Ferdinand's new job and his son gave him renewed vigor, temporarily. He felt more able to meet his professional duties, and he told Fichte, "My Hermann gives me only blessed moments." Yet he longed to leave Potsdam, where he felt isolated. He lacked enough money even to travel to Berlin. He found some relief in Christian piety. Only myopic man, he lectured Fichte, thought life was endless; "more beautiful days" lay in the beyond. His letters to Fichte overflowed with romantic and (at points) incoherent thoughts about art, religion, God, and life. There was an infinite world of spirits (*Geisterwelt*), he maintained, just as there was a physical world; both worlds ultimately depended on the infinite universe. In good Fichtian manner, Ferdinand explained that it was man's inner, spiritual, intellectual life that distinguished him; if there was to be any "harmony and self-cultivation [*Bildung*]" in man, it had to be here. He advised Fichte to "do God's will." To these philosophical musings he attached a deadly conclusion: "Oh, how often I long for the stillness of the grave, how often I long: 'Oh lay me down quickly into the cool grave, because my condition, my intellectual and bodily weakness, do not let me come down to quiet, rest, love, and salvation.'"[16] Hermann soon enough heard his father's piety, experienced his romanticism, and, it may be presumed, saw his depression firsthand.

At Christmas 1822, when Helmholtz was sixteen months old, Ferdinand (once again) wrote Fichte of his feelings of spiritual emptiness, of the banality of his existence, and to report that his and Lina's great hope lay in their son. The boy renewed his father's strength for work: "A sweet child," he "smiles at me full of hope, . . . has opened for me the ways to the deepest wisdom." Still, he felt deeply frustrated as his ambitions remained unfulfilled: when he read a Shakespearean tragedy alone, it was "boring and unbearable," no matter how much "passion" and "struggle" it contained; he needed people for pleasure and encouragement. Art was an important part of Ferdinand's life, and so of his home. Those individuals who lacked insight into art, Ferdinand opined, could do no more than follow mechanical rules. His pedagogical cum epistemological opinions prefigured some of his son's own (namely, those on unconscious inferences) in maturity. In teaching the basic levels at school, Ferdinand thought it

> absolutely necessary . . . that the first bits of knowledge be so impregnated into the mind that they require not the least reflection and freedom in their subsequent use. Think about how many judgments are required to understand the simplest Latin sentence, and yet doing so is the work of an instant. You cannot even know all the functions involved; you complete them with incredible speed. In fact, this action, this knowledge, is the result of an artistic reflection, which you only need to rediscover again. For this, however, one needs solidity and sureness in the initial elementary judgments, by which alone, step-by-step, secure progress is possible and by which alone the child can feel well.

Yet in unmistakable contrast to his son's future epistemology, Ferdinand wrapped his pedagogical views in strong Christian dogma: "the real truth" was God's revelation. "God is love," he declared. "Only he who always keeps in mind . . . this single revelation of Christ, that God is love, and is completely convinced by this conviction, to him is every story (on the outside) a rapturous present, to him the way to Heaven may be opened."[17] Piety as well as culture enveloped the Helmholtz home.

Religion, philosophy, and Hermann were not the only sources of solace for Ferdinand. By now he had come to love the city and its beautiful environs, to which he introduced his son. He enjoyed the king's gardens (from the outside) as well as his own little garden. Teaching, by contrast, offered him little solace: "He alone may be a teacher who is unfit for any other business." He thought that few of his fellow teachers were committed to "higher, freer self-cultivation" and that "solid [intellectual] enthusiasm" was "so rare." He had bad relations with his colleagues, he said, and his salary was inadequate. He felt stultified intellectually. His pleasures came in good measure from being with his eighteen-month-old son: "My little son is beginning to chatter. He naturally seems to me like a little Christ. I am of course his father. Yet he also finds friends among uncivilized strangers. He frequently gets candies as signs of human kindness from them in the middle of the street, at the post office, and at other public places, so that people greet him sweetly." Aside from his pleasure in Hermann, Ferdinand longed only to be in his "garden with the flowers and fruits of dear nature." A few months later he repeated the same points to Fichte but concluded: "My life is rotten; I am sick in body and spirit."[18]

Lina saw their world differently. She said they had an active social life: they gave and attended punch fetes, played cards, and had a few friends. The one thing that they certainly agreed about was their baby son. She wrote: "Hermann gives me much, much joy. He answers 'Yes' and 'No' quite properly. So one can already find out quite a bit from him. He is now making an effort at speaking, and takes pleasure in naming things whose names he can speak. He already knows the [letter] 'i' and looks for it himself on each piece of printed paper, and takes pleasure in each 'i' that he finds. Since he can again enjoy fresh air, which he had to do without because of the great cold, he now, once again, has red cheeks."[19]

In stark contrast to Ferdinand, moreover, Lina was sober-minded, concrete, and optimistic, traits that Helmholtz inherited. A family acquaintance claimed that Helmholtz's head was shaped like Lina's and that his nature was "the same" as hers, meaning that she willingly tackled problems without worrying beforehand about the final details and that she spoke in a simple manner.[20] Helmholtz himself later certainly displayed these qualities. Hence it appears that he got his first interest in art and his early education from his father, while he got his optimism, and perhaps even a good deal of his intelligence and problem-solving abilities, from his mother. Certainly he got his enormous self-confidence from her; that was something the self-loathing, self-pitying Ferdinand completely lacked.

Ferdinand and Lina's union was fruitful beyond Hermann: all told, they had six children. The birth of their second child and eldest daughter, Marie (1823–67), forced Ferdinand to spend most of his free time with the family, leaving him little time to read and additional concern over money, about which he worried constantly. Lina's pregnancy with Marie left her quite ill. Her troubles were compounded when Hermann contracted a serious case of the measles. All this depressed Ferdinand still further; by his own account, he suffered from "melancholy." Moreover, his relationship with the Gymnasium's rector was (at least then) not good. Further, he meddled to prevent a marriage between Fichte and Bertha Leithold, one of Hermann's godmothers and a distant relative of the Helmholtzes, and his effort led to strained relations between Ferdinand and the Leitholds in Berlin; it also affected Helmholtz when he became a medical student there.[21]

At age two, Helmholtz had a serious accident—he fell against the edge of the kitchen stove—that, however, ultimately affected his mother more than him: Ferdinand and their friends thought Lina might die from the fright. After she recovered her health, the couple had four additional children: Julie Caroline Louise (1827–94); Ferdinand Carl Ludwig (1831–34); August Otto Karl (1834–1913), known as Otto; and Johannes Heinrich (1837–41). From Hermann's birth in 1821 until the late 1840s, Lina devoted her life to birthing, nursing, and rearing her six children, two of whom died before the age of four. She kept an immaculate house and ran a frugal household. Ferdinand, for his part at least, felt that there was "nothing more beautiful, more delightful, more charming, and more dear than a healthy, happy child." His children, he said, were apparently God's way of compensating him for his situation in life.[22]

In short, Ferdinand and Lina built and provided a strong and stable family life. They loved their boy Hermann deeply, and he became well attached to them. It may be that, as the firstborn, his parents gave him extra attention and resources, at least until his five siblings starting arriving. This too may have contributed to his intellectual development and sense of confidence. As the firstborn of the six, he perhaps felt a sense of responsibility for his siblings, and this in turn may have contributed to his strong sense (as an adult) of responsibility for family and others.

The couple faced their greatest challenge with Hermann when he was age five and suffered a life-threatening illness, probably "a light hydrocephalus," as he later told his doctors. It took him two years to recover, which he did thanks to "the goodness of God almighty" and the care of his parents. He regained his health and strength "above all through the use of the sports gymnasium and the baths." All in all, he reported that he had been "a sickly boy" during his first seven years, "long bound to his room, often enough to the bed." Still, he had a "lively drive to talk and be active." If nothing else, this extended childhood illness gave him a lifelong and constant concern about his health. His parents devoted much time to him, at least when he was not alone and involved with "picture books and games, principally little wooden blocks." They gave him piano lessons, though the lessons left him without any feeling for music itself, he said. He thought himself "a very obedient boy," except for the time when his piano teacher became "so insufferable" that one day he "threw the piano notes down at his feet and thus brought the lesson to an early end."[23] That was his sole act of rebellion in life.

Already as a child, Helmholtz gained firsthand knowledge of geometry by playing with stereometric bodies. The wooden blocks provided him with his first notions of proportionality and shape, while geometry gave him his first sense of law. "From my childhood games with the wooden blocks," he wrote, "the connections of spatial relations to one another were well known to me and came intuitively." This youthful playing with blocks and reasoning about geometric relationships may also have instilled his first, inchoate and intuitive understanding of epistemological matters. The hands-on experiences also gave him an indelible memory of his first sense of perspective. For example, he later noted how young children sometimes misjudge distances:

> I still remember clearly the moment when the law of perspective first appeared: that distant things seem small. I went past a tall tower. Some people stood on

the top balcony and I called out to my mother to reach out for the pretty little puppets—since I was thoroughly convinced that, if she would just stretch out her arm, she would be able to reach up to the tower's balcony. Later, I quite often looked up to the tower's balcony when people were there, but they did not, to the more-practiced eye, any longer want to become sweet little puppets.[24]

He thus grew up with geometry (and in a physical sense) and an awareness of problems of spatial perception; his mother, in her own way, contributed to that sensitivity.

Helmholtz acquired these first geometric experiences and sense of law before he attended the local elementary school (from about 1826 to 1829). Even as they continued to inculcate conservative values, such schools as Potsdam's were greatly expanded in number and their curricula reformed in the wake of the Prussian defeats at the hands of Napoleon. Ferdinand may have owed his appointment partly to this expansion, and Hermann was doubtless a beneficiary of these reforms. Nevertheless, as Helmholtz later claimed, he knew the facts of geometry quite well before he had learned any formal geometry theorems, and this surprised his teachers. He also learned to read early in life, though he found that his memory "for unconnected things" was weak and said that as a child he had difficulty distinguishing left from right—a fact that may also help account for his preoccupation with the problem of spatial perception during adulthood. In elementary school he found the study of languages more difficult than his fellow pupils did; history was more difficult still; and memorization of prose was virtually impossible. But the school's teachers taught him "the rigorous method of science, and with their help I felt the difficulties disappear that had slowed me up in other areas."[25] When he left the school, at about age nine, he was more than prepared for the challenges of a rigorous German Gymnasium.

2

At the Gymnasium: Father and Son

The Potsdam Gymnasium

The Potsdam Gymnasium's development reflected the educational needs and the growth of the city's *Bildungsbürgertum*. It offered all nine grades of Gymnasium instruction: the three lower (Sexta, Quinta, and Quarta) and the six upper (Tertia, Secunda, and Prima, each of which included a lower and an upper division). In 1831, the year after Helmholtz entered, it enrolled 299 pupils, all boys; in 1838, the year he graduated, 306 were enrolled. It had a staff of nine teachers, along with two or three instructors in penmanship, drawing, and singing. It housed a large lecture hall, a drawing room, a physical laboratory (*Saal*), seven lecture rooms, one library each for teachers and pupils, a conference room, and a director's residence. The pupils' library contained more than seven hundred volumes. The laboratory, with more than one hundred pieces of physical and mathematical apparatus, allowed pupils to conduct experiments in conjunction with the lessons they learned in class. In addition, there was a collection of natural science literature and a mineral collection.[1] These modern facilities were probably among the German states' best. They attested to Prussia's greatly enhanced commitment to education in the post-Napoleonic era. For its instructors and pupils alike, the Gymnasium was the breeding ground of German *Kultur*.

Like all classical Gymnasia, Potsdam's aimed ultimately at imparting the essential elements of *Bildung* (self-formation or self-cultivation). It was a neohumanist institution that sought to form the nation's elite: to educate its wards for the nation by developing in them a national consciousness. Its officials were state officials, but sans republicanism or any interest in industry, technology, or practice generally. It sought to develop in its pupils moral and spiritual, as well as intellectual and physical, character and to make them into ethical human beings who knew how to learn and use knowledge in ways that would ultimately develop them most fully. It sought to accomplish these lofty ideals by the required study of languages and literature (Greek, Latin, German, and French) as well as the *Wissenschaften* (i.e., the "sciences": not only mathematics, physics, natural history, and geography, but also history and philosophy); by instruction in drawing, singing, and gymnastics; and by instruction in religion (i.e., Christian dogma and history). For in Potsdam as elsewhere, the Gymnasium's overall purpose included imparting a Christian moral orientation as well as a classical or humanistic education. Helmholtz and his fellow pupils sang religious songs like "Vater unser" ("Our Father") and "Der Auferstandene" ("The Resurrected") as well as the school's Christian prayer song. The school thus added to whatever religious instruction Helmholtz's parents and others provided. Finally, it also propagated a code of behavior intended to inculcate a sense of civil order and moral rectitude. The aim, stated one of the rectors, was to give pupils a sense of "punctuality, external order, and strong lawfulness," to encourage humility and industry, and to inculcate these moral virtues along with a "Christian sense of nobility," and so give them moral independence.[2] The school, like the Helmholtz home, gave Hermann a strong sense of Christian piety and civic propriety.

Ferdinand Helmholtz as Pedagogue and Philosopher

Ferdinand Helmholtz was an integral part of the Potsdam Gymnasium. After an initial probationary year, he served five years as a teacher in the three upper grades. Thanks to his "solid erudition and proven loyalty to office," he was promoted in 1827 to subrector; the following year he received the title of "Professor" and became head of the Secunda.[3] He served in this position and at this grade for the next three decades. Like most of the new mass of schoolteachers of his day, he became a moral model and a pedagogical facilitator for all lower- and middle-class students and families who sought to move ahead and up.

During his first professional years, Ferdinand taught a very wide variety of courses. After 1830, however, he no longer taught mathematics or physics. He told Fichte, "On the one hand my main education has certainly not been mathematical but rather philological, while, on the other, philological instruction, due to its versatility in developing intellectual powers, has so far been more interesting than the mathematical." He concentrated instead on teaching language, literature, philosophy, and religion, and on his administrative duties. Pupils recalled that it was particularly pleasurable to hear him read poetry, plays, and the like.[4]

Ferdinand's supervisors praised him throughout his career. They noted his excellent aesthetic education, wide reading in literature, inspired teaching, school loyalty, faithfulness to duty, and, in general, his *Bildung*. During later years he was reportedly increasingly lenient with pupils, and during the last phase of his career, gossip had it that the quality of his teaching had slipped. There is only one blemish (if it was a blemish) on his record: in early 1848, on the eve of the revolution in Prussia and at the request of his pupils, he canceled his German-language classes for three hours and spoke about his experiences in the Wars of Liberation, about how king Frederick William III had followed rather than led the war effort, and about the Restoration. In doing so, he touched on the rawest of nerves in recent Prussian political history: To what extent were the wars patriotic and to what extent nationalistic? Was it the state or the nation, the Hohenzollern dynasty or the Prussian people, who led the way? Were they wars of liberation or liberty? The Gymnasium director reprimanded him for this insubordination and these indiscretions, threatening him with immediate dismissal should he repeat such actions. He never did, though he remained enthusiastically patriotic and often inspired his pupils by his patriotism. Helmholtz himself portrayed his father as a teacher who had "a strict sense of duty but who was also an enthusiastic person, impassioned for poetry, especially for the great age of German classical literature." Until ill health forced him to retire in 1857, he served, according to the school's eulogy, "with true devotion."[5] Here was a reminder, writ small, of the subordination of culture to politics in the German lands.

During the course of his long career, Ferdinand gave four talks before the Gymnasium's plenary sessions, and they constituted the sum total of his publications. The most pertinent of these in regard to his son was the one of 1837, on aesthetic education. Its subject suggests that aspects of Helmholtz's mature intellectual interests originated in his father's own concerns and that Ferdinand's assessment of contemporary cultural life and recommendations on education bore strongly on his son. Helmholtz himself said as much: "Happy is the child who grows up with parents who cultivate the arts."[6] When Ferdinand gave his 1837 presentation, Hermann was in his final year at the Gymnasium and, in all likelihood, sat in the audience as his father spoke or read the address.

Ferdinand believed that the material and spiritual life "of the civilized European peoples" was progressing and that "civilization" was expanding. But he also thought progress and civilization had led to a certain one-sidedness in outlook: the spiritual dimension of man had been neglected and the still-unknown was considered merely an unsolved conceptual problem. In his own view, human existence was a combination of an outer and an inner world, of the temporal and the eternal, of science and spiritual life. Life was a "unity" of these binary pairings. The Greeks had shown the way for the former of them, he said, and Christianity for the latter. He thought science had value only insofar as it allowed people to enjoy life. He did not believe in science (including scholarship) for its own sake or as separate from practical life. Science, technology, and human values had to serve one another. In his eyes, contemporary science was unchristian. He claimed that life had a "higher unity," that there was a "lawfulness to all life and all phenomena." He thought the gaps in contemporary *Bildung* were responsible for the "troubles of the age."[7]

Ferdinand maintained that the modern world was one of freedom and of Western culture. It arose out of man's "struggle with nature and with himself" and so had led to man's freedom. Here "the development of the mind as freedom forms the peculiar nature of Western life." It had begun with the Greeks, who created abstract concepts and laws of nature. As a consequence, man had become increasingly alienated from nature, even as he believed that he was observing reality and discovering laws. This analytical phase had destroyed life's real unity and replaced it with an abstract unity. Man came to believe that he had a free will. Belief in reason and the individual ruled. But this, Ferdinand argued, represented only a limited sort of progress. True freedom was not freedom for its own sake but rather for "the revelation of God." Reason was nothing if it was not used for "revelation." This meant that the development of feeling and intuition had to proceed from the development of reason. Hence, now that "intellectual freedom" had been achieved, it was time to turn away from an abstract, "empty understanding of life" and toward "realizing the Christian task of living and loving God in freedom." That required the "development and formation of the feelings," which constituted "the next and most important task of our pedagogy."[8]

Art, he maintained, shaped feeling and did so "according to the laws of beauty." The development of these laws was necessary for personal, familial, and societal reasons. The era's principal task was to unite feelings and freedom, to cultivate the former for the benefit of the latter. Like truth, beauty was an absolute that had to be discovered, not invented. Hence it was the task of art to express the *ought* in the *is*, to educate the senses. Beauty was a necessity "to a truly Godly life," that is, a Christian life. In this way, aesthetic education helped found morality and ground compassion. Poets, musicians, and dancers were the founders of morality, because they emphasized harmony, sounds, and form. Art led to spirituality, freedom, and the noblest (the inner) mode of existence. At their highest levels, the arts satisfied "a more serious religious need."[9]

Ferdinand further held that while reason produces itself from within, feelings are immediate responses to external objects. He thought every people had its own art and that the more developed a people was, the more developed its art; the inner nature of a people and its type of art were closely connected. For Ferdinand art was the measure of *Bildung*. He thought the German Gymnasium's aesthetic education was in a poor state, and he called for its improvement. He thought his entire era displayed tastelessness in dress and fashion and showed inadequacies in architecture, speech, and theater.[10]

For Ferdinand, the highest beauty was the full achievement of unity. From such unity came "peace, harmony, and ease of mind," which would spread "beauty over the soul." Art thus brought people beyond a one-sided rationality. It did precisely what science did not do: it resolved disparate elements into a unified whole that included inwardness and outwardness, freedom and law, absolute and relative, *is* and *ought*, the temporal and the eternal, the finite and the infinite, the individual and the whole. Art unified all. It brought "ever-more peace and ever-more ease of mind and bliss." Beauty thus had a powerful effect on "the civilization of peoples," and so man had to perceive "beauty as a form of the Godly life." Paradise, the Golden Age, would be living life "in the unity of mind and nature," in the unification of self and not-self, of body and soul, of becoming and being, of life and death.[11]

Ferdinand's romantic, holy vision of life claimed that the individual's sense of beauty had developed through a series of historical stages, the last of which embodied his highest ideals in art. These included the literary works of Aeschylus, Aristophanes, Sophocles, Shakespeare, Molière, Calderon, Schiller, and Goethe, as well as the music of Beethoven, Handel, Gluck, and Mozart.[12] With the exception of the Frenchman Molière, his son Hermann in due course regularly cited these very names as touchstones of great literature and music; their works provided Hermann with some of his greatest hours of literary and musical pleasure.

Ferdinand's long, turgid, rambling, and highly romantic essay of 1837, a virtual philosophy of life, also listed several practical rules for cultivating feelings in young people. Children, like adults, must learn to control their animal passions and so to a certain extent become godlike. They must develop good manners and social behavior in general. Here, too, control over the will was essential. They must learn to eat and drink gracefully and without inordinate desire and to walk with a natural stride. Hence the importance of participation in orchestra and gymnastics and of careful attention to dress. They must learn to beautify life through refinement and graciousness, while avoiding "arrogance, lust for power, pride, and ill-bred indifference." They must seek to create or find artifacts that fully exhibited unity and naturalness. They must cultivate beautiful language as the expression of the soul's beauty and of civilization. They must make sure every individual action was "an act of freedom," one that was simultaneously an individual act and in accord with reason in general. Finally, they must seek to unify "reality and the ideal, this life on Earth and eternity, in short, the resolution of all antitheses [*Gegensätze*] of man as an intellectual and material being." Ferdinand believed that "the need of beauty is as deeply rooted in the human soul as that of truth" but that aesthetic education occurred "only through intuition, not through concepts."[13]

Though in his own eyes Ferdinand could never live up to such ideals, he nonetheless sought to instill them in his children and his Gymnasium pupils. Helmholtz related that once, when he was "standing before the sculpture of Athena in the large fountains of Sanssouci," and "wanted to begin criticizing details of the hate-filled goddess," his father once told him, "My son, you must first try to absorb into you the sense of the whole and let it take effect on you. Only when you've succeeded in doing this can you allow yourself to break it into parts. In this way you protect yourself from letting the small matters that you don't like ruin the joy of the entire phenomenon." Otto Helmholtz likewise noted that his father always sought to inspire his sons through art and nature. This suggests that Helmholtz's life with his father may not have been easy. Anna von Mohl, Helmholtz's second wife, indicated as much, adding that Ferdinand apparently "had the air of being a serious pedant who believed himself to be the wisest of mortals."[14] Ferdinand, who may well have been overbearing with his children, longed for an integrated, holistic world. His eldest son strove after similar goals, but he did so with the findings and tools of natural science and with a realistic outlook, not with religion and romanticism. The tensions between the father and the son were also those between one era and another. Nevertheless, the father's emphasis on the importance of aesthetic education in children contributed to the son's overall development, including that of his imaginative powers.

A Model Pupil

The Potsdam Gymnasium had the advantage of building on Helmholtz's self-education, his father's home instruction, and the elementary school's teaching. From start (spring 1830) to finish (fall 1838), Helmholtz was a very good if not quite outstanding Gymnasium pupil. Already during his first semester, his teachers judged his conduct as "quite good," his attention and participation during instruction as "independent and good," and his diligence and overall progress as generally praiseworthy. The only minor complaints concerned his reckoning, penmanship, and orderliness. Ferdinand's colleagues knew, of course, that they were dealing with his son, and that may have influenced their assessments.

Ferdinand himself judged Hermann's behavior and talent similarly. When Hermann was twelve years old, Ferdinand wrote Fichte: "My Hermann is a very talented boy who is happily moving ahead without being industrious (due to his health I don't demand too much of him). People find him too serious and quiet, however; only with other children is he communicative and happy." The boy had talent and self-control; both turned out to be essential for his future scientific work.[15]

At the end of Helmholtz's Gymnasium education, his teachers judged him in essentially the same way as at the start. His conduct and attention were considered "good, always serious and reasonable, and showing lively participation in all subjects of instruction"; his attendance was perfect; and his "diligence and order in written work" were deemed "praiseworthy." His gymnastics teacher reported him to be enthusiastic and described him as "reserved, serious, and very kind toward the younger pupils."[16] Already as a teenager, he demonstrated habits and virtues of discipline, hard work, and stamina. He knew how to get things done and get them done on time.

Like all other Gymnasium pupils, Helmholtz studied Latin, Greek, French, and German; unlike most others, he also studied Hebrew. His performance in Latin, which he had begun to learn at home under Ferdinand's direction, was deemed "praiseworthy," with only his written work in need of improvement. His progress in translation was judged "very good." Likewise in Greek, he was considered "very good," and his progress "delightful." In French, too, his teachers found his diligence "very good" and his progress "good," and in German both were "keen" and "very good." As for Hebrew, his teacher noted Helmholtz's dedication and that he took the final examination even though he planned to study medicine; overall, he judged Helmholtz's performance "a praiseworthy effort." Nor did his language learning end there. He also studied English and Italian privately at school, and even a little Arabic. As for "the sciences" (*Wissenschaften*), his teachers judged both his diligence and his progress in religious doctrine as "good," in history and geography "very good," and in mathematics and physics "praiseworthy" for diligence and "excellent" for progress.[17] All of this indicates much and widespread intellectual talent and ambition. His strong language skills and his equally solid knowledge of the sciences proved essential to his success in medical school and as a scientist. His Gymnasium education was broad and his overall performance was close to excellent.

Three points about this record of achievement merit special notice. First, although he was the son of a pious father who constantly stressed the importance of religion in general and Christianity in particular, Helmholtz received his lowest marks in religion. Nonetheless, his religious education as a young man was quite extensive, including eight years' worth of instruction at the Gymnasium in Christianity. The bishop of Potsdam, Helmholtz wrote, "taught me the precepts of our divine religion and received me into Christian society"; he would "always preserve [the bishop's] memory" in his "most grateful soul."[18] By age seventeen Helmholtz had a thorough understanding of Lutheran dogma and practices, and he appears to have been a believer.

Second, his apparent facility in ancient languages and literatures as well as in history does not quite square with his later report that he found it hard to learn foreign languages, that history was exceptionally difficult for him, and that to remember prose was "martyrdom." At school he acquired a "little mnemonic technical aid" in order to remember poems. He knew some songs from the *Odyssey*, many of Horace's odes, and much German poetry. He became "a great admirer of poesy." In this regard, he described his father as "a dutiful but enthusiastic person, inspired by the art of poetry, especially for the great age of German literature," and he noted that Ferdinand encouraged his efforts in language and literature. Ferdinand thus helped him and his schoolmates learn to express themselves. That instruction in German redounded to Helmholtz's benefit, especially when composing his popular and philosophical addresses. He judged his instruction in Latin to have been useful in developing "the training of systematic feeling" but said Greek did little "for the fine formation of taste, not merely for spoken but also for moral and aesthetic things." He and other pupils spent much time reading classical authors, often in after-school hours. Although it was doubtless partly rhetorical on his part, Helmholtz praised his teachers in the humanistic disciplines.[19] They and Ferdinand sensitized him to other civilizations and cultures and their evolving natures. Together, they helped instill in him *Bildung*.

Third, Helmholtz's real intellectual interest and ability lay in physics. Already in childhood, he reportedly declared, he had felt a strong need to focus on "reality" (as opposed to adopting philosophical theories like those of Fichte or Hegel) and, in order to compensate for his poor memory, on laws that connected things together. For him, physics, with its lawlike structure, best satisfied these needs. Physics helped him comprehend "reality" and minimized the burden of memory. "The most complete mnemonic aid available," he later wrote, "is knowledge of the laws of phenomena." Mathematics held second place for him, not least since he thought it played a subordinate role to physical reality. What interested him about mathematics was its application to understanding physical reality. He preferred the "full reality" of physics to the mere abstractions of mathematics. He judged his mathematical competence as merely comparable to that of his fellow pupils at the Gymnasium and, later, at medical school.[20] That seems far too modest an assessment.

"From early youth on," he later wrote, "my inclination and my interest were devoted to physics." This was his dream: to understand the physical laws of nature. His love of physics, and of natural science in general, might not have blossomed fully without his parents' indirect support. "I plunged then with great zeal and pleasure into the study of all the books on physics [that] I found in my father's library," he reported, while noting that some "were very old-fashioned" (e.g., with discussions of phlogiston). Along with reading came home experiments: he and a friend investigated the effects of acids on their mothers' linens, leaving stains on the sheets and guilt on his conscience. He constructed optical instruments out of old eyeglasses from the local optician and used his father's botanical lens. His limited experimental means during his boyhood perhaps led him later, as a professional scientist, to maximize whatever limited means came to hand.[21] He never experienced any "identity crisis" about himself: quite simply, from youth onward he was sure that all he wanted to be was a scientist, that is, a physicist. Though financial reasons forced him to take a long, circuitous road in becoming one, he did just that in due course.

In becoming a scientist, he also owed to his parents the general if intangible benefits of growing up in Potsdam. During his youth he loved to walk on Potsdam's hills and in its woods, fields, and gardens. He adored the Havel River and Potsdam's many lakes. He loved being in and around water, and later, as an adult, he analyzed its flow. From his teenage years onward, he also walked and hiked in part to restore his body and mind. His first such trip was taken in July 1837, when he and some friends went to the Harz mountains. Along the way, they visited various cities and sites. He admired the flora, the landscapes, and the countryside. The walks in Potsdam and beyond gave him a love of nature and a desire to understand it. These physical experiences sharpened and deepened what he learned through reading: "In fact, the first and foremost thing that excited me was the intellectual mastery of nature that stood as something foreign before us, doing so through the logical form of law. Naturally, I soon realized that knowledge of the laws of natural processes may also be the magic key that may give its holder power over nature. I felt at home in such thoughts." Understanding the laws of nature gave him a sense of power and was, he said, the motive behind his drive to become a scientist: "This drive to conquer reality through concepts, or, to say what I mean in another way, to discover the causal connections of phenomena, has guided me throughout my life. Such intensity of feeling has also very much left me with a guilty feeling that I shall find no rest in the apparent solution of a problem so long as I feel that it still contains unclear points."[22]

Finally, Ferdinand also gave Hermann his first taste of philosophy: "The interest for questions of the theory of cognition had been implanted in me in my youth, when I had often heard my father, who had retained a strong impression from Fichte's idealism, dispute with his colleagues who believed in Kant or Hegel." He later came to take little pride in such disputations or studies.[23]

It was his mathematics and physics teacher, Karl Ferdinand Meyer, who most inspired him in these areas. His link to Meyer was also a link to two seminal figures in German science and mathematics: the astronomer Friedrich Wilhelm Bessel and the mathematician Carl Gustav Jacob Jacobi, who had also attended the Potsdam Gymnasium. Under their tutelage, Meyer became a rigorously trained secondary-school teacher in mathematics and natural science. Meyer's superiors, including Ferdinand Helmholtz, considered him to be an excellent teacher. In 1838 he published an essay on optics; that subject, and his training at Berlin and Königsberg, prefigured Helmholtz's own interests and career. For his physics classes, Meyer used Ernst Gottfried Fischer's *Lehrbuch der mechanischen Naturlehre*. It placed a stronger emphasis on mechanics, on the mathematical treatment of physical problems, on relating experiment to theory, and on the role of measurement in experiment than was to be found in other textbooks. While still in the Tertia, Helmholtz wrote an essay on Benjamin Thompson's (Count Rumford's) and Humphrey Davy's experiments on the heat of friction. During his school years he heard often of the impossibility of creating a perpetual motion machine. Taken all together—the excellent teacher, the impressive physical laboratory, and the use of an excellent physics textbook—this complex of pedagogical resources suggests that Potsdam's Gymnasium was among the most advanced and intellectually challenging schools of its day. It certainly offered Helmholtz an excellent grounding in physics. "I must confess," he wrote a half century later, with a lingering sense of guilt, "that sometimes when the class read Cicero or Virgil, both of which I found extremely boring, I calculated (under my desk) the path of rays through a telescope and so already discovered some optical theorems not then found in textbooks, but which I later found useful in the construction of the ophthalmoscope."[24]

Finals

As a teenager, Helmholtz established a lifelong pattern of stopping work for several weeks in the summer and taking a walking tour. In July 1838, having completed his Gymnasium coursework but before taking his leaving examinations, he sought relaxation by visiting relatives in and around Berlin. In August he was back in Potsdam, ready for his last act at the Gymnasium: three and a half days' worth of demanding examinations. His father initially tried to prevent him from taking his examinations, because he did not see Hermann sitting at his desk and studying—instead, he was wandering in the forest—and so he thought Hermann was not properly preparing himself. (Helmholtz, it turned out, had a lifelong aversion to sitting down at a desk.) Ferdinand finally relented, and Hermann proceeded to take his examinations.[25]

The examinations called for solving four problems in mathematics and one in physics, writing an essay on a German text, and translating Greek, Latin,

French, and Hebrew texts. In mathematics, he was asked to solve problems from geometry and algebra, and in physics he was to discuss the law of free fall. Meyer judged his work as showing "great clarity and soundness" in mathematics; all in all, he declared, "The work is excellent." (He later reportedly told Ferdinand that Hermann was one of the best pupils he had ever had.) Helmholtz himself judged his Latin essay to be the poorest of his language essays. For the Hebrew examination, he had to explain, in Latin, selected Hebrew terms from Deuteronomy. For the Greek, he had to translate a passage from Euripides, and his translation was judged as more than satisfactory. As for French, his teacher judged his translations to be virtually free of grammatical errors, idiomatically *juste*, and generally well done.[26]

By far the most instructive of all these examinations was his German essay, for it gives the first and strongest indication of his general outlook on life. His topic was Gotthold Ephraim Lessing's *Die Idee und Kunst in Lessings Nathan, der Weise*. He "admired" Lessing and thought him "excellent both in clarity and clear-sighted knowledge as in the lively and artful presentation of the poetry." He also generally admired the way "the characters are masterly true and deep" and thought that in *Nathan the Wise*, Lessing had outdone himself. Helmholtz portrayed the Jew who stood at the heart of Lessing's play as "the noblest, wisest of his people, who . . . recognized the great truth that all men are brothers, that, in spite of differences of religion, there exists in all of them, without any distinction, the love of a Jew who, moreover, through his spiritual ideal wins the Christian's and the Muslim's heart and who unites them together into a loving family"; that, he said, was "the heart of the poem." He disapproved of those characters who displayed religious prejudice. The play's principal idea, he wrote, was this:

> Among the peoples of all beliefs there can be good men who believe they can find the single way to Heaven, and who do so not through pursuing positive prescriptions of religion, but rather freed from these prejudices, encompassing all men in the same bond of love and thus reaching true virtue. Lessing thus develops in Nathan the deepest, most Godly lesson of our holy religion, which unfortunately has too often been forgotten. His Nathan is no longer a Jew, his Saladin no longer a Muslim. Instead, according to their inner convictions, they are Christians, true Christians. The poet has thus been very wrongly reproached with seeking to say that all positive religions are to be rejected and, in their place, to substitute reason alone.

On Helmholtz's reading, his own religion remained holy; Nathan and Saladin had evolved from being a Jew and a Muslim, respectively, into Christians—and Lessing's deism dissolved altogether. Above all, he admired Nathan's "clear understanding" and "deep, inward feeling." These were his "virtues." Ferdinand strongly criticized the essay as showing "more training in the ability to apprehend than to reflect." He found its conceptual analysis of the play's main themes weak, though sensitive in its understanding of literary character. However, in the end he judged it to be very satisfactory.[27] Certainly the essay suggests Helmholtz's sense of tolerance and his intellectual roots in the Enlightenment.

In September the Royal Examination Commission also assessed Helmholtz. It found that he showed "the highest becoming and modest demeanor" and noted "his outwardly quiet and peaceful nature," which was "united with great intellectual versatility." It emphasized his clear reasoning and "deep kindliness." "His morals display a true, rare purity, and a truly childlike purity." In general, the commission found him to be intellectually mature: its members harbored great hopes for his future and noted his ambitions to develop his talents. They said he was hardworking and orderly and had attended school regularly and completed his written work.[28]

Those were the generalities. As to specifics, the commission found that he could read and translate Latin authors without difficulty and that his Greek was distinguished "through thoroughness and considerable range." It praised his great efforts in Hebrew and judged his German as quite good. His reading ability in French displayed "praiseworthy accomplishment." He could also easily read English, not only the modern authors but Shakespeare and other poets. The same was true for Italian authors. Furthermore, the commission judged his knowledge of history and geography as good and said he was well grounded in Christian dogma and ethics. He knew the elements of logic and rhetoric and had studied some psychology. His knowledge of mathematics was found to exceed that offered by the Gymnasium, and his confident self-study in this field was noted. The commission determined, too, that he had a secure knowledge of physics. As for his drawing ability, it was no more than elementary. When it came to singing, he exerted much effort and had some accomplishments. By way of comparison, the intellectual level and difficulty of Helmholtz's Gymnasium coursework in its final year and the accompanying examinations were higher and more rigorous than Charles Darwin's examinations at Cambridge University circa 1830. In the autumn of 1838, eight and a half years after entering the Gymnasium, the seventeen-year-old Helmholtz graduated (early) and had his leaving-certificate in hand, which in principle allowed him to attend a German university.[29] His success in all his subjects went far toward ensuring his future academic success.

Medical School as Ersatz

From his parents and the Gymnasium, Helmholtz had acquired a culture that advocated and exemplified the values of work, discipline, and diligence; helped him find purpose in life; and helped give him the intellectual tools and moral character essential to achieving his scientific and other goals. Yet Ferdinand's limited means left him unable to finance a university education, especially since he believed Hermann had little if any chance of gaining employment as a physicist. Hence Ferdinand's suggestion that Hermann study medicine instead, in particular at the Friedrich-Wilhelms-Institut in Berlin, the Prussian military medical school, where he could, in effect, study largely "for free." Helmholtz later explained: "My father, a Gymnasium teacher who lived in very modest circumstances but a man who had kept himself intellectually alive through the highflying, scientific inspiration of Fichte's philosophy and the Wars of Liberation, explained to me that, sorry as he was to say it, physics was not a science that one could earn a living by—in fact, it then was not—but that if I wanted to study medicine I would also be able to pursue the natural sciences." He repeated this explanation on several occasions. For financial reasons, he chose medicine as a career; he had no particular intellectual or moral interest in it. The study of medicine would at least allow him to study the life sciences and eventually earn a living. He readily accepted this situation.[30] Helmholtz's acquiescence in his father's judgment suggests his strong sense of obedience. He never openly expressed resentment toward his father in this matter. Nor did he ever second-guess his own decision; that decision is of a piece with his subsequent decisions in choosing scientific problems: he opted for the concrete and the doable, choosing problems that could be solved. His decision was that of the dutiful son, the impoverished realist, and the self-confident student. This discouragement from his father to pursue the career of becoming a professional physicist may perhaps (and ironically) have fueled Helmholtz's drive to become one; at the least, it may help explain why he continued to conduct research in physics and to publish his results until the very end of his days.

Already in 1835, Ferdinand had attempted to secure a place for Helmholtz at the Friedrich-Wilhelms-Institut. It declined to admit him because it had too many applicants that year and because Helmholtz still had three years remaining at the Gymnasium. But it encouraged him to reapply early in his final year. In late March 1837, Hermann went to Berlin for several days to take the institute's entrance examinations. He was received in a "very friendly" manner by Friedrich August Schulz, the subdirector, who knew Ferdinand personally. Schulz had served at the Charité Hospital when the late Christian Ludwig Mursinna, Hermann's great-uncle by marriage and one of his godparents, was a leading figure at these institutions and a former surgeon-general of the Prussian army.

Thus, when Helmholtz arrived for the examinations, he had strong family connections to the institute. Schulz admonished him to emulate his great-uncle and not to worry about the academic examination, which, he said, was largely an aptitude test. As for the accompanying physical examination, it would have shown that Helmholtz stood about five feet six inches (1.66 meters) tall and was of strong build.[31]

Even before he left Berlin, he was admitted to the institute. Along with his excellent academic record, upright behavior, and satisfactory performance on the institute's academic entrance examination, Helmholtz also owed his admittance to his connections to Mursinna, a fact he never sought to hide. His relationship with Mursinna, "the only influential man" in his extended family, he said, "distinguished" him from other applicants to the military teaching institute, the Friedrich-Wilhelms-Institut, "which very largely made it possible for impoverished students to study medicine." In the Prussian military, as elsewhere, pedigree counted. Before he left Berlin, Helmholtz managed to see Ludwig Rellstab's tragedy *Die Venezianer* and intended to see Heimbert Hinze's comedy *Oben und Unten*.[32] Like his father, he had already become a *Kunstfreund*.

Becoming a Medical Doctor

The Institute and Berlin

In the autumn of 1838, Helmholtz enrolled at the Prussian army's medical school, the Friedrich-Wilhelms-Institut, also known as the Pépinière (fig. 3.1). Located near Berlin's center, it was closely associated with both the nearby University of Berlin's Medical Faculty and the Charité Hospital. The institute did not charge for room or instruction, but its students, who came from families of insufficient means, had to agree to serve as army medical doctors and surgeons after their training. For each year at the institute, a student was obliged to serve two in the Prussian army. Thus, after completing a year-long internship at the Charité and taking the state medical examinations, a student would complete eight years of military service.[1]

Each student received a small allowance from the government for meals, general needs, and the future purchase of uniforms and surgical instruments. Helmholtz found the meals at the medical school to be not as bad as many thought, but less nutritious than home-cooked meals. The institute required that parents guarantee a small allowance, and Ferdinand gave Hermann eight, later nine, talers per month. This allowed him a few small pleasures, including attending the opera once

Fig. 3.1 Courtyard of the Friedrich-Wilhelms-Institut on Friedrichstrasse in Berlin.
Bildarchiv, Institut für Geschichte der Medizin, Freie Universität Berlin.

or twice a month. He led a simple, modest life. But however constrained the family budget was, he could afford to have a piano and could pay someone to polish his boots.[2]

The institute sought to instill military discipline. During the summer semester, students arose at 5:00 a.m. (in the winter semester, at 6:00 a.m.), had to be in the institute building by 10:00 p.m. (in the winter semester, by 9:00 p.m.), and went to bed shortly thereafter. Entrance and exit, leaves, and vacations were all strictly controlled. Instruction began at 6:00 a.m. in summer and 7:00 a.m. in winter and, except for a one-hour lunch break, continued until 8:00 p.m. Students and staff were also expected to attend a weekly, common lecture every Saturday evening from 6:00 to 8:00. Rudolf Virchow, a future pathological anatomist, anthropologist, and liberal politician, who was Helmholtz's junior at the institute by one year, had "scarcely an hour for relaxation." He found that all the classroom hours plus private study were "nearly too much." Often enough, some students failed to follow regulations. Virchow said the majority of students in his section skipped classes, played cards, drank beer, and so on. Like Virchow, Helmholtz disapproved of such behavior; he was a very serious student, and he complained that some of his fellow students disturbed him. The institute aimed not only to teach medicine but also to inculcate duty, obedience, and respect for law and order.[3] It thus continued what Helmholtz's parents, elementary school, and Gymnasium had begun, further honing his work habits and giving him a sense of belonging. To a certain degree, his values and interests became those of the Prussian military.

Helmholtz moved into the institute dormitory in late October 1838. He described it in full detail to his father, including the location of his piano, and he spoke of his roommate's "truly amazing dexterity in piano playing," though he disliked having his roommate play "pieces with a lot of color along with modern Italian music." He cared still less for the crude noises made by some fellow students.[4] His piano playing was the expression of his spiritual self, and in this he always favored the classical, not to say the subdued.

Helmholtz's first weekend at the institute marked the start of his regular visits to relatives and friends in Berlin—then a city of some three hundred thousand—along with his exploration of its cultural offerings. He visited his aunt, Julie von Bernuth, the daughter of Mursinna and wife of Louis von Bernuth, a senior figure in government finance. They lived near the Tiergarten, the city's most elegant public park, in one of its most elegant residential quarters. His aunt fed him so well that afterward he "could hardly climb the stairs" to his room. She also gave him unsolicited lessons in table manners: "Each time that I get up from the table she lists everything that I did wrong, but she does say that I've improved somewhat." Ferdinand approved of Julie's instructions in etiquette. Helmholtz also visited the homes of Friedrich Gottlob Hufeland and Emil Osann, where he was "received in a very friendly manner." Both men were professors at the institute and the university. Though he received a standing invitation to their homes for Sunday evenings, he did not go to the Osanns that first Sunday because he wanted to see a performance of *Don Juan*. He also visited the family of Johannes Wilhelm Rabe, a portraitist and drawing instructor, and Justus Friedrich Karl Hecker, who was a leading medical historian, the university's first professor of medical history, and dean of the Medical Faculty (1839–40). The Helmholtz name gave him entrée into Berlin's *Bildungsbürgertum*. That same first weekend, he also attended an art exhibition, where he saw "some new paintings"; but, he continued, "There isn't much there. The only thing that I liked is a Jephta." Despite this busy first weekend, he was homesick: "I think about everyone at home a lot. Don't forget your loving Hermann." They did not: within days his family wrote to express their love and support for him.[5]

When classes began the following week, he was glad in part because they brought relief from having too many visitors: "These guests are usually burdensome; especially when I'm playing [the piano;] they often demand that I play dances and the like for them." He largely avoided the other students and therefore said, "I've gotten the reputation of being unsocial." He was impatient with vulgar, especially nonintellectual, students from the start. The next semester he got a new roommate, whose intellectual, cultural, and artistic efforts impressed him. This was the sort of person he liked and respected. He told his parents not to worry that he would "forget about playing music," since, he wrote, "the more modern music that my companion loves so much doesn't satisfy me. To hear something deeper I have to play myself. Also, it's rare that someone else's [musical] expression satisfies me. I find music far more satisfying when I play myself." He played at Aunt Julie's, too, where he was invited for lunch on the weekend. Yet he was not at home with her, either; he preferred the more cultured Osanns, where he spent much time discussing "literature, study, and all sorts of things." He was especially impressed with Mrs. Osann, who was "excellently experienced in all matters of *Bildung*." He liked talking about art. His mother, however, tenderly complained that he and other men were "silent, uncommunicative people!" She wanted to know all about his feelings as well as his studies, and added: "May God inspire you to do the right thing and to avoid the wrong."[6]

A Medical Education and Beyond

The medical curricula at the institute and at the university were largely the same. Indeed, institute students normally sat with university students in the same lecture rooms, and clinical training for both groups occurred at the Charité. The social distinctions were not great, either. The institute was a very high-quality medical school; its students received training that at least equaled that of university medical schools. It stressed clinical, practical medicine as well as instruction in the basic sciences. The main difference between the two types of institutions (in terms of medical education) was that the university allowed its students far more flexibility—consonant with the German university system's general educational principle of *Lernfreiheit* (freedom to learn)—while the institute prescribed a rigid, set curriculum that its students had to follow. The institute was the very antithesis of the Humboldtian vision of *Lernfreiheit* and *Lehrfreiheit* (freedom to teach) that led to *Bildung*. Moreover, the university was far larger: it enrolled some 1,600 students, of whom there were about 350 in the Medical and 320 in the Philosophical (arts and sciences) Faculty; by contrast, the institute had only around 90 students.[7] Karl Marx, Ivan Turgenev, and Michael Bakunin, for example, all studied at the University of Berlin circa 1840; Helmholtz apparently never met any of them or their like. See figure 3.2.

Like all other medical students, Helmholtz spent his first two years taking courses in the basic sciences and, to a much lesser extent, the humanities. Included were logic and psychology; physics and meteorology (with Karl Daniel Turte and Heinrich Wilhelm Dove); chemistry (with Eilhard Mitscherlich); botany and natural history; encyclopedia medica and the history of medicine; osteology, syndesmology, and splanchnology; anatomy and physiology (with Johannes Müller); and embryology.[8] Mitscherlich and Müller were among Europe's leaders in their respective disciplines.

Fig. 3.2 The University of Berlin, on Unter den Linden, 1840.
Bildarchiv Preussischer Kulturbesitz.

Helmholtz had mixed reactions to his courses and teachers during his first year. On the whole, he reported home, "Our courses are moving merrily along," but he found them to be very time-consuming. He complained about the tedium of "often sitting there in the evening learning muscle upon muscle," saying, "That makes the head reek." He never warmed to anatomy, since he thought it difficult to maintain order "in this deluge of facts." He found Mitscherlich's chemistry courses "very interesting" but also "just a little bit boring" as well as "stinking full." He never became attracted to chemistry, either. He thought that Heinrich Friedrich Link, the botanist and natural historian, suffered from "a superabundance of intellect"; six weeks into natural history, Link was "still at the philosophical introduction (Oh God!)." He found Müller's course on physiology, by contrast, "excellent." Dove, an experimental physicist and meteorologist, impressed him very favorably; he found his teaching insightful. All told, Helmholtz spent no less than forty-two hours a week in class: "That's military order!," he exclaimed.[9]

Despite this heavy, demanding course load, he found time for his aesthetic, philosophical, and social interests. Although he needed no reminder from his parents to keep up with his piano playing, he got one. He played about an hour a day during the week and still more on the weekend. He favored Mozart and Beethoven sonatas. He was also a keen admirer of Gluck. He attended the theater and saw performances of *Hamlet* ("performed terribly badly"); Carl Maria von Weber's popular romantic comedy *Euryanthe* ("excellent"); and *Faust*. He thought the performance of Goethe's masterpiece "made . . . a powerful impression on everyone, in part due to the godly poetry itself and in part due to the excellent presentation of Mephistopheles ([Karl] Seydelmann) and Gretchen (Clara Stich)." They were presentations like he'd never seen before, "the former just as satanic and funny as the latter was fragile and simple." He had been studying the second part of *Faust* "since some time" and had concluded, "The thing is colossally wild." He asked Ferdinand to obtain Johannes Falk's study of Goethe from the Gymnasium's library and have it ready for him when he returned home shortly. In the meantime, he and a friend were writing a play together.[10]

Helmholtz read broadly, including works by Homer, Kant, Goethe, Byron, and the French physicist Jean-Baptiste Biot. He conceded that he had lately lost the feel for some of them, especially Kant, and that he had had to (and did) work his way into them again: "Once that's happened," he told his father, "they're even more captivating; in particular, I can barely tear myself away from Homer: I consume two or three songs in an evening, one right after another." For a change of pace, he pursued integral calculus.[11] His intellectual ambition and energy were enormous.

He also continued to receive social invitations to the homes of various Berliners. At Privy Counselor Langner's, for example, he met several law students and played whist. "It was a terrific party, but also nonsense," he reported. Aunt Julie, at a visit one cold day, gave him a pair of gloves, "which were very useful . . . given the current, rather delicate weather." He explained, "Every morning we have an hour long anatomical lesson in an unheated room and, by the end of the lesson, it's rather cold without a topcoat." He even wore the gloves in his dorm room, where it was so cold that he could neither write nor play the piano. Although he continued to visit Aunt Julie, he also sought to keep his distance, since there were tensions between her and the Helmholtzes. He saw other family friends and relatives as well: the Rabes; August Spilleke, the director of Berlin's Friedrich Wilhelms–Gymnasium and the Realschule; and the Hamanns. From time to time he would return home for part of a weekend, either by taking the train or by walking, which took him about five hours.[12]

He found time for physical fitness, too: he rowed, swam, and fenced. He was "a very able swimmer." During his second semester, he was asked to assist the institute's librarian. Though he "lost"—the word is his—two hours per week doing so, it was also "the only way of learning what's in the library that's good amongst the endless number of old, trashy books." Snooping around in the collection, he discovered "the works of Daniel Bernoulli, [Jean le Rond] d'Alembert, and other mathematicians of the previous century." His readings in these works raised some fundamental questions in physics for him. "I thus came upon the question: 'What relations must hold between the different types of natural forces when, in general, no perpetuum mobile can possibly exist'; and the additional question: 'In fact, do all these relations indeed hold?'"[13] Within a decade, he provided answers to these questions in his essay on the conservation of force.

As his first year of studies drew to a close, Helmholtz became ill: for several weeks he suffered seriously from diarrhea and related bleeding problems. While visiting the Bernuths, he drank too much wine — "beyond my usual amount of wine, two extra glasses that evening" — and his condition worsened. In Mitscherlich's chemistry class the next day, he felt faint and had "terrible headaches." He tried self-doctoring, but it failed; so he went to the institute doctor, who told him to stay in his room for two days. He felt "very distressed [*sehr molestirt*], was very weak, and looked completely green and yellow." He was exhausted and yearned for a vacation. This he took, spending about two weeks with the family of an uncle, August Helmholtz. While on vacation, he read Schiller, the great German poet of and thinker about freedom, for the first time in several years, as well as essays and biographies by Ludwig Rellstab, a music and theater critic. He went to the theater, and he played the piano (Mozart, Strauss, Lanner, Czerny, Hünten, Auber, Ross, Bellini, and others, though he always returned to Mozart and Cramer, "so as to strengthen somewhat" his "spiritual stomach"). He danced with young ladies and played the piano at an impromptu party. He even did a bit of sailing. While at Uncle August's, he also performed in a comedy and joined his uncle on a business trip to Stettin and Swinemünde. He was impressed with the northern German landscapes and rivers, especially around the Oder and the Swinemünde, in which he sailed and swam: "The sea especially enchanted me through its ever-changing colors that originate from the light pressing through the different layers of clouds. I was completely enraptured one evening when I went to the end of one of two harbor entrances that had enormous stone piers extending well into the sea. From there I looked at the surf as it rose so high, yet one could stand on the piers and keep one's feet dry. To be sure, the breaking waves weren't strong enough for the swimmers and other guests; but on me these waves made a powerful impression."[14] Bodies of water often left him spellbound. A pattern of overwork and illness late in the academic year, followed by rest and relaxation in the summer, recurred regularly in his life.

Yet he could relax only so much. He had brought along a zoology textbook to prepare for his upcoming qualifying examinations. This was the *tentamen philosophicum*, the first of three sets of examinations — the other two being the faculty and the final state medical examinations — that all medical students had to pass in order to continue their studies.[15]

When he took the *tentamen* (10 December 1839), he had more on his mind than academics, however. His mother was sick — he was particularly concerned about her "anxiety and strain" — and his siblings were suffering from scarlet fever. The siblings soon improved, but his mother recovered more slowly. He reported home that he had passed the *tentamen*, receiving a "rather good" in mineralogy; a "very good" each in logic and psychology, physics, zoology, and botany; and an "outstandingly good" in chemistry. His overall grade was a "good," which was the highest possible. He bragged, "My certificate [of grades] was the best of the four of us."[16] His family was no doubt proud of him and cheered up by this good news.

After the *tentamen*, Helmholtz rarely mentioned course work again, an omission that suggests a growing sense of personal or professional confidence and independence, or perhaps simply less interest in course work. Nearly all of that work now involved medicine, though principally in its theoretical, nonclinical form. Some of the lecture courses were no more than dictations: the professor read and the students recorded; others included some experimental demonstrations (only occasionally was there a microscopic demonstration). There was no individual student work in physiological or physical laboratories, though there were anatomy exercises.[17]

If the curriculum was rigid, it was not backward. During the 1830s and 1840s, German medical professors, especially those in Berlin, gradually reformed medical instruction. Helmholtz later characterized these decades as "a time of ferment and struggle between the learned tradition and the new natural scientific spirit, which believed in no tradition and wanted instead to be based on its own experience." If his argument was quite a bit rhetorical and quite Whiggish, it was also not so far off the mark: movements and theories such as *Naturphilosophie*, Brownianism, Cullenism, Albrecht von Haller's irritability theory, and vitalism had held center stage at various times in the late eighteenth or early nineteenth centuries, but during the period 1830 to 1850, they gradually disappeared. The great shortcoming of these earlier medical movements, Helmholtz thought, was their "one-sided, false overestimation of deductive methods"; romantic medicine, as he and others called it, was "only the leftover ruins of the old dogmatism." By the time he entered medical school, not only had romantic medicine (especially *Naturphilosophie*) long been under attack; so, too, was its successor, an (exaggerated) empirical medicine. His student years thus came at the start of the eventual triumph of theory-oriented medicine over narrow empirical medicine.[18]

He probably had little interest in the medical and clinically oriented courses. He now began writing home mainly about his personal and social life. At the end of his second year of study, he took a summer holiday trip to Silesia by coach with several friends. It was hot and crowded on board, and he did not find all of his fellow passengers especially congenial: "Two Jewesses from Breslau and a Jewish Secundaner sat across from us. He looked and behaved like a Sextander, however: he could not sit still for a moment and, while one of the Jewesses quarreled with him, he chattered away loudly during the entire trip. He thrust his arm around hers." By contrast, Helmholtz found the Silesian countryside, the local food, and the people most pleasant. He was especially taken with the Silesian mountains and the Hirschberger valley, which "everywhere give off a romantic, alluring, vivacious image."[19] The long walks with his traveling companions did him much good, physically and spiritually: he relaxed and became more open to the world around him.

After spending the Christmas holiday of 1840 with his family, he returned to find "the mood . . . in Berlin . . . noticeably worse." The Prussian monarch, Frederick William III, had died in June 1840, and Berliners, like others, had expected that his successor, Frederick William IV, would be a liberal-minded monarch who welcomed change. Instead, they got a romantic, nationalistic, and highly conservative ruler who sought a Christian-German state as an alternative to the revolutionary forces that had swept over Europe after 1789. They got "the romantic on the throne," as he became known, and many were bitterly disappointed. Helmholtz reported the "scandal" over the king's request to have a revised version of Racine's religious drama *Athalie* (1691) performed, which Frederick hoped would help instill piety. The king was enchanted with the new version, but the audience hissed through the opening performance, and most if not all audiences considered the play pietistic propaganda. The result was greater mistrust of Frederick. Helmholtz noted that a parody, one modeled on Goethe's refrain (from his "*Nachtgesang*": "Sleep, what more do you want?") was making the rounds at the king's expense: "He prays, what more do you want?" Helmholtz shared the general scorn for the new king. He also used some free time to attend lectures by Eduard Devrient, a Berlin singer, librettist, and theater historian.[20]

Late in his third academic year, Helmholtz contacted Immanuel Herrmann Fichte, who had become professor of philosophy at the University of Bonn. The two had never met (or even corresponded). It was Ferdinand who initiated the contact. He told Fichte that his own temperament had changed little over the years. He was his old, inharmonious self and had grown "very old." He proposed to Fichte that he either send Hermann to see him in Bonn, "so that, at least for a few weeks, he gets to know his godfather and makes his seed his own," or that Fichte come to Potsdam. He recalled "the joy with which you once greeted my first-born and . . . at his baptism promised your fidelity." Hermann wrote that, notwithstanding his infant baptismal screamings, he had never before addressed Fichte, but he felt nonetheless as if he knew him: "Many times Father has told me a great deal of his dearest friend and of his kindness; and for a long time I've imagined the Rhein to be the most beautiful region of Germany. So I've always longed to meet both together; all the more so, as I've recently come to understand some of the works of your great father, about which I nonetheless still hope to receive some instruction." He expected to meet Fichte either on the Rhine or on the Havel and Spree. Nearly two months later, he was still hoping

Becoming a Medical Doctor 47

to visit Fichte in Bonn—as Fichte had invited him to do—and "also to hear the lectures of some of the famous names there [i.e., at the University of Bonn]."[21]

Typhoid and a Microscope

Nothing came of the proposed meeting, however, because late in July Helmholtz contracted typhoid fever. It started with a cough and hemorrhoids; he reported sick to the institute authorities and was given medication and told to stay in his room and rest. He feared that his mother would be very worried about his health: "*On my word of honor* you can assure Mother," he told Ferdinand, "that it is absolutely nothing other than what was reported and that she should not be in the least concerned." They were probably quite concerned, not least since he confessed that he was in a "dangerous condition." His cough got better, but he contracted a serious gastric fever and became further exhausted through "a lot of sweating and, especially, nosebleeds." He again stressed that his parents had no reason to worry. Yet in fact they did. An institute physician wrote Ferdinand that while Hermann's fever had gone down, his appetite had improved, and a bloodletting had helped, he could get no rest at the institute and so had to be moved to the Charité. Another reported that Hermann had a "gastral-catarrhic fever with congestion in the chest and head." He had let his blood that very morning and "had the best success"; the prescribed medicine had helped, too. He thought the Helmholtzes should not be unnecessarily alarmed. Yet a week later Hermann was still in the Charité, still had a fever, and still slept poorly, though he was slowly regaining his appetite. He continued to suffer from nose-bleeds, which often occurred even when he was healthy. His doctor thought Helmholtz's illness would soon reach a turning point, and so he asked Ferdinand and Lina to delay their planned visit to see him. Helmholtz concurred. Ferdinand wrote Fichte that Hermann was in "risk of life," too sick even to return to Potsdam. Ferdinand would be happy if he could bring him home in three weeks and if Hermann could begin the next semester on time. His "poor young boy," he wrote, lay "in a heavy fever instead of having a new day of life on the beautiful Rhein." He continued: "We poor parents beseech nothing more from God than that He may want only to leave him with us and not take him away, he to whom we've given all our love and understanding, and who seemed to turn out so magnificently. My poor wife had scarcely recovered from her earlier loss [i.e., the death of their four-year-old son, Johannes Heinrich Helmholtz, earlier that month], and now she is again brought back [to her worries]. We both fear the worst." The fever did not fully pass until mid-September. Helmholtz remained in the hospital and could do no more than take short walks.[22]

The illness thus lasted two months and cost Helmholtz his vacation and the meeting with Fichte. Yet it proved of some moment in his career. Since institute students received free medical care, he was freed of his normal expenses, and he displayed his usual frugality: he saved enough money to buy a microscope that he intended to use in his upcoming dissertation research. (Not until the 1830s and 1840s did leading biological researchers and institutions begin purchasing their own microscopes.) His purchase indicates the seriousness of his commitment to science. The instrument wasn't pretty, he thought; "Yet with it I was in a position to recognize the nerve extensions of the ganglia [i.e., nerve] cells in invertebrate animals described in my dissertation and [later, in 1843] to study the vibrions [*Vibrionen*] in my work on putrefaction and fermentation." Ferdinand thought he had wasted his money and reprimanded him.[23]

A Dissertation with Johannes Müller

Helmholtz spent most of his final academic year following the standard set of clinical courses. As the year drew to a close, he confronted three important sets of events. The first was the faculty examinations, which covered, in Latin, eight areas of instruction: anatomy and physiology, nosology, pharmacology, practical medicine, surgery, obstetrics, state drug regulations and forensics, and the literature and history of medicine. He readily passed on 25 June 1842, and so was admitted into clinical work. Second, on 2 August the institute celebrated its annual Founding Day, to which the prince of Prussia, other royal and aristocratic figures, and various institute friends were invited. The institute's director used the occasion to recall its purpose, relate its accomplishments, and reward its students. Each year he chose one student and one professor to deliver a lecture on some scientific topic; in 1842 he chose Helmholtz. Helmholtz's topic was "operating on tumors in blood vessels." He thought his address was rather well received, and as a reward and memento he was given several books, which he long cherished. Yet in fact he had never seen such an operation; his knowledge came "only from books." Indeed, his entire medical education had been based essentially on book learning and only occasionally on lecture demonstrations, certainly not on laboratory instruction. Throughout his career he looked down on mere book learning in the sciences; it was contact with experience, he believed, that brought more trustworthy knowledge. Medicine during his years of study, he later claimed, lay in such a theory-dependent state that it often led to the dismissal of facts. However, he valued his medical education insofar as it was united with the natural sciences. Indeed, he thought medicine had provided him with "the everlasting foundations of all scientific work." It gave him "a far broader knowledge of all the natural sciences than is usually given to students of mathematics and physics."[24] It also allowed him to study natural science while still earning a living.

The third set of events concerned his dissertation. He wrote his father on

Fig. 3.3 Johannes Müller. Bildarchiv, Institut für Geschichte der Medizin, Freie Universität Berlin.

1 June 1842 that he was hard at work on it and had "already found a very important result" but that he wanted "to examine the matter still more closely." So he worked on it through July and then went to see Müller, who, though a professor at the University of Berlin, was his adviser. Müller received him warmly.[25]

It was one of the great pieces of luck in Helmholtz's career that Müller became his teacher, mentor, and patron. Müller was a leader in the fields of anatomy, physiology, zoology, embryology, and pathology at midcentury, during an era in which the Germans decidedly led other national efforts in physiological research (see fig. 3.3). (Humboldt called Müller "the greatest anatomist of our age.") He was noted for his extraordinarily exact and prolific research, and he emphasized to students the importance of using physics and chemistry (and their associated instrumentation) for studying biological phenomena. His arrival at the University of Berlin in 1833 helped shift medicine there from a strict adherence to empirical phenomena toward a greater use of theory.[26] The additional acquisitions of the gifted and innovative clinician Johann Lucas Schönlein in 1840 and the equally gifted surgeon Johann Friedrich Dieffenbach further strengthened Berlin's reputation. Medical education in Berlin, especially under Müller's and Schönlein's leadership, became more scientific, and in this regard (at least) Berlin was the leader in the German lands if not in most of Europe.

Müller was especially known for his epochal work on the physiology of vision, in particular his law of specific nerve (or sense) energies. His *Handbuch der Physiologie des Menschen* (1833–40) was widely used by students and professionals alike, and Helmholtz doubtless studied it carefully. He attended four of Müller's lecture courses: general anatomy, comparative anatomy, pathological anatomy, and physiology. His notebooks from Müller's courses on comparative and pathological anatomy suggest that the lectures were clear and orderly. The courses attracted large numbers of students. Most of the 150 to 200 (or more) students who normally heard him lecture on human anatomy and attended his dissection section each semester had no contact with him, however. He remained distant from nearly all of them. He spent scarcely a half hour overseeing the section; most students received no instruction, not even an introduction, to dissection. His lectures on physiology, for their part, had virtually no demonstrations, though those on comparative anatomy did.[27]

Yet Müller inspired many students, and he spent just enough time among the masses of them to spot potential talent. He invited a small number of the most able to work in his two-room, sparsely equipped quarters—which in the mid-1830s contained only one microscope—in the Theatrum anatomicum (see fig. 3.4). In addition to Helmholtz, his best-known students included Theodor Schwann, Jacob Henle, Robert Remak, Emil du Bois–Reymond, Ernst Brücke, Rudolf Virchow, and Ernst Haeckel. All of them became major figures in science and medicine. Some studied with Müller before Helmholtz's day, some after, and it is uncertain just when Helmholtz first met several of them. (For example, he was completing his dissertation just as du Bois–Reymond, who was then working in Müller's laboratory, was having his first successes in electrophysiology; still, the two had yet to meet.) Müller drew these and other promising students into his circle. His undogmatic nature especially attracted them. He treated theories as mere hypotheses and, according to Helmholtz, let the facts alone decide which hypotheses were correct. To be sure, Helmholtz found unacceptable Müller's vitalistic viewpoint and his associated belief in a *Lebenskraft* ("life force") that supposedly served as an organizing principle of the physical body and its functions and dissolved upon death, as well as his belief in a conscious soul. Yet even in these beliefs he considered Müller undogmatic and open to factual analysis. As a leader of the new, the experimental direction in physiology, Müller advocated the use of chemical and physical methods in anatomical and physiological investigations, and this Helmholtz found seductive. He especially liked Müller's idea of the law of specific nerve energies, later deeming it a scientific accomplishment whose value he was "inclined to equate with the discovery of the law of gravitation."[28] The overestimation is telling: for all his talent, Müller was no Newton.

In late July, Helmholtz went to see Müller to present his preliminary research results on the origins of the nerve fibers in several higher animals and to discuss the status of his dissertation. His topic involved an aspect of cell differentiation, and it seems likely that, in light of the recently proposed cell theory of

Fig. 3.4 The Theatrum Anatomicum, University of Berlin, 1841. C. E. Geppert, *Chronik von Berlin von Entstehung der Stadt an bis heute*, 3 vols. (Berlin: Ferdinand Rubach, 1839–41), vol. 3, unpaginated foldout.

Schwann, Remak's dissertation work on cell-fiber relations, and the new emphasis on using the improved achromatic microscope for neuroanatomical studies, Müller may have suggested to Helmholtz that he work on the origins of nerve fibers. It is also possible that Helmholtz came across this topic independently in Müller's *Handbuch*. In any case, Müller assessed Helmholtz's results and evidence and declared that his work was "of great interest." He noted that others before him had only surmised what Helmholtz had proven, but he also recommended that Helmholtz extend his observations beyond the three or four animals he had studied to date—"in order to provide a rigorous proof"—and he offered the use of his own instruments in the Anatomical Museum (Anatomisches Museum). He suggested that, if Helmholtz was not in too great a hurry to complete his promotion to doctor of medicine, he use the upcoming vacation period to do the additional research, "in order to bring a fully developed child into the world, something that would have to fear no attacks whatsoever," as Helmholtz gently put it to his parents. This was a very positive yet also slightly discouraging meeting for him. Müller showed himself to be a helpful but demanding adviser; in truth, he said little that Helmholtz himself had not already thought. Helmholtz's only hesitancy concerned the consequent delay in his promotion, since that might disappoint his parents. He wrote home: "Should that cause the two of you too much pain, then write to me and I'll revise the talk that I held here at the institute at Whitsuntide, and a week later I'll be a doctor. The petty folks in Potsdam will perhaps think that I failed the exam; those in Berlin, that I want to abscond with the doctor's banquet. In time, both will come to understand. Actually, it was rather surprising to me as well, and not quite right, but, as I said, I have nothing reasonable to object against it."[29]

While he did not want to proceed definitively without his parents' permission, and while the delay pained him too, he knew what he needed to do to produce a first-rate piece of scientific work. Yet he could not and would not dismiss his parents' concerns. Two months earlier, in his curriculum vitae, he had written, "God Almighty has kept [his parents] alive and well." He remained the loving son, even as he tested his mettle at the highest scientific level. His decision to conduct further research was ultimately an act of personal self-denial and a vigorous assertion of his scientific self.[30] It was also one of the first and strongest indicators that, as regarded scientific research, he had fire in the belly.

Building on earlier neuroanatomical work, Helmholtz observed in his dissertation that in the ganglia of invertebrates, the axons originated in the cells. As such, his study was one element in an emerging neuroscientific context that showed that nerve fibers arose from nerve cells. He wrote and defended his dissertation results in Latin; by the time he had finished, he was already an intern at the Charité. Like nearly all candidates, he had had little chance of failing: dissertation defenses were usually announced and available only the day before the defense, and the "opponents" were chosen from among one's friends. On 2 November 1842, Helmholtz successfully defended his dissertation, received his diploma, and became a medical doctor.[31]

Yet he had also become something more: a physiologist. He had transformed the final routine requirement (the dissertation) for becoming a doctor into a noteworthy piece of science, and thereby he also revealed his scientific ambition. He owed something here to Müller, whom he revered ever after. He was "a preeminent man who gave us the enthusiasm to work in the right direction" and his "great teacher, the powerful Johannes Müller."[32] This was not mere pious rhetoric.

Undiscovered

Intern at the Charité

By age twenty-one, Helmholtz had become a medical doctor, a physiologist, and a man of *Bildung* who relished the arts. In early October 1842, he moved from his institute quarters into those of the Charité Hospital.[1] He spent the next seven years living in Berlin or Potsdam, laying the foundations for his scientific work, and falling in love.

Although the Charité was Prussia's oldest institution for medical education (founded in 1710), its facilities had recently been rebuilt, making it a modern teaching hospital (fig. 4.1). It provided free care for the destitute, and so diverse types of patients offered its doctors a wide variety of medical cases to examine and treat. It had clinics for internal medicine, medical surgery, clinical surgery, ophthalmology, obstetrics, pediatrics, syphilis, and psychiatry, and its interns received bedside instruction as they rotated from one clinic to another during the course of a year.[2]

A week into his first clinic, Helmholtz was seeing patients, prescribing medications, performing postmortems, keeping records, and the like. He found most patients to have incurable (and thus boring) illnesses, against which he could only prescribe opium. He complained

Fig. 4.1 The Charité Hospital, Berlin. Bildarchiv, Institut für Geschichte der Medizin, Freie Universität Berlin.

that what little free time he had—he worked from 7:00 a.m. to 8:00 p.m.—was so short and irregular that he could not undertake very much of consequence. Still, he found time for social and cultural life. He visited the Bernuths and his friend Theodor Rabe, a painter. He attended an art exhibit, where he saw "masses of the most horrible trash and very little good stuff," though he was impressed with the work of Carl Friedrich Lessing, a historical and landscape painter: He thought Lessing's *Johann Huss zu Konstanz*, an anti-Catholic painting, the best piece he had ever seen in Berlin, and he strongly urged his parents to see it. He attended a ball and a performance of Christoph Willibald Gluck's *Iphigenie in Tauris*.[3]

When his mother saw him that Christmas, she was shocked by his appearance—both parents suspected a bad love affair—and gently admonished him to take more care with his dress. He promised to do so and to be more social. In self-defense, he pleaded that his duties had so absorbed his time and attention that he saw virtually no one in Berlin and that this made him forget to send his dirty clothing home for his mother to wash. Relief finally came in February, when he rotated into the pediatrics clinic, where he had little to do. He socialized more—attending a ball at the Rabes', a symphony, and a quartet, and visiting friends—and he hoped that this activity would ease his parents' concerns.[4]

When his mother became gravely ill and his hospital duties prevented him from visiting her, he begged his father for news of her condition, gave medical advice, and worried about her. When he finally did see her, he became "extremely alarmed" and depressed. "Human help here is frankly at an end," he told his father. He was completely uncertain as to the outcome of her illness but feared the worst. Her condition forced him into a rare reference to a supernatural, higher power: "We have to leave the decision up to a higher will and know how to find thanks and resignation in his [God's?] decision." He longed to see her, and he thanked his siblings for the love they had shown her. She made a partial recovery, and on her birthday he wrote her, "All of Potsdam should become an Elysium for you."[5] He loved her deeply.

In the spring of 1843, Helmholtz and his family began lobbying for a position that would bring him back to Potsdam as a general doctor with the Royal Guard Hussars Regiment (Königliches Garde-Husaren-Regiment). He spoke with Johann Karl Jakob Lohmeyer, the army's surgeon general, who received him "in a very friendly manner." Lohmeyer anticipated that a position with the regiment would soon open up and said that, in principle, Helmholtz could have it, since he had heard only "the best reports" of his work. Lohmeyer also expected that Wilhelm Puhlmann, the regiment's long-serving physician and an old family friend, would want him. Puhlmann had clout, and Helmholtz asked his mother to speak with him. In July Lohmeyer, who had the power to fulfill the family's "wish," told him he need only say what he wanted and it was his. Family connections helped him in yet another way: he told his father that he was about to write up his experimental results on fermentation and that he needed to see Mitscherlich's latest article on the subject. He asked him to ask Meyer, his former Gymnasium teacher, for a copy; failing that, he wanted to come and see Meyer to discuss Mitscherlich's findings.[6]

In late September he completed his last clinic. While he was more than satisfied with the quality of instruction he had received, to judge by later remarks, the resistance of some medical men to the use of mechanical instruments for clinical purposes disturbed him. Most doctors only measured pulse with a watch that had a second hand, listened with a stethoscope, and tapped body parts. No one even considered taking a patient's temperature. He found these and similar unscientific attitudes deplorable. On 1 October 1843 he was promoted to assistant doctor and entered active military service as an army physician and surgeon, assisting Puhlmann in the Guard Hussars.[7]

"Postdoc" with the Hussars

The Hussars regiment had its rather new barracks in the Neue Königstrasse (later Berliner Strasse), and Helmholtz moved in there (fig. 4.2). The bugle sounded at 5:00 a.m. each day, forcing Helmholtz out of bed. Yet the Hussars, like the entire Prussian army during the peaceful years from 1815 to 1864 (except 1848–49), had little to do. There were no war-wounded and few ill soldiers for Helmholtz and others to treat, and so Helmholtz gained little in the way of

Fig. 4.2 The barracks of the Königliche Garde-Husaren in Potsdam. Hans Kania, *Potsdam Staats- und Bürgerbauten* (Berlin: Deutscher Kunstverlag, 1939), 115.

practical medical experience. Though the senior officers treated him in a "very disdainful" manner, his position left him considerable free time to pursue scientific research, and his barracks provided enough space to set up a small laboratory. In effect, he had garnered state support for his scientific work—a sort of "postdoc"—even if he earned (as he had as an intern) only 210 talers annually.[8]

He made the most of these favorable circumstances during his Potsdam years of army service (1843–48) by doing research and publishing a series of articles on physiological and physical topics. For the first two years, he worked alone in his military quarters and, apart from du Bois–Reymond, who visited him there, and Müller, he had little if any contact with Berlin scientists. His first publication reported his results on fermentation and putrefaction, work he had begun as an intern in Berlin. His topic belonged to an ongoing, broad set of arguments among chemists, physiologists, and others variously concerned with vitalism versus reductionism, the relations of organic and inorganic phenomena, and the origins of life. His outlook owed much to recent work by Schwann, Mitscherlich, and Justus von Liebig. Physiologists and chemists, he noted, disputed the causes of the decomposition of organic bodies (i.e., fermentation and putrefaction), and he chided many of the "greatest chemists" for ignoring the facts and speaking instead of "physiological fantasies." (Liebig thought decomposition fundamentally arose out of the motions of atoms [i.e., was chemical], while Schwann attributed it to microorganisms.) Helmholtz emphasized what the microscope and experiment could reveal, and his topic and approach suggest the influence of Mitscherlich and Müller (he published in Müller's *Archiv*). Yet his conclusions about decomposition were ambiguous: it was not solely a chemical phenomenon, he found, but one that could occur independently of living phenomena and yet provide the basis for life. If he had perhaps helped clarify the debate on decomposition, he had certainly not resolved it or the problem of spontaneous generation. (Liebig's chemical explanation reigned until Louis Pasteur overthrew it definitively in the mid-1860s with his microorganismic explanation.) But Helmholtz gained greater experimental skill from doing these investigations, and his stance left him firmly in opposition to Liebig, Germany's foremost chemist. The two remained cordial and respectful but never became close.[9]

In 1843–45, Helmholtz turned directly to the issue of the role of a "life force" (*Lebenskraft*) in physiology. Did organic life come from some force within itself, or did it emerge out of inorganic phenomena? The question here, as Helmholtz, following Liebig, put it, was partly "whether or not the mechanical force and heat produced in organisms are derived completely from metabolic changes." He sought to determine whether chemical reactions occurred when mechanical effects were produced, and he used frogs—"the old martyrs of science"—as his experimental victims: he stimulated them electrically to show that a chemical transformation occurred during muscle contraction. He thought his results were preliminary and that he had certainly not shown that metabolism caused muscular force and heat, let alone resolved the big issue of *Lebenskraft* in physiology. Yet his study of chemical changes in muscles became a seminal one: it marked the start of a forty-year study by others of the oxidation-reduction process in muscles. His study of metabolism in muscles (and associated analyses of physiological heat) also set the stage for his own theoretical work on the conservation of force. Finally, and in a rather paradoxical way, it showed that as he mastered the use of instruments, quantitative methods, and precise measurements in physiology, he simultaneously began to realize the limitations of the latter two.[10]

State Medical Examinations

Shortly before October 1845, Helmholtz completed a long essay review on physiological heat for an encyclopedic reference book on medicine published by Berlin's Medical Faculty (its venue again suggests Müller's involvement). He critically reviewed dozens of studies concerned with temperature measurements in animals, including the origins of animal heat; he found the methodological procedures in most of these studies to be wanting. He showed concern with the physical nature of heat itself and expressed his disbelief in a *Lebenskraft* (since that notion "contradicts all the logical laws of the natural sciences"), though he admitted that he had nothing to offer physiologists in lieu of it.[11]

Then in October 1845, his regiment granted him leave to prepare for and take his state medical examinations. Officially, he became a so-called attached surgeon to the institute in Berlin, living with a friend in "a very pleasant little apartment" and eating "very well and plentifully" in the Café Belvedere next to the Opera.[12] Unofficially, he became a researcher in the private laboratory of the physicist and chemist Gustav Magnus.

The examination process involved clinical as well as theoretical testing and lasted four to six months. It placed heavy demands on the examinee's time and financial resources and created the usual anxieties about passing. The examiners were members of a Berlin commission; most came from the university's Medical Faculty, and Helmholtz knew some of them quite well. The examination proper consisted of five parts: anatomy, surgical technique, clinical surgery (a week-long examination that concerned pathological, therapeutic, and operative aspects, with good performance in this last bringing the examinee the title of "Operating Surgeon"), clinical medicine (a fourteen- to twenty-one-day examination, in Latin, of two patients at the Charité), and an oral final examination over all fields of medicine.[13]

The process began with a preliminary examination that the candidate had to pass in order to be admitted to the examinations proper. The examinee was tested in the chief branches of medicine and had to provide evidence that he had attended a clinical-practical course. As part of the preliminary examination, Helmholtz also attended lectures. In late October he told his anxious parents that he had successfully completed two days of a challenging anatomical examination. To relax, he saw a "splendid performance" of Friedrich Schneider's *Das Weltgericht*, attended a party given by a former roommate, and spent an evening at the Hamanns'. The preliminaries would not be over for another ten days, after which he intended to go to a ball in Potsdam.[14]

The clinical examinations went well, too. For one of the tests, each examinee was assigned four Charité patients, whom he had to diagnose before the examiners. Examinees were usually given unambiguous cases, but one of Helmholtz's turned out to be of such a "rather tricky nature" that even the examiners were uncertain about it. He awaited their judgment and looked forward to seeing his family at Christmas. In late January 1846, he heard unofficially that he had passed the clinical examinations. He had crammed down a "flood" of information about prescribing dosages of medication, he told Ferdinand, explaining, "Our mnemonic technique has served me well." Still, he was not at all confident he had passed; it was by no means rare for a candidate to fail the state medical examinations. In fact, he passed with a grade of "very good" on the examinations for doctor and surgeon but did not receive the title of "Operating Surgeon," which suggests he may have lacked sufficient manual dexterity. Still, he was now a licensed medical doctor.[15]

A Network of Supporters

Heinrich Gustav Magnus stood at the center of Berlin science at midcentury. He was not only a chemist and physicist; he was also keenly interested in applied science and promoting the economy. At various points in his career, he held the titles professor of chemistry, of experimental physics, and of technology at the university. His predilection and ability lay in precision and experimental work. In 1843 he initiated the Physics Colloquium (Physikalisches Kolloquium) at the university, and he used his own money and organizational skills to create and sustain a private physics laboratory in his home (next to the university), where he invited promising younger scientists to use his excellent instrument collection for their research.[16]

It is scarcely surprising that Helmholtz and Magnus became acquainted. Magnus was Müller's neighbor, friend, and colleague, and Magnus's brother Eduard was a well-known painter and a friend of Puhlmann's. Moreover, Magnus probably knew Helmholtz's publication on fermentation, a topic that interested him greatly. For about three months, at Magnus's invitation, Helmholtz worked in his laboratory "almost daily." His fermentation experiments there turned out successfully, and he began some new work on this topic. So instead of returning to his regiment immediately, he remained in Berlin for two weeks to extend his experimental work in Magnus's laboratory and "to read up on some studies" for his additional investigations, for which there were "no books available in Potsdam."[17]

Helmholtz now also became acquainted with several young scientists around Magnus, including du Bois–Reymond, Brücke, Wilhelm von Beetz, Gustav Karsten, Carl Hermann Knoblauch, and Wilhelm Heintz. In January 1845 these individuals established the Berlin Physical Society (Physikalische Gesellschaft zu Berlin); it emerged (if only indirectly) out of Magnus's colloquium, yet Magnus himself did not join, and the early society had a nonacademic, intellectually progressive, and reformist whiff to it. The name was, moreover, a trifle misleading, since its founders and earliest members included not only physicists but also chemists, physiologists, medical doctors, astronomers, instrument makers, engineers, and various army officers. They discussed and presented new research in physics but also emphasized the physical foundations of other disciplines (e.g., physiology) and related physics to other sciences; the society was "interdisciplinary" in its nature if not in its outlook. At first, only younger Berlin scientists belonged; many of them wanted academic, economic, and political reform in Prussia and expected that science and technology would play important roles in Germany's future political economy. Meeting fortnightly on Friday evenings in the Cadet House (Kadettenhaus), and with du Bois–Reymond as its chairman, the society often attracted thirty or more members per session.

In time, physicists like Rudolph Clausius, Gustav Robert Kirchhoff, and Gustav Wiedemann also joined, as well as the physicist-technologist and budding industrialist Werner Siemens. Helmholtz became a member by the end of 1845, and between then and 1850 he lectured before the society on three occasions and contributed three review articles (on the physiological theory of heat) to its journal, the *Fortschritte der Physik*. Edited by Karsten, the journal was devoted to reviewing the physics (and related) literature; it and holding meetings were the society's most important purposes. The *Fortschritte* soon became essential to all physicists, but it also soon fell a year, and then multiple years, behind its scheduled appearance. Nevertheless, the society, its members, and its journal constituted Helmholtz's first sustained contact with physicists. For example, he and Wiedemann studied mathematical physics together, concentrating especially on Carl Friedrich Gauss's magnetic investigations and Siméon Denis Poisson's theory of elasticity.[18]

Helmholtz attended society meetings, and du Bois–Reymond introduced him around. He became close friends with Brücke and du Bois–Reymond. He had known Brücke since 1841; Brücke served (1843–46) as Müller's assistant at the Anatomical Museum, until he became a lecturer in anatomy at the Academy of Arts (Akademie der Künste). Helmholtz first met du Bois–Reymond in December 1845. When physics problems stumped du Bois–Reymond, he made a pilgrimage to Helmholtz in his barracks "to get his advice." Helmholtz himself was an autodidact (beyond the elementary level) in mathematics and physics. His brother Otto noted that while serving in Potsdam, Helmholtz sometimes lunched at his parents' home and afterward would lie on the sofa reading a mathematical text, for example, one on Jacobi's theory of elliptical functions; this observation suggests that already as a young man, Helmholtz undertook his own study of advanced mathematical topics, not least those, like Jacobi's, that were highly pertinent to mechanics and other parts of physics. Du Bois–Reymond wrote to a friend about his early impressions of Helmholtz: "In the meantime I've made Helmholtz's acquaintance, and it's given me much pleasure. He is, along with Brücke and (*sauf la modestie*) my humble self, the third organic physicist in our alliance. He's a fellow who has devoured chemistry, physics, and mathematics with a spoon, who completely holds our worldview, and who has a wealth of thoughts and new ways of looking at things."[19]

The "alliance" referred to the oath that du Bois–Reymond and Brücke had taken in 1842 to explain all organic phenomena in terms of physical and chemical forces, using physical, chemical, and mathematical methods or instruments and avoiding the use of the vague *"Lebenskraft"* as an explanatory term. They sought to understand living phenomena as material matter in mechanical motion. In the coming years, they also stood for and embodied (individually and institutionally) a more or less sharp distinction between morphology and physiology. They helped give "physiology" a new, far less anatomical and much more physical meaning. Helmholtz and Carl Ludwig joined with them to form a group of so-called organic physicists seeking to rid physiology of the romantic, non-material notion of a "vital" force (i.e., the *Lebenskraft*) that supposedly gave life to all organic phenomena. For them, organic phenomena were based on inorganic phenomena. They built on the recent emergence of physiological chemistry, on new experimental methods within physics, and on associated developments in precision measurement. They were a younger, more materialistically and mechanistically inclined set of scientists who challenged the older, romantic generation that to one degree or another had come under the spell of the now long-discredited *Naturphilosophie*, the speculative, idealistic philosophy of nature promoted by Lorenz Oken, Friedrich Schelling, Goethe, and others who believed in the fundamental unity of the forces of nature. The organic physicists, who themselves at some points displayed philosophical tendencies that had their own romantic inflections, invigorated or launched sensory physiology, electrophysiology, and experimental psychology, while also holding strong, related interests in art and aesthetics.[20] Kirchhoff, Siemens, du Bois–Reymond, Brücke, and Ludwig remained central figures in Helmholtz's life and career. He had found a network of supporters—and they had found him.

Courting Dot

Helmholtz returned to the Hussars in early February 1846 and spent the next eighteen months engaged principally in courting a young woman and writing an essay on the conservation of force. He worked in his barracks every morning until eleven, in the small laboratory that he had fashioned; his medical duties were minimal. He occasionally visited Berlin to consult with an instrument maker, Johann Georg Halske, or to use the library. He corresponded and exchanged visits with du Bois–Reymond. "During the next quarter," he wrote him late in 1846, "I'm on hospital watch; hence, I'll work above all on force conservation." That is, he anticipated not having much time for experimental work, and so he planned instead to work on his theory of force conservation during his free time. Less than two months later, he sent du Bois–Reymond the draft of his introduction to *Ueber die Erhaltung der Kraft* (On the conservation of force) and asked for comments and criticisms. He wanted to know whether he thought the type of exposition would be acceptable to physicists, explaining, "By the last revision I made an effort to get rid of everything that, insofar as was not absolutely necessary, smelled of philosophy. For that reason there may be some intellectual gaps."[21]

His other, simultaneous pursuit was Olga Leopoldine von Velten (1826–59), a young woman from a Prussian family of some distinction. It is hardly surprising that they met. Her maternal grandfather, Johann Gottlieb Puhlmann, was the affluent director of the state art gallery in Berlin, a court painter and conservator, and Wilhelm Puhlmann's father. Olga's father was Leopold von Velten, who, like Helmholtz, was a military physician; her mother was Julie Puhlmann. While serving in West Prussia, Leopold and Julie had two daughters, Sophie Julie Betty and Olga; when Leopold died in 1828, Julie returned with her daughters to her native Potsdam, where her brother, Wilhelm Puhlmann, lived. There she raised her daughters, emphasizing their language, literary, and musical education; stressing patriotism and living for ideals; and teaching the importance of good manners and enjoying life. Betty thought Olga "not beautiful, but fine and charming, not one who was outgoing but rather one who reasoned attentively and observed carefully. She was quick-witted, amusing, and witty to the point of being sharply sarcastic. Above all, and what left an immediate and the most-lasting impression, there surrounded her, like a sweet wonder, a whiff of woman-liness and a simple, unpretentious purity, something that was completely irresistible." She sang and was greatly interested in the fine arts (especially music) and literature. She possessed precisely the sort of talents and culture that attracted Helmholtz. He fell in love with her—reportedly at first sight—while creating his theory of the conservation of force. Both parties came from Potsdam's *Bildungsbürgertum*, and they had Wilhelm Puhlmann as their respective uncle, friend, and superior. Since at least the spring of 1846, they moved in the same social circle. Then on 11 March 1847, they secretly became engaged. "Your future is entrusted to my industry," he wrote his "dearly beloved." Yet he earned so little that marriage was out of the question at that time.[22]

He called her "Dot" (short for "Dötchen"). Her presence, or merely the thought of her, enraptured him. He suffered when he was not with his "beloved sweetheart." He took a walk with Ferdinand and one of the latter's colleagues— "we did a lot of disputing"—and in the evening saw a play and an opera (Daniel-François-Esprit Auber's *Der Maurer und Schlosser* [*Le maçon*]). He found the double-headed artistic evening most amusing but bemoaned being without his "little, good queen." When he returned to his scientific work on conservation of force and other matters, he was all reason; when he thought of Dot, all passion:

> You will not find, dear, dear Olga, much reason in what I've written. When I talk to myself about you, reason always flies away completely. In other respects, I'm always frightfully reasonable. I don't believe that anyone can see what infinity of bliss I bear in my heart. The people and the houses that I see are still always the same old ones, but they look to me to be so much fresher and clearer. If my

entire, long life had given me no other joy than the love of my Olga, it would have already been worth the trouble of living.

She had a cough, and he worried about it.[23] It was not the last time that either did so.

His letters to her overflowed with expressions of love, and he was occasionally inspired to write (undistinguished) poems. At the ten-week celebration of their engagement, he was *hors de soi*: He hoped his "angel" would love him "through all eternity," as he loved her. However, he returned to his senses enough to tell her that he was studying the heating of frog muscles and that he had gone to Berlin to get Halske to build him an instrument for his work. He was so madly in love with her that he first entitled his manuscript "On the Conservation of Force, a Physics Essay for the Instruction of His Beloved Olga," thereby mixing reason with passion (fig. 4.3). When reason returned fully, he deleted the reference to her. The essay was certainly far too scientific to be appropriate for her instruction. Still, he viewed it in part as a means to advance his career, and so perhaps eventually to accelerate their wedding date. They revealed their engagement to his parents, and he told his "sweet angel" that Ferdinand "sprang . . . to the ceiling with joy" and congratulated him "unusually warmly." (Obtaining his father's approval was never a small matter.) He pined after her: "Although it's so beautiful outside in nature (as it otherwise so rarely is), without you I don't have the right love of springtime, at least not the sort of joy I often had last spring when I strolled around with you in the beautiful evenings."[24]

That spring, too, he left the Hussars and on 1 June joined the Royal Regiment of the Corps Guard (Königliche Regiment der Gardes-du-Corps). His transfer meant a promotion to senior doctor and a modest salary increase (to 315 talers). In his new barracks, he again established a small laboratory. His new superior was Friedrich Wilhelm Branco, whose duties included treating the royal family and who, like Puhlmann, became one of his personal friends. His son, much later, became one of Helmholtz's sons-in-law. He reminded his "infinitely loved Olga" that they had been engaged for a full quarter of a year. He also wrote that he was finishing the draft of his essay on conservation of force and that he longed to see her: "I would not be able to hold out longer except that, when I have plenty to work on, I can keep thinking that I'm working for both our futures. But in idle hours I'm seized by very hard and heavy feelings of loneliness and longing for the one who unites for me all love, all hope, all rapture."[25]

He was ready to announce his thoughts on force conservation, since he told du Bois–Reymond that he would come to Berlin on 23 July to lecture on the topic before the Physical Society, which he did. He found Potsdam without Olga a "lonesome" place. But he kept busy swimming every morning and visiting the

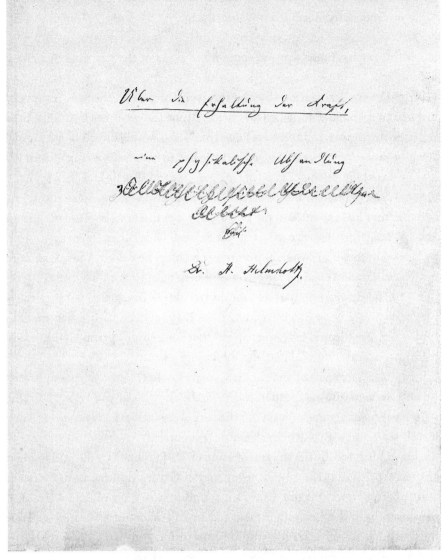

Fig. 4.3 Original title page of the manuscript "Über die Erhaltung der Kraft, eine physikalische Abhandlung zur Belehrung seiner theuren Olga bearbeitet von Dr. H. Helmholtz," 1847. Berlin-Brandenburgische Akademie der Wissenschaften, Akademiearchiv, Helmholtz *Nachlass* 598.

Puhlmanns; sending Olga novels he read (by Henrik Steffens, a *Naturphilosoph* who was one of his former teachers, and by Charles Sealsfield [Karl Anton Postl]); ordering a book by Ida Gräfin von Hahn-Hahn (*Der Rechte*, 1839); and buying musical scores.[26]

On the Conservation of Force

Helmholtz's interest in force conservation stemmed from multiple sources and contexts: philosophical, physical, physiological, cultural, technological, and political-economic. General efforts, partly of a philosophical nature, to understand the conservation of motion (*vis viva*) and to refute the notion of perpetual motion reached back to Gottfried Wilhelm Leibniz and others in the early modern period. During Helmholtz's teenage years, as already noted, he became convinced of the impossibility of a perpetual motion machine, and as an assistant at the institute's library he discovered the Enlightenment works of, inter alia, Daniel Bernoulli and d'Alembert, which contained pertinent physical principles and methods of analysis. In time, he acquired a thorough understanding of conservation mechanics. Sometime before his fourth year at the institute, moreover, he formulated the issue into "a precise question," reasoning that *Lebenskraft* was simply another form of a *perpetuum mobile*. The topic of force conservation fit his ambition to tackle the most challenging scientific problems. "Young people," he later confessed, "prefer right from the start to tackle the deepest problems. So it was with me about the question of the puzzling nature of *Lebenskraft*." To eliminate it was not only central to the organic physicists' program of accounting for biological phenomena in terms of physics and chemistry; it was also relevant, he later claimed, to generally fighting against *Naturphilosophie* and its mystical roots, though by then he and they were whipping a dead horse. Then, too, force conservation interested Helmholtz as part of his ongoing effort to understand heat as a physiological phenomenon.[27] His essay on the physics of force conservation extended the organic physicists' program beyond physiology and nicely complemented du Bois–Reymond's own seminal work on the biophysical foundations of physiology, *Untersuchungen über thierische Elektricität* (1848).

The essay also complemented Alexander von Humboldt's ongoing publication of *Kosmos* (1845–50), then a three-volume work devoted to understanding nature as a whole. Yet, whereas Humboldt worked in a largely descriptive, semipopular manner, Helmholtz articulated a causal principle in an explanatory, mathematical mode that governed all natural phenomena. More generally, Helmholtz's search for a unifying principle in nature belonged to a German tradition that, in its search for unifying forces of nature, included not only Humboldt but also Kepler, Leibniz, Kant, Hegel, Goethe, the romantic *Naturphilosophen*, and the organic physicists themselves; it later included such figures as Ernst Haeckel, Wilhelm Ostwald, Max Planck, and Albert Einstein, each of whom emphasized the unity and interconnectedness of all nature and sought to provide a unified world picture (*Weltbild*) of nature's phenomena and laws. Even as a young student, Helmholtz had harbored such philosophical ambitions and interests, as his study of Kant (especially) and Fichte suggests, and for him as for numerous German thinkers, the concepts of unity, causality, and world picture played an important role in his thinking. However, in his essay of 1847, Helmholtz neither mentioned Kant by name nor referred to any of his writings.[28]

Helmholtz's sources and methods were, lastly, also technological and political-economic. His interests here were informed by French engineering mechanics (e.g., that of Sadi Carnot and Emile Clapeyron) and the central concept of work, which by the 1840s had become a well-studied topic. Moreover, living in Potsdam and Berlin during the 1830s and 1840s, as the railroad made its appearance around him and as steam engines were placed even in Potsdam's luxuriant public gardens and other areas, he (like many others) became increasingly aware of steam engines and their associated concepts and technology (e.g., work and indicator diagrams).[29]

Hence Helmholtz's awareness about force conservation as a general scientific and philosophical topic in the mid-1840s was anything but singular. In fact, between the 1820s and 1847, a series of European figures—including Carnot, Carl Friedrich Mohr, Marc Séguin, Michael Faraday, Julius Robert Mayer, William Robert Grove, Liebig, James Prescott Joule, and Ludvig Colding—had in one way or another argued that forces could not be created or destroyed but were, instead, subject to conversion or correlation, and in particular that heat and work were convertible. Helmholtz knew firsthand the work of at least some of his predecessors and contemporaries (though he did not then know Mayer's). His later claim that, apart from Johann Christian Poggendorff's *Annalen der Physik und Chemie*, he had little access to the relevant literature, seems suspect.[30] He certainly borrowed elements of his theory from many physicists, chemists, engineers, and philosophers before him; by the mid-1840s their ideas and findings had become commonplaces. The novelty of his essay lay, rather, in both the general applicability of his principle of the conservation of force throughout physics (and, he hoped, eventually beyond it, to physiology) and in its quantitative specificity to particular physical problems. While physiology was the immediate context for his essay, his analysis of force conservation was intended to show his competence as a physicist. Above all, however, it was Helmholtz's ambition, pursued intermittently since his teenage years, to show the impossibility of perpetual motion (i.e., it was the physics) that motivated him.

The essay thus addressed two audiences. First came the physicists, a group with whom he had no professional accreditation and to which he did not belong, but whose recognition he sought. He saw his work as a critical analysis, generalizing and ordering what others before him had done from the eighteenth century to his own day.[31] Second, he addressed his fellow physiologists, who he hoped would employ force conservation to explain physiological phenomena and so give a rigorous foundation to organic physics.

Helmholtz began by acknowledging that his philosophical introduction could be read independently of the essay itself (i.e., the physics). He argued that all the physical theorems he proposed could be derived from two seemingly disparate but ultimately identical viewpoints: either from the impossibility of a perpetual motion machine (i.e., creating an infinite amount of work from a finite amount of force) or from explaining all nature's effects in terms of attractive and repulsive forces (point masses interacting via central forces). The task of physics, he proclaimed, was to find the causal laws "by which the individual processes in nature can be traced back to general rules and can again, in turn, be determined from the latter." Such laws encompassed all relevant phenomena. He believed that nature was, in principle, completely comprehensible by means of laws that reduced all phenomena to a set of unchanging, necessary, and sufficient causes.[32]

In good Newtonian and Kantian fashion, he distinguished matter and force as the fundamental abstractions by which science conceived the external world. In itself, matter has no effects, he argued; it informs us only of the spatial distribution and quantity of mass. Hence the only change that matter can undergo is a spatial one (i.e., motion). Since we can know nature only through its effects on our sense organs, the abstraction "matter" requires a second abstraction, namely, "force," for it is force that elicits all effects. The two concepts are always and necessarily linked together. Both are mere abstractions "from the real" and, taken individually, have no operational meaning.[33]

Finding the ultimate causes, he continued, meant finding unchanging forces over time; this in turn required reducing natural phenomena to the motion of matter according to the mechanical (force) laws of point masses. Mechanical forces are either attractive or repulsive. He thus concluded that "ultimately the task of the physical sciences [is] to reduce natural phenomena to unchanging attractive and repulsive forces whose intensity depends on distance. The resolution of this task is simultaneously the condition of the complete comprehensibility of nature." The "business" of theoretical natural science would be completed, he believed, once such forces were discovered and shown to be unique and to account for all phenomena. Such forces would constitute "objective truth."[34] Twenty or so years later, he effectively abandoned this reductionist, causal approach in favor of a more phenomenological one.

With the philosophical introduction behind him, Helmholtz plunged into force conservation in scientific detail. He assumed that no combination of bodies could continuously create forces from nothing, or, in the language of mechanics, that the quantity of work gained when a system changes from its initial into a second state always equaled the quantity of work lost when the system changed from the second back into the initial state, no matter how or by what path or at what speed the change was effected. For otherwise the quantity of work would somehow have increased, which in effect meant the creation of a perpetual motion machine. He sought to show the truth of this assumption in all branches of physics and its use as a guide to experiment in areas where the laws of physics had (to date) been only partly tested. He then mathematized and generalized the point, calling it the principle of the conservation of force: "In all cases of the motion of free material points under the influence of their attractive and repulsive forces, whose intensities are dependent only on distance, the loss in the quantity of tensional force [*Spannkraft*] is always equal to the gain in living force [*lebendiger Kraft*], and the gain of the first is the loss of the second. Thus, the sum of the living and tensional forces is always constant." (Between the early 1850s and 1862, to anticipate, this "tensional force" became known and understood as potential energy, and this "living force" as kinetic energy, but in 1847 that was language and a conceptualization unknown to Helmholtz and others. They inherited the ambiguous eighteenth-century notion of "force" as either, following the Newtonians, "dead force" or, following the Leibnizians, "living force.") Helmholtz then applied his principle to a variety of mechanical theorems and analyzed the force equivalent of a variety of thermal, electrical, and magnetic processes. Toward the end of the essay he briefly mentioned his hope for the principle's eventual application to analyzing organic life to explain the development of heat in plants and animals.[35]

He had shown, he maintained, that the principle of conservation of force was confirmed "by a large number" of facts in natural science and that it was consistent with numerous, well-known laws in mechanics, electricity, and magnetism. The principle confirmed these laws in terms of known observational and experimental results. He considered "its complete confirmation as one of the main tasks for physics in the near future."[36] He had plotted out a future research program for his (would-be) fellow physicists and physiologists. At age twenty-six, just five years after completing his medical education, he had presented his solution to the vague, partly philosophical issue of the impossibility of a perpetual motion machine and had done so by turning it into a concrete, rigorous scientific analysis.

Rejection and Neglect

To get published in science, it helps to have connections. In physics, Helmholtz's were secondhand. He turned to Magnus (via du Bois–Reymond) for aid in getting his manuscript published in Poggendorff's *Annalen*, Germany's premier physics journal. Magnus was a close friend and colleague of Poggendorff. Poggendorff reviewed the manuscript promptly and wrote Magnus, not Helmholtz. While he vaguely conceded its importance, he found it too long an article for his journal and, more importantly, not devoted to experimental results, which was what, he said, the *Annalen* principally published. He claimed that to publish the work of "the theorizers, . . . to whom, by the way, I absolutely do not deny my respect and appreciation for their usefulness," would be tantamount to excluding much work of an experimental nature. He thus declined to publish Helmholtz's manuscript and recommended that Helmholtz publish it as an independent monograph. He sought to soften the blow by leaving the door open to a potential future relationship, if and when Helmholtz chose "to verify or even merely to test his stimulating speculations through experiments."[37]

His rejection of Helmholtz's manuscript bids well to be among the most egregious rejections of a manuscript in the history of scientific publication. Magnus (again via du Bois–Reymond) sent Helmholtz Poggendorff's rejection letter and said he regretted Poggendorff's decision: "In my view the essay can have a very useful effect, and, moreover, gives a rare example of the broad-range of knowledge as well as a new proof of Dr. H's acumen and talent." He too recommended the manuscript be published as a monograph.[38]

It was Helmholtz's first professional setback. The experience of rejection wounded him deeply and left scars that remained forever. In 1859 he told a colleague that his theory "was then [in 1847] very unpopular with the physicists." In 1868 he told Peter Guthrie Tait, a Scottish physicist, how the German physics community rejected both his and Mayer's work on force conservation. "It's difficult now," he added, "to imagine oneself back in the state of mind of that time and to realize clearly how absolutely new the matter then seemed." He was stunned over the "resistance" that he encountered from experts. He noted that at the Royal Prussian Academy of Sciences (Königliche Preussische Akademie der Wissenschaften), only Jacobi had shown sympathy for his work. As late as 1891, three years before his death, he publicly declared, with exaggerated rhetoric, that the (unnamed) "physical authorities" thought his work was "a fantastic speculation" in the tradition of Hegel's nature philosophy.[39] Though his manuscript's rejection scarred him for life, it also served to fuel his ambitions to become a physicist and, in general, to see his name in print.

While still awaiting Poggendorff's decision, Helmholtz wrote Olga to say that he thought of her "always." He did little else and wanted little else than "to think of her." He was lonely again and lacked blocks of time to get real work done. When she was around, his little frustrations disappeared because she united in herself "the greatest interests" for him. He could talk to her as to no other: "It is indeed good . . . , sweet Olga, when one has so good a human being as you to whom one can confidentially confess everything that weighs on one and that one has no true inclination to overcome." He revealed himself only to her: with her he did not worry about appearing self-absorbed or even downright foolish. He stole time from his army duties to study (especially mathematics), to conduct (unsuccessful) experimental work on heat formation in frog muscles and nerves, and to write. He played piano pieces by Otto Tiehsen to see if Olga might be interested in them, though he himself did not care for them. He wrote her, too, of "a totally special harmony" of their tempers and of his longing for her: "How infinitely, how fervently, and how submissively I love you." He guiltily reported that he had spent one afternoon reading a novel, Samuel Warren's *Ten Thousand a Year*. He did not like its "crude sentimental morality and religiosity." Nevertheless, he had read two of its three volumes and planned to read the third as well.[40]

Du Bois–Reymond was angered and saddened by Poggendorff's rejection. He tried to turn the situation around by arguing that a monograph had several advantages over an article: it would bring an honorarium; it was more impressive; Helmholtz could resurrect the philosophical introduction, which he had scuttled, "wherein many splendid things are said"; and he would get more readers. The omniscient du Bois–Reymond recommended Georg Ernst Reimer, a well-known Berlin publisher of politically liberal sentiments who had published books by himself, Brücke, and others known to Helmholtz, and who published works of the academy as well as the Physical Society's *Fortschritte*. Finally, he reproached and advised Helmholtz that he might have avoided altogether this rejection by Poggendorff had he spent more time in Berlin meeting with him regularly. It was not enough to do good scientific work, du Bois–Reymond was in effect saying; one also had to network. That was precisely the sort of thing that the politico du Bois–Reymond excelled at and that Helmholtz had yet to learn. Yet neither Helmholtz's circumstances nor his personality would have allowed him to hobnob much with Poggendorff in Berlin. Besides, his scientific work was something that Helmholtz did alone and did not discuss much with others. Furthermore, his geographical and social distance from the community of Berlin physicists may have strengthened his intellectual independence and encouraged him to pursue big issues like force conservation. He appreciated du Bois–Reymond's support and agreed he should publish the essay as a monograph. The only change he made was to resurrect the introduction. He valued du Bois–Reymond as an adviser and referred to himself as "a young beginner." At the same time, he rejected Poggendorff's stated reasons for not publishing his essay, pointing out that he had in fact previously published long as well as theoretical pieces.[41]

He sent Reimer the manuscript and asked if he might publish it. He described its purpose as providing a general, fundamental law of mechanics that concerned all branches of physics. He claimed his essay had been well received so far and was of general interest; he dropped Magnus's, du Bois–Reymond's, and Brücke's names and said they were ready to speak for his work. His pitch conveniently left out their partiality and Poggendorff's rejection. He also said that Müller could attest to the scientific quality of his previous work, which he listed. Reimer quickly offered to publish the manuscript, and Helmholtz quickly accepted. It was clearly not refereed independently, and, in fact, du Bois–Reymond had intervened on Helmholtz's behalf. It appeared within weeks (on 3 November 1847).[42]

In early September, with the manuscript's publication finally secured, Hermann and Olga publicly announced their engagement. He now felt that he "frequented an entirely new world of ideas" that had little connection with the previous one. Until then, he had not realized what a difference this announcement would make not only for their "relationship vis-à-vis other people," but also for themselves. When they walked together, she felt like his bride. He became her "protector." He felt an "ecstasy" that he had not felt before "so stirringly." On the day of the announcement, he finished correcting the proofs of *On the Conservation of Force* and began writing up his research on the heating of frogs' muscles, "in order to inundate the literary market" with his products, he self-mockingly yet most revealingly told his fiancée. He socialized with the Puhlmanns and started reading Hahn-Hahn's *Cecil*. Everyone in Potsdam congratulated them on their engagement. The Puhlmanns invited her to stay with them for the coming weekend, which meant they could be together. He was reading August von Kotzebue's well-known comedy *Die Indianer in England*, which concerned a wedding engagement and its postponement, but he gave it up: "These endlessly mere matters of love, most of which are of a rather twisted sort, now bore me since I myself know it [love] better, much better than before." He missed her voice and impatiently awaited singing with her.[43] Sound, in all its manifestations, was always important to him.

During the next four years, Helmholtz's essay was virtually ignored by the physics and physiology communities. It was (initially) a scientific dud. He found little or no support for it, and Clausius was outright hostile. In the German lands, it was only his fellow organic physicists, certain members of the Physical Society, and Jacobi who accepted his viewpoint or found his principle of the conservation of force to be interesting and even pathfinding. Du Bois–Reymond, who was then having his first success in demonstrating signals in nerve currents (negative variation), told Ludwig that Helmholtz's essay "simply cannot be praised enough." Ludwig complained, however, that he could not get the essay from his Marburg bookseller, and even if he could have, he feared it would be too difficult for him. Indeed, few contemporary physiologists could follow it — too much physics and mathematics — and perhaps fewer still thought it relevant to their work. One of Helmholtz's few German readers was Mayer, who judged Helmholtz's theory to be "absolutely the same" as his own. At least Helmholtz's military superiors were impressed with his essay. Or at least, as du Bois–Reymond reported, they praised the practical nature of his work, because they managed to confuse his notion of physical force with theirs of military force! In France, his work remained largely unknown until the late 1860s. In Britain, too, his essay was at first unknown or ignored, though by 1851–52 a group of British physicists came to appreciate the seminal nature of his principle. In the meantime, and perhaps as a result of the stress brought on by overwork and personal concerns, what he got in 1847 was not recognition but hay fever. He suffered from it annually every spring for the next twenty-one years.[44]

Assistant, Lecturer, and Bystander

While working on the essay on force conservation, Helmholtz also conducted experimental work and drafted an article aimed at understanding heat formation in muscles. His research effort here—indebted partly to du Bois–Reymond's own effort on animal electricity—greatly enhanced his understanding of and skills with delicate thermoelectric apparatuses and multiplicators (for detecting small current differences), as well as the associated precision-measurement techniques and error analysis. He fretted a great deal over, and improved upon, the quality of his instrumentation and the standards of precision and accuracy, and in so doing he helped experimental physiology approach the more exacting standards of experimental physics. He again used frogs to experiment with, this time to gain data on heat formation in muscles (and nerves). If his article was long on rigorous methodology and short on definitive numerical results, he nonetheless showed that muscles produced heat—just as they produced work—and concluded that the differences between muscles and nerves in heat formation were vanishingly small. Already by 12 November 1847, two weeks after the appearance of his essay on force conservation, he was again lecturing before the Physical Society, this time on the results of heat formation in frog muscles. His results appeared in print the following year, again in Müller's *Archiv*.[45]

A professional break now emerged for him. Brücke was to be offered the extraordinary professorship of physiology and pathology at the University of Königsberg. This meant that he would relinquish his positions as assistant to Müller at the Anatomical Museum and as instructor at the Academy of Arts. In mid-December Helmholtz met with Johannes Schulze, the official responsible for educational matters (Gymnasia and universities) within the Prussian Ministry of Culture (known formally and in German as the Ministerium der geistlichen, Unterrichts- und Medizinalangelegenheiten). He conveyed his hope of becoming Brücke's successor in Berlin. Du Bois–Reymond told Ludwig that Helmholtz might move there after Easter. Helmholtz felt expectant: "Since hopes of a better turn in my career have gained greater foothold," he teased Olga, "the sense of luxury has powerfully begun to work its way into me."[46]

They spent Christmas in Dahlem with her family. His future sister-in-law described him as at first "very serious and inward looking, somewhat awkward and oppressed among the young men who were lively, animated, and worldly." Yet he soon fit in well with the von Veltens and the Puhlmanns and in fact became the center of their household, where there was always a lot of music. Under his tutelage, Olga's family came to love Beethoven and Shakespeare. They read books aloud together, each taking the part of a different character. (Helmholtz read especially well and favored comedic parts.) He said his sensory organs and soul had become dependent on her, and when she could not go to a symphony, as they had planned to do together, he said his hearing went bad: "To me it was as if until now I had let the [musical] harmonies into my understanding only through your soul, with its deep musical inwardness. My ears heard only musical figures and my soul heard nothing at all." He listened to Mozart and felt lonely, though Beethoven's *Overture to Collin's Coriolan* cheered him up: he considered it "an unsurpassed masterpiece."[47] The von Veltens and the Puhlmanns became his new, enlarged family.

As he awaited the decision about a possible position in Berlin, the revolutions of 1848 broke out. The one in Berlin began in mid-March and lasted until early December; it included heavy street fighting and riots. While the Prussian throne and bureaucracy were shaken to their foundations, they were not overthrown. The monarch and his troops decamped to Potsdam, which, though itself in a shaky political state, nonetheless remained true to its quietist and conservative traditions; Potsdam became home to the counterrevolutionary forces. In September Helmholtz's regiment conducted crowd control. Some revolutionaries sought to persuade his unit to join them, but none of its members did.[48]

Nor did Helmholtz. Unlike many young scientists in Berlin (du Bois–Reymond, Beetz, Heintz, Wiedemann, Kirchhoff, Halske, Clausius, Knoblauch, and Remak), he did not sign a petition initiated by the Physical Society calling for greater openness on the part of the Academy of Sciences toward nonmembers. (The academy rejected the petition.) Moreover, he saw that his father was reprimanded that March for insubordination and indiscretions and was threatened with losing his position. He perceived that signing petitions and similar sorts of mild political activity would scarcely help him get the new positions he coveted, especially since his former teacher and academic patron, Müller, was a firm antirevolutionary. Any oppositional activity on Helmholtz's part that came to Müller's notice might well have injured his standing with him. In any case, unlike Virchow, Helmholtz was no revolutionary or radical democrat. Ludwig dubbed him "an unflinching member of the liberal party," a point that had more intellectual than political resonance. He remained loyal to the Hohenzollern regime and Prussia's institutions throughout the revolutionary period. He came no closer to a political remark that revolutionary year than to tell Olga he was reading Johann Wirth's *Geschichte der Deutschen* (1842–45), "a highly interesting work" by a radical liberal journalist and republican. He was impressed with but critical of the constitutional structure of the ancient Teutons, since nearly all of them lived under tyranny. His (few) extant letters to Olga that year are filled with amorous expressions for her. It was the sound of her voice, not Prussia's

Fig. 4.4 Helmholtz in 1848. *Nachlass* Emil du Bois-Reymond. Courtesy of the Deutsches Museum, Munich, Porträts-Sammlung 1932 Pt A 35/2.

political system or its opponents, that moved him. Other than his love, he talked not about politics but mainly about such things as walking through Sanssouci in the afternoons and reading Charles Dickens's *Martin Chuzzlewit* on his roof near the Havel River "by the most beautiful illumination at sunset." With his highly refined senses, he often stood at a fountain in Sanssouci, where frogs ate and squirted water, and listened "to the splashing and rippling." He heard melodies in the fountain's noise though his friends could hear none. He was busy with his myograph and seeing patients (not always successfully: one of them died, and, to his shame, he had confused him with a sibling).[49] He was, as the earliest known portrait of him (March 1848) shows, a soigné young man who took much care with his dress. He looked like a middle-class *Bürger*, a professional medical doctor, and an army officer, not a revolutionary (see fig. 4.4). He stayed on the side of the conservative forces.

That June, Müller offered him the appointment as his assistant at the museum. He praised him greatly to the ministry and, as in the past, supported him warmly. But since the appointment paid so little (two hundred talers per year), he needed an additional appointment of lecturer in anatomy at the Academy of Arts, which paid four hundred talers. In late July, a trial lecture was scheduled at the academy. Although the ministry had no doubt as to Helmholtz's scientific competence, since Müller had vouched for his "thorough knowledge of anatomy" as well as "his versatile education [Bildung] for the teaching position in question," it sought assurance that Helmholtz could teach anatomy to art students.[50] The lecture took place on 19 August, before the academy's senate and faculty, in its building on Unter den Linden, in the heart of Berlin and in the midst of the revolution.

Helmholtz's lecture aimed to demonstrate the outlook and methods that an anatomy teacher at an art institute should adopt in instructing such students. He thought that perhaps the greatest challenge lay in turning the numerous and dry facts of anatomy into "a living picture" useful to artists. In contrast to anatomical instruction for medical students, who needed well-defined anatomical details, art students required "insight" and a sense of particular muscles and their relationships with other anatomical elements.[51]

How then, he asked, can anatomy help the artist, and why is it necessary? Ancient Greek artists and others had done superb artistic work, he explained, even though they had only a limited, and in some cases even false, knowledge of anatomy. He cited Cleomenes the Younger's statue Germanicus and the shooting Apollo in the Berlin Museum as having obviously flawed muscular representations. Yet the creative artist, he maintained, is far more concerned with the "the sense for the ideally beautiful" than with isolated anatomical details:

> The artist's genius is precisely the secret power of finding and presenting in original intuition and without calculated reflection that which subsequent, pondering reflection must also recognize and justify as true and complete. The more the sensitivity of the impressionable spectator is stimulated, the richer and the truer the creating artist has known how to conceive and render the ideal content of his work; similarly, and just as certain, it [the spectator's sensitivity] will also find every deficiency as an injury to the life and the beauty of the whole, even if it cannot be determined where the error may lie and what its cause may be.

He thought even the great ancient artists, despite their insights into beauty and truth, could have profited from greater knowledge of anatomy.[52]

Anatomical instruction for modern artists, he continued, boiled down to a matter of how knowledge of the body's inner structure could help them get beyond knowledge of its external surface as acquired by the observation of living models. There were three ways. First, it aided the artist's understanding of the body's arrangements of individual parts by providing him, when he had no model to work with, "an anatomical mechanism." Second, it taught him to differentiate the essential from the nonessential. The artist should not seek to be merely an imitator of the real human body with all its flaws; rather, he must find and express its "spiritual [*geistigen*] content." Finally, it could help the artist represent the body in motion—something that static models could never do—by helping him to know what muscles looked like when the body was in motion.[53]

Yet Helmholtz warned that instruction in anatomy, like all artistic instruction, could never substitute for "the insight into these [bodily] forms and the artistic sense of beauty." Instead, it was "a means which eases for the artist the spiritual mastery of the eternally changing multiplicity of earthly objects, including the human form, which sharpens for him the view of the essential nature of the whole, which makes the entire form, as it were, transparent, and which should arm him with the resources of testing criticism for the work created. . . . The artistic spirit first shows itself in the wise application of forms, their context, and the simple principles which anatomy has taught, in the distinctive characteristics of the form." Even the greatest artists, like Michelangelo, who use anatomy to be showy, end up producing "unpleasant and untruthful" figures; conversely, those who neglect anatomical truths produce "lifeless and distorted figures."[54]

The essence of an anatomy lecture for art students, Helmholtz concluded, was to portray a sense of the "living, uninjured form"; later, the student could compare this ideal with a live model as well as with artistic works. Since anatomy for art students concerned only those bodily parts that influenced the external form, it must offer lessons about bones and externally visible cartilage, joints, and ligaments (to understand how to represent motion), as well as the theory of muscles.[55]

The examiners found Helmholtz's lecture to be "thorough" and judged that he had "a sound scientific education" in anatomy. They thought his lecturing ability satisfactory and expected that it would mature as he gained some experience; they apparently thought he was a no-more-than-adequate lecturer. (Helmholtz later confessed that as "a young military physician" his hands had shaken before his first lecture.) He got the position.[56]

To finalize his appointments, Prussia had to release Helmholtz from his army medical obligations. This was no routine matter, and it took no less a figure than Humboldt, the doyen of Prussian science, to help make it happen. Humboldt was, in effect, a scientific headhunter. As a scientist of broad experience and international reputation, as the monarch's chief science adviser, as a member of the state council, and as adviser to the minister of culture, Humboldt was the most politically influential figure in Prussian science. He did a great deal to advance science in Berlin specifically and Prussia in general. He was also a principal figure in German liberalism, especially in cultural and educational matters. Like most scientists of his day, he believed both in the value of scientific research for its own sake and in its use for social purposes.[57]

Du Bois–Reymond introduced Helmholtz to Humboldt, who had long admired Müller and was also friendly with several of Helmholtz's other teachers. Humboldt's library contained Helmholtz's dissertation as well as his articles on fermentation and putrefaction, on muscle metabolism, and on heat development in muscles and his essay on force conservation (though he was skeptical about Helmholtz's law of force conservation, which he apparently could not understand). Helmholtz, like many other young scientists, visited him at his home occasionally to seek his advice and aid. Thanks in part to Humboldt's intervention, the army released him from his obligations three years early (on 30 September 1848) and the Ministry of Culture authorized his new appointments.[58] The two appointments together gave him a total annual income of six hundred talers.

Counting his four years of medical school, his year-long internship, and his five years as an army medical doctor and surgeon, Helmholtz had spent a decade in the Prussian military. It had given him not only a tangible medical education but also several intangibles. It reinforced in him the values of order and discipline and his already high sense of duty. He was a soldier and a patriot, and when he became a civilian, he carried a sense of civic responsibility into the world of science. His military career also increased his social status, since as a Prussian army officer (albeit a medical one), he belonged to nineteenth-century Europe's elite, model military institution, one held in high regard within and beyond Prussia. His extended military service enhanced his credentials with the state by demonstrating his commitment to Prussia as a state. Few, if any, German scientists could boast as much. Still, when it proved advantageous, as it did now, he readily broke his military tie.

He spent the academic year 1848–49 in Berlin lecturing at the academy and assisting Müller at the museum. He deepened his relationships with Berlin scientists. Du Bois–Reymond, Kirchhoff, and Siemens became his closest friends, while Müller, Humboldt, and Magnus continued as his senior colleagues and backers. He saw Clausius and Wiedemann "almost daily," since they frequented the same restaurant. Du Bois–Reymond told an old friend that Helmholtz had taken Brücke's place as his closest friend in Berlin: "In scientific matters . . . his talent is truly limitless and it's difficult to find his equal in matters of knowledge. But he isn't like Brücke, a deep and richly lived person, or perhaps it's only that I wasn't young with him, as I was with you and Brücke." In May 1849 Helmholtz joined Müller, Humboldt, and du Bois–Reymond in the latter's quarters to observe a pathbreaking experiment: du Bois–Reymond demonstrated, via the deflection of a galvanometer needle, that tetanic current was present in living human beings. He often visited du Bois–Reymond in his apartment. He became a full-time and important participant in the Berlin scientific community, one that Benjamin Silliman, the Yale chemist who visited Humboldt in 1851, thought was "probably the finest in Europe" and whose figures represented "the highest result of the refining influence of modern civilization."[59] Helmholtz helped strengthen that community.

Appointment to Königsberg and Marriage

Yet his new positions were no more than temporary; they could hardly satisfy his ambitions and need for more income so he could marry. Ultimately, he hoped for a permanent academic appointment. This was not unreasonable, because he was considered to be among Germany's best younger physiologists. Ludwig told Henle that, with the emergence of such young figures as du Bois–Reymond, Helmholtz, Brücke, Karl von Vierordt, Ludwig Traube, and Eduard Weber, German physiology, not to mention Liebig and his school, had great talent and potential. With eight publications to his name and with his growing reputation, Helmholtz had become a player in academic physiology. His second professional break came in December 1848, when Brücke at Königsberg received a call to the University of Vienna. Brücke considered du Bois–Reymond, Ludwig, and Helmholtz, in that order, as his possible replacements. He thought that for Helmholtz the position would mean "an essential improvement . . . and a means for him to get married." But he told du Bois–Reymond that, should he leave and should the latter want the Königsberg position, he would recommend only him; if not, then only Ludwig and Helmholtz. In any case, he wanted someone from their crowd, an organic physicist, to get the job. The position became formally available in the late spring. Du Bois–Reymond tried to finagle it for Ludwig—du Bois–Reymond himself much preferred to stay in Berlin—but then Müller, in du Bois–Reymond's view, "ruined" everything: he not only praised Ludwig but also recommended both Helmholtz and (he believed) Remak. Du Bois–Reymond thought that Remak was no competition but that Helmholtz certainly was.[60]

The Königsberg Medical Faculty, however, recommended du Bois–Reymond, Helmholtz, and Ludwig, in that order. Du Bois–Reymond thought Helmholtz would be "the most appropriate," but he also thought "he had little inclination to accept that position," and so he suggested Ludwig, who, he assured the ministry, was not a radical but rather "a conservative liberal." In fact, in 1848 Ludwig had been an open democrat.[61]

Müller, who had always been much impressed with Helmholtz and was his great backer, wrote the minister about the best candidates for physiology, in particular experimental physiology. He said that the discipline had become more closely tied to physics and chemistry; its leading edge consisted of individuals who understood both physical methods and physiology. He considered Brücke, du Bois–Reymond, Helmholtz, and Ludwig as "the most-promising younger talents . . . in Germany." Since Brücke was leaving Königsberg for Vienna, that left the other three as potential candidates. Du Bois–Reymond's work in electrophysiology made him the highest qualified, Müller said, but he was in the midst of completing that work and so was not now ready to leave Berlin. As for Helmholtz, he described him as "one of the most important talents in physiology" and harbored no doubts about his teaching qualifications. He held Ludwig in as high regard as the others. He also mentioned Remak but said his approach was not that of a physical physiologist but rather of a microscopist and pathologist, and he noted that Martin Heinrich Rathke at Königsberg already did this type of work. In effect, Müller's letter left the minister with a choice between Helmholtz and Ludwig.[62]

Helmholtz's impressive scientific results and strong publication record constituted the essential basis for his getting the appointment. Yet outside Müller and his circle, he was scarcely known; none of his work through 1848 had made much of a noticeable impact on physiology, and it had had no impact on physics. Müller's recommendation must thus have been vital in the ministry's eyes. Helmholtz's military service and nonparticipation in the revolution of 1848 doubtless further favored him; unlike Ludwig, he was no political troublemaker; and unlike Remak, who was Jewish, he had no religious hurdle to overcome. Ludwig complained bitterly to Henle that the ministry preferred Helmholtz and Remak to him because of his "suspiciously democratic opinions" and that Müller favored Helmholtz and Remak because, unlike himself, they were Prussians. In early June, Helmholtz was appointed to the post with a salary of about eight hundred talers. His selection showed that the Prussian educational authorities, at a moment when political reaction had set in, trusted him. The ministry wanted him to go to Königsberg immediately and begin lecturing that very summer semester. But completing the paperwork made it impossible to do so. He visited Königsberg for the first time in midsummer, and liked it.[63]

After a nearly two-and-a-half-year engagement, Hermann and Olga could now afford to marry. He invited du Bois–Reymond to the wedding: "I cannot in good conscience exempt you from participation in this festivity because I would be sorry to see it if my best friend didn't want to have any part in this, the most joyful day of my life and the goal of many years of effort. So overcome your shyness against all people who aren't physiologists—and come." He also asked him to fetch a book at the library and place an order with Halske for some magnetic rods. His best friend was always his useful factotum in Berlin. Hermann and Olga married on Sunday, 26 August 1849. He was twenty-eight, she twenty-three. The ceremony took place at the picturesque village church of Sankt Annen, on the Dahlem estate of Betty and Emil Puhlmann, "under the old trees" amid sunshine and flowers, in the presence of the bride and groom's families and friends, "everyone filled with certainty of this happiness." Yet even here, science and technology were not quite absent, for the church also served as one of sixty-one stations—it was No. 4, known as the Telegrafenberg—of the Berlin-Koblenz optical (later electromagnetic) telegraph line. The newlyweds left for Königsberg immediately after the wedding.[64] He now belonged fully to her and to science.

PART II

Gaining Scientific Renown

Scientific Networking

Königsberg by the Sea

Königsberg was the principal city of Prussia's Wild East (once Polish territory) and the provincial capital of East Prussia. It was dominated by governmental, military, and business figures. Once the home of the Teutonic Knights, it had a heritage as a seat of regional power, and on several occasions it witnessed the coronation of a new Hohenzollern monarch. Located on a bay of the Baltic Sea at the mouth of the Pregel River in the Samland, it constituted an entrepôt that boasted commercial ties with the Prussian, Russian, and Lithuanian interiors as well as with more-distant regions. The Junkers, who resided on their large landed estates, ruled its hinterland. The production of grain dominated the East Prussian economy.

While the coming of the railroad to the German states during the 1830s and 1840s made Königsberg less remote from the surrounding Germanic and Slavic worlds, the sea remained vital to it. Helmholtz thought that for East Prussians the sea took the place of the Alps. During his six years there, he passed many an hour contemplating "the steep, well-wooded coasts of the Samland" and often became transfixed by the sea's waves and their combinatorial, manifold nature. Such a scene, he thought, "grips and raises the mind, since the eye easily learns to recognize law and order" in the trains of wave.[1] He loved being around bodies of water; they relaxed and inspired him. The power, beauty, and peace of Königsberg's coast and sea helped to compensate for the loss of Potsdam's gardens, forests, and lakes.

Königsberg was also Kant's town: he had spent his entire life there, and within the sphere of culture, his name and the city's name became synonymous. Kant's philosophical works established Königsberg as a center of the Enlightenment. His idealist theories of pure and practical reason and of judgment, as well as his analyses of the nature of science, morals, religion, man, and peace, made Königsberg into an intellectual and liberal stronghold. But neither then nor later did the university attract noticeable numbers of students beyond its own region. Helmholtz, like many others, thought Königsberg a "remote" city that had in part "limited" Kant.[2] Certainly the university and its professors remained far less important to the city than their counterparts in small German university towns like Göttingen; the political and business classes overshadowed them. Nor did the fine arts flourish there to the extent that they did in many central and western German cities. At midcentury, Königsberg remained a distant, isolated, and (to many) unknown place—cold, damp, and Lutheran.

Nevertheless, the university, affectionately known as the Albertina, constituted perhaps the most important element in the city's cultural life. It was one of the oldest German universities, with strong Lutheran ties and a strong mission to civilize the reputedly barbarian East. It was known as a liberal and reform-minded place. Its enrollment numbers placed it in the mid range of German universities: when Helmholtz arrived for his first semester (1849–50), there were some 300 students enrolled; when he left at the close of the summer semester 1855, there were about 350. Yet because most of the students were poor, they provided little in the way of seminar fees that might have increased a faculty member's income. The university's facilities were outdated and limited in scope.[3] See figure 5.1.

On the other hand, from about 1830 to the mid-1850s, Königsberg was one of the best German universities in mathematics and the natural sciences. Its reputation rested above all on the innovative teaching and research of the mathematicians Carl Gustav Jacob Jacobi and Friedrich Julius Richelot, the astronomer Friedrich Wilhelm Bessel, the physicists Franz Ernst Neumann and Ludwig Moser, and the anatomists and physiologists Karl Friedrich Burdach and Karl Ernst von Baer. By the time Helmholtz arrived, only Richelot, Neumann, and Moser were still there or alive. Especially Neumann, with his innovative mathematical-physical seminar, and Bessel, with his rigorous error analysis, led the way in showing the importance of and setting a tone for demanding precision, exactness, and empirical care in scientific investigation.[4]

Fig. 5.1 The University of Königsberg at its 300-year jubilee (1844). *Die Albertus-Universität zu Königsberg: Eine Denkschrift zur Jubelfeier ihrer 300 jährigen Dauer in den Tagen vom 27sten bis 31sten August 1844* (Königsberg: H. L. Voigt, 1844), facing title page, in Niedersächsische Staats- und Universitätsbibliothek Göttingen.

The city's cultural life was also enriched by several scientific, literary, and cultural societies, three of which Helmholtz became closely associated with. The first of these, the Physical-Economic Society (Physikalisch-ökonomische Gesellschaft), was a town-and-gown organization founded in 1789, one of the hundreds of patriotic and economic societies established in central Europe during the Enlightenment. Its original concern had been to advance agricultural-economic issues, and so it naturally favored the interests of the local (and conservative) landed elite. By Helmholtz's day, however, it had largely evolved into a general science society devoted to the popularization of science. He joined shortly after he arrived in town and became an active member: he gave an invited presentation on Brücke's dissertation on the physical-chemical foundations of osmosis. In 1852 he became a director of the society and served as its president in 1853–54. The second was the German Society of Königsberg (Deutsche Gesellschaft zu Königsberg), a literary organization also founded during the Enlightenment (1741), but one with a strongly patriotic-monarchical bent, devoted to encouraging the appreciation of German history, language, and literature, as well as the fine arts and related matters, above all as these pertained to Germany and

Prussia. Reasoned discussion was its hallmark. The third was the Association for Scientific Medicine (Verein für wissenschaftliche Heilkunde), founded in late October 1851 by a small group of Königsberg medical doctors. It, too, was a town-and-gown organization; it aimed to promote medicine as a science and an art and to gather together Königsberg's medical practitioners. Helmholtz stood at its center: he was a founding member and was elected its first chairman (a position he held until he left the city permanently).[5] These three societies helped give Königsberg a liberal ambiance and reputation, and they served as vehicles for Helmholtz to develop relationships with Königsberg's nonacademic cultural and professional elites. They helped him cultivate his own career, while serving the public weal. His participation in them paralleled that of his father in similar associations in Potsdam.

In August 1849 Hermann and Olga moved into the first of their three successive residences in Königsberg. Shortly after they had settled in, Olga's mother also moved there, to live in or near their residence. Hermann and Olga filled their home with music. He loved being married and gave this unsolicited advice to du Bois–Reymond:

> After we got our household set up, it was quite nice and comfortable. We could thoroughly enjoy the happiest time of life without disturbance. In all good conscience, I recommend that you find as . . . charming a wife as I have found. For apart from the existential comforts that a young man can never provide and the elimination of many things that one must otherwise necessarily worry about, it [married life] makes the mind feel so completely satisfied in the here-and-now and gives such a calm security of possession that even my ability to work has increased considerably.

On 22 June 1850, Olga gave birth to their first child, Katharina ("Käthe") Caroline Julie Betty Helmholtz. The newborn was a "well-formed and healthy little girl," and both daughter and mother did well at first. But shortly after Käthe's birth, Olga started coughing a great deal.[6] The cough, which may have been brought on as much by Königsberg's raw climate as by the stresses of childbirth, became chronic and fateful for her, her daughter, and Helmholtz himself.

The couple made friends easily. Among them were the anatomist Martin Heinrich Rathke and the physiologist Wilhelm Heinrich von Wittich. Anatomy and physiology had become separate disciplines at Königsberg in 1826. Rathke, a developmental biologist, was a former student of Müller's and was von Baer's successor. Wittich, a histologist, studied with Helmholtz and became an unsalaried private lecturer (*Privatdozent*) in 1850; he succeeded Helmholtz (in 1854) as an extraordinary professor of physiology. At first, Helmholtz was not especially impressed with either man. But he later revised his opinion of Wittich, characterizing him as "a talented and skilled microscopist." He saw more of him than of anyone else in Königsberg and considered him "especially useful" because he willingly taught histology, which Helmholtz did not want to do.[7]

When Helmholtz first arrived, he anticipated having closer collegial relations with the "mathematicians" (i.e., the physical scientists) than with the medical professors. Though he came to know the astronomer August Ludwig Busch slightly, he never became close with Neumann, "the most important" physical scientist there. Neumann, he said, seemed "somewhat difficult to get close to." Though he studied Neumann's electromagnetic papers carefully, and though Neumann provided him with some help in his investigations, the "difficult" Neumann ignored Helmholtz's work, kept his seminar separate from the newer and younger scientists at Königsberg, and did not incorporate Helmholtz's work into it.[8]

By contrast, the Helmholtzes socialized with Wilhelm Friedrich Schiefferdecker and with Richelot (and their wives). Schiefferdecker, a public hygienist, was a cofounder of the Association for Scientific Medicine and a leading figure in the Physical-Economic Society. As for Friedrich Julius Richelot, Helmholtz characterized him this way: "a local, brave mathematician, who, however, is somewhat confused in regard to non-mathematical logic and who lectures here on mechanics. After a hard struggle I've finally converted him to the conservation of force . . . so that the latter has in fact become official at this university." That was still a lonely battle, however: Helmholtz lamented that his essay had yet to be reviewed in the *Fortschritte der Physik* and thought he would have to convince his colleagues about force conservation on a one-by-one basis, as he did with Richelot. (He himself eventually reviewed his own essay in the *Fortschritte*; before 1851, no one else seemed interested in doing so.)[9]

There were other new friends as well, including August Werther, the professor of chemistry, with whom Helmholtz walked to the university daily; Ludwig Friedländer, the professor of history; and the family of Friedrich Carl Ulrich, a local judge and relative. Along with Wittich, these were his (their) closest friends in Königsberg. By the time of his first spring there, he judged Königsberg to be "a splendid place to work precisely because it doesn't offer much else in the way of temptations and yet keeps intellectual life sufficiently well astir." Their other friends or acquaintances included Moritz von Adelson, the Russian general counsel; Eduard von Simson, a professor of law and an important liberal political figure, who became Königsberg's representative to the Prussian parliament (Landtag) in 1849; and Theodor von Schön, once a leading reformer and liberal figure of Prussia, who had retired to his estate outside Königsberg. Schön was, however, still a well-known and influential man in the city and province. He was a former student and disciple of Kant's and a friend of Johann Gottlieb Fichte's. An Anglophile who advocated Smithian economic principles and industrialization for East Prussia, he was a former liberal minister of state, governor of East Prussia, and adviser to Frederick William IV. He was a strong supporter of the university in general and a warm friend to many of its academics. He was, furthermore, the "protector" of the Physical-Economic Society and president of the Association for Science and Art (Verein für Wissenschaft und Kunst). To be in Schön's set was to be in the world of liberal, progressive, and Enlightenment values and practices. Helmholtz visited him at his country home.[10]

He found that politics was widely discussed in Königsberg. "[Johann] Jacoby is currently the idol of the democrats," he told his father, "while the others proclaim the most excessive abhorrence against him." Jacoby was a Königsberg medical doctor, a radical democrat, and a sometime parliamentary representative in Berlin, who in 1849 was charged (for the second time) with high treason and later acquitted. This was not someone Helmholtz could support. "The democrats talk about him with the most pompous expressions," he sneered: "how he eats and drinks, and who first said 'Cheers,' as if Emperor Napoleon stayed in Krähwinkel."[11] Helmholtz himself avoided all overt political activity.

As *homo academicus* Helmholtz at times complained about his situation, as he did to Carl Ludwig about his Medical Faculty colleagues. Ludwig, who in 1849 had managed to overcome his liberal political handicap in Prussia by getting an appointment as professor of anatomy and physiology at the University of Zurich, counseled him that he would encounter difficulties within all medical faculties. He thought Helmholtz's position was an enviable one, since it allowed him to "get into the superior circle of the physicists and mathematicians." Nor did Ludwig, who considered Helmholtz "an unshakable member of the liberal party," hesitate to express his liberal political opinions to Helmholtz. He deplored Prussia's "loathsome egoism" and nationalism; he said its standing among Swiss politicians was as low as it had been in 1806, when it was decisively defeated by French forces, and predicted that it would eventually have to pay for responding only with force to any and all calls for political change. Ludwig deplored the politically reactionary regime that had enveloped Prussia since 1849 and felt comfortable expressing such sentiments to Helmholtz. He hoped Helmholtz would visit him soon, saying his house was "always open" to him and "It will be an honor to have a guest like you."[12] Helmholtz did indeed soon grace it. In the meantime, he was preoccupied with two scientific issues of great moment to them both.

Slow Nerves

It was (and is) a central ideological tenet of the German university system that teaching and research mutually reinforce one another. Whatever the truth of this claim, it was the reality for Helmholtz during his second academic year at

Königsberg. The Ministry of Culture had called him there to teach physiology and general pathology, and these pedagogical responsibilities brought him "above all two valuable pieces of fruit," by which he meant his discovery of the speed of nerve impulse and his invention of the ophthalmoscope.[13] The former achievement overlapped the latter, and the two must be seen within both their local pedagogical and their European research contexts. In hiring Helmholtz, as in hiring Brücke the year before, the Königsberg Medical Faculty implicitly indicated their strong commitment to conveying to their students an understanding of the physiological (i.e., natural scientific) foundation of medicine. They wanted their students to have the sort of intellectually rigorous education in medicine that students in the physical sciences were getting from Neumann and his colleagues.

When Helmholtz arrived for his first semester, there were only some forty-three students enrolled in the Medical Faculty. That first semester only seven registered for one of his classes and only three to five of them showed up, "depending on the weather." His teaching duties were light (though he found preparing lectures demanding), and his material resources were limited (one room for storing instruments and doing experiments and a small annual budget for the same). But the situation was better than what he had had in Berlin. He taught physiology annually and general pathology every winter semester. His older colleagues considered general pathology "the finest blossom of medical science." Helmholtz himself took little pleasure in teaching it.[14]

His pedagogical abilities appear to have been no more than satisfactory. Brücke harbored doubts about Helmholtz's lecturing abilities, but he thought Helmholtz's "solid knowledge" and the small number of students in Königsberg meant that he would do well there. Ludwig thought him a sufficiently decent teacher, though one who certainly lacked du Bois–Reymond's gift for lecturing: "Given the clarity of his mind," he said, "Helmholtz can never speak badly; perhaps, however, this very clarity makes his speech appear dry." After Helmholtz had taught for two years, his Medical Faculty colleagues declared, "He has proven to be a very ardent and effective teacher, and has instructed his listeners with much success."[15]

If it is difficult to judge how effective he was as a teacher, it is much less difficult to judge his effectiveness as a researcher. During this period he produced a pathbreaking analysis of the finite speed of nerve impulse, which had its origin in his work of 1848–50 on muscle contraction and also reflected a broader research program and methods. It was a variation of Eduard Weber's recent study of muscle contraction, but unlike Weber's, it was concerned with the mechanical work performed by muscle, and thus it was probably meant to show the application of his own law of conservation of force to physiology. It employed a variation on a measuring device recently devised by Ludwig to yield a graphical method for relating the muscle's work to its energy. Finally, it adapted a galvanometric method, developed for other purposes by Claude Servais Mathias Pouillet, for measuring time intervals more accurately than the graphical method could do. (Making instruments or having them made was always one of Helmholtz's primary concerns: after he saw what little the local Königsberg instrument maker had to offer, he came to value Halske in Berlin even more.) His research also depended on Olga, who, "as the protocol coordinator of the measuring scale that must be observed," stood faithfully by his side during the experiments. "This is something that is absolutely necessary to do since, were I working alone, I would become completely confused as I should have to pay simultaneous attention to all the many things that are there."[16] As a result of his careful measurement, he discovered a time difference between the moment a frog's muscle was stimulated and its subsequent action.

Most of the work on measuring the speed of the nerve impulse in frogs was done during the winter holiday break of 1849–50. His preliminary results indicated that the nerve's speed was variable and in the range of approximately 24 to 38 meters per second; the precise speed—around 24 to 27 meters per second—was less important (and indeed was long disputed) than the fact that he had shown it to be finite and measurable, though unexpectedly quite slow, far too slow to be identified with an electrical current. Helmholtz asked du Bois-Reymond to convey his results "as priority protection" before the Physical Society, which the latter did on 1 February 1850. Helmholtz then rapidly produced five papers, including one in French for the Paris Academy of Sciences and one for Poggendorff's *Annalen*, while Müller read a report on Helmholtz's work to the Prussian Academy of Sciences on 21 January 1850.[17]

Müller enthusiastically told his "Dearest Friend" that he was greatly excited by Helmholtz's manuscript and thought this work had greatly advanced nerve measurement. He not only immediately announced it before the academy; he also arranged for publication in the next issue of its *Monatsberichte* and told Helmholtz that if he intended to publish further on this topic, he would publish his manuscripts "immediately" in his *Archiv*. Ludwig was equally enthusiastic. Du Bois-Reymond, by contrast, checked Müller's enthusiasm. He claimed—adding, "I say this with pride and sorrow"—that Helmholtz had written the nerve-impulse paper "in such an immeasurably obscure manner" that it was "appreciated and understood . . . in Berlin" only by him. He had to explain the paper, he claimed, to Dove, Magnus, Poggendorff, Mitscherlich, Müller, Peter Theophil Riess, and other academicians, and he had to rewrite it before he could get Humboldt to send the French version to Paris. Humboldt also needed a clarification of the work of the "admirable Helmholtz." Nevertheless, within two months he had arranged for its French-language publication in the *Comptes Rendus*. Müller's letters to Helmholtz suggest that he understood Helmholtz's work well enough. Those who supposedly needed du Bois–Reymond's tutorial were physicists and chemists of an older generation, most of whom had also failed to appreciate Helmholtz's essay on force conservation three years earlier. Still, Helmholtz was grateful to du Bois–Reymond for his editorial assistance, promotion of his work, and other services.[18]

Du Bois–Reymond went to Paris that spring to lecture on his own work in electrophysiology and found his French confreres initially skeptical if not hostile to it. He also tried to promote Helmholtz's work on the speed of nerve impulse, but in vain. Helmholtz's work was "scoffed at" in the Paris academy. At the Société Philomatique, where du Bois–Reymond also lectured on it, the reaction was only marginally better: "Although no one wanted to raise objections publicly, I was nonetheless plagued by them privately with the stupidest doubts and objections." He told Ludwig that, as regarded Helmholtz's work on nerve impulse, the French thought Helmholtz "a madman." That perception may have owed something to the organicist (i.e., nonreductive) orientation of French physiology. These practitioners were not likely to appreciate the importance of the new physiological instruments that du Bois–Reymond, Ludwig, and other young German physiologists were developing. Instead, leading French physiologists like Claude Bernard had a completely different orientation in biology: they emphasized the importance of the environment and dismissed efforts at physicalist reductionism. Their outlook may have also owed something to a more general midcentury French inability to respond to German challenges in scientific research by committing adequate resources for research, and to a measure of complacency about novelty in science (instead stressing oratory about known matters). In the end, du Bois–Reymond thought he had convinced the Parisians of his own work but not of Helmholtz's. He disliked their collective attitude, though: "In general, we have no idea of their mixture of stupidity, arrogance, ignorance, and vileness." Helmholtz naturally gave his full moral support to his friend and agent: "I was in part already previously convinced that they [the Parisian scientists] richly deserve the pleasant epithets that you give them."[19]

In the spring of 1850, in between semesters, Helmholtz began working on the propagation of nerve impulses in humans. He told an uncle that he had performed experiments on humans (as well as on frogs) that showed a propagation speed in humans of roughly 150 to 180 feet per second, "so that information from the big toe arrives in the brain roughly $1/30$th of a second later." He added a personal note: "We are very happy living here, even if Königsberg itself does not make a very great contribution to our happiness." He also told his father of his recent work and that he had determined that the more distant an impulse's point of bodily stimulation was from the brain, the longer the time interval between the two became. "I figure this finding to be a very great success; it will not fail to gain attention." Yet he also confessed to his parents that his report to the French academy had, unsurprisingly, not been well received: "Do not worry about it any further. It is simply not possible for the French to show an attitude of goodwill in such matters [of scientific discovery] by Germans. Besides, the provisional goal, that of generally drawing attention to the matter in the first place, has already been achieved." Ferdinand picked up immediately on the philosophical implication—it was a positive contribution to the age-old mind-body problem—of his son's findings: he was astonished to learn that thought and bodily action were serial, not simultaneous, events, as he had always assumed. He hoped Hermann could better clarify the meaning of his findings in terms a layperson could understand. Helmholtz sought to do so by suggesting physical analogies and explaining the inherent, unavoidable uncertainties in the measuring process (i.e., the so-called personal equation that Bessel had first explained in astronomy). Giving such explanations helped him better appreciate that different audiences, lay and professional (including medical colleagues), required different types of explanations (e.g., using simple graphs versus complicated precision measurements), just as he came to appreciate better the epistemological problems or uses of data, protocol, and error analysis in both the discovery process and the interpretation of results. His overall efforts between 1849 and 1851 show that scientific discovery and persuasion were neither singular nor isolated events but rather were continuous, closely related processes.[20] They were also a nice instantiation of the organic physicists' program and marked an important step in the beginnings of neurophysiology.

In July 1850 Helmholtz sent Müller the final version of his manuscript on the speed of nerve impulse in animals. By late August he found that "work and heat" had left him "somewhat tired." He said, "I had headaches more often than usual, and so I decided to give my liver a few days of walking on the beach and to scrub my brain clean with seawater." He spent time "with great success" in the company of Kirchhoff, who was a Königsberg native and who had studied mathematics with Richelot and physics with Neumann; Kirchhoff taught in Breslau. But he could not rid himself of his hostility toward the French. He told du Bois—Reymond that the Paris academy's report on the latter's electrophysiological works was filled "with the greatest possible ill-will." He thought it "perfidious" that it had failed to acknowledge fully du Bois—Reymond's earlier work. "And then, in an attempt to find a contradiction, they [the French] get entangled in the most sterile theoretical discussions, while all the while they themselves could not perform the experiments." They did not even debate du Bois—Reymond's work on frog-muscle currents. He then gave full vent to his anti-French rage:

> The doubts on the connection of the experiment on humans with those on
> frogs can only be expressed in this way by those who refuse to see how it has

worked. They show themselves at their most impertinent in regards to your theory of the phenomena, which they do not understand, and in their good advice try to burden you with ever-more rigorous methods. I have become extremely angry over this way of proceeding. It would be very good to roundly blame these fellows, since those in Germany who are ignorant of the matter could then find in their judgments some nourishment for their own glosses. When one considers that, apart from the withered [François] Magendie, there does not exist a single noteworthy physiologist in France, and that the way they speak when awards are announced, with which they crown such a scoundrel as Bernard for his wretched work on pancreatic juice — "a work that in any century can occur but rarely" etc. — so one almost wishes to conclude, for the moment, to completely ignore these gangs in their nullity until they have finally learned to see how bad things are with them. Let's wait and see whether they will have the honor to recognize you with the next physiological prize. It is beyond all doubt that there is no one who can justly rival you.[21]

Such venom directed against his colleagues beyond the Rhine long remained within him.

Measuring the speed of nerve-impulse transmission within the animal and the human body was a historic development in science, one comparable to, say, the first measurements of the speed of light in 1676 by the Danish astronomer Ole Christensen Rømer. Yet Helmholtz's results and technique were not only too revolutionary for the French; they were also too much for the older generation of scientists in Berlin to fully appreciate. Only young friends and associates like Ludwig, who knew Helmholtz and shared his intellectual orientation in physiology, needed no special convincing before discovering the value of this accomplishment. Ludwig wrote him: "Your works on the speed of so-called nerve forces have filled me, as an adherent of your person and viewpoint, with great pride." Helmholtz thought others would in time come to understand his work, too. Yet his work at persuasion was hardly finished. On 13 December 1850, he lectured before the Physical-Economic Society on methodological issues in measuring nerve impulses and on his measurements in humans. A week later, he had du Bois–Reymond deliver essentially the same lecture before the Physical Society in Berlin. Now, as in the past, that society was an important outlet for stating his views. Though his work on the speed of nerve impulse proved fundamental to the future of neurophysiology and, much later, to the emerging field of neuroscience, few beyond his immediate circle initially found it credible. The very idea of a noninstantaneous, finite, measurable time-lapse "between the infliction of a wound and the feeling of the injury," as John Tyndall later so vividly put it, was initially considered "preposterous." No wonder that Helmholtz soon put on a roadshow to promote his work in person.[22]

Seeing the Living Retina (or Inventing the Ophthalmoscope)

While conducting research on the speed of nerve impulse, Helmholtz also had to teach. In particular, his course on sensory physiology in the winter semester of 1850 led him to present students with a theory of the eye's illumination and to demonstrate such illumination. From the late eighteenth century onward, many investigators had noticed the pupil's luminous appearance. Bénédict Prevost, for example, argued in 1818 that such luminosity was reflected light that had entered the eye, while William Cumming in 1846 and Brücke in 1847 independently showed that healthy human eyes could be made luminous. According to Helmholtz's own account, Brücke came within "a hair's breadth of the invention of the ophthalmoscope." It was only Brücke's failure to inquire about the return path followed by the light sent into the observed eye and how an optical image is formed in the observer's receiving eye that prevented him (in principle) from becoming the ophthalmoscope's inventor. In fact, Helmholtz thought there were five to ten other German scientists "who doubtless would have performed in a completely logical way exactly the same as I did if they had been put before the same tasks and under the same conditions."[23]

Building on Brücke's finding that reflected light could be emitted from the eye, Helmholtz, ever the physically minded physiologist, recognized that light rays entering the eye were reflected back precisely along the same path they had entered. The novelty of his contribution lay in explaining the physical optics behind the eye's luminosity and in creating an instrument that could capture the returning light for view. It took him eight days to construct his first, crude ophthalmoscope: it was a jerry-built combination of fragile cardboard, eyeglass lenses, and a covering glass used in a microscope. Yet it was good enough to make him the first person ever to see the living human retina. He called himself a "diletantte in practical mechanics" but one who, in seeking a new way to investigate a topic, had a penchant for building a crude model instrument. He would probably have given up constructing such an instrument, he said, except that he was convinced on theoretical grounds that it had to work.[24] Though it was difficult enough to use, the pleasure he received in being the first to observe the living retina was probably akin to Galileo's when, in 1609, he became the first to view the moon and other heavenly bodies systematically through his own improved instrument. And like Galileo, Helmholtz immediately foresaw the potential material and professional benefits that might arise from his invention.

He explained the invention to his father in a letter. He also planned to send him a copy of the article on nerve and nerve-reaction times, but in the meantime he sent the manuscript of a popular lecture on this topic that he had recently given at the Physical-Economic Society. He wanted it back, however, because various people in Königsberg, including His Excellency von Schön, wanted to read it. He was also constructing a new apparatus for his work on time intervals, but at the moment he had no new results to convey. Then he got to the main point:

> I have . . . made a discovery that can possibly be for ophthalmology of the greatest importance. It was obvious; it demanded no more knowledge than what I had learned about optics in the Gymnasium. It now seems to me laughable that other people, and myself as well, could be so wooden-headed as not to have found it. To wit: it is a combination of glasses through which it becomes possible to illuminate through the pupil the dark background of the eye, and to do so without applying any blinding light and simultaneously to see exactly all details of the retina, indeed even more exactly than one sees the exterior parts of the eye without enlargement. . . . One sees the blood-vessels most elegantly, the branched arteries and veins, the entrance of the optical nerves into the eye, etc. Until now, there was a series of extraordinarily important eye diseases that combined together under the name "black star" [i.e., an ill-defined group of eye diseases, including blindness], a terra incognita, since neither in life nor, for the most part, even in death did one learn something about the changes in the eye. Thanks to my invention the most specialized investigation of the interior structure of the eye will become possible. I have treated my invention very carefully as a Columbus's egg [i.e., a simple, clever way of solving a difficult problem] and have immediately let it be announced as my property in the Physical Society in Berlin. I am at present having another such instrument built, one which is better and easier to use than the sticky-paste version that I've used so far. If possible, I'll do some investigations on sick patients with our local ophthalmologist and will then publish the matter.

Helmholtz's vivid description of his invention indicates his enduring debt to his Gymnasium and shows that from the very start he viewed the ophthalmoscope as his intellectual property and took immediate steps to so claim it publicly. He wrote up a brief communication of his method and invention, which du Bois–Reymond read before the Physical Society. Even the otherworldly, romantic Ferdinand understood his son's proprietary intent. He wrote Hermann that since, as he had heard, it would not be possible to publish his description of his instrument in Müller's journal, it was "all the better that you now want independently to bring out your treatise under your own name." He was eager to read it and planned to get Puhlmann to communicate it to the Literary Society. "The discovery on the observation of the eye, even if it did not presuppose so much knowledge," Ferdinand foresaw, "will probably rapidly create a name for you because it seems immediately practical. One wonders whether you shouldn't get a patent for the instrument of observation."[25]

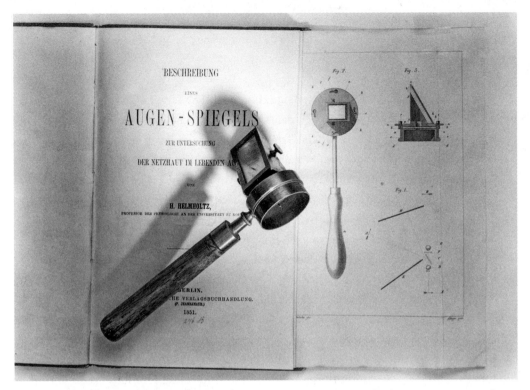

Fig. 5.2 Helmholtz's ophthalmoscope, resting on his booklet *Beschreibung eines Augen-Spiegels zur Untersuchung der Netzhaut im lebenden Auge* (Berlin: A. Förstner, 1851). Deutsches Museum, Munich, No. 30073.

The formal essay itself appeared in the autumn of 1851. It was a forty-three-page pamphlet that laid out in detail the relevant laws of physical optics and the workings of his ophthalmoscope. Helmholtz noted that an observer cannot normally see any light from the observed eye because, in attempting to do so, the observer necessarily cuts off the incident light to that eye. To avoid that problem, he sent light indirectly to the observed eye from a lamp (or sunlight) through reflection from a small, half-silvered plane glass plate. The observed eye thus sees only the light's mirror image, thereby allowing the observer simultaneously to view it. The reflected light strikes the retina (the eye's dorsal region) and there produces an inverted optical image. The light is in turn reflected from the retina and the image is erected so as to allow the observer to see it upright. Seated within a darkened room illuminated only by a single source of directed light, the observer could see the living human retina in great detail. Helmholtz's ophthalmoscope—his original German term, *Augenspiegel*, is more suggestive (see fig. 5.2)—was thus fundamentally the creation of an instrument by which reflected rays were formed into a distinct image in the observer's eye. His essay, like the essay on the conservation of force, appeared as a separate publication, not as an article in a scientific journal. As compensation, he received the modest sum of eighty talers.[26]

There was no patent and there were no royalties. Helmholtz later conceded that during the first half of his life, when, in his words, "I still had to work for my position in the world," his effort was motivated not only by a combination of "higher ethical motivations," "curiosity, and a feeling of duty as a state official [i.e., a university professor]," but also by "egotistical motives." He thought other scientists were subject to the same material pressures, at least until the later phases of their careers, when, for those who had not given up science altogether, there "comes into the foreground for those who want to work further a higher view of their relationship to humanity." Yet his invention was also a matter of "luck."[27] Beyond the pedagogical task of having to lecture to medical students on the eye's illumination, there was a conjunction of diverse developments that led to his becoming the inventor of the ophthalmoscope.

The combination of luck, broad training (in physics, physiology, and medicine), and pedagogical needs led Helmholtz later to declare that "the ophthalmoscope was more a discovery than an invention." Indeed, his fortunate discovery (invention) embarrassed him. He too modestly described his achievement as "merely the work of a schooled worker who had learned to apply correctly the knowledge and the means already created by his predecessors." He considered his achievement an act of applied science. He did not hesitate in the future to cite it as an instance of how science pursued for its own sake may someday bring great practical consequences for society. It brought him more fame than anything else he did before or after.[28]

The Roadshow

Like corn and soybeans, scientific results are a commodity; they too exist within a market. Scientists discover, invent, trade, purchase, and advertise their scientific goods and services—facts, concepts, theories, instruments, equipment, conversation, and more—and they do so both for themselves and for nonscientists. Helmholtz well appreciated this view of science as a marketplace, and he knew, too, that those who want to achieve distinction sometimes have to bring their products to market even before they are fully developed.[29]

In the summer of 1851, Helmholtz capitalized on his recent discoveries and his invention by undertaking a tour of German-speaking universities. His ostensible purpose was to visit German physiological institutes; the trip's principal purpose, however, was to display what he had to offer his colleagues. He was marketing his work on measuring the speed of nerve impulse and the ophthalmoscope—as well as himself. He proved to be an exceptional salesman.

He traveled for nearly two months (early August to late September), mostly on his own and mostly in the German states, but occasionally with others in Switzerland and Italy and also briefly in France and Austria. The Prussian Ministry of Culture underwrote the trip with two hundred talers in travel-expense money. He left Olga and their one-year-old Käthe behind in Dahlem; it was the first of what became annual separations from his wife and family for several weeks or more each summer. (The trips were far too strenuous for them.) He himself felt "rather fatigued" at the trip's start—he had been ill with gastric fever that spring—and hoped the trip would help him "again re-start the matter lost" from his "life machine." To have spent his summer in Königsberg, as he heard Kirchhoff was to do, smacked of "tastelessness."[30]

The first stop was Halle, where his old friend the chemist Wilhelm Heintz, and his family, received him warmly; with them he spent two pleasurable days. Then he headed for Göttingen. The railway journey through the Thuringian countryside much impressed him, as did "a series of very romantically situated cities." He arrived in Göttingen at 3:30 a.m. and wrote Olga: "First of all, a hundred thousand of the truest greetings from the heart from afar, and the report that more than ever I love only one who was once named Dötchen von Velten." He teased her that on his trip (so far) he had yet to find anyone better than her: "If everyone had as much reason to be satisfied as I do, then the world would look like a better place."[31] Distance made his heart grow fonder.

He found the University of Göttingen to be "very rich." It had more than six hundred students, twice as many as Königsberg, and they were "most well-to-do." The professoriate was "the city's aristocracy." They felt superior to the (other) townsfolk. Unfortunately, the king of Hanover, Ernst August II, was in town for several days to dedicate a new hospital clinic, visit the university and other sites, and receive faculty and students, and his visit made it more difficult for Helmholtz to meet Göttingen's professors. Though the king had initiated anticonstitutional measures in 1837 that led to the infamous dismissal of seven professors—the "Göttingen Seven," whose numbers included the physicist Wilhelm Weber—and though he had opposed the Frankfurt Parliament in 1848, he had nonetheless treated his subjects liberally during the revolution. "The Hanoverians," Helmholtz wrote Olga, "are on the whole satisfied with their political situation. They are almost the only Germans who since 1848 have not had any of their political rights violated, and in recognition of this they seem to depend very much on their king, in spite of his perverse character."[32]

He first visited August Ritter, the professor of philosophy, and his wife. They received him "very warmly," since Ritter was an old and dear friend of Ferdinand's and Helmholtz himself had known Ritter and his family from Berlin. The Ritters arranged a party for him and invited all the professors he wanted to speak with. "It's extremely pleasant to travel around the world and go to parties that are arranged to honor us." He met, among others, the physiologist Rudolph Wagner, "an older man, about whom one notes a somewhat self-conscious importance." He thought that Wagner, like Alfred W. Volkmann, the professor of physiology at Halle, was "not completely at the level of necessary physical knowledge" that he should have had. Wagner "feels it, and is cautious enough not to get lost" (in Helmholtz's scientific explanations), he said. Yet in fact Wagner was one of the more physically oriented physiologists of the older generation, a group that included figures like Müller, Jan Evangelista Purkyně, and Magendie.[33]

In Göttingen he also met Weber, whom he considered "after Neumann, the first mathematical physicist of Germany." Weber was one of Europe's leading electromagnetic theorists, and Helmholtz already knew his work: he had studied Weber's book on electrodynamics (*Elektrodynamische Maassbestimmungen*, 1846) while preparing his essay on the conservation of force, for example. Weber showed him, "with somewhat less-than-smiling friendliness than his brother in Leipzig did" (i.e., either Ernst Heinrich or Eduard, both of whom were physically oriented physiologists), "many interesting pieces of physical apparatus that were largely completed." He also met Carl Georg Bergmann, an extraordinary professor of anatomy and physiology, who, like Helmholtz, was much interested in the problem of animal heat, though Helmholtz was not impressed with his work. Furthermore, he met Christian Georg Theodor Ruete, a professor of ophthalmology, who had done "valuable work on the physiology of the eye"; Wilhelm Baum, a clinically oriented professor of surgery and also an ophthalmologist; and Johann Benedict Listing, the professor of physics, who worked in mathematical optics and who was "very worthy," though Helmholtz had not been aware of his work until now. He demonstrated his ophthalmoscope before all of them. Ruete, who the following year made noteworthy improvements to Helmholtz's ophthalmoscope (by replacing the plain glass plates with a concave mirror), wrote of Helmholtz's visit: "When, under his guidance during his appearance in Göttingen, I first looked with the help of that instrument at the optical nerve with the Arteria centralis re[t?]inae, it immediately became clear to me that in this way much could be gained for the diagnoses of the diseased eye of this organ." Helmholtz had little doubt about the importance of his demonstrations here (and elsewhere): "The ophthalmoscope is truly excellent for my tour. I demonstrated it this morning, and here too created a kind of sensation." It impressed people, he said, and at least two ophthalmologists placed orders with Egbart Rekoss, his instrument maker in Königsberg. "I also demonstrated my frog curves everywhere," he added. He drove the point home to Olga: "All these people also show me great respect and friendliness; they all give me whatever time they can. It was very pleasant to see that they have worked their way into my somewhat difficult nerve work [i.e., measuring the speed of nerve impulse] and agree with it or, so at least it seems according to Weber's judgments, that they have sufficient trust in my physical knowledge to believe in my results." Yet the Göttingen scientists apparently did not have a word to say about his work on the conservation of force. Finally, he also met the philosopher Rudolf Hermann Lotze, who had "worked a great deal on the principles of pathology and physiology, but who, unfortunately," was "hypochondriacal and introverted." "It may not be possible to have an intellectual exchange with him in such a brief period of time."[34]

He left Göttingen at 1:30 a.m. by train; returned for a day to Kassel, went briefly to Marburg, and then stopped in Giessen for a day. The pace was already grueling, and his family's health (both Olga's and Käthe's) and money matters were, as always, on his mind. He adored the setting of the University of Marburg, which lay "beautifully between high and steep mountain forests." There he visited Karl Hermann Knoblauch, an extraordinary professor of physics, and Hermann Nasse, the professor of physiology. Knoblauch, whom Helmholtz knew from Berlin, had worked in Magnus's laboratory, had habilitated (i.e., published a new, special research dissertation or book beyond the doctorate and so effectively received a second doctorate and was qualified to teach [receiving the *venia legendi*]) at Berlin, and had cofounded the Physical Society. The well-to-do Knoblauch invited him to his home for coffee, and Helmholtz commented, "He showed me his expensive instrument collection (totally his own property) and also his own interesting experiments, whose invention certainly was not his but which nonetheless have scarcely been repeated in Germany." But he thought that Knoblauch had no ideas of his own, that he only tested or tweaked other scientists' ideas. He thought little more (and perhaps even less) of Nasse, who belonged to the school that "gladly reduces life to the most mystical ideas and stands in direct opposition" to Helmholtz and his scientific circle. He had done no important work, though some of it was "diligent and valuable." "So he receives me only courteously, avoids everything scientific, and in the end expressed doubts about my nerve-conduction studies." When Helmholtz explained his frog-muscle curves to him, however, Nasse's attitude suddenly changed: "Then he became completely different. He begged me to stay longer in Marburg. I had no desire to do that, however, and so we parted the best of friends."[35]

He went to Giessen in part, as many others did, to meet Liebig, "the king of chemists . . . as he himself and his students consider him." But Liebig had gone to London to attend the 1851 Exhibition "and to let himself be feted by the English. I would have liked to meet him," Helmholtz conceded to Olga. So he settled for Liebig's son, Georg, a medical doctor who had worked in Magnus's laboratory, who showed him around his father's "empty laboratory . . . , to which flock students from all of Europe and America in order to do practical [laboratory] exercises." This taught Helmholtz an important lesson about laboratories: "I was astounded that there were no especially important furnishings; on the contrary, everything had dirt on it. There were few laboratory workers there. It made a strange, precise contrast to the more purposeful, much better equipped, well-ordered, and cleaned laboratory of Heintz and others. One sees that it isn't the external things that make [a laboratory]. For in spite of all the vanity and arrogance, Liebig is indeed the most important of living chemists, and a teacher of colossally widespread influence."[36] Even the best-equipped laboratories were of little avail if their leaders had few or no ideas.

Helmholtz had also gone to Giessen to see Theodor Bischoff, the professor of anatomy and physiology and a former student of Müller's, and Bischoff's prosector, Konrad Eckhard, who had previously served as Ludwig's assistant in Marburg. Eckhard had there performed experimental work on nerves and muscles. Helmholtz considered him an ally of the organic physicists and a man of great promise. Eckhard arranged a dinner party to introduce him to Giessen's younger scientists, and Helmholtz much appreciated getting to know him. By contrast, he portrayed Bischoff unflatteringly as "a fat man," with a weird hairdo and "a nose whose bridge projects out more than its tip"; he was as little impressed with Bischoff's anatomical research as he was with his anatomy and coiffure. The best he could say about him was that he had "been driven out of his earlier mystical outlook by Liebig." "He is now completely inclined to our investigations, yet he has difficulty understanding and judging them." Helmholtz instructed him on the speed of nerve conduction and apparently converted him. But he "was very unskilled with the ophthalmoscope and could see but scantily; his wife saw more easily than he did." He thought Bischoff's wife—she was the daughter of Friedrich Tiedemann, the former professor of physiology at Heidelberg—was "of a more important nature than her husband." He discussed politics with her. He found that the people of small German states like Baden and Thuringia remained deeply shaken by the events of 1848–49. Yet the Thuringians with whom he spoke welcomed the steadfastness of the Prussian military during this politically uncertain period, and, despite certain misgivings, preferred to be "Prussian subjects."[37]

From Giessen he headed toward southern Germany. It was his first trip to the south, and he was struck by its inhabitants' drinking of light wine rather than water: "I've consumed a bottle of the stuff daily, but don't notice the least discomfort, whereas at home I can't bear regularly drinking even a glass without getting headaches." He was astounded to see that women in the south could skillfully balance loads on their heads and that donkeys were widely used to carry goods. Yet he judged the women harshly and hastily: he thought the national dress of the Hessian farm women was awful, "all the more so since the women themselves" were "almost generally hideous."[38]

In "magnificent" Frankfurt am Main, he saw old buildings, including its famous dome and city hall, in serious decline and new ones that suggested modernity. "Frankfurt is now the city of uncultured moneybags and their palaces; by contrast, Berlin's best sections have to hide their faces in shame." He lived "sybaritically" in a middle-class inn. He told Olga, "A Frankfurter hotel is really something worth seeing." Though never a gourmet, he was impressed by Frankfurt's culinary offerings. He took a swim and then went to the Paulskirche, which made him "very melancholic." He visited the municipal museum, where he "again reveled in" Carl Friedrich Lessing's *Johann Huss zu Konstanz* and *Ezzelino da Romano* "and two small landscapes by the same." He clearly enjoyed seeing Lessing's works, which were widely perceived as anti-Catholic. He reported further on the museum's holdings, which included a few "pretty landscapes" and "lots of unimportant older and new items." The "very much praised" allegorical pictures by Friedrich Overbeck and Friedrich Wilhelm Schadow, which Frankfurt's Catholics seemed to have brought in "to form a counterweight to the *Huss*," he "didn't like at all." But he did like the "quite beautiful gypsum castings from Antiquity" and "a very beautiful shield of Hercules according to Hesiod's description by [Ludwig Michael] Schwanthaler." "I saw the Laocoön here for the first time in a complete group; it compared however to other objects of Antiquity like Victor Hugo to Sophocles."[39] He knew what he liked and disliked. He left for Heidelberg that afternoon.

Heidelberg's countryside and the town itself also impressed him. The famous castle, where he "enjoyed sunset and wished that Dot were" with him, "surpassed everything." He was surprised to find many English and French, as well as German, families there, all of which made the town a still more "wonderfully romantic" place.[40]

He devoted a day to seeing the physiological institute, which meant meeting Jacob Henle, the professor of anatomy and physiology and a close friend of Müller's. Henle, who had been a student of Müller's, was "somewhat Jewish"; du Bois–Reymond had unjustly made Helmholtz suspicious of Henle. But Helmholtz had good reason to like Henle, who, as he excitedly told Olga, "opened up" for their future "a potential success," to wit: Henle and other younger members of the Medical Faculty wanted to bring him to Heidelberg. The situation was complex, however. Henle had originally been responsible solely for physiology, while Tiedemann had been responsible for anatomy. But then "furious frictions" occurred between the two; the faculty asked Tiedemann to apologize, but instead he resigned, and Henle had to assume responsibility for both anatomy and physiology. Henle and his like-minded colleagues now wanted Helmholtz as their physiologist. Tiedemann's former supporters wanted Bischoff, who just happened to be Tiedemann's son-in-law. More precisely, they did not so much want Bischoff as his patron, Liebig, who said he might come to Heidelberg if Bischoff and others of his choosing were also brought there. Various university officials believed that, for political reasons, the University of Heidelberg had "terribly declined," and they wanted to return it "to its previous brilliance." Hence they sought chemistry's king. But it appeared that the king would not come, which in turn meant that bringing in Bischoff—a man, as Helmholtz saw it, of limited scientific abilities, who was under Liebig's sway and who could only cover the very same fields already covered by Henle—served no good purpose. Henle told Helmholtz that he had already recommended him for the position, even before meeting him in Heidelberg. "And, since we got on excellently while I was here, I was in a position to say all kinds of flattering things to him, for which he is not insensitive, and so he will not take these recommendations back." He wrote Olga: "So let's see what happens. The Heidelberg region is not bad. The Germans have become somewhat accustomed to staying away because, at present, there is a lack of teachers, but students from North America, Brazil, England, France, Greece, and Russia still come. Life is ridiculously cheap."[41]

Henle's idea of bringing him to Heidelberg may not have been his own. A month before Helmholtz's visit, their mutual friend Ludwig wrote Henle to say that Heidelberg only needed a chemist—he strongly recommended Adolf Strecker, his own teacher and a minor student of Liebig's, over Liebig himself!— and a physiologist to recover its standing in science and medicine and so become competitive with Vienna, Prague, Berlin, or Würzburg. Ludwig thought Heidelberg needed someone who was as scientifically "virtuous" as Henle, and he said there were in principle three possibilities: Brücke, who was not available; du Bois–Reymond, who would not leave Berlin; and Helmholtz, whom he called a "truly noble human being." He noted that at Königsberg Helmholtz had "no listeners. So no great pecuniary sacrifice should be required to get him. And what a win that would be." Ludwig proved himself to be an excellent friend and member of Helmholtz's small but growing network, and he strongly recommended Helmholtz to Henle. He was perhaps an even better friend than du Bois–Reymond, who told Henle that he favored Ludwig over Helmholtz for the appointment, not because he thought Ludwig a better (or worse) physiologist than Helmholtz but because he thought Ludwig would make a better fit in "little Heidelberg, with all its intrigues and splits," than Helmholtz would.[42]

Henle showed Helmholtz around the institutes of anatomy ("excellent") and physiology ("extremely poor"). They strolled around Heidelberg together. Henle had him over for tea that evening, and Helmholtz met his wife and children; he thought the Henles had a "charming relationship." As for Heidelberg, its location and environment were "wonderful, and very easy to enjoy."[43] He sought to sell Olga on it just as much as he sought to sell Henle on himself.

With Ludwig in Zurich

On Henle's advice, he visited nearby Baden-Baden, and along the way he spent a few hours enjoying the Schwarzwald, dined in an old castle, and "ate ice cream in the garden of the Conversation House, so as to watch the bathers." But he could not fully enjoy the region's magnificent landscape, he confessed, without Olga, and so he soon went to Kehl and crossed the Rhine into Strasbourg. It was his first step into the "*République française*," as he sneeringly called it: "There's a lot of amusing things here. Everywhere *Liberté, Fraternité, Egalité* is displayed on every public building, *Propriété de la nation*, and on many private houses other terrible democratic mottos. The peasantry and the lower urban classes seem to be just as in Baden, only they seem to be more stupid. The better parts of the city look completely French." He visited Strasbourg's renowned cathedral and noted its disrepair but admitted that it was "in part extremely imposing and noble." He watched in amazement as crowds gathered at noon to hear the cathedral's mechanical roosters crow, "to see the 12 apostles of Christ pass by, etc." He continued, "The peasantry seems to consider it a sort of daily, renewed wonder; they make a pilgrimage to see it. I too naturally took a look at the joke." He also had to put his French to use, which included talking to two salesladies. The first, a "very fine counter lady," courteously endured his schoolboy French, and they "understood one another." The second was "so flattering as to wonder" that he was not French. (Her salesmanship skills were excellent: Helmholtz made a purchase.) "In short, I'd no longer any doubt that I could fight my way through France, something that until now I really had no hope of doing."[44] Not until fifteen years later did he again set foot in *la grande nation*.

That morning he went first to Freiburg, then "through the extremely romantic valleys of the Schwarzwald," and on to Schaffhausen, Switzerland. (It was his first of many trips into Switzerland.) He arrived late and stayed outside the city in a hotel whose terrace gave him a good view of Schaffhausen's waterfall, this "*Weltwunder*," as he called it. He found its effects that night to be rather unimpressive, but the next morning he "saw it completely differently." Its "most magnificent effects" could only be appreciated during the daytime. He walked onto a trestle to see the froth and the clouds of spume more closely, and he found the entire experience exhilarating and spellbinding. As always, water in motion impressed him. Then he had "a luxurious lunch" that was cheaper than any that could be had in Berlin.[45]

After lunch he traveled to Zurich, where he stayed with Ludwig, "who received me very warmly." Ludwig, who adored Helmholtz's "lively eyes," had sought to entice him to visit by telling him how beautiful Zurich's nature was. Du Bois–Reymond, who knew of the visit ahead of time, wrote Ludwig some ten days earlier: "You'll also be seeing Helmholtz, that giant of an intellect, there soon. One should only become a scientist when one has talent like him, that in such a quiet and relaxed way brings great things to light without effort." Helmholtz told Olga that he considered Ludwig to have "a truly noble and charming nature" and that probably Ludwig's wife had rid him of his "formerly boisterous nature." The next day he visited the University of Zurich's chemistry laboratory and went with Ludwig and his prosector, Georg Hermann Meyer, to the "Uetliberge." There, he exclaimed, "I occasionally saw the glacier with its majestic fields of snow. Compared to these mountains, all others are molehills."[46]

Three weeks into his trip, he got the first of only two letters from Olga. She told her "Dear Angel," "A wife without a man and household is like a dot without an i." She could be "reasonable" and hold out for the remaining five weeks, she said, but no more. She was deeply moved that her husband missed her so much but thought the beauty of southern Germany and Switzerland would take his mind off her. She wanted only what was good for him: he should enjoy life to the full, avoid danger, and not worry about her (except when he attempted some "risky undertaking"). His father had visited her and, though talkative and in a good mood, "was nonetheless somewhat piqued" that Helmholtz had not asked him "for more extensive instructions concerning your trip to Switzerland." She let him read (some of) Hermann's letters to her; he was pleased with all he read, though he became "enraged" over his son's "judgment about the Laocoön." Father and son would have much to talk about on this topic, she warned. She also reported that his sister, Julie, had come to Dahlem and painted "a very sad picture of their family life," in particular saying that Marie, the oldest of his sisters, ran the household and treated his mother "often in a very insulting, disdainful manner." Marie seemed "extremely egotistical." Olga said she was more than ready to move anywhere at any time with him and Käthe, "but above all to Heidelberg." By contrast, she liked Königsberg only because it had been good to her. She did not breathe a word of this possibility to his parents, since people like them, "who have so few interests in the larger world," tend "to talk more and with greater confidence to other people [about such matters] than it's good to do." She wrote, too, of Käthe, and sent "thousands of kisses."[47]

Helmholtz spent eight days with Ludwig in Zurich, where he was much taken with the region's beauty. He thought Ludwig a very warm-hearted person whose excellent image of Helmholtz's abilities was due partly to du Bois–Reymond. "If you would have heard all the encomia he said in his honest way to me," he wrote Olga, "you would doubtless have been very pleased with him." Ludwig constantly sought to maintain conversation, keep it à deux, and keep Helmholtz away from his Zurich colleagues, so that he could speak with Helmholtz "alone." They had broad-ranging discussions "concerning every possible physiological and physical subject." Ludwig worked indefatigably "in the best direction," and his students "enthusiastically" loved him. Helmholtz judged him a man of achievement and of "still greater" promise, yet thought him subdued and hypochondriacal.[48]

The two spent the mornings in Ludwig's institute, where Helmholtz saw his facilities, instruments, and experiments and met his junior colleagues, one of whom, "due to his arrogance and contempt for the world, was of no use." They spent the afternoons walking around the Zurich region. Helmholtz loved Lake Zurich and the surrounding mountains and was impressed with Swiss reliability in mundane matters. "In general, as far as the lower virtues of the human race are concerned—diligence, activity, prudence, and a not-all-too-strict honesty—the Swiss are perhaps the most excellent people that one can imagine. The canton of Zurich is a model of culture, without poverty, and all that under conditions of nature that are in no way favorable." On the other hand, Zurich "completely lacked all the nobler characteristics, so that life with them" appeared to be "a plague." He thought Switzerland's political structure effectively robbed its people of their ambition. "And their main support is a kind of craftiness, of which every decent person in Germany would often enough be ashamed of." Within the Swiss family, he further opined, wives were "always supposed to play a subordinate role," one that made them largely "more intimate with the maid than with the husband, even in the wealthiest and best-educated households." He thought Swiss women looked rather "crude" and sometimes acted "uncouth." Still, he was surprised to learn that they were allowed to attend university lectures, something that did not generally happen in the German states until after 1900. What surprised him most, however, was that Swiss marriage contracts allowed most "city ladies" to spend three or four weeks during the summer in the mountains.[49]

After Helmholtz left Zurich, Ludwig wrote Henle that he found Helmholtz well versed in every subject they discussed (with the partial exception of comparative anatomy) and that he was "at home" with and could discuss developmental biology, general and pathological anatomy, chemistry, and general botany "with whomever he wanted." Du Bois–Reymond thought Helmholtz's broad knowledge issued in part from his "indefatigable and . . . immense feeling of duty"; he could "do nothing that leaves a trace of incompleteness." Ludwig replied that Helmholtz's days with him were "a stage" in his own scientific development, and he would "long remember" them.[50]

In the Alps and Italy

He left Zurich to wander, first, in the Rigi, where by chance he met Peter Leopold von Schrenk, a zoologist who had studied in Dorpat and Berlin before taking his PhD degree at Königsberg. The two hiked together. This was the first of Helmholtz's untold number of hikes in the Alps. He found the Rigi stunning. He described it, especially its glaciers, to Olga in much detail. Its views inspired him. Sunset was "extremely imposing." Words could not describe the mountain and glacial setting and the effects of sun and water on them. The only disappointment came when he got to the top, where he found "a large number of people, for the most part a rotten pack of tourists." After he and Schrenk descended, they took a steamer to Lucerne. As with Zurich, he was thrilled with Lucerne's landscape and environs; the region reminded him of Schiller's description in *Wilhelm Tell*. From Lucerne they voyaged to nearby Flüelen am Vierwaldstätter See, whose beauty he said Olga could only appreciate if she came there herself.[51]

Then from the Berne countryside he detailed with gusto the features of its mountain landscapes, valleys, and glaciers. This romantic area was just his taste. For four days he and Schrenk had been on a "wild mountain party." They visited the villages that Tell had lived and shot in. In Realp, they drank two bottles of Italian wine with a Capuciner innkeeper. He found the people of canton Berne to be completely different from those of Zurich and was forced to "completely recant" his earlier, hasty generalizations about the Swiss. "It's remarkable how in the cantons races of man, clothing, and customs are different from one another." They walked a great deal, often in the rain; along the way, he lost the umbrella Olga had given him and begged her forgiveness.[52]

They hiked to Rosenlaui-Bad, whose glacier Helmholtz characterized as Switzerland's "purest piece of ice." It was like one of those "fairytales of ice palaces." He sketched it rapidly, using his "deficient talent for painting." They climbed "like a turtle" to the top of the Faulhorn (some eighty-eight hundred feet above sea level): "It's the principal observation post for the Alps chain, and the view is absolutely tremendous and powerful." Although he walked in the hot sun "without coat and hat," his health was excellent. Yet the stay on the Faulhorn's peak was "extremely unpleasant" because it was "so cold and windy" and because he was exhausted from lack of sleep the night before (having drunk so much coffee to keep warm). The English ladies staying in their inn did not even emerge from their rooms. He and Schrenk were all too glad to descend the mountain the next morning.[53]

But a snowstorm forced them to stay indoors. So he rested, played chess, and worked on his ophthalmoscope manuscript. He and Schrenk finally left, out of fear that they might otherwise get stuck there, and they walked through snow and rain until they found a carriage that took them to Interlaken. The region's beauty meant far less to him than the letter from Olga that awaited him there, "which told me that she was well." Also, "She still keeps my heart dear." He learned of Käthe's doings and took pride in them, but he feared that when he returned, father and daughter would not recognize one another. The day was 31 August, his thirtieth birthday, and he congratulated his wife on it and on their wedding anniversary, which in fact had occurred five days earlier. He shamefully confessed that he had lost track of time and had failed to celebrate their anniversary. Yet he thought much about her, he claimed: "What a dear soul she is, and how happy she makes her husband."[54]

His life at Interlaken was "very noble." Most of his fellow tourists were English, but they had little interest in others. He ate well. There were three dining tables, two for the English and one for the Germans. At the latter sat "an old Frankfurt banker (Jew) with his cultivated Jewish lady," the family of Nikolai Baron von Dellinghausen from Estonia, and two single ladies. He was astonished to learn that Dellinghausen's son "eagerly pursued difficult philosophical and physical studies." That same summer, Helmholtz exclaimed, the young man "also had a piece of writing on speculative physics published in which he honored me and wherein appear some very good individual thoughts along with a lot of nonsense from Hegel's system." Since it rained endlessly, the guests prolonged their meals and "talked a lot." The inn had a reading room, and, while the English kept to themselves around the fireplace, he read the third part of Humboldt's *Kosmos* (1851), devoted to astronomy. He also climbed a nearby mountain and visited, but only half intentionally, "a sanatorium and educational establishment for cretins."[55]

From Interlaken, Helmholtz and Schrenk took a steamer over Lake Brienz to Giessbach, where they continued hiking. In Thun he capitulated and bought a new, good umbrella: "The cheaper ones were too indecent." As always, he noticed prices and hated to overpay. Then the two went to Kanderthal, where rain forced them to stay in a village pub; they passed the time playing chess with a makeshift set. When the weather cleared the next day, they climbed over the Gemmi, the most impressive pass they had seen to date. At its other end was Bad Leuk, which contained large, warm baths and where everyone wore special bathing suits, took four-hour baths twice daily, played on swimming boards, and the like. "It's extremely ridiculous here."[56] He was relaxed.

At Bad Leuk, Helmholtz and Schrenk parted, and he now hiked alone through the Rhône Valley. Five hours into his hike, he realized he had wounds and lesions on his feet; by the afternoon he was in too much pain to walk further and had to travel on horseback (with a horseman). They rode for six unpleasant hours over hilly terrain. The next day they climbed Monte Rosa, near the Swiss-Italian border. By the time they got to the top (ninety-six hundred feet), he was "tired, freezing, with wet boots," and because of the heavy snow he could see neither Italy nor Switzerland. Back down in the valley he watched a parade, "where the women appear in the white Italian veils." He headed to Lake Maggiore.[57]

He rode on horseback part of the way to Ceppo Morelli, but once there he decided he had had enough of the horse and again began walking (in the rain). He had two seven-hour days of walking before him. The area was "very Italian," and he found the girls "strikingly pretty, but of a more German type in face and dress." He tried speaking with the locals. "I speak Italian with the greatest lack of constraint, and find my pronunciation better than that of the natives. They understand me, and don't know whether my strange pronunciation may be not exactly good Italian. I understand them, however, just as little as the German of the Welshmen." They took him for a Swiss, because he spoke German so fluently. "That someone could do that in other countries, went beyond them." The region's culture was "arch-Catholic," with many biblical holy scenes and crucifixes on the walls. He heard pleasant evening church music and the impressive ringing of church bells, though he was perplexed that they were partly out of tune with one another. He also heard Swiss music: "Fortunate, he who has not heard it."[58]

In the countryside along the way to Lake Maggiore, he saw several villages and churches, each of which he judged "more poetic" than any he had seen in Switzerland. He was stunned by the masses of snow in the countryside and their "absolutely brilliant light effects," but he did not like the desolate mountain valleys. His journey sensitized him to a problem of perception: "The [mountain] masses by themselves are impressive for at most a moment. They are so tremendously large that we can't at all comprehend them by judging with our eye. Rather, we first do so when we climb up to the top."[59] Nature gave him food for future scientific and philosophical thought.

He could now barely walk. "Unfortunately, the last mile, a semi-completed roadway strewn with stones, was poison for my feet." Toward the end of his tour, he got a hole in one of his boots, which caused a new blister on his left foot (older ones were still there, too). He (temporarily) lost his youthful, bourgeois, professorial self during the tour. He looked "like a lame old man." He had not shaved since Interlaken, his hands were tanned brown as coffee, and his hair was uncut. "In short, I look very seductive."[60] The changed appearance reflected his now relaxed body and mind. He had become the modern tourist from a dark, northern European city, one who visits sunny southern climes to tan and rest his body. He sought sun and mountains to gain bodily renewal, beauty, and freedom. The travels through the Alps relieved the stress from the past, intellectually creative academic year and brought out the romantic in him. The mountains and lakes renewed his sensitivity to natural phenomena: to light, color, and air, to cloud, mountain, snow, and ice formations, to changing weather conditions, to waterfalls, and to glaciers. He started to think more about optical illusions and the changing nature of perspective.

Italy itself simply overwhelmed him: "There's frightfully much to experience." When two Milanese couples failed to agree on a price with a carriage driver for a trip from Luino to Lugano for themselves and Helmholtz, all five simply starting walking: "It was a gorgeous evening moon. I went along. They had charming manners. We spoke with one another as well as my Italian and French would allow (the prettier of the young women spoke the latter)." Then they chanced upon "a very cheap carriage" that brought them to Lugano. "I really wished that my father could walk here one time; the man would completely break into delight."[61] He had left his daydreaming father far behind.

He took a boat across the "very beautiful" Lake Lugano to Como, and from there a steamer to Bellaggio. He thought Lake Como's beauty, the luxury villas, and the sculptures made it "the most beautiful summer spot . . . that one can imagine." He was above all impressed with the sculpture gallery in the Villa Carlotta, including Albert Thorvaldsen's famous marble frieze, the *Triumph of Alexander* (in Babylon). But "the pearl among them, and in general of all the more modern sculpture" that he had seen, was "[Antonio] Canova's Amor [*Cupid and Psyche*], which awakens the psyche, a work of the most wonderful beauty and intimacy." After sunset, he saw the lake's three sections from the Villa Serbelloni near Bellaggio: "I began to feel the enchantment of Italy."[62]

Early the next morning he took the steamer back to Como, and from there the train to Milan, which he found to be "a large, sumptuous city with all the luster and noise of Italian life." While he thought the Duomo was "Milan's climax," he also thought that "in terms of beauty of form," it was "by far second to the Gothic domes of Germany." Its Gothic forms were "arbitrary," if "tasteful." Besides, there were many spires, flying buttresses, and statues, all done "in white marble against the blue heaven." It is "a sight whose splendor one simply can't imagine." At sunset he and some fellow travelers climbed to the top and saw "the immense plain of Lombardy," the Tyrols, the Berne Alps, Monte Rosa, Mont Blanc, and the Apennines. They ate ice cream—"ridiculously cheap in Italy"— and went to the theater, where he saw an opera, *Atala*, and a ballet, *Le Corsaire*. "I've certainly heard better individual singers, but never a more beautiful *ensemble* and such beautifully cultivated voices as here. The song was so fresh and full that it was a joy [to hear]. The ballet was less bad than that in Berlin; it had the advantage, however, that everything went very quickly."[63]

In Milan, too, he saw Leonardo's *The Last Supper* and the Brera Palace's art collection. He was disappointed that the Brera was only showing its modern, not its classical, collection of paintings. Though there were a lot of them, he thought that, with the exception of the portrait collection, Königsberg's most recent exhibition of modern works was definitely superior. "The paintings were terribly shabby; the sculptures, by contrast, were for the most part excellent. Of the old paintings, one saw little that is beautiful."[64]

He left for Venice early the next morning, observing Lombardy's rich vegetation and flourishing farmland along the way, but it bored him. In Verona, he saw the old amphitheater and the Casa di Giulietta (but could not figure out which balcony was Juliet's). Venice was "the city of wonders, a living fairyland."

He claimed, "The sight of it goes beyond all that one has seen in pictures and heard in descriptions. Saint Mark's Square, with its mosque-like, colored church lies amongst a series of palaces, with numerous gas lights, a deep-blue moonlight heaven, and some steps into the deep-blue sea, and in addition there's the waving crowd of people, gathered like a fist—that is an indescribable scene." He spent the day sightseeing and being overwhelmed by Venice. "The historical memories, the tremendous wealth that Venice has brought together from around half the globe, these art treasures, which for the most part still glitter in full, fresh colors, can't be missed." He decided to stay for a third day in order to "study thoroughly, with meditation, the local art gallery." "In Germany one can only get a rough idea of the bloom of Italian art; here one can soak it up fully." During the remainder of the day and the evening he indulged himself by drinking wine, eating oysters, ice cream, and more, though he avoided fruit, which in general he apparently could not or would not eat. (None of this feasting stopped him from noticing prices.) He planned to head the next day to Trieste by boat.[65]

By mid-September Olga's patience had worn out: "I must soon have my dear own husband back again. For I often feel strange and alone, and yearn for my independence, for my house, for you," she wrote. She pined after him and longed to be with him and Käthe in Königsberg. She worried about the holes in his boots and his wounded feet. She had visited his parents in Potsdam to report about him. She showed some of his letters from Italy to Ferdinand, and he exaggerated their contents to others. "He is so proud of you that he recently even said that when the Berlin gentlemen like [the ophthalmologist Johann Christian] Jüngken, etc. want to speak to you, they can come to you in Potsdam." Yet everyone, especially Ferdinand and Marie, had been quite annoyed that she had not come sooner; however, she stayed for eight days and that seemed to calm the family tensions. Olga confessed to finding it difficult to cope with his family, in particular with their advice on how to treat Käthe, who had come down with a fever and was difficult to handle. She wondered how Lina, who now spent all her time cooking, could have raised six children. Still, she held her tongue before her in-laws, recognized their "unhappy peculiarity," and in the end forgave them.[66]

Helmholtz was grateful for the diplomatic, restrained way Olga had handled the delicate situation with his parents: "I thank you very much that you have held out so valiantly in Potsdam, and believe that they made it plenty uncomfortable for you." He perhaps felt guilty that she had to deal with his family in Potsdam while he was in Venice looking at paintings and feasting; nonetheless, he could not refrain from describing some of the art:

There's a collection here of masterpieces of older Venetians. The principal piece is Titian's *Assumption of the Virgin*, of which I had already previously seen engravings. . . . The indescribable beauty of the work lies in the orderly, intoxicating glow of colors and of light. I had never seen anything similar, nor can one imagine it without having seen it, since this kind of beauty is so completely different from our German pictures. . . . Many of them have a wonderful degree of joyful color, and one encounters a large number of the most radiant and most ideal human heads that one can imagine.

Before leaving Venice, he bought Olga a little gift and in the evening went to St. Mark's Square, where he ate ice cream, heard music, and took a gondola to a steamer headed to Trieste. On board he ran into none other than Jüngken, a leading but conservative ophthalmologist. Jüngken told him that he had heard of his ophthalmoscope, and he "acted extremely friendly." Word of the invention was spreading. In Trieste, he sought out Rudolph Wagner and Johannes Müller, both of whom were supposed to be there, but he could find neither man. Trieste's market and modern commercial life much impressed him. He boarded a train for the sixteen-hour journey to Vienna, arriving at 6:00 a.m., "very bruised and worn out."[67]

With Brücke in Vienna

After going to an inn to clean himself up, he went to Brücke's, where he got a warm reception. Wagner arrived shortly after, and a day later so too did Robert Bunsen, "one of the most highly gifted" of chemists; "We here formed a very learned society."[68] This serendipitous meeting in a relaxed setting with Bunsen, who was then teaching in Breslau but who moved the following year to Heidelberg, proved most useful for Helmholtz's future. His sightseeing and hiking in Switzerland and Italy were behind him, and he was again promoting his ideas and networking.

Brücke had keenly wanted Helmholtz to visit him. They were close friends, not merely because they were both medical doctors and physically minded physiologists trained under Müller, but also because they shared a deep, abiding interest in the scientific basis—that is, the anatomy, physiology, and physics—of painting (aesthetics) and speech, which in turn led them into psychological and philosophical issues. Both were savants, not *Fachidioten*. Helmholtz found Brücke to be the same as ever, and he found his wife "rather pretty and similarly of a very pleasant, serene nature." He and Brücke slept in the latter's study; he lived intimately with the family. He got to know Brücke's working quarters and adjoining physiological institute with its "quite pretty equipment" very well, especially since it rained constantly and, as a result, they mostly stayed indoors and talked science. He showed Brücke his ophthalmoscope. Brücke had doubtless been keen to see and study it, and he was certainly not surprised by Helmholtz's invention, since he himself had unsuccessfully tried to construct a similar device in 1846.[69]

Helmholtz also managed to see others in Vienna. He went to Vienna's renowned General Hospital, in particular to its morgue, to introduce himself to Karl Rokitansky, the professor of pathological anatomy at the university and one of Vienna's premier medical figures. Afterward, he showed Wagner, Bunsen, and others his ophthalmoscope, and later in the day he explained his induced electric current work to Brücke. He spent the evening at the home of the philosopher Franz Karl Lott; Wagner and various Viennese professors were there, too. "It was a pleasant, friendly atmosphere, but from time to time there was a series of somewhat trivial anecdotes."[70]

He paid a second visit to Rokitansky to see (with others) his "very excellent collection of pathological-anatomical preparations," including his "world-famous" wax preparations. Later he went with Brücke and Wagner to see "two famous paintings by Canova, a tomb of a princess [the Archduchess Christine] in the Augustine church, and the statue of *Theseus* in the Volksgarten." He thought these works much inferior to what he had seen in Italy. They walked around the city's walls, where it was "rather pretty," and spent much time talking to Wagner. Helmholtz later told Ludwig that Wagner wanted to know his opinion on the relationship of the body to the soul "and other dark points of physiology." He thought that, at least given the present state of physiology, there was nothing to say about such matters. Wagner found that Helmholtz's "total personality" left him with "an extremely pleasant impression." Brücke was similarly stimulated by Helmholtz's visit.[71]

Apart from scientific Vienna, Helmholtz thought the city "terribly boring and oppressively poor in spirit," that it suffered, in physical and architectural terms, in comparison to Berlin. Most of its streets were short and narrow, dotted with "tasteless surprises of art." In general, he found that things were simply too far apart from one another. He passed up opportunities to see Giacomo Meyerbeer's grand opera *Le prophète* and Vincenzo Bellini's *Norma*, though he did hope to visit three art galleries. But he had also lost his patience for sightseeing and wanted to be home with Olga: "Keep it together for this week, then I'll be with you again—and will not leave you again so soon." By late September 1851 he was back with her and Käthe in Dahlem.[72]

His eight-week tour was a watershed in his life and career. His predilection for travel, evident already in his briefer, more local trips during his Gymnasium days, had now expanded onto a wider European stage. Thanks to Europe's burgeoning railroads, he networked with scientists in the German-speaking world, solidifying old friendships and making new ones, unostentatiously promoting his neurophysiological findings, his ophthalmoscope, his scientific viewpoint and general ideas, and, not least, himself. His ophthalmoscope and his tour in the summer of 1851 made him the cynosure of physiology and medicine. He saw firsthand the physiological institutes of leading universities. He advanced himself and science internationally. Along the way he visited several of Europe's finest art galleries and museums, gaining further firsthand experience of European art. In the coming years he put all this networking and sightseeing to further use. He was a rising star, one who knew not only science but, now, how to network with scientists.

6

In the Private and the Public Eye

Coming up Empty-handed

With the public announcement of Helmholtz's ophthalmoscope in the autumn of 1851, Prussia moved fast to reward him. Königsberg's Medical Faculty declared their intent to do all they could to retain him, asking the ministry to recognize his achievements and promote him to ordinary professor of physiology. Their request emphasized teaching needs—only a full professor could conduct examinations, they noted—rather than his research accomplishments, probably because the latter were all too obvious: he already had seventeen publications to his name. Du Bois–Reymond, for his part, was astounded by Helmholtz's "tremendous capacity to work" and the extent of his knowledge, as well as by his ability to "simultaneously teach new classes and accomplish so much research." He seemed to think Helmholtz's ability lay in pure brain power: he believed Helmholtz could understand more in fifteen minutes than he himself could in a week. Brücke declared that du Bois–Reymond worked a lot, but Helmholtz still more. At the ministry, Johannes Schulze, in charge of the universities, pointed not only to Helmholtz's strengths as a teacher and researcher, but also to his "moral and political conduct."[1] By January 1852 Helmholtz was a full professor.

His rapid promotion showed that the ministry rewarded research and recognized the public benefits that might issue from it.

His parents took great pride and joy in his accomplishments. Ferdinand was thrilled when Potsdam's medical men and other respected citizens complimented him on his son's achievements. He took pride, too, in Hermann's becoming a university professor and recognized that his son had achieved incomparably greater professional recognition than he had. Yet tensions in the father-son relationship remained. While Ferdinand longed to be a part of his son's work, his frequent recourse to highly speculative philosophy and his judgments (occasionally negative) about Hermann's work on the basis of what was and was not philosophically sound, made it increasingly difficult for Hermann to deal with him. So Hermann adopted the strategy of simply not discussing his work with his father, which, according to his brother Otto, at times led to bitter feelings on Ferdinand's part. After Hermann became a full professor and a leading scientific figure, such feelings slowly subsided and the irritations largely disappeared. By Ferdinand's own account, Hermann's scientific achievements gave meaning to his own life.[2]

When the ministry promoted Helmholtz, it doubled his research budget (to 200 talers) but gave him no salary increase, which left him feeling dissatisfied. Müller tried to mollify him by declaring that the ministry would "certainly not hesitate" to satisfy his wishes since, in this regard, it had "never been wanting." Ludwig foresaw a possible professional move: "Absolutely no physiologist can be given preference [to you]; you'll soon come into the heart of Germany. . . . It's obvious that you'll play one of the outstanding [scientific] roles in this country." Like du Bois–Reymond, Ludwig felt depressed when he compared his own recent record of achievement to Helmholtz's—forgetting that he had done major work on the theory of blood pressure and its measurement, on the processing of urine, and on nerve secretion, not to mention inventing the kymograph. Nonetheless, he took comfort in being Helmholtz's friend and hoping for a visit. "What do you have to say about Helmholtz's tremendous works?," du Bois–Reymond wrote Ludwig. "It's a fortunate man who, without exertion or hesitation, achieves the greatest of things, just in the way a tree grows! My only solace is the thought that whoever is so calm and without deep suffering doesn't know its opposite— burning, full pleasure." Yet Ludwig (still) thought du Bois–Reymond was Helmholtz's scientific superior, since, unlike du Bois–Reymond, Helmholtz could not stick to one scientific direction.[3]

The possibility of transferring to Heidelberg, at least as dangled by Henle the previous summer, still remained. Yet six weeks after his promotion, Helmholtz had heard nothing. Henle recommended both Helmholtz and du Bois–Reymond for the chair in physiology and thought a decision would be made shortly. He himself was considering leaving Heidelberg, and he made the long-sought appointment to the chair one of his conditions for staying. Yet Helmholtz's chances of getting the position soon decreased, and when he sensed that it would not happen, he became disappointed. A host of struggles—ideological, disciplinary, financial, and political—internal to Heidelberg and between it and the Baden government brought an end to his candidacy. In July Henle left (for Göttingen) and Friedrich Arnold, an anatomist and physiologist, replaced him. Henle despaired that Heidelberg had failed to secure a (genuine) physiologist, first passing over Ludwig, then Carl Theodor Ernst von Siebold, and now Helmholtz and du Bois–Reymond.[4]

Some consolation was to be found at home: about nine months after he came up empty-handed professionally, Olga gave birth to their second child, Richard Wilhelm Ferdinand Helmholtz, a "very splendid boy," whom they had baptized in January 1853. She paid a heavy price, however: she was constantly ill during the pregnancy, and after Richard's birth she became seriously ill with "a gastric-nervous" fever and bronchial catarrh (inflamed mucus in the nose and air passages); her chronic cough worsened. Käthe became ill, too, and Helmholtz spent the winter filled with "anxiety and concern." He lacked the peace of mind, he said, to conduct his time measurements of nerve impulses, though he did work on accommodation of the eye. Yet Richard's birth made him still more family-oriented. He praised the blessings of marriage. "I know it from my own [past] experience and I still experience it daily," he preached to du Bois–Reymond, who had finally found his own soul mate, "that it [marriage] is [one of] the most important and most beautiful steps in life. May it bring you the wealth of flowers and fruits it has brought me." Among those "fruits" was also better science: "If through the satisfying pleasures of marriage your diligence and capacity to work increase in the same proportion as has happened with me, then the world can expect wonders." Marriage and family were costly, however; Helmholtz needed more money, which he finally got in early 1853 in the form of a raise to one thousand talers after he declined a call from the University of Kiel.[5]

Revolution in Ophthalmology

Physiologists and ophthalmologists first learned of Helmholtz's ophthalmoscope through word of mouth, through his presentations during his summer tour of 1851, and then through reading his essay that autumn. Albrecht von Graefe, for example, who had opened a small ophthalmology clinic in Berlin in 1850 and who had yet to meet Helmholtz, became keenly interested in the ophthalmoscope; he read the essay and wrote Helmholtz a laudatory letter about it, asking to have one built for him. He also told his fellow ophthalmologists William Bowman in London and Louis-Auguste Desmarres in Paris about it. The number of orders to Helmholtz's instrument maker soon became so large that

Helmholtz had to refer them elsewhere. Shortly after the essay's appearance, moreover, he gave his first public lecture (November 1851) on the ophthalmoscope, at the initial meeting of the Association for Scientific Medicine, where he also demonstrated its application to his fellow medical doctors. Ruete in Göttingen also gave a lecture on it, and in 1852 he demonstrated an improved version with an innovative, practical lens system that created an inverted image of the retina (the so-called indirect method). Richard Liebreich, an assistant of Graefe's, built an improved indirect ophthalmoscope. He advanced knowledge of the instrument by making connections with key figures in Berlin, Utrecht, Paris, and London.[6]

In the 1850s, then, interest in and further development of the ophthalmoscope grew rapidly. It joined other new, nineteenth-century technological innovations in medicine—for example, the stethoscope, the endoscope, the clinical thermometer, the laryngoscope, and the sphygmomanometer—for diagnosing the internal state and workings of the living human body. Together these instruments gradually changed the character of the doctor-patient relationship. In early 1852 Graefe ordered several ophthalmoscopes from local Berlin opticians, and he promised to send one to the Dutch ophthalmologist Franciscus Cornelis Donders in Utrecht, though he said it required changes before practitioners could make use of it. Nevertheless, he told Donders, who was a leading advocate of science-based medicine and who cultivated relations with several of Müller's students, "For diagnosing the onset of cataract the instrument seems to me to be capital," adding that he had already made many diagnoses of patients' eyes with it. Brücke too had Helmholtz's instrument maker construct several ophthalmoscopes. Ludwig begged du Bois–Reymond to have one made for him in Berlin: "I absolutely must have an ophthalmoscope," he declared. Du Bois–Reymond himself found the original form of Helmholtz's ophthalmoscope difficult to observe with and more complex than it appeared to be, and he thought only a handful of mechanics had the equipment to build and test one. Helmholtz, for his part, stimulated by Ruete's indirect method, devised a newer, simpler ophthalmoscope, one that employed a convex lens, and announced that his instrument maker had attached two rotating disks (each holding four concave lenses) to it to further ease its use. Thus, virtually everyone who wanted an ophthalmoscope had to turn to a professional instrument maker to get one, especially since constant innovations made the device increasingly complex technologically. Only a few creative individuals, like the Scottish physicist James Clerk Maxwell, could construct their own on the basis of Helmholtz's description. A few ophthalmologists simply refused to adopt the ophthalmoscope at all. One surgical colleague told Helmholtz that it was too dangerous to use; another said it was simply unnecessary.[7]

A year after inventing the ophthalmoscope, Helmholtz turned his attention to developing a theory of the eye's accommodation, that is, how its lens focuses on objects at various distances. This old issue—investigations reached back to the Middle Ages—was of keen interest to several modern investigators. Accommodation was effectively about light's refraction in the eye, and Helmholtz's analysis of how the eye focuses largely concerned the role of the lens's changing geometry and mechanics, rather than its anatomy or biochemistry. Typically, he created a new instrument—in this case, the ophthalmometer—to measure the cornea's changing curvature as well as other changing surface curvatures in the eye. His measurements aimed—also typically—more at illustration than exactitude. Though his analysis was mathematical and mechanical, he also pointed to the use of his theory and his instrument in the practical understanding of eye conditions, functions, and ailments.[8] The ophthalmometer, the ophthalmoscope, and the theory of accommodation together revolutionized the understanding and diagnosis of the eye and aided in ophthalmic surgery. They became to ophthalmology what the telescope and the graduated scale were to astronomy.

His work strengthened his ties to Berlin. He had begun his accommodation studies in the winter of 1852–53, but because of family illnesses he had a hard time working. As had happened before, he asked du Bois-Reymond to help him get a preliminary manuscript on accommodation published. Du Bois-Reymond, recently elected an ordinary member of the Academy of Sciences, told Helmholtz that he was in a stronger position than before to help him get published there. In mid-January 1853, Helmholtz himself went to Berlin to participate in the annual meeting of the Physical Society, where he saw his close friends du Bois-Reymond, Karsten, Brücke, Knoblauch, and Kirchhoff. Later that month, he sent du Bois-Reymond a short report for the academy's *Monatsberichte*, adding that it would be a while before he could finalize his work into a fully complete article. He returned to Berlin in late February to lecture on accommodation at the society.[9] Not until nearly two years later did he finalize his manuscript on the subject.

In the meantime, priority claims were made against his preliminary results. Donders (and soon Brücke) told him that Antoine Cramer, a Dutch medical doctor, had just finished similar work, and he begged Helmholtz to mention Cramer's name: "Otherwise his name will be crushed under the weight of yours." Others (including Brücke himself and Max Langenbeck) were also at work on accommodation. The issue of priority was ever present. It could come from his closest friends as well as from strangers. Helmholtz freely conceded to du Bois-Reymond, for example, priority in comparing the human nerve-conduction system to the new electrical telegraph systems. This was a widely noted comparison (along with its associated metaphorical notions of web or network) at midcentury that was to be found in imaginative literature as well as in science and technology.[10]

During the 1850s and 1860s Helmholtz's achievements in physiology and ophthalmology were joined with the innovative work of others to usher in a revolution in ophthalmology both as science and as clinical practice. A broad mix of scientific, technological, and institutional developments were interwoven. Donders did fundamental research on refraction and accommodation of the eye, as well as associated clinical work on prescribing corrective eyeglasses. Graefe's clinic became renowned for its surgical and organizational innovations, including testing, using, and teaching about Helmholtz's ophthalmoscope. Through Graefe, Berlin became the center of ophthalmology, and he led a new generation of clinical ophthalmologists. In 1854 Graefe founded his *Archiv für Ophthalmologie*, the first and leading journal devoted to the subject. In doing so he cited above all the role played by Helmholtz's ophthalmoscope in diagnosis and therapy. By the mid-1850s, leading ophthalmologists had introduced it into their practices. The doctor's observation of the eye began to replace the patient's report about it. Moreover, academic ophthalmology gradually split off from surgery and became a distinct discipline. In 1852 Ruete at Leipzig became the first ordinary professor of ophthalmology; by 1873 every Prussian university had an ordinary professor for ophthalmology. As ophthalmology became an independent academic field and more dependent on physiological knowledge, Helmholtz's subsequent physiological work—studies on color and space perception and on eye movement and coordination—also advanced the field's general development and influenced clinical research and practice. Graefe reportedly claimed of Helmholtz: "A Luther has arisen among us." The invention of the ophthalmoscope, Donders told Helmholtz, played a role in the establishment of a Dutch hospital for eye patients. When the Dutch king, William III, learned of all this, he made Helmholtz a knight in the Order of the Netherlands Lion.[11]

The German lands led, but were not alone, in this revolution. The ophthalmoscope spread to Holland and France—as in so many other areas of science and medicine, in ophthalmology the German lands now surpassed France as the innovative center—and with it Helmholtz's reputation. Donders told him that no one wanted to compete with him in scientific matters. News of the ophthalmoscope also spread to America, and young American ophthalmologists started flocking to German-speaking universities, especially Berlin and Vienna, to learn about it. News spread to Britain as well. Henry Bence Jones, a well-known London physician and medical chemist who had studied with Liebig and who followed German science closely, asked du Bois–Reymond to bring one of Helmholtz's ophthalmoscopes with him when he came to London: "I want to give it to an occulist here." Du Bois–Reymond also brought it to David Brewster's attention, and the two men sought to arrange for an English-language translation of Helmholtz's essay. Helmholtz declined to cooperate, however, since he was about to publish a devastating critique of Brewster's theory of colors. Bowman, a leading English physician who soon became noted for his ophthalmic surgical work, recognized early on the importance of Helmholtz's ophthalmoscope, and he introduced it into his practice; in 1857 he became the first in England to perform Graefe's iridectomy procedure for glaucoma. That year, too, Graefe organized the first congress of ophthalmologists in Heidelberg, and informal meetings were held there annually from then on. In 1863 the congress's members formally established the Ophthalmological Society (Ophthalmologische Gesellschaft), which continued to meet in Heidelberg.[12]

The improvement and widespread adoption of Helmholtz's inventions and his analyses of the eye meant quick and lasting fame for him among medical men, scientists, and government officials. Du Bois–Reymond told his fiancée in 1853 that in the natural sciences Helmholtz was "the top thinker alive and, in general, one of the top" who had ever lived. "But in social terms," he added for good measure, "he's somewhat dull." In Würzburg, according to Ernst Haeckel, then a young student, Helmholtz was regarded as "the most exact of contemporary physiologists, someone who has treated everything with great mathematical exactness in the smashing field of physical physiology." Helmholtz himself later conceded: "The construction of the ophthalmoscope was very decisive for my appearance before the world. I now found the most eager recognition and favor among state officials and colleagues for my wishes, so that henceforth I could follow much more freely the inner drives of my curiosity."[13] He owed his first, broad, and sustained recognition not to his law of the conservation of force but to his ophthalmoscope.

Beyond ophthalmology, he was simultaneously involved in other, diverse scientific activity. Between 1847 and 1854, he developed two different methods for measuring and displaying the speed of nerve impulses. Along with the method based on Pouillet's ballistic galvanometer, he also discussed a second, graphical method resulting from his new myograph, an instrument that he invented by adapting Ludwig's kymograph (for recording pulse and blood pressure on a rotating drum). Although the myograph's results were seemingly less exact than those based on the Pouillet method, it had the advantages of tracing curves of muscle contraction that were displayed graphically (and therefore easily visible) and of being simpler to use. Its persuasive power in demonstrating muscle contraction and the speed of nerve impulses was thus greater.[14]

Intimately related to these pathbreaking studies in electrophysiology was a set of studies on induced electrical currents, currents in human bodies, and animal electricity. Helmholtz's study of induced currents built on the work of Georg Ohm, Kirchhoff, and others to establish the concept of equivalent circuit.

The work was largely physical (as opposed to physiological), but Helmholtz emphasized the pertinence of the one approach for the other. Two of these papers appeared in Poggendorff's *Annalen*; the two men had overcome any lingering after-effects resulting from Poggendorff's earlier refusal to publish Helmholtz's essay on conservation of force. Further, Helmholtz also published a "popular" exposition on animal electricity, one largely concerned with and critical of du Bois–Reymond's recent results on "negative variation." It appeared in the *Allgemeine Monatsschrift für Wissenschaft und Literatur*, a new journal edited by Karsten in Kiel, which sought to popularize science and literature and to underscore the presence of Germanic culture in northwestern "Germany" as Prussians and Danes struggled for mastery of Schleswig-Holstein. Taken all together, the series of essays that Helmholtz produced on the speed of nerve impulse—plus his earlier work on fermentation and putrefaction and on the chemical changes that muscles undergo while contracting—created new subspecialties within experimental physiology that physiologists pursued during the remainder of the century. He thus played a major role in opening up the new field of bodily time measurements (e.g., reaction and association times) for the emerging interdisciplinary field of psychophysics, as led by figures such as Gustav Theodor Fechner and Ernst Heinrich Weber.[15]

Critique of Goethe and Color Mixing

Helmholtz's ascent to the top of the German scientific world included his debut as an essayist and public spokesman for German science. In the early winter of 1852–53, he was busy preparing a lecture on Goethe as a scientist. He dictated his draft presentation to Olga and then read it aloud to her, using her understanding to represent that of the generally educated listener. He thought he gave his best lectures when he had her help. He delivered his lecture, "On Goethe's Scientific Researches," on 18 January 1853 before the German Society of Königsberg. It was a moment when Goethe's literary reputation probably stood at its lowest point. His talk on Goethe's scientific work, at a time when the popularization of science in Germany was in its earliest stages, did nothing to help and something to damage Goethe's general intellectual reputation.[16]

Helmholtz's lecture concerned much more than its title suggests. He analyzed Goethe's scientific work—botany and comparative anatomy (osteology), on the one hand, and optics and the theory of colors, on the other, leaving unmentioned Goethe's studies of geology, mineralogy, meteorology, and other topics—and sought to assess it in the light of contemporary natural science and to relate its "common guiding ideas." But in doing so, he also addressed the nature of science in general, comparing and contrasting it to the arts. This became his first attempt to explain to a broad audience his own philosophy of science and epistemology, and to do so within the broader context of his long-standing interest in the arts. His strategy was less to explain things at a level that all could understand than to stimulate his listeners' imaginations, even if that meant leaving some of them with unresolved questions.[17]

The "more intellectual part" of the descriptive sciences, he explained, only began and became interesting when the scientist united unrelated findings into lawlike statements. He thought this sort of work, and the historical moment, was perfect for "the ordering and surmising mind of our poet." Goethe had two guiding ideas in the descriptive sciences, he argued. The first was that the diverse anatomical structures of all animals were variations of an archetype. The variety of behavioral habits, geographical locations, and food supplies had led over time to various anatomical structures. There was, however, unity between man and animals. Goethe found features in human skulls and animals' skulls that led him to claim that both had derived originally from the intermaxillary bone, which itself later disappeared. Helmholtz considered this variation from a general type as Goethe's "guiding idea" in comparative anatomy, and he said no one had expressed it "better and clearer" than Goethe had.[18] That was a generous interpretation, since Goethe may not have been either the first or the only source of the idea that the human skull originated from one or more altered vertebrae, while the broader notion of the transformation of all life over time was well known (if only rarely accepted) to natural historians from the late Enlightenment onward.

Goethe's "second guiding idea," according to Helmholtz, was to see an "analogy between the different parts" of an organic being. In particular, Goethe's theory of the metamorphosis of plants maintained that the seemingly different forms of parts of plants had developed from—were transformations of—an underlying, foundational structure (the ideal leaf). By 1850, Helmholtz claimed, botanists fully accepted this point. However, he said, Goethe's views on an original, common type of animal kingdom had found little or no favor, or at least not until scientists arrived at the same views independently. Nevertheless, he credited Goethe as the first to have thought of these "guiding ideas," which botanists and anatomists now pursued.[19]

The tone of Helmholtz's lecture changed noticeably when he turned to analyzing Goethe's optical ideas and his theory of color. Here he spoke with first-hand experience and authority. During the previous two years, he had done much work in optics and color research, and he probably chose to lecture on Goethe's scientific research in part because he had just finished critically reviewing the work of many others on optics and color and had begun to develop his own ideas as well. Every winter semester at Königsberg, he lectured on color as part of his sensory physiology course. In contrast to much of his previous physiological work, however, his color research was largely synthetic in nature; he did not here create new instruments or make new findings, but rather criticized, borrowed, and synthesized the work of others. Among those others was the reigning master of experimental optics and color theory, David Brewster. In the 1820s and 1830s, Brewster developed a theory of the solar spectrum according to which three primary colors produced white; if not quite anti-Newtonian, like Goethe, he nonetheless challenged Newton's color theory. Newton showed that there were seven spectral colors decomposable (by a prism) from white light. Brewster claimed that there were only three, with some white light remaining nondecomposable. Although by no means the first to criticize Brewster's triple-spectrum theory, Helmholtz did so most effectively because he had several advantages that others lacked: unsurpassed understanding of the eye's physics and physiology; Königsberg colleagues to whom he turned for help (Moser for his work on the effects of color intensity, i.e., the Purkyně effect, and Neumann for optics in general and the loan of his prism); his own ongoing work on the mixing of lights; and close familiarity with Brücke's study of subjective colors. In a coup-like article of 1852, he refuted and dethroned Brewster's solar-spectrum theory and strongly defended Newton's color theory to physicists.[20]

Helmholtz published his anti-Brewster lecture on the theory of compound colors in both Poggendorff's *Annalen* and Müller's *Archiv*, thus indicating its dual physical and physiological character and reflecting his own growing reputation as a physicist as well as a physiologist. Before publication, he sought du Bois–Reymond's aid to make sure it would appear in Müller's journal and, typically, to minimize his costs for the two hundred copies he needed to habilitate with it. He also pettily complained that his most recent article in the *Archiv* had been only loosely attached to a recent issue and placed at the end of the issue; this false sense of mistreatment reflected his ambition and drive. He even peevishly wondered whether the *Archiv* was still interested in publishing on experimental physiology and said it was only his respect for Müller that prevented him and Ludwig from founding a new journal devoted to experimental physiology. Du Bois–Reymond told him he had completely misinterpreted things: the *Archiv* had not mistreated him; rather, in its eagerness to publish anything that had his name on it, the editor had decided to place his article at the end of an issue that would appear soon rather than hold it back until the next issue was ready to appear.[21]

In his habilitation essay as ordinary professor, Helmholtz reviewed earlier theories of colors, discussed (and rejected) Thomas Young's tripartite theory of color perception, and then concluded, inter alia, that colored lights or pigments could be mixed by "adding" or "subtracting" (i.e., by illuminating rays of light on the same retinal spot or by combining different pigments). Color mixing, from Newton onward, had been a subject in disarray. Some workers (not all were scientists) used pigments to mix colors, and others used lights; some came from a painterly tradition of color mixing, others from a physical tradition that mixed colors via spinning tops or wheels with colored sectors; some used prisms and lenses, and still others mixed colors on a physiological basis. Helmholtz's theory and demonstration of color mixing in terms of adding and subtracting colors brought conceptual order to color mixing. Yet he could not explain precisely how the mixing occurred, and there were many (not least Joseph Grailich and Hermann Günther Grassmann) who criticized his work, in terms of both its priority and its substance. He also had his admirers. Not just friends like Ludwig or Donders, who told him that no one any longer believed in Brewster's theory of colors, but also, after the essay appeared in English in the *Philosophical Magazine*, strangers like William Barton Rogers, a young American chemist and geologist.[22] Helmholtz continued developing this work through 1855.

Furthermore, in late June of 1852, he delivered his habilitation lecture, despite suffering from "a severe catarrh that always accompanies me in the summer heat." His subject was human sense perception. He concentrated on the nature of light, colors, and their perception. Only toward the end did he turn to epistemological matters, briefly declaring that sensations of light and colors were only "symbols for the relations of reality." "They have as little in common, and as much similarity or relationship, with [reality] as the name of a man has, or the handwriting for the name has, with the man himself." These symbols, he said, do not tell us about nature itself.[23] His viewpoint, which was only the kernel of a developing epistemological position, expressed the importance of both the material (physical and physiological) dimension and the immaterial dimension (symbolic relations) in analyzing a scientific issue.

Thus, when he turned to Goethe's theory of light and colors in his January 1853 Goethe lecture, he did so with firsthand experience of these subjects, both experimental and theoretical, and as part of his larger critique of Brewster and others who had worked in these fields from the time of Newton onward. In contrast to his positive assessment of Goethe's research on comparative anatomy, Helmholtz offered a scathing critique of his work on optics and color and, more generally, of Goethe's methodological procedures in the physical sciences. He portrayed Goethe as a complete amateur in physics. He said Goethe's original interest was in aesthetics, not physics; that Goethe had to borrow a prism— apparently forgetting that he himself had borrowed one of Neumann's—which he (Goethe) long failed to use; and that (at least initially) he had only poorly understood Newton's theory of light (yet passionately scorned and polemicized against Newton). In contrast to Newton's analysis of white light as composed of the spectrum of individual monochromatic rays, Goethe argued, *grosso modo*, that it could not be further decomposed and that colors belonged to a "plus" or "minus" category. Colors for Goethe, Helmholtz said, emerged through a prism at the intersection of white and black boundaries, and Goethe believed that Newton had failed to see this phenomenon and so his theory had failed to account for it. As for so many other romantics of the day, Newton was Goethe's bugbear. Physicists responded to Goethe by countering that they knew all this quite well and that they could explain it using Newton's theory. In the end, Goethe's *Zur Farbenlehre* (*Theory of Colors*, 1810) convinced virtually no one.[24]

Helmholtz was by no means the first physicist to reject Goethe's optical work, which had received a critical, largely hostile reception from its first appearance in 1791. Goethe became extremely, and uncharacteristically, irritated by this criticism, even driven to famously assert that his theory of colors was more important than his poetry. He saw himself as a scientist as well as a poet. But at least by 1850, no scientist took any serious interest in Goethe qua physicist, as someone who had a theory or data that was worth developing scientifically. Helmholtz later said that Goethe's theory of colors had gotten "much attention" in Germany because "in part . . . the broad public [was] inexperienced about rigorous scientific investigations" and so preferred an artistic account to one of mathematical physics. But the public also welcomed Goethe's account partly because it accorded nicely with Hegel's *Naturphilosophie*.[25]

To be sure, Helmholtz credited Goethe with giving clear, lively, and well-ordered descriptions of his optical observations. He was "rigorously true to nature" and appeared, "as everywhere in the realm of the factual, as the great master of description." But he had "obviously" insufficiently appreciated Newton's optics. Even after Goethe had received instruction in Newton's views and had indeed come to understand them, he nonetheless continued to maintain that the facts showed Newton to be wrong. Goethe's writings on optics, Helmholtz argued, remained "polemical" and vague. His *Theory of Colors* gave Helmholtz "an uneasy, anxious feeling." He declared that Goethe's work in optics remained baseless and his theory undeveloped and that all physicists—he conveniently forgot the anti-Newtonian Brewster—agreed that the optical facts in question could be explained in terms of Newton's theory. Against them stood Goethe, "a man whose rare intellectual greatness, whose special talent for the apprehension of factual reality," must be recognized "not only in poetry but in the descriptive parts of the natural sciences."[26]

But then, in the second half of the essay, it got truly ugly. Helmholtz analyzed the sources of Goethe's strengths and weaknesses in science (and in art) and presented some of his own views on aesthetics, the philosophy of science, and the relations between science and art. He maintained that the essence of art is to give an "immediate expression of the idea." It operates not by concepts but by "immediate intellectual intuition." The idea itself lies in the work of art. He thought that Goethe's key failure, like Schelling's, Hegel's, and all the *Naturphilosophen* in natural-scientific matters was his (their) inability to analyze nature by means of concepts and to do experiments. Goethe believed the scientific observer could reveal all of nature, and he mocked Newton's use of experimental apparatus and spectra. Rather than advocating the importance of controlled experiment in science, Goethe privileged direct observation. This approach worked well enough in Goethe's morphological work, Helmholtz acknowledged. Goethe's "greatness" in science came from his intuition that some sort of law was near at hand and from his pursuit of it. Yet Goethe himself neither specified nor pursued that law, for it was "not in the line of his activity" to do so. Goethe treated nature as he treated art.[27]

Helmholtz further argued that Goethe believed that observations could be easily arranged in science such "that one fact always explains the other," that context gives insight, and that sense perception alone was enough. Helmholtz thought this mistaken. To fully explain natural phenomena, he argued, contra Goethe, means ultimately to reduce them to the natural forces on which they are based. But scientists and others only perceive the *effects* of natural forces, not the forces themselves. To explain natural phenomena ultimately means leaving the realm of the senses and entering that of concepts. To understand the causes of phenomena thus requires determining the forces behind them and how the phenomena depend on those forces.[28] Goethe had an organic conception of nature that stood in opposition to the mechanical tradition advocated by Helmholtz, one that reached back to Descartes, Newton, and others in the seventeenth century.

Goethe feared, in Helmholtz's view, taking "this step into the realm of concepts, something which must be done . . . to ascend to the causes of natural phenomena." The source of his success as a poet and his failure as a scientist was that he did not use concepts. Helmholtz explained that, in contrast to Goethe's way of proceeding, the physicist "leads one into a world of invisible atoms, motions, of attracting and repulsing forces, which, in a lawlike but scarcely comprehensible maze, work through one another. The sense impression is for him no irrefutable authority. . . . The sensory organs do indeed inform us about external effects, but these are brought to our consciousness in a completely different form, so that the ways and means of sense perception depend less on the peculiarities of the perceived object than on those of the sense organ through which we receive the information." This was precisely Müller's theory of specific sense energies. Helmholtz's critique of Goethe now went beyond merely arguing that Goethe had mistaken views about light and colors; his larger critique was that Goethe did not understand the very nature of physics and that he had made epistemological errors. "For us, the sense impressions are only symbols for the objects of the external world," Helmholtz maintained, "and these correspond roughly as handwriting or wording corresponds to the thing designated. They indeed give us information about the peculiarities of the external world, but they do so no better than we can do by giving a sense of color to a blind person through a description in words." Goethe's optical theory was, in effect, no science at all; rather, it was an attempt "to save the direct reality of the sense impressions against the attack of science." His emphasis on the importance of immediate sense impressions in optics, as in poetry, meant that his entire approach to optics was an "embarrassment."[29]

True physics, by contrast, sought "to discover the levers, cords, and pulleys which work behind the scenes and direct things. The mechanical view clearly destroys the beautiful appearance." Goethe mistakenly sought to mark "reality itself" with the image of poetry. This was the uniqueness of his poetic gift, and this was why he opposed the use of intervening mechanisms, which threatened to disturb "his poetic pleasure." Helmholtz warned that the physicist cannot proceed as Goethe did: "We cannot . . . be victorious over the mechanism of matter by denying it, but only through it, by subjugating it to the purposes of moral intelligence. We must learn to recognize the pulleys and cords . . . in order to be able to control them according to our own will. Therein lies the great meaning of physical research for the culture of the human race and therein is its full justification established." Goethe's approach to natural science, like his approach to poetry, brought him great success in the descriptive sciences of botany and osteology but "failure" in physics.[30] In effect, then, Helmholtz was portraying science and art as two distinct realms.

His critique, moreover, differed from those of his predecessors partly in that he presented it to a broad public and gave it in a language that laypersons could understand. His lecture, published later that year as an essay in a journal devoted to science and literature, was particularly effective because his growing stature within German science made him an authority, because he put Goethe's optics within a broader intellectual framework, and because the times had changed. In criticizing Goethe publicly and in linking him to other (now largely dead) romantic philosophers ("scientists"), Helmholtz became a public spokesman for a hardheaded, realistic, and mechanistic view of nature, a post-1848 outlook that stood in opposition to the romanticism of *Naturphilosophie* that had reigned during the first one-third of the century. If he did not write Goethe out of the physical science community entirely, that was largely because Goethe had never really been in it. In any case, Helmholtz showed some of the philosophical differences between science and art. His essay, finally, was easy to read and became widely available. As a result of it, Goethe's reputation as a scientist (especially as a physicist) suffered still more. The essay decisively shaped future interpretations of Goethe's writings on optics until at least the 1920s, when a more sympathetic attitude toward organicism again emerged in German science and culture.[31]

Networking among the Germans Again

In the late spring of 1853, Helmholtz began preparing for a trip to Britain. He also had to attend to Olga's medical needs: she had been sick all winter with an inflamed throat and, though she had improved, he was still treating her by using fresh cod-liver capsules. The climate in Königsberg was highly unfavorable to someone with her medical problems, and she had overexerted herself in caring for the children and tending to the household. Moreover, since Easter she had again suffered from catarrh, and he was very worried about her. He himself rested during the four weeks before his trip and drank Marienbad spring water for his "frequent and severe attacks of colic."[32]

His trip was emblematic of two broad social changes occurring in science. First, it exemplified the emerging internationalization of science, a development due largely to the expanding railroad systems. Scientists from different countries or distant places within a country could now meet far more easily for conferences or for other reasons. Second, Helmholtz's trip to Britain, rather than to France, reflected the declining importance of France in science at midcentury. In previous generations, many young German scientists, such as Humboldt and Magnus, considered a trip to France (i.e., Paris) de rigueur for their future careers. Visiting Paris helped them become au courant with the latest, if not the best, work in science; make personal contacts with leading and promising scientists; and gain entrée to the institutions where scientists worked. By midcentury all this was passing, and French science was in relative decline.

Helmholtz felt no need or desire to visit France, and indeed he had a hostile, derisive attitude toward French scientists and science. He told du Bois-Reymond that he "was angry that such a shabby fellow as Bernard has again received the great physiological prize and was almost elected into the Academy. It appears that he sweetens Mr. [François] Magendie and Mr. [Pierre] Flourens as well with flatteries." Du Bois-Reymond egged him on: "I was very much amused that the stupid rabble in Paris poured prizes upon Mr. Bernard and Mr. [Julius] Budge and Mr. [Augustus] Waller, but had nothing for you—even though you have done nothing to them. As concerns me, they have their good reasons." In Helmholtz's view, "The French Academy now so disgraces itself in physiology that one has very much to reconsider whether or not to even send it physiological pieces."[33] He had inherited some of this hostility toward the French from his father, but some was also rooted in a lack of recognition from his French colleagues—none of whom he had ever met personally—of the new, physically oriented physiology that he and the other organic physicists in the German lands were pursuing. Better, then, to cross the North Sea than the Rhine.

One of Helmholtz's principal aims in visiting Britain was to attend the meeting of the British Association for the Advancement of Science (BAAS) in Hull, where he hoped to meet John Tyndall. Tyndall, who was elected professor of natural philosophy at the Royal Institution in 1853 (an appointment that soon made him a highly visible, public scientist), had studied in Marburg with Bunsen and Knoblauch. After receiving his PhD in physics there in 1850, he had spent the spring of 1851 working in Magnus's laboratory, where he became friendly with Helmholtz's crowd. Du Bois–Reymond gave Tyndall Helmholtz's essay on force conservation, saying it "was the production of the first head in Europe since the death of Jacobi" and suggesting that Tyndall have it translated into English.[34]

Helmholtz began his trip in early August with a stopover in Berlin and Potsdam to see family, friends, and colleagues. He first went to Müller—"He was very friendly towards me"—and then to Poggendorff, whom he found "very obliging" toward him, as well as to August Karl Krönig, a former student of Magnus's, the new editor of the *Fortschritte der Physik*, and (after 1856) one of the cofounders of the kinetic theory of gases. Helmholtz was preparing a report for Krönig on recent work on acoustic theory, and Krönig asked him to prepare another on the theory of heat. He published eleven such reports between 1847 and 1859 on these topics and was one of the journal's (and hence the Physical Society's) most important contributors. He visited his tailor—"I ordered a fine, black topcoat and a white vest for myself; both turned out very well"—and ordered calling cards, all of which he needed for Britain. For relaxation, he attended Johann Nestroy's play *Kampl oder das Mädchen mit Millionen und die Näherin*.[35]

He dined at Magnus's in Berlin—"The dinner was very elegant, with champagne"—where the other guests included the Berlin chemist Heinrich Rose and none other than Tyndall. He found Tyndall to be "a very talented young man" and the most interesting of the guests, but he was disappointed to learn that Tyndall would not be in England while he (Helmholtz) was there. "I had actually counted on his help. Still, I received instructions and recommendations from him that will be useful to me." After dinner, the two walked in the Tiergarten, and although Tyndall spoke "rather good German," they conversed in English. "It was very agreeable that I understood him, when he did not speak too quickly. He praised my pronunciation as very good." In turn, he analyzed several physics topics for Tyndall.[36] A friendship was born.

The next day he visited Dove, "who, of all the Berliners," could give him "the best letters of introduction, which he was immediately ready to do." That afternoon, he tried to see Tyndall again but instead ran into Graefe, who showed him his latest ophthalmoscopic results "and said many flattering things" to him about the ophthalmoscope. He also saw Wiedemann, who invited him to dine at his home. There they drank a beer and ate "warm fricassee, cold pudding, and bread and butter" and saw lots of interesting people (including "a number of well-educated, tastefully dressed, and self-confident women").[37]

The next morning he called upon Karl Otto von Raumer, the conservative head of the Prussian Ministry of Culture, and then went to the Kunstverein's painting exhibition, where he was again very impressed by Adolph Menzel's *König Friedrichs II: Tafelrund in Sanssouci* and Hans Frederik Gude's *Sommerabend auf einem Norwegischen Binnensee*. Afterward he visited Schulze at the ministry, who received him "extremely favorably" and said he would get a bonus immediately (if the ministry had more money) and that he hoped Helmholtz would remain in Königsberg. "When I complained about the climate, he told me that, given my appearance, he had to conclude that, on the contrary, it must be splendid." In the afternoon, he visited his uncle and sang duets with his brother Otto, who had "a very good voice" and sang "some things very sweetly." But he complained, "However, he has, or so it appears, gone in completely for sentimental songs."[38]

Before leaving Berlin, he turned in his report to Krönig for the *Fortschritte*; saw Dove, who gave him two letters of introduction; Magnus, who gave him two more; and Rose, who gave him a certificate to deliver to August Wilhelm Hofmann, the German director of the Royal College of Chemistry in London. He also visited the Neues Museum, where his main interest was Wilhelm von Kaulbach's *Die Blüte Griechenlands*, one of six large, historical wall paintings that he created for the museum. Finally, he saw Tyndall; they lunched and strolled through the Zoological Garden together. "We spoke English the entire time, concerning all the toughest issues in physics. He is a very charming, modest, and clear thinker." Tyndall gave him several letters of introduction.[39]

He also experienced painful toothaches, which became even more painful as he traveled by railway through Germany during the next ten days or so and even after his arrival in London. From Berlin, he went to Halle to see, among others, the physiologist Volkmann. A year or so earlier, Helmholtz had had difficulties with him. Now things had changed, and he liked him, "although he had gotten himself entangled in false mechanical ideas," and several times he had put Helmholtz "in the embarrassing position of explaining things to him without saying falsehoods and offending him." Nevertheless, Helmholtz spent a pleasant evening with him and his family: "The wife seems to be more intelligent than he is[;] she teases him about table-turning, for example, about which he had let himself become a bit deluded." Talk of table-turning seems to have been widespread: Helmholtz had recently told du Bois–Reymond that work on animal electricity had somehow become associated with it and that "a learned lady" had told him that results on the negative variation of electrical currents in nerves were being used as "an extremely foolish proof for the immortality of the soul."[40]

That night he went to Frankfurt, but it was so cold that his toothaches returned. "I wrapped myself up in a shawl and top-coat and sat on the sunny side of the train car until I was in a steady sweat. Then they [the toothaches] stopped."

He dined at the elegant Landsberg hotel: "I'm not normally very susceptible to hotel luxury. But when it's so well displayed, as it is here, it's indeed not unpleasant," he told Olga. Following a "most boring concert," he heard "a terrific fragment" of Felix Mendelssohn's unfinished opera *Die Loreley*.[41]

On the train to Mainz early the next morning, he met two English tourists; though their female guide was well educated, they themselves "did not belong to the *crème*." From Mainz, he took a steamer along the Rhine ("terribly beautiful"). It was "the height of romanticism," and his low expectations for the Rhine region were easily met. In Bonn he found a hotel room with a beautiful view of the river and was impressed by the passenger and coal-carrying steamships. He sought in vain to see the physiologist Julius Budge at the university and called unannounced at the home of its physicist, Julius Plücker, who was also a mathematician. Plücker was not in, but his attractive and intelligent wife—"not so that you should become jealous," he assured Olga—was. Just as Helmholtz arrived, so did Plücker's former student August Beer, a young physicist and mathematician whom Helmholtz (rightly) thought might be Jewish and who had "done very good optical work." When Plücker arrived home, exhausted from the day's work, he opened a bottle of good white wine in Helmholtz's honor, and Helmholtz was forced to exchange his red for the white. This led to Plücker's revival but to a recurrence of Helmholtz's toothaches (and so a bad night's sleep). The next morning, Plücker showed him his laboratory. Helmholtz's toothaches forced him to delay his departure for Cologne by a day.[42]

From Cologne he traveled first to Ostend, meeting two "fashionable" young English sisters on board. Then he boarded a ship for six rainy and windy hours from Ostend to Dover. From there he took a fast train to London—"The English travel like the devil"—arriving at 6:00 a.m. The charm of English lawns overwhelmed him. Later that day he dined at the home of Christian K. J. Freiherr von Bunsen, the well-connected Prussian diplomat, scholar, and liberal who headed the London legation. London was very far from Königsberg, and because at that distance he could not keep up with the details of Olga's health, he decided, for once, that he should not give her further medical advice.[43]

In London Town

London was a "great Babylon"; compared in size and culture to London, Berlin was "like a village." He toured London alone on foot. He was much taken with the expanses of lawns and saw lords riding horses and driving carriages on fenced-off paths. He was impressed, too, by the healthy, well-dressed look of everyone he saw in the beautiful parks; he judged even the nannies to be so well-dressed that "some Prussian countesses" might envy them. London's "wonderful parks" and greenery were compensation, he found, for its disadvantages as a big city. While he thought that neither London nor West End homes were attractive by day, by night they were elegant. And he appreciated that, in contrast to Berlin, there were many comfortable and constantly running buses. He felt at home.[44]

To save money, he moved out of his hotel and rented a pair of small rooms. Along with two German orientalist scholars, he lunched that day at Freiherr von Bunsen's, where everything was of the finest quality. He found Bunsen a touch pretentious—he sought to give the impression of knowing Michael Faraday well, whereas Helmholtz believed he did not—but very cordial toward him personally; he even wrote an unwanted letter of introduction to the zoologist Richard Owen.[45]

That evening he saw Byron's *Sardanapalus*, a popular dramatic tragedy about an Assyrian emperor. The costumes, set decorations, and other mechanical aspects of the theater production particularly impressed him, though he understood only parts of the play. But "in spite of the exaggerated, screaming, and screeching pathos of the English," he judged it to be "not badly acted." He thought the audience absolutely tasteless in its appreciation of music, drama, and the performing arts: taste here was merely a matter of utility.[46]

At the British Museum the next day, he saw Austen Henry Layard's archaeological excavations from "Nineveh" and Lord Elgin's Parthenon findings. The large Assyrian bulls had humanlike heads with "the most beautiful Jewish faces." He thought them far more impressive stylistically than Egyptian art and equal to the better ancient Greek art pieces. The Parthenon items, by contrast, made a poor impression, since they were so damaged.[47]

The next day he sought out Owen in vain, but he did find Faraday, whom he described as "at present . . . the first physicist of England and Europe," at the Royal Institution. Helmholtz was charmed and delighted by him: "He is simple, charming, and unpretentious, like a child. I have never seen a man with such a heart-winning nature." Faraday was "extremely obliging," showing Helmholtz his entire laboratory setup. Helmholtz, however, thought "very little" of this, "since it seems to him that some old pieces of wood, wire, and iron will be enough for the greatest discoveries." Afterward he attended an exhibition featuring a giant, three-story-high globe, which had the advantage that one could walk inside and get a feel for the earth's dimensions, though he thought this more a means to attract audiences than to teach. In the evening he went to the Zoological Garden, where he was much taken with the collection of exotic animals and perhaps even more so with a house filled with glass vats containing sea animals: "One sees here these beasts, which one otherwise only sees as corpses, in their most secret doings."[48]

After eight days in London, he took a steamer up the Thames to Hampton Court, Cardinal Wolsey's and Henry VIII's old castle. Many of his fellow travelers were working-class people, and he thought the maids on board were both pretty and relatively well-dressed. He also thought that in London clothes gave no indication of a woman's class, since everyone seemed to dress so finely and since English women were marked by good skin color and richly colored brown hair. The painting collection at Hampton, he reported, was strongest in Dutch items, including Benjamin West's fine *Death of General Wolfe*; on the whole, however, he found the collection "unimportant" and disappointing. The windy return trip down the Thames renewed his toothaches, and so the next morning he doctored himself "with chloroform on the gums and binding of the face." The pain ceased, though he stayed at home to write his lecture on his optical experiments for the Hull meeting.[49]

The following morning he went to see the book and journal publisher William Francis. Francis, who had trained as a chemist, was the illegitimate son of Richard Taylor, himself a chemist and book publisher. In 1852 Francis took over his father's publishing firm, which he rechristened Taylor & Francis and which he helped make into Britain's leading scientific publisher. Among its many publications were the *Annals and Magazine of Natural History* and the *Philosophical Magazine*, which had previously included chemical and biological papers but during the 1850s became largely devoted to physics and mathematics. Helmholtz learned that Francis had studied science in Berlin in 1839–41 and then took his PhD degree in chemistry with Liebig in Giessen in 1841–42. While there, Francis met other young British men of science, including Tyndall, the mathematician Thomas Archer Hirst, and Edmund Atkinson. Francis knew the British scientific community well, and he encouraged these and other young men of science to look for important German- and French-language papers that might be translated into English. As a result, in 1853 one of Helmholtz's papers on the time-measurement of nerve impulses was translated into English and appeared in the *Philosophical Magazine*. As a close friend of Francis's and as co-editor of the journal, Tyndall became a key figure linking German and British science. Helmholtz thought Francis "a young, modest, and pleasing man who is very well-instructed in all scientific matters and London personalities." Francis took Helmholtz around to meet several leading British men of science.[50]

In the afternoon he saw the National Gallery's picture collection. He enjoyed its "beautiful Rembrandts, modest Rubens and Italians" but became especially enamored of two sentimental religious pieces: Bartolomé Esteban Murillo's *St. John and the Lamb* and *The Two Trinities*. Later he called upon Charles Wheatstone, the experimental physicist who had helped invent the electric telegraph and who, like Helmholtz, keenly pursued optics, acoustics, and electricity. Wheatstone was away, but Helmholtz still hoped to meet him in Hull. He spent the remainder of the day and the evening at the theater, where he saw no fewer than three plays—Dion Biocicault's *Genevieve, or the Reign of Terror*, a historical drama about the French Revolution; a parody of Byron's *Sardanapalus*; and

Richard Brinsley Sheridan's comedy *The Camp*.[51] These ongoing visits to art museums, as well as to various cities and countrysides, themselves suggest the unceasing role of observation in his life and his fascination with color. In the museum or wandering in the city or countryside, as in the laboratory, he never ceased to observe and be fascinated by color.

The following day he took a steamer to Woolwich to see one of the grandees of the British scientific community, Colonel Edward Sabine, a geophysicist who had promoted the "Magnetic Crusade" of the 1830s, managed the Kew Observatory, and held high office in the Royal Society of London. In 1852 Sabine had served as president of the BAAS, and in his presidential address he referred not only to new British work on the theory of heat but also to Helmholtz's. Helmholtz had come armed with a letter of introduction from Dove, Sabine's fellow meteorologist and long-time correspondent. Dove portrayed Helmholtz as both a physiologist and a physicist: "He is considered by us in Germany as one of the most-gifted and sound of the younger natural researchers." It turned out that Sabine, too, was away. Nonetheless, later that day Helmholtz received "an extremely courteous letter" from Sabine enthusiastically inviting him to return and offering to send his carriage to meet him at the station. So Helmholtz, the former Prussian army medical doctor, used the afternoon to tour Woolwich military barracks and to view an impressive collection of military equipment. Woolwich, and the many large ships in London's docks, brought home to him England's commercial and naval power.[52]

He worried about Olga because he had not heard from her for three weeks. He facetiously discounted a kidnapping, and, since he knew she was "reasonable," feared that her letters had simply gone astray. He thought about telegraphing but judged it expensive, and so for the moment he limited himself to doing no more than writing again. He reported that his health was good (the toothaches had ceased); he had met du Bois–Reymond's new wife, Jeannette. (Du Bois–Reymond did not invite Helmholtz to his wedding, explaining to his bride, "Helmholtz . . . is, in my opinion, one of the truly best living scientists as well as one of the best who has ever lived. But in company he's a bit stiff.") Helmholtz thought Jeannette taciturn and only "moderately pretty." It rained a lot in England, he continued, and his English was improving: "The English have understood me from the very beginning, and I also understand educated people who speak to me very well, though not the waiters, artisans, etc." Two letters from Olga finally arrived, which he called his "birthday present." (It turned out that he had sent his wife the wrong address. "Now one of the two of us, let's admit it, was a silly ass. Will you be immediately able to decide who?" In fact, at first he, the thirty-two-year-old rising international star scientist, thought she was, but a week later he had to admit that the mistake was his.) The publisher Leopold Voss had sent him a book contract, and he asked Olga to forward it to Potsdam.[53]

In London, too, he went to the Museum of Practical Geology to see various chemists and geologists. He was accompanied by William Francis and Axel Erdmann, a Swedish chemist and geologist whom he unflatteringly characterized as "a very dull companion, boring, gauche, and indecisive." He had hoped to see Hofmann there, but Hofmann had gone to Germany. The trio then went to the British Museum, where Francis made introductions and got them special privileges for using the library and the museum. Though the library impressed him — he saw the original Magna Carta as well as other original manuscripts signed by various English kings and Protestant reformers — he knew he would never avail himself of the privileges Francis had obtained for him.[54]

He returned to Woolwich to visit Sabine, "the director of meteorologic and magnetic observations in England, an old, dignified gentleman." Sabine's wife, Elizabeth, had done, under his direction, an authorized translation into English of the first three (of five) volumes of Humboldt's *Cosmos*. Later he sent Helmholtz some of his publications as well as more letters of introduction than Helmholtz cared to have; he offered to help him however he could. "I thus discovered that he was a much more important scientific thinker than I had originally thought and was very well informed in all matters." The next morning he met up with the du Bois–Reymonds and then went again to the British Museum, this time to see the Aztec exhibition and its alleged human remains. In the afternoon he went (again) to the Zoological Garden with Erdmann to hear music (he judged the military music he heard there to be no better than what one heard in Königsberg) and induced Erdmann to continue with him to the Sadler Wells Theatre, "where according to English judgment the best performances in London of Shakespeare's plays" were given. He saw *Macbeth*. "It was terrible! Endless fits, shouting, screaming, shrieking, even at the most indifferent words." He was stunned by the endless applause and cheering. Afterward he enjoyed a comedy, *Brother Ben*, that he "almost understood completely."[55]

Francis invited Helmholtz and Erdmann to his home in Richmond. Helmholtz loved its hilly, parklike setting. They dined with Francis, Taylor's wife, and the wife's elder daughter. Though Taylor himself was mentally ill, he impressed Helmholtz, who found that one could talk to him and not know he was ill. But when he started to speak of his idée fixe — the dangers of Protestantism in Germany — the family sent him off to bed. "Social intercourse in English families is very beneficial," Helmholtz concluded. "They are considerate and yet free and easy." Erdmann and Helmholtz also went to the nearby Kew Gardens, which greatly impressed him.[56]

The next morning he returned to the British Museum to see its natural history collection, in particular its fossil skeletons. He was unimpressed with the ethnographic collection, however. He met the du Bois–Reymonds there, and that afternoon they went to the houses of Parliament and to Greenwich Park. Armed with a letter of introduction from Sabine, he returned to Greenwich Park the next day to see George Biddell Airy, the British Astronomer Royal and head of the Royal Observatory there. But Airy's tight schedule meant their meeting had to be short, so they rescheduled for a few days later. Olga could write to him at Hull under the title "Fellow of the British Association," he boasted.[57]

After visiting the famous, large Barclay and Perkins brewery and Hunter's Anatomical Museum, where he was impressed by "the most complete skeleton of the giant, primeval sloths and other beasts," Helmholtz dined that evening at Bence Jones's (with the du Bois–Reymonds). The following evening he again dined at Bence Jones's, and Bence Jones insisted that he come to his beach villa in Folkestone for several days.[58]

Helmholtz had heard that Airy could "supposedly be very unpleasant," and at their first visit (in Greenwich) he had found him "rather formal." But now he seemed "extremely charming." "Since I paid attention to his expositions, and praised much and contradicted only a bit, he could hardly come to an end with all the showing around, so that I've seen a great deal more of the observatory than perhaps anyone else." He admired Greenwich's equipment, including the magnetic and meteorologic apparatus; the automatic, continuous photographing of heavenly objects; the impressive electromagnetic time-measuring apparatus for star transits; and the electric clocks, which were "supposed to give the time simultaneously in London and at the mouth of the Thames and at all London railroad stations." The Airys lived in grand style. (Helmholtz judged that most English professors lived well.) Airy's wife was "somewhat formal, well-preserved and pleasant in conversation." He commented, "The English ladies are always very much interested in their husbands' pursuits, and she too was very well informed about everything." Nevertheless, Airy was haughty, and that led to difficulties with his colleagues. Thanks to his "favored position," he signed his articles under his title ("Astronomer Royal") rather than under his own name, and this too put distance between himself and his colleagues. Helmholtz also thought that Airy, unlike most of his fellow English physicists but like their French counterparts, had "methodical training" as a man of science, which made him "superior." Most English physicists who had done "great work" did so, he said, simply "through pure instinct," though "often enough" their lack of basic, common knowledge could ruin things for them. Whatever arrogance he and others detected in Airy, he left with the feeling that "the afternoon in Greenwich belonged to the most interesting and most pleasant" of his trip.[59]

He then spent the weekend at Bence Jones's place in Folkestone (the du Bois–Reymonds were there, too). He thought Bence Jones to be "so much like a German, except for the assured, distinguished formality" that he always had. He was pleased that he could (just) understand Bence Jones's English, and he proudly reported that others said his own had grown more fluent (if still accented) and that he was praised for it. He still found Jeannette du Bois–Reymond unappealing: "Even if there were no Dötchen, I wouldn't choose her for me." Saturday afternoon he and Bence Jones walked along the Dover cliffs. When the weather was good, one could see the French coast from his house.[60]

Early the next morning he went sea bathing, "*au naturel*," near the house; "that's allowed here because, before 9 a.m., no respectable person is to be seen." Then he was forced to attend church with his host. "The Anglican liturgy is, however, really deadly long and boring, and is carried off as much as possible without expression by both the minister and the congregation. The sermon was, according to our ideas, silly and tedious." That afternoon, Bence Jones, Helmholtz, and du Bois–Reymond took a long walk on the beach and saw the "pretty valleys and ravines." He liked the comforts, cleanliness, customs, and atmosphere of English landed life, in which, "notwithstanding all the formalities, there ruled a great lack of constraint, so that everyone speaks his mind just as he wishes." It was still embarrassingly difficult for him to figure out which piece of silverware to use when, which suggests that neither his mother nor his aunt Julie had educated him sufficiently in table manners for the worldly life he was now leading.[61]

He returned to London the following morning, visiting the Tower of London, where he saw weaponry, the Crown Jewels, prison cells, and more, but scarcely thought it worth the visit. By contrast, he loved the Catherine and London docks with their flourishing commercial life; they left Königsberg's far in their shadow. As always, he admired commercial activity. That evening he and Moritz Alberts, a secretary at the Prussian legation, visited a debating club, "where some comical lawyers in judges' costumes and with very strict observance of the courts' forms" presided over "imaginary trials." It amused him very much.[62]

The next day was spent with Wittich, his colleague from Königsberg who was also visiting Britain. In the morning they visited Westminster Abbey. Helmholtz was distinctly unimpressed with its size and architectural forms. He said, though:

> The series of memorials to the famous dead is indeed imposing, and must, I believe, excite the pride of the English to the highest degree. To have had such men, and to see them so honored, is a great thing. The professors of physics and chemistry lie there between the kings, commanders-in-chief, [and the] artists. And even actors and actresses of the first rank have found their places and memorials here. Newton, James Watt, Humphry Davy, Thomas Young, Shakespeare, Milton, [David] Garrick, Mrs. [Sarah] Siddons, Henry V, Richard II, the sons of Edward, Warren Hastings, both Pitts, Maria Stuart, and Elizabeth are all together in the same place.

In effect, he was saying that being a scientist was something to be proud of, like being a statesman, soldier, artist, or actor. That evening he and Wittich went to Drury Lane, where they saw *The Merchant of Venice*.[63]

The BAAS at Hull and Energy Conservation

After nearly three weeks in London, Helmholtz left by train for the BAAS meeting at Hull. There he stayed with a medical doctor, Henry Cooper: "I reside very fashionably and am cared for." Plücker, Helmholtz, and a Russian colleague were the only foreign scientists at the meeting, and they were "treated with the most exquisite courtesy." Some 600 people, including 175 women, attended the first day's meeting. At the opening, plenary session that evening, William Hopkins, the much-respected mathematical coach at Cambridge, geologist, and current president of the BAAS, reviewed the principal "progress" in science during the past year. The secretary (Sabine) read aloud the names of all foreigners present, including Helmholtz's, saying that he had "made some of the most important progress of continental science." Helmholtz seemed quite surprised to learn that his conservation of force essay was better known there than in Germany and better known than his other works.[64]

How *did* his essay become better known in Britain than in Germany and (if true) better known there than his physiological studies? For nearly five years (1847–52) it had lain fallow. The German scientific community, it will be recalled, had virtually ignored it; only a handful of scientists (mostly Helmholtz's friends) had reacted positively to it, and a few had reacted negatively. But things began to change in 1852. This was due in part to the attention that his physiological studies had brought to his name and to his invention of the ophthalmoscope.

There was another, more direct reason why the essay suddenly started to draw attention, one perhaps related to why Helmholtz had chosen to visit Britain and meet some of its men of science: namely, their recent recognition of his essay. In late January 1852, the young William Thomson, who had recently begun thinking in terms of "energy," read Helmholtz's essay, and by March he had publicly expressed his admiration of it. During the next two decades he led a group of northern British physicists and engineers—including his brother James Thomson, James Prescott Joule, W. J. M. Rankine, James Clerk Maxwell, Peter Guthrie Tait, and Fleeming Jenkin—into transforming Helmholtz's law of conservation of force into the law of conservation of energy. By the spring of 1853, an English-language version of Helmholtz's essay had appeared. Its translator and coeditor was none other than Tyndall; the other coeditor, and the publisher, was William Francis. Helmholtz's essay was reborn in Britain. Moreover, its endorsement by Thomson and other British physicists represented some of the first real recognition of Helmholtz qua physicist. However, there were important philosophical differences between him and these northern British physicists and engineers, several of whom soon became his friends. Both the northern British and Helmholtz constructed deterministic, if different, accounts of the role of force in a mechanical universe; but one (Helmholtz's) was rooted in eighteenth-century Laplacian physics and centered on attractive and repulsive forces, while the other (the British) posited a Creator who had established a continuum of matter and energy and who left room for human will to intervene. Both Helmholtz and his British supporters largely glossed over these differences as they all promoted what suddenly became the law of conservation of energy. It was thus more than semantics ("force" versus "energy") that separated him from them, even as he and they allied to promote their common cause and to move it beyond philosophical battles within physiology and beyond its British industrial and professional contexts.[65]

On Helmholtz's first morning in Hull, Charles Frost, a wealthy local solicitor, geologist, antiquarian, genealogist, and a vice president of the BAAS meeting, invited him to breakfast with several people, including George Gabriel Stokes, "a young man, however one of the highest, most excellent abilities." Meeting one of the leading mathematical physicists of the era—Stokes was the Lucasian professor of natural philosophy at Cambridge—was only one of many rewards that Helmholtz received for attending the Hull meeting. At Frost's breakfast table, he also met several well-to-do Englishmen and their wives, including "a very rich Lord Landsborough [also a vice president of the meeting] with a stately, young, and rather beautiful wife." Unfortunately, Helmholtz sat before the ham and was forced to cut pieces for everyone. Landsborough invited Helmholtz to join him on his yacht, and that pleased him.[66]

As for Dr. Cooper, he was a well-educated man and seemed to be "a lively doctor of scientific passion." He provided Helmholtz with his own room and treated him "like an old friend of the house." Helmholtz found Cooper's furnishings not quite as elegant as those of a well-to-do German. He thought, however, "All these carpets, curtains, waterpipes, bellpulls, other pulls, etc., which serve the purpose of making things as thoroughly comfortable as possible—given that one knows in the first place how to use them all—cost perhaps twice as much" as in Germany. Here and elsewhere in England he tasted upper-middle-class if not upper-class life. Twice he had to have lunch—at his own expense—with lots of colleagues at a large table. He found the serving protocols unwieldy, the toasts too many, and the speeches too long, too poor, and too meandering. Even most English men of science conceded that Continental lunch meetings were more pleasant than their own.[67]

BAAS meetings served largely to popularize science in general and to educate the British public, above all those who attended, about the latest results of science. There was a premium on showmanship. During the Hull meeting, attendance climbed to 850, including 236 women. "The latter seem here in England to be in fact frequently rather informed in the natural sciences," Helmholtz wrote Olga. He thought they came variously to be seen, to learn, "and to be amused." They were generally attentive and apparently did not fall asleep, "even when the temptation was there." The association's six (or seven) sections met each day from 11:00 a.m. to 3:00 p.m. (Beforehand, from 10:00 to 11:00 a.m., there were committee meetings; Helmholtz was placed in the physics section.) Audiences moved from one section to another, "in order to hear the most famous and most popular speakers." It was the embodiment of science as a marketplace of ideas and personalities. The quality of the talks ranged from the scientifically serious to presentations by fools who thought they had made important new discoveries. (The chairmen got rid of the fools quickly.) The leadership of British physics, chemistry, "and such like sciences, in which a man must work by himself," was kept in the background. "For other sciences, however, such as meteorology, ethnology and geology, where there must be co-operation among many observers, the meetings (of sections) were of great importance." Helmholtz most appreciated those talks that summarized past work and suggested future areas for investigation that should be pursued by a team of scientists, something that the English had a particular gift for doing, he said. He noted that there was currently a team of astronomers (including amateurs) studying the moon's surface and comparing its geology with the earth's. Another team was preparing to send a large telescope to the southern hemisphere for study of the skies from there.[68]

Geology, geography, and ethnology were "the public's favorite sections." They were also the sections that contained "the most famous scientists," since for these disciplines it was important to instruct a large number of people to work together as a community. "For that," he said, "the Association is a very appropriate tool." However, Helmholtz noticed that many of the most important chemists, physicists, and astronomers did not attend the meeting—he named Airy, Faraday, and Wheatstone as examples. Nonetheless, he met several big names whose acquaintance he "very much wanted [to make]": William Robert Grove, "a lawyer and important physicist from London," best known for publishing one of the first versions of the conversion of forces as well as for his work in electrochemistry; Thomas Andrews, the professor of chemistry at Belfast who specialized in thermochemistry; and Stokes. Not a single physiologist attended, which probably reflected the low state of physiology in England at midcentury.[69]

While he found that some lectures were "excellent in regards to clarity and popularity, and yet scientifically important," including presentations by the geologists John Phillips and William Hopkins and by the ethnographer Robert Latham, he said, "There were also many ponderous ones, and many—and this surprised me here with the English—were muttered and so poorly delivered that they were not to be understood." To his ever-sensitive German ear, spoken English in public lectures sounded "worthless, because of its too great similarity to Low German, and hesitant, because they push forward the accentuated syllables very sharply and let everything else fall weakly and quickly by the side." He could now easily understand the better-spoken presenters. He asked Dr. Cooper to explain several provincial English expressions, and as they both discovered, these turned out to be abuses of English. In this context, as in musical matters and in experimental acoustics, his ear was acute. In one section, he participated extemporaneously in a discussion about the optics of the eye. He thought he had spoken clearly and was completely understandable; and though his colleagues praised his performance, he admitted to Olga that he had made some linguistic mistakes. He also took pride and care in his own oral presentation (on color mixing). He held a dry run of it before Francis, who corrected his linguistic mistakes and praised him. At the actual presentation, his listeners were "very satisfied" with his enunciation. "I received lots of compliments at the expense of Prof. Plücker, who, despite his many and long visits to England, articulates [English] very badly." "I was also honored both times with applause," he reported home with further pride, though he was surprised that, instead of clapping, the English rapped their fists on tables. They did this "whenever a speaker uttered something clever or any sort of a favorite catchword or a joke," and it was "relatively easy to reap applause." Though the BAAS was somewhat disappointed that only three foreigners attended, it nonetheless said this redounded to its benefit.[70]

On three separate evenings there were major lectures, one by Hopkins reviewing the "progress" in science during the previous year, one by Phillips on the geology of Yorkshire, and one by Robert Hunt on the present state of photography. On three other evenings the city of Hull sponsored (overcrowded) soirées. That left Sunday morning free, which presented Helmholtz with a little problem: he was forced to attend a local German Lutheran church, but he got out of doing so because someone had given him the wrong starting time. In the afternoon, Dr. Cooper showed him the Hull docks, which were packed with German, Russian, and Danish ships. He saw more Prussian shipping activity here than he had ever seen in Königsberg. The meeting's penultimate day was devoted to a general session for planning next year's meeting, which held "no interest" for him. Nor was he interested in the final day's offerings of local excursions. So he skipped these and instead gained more time and saved more money for a visit to Scotland. He sailed for Edinburgh.[71]

Scotland

He experienced no seasickness on the unexpectedly long, twenty-eight-hour voyage, and he slept well on board, though he did not like his fellow passengers, who, like himself, traveled by ship because it was less expensive than the train. He found the Scottish coastline "very romantic" and in certain respects similar to that of the Rhine. Along the way he saw Lammermoor Hills and expected to see the Lady of the Lake area, both of which reminded him of two of Walter Scott's novels. He wanted to know how Olga's financial situation stood, since that would affect his plans. He "spent eight days in Scotland, to feast on nature."[72]

Edinburgh delighted him. (It "is a jewel among cities," he told Ludwig.) He thought it picturesque and "absolutely, extraordinarily beautiful." He noted its location on a series of steep hills, where houses were built seemingly one on top of another. The city as a whole appeared to him like the side tower of a gothic cathedral. He climbed to the top of Carlton Hill to view the city from on high, noting the Roman-built observatories, other monuments that were copies of Athenian buildings, and a partially completed building that appeared to be a copy of the Parthenon. Then he climbed down to the city's center and up to the castle. He liked the beautiful "Saxonian" architectural style of many buildings. The newer, more elegant part of the city contained houses done more in a German or Italian style than in an English one; this gave it "a much friendlier and more elegant look than the English cities." The homes in the city's older sections, by contrast, had a dirtier, more decayed look and were "resided in by a poor, ragged people with a profusion of red hair." But on the whole, he thought Edinburgh's homes, castles, and art collections offered him nothing that he could not find elsewhere, and so he soon left.[73]

He went to Glasgow, a large, smoke-filled, busy, commercial city that he found also filled with red-haired workers who looked unhealthy, dirty, and impoverished. His reason for going there, and his first order of business, was to meet William Thomson, who had "worked a great deal on the subject of conservation of force." But Thomson was at the seaside. And so early the next morning Helmholtz took a steamer to the port of Oban in western Scotland, where he hoped to meet up with Wittich. Along the way the sea traffic and wharfs of the River Clyde impressed him just as much as those in London had. As in Edinburgh, he found the Clyde itself "extremely picturesque and rich." The region's steamers were "all very elegantly equipped." He also found the local people easy to talk to and especially solicitous of foreigners. He liked the Scots a lot and could understand anyone who spoke to him (whereas in England he could usually understand only "educated people"). But there were not as many well-educated gentlemen in Scotland as in England, he thought, and he was stunned by the level of alcohol consumption ("Young and old drink whisky by day and by night"); those of Saxonian descent could tolerate this well, he claimed, while their Highlander countrymen "seem to become confused by the first glass."[74]

Helmholtz and Wittich walked around Oban and its environs. They visited Dunollie Castle, a medieval ruin and the original residence of the Scottish kings. They saw the grave of King Fergus MacRoich of Old Gaelic legend. The Scottish countryside here had all too few trees and plenty of brown heaths and moors, he judged; it made "a very desolate impression." The next day they took a steamer north to Glencoe, and from there Wittich headed up north alone and Helmholtz returned to Oban. He toured the peninsula of Morvern, which he noted was supposedly once the kingdom of the semimythical Fingal (Fionn MacCumhal). He sailed around the island of Mull and to the Atlantic islands Staffa and Iona. As nearly everywhere he went, he found traveling companions along the way; he was congenial and good at meeting people. He noted that Scott, in *The Lord of the Isles*, had written about the sagas of the ruins in this region. On Iona, he saw old churches, cloisters, and graves, including Macbeth's grave.[75]

Bad weather forced him to return to Glasgow, where he expected to meet up with Wittich again. But Wittich had gone to Edinburgh, and so Helmholtz decided to cut short his Scottish excursion and take the train back to Hull, where he stayed for barely a day. A letter from Olga awaited him and told him she had been sick. This depressed him and made him anxious. He told her that if she got sick again, or if she simply wanted, he would immediately eliminate his planned stopover in Potsdam and Berlin and return directly to Königsberg. She needed only to write to him in Hamburg or Potsdam.[76]

He sailed for nearly three days from Hull to Hamburg. Along the way, the boat hit strong winds and swayed from side to side, and Helmholtz, who had just finished his lunch (fish and roast), suddenly got seasick. After vomiting up half the lunch, he ordered two fresh portions of roast. The stormy seas limited his activities to reading, made him dizzy, and made it difficult to eat and digest his meals. But he also suffered from headaches, which in fact had begun before the sea voyage and which he thought were "a consequence of the strained trip."[77]

He had no desire to see Hamburg, and he feared cholera and the expensive hotels. So he left forthwith for Berlin, where he arrived at 5:30 a.m., had coffee at the railroad station, slept on a sofa until 7:00 a.m., and then went to Potsdam. He found that his family was well. By 6 October he was back in Königsberg, after two months of travel through the German lands and Britain. He had done a great deal of networking and had made new friendships. These included more than a few of the leaders of the British scientific and medical communities. He had presented a paper before their national scientific organization and had spoken with dozens of them on a personal, individual basis. He had promoted his ophthalmoscope and his law of conservation of force. In effect, and certainly unintentionally, he had become a representative of German science, a nonappointed, unofficial statesman. He wrote Ludwig: "England is a great country, and one feels here what a terrific and magnificent thing civilization is when it penetrates into all life's tiniest relationships." He thought that Berlin and Vienna, in comparison to London, were "mere villages." It was simply indescribable. "A visit to London marks an epoch in one's life; after such a visit, one learns to judge human actions on a scale hitherto unknown." He regretted not having been able to see Rankine, Brewster, Joule, Thomson, and Wheatstone, though he had "had better luck" in meeting Faraday, Stokes, Sabine, Grove, Airy, Bence Jones, Andrews, William Rowan Hamilton, "and many others of lesser importance." He was glad he had gone but also glad to be home, though "with a very empty purse." "My health has greatly benefited by the trip," he said, "but my teeth troubled me on the journey, and made for a time my physiognomy asymmetrical."[78]

7

The New Dispensation

The Thermodynamics of History

By 1852–53 Helmholtz began receiving recognition both as a physicist
and as a physiologist. The British were reinterpreting the meaning of
his conservation of force. In Germany, Rudolph Clausius, who generally much admired Helmholtz's essay on force conservation, nonetheless challenged its analysis altogether. The criticism and response redounded to Helmholtz's benefit, however, since it meant that force
conservation was finally being taken seriously by a leading German
physicist and since the exchange occurred in the pages of the *Annalen
der Physik und Chemie*. Clausius, who in 1850 had first presented a version of what became in the ensuing two decades one form of the second
law of thermodynamics, criticized Helmholtz's understanding of potential theory and the concept of work, as well as his view of the role that
central forces played in conservation. Yet Helmholtz skillfully defended
the conceptual thrust and general approach of his 1847 essay, conceding
only minor points (though three decades later he conceded major ones
as well).[1] Further development of the principle of energy conservation
now became a matter for physicists, not physiologists.

Helmholtz also completed his definitive study of the eye's accommodation. However, he published it in a medical, physiological journal that few physicists read. Magnus, who was greatly impressed by it, suggested that Helmholtz write an article for Poggendorff's *Annalen* giving the essence of his results and emphasizing points of special interest to physicists. (He assured Helmholtz that Poggendorff would publish such an article.) Further, Kirchhoff asked Helmholtz to write him a letter of recommendation for the available professorship of physics at Heidelberg. In this connection, Wilhelm Weber wrote to acknowledge Helmholtz's "beautiful, physically as well as physiologically important and extremely interesting, essays."[2] Germany's leading physicists now considered him a colleague.

He spent the winter of 1853–54 constructing "new apparatus for human time measurements" but expected to have little time that summer to do any experiments with them, since he had just moved into a new, unfinished laboratory. His working conditions were difficult, and besides, he was the dean of the Medical Faculty for the academic year 1854–55. "Unfortunately, the Königsberg deanship doesn't compare in monetary terms with that at Berlin," he told du Bois–Reymond. Beyond completing his work on accommodation, however, he did manage to send du Bois–Reymond a short notice on the speed of nerve impulses for the academy's *Monatsberichte* and to lecture before the Berlin Physical Society (on 30 June 1854) on measuring processes in nerves and muscles. He ordered a myograph to be constructed for the physiological institute in Giessen, but he feared that if the physiologist Eckhard got his hands on it, Eckhard might "take away" what Helmholtz had discovered.[3] He wanted property protection for his science.

Helmholtz also now gave the first of numerous popular lectures on the conservation of force (energy), thus bringing the subject to the educated world and beyond. On 7 February 1854, as president of the Physical-Economic Society in Königsberg, in his address "On the Interaction of the Natural Forces and the Most Recent Investigations of Physics Related to Them," he declared that there was "a new, general law of nature" in physics, one "of very general interest." It concerned the mutual relationship of all forces of nature with one another and was thus of importance for understanding nature theoretically as well as — this was a point that his nonacademic listeners and readers would readily welcome — for its "technical application."[4] In the very year that Thomson christened the subject with a new name, "thermodynamics," Helmholtz began to popularize it.

Helmholtz recalled that inventors had long hoped to create a perpetual motion machine; it was a sort of modern Holy Grail. Such enthusiasts dreamed of a machine that would replace animal and human labor, requiring no (or minimal) consumption of material resources, a machine that would operate indefinitely, thus becoming an "inexhaustible force of work." They wanted "to make money from nothing." "But work is money," Helmholtz reminded his listeners and readers, and he declared that those who sought to create such a machine were "confused and poorly instructed heads."[5] His general analysis of machinery, work, force, and money came as the German economy was industrializing. It was just the sort of talk that listeners in commercial Königsberg and readers throughout the economically developing German lands could well appreciate.

He laid out for listeners and readers the central concept of a machine, its "motive power or force of work," and explained that its "work" was equivalent to its "expenditure of force." He facilitated understanding of these and other abstract, mechanical concepts by referring to the internal motions of concrete, familiar objects—water mills, iron hammers, wall clocks with sinking weights, and pocket watches with tension springs—that, again, his listeners and readers could readily appreciate.[6]

He declared, further, that the era's novelty lay not merely in showing the impossibility of a perpetual motion machine but in recognizing that conservation applied to all forces, not just mechanical forces but also thermal, electrical, magnetic, light, and chemical forces. It was in these other, nonmechanical forces that "the progress of modern physics" lay. A series of natural philosophers and physicists—he named Mayer, inter alia—had in one form or another recognized the proposition that all forces could be transformed into one another. He reported that in England he had found growing interest among scientists in the law of conservation of force (soon to become known as the conservation of energy and, later, as the first law of thermodynamics). Joule and others had confirmed the theory experimentally, while Henri Victor Regnault, "the most important of the French physicists," had used it as the basis for investigating the specific heats of gases. While he conceded that more proof was still needed, he nonetheless thought the theory was sufficiently confirmed to merit presentation before "a non-scientific public."[7]

The new issue, then, was to understand the relationships between forces. Nature as a whole contained a store of force, he explained, that could be neither increased nor decreased; its total quantity was "eternal and unchangeable, like the quantity of matter." He presented nature as something to analyze and understand but also to enjoy and exploit commercially: "The woodland stream and the wind that drive our mills, the forest and the coalbed that feed our steam engines and heat our rooms, are only carriers for us of a part of nature's great store of force, which we exploit for our purposes and whose effects we try to bend to our will." The windmill owner considered the downward flow of water and wind power "as his property."[8] Helmholtz loved nature as much as anyone, but he was no modern environmentalist.

William Thomson (following Sadi Carnot) had recently conceived a law, Helmholtz further explained, that showed that the transformation of heat into mechanical work occurred only (and only in part) when heat flowed from a warmer to a cooler body. While the law—which eventually became known as the second law of thermodynamics—had yet to be proven, Helmholtz thought it was probably true. (He gave scant recognition to Clausius here.) Left to themselves, the forces of nature would yield a temperature equilibrium, nature would come to a "complete standstill," and the entire universe would be in "eternal rest." In the infinite span of time, the world was threatened "with eternal death," or, as he now famously put it, a "heat death."[9] This was the first time, and one of the few times, that he mentioned the second law. Perhaps it was because his own emotional disposition was too optimistic for this "negative" law centered on notions of disorganization, irreversibility, decline, and death?

The interaction of nature's forces also led Helmholtz to speculate on other cosmological matters. Force conservation helped in understanding "the household of the universe." He rehearsed the Kant-Laplace nebular hypothesis of the formation of the solar system by means of the action of gravitational forces upon dispersed celestial matter, taking pride and showing diplomatic skill in noting that Kant had formulated this hypothesis "within the walls of this city," yet without denigrating Pierre-Simon de Laplace, "the great author of the *Méchanique céleste*." The hypothesis provided a rational explanation of the origin and structure of the earth, the solar system, and beyond; it brought humankind "out of the dark of hypothetical ideas and into the brightness of knowledge." Yet it was no more than a hypothesis and an incomplete cosmological idea. He recognized that on this issue science stood all too close to ancient human legends and the premonitions of poetic fantasies, and he quoted from both the Hebrew Bible and Goethe's *Faust* to suggest (presumably) that there were structural similarities between ancient and modern cosmologies.[10] He appealed to intellectual tolerance and the use of the imagination.

Nor was that all. Helmholtz drew implications from force conservation for understanding such phenomena as movement and work in organic beings, the effects of the sun on the earth's atmosphere, and geological formations. He argued that in terms of food consumption and processing, organic beings were like steam engines. In principle, one could calculate how much heat (or its work equivalent) was produced from the combustion of a given amount of material in an animal's body. In effect, he provided his listeners and readers with an instance of the organic physicists' program. He reminded them that the plant world (or at least the meat of plant-eating animals) constituted the source of food that provided animals with energy: plants processed the food and nutrients that animals then ate and burned up, and in turn plants lived off the animals' combustion products. There was "a circle" of energy. If all this proved to be true, he argued, then all force ultimately originated in the sun, and "none of us are inferior to the noble descent of the great monarch of the Chinese Empire, who called himself the son of the sun." The point also held, he said, for "all our lower fellow creatures, like the toads and the leeches, the entire plant world and even the fuel, both the primeval and the recently planted, with which we heat our ovens and machines." Though experimental proof of such biochemical energy cycles had yet to be given, he acknowledged, his account, coming five years before Darwin's *Origin of Species* (1859), contained a similar general message: all life was netted together. For Helmholtz, the sun, through its illuminating and warming rays, constituted the dynamics behind this system. All planets had atmospheres; and Mars, he averred, even showed signs of water and ice.[11]

Helmholtz reasoned through all of the above largely on the basis of the laws of force conservation and universal gravitation. "Physical-mechanical laws," he declared, "are like telescopes of our mind's eye; they press into the most distant night of the past and future." Such laws, in conjunction with empirical data on the spread of cultivated plants as a function of change in mean annual temperature, could indicate the earth's temperature in the distant past and the distant future. Earth's history "shows how brief the moment of time has been in which the existence of the human race has occurred." The human race, he opined, had existed for something like six thousand years; and many species were now extinct. Experiments suggested that the earth, once a hot ball of molten rock, had required roughly 350 million years to cool down to 200 °C. Geologists estimated that the earth's geological formation and the subsequent formation of organic (but not human) life had lasted anywhere from 1 to 9 million years and that life today was distributed geographically very differently than it was in the distant past. The human race "was thus only a short wave in the ocean of time." Nevertheless, he assured his listeners and readers that life would continue for many generations to come and that they had nothing to fear. Yet he also told them (as if quoting from James Hutton or Charles Lyell) that the same geological forces at work today had been at work in the past; as these forces had transformed the earth and its life forms in the past, they would continue to do so in the future. The law of conservation of force, he declared, had given humankind a long but not an immortal history. "Like the individual, the species must also bear the thought of its death. But unlike other extinct forms of life, it bears higher moral tasks whose completion fulfills its destiny."[12] Pessimism was not his style. He offered a naturalistic, evolutionary, and non-Christian analysis of the earth's and humankind's history. It was one that would scarcely have appealed to his supporters in physics in northern Britain.

A Dispute over Materialism

Helmholtz's essay on the interaction of the natural forces impressed many, among them du Bois–Reymond, who was seeking to have him appointed a corresponding member of the Prussian Academy of Sciences. The academy viewed

Helmholtz favorably, and du Bois–Reymond hoped to achieve his goal as soon as possible. In fact, they had to wait two and a half more years, until January 1857, before Helmholtz was appointed.[13]

Among the essay's other admirers were Ludwig Büchner, the philosopher Friedrich Albert Lange, and Karl von Vierordt. Ernst Haeckel, while studying in Vienna in 1857, thought he should read the speech along with works by Jacob Moleschott and Büchner. It even impressed Ferdinand Helmholtz, who praised its contents and, for once, his son's style, though he preached that "nature and history are the expression of godly life" and objected to his son's inclusion of the "mosaic history of creation" as being a result of the universe's thermodynamic processes. In Britain, Hirst read Helmholtz's "excellent lecture"; he thought it "very speculative" yet also thought it abided "very closely to facts." He commented, "I must say that for me it has quite peculiar attractions, and I shall read it again several times." Tyndall translated it into English for the *Philosophical Magazine*. Thomson read it carefully, being especially interested in the issues of the sun's heat and heat death. But he rejected Helmholtz's hypothesis that the sun's source of heat (energy) was due to gravitational contraction rather than to meteoric impact (as he and Mayer were inclined to believe), since Helmholtz's view relied on the Kant-Laplace nebular hypothesis and its evolutionary nature. In 1861 Thomson revised his views in favor of Helmholtz's hypothesis, which, with its argument for the importance of gravitational contraction, had now become and long remained the dominant view of the source of solar energy.[14]

Helmholtz's essay helped herald a new moment in German intellectual life. Since the deaths of Hegel (1831), Goethe (1832), and Schelling (1841), both the substance and the tone of German intellectual life had changed markedly. The Age of Goethe ("Goethe-*Zeit*," 1770–1830), or, philosophically, the age of absolute idealism and romanticism, was definitively over. In the 1830s and 1840s, as German industrialization commenced and the natural sciences began to flourish in the German lands, Hegelianism and *Naturphilosophie* came under attack from and were swept away by a younger generation of scholars and scientists, who saw these and related philosophical orientations as too speculative, too unempirical, too romantic. The new intellectual trend emphasized historicism, empirical science, materialism, mechanism, and a renewed interest in Kant.[15] Helmholtz's essay, relying as it did on what were becoming the first two laws of thermodynamics for understanding the sun as the source of all energy, helped show the importance of science for understanding man's place in nature, the sources of energy, and the age of the universe.

One manifestation of this changed cultural atmosphere was the rise of the Young Hegelians, scholarly critics of Hegel. These included David Friedrich Strauss, whose *Das Leben Jesu, kritisch bearbeitet* (1835) questioned the miraculous, unhistorical portrayal of Jesus in the Bible and pointed to the mythmaking function of Christianity, and Ludwig Feuerbach, whose *Das Wesen des Christentums* (1841) portrayed Christian theology as a projection of human intellectual needs and, instead, emphasized physical sensations and experiences. Though neither critic, unlike their fellow Young Hegelian Karl Marx, was a materialist, their writings pointed in the direction of naturalism. This was true as well for the biological reductionists (including the organic physicists) who emerged on the scene in the mid-1840s and who sought to reduce the understanding of physiological life to the laws of physics and chemistry. They (Helmholtz included) were of course natural scientists, and mechanists, but certainly not materialists.

A second manifestation of the changed cultural atmosphere was the general rise of the natural sciences and technology and, with them, the emergence of a variety of new educational institutes and institutions for their advancement. In many quarters philosophy was now viewed either as irrelevant or in opposition to the natural sciences. The natural and the social worlds were seen as the objects of detailed empirical, ever-more-specialized study, not of general philosophical speculation, and especially in the German lands, empirical research was pursued with unprecedented passion and scope. Empiricism, naturalism, realism, and materialism became the hallmarks of contemporary science; philosophy was ignored by the sciences, attacked by them, or subsumed under them.[16]

Yet a third manifestation of the changed cultural atmosphere was the emergence of a group of scientists and medical doctors aptly dubbed the "scientific materialists." Three of them in particular—Karl Vogt, Jacob Moleschott, and Ludwig Büchner—were prolific polemicists and popularizers of science who helped shape the public's understanding of science through their materialist outlook as well as their general intellectual and political radicalism, which they considered to be based on natural science. Vogt was a physiologist who became best known for his *Physiologische Briefe für Gebildete aller Stände* (1845–47) and for his *Köhlerglaube und Wissenschaft: Eine Streitschrift gegen Hofrath R. Wagner in Göttingen* (1855). Moleschott had trained in anatomy and physiology—he read Helmholtz's studies on physiological heat—before devoting himself largely to physiological chemistry. He became especially known for his *Die Lehre der Nahrungsmittel: Für das Volk* (1850) and *Der Kreislauf des Lebens: Physiologische Antworten auf Liebig's Chemische Briefe* (1852). Büchner was the most famous of the three and indeed was arguably the century's leading German materialist. He was also a political republican, and he became famous above all for his *Kraft und Stoff: Empirisch-naturphilosopische Studien* (1855). Its publication met great hostility from the clergy and conservative elements in German society. He was forced to resign his teaching position at Tübingen. These and other works of the "scientific materialists" became widely known in the German-speaking world and beyond and contributed greatly to the generally materialistic and naturalistic atmosphere that became increasingly dominant in Europe, especially in the German lands, in the post-1848 era.[17]

The changed cultural atmosphere also manifested itself sharply at a meeting of the Association of German Naturalists and Physicians (Gesellschaft Deutscher Naturforscher und Ärzte), held in Göttingen in mid-September 1854. The association, known more colloquially as the Naturforscherversammlung, was the institutional vehicle that had brought German scientists and medical doctors together on an annual basis since 1822 to present their work, discuss the state of science, and propagandize for the benefits science might offer society. It had served as the model for the BAAS (1831). The historic nature of the 1854 meeting stemmed from a plenary address given by Rudolph Wagner, the professor of physiology (as well as zoology and comparative anatomy) at Göttingen, where Helmholtz had met him three years earlier.[18]

Like Liebig and Vogt before him, Wagner sought to popularize knowledge of anatomy and physiology by publishing a series of article-like newspaper letters (1851–52) that were then gathered into book form. Writing in light of the failed revolutions of 1848, Wagner, a political conservative, wanted to compensate for the unachieved political unity by arguing that Germans had a higher mission, namely, to make intellectual and artistic contributions to the world. He was a friend and colleague of Johannes Müller, and he noted the accomplishments of Müller's physically oriented students, including Helmholtz. He said it was rare for physicists to have knowledge of physiology. "It is thus so much the more gratifying that Helmholtz in Königsberg, a younger researcher, one versed in all branches of physics, has completely dedicated himself to physiology." He especially praised Helmholtz's essay on conservation of force and his measurements of the speed of nerve impulses. (Helmholtz read these flattering letters.) By contrast, he strongly opposed Vogt's materialism and political radicalism, and he declared that science and religion were two separate worlds. In turn, Vogt attacked Wagner. Neither avoided using ad hominem arguments.[19]

Wagner advanced these same points at the Göttingen meeting, where he spoke before some five hundred registered attendees on "Human Creation and the Soul's Substance." He maintained that, in general, there was no conflict between science and biblical belief; he rejected the use of science to undermine religion, as Strauss had done in his historical account of Jesus's life and early Christianity. Wagner said, in particular, that contemporary science had nothing to say about the claim that all human beings descended from one original pair of human beings (that was solely a matter of faith), while the recent emergence and spread of a materialist outlook among natural scientists, especially physiologists (not least Vogt and Ludwig), had undermined belief in the soul's existence, freedom of the will, and more. Wagner urged his fellow scientists not to misuse science by attacking such concerns as human creation and the soul's existence; neither should science be used to undermine the moral foundations of society. He also proposed a public discussion of these issues at the next day's session on anatomy and physiology, and he invited Ludwig in particular to participate. Ludwig did just that, but Wagner himself failed to show up. A few weeks later, three thousand copies of Wagner's address became available for the public to read. Early in 1855 Vogt, one of the main culprits in Wagner's eyes, published a reply. The entire matter had now transcended the assembly and impersonal, academic debate. It was now at least as much about politics as about epistemology or science; it was a matter of defending religious (Christian) belief and the political status quo as opposed to advocating a rationally ordered, progressive state and society.[20] Büchner's *Kraft und Stoff*, for its part, was written with Moleschott's recent pamphlet and the entire assembly dispute in mind. He argued that all knowledge should be based on observation and experiment, which became the basis of his materialistic worldview, and emphasized the role of force in the causes and unchangeable laws of nature. Wagner, for his part, became known as a spiritualist.

Helmholtz soon heard of the Göttingen *Materialismusstreit* (materialism dispute), as it became known. Ludwig wrote him that everyone at the Göttingen meeting praised him. "Of all contemporary physiologists, you have entered into the hearts of practitioners, above all thanks to your magnificent ophthalmoscope, which, as is generally said, marks a new era in ophthalmology." Helmholtz told Ludwig there were rumors that compared Ludwig and Wagner "like once upon a time Dr. Eck and Dr. Luther, who wanted to hold a public disputation on the nature of the soul, wherein Wagner naturally argued with the Bible in hand and you supposedly championed the cause of the Devil, of atheism, etc." He did not understand that Wagner had failed to appear at the open forum that he himself had proposed. But he also thought, quite rightly, that "Wagner's denunciation at the Naturforscherversammlung had injured him [Ludwig] with the government."[21] The materialism dispute was anything but over, either in the German lands or, for that matter, in Britain.

On Philosophy and Human Perception

In the early 1850s, the city of Königsberg sought to strengthen its legacy with Kant. In 1852 Karl Rosenkranz, a liberal-minded Königsberg patriot and Hegelian philosopher, established the Königsberger Kant Memorial, and in 1855 a bronze statue of Kant was completed (though it was not erected until 1862). They and others asked Helmholtz to deliver a Kant Memorial Address.[22] He was doubtless chosen because he had become Königsberg's most important intellectual figure since Kant. His lecture, entitled "On Human Perception," concerned not only his and others' recent understandings of human perception and his assessment of Kant's contributions to this topic; it also reflected the ongoing philosophical and cultural changes in Germany, and it came (February 1855) hard on the heels of the recent meeting in Göttingen and its materialism dispute.

Ever the diplomat, Helmholtz noted that it was the memory of Kant that brought his audience together that day and that Kant, "perhaps more than anyone else, had contributed to linking indissolubly" Königsberg's name with the cultural history of humanity. He sought to strengthen tradition, saying, "Our age and city has a grateful and honoring memory for men to whom it owes thanks for scientific progress and instruction."[23]

Helmholtz first outlined his relationship to recent philosophy. He supposed that some might question whether a natural scientist could honor a philosopher (i.e., Kant), thereby alluding to the current cultural struggle between natural science and philosophy: "Doesn't one generally know that at present the natural scientist and the philosopher are not exactly good friends, at least in their scientific work? Doesn't one know that for a long time both sides have conducted a bitter struggle? To be sure, the struggle seems to have ended recently, though not because one party may have convinced the other but because each has doubted being able enough to convince the other." Natural scientists spoke "loud and gladly" of "the great progress of science in the most recent times." This occurred precisely as science "thoroughly purged its realm of the influences of *Naturphilosophie*." During Kant's lifetime, Helmholtz said, there had been no split between philosophy and the natural sciences. Kant had held the same views as natural scientists concerning the foundations of natural science: that is, he was a Newtonian. The natural sciences still had the same foundations as in Kant's day; but philosophy had since changed its attitude toward them. The aim of Kant's philosophy, Helmholtz said, was not to increase knowledge through pure thought—since Kant believed that knowledge of reality could come only through experience—but rather "to investigate the sources of our knowledge and the degree of its justification." That was what philosophy must always do, he said; it was about epistemology, not metaphysics. Along with Kant, Helmholtz also appropriated Fichte, his father's hero, for the natural sciences. On his view, Fichte, who had himself briefly taught in Königsberg, was another friend of the natural sciences, even if from another era.[24]

There were two philosophers to whom Helmholtz wanted absolutely no linkage: Schelling and Hegel. He claimed that after Fichte's death Schelling and Hegel "began the quarrel" with the natural sciences. These two thought that Kant's view of the role of philosophy vis-à-vis science was too limited. They sought to transcend him by claiming that pure thought alone could yield new scientific results without resorting to the experimental sciences, and they thought philosophy could address all issues. (Contrary to Helmholtz's opinion, this was not true for the early Schelling, who indeed emphasized the importance and necessity of experience and experiment and who contrasted his view to Fichte's.)[25] Hegel and his students polemicized "against the scientific foundations of natural research," above all "against Newton and his theories." His philosophical opponents, Helmholtz said, first identified nature with the human mind and then attempted to identify the laws of the mind with the laws of external reality. (As did Kant, he might have added.) Hence they sought to show "the identity" of human sense perceptions "with the actual properties of perceived bodies." This made them "defenders of Goethe's theory of colors." During the first third of the nineteenth century, when the natural sciences were, Helmholtz maintained, not much cultivated in Germany, philosophy had the upper hand. Apart from a few "honorable" exceptions, such as Humboldt, most German natural scientists had succumbed to *Naturphilosophie*, "until finally the great rise of the natural sciences" among Germany's European neighbors "also pulled Germany along with them." (That was a good bit of rhetoric on his part.) The imperialist *Naturphilosophen* "had wanted to claim everything," and as a consequence, some now thought philosophy should be shunned or eliminated altogether. He begged to differ; he thought it had a legitimate, though limited, role to play, and he urged that *Naturphilosophie* not be equated "with philosophy in general."[26] He did not consider the possibility that part of the contest between philosophy and the sciences arose from the recent rise of the sciences in Germany; that the introduction of many new professorships and other teaching positions, along with new institutes and laboratories, was perhaps perceived as a threat by philosophers and other humanists.

Having dealt with the philosophers, he focused on showing how the latest natural scientific findings bore on the understanding of human sensory perception, for it was here that "philosophy and the natural sciences came closest to one another." He characterized the human eye as "an optical instrument developed by nature, a natural camera obscura." He thought daguerreotypes and photography had been around long enough for his listeners and readers to appreciate the comparison of the eye to a manmade camera. The only substantial difference between the two, he said, was that, while the camera used light-sensitive glass plates to receive light, the eye had a sensitive retina by which it conducted light through a complex system of nerve fibers to the brain. He then explained the process of accommodation and of observing the retina by means of an ophthalmoscope. Physically speaking, the eye was a *flawed* optical instrument, he said, so flawed that when he took measurements of the cornea's curvature for his study of accommodation, he did so for illustrative purposes only: he thought it useless to waste too much time trying to obtain highly precise results. He then explained how, as a physical process, the eye saw: Every image point in the retina corresponded to a point of light emanating from the external world. Light rays reflected off objects and passed through the cornea and aqueous humor to the crystalline lens, which focused them and transmitted an image through the vitreous humor onto the retina. The eye's nerve apparatus discerned the brightness of various objects. Our images of things, he explained, are formed by the eye's crystalline lens, which cast the image onto the retina.[27]

Moreover, he argued that there are several different ways (for example, mechanical means or electrical excitation) for the sensation of light to stimulate the eye, and when such sensations affect the optical nerve, they always yield an optical sensation. If the same stimulus is directed to another type of nerve, some other type of sensation (not light sensation) results. If exactly the same stimulus is directed to the auditory nerve, sensations of sound are produced; to the nerves of the skin, sensations of taste or heat; and to nerves in the muscles, no sensations at all, only muscle spasms. In all these cases, one and the same stimulus produces different sensations. The quality of a sensation did not depend on the external object from which it issued, he maintained, but rather on the sensory nerve that received it. This was of course Müller's law of specific nerve (or sense) energies, first announced in *Über die phantastischen Gesichtserscheinungen* (On phantastic phenomena of sight, 1826). In Helmholtz's judgment this constituted "the most important progress that the physiology of the sensory organs" had made "in modern times." Similarly, he argued that when light is modified, it produces different colors. Depending on the vibration speed, one obtained the colors running from violet to red. Also, mixing light of different colors produced a new, mixed color.[28]

These scientific facts about the production of sensations led him to an epistemological point "of the greatest importance" about perception: our perceptions are as dependent on our senses as on external objects. According to Helmholtz, modern sensory physiology, an experimental science, had demonstrated what Kant had once sought to prove concerning the role of innate laws and the organization of the mind in the formation of ideas. Helmholtz thought that modern science brought philosophical understanding; that there was a parallel nature between a modern, physiologically based theory of perception and Kant's epistemology; and he expressed, without ever quite saying so, his view that Kant's theory of the mind was no longer valid. He invoked Kant's name yet effectively rejected his viewpoint.[29] Kant was back, if not quite all of him.

Yet for Helmholtz the physical and physiological process of light's reception was not itself "seeing," but rather a preliminary to it. "Seeing" occurred only subsequently, "in the understanding [*Verständnis*] of the sensation of light." He thus introduced a psychological dimension into his analysis of human perception, and in doing so he also, and not so subtly, distanced himself from the views of the materialists. Upon determining the presence of light sensations within its field of view, he argued, the eye seeks to find the sensations' location (i.e., direction). This provided perspective on the depth perception of objects.[30]

To help his listeners and viewers understand depth perception, that is, to help them appreciate the difference between viewing a perspectival drawing and viewing an object itself, he now introduced Wheatstone's stereoscope. Our two eyes, he said, give two constantly different perspectives on the world. Yet when the eyes view a perspectival drawing of an object on a flat surface, they yield only one perspective. That perspectival view enables them "to distinguish the real object from its image." If one were to take two perspectival drawings—one for the left eye, one for the right—of the same object, "and then show each eye the corresponding figure in its proper position, then the essential difference between the view of the object and its image ceases, and we now in fact believe we see, instead of the figures, the objects." The stereoscope presented two drawings of the same object from two slightly different perspectives, and it could be so adjusted as to make the two drawings appear to be in one and the same place. This is precisely what individuals do when they squint at two objects lying next to each other until they finally appear to be one. In both cases, Helmholtz argued, they create an optical illusion.[31]

Though he had dedicated his address to Kant's memory, his argument about spatial relations was rather un-Kantian. Instead of spatial relations being an a priori given, as they were for Kant, Helmholtz considered them as something constantly constructed out of the eyes' two perspectives and their individual motions, that is, by their empirical activities. Furthermore, a psychological element—that of human will—entered into the mental formation of spatial relations. He cited various illusions of motion experienced by individuals, including dizziness, high fever, circular rotation, sitting in a moving railway car, and traveling by boat for a longish period of time, that result from bodily conditions or changes in orientation that can be largely explained by not properly interpreting the actions of the eye's muscles. He thought science explained many otherwise inexplicable illusions. For him, seeing was in large measure a learned behavior.[32]

Helmholtz further maintained that some mental acts are unconscious and occur independently of our will, intelligence, and convictions. He even wondered whether they could be called "thought" as such. He readily conceded that it is difficult to understand the psychological process by which physical sensations became perceptions. The act of perception, he argued, involves an unconscious act of inference, based on past learned behavior. To illustrate his point, he considered a skilled actor—given his many visits to the theater, his choice of this example is hardly surprising—who wears the clothes and adopts the movements and ways of someone whom the actor is portraying on stage. No matter how good the performer, the audience is always aware that it is the actor, and not the real person, who is there on stage. The audience continues "involuntarily" to expect the actor's feelings and actions to correspond to the role being played. The best actors, Helmholtz thought, never even allow the audience to realize they are acting; rather, they seem "completely natural." It is only when one compares them with less skilled actors that one realizes how outstanding they are and is reminded that they are, in point of fact, acting. A similar situation obtains with optical illusions, he argued. Even though we know that the idea or mental image evoked in us by certain sensations is false, we still retain that idea or image "in all its liveliness." And just as the actor's skills evoke and maintain an illusion, so sensory perceptions are linked to conceptions conditioned "through the nature of our senses themselves." Individuals repeat this process millions of times throughout their lives; for example, "when light stimulates certain nerve fibers in our eyes we stretch our arm or walk a certain distance in order to reach an object. Hence thereby is produced the involuntary connection between certain facial impressions and the distance at and direction in which the object is to be sought." The same holds for optical judgments about distance; that, too, is a learned process. Helmholtz vividly recalled here that as a boy he had his first understanding of "the law of perspective" while walking with his mother past a high tower with people on its upper level. (When, a year after his address, he congratulated the du Bois–Reymonds on the birth of their first child, now three months old, he only half-jokingly said that their child "is probably already very much involved with difficult but practical questions like the formation of spatial and temporal ideas, and so already knows more about that than all the world's learned physiologists.") For Helmholtz, then, the sense organs became educated, and this explained "the beauty and precision of our eyes' construction of space." Like skilled jugglers or billiard players, he believed, humans learned to make judgments about visual objects. They learned to see, which for Helmholtz meant they learned to associate the idea or image of a certain object with certain past and present sensations experienced by them.[33]

He concluded his address by returning to Kant. If there was to be a connection between sensations and images of a particular body, he argued, an idea of bodies as such was needed. He claimed that our senses never lead us to perceive directly the objects of external reality. Instead, from birth onward we only receive their effects on our sensory apparatus. We make the passage from the world of sensations in our nervous system to external reality "only through a conclusion [i.e., an inference]. We must presuppose the presence of external objects as the cause of our nerve sensation. For there can be no effect without a cause." This law of no effect without a cause is not an empirical law, he maintained, but rather a necessary one for knowledge of objects in the external world. It did not come "from the inner experience of our self-consciousness," "for we consider the self-conscious act of our will and thought as free." Thus the analysis of sense perceptions brought him back to Kant's theorem of "No effect without a cause," an a priori theorem of thought. He ended as diplomatically as he had begun, with praise for Kant, who had contributed "the most extraordinary progress" ever made in philosophy by bringing and providing this law "and the other innate forms of intuition and laws of thought." In fact, according to Helmholtz, Kant did "for the theory of the mental representations in general" what Müller did for the study of perceptions through the specialized, empirical discipline of sensory physiology. Hence, "Kant's ideas are still alive, and will always develop richly, even in fields where one perhaps would not have expected any fruition." Hence, too, the conflict between philosophy and the natural sciences was due "only to certain newer systems of philosophy" (i.e., the Hegelians and their ilk), not to "philosophy in general." He had sought to show his audience "that the common thread which should bind, and which is supposed to bind, all the sciences is in no way torn asunder by modern natural science." Kant was back, and Helmholtz became one of the early and principal leaders in bringing him back.[34]

That said, and though he had not explicitly criticized Kant—and his implicit criticism was certainly not as severe as his criticism of Goethe on optics—he had nonetheless subtly revised him. He had also irrevocably linked his own name to modern Germany's two greatest cultural figures. Along with Humboldt, he had now arguably become the foremost representative of culture among German natural scientists. His address on Goethe and the one on human perception further weakened romanticist and idealist viewpoints in German philosophy and culture in general; together, they helped reset the general European intellectual scene and tone, making it more naturalistic, just as Darwin soon did with his *Origin of Species* (1859) and *Descent of Man* (1871) and as Marx did with his writings on political economy. For many European thinkers, the world increasingly became a place to be explained in naturalistic, not supernatural (religious), terms. There was a new dispensation abroad.

Launching a Science of Colors

Less than two months before Helmholtz's address on human perception, Maxwell characterized Helmholtz's theory of compound colors as "the most philosophical inquiry into" the subject that he had seen. He classed Helmholtz as a creator of color theory alongside Newton, Thomas Young, James David Forbes, and Hermann Günther Grassmann. Maxwell's praise notwithstanding, Helmholtz was not without his critics on color mixing. In an attempt to respond to criticisms leveled by Grassmann, a mathematician and scientist, Helmholtz revised and expanded a version of his earlier presentation on color mixing at the BAAS in Hull. Based on a new instrumental color-mixing technique, and using his own eyes' color perception as his standard, he variously criticized, adopted, and adapted Grassmann's mathematical (geometric) laws and empirical results on color mixing, arguing that, instead of a circle or barycentric figure (as Newton and Grassmann had proposed) to represent color mixing, a triangle-like curve worked best. He showed graphically and empirically how the eye mixed colors additively. Moreover, he remained skeptical about Young's three-color hypothesis and, equally important, again introduced what became one of the major distinguishing (and controversial) features of his own work on color perception: according a role to human judgment. Before 1855 ended, Maxwell presented his own set of color equations, which itself was probably indebted to both Grassmann's and Helmholtz's results. Together, within the space of three years, the three men effectively launched the science of colorimetry.[35]

Jockeying for a Position at Bonn

Despite the patriotic public face that he displayed before his fellow residents of Königsberg, Helmholtz had wanted out of Königsberg for quite some time. It was too distant from Berlin and from many colleagues and potential students in central and western Germany, not to mention family. His mother died in 1854, and he did not attend her funeral, presumably because the semester was about to start and Königsberg was too far away from Potsdam for him to arrive in time for it. She had suffered in life, he said, and had lived for her loved ones. While her influence on him appears less transparent than that of his father, the great value that he placed on having a strong family life, his admiration for women of culture, and the apparent fact that he felt more comfortable in their presence (with them he did not have to compete professionally)—were all owed in some measure to his mother. Above all, he wanted to leave Königsberg because it was too dangerous for Olga's fragile health. He was determined to find an academic position in a warmer climate that offered her a more congenial setting; he regretted, especially, that Heidelberg had eluded him, just as Bunsen lamented that he could not bring him there.[36]

Then in 1854 a job possibility opened up at Bonn. The Bonn Medical Faculty needed a successor to a retiring anatomist and, more generally, also wanted to expand the faculty's personnel and offerings in several disciplines. It recommended that Julius Budge, who since 1847 had served as extraordinary professor of anatomy, physiology, and zoology, be promoted to ordinary professor of physiology. The ministry did so, but it was otherwise not ready to move as rapidly as Bonn wanted. It was a fluid situation there.[37]

Helmholtz undertook a serious effort to obtain the evolving position at Bonn. Doing so, however, meant potential conflict with du Bois–Reymond and Ludwig, both of whom were also interested in the position. Helmholtz thought Bonn was a better place for du Bois–Reymond's scientific work and his "own feeling of self-esteem" than his current extraordinary professorship in Berlin. The ministry, he believed, intended to bring several able scientists to Bonn "and so raise the Medical Faculty out of its current swampiness." He advised du Bois–Reymond to take the position if it was offered to him. But if he did not want it, Helmholtz asked that he notify him "as soon as possible" so that he (Helmholtz) could discuss it with the ministry. If the salary was equivalent to what he got at Königsberg, then he preferred Bonn, since there he would have "a larger sphere of activity" and earn more money, and since it would provide Olga with a more hospitable climate than Königsberg, where her health seemed to be "constantly endangered." He claimed that Königsberg's climate had made him lose "a good part of my ability to do work" during the previous eighteen months; Königsberg's climate had supposedly led to his "dangerously" acute abdominal pains, which in turn were caused, he said, by bowel problems. However, he thought his reasons were "not as pressing" as du Bois–Reymond's, and he hoped the latter would get the position. If not, he asked du Bois–Reymond to inform him of the terms of his offer; he would then go to Berlin to negotiate directly with the ministry.[38]

But du Bois–Reymond did not keep him up to date. So he asked Schulze (at the ministry) to reconsider him for the post. He knew nothing about Bonn, he said, and felt he even had to enquire as to whether the position was in fact still vacant, whether Bonn had a physiological institute, and what kinds of lectures would be expected of him. He preferred Bonn for several reasons, he said. It lay relatively close to England and Holland, where he had close relations with English and Dutch colleagues. Julius Budge was there. Königsberg was isolated. Finally, and most importantly, there was the deleterious effect of Königsberg's harsh climate on Olga's and his own health. He thought her life endangered, and so he set aside all other considerations as he asked the ministry to transfer him to a university with a warmer climate. Still, he was worried that Bonn was expensive, and he hoped to earn there at least what he was earning in Königsberg. All this, he explained, was "a matter that lay very close to my heart." If it proved difficult to find someone to replace him, then the ministry should know that his colleague Wittich could do the job.[39] He did not even bother to lay out his record of academic research: the ministry was well aware of his many accomplishments and growing reputation, and it could easily have learned that during his thirteen-year professional career to date he had thirty-four publications to his name, twenty-six of which had appeared since he arrived in Königsberg. In both qualitative and quantitative terms, his record was probably unmatched by any of his peers.

Within a fortnight, he received a positive, but conditional, response from the ministry. He replied in turn that he would gladly exchange teaching pathology for anatomy (the condition), since the latter was closer to his interests than the former. He said he was becoming increasingly distant from developments in pathology and that he had taught anatomy at the Academy of Arts. He noted that his earliest publications had concerned microscopic anatomy and that, although a superficial review of his work might lead one to think he had little to do with anatomy, his interests were "in no way so far from the anatomical sciences."[40] That was spin.

He had three concerns, not to say conditions, of his own, however. He wanted appointment as professor of anatomy *and* physiology, since thus far he had concentrated his efforts in the latter discipline. He had learned that Bonn had no physiological institute, and he asked if funds might be made available for purchasing physiological instruments, "since the very possibility of doing scientific work in the most appropriate field for me is in fact tied to having such a collection." In addition, to match his salary of 1,000 talers at Königsberg, he needed 1,200 at Bonn; in fact, he thought 1,400 would be fairer, since Bonn had more (about eighty-five) medical students than Königsberg, which meant he would have more work to do there but would scarcely earn more in the way of honorary income. Under these conditions, he would gladly accept the position.[41]

As for Ludwig, he had few expectations of getting a position in Prussia, and he appeared unperturbed when Helmholtz bested him for the Bonn position. Ludwig told Henle that Helmholtz had written to explain his negotiations with Schulze and that he had learned from a well-informed friend of his at the ministry that his (Ludwig's) candidacy had been rejected (only) by the minister himself, the archconservative Karl Otto von Raumer. He suspected that someone, probably Wagner, had let the minister know of Ludwig's "all too materialistic" views, and "the disagreeable Göttingen affair." It was a matter of piety versus materialism, and Ludwig thought it was impossible for Raumer to recommend that the king appoint him to Bonn. Any hopes that he may have entertained had fallen victim to the reactionary political spirit of Prussia in the 1850s. In 1855 Ludwig left Zurich to become professor of physiology and zoology at the Medical-Surgical Military Academy (Medizin-chirurgische Militärakademie, known as the Josephinum) in Vienna.[42]

In mid-March 1855, the Bonn appointment entered its final, decision-making phase. The politics were complex. Humboldt told du Bois–Reymond that he was being pressured into supporting Helmholtz, "for whom I have as much affection and esteem as you do," but that he did not want to do anything until he first heard from du Bois–Reymond as to whether or not he indeed wanted to go to Bonn. He assured du Bois–Reymond that he had "a great future" in Berlin. Raumer, for his part, wanted Helmholtz, but Schönlein wanted someone else and had stymied him. Du Bois–Reymond told Humboldt to support Helmholtz, whom he wished well; Humboldt now drew up a list of Helmholtz's scientific achievements. Helmholtz appreciated du Bois–Reymond's support and wrote Humboldt to ask for his own support and to learn about the status of the appointment. He worried that the possibility of war (presumably in the Crimea) might force the government to delay the appointment. He was also in direct contact with Schulze, and a chemist friend of his wrote a friend at the ministry to find out what he knew. He spared no effort, especially since Olga's condition had worsened during the second half of the winter: she had gotten the flu—there was an epidemic in Königsberg—and spent two weeks in bed, followed by a month when she could not speak. Although she had partially recovered, her voice was still not fully normal and she still had a cough.[43]

Humboldt wrote Raumer and Helmholtz immediately. He told the former that he had been much pleased when Helmholtz had been promoted to ordinary professor and that he had "friendly relations" with "the young man." He said that Olga's "very seriously endangered state of health" had effectively forced Helmholtz to apply for the Bonn position. He gladly supported "this very talented, exceedingly active and ambitious scholar," referring almost exclusively to Helmholtz's work in anatomy (such as it was). He said Helmholtz's dissertation on ganglia cells was among "the finest works of modern microscopic anatomy," that his anatomical work on the eye and the theory of accommodation were "not less important," and that everyone agreed how useful his ophthalmoscope was. "In the current state of knowledge," he added, "there is no other single individual who shows equally great strength in anatomy and physiology."[44]

Humboldt told Helmholtz that he had given his support even before Helmholtz had requested it. Once he had learned that "our mutual friend du Bois" did not want to leave Berlin, he was, given his "older friendship with du Bois, free to act." He based his judgment and warm recommendation to the ministry on the publications Helmholtz had sent him, he said, but additional motives were the state of Olga's health, his friendship with Helmholtz, and Helmholtz's "outstanding talents and ingenious activity." He thought the move would be good for him and welcomed the opportunity to support him.[45] Thus Humboldt, by far Prussia's best-connected scientist at court and with the Ministry of Culture, its most internationally renowned figure in science, again gave him a ringing endorsement.

Raumer was not fully satisfied, however: he wanted further assurance from Helmholtz that he could and would teach anatomy. Helmholtz declared that he was fully able to do so: that his teaching of both human and animal anatomy at the Academy of Arts in Berlin, and his research from 1843 onward, proved this; and that he had been an assistant at the Anatomical Museum, where he had helped with the collections. He added in a pleading tone: "If your Excellency should kindly grant me this position, I will approach my new task with full passion."[46] He was desperate.

The long-sought appointment finally came in early May: he was to become professor of anatomy and physiology at Bonn on 1 October 1855 and would be placed in charge of the anatomical institute and museum. He was given a salary of 1,200 talers along with 300 in moving expenses. On the whole, he had gotten the compensation he needed, though he received a measly 50 talers for apparatus. That would make doing experimental research at Bonn difficult at best. Moreover, he had to leave his instruments behind in Königsberg; that, he told Ferdinand, "is the most unpleasant loss concerning my resettlement." Moritz Naumann, the dean of Bonn's Medical Faculty, informed him that he did not need to give a habilitation lecture, since he was already an ordinary professor at another Prussian university. The faculty looked forward to meeting him, Naumann added, and recognized his "excellent accomplishments." Members of the local Lower Rhine Association for Natural History and Medicine (Niederrheinische Gesellschaft für Natur- und Heilkunde) had "often" discussed Helmholtz's "penetrating works," he said, and he asked him to decide quickly about which anatomy courses he would teach next semester. In this connection, Helmholtz immediately wrote to Moritz Ignaz Weber, the second ordinary professor of anatomy at Bonn, that he agreed to the latter's request in regard to teaching anatomy. Weber was extremely grateful for Helmholtz's goodwill. He spelled out the details of his own schedule and gave him a four-page description of the anatomy and physiology institute, including its apparatus and instruments, its library, its number of students (43–60 in anatomy), and its budget.[47] The deal was finally done.

Goodbye Königsberg

The news spread rapidly. Ludwig heard that Raumer had declined to choose him because his views were "all too materialistic" and that behind the ministry stood someone who had the king's ear. He said that as recently as the previous October it had looked as if he would get the post. He was told that he "may not be worse than Helmholtz and du Bois." He thought the ministry was "misusing" Helmholtz's talent by making him teach anatomy. "They should set you free, like a Parisian *savant*, and naturally leave you with your instruments in Bonn as professor of physiology." That Helmholtz and not Ludwig got the appointment was due in some measure to Helmholtz's political quietism and trustworthiness (and Ludwig's lack thereof). Brücke was equally congratulatory. Although Helmholtz knew he would find some colleagues at Bonn who would not welcome his appointment, he thought that in time they would change their minds.[48]

Helmholtz spent much of June and July preparing to leave Königsberg. In mid-June he resigned as president of the Association for Scientific Medicine. His departure was "a heavy loss" for the association, whose members respected and loved him; they honored him with a banquet at which they gave him a silver votive tablet and a diploma declaring him an honorary member. In mid-July,

Olga and the children left for Dahlem, where she underwent medical treatment and tried to regain her health. Helmholtz remained behind in Königsberg to administer some final doctoral examinations (from which he had already earned about 100 talers), to conclude their personal business, and to say farewell to friends and colleagues. He continued to give Olga medical advice, reminding her to take her medication. He warned her to guard against potential colic and loss of appetite, and he wanted her to describe precisely "how things stood" with her "appetite and energy." He was also worried about the behavior of Julie, his younger sister, who had done something—he gave no specifics—that, were it to become known, would destroy her reputation, or so he thought. He was angry with and ashamed of her. He planned to leave Königsberg on 29 July for Berlin and Dahlem; he would see his family and leave two days later for Bonn to find housing for them and generally assess the situation firsthand.[49]

He spent nearly two weeks saying goodbye to friends and colleagues. During the second half of July, he scheduled no fewer than sixty individual farewell visits. On the afternoon of 18 July, thirty-seven of his colleagues and friends gave him "a great, final dinner in the stock-exchange garden," music included. Most were faculty members (more than half of the university's total faculty attended), while a few were university officials or clinical doctors. They each paid two talers to attend—"I for free, naturally," he happily and characteristically noted. He sat between the university curator, His Excellency Franz Eichmann, and Eduard von Simson, the professor of law. Eichmann led a toast to the king; Simson to Helmholtz. "Apart from the endless flatteries, the speech was very beautiful, warm, and touching," he wrote Olga. "That I myself was thereby rather touched, was only natural; but there were also others, as they came up to me and toasted, who had some very teary looks, Rathke, [Justus] Olshausen, Wittich, [and] Richelot were among them." So too were Neumann and Moser. After dessert, Helmholtz toasted the citizens of Königsberg and the Albertina's professors. He claimed that his decision to leave was a difficult one. "Within these [the university's] walls I have experienced beautiful, rich years that elevated my mind and heart; I have found here a circle of colleagues who are not inferior to any other German university in the realm of knowledge and intellectual creative power; and who are perhaps ahead of all German universities in terms of undisturbed harmony of collegial relations, selfless recognition of the displaced individual, and ready support for the fields of work of every colleague." He was leaving Prussia's far northeastern border for its far western one, and he would never forget Königsberg's "sensible and gallant residents," he said. He hoped his colleagues would always remember him and that some would visit him, and he said he had taken Königsberg into his heart. He praised the university's "serious, austere," even "outstanding, critical scientific spirit." There was, all in all, probably as much truth as diplomatic pleasantry and professional courtesy in his pious public toast. It had been, he told Olga, "a very animated party, and the people were extraordinarily warm" toward him. That evening he and Adolf Sotteck went to his good friends the Olshausens.[50]

He spent the evening of 20 July at the Richelots'. He was taken with Frau Richelot's singing of Beethoven *Lieder*, and he hoped that Olga, after her treatments, might perform one of them. The next day he paid upward of fifteen additional farewell visits and spent the remainder of the following week making still more visits and (further) packing boxes. By the twenty-sixth he had completed the last of his visits. Olga felt better, but he was irritated with himself for not controlling her diet better, since she had eaten French beans, which were hard for her to digest. He was less worried about her sniffles than her possibly getting the chills.[51]

There was a farewell party for him on his final day in Königsberg (28 July). The *Ostpreussische Zeitung* reported that he was loved and honored by his university colleagues and students. He was charming and modest, it said. It noted that he had made a name for himself not only in physiology but also in "practical medicine." He had even come to know the king of Prussia personally. (When the monarch was last in town, he met all the faculty members, including Helmholtz, who was then serving as dean of the Medical Faculty and who thus wore the dean's official red gown. The king told him: "It wasn't necessary for you to wear the red gown; I would have recognized you without it!") That evening, just before he and Wittich—his friend, colleague, and now successor—departed together by train for Berlin, the Medical Faculty gave him a silver tablet inscribed with all the faculty members' names. His students, who knew his special love "for the plastic art [i.e., sculpture]," gave him a set of copperplate engravings of famous Raphael drawings, and he also received a picture of the newly completed Anatomical Museum. He and Wittich were expected to arrive in Berlin the following evening at 9:15 p.m.[52] Helmholtz never returned to Königsberg.

Unhappy Intermezzo in Bonn

In Beethoven's City

Bonn certainly had advantages over Königsberg. The city had experienced important changes in the forty years before Helmholtz arrived in 1855. In 1815, as a result of the Congress of Vienna, the Rhineland, and with it Bonn, became part of Prussia. By the mid-1850s, Bonn's population had reached about twenty thousand, and the city, though otherwise isolated, was networked into the expanding German railroad system, which greatly facilitated travel. Above all, located on the Rhine and having a mild climate, it became a cultural center, with the Rheinische Friedrich-Wilhelms-Universität Bonn (Frederick William University, established in 1818 by Frederick William III of Prussia) as its single greatest cultural asset. That linked the Rhineland to the Hohenzollern regime. Though Bonn was the youngest of all German universities, by midcentury it had become a midsized university of some renown (mostly for its historical and philological disciplines). Bonn also served as the seat of the Lower Rhine Association for Natural History and Medicine, where Helmholtz eventually lectured several times; he also published in its transactions and served as its chairman. Finally, in an effort to enhance its image in 1845, Bonn claimed to be the City of

Fig. 8.1 Helmholtz in 1857. Lithography by Rud Hoffmann (1857) from a photograph by Schallenberg in Bonn. Courtesy of The Bakken, a Library and Museum of Electricity in Life, Minneapolis.

Beethoven—it was the composer's birthplace (1770) and longtime home (until 1792)—and it erected a statue of him and initiated an annual three-day musical festival. Bonn seemed perfect for Helmholtz.

And so in late July 1855, as he left Königsberg for Bonn, his future looked brilliant. His academic achievements had transformed him, just shy of his thirty-fourth birthday, into a leader of German science and had given him an international reputation. He had a loving family and good reason to believe that the mild Rhineland climate would ameliorate if not cure his wife's illness. He had no reason to expect, therefore, that the coming three years in Beethoven's city would be the most trying of his professional life. (See figure 8.1.)

He spent the first half of August visiting Bonn, Kreuznach, Heidelberg, and Berlin, all the while worrying about Olga's cough, Käthe's sniffles, and his family in general. In Bonn he looked for housing for his family, met two (of three) colleagues in anatomy, and assessed firsthand the state of anatomy and physiology there. The leafy, secluded residence that interested him, known as the Vinea Domini, sat on a high terrace above the Rhine, before the Coblenz Gate and in a garden in Coblenzer Strasse; he considered it a "paradise-like" villa and rented one-half of it. He thought it would offer Olga the best view she had ever had. It was an elegant, nine-room villa with plenty of space for them, their children

Käthe and Richard, and Olga's mother. He negotiated the annual rent down to 250 talers plus two additional rooms. He drove a hard bargain in part because he needed his cash to buy new furniture. He did comparative furniture shopping—he favored oak wood—at shops in nearby Cologne (where he also saw the renowned cathedral), in Bonn, and in Berlin. By mid-October the family was settled in and enjoying its charming view of the Rhine (see fig. 8.2). During their first months there, Olga's health did not seem to improve, however. Richard fell greatly under the spell of the nearby railroad, and this childhood passion for what was the era's most exciting technology and the pride of all liberal enthusiasts eventually became his profession. The family remained in this residence until 1857, when they moved into a house owned by Helmholtz's colleague the anatomist Moritz Ignaz Weber, located in the university garden. All in all, Helmholtz found Bonn "stimulating."[1]

During his initial visit, Helmholtz met his new colleagues Weber and Budge and saw their institutional setup in anatomy and physiology firsthand. Budge, as the ordinary professor of physiology, had written him in advance to say he felt isolated at Bonn and that he thus greatly looked forward to working with him; that they should work with, not against, one another; and that he wanted to help him in his move to Bonn. At first Helmholtz found Budge "very reasonable" and approachable, though he also felt "dragged around like a crazy poodle." They

Fig. 8.2 At Bonn (1855–57) Helmholtz's residence was the Vinea Domini (*the building on the left*); the upper four windows to the right of the central tower are those of the Helmholtz family apartment. Stadtarchiv und Stadthistorische Bibliothek Bonn.

Fig. 8.3 The Bonn anatomy building. Archiv der Rheinischen Friedrich-Wilhelms-Universität Bonn.

planned a joint course in physiology (but in fact never taught it), and Helmholtz gladly ceded microscopic anatomy, developmental biology, and comparative anatomy to him. He found the anatomy building to be much wanting, but preparations for a new building had been completed; it was only a matter of purchasing land for its location (see fig. 8.3). The institute was "in a gruesome condition," and Helmholtz thought Budge had failed to properly conserve even its few, inadequate instruments for physiology. (The institute budget allowed a mere 54 talers for its course in experimental physiology, whereas Helmholtz thought it required about 200 to meet its pedagogical needs.) On the afternoon of his first visit, he accompanied Budge to a party.[2] It had been a good start with Budge.

Before retrieving his family in Dahlem, Helmholtz hoped to meet up with William Thomson, who had brought his ailing wife, Margaret, to Kreuznach, a spa in southwestern Germany. Although this first Helmholtz-Thomson meeting came at Thomson's initiative, Helmholtz had sought him out (in vain), it may be recalled, two years earlier in Glasgow. Thomson "regretted extremely" that he had not gone to the BAAS meeting in Hull to meet Helmholtz and that he then missed Helmholtz again when the latter tried to visit him in Glasgow. He knew that Helmholtz had received an official invitation to attend the upcoming Glasgow meeting of the BAAS, and he invited him "personally" and offered to arrange accommodations for him: "I should consider your presence as one of the most distinguished acquisitions the meeting could have." He had "been anxious" to make his acquaintance ever since he "first had the 'Erhaltung der Kraft'" in his hands. No one, not even Helmholtz's greatest admirers in Germany, had previously singled out that essay as his most distinguished piece of work. Like Clausius in 1852, Thomson gave Helmholtz real recognition qua physicist, treating him as a highly valued colleague working on abstract and theoretical matters in physics. Thomson had hoped soon to be in Germany and to meet Helmholtz there.[3]

Helmholtz had received Thomson's letter shortly before leaving Königsberg. He had replied that since he was in the process of moving, he would find it impossible, as much as he wanted to, to attend the BAAS this year in Glasgow, "which is so distinguished through the most-terrific development of its industry." Instead, he hoped to meet Thomson in Kreuznach.[4]

Helmholtz arrived in the picturesque town of Bingen—"it looks just like in a stereoscope"—on the evening of 6 August, and from there he took a carriage to nearby Kreuznach. Thomson's youthfulness surprised him: "I expected him, one of the top mathematical physicists in Europe," he told Olga, "to be a bit older than me, and was not a little astounded when I found someone facing me who was very youthful, with light-blond hair, and a very girlish look." He met Margaret Thomson only briefly, finding her "a rather pretty, very charming, and witty young woman, but in a terrible condition" (she could not walk, sit up, or stand up without pain).[5] The two men had both physics and wives with debilitating conditions in common. Little wonder that they developed strong bonds of affection for each other.

Helmholtz had planned to leave Kreuznach the following morning, but the two men had so much to talk about that, begging for Olga's understanding, he stayed an extra day. He was awed by Thomson's intellectual abilities: "In terms of penetration, clarity, and intellectual flexibility," Helmholtz wrote: "He exceeds every scientifically great figure that I've known personally, so that at points I myself appear somewhat dull beside him." Thomson inquired about electrical wires for resistance standards, a topic they would discuss on and off for the next forty years. Helmholtz told him that he had spoken about them with Kirchhoff, who said he gave the wires he had used for such work to Wilhelm Weber in Göttingen and that Weber had used them when in Leipzig while working on the electrodynamic determinations of mass. He promised to report back to Thomson on the results out of Leipzig. This Kreuznach visit marked the beginning of a scientifically stimulating, professionally advantageous, and personally warm forty-year relationship between the two men. From Kreuznach, Helmholtz went to Heidelberg for several days to visit Bunsen and Kirchhoff; then to Berlin, where he planned to call on ministry officials in the hope of eventually improving his and the institute's situation in Bonn; and finally to his loved ones in Dahlem.[6] From there they left, *en famille*, for their new home in Bonn.

As in Königsberg, the Helmholtzes built a close circle of friends in Bonn. Most were historians and philologists. Helmholtz believed that the "most important" faculty members at Bonn were in those disciplines and that, with the exception of geology, the natural sciences there were treated "rather shabbily." Their friends included Otto Jahn, a philologist, archaeologist, art and music historian, and the author of the four-volume *Mozart*; Ernst Moritz Arndt, a liberal, nationalist Prussian patriot, historian, and poet; Klaus Groth, a poet, man of letters, and linguist; Friedrich Christoph Dahlmann, the historian and political scientist; Moritz Naumann, the university's specialist in clinical medicine; Karl Otto Weber and Wilhelm Busch, both surgeons at the university; and Heinrich Eberhard Heine, an extraordinary professor of mathematics. Busch was the sole medical doctor or scientist at Bonn with whom Helmholtz became close. The Helmholtzes also received visitors from abroad, including various English families and the Dutchman Donders, who became a close friend of Helmholtz's. Their home was regularly filled with intelligent, educated guests and good music.[7]

Helmholtz also built one other relationship while at Bonn, an exceptional one. In November 1856 the young mathematician Rudolf Lipschitz, who had studied at Königsberg (with Neumann) and Berlin (with Peter Gustav Lejeune Dirichlet) and had then become a Gymnasium teacher in Elbing, asked him for advice about becoming a private lecturer at a university, perhaps at Bonn or Königsberg. Helmholtz supported him in this effort, in part because he and Olga knew him from Königsberg—Lipschitz, too, was a great lover of music and counted various artists among his friends—and in part (probably) because Lipschitz had studied and maintained close relations with Berlin mathematicians. He had broad mathematical interests: he was a specialist and soon a leader in linear algebra, differential equations, differential geometry, and mathematical physics. He became a notable representative of the Berlin mathematics school, which stood in opposition to its archrival Göttingen, where Gauss and his successors ruled.[8]

Helmholtz explained to Lipschitz that mathematics at Bonn had two ordinary professors: Plücker (for mathematics and physics) and Beer (for mathematical physics). He thought Plücker an able but limited mathematician, driven principally by career ambitions and fearful of his more able and analytical colleagues. He thought Beer more pleasant and honorable but without great ambition and, like Plücker, a geometry-oriented (that is, not analysis-oriented), mathematician. For these reasons he thought Bonn could use an analysis-oriented mathematician like Lipschitz, and he encouraged him to apply there (and to Halle). He told him that Bonn's setting was "charming," the faculty "friendly and noble," and its "high society inflexible and luxurious"; in short, the Helmholtzes would very much welcome Lipschitz's coming to Bonn and joining their "circle of north German souls." Lipschitz decided to habilitate at Bonn, and he asked Helmholtz to let Beer know that he hoped to start lecturing there after Easter. In late March 1857, Helmholtz gave him a detailed list of the steps a candidate had to follow and told him that his application for habilitation at Bonn was going well. However, he warned him that he thought Plücker was "mistrustful" of Helmholtz because he had "visited a lot with" the mathematician Heine, formerly at Bonn and now at Halle. He looked forward to seeing Lipschitz in Bonn in part because of their old friendship and in part because he needed "a mathematical advisor." He also looked forward to seeing him as a guest in his home, but he said it would be best for Lipschitz if they did not let Plücker know of their friendly relations. In April 1857 Lipschitz successfully habilitated at Bonn. Helmholtz's support, he said, was "extraordinarily valuable" for him. No one else did for him what Helmholtz had done. Plücker could not or would not help, though Beer was somewhat better; in contrast to Helmholtz, they were uninterested in his work. Lipschitz soon enough decided to apply for an open ordinary professorship in Zurich, in part because he saw little chance of getting an extraordinary professorship at Bonn, and in this regard he again sought Helmholtz's advice. To be sure, he did not want to leave Prussia, which, he noted, had a promising scientific future. In 1862 he left to become extraordinary professor of mathematics in Breslau, but he returned to Bonn two years later as ordinary professor of mathematics.[9] Helmholtz and Lipschitz developed a solid friendship in Bonn that remained important to both during the next thirty-plus years.

Anatomists vs. Physiologists

The university's natural science institutes (botany, mineralogy, zoology, chemistry, and physics) were located in the Poppelsdorfer Castle and were modest in size and outfitting. Helmholtz's office was elsewhere, in the Anatomical Theater (Anatomisches Theater), an old building in the university garden. It functioned as the home of anatomy, physiology, and pathological anatomy. Helmholtz divested himself as much as possible of his anatomical duties. He left anatomy exercises largely to the elderly Moritz Ignaz Weber, thereby maximizing Weber's academic pleasures and minimizing Helmholtz's pain. During Helmholtz's time at Bonn, a chair for pathological anatomy was also separated off from anatomy proper; it went to surgery. He was thus effectively left responsible mainly for physiology. He also sought to get Georg Meissner, a young anatomist at Göttingen, to join him and further relieve him of some of his responsibilities in anatomy, but Meissner feared that, for all its advantages, becoming Helmholtz's assistant or prosector would effectively extend his student years and make him dependent on him. Meissner therefore declined the offer. All told, at Bonn Helmholtz offered classes in sensory physiology and anatomy every winter semester from 1855 to 1858, as well as microscopic anatomy and experimental physiology (summer semester, 1856); the physiology of generation (summer 1857); and experimental physiology (summer 1857).[10]

Helmholtz soon concluded that Budge had become an ordinary professor because he had joined the Evangelical Reform Movement (Innere Mission), a sort of Christian socialism, of which Frederick William IV was a strong supporter. He scorned Budge's recent conversion from Judaism to Christianity since, he implied, Budge had converted (merely) to advance himself professionally. His judgment was less than generous, however. After promoting at Bonn, Budge had served for nearly a decade as a practicing physician and had turned himself into a specialist in nerve physiology. In 1846 he discovered that stimulation of the vagus nerve inhibited the heart's operation. Though the discovery was an important one in pathology and clinical medicine, the Weber brothers had also discovered it simultaneously, and it was largely their name that became associated with it. Similarly, in 1853 Budge demonstrated the dependency of the pupil's movement on the nerve center of the medulla oblongata, for which he received the Prix Monthyon for experimental psychology from the Paris Academy of Sciences. However, his assistant, the Englishman Augustus Waller, showed that a nerve bundle degenerated when separated from its origin, and that discovery overshadowed Budge's. Budge thus had substantial professional accomplishments to his name, and his advancement was arguably due to more than his religious affiliation (or lack thereof). As for Weber, Helmholtz assessed him a touch more generously: "an old, skilled prosector, like Friedrich Schlemm, but neglected and sick, and so he has no great motivation to work."[11] He set high standards.

The relationship between Helmholtz and Budge soon deteriorated, though du Bois–Reymond believed Helmholtz was in fact too conciliatory toward him, and Ludwig apparently shared that view. Du Bois–Reymond stoked the fire by telling Helmholtz that Budge was spreading a rumor within the ministry that Helmholtz was incompetent at teaching anatomy, and he intended to speak with the minister in order "to protect" Helmholtz. (Brücke had also heard "that in Bonn one bears enmity" to Helmholtz.) The rumor angered Helmholtz, who said it was "a pure fabrication . . . one that throws no good light on the intention of whoever has brought it forth." Still, he thought he could improve his anatomy lectures and intended to do so. Several ill-willed auditors scrutinized his lectures, and, as a result, he received hints that his inclusion of physiology and chemistry in them was coming at the expense of general anatomy. Some students even laughed when he introduced a cosine into his lectures on physiological optics. However, a number of his colleagues and older students valued his lectures. Helmholtz claimed that virtually all the students who wanted to learn physiology had abandoned Budge for him. "This result, I think, proves decisively that in the anatomy lectures I can have neither disgraced myself nor dissatisfied [the students] through my lecture style." So he thought it unnecessary to try to squelch the rumor. He did, however, hope that du Bois–Reymond would speak with the minister or Schulze, since preliminary steps had been taken toward obtaining a new building for anatomy and physiology, and he did not want to endanger that undertaking. As for Budge, "I prefer not to depict him further; in him the weak sides of the Semitic nationality are too powerful, so that he was against me because I've wounded him in his pecuniary interests. Thanks to his success, his immeasurable self-conceit has perhaps received the first strong push to think and act straight—and he could be inclined to so act." They remained on speaking terms, but he hoped Budge might be offered another position elsewhere.[12]

In fact, when he returned from a late-summer trip in 1856 to Switzerland, he learned that Budge had accepted an offer from Greifswald. That naturally eased the tension in Bonn, Helmholtz confessed, but he regretted "that the ministry, although convinced of his scientific incompetence, nonetheless values him sufficiently to give him such a position thanks to his service for the Inner Mission." He had in effect driven Budge out of Bonn. His departure had exceeded Helmholtz's hopes, since he was convinced that no non-Prussian university would employ Budge and that Prussia did not like to move professors internally. "Only the Inner Mission could happily solve this puzzle."[13] In a sense, religion had saved Helmholtz.

His problems with Budge probably affected his judgment about the Bonn professoriate in general, which he thought displayed "a curious commitment." "They seek to shine less through science," he said, "than through their connections with the princes, a point that is all the more noticeable when one comes from stringent and honorable Königsberg with its simple morals." Conversely, though the Bonn faculty recognized Helmholtz's talents and reputation, he had in fact been imposed upon them by the ministry, and that was something that not everyone appreciated. As for Weber, he sardonically dubbed him "Bone Weber." He thought him an able enough teacher, and he had a good relationship with him, but he also thought he lacked research talent. Weber recognized this, Helmholtz said; "He can never reach higher goals. He's a hardworking coach of absolutely basic anatomy, and always a useful member of the university." He characterized his other (third) colleague in anatomy, Hermann Schaaffhausen, an anthropologist who was extraordinary professor of anatomy, as someone who "considered science from the standpoint of a wealthy man, as only a means of amusement and as an object concerning which fine-spoken phrases are uttered." Helmholtz respected him as little as he respected Budge. Neither man, nor Weber for that matter, came close to his vision of a true (professional) scientist.[14]

Hence, a year after arriving in Bonn, he found himself as "the principal representative of physiology and human anatomy . . . both of which I enjoy lecturing on." This meant the ministry could no longer expect him to teach microscopic and comparative anatomy, "which would indeed have become burdensome." He had taught the former the past summer "for my own instruction," but he planned to teach it again only as part of a course on human anatomy.[15]

On the institutional level, Budge's departure did not improve anything: in mid-1857, Helmholtz judged the institute to be an "old, dirty hole," and he deplored "the endless negotiations over providing corpses, concerning which the Catholic clergy secretly intrigues and lets no opportunity pass where they can treat us as indecently as possible." He had no more (and perhaps even less) institutional capacity to do research in Bonn than he had had in Königsberg: neither had a physiological institute, and Bonn lacked physiological instruments. He conducted nearly all of his experimental work at home, and the students used the auditorium for theirs. The state of the anatomical institute so embarrassed him that he constantly sought to keep visitors away: it was "a pigsty, out of which not even dirt and disorder can come."[16]

Nevertheless, student numbers increased during his tenure. In the winter semester 1855–56, he had forty-four students (and fifty corpses for those in the dissection exercises to practice on); a year later (winter semester 1856–57) he had fifty-two (but only forty-five corpses). He experienced "extremely irritating negotiations" with both the Catholic clergy's and the city of Cologne's municipal representatives in obtaining corpses for the institute. By his last semester at Bonn, he boasted that student numbers had reached sixty, though the number of senior medical students was "pitifully small." (When he arrived in 1855, there were 78 students enrolled in the Medical Faculty; when he left, there were 88.) He believed the medical students were more interested in anatomy than physiology: "To the young medical students, physiology is, foolishly, always a side issue." And while he pointed to the rapid increase in student numbers, he also recognized that "by no means do they all come to listen to physiology with me, since Budge and Schaaffhausen's winter semester course has remained rather full, [even] more than in the last winter semester. To be sure, in the summer semester, listeners come to me." He thought many came to him to learn anatomy, not physiology, which last they thought "too unimportant."[17]

The *Handbuch* (Part One)

Helmholtz normally did not announce to others what he was working on. Ludwig told du Bois–Reymond in May 1854 that he often received "pages-long" letters from Helmholtz. "Still," he added, "one never entirely knows what he's doing." Donders heard that Helmholtz had a study on physiological optics under way, but Ludwig knew nothing of it. Then in March of 1855, Helmholtz told du Bois–Reymond that he was writing a book on physiological optics and that the first half would be ready for the press the following month: "It'll be rather bulky." Indeed it was: the work in question was his *Handbuch der physiologischen Optik*, part one of which appeared in 1856 (part two was published in 1860, and part three in 1866). Together the volumes reviewed, expanded, and synthesized the modern, scientific study of vision as it had developed under Helmholtz's leadership.[18] Helmholtz's work on the *Handbuch*, and on the associated parts that he published in journals, marked his move away from electrophysiology and into sensory physiology.

The book's inspirations and origins were at least fourfold. First, Helmholtz probably aspired to do something similar to what his teacher Müller had done in his seminal two-volume *Handbuch der Physiologie des Menschens* (1838–40). Similarly, Ludwig had published the second volume of his own enormously influential *Lehrbuch der Physiologie des Menschen* (1852 and 1856), dedicating it to Brücke, Helmholtz, and du Bois-Reymond. And du Bois–Reymond had published the first volume of his two-volume *Untersuchungen über thierische Elektrizität* (1848–84). Helmholtz was a part of this group, and these major publications — to which should be added the work of the younger, highly mathematical physiologist Adolf Fick, a student of Ludwig's and the author of *Die medizinische Physik* (1856) — gave organic physics standing in a way that it never had had before: all together, they helped make organic physics the prevailing outlook in physiology. Theirs was now the leading school; they sought to make physiology an exact science and to base the understanding of physiological phenomena, especially human physiological phenomena, on the foundations of physics and chemistry, refusing recourse to empty words like "vital forces." This was no longer merely a program; it was now a school of thought and practice.

Second, Helmholtz had been urged to write his large-scale handbook of physiological optics by his old Berlin colleague and friend Gustav Karsten. Karsten was Kiel's physicist and had close ties to the Physical Society, but it might be more telling to say he was a science editor: he hoped that Helmholtz's book would be one of a series he was editing. In the end, it appeared as part of another series edited by Karsten: it became volume nine of the *Allgemeine Encyklopädie der Physik*. A main intended audience (if not the primary one) was medical doctors. Yet the volume was not a mere summary of the extant literature on the dioptrics of the eye (its subject). While Helmholtz wrote parts of it that were not too scientifically technical for medical doctors, he nonetheless kept it at the highest scientific level, checking the measurements of previous research and conducting new measurements himself. All that meant that he devoted a good two years (1854–56) to work on it, more than he had intended. Third, Helmholtz's work in color theory, accommodation, and sensory perception finally led him "to the decision to re-work the entire field of physiological optics," which is what he said he did in the *Handbuch*. Indeed, in agreeing to write a handbook, he committed himself to covering the entire field of physiological optics, a field that, in the 1850s and 1860s, exploded in terms of its range and numbers of publications.[19] He thus originally had only a limited idea of what his true commitment in terms of research time for this handbook would be. Fourth and finally, the impoverished state of Bonn's anatomical and physiological facilities further induced him to write the *Handbuch*: he could do synthetic work there, but not much experimentation.

Part one of the *Handbuch*, which alone required about 190 pages, was devoted to the eye's dioptrics. Helmholtz began by describing the eye's anatomy, including its general structure, the sclerotica and the cornea, the uvea, the retina, the crystalline lens, the aqueous and vitreous humors surrounding the lens, and the eye's supportive structures. He then turned to the eye's physiological optics proper. After briefly discussing the general physical properties of light, he focused on his main subject: the eye's dioptrics. Here he laid out in eight sections—using dozens of illustrative line drawings (figures) and numerous mathematical equations—the laws of refraction in systems with spherical surfaces, the refraction of light rays in the eye, dispersion images on the retina, the mechanism of accommodation, chromatic aberration in the eye, monochromatic deviations (astigmatism), entoptical phenomena, and, finally, the ophthalmoscope and illuminating the eye. He assured his readers that he had personally observed or experienced or conducted experiments on all of the physiological-optical facts and phenomena that he reported in the book.[20] With its extensive citation of the pertinent literature, the *Handbuch* constituted a synthesis of modern physiological-optical studies as concerned the eye's dioptrics, and in conjunction with parts two and three it became the bible of physiological optics, ophthalmology, sensory physiology, and experimental psychology—though it did not go unchallenged, above all concerning the theory of vision proffered in part three. Part one was the least novel and certainly the least controversial of the three parts.

Hiking in the Alps

By June 1856, Helmholtz had completed part one of the *Handbuch*. Thomson had missed seeing Helmholtz as he passed through Bonn; he invited him to Kreuznach, but Helmholtz could not get away from his teaching, and he hoped Thomson would come through Bonn again on his return trip. He thought the Thomsons would enjoy Bonn's "very beautiful" setting. He reported that he was doing well and that, most importantly, Bonn had spared Olga further suffering from her throat problems. He complained that he had lost time in moving to

Bonn and by having to teach anatomy, which he "had not taught for 6 years," and that all this had slowed down his research.[21]

The Thomsons had returned to southwestern Germany (Schwalbach) that summer so that Margaret could again take spa treatments. He thanked Helmholtz belatedly for sending the "wire of measured galvanic resistance," which he had put to use. Thomson intended to visit Helmholtz in Bonn, and Helmholtz, whose last lecture was on 8 August and who was undergoing medical treatments by drinking Marienbad spring water, advised him to do so before 12 August, because afterward he would be going to Switzerland. Thomson visited Helmholtz in Bonn for two days in early August, and the two planned for Helmholtz to stay with the Thomsons in Schwalbach a week later.[22]

Schwalbach thus became Helmholtz's first stop on his summer holiday. Thomson awaited him at the station, and he was welcomed to a place at the Thomsons' table: "In short," he told Olga, "I was again, like last year, his guest." Thomson "was again very charming and lively," and the two men spent the day together without interruption. Helmholtz, who noticed that the female guests in Schwalbach sought to improve their color, twice managed to drink the local spring's "ice-cold carbonated water." Thomson accompanied him to the station early the next morning, as they talked acoustics (experiments with sirens). He wrote his father about Thomson: "He is certainly one of the first mathematical physicists of the day, with powers of rapid invention such as I have seen in no other man." They conducted experiments together: one day with a siren, the next on combination tones. Helmholtz soon incorporated one of Thomson's suggestions for making combination tones audible.[23]

From Schwalbach Helmholtz went to Frankfurt am Main to meet up with his Bonn colleague and friend Karl Otto Weber, who was accompanying him to Switzerland. They first went to Heidelberg, but Kirchhoff was away and Bunsen about to depart; his networking failed him for once. So they climbed up to the castle, "which shone in its renowned beauty," and then walked the forested hills and fields, whose beauty and freshness he appreciated more than before. He regretted not having taken Olga along, since the climbing was not so difficult. "Through its closeness to mountains and the beautiful forest," he told her, "Heidelberg is even more beautiful than Bonn, though it lacks a river, and is probably much less livable than Bonn, so that one doesn't need to be envious of the local beauty here." The next morning he and Weber left for Basel.[24]

There they visited the art museum, since Weber, too, was an enthusiast for the arts. They saw drawings by Hans Holbein ("of really excellent execution"), which Helmholtz valued for their extraordinary "force, character, and dramatic life," though he thought they lacked "grace." He also viewed drawings by Albrecht Dürer ("of the most excellent kind"), and it was only now that he paid "great attention to these masters." On the other hand, he found the museum's oil paintings to be "of the well-known barbaric type." Afterward he and Weber visited the medieval cathedral and went swimming. Basel left "a very pleasant impression." They traveled that night to Lausanne, "through a terrifically romantic valley of the northern Jura." Weber provided geological instruction along the way.[25]

On this trip he preferred the French Swiss to the German Swiss. He liked that the former were "a very courteous, educated, and obviously very industrious and orderly people." He thought they did not exploit foreigners as much as their countrymen in Zurich and around Berne did. "The people here completely give the impression that they value social development [*gesellige Bildung*] and honesty more highly than moneymaking, and [acted] as if they despised indecent types of profit." If he valued *Bildung* over *Gelderwerb*, he nonetheless did not fail to note the exact prices at the "best and cleanest inn," where he stayed and ate food "of excellent quality."[26]

He was equally taken with Lake Geneva, with the Rhône valley: "The most notable thing in this trip was a waterfall whose name in German is, to translate indecently, Pissevache [i.e., "Cow piss"]," though he thought the name unworthy. He ironically bragged that he spoke "fluent French, and have, for the moment, happily not again (yet) come into an embarrassing situation." He knew his limits: "That's how things are with a husband who doesn't take along his little, French-speaking wife."[27]

He and Weber then walked from Martigny to Chamonix (at Mont Blanc), climbing the intervening mountains, and got thoroughly soaked for their efforts. Weber proved to be the sturdier and swifter climber; Helmholtz could barely make it. When Weber refused to slow down, Helmholtz simply trailed behind him, eventually reaching Chamonix "with a lot of sweat." From there he looked out onto Mont Blanc and its glaciers, "thus in the midst of the greatest Alps." It was rainy and cloudy; their hotel was freezing and they were thoroughly chilled; but at least the food was decent—they ordered "a very good, strong red wine"—and the two talked as Helmholtz smoked a cigar to celebrate their journey.[28] He enjoyed simple pleasures.

The rest did them both good, and with the sun's return the next day, they saw the top of Mont Blanc and the nearby Savoy and Berne mountain ranges, "so that one saw the highest mountains of Europe distributed across the horizon." They hired a guide to lead them through the Mer de glace, "Europe's largest glacier," which Helmholtz described in great detail while noting that Goethe had mentioned it too. They crossed it, which he found to be quite dangerous. Had he realized just how dangerous the crossing was, he said, he would not have attempted it: "I think it is irresponsible that they have made a route for tourists out of it, and especially that they allow women to go over it." After they reached their destination, exhausted and in a sour mood, they dined "between some Märkian and Mecklenburgian landed nobles, to whom I would have preferred, to be sure, the worst sort of English tourists." The next day they rested, "in order to heal [their] feet."[29]

On his thirty-fifth birthday (31 August 1856), he found himself in Interlaken, from where he sarcastically congratulated Olga for having a husband who "from time to time runs away from you and therefore doesn't always pain you, but allows you some vacation time too." His humor sought to ease the ticklish fact that Weber was about to return to Bonn but that he (Helmholtz) wanted to take an extra day to make "a scientific trip to Zurich." Both men were exhausted from climbing the Faulhorn and then walking to Giessbach; it proved to be their most difficult walk to date. After having seen Mont Blanc, he thought the Berne mountain chain less impressive; still, it was "more diverse and in terms of countryside more beautiful than the Mont Blanc chain, and so always pleases and refreshes anew." His feet pained him greatly—he called himself Olga's "marriage cripple." A few days later he was back in Bonn.[30]

That autumn brought another honor and also a problem with du Bois–Reymond, who nominated him to be a corresponding member of the Prussian Academy of Sciences. "You can consider it, moreover, a done deal," du Bois–Reymond wrote. In thanking him, Helmholtz spelled out explicitly the honor's (different) meanings to himself and to his Bonn colleagues: "If the result comes out favorably, it will also be, apart from the inner satisfaction that every recognition gives to people competent to judge, of importance for my social position here in Bonn. In this latter regard, it would have had less influence in Königsberg; but here, unfortunately, most colleagues are prouder of superficialities than of scientific importance, and they judge the latter only in terms of success." The academy unanimously elected him a corresponding member for anatomy and physiology. But du Bois–Reymond became miffed at what he perceived as Helmholtz's ingratitude. He reproached him for supposedly not thanking him for nominating him. Helmholtz replied immediately and humbly—"I solemnly and regretfully beg your forgiveness"—but thought that he had already thanked him. "I wrote to you earlier," he said by way of exoneration, referring to his previous statement in regard to Bonn. He explained further: "But such a recognition is, however, also a strengthening of and encouragement for one's own conviction, particularly if one has to live amongst people who for the most part have no sense at all of scientific effectiveness. At least this is very much the case amongst our natural researchers [in Bonn]. Amongst the philological-historical people, to be sure, there are respectable heads and characters, but their studies lie far afield from mine."[31] He never felt a sense of belonging or respect for his natural science colleagues at Bonn.

From the other side of the Rhine there were those who, for their part, were less than enamored of Helmholtz as a scientist. Claude Bernard, in particular, believed that, in contrast to himself, a self-described seeker after new facts (at least until the later 1850s), Helmholtz sought only laws of known phenomena: "I have thus had the objective of extending the analysis," Bernard praised himself in his notebook, "because there are those who are solely concerned with looking for the law of things that are known, but who do not seek to learn anything new. Helmholtz, du Bois Reymond, etc. are in this category. I have wanted to represent the side of invention." "I have been told," he further recorded, "that I find what I am not looking for, while Helmholtz finds only what he looks for. It is true, but the second direction is bad if [it is] exclusive."[32] There was no shortage of envy on either side of the Rhine.

How We Hear: On Combination Tones

Part one of the *Handbuch* appeared in the autumn of 1856. Several scientists keenly anticipated its appearance. Already in mid-December Maxwell wrote Thomson: "Where is Helmholtz on the Eye [*sic*] to be found?" Two weeks later, Thomson asked Helmholtz: "When will your book on the eye be completed, or is it so already? I find people greatly interested in it, especially regarding the adjustments." He also informed him that he wanted to learn all about Helmholtz's galvanometer, manufactured by Siemens and Halske, and buy it; and that the Atlantic telegraph was currently being manufactured and would be laid next May, thus permitting, he hoped, the first transatlantic (Ireland to Newfoundland) telegraphic messages by July. In his reply, Helmholtz discussed his most recent work and publications in physiological optics and acoustics. But he had not made much progress with part two of the *Handbuch*, he said, because he was now so involved with physiological acoustics. As early as 1848–49, he had reviewed (in the *Fortschritte der Physik*) recent work in theoretical acoustics and acoustic phenomena. (The journal was several years behind in publication, and so his two reviews did not appear until 1852 and 1854.) Sometime in late 1855 or early 1856, he began turning his attention to acoustics in a sustained manner aimed at publication. For example, he ordered a custom-designed double siren (for producing sounds) from Ferdinand Sauerwald in Berlin; he was very pleased with it as he investigated Tartini tones. In early May 1856, he was finishing his piece on Tartini tones, one that he thought would simplify the relationship between hearing, on the one hand, and the consonance and dissonance of sounds, on the other; it appeared before the month was out. In early June he lectured on that topic before the Natural History Association of the Rheinland and Westphalia (Naturhistorischer Verein von Rheinland und Westphalen). He did not bring out his full-scale exposition until later that year, however, in a long piece on combination tones in the *Annalen*. As for working on part two of the *Handbuch*, he did not think he would be able to do so for at least a year.[33]

While the broad appreciation of a mathematical (harmonic) relationship between sounds and music reached back to the ancient Greeks, it was not until the creation of harmonic analysis between the seventeenth and the early nineteenth centuries that a mathematical-physical tool became available for analyzing acoustic phenomena. The culmination in the 1820s of Joseph Fourier's method of harmonic analysis, aimed at understanding thermal phenomena, established the mathematical basis on which acoustic theory could first restructure itself in the 1840s. (To be sure, Fourier and several of his predecessors were aware of the relationship between their mathematical analysis and that of acoustic phenomena.) Ultimately, acoustic theory found its most brilliant applications in Helmholtz's theory of combination tones of 1856.[34] That theory laid the foundations of physiological acoustics in the modern sense, both for Helmholtz's own later work and for the field at large.

His analysis began with a reexamination of an older (1839–49), otherwise forgotten dispute between the physicists Georg Ohm and August Seebeck over the mathematical definition of tone.[35] Ohm had become best known for his electrical law expressing the relationship between current, voltage, and resistance in a conductor (namely, that the voltage was equal to the current times the resistance). But he had also become known for his acoustics law that was central to auditory perception: musical tones are periodic in nature, with the ear perceiving a collection of tones consisting of pure harmonic tones; through Fourier analysis, he maintained, the ear resolves these into simple or compound tones. Seebeck, for his part, had previously made significant improvements on Charles Cagniard de la Tour's siren (1819) for producing tones of a given frequency, and he made German and other scientists aware of its potential use in acoustic analysis. Ohm employed Fourier analysis to represent wave vibrations in the ear generated by a tone of a given pitch: in effect, he turned the ear into a harmonic analyzer. Seebeck, however, showed that Ohm's analysis contained mathematical errors that vitiated his law and that the law depended on specific empirical facts, including those of a physiological as well as a physical nature. Seebeck's analysis claimed that the so-called higher harmonics of a sound produced "combination tones"; that was why, he said, the fundamental was heard more loudly than the higher harmonics. It appeared that Seebeck had bested Ohm in this dispute, because Ohm could not explain why a siren's fundamental tone was heard much more easily than its higher harmonics. Furthermore, Seebeck suggested that a tone's timbre was somehow controlled by its upper partial tones. All together, these issues and their respective analyses set the scene for Helmholtz's own analysis. Then, too, his lifelong passion for music and the recent completion of his work on the eye (that is, part one of the *Handbuch*) were also parts of the context of his turn to physiological acoustics. He may have also been stimulated here by his recent philosophical analyses of human sensory physiology (1852), proper method in science (Goethe, 1853), and human perception (1855).

Helmholtz championed Ohm's law. Although he may have had some doubts about it and its experimental basis, he nonetheless thought it provided him with the means for understanding combination tones. Using a siren for producing tones, tuning forks for setting frequencies, and resonators for listening, he was able to separate the fundamental from the upper partial tones and produce first-order combination tones. From these he arrived at a definition of tone. He saw the centrality of the higher harmonics for his beat theory of consonance and for understanding timbre. His experimental work led him to recognize the importance of harmonics (that is, the upper partial tones) in acoustic phenomena and in their mathematical representation. With his resonators, he could hear either his newly discovered "summation tones" or the already known "difference tones." These combination tones, he found, originated in the middle ear's auditory ossicles and were due to the nonlinearity of the ear's response; and that, he later concluded, made them largely objective, even if they (as signs, not images, of external reality) required constant interpretation by the mind through a process that he called "unconscious inference." Altogether, he showed himself to be part philosopher, part empirical investigator, and part mathematical scientist. He returned to these and related acoustic and philosophical issues and further developed them in *On the Sensations of Tone* (1863) as well as in "The Facts in Perception" (1878).[36]

Market Value: Prussia Trumps Baden

After the turmoil of 1848, Baden, like a number of other German states, increased its commitment to higher education and looked to the natural sciences for potential help in matters pertaining to agriculture, medicine, and industry. In so doing, it attended to long-standing, liberal-minded, middle-class concerns about modernization. It built costly new chemical institutes at the Polytechnische Schule at Karlsruhe (1851) and at Heidelberg (1855), the latter for Bunsen. Heidelberg also got a renovated anatomy institute in 1849; some of its faculty also argued for the increasing importance of experimental physiology for medicine, though the state did not commit to building a new physiological institute or splitting its joint professorship of anatomy and physiology until 1858. Nonetheless, Baden's desire to reform medicine throughout the province and to strengthen its medical offerings at Heidelberg led to a readiness to commit extraordinary resources to attracting an outstanding physiologist. So too did its desire to rival Prussia in cultural terms.[37]

The 1850s and especially the 1860s were boom years in the German lands and in central and western Europe. Thanks to the combined growth of the railroads; iron and steel manufacturing; machine-building; and coal, textiles, and heavy industries generally; as well as similar growth in agriculture and the discovery of gold in California and Australia, the German economy experienced sustained growth until the depression of 1873. Governments had money to spend. By the mid-1850s, as Baden was increasing its investment in higher education and as Heidelberg sought to enhance its commitment to physiology, Helmholtz became a highly sought-after commodity. In May 1857, Bunsen informed him that Heidelberg would get a new physiologist and he hoped it would be Helmholtz (as opposed to Brücke, du Bois–Reymond, or Ludwig). Heidelberg was prepared to pay top dollar to bring Helmholtz there, and as the professor of physiology he would head a well-funded institute. There were about 120 medical students as well as about 40 others who might take medical courses. Heidelberg valued him greatly, Bunsen emphasized, and he wanted to know if Helmholtz was interested in the position and, if so, what his conditions were for coming. He told the Baden Ministry of the Interior that, of the four nominees (Helmholtz, Brücke, du Bois–Reymond, and Ludwig, in that order), Helmholtz was "doubtless the most ingenious, most gifted, and most versatilely educated," as his accompanying list of publications made clear.[38]

Helmholtz told du Bois–Reymond the essence of Bunsen's letter. He declined to be a candidate, he said, because he thought du Bois–Reymond wanted to go there and because Heidelberg offered him "no fundamental advantage" compared to what he already had in Bonn. He did not want to block du Bois–Reymond and told him that he had asked Bunsen to negotiate only with him. He added that he still owed the Prussian ministry "a certain duty of personal gratitude" because it had brought him to Bonn on account of Olga's health.[39] It was an act of self-interest, friendship, and noblesse oblige — or so it seemed.

His decision left him feeling depressed, however, since he found the Bonn academic scene "unfortunately very lazy." He saw no prospect of getting a new anatomy building and rightly thought this was due principally to the noncooperation of the university senate, which for several years had opposed the sale of university land to raise the necessary capital to construct new institute buildings. So he sat in his "old, dirty hole" of anatomy, without an assistant, and did nothing for the subject. He felt as offended by several of his Bonn science and medical colleagues as by the ministry. He found the "lower motives" (i.e., the pecuniary ones) displayed by some Bonn faculty members shameful and their behavior unworthy. He was especially offended by senior faculty who manipulated the types and scheduling of courses to be taught by private lecturers or himself. He explained that he and Busch and (in the past) Budge "constituted the progressive party," but now that he and Busch stood alone against the "party of reaction," they had no chance. In both the medical and the philosophical faculties, he knew of cases in which, to be granted a habilitation, young natural scientists had to agree formally beforehand not to offer courses that might infringe upon those already being taught by senior faculty. He found such behavior morally "scandalous" and was equally astounded when others did not also think so. His colleagues irked him further because they had scheduled his physiology course for medical students at the same time as inorganic chemistry. He felt that they showed no goodwill toward him. He could apparently do nothing but tend his own garden. He threw himself with renewed vigor into his work: "I've slowly accumulated rather a lot of stuff for reforming physiological acoustics," he told du Bois–Reymond, and he was awaiting some instruments so that he could complete the work.[40]

From Berlin, du Bois–Reymond responded that for the past six months he had been negotiating with Kirchhoff about going to Heidelberg and that, like Helmholtz and Brücke, he had recently heard from Bunsen, who asked him his conditions. If Heidelberg met his demands, he would accept, because, he regretfully reasoned, if Helmholtz got the call and accepted, then he (du Bois–Reymond) would in effect be forced to assume Helmholtz's position in Bonn. Du Bois–Reymond had asked Berlin to meet his demands, though he knew they would not. Hence he expected (but did not want) to go to Heidelberg, which he considered "a kind of defeat for me." Still, the only substantial difference between Heidelberg and Bonn, he thought, was the advantage the former offered as a purely physiological position. He badly misjudged Heidelberg's interest in him. He believed that his academic fate lay in decisions concerning Helmholtz. Helmholtz immediately updated du Bois–Reymond that the Heidelberg Medical Faculty had just dispatched Friedrich Wilhelm Delffs, its professor of pharmaceutical, physiological, and organic chemistry, and the faculty's associate dean, to negotiate with him. He had told Delffs that he would not consider moving for a salary of less than two thousand talers and that he could not imagine that Baden would pay that much for a professor of physiology nor that he could ask Prussia to allow him to go to Baden.[41]

Helmholtz did hope, however, that Baden's expression of interest would give him leverage against Prussia so as to improve his situation in Bonn. He asked du Bois–Reymond to conspire with him: not to let the ministry know of his intentions, or at least to say it was unlikely he would go to Heidelberg. He thought that Arnold considered the professorship of physiology and its institute as his own but would yield them if he liked the new physiologist. He also let du Bois–Reymond know that the Heidelberg faculty had asked a Russian student to attend both his and du Bois–Reymond's classes and report back on their teaching practices and abilities. As expected, the Russian was critical of their physically oriented physiology lectures. Du Bois–Reymond thought him a liar.[42]

In fact, Helmholtz did more than tend his garden. He showed Bunsen's letter to Bonn officials and reported that Delffs had come to Bonn to negotiate with him and that the latter said he would either be the only candidate or "*primo loco*," and that the offer included a new institute of physiology. He informed Bonn of these developments, he said, because he hoped it and the academic senate might change its mind and sell some university land to raise capital for a new institute building. He thought his colleagues (including his dean) had too little faith in his scientific work and were not ready to support him sufficiently. He saw no hope of ever getting a new institute building; all in all, it was "a very depressing outlook" for him. The lack of proper institutional facilities prevented him from accomplishing much of his planned work, he said, and further depressed him because he (and Busch) had no influence on the faculty majority, who rejected all their suggestions and even put many minor obstacles in their way. All this explained why he had yet to decline Baden's offer, even as he continued to be grateful for the ministry's past support in moving him from Königsberg to Bonn. He had two conditions for staying: agree to build the institute and increase his salary by 400 talers to 1,600. A Bonn administrator noted: "It would be a terrible blow if we lost Helmholtz, and I want to prevent precisely this from happening. If he were to leave us, our university would be robbed of a man who already today, and doubtless still more so in the future, must be considered one of the university's principal attractions, thanks not only to his ingenious scientific results but also to his natural and fortunate teaching abilities and moral value." Moreover, he added, Helmholtz, together with Busch, stood against the older (or scientifically weak), nonprogressive members within the Medical Faculty, and if the latter were to be overcome, it would have to be through the efforts of Helmholtz and Busch. Helmholtz represented "the Prussian banner of science in the west [of Prussia]," and Bonn's reputation would decline if Helmholtz left. He thought the ministry shared his opinion and asked that it do all it could to meet Helmholtz's conditions, all of which he thought were reasonable. The ministry responded with a salary increase of 400 talers, hinted at additional increases, and promised a new anatomy building as soon as its financial circumstances allowed.[43] The Prussian ministry had moved expeditiously.

Ten days later, Kirchhoff confidentially told Helmholtz that he alone had been recommended for the position, that the (Baden) ministry wanted him "in the most urgent way," and that a formal offer would soon be forthcoming. He added that everyone in Heidelberg, where Helmholtz already had friends, wanted him to come. In mid-July, in response to the enhanced compensation from Prussia, Helmholtz formally declined Heidelberg's offer. He told du Bois–Reymond his decision was "definitive."[44]

He showed strategic negotiating skill in exploiting the bidding between Prussia and Baden for his professional services, and he did so without offending anyone. He had enhanced his reputation and the perceived importance of physiology to gain a 25 percent salary increase and a renewed promise of a new institute building. Even little things were going his way. Ludwig reported that the proprietor of a Viennese scientific instrument firm was compiling a photograph album of the greatest contemporary natural scientists and "naturally" wanted a photograph of Helmholtz. And he collected past-due lecture fees from his Königsberg days. Justus Olshausen, who forwarded the money, thought Helmholtz had done well in choosing not to go to Heidelberg, "which doesn't have as great a reputation" as Bonn.[45]

In mid-August, with the Heidelberg negotiations behind him, and after a visit to Bonn by his father and his sister Julie, Helmholtz undertook (again without Olga) another hiking tour in Switzerland. He passed through Zurich so quickly that he did not have time to see his friends there as he headed for the Rigi. There he met up with friends from Bonn as well as with his friend Heine, the Halle mathematician. Then he and a Russian physician headed for Riffelhaus and the Monte Rosa, seven thousand feet above sea level. They trekked through the Rhône Valley, whose glaciers he thought stunning and like a "crystal palace." He wore out his leather boots from so much walking and had to buy a pair of Alpine shoes to march up Monte Rosa. He climbed so high that his nose bled. He climbed slowly and without a backpack. He and the Russian planned to cross over to the Italian side, and from there he intended to go to Lake Maggiore before heading into the Rhine Valley and then back to Bonn.[46]

A Philosophical Dispute *en famille*

As he worked on physiological acoustics, Helmholtz also had to attend to some "philosophical" issues that were as much personal as they were intellectual in nature. One concerned his father, whose health continued to decline. Ferdinand complained to his old friend Fichte about his aging and the decline of his memory, "pressure on the brain, and the like." In the autumn of 1856, after thirty-six years of teaching, Ferdinand was forced by poor health to retire. Fichte had become a prolific philosopher since he had last seen Ferdinand in 1821, at Hermann's baptism; his writings ranged through theology, ethics, and metaphysics to anthropology and psychology. His general orientation was antimaterialist and pro-Christian, yet also liberal. He had become, too, a keen student of Hermann's writings and a great admirer of his godson. He read him on the conservation of force and "applied" the principle to his own psychological system; he read him on Goethe; and he sought to use Helmholtz's sign theory of perception in his own *Psychologie* (1864, 1873). He noted in his diary that Hermann had just sent him "an important letter . . . with insightful remarks" about the relationship between natural science and philosophy. He wrote Ferdinand about Hermann: "You know how greatly I estimate his [Hermann's] mind, and that his solid judgment illuminates each and every one of his [written] lines. I believe I'm able to come to a full [philosophical] understanding with him." He thought that Hermann's philosophical standpoint—he called it Kantian—was also pertinent to Ferdinand's philosophical outlook, and he told Ferdinand that Hermann's (and others') empirical findings in and understanding of sensory physiology had aroused "attention" in psychology. He found that all of Hermann's recent physiological and psychological findings accorded with his (Fichte's) *Anthropologie* (1856) as well as with the book that he was then writing on psychology. "Nothing would be more desirable for me personally than to hear a full lecture on physiology by your son."[47] Ferdinand was doubtless prouder than ever of his son's intellectual accomplishments and standing.

Not all German philosophers were so adoring, however. Although his influence lay principally outside the academic realm, Arthur Schopenhauer was among the most influential and widely read figures in nineteenth-century philosophy, especially as the debate about materialism heated up and as neo-Kantianism came to the fore. His emphasis on the emotions, the will, the irrational in life, and asceticism, to say nothing of his critiques of optimism and progress, held no interest for or appeal to someone like Helmholtz. Yet Schopenhauer and his followers had a quarrel with him: they believed that the publication of Schopenhauer's *Ueber das Sehn und die Farben* (first edition, 1816; second edition, 1854), which vigorously supported and extended Goethe's theory of colors, qualified him as a color theorist and vision scientist. Like his fellow amateur Goethe, he was very frustrated by what he considered the academic professionals' neglect and lack of recognition of his outstanding work in color theory and philosophically insightful understanding of vision.[48] Though Helmholtz already had several scientific opponents, Schopenhauer became his first (and not the last) enemy. He represented a current of antirational German philosophical and cultural thought that at some points masqueraded as natural science; it remained alive through the end of the Nazi era.

As early as June 1853, Helmholtz's name had come to Schopenhauer's attention. At first, Schopenhauer dismissed him as part of an alleged self-perpetuating academic clique, and he considered Helmholtz's essay on Goethe to be an attempt to curry favor. (He also dismissed Ferdinand, his former friend; the two had, in fact, attended Fichte's lectures together in Berlin in 1811–12.) Schopenhauer found Helmholtz's essay on Goethe "tasteless" and criticized his own friend and defender Julius Frauenstädt for regarding Helmholtz as his (Schopenhauer's) equal in matters of color and vision (that was like comparing Mont Blanc to a molehill, he said). By January 1856, Schopenhauer had classified Helmholtz as "a scoundrel," claiming that, in his Kant Memorial Address, he had failed to cite Schopenhauer's contributions to color theory and allegedly plagiarized from him and others as well. Helmholtz's essay on Goethe was worthless, Schopenhauer thought, and his essay on the interaction of forces scarcely better.[49]

In September of that year, after Hermann had sent Ferdinand a copy of the

Handbuch (part one), Ferdinand drew his attention to Francis Bacon's epistemological views on the relationship of empirical and a priori knowledge, as well as
to Schopenhauer. He included the charge by Frauenstädt that, in the Kant Address, Hermann had plagiarized Schopenhauer's epistemological views. Ferdinand naturally defended Hermann, declaring that Schopenhauer's own views
were already to be found in Kant and in Johann Gottlieb Fichte. Hermann replied that he was "very pleased" that Ferdinand was doing so well and that Ferdinand continued to write (though not publish) philosophy. He thought it the right
moment for older men like his father to reexplain to a younger generation the
meaning of Kant and Johann Gottlieb Fichte, "since the philosophical drunken-
ness and the attending hangover of Hegel's and Schelling's natural-philosophical
systems" seemed to be over, and people were again "beginning to be interested
in philosophy." Hegel's idealistic philosophy was worse than useless, he said,
since it tried to substitute itself for science. Nor was he much impressed with
the work of Immanuel Herrmann Fichte, who in his own way did the same thing.
Schelling, Hegel, and Fichte junior all failed to provide an empirical basis for
their philosophical outlooks, he said, and so their enterprises were necessarily
doomed. (In fact, he was distinctly unimpressed with the philosophical work of
Fichte junior, and told his father so.) He believed that philosophy was above all
about testing the sources of scientific knowledge, that is, about epistemology.
Kant, and perhaps Fichte senior, he added, had understood this. As for Schopenhauer, Helmholtz attributed his recent success to his building on Kant.[50]

Ferdinand vigorously defended his old friend Immanuel Herrmann Fichte.
Though Ferdinand was in failing health—he still felt pressure on the brain, his
hand shook so much he could barely write, and his eyes were failing—he was
determined to "catch up" philosophically, to find inner peace, and to bring some
sort of "unity" to his life before death, as he now sensed, caught up with him.
But he felt strong enough to vigorously criticize his son for what he referred to as
Hermann's insufficiently considered thoughts on philosophy and his "prejudice"
vis-à-vis Fichte junior's work. He thought Hermann had perhaps been driven too
far by his interest in natural science (i.e., empiricism) and thus had forgotten the
place and importance of a priori ideals and their interaction with objects in the
world. To produce true knowledge, Ferdinand believed, thinking and observation needed constantly to interact with each other. The problem with Schelling,
Hegel, and their followers, or so the now retired, unpublished, amateur philosopher lectured his son, was that they had forgotten the importance of the empirical in producing knowledge, while the problem with men like Moleschott and
Vogt was that they sought to reduce everything to mere matter and its interaction. Fichte, by contrast, did not have the shortcomings of either these ideal-
ists or materialists; instead, Ferdinand argued, he knew how to exploit genuine
scientific knowledge for philosophical purposes to gain "knowledge of the soul,"

which was something else that Hermann should not forget. Ferdinand further lectured him on Fichte's understanding of the nature and properties of the soul, even as he himself ever so gently criticized Fichte for going too far. Most importantly, at one point or another philosophy had to absorb the results of natural science in order to expand "self knowledge" and hence understanding of the soul. Not even the applications of the sciences to man's social life was as important. As for Schopenhauer, Ferdinand dismissed his charges against Hermann and thought him guilty of seeking in the public arena the sort of respect he could not otherwise attain among philosophers.[51]

Hermann graciously and piously replied to his aging and ill father that he largely agreed with him and that he would now reconsider if not reread Fichte's *Anthropologie*. The problem with "mathematical natural researchers" like himself, he acknowledged, was that they were so concerned about testing hypotheses with facts that they perhaps had "too great a fear before a bolder use of scientific facts, which, on other occasions, can indeed be legitimate." He protested against any charge of being a follower of Moleschott and Vogt, neither of whom he considered to be a scientist. No matter how deep the discoveries of a natural scientist may be, he said, this did not justify making pronouncements about the soul. Hence, this also did not justify Ferdinand's remarks about the hostility of many natural scientists toward philosophy, most of whom were, in any case, "indifferent" toward it. Besides, most of the problems between philosophers and natural scientists were due to men like Schelling and Hegel. Nor did he think much of Lotze, either. It was Kant and epistemological questions concerning science, he said, that attracted him. As for Schopenhauer, he agreed with his father: what he had read of him, he did not like. Only much later, and then with only a brief, unflattering glance, did Helmholtz again mention Schopenhauer in print. This peaceful philosophical exchange between Ferdinand and Hermann had been brewing for decades, and it perhaps represented, if only on a rational, intellectual surface, unspoken emotional tensions that lay below. In one sense, father and son had a sort of materialism dispute *en famille*.[52] In another, they had scarcely begun to talk.

The Physiology of Musical Harmony

In June 1857, in a presentation before the Natural History Association of the Rheinland and Westphalia, Helmholtz unveiled a new invention, the "telestereoscope." It was an enhanced stereoscope that gave the viewer clearer images (via mirrors) of objects whose lines or surfaces were not smooth (for example, landscapes). It was easy to construct, he said, and useful not only for observing near and distant objects, but also as "an amusing optical trifle." A local newspaper picked up the story, which was now making "the rounds of the political newspapers," and so he needed to write it up, at Poggendorff's request, for the *Annalen*.[53]

He also presented his telestereoscope at the meeting of the Naturforscherversammlung held in Bonn (September 1857). Around one thousand people attended, although Helmholtz thought that "for the most part the most important scientists failed" to attend. He found it both "very interesting indeed" and "also truly a great rush" for him, even though he avoided most of the social gatherings. Dove and Wittich stayed with the Helmholtzes. He and Olga had people over for only one meal ("made me *caput* for the following day") and served lots of coffee and tea between sessions but otherwise minimized having guests. He thought the deleterious state of his anatomy institute hurt his reputation among those who visited it during the meeting. This was the first time he spoke at a Naturforscherversammlung meeting, and he did so on only four other occasions (1858, 1869, 1872, and 1889). The telestereoscope's utility—he presented his lecture before the entire association—resonated with a number of users, for example, Brücke, who wrote to say that it was "very clever," and Anton Danga, a "simple man" from the Palatinate, who said he was fascinated with the different pictures resulting from Helmholtz's telestereoscope. Helmholtz did more than merely lecture on the telestereoscope, however. He also showed visitors his myograph for taking measurements in nerves, gave a brief lecture on the movements of small bones in the ear, and spoke on combination tones. He thus lectured in both the Anatomy Section (twice) and the Physics Section. (All told, he gave four lectures at the meeting.) Furthermore, he discussed with others the need for money in science, in particular sums that went beyond the financial resources of any individual scientist. Later, to his surprise, he received an (apparently unsolicited) annual sum from King Maximilian II of Bavaria for precisely this purpose. (The king was much interested in Helmholtz's acoustic studies and asked in return that Helmholtz briefly report his results to him on a quarterly basis.) Finally, he was also presented to Princess Elisabeth of Prussia as "the discoverer of the ophthalmoscope."[54]

During the winter of 1857, "in the hometown of Beethoven, of the most powerful among the heroes of the tonal art," Helmholtz delivered a popular science lecture on the physiological causes of musical harmony. He sought to explain what physics and physiology had to offer for understanding music and musical phenomena. He thought that music, more than any of the other arts, had up to then escaped scientific analysis; the other fine arts, by contrast, used concrete materials and represented natural and human objects, thus making them more accessible to scientific and aesthetic analyses. Music employed no such materials and made no such representations; hence, to some its effects appeared "incomprehensible and wonderful." But music did indeed have a sort of material, he argued, that of "the tones or tone sensations." While it appeared to be "the most immaterial, the most fleeting, and the most fragile creator of incalculable and indescribable tempers," in fact it was subject to mathematical analysis. Though mathematics and music appeared worlds apart, they were actually related: mathematics could give insight into music. More specifically, Helmholtz sought to show his audience and readers how physical and physiological acoustics pertained to the understanding of music, focusing attention especially on the reasons for musical consonance. He thus sought to use modern science and mathematics to answer the ancient question that Pythagoras and many others since had never satisfactorily answered: What do the frequency relationships of small, whole numbers have to do with consonance?[55]

A musical tone, he explained, issued from the repetition of rapid, regular impulses in equal periods of time. The tone's height was a function of the number of such impulses; more impulses per time period resulted in a higher tone. It mattered not the slightest how the tone was made or by what type of instrument, including the human voice. To show the dependency of tones on frequency relationships, he illustrated his point using a siren.[56]

What makes a sound vibration a tone sensation, he continued, had to do with its reception by the ear (and not, for example, by the skin) and the nature of the ear's nerves. In the first instance, this was a matter of the physics of wave motions as these propagated through the air. He explained the simplest properties of sound waves: their motions and their interactions, in particular the relationship among wavelength and tone height, strength, and intensity. The variations in the form of sound waves caused variations in tone quality (or timbre). He explained that he had often spent hours on the East Prussian coast watching and being fascinated by sea waves and their interacting systems. Doing so, he found, "captivates and raises the spirit [Geist], because the eye easily recognizes law and order in it." For sound systems and their perceptions, a concert or dance hall was analogous to the behavior of waves at sea and their perception by the viewer. Indeed, he pointed to similarities or close analogies between sound and sight perception.[57]

He then explained the relevant anatomical and physiological (including nerve) properties of the ear and its perception of sound waves. Thanks to recently enhanced understandings of the ear's anatomy and physiology, it was now clear how the ear analyzes sound waves into their individual component tones, a phenomenon very similar to Fourier analysis in mathematical physics. The analysis of tone by the ear leads, for example, to recognition of the differences between individual voices and musical instruments. For Helmholtz, the ear was a mathematical instrument for analyzing sound waves, just as the eye was one for optical information.[58]

He also emphasized that tone perception was due not only to "the corporeal ear of the body" but also to "the spiritual [geistige] ear of the power of imagination." The former operated automatically; its action was similar to how a mathematician does Fourier analysis of wave systems, allowing distinction of the individual upper partial tones from fundamental tones, and he emphasized how much attention it required, even for an experienced listener, to perceive such tones. To the listener, by contrast, "still belongs a peculiar activity of the soul in order to get from the nerve's sensation to the representation of the given external object that has stimulated that sensation." Here he invoked his sign theory of epistemology, that sensory perceptions only give us signs that external objects are present; experience gradually teaches us to conclude that such internal, mental signs are indicators of certain external, material objects. Our perceptions are dependent on practical experience, he argued. We are rarely aware of the upper partial tones, and it requires the soul (*Seele*) as much as the ear's nerves for us to become aware of them.[59]

The upper partials produced the timbre ("Klangfarbe"), which depended on the wave form. The upper partials produced by clocks were the easiest to hear. Helmholtz emphasized that the upper partials played an "important role . . . in the artistic effect of music." Though they were not so easy to hear, his invention of the Helmholtz resonator (as it since became known)—a glass or metal sphere with two openings, one leading to the sound source and the other to the entrance to the ear—provided a way for listeners to hear the upper partials.[60]

He then explained the phenomena of musical beats in terms of the superposition of two sine waves. Beats that occur rapidly enough become (to the ear) indistinguishable from one another, producing what he called "a continuous tone sensation," or a sensation of tone. Such beats contrasted with sensations that were discontinuous and so produced a sense of dissonance. Moreover, the superposition of two loud tones could induce the hearing of further tones, called combination tones.[61]

In the light of these explanations of the elements of physiological harmony, Helmholtz declared that the distinction between harmony and disharmony was "that in the former the tones flow by one another regularly, each individual tone for itself, while in disharmony there occurs an incompatibility, and the tones disintegrate as they mutually push against one another." Musical beats depend on the interference of wave motions, which is to say that sound is a wave motion. The ear is able to distinguish upper partials and analyze compound wave systems thanks to its ability to do Fourier analysis; indeed, Fourier's theorem helps determine the relative amplitudes of the partials. In all this, he concluded, we see a strong comparison between the ear and the eye. Light, too, is a wave phenomenon and hence shows interference properties, while its various frequencies produce various colors perceived by the eye. Yet the eye cannot distinguish among compound colors, which is to say that "it has no harmony in the sense that the ear does; it has no music."[62]

Despite all this anatomical, physiological, physical, and mathematical analysis, in the end Helmholtz maintained that "aesthetics seeks the essence of artistic beauty in its unconscious reasonableness." While upper partials are central to the musical experience and dependent on the ear's processing of waves, this processing nonetheless occurs on an unconscious level. He maintained that musical beauty, that is, harmony and disharmony, was a matter of degree, something that rose up a continuous scale of beauty from the lowest sensual satisfaction to the highest intellectual satisfaction. As in the sea, he believed, there were currents in the artist's soul, currents that, though the artist himself could not explain them, connected him spiritually to similar currents "in the soul of the listener, finally lifting him upward into the peace of immortal beauty, of whose prophet among men the Godhead has chosen but a few favorites." Helmholtz concluded that, at this point, natural science reached its limits.[63] Here was his own touch of romanticism.

Helmholtz thus grounded his theory of musical harmony in the ear's anatomy and the physics of waves, just as he grounded his theory of human perception in the eye's anatomy and physiology and the associated physics of waves. Yet he did not "reduce" his theories of hearing and sight to such bodily and physical phenomena. For he always maintained that there was a psychological component in auditory and visual perception, one that he vaguely referred to as being part of the "soul" or the creative spirit in human beings. To be sure, that element was itself a function of human behavior and experience over long periods of time. But Helmholtz never thought the "hard" sciences of anatomy, physiology, and physics could themselves fully account for human visual and auditory perception.

Flowing Water

In November 1857, the editors of the *Illustrirte Deutsche Monatshefte* asked Helmholtz to send them a popular scientific essay for publication in their magazine. He had none to send. "My time is already so taken up that I can't undertake special work for popular purposes. However, should an occasion arise that leads me to write such an essay, and should it seem right for the *Deutsche Monatshefte* (something that can indeed easily happen), then I'll send it to you." At the same time, Rudolf Haym, the founding editor (1858–64) of the *Preussische Jahrbücher*, also wrote him to explain the sorts of articles he hoped to get for his new magazine; included were articles on the history of science, on the use of science, and on developments in the latest scientific fields, especially as these related to bigger issues concerning the "nation's cultural life [*Bildungsleben*], etc. etc." Haym thought Helmholtz understood what he was talking about, since he had already written such essays, some of which Haym had read. He continued:

I would consider myself very fortunate if I could soon see the *Jahrbücher* adorned with an essay like that on Goethe's attitude vis-à-vis the natural sciences. I can hardly but believe that such an involvement of natural science in the *moral* development of the nation's culture—for that is what our magazine is about—must be close to your heart, too. Don't reject the hand from this side that is here offered you, a hand which otherwise tends to ignore the natural sciences completely. In an age when philosophy has become powerless—and thus the dualism of the historical and the physical sciences naturally ceases—do help us so that the dualism doesn't gain further ground.

Haym also asked Helmholtz to suggest names of other natural scientists who might also write for the *Jahrbücher*, and in particular to recommend someone to write a biography and appreciation of Humboldt. Helmholtz did not then agree to write an article (as he eventually did), but he sent Haym several names. Haym now considered himself to be in an "alliance" with Helmholtz, an alliance, Haym said, "whose conclusion a later generation will experience and will enjoy in a freer state and national life than ours."[64]

Rather than thinking about writing popular scientific essays, in 1857 at least, Helmholtz was much preoccupied with the scientific analysis of the flow of water. The following year he published what became one of the foundational papers and theorems in the study of the flow of water (and other physical and mathematical phenomena). Since the Renaissance, and above all since the mid-eighteenth-century work of Daniel Bernoulli and the mid-nineteenth-century work of a variety of British and French natural philosophers and engineers, the subject of the flow of fluids meant both hydrodynamics and hydraulics; it meant both the ideal and the practical, both mathematical analysis and empirical data-gathering and formulas, with or without theory. Helmholtz now joined and indeed reinvigorated this tradition.[65]

His work on hydrodynamics and its place of publication are at once unsurprising and surprising. To understand why this is so, one must look far beyond the highly mathematical and physical content of his epochal paper of 1858 and toward its cultural and personal meaning. As to the unsurprising part, Helmholtz, it will be recalled, had always wanted to do physics. By the mid-1850s, he had accomplished enough professionally within physiology proper that he could devote more time to strictly physical topics. Nevertheless, the hydrodynamics paper of 1858—highly mathematical and technical as it was—emerged from his recent work in physiological acoustics. His general scientific orientation was beginning to change, and it distinguished him even further as a physicist and, for the first time, as a mathematician.

Perhaps equally important, Helmholtz had always loved being near water and was fascinated by wave formations and surfaces. As already noted, he took great pleasure in watching the rivers, lakes, and seas in and around Potsdam, Berlin, Königsberg, and Bonn; it was later the same with the Neckar in Heidelberg. He had also greatly enjoyed the rivers of Scotland. Water, like the mountains, made him feel a part of nature and was psychologically restorative:

> Water in motion, as in cascades or sea waves, has an effect in some respects similar to music. How long and how often can we sit and look at the waves rolling in to shore! Their rhythmic motion, perpetually varied in detail, produces a peculiar feeling of pleasant repose or weariness, and the impression of a mighty orderly life, finely linked together. When the sea is quiet and smooth we can enjoy its coloring for a while, but this gives no such lasting pleasure as the rolling waves. Small undulations, on the other hand, on small surfaces of water, follow one another too rapidly, and disturb rather than please.

Again speaking of water surfaces and their associated, multitudinous wave formations, he wrote: "I must own that whenever I attentively observe this spectacle it awakens in me a peculiar kind of intellectual pleasure, because it bears to the bodily eye what the mind's eye grasps only by the help of a long series of complicated conclusions for the waves of the invisible atmospheric ocean."[66] His paper of 1858, it is also worth noting, was related to his ongoing, intense interest in music, that is, to the physics of sound waves.

On the other hand, the paper's appearance is rather surprising and is related to the institutional state of physiology at Bonn. The dilapidated condition of Bonn's limited anatomical and physiological facilities encouraged Helmholtz to take up more nonexperimental, abstract work. As he wrote du Bois–Reymond: "That's why here [in Bonn] I've thrown myself into mathematical work, of which a piece on vortex motion is already in press in Crelle [a journal]; some acoustic work still remains to be done." He apparently did this hydrodynamic work in 1857, because in early January 1858 he sent Carl Wilhelm Borchardt, the editor of (Crelle's) *Journal für die reine und angewandte Mathematik*, the first half of his manuscript. He saw its strength in the clarification of the mathematical equations of hydrodynamics and in their eventual application to understanding the motion of water. Borchardt warmly welcomed it both because he did not want his journal to be solely for pure mathematics and because Helmholtz's piece would strengthen the journal's connections with the mathematical sciences. Borchardt immediately initiated the publication process, and by early February Helmholtz was correcting the proofs. Beyond all of the above, Helmholtz's piece on hydrodynamics also proved seminal both for William Thomson's vortex theory and for the development of knot theory by Peter Guthrie Tait and others.[67]

The paper also emerged in part out of Helmholtz's interest in physiological acoustics, in particular his analysis of how organ pipes produced sound.[68]

As he sought to improve the theoretical understanding of the organ, he was led to rethink earlier work on it by Daniel Bernoulli, Leonhard Euler, Joseph-Louis Lagrange, and Poisson concerning the air's internal friction and damping effect and related observational studies by Guillaume Wertheim and Friedrich Zamminer. Fundamental to Helmholtz's hydrodynamics work of 1858 was his creative borrowing of Bernhard Riemann's mathematical notion of "cut surfaces" or "intersections" ("Schnittflächen"). Using Green's theorem, he concentrated his analysis on the motion of air in open-ended pipes. In turn, he applied a similar sort of analysis to fluids and the role of friction. His papers on organ pipes and hydrodynamics showed that he had mastered the use of differential equations and, with them, mathematical physics; his sources were at turns French, British, and German, and his reasoning at turns analytical, geometric, and analogical (here comparing electromagnetism and continuum mechanics). (Maxwell, ever a keen reader of Helmholtz, rightly thought the latter's paper suggested a strong physical analogy between electromagnetism and hydrodynamics.) Moreover, as a leading articulator of the law of conservation of force (energy), he naturally employed that too. He borrowed eclectically from all and sundry to produce an innovative, foundational mathematical theory of the behavior of fluids in terms of vortex motion; he also provided a simple illustration for observing the behavior of vortex rings: dipping a spoon into a calm water surface and quickly withdrawing it. His analysis brought quick, positive responses from mathematicians and physicists, several of whom (Rudolph Clebsch, Riemann, Kirchhoff, and Hermann Hankel) soon improved upon parts of his analysis or extended it to other domains. In the late 1860s, Helmholtz himself, along with Thomson, Maxwell, and Tait, made still further improvements.[69] His paper on hydrodynamics became as seminal as the one on force (energy) conservation; only now he found immediate and positive reaction from both physicists and mathematicians.

Market Value: Baden Trumps Prussia

Despite Helmholtz's choice of Prussia over Baden in July 1857, Heidelberg (Baden) never really ceased its effort to hire him; in October there was renewed talk of doing so. Bunsen began reexploring the possibility of Helmholtz's moving to Heidelberg. Helmholtz replied that he owed Prussia too much to move so soon and that, moreover, du Bois–Reymond deserved the ordinary professorship at Heidelberg more than he (Helmholtz) needed to switch from one ordinary professorship to another. Later that year Baden restated its desire to acquire him for Heidelberg, saying that it placed a "high value" on doing so and would do everything possible to achieve that goal; and, in the person of Bunsen, Baden again explored such possibilities with him. On his own authority and to accelerate the process, Bunsen told the ministry that Helmholtz was interested. In early February 1858, Helmholtz indicated to Kirchhoff that he might indeed move to Heidelberg. Kirchhoff told him that the anatomy building was large but in poor condition and would be of little use to him. He advised him to come see for himself when he went to Karlsruhe (Baden's capital) for discussions with the authorities, and he said Bunsen was convinced that Baden would give Helmholtz "everything" he wanted so long as he asked for things before making a commitment to them. Helmholtz scheduled a visit to Heidelberg, and by early March he had accepted Baden's offer. He told the Prussian minister of culture that he had turned down Baden's offer the previous summer because Prussia had met his conditions; but then subsequent actions by the Bonn senate had made it clear that it would not in fact cooperate anytime soon in helping him get a new institute building there. When Heidelberg renewed its offer in October, he had hoped that the senate would reconsider and make a favorable decision. He said that Bonn's poor facilities "almost totally hindered" his (and his students') scientific efforts. When the senate still did nothing, he accepted Heidelberg's offer. He regretted leaving his native Prussia, he said, but felt duty-bound to accept Heidelberg's offer, which granted "the most favorable conditions" for his "scientific effectiveness." Heidelberg wanted him to begin in April, at the start of the summer semester, and he intended to do so if Prussia could find a replacement for him in time; if not, he would leave in the autumn. (Nonetheless, the deal was not done.)[70]

The local politics of Bonn had defeated Helmholtz's (if not Berlin's) modernization plans. Bonn's curatorium told the Medical Faculty that its loss "of an outstanding member" was "extremely deplorable." A local medical journal reported that "everybody" in Bonn was talking about his leaving and wondered why he was doing so since he had asked to go there and had arrived only recently. There was speculation that Bonn had not met his expectations. At the Naturforscherversammlung meeting in Bonn in September, everyone had hoped that Bonn would act soon and that this would increase its scientific standing. But nothing happened, and Helmholtz's departure left the medical community there feeling discouraged and depressed. It was, they said, "a bitter loss for our medical faculty."[71]

Helmholtz immediately informed du Bois–Reymond about the nature and extent of his negotiations with Baden and that he had accepted Heidelberg's offer and requested permission from Prussia to be released from its service. He felt guilty that he may have interfered with du Bois–Reymond's own hopes for an offer from Heidelberg but made it clear that Heidelberg, not he, had reinitiated possibilities. He added that the Heidelberg Medical Faculty largely opposed "physical physiology" but nonetheless awarded him "the very dubious distinction" of being the least physical of any such candidates. As for the Bonn senate, he had been "miserably swindled" and they had left him hanging. After this "scandalosum" became known to Karlsruhe, the ministry then again offered him the position. By the time the Bonn senate and the ministry in Berlin had finally approved the sale of the land (in February), it was too late: at Christmas of 1857, Baden offered him the position for a third time. It gave him everything he asked for "so as to secure for [him] undisturbed [scientific] effectiveness."[72]

In an effort to help secure his position for du Bois–Reymond at Bonn, he told him that he taught two courses per semester and earned a salary of 1,600 talers as well as (last year) honorary income (presumably student lecture fees) of 1,060 talers and examination fees of 100 talers, making a total of about 2,760 talers. He advised him not to accept an offer from Bonn unless he was promised a new institute building. He did nearly all of his experimental work in his own home, he said, while his students used the auditorium. This lack of institutional facilities, he explained, was why he had concentrated his efforts on mathematical topics. The anatomy institute was "a pigsty." Bonn itself, however, was very stimulating.[73]

In response to a request from du Bois–Reymond, Helmholtz further declared that Heidelberg would be paying him a salary of 3,600 guldens (2,057 talers) along with financial support for an assistant (300 guldens, or 171 talers), an institute servant (150 guldens, or 86 talers), a research budget (600 guldens, or 343 talers), and so on. Moreover, Baden had promised him a new institute building. His salary at Heidelberg was thus 25 percent higher than what he was earning at Bonn (and was the second-highest at Heidelberg and virtually the same as Virchow's at Berlin). While he thought Heidelberg's setting was more attractive than Bonn's, this was not an important factor in his decision. What above all attracted him to Heidelberg, as he later explicitly said, was getting a proper, well-equipped scientific institute. Moreover, he already had good friends there in the persons of Bunsen and Kirchhoff; by contrast, in Bonn he had made virtually no friendships among his science colleagues. Ludwig congratulated him on his decision and was especially pleased that Helmholtz no longer had to teach anatomy and that he would live "in glorious Heidelberg." Kirchhoff was already looking for a residence for him.[74]

Du Bois–Reymond regretfully accepted Helmholtz's offer to try to get him nominated as his successor: he would have preferred Heidelberg by far to Bonn. But even with Helmholtz's support, the Bonn Medical Faculty was not really interested in him. However, after Johannes Müller died unexpectedly that April, du Bois–Reymond (quite correctly) anticipated that he would become his successor at Berlin.[75]

Self-Justification

Yet Bonn and Prussia would not consider Helmholtz's transfer to Heidelberg a done deal; instead, they attempted to make him renege. Prince William of Prussia himself expressed his disappointment at Helmholtz's leaving and said he intended to visit Baden in order to get him released from his agreement and to renegotiate his contract with Prussia. The ministry seemed not to share the prince's concern and portrayed Helmholtz "as an ingrate." In late May the rector of Bonn asked Helmholtz to meet with Ferdinand Knerck, a senior official in the Prussian Ministry of Culture: the ministry hoped that Helmholtz might yet be released from his agreement with Baden. This would be arranged, it told Helmholtz, "with your doing." The Baden authorities would understand. Helmholtz now agreed to stay at Bonn if and when Prussia (on its own) got Baden to release him from his promise to go to Heidelberg. Knerck said Prussia would match Baden's offer. He replied that he was bound by his word to Baden, and he refused to himself try to undo his agreement with Baden. This proved enough of an opening to initiate diplomatic negotiations (and so proceed without him). But Helmholtz did not expect that Baden would release him.[76]

Simultaneously, Helmholtz told Baden that Prussia had delayed his release from its service in an attempt to keep him there; they wanted to use their diplomatic relations with Baden to undo the agreement. He reported Knerck's coming to Bonn and his proposals. He said that he in no way sought a counteroffer and that he could not consider one because he had already given Baden his word. He acknowledged that, as a Prussian native and civil servant, he still had "some obligations." But while he said it was up to Baden to decide, his tone throughout was of someone who intended and wanted to go to Heidelberg. Baden responded that it had informed Prussia that it would not release him from his promise. Helmholtz felt bound to Baden unless it annulled the agreement. He twice again asked Prussia to release him from its service.[77]

Du Bois–Reymond wrote his father: "The Grand Duchy of Baden is having a good laugh at Prussia." He also wrote Helmholtz that the latter had made a real "faux pas" with the Prussian authorities (or else, and more probably, some negotiator had erred in his understanding). He defended Helmholtz before several senior figures in the ministry. But they remained unconvinced and intended to make him pay a price for his actions. Still, they hoped to have him back at Bonn by next Easter. Though du Bois–Reymond conceded that he was disappointed that he did not get the call to Heidelberg after he had for so long expected it, what irritated him most was that Helmholtz had not kept him au courant. He knew that what bothered people the most was their envy of Helmholtz's "serene scientific brilliance," so now they tried to find fault with his person.[78]

Helmholtz immediately wrote the minister: "I'm told that you are reportedly angry with me for deciding to leave Prussia." He intended, when he next came to Berlin, to call upon him personally to explain all that had happened over the past several years and to defend himself before the minister: "It would be disturbing for me to leave you with anything other than a good impression of myself." As for du Bois–Reymond, Helmholtz wrote a long, exculpatory letter again recounting the history of the negotiations with Baden and Prussia, not least to protect himself from disinformation and false rumors that he feared the Prussian ministry would now spread. He asked him to let their "scientific friends in Berlin" know of the letter's contents. He remained uncertain which of the two governments "had most bent and twisted the truth." His greatest fear, however, was that all that had happened might endanger his friendship with du Bois–Reymond or that it might have injured the latter's career. He assured du Bois–Reymond that, rumors notwithstanding, he had never had any intention of competing with him over Berlin, a subject that he had never even mentioned, let alone discussed, with the ministry.[79]

It was not until July 1858, then, that it became absolutely certain that Helmholtz would move to Heidelberg. He sought to justify his motives and actions before his father as well as before du Bois–Reymond, giving both men essentially the same account of Prussia's attempt to undo the deal. In the end, he said, he had decided for Heidelberg over Bonn because Baden, in contrast to Prussia, had made it clear it would do everything needed to provide him with a situation that permitted his future "scientific success." The Prussian ministry, he said, had acted deceitfully and arrogantly.[80]

That July, too, brought news that Helmholtz had been chosen a corresponding member of the Bavarian Academy of Sciences, which judged him to be "one of the most talented younger researchers in the so-called physical-physiological school." It noted that he had distinguished himself not only among physiologists but also among physicists, mathematicians, and others and had devised and creatively used new scientific instruments. It referred to his invention of the ophthalmoscope as an instrument of "epoch-making influence" for ophthalmology and of importance for physiology, too. As the month drew to a close, he finally received his release from Prussia's service. A month or so later, he moved his family to Heidelberg and then spent a few days in Karlsruhe, where in mid-September that year's Naturforscherversammlung was being held. There he delivered two different talks, one "on afterimages" before the Anatomy and Physiology Section and another "on the physical causes of harmony and disharmony" before the Physics Section.[81]

In late September Bonn again solicited his opinion on the matter of his replacement there. He supported Weber as the nominee for the professorship of anatomy (now separated from physiology) and director of the Anatomical Museum, as he had done once before. As for physiology, he believed that Eduard

Pflüger, a student of Müller's and an assistant of du Bois–Reymond's, was "by far the most-talented and promising of the younger, pure physiologists" and a practitioner of physical physiology. While Pflüger got the position for physiology, the sixty-three-year-old Weber did not get it for anatomy; instead, it went to Max Schultze, the associate professor of anatomy and prosector at Halle. Both Pflüger and Schultze became friendly colleagues of Helmholtz's. Pflüger held the professorship of physiology there for the next half century and became one of the outstanding physiologists of his generation. Du Bois–Reymond pridefully noted in the late autumn of 1858 how physical physiology had overcome many of its opponents. He now held the chair in Berlin, Brücke and Ludwig the chairs in Vienna, Helmholtz the chair in Heidelberg, and (presumably) Pflüger the chair in Bonn and Rudolf Heidenhain the chair in Breslau. He told Ludwig there was no shortage of people in Berlin who sought to ascribe the failed negotiations with Helmholtz "not to the ineptitude of the ministry's negotiator, but rather to the lack of moral principles of such men of science who imagine that organisms are composed of atoms directed by central forces."[82]

The Turning Point

Heidelberg on the Rise

The city of Heidelberg lies nestled in a valley between the Neckar River and the Königstuhl hill, tucked into the southwestern corner of Germany within the Rhine's broad plain. By midcentury Heidelberg's population (largely Protestant) stood at about fifteen thousand. One of its two principal attractions (and one of Europe's great sights) was its restored baroque castle on the hill that rises behind the town (see fig. 9.1). When the Yale chemist Benjamin Silliman visited there (in 1851) he was enraptured by Heidelberg Castle, the beautiful views from the forested hills, the cultivated fields, the Neckar, the Rhine Valley, the cleanliness of the "handsome" city itself, and, on the north shore, opposite the university and the town, the famous Philosopher's Walk. The city appealed especially to enthusiastic walkers like Helmholtz, who enjoyed the nearby hills, forests, and river. For him such walks were partly physical exercise and partly an opportunity to think about scientific problems; he supposedly got his best ideas while walking, then later worked them out in detail at his desk. As with all poets and scientists, he said, his inspiration did not come from merely working, though that was certainly a necessity; instead, ideas sometimes sprang forth unexpectedly, as if

Fig. 9.1 A view of Heidelberg and the Heidelberg Castle. Probably around 1850. Theodor Verhas, *Heidelberg von der Schloss-Terrasse*, drawing, Handschriftenabteilung, Universitätsbibliothek Heidelberg II, 48. Courtesy of Kunsthistorisches Institut der Universität Heidelberg.

from nowhere, and for this one needed to refresh the mind from time to time through walks in peaceful surroundings.[1] Heidelberg was his kind of place.

Heidelberg's other main attraction, and its predominant institution, was the university. Founded in 1386, it was the third-oldest in the German-speaking world, though it had been nearly destroyed during the Napoleonic wars. When the Holy Roman Empire was dissolved in 1806, Baden enlarged its territory and Grand Duke Karl Frederick saved and took control of the university: he claimed it (and another in Freiburg) for Baden; renamed it the Ruperto Carola University (i.e., jointly after its founder, Ruprecht I, and himself), reorganized and refinanced it, and assumed the formal title of rector (letting the university select one of its members to serve as the so-called prorector for one-year terms). Governmental responsibility for the university's affairs was given to the Ministry of the Interior in Karlsruhe, Baden's capital. Though the relationship between the grand duke and his officials in Karlsruhe, on the one hand, and between him and the state's universities in Heidelberg and Freiburg, on the other, was at times problematic, Karl Frederick was liberal-minded toward the universities and solicitous of their needs. Then in 1856 Prince Frederick became Baden's new grand duke and married Princess Louisa of Prussia, whose father subsequently became the Prussian king and German emperor (William I), thus establishing a familial, royal bond between the grand duchy and Prussia.[2]

By the late 1850s, Heidelberg had become more a German-wide than a provincial university. It enrolled a large number of "foreign" students from the other German provinces as well as from beyond, especially Congress Poland and Russia. The "foreigners" far outnumbered the Badenese. During Helmholtz's time (1858–71), Heidelberg averaged about 660 students per semester, making it a medium-sized German university. The student body and the university as a whole had a decisive economic and cultural impact on the city proper, which nevertheless retained the small-town, relaxed atmosphere characteristic of the region.[3]

At midcentury, law was still the university's premier faculty. Fully one-half to two-thirds of Heidelberg's students were enrolled in it. The leading law professors included Karl Adolf von Vangerow, Robert von Mohl, and Johann Kaspar Bluntschli. Philosophy's leaders included Friedrich Schlosser (history), Ludwig Häusser (history), Georg Gottfried Gervinus (history and literary history), Eduard Zeller (philosophy, after 1862), Heinrich von Treitschke (history, after 1867), Heinrich Georg Bronn (zoology and paleontology), Reinhard Blum (mineralogy), Robert Bunsen (chemistry), and Gustav Robert Kirchhoff (physics). Medicine included such prominent names as Friedrich Arnold (anatomy), Wilhelm Delffs (medical chemistry), Nikolaus Friedreich (pathological anatomy), and Helmholtz (physiology). The state added several new professorships and new institute facilities during the 1850s and 1860s, but often enough (especially as concerned medicine) it did so against the wishes of the faculty. In many ways, it was the ministry in Karlsruhe, not a university faculty, that served as the dynamic force behind change at Heidelberg. It filled many chairs with outsiders, often first-class scientists and scholars, paying top dollar to attract them, and built them modern institutes in which to teach and do research.[4] The appointments of Bunsen, Kirchhoff, and Helmholtz in particular helped make the university among the best in the German-speaking world, perhaps slightly behind the big universities of Berlin, Leipzig, and Munich, but roughly equal to Bonn.

While most of the faculty was politically conservative, there was a substantial and outspoken minority who promoted liberalism and helped give Heidelberg a reputation as a center of liberalism. The contrast between these two forces became especially pronounced between 1848 and 1854, when the university experienced a chilling effect on its academic freedom: several unsalaried private lecturers lost the right to lecture (*venia legendi*). Gervinus, who had previously taught at Göttingen, where in 1837 he became a member of the Göttingen Seven and was well known for his strong liberal views, not only lost his right to teach but was tried for high treason. The philosopher Kuno Fischer lost his right because of his "allegedly pantheistic and Spinzoistic viewpoint." Jacob Moleschott, a physiological chemist, philosopher, polemicist, and popularizer who openly taught materialism and atheism, received a warning that his radical views in scientific and political matters might cost him his lectureship. He anticipated the authorities by resigning in protest in 1854. Despite the governmental assaults on the freedom to teach and the numerical minority status of the liberals, throughout the 1850s and 1860s Heidelberg was widely perceived as carrying the banner of liberalism and nationalism for the German universities. Its vocal liberal leaders, including Bunsen, Kirchhoff, and Helmholtz, called for German unification in its *Kleindeutsch* (i.e., non-Austrian) format.[5]

During the 1850s and 1860s, medicine and philosophy challenged law as the leading Faculty. A series of exceptional new appointments in the natural sciences made Heidelberg an international science center. When Helmholtz arrived to take over physiology, medicine's other ordinary professors included Arnold, Friedreich, Delffs, Maximilian Josef von Chelius (surgery), and Wilhelm Lange (gynecology). In 1865 Helmholtz helped win the appointment of Karl Otto Weber, an old friend and colleague from Bonn, to the ordinary professorship in surgery (which still included ophthalmology). Yet even as the quality of the Medical Faculty improved, its student enrollments actually decreased.[6]

At Heidelberg, as elsewhere, the natural sciences had traditionally played a subordinate role to medicine: they served in the first instance to give instruction in the basic sciences to medical students. This relationship changed after the revolutions of 1848, when Baden, hoping to avoid or at least mitigate the sorts of famines that contributed to social unrest, made a fundamental policy decision to give greater support to the natural sciences as a means of developing its substantial agricultural and tiny industrial sectors. It reformed medicine at Heidelberg and sought to rival Prussia and other German states in cultural matters. To do so, it had to attract high-quality scientists and provide them with the institutional and financial means they needed to pursue their research.[7]

The appointments of Bunsen in chemistry, Kirchhoff in physics, and Helmholtz in physiology served precisely these ends. Bunsen, an inorganic chemist, came from Breslau in 1852 as the new ordinary professor of chemistry. Baden made him (after Vangerow) Heidelberg's second-highest-paid professor and built him an entirely new chemical institute, which, when it opened its doors in 1855, was Germany's most modern such institute. He was an excellent lecturer who attracted many introductory students, but he had few assistants or advanced students. However, he also attracted quite a few English students or colleagues, including Henry Enfield Roscoe and John Tyndall, who came to study or work with him; thus he helped build friendly relations between Heidelberg and British scientists.[8]

Kirchhoff, a specialist in mechanics and electromagnetism, exemplified a new type of physicist—the theoretical physicist—that was now emerging on the German scene. Thanks largely to Bunsen's efforts, Baden attracted him to Heidelberg in 1854 as its new ordinary professor of physics. Kirchhoff's twenty-year tenure there proved to be the most creative period of his career. Between 1859 and 1862, he and Bunsen collaborated in the discovery and development of the new field of spectrum analysis. In 1859 Kirchhoff explained the Fraunhofer (or dark) lines in the solar spectrum, thereby devising a means for understanding the chemical composition of both celestial and terrestrial phenomena. During the next three years, Bunsen used spectrum analysis to discover two new elements (cesium and rubidium).[9]

In effect, Bunsen and Kirchhoff launched a new scientific discipline, at first variously known as celestial chemistry, solar chemistry, the new astronomy, or physical astronomy and, eventually, as astrophysics. Helmholtz thought spectrum analysis was the "most brilliant discovery of the last few years," though not quite everyone agreed it was Bunsen and Kirchhoff's discovery. When he told Thomson of Kirchhoff's research on the solar spectrum, Thomson replied that he had just been teaching about the known components (other than sodium) of the sun's atmosphere and that he had read aloud before his class the relevant parts of Helmholtz's letter concerning sodium. He also made a priority claim for his friend Stokes as against Kirchhoff. Yet Thomson's claim on Stokes's behalf did not even convince Stokes himself, who thought Thomson had jumped to conclusions that he (Stokes) had never reached. Bunsen and Kirchhoff's work, along with the latter's subsequent law of heat and light radiation and emission, enormously enhanced Heidelberg's (and their own) international reputations. Helmholtz thought it was probably no accident that spectrum analysis had been conceived in the hills of Heidelberg. Together with Helmholtz, Bunsen and Kirchhoff provided Heidelberg with its period of greatest fame and allowed it to display science as a cultural ornament.[10]

The enhancement of the natural sciences and medicine in Heidelberg could also be seen in the founding (October 1856) of the Natural History and Medical Association (Naturhistorisch-medicinischer Verein), a group consisting of local ordinary professors of natural science and practicing physicians. The association served as a cultural center for local medical and scientific figures, who lectured on a variety of topics, the lectures later being published in its transactions, *Verhandlungen des naturhistorisch-medicinischen Vereins zu Heidelberg.* Helmholtz joined as soon as he came to Heidelberg. He became a regular, very active participant in the association's biweekly Friday sessions. He lectured there several times each year, giving some thirty lectures all told. In December 1858 he was elected its director, an office that he held until he left Heidelberg. The association served him professionally as well as socially: he used it as a forum for testing some of his new, unfinished ideas or results. He especially liked its broad orientation and later came to feel that similar associations in Berlin lacked a comparable function. The Heidelberg association kept its participants apprised of the latest developments in natural science.[11]

The city and the university were inseparable, and they enhanced each other's attractiveness. At midcareer, Heidelberg gave Helmholtz the sort of peaceful yet stimulating environment that Potsdam had once given him. He had every reason to stay, though he did think Heidelberg's music scene was unimpressive and certainly not as good as Bonn's. At Easter of 1859, Rudolph Wagner sought to make him his successor at Göttingen. He thought that if he could get Helmholtz (and Rudolf Leuckart), Göttingen would become the leading German institution in physiology, zoology, and comparative anatomy. From his sickbed, he opened negotiations with Helmholtz and persuaded the university authorities to make a formal offer. But Helmholtz quickly declined Göttingen's offer. Even so, late in 1859 the Physical Science Class (physikalische Klasse) of the Royal Society of Sciences (Königliche Sozietät der Wissenschaften zu Göttingen), Göttingen, made him a corresponding member. It characterized him as being "indisputably amongst Germany's most outstanding physiologists." Like the appointment the previous year as a member of the Halle Natural Research Society (Naturforschende Gesellschaft), Göttingen's served as a useful piece of career enhancement.[12]

An Old Institute and a Young Assistant

Although Helmholtz got a far warmer welcome from his colleagues in Heidelberg than he had received in Bonn, in a sense Baden treated him little differently than Prussia had: he had to wait a long time (five years) before getting his promised new institute. This also meant that he lived in a residence apart from his working quarters. In the autumn of 1858, he and Olga, their children Käthe and Richard, and his sister-in-law Betty Johannes took a flat in a new, three-story building on the town's main promenade. Olga's health remained fragile: although the pains in her breasts and throat lessened at times, and although she and Hermann reveled in wandering "in the wonderful mountains," she continued to be in ill health and led a curtailed social life.[13]

Helmholtz's institute was located in a three-story, eighteenth-century building known as The Giant's House (Haus "zum Riesen") on Heidelberg's main street, almost directly across from Bunsen's chemistry building. It housed the anatomy, physiology, zoology, and physical institutes (better: "cabinets"). His physiology institute was on the first floor, Kirchhoff's physics institute on the second. Bunsen, Kirchhoff, and Helmholtz came into close, daily contact with one another. During the next five years, Helmholtz's institute remained provisional: his facilities consisted of a lecture hall, which also served as a laboratory and an instrument-storage and work area; a small room for chemical work; two small cabinets to hold animals; holding areas for dogs and other animals; and two frog basins. His annual budget amounted to 1,050 guldens (600 talers). Along with his salary of 3,600 guldens (2,057 talers), the institute as a whole cost 4,250 guldens (2,429 talers) to support annually. The budget allowed him to purchase at least some new instruments and apparatus, such as the optical apparatus and instruments he ordered from Karl Steinheil in Munich.[14] Nonetheless, the limited facilities were not much better than those he had had at Bonn; here, too, laboratory limitations helped set the agenda for his research during the next five years.

In January 1859, after he had gained a firsthand feel for the institutional situation at Heidelberg, Helmholtz submitted his own plans for a new institute. Given Baden's new medical training requirements, his institute enrolled about thirty laboratory students per semester. His plans called for, inter alia, an auditorium, a student laboratory, a smaller room for a limited number of students to do precision-instrument physical work, a director's office and workroom, a residence for the assistant and another for a married custodian, and, he hoped, a director's residence as well. He offered to pay rent for the latter, which would, he said, generally make it easier to do his own work.[15] He then waited for Karlsruhe to assess his plans and those of related sciences at Heidelberg.

He needed an assistant to help run the institute. About a year earlier, in February 1858, during his visit to Heidelberg while considering Baden's offer, he had met a young physiologist named Wilhelm Wundt, who asked to become his assistant. Wundt had strong roots in Heidelberg: he had studied medicine there as well as at Tübingen, had done research in Bunsen's laboratory, and was the nephew of Friedrich Arnold. By 1856 he had decided to become a research physiologist, and so he went to study with Müller and du Bois–Reymond in Berlin. A year later he habilitated at Heidelberg, where he offered lectures on experimental physiology.[16]

Wundt made a good first impression on Helmholtz, as did his book on muscular motion. Helmholtz declared him to be "completely reasonable and talented." But before hiring him, he wanted du Bois–Reymond's opinion. The latter judged Wundt very positively, but he also detected shortcomings in him qua experimentalist and said he would hire him only "faute de mieux," since Wundt's abilities did not complement his own, though that should not constrain Helmholtz, since Helmholtz was "at home everywhere." Helmholtz regretted that he could offer Wundt only a salary of 300 guldens (about 171 talers) annually—a mere one-twelfth of his own salary—though Wundt would have the right to use the institute's equipment for his own research. Wundt's principal responsibility, Helmholtz explained, would be to supervise the students' physiological exercises; this required two to three hours per week. He also needed to be present daily to advise students or supervise those who wanted to use the laboratory. Helmholtz imagined that Wundt could probably use most of his laboratory time for his own purposes. He expected that he himself would only occasionally come into the laboratory to speak to students and that he would only occasionally require Wundt's help in giving lecture demonstrations. Wundt did not even need to attend the lectures, though Helmholtz hoped he would give the lectures on microscopic anatomy, "because that would very much ease the effectiveness of the laboratory." He continued, revealingly, "I myself do not want to give these lectures, because, when doing extended microscopic work, I easily get headaches, and so I'm not exactly up-to-date with all the developments in histology, as it is necessary to be in giving a course of lectures on it." If Wundt wanted the position under these conditions, Helmholtz would recommend him to the authorities.[17] Wundt accepted.

Wundt assumed his position in October 1858 and remained as Helmholtz's assistant until March 1865. The two shared several research interests (especially nerve and sensory physiology), but they never became true or equal colleagues, let alone friends; Wundt was very much the junior figure in this relationship, and though Helmholtz treated him cordially and respectfully, the relationship never became close or warm. Wundt admired Helmholtz enormously, but they rarely if ever had substantive scientific discussions, not least, according to Wundt, because Helmholtz was so taciturn. He never told Wundt what he was working on. Moreover, as if to confirm du Bois–Reymond's judgment, Wundt lacked precisely the sort of experimental skills in microscopy that Helmholtz wanted in an assistant. What he needed was someone who could supervise the flood of medical students, who were required to spend one semester doing exercises in a physiology laboratory to be admitted to the state medical examination. Being Helmholtz's "assistant," therefore, largely meant supervising the medical students seeking to meet this requirement.[18]

Helmholtz taught only one course—physiology—that all medical students were required to take. He and Wundt faced problems aplenty in getting his laboratory established during the first semester, especially since it was "only a modestly adequate locale" and the instrument collection was also equally "very modest." Even with Wundt's assistance, Helmholtz found the load of medical students burdensome. There were twenty-six students, each of whom had to take an examination (though Helmholtz hoped they would not be overeager to do so). There were also three students interested in "pure science." Helmholtz thought this first semester would be his worst. Moreover, he judged Heidelberg students to be particularly poor at mathematics and uninterested in it. He attributed this to Hegel's bad influence when he had been at Heidelberg (1816–18)![19]

On the other hand, Helmholtz's European-wide reputation and Heidelberg's pleasant location enabled him to attract a few research students, something he had been unable to do in distant Königsberg or during his brief stint in Bonn. He attracted the physiologist Hugo Kronecker, the ophthalmologist Hermann Knapp, Gustav von Piotrowski, and the histologist (and future pioneer of the use of photography in biological research) Gustav Fritsch. Yet few if any of these were exclusively his students. Nearly all had previously worked with others, including Ludwig, du Bois–Reymond, and Brücke, or later worked with them; several also came to work with Bunsen or Kirchhoff. For most, contact with Helmholtz almost certainly meant not so much hearing him lecture as the opportunity to get direct, individual advice and, most important of all, to see the example he set as a researcher. Moreover, he could offer them a place of publication for their work. Following Müller's death in 1858, du Bois–Reymond assumed the editorship of the *Archiv für Anatomie, Physiologie und wissenschaftliche Medizin*, and he asked that Helmholtz and his students send him their manuscripts, declaring that there would be no difficulties in getting them published.[20]

Heidelberg at midcentury also became a mecca for Russian science students, many of whom came to study with one or all of the trio Bunsen, Kirchhoff, and Helmholtz. These students included the ophthalmologist Eduard Andreevitsch Junge, the pharmacologist I. M. Dogel, and the Moscow physicist Constantin Raczinsii. Junge, who came to do ophthalmological research on blood pressure as measured in the eye, was especially grateful for the support and guidance Helmholtz gave him. Ivan Mikhailovich Sechenov was one of Helmholtz's first and more advanced Russian students: he later became the intellectual leader of Russian physiology, known especially for his work on reflexes and neurology. After working with Müller, Magnus, du Bois–Reymond, and others in Berlin, as well as Ludwig in Leipzig, he went to work with both Helmholtz and Bunsen for six months in 1859. He thought Helmholtz such "a great physiologist in the eyes of the whole world" that he "went with trepidation, carrying in my head the whole plan of the conversation." He proposed four research topics, and Helmholtz approved one of them (on the fluorescence of refracting media in the eye). Helmholtz let him work in his laboratory and arranged for him to work with Bunsen. Sechenov, however, felt so scientifically ill prepared that he failed to get close to Helmholtz, and he "always remained quiet in his presence." He found Helmholtz to be a "quiet figure with thoughtful eyes [who] breathed a certain peace, as if he were not from this world." Although he had little direct involvement with Helmholtz, he nonetheless learned (from Ludwig) to his surprise that Helmholtz said he was pleased with him.[21]

Of Air Vibrations and Vowels

Between 1842 and 1859, Helmholtz published no fewer than sixty-three items; in quantitative terms alone, this would have made an outstanding, career-long research record for any scientist. Yet he was only thirty-eight-years old, and his best, most productive years still lay ahead of him. By the mid-1850s, his general field of research had become sensory physiology—as the publications of the first two parts (1856 and 1860) of the *Handbuch* and his results on combination tones show—but he had also published his first important work on hydrodynamics. He was planning the third and final part (1866) of the *Handbuch*, as well as a volume on the physiological basis of music (1863). He achieved all this despite the move from Bonn to Heidelberg and daily concern about Olga's fragile health.

It was this larger research framework that shaped his relatively minor studies on air vibrations in opened-ended tubes and on vowels. In March 1859, Borchardt, the editor of the *Journal für reine und angewandte Mathematik*, decided immediately upon receipt to publish Helmholtz's manuscript on the theory of air vibrations in open-ended tubes; he thought it an "extensive and valuable contribution to the *Journal*" and intended to make it the lead piece in the next issue. At the same time, Helmholtz also presented a shorter version for the *Heidelberger Jahrbücher der Literatur*.[22] It was a scientific theory of potential interest to students of the fine arts as well as mathematics and the physical sciences.

Helmholtz was also busy preparing and giving two extraordinary lectures. One was at the Karlsruhe Museum Lecture Hall (Museumssaal) and was part of a series of lectures instituted by the Grand Duke of Baden. His topic was musical tones and sensations of tone; it was probably a version of his talk on a nearly identical topic of 1857. The other was on the tonal qualities of vowels. He gave it in Munich, and the occasion was his appointment in 1858 as a Foreign Member of the Bavarian Academy of Sciences (Bayerische Akademie der Wissenschaften), which was then also celebrating its centennial. Ludwig I appeared at the first session, though Prince Max Luitpold represented him thereafter. For the festivities, Helmholtz joined a company of nearly one hundred members. While there, he worried about Käthe's and Olga's health and expected that Käthe's current illness would further strain Olga's own health.[23]

When both improved, he decided to extend his stay in Munich, and he went to see Terence's *The Brothers*. He also went, with Wilhelm Eisenlohr, the professor of physics at the Karlsruhe Polytechnic, and Philip von Jolly, the professor of physics at Munich, to visit Wilhelm von Kaulbach's atelier. They found him working on his *Schlacht bei Salamis*, an early version of which he displayed at the academy's centennial. Helmholtz thought Kaulbach quite a talented artist with broad interests in everything related to art. Jolly invited Helmholtz to his home for dinner, and along the way they looked at Ferdinand von Miller's statue

Bavaria. The next morning, following "divine service with a fully stout-hearted sermon in the Protestant church," Ludwig I showed up again, this time to meet some of his academicians, including Helmholtz, personally. "This man is pure caricature," he wrote Olga; he painted an unflattering picture of the monarch's dress, face, gait, and forthcoming marriage to a seventeen-year-old girl. Ludwig, for his part, mistook Helmholtz for a Dutchman, perhaps because Helmholtz may have been wearing the medal (the Knight's Cross of the Royal Dutch Order of the Lion) that the Dutch had given him the previous year for his invention of the ophthalmoscope. Then came a speech on the academy's history by an orientalist who rabidly attacked the Jesuits; Helmholtz could not believe his ears. Perhaps he got some relief from the Bavarian beer, which he praised and found better "by far" than foreign beer. He spent an "amusing" time with Christian F. Schönbein, the professor of chemistry and physics at Basel; Theodor Bischoff; Liebig; and Kaulbach. He and Eisenlohr visited Steinheil's optical workshop. Later he went to dine with King Maximilian II, with whom he had "a very long and extensive audience." The king was friendly and well-spoken, but he had apparently inherited his father's (Ludwig I's) "bad malnutrition." The king hoped Helmholtz could give him some insight into the auditorium's architecture, but Helmholtz had little to offer. Afterward he went to the theater, where he saw a production of Sophocles's *Oedipus at Colonus*, with music by Felix Mendelssohn, though he found the music less flowing than that in a recent production of Sophocles's *Antigone*. The next evening there was a big party and meal at the City Hall. The king was there, and a select group of Munich savants surrounded him at his table. The king asked questions of various scientists and scholars, which promptly turned into a set of minilectures: the Egyptologist Richard Lepsius on Egypt, Liebig on the "non-nutritiousness of beer," and Helmholtz on tone qualities.[24]

A few days later he read his paper on "the tonal qualities of vowels" before the academy. Like Brücke and Donders, Helmholtz had for quite some time been concerned with analyzing the production of human speech sounds. Building on his results for combination tones, and emphasizing the distinction between the sensations experienced in the ear by its nerves and the psychological image produced in the mind, he defined a tone as any "simple sensation as it is produced by a simple, pendulum-like movement of air." The ear, he emphasized again, was a harmonic analyzer. Here and in a subsequent piece for the Natural History and Medical Association, however, he analyzed the vowels of human speech by means of tuning forks and specially made glass resonators used for distinguishing between combination tones and upper partials. The mouth, he argued, was a cavity that acted like a variable resonator. Magnus, for one, was much impressed with Helmholtz's study, and he intended to visit him in Heidelberg to see his machine for producing vowel sounds.[25]

Olga's Death: Loss, Grief, Depression, and Work

In the spring of 1859, Thomson invited Helmholtz to attend the BAAS meeting in Aberdeen in September. He sought to entice him by offering him private quarters with a relative, by noting that Queen Victoria and Prince Consort Albert would be in attendance (with the prince serving as the BAAS's president), and by inviting Helmholtz to join him on the Isle of Arran. Thomson had not seen Helmholtz for "a long time," and they had much to talk about. In the meantime, he had studied Helmholtz's paper on rotary motion in fluids "with great interest" and had made progress on improving resistance standards and on his marine and land galvanometers.[26]

Helmholtz, however, had several reasons for delaying a positive reply to Thomson. For one, his father was dying. He traveled overnight to Potsdam—and got a headache and nosebleeds for his effort—but still arrived several hours after his father's death in the early morning of 4 June 1859. (Both of his parents died of what he referred to as "softening of the brain.") Relatives and old friends, including Betty Johannes and Wilhelm Puhlmann, paid their condolences. The Wilkens family invited the Helmholtz children (Hermann, Marie, Julie, and Otto) for lunch, and afterward the four siblings walked through Sanssouci. Hermann invited Julie, who had lived with their father in Potsdam, to stay with him in Heidelberg if she had no alternative. (She was not penniless, since Ferdinand left his four children a total estate of 3,566 talers in assets.) Before returning to Heidelberg, Helmholtz planned to visit Dahlem and Berlin for a day each and to stay overnight in Frankfurt an der Oder so as to avoid more migraines. He asked Olga to tell Wundt that he would recommence his lectures in about two weeks and that in the meantime Wundt should continue supervising the laboratory work.[27]

The news of Ferdinand's death left Immanuel Herrmann Fichte shaken and filled with regret that since leaving Berlin as a young man he had never again visited his old and dearest friend. So he decided to visit Helmholtz, whom he had never met, in Heidelberg; but when he arrived, he learned that Helmholtz had not yet returned from Potsdam. He left behind a condolence letter for Helmholtz, calling him "the fortunate son of a less-fortunate father," and said it was his "duty as a son to establish a memorial to him" by editing and publishing Ferdinand's collected essays. "Your own famous name," he continued, "makes this easy for you. And the feeling that you have brilliantly attained what your father—despite his deep learning and a not limited intellectual capacity, hindered by a sick body and demanding professional activity—did not so completely succeed at will let this seem to you, along with your filial piety, like a holy duty." Helmholtz did no such thing. But nine months later, in mid-April 1860, he visited Fichte in Tübingen. Fichte liked Helmholtz's "clear, reasonable nature," and he again asked Helmholtz to edit the volume of Ferdinand's essays, but Helmholtz again declined to do so.[28] That was their first and last meeting.

Helmholtz could also not think of traveling to Scotland that autumn because of the outbreak of war in Europe, namely, the struggle for Italian independence that simultaneously rekindled German national sentiments. In the spring of 1859, war broke out between Austria, on the one side, and France and Italy, on the other, over control of Lombardy-Venetia, which, as a result of the accords reached at the Congress of Vienna, belonged to Austria. The southern Germans, including the Badenese, strongly supported the Austrians, and to a certain extent so did the northern Germans. Anti-French feeling was strong throughout the German lands, not least since a war led by Napoleon III served as a vivid reminder of Napoleon I's treatment of the Germans earlier in the century. The war in northern Italy was too close to Baden for its residents to feel comfortable about leaving home. Even the armistice that was reached in mid-July did not allow Helmholtz to think of traveling too far from home that summer. While Thomson appreciated the situation, he again invited both Hermann and Olga to come to Scotland. A month later he was still hoping that the quieter political situation would allow the Helmholtzes to visit.[29]

If Ferdinand's death and the threat of war were not themselves enough to stop Helmholtz from visiting Scotland, then Olga's poor and declining health was. Helmholtz told Thomson, "I can't leave her for too long nor go too far away, as originally planned." He thought he might have to take her to the southern Alps for convalescence next winter. He politely excused himself for not being able to visit Scotland. "I had been very happy about making this trip and the prospect of talking about lots of things with you, and I'm still hoping to come to England in the not-too-distant future and see you in Glasgow." He hoped this might occur during next year's Easter holidays. The Thomsons were disappointed. He told much the same thing to Gustav Michaelis, a mathematician at Bonn, about Olga's poor health and his fear of traveling too long and too far away. He did, however, intend to make a short trip to Switzerland.[30]

So instead of taking Scotland's mountain air, that September he took Switzerland's and Italy's. He traveled without Olga and mostly alone. He and Friedreich climbed the Rigi. They met up with several colleagues as well as with Gisela von Arnim, the dramatist and daughter of Achim and Bettina von Arnim. He and his traveling companions viewed monuments and pictures that brought to mind "[William] Tell's shot at the apple," and then walked through the valleys that Tell had once traversed and to Tell's birthplace. After visiting Chur and St. Moritz, he went to Pontresina to take the spa, because he was much concerned about his own health. He and Friedreich had visited the Wallensee ("small but very romantic, of very beautifully clear color"). He climbed to the top of Pontresina (some six thousand feet); it was cold, and the air was "splendid." He liked the locals: "The people here are very solid, friendly, instructed, and not yet focused on tricking strangers." He was in and around glaciers. "These ice formations are always magnificent and interesting, no matter how often one sees them," he reported. It was a good day's work for a tourist to climb the Rigi, and he thought the fitness training he had received by climbing Heidelberg's lower mountains helped him greatly: "I arrived at the top very comfortably and quickly enough. But today I was gasping as heavily as a dog who has just run a mile." The next day they were going to walk over to Italy. He asked Olga for photographs of herself and the children, and in exchange he sent her two little flowers as a greeting. In Italy, he visited Milan, Lake Como, Lugano, and Bellinzona. The trip did him good: he felt "quite strengthened and refreshed." But he learned from Olga's letter that her own health was still not good.[31]

In fact, shortly after his return in late September, her illness worsened dramatically. Early that month a serious catarrhal epidemic had struck the Heidelberg region; for those like Olga who already suffered from chronic respiratory problems, this was especially dangerous. She was struck by the catarrh, and the pains in her breasts worsened noticeably. Helmholtz wrote Thomson: "She is constantly in an extremely wretched condition, one that has already given me the greatest anxiety about the immediate future." He was so preoccupied with her illness that he had neither the time nor the peace of mind to respond promptly to Thomson's efforts concerning an essay by Stokes. Helmholtz was doing some work that involved using Stokes's theory of the pendulum in air; it caused him "rather a lot of headaches." He had already sent Thomson his paper on vowel sounds and another on the mathematics of sound motion in open-ended tubes, and Thomson had asked permission to have the former translated into English. Helmholtz authorized him to do this. He also told Thomson that that the second part of his *Handbuch* was in press and that after its appearance he intended, time permitting, to write on physiological acoustics.[32] Work helped keep his mind in a positive state, and that became his salvation.

The moves to the warmer, milder climates of Bonn and Heidelberg had not helped Olga's health much, if at all. Since at least early December, Helmholtz surmised, she had been dying. Their friends may have sensed this, too, for during her sixteen months in Heidelberg, they saw only a shadow of her former self. During her final year, she gradually lost interest in life, her sister claimed, and Helmholtz became a lonely husband; he needed her companionship and conviviality. In all that counted in his life beyond science, she was his full partner. She died on 28 December 1859, "clear, strong, and unpretentious, just as she had lived," with Helmholtz "at her side, without trepidation, and as always in life aiming at what is highest." She left him with their two small children (nine-year-old Käthe and seven-year-old Richard) to raise and a mother-in-law who, at sixty, helped as much as she could and remained with him and the children until 1865. They buried Olga on New Year's Eve on the Bergfriedhof in Heidelberg. "I enjoyed the purest and greatest happiness that marriage can offer," he wrote his friend Carl Binz, a pharmacologist at the University of Bonn. "It was too beautiful for this earth."[33]

Friends, like Heintz, the du Bois–Reymonds, and the Thomsons, sent their condolences. So, too, did Magnus, who believed that rededicating himself to science would be the only way for Helmholtz to ease his pain. Lipschitz said that, as bad as the blow was for Helmholtz, it would not stop him from living for science: "Nature has clearly ordained that you be a lamp that shows the path of science for many. In such an act you serve God; that is your whole life, and for such acts you will always have the needed strength." Like Magnus, Ludwig sought to strengthen Helmholtz's resolve to face the future. He said that the death of a loved one was "a turning point in life," and he believed that Helmholtz's having his great love torn away from him meant that he would be "all the more sympathetic for what to date has always and already" filled his life "with happiness." He predicted, "Science will give you a high and clear spirit, stability, and peace, and will transform the painful longing into a tender and blessed memory."[34] These friends saw doing science as his nature, destiny, and salvation.

By the end of the 1850s, Helmholtz's life had indeed reached a turning point, both personally and professionally. His parents and Olga were gone, and he was a widower with two young children who needed his care. He went into mourning and, for the next two months or so, depression. He suffered in body and soul. "It was a very difficult time," he told his friend Donders. "Because I couldn't work I also lost the main means of resistance against the feeling of loneliness and loss of interest in the world. So for two months I had very long days and endless nights. Since the beginning of March, I could again help myself by working." Years later, when Knapp's wife died, Helmholtz confessed to him, "I know from my own experience what it means to see one's dearest hopes in life destroyed, and to have a future before one that, for the moment, looks only like a barren desert. What kept me together then was the duty that I still had to care for my children as a legacy and a part of the very life of the deceased, and to do so in her way." Knapp managed to overcome the worst, and Helmholtz wrote him again: "Under such circumstances the compulsion to work is a real help, even if, at first, it isn't always felt as such. You know that I [too] have gone through a similar period of trial."[35]

10

The New Angel

Recovery

The late 1850s marked the onset of Helmholtz's accrual of honors and awards. Having already been honored by the Dutch, the Bavarians, and the Hanoverians (Göttingen), he was in 1860 nominated and elected a foreign corresponding member of the mathematical-natural scientific section of the Vienna Academy of Sciences, his supporters declaring that his "high services for the sciences" were well known to a wide variety of scientists, not only physicists and physiologists. The Baden authorities took notice. That spring also brought election as a Fellow (Foreign Member) of the Royal Society of London, an honor that Helmholtz regarded "as the most valuable reward and the most powerful encouragement for the pains and the continued efforts of scientific work." The following autumn, to help celebrate the fifty-year jubilee of the founding of the University of Berlin, he was one of twenty-two individuals selected to receive an honorary doctorate; Magnus played a key role here.[1] He appeared in Berlin in mid-October as one of two representatives of the University of Heidelberg: he was serving as dean of its Medical Faculty. The honorary degree was meant in part to rebuild his ties to Prussia and Berlin, whose officials may have had an eye on eventually returning him to the Prussian fold. With the deaths of his former patrons Müller (1858) and Humboldt (1859), he had arguably become the most prominent figure in German science and one of Germany's leading cultural figures.

Six months after Olga's death, he had to a considerable extent recovered from his mourning and depression. In August 1860 he visited Thomson on the Scottish Isle of Arran. The sea and the air helped his health, not least by largely ridding him of the headaches "and other little evils" that had plagued him the previous winter. He visited Walter Crum, an industrial chemist in Glasgow who was also Thomson's father-in-law, and he met William Barton Rogers, who the following year became the founding president of the Massachusetts Institute of Technology.[2] He was back.

Anna von Mohl

The principal source of his recovery was not receiving scientific honors or visiting the Scottish seaside, however; it was meeting a new woman, Anna von Mohl. Anna was born in Tübingen in 1834 into a Württemberg family of highly distinguished public officials and scholars; she was the great-granddaughter of Johann Jakob Moser, a prominent legal and constitutional theorist during the German Enlightenment. Her father, Robert von Mohl, was a well-known academic, parliamentarian, and statesman and one of the leaders of German liberalism. His three brothers also became distinguished individuals: Julius was a leading Orientalist, professor of oriental languages at the Collège de France, and president of the Société Asiatique and of the Académie des Inscriptions et Belles-Lettres; Hugo was a professor of botany at Tübingen; and Moritz was a politician and economist. The family was imbued with learning and greatly valued education. From Robert's first marriage came his eldest, Ida, who married Franz Freiherr von Schmidt-Zabiérow, an Austrian government official, and from his second (to Pauline Becher) came Anna and her three brothers.[3]

In 1847 Robert von Mohl moved to Heidelberg, where he became professor of political science. He represented the university in Baden's parliament in Karlsruhe and became a close adviser to the grand duke. His was more a practical than a scholarly figure; he was a political professor who became best known for championing the liberal notion of *Rechtsstaat* (rule by law). He advocated (for the most part) a free-market economy and called for equal representation for the lower and middle classes alike. He was active in the Frankfurt parliament of 1848, where he called for a united Germany, and he later represented Baden in Munich and, after 1874, in Berlin. He was one of the most prominent liberals in Baden, and while at Heidelberg his circle included Gervinus, Häusser, Fischer, Vangerow, Henle, Chelius, Bunsen, Kirchhoff, and Helmholtz. His wife Pauline was a woman of social grace and aristocratic disposition, and she made their home into an elegant salon, a relaxed and intellectually refined center. In time, his daughters Ida and Anna also became *salonnières*. Cultural figures of all sorts visited them, including the English chemist Henry Enfield Roscoe, who came first in 1853 (to work with Bunsen) and then regularly from 1857 to 1863.

Anna's parents paid close attention to her education. Like her mother, she was musical—she played the piano, especially favoring Beethoven—and was keenly interested in the arts. She spoke excellent French. The values she acquired at home and at school—cosmopolitanism, elegance, cultural refinement, and the importance of learning—were reinforced by a long stay in Paris and many visits there, at the home of her uncle Julius and his extraordinary wife, Mary Clarke, an Englishwoman who became one of Paris's leading *salonnières*. Mary Clarke's salon was the meeting place of the liberal-minded elite in Paris, also attracting leading English, German, Italian, and other men and women. Mary considered Ida and Anna her adopted daughters. In Paris, Anna met the geologist Elie de Beaumont, Count Casanova, Victor Cousin, Leopold von Ranke, Leopold von Buch, Lady Elgin, Florence Nightingale, François Arago, and other leading figures. In the spring of 1853, Mary took Anna to England. In addition to introducing her to a good number of Europe's leading intellectuals and liberal thinkers, Mary guided her in how to dress and arrange her coiffure and what to read and taught her proper salon etiquette. Anna was deeply affected by Mary and her home.[4] By the mid-1850s, Anna was considered one of the most eligible young women in Heidelberg, not only because of her excellent education, cultural refinement and elegance, but also because of her beauty. She and Helmholtz met while he was putting the finishing touches on part two of his *Handbuch*.

The *Handbuch* (Part Two)

Part two of the *Handbuch*, which at some 236 pages might have been better called volume two, appeared in July 1860, eight years after Helmholtz had begun systematic research for it. Its general subject was the theory of visual sensations (*Gesichtsempfindungen*). He systematically reviewed and analyzed work on nine different subtopics: the stimulation of the nerve apparatus of the organ of vision, stimulation by light, simple colors, compound colors, the intensity of light sensation, the duration of light sensation, changes of sensitiveness, contrast, and, finally, a grab bag of subjective phenomena. For all but two of these, he also tacked on a brief historical discussion.[5] With its extensive citation of the pertinent literature, part two constituted a synthesis of studies on visual sensations, and, as already noted in regard to part one, the entire work (parts one through three) was going to become the bible of physiological optics, ophthalmology, sensory physiology, and experimental psychology.

Part two provided innovation in color theory by synthesizing previous work, largely that of others; its strength lay in its theoretical explanations of various color phenomena, not in its reporting of new observational or empirical results, of which it offered relatively few, though that did not stop Helmholtz from sometimes claiming too much originality.[6] In the early 1850s, he had rejected Young's trichromatic theory of color vision; here, however, he became a convert, revising it and placing it at the center of his theoretical analysis, which synthesized a wide variety of color phenomena. Like Young, Helmholtz assumed that the eye contained three types of nerve fibers that were excited (to one degree or another) by light and thereby produced the entire range of spectral colors. Yet he considered his revised Young-Helmholtz theory—as it became known, notwithstanding the very similar viewpoint and results that Maxwell had already developed—as itself only an instance of Müller's law of specific sense energies (that is, one kind of fiber, one kind of sensation). Helmholtz certainly knew Maxwell's work on color (and color blindness) before he wrote the *Handbuch*, just as he knew of and defended Fechner's theory of afterimages (on which he had given two speeches in 1858). But both his habilitation lecture of 1852 and his Kant Memorial Address of 1855 suggest that he mainly drew on Müller's law as the foundation for his work in color theory. Still, the issue of priority here is not unambiguous.[7]

In addition to the theory of color vision, Helmholtz also used Young's hypothesis to explain other color phenomena, such as color blindness, the Purkyně shift, color harmony, and subjective colors. In analyzing subjective colors, he adopted not only physiological (i.e., mechanical) processes to explain such phenomena, but also invoked psychological judgments. In doing so, he shattered the methodological unity and simplicity that was otherwise to be found through much of part two of the *Handbuch*. He probably drew on the work of Fechner and Brücke to advance his explanation of subjective colors. That said, it was his extension of Young's hypothesis to all these other phenomena that justified what became the novel descriptor of the "Young-Helmholtz" hypothesis and that distinguished his own work from Maxwell's much more limited work on colors. Finally, he also took up the issue of explaining contrast, which likewise forced him to emphasize the role of psychological judgment. In his hands, color became a thoroughly subjective phenomenon: the study of color became the study of color vision.[8]

While discussing these major issues about color, and while employing the Young-Helmholtz hypothesis and invoking the role of psychological judgments, Helmholtz also discussed such topics as the retina, indirect vision, comparison of the color and musical scales, color mixing, pigments, complementary colors, color tables, methods for mixing colored lights, the trichromatic and dichromatic color systems, Ewald Hering's theory of colors, principles of photometry, rotating and stroboscopic discs, and color tops, to name but a few.

The publication of part two of the *Handbuch* amounted to a confirmation of Helmholtz's advancing status; a retired doctor from Potsdam said, "Helmholtz is our most active and practical master of physiology." The Senckenberg Natural Research Society (Senckenbergische Naturforschende Gesellschaft) in Frankfurt am Main said as much in 1861 when it honored him with its Sömmerring Prize in recognition of his overall work in physiology and, more particularly, for the new installment of his *Handbuch*. As for Brücke, he read part two to see what if any gleanings Helmholtz might have left behind for him. Du Bois–Reymond could only complain that every time he opened the *Handbuch* he felt angry that Helmholtz had published it in Karsten's series, because he thought it had limited his readership and because the font size was too small ("and when you write something, there's absolutely no reason to print it small"). In short, with part two Helmholtz had synthesized and ordered the literature on color theory from Newton down to his own day. He became its leading figure and gave direction to the subject (until the 1920s). And yet, ironically, he scarcely published again on color theory; thereafter the field opened up to a wide array of new but intellectually disordered studies on color vision.[9]

Courtship

By the summer of 1860, Anna von Mohl was twenty-six years old, and Helmholtz was thirty-nine. She first learned of him from a newspaper article describing his invention of the ophthalmoscope. And since the circle of Heidelberg professors, especially those with a liberal bent, was small, it is hardly surprising that they met. She already knew Bunsen, whom she visited in his laboratory and through whom she met Roscoe. They showed her some of their experiments, and she learned of Bunsen and Kirchhoff's recent discovery of spectroscopy and other news concerning science, of which she was otherwise ignorant. She also told her aunt about Helmholtz: that he was a widower, an outstanding physiologist, and an "excellent musician and a very pleasing man." "Coming from the North, he doesn't like beer, an advantage that you'll know to appreciate with me." She considered him one of Germany's leaders, and she now began to see him within her father's circle. It was their common love of music that first brought them together that summer of 1860.[10] She had her eye on him.

Just as he had, if a touch fearfully, his on her, because in early 1861 they announced their engagement. He told Thomson that in the previous winter it had become increasingly clear that his mother-in-law could not take care of his children and oversee the household for much longer and that he needed to decide soon about an alternative arrangement. His relationship with Anna developed more quickly than he had anticipated, he said, "for when love is once allowed to develop, reason doesn't afterwards ask for permission as to how fast it may

Fig. 10.1 Helmholtz in 1861.
Deutsches Museum, Munich,
Porträts-Sammlung, No. 1439/5
(= 22194).

grow." Anna was, he said, "a richly talented, relatively young (vis-à-vis me) girl, and one who is considered, I think, to be among Heidelberg's beauties. She has a very quick understanding and wit; she is very skilled in societal matters." She spoke fluent French and English, certainly better than he did, he added. "Moreover, her fashionable education hasn't affected her quiet, good, and pure nature. In short, you see, I think her an angel, just as every fiancé does of his betrothed, and so you will thus perhaps not want to believe everything that I say about her." He thought that the next time he visited Thomson and the sea—"when I must once again try to recover my health in the waves of the sea"—Anna would accompany him, and then Thomson could see for himself the truth of Helmholtz's portrayal of her. (See figs. 10.1 and 10.2.) Meeting and courting her had consumed so much of his time and thoughts lately that he had made little progress with his book on physiology and music. Moreover, he had scheduled lectures in Germany and in London on the physiological foundations of music, and instead of writing the book itself he had been composing his lectures, preparing diagrams and experimental apparatus, and so on.[11]

He also wrote to du Bois–Reymond of his engagement, expressing himself more frankly than he had to Thomson. He said that "from the very beginning" Anna had struck him "as a very quick-witted girl," though he initially did not see her all that much. After telling him of her highly cultured background and nature, he then spoke of the difference in age between them and his concerns for his children: "I must say that last summer I sought to avoid Anna von Mohl because I felt that a girl of her type would be dangerous for me. Besides, I never really imagined that, as a widower with two children, and as someone who was far beyond a youthful age, I might ask for the hand of a much younger lady who had all the qualities for playing a distinguished role in society." But as he had explained to Thomson, he realized during the past winter that his aging mother-in-

Fig. 10.2 Anna Helmholtz in 1869. Siemens Forum, Munich.

law would be incapable of running his household and caring for his children for too much longer. Anna had agreed to marriage, and he now saw the future "with a new, happy outlook." They married that spring.[12]

Anna saw things similarly. She wrote her aunt Mary that she hoped she was up to her chosen task. She explained that Helmholtz was a forty-year-old widower with two children, that he held "a beautiful position in the world of science," that he was "a charming, pleasing man, with a sweet, balanced personality, filled with *esprit*, with a physiognomy that you'll like, I'm sure, and one who has dedicated a passionate and inexplicable attachment to me, of which until a few weeks ago I myself doubted." She loved everything about him: "His tastes, his person, his mind are infinitely pleasing to me and have won me over, above all as he pleaded his cause in a very elegant manner." She at first tried to postpone making a decision; however, "Mr. Helmholtz was so unhappy, and then became so much more eloquent—and because my reasons, which at bottom were limited to my lack of courage—I finally said yes—and, Dear Aunt, did so with a full heart." She believed "that one would not find a second man like him." Her father was extremely pleased by her decision, and the two men already understood one another quite well, although the one was a natural scientist and the other "a man of State." She had only seen the children once: "They are a little pale, but very intelligent and with large black eyes. They say I'm too young to be their mother, about which they're not wrong." She could only hope to do her best, she said. Helmholtz was scheduled to go to England in March, where he would lecture at the Royal Institution. She was trying to persuade him to visit Paris afterward.[13]

That was not so simple. Anna also learned that Helmholtz disliked the French physiologist Pierre Flourens and that he thought the French physiologists were "always distinguished by their lack of sympathy for animals." Helmholtz, by contrast, was going to devote all of his future physiological studies to living human beings, suggesting that he was opposed to experiments on animals, which neither accorded with his early work on frogs nor his general opposition to the antivivisection movement. In any case, she adored him: "To see his eyes and ears, and how they work: that's what has made his great reputation." She claimed, incorrectly, if understandably for a fiancée who knew little about scientific awards and honors, that he was a member of all the German academies and their English and Dutch counterparts. "Every day I find that I have reason to be even happier about my future." She and her father posed scientific questions to Helmholtz: "We've acquired marvelous knowledge." Helmholtz's mother-in-law would go to Potsdam in April to be with her daughter (Betty Johannes), and the children would accompany her and stay at Betty's place for the summer. Anna and Hermann would be married in mid-May, at Whitsuntide. All this meant that Anna had to start their (i.e., reorganize his) household alone, which was fine with her. In August they would travel, and then they would collect the children and return to Heidelberg. In regard to Helmholtz's upcoming trip to England, she was asked to provide a picture of herself for his British friends and colleagues, who of course had yet to meet her and were keen to see "the lady who is happy enough to be Mrs. Helmholtz." She was "very proud" that everyone was filled with praise for her fiancé: "Although I've never had any doubt, some tell me that they hope I'll be worthy of his position, others that they're happy for him to have such a charming wife. In short, our self-love has much to feed on." She conceded that it was unlikely that Helmholtz would go to Paris after visiting England.[14]

Anna was awed by his knowledge of things. "He knows the reason for everything, and he loves to talk about it [the natural world], provided that the questions aren't too stupid. He has a vast terrain to cultivate in me, because certainly few people are as ignorant in natural history; but that seems to amuse him, and I assure you, my dear aunt, that our daily walks are extremely instructive." Her father was reading Helmholtz's essays on vision and the eye, and he was "filled with admiration" for them. She assured her aunt that she would not try to turn Helmholtz into "a dandy . . . but I will try a bit more to make him a man of the world." She and he agreed completely that they would not emphasize the "pleasures of the table," as did so many others in Heidelberg, or spend their money on serving salmon to their late-night guests. "They will have a cup of tea with sandwiches, a gay *salon*, the freedom to move around and say 'Hello,' and the music they want to hear."[15]

In early February, Helmholtz lectured in Karlsruhe on the conservation of force (energy). Among his listeners was Eduard Devrient, a Berlin singer, librettist, and theater historian. Twenty years before, as a student in Berlin, Helmholtz had heard him lecture there. Now the two met for the first time, with Devrient as the director of the Court Theater (*Hoftheater*) in Karlsruhe. He found Helmholtz's lecture, like similar popular lectures by Wilhelm Eisenlohr and Ferdinand Redtenbacher, to be "learned" but guilty of using "disconnected language, [and] assuming too much knowledge." He thought that all three erred in "drawing too few clear results," which would interest the layperson.[16]

Helmholtz's old friends congratulated him on his forthcoming marriage. Brücke, who had an avid interest in language, asked Helmholtz to inquire of his future brother-in-law, Julius von Mohl, about the Prix Volney—an award proposed by the Académie des Inscriptions et Belles-Lettres but given by the Institut de France for distinguished philological work—on the topic of the "general alphabet," since Brücke thought he might well submit a paper for it. Olshausen, who was now with the university division of the Prussian Ministry of Culture in Berlin, had heard "from all sides" that Anna was an "outstandingly fitting assistant." As an orientalist and philologist, Olshausen was already friends with his colleague Julius von Mohl, with whom he had in 1829 coauthored a book on Zoroastrianism. Magnus hoped that Helmholtz's new marriage would provide him with "new strength in order to delight the world" with works like those through which he had "contributed so much" to the scientific community's "enlightenment." From Britain came congratulations not only from the Thomsons and Henry Bence Jones, but also from William Benjamin Carpenter, a London physiologist and naturalist, who invited the Helmholtzes to stay with him.[17]

However, some of Helmholtz's friends had difficulty approving of his engagement and marriage a mere eighteen months after Olga's death. Betty Johannes thought these (unnamed) people did not appreciate his situation and that they wronged him. She herself recognized that Helmholtz had, in effect, already lost Olga during the final year or two of her life, and that he had the children, not to mention their aging grandmother, to think of. They needed a mother as much as he needed a wife.[18]

As Hermann and Anna passed the days before their wedding together, her respect for him only deepened—both in terms of his character and his mind. "He always sees the greater side of things," she told her aunt, and at the same time "is sweet and kindly, and scorns no one." Though her father and Helmholtz respected and befriended each other, her father could not understand Helmholtz's explanations of acoustic phenomena, in part because he had a limited understanding of music; Helmholtz, for his part, had "only a moderate interest for politics." While on a walk together, Helmholtz and Anna ran into Kirchhoff, and he asked them a question about light reflecting off the sea. The two men spent a half hour pondering the topic, while Anna, with her "paws in the water," sat there as a sudden downpour of rain came upon them. Scientists, she thought, were odd ("drole") people who represented a "side of human nature" that was new to her.[19]

Lecturing at the Royal Institution

After 1826 the Royal Institution became the central theater in Britain for popular lectures on science, thanks above all to its Friday Evening Discourses. Located in fashionable Mayfair, the Discourses attracted London society's intellectual and social elites, not least for the style and dignity of its evenings. Lecturers apprised the elites of the latest, most interesting developments in science while also serving as a sort of entertainment. At midcentury the Royal Institution was experiencing a golden age. It was led by old Michael Faraday and young John Tyndall, who since 1853 had held the Institution's chair of natural philosophy and was a manager of the Institution and its secretary, as well as by the chemist Edward Frankland. The well-attended lectures were published and distributed throughout the world. Many notable figures in mid-Victorian scientific life—not only Faraday and Tyndall, but also Charles Lyell, Richard Owen, Thomas Henry Huxley, Lyon Playfair, William Thomson, and J. Norman Lockyer, to name only some of the most prominent—delivered the lectures, as did a few select foreigners, Helmholtz among them. By the 1860s, attendance generally ranged between four hundred and five hundred ladies and gentlemen.[20]

The forces behind Helmholtz's visit were Faraday, Bence Jones, and William Benjamin Carpenter. (Carpenter was Fullerian professor of physiology at the Institution and supported the popularization of science.) Bence Jones and Faraday wanted Helmholtz to lecture on conservation of force; he agreed to give three lectures in mid-April 1861, one of which, the Friday Evening Discourse, was expected to attract as many as one thousand attendees; for this lecture the Institution paid nothing! Faraday especially wanted the Friday Discourse to be on the law of conservation of force "in its application to life." That topic, he said, "will interest us all and get us out of the metaphysical groove into which this subject has got here." The other two lectures would be more specialized, and Helmholtz would be compensated for them (£20); an audience of about three hundred was anticipated. In truth, Helmholtz preferred not to lecture on the conservation of force (energy); Bence Jones and Faraday, however, insisted, and since they were paying for his other two lectures (both on the physiological foundations of music), he was forced to agree. He took solace for this English insistence from the fact that his recent article on the motions of a violin's strings had just appeared in English.[21]

He spent around a month (mid-March to mid-April) in London, and everybody there and beyond wanted to see him. Carpenter gave an evening party for him, which Maxwell attended, as did Thomas Archer Hirst, a minor but well-connected mathematician. Like his mentor and friend Tyndall, Hirst had studied in Germany; in addition to teaching and holding various administrative appointments, he translated scientific pieces from German (and French) into English. Hirst "had long conversations" with both Helmholtz and Maxwell that evening: "The former is a little reserved, the latter talkative with a Scotch brogue."[22]

The Thomsons, for their part, wanted Helmholtz to visit them in Scotland, since an illness prevented Thomson from attending Helmholtz's talks. Helmholtz wanted to, but because his London lectures were only half finished, he could not come. He told Thomson, "In the last few years I've gotten used to moving around using Europe's railroads," yet there was still not enough time for him to travel up north to Thomson.[23] Helmholtz embodied the new and coming type of international scientist, one who traveled abroad by means of the new, rapid railroad systems or steamships to present his latest work, meet foreign colleagues and learn of their work, and generally promote his career. He was a modern professional.

Furthermore, he told the Thomsons, he planned to hear a lecture that evening by William Barton Rogers at the Institution and also to meet Rogers. Roscoe, who intended to see Helmholtz in London, also invited him to stay with him in Manchester for several days. William Sharpey, professor of anatomy and physiology at University College, London, and a secretary of the Royal Society, also looked forward to Helmholtz's London visit and wanted to know whether Helmholtz, a foreign member of the society, would also be willing to give this year's Croonian Lecture. The lectureship, he explained, was founded in 1738 and was originally meant to advance understanding of "the nature and laws of muscular motion," but Helmholtz could speak on any topic in physiology or anatomy that he cared to. (Sharpey preferred that he speak on his "experiments on the excitation of muscular and nervous excitement" or on vision, but he left the choice up to him.) The pay, he shamefacedly confessed, was "a miserable pittance, not quite three Pounds."[24] Helmholtz declined the offer, for now.

Anna, meanwhile, put their new household in order. She missed him terribly; she worried that he would not get enough rest; and she reported that a friend of hers had said that one day people would speak of Helmholtz the way they now spoke of Gutenberg and Luther. She tried to clean up his desk: "in itself no easy task, given the quantity of impediments in the form of papers." "You see, Hermann, I'm in your place, and so my better nature comes out and peace comes into my soul." She had a strong penchant, not to say obsession, for orderliness. If she had not been raised on the principle of "learned disorder," she said, she would have separated the fresh pieces of paper from those with notes on them, put his letters in a drawer, and then, "according to Miss Nightingale's principle," dusted with a damp cloth. She worried that she lacked the skills to be a good wife and mother.[25]

Helmholtz's first two lectures at the Institution were on the physiological foundations of music, while the third, the Friday evening talk, was "On the Application of the Law of the Conservation of Force to Organic Nature." Hirst attended the second lecture (on vowel sounds) and again met Helmholtz at a dinner party. The guests included Roscoe; Thomas Graham, a chemist and the Master of the Mint (1855–69); and the Carpenters. Hirst also attended Helmholtz's lecture on the conservation of force, which he found "instructive but not as interesting" as he had anticipated. Bence Jones saw it somewhat more positively. He told his good friend du Bois-Reymond: "Helmholtz did very well [but] not splendidly[;] his Friday evening was very good & I had it reported & will send you a verbatim report. His acoustic lectures were not so taking[;] he said too much that was known in the first & this filled his 2d lecture too full. He ought to have given half a dozen—Those who knew most said it was very good; but some of the inveterate grumblers amused themselves—However I do not mind a little grumbling at the Institution. Another year I expect a greater success if I can persuade him to come."[26] Perhaps the best proof that the lecture was successful is that less than two years later Helmholtz was invited back to the Institution to lecture again on the conservation of force (energy) and its applications (though this time he gave six lectures on the topic).

In his Friday Evening Discourse, he argued that the law of conservation of force was very general, that it "embraces and rules all the various branches of physics and chemistry." It was important not only for understanding "the nature of forces" but also for understanding "immediate and practical questions in the construction of machines." (He refrained from discussing this latter point, however, saying, "You will hear these results better explained by your own countrymen.") Following Rankine, he thought it might as well be called "the conservation of energy" as "the conservation of force"; this comment, if nothing else, is a mark of the law's changing nature and status in Helmholtz's thought. He explained that while "it is the nature of all inorganic forces to become exhausted by their own working, the power of the whole system in which these alterations take place is neither exhausted nor increased in quantity, but only changed in form." He said gravity was an example of this process and cited weight, velocity, elasticity, heat, and chemical fuel as examples of "motive power." He also explained that Joule had determined the mechanical equivalent of heat and that heat, as a motive power, could be converted into mechanical power, such as in a steam engine through the chemical process of burning fuel.[27]

Thus the amount of energy (or "working power") of the entire universe must be constant, he argued. Assuming Laplace's nebular hypothesis to be true—that the universe in its initial state was constituted by chaotic nebular matter spread throughout all space, which then, under gravitational forces, formed into the present planetary system—an "immensely great" amount of heat must have also been produced, and in particular the sun was heated to 28 million degrees. The sun's contraction converted matter into (mostly) heat energy. This contraction hypothesis thus led to the result that the earth was far older than many geologists, physicists, theologians, and others had believed. Recent spectroscopic analysis of the sun by Kirchhoff and Bunsen showed that it is hotter than any body on earth and that its atmosphere contains vaporous iron, among other metals. Most of the earth's heat, by contrast, is in its interior, and it cannot get to the surface (except by volcanic activity). Therefore, except for the tidal forces, all changes produced on the earth's surface are ultimately due to solar radiation. Temperature differentials on the surface are in turn responsible for creating the earth's atmosphere (including winds, vapors, clouds, and rain). In short, he argued that the earth's meteorology is essentially due to the sun's heat. Similarly, the sun causes plants to grow. It was "very remarkable and curious," he noted, that first Mayer and then himself, both physiologists, "should come to such a law." He continued: "It appears more natural, that it should be detected by natural philosophers or engineers, as it was in England; but there is, indeed, a close connection between both the fundamental questions of engineering and the fundamental questions of physiology with the conservation of force. For getting machines into motion, it is always necessary to have motive-power, either in water, fuel, or living animal matter." That was why, previously, many machine builders sought to create some sort of perpetual motion machine. They even thought that an animal's body was one such, because they failed to appreciate that the food eaten by the animal was, in fact, its fuel. There was a very strong analogy between the living body and a steam engine. Animals consume food (fuel) and breathe oxygen from the air, which is similar to the sources that give steam engines their power. An animal can do a certain amount of mechanical work, expressed either in the form of heat or as "real mechanical work" (i.e., muscular exertion). But the analogy between the body and a steam engine remained just that because, in terms of the amount of work performed, "the human body is . . . a better machine than a steam-engine[;] only its fuel is more expensive than the fuel of steam-engines." Hence Helmholtz concluded: "The laws of animal life agree with the conservation of force, at least as far as we can judge at present regarding this subject. As yet we cannot prove that the work produced by living bodies is an exact equivalent of the chemical forces which have been set into action." Still, he thought it "extremely probable that the law of the conservation of force holds good for living bodies." What was not open for discussion, he said, was that the agents or inorganic forces operating within the body did so by necessity, "and that there cannot exist any arbitrary choice in the direction of their actions." His essentially deterministic and reductionist outlook stood in opposition to the role allotted to free will in nature by his Scottish friends and colleagues (the Thomsons, Maxwell, Tait, et al.). By contrast, Tyndall, whose views and efforts were repeatedly attacked or belittled by Thomson and Tait, expressed scientific naturalist views.[28] In this regard, Helmholtz stood much closer to Tyndall than to his good friend Thomson.

A Second Marriage, a Salon, and Family Pain

By late April, Helmholtz was back in Heidelberg, where du Bois–Reymond visited him. Helmholtz awaited his close friend at the railroad station. They dined with Bunsen at du Bois–Reymond's hotel and afterward Kirchhoff joined them; they went to see Bunsen's laboratory, and Kirchhoff showed them a few experiments. They climbed up to the castle and then returned to Kirchhoff's for tea. It was there that du Bois–Reymond first met Anna. "The bride is nothing less than pretty, but very lively and *knowing*. I should think she brings as much to the party as Helmholtz," he told his wife Jeannette. He envied the close relationship that Helmholtz, Bunsen, and Kirchhoff had developed among themselves; he thought it gave them a "totally incalculable advantage." At one of their meals together, the "Fish," as he called Helmholtz, gave him his bowl of pineapples, "which he disdains however, since (as is well known) he eats no fruit." He might have mentioned, too, that Helmholtz did not smoke and indulged in only small amounts of alcohol, since he thought the latter hindered his intellectual processes, and so his ability to develop scientific ideas. In general, he was temperate in all that he did and in what he consumed. All told, Helmholtz's scientific, collegial, and environmental riches left du Bois–Reymond depressed. By contrast, Magnus, who visited Helmholtz and met Anna in May, apparently felt no envy.[29]

Hermann Helmholtz and Anna von Mohl were married on 16 May 1861 at 11:30 a.m. in a church wedding (performed by a municipal cleric) in Heidelberg. His children were not present; they remained in Dahlem for the time being.[30] Though she lost her "von," she gained one of the grandest surnames in contemporary German culture.

Anna moved into and redecorated his residence. If she did not know it already, she now learned that generally he arose early; that mornings were "his best working time"; that around 1:00 p.m. he took a break that lasted until 4:00 p.m.; and that he then went back to work, reserving the evenings for relaxation and family. In early August, they honeymooned in Switzerland and northern Italy. He wrote Roscoe: "In every way things have gone well for us so far in this first period of our marriage." They were very compatible. Anna wrote her aunt Mary that the honeymoon had been "a charming voyage" and that Hermann was the perfect traveling companion.[31]

She already had her household running smoothly. Speaking probably of his book-in-progress on physiology and music, she said, "When Hermann is finished working, unless it's on mathematics, of which I have a horror and which he adores, I try to get myself up-to-date on what he's done, and when he's finished any paragraph, he lets me read it — and I like that a lot." She said, too, that he kept his daily and personal needs simple. She turned their home into one of Heidelberg's social and intellectual centers. It became known as a place of stimulating conversation and good music, while deemphasizing food and drink; it became a salon, like the ones her mother and her aunt had created in their homes. Their regular guests included the scholars Häusser, Vangerow, Gervinus, and Zeller, as well as Bunsen and Kirchhoff. (Zeller, a learned theologian and scholar of the ancient Greeks, became good friends with Helmholtz in part because he too was one of the leaders of the back-to-Kant movement and in part because they shared similar liberal sentiments and views.)[32] Anna brought *savoir vivre* and joie de vivre to Helmholtz and his household. She thrived in their home and in her new role as Frau Professor Helmholtz.

Their temperaments were quite different, however. Anna was lively and energetic, warm-hearted, headstrong, and with a biting sense of humor that often became sarcastic and led to a reputation for arrogance; she was an excellent judge of people; she needed to create order and beauty in the household and, sometimes, beyond; and she was religious though not dogmatic.[33] Hermann, by contrast, was quiet, peaceable, and (sometimes) stolid. Their personalities complemented each other, and that perhaps helped make their marriage work well. While Hermann's marriage to Olga had been an *affaire d'amour*, his relationship with Anna was also an *affaire d'intérêt*. His letters to Olga showed a greater depth of passion beneath his cool scientific persona than those to Anna. Yet in both cases there was love: even though neither Olga nor Anna had any higher education and did not understand science, each was effectively his equal in marriage. He felt comfortable around them, as around women in general, indeed he was perhaps even more comfortable around women than around men. He had a strong sense of wanting to be married—and, after Olga's death, of needing to be married—and he had a strong sense of home and family. Marriage gave him a greater sense of stability and responsibility. It suited him.

Despite the blessings of his new marriage, the winter of 1861–62 proved to be a difficult one for Helmholtz. He suffered again enormously from migraine headaches, to the point where he was sometimes unable to work for days on end. (He found it "very useful" to rub veratrine salve into the sides of his aching eyes.) However, he managed to complete the manuscript of his book on physiology and sound. He also received a visit from Alexander Crum Brown, an organic chemist from Edinburgh, whom he quite liked and who was involved in a controversy with Brewster, Carpenter, and Wheatstone over whether a set of drawings by a mid-seventeenth-century Italian artist that Crum Brown had found in a Lille museum were stereoscopic or not. Crum Brown sent a copy of the drawings to Helmholtz for his opinion.[34]

For much of March 1862, however, what preoccupied Helmholtz was less his own health than family matters. On 3 March 1862, less than ten months after marrying, Anna gave birth to their first child, Robert Julius Helmholtz. The long and difficult birth for both mother and son—it took forceps to get him out—was a portent of Robert's future. From the start the couple was much worried about their newborn's health. Anna herself suffered a life-threatening inflammation. They thought they might lose Robert, but two weeks after his birth, Helmholtz, with great relief, pronounced his son as healthy as could be. "Since father and mother have thick heads," he joked to du Bois–Reymond, "he must, according to Darwin, have a still thicker one," and he asked him to relay all this privately at the next academy meeting "to my friends Magnus, Dove, Olshausen, Weierstrass, [and] Borchardt." But his sense of relief proved to be slightly premature; it took another month for Robert to recover; by May, both mother and son had recovered fully. It appeared that the worst was over and that they could think of having another child. In August and September, Helmholtz went first to Bad Kissingen for a short-term health treatment and then mountain climbing. His headaches the previous winter had been so bad that his doctors insisted he go there to drink the local mineral water. He would have preferred the sea, but he later conceded that the treatment had helped him get through the following winter. This was Hermann and Anna's first separation since their wedding, and Anna felt melancholic and lonesome without him. In September, the two went together to the mountains around Salzburg. That month, too, he planned to visit Brücke near Vienna. In October the congress of ophthalmology met in Vienna and toasted him in recognition of his contributions to the field, even though he was not there.[35]

During the winter of 1863–64, when Anna was again pregnant, it became apparent that Robert was not a normal, healthy child. The Helmholtzes realized that he had serious developmental problems; over time it became clear that he suffered a congenital osteological malformation in his hip and back. (It appears that bones near the hip never fused properly.) Though Robert was mentally healthy, he could not walk or run normally. This pained Anna deeply; she believed, rightly as it turned out, that he would be disabled for life. His right knee was also particularly bad, and he was in much pain. Helmholtz said that his family was doing fine but added, "Our little Robert gives us a lot of concern because, for some time now, he has been lame in his hip, and now cannot move." He asked his friend and colleague Wilhelm Busch, a professor of surgery at the University of Bonn, for advice. After Busch came to Heidelberg and examined Robert, he had him wear a custom-made orthopedic device. He thought Robert would recover, and he advised the Helmholtzes to be patient. Anna, however, remained extremely anxious about his condition. "Naturally everything [we do] revolves entirely around his existence," she told her uncle Julius. In the course of time, the Helmholtzes turned to numerous doctors for diagnoses and treatments. Though Helmholtz certainly worried about Robert, it appears that it was Anna, the mother, who mostly cared for him on a daily basis.[36] The birth (on 24 April 1864) of their second child, Ellen, a healthy child, perhaps gave them some consolation.

Helmholtz told friends around the beginning of 1865 that during the past two months all of his four children had consecutively contracted measles and that Robert was still confined to bed. "Our family life went through a very uncomfortable period and, in the end, nothing helped." The worry particularly affected Anna, who herself had a fever. A month later things had largely improved around his household, but not for Robert, whose condition seemed worse than ever. "We've had a lot of stressful situations and illness in our family," he reported. They "longed for more peaceful and healthy times." Eight months later he told Jacques-Louis Soret, a Genevan chemist and physicist, that Robert and Ellen were well but that his son's leg was "stiff, if without pain," and he was just again beginning to get some use out of it. In the meantime, he was now getting around by supporting himself with his hands. Anna had Robert examined by three new surgeons. She wondered whether, suffering as he was, he might not be better off in "Heaven." She thanked God daily that he had blessed them, and she hated disturbing Hermann with all this trouble and worry as he sought to recover his own health. Robert's condition became better than the doctors had expected. Still, he had been in this state for eighteen months now, and as a consequence, she had aged markedly.[37]

The Helmholtzes thought Robert would not survive the winter of 1865–66, but he did, though his various medical and surgical treatments were without any positive result. While their other children remained healthy, they continued to feel "much distress with our little, lame Robert." He was lively and energetic, but, Anna said, "Frankly he can't use his sick leg at all, and he moves forward only by creaking along the ground." He was so badly disabled that even at five years of age he could not yet use crutches. Anna saw clearly what lay before them. She wrote Hermann, who had to travel to Berlin without his family: "The dear Lord has now denied the little lad a normal existence, and so he must be and remain our first concern. Neither for him nor for us will the future be any easier; that I know more and more each day, even if it also doesn't help to talk about it." Nearly six months later she told her aunt in Paris much the same thing. Yet he was a happy child, spending much of the day drawing and playing while lying on the couch. Everyone seemed to agree that there was little or nothing to be done about his medical condition. In the meantime, he had learned to read quite well, had taught himself the French letters and some numbers, and "ardently" wanted "to have lessons from a true master." People liked him and he had playmates. Helmholtz told du Bois–Reymond: "Little Robert's health is much stronger, but unfortunately he has a severe curvature of the spine, though he is an intelligent and lively lad." His condition remained more or less the same until his death at age twenty-eight.[38]

The Relations of Science

The Sciences and the Modern Nation-State

During the 1860s, Baden enacted several liberal reforms, including some affecting higher education. At Heidelberg in late 1861, a majority of the academic senate, which consisted mostly of professors of a very liberal persuasion, advocated reforming the university's governance and, among other things, giving more rights to those who were not ordinary professors; Helmholtz favored such reforms. In January 1862 the ordinary professors overwhelmingly elected him as their new prorector. His basic responsibility was to oversee the university's daily business.[1] He assumed office that April and held the position for the next two semesters. His election and subsequent prorectoral address were of a piece with larger changes in Baden.

As prorector, Helmholtz reported to the Ministry of the Interior, which in turn reported to the grand duke. From the start Helmholtz realized that the office would demand much of his time. As the former dean of the Medical Faculty at Königsberg, he had acquired some administrative experience as well as a sense of the demands on his time that went with such appointments. He enjoyed widespread trust among the more than seven hundred students enrolled at Heidelberg.[2]

Already in May, Anna reported that he was "almost never at home" because he was "either enrolling students or doing other busy work." Helmholtz himself told Thomson that nearly all his time was consumed by administrative duties. He planned to go to the sea in the autumn in the hope of getting relief from his headaches, but he was uncertain that he would have enough time to visit London and see an exhibition there because, as prorector, he could not leave Heidelberg without first getting someone to substitute for him. Then, too, he had to attend to the endless requests for letters of support for people, such as a letter for Nathanael Pringsheim for the professorship of botany at Heidelberg, which Pringsheim, who was the foremost research botanist of his day as well as an excellent teacher, and who was Jewish, did not get.[3] Even as prorector, Helmholtz's influence went only so far at the university.

He thus had little time to do scientific work that academic year; in addition, he was busy seeing his book on physiological acoustics and music through the press as well as overseeing the construction of his new physiological institute. Still, he managed to offer a new course in the winter semester 1862–63 on the general results of the natural sciences. It was, in effect, a series of popular lectures, and their preparation demanded a lot of work. He invoked the law of conservation of energy and its consequences as his guiding theme to present and connect most everything in science with the whole, including "the astronomical, geological, climatic, and meteorological phenomena, as well as the life of plants and animals," not to mention cosmological and anthropological topics. This meant that he "had naturally to do a lot of studying about" things that he "no longer knew so exactly." In this connection, he was particularly interested in Thomson's recent results on the earth's cooling and the changes in form of elastic spherical shells with reference to the earth. Most of the attendees at these lectures were not science students. Among them was Alfred Stern, who was studying history and law and who found Helmholtz's lectures "loosely structured but always stimulating and ingenious, treating a wide array of subjects, like sensations of tone, the structure of the human eye, spectrum analysis, [and] the Neanderthals." While attending them he met Aaron Bernstein, a Jewish bookseller, theologian, liberal politician, journalist, and author. Among Bernstein's many writings was an immensely popular set of books on natural science for a popular audience (*Naturwissenschaftliche Volksbücher*, first edition, 1853–57). These books in turn inspired the young Albert Einstein, who as a student at the Zurich Polytechnic, where Stern had become a professor of history, was befriended by him.[4]

In late November 1862, Helmholtz delivered his obligatory prorectoral address. His subject was the relationship of the natural sciences to all of science within the organizational framework of the German university and society. He began with a symbolic bow to Karl Frederick, "an enlightened prince" of the province, who had refounded and rebuilt the university during the Napoleonic era. He declared that Karl Frederick understood that the university's well-being constituted a key component of Baden's future well-being. Education, Helmholtz believed, led to social improvement, and as leader of the university he thought it fitting to look at the larger societal context of the sciences and their study as a whole, even as he recognized the limitations of any single individual in doing so.[5]

Helmholtz suggested that in the past half century or so, the relations of the sciences to one another had loosened, since modern scholars and scientists had become thoroughly involved in the detailed, specialist research issues of their individual disciplines. No one could possibly master anything more than a tiny portion of contemporary science and scholarship. Moreover, while some scholars vigorously and endlessly pursued empirical evidence wherever it might lead them, even "in an unknown quarter of Hungary, Spain, or Africa," others sought to systematize and compare what their colleagues had already discovered.[6]

His address thus sought to portray the Big Picture, to see the connections among the various branches of knowledge. The modern natural sciences, he said, stood accused of creating a situation of hyperspecialization and of distancing themselves from humanistic studies. Although this had once been true, he acknowledged, it was the Hegelians, not the scientists, who had been responsible for the development. (This was neither his first, nor his last, bit of anti-Hegelian rhetoric.) The separation of the disciplines could not have been due to Kantianism, he maintained, since it "stood much more (on exactly the same ground) with the natural sciences." Kant's own natural scientific work had been intimately involved with Newton's and Laplace's, and his critical philosophy aimed to explore "the sources and the justification of our knowledge." By contrast, in Hegel's "philosophy of identity," nature and human life found their origins in *Geist* and, in turn, in the human mind. Hence Hegelians believed they could gain knowledge of the world through thought alone, through a priori reasoning, without empirical investigation. Helmholtz admitted that this approach could yield some fruit with respect to understanding the "*Geisteswissenschaften*" (the human or moral sciences—that is, disciplines such as religion, law, government, language, art, and history), whose objects of study essentially developed from a psychological foundation that aimed at satisfying human intellectual or moral needs. He believed that, given the randomness and accidental character of human activity, in these fields it made sense to begin with a priori abstractions and then to connect empirical results to them. Even Hegel had some successes in this regard, at least in terms of amassing a following of like-minded philosophers and constructing a systematic philosophy with its own language.[7] That was the kindest thing he ever said about Hegel.

By contrast, he thought that in the natural sciences, where "the facts of nature were the deciding means of testing," it made no sense to begin with an a priori approach. The natural world did not follow the mind's thought processes;

this explained why Hegel's philosophical approach failed "completely." Hegelians criticized especially Newton as the great representative of an alternative philosophical approach, and they characterized natural researchers as narrow-minded. And so, Helmholtz maintained, natural researchers began to distance themselves from philosophy altogether, seeing it as "useless, indeed even condemning it as harmful day-dreaming." Hegelianism's attempt to order and direct all disciplinary knowledge had led natural researchers to discard even "the justified demands of philosophy: namely, to conduct criticism of the sources of knowledge and to establish the standards of intellectual work."[8] His arguments against Hegel in 1862 strongly resembled those he had made against Goethe in 1853.

This sharp and bitter distinction between the natural and the human sciences—one that had been under debate since the 1830s and 1840s—soon helped lead the natural sciences to make many "brilliant discoveries and applications," and so to gain the respect due them. In the human sciences, too, the search for and collection of facts soon worked "against the all-too-bold Icarus flight of speculation." But the Hegelian system had left traces of itself within the human sciences. While Helmholtz maintained that, insofar as empirical work had penetrated into the human sciences, the contrast between them and the natural sciences had receded; but there were limits to their similarities. He believed that the human and natural sciences were different kinds of intellectual (mental) work and that their individual disciplines naturally had different contents. As in his address on Goethe, he maintained that the physicist, for example, abstracted from phenomena and employed geometric and mechanical reasoning; this was something that a philologist or a jurist, for example, could or would not do. By contrast, the aesthetician or theologian, for example, might consider the natural scientist's "mechanical and materialistic explanations" to be trivial. He argued that it must ultimately be remembered that the human sciences were involved directly with the most important human interests and with the conceptual order they brought to the human world; the natural sciences remained limited to a concern with inert matter, which, while they might have practical use, apparently had "no direct interest for the development of the mind."[9]

Thus the gradual but dramatic rise of the natural sciences in the German lands after 1830 challenged the idea of *Bildung*, which in turn challenged the organizational structure of the Humboldtian-style university. And since no one individual could possibly grasp even one significant part, let alone the entire range of the sciences, Helmholtz rhetorically asked if it still made sense to retain the human and the natural sciences together in a single institution. His interest in this issue was focused on the present. The State of Baden was contemplating such organizational changes. At Tübingen in 1863 and later at Strasbourg in 1873, moreover, the Philosophical Faculty was dissolved and replaced by separate faculties for the natural and other sciences. Was the notion of the four faculties (Law, Medicine, Theology, and Philosophy), Helmholtz asked, merely a medieval organizational vestige? Why not put the Medical Faculty in with the big-city hospitals? Why not attach the natural scientists to the polytechnic schools? And why not develop special seminars and schools for theology and law? He hoped this would not in fact come to pass, for if it did, it would sunder the connections among the various sciences. Here, on the organizational and institutional plane, was his old intellectual program of the conservation of force's unifying function for all branches of physics and, potentially, all natural knowledge. It was the widespread dream of unity in the German-speaking cultural world, one that found relatively little resonance elsewhere.[10]

Helmholtz thought there were formal (structural) and material reasons for keeping the four faculties together and retaining the existing university structure. From the formal viewpoint, doing so would "maintain the healthy equilibrium of intellectual forces," since each science had its own demands on (and so strengthened) different intellectual abilities in individuals. Moreover, there were differences within each of the sciences themselves to consider: no matter how talented an individual scientist might be, there were always areas and topics that someone else could perform better. A too narrowly conceived organizational structure lessened the ability to see the larger context of knowledge and life. Specialization led "easily to self-overestimation," and the latter was "the greatest and worst enemy of all scientific activity." Every scholar and scientist, he said, must be self-critical if he is to practice scholarship and science successfully. That meant searching for and collecting facts, no matter how trivial and unworthy they may appear, and avoiding empty theorizing. For Helmholtz, science was as much about character as it was about intellect: it required strong moral as well as cognitive training.[11]

He also argued that the sciences required good order and organization: the more facts each branch of knowledge acquired, the greater became the need for order and organization. That meant, in the first instance, a need for such banal things as catalogs, dictionaries, indexes, literature reviews, annual reports, systems of natural history, and the like. He compared them, metaphorically speaking, to investment capital. Even though such printed works might appear to be boring and dry, they constituted collections of facts, and every individual fact, he said, had to be found, studied, tested, and compared with others, while the important ones had to be separated from the unimportant. To do all this required that the scientist or scholar be able to see the larger purpose and structure of his research program.[12]

But all that was only the preliminaries to science: "Science first originates when laws and their causes are revealed." The logic of science required determining the similarity of things and placing them within a general concept or, if it concerned processes or events, a law. Concepts and laws allow scientists to store, manipulate, and extend knowledge, he maintained. The quintessential difficulty of the human sciences was that of discerning similarities and forming sharp concepts. How, for example, can we discern, conceive, and measure human ambition? In such cases, scientists are dependent on "a certain psychological tact," that is, an unconscious mental conclusion. He argued that this type of induction—he called it "artistic induction"—lacked the full logical structure and the well-defined theorems of normal induction, that is, logical induction, and that it did not lead to laws without exceptions. Yet all sensory perceptions, he held, are dependent on precisely such induction, widespread in human life. Furthermore, "because of the extraordinary intricacy of influences which condition the formation of human character and of a momentary frame of mind," such induction lies at the heart of psychological processes in general. Indeed, precisely because human beings have free will—are not subject to a rigid, inflexible law of causality—there is no possibility of reducing the soul's utterances to strict laws.[13]

In making this distinction between types of induction and sciences, he drew on John Stuart Mill's *System of Logic, Ratiocinative and Inductive* (1843; first complete German translation, 1862), though he by no means agreed with all that Mill said. Helmholtz thought that human beliefs and actions in the moral and juridical realms were subject to commands, not laws, and while he thought there was a great deal of logic in the human sciences, it was of a different kind from that of the natural sciences. The human sciences did not have as high a degree of consistency and completeness as the natural sciences, and they often required inquiry into intentions.[14]

The natural sciences, by contrast, concerned rather different aspects of intellectual work. To be sure, they too involved a certain degree of "instinctive feeling for analogies and a certain artistic tact"; this was particularly important in matters of natural history, as Goethe's work in comparative anatomy showed. But in comparison to the human sciences, the natural sciences displayed much sharper and more general concepts and propositions. These latter are more fully realized in the experimental and mathematical sciences, and completely so in pure mathematics. In Helmholtz's view, the experimental and mathematical sciences were best able to turn individual observations and experiments into generally valid laws of considerable scope. It was precisely this that the human sciences could not do. In mathematics, the mind operated "in its purest and most complete form," that is, most logically. Similarly, mathematical physics and physical astronomy employed abstract concepts and logic. Newton's law of gravitation, with its manifold consequences, was, in Helmholtz's estimation, "the most impressive achievement of the logical force of the human mind that has ever been made." As for the experimental sciences, they had the distinct advantage over the observational sciences of being able to manipulate arbitrarily the conditions leading to anticipated results, and thus of formulating general laws. This was why, Helmholtz explained, the physical sciences, "after the right methods were once found," had made such rapid progress. They allowed physical scientists to look back in time to the very beginnings of the universe, to determine the very composition of the sun's atmosphere, and to understand nature's forces such that people could exploit these and "make them serviceable to our will." The physical sciences concerned laws characterized by "a completely rigorous causal connection"; they allowed "no exceptions." It was precisely this characteristic that pushed physical scientists to continue their work until such laws were found, until scientists ultimately conquered nature. Such work proceeded slowly and demanded "great obstinacy and caution." Only rarely did "rapid intellectual insight" occur. Although he held that the mathematical dimensions of the natural sciences had progressed further than the other sciences had, he also emphasized that he did not mean to deprecate the human sciences, since, dealing as they do with man, his mind, his feelings, and his actions, they "have the higher and more difficult task." Still, he thought it was important that they not forget the methodological and substantive accomplishments of the more advanced formal sciences. The natural sciences could teach the human sciences respect for the facts and the search for causal connections among them.[15]

Helmholtz then turned to the applications of the sciences and, as he saw it, to the larger purpose of the sciences as a whole and their place within the larger framework of human life and society. Like his father before him, he believed that mankind's ultimate purpose could not be knowledge for its own sake; the sciences cannot fulfill man existentially. He differed with those who believed that the highest nobility lies in devoting one's life to increasing knowledge and cultivating one's mind; in his view, it was action that made a man's life worth living, and by "action" he meant either applying known knowledge or increasing science itself, since that too "is an action for the progress of humanity."[16]

Quoting Bacon's platitudinous declaration "Knowledge is Power," he claimed that no previous era had made that point so obvious. Scientists had learned how to use the forces of inorganic life for human purposes, for example, in the use of steam power. But he also noted that what made a nation strong—he perhaps had in mind Prussia, then in the midst of a constitutional crisis and attendant politico-military tensions—was not only military goods, or machinery, or adequate food and money supplies. He thought, rather, "Even the proudest and most unyielding of the absolute governments of our day must consider unchaining industry and conceding to the political interests of the working middle classes a justified voice in its [governing] council." For the modern nation-state also required solid "political and legal organization" as well as "the moral discipline of the individual." It was these characteristics that made "the superiority of the educated nations over the inferior ones." Those states that failed to develop culturally would inevitably collapse: "Where there is no rule of law and where the interests of the majority of the people cannot be expressed in an orderly way, there is also no possibility to develop national wealth and the power upon which it rests. An individual can only become a proper soldier if he has trained under legal laws and has the feeling of honor of an independent man, not that of the slave who is subjugated to the whims of an arbitrary master."[17] He spoke to the spirit of the times: in 1862 Baden ended enforced membership in guilds and introduced freedom of trade, and it passed a law on the civil equality of Jews. His lecture had a strong nationalist and liberal message.

His two themes—the nature of science and the nature of the modern nation-state—were closely linked in Helmholtz's thought. For its own preservation every modern nation that sought to be "independent and influential" had to support all the sciences, the natural as well as the human. Its interest lay not only in the "higher, ideal demands" and the training of scientists and technologists, but also in preparing civil servants, lawyers, and scholars of the moral, historical, and philological sciences. "The cultured peoples of Europe" recognized this, and it was reflected in the unprecedented level of state support for schools, universities, and scientific research in general. "In fact, the men of science form a sort of organized army," he declared. "They seek the best for the entire nation and to increase knowledge (almost always under its mandate and at its costs), which in turn can serve to increase industry, wealth, the beauty of life, and the improvement of the political organization and moral development of the individual."[18] His address reflected the industrialization taking place in the liberal state of Baden and elsewhere. Its liberalism, nationalist fervor, and sense of progress reflected and embodied widespread sentiments in Europe in the 1860s.

But then suddenly, he reversed course and declared that one should not expect science to pursue "immediate uses." Instead, he declared, all scientific and scholarly research that gave knowledge about nature's forces and the powers of the human mind was useful, quite often where one least expected it to be so. He cited as examples the distant linkage between Galileo's analysis of a swinging pendulum and the development of knowledge about longitude as well as that between Luigi Galvani's discovery of "animal electricity" and the development of the electric telegraph. "Science can only strive after complete knowledge and complete understanding of the rules of the natural and intellectual forces," he maintained. Each new discovery by each individual researcher represented a "new victory" against nature's tendency to hide its properties and an increased sense of "the aesthetic beauty" that came through greater understanding of the order of the parts within knowledge as a whole and understanding that "everything shows the traces of the rule of intellect [Geist]." Hence one type of reward for the researcher was the knowledge that he had contributed "to the increasing capital of knowledge, upon which rests the rule of humanity over the forces hostile to the intellect." The fact that public opinion gave ever greater recognition to scientific discovery showed that there was a second type of reward for scientific work: "Governments and peoples as a whole have become conscious of the duty to reward outstanding performance in science, either by offering appropriate positions or through specially announced national rewards." That year he received the title of court counselor (*Hofrat*) to the Grand Duke of Baden.[19]

His prorectoral address exemplified his leadership abilities. As a philosopher of science and as prorector, he aimed to include all scientists and scholars. The "common purpose" of the sciences was "to make the intellect the master of the world." The very nature of the human sciences did so directly, thereby enriching life, while the natural sciences did so indirectly, constantly seeking to liberate mankind from the forces of the external world. Every scientist and every scholar had his individual part to play in what was a common effort of the scientific community. Individual scientists, he emphasized, worked within this larger communal context, and thus each had to make his results as complete and as readily available to others as possible. In this way, scientists and scholars supported one another: calendar makers depended in large part on astronomical calculations, linguistics on physiology, historical linguistics on historical knowledge, sculptors on anatomists, and so on. Similarly, the physics of sound and the physiology of tone sensations helped in understanding the foundational elements of music, "a problem that essentially brings us into the specialty of aesthetics." More generally, each science depended on one or more closely affiliated sciences. He declared that every individual worker works for "the noblest interests of all humanity, not as an individual worker as such." He believed that this conscious appreciation of the relations of the sciences to one another and of the place of the individual scientist within the larger communal framework and effort of science was "the great task of the universities," and to accomplish it the four faculties needed to remain together.[20] In effect, liberal as the address may have been, he called for retaining the status quo, for not changing the university's structure.

The address first appeared in written form as a university publication, which meant that only a few had access to it. Brücke was one of them, and he was much impressed with it. Helmholtz told the publisher Friedrich Vieweg that the address had been "very favorably received," that he had received many personal requests for copies of it, and that his own supply of offprints was now exhausted. He wondered whether Vieweg might be interested in publishing it and might offer him an honorarium. Vieweg did just that: he published it two years later in the first collection of Helmholtz's *Populäre wissenschaftliche Vorträge* (1865).[21]

Fig. 11.1 The Friedrichsbau, home to the University of Heidelberg's natural sciences institutes (except for chemistry), as of 1863. Universitätsarchiv, Ruprecht-Karls-Universität Heidelberg.

A New Institute and Its Assistants

It took five years for the Baden authorities to approve and construct the institute they had promised in 1858. Part of the delay stemmed from the fact that Helmholtz's new institute became a part of a new natural science complex. Early in 1859 he sent the university construction plans for his proposed new institute. The complex included institutes for physiology, physics, mathematics, and medical chemistry and rooms for scientific collections. Helmholtz was also provided with his own residence within the institute, he explained to Thomson, "so that the use of the laboratory will be very convenient."[22]

The new building, known as the Friedrichsbau and located in the center of town, consisted of a three-story main front with a Renaissance facade and two two-story wings (fig. 11.1). It became available for use in September 1863, at which time Helmholtz and Kirchhoff moved their respective families into their new residences in the building. The physiological institute was located in the western half of the second floor, the physics institute in the eastern half. The institute was Helmholtz's first true, working institute. He no longer had to work out of his own private home or apartment, but instead had a public space for doing so. This was emblematic of the evolving nature of German scientific institutes. The French chemist Adolphe Wurtz, who toured German science facilities in 1869 on behalf of the French government, judged Helmholtz's institute to be one of the leading and best such institutes and thought the intimacy of his and Kirchhoff's was most fortunate, since it brought together "two savants of whom all Europe knows the names." Yet he also judged the facilities to be inferior to the men. Each had only a few rooms; and, all in all, the physiological institute was "a modest establishment."[23] Nevertheless, this complex of scientific institutes in their new building, plus Bunsen's own new building, signaled that the natural sciences now lay more than ever at the center of the University of Heidelberg and were rivaling law and history for support and distinction.

In March 1865 Wundt resigned as Helmholtz's assistant and became an independent scholar in Heidelberg. He had become quite dissatisfied with his position and wanted to move up professionally. For about six years he had supervised student laboratory work and related teaching matters, but he had not otherwise assisted Helmholtz in his work. Wundt was independent-minded and did not let Helmholtz or anyone else guide him; he worked alone by preference. Much the same was true of Helmholtz. They had a good working relationship, though, and Wundt greatly admired Helmholtz.[24]

Helmholtz, for his part, had a quite positive view of Wundt, both as a teacher and as a researcher, and sought to help him obtain professorships of philosophy (i.e., psychology). In late October 1863, he wrote to Kiel in support of Wundt's application for a professorship there. He characterized Wundt as a good lecturer, pointing in particular to his lectures on reproduction, anthropology, and psychology. He explained, too, that Wundt supervised the students in his (Helmholtz's) laboratory, giving them instructions and generally watching over them with care, and that students were "always appreciative" of this. Moreover, he noted Wundt's publications in nerve, muscle, and sensory physiology, among other fields. He said Wundt had made important factual discoveries, worked hard, and had "a broad, general education" and a solid knowledge of physical and mathematical methods. He also praised Wundt's moral character. He thought Wundt a talented young physiologist worthy of a position on Kiel's faculty.[25] But Wundt did not get an offer.

Helmholtz also tried to help Wundt get Heidelberg to name him extraordinary professor of anthropology and medical psychology in the Medical Faculty. He indicated that he had every reason to warmly favor Wundt's request and that he was not concerned about any clash of interests between Wundt as his assistant and Wundt as an independent extraordinary professor.[26] Wundt received the extraordinary professorship the following year.

Nearly a decade later, in early 1872, when a professorship of physiology became vacant at Marburg, Helmholtz was asked for his advice about Wundt (and others) for the position. He again recommended Wundt, who came in second on the short list. Also that same year he wrote on Wundt's behalf for the professorship of philosophy at Giessen. He again declared that Wundt had been "beloved" as a teacher at Heidelberg and, until Zeller arrived in 1862, had also given well-received philosophical lectures. He noted that he had written on Wundt's behalf for the professorship at Marburg, adding, "But our government [that is, the Prussian] was not yet prepared for such an unusual step." While he did not always agree with Wundt about details—and it could not be otherwise, he said: "In so new a field [as sensory physiology] the explanations of many processes are doubtful and judgment of their probability has broad, individual latitudes"—he nonetheless thought that Wundt was "on the right path." Wundt failed to get the position at Giessen, just as he failed to get the professorship of philosophy at Würzburg, for which Helmholtz also wrote on his behalf.[27]

Above all, Helmholtz admired Wundt's attempt to pursue philosophy (that is, epistemology) through the study of sensory physiology (that is, using an empirical approach). What he did not like about Wundt was his politics. He scorned Wundt's democratic sympathies and activities on behalf of the working class. He wrote du Bois–Reymond privately that Wundt was "now mostly doing politics, workers' unions, etc."[28] Despite such political differences or judgments, he gave Wundt his full support.

After Wundt left his position at Heidelberg, he was replaced by Julius Bernstein, the son of Aaron Bernstein. Julius had previously studied with Rudolf Heidenhain and, especially, du Bois–Reymond. He took his medical degree in Berlin and in the spring of 1865 became Helmholtz's assistant. Like du Bois–Reymond and Helmholtz, Bernstein emphasized exacting methods of research, employed thermodynamics in physiology, was antispeculative, and argued for a mechanistic understanding of organic phenomena. Building on their work and approaches, he became one of the leading electrophysiologists of the next half century. His exact measurements in 1868 of the "action potential," by means of his differential rheotome, gave highly accurate measurements of electrical nerve and muscle activity. Even more important was his later work (1890–1902) that led to the development of the membrane theory of action potential in cells and tissues. Like his teacher du Bois–Reymond and his mentor Helmholtz, Bernstein's work proved fundamental to the future of neurophysiology and the emerging field of neuroscience. With his use of concentration cells, ion theory, and electrolytic dissociation (part of which he owed to Helmholtz's work in electrochemistry and chemical thermodynamics during the 1870s and 1880s), he helped bring the "organic physics" of du Bois–Reymond and Helmholtz into the next generation.[29]

Physiology Teacher and Foreign Attraction

Most German university professors taught two fundamentally different types of students: those in their elementary (or introductory) courses and those in their advanced courses. Being able to teach one group well did not necessarily mean being able to teach the other equally well. It seems that, as a lecturer before university medical and other types of science students (for example, physiology) in introductory courses, Helmholtz was not very talented or motivated. Theodor Wilhelm Engelmann, who studied only briefly with him but was one of his best students, maintained that Helmholtz never prepared his lectures in detail beforehand, but instead spoke rather freely. He added that Helmholtz "spoke slowly, in a measured way, and sometimes a little hesitantly." Engelmann spotted other deficiencies in Helmholtz as a lecturer. He did not make eye contact with his listeners but instead gazed off into the distant horizon. Nor did he assess the varying levels of his students' preparation for a course and adjust his lectures accordingly. Bernstein said much the same thing. After noting that Helmholtz taught a general survey course on physiology and a special course on sensory physiology, he euphemistically indicated his view of Helmholtz's shortcoming as a pedagogue: "He did not belong to those academic teachers who shine through oratorical eloquence." The ophthalmologist Albrecht von Graefe reached essentially the same conclusion without ever entering one of Helmholtz's classes. While Graefe urged Johann Friedrich Horner, a well-known Zurich ophthalmologist, to try to attract Helmholtz to the University of Zurich, he also warned him that as a lecturer Helmholtz had his limitations. "It would be [worth it] for the name and the man alone; to be sure, it is to be considered whether he fits [Zurich's] teaching needs since he is in no way keenly interested in all branches of physiology, and, correspondingly, lectures fully well only on individual parts."[30]

Others also found shortcomings. Wundt claimed that when Helmholtz described his ophthalmometer during a lecture to students, he did not say what its practical purpose was. Sechenov reported that Helmholtz had only minimal contact with students, not least since it was Wundt who supervised the physiology exercises. The students caught "only a glimpse" of him, though he did appear once a day in the laboratory to review what the students were doing and to see if they needed any help. Sechenov found Helmholtz's lectures to medical students rather plain, elementary, and more or less boring. On the other hand, he had heard Helmholtz lecture "gaily" one evening at the Heidelberg Association on the analysis of sound by means of resonators.[31]

Helmholtz's student, junior colleague, and friend Hermann Knapp, a clinical ophthalmologist who habilitated at Heidelberg in 1859, had a more positive view of Helmholtz as teacher. He claimed that Helmholtz liked teaching and that he discharged his duties punctually: Helmholtz "never missed a lecture, never was late." He stuck to his subject, avoiding jokes and amusing stories; instruction as entertainment was not his mode of delivery. He was patient with his pupils, spending hours helping them in the laboratory. Knapp was probably referring to the instruction of advanced research students, for he added that Helmholtz helped such students with their experimental and mathematical problems and in the revisions of their manuscripts for publication. They became his friends; they took walks together on Saturday afternoons in the Heidelberg countryside and in the evening took a meal together.[32]

Bernstein also acknowledged certain of Helmholtz's pedagogical assets: he lectured simply and clearly; students very much felt "his imposing personality"; and he impressed those whom he visited while they worked in the laboratory. The advanced students who came into personal contact with him found that he taught in good measure by example. He explained: "Whoever has seen Helmholtz experiment, will have been amazed at the calmness and patience that ruled over him, and which could not be shaken by any sort of mishap. Helmholtz's happy temperament, in which seriousness and clear peace shared a place, also made him a born experimenter." That example benefited the advanced students who came to work with him; he gave them his time and advice.[33]

His shortcomings as a classroom lecturer in physiology were in part compensated by what appear to be the highly successful, popular lectures he gave to students in all four of Heidelberg's faculties on the most general and important results in the sciences, and by the many well-received lectures he gave on his research at Heidelberg's Natural History and Medical Association, where for years he lectured, chaired the organization, and partook in social events. Not everyone was enamored of or shared his viewpoint, however. For example, the young John Theodore Merz, a future British historian of science, philosophy, and religion, who studied privately in Heidelberg between 1862 and 1864, attended one of Helmholtz's public lectures and "heard for the first time an exposition of his celebrated theory of the contraction of the sun through gravitational forces as the source maintaining the sun's heat." Merz found Heidelberg and other places in Germany too rationalist for his taste: he said there was "an anti-religious spirit" to the place that opposed his own sympathy for religious thought. Indeed, he did not like the whole spirit of modern philosophy, including that represented by thinkers like Zeller, Strauss, and Helmholtz (whom he nonetheless respected enormously).[34]

Beyond general audiences, Helmholtz's pedagogical strength was with the advanced, research-oriented students. He had his share of distinguished students of physiology, though perhaps not nearly so many as Ludwig, Brücke, du Bois–Reymond, and others. German students often attended two or more universities during their years of study; often enough, they had two or more intellectual teachers; and sometimes they received a degree from a university where they had spent only a semester or two. Thus "Helmholtz's" students also studied with physiologists and others at other universities. "His" German or Austrian physiology students included Engelmann, who succeeded du Bois–Reymond at Berlin in 1897; Sigmund Exner, who initially studied with Brücke in Vienna and, after working with Helmholtz, succeeded Brücke there in 1891; and Hugo Kronecker, who studied with du Bois–Reymond and Helmholtz, later became Ludwig's assistant, and in 1884 became professor of physiology at Berne.

Beyond physiology proper, Helmholtz's very presence at Heidelberg ipso facto turned it into a center for ophthalmology, even though Helmholtz himself never taught the subject nor directed a clinic. Yet his invention of the ophthalmoscope and the ophthalmometer, his fundamental research in physiological optics, and even his mere membership in the Medical Faculty all made him and Heidelberg an attraction for students of ophthalmology. His presence there led Graefe, Donders, and Carl Ferdinand Arlt, the other doyens of the field, to choose Heidelberg as the meeting place for their ophthalmology congresses each autumn.[35]

Helmholtz's two best students in ophthalmology were Theodor Leber and Hermann Knapp. Leber came to Heidelberg about the same time Helmholtz did. Helmholtz inspired and encouraged him and drew him to experimental physiology. He suggested that Leber work on a prize essay in the Medical Faculty. He did, and won it (his being the only entry). Still, Leber found that while Helmholtz could show him where new, unexplored scientific territory lay, he could not show him how to explore it. So he went to study with Ludwig in Vienna. In September 1864, at the annual meeting of the ophthalmological congress in Heidelberg, Leber gave a stunning presentation titled "The Eye's Blood-Vessel System." This lecture brought him immediate fame and effectively marked the beginning of experimental ophthalmology. He assumed a professorship at Göttingen in 1870, set up the first laboratory for experimental ophthalmology, and became one of the leading experimental ophthalmologists of his generation.[36]

As for Knapp, he studied with Graefe in Berlin before coming to Heidelberg to be an assistant with Chelius and to work with Helmholtz. For eight years (1859–67), he lived near Helmholtz and evolved from student, to junior colleague, to lifelong friend. He did his original work in Helmholtz's laboratory, where he wrote a thesis titled "The Optical Constants of the Eye." He habilitated at Heidelberg, founded the highly successful University Eye and Ear Hospital (1862), and became an extraordinary professor of ophthalmology there. He saw Helmholtz almost daily, both at the institute and socially. He found that, despite Helmholtz's profound scientific originality, he could also be "a believer in [scientific] authority," citing Helmholtz's trust in the work of Young. He said Helmholtz investigated only those problems that, after due consideration, he thought solvable. This made his work much more effective. And while he had little time for those who "were selfish, vain and frivolous," he would patiently explain to even an uneducated layperson (most definitely women included) any scientific or philosophical topic, as long as he detected a genuine interest in the matter. But Knapp felt frustrated by the firm resistance of Heidelberg's Medical Faculty—where Helmholtz was his backer—to his request for an independent, ordinary professorship and a clinic for ophthalmology. And so in 1868 he emigrated to New York, leaving behind a disappointed Helmholtz. The incident suggests that, despite everything, Helmholtz's influence went only so far.[37]

Helmholtz's largest numbers of foreign physiology students were Americans and Russians. After the American Civil War, the Americans' principal foreign destinations were Berlin, Leipzig, Göttingen, Heidelberg, Strasbourg, and Vienna. For young American medical doctors, physiologists, and others, the German states embodied the ideals of research; they were perceived as the leading edge in basic and clinical medicine. Jeffries Wyman, a Harvard comparative anatomist, naturalist, and archaeologist, wrote his brother in 1870: "Ludwig at Leipzig & Helmholtz at Heidelberg are got up in a most expensive manner & with a great outlay for apparatus." It was all needed, he said, to do experiments. The young medical doctors and physiologists returned to America with an enhanced belief in the importance of research in medicine and science generally, and in physiology in particular. In 1870 Heidelberg alone had thirty-three Americans enrolled in its Medical Faculty, some if not many of whom came into contact with Helmholtz.[38]

Henry Pickering Bowditch and William James were among them. Bowditch was a physiologist and the future dean of the Harvard Medical School. During the period 1869–71, the young Bowditch spent several months studying with Max Schultze at Bonn, then a year with Ludwig in Leipzig, and, finally, another year divided between working with Helmholtz and Virchow. James, despite his seemingly good intentions, failed to make contact with Helmholtz in Heidelberg. In the autumn of 1867, he wrote a friend: "I am going on to study what is already known, and perhaps may be able to do some work at it [experimental psychology]. Helmholtz and a man named Wundt at Heidelberg are working at it, and I hope I live through this winter to go to them in the summer." James also wrote Bowditch and again mentioned his intention to study with Helmholtz and Wundt. "The immortal Helmholtz is such an ingrained mathematician that I suppose I shall not profit much by him." From Dresden he wrote Bowditch again: "Helmholtz, who is perhaps the first scientific genius now above ground, is said to be a very poor teacher though he has the finest laboratory." Yet when James finally reached Heidelberg in late June 1868, he became so depressed that he soon left town, apparently without ever meeting Helmholtz or hearing him lecture. James did manage, however, to acquire a photograph of Helmholtz, which he sent to his parents, "begging you to notice how mean is the lower part of his immortal face. He is probably the greatest scientific genius extant notwithstanding, and in his company your despised child can well afford to let rebound the shafts of your ridicule."[39]

Like the Americans, the Russians sought out the excellent institutional facilities and intellectual leadership that German scientists could provide. Russian science in the 1860s was at least as undeveloped as American science. Russian students of science and medicine flocked to the German lands. In physiology they tended to go especially to the laboratories of Müller, Purkyně, Bernard, Helmholtz, du Bois–Reymond, and above all Ludwig. Helmholtz's name and work was certainly widely known in Russia. After Zurich, Heidelberg was the principal foreign university that Russian students attended. Of Helmholtz's Russian students, the most important were Sechenov and Junge (Junge became an important figure in Russian ophthalmology).[40] Other important Russian students included Kliment A. Timiriazev, an outstanding plant physiologist, a Darwinist, and a highly effective popularizer of science; and the physiologist I. F. Tsion, who became Ivan Pavlov's most-adored mentor.

These foreign students and colleagues who came to work with Helmholtz in the field of physiology—both the fact that they came and their eventual stature—are indicative of the leadership he gave to the discipline. From 1861 onward, his home became the virtual cultural center of Heidelberg. Advanced research students, especially the foreign ones, along with visiting colleagues and friends from around Germany, were often invited there. Every fortnight Anna and Hermann held a *Musikabend*, where "artist friends and dilettantes" performed; Helmholtz himself often played Bach fugues as well as other classical works on a harmonium that he had built. Always conscious of expenses, the Helmholtzes made sure that an evening cost them no more than five talers in food and drink. Anna, not Hermann, animated these evenings, yet Helmholtz was the center of attraction and enjoyed them fully; they were even essential for his well-being. He also continued to take great interest in the theater; often enough, the Helmholtzes devoted an entire evening to home theatrical performances. He loved to perform, and on occasion played the role of an Englishman speaking mangled German. Wundt reported: "He played the role with such an overpowering comic effect that one would have had a hard time distinguishing him from a [professional] comedian." Such performances helped him to relax. He preferred not to talk about scientific or even artistic matters on such evenings. He also loved to do readings from classical plays, especially *Don Carlos*, whose leitmotif is freedom of thought. Wundt claimed that Helmholtz allotted the most important roles to the socially most important guests: Hamlet, for example, might be played by a *Geheimrat*, Rosencrantz and Guildenstern by *Privatdozenten*. Wundt also thought that Helmholtz favored (in order) English, American, and Russian colleagues and students of science.[41]

12

The Relations of Music

A Musical Life and a Publisher

Helmholtz formally began working on *Die Lehre von den Tonempfindungen, als physiologische Grundlage für die Theorie der Musik* (*On the Sensations of Tone as a Physiological Basis for the Theory of Music*) in 1854. But in effect he had been at it much of his life. From youth onward, he performed, listened to, and analyzed music on a daily basis. Having learned to play the piano as a child, he continued to do so throughout his life. (He apparently also learned to play the organ and perhaps one or more other instruments, but the piano remained by far his preferred instrument.) The composers he admired most included Palestrina, Bach, Mozart, Handel, and Beethoven. He considered Beethoven "the most powerful and deeply affecting of all composers," and when alone he played "almost nothing but" him. In terms of "harmoniousness and the fine artistic beauty of harmonic flow," though, he thought Mozart the finest of composers, "even if he doesn't move us so powerfully." As he grew older, as he had ever more "scars" on his soul, he came to prefer Mozart's gentleness to Beethoven's power. He also liked to sing, though his voice was undistinguished. Late in life he wrote: "Someone who loves and feels music, but can't make it, happily extols others who love, feel, and make it." He sometimes performed with friends and their families in his and their homes. Music drew him to others. He firmly believed that there was a relationship between his scientific study of physiological acoustics and his love of music.[1] He continually cultivated his senses and sought aesthetic standards of judgment long before he philosophized about music (and painting). The fine arts as well as the natural sciences stimulated his intellect and imagination.

In the later 1850s, Helmholtz was hampered in working on his book on physiology and music as he devoted his efforts to completing part two of the *Handbuch*. After finishing part two, he returned to the physiological foundations of music. He told Donders that the book was meant to be a popularization so that "Musikliebhabern [music-lovers]" could understand the physical and physiological foundations of harmony. By March 1861, some seven years after he had begun, he had virtually completed the book's first draft. During the winter of 1861–62, he was also hampered by painful migraine headaches. By the spring of 1862 he was completing the book.[2] He now faced the problem of finding the right publisher; and here he had choices.

In the summer of 1861, Eduard Vieweg, the head of Friedrich Vieweg, asked him for a second time to write a textbook on medical physics. Vieweg sought a "literary connection" with Helmholtz, he said, and he was open to other ideas if Helmholtz did not want to do a textbook. Vieweg cultivated personal relations with especially important scientists, including Liebig, Poggendorff, and Helmholtz. Under his leadership, the firm began emphasizing publications in technology and science that stressed the practical uses of theoretical work. It favored works that showed the relationship between science and practical (especially economic) life and was particularly strong in chemistry. For example, it published works by Friedrich Wöhler, Bunsen, and Virchow.[3]

Helmholtz told Vieweg that he was currently too busy "with literary works," by which he meant his manuscript on acoustics and music, originally entitled *Studien über physiologische und musikalische Akustik*. "I've sought as much as possible to make it popular and generally understandable, like the book by [Friedrich] Zamminer on music and musical instruments, because I've wanted to make it accessible to musicians and music aficionados as well as to physiologists and medical doctors." His manuscript developed a theory of harmony based on physical and physiological principles. It would require lots of illustrations to help enlighten readers about the scientific theory and experiments discussed in the text. He said Leopold Voss in Leipzig wanted to publish it if they could come to an agreement about length and the number of illustrations. Once this book was finished, Helmholtz intended to return to his suspended work on physiological optics, "because I must maintain my priority in regards to the facts and views developed therein." That work would take him at least a year to complete, he told Vieweg, so that he could not now make any promises. Moreover, he had

had enough of writing textbooks or reference books, though he greatly valued "the service which you have done in this regard for the spread of science." Vieweg wanted Helmholtz's manuscript on acoustics and music, and he asked for details as to length, illustrations, and honorarium. He would feel "a great joy and honor" in being Helmholtz's publisher.[4]

Helmholtz asked for 600 talers. This made the book more expensive, however, and so Vieweg asked him to lower his honorarium to 500 talers, in which case Vieweg would produce 1,800 to 2,000 copies. If the book reached a second printing, he promised Helmholtz another 400 talers. Helmholtz agreed, and he signed with Vieweg. By February 1862, Vieweg was printing the proof pages. Helmholtz told his old publisher, Georg Reimer, who belatedly also sought to publish the manuscript, that its production had already begun and that he had accepted Vieweg's offer because it was "the most favorable for me." His signing with Vieweg marked the first in a long series of publications with the firm, and he also brought several works by his British colleagues to Vieweg.[5]

As he was completing his manuscript, he gave Thomson's wife, Margaret, a sense of his creative process and working procedure. He wrote:

> With my physical theories I've penetrated rather far into the theory of music, at least farther than I myself had thought possible at the outset. The work became extremely amusing for me. When one develops, from the right general principle, the consequences in individual cases of the principle's applications, then new surprises always come forth that one had previously not imagined. And since the consequences don't develop from the author's arbitrary will, but rather according to laws, then it often made the impression on me as if it was not even my own work that I'm writing down but rather as if I had first studied up the work of somebody else.

On this account, it was as if Helmholtz were merely a conduit for some higher power. Yet the work scarcely flowed effortlessly from first principles; in addition to general principles (theory), it involved much observation, experimentation, instrument making and use, and scholarly historical research. With interruptions, it had been an eight-year-long struggle: he did not sign the preface to the volume until October 1862. Anna considered the book Helmholtz's "spiritual child." He told Thomson that writing it was "very much worth while, but frankly also very laborious."[6]

Acknowledging Debts

Leaving aside the array of physicists and physiologists whose work he drew upon and synthesized, Helmholtz had incurred two special debts while working on

acoustics and music during the previous eight years. The first involved his creation, development, use, or simple purchase of instruments, both musical and
scientific. He acknowledged that his work had depended on constructing new
instruments—resonators, sirens, the vibration microscope, and a harmonium,
not to mention tuning forks. These were otherwise unavailable to him and to
his colleagues in German physiological institutes, and constructing them far
transcended the financial resources of any individual scientist. He had been in
frequent contact with instrument makers. He could not have done his research
without state patronage. In particular, he owed King Maximilian II of Bavaria a
debt for the financial support that enabled him to order a custom-made vowel
synthesizer for analyzing vowels and timbre. He used the Sömmering prize
money that he had won to have a harmonium constructed, which he used for
studying just intonation. Relatedly, his analysis of acoustics and its relationship with music and instrument-making drew on a theoretical tradition (especially strong in the German lands) that reached back to Ernst Florens Friedrich
Chladni in the late eighteenth century but also included a series of nineteenth-
century figures. Physical (acoustic) theory contributed much to the understanding and making of musical instruments, just as the latter did to the former.[7]

In the 1850s Helmholtz created his resonator to analyze sounds, in particular to hear the upper partials (overtones) that lay at the heart of his theory of
sound. Each individual resonator—a sphere with a short neck and opening that
was inserted into the ear—was tuned to a particular, individual frequency (depending on the sphere's volume, the height of the neck, and the size of the opening). Helmholtz's own first resonators were effective but rather crude glass devices for detecting the tricky-to-hear upper partials. But, as he had done in the
ophthalmoscope's development, he turned to a professional instrument maker
to improve upon his early, crude version of the resonator, as well as to provide
him with better tuning forks for generating tones and beats.

He turned especially to Rudolph Koenig, who in 1858 opened his atelier
in Paris. Koenig's instrumentational work—he specialized in acoustic instruments—greatly advanced Helmholtz's own work. For example, he made the
resonators of brass, so that they were much sturdier and more precise; he made
more reliable and precise tuning forks; he created a tonometer (a combination of
dozens or even hundreds of tuning forks) for demonstrating combination tones;
he rebuilt the double siren (first built for Helmholtz by Ferdinand Sauerwald) for
generating sounds that Helmholtz used to study combination tones and interference effects; and he created a so-called manometric flame apparatus for visually representing sound waves. Koenig's instrumentational advances also helped
convince others of the truth of Helmholtz's musicological analysis: the tuning
forks and resonators, which produced and detected the tones and beats, in effect
became the embodiment of Helmholtz's theory—even though Koenig later criti-

cized it. Koenig's Parisian workshop became a demonstration and sales center for Helmholtz's theory for visitors from Europe, North America, and elsewhere. Moreover, "Helmholtz resonators" became a standard teaching and research instrument in every physics and (future) psychology laboratory.[8]

Helmholtz also further developed a vibration microscope and an associated optical method for observing vibrations, in particular those of violin strings. The original method and microscope had been conceived by the French physicist Jules Antoine Lissajous, known for his "Lissajous figures." Helmholtz extended the method and varied the instrument so as to observe the vibrational forms of the parts of a string (in particular its stick-slip motion during bowing, subsequently known as the Helmholtz motion) and, in turn, to calculate the string's motion and determine the intensity of its upper partials. His analysis of the motions of a violin string provided the classical theory (later confirmed and extended by C. V. Raman and others). He also studied the vibration and plucking of strings more generally and piano strings more particularly. Building on the work of Wilhelm Weber, he also developed novel and lasting analyses of the theory of flute pipes, reed pipes, and wind instruments (clarinets, oboes, bassoons, trumpets, trombones, and horns).[9] These studies were expressly designed to give observational evidence for the different concepts and theories that he advanced in the *Tonempfindungen*, especially part one. Like Darwin's *On the Origin of Species* (1859), Helmholtz's text regularly gave concrete, evidentiary illustrations of his theory. This was a new sort of book for the musically educated, or those who hoped to be so.

Helmholtz also became indebted to two musical women. One was Anna, who provided a sounding board for the writing of the book's nonmathematical chapters. When she did not understand something clearly, he simply rewrote the text until she did. She became his standard for addressing the intelligent, educated, musically interested but scientifically illiterate adult. The other woman was Emma Seiler, a student of voice who in the 1850s gave singing lessons in Heidelberg, where she became a close friend of Bunsen and Kirchhoff and met Helmholtz. Seiler sought out Helmholtz to help her investigate the human voice scientifically, so as to improve the quality of singing. She joined him in his investigations of singing and thanked him for teaching her how to use the laryngoscope to study the physiology of the larynx as tones were produced. She felt deeply indebted to him for helping her appreciate the relationship between the scientific understanding of human sound production, on the one hand, and human speech and voice improvements, on the other. He also owed her. While working on his book, he visited her "almost daily for several months for advice and for verification of his calculations by her experiments." In the *Tonempfindungen*, he referred to her work on hearing the upper partials of a watchman's voice and on dogs' sensitivity to the violin's E-double-sharp. Seiler subsequently moved to Philadelphia, where she established a singing academy and was elected a member of the American Philosophical Society, one of only six women through 1891 to have been so chosen. She became well-known for her books *The Voice in Singing* and *The Voice in Speaking*.[10]

On the Sensations of Tone

Helmholtz divided his book into three general parts, paralleling the organizational structure of the *Handbuch der physiologischen Optik* (even with part three still to be written): the physical foundations; the physiology; and the introduction of a psychological component to help provide an explanation. He acknowledged the recent, important contributions by music aestheticians such as Eduard Hanslick and Friedrich Theodor Vischer but noted that neither had dealt with physical motion as it pertained to music. By contrast, he aimed to give musicology its "proper origin and foundation," to discuss the "scientific foundation" of the "elementary rules relating to the construction of scales, chords, keys and modes." He suggested, "Music stands in a much closer connection with pure sensation than any of the other arts." Poetry and the "plastic arts" seek to generate images in the mind, he said; but the sensations of musical tone constitute the very "material of the art" and do not produce "images of external objects or actions." He thus thought that understanding musical sensations was central to understanding musical aesthetics. Moreover, unlike the other fine arts, music did not seek to represent nature; rather, tones and their sensations existed for their own sake.[11]

He began by explaining the acoustic processes that occur "within the ear itself." The ear, like the eye, generally required three areas of study: the physical, or "how the agent reaches the nerves to be excited"; the physiological, or how the nerves produce sensations; and the psychological, or how perceptions are produced out of sensations. The physiological and the psychological parts, he said, required the greatest investigatory effort for understanding the scientific foundations of the theory of music. The *Tonempfindungen* thus aimed to synthesize all past research on physiological acoustics and add Helmholtz's own, new findings. Even so, he considered it only "a first attempt" and thus "somewhat imperfect." He conceded that he could do no more than give "the elements and the most interesting divisions of this subject."[12]

Part one, to return to the book's overall structure and contents, was largely concerned with physical and physiological acoustics (the "composition of vibrations"), more particularly with harmonic upper partial tones, including their nature and relationship to the quality of tone. In contrast to his predecessors, he argued that harmonic upper partial tones are essential musical phenomena. He analyzed the ear's sensation of them, which led him "to an hypothesis respecting the mode in which the auditory nerves are excited, which is well fitted to reduce all the facts and laws in this department to a relatively simple mechanical conception." In part two he discussed music proper (the "interruptions of harmony") by giving an analysis of combination tones and beats as the basis of musical consonance and dissonance. He sought to provide a physiological grounding for the numerical relationships that many had remarked on or utilized ever since Pythagoras. Part three treated the nature of musical scales and notes (the "relationship of musical tones"). Here Helmholtz offered his aesthetic analysis. Whereas parts one and two were based upon mathematical, physical, and physiological facts and laws—in a word, science—the final part concerned historically contingent matters of taste, both of a national and of an individual sort. Here he demonstrated both his long-standing practical experience with musical phenomena and his scholarly abilities, giving abundant, erudite references ranging from Aristotle to modern music historians and aestheticians. He also showed a sympathetic, and for the most part nonpatronizing, appreciation for both premodern and non-European (Persian, Arabic, Indian, and Chinese) music. While he thought music could to a certain degree be understood by natural science, that hardly exhausted the matter; physics and physiology were only the necessary, not the sufficient, conditions that made further analysis possible. Nevertheless, while he emphasized that his investigations were "confined to the lowest grade of musical grammar," he feared that some music theorists would see his approach as "too mechanical and unworthy of the dignity of art."[13] His fear proved to be all too prescient.

As Helmholtz himself emphasized, and as many have since noted, he did *not* believe there was a basis in nature for music. Yet some musicians and theorists had (or have) failed to appreciate his crucial distinction between nature as setting the conditions for music, on the one hand, and musical style as an evolving historical phenomenon, on the other. Or they saw it but neglected it, as they sought to establish a basis in nature for music. By contrast, Helmholtz viewed music as a historically evolving, stylistic phenomenon. Nature only laid down the conditions for music; in his view, nature was not the principle behind music.[14]

Helmholtz began his discussion of the composition of vibrations (in part one) by describing the sensation of sound in general, crucially distinguishing noises from musical tones. "The sensation of a musical tone," he explained, "is due to a rapid periodic motion of the sonorous body; the sensation of a noise to non-periodic motions." He further distinguished musical tones (more simply, tones) by their force, pitch, and quality. He explained that the ear hears an entire range of "higher musical tones," which he designated as the harmonic upper partial tones (more simply, upper partials, or partials, or overtones), in contrast to the fundamental or prime, "which is the lowest and generally the loudest of all the partial tones." A musical tone is thus the periodic vibration of the air sensed by the ear. It is a compound of a series of partial tones, beginning with the prime followed by the upper partials. Tones can be either simple (that is, a partial) or compound (that is, a note).[15]

Using layman's language, Helmholtz described the physical nature of waves and explained that musical tones were like such waves. Ohm's law and Fourier's theorem could be used to analyze them into simple vibrations. He explained, too, how resonators could detect otherwise barely perceptible tones, allowing even the uninitiated to hear them; such resonators were the sine qua non of his entire musicological investigation. He explained how upper partials were "an essential condition for a good musical quality of tone." Learning to hear them, he argued, was like learning to recognize the different ingredients and flavors in food or wine and distinguishing good from bad: it was simply a matter of practice or cultivation. He described in great anatomical detail the ear's fine structure, in particular the membranous cochlea, which led him to propose his resonance theory of hearing: sound waves put into sympathetic vibration the auditory nerves containing fibers of differing lengths and densities whose ends are spread out within or on the basilar membrane in a liquid-filled cavity. (The germinal idea behind his theory could be found as far back as the eighteenth century in work by Domenico Cotugno, Giordano Raccati, Jean-Philippe Rameau, Albrecht von Haller, and others.) As with the tuned strings of a musical instrument (for example, a harp, a cello, a violin, or a piano), the basilar membrane contained a varying set of transverse fibers that acted as resonators set for tones of different frequencies. Hence, underlying this analysis of sensory (auditory) physiology lay an analysis of nerve fibers, that is, Müller's old law of specific sense energies. Helmholtz's results, theory, and synthesis of work in a wide variety of fields became the dominant view of the ear's function, of otology, and of auditory perception for much of the next half century or so. Though his resonance theory of hearing did not go entirely unchallenged (by Ernst Mach, for one), it was not until the appearance of Georg von Békésy's modification of it through his so-called traveling-wave theory of hearing (1928), for which he much later (1961) won the Nobel Prize for Physiology or Medicine, that a superior explanation appeared on the scene.[16] In the meantime, Helmholtz's theory found applications in Alexander Graham Bell's telephone and in medicine (otology) and use in musicology.

Intimately connected with the issue of the ear and hearing was that of vowel qualities. Helmholtz investigated the vowels of speech not out of an interest in phonetics per se but rather to understand the vowel qualities of musical tones, both of instruments and of the human voice. This meant understanding the vocal chords "as membranous tongues" producing "pulses of air" that the ear received as compound tones and then analyzed into partials. Here, too, he made extensive use of tuning forks and resonators.[17]

Part two was devoted to the interruptions of harmony, in particular combination tones and beats, and to consonance and dissonance. Combination tones occur, Helmholtz explained, when "two musical tones of different pitches are sounded together, loudly and continuously"; this normally results in a new pitch. He distinguished combination tones from beats, noting that the ear resolves combination tones "into a series of simple tones" without affecting the sensations, while in beats the sensations are disturbed. Consonance occurs when, with certain ratios of pitch numbers, two musical tones are sounded simultaneously and united but no beats are formed; it invokes a continuous sensation of tone. Dissonance occurs when beats of the fundamentals and beats involving upper partials produce an unpleasant disturbance; it invokes intermittent sensations. In "classical art," the aim was to avoid all unpleasant sounds; in "modern art," by contrast, where novel (and an increased number of) instruments came into play to produce greater musical expression, a greater number of unpleasant sounds were necessarily produced.[18]

His discussion of upper partials, combination tones, and beats had been standard fare to all his musicological predecessors; he maintained that, as in his work on the ophthalmoscope, he had merely extended their results. He did this through the use of the resonator to hear upper partials and so to show that virtually all tones "were compounded of partial tones" and that partial tones were especially important in yielding positive musical effects. He emphasized that his work was not "empty theoretical speculation" but rather was largely empirical and stood in conjunction with certain well-known physical laws. In the hands of a practiced listener, he maintained, the resonator was essential to musicological analysis. Though he emphasized that there was still much empirical work to be done, he felt confident enough to declare that he had exhibited "the true and sufficient cause of consonance and dissonance in music." Although various civilizations had developed different tonal systems, he respectfully noted in passing that only the modern, that is, the European, form used the system of harmonic chords. "It was only in this system that a complete regard was paid to all the requisitions of interwoven harmonies."[19] He clearly thought it superior.

The ear's power to resolve complex sounds via "the laws of sympathetic vibration," and the associated notion that harmony consists in continuous nerve stimulation, explained Pythagoras's ancient numerical findings, he argued. This scientific understanding of musical tones gave him grounds to criticize any and all numerological and mystical claims about the source of harmony. Such claims were "the foundation of extravagant and fanciful speculation." Included here were the ancient Greeks and Orientals, "the musical writings of the

Arabs," medieval claims about the harmony of the spheres, and Athanasius Kircher's claim that the macrocosm as well as the microcosm was musical. All were mere musical imaginings, he said. "Even Keppler [sic], a man of the deepest scientific spirit, could not keep himself free from imaginations of this kind. Nay, even in the most recent times, theorising friends of music may be found who will rather feast on arithmetical mysticism than endeavour to hear upper partial tones." All this wild, irrational speculation was anathema to Helmholtz. Instead, he pointed to the achievements of mathematicians and natural philosophers like Euler, Rameau, d'Alembert, and Giuseppe Tartini, who had led the way to a theory of consonance. Though he made clear what he saw as the shortcomings of their work and their lack of sufficient knowledge in certain areas, he made it equally clear that his own work built on theirs and that of others.[20]

Part three concerned the relationship of musical tones. Here he turned from the purely scientific, largely "mechanical" parts of his musicological analysis to examine such issues as musical style, tonality, and aesthetics. It was "the duty of science," he said, to continue its investigation until everything arbitrary was eliminated from scientific laws before then entering, as he did here in part three, into "the domain of aesthetics."[21]

He thought "historical and national differences of taste" helped shape the changing "boundary between consonances and dissonances," scales, and modes, not only "among uncultivated or savage people, but even in those periods of the world's history and among those nations where the noblest flowers of human culture have expanded." Musicologists and music historians had not sufficiently appreciated this point, he claimed. That did not mean that the elements of music were "arbitrary" or that music did not follow "from some more general law." Rather, it was the business of science "to discover the motors, whether psychological [that is, aesthetic] or technical [that is, scientific], which have been at work in this artistic process." Older musical elements and styles should not be judged by later ones. He discerned three main principles and eras of musical style: the homophonic of the ancients (to which also belonged the extant music of the Oriental and Asiatic nations), the polyphonic of the medieval period, and the harmonic of the modern, post-1500 period. He showed great admiration for the musical achievements of the Greek, Persian, Arab, Indian, and medieval Latin worlds, even as he pointed to their limitations. Yet he also spoke vaguely of "the less civilised nations." He judged that, though the roots of harmonic music issued from the ancient and medieval worlds, and though the pace of development had quickened after 1500, harmonic music in its full, modern form was barely two hundred years old and was "limited nationally to the German, Roman, Celtic, and Slavonic races." He considered modern, Western, harmonic music "the best of all," yet he repeatedly emphasized its multicultural roots and historical dependency on earlier systems.[22] This historical, evolutionary approach to music was entirely of a piece with nineteenth-century general intellectual discourse: music, like organisms and people, and like ideas and societies, had evolved and had a history. Helmholtz sprinkled his text with references to Aristotle, Boethius, Plutarch, and a variety of modern music historians and musicologists, revealing the breadth of his learning.

He also adumbrated a musical aesthetics but did little more in that direction. Instrumental music, he argued, "expresses the kind of mental transition which is due to the feeling." Different listeners to the same music are differently impressed: "They often adduce entirely different situations or feelings which they suppose to have been symbolised by the music." There is no true, unique impression here, he emphasized, "because music does not represent feelings and situations, but only frames of mind," which vary within individuals over time and from individual to individual. His analysis ultimately rested on a sense of tolerance and understanding of others.[23]

In considering aesthetic relations, Helmholtz argued that compound tones containing harmonic upper partials were "preferred for all kinds of music," melodic as well as harmonic, and that this "is subjective and conditioned by the construction of our ear." "On the other hand, the construction of scales and of harmonic tissue is a product of artistic invention," he said, "and [is] by no means furnished by the natural formation or natural function of our ear, as it has been hitherto most generally asserted." He here again warned that music was ultimately a cultural phenomenon, not one of nature per se. This did not mean that "the laws of the natural function of our ear" were not very important in influencing the type of music produced; rather, they constituted (only) "the building stones with which the edifice of our musical system has been erected, and the necessity of accurately understanding the nature of these materials in order to understand the construction of the edifice itself, has been clearly shewn by the course of our investigations upon this very subject." Music history showed, he said, "that the same properties of the human ear could serve as the foundation of very different musical systems." Music was "the work of artistic invention, and hence must be subject to the laws of artistic beauty."[24]

In addressing "artistic beauty in general," Helmholtz sought to "illustrate the darkest and most difficult points of general aesthetics," which he thought was "closely connected with the theory of sensual perception, and hence with physiology in general." He believed there was widespread agreement "that beauty is subject to laws and rules dependent on the nature of human intelligence" but that neither artists nor their listeners (or viewers) were consciously aware of these and that they remained unarticulated. Art only gave the appearance of being without design; its sources seemed to lie in the unconscious. He thought that any art produced by conscious laws and rules was poor art. Yet art should be "reasonable," by which he meant open to critical discussion, suitable, well arranged, and internally balanced; in short, it should be harmonic. The measure of a work of art's greatness was that the continued observation and intellectual analysis of it led to its being seen as ever "richer" and ever more reasonable. A person with "cultivated" artistic taste knew what was "pleasing or displeasing, without any comparison whatever with law or conception." A work of art was beautiful not through its individuality but rather because it was "in regular accordance with the nature of mind in general." That meant that others must agree that something in it is beautiful. To be sure, he acknowledged that individual and national tastes, education, and experience played some role in judging a work of art, but he thought these were all secondary considerations. The problem, then, was to determine how intuition succeeded in grasping a work's unconscious "regularity." That apprehension gave a sense "that the work of art which we are contemplating is the product of a design which far exceeds anything we can conceive at the moment, and which hence partakes of the character of the illimitable." Here he quoted Goethe (*Faust*), using one of his favorite quotations: "*Du gleichst dem Geist, den du begreifst*" (You are like the spirit you conceive). Both the artist and the listener or viewer were guided by "tact and taste," unconscious of the laws and rules at play in the work of art. To a certain but ultimately limited extent, he maintained, critical analysis could help one grasp those laws and rules of harmony and beauty.[25]

For Helmholtz, artists were, mentally speaking, just like nonartists, except that to one degree or another they had "a spark of divine creative fire, which far transcends the limits of our intelligent and conscious forecast." Since all human beings were similar mentally, he believed, a work of art gave the nonartist a feeling of moral uplift and "ecstatic satisfaction." Art thus allows people to see that "there slumbers a germ of order [in them] that is capable of rich intellectual cultivation, and we learn to recognise and admire in the work of art, though draughted in unimportant material, the picture of a similar arrangement of the universe, governed by law and reason in all its parts. The contemplation of a real work of art awakens our confidence in the originally healthy nature of the human mind, when uncribbed, unharassed, unobscured, and unfalsified."[26] Thus art, like science, was rational for Helmholtz, but there was a difference in that art did not allow any articulation of the unconscious laws that led to its creation and its aesthetic effects.

Understanding aesthetic considerations was thus like understanding sensory perception: it was "the apprehension of compound aggregates of sensations as sensible symbols of simple external objects, without analysing them." Though music and all art can imitate nature, Helmholtz maintained, it in fact goes "far beyond" such imitation. He ended his masterly, epochal book by noting that he could not go beyond these elementary musicological considerations and their relationship with physiological phenomena. For to do so, to discuss the role not only of harmony but also of rhythm, compositional form, psychological motivation, and so on, was, he thought, to go beyond natural philosophy. Though tempted, he left such deeper considerations to others and instead chose to "remain on the safe ground of natural philosophy," where he was "at home."[27]

Responses

The German Lands

Reactions to Helmholtz's book came almost immediately and, especially as new editions appeared, continued into the early twentieth century. One measure of the book's general impact among German readers (or those who read German) is that between 1863 and 1913 Vieweg published no fewer than six editions of the *Tonempfindungen*. Already in 1865, a second edition appeared. A third, revised edition appeared in 1870, and the fourth, and final revised edition appeared in 1877. After Helmholtz's death in 1894, two more editions were published by Vieweg (the fifth in 1896 and the sixth in 1913).[28] Thereafter, several other German presses have reprinted the book several times down to 2007.

Good friends naturally had good things to say about it. Brücke stopped reading it "only intermittently, in order to sleep, to drink coffee, and to write you [Helmholtz] these lines." Helmholtz had opened up a new intellectual path for him, and in 1871 Brücke published a small volume on speech that built on Helmholtz's work on the pitch of musical tones and on vowels. Helmholtz's book addressed Brücke's "inner self," he said. Ludwig, another keen scientific student of the arts, thought Helmholtz had doubtless heard "so many beautiful things" about the *Tonempfindungen*. In the autumn of 1863, the book became Ludwig's "constant companion"; he compared it to the work of a master Renaissance painter. But he also thought that, with the exception of Hanslick, "who very much raves about you," Vienna's musicians were "too immature" to understand the book. Helmholtz himself told Ludwig that he thought, "On the whole [the book] has had more *succès d'estime* than it has convinced people. Incidentally, I never had any illusions that it would be otherwise." Ludwig begged to differ, and when the book's second edition appeared, he claimed that the public proved him (Ludwig) more right than Helmholtz. Helmholtz's old friend Klaus Groth of Bonn, a poet, became fully convinced of the book's importance, but he too thought that musicians would not be able to understand the scientific parts and that physicists and physiologists did not have enough interest in music to appreciate the musicological part.[29]

The nonfriends were somewhat less kind. The philosopher Lotze enthusiastically received and wrote about Helmholtz's musicological work. Like many others, he was especially impressed with the many new and interesting acoustic and physiological facts that Helmholtz presented concerning music. He was less impressed, however, with the "aesthetic meaning" of these facts. Still, he declared that what counted here musicologically was the effect that the sensations of tone had "on the soul," and so he understood (as not quite everyone else did) one of Helmholtz's biggest philosophical points.[30]

The young Ernst Mach was also an early reader and (initially) an enthusiast. At Vienna, where he habilitated in physics in 1861, Mach studied under Brücke, met Ludwig, and, thanks in no small part to Helmholtz's works, developed a deep, lifelong interest in sensory physiology. Living in a tight financial situation as a private lecturer in physics, he gave private lessons to help make ends meet, including lectures on the *Tonempfindungen*. He published these as *Einleitung in die Helmholtz'sche Musiktheorie: Populär für Musiker dargestellt* (1866), giving a sympathetic and well-received account of Helmholtz's theory. He sent a copy to Helmholtz, who belatedly sent the young, then unknown lecturer his appreciative thanks for Mach's help in making his book more understandable.[31] Although Helmholtz's book strongly influenced Mach's thinking during this early phase of his career—for example, it led him to analyze mathematically the ear's processing of tones—he later developed major differences with Helmholtz about hearing, musicology, and epistemology.

Some nonscientists could not or would not understand the *Tonempfindungen*'s scientific parts. The *Niederrheinische Musik-Zeitung* found Helmholtz's book "important," "epochal," and "excellent" and thought everyone recognized it to be so, but nonetheless pointed out that musicians did not generally recognize it or fully appreciate its results. It was so scientific (too much physics) that even many individuals with an interest in musical theory could not understand it. Oskar Paul, a leading music scholar and musicologist, called Helmholtz, with reference to the *Tonempfindungen*, "the greatest acoustician of our time." Instrument makers now had to learn Helmholtz's acoustic theory, he said, not least since his work had already influenced them, as could be seen clearly at the World Exposition in Paris (1867).[32]

Music journals and works were generally very positive about the book. The *Allgemeine Musikalische Zeitung* lauded its value for musicians, not least since it brought art and science together. The *Musikalisches Konversation-Lexikon* provided an individual, highly admiring entry on Helmholtz, calling him "the most important contemporary researcher in the field of acoustics," and said that his book shone "light into the darkest parts of the science of music" and solved "in the simplest way" puzzles before which philosophers and musical theorists had until now "stood helplessly by." In the *Musikalisches Wochenblatt: Organ für Tonkünstler und Musikfreunde*, Gustav Schubring reviewed Helmholtz's life and scientific work, noting that since 1863 his name had appeared "frequently enough" in music journals and observing that though his book was widely discussed among musicians, attitudes toward it certainly varied. Still, it became the marker for those who generally favored the "scientific" approach to musicology, and he thought it had import for practicing musicians. Helmholtz's scientific instruments helped make various musical principles more understandable and, especially, allowed much better tuning of instruments and voices. *Die Musik*, finally, praised Helmholtz and his views of music but noted that his book "was much read but little understood."[33]

After studying the *Tonempfindungen*, the psychophysicist Fechner wrote Helmholtz of "the justified wonder which the world, myself included, renders to your works, and which can only be increased by your *Tonempfindungen*." He warned Helmholtz that certain philosophers of aesthetics, in particular Robert Zimmermann, who favored an abstract formalist aesthetics, was misusing Helmholtz's views in his own book. Meanwhile, William Thierry Preyer, professor of physiology at Jena and a friend of Helmholtz's, arrived at some new acoustic results that disagreed with Helmholtz's. But Preyer assured him, "Everyone recognizes on every page of my writings their dependency on your fundamental works," "without which my experiments could not have come about."[34]

Indeed, although there were many individuals, like Arthur von Oettingen, a physicist at Dorpat, who thought that Helmholtz's book marked the beginnings of modern music theory and that it had the great merit of bringing science and art together, there were few like him who could criticize it intelligently, let alone propose an alternative. Oettingen himself sought respectfully to do this with his so-called dual-harmony system, which was different from but built upon Helmholtz's. The *Allgemeine Musik-Zeitung* declared that very few musicians could understand the physics and the physiology presented in the *Tonempfindungen*, and so to a greater or lesser extent they remained skeptical of the book's contents. Selmar Bagge declared in the *Leipziger Allgemeine Musikalische Zeitung* that most musicians simply could not understand Helmholtz's book. Similarly, the physicist Felix Auerbach, who had studied with Helmholtz, wrote a piece in 1881 explaining and praising Helmholtz's book to readers of *Nord und Sud*, which catered to a generally educated but nonscientific audience. Auerbach conceded that Helmholtz had his opponents, particularly among musicians.[35]

Among the outright critics, perhaps the harshest were Moritz Hauptmann and his followers. Until 1863 Hauptmann's *Die Natur der Harmonik und der Metrik* (1853) had constituted the leading music theory. It was written from an idealist, not to say Hegelian, point of view (as Helmholtz, among others, thought). Hauptmann, who confessed that he had only read in, rather than thoroughly through, Helmholtz's book, thought it failed to portray the overall architecture of music, even as he recognized that it had some valuable things to say about individual parts of music. It lacked a theory of music even as it showed some interesting empirical findings. Bagge, who favored Hauptmann's views, recognized, however, that Helmholtz's book was of a relatively more popular nature than Hauptmann's. Moreover, scientists were extending Helmholtz's work, something that did not seem possible with Hauptmann's.[36] Helmholtz's highly popular book swept Hauptmann's from the market.

Helmholtz's most powerful and informed critics appeared only after 1873, however. Mach now strongly criticized Helmholtz's beat theory as the basis of musical consonance, and thus his harmony theory, and he offered an alternative model. Koenig, who had done so much to promote Helmholtz's theory and who had greatly helped him by making customized resonators for him, came to doubt Helmholtz's idea of combination tones and his explanation of their production, arguing instead for the importance of beat tones. Hugo Riemann, who as a young musicologist and music teacher largely accepted Helmholtz's approach, thought Helmholtz had created the "cultural" foundation on which Riemann developed his own work. He was in agreement with him on many principal points, even if he did not always agree with Helmholtz's understanding of upper partials. He asked Helmholtz for a letter of recommendation to help him become professor of music at the University of Bonn; he thought such a letter would be enormously helpful. (He got the letter, but not the position.) Riemann ultimately proved, however, to be a far more formidable opponent to Helmholtz than Hauptmann had been. For unlike Hauptmann, and like Oettingen, he claimed that his own approach was "scientific," and so based on nature. He developed Hauptmann's and Oettingen's notion of "harmonic dualism" (roughly, that both undertones and overtones were to be found in all harmonic series). This ever-evolving notion led him to become the leading German musicologist by the turn of the century. Even so, both he and Oettingen built upon Helmholtz's nascent musicology, even as they differed with him in central ways. Much the same also held for Carl Stumpf, an experimental psychologist and philosopher who, though very sympathetic to Helmholtz's work and empirical approach, thought that consonance was due to a perception of tones that were blended (or fused) together. Whereas Helmholtz emphasized the sensations of tone, Stump emphasized their perception.[37]

In 1863 Helmholtz was appointed a privy counselor to the Grand Duke of Baden, probably thanks in part to the enhancement of his reputation through the *Tonempfindungen*. The publication also helped spread his name among other educated but nonscientific people in the German lands and beyond. Nietzsche, for example, had a copy in his library. In 1875 Helmholtz's picture appeared on the front page of the Sunday weekly *Über Land und Meer: Allgemeine Illustrirte Zeitung*. An accompanying article informed readers that the *Tonempfindungen* was a work that every layperson should read. In 1876 *Daheim*, another general, middle-class magazine, also published Helmholtz's portrait and referred to the *Tonempfindungen* as "this most important work," for its outstanding science but also for the help it gave to laypersons in appreciating the scientific foundations of music.[38] In short, Helmholtz's book contributed to music education among the *Bildungsbürgertum* in the German lands.

France

The *Tonempfindungen* enormously enhanced Helmholtz's reputation in France and the French-speaking world. The year 1863 marked a profound change in his reputation there. Before then, his physiological work had not been particularly well received in that world, and his physics papers had drawn little notice. His reputation now underwent a reversal of fortune, particularly after his first visit to Paris (1866). Rodolphe Radau declared that research on the scientific theory of music was essentially something new and that Helmholtz held "the key" to explaining music scientifically to his readers. The "illustrious physiologist," Radau explained, had built on the work of such Enlightenment predecessors as Rameau and d'Alembert and had shown the physiological foundations of music. Privately, however, in a friendly but respectful manner, Radau also told Helmholtz what he thought were the book's minor weak points.[39]

Publication of the *Tonempfindungen* also significantly boosted Helmholtz's international reputation as a man of high culture. Mary von Mohl, who had yet to meet her niece's famous husband, wrote to a friend that she could not understand how physiology and mathematics went together, but she said, "He has a great reputation, so I bow my long-ear'd head and bray." In April 1866 Helmholtz spent two weeks in Mary and Julius von Mohl's home in Paris. She invited many people to meet him, including Auguste Laugel, a writer who had trained as an engineer and did a good deal of science writing, "because he makes a great talk about Helmholtz."[40]

Helmholtz had gone to Paris to meet other scientists and almost certainly at the urging of his Francophile wife. His first stop was the Louvre, to see its Italian paintings. However, he had to leave in midmorning to breakfast with the mathematicians Charles Hermite and Henry J. S. Smith (the Savilian Professor of Geometry at Oxford). Smith told Helmholtz that Oxford wanted him as its professor of physics. Friedrich Max Müller, a German who taught at Oxford, had assured them that Helmholtz would never accept, and so no formal offer was extended. Helmholtz found Hermite "very flattering" toward him. Hermite, then France's leading mathematician and an extremely well-connected one in Paris, introduced him to Louis Grandeau, an agronomist who was an assistant first to the chemist Henri Sainte-Claire-Deville and then to Claude Bernard. Grandeau brought Helmholtz to the Ecole Normale Supérieure, where he met Sainte-Claire-Deville, who received him most warmly and showed him his physics cabinet. Helmholtz was applauded by students as the two men passed through a classroom. "You thus see," he wrote Anna somewhat ironically, "that I'm already a sort of popular character here." At the Ecole, he also met the ophthalmologist Louis-Emile Javal and Soret.[41]

Helmholtz began to appreciate Paris's beauty: cracks appeared in his old prejudices against France. Grandeau and Laugel brought him to see the atelier of Aristide Cavaillé-Col, one of the era's foremost organ builders; Helmholtz was much impressed. They continued on to Saint Sulpice to see the church's organ, the largest in Europe, which had been built by Cavaillé-Col. Grandeau also took him to a conservatory to hear a Haydn symphony, part of Beethoven's *Prometheus*, the entire *Sommernachtstraum*, a Bach choir, and Handel's *Halleluja*. Though he thought the chorus in Germany was better, he also thought the orchestra in Paris was unique and certainly far better than what Heidelberg had to offer. Its performance of Mendelssohn's *Ouverture* was superb. The concert and seeing the Venus de Milo were "life experiences." He told Mach how impressed and pleased he had been with musical Paris and exclaimed about how well the instrument makers there knew his theory. Moreover, he thought, "In France mathematical-physical education is more widespread than unfortunately is the case in Germany, and at the same time there is much understanding of music."[42]

He visited the Jardin d'Acclimatation (a zoological park where animals could more easily acclimatize themselves), the aquarium, the Bois de Boulogne, and Passy and then went to the Institut de France, where he experienced a working session and met several of its leading members. "The Académie looks far better in print with its *Comptes rendus* than in person," he said. His unflattering remark reflected the decreasing importance of academies of science to the advancement of science. That evening, he dined with the historian François Mignet, the savant and statesman Jules Barthélemy-Saint-Hilaire, the chemist Michel Eugène Chevreul, the jurist Louis Renault, the Oxford zoologist Lovell Reeve (and family), and the Danish aesthetician, literary critic, and historian Georg Brandes.[43]

Shortly thereafter he went to the Jardin des Plantes, where the physicist Edmond Becquerel (son of physicist Antoine César Becquerel and father of physicist Henri Becquerel) had shown him "very pretty and interesting" experiments concerning bodily phosphorescence. He then went to the Galérie d'Anthropologie and made another trip to the Ecole Normale to meet Grandeau and Sainte-Claire-Deville. Then, to his amazement, Victor Duruy, the minister of public instruction, and one of his advisers unexpectedly appeared. Duruy asked him to give an impromptu lecture on the analysis of vowels, which he did, "without getting stuck." Duruy strongly supported more research in France and so had come to meet Helmholtz and question him "extensively," especially about the facilities in Germany for medical instruction. The minister "spoke badly enough about French medical men" and explained that he would be very happy if he could "win over" Bunsen, Kirchhoff, and himself (Helmholtz) "for France." Duruy's visit reflected deep French concerns about the state of scientific research there in the 1860s and the scarcely latent rivalry between the Germans and the French.[44] Helmholtz's remarks suggest that, in science and medicine, some French leaders felt themselves ever weaker relative to the Germans.

That evening he dined at Javal's home, where other guests included several National Assembly deputies as well as the political economist Michel Chevalier and the journalist Adolphe Guéroult. They discussed "literary property," since the government was then proposing a new law in this area. One of their company "proved himself to be a rather red socialist," but Helmholtz enjoyed the evening nonetheless. He also visited the Luxembourg Museum, though he thought most of its contents were "repulsively crass or lifeless, or both," and he again visited the Louvre, this time to see various sketches. The Mohls again invited people to their exclusive salon to meet him; among them were the historian and politician François Guizot and the natural historian and anthropologist Jean Louis Armand de Quatrefages de Bréau. He breakfasted the next morning with Barthélemy-Saint-Hilaire.[45]

Helmholtz soon reaped concrete results from this trip to Paris. Two months afterward, a French translation of his essay on ice and glaciers appeared, although his essay on the conservation of force did not appear in French for another three years. The following year a French edition of his *Handbuch* came out, as well as French translations of his essays on the relation of the natural sciences to science in general and the essay on the physiological origins of musical harmony. A French translation of the *Tonempfindungen* was also begun; it appeared in 1868. Helmholtz reviewed it while in press. He said that his fundamental theoretical views remained the same, even though he acknowledged that some critics had attacked him over the issue of the difference between major and minor chords. However, he was no longer so confident about what he had said about the history of music, in part because of his lack of expertise and the limited sources he had to work with, and in part because the field was largely undeveloped. The French and the Belgians had done more for music history than the Germans, he said. Finally, he noted that during two recent trips to Paris—the second one in August 1867—he "found amongst a large number of French *savants* and musicians a more favorable reception" than he had dared to hope. He said publicly what he had told Mach privately: the French had a better combined understanding of music and the use of scientific thinking to understand it than other peoples of Europe did. A second printing of the French edition of the *Tonempfindungen* appeared in 1874.[46]

Helmholtz came into deeper contact with French culture in other ways. In 1867 Emile Alglave, the editor of the *Revue des Cours Littéraires et Scientifiques*, wanted to republish several essays on the mechanical theory of heat, including something by Helmholtz (on conservation of force). But he hesitated to publish some of them since they were a little old, "even if they have remained unknown in France." In 1877 he published Pietro Blaserna's *Le son et la musique* along with Helmholtz's lecture on the physiological causes of harmony in music. (A decade later, Alglave published the French translation of Helmholtz's lecture on the foundations of geometry as well.) Blaserna was an Italian physicist who was educated at the University of Vienna. In 1872 he became professor of experimental physics at Sapienza University of Rome. Like his good friend Helmholtz, he had broad interests in both physics and the arts (especially music). In 1875, working within a completely Helmholtzian framework, he published *La teoria del suono nei suo rapport con la musica*. He was an admirer of German (and French) science in general and was a major institutional officeholder in Italian physics and science in general. For example, he became a member of the Accademia dei Lincei in 1873, its assistant secretary from 1877 to 1879, and eventually its president (1904–1916). He referred to Helmholtz's *Théorie physiologique de la musique fondée sur l'étude des sensations auditives*, to use the French title of the *Tonempfindungen*, as a "classic" and said his own study followed Helmholtz's example of trying to unite the science of sound with that of music. In 1868 Louis Pérard, professor of physics at Liège (Belgium), did the first French translation of Helmholtz's essay on the conservation of force. It was now so rare that he had been unable to purchase it and had to borrow a copy from a colleague. Pérard said that Mayer's and, especially, Helmholtz's essays were not as well known beyond Germany as they should have been, a comment that said far more about the French-speaking scientific world than it did about Helmholtz. Pérard also translated Helmholtz's essay on the interaction of the forces of nature.[47]

Moreover, Laugel brought attention to the *Tonempfindungen*. He thought Helmholtz's writings on acoustics the most impressive he had ever seen and believed they were of interest to students of physics as well as aesthetics. He called Helmholtz "the indefatigable professor from Heidelberg" who had renewed acoustics by building on the foundations laid by Newton, Euler, Laplace, and Poisson. Helmholtz had turned acoustics, "this arid and banal science," into "a branch of universal dynamics as well as of aesthetics." Helmholtz had completed the work begun by Rameau and declared, "His instruments furnish to harmony some sure guides; the analysis of sounds becomes as easy and precise as it was formerly vague and difficult." Helmholtz had also made a penetrating analysis of the ear's operation. For Laugel, Helmholtz had revolutionized acoustics and had unlocked the secret of musical harmony.[48]

Not everyone shared this panegyrical view. In a review of Laugel's book *La voix, l'oreille, et la musique* (The voice, the ear, and music, 1867), Gustave Bertrand was quite critical of Helmholtz's musicological analysis. To be sure, he thought Laugel's book, like Radau's article, helped students of musicology understand

Helmholtz's book, and he even praised some of Helmholtz's accomplishments. But he had doubts as to how well Helmholtz had succeeded in explaining melody, harmony, and musical aesthetics in terms of acoustics. He also praised Helmholtz's analysis of timbre, but he believed that Helmholtz had sought to do too much when, like Rameau, he built his entire musical system around the phenomena of harmonics. While Helmholtz was "the eminent physicist from Heidelberg," Bertrand did not like the idea of explaining music in terms of harmonics. Helmholtz had tried to create a utopia. He was a creator of systems, and Bertrand objected to this. Nevertheless, he had explained the theory of timbre, and that was a very big accomplishment.[49] Such critiques could be found on both sides of the Rhine.

Still, Helmholtz's reputation and influence in France experienced a complete turnaround from that of circa 1850. Small wonder that in January 1864 he was elected an honorary member of the Académie Royale de Médicine de Belgique and that in 1870 he became a corresponding member of the Académie des Sciences.[50]

Britain

Helmholtz's solid anchoring in the British scientific scene since the early 1850s seemingly assured that the *Tonempfindungen* would get attention in the English-speaking world. Yet the matter proved to be not so simple. Tyndall thought the book an "excellent work" and believed there was a market in Britain for reading Helmholtz on music, but only in English. He approached Longmans about publishing a translation; but Longmans thought it would be unprofitable and refused. The firm did, however, agree to publish Helmholtz's *Populäre wissenschaftliche Vorträge* in English.[51]

Tyndall was not alone in wanting to see an English translation, but Longmans was not alone in its market assessment. Within six months of its appearance, Vieweg declared Helmholtz's book to be a scientific classic but said its readership was largely limited to the German-speaking world. He did not want to see an English translation now, since he thought the market among the German-reading public in England and elsewhere would disappear once an English translation was available.[52] But in any case, who would translate and publish it?

The catalytic figure here was Max Müller, who soon became Corpus Professor of comparative philology at Oxford. Shortly after the *Tonempfindungen* appeared, he wrote Helmholtz, whom he did not yet know personally, to say that he had just finished reading part one "with great interest" and to note "the great instruction" that he had received from it. He had just given two well-attended lectures at the Royal Institution on linguistics, and so he was especially interested in Helmholtz's work on vowels. One of his listeners was Alexander John

Ellis, a "Cambridge man," who worked on phonetics and wanted to translate Helmholtz's book. Ellis had a thorough command of German, Max Müller said, and was also well instructed in mathematics and acoustics. He thought Ellis would be unsurpassed as a translator of Helmholtz's book. Ellis had asked Max Müller to discuss the matter with Helmholtz. Max Müller, like Tyndall, thought the book would enjoy good sales in Britain. He told Tyndall: "I am much pleased with Helmholtz's book, and should give a great deal to be able to hear your lectures on Sound, and to see some of the experiments which, though so well described by Helmholtz, are yet imperfect and unsatisfactory on paper."[53]

Ellis began translating Helmholtz's book in 1863. He asked Tyndall, who had just been lecturing on sound at the Royal Institution and, in the process, often mentioning Helmholtz's book, to write a letter to potential publishers. Tyndall's letter of support spoke of Helmholtz's "excellent" book, calling it "incomparably superior . . . to anything hitherto written on the same subject." It was "written so clearly as to be quite intelligible to the general reader." He characterized Helmholtz as "a man of the very highest scientific eminence" and said, "His name alone is a guarantee of the superior character of any work to which it is appended." But the book still could not find an English-language publisher. Ellis's proposal was turned down by no fewer than seven English publishers. In general, they thought such a publication too costly and the market for it too small, even as they recognized Helmholtz's reputation and the work's quality and general cultural importance. Ellis simply could not get the book published in English. Helmholtz himself entered the fray by asking Henry Wentworth Acland, the Regius Professor of Medicine at Oxford and the dominating figure there in medicine in the second half of the century, to try to "persuade" any of his booksellers to print it. As late as 1871, Tyndall again sought to secure an English-language publisher; Longmans again refused, though this time the company expressed an interest in publishing a condensed version.[54]

In the meantime, for those who could not read German or (until 1868) French, there were two principal vehicles for learning about Helmholtz's musicological thought in English. One was Tyndall's recent lectures published as *Sound* (1867); the other was a book by Sedley Taylor, of late a fellow of Trinity College, Cambridge, and a former president of its Musical Society and Musical Club, who presented Helmholtz's chief musicological results in semipopular form as *Sound and Music* (1873). Ellis told Helmholtz that Taylor's book "introduced" his "theories and discoveries to English readers." Taylor had long been a partisan of Helmholtz's but at odds with Tyndall over the latter's presentation of Helmholtz's music theory. He told Helmholtz that Tyndall's *Sound* presented Helmholtz's "theory of musical consonance" in a way that was "entirely erroneous." Tyndall disagreed, and Taylor wanted to know if Helmholtz thought Tyndall's reply was adequate. Helmholtz agreed with Taylor that Tyndall had not given "a *complete* exposition" of his "theory of Harmony," but he added, "I suppose, he never intended to do so." Helmholtz thought, however, that Tyndall had "touched on the principal point" of his theory, which was all that could be expected in a brief lecture. "I think, a popular lecturer ought to have the liberty, to choose, on what he thinks convenient to speak," he said. "Popular lectures have a great degree of real utility, if they give to the reader a vivid impression of a number of simple facts, but of course, they cannot be at the same time scientific manuals." Taylor thought the issue was not that of a popularizer's freedom of choice, but rather "that Prof. Tyndall's statements on the theory of harmony" were "not merely *incomplete* but *fundamentally erroneous*." In 1875, moreover, Taylor defended Helmholtz's theory against a long, highly critical review of it, indeed an attack, by William Chappell. Helmholtz had little patience for Chappell, whom he thought incompetent in this and other matters. He certainly thought Taylor's brief rendition of his musicological work was "very useful," and he hoped to see a German-language version of it. As for his own book, it had "indeed been much used by German musicians," as he told one potential expositor. However, many of his readers simply took the individual points they needed from the book and adapted them to fit their own needs, thereby distorting the larger context in which he discussed those particular points.[55]

But neither Tyndall's nor Taylor's accounts of Helmholtz's book were sufficient for those who wanted a firsthand account. And besides, Ellis refused to abandon the project. Between 1863 and 1873 he intermittently translated much of the book and discussed his project with two publishers. He was on a mission. In February 1873, Longmans, which was then publishing the English edition of Helmholtz's *Populäre wissenschaftliche Vorträge*, finally showed some interest. Ellis did not finish the translation until August 1874: it was a labor of love, not least since it contained many long editorial remarks by Ellis. To Helmholtz's own many footnotes and nineteen appendices, Ellis inserted his own comments bracketed in the text, added his own set of footnotes, and included an additional appendix of 126 pages that clarified and commented on various points. He thus did far more than translate the book, which finally appeared in July 1875, twelve years after he had first broached the idea.[56]

Publication of the translation further enhanced Helmholtz's reputation abroad. In his preface Ellis noted "the high scientific position" the volume had "assumed immediately on its first publication in 1862 [sic], and which, after passing through three editions in Germany, it now occupied as the indispensable preliminary to all scientific investigations on the nature of music and our sensations of sound in general." Helmholtz's work contained beautiful explanations of phenomena that, Ellis wrote, had "perplexed philosophers and musicians from the time of Pythagoras in Greece, and thousands of years before Pythagoras in China, to the year 1862, when Prof. Helmholtz first gave his results to the world in a connected form." Here was "a master's hand directed by a master's mind." In the second English edition (1885), Ellis noted that Helmholtz's book had "taken its place as a work which all candidates for musical degrees are expected to study." He added that Helmholtz's "theories . . . were quite strange to musicians when they appeared in the *first German* edition of 1863," but that in the twenty-two years since, they had been received "as essentially valid by those competent to pass judgment."[57] Subsequent English-language editions appeared, the last in 1948, and the most recent reprint in 2010.

Helmholtz found another English expositor, along with Tyndall, Taylor, and Ellis, in William Pole, a professor of civil engineering at University College London and a musician. At the request of William Spottiswoode, a London printer who was also an amateur mathematician and a physicist of some repute, as well as a cofounder of the British Musical Association, Pole gave a series of lectures at the Royal Institution in 1877 on the philosophy of music. He told Helmholtz that he "had often spoken" with Spottiswoode "in admiration of the *Musical* part" of his "great work, as distinguished from the Physical part." Then he reported, "He wished me to endeavour to give some popular account of its main features, from a musical point of view." Pole sent Helmholtz an outline and an announcement of these lectures, as well as the page proofs. His admiring exposition of Helmholtz's book appeared as *The Philosophy of Music* (1879).[58]

Helmholtz's work thus became widely known among British readers. Many leading British intellectuals read him on music, either directly or indirectly. They learned of his work through his several British interpreters as well as in the pages of *Nature* (founded 1869) and of *Mind* (established 1876). The articles in the fourth volume of *Mind* (1879) in no small measure referred to Helmholtz's acoustics, optics, or aesthetics. The poet Gerard Manley Hopkins, for example, was a keen reader of science in general, especially the works of Tyndall and Helmholtz, both of whom had important influences on his notions of meter, sound, and color. George Henry Lewes, a widely read man of letters, critic, and amateur physiologist, was another keen reader of Helmholtz. In 1868 Lewes, who lived apart from his wife and was the lover of the novelist George Eliot, traveled with Eliot to Germany, where he (but not she) met Helmholtz in Heidelberg. His diaries indicate that he read both Helmholtz's *Handbuch* and the *Tonempfindungen*, as well as Helmholtz's popular and philosophical essays. Eliot herself was a keen student of science and an erudite scholar as well as a gifted creative writer; she was then working on her masterpiece *Middlemarch* (1871–72), a work that often touched on the natural sciences. She recorded in her diary in 1869: "I am reading about plants, and Helmholtz on music."[59]

Another major British intellectual figure who followed Helmholtz's musicological (and other) work was Darwin. In his *Expression of the Emotions in Man and Animals* (1872), Darwin cited Helmholtz on vowel formation, which he sought to relate to expressions like pain and laughter. To a correspondent who asked for help in understanding the relationship between sound and the ear, Darwin advised: "I presume that you have of course studied Helmholtz's work." He first discussed Helmholtz on harmony and disharmony in the *Descent of Man* (1871):

> Therefore, as far as the mere perception of musical notes is concerned, there seems no special difficulty in the case of man or of any other animal. Helmholtz has explained on physiological principles why concords are agreeable, and discords disagreeable to the human ear; but we are little concerned with these, as music in harmony is a late invention. We are more concerned with melody, and here again, according to Helmholtz, it is intelligible why the notes of our musical scale are used. The ear analyses all sounds into their component "simple vibrations," although we are not conscious of this analysis.

Darwin knew who to lean on.[60]

Helmholtz's most profound British reader (at least in acoustics), however, was John William Strutt (Lord Rayleigh), a distinguished physicist and well-to-do aristocrat, who both drew on and surpassed Helmholtz as a scientist of sound. The *Tonempfindungen* inspired Rayleigh to work on resonance. Maxwell, for his part, told Rayleigh that he thought it was the best book on sound. He wrote Rayleigh: "You speak modestly of a want of Sound books in English. In what language are there such, except Helmholtz, who is sound, not because he is German but because he is Helmholtz." In his Rede Lecture ("On the Telephone") of 1878 at Cambridge, Maxwell also said that Helmholtz's book and his work in acoustics and music in general constituted "a series of daring strides . . . over that untrodden wild between acoustics and music—that Servonian bog where whole armies of scientific musicians and musical men of science have sunk without filling it up." In his view, Helmholtz's theoretical and instrumentational work in electromagnetism and in sensory physiology served as an important source upon which Bell (and others) drew as they invented the telephone. It is a measure of just how aware two of the era's leading physicists were of technological developments and possibilities that both Maxwell and Helmholtz independently published articles in 1878 on the telephone. In 1877 Rayleigh published his own masterpiece, *The Theory of Sound*. In it he maintained, "A large part of our knowledge upon this subject is due to Helmholtz, but most of the workers who have since published their researches entertain divergent views, in some cases, it would seem, without recognizing how fundamental their objections really are." He urged his readers to study Helmholtz's views themselves. "Only one thoroughly familiar with the *Tonempfindungen* is in a position to appreciate many of the observations and criticisms of subsequent writers." He referred to Helmholtz's "great work" and noted that the last chapter of his own book, "Facts and Theories of Audition," was built on Helmholtz's work: "Constant reference to the great work [the *Tonempfindungen*] is indispensable. Although, as we shall see, some of the positions taken by the author have been relinquished, perhaps too hastily, by subsequent writers, the importance of the observations and reasonings contained in it, as well as the charm with which they are expounded, ensure its long remaining the starting point of all discussions relating to sound sensations." The relationship between Helmholtz and Rayleigh, the two leading acousticians of the era, became (on the intellectual plane) one of mutual self-help: When Rayleigh's book appeared in 1877, Helmholtz praised it in *Nature*; then he got one of his assistants, Friedrich Neesen, to do a German-language translation of it (1878).[61]

At Cambridge, where first Maxwell (1871–79) and then Rayleigh (1879–84) was the Cavendish Professor of experimental physics, acoustic science was taught and researched in an up-to-date manner, and musical studies were taught in the light of science. The Board of Musical Studies assured the traditional place of musical study within a liberal education. It did so on the scientific side through the teaching of Rayleigh's *Theory of Sound* and "on the aesthetic side" through the University Musical Society as well as through lectures by Sedley Taylor, "where the wail of the Siren draws musician and mathematician together down into the depths of their sensational being," as Maxwell had it. Every candidate for a musical degree at Cambridge was "expected to study" the *Tonempfindungen*. There and elsewhere their path was eased after 1881 by John Broadhouse's textbook *Musical Acoustics; or, The Phenomena of Sound, as Connected with Music*, or, as on the title page, *The Student's Helmholtz*. The book aimed to prepare students at Cambridge and London for the acoustic portion of their set examination papers in music and, more generally, to teach "cultured people" about musical acoustics. Its outlook and substance were completely Helmholtzian. Broadhouse quoted liberally from the *Tonempfindungen* as well as from the major acoustic (Helmholtzian) works of Pole, Tyndall, and Taylor, among others. The appendix provided a set of "Examination Questions" that further revealed Helmholtz's influence on musical acoustics for Cambridge, London, and other students. At a more popular level, John Curwen, the British inventor of the Tonic Sol-fa method of musical instruction, employed Helmholtz's observational work on voice to support his own understanding of quality of voice, citing Helmholtz as an authority on how vowels function in singing.[62]

America

As in Britain, *On the Sensations of Tone* found a large number of readers and converts in America. In 1866 the young Alexander Graham Bell began studying vowel sounds as part of his quest to understand speech and to aid the hearing-impaired. He wrote in this regard to Ellis, who was then repeating Helmholtz's work on the use of tuning forks to determine musical vowel tones. Ellis advised Bell to study Helmholtz's book, which he did, taking the book with him when he departed Britain for Canada in 1870. Bell became a keen student and a great admirer of Helmholtz's acoustic works, especially his vowel theory. His belief that speech could be transmitted via the telegraph in turn inspired him to study electricity. His work with tuning forks and electromagnets for studying and transmitting speech became key elements in his route to the telephone in 1876. Thus his study of Helmholtz's book helped lay the foundation for the telephone's development.[63]

At virtually the same moment, in July 1875, Thomas Alva Edison began experiments in acoustic telegraphy for Western Union. To learn something about acoustics, he turned, inter alia, to Helmholtz, acquiring the just-issued English-language edition of the *Tonempfindungen* and reading it carefully, as his extensive marginal notes show. His interest in transmitting articulate speech led him later that year to construct a so-called water telephone: a tuning fork with a resonant chamber for speaking; the chamber was attached to the fork's upper arm, while the lower arm held a cup containing mercury (labeled "mercury like Helmholtz"). Edison also redesigned Helmholtz's proposed electromagnetic tuning fork to use it as a telegraph instrument. His work employed a Helmholtz resonator, too. In addition, Helmholtz's aesthetics, with its rational, orderly approach to sound transmission, sensation, and perception, had special appeal to the hearing-impaired and music-loving Edison as he worked on the telephone, the telegraph, and the phonograph. He also bought a copy of Helmholtz's *Popular Scientific Lectures* soon after the book appeared. In turn, Edison's invention of the phonograph (1877) itself soon allowed the British engineer Fleeming Jenkin and the British physicist James Alfred Ewing, on the one hand, and the American physicist Charles R. Cross, on the other, to test Helmholtz's theory of vowels.[64]

Another major American debt to Helmholtz and the *Tonempfindungen* issued from the piano-making firm of Steinway & Sons. The firm's emergence at mid-century reflected the new understanding of acoustic phenomena, the wider availability of precision instrumentation, and, more broadly, the increased role of science in industry, the emergence of the middle classes in Germany and America, and a desire for higher culture among the elites of those and other nations.

In 1850 Heinrich Engelhard Steinweg and four of his five sons emigrated from Germany to New York, where they refashioned themselves as "Steinway" and established a new piano factory. They found extraordinary commercial success through a combination of business acumen, the ability to mix with artists and other professionals, and constant technological innovation. The firm became distinguished for the high quality of its piano construction, and it discovered or invented nearly all the major technological changes that modern grand and baby grand pianos came to incorporate. Among its major innovations were the use of the principle of cross-stringing and using strings that were strung in a fan-shaped manner so as to greatly increase the piano's general tension. Another was to relocate the sounding-board bridge near the middle and introduce the duplex scale. From the late 1860s onward, Steinway & Sons gradually reduced its competitors to insignificance. It produced pianos whose tones were strong and sensitive and of unprecedented volume and sound quality. The Steinway name became synonymous with excellence in pianos.[65]

Henry E. Steinway and, especially, his son Theodore were the principal leaders in the firm's technological innovations. Theodore had studied physics, including acoustics; he knew Helmholtz personally and knew his writings on acoustics. Under Theodore's direction "science became the guide" in piano-making at the firm. The firm patented several of its technological improvements, including (in 1872) that of "the important double duration of a note [*Doppelmensur*]," whose discovery it attributed to Helmholtz as part of his investigations on tones and strings. Theodore's study of Helmholtz on tone supposedly led to his decision to restring the piano so as to include the entire string. The result was his duplex scale, which gave the string a double vibration and hence a richer sound.[66]

As with the work of the German-American student of voice Emma Seiler, Helmholtz's work on sound penetrated the developing American culture. His book and his views had a practical influence on the understanding and development of the piano: to say the very least, the Steinways (like Bell and Edison) studied his book, there was contact between the Steinway firm and Helmholtz, and the introduction of resonators was used to determine harmonics. Helmholtz's musical acoustics helped musicians understand the piano from a mechanical point of view, including design, sound, pedaling, and touch. His book became widely known among piano manufacturers, teachers, and performers. Rudolph H. Wurlitzer, the son of the founder of the Wurlitzer musical instrument company and "the artistic genius" of the family, went from Cincinnati in 1891 to Berlin to study acoustics with Helmholtz, experimental physics with August Kundt, and music with others. He became a leading authority in understanding the violin as an instrument and in understanding and measuring tone production. He (with his brothers) helped lead Wurlitzer into becoming the leading manufacturer of automated musical instruments.[67]

Coda

During the late nineteenth and early twentieth centuries, Helmholtz's book continued to be read by various leading figures in the world of music. The Czech composer, musical theorist and editor, and folklorist Leoš Janáček; the Austrian composer and conductor Gustav Mahler; and the Franco-American modernist composer Edgard Varèse, to name but a few, were all keen students of Helmholtz's book. In a very different way, so too was the sociologist Max Weber. By emphasizing the historical nature of musical development and by distinguishing between natural laws and aesthetics, Helmholtz's book effectively laid (for Weber) the basis for the sociology of music. Weber turned to Helmholtz as one of his principal musicological and music historical guides, even though, on some points, he disagreed with him.[68]

Down to his final years, Helmholtz continued to receive inquires from the general public about musical matters or simply responses to his book. A century after the appearance of the *Tonempfindungen*, the *New Grove Dictionary of Music and Musicians* called the era from 1862 to around 1900 "The Age of Helmholtz." Helmholtz's work on understanding sound—its production, transmission, amplification, damping, and detection (by the ear)—found its way, sometimes more indirectly than directly, and sometimes more through his students than through him, into a series of technological inventions and innovations: the telephone, piano stringing, the microphone, the loudspeaker, amplifiers in general, vacuum tubes and their amplifiers, the oscillator, the oscilloscope, acoustic impedance, and architectural acoustics.[69] The *Tonempfindungen* mixed natural science and musical culture in a way that showed how science could illuminate human culture without at the same time reducing the sources of that culture to mere materialism and mechanism. It used the *Wissenschaften* to advance *Bildung*, and it showed that Helmholtz was among the most *gebildete* figures of the rapidly emerging scientific-technological age.

13

Popularizing Science in Britain and Germany

Britain as an Intellectual Spa

No foreign scientist had a stronger reputation in mid-Victorian Britain than Helmholtz, not least because he reinforced it by visits and personal friendships with leading British men of science. In the spring of 1864, he returned to Britain for six weeks, this time to deliver a series of invited lectures at the Royal Institution and the Croonian Lecture at the Royal Society. Along the way he visited friends in London, Oxford, Glasgow, and Manchester.

The entrepreneur behind the Royal Institution lectures was Henry Bence Jones, who had heard about Helmholtz's recent lectures (1862–63) in Heidelberg and Karlsruhe on the general results of the natural sciences. Bence Jones wanted Helmholtz to deliver a similar set of lectures at the Institution, and Helmholtz accommodated him with a set of potential lecture topics: "1) Exposition of the Principle of Conservation; 2) Mechanical Energy of the Stellar System; 3) Stellar and Solar Heat; 4) Formation of the Earth; 5) Motions of Atmosphere and Ocean; 6) Circulation of Water on the Earth; 7) Food of Plants and Animals; 8) Mechanical Energy of Plants and Animals." Bence Jones agreed that the general subject and theme would be the conservation of energy; it would be a set of eight lectures given at the Institution in April 1864, and Helmholtz would receive £80 for his efforts. The desire of the Institution's leadership to have Helmholtz lecture on energy conservation shows that, although "energy physics" was now more than a decade old, in 1864 it was a subject that the general public was still very much just hearing and learning about. Helmholtz had recently lectured on conservation of energy at Karlsruhe partly because he needed money to help pay for equipping his new physiological institute; he "treated conservation as a nourishing cow," as he told du Bois–Reymond. By contrast, he envisioned the London lectures "mainly as a cheap and advantageous way of getting in close connection with London *savants*." He expected England to offer him "a sort of intellectual spa for shaking up the activity of the mind," an antidote to sleepy southern Germany.[1]

At Heidelberg in 1862, with his lecture on the relationship of the natural sciences to the sciences as a whole, and at Karlsruhe during the winter of 1862–63, with his series of lectures, Helmholtz sought to present his understanding of the central features of the natural and human sciences and the differences and relationships between the two; above all, he wanted to provide "insight into the peculiar character of those sciences" whose study had been his "life's task." As to the Karlsruhe series, only the first (on conservation of force) survives. It effectively recapitulated parts of his 1854 lecture at Königsberg on the interaction of the natural forces and parts of his Heidelberg lecture of 1862. He declared in Karlsruhe that since the Renaissance the natural sciences had greatly reshaped "all the relations of the life of civilized nations," in terms of both their practical applications and their intellectual influence. They had facilitated an increase of wealth, greater enjoyment of life, better health, and greater industrial and political power. Hence any educated person who sought to understand the forces shaping the modern world had to be interested in modern science.[2]

The laws of natural science had become, he said in Karlsruhe, humankind's "obedient servant." The human sciences, which were without laws (at least so far), were by contrast about the study "of a civilized life." Science as a whole was worthwhile not simply because it gave intellectual satisfaction or an ability to harness nature's phenomena to humankind's purposes, but also because it gave "artistic satisfaction" to understanding nature's complexity as a "lawlike ordered whole." His lectures aimed at imparting some sense of the recently discovered general law of all natural phenomena, that of conservation of energy. He appositely employed a metaphor from capitalism to express his point about the impossibility of a perpetual motion machine: to create such a machine would in effect be to create money. "Work is money. A machine that could create work out of nothing was as good as one which made gold." Conservation of energy, he declared, applied not only to phenomena in laboratories and factories but also "to the large processes in the life of the Earth and the universe." It applied to climatological processes as well as to those of plants and animals (including human beings). Humankind put such processes "in our service."[3] In truth, he said little in Karlsruhe that he hadn't already said in Königsberg in 1854 and in Heidelberg in 1862. He had, as he told du Bois-Reymond, "treated conservation as a nourishing cow." He nourished it further in Britain.

Late in the autumn of 1863, after advertising Helmholtz's lecture series for the following spring, Bence Jones learned of a possible conflict between Helmholtz's lectures and a set of lectures to be given by the chemist Edward Frankland at a London university. While Bence Jones judged the Royal Institution audiences to be different from—he surely meant more fashionable and less educated scientifically—a university audience, he and Faraday nonetheless decided that the wisest course was to avoid a conflict of dates with Frankland's lectures. He also began making arrangements for Helmholtz to become a member of the Athenaeum Club and, assuming Helmholtz agreed, for a medical journal to dispatch a reporter to take notes on the lectures for press publication.[4]

With the London lecture series now set, with the *Tonempfindungen* now being widely noticed, with his new physiological institute now under construction, and with a year's service as prorector now behind him, Helmholtz needed a long rest from work. In September he went hiking in the Alps with friends. The climbing went well, and he bragged that he walked faster than his guide—except that the latter carried a thirty-five-pound backpack! He took notice (as usual) of pretty women and felt comfortable mentioning such things to Anna. His group hiked into the Italian Alps, but in four hours of hiking they saw not a single soul. He felt lonely and sad there.[5]

His visit to Britain was much anticipated, Tyndall told him. Helmholtz planned to arrive in London in mid-March and spend much of the next thirty days alternately residing with Bence Jones, the Enfields (Henry Roscoe's sister), and his old friend William Benjamin Carpenter. Among other things, he was keen to participate in meetings of the Royal Society. The Thomsons also invited him to their place in Glasgow for late March. Thomson wanted to have "a great deal of conversation" with Helmholtz "on many subjects." To make that trip work, Helmholtz had to abandon his original plan to visit Cambridge, above all to see Stokes, its mathematical physicist.[6]

Helmholtz left Heidelberg on 11 March, with Anna accompanying him to the station and returning home in tears: she wrote to him, "Then I went into your room, cleaned up, and cried again, at which point Robert appeared and comforted me." He traveled to London via Utrecht, where he briefly visited Donders and discovered that everyone (including musicians) was reading the *Tonempfindungen*, in part because Donders himself was lecturing to the public on acoustics. The two attended a concert together. Once in London, that "gigantic Babel," he went first to the Royal Institution, where he sought in vain to see

Tyndall but managed to find Faraday, who "as in the old days was extremely charming." Faraday told him that, owing to memory loss, he himself had given up lecturing. Helmholtz dined that evening at the Philosophical Club in St. James' Hotel, "where rather the most interesting members of the Royal Society gathered." After dinner the group went to hear Tyndall lecture at the society. He found its quarters to be "extremely impressive"; he was most impressed by "the powerful golden mace" that symbolized the society's royal privileges. Late that evening he and Bence Jones called upon William Gladstone, one of the leaders of the Liberal Party, the chancellor of the Exchequer, and a future prime minister. Gladstone knew many leading British men of science, including Huxley; Joseph Dalton Hooker, a botanist and the director of Kew Gardens; and Darwin, and had himself published on human color vision. He had numerous scientific interests, including the sounds of vowels, and had studied Helmholtz's work on that topic.[7]

Helmholtz encountered a wide swath of London's cultural life. He admired the "beautiful materials and clothing" worn by various ladies, though he made fun of their "pompous headpieces." He met the Duke of Argyle and various diplomatic envoys, visited the liberal Scottish politician M. E. Grant Duff, and went to the Kensington Museum to see some pictures. (He prized the museum's showing of William Mulready's oil paintings and aquarelles, liking especially their warm colors.) He attended a large dinner party at Bence Jones's well-appointed place. Bence Jones had given up scientific work since Helmholtz's last visit and had become an extremely busy and fashionable London doctor. At dinner Helmholtz was seated next to a musically talented woman; speaking of her to Anna, he said, "[Miss Gabriel] plays very beautifully, is very lively and smart," and that she had declared "Under your care I've become much younger." She also composed cantatas for orchestras: "Her compositions," he went on, "have had many performances. At table I had a deep discussion with her about music." After dinner the guests went to hear Tyndall give a popular lecture ("based upon very brilliant experiments") at the Royal Institution. The Prince of Wales, "a very good-looking young man," also attended.[8]

The next morning he breakfasted with Grant Duff and "his very pretty wife"; some other Scottish politicians; Arthur Russell, a writer and politician; the Berlin military historian and diplomat Theodor von Bernhardi; and others. Crown Princess Victoria had dispatched Bernhardi to England to explain Prussia's heavy-handed threat to seize (with Austrian help) the duchies of Schleswig and Holstein from Denmark. Prussia's plan was opposed by many in Baden and elsewhere. Helmholtz told du Bois–Reymond that the current political situation was "like the vulture of Prometheus": "It eats up one's liver every morning." The remark reflects Helmholtz's own distance from and distaste for actual political power. Bernhardi, he said, had offended people in England, and Helmholtz was depressed by the quality of Prussia's diplomats. He was impressed, however, by the Scottish politicians who could discuss, inter alia, the work of such well-known intellectuals and liberals as David Friedrich Strauss and Ernest Renan. Like Strauss, Renan sought to explain away (in his *Vie de Jésus* [1863]) Jesus's alleged miracles and show how his disciples had transformed his historical activities into legends. No God here. Renan's book shocked and scandalized many, especially among the bourgeoisie and within the church; as a result, he lost his chair at the Collège de France (though he was reinstated in 1870). Helmholtz himself was familiar with this so-called Higher Criticism of biblical scholarship. All in all, he found the breakfast company "very interesting." Afterward he went to Hyde Park to watch ladies ride horseback and then to Regents Park to visit Wheatstone, who, he said, had "set up some fine recording instruments; like everything he does, very able." He found London quite exciting.[9]

Things quieted down soon enough. He worked on his Croonian Lecture and met with Tyndall at the Institution to discuss preparations for his first two lectures there. He took a break from science by doing a bit of painting, seeking to "compete with [J. M. W.] Turner in using bold colors and clouds" and by seeing a play about the mythical ancient Greek king Ixion. He suddenly found London to be dark and "melancholic." Before leaving the city, he went to the College of Surgeons to see Huxley; he described Huxley: "a young, very intelligent man," who was now "the main fighter for the Enlightenment here" and who opposed "biblical natural history." He also visited the British Museum to see its "very rich, instructive collection of primeval animals."[10]

In late March he spent two days at Oxford, staying with Max Müller. Everything about Max Müller, from his intelligence to his dress, not to mention his beautiful, charming, and knowledgeable English wife, impressed Helmholtz, as did nearly everything at Oxford. "One can have no idea of [Oxford] before having seen it. Now I understand the love of the English for their university." However, he did not lose his critical sense about Oxford: "For educating a gentleman, [the university] is absolutely perfect; for science, however, much can't come from it: for a fellow not to sink into inertia, an unusually strong interest for science is obviously needed. At the moment, Max Müller is perhaps the only man here who is working." More than a few scientists at Oxford and elsewhere in Britain shared his skeptical view about the university. He visited the new natural science institutes (in the University Museum) of George Rolleston (the first Linacre Professor of Anatomy and Physiology), George Griffith (a lecturer in natural science and the Deputy Professor of Experimental Philosophy), and Benjamin Brodie (the Waynflete Professor of Chemistry). He dined one evening at All Souls, Max Müller's college, which also much impressed him.[11]

Helmholtz then spent two "very interesting" days in Glasgow with the Thomsons, who lived in University College. Two of the many topics that Helmholtz and Thomson presumably talked about were electrical standards (especially concerning resistance) and vortex atoms. He saw Thomson's extensive collection of new physical apparatus, including many self-constructed, clever pieces of electric measuring instruments; his current experiments (the two men spent most of their time experimenting together); and the many students at work in the laboratory. He was above all immensely impressed by Thomson's intellectual quickness. The Thomsons treated him very well, and he felt "still more at home in Scotland than in England." He met James Thomson, William's brother, who was a professor of engineering at Belfast. He thought him "a level-headed fellow, full of good ideas," but intellectually narrow in that he could only (and "ceaselessly") talk of engineering.[12]

He returned to London via Manchester, where Roscoe had invited him to stay in his home. He had been keen to experience Manchester's urban, commercial life. At Roscoe's country house he dined with Joule, the "principal discoverer of the conservation of force," and with Robert Bellamy Clifton, a former student of Stokes's and the first (and young) professor of natural philosophy at Owens College. In 1865 the promising and socially well-connected Clifton was elected the new professor of experimental philosophy (that is, physics) at Oxford. His only definite rival for the post was Griffith, though Helmholtz himself was considered as a possible nominee. During a fifty-year reign there, Clifton devoted himself to teaching but did virtually no research, though he did oversee the construction of Oxford's new Clarendon laboratory. Research physics failed to advance at Oxford, which was increasingly eclipsed by other British and Continental physics centers. Helmholtz noted that the Roscoes, unlike the Thomsons, had completely done away with premeal blessings, and he judged that England was rapidly becoming liberal in religious matters. He and Roscoe discussed English universities, "about which we agreed completely." He was impressed with Roscoe's "well-equipped laboratory." After two days, he returned to London. He stayed at the Athenaeum, visited Carpenter, and talked with Faraday again at the Institution as he prepared for his forthcoming lectures. He would have enjoyed it all much more, he said, if Anna had come along.[13]

Lecturing at the Royal Institution and the Royal Society

It turned out that Helmholtz gave six (not eight) lectures at the Institution on conservation of energy and its ramifications into fields beyond physics. His first lecture was well attended but (owing to rain) not quite full. He told his London audience that because the natural sciences occupied at present "so prominent a place among the different branches of human knowledge" and because "they have exerted so great and all-pervading an influence on the whole state of modern society and civilized life," it was only natural to ask why this state of affairs had developed in the past two or three centuries. The humanities and social sciences, by contrast, had progressed little beyond "that of ancient times." The reason for the development of the natural sciences in modern times, he said, was "the discovery of the strict order and lawfulness pervading all nature." The discovery of physical laws made the difference between the growth of the natural sciences and that of other branches of knowledge. By contrast, "Wherever the faculties and instincts of the human mind, the energy of the human will intervene, we are not able to determine the consequences with certainty and without doubt." He doubted that even in the course of time scientists could discover "definite laws of the human mind." There were simply too many factors to take into account to make the humanities and the social sciences lawlike. "Mental and moral science" was vague and uncertain in comparison to, say, "the great and admirable system of modern Astronomy," which could be considered "the ideal of scientific perfection" and which was based on Newton's law of gravitation. It was "a few simple laws" of physical science that allowed humankind to order and control knowledge and the natural world itself. For Helmholtz, science and industry marched closely together. He urged his audience to "avoid falling into the error of the philosophers of ancient times and the middle ages, who believed that they could find laws of nature by metaphysical speculations, transferring the laws of thinking to the external world." Beyond referring to "the renowned logician, Mr. Stuart Mill," he made no further reference to any philosopher. Instead, he advised that finding the laws of nature meant conducting observations and experiments. He hoped to demonstrate some of these claims by discussing the law of conservation of energy—and that was the word he now but not always used before his British audience—and its various, rich consequences for other fields of science and for practical purposes. He devoted the remainder of this first lecture to explaining the law, including two demonstration experiments. He never (in any of these lectures) used the term "potential energy" or "kinetic energy" or "thermodynamics." Even in 1864, before a British audience, there were residues of his language (if not thinking) from the late 1840s.[14]

The second lecture, two days later, was devoted to the universe's overall energy. The law of conservation of energy, Helmholtz explained, implied that energy needed for human purposes, as well as that on the earth's surface, had to come from the universe's general supply of energy. This largely amounted to understanding solar energy and its effects on earth, including the formation of fossil fuels. Similarly, it was solar and planetary energy that was responsible for the mechanics and dynamics of the heavenly bodies, as heat was converted into mechanical power. Hence his main concern in this lecture devolved into "the nature of the Sun." He showed a photograph of sunspots to indicate that the sun's surface was characterized by "violent motion" and an immensely high and dense atmosphere. Then he brought this static, empirical knowledge of sunspots into conjunction with Bunsen and Kirchhoff's recent discovery of spectroscopy, explaining that each chemical element has a particular spectrum and pointing out that their analyses showed, inter alia, the chemical composition of the sun's atmosphere. He used slides to present various line spectra and noted that the sun's spectrum was continuous; and he discussed Fraunhofer lines (the dark lines in the spectrum) as well. Moreover, he explained that Kirchhoff had given an analysis of the absorption and transmission of solar gases, while Bunsen (and Kirchhoff, too, he should have mentioned) had used spectroscopy to discover two new elements (cesium and rubidium). Thanks to these and other researches (for example, by Roscoe), it was now known that the sun's atmosphere contained a great deal of iron as well as other elements. The principal conclusion from all this solar astrophysics was that the sun was very hot; its temperature "must be much higher than any temperature we can obtain by our terrestrial means." That point stood in contradiction, he explained, to the earlier astronomical results about sunspots, which declared that a "photosphere" gave off the sun's light and "that the body of the Sun itself was dark." Rather, according to Kirchhoff's analysis, clouds of various density resided above the sun, and the changing nature of sunspots issued from their varying distances from their clouds.[15] The lecture was thus in part a presentation about a new subdiscipline, astrophysics, which Bunsen and Kirchhoff's creation of spectrum analysis greatly enabled and which emerged in the 1860s and 1870s.

This second lecture was better attended, and, thanks to Tyndall, Helmholtz's experimental demonstrations came off well. Helmholtz was pleased that *Punch* magazine reported on the lecture, but he thought he had again spoken too slowly; the lecture had lasted well over an hour. The next day he visited Thomas Graham, not only "one of the great chemists of England" but also Master of the Mint. Graham gave him a tour around the Mint. Helmholtz had a break of several days before his third lecture, and he spent one morning further preparing it. He noted, "Mr. Faraday always dispensed coffee as encouragement."[16]

The third lecture aimed to show that even the relatively small amount of heat that the sun gives off to the earth is enough to induce meteorologic change, circulation of water, and so on. Speaking a mere five years after the appearance of Darwin's *On the Origin of Species*, Helmholtz noted that the sun had been emitting heat for far longer than six thousand years, and thus he concluded "that chemical combinations are not at all adapted to produce such an amount of heat as the Sun gives out every year." According to Kirchhoff and Bunsen's analyses, the sun's composition was similar to the earth's. The sun, he argued, had not perceptibly cooled during the period of human history. In all likelihood, he thought, the sun had a "means to produce new heat," and he suggested that this was based on a combination of its contraction and the gravity at work during such contraction. He disagreed with several scientists (Mayer, the Scottish physicist John James Waterston, and especially William Thomson) who believed that the sun's heat had originated in massive amounts of meteors that crashed onto the sun's surface and so produced solar heat. Urbain Le Verrier's observations of the precession of Mercury's orbit showed that this hypothesis was not tenable.[17]

Following Laplace, Helmholtz suggested that balls of mist in interplanetary space gradually condensed. Those "balls" closest to the sun had greater rotational motion. While Laplace had invented this nebular hypothesis "only to explain the agreement in the direction of the motion of the planets," it also, unbeknownst to Laplace, explained "the origin of the solar heat and the interior heat of our Earth." For as the masses formed together under gravitational influence, heat was given off.[18]

After the third lecture, Helmholtz went to the Athenaeum, where he worked on his Croonian Lecture until 4:00 p.m., seated constantly at a window and hoping to catch sight of a parade honoring Giuseppe Garibaldi. The Italian revolutionary was in Britain that April, and he was the talk of all London. His appearance and addresses brought out enormous crowds. Helmholtz waited until 6:30 p.m. for Garibaldi to appear; then a dinner engagement forced him to leave. He finally did see him, apparently by chance, when Garibaldi's carriage pulled up to the front of the Royal Institution. Helmholtz felt "envied," because he "replied in a friendly way" to Helmholtz's greeting. The dinner was a benefit for the University of London Hospital. The chairman of the benefit and the main speaker was Charles Dickens. Helmholtz actually spoke with Dickens—in contrast to his brief "encounter" with Garibaldi. Dickens was in London that spring working on *Our Mutual Friend*. Helmholtz "assured him of the admiration" that Germans had for his books. "The entire [evening] was very interesting," he wrote Anna. The following evening he dined with Huxley, Carpenter, Tyndall, the philosopher Herbert Spencer, the naturalist, archaeologist, and banker John Lubbock, and an Australian named Wallace. "They were all very witty, lively people, who frequently provoked one another; the evening was very amusing." An exception arose when Lubbock started arguing somewhat heatedly about the Schleswig-Holstein affair, something that "the other scientific people" remained rather cool about.[19]

Helmholtz spent the morning of 14 April preparing the experiments for the fourth lecture later that day on wind, rain, and the growth of plants.[20] He began with a brief account "of the influence which heat produces on the inorganic processes on the surface of our globe." He noted that temperature increases with increasing depth in the earth, and that the earth has been cooling very slowly. Thomson had calculated that it would have required 24 million years to cool the earth from 2,000 degrees to its current approximate temperature. Helmholtz described the probable changes in the earth's geological structures that were due to cooling, above all "that the surface of the Earth has been broken many times, that new masses have come out from the interior; that some parts may have been raised and others depressed, not only once, but many times," all in all producing an extremely complicated geological structure. (Here he referred to Darwin's geological work on the Andes.) In short, there were changes in the earth's surface level, made "either quietly" or in the form of earthquakes, and in both cases he supposed these were due to the "slow cooling of the interior of the Earth, by which it contracts more and more, and becomes less adapted in its form to its solid covering."[21]

He summed up his geophysical survey: "All the motions of the Earth, the winds, the rain, the weather, the snow, the phenomena of glaciers and rivers, the geological deposition of new layers of earth, even the electric processes in our atmosphere, are produced in this way by this one cause—by the heat of the Sun coming down to the surface of the Earth." Solar radiation also affected organic life. Plants, he explained, were principally composed of carbon, hydrogen, and oxygen, along with a small (but crucially important) portion of nitrogen as well as other elements. By absorbing solar light, he explained in some detail, the plants are effectively storing energy.[22]

That evening he gave his Croonian Lecture at the Royal Society: on human eye movements and their relations to perceptions. His subject was the motions of the human eye, a subject he thought of interest not only in terms of the eye's physiology but also in terms of voluntary muscular movement in general. He hoped that his results might interest physiologists, medical men, and "every scientific man who desires to understand the mechanism of the perceptions of our senses."[23] He presented to the society a brief, simplified English-language version of a complicated series of theoretical, mathematical, and experimental studies of eye movements that he had recently published in German.

The eyeball, he explained, is roughly a sphere centered on and moving around a fixed point. It has six muscles attached to it, which form antagonistic pairs. The eye's motion could thus be understood as a system of rigid pairs of muscles, and he presented various empirical results and Donders's and Listing's laws of eye motion. He discussed stereoscopic vision briefly, as well as both monocular and binocular vision and the horopter, and said that details of all these topics could be found in his *Handbuch*.[24]

In contrast to his popular lectures at the Royal Institution, at the Royal Society he spoke without notes for more than an hour; he thought that approach made his lecture noticeably better. His talk was apparently well received, since Edward Sabine, the society's president, who praised his English, insisted that he continue. This he did for another half hour, during which several listeners made comments that supported his views. Among these were Hirst and the science writer and editor Edward William Brayley, who was with the London Institution. Brayley had written the article on the correlation of physical forces for the *English Cyclopedia*, reporting on Helmholtz's law of conservation of energy. The following morning, Helmholtz took a train to visit Alexander John Ellis; Ellis had attended at least one of Helmholtz's lectures, and he had invited him to his place to dine and talk (presumably, inter alia, about translating the *Tonempfindungen*). At this time he still had not been able to get Helmholtz's book published in English. Ellis also took Helmholtz to an English school to hear schoolchildren sing.[25]

As his stay in Britain began to draw to a close, Helmholtz became busier than ever in London. He attended a lecture at the Institution one evening on guncotton. He met James Martineau, a Unitarian minister, liberal moral philosopher, and close brother of the writer and feminist Harriet Martineau; James Joseph Sylvester, the distinguished mathematician; and Stokes, "one of the best mathematical physicists." On one day, after preparing his Croonian lecture for publication, he went to Maxwell's place in Kensington for lunch. (Maxwell chose lunch so that the two could have "light to analyze [by].") Maxwell had finally gotten his color-mixing instruments in sufficient order to be able to invite him for lunch, as well as William Pole, who was color blind. Maxwell showed Helmholtz his "beautiful apparatus [a color wheel?] for [understanding] the theory of colors," and the two performed experiments on the cooperative Pole. "We had there," Helmholtz said, "a splendid lunch with champagne and every imaginable delicious item." The following morning, a Sunday, Helmholtz went to the Unitarian Chapel to hear Martineau preach: "very beautiful, completely clear, and yet warm and deep, so that I understood the enthusiasm of the Unitarian women for him. (The liturgy and singing were less beautiful.)"[26]

The following Tuesday, in his fifth lecture at the Institution, Helmholtz focused on the importance of the law of conservation of energy for physiology. Here the difficulty lay, he said, in the fact that physiological processes were so complex and sensitive to influences that understanding the effect of chemical and physical forces seemed impossible. He explained that physiologists before his generation often spoke of "the vital principle" or "the vital force," some sort of soul-like characteristic that regulated the whole body, with its chemical and physical forces. This notion, however, had hindered progress in physiology, Helmholtz declared, because it introduced an element of arbitrariness and "mystery" rather than providing a "real explanation." Just as natural philosophers had discovered that the same laws held for celestial and terrestrial phenomena, so "the modern great development of physiology depended principally upon the supposition, that in the interior of living things those same forces were acting which we know already to be acting throughout organic nature." Mayer and Helmholtz had first pursued this idea of explaining physiological processes in terms solely of inorganic forces and trying to establish whether other types of forces might be at work here. That was what the law of conservation of energy was intended to achieve. But he conceded that the law could "not yet . . . show that there exists an accurate equivalent proportion between the power or energy of these solar rays which are absorbed by the green leaves of the plants, and the energy which is stored up in the form of chemical forces in the interior of the plants." All that could be said so far was that there was enough solar heat "to produce the effects of vegetable life." Essentially the same point held for animals, which either directly or indirectly lived off of plant life. Much of the rest of his lecture became a discussion of the biochemistry of food.[27]

If the fifth lecture was mainly devoted to analyzing the intake of energy by the human body, the sixth and final one was devoted to discussion of the body's output of energy, that is, to the mechanical effects and heat. The higher classes of animals, Helmholtz noted, produce their mechanical effects through the contraction of their muscles. Muscles have elasticity, just as the air and steam have, and so there can be a change in muscular forces. He then showed several of du Bois–Reymond's electrophysiological experiments. He discussed a series of experiments demonstrating how heat is produced by the contraction of muscles and described the chemical processes occurring within them. Here, too, he referred to the work of his "friend, Du Bois Reymond" as well as that of Ludwig, and especially the invention of a new physiological instrument ("a little portable gas-meter") by Edward Smith of London. This last was for measuring the intake of air. Smith tried out his instrument while sleeping, and Helmholtz did likewise: "I found that when awake, and in a state of repose during the day, I inspired nineteen times in a minute. Going up a mountain as fast as I could, I increased the rate of respiration to fifty times in a minute. The number of pulsations of my heart was seventy in the minute in a state of repose, and they were increased to nearly 200 when I was in a state of bodily exertion." Increased respiration, he concluded, led to increased chemical processes: "You see, therefore, that no mechanical power can be developed by our muscles without a great increase of chemical processes." Moreover, "in a state of repose the quantity of heat developed by the human body must be like the difference between the heat of combustion of the food introduced and the heat of combustion of the excretions." Though he could not prove it, his broad point was that "the quantity of heat developed in the interior of the body is the same as the heat of combustion. When work is being done by the human body, then work must be done at the expense of a part, at least, of the heat given out by the body." With that, he brought his survey of the law of the conservation of energy and its various ramifications in the inorganic and organic worlds to a close.[28]

While still in London, the question arose of electing du Bois–Reymond a foreign member of the Royal Society. Helmholtz naturally argued strongly in his favor, as did Bence Jones. But several fellows favored Carlo Matteucci "as a counter-candidate . . . , not in order really to elect him, as B. Jones meant, but rather," Helmholtz explained to du Bois–Reymond, "to make it appear as if your work is too contested." There was widespread conviction that du Bois–Reymond's experimental work was "too subtle in order to succeed securely." Helmholtz sought to counter that view. Where some fifteen years ago du Bois–Reymond had played the role of Helmholtz's advocate abroad (in Paris), now it was Helmholtz who played that role for du Bois–Reymond (in London). However, before he was finished, Helmholtz lost Matteucci as his one-time admirer. For Matteucci read at least one of Helmholtz's Royal Institution lectures as given in the *Medical Times and Gazette*, and he saw that Helmholtz referred to du Bois–Reymond's work on electrophysiology without also mentioning his (Matteucci's) own. Matteucci was angered by this lack of recognition and in two long letters told Helmholtz of his disappointment in this regard. Helmholtz tried to clarify his remarks, but that seems to have had little effect.[29] Matteucci never wrote to Helmholtz again.

Helmholtz arrived back in Heidelberg on 23 April 1864 at 4:00 p.m.; the following morning, at 2:30 a.m., his daughter Ellen Ida Elisabeth was born. His trip to Great Britain had lasted six weeks (11 March to 23 April). He reported to du Bois–Reymond that as for the popular lectures at the Royal Institution, he had adopted du Bois–Reymond's view: he would think twice before again agreeing to give such lectures. While satisfied with his lectures' "external success" (he had attracted about three hundred attendees per lecture), he was dissatisfied with the lectures themselves. There was so much competition in London among popular science lecturers, he said, that these had a tendency to "sink down into the most general sensationalism." English physicists charged that Tyndall in particular showed such tendencies and that this threatened to bring the Institution down to the level of the Polytechnic Institute, "though at the latter," said Helmholtz, "one can nonetheless bring forth still greater means for creating brilliant optical experiments and electric fireworks." Helmholtz had heard Tyndall give a Friday Evening Discourse, and he was in agreement with his English colleagues on this point. Bence Jones made the same charge about Tyndall before the Prince of Wales. On the other hand, Helmholtz recognized Tyndall's "excellent talent for popular lectures" and the public recognition that he received. "A medium for door-knocking by ghosts recently had his [Tyndall's] name spelled out in heaven, namely: Poet of Science." Du Bois–Reymond's own view was even harsher than Helmholtz's and their English colleagues' about Tyndall: "He finds himself in a very false position. He wants completely to be, he must be, a great man, something that he in no way is."[30]

Even before Helmholtz had left for Britain he had not expected to learn much there about experimental physiology. So it is perhaps unsurprising that, when he returned to Heidelberg, he told du Bois–Reymond that, generally speaking, he thought little of English physiology. He included here the work of Sharpey, whom he considered one of the better practitioners. He thought Lionel

Smith Beale, a medical doctor and microscopist who was professor of physiology and anatomy at King's College, London, "a principled obscurantist"; and he said Rolleston at Oxford was "a shy and intelligent man, but . . . for the present" was "still without his own ideas." He knew "to do nothing other than investigate the Pes hippocampi in ape brains." As for Glasgow's poor Joseph Lister, "likewise a shy and able man," Helmholtz said he had to "earn his bread by means of medical practice." Helmholtz was not optimistic about the future of British physiology, either. But thirteen years later he surprisingly offered a more generous assessment of English science to Alexander William Williamson, professor of chemistry at University College, London, and the foreign secretary of the Royal Society. He wrote Williamson on the occasion of his being awarded the society's Copley Medal, the society's highest honor, which may help explain his newfound generosity:

> I have seen in your country a good deal of strong enthusiasm and strong intellectual energy, devoted to scientific labour arising among men of different classes of society and of the most different occupations. I perceived that herein was the source of the strong individual originality, which is a characteristic of English science, and also the source of its practical fruitfulness. On the Continent the conditions of life for scientific men have been different; the great part of them belonged always to a peculiar class, more isolated from other classes of men, more connected by its interests and occupations. These conditions are more favorable to develop scientific schools, with all the advantages and disadvantages which tradition and the discipline of such a school tend to produce. Frenchmen turned more to the methodical and refined elaboration of detail. We Germans, partly driven by the native tendency, partly by the social and political consequences of our long religious struggles, turned more to the first principles of knowledge in general, and of scientific theories especially. I cannot disown this national tendency for myself; my own exertions have been devoted partly to the great natural law of the conservation of energy which lies at the root of all questions about the nature of force, partly to the physiological theory of nervous action and sensation which leads to that of perception, the source of all other knowledge. But I owe a great deal to England for my own intellectual education. Grown up among the traditions of high-flown metaphysics, I have learned to value the reality of facts in opposition to theoretical probabilities by the great example of English science. It was, for the greater part, this example which has sheltered me against losing myself in overstrained theoretical speculations.[31]

Helmholtz had met virtually every leading British man of science of the day (with the important exception of the reclusive Darwin) and had visited several

British scientific institutions. For six weeks the British had wined and dined him in London, Oxford, Glasgow, and Manchester. The old Faraday, the leading British chemist and physicist of his generation, had served him coffee as he wrote his lectures. Thomson and Maxwell, the leading British physicists of their generation, invited him into their homes to socialize and experiment. Between giving lectures to hundreds in London, being reported on in newspapers, and touring the country, he had come to know personally practically everyone of scientific consequence in Britain. Perhaps no other foreign scientist had ever received such a warm reception in Britain as Helmholtz had in the spring of 1864. All this also helps explain why in the 1860s and 1870s an array of leading British writers and intellectuals — Spencer, Eliot, Lewes, and others, including such widely read scientists as Darwin, Huxley, and Tyndall — read Helmholtz. He was soon invited there several times again, but he did not return for another seven years.

The Thermodynamics of Beauty: Ice and Glaciers in the Alps

During the mid-nineteenth century, mountaineering spread as a means of relaxation and as sport. Tyndall helped make it more comprehensible for all through such writings as *The Glaciers of the Alps* (1860), *Mountaineering in 1861* (1862), and *Hours of Exercise in the Alps* (1871). Helmholtz's frequent if not quite annual visits to the Alps stood in contrast to those of mountaineers who sought extreme or overly strenuous physical challenges; he walked and hiked in the mountains to restore himself physically and mentally, to forget the intense academic and intellectual work that he pursued daily during the academic year and that manifested itself in his various bodily ills. His hikes also gave him enough first-hand experience to offer popular scientific lectures in Heidelberg and Frankfurt am Main: "On the Properties of Ice" and "The Movement of Glaciers." These soon appeared in essay form as "Ice and Glaciers" (1865).[32]

For Helmholtz the snow and ice of the Alps held a "completely special magic." They revealed part of the earth's history. So he took the Alps' beauty for granted and instead sought to present listeners and readers with the latest scientific analyses of their climatological conditions and causes. As in the lecture on Goethe and in the *Tonempfindungen*, he held that to understand nature, one must appreciate its mechanics as well as its surface phenomena. Beauty, whether natural or manmade, was subject to scientific analysis.[33]

He proceeded from the individual "small peculiarities of ice" to understanding the "most important processes in the glaciers." Indeed, the analysis of glaciers could inform human history because it reached "into the darkness of primordial time." He treated the earth's atmosphere as a thermodynamic system. The temperature decreases as one climbs higher up a mountain, because the atmosphere allows nearly all the sun's illuminating rays to pass through but blocks much of the heat radiation that issues from the earth. Air masses heat up, and heat is eventually given off. "The exhaust of heat is thus delayed in relation to its intake, and thereby a certain heat supply is maintained along the Earth's surface." (He also here gave an explanation, but only sketchily and en passant, of the much-disputed origins of the foehn wind, a warm, dry wind in the Alps; his thermodynamic explanation of it as a local phenomenon was ignored, but only a year later, in 1866–67, the Austrian meteorologist and climatologist Julius Hann more extensively and successfully defended a similar explanation in terms of adiabatic temperature change.) But since the protective atmosphere higher up in the mountains is thinner, heat radiation escapes more quickly into space, while the heat supply and temperature are both lower.[34]

Expansions and contractions of air masses at different temperatures and pressures, and the associated moving systems of air, manifested themselves meteorologically as rain or snow, heat or cold, wind or no wind, depending on the particular geographical configurations that they encountered. Owing to western Europe's damp climate, Helmholtz explained, only a few places on earth (the Himalayas, Greenland, northern Norway, Iceland, and New Zealand) could compare with the number and size of the Alps' glaciers. There was a limit to the height of the snow peaks that formed in the mountains because they could bear only so much weight; if the peaks became too steep, they could receive no new snow. That forced some of the snow to go lower (that is, below the snow line), where, owing to the warmer air it encountered, it gradually melted and flowed down into the valleys. Sometimes, however, the downward movement of snow happened suddenly (as in an avalanche) and sometimes very slowly (as in the progress of a glacier). Hence, Helmholtz further explained, there are two different parts of the ice field to consider: the part above the snow line, which concerns the newly fallen snow and covers the snow peaks, and the part below, the glaciers, which consists of solid ice that has come down from above the snow line. The surfaces of most glaciers, he noted, were filled with an odd assortment of rocks and other objects and were dirty; underneath, however, the glaciers showed "in their purity and clearness a glorious blue." Here was real beauty, he thought, "a play that is strongly spiced with the stimulating interest of danger."[35]

He compared the internal and external forms of glaciers to a slowly flowing current (for example, of water, honey, or tar). Although this flow may not always be visible, he explained, it was measurable. Geologists had proved that a series of Alpine glaciers had descended into the valleys of Switzerland, France, and Germany and that similar glacial movements had occurred on the British Isles, the Scandinavian peninsula, and elsewhere. He referred to the empirical findings and theoretical analyses of such researchers as Horace-Bénédict de Saussure, Louis Agassiz, Forbes, Tyndall, James and William Thomson, Clausius, and others.[36]

These and other researchers also asked how such large masses as glaciers could possibly move over solid ground. To answer this question, Helmholtz explained, they had turned to the mechanical theory of heat. The temperature of glaciers was more or less always at the freezing point of water. However, James Thomson had proved that water's freezing point was lowered by strong pressure, and shortly thereafter Clausius independently showed, using the mechanical theory of heat, that "strong pressure" lowered water's freezing point and so changed the glacier's temperature. William Thomson then confirmed this finding, namely, that an increase in pressure leads to a lower temperature. Helmholtz explained that the changing pressure and temperature of mixtures of ice and water led to changes in the volume and state of a mass, for example, that of a glacier. He thus concluded that although glaciers appeared as immobile masses laced with "barren, stony, and dirty ice surfaces," in fact they were "majestic currents" that flowed "peacefully and regularly" and reshaped themselves "according to definitely determined laws." Indeed, the water in the Alpine lakes (Geneva, Thun, Constance, Como, Garda, and so on) was principally glacial in origin: "The clarity and the wonderfully beautiful blue or blue-green color" of their water "is the delight of all travelers." Moreover, the gradual melting of the glaciers into water and the various chemical constituents that they transported or helped expose to the environment through the mechanical processes of grinding down the grosser organic phenomena were essential for vegetation. Hence, what at first glance appeared as little more than "wild, dead ice deserts" turned out to be the source of some of Europe's most beautiful and life-giving brooks, lakes, and rivers.[37] The lecture offered a good example of how thermodynamics might be applied to understanding the environment at large.

Popular Scientific Lectures

In 1865–66 Helmholtz received several new honors or other forms of public recognition. He was nominated a foreign honorary member of the Vienna Academy of Sciences, of which he had been a foreign corresponding member since 1860. Brücke inquired whether Helmholtz already had an honorary doctorate of philosophy from the University of Vienna, which was interested in making such an award to him; in fact it did so on the occasion of the university's five hundredth anniversary. Helmholtz, along with twenty-three others (including Purkyně, von Baer, Liebig, Bunsen, and Bernard), was also elected an honorary member of the Viennese College of Medical Doctors (Medicinisches Doctoren Collegium) and received an honorary doctorate. In 1865 he was appointed Privy Counselor

Third Class in Baden and also received the Imperial Russian St. Stanislaus Order (Second Class) as well as the Royal Bavarian Maximilian Order for Art and Science from King Maximilian II. Finally, he was offered an appointment at Vienna's Josephinum when Ludwig, Brücke's colleague in Vienna, moved to Leipzig in 1865. He declined, saying it represented no material gain for him, though he did receive an increase in salary (to 2,286 talers) and more expense money to run his institute. (Ewald Hering, Ludwig's assistant at Leipzig, who soon became Helmholtz's archrival in matters of perception theory, was then offered the position in Vienna and quickly accepted.)[38]

Meanwhile, in late 1864, Eduard Vieweg told Helmholtz that his firm, Friedrich Vieweg, wanted to publish his recent prorectoral address. Vieweg also soon obtained the right to republish Helmholtz's essay on Goethe and wanted to bring out a collection of Helmholtz's addresses. (He also hoped, in vain, to obtain the publication rights to Helmholtz's *Handbuch*.) In response, Helmholtz partly rewrote the two addresses along with the lectures on the physiological causes of harmony in music and on ice and glaciers. Vieweg made his move just in time, because just as these essays appeared as *Populäre wissenschaftliche Vorträge* (1865), the firm of Georg Reimer also asked to publish a collection of Helmholtz's essays.[39]

As these offers from Vieweg and Reimer suggest, Germany had a vigorous and broad-based market for the popularization of science. Although popularization had its roots in the Enlightenment, it grew markedly during the nineteenth century, especially after 1848, not only in Europe but also in the United States. Popularizers aimed at different market sectors and socioeconomic groups. Books by Liebig and Humboldt, for example, generally appealed to the better educated, upper end of the market; those by Vogt, Moleschott, and Büchner generally appealed to a less-educated sector. However, all levels of popularization contributed to and depended on growing literacy and an increase in the number of public libraries, newspapers, and general reference works, as well as untold numbers of associations, societies, and the like that were dedicated to or supportive of the popularization of science through lectures and discussion groups. Many of these institutions of popular science were short-lived, and the works of many popularizers often disappeared from the scene within a few years (or even months) of their initial appearance. The popularizers sought not only to educate their audiences or readers in science; many also saw science as the principal source for securing Germany's technological and, hence, economic development, and they called for an increase in state support for science. Still other writers emphasized the philosophical dimensions and consequences of science. Humboldt's *Cosmos* attempted (in vain) to synthesize all scientific knowledge; it became, after the Bible, the most-read book in Germany. As for Helmholtz, his volume aimed at an educated readership that lacked special knowledge of the natural sciences. The essays had no one particular theme—they were a collection of occasional pieces—yet, taken as a whole, they sought to instruct readers about the nature and results of recent scientific laws and to show something of their bearing on other parts of intellectual life. His essays demonstrated what a cultivated man of science was—or at least could be—in the nineteenth century. As lecturer and author, Helmholtz became one of the leading public voices of the liberal *Bildungsbürgertum* of nineteenth-century Germany. With his volume's publication, he joined older German scientists like Liebig and Humboldt as a leading public scientist.[40]

A few months after the appearance of Helmholtz's collection, Eduard Vieweg reported that the volume had found "such a great approval" that he had just increased the print run by five hundred and he thought it had a promising future. The good sales figures led him to ask Helmholtz to produce a small but similar volume devoted to physiology, one that would reach the same market that the *Tonempfindungen* had reached. He believed there was an "educated" ("gebildete") public that was looking for further instruction "by the master [*des Herren*] of science."[41]

Certainly Helmholtz's friends and colleagues were members of that educated public. Ludwig discovered that even though he had previously read the essays as individual pieces, he now "saw by repeated readings" that there was much he had "overlooked the first time." Du Bois–Reymond felt both envy and admiration and thought Helmholtz's popular essays were far better than Tyndall's. Tyndall himself told Helmholtz that he "read these lectures with extreme pleasure, and more than once cried Bravo! to the sentiments therein uttered." In his famous Belfast Address—or infamous in the eyes of those repelled by its support for methodological or philosophical materialism and its urge to separate the realms of science and religion—before the BAAS in 1874, Tyndall declared: "It has been said that science divorces itself from literature. The statement, like so many others, arises from lack of knowledge. A glance at the less technical writings of its leaders—of its Helmholtz, its Huxley, and its Du Bois–Reymond—would show what breadth of literary culture they command." When Joseph Henry, an American physicist and the secretary of the Smithsonian Institution, heard about the volume's publication, he ordered a copy and tried to arrange for a translation; he felt sure that Helmholtz's reputation as well as his own knowledge of Helmholtz's work meant that a translation would appear. Darwin, too, became a keen reader of Helmholtz's popular scientific essays; he was especially taken with Helmholtz's discussion of the imperfections of the eye.[42] Eliot, Edison, James, Dilthey, Nietzsche, and Freud likewise became keen readers. The last two merit special attention.

Nietzsche began reading about natural science (principally in the areas of Darwinism, physiology, and physics) after 1870. He turned to authors like Helmholtz for instruction. His library included copies of *Über die Erhaltung der Kraft*, "Über die Wechselwirkung der Naturkräfte," and the *Tonempfindungen*. He read Helmholtz on the loss of solar energy, a topic that interested him because he (presumably) thought it showed that even in the astronomical realm there was nothing that did not change, that "Becoming" was as much an issue for the natural sciences as it was as for philosophy. Nietzsche of course also put the theme of the "Will to Power" at the heart of his writings, and for this he undoubtedly drew on the midcentury development of the law of conservation of energy (perhaps more through Mayer than through Helmholtz) and the more general emergence of energy physics. He also tied energy conservation to another of his favorite themes, that of "Eternal Recurrence." However, Nietzsche rejected Helmholtz's theory of perception, in particular the hypothesis of unconscious inference.[43] His skepticism about the epistemological status of science in general—its claim to represent objective reality—meant that any influence by Helmholtz or any other scientist on him had at best to be quite limited. He operated on the principle of "Know Thy Enemy."

In this he stood in contrast to Freud, who in 1873 enrolled at the University of Vienna as a medical student with serious scientific interests and who had Brücke as one of his principal teachers. As a student, Freud studied the law of conservation of energy and became an adept of the organic physicists' program. (He considered Helmholtz's lectures as among the most valuable of his readings, along with those by Aristotle and Thomas Carlyle.) In his early career, as he himself stated, he emphasized understanding the biophysical basis of mind and the importance of "energy" (as he understood it) within psychology. His intellectual debt to Helmholtz was profound.[44]

Tyndall and the Helmholtzes: Translating Popular Science

Helmholtz and Tyndall were friends and admirers of each other. They read and drew on each other's accounts of the formation of glaciers. Helmholtz climbed in many of the Alpine regions described in Tyndall's books. Tyndall lauded Helmholtz by, inter alia, devoting an entire essay to "Helmholtz on Ice and Glaciers" in his *Hours of Exercise in the Alps*. When Helmholtz was in London, he saw Tyndall there, as he occasionally did in the Alps. When he read in the *Philosophical Magazine* that a lecture by Tyndall at the Royal Institution had demonstrated the composition of sounds using electrical light, he expressed his gratitude that Tyndall "had in this way made the foundations of the sensations of tones understandable to people."[45]

Yet the nature of Helmholtz's friendship with Tyndall was different from that with Thomson. Helmholtz admired and valued Thomson as a creative scientific genius with an astoundingly quick mind; he admired Tyndall as a talented scientist, but above all as a gifted popularizer, not to mention as a colleague who promoted his (Helmholtz's) work in Britain. He even wrote, at the urging of the editor Norman Lockyer, a flattering portrait of Tyndall for *Nature*. It was their common interest in promoting popularization that especially sustained Helmholtz's (and Anna's) relationship with him. Helmholtz proposed that Anna translate into German, under his supervision, Tyndall's *Heat Considered as a Mode of Motion* (1863), a book that Helmholtz quite liked, and he asked Vieweg to publish it. Anna, who knew no science, found the translation difficult to do. So it became something of a family affair, with Helmholtz's daughter Käthe and Anna's brother Ottmar helping her with parts of the first half of the book (Helmholtz himself reviewed and corrected the translations for scientific accuracy), and Gustav Wiedemann translating the second half. As thanks, Tyndall sent her a small brooch with inlaid gold and pearls. The translation proved a success: following its original appearance (1867), Vieweg issued a second, enlarged edition (1871), then reprinted it thrice more before bringing out a third enlarged edition (1875) and then a fourth enlarged edition (1894), by which time it had gone through nine printings and the names of Anna von Helmholtz and Clara Wiedemann—Gustav's wife, who was the daughter of Helmholtz's old chemistry teacher, Eilhard Mitscherlich—had replaced those of their husbands on the title page.[46]

Immediately after the translation appeared, both parties, Tyndall and the Helmholtzes, wanted more. When Tyndall's *Sound* appeared (1867), a German firm offered to publish a translation of it. The book was originally a set of Tyndall's lectures at the Royal Institution. But before agreeing to a contract, Tyndall cautiously asked Helmholtz whether he thought the book might "interfere in any way with the *Tonempfindungen*." Moreover, he had already written to Vieweg to inquire whether he might publish a German edition. He also told Helmholtz that he was working feverishly at finishing his essay on Faraday (who died in August 1867) and that he hoped to see it, too, published in a German translation. Helmholtz replied forthwith that he did not think there would be any rivalry between Tyndall's *Sound* and his own *Tonempfindungen*; quite the contrary, he thought Tyndall's book would help increase knowledge of his (Helmholtz's) own views "for the public at large and for musicians, who here in Germany move around sluggishly." *Sound* appeared in its first German edition (with the translation again by Anna and supervised by Helmholtz and Wiedemann), published by Vieweg, in 1869, followed by a second edition in 1874 and a third in 1897. Anna reluctantly agreed to accept an honorarium that Tyndall sent her for this translation, but she thought it out of proportion to her effort. She again invited Tyndall to visit them: "It's not good that we haven't seen one another at all."[47]

As for Tyndall's essay *Faraday as a Discoverer* (1868), Helmholtz and Anna wanted to translate it themselves; that is, Anna would do the translation under

Helmholtz's supervision. He wanted to be associated with the book's appearance in German, he said; "I myself still have to pay off a great debt of thanks to Faraday, who was always extremely charming and obliging towards me." He thought the book would do well in Germany. Tyndall was naturally very pleased by this, and he sent part of his essay to Helmholtz even before the entire manuscript was completed and promised to send him the rest soon.[48]

But Anna's pregnancy and the birth of Friedrich Julius ("Fritz") on 15 October 1868 delayed her translation work. By mid-November, she had finished only about half of the work, which Helmholtz then corrected. In early 1869 Helmholtz told Tyndall that Anna had finished the translation but he had not yet finished reviewing it and would not write the preface until May. That October, he sent Tyndall a copy of the German version of the book. Tyndall was most pleased: "It is clear and forcible. In fact I cannot congratulate myself too much on the fact of coming before the German public under such auspices."[49] As still other translations of works by Tyndall and associated political polemics within science during the 1870s showed, Helmholtz's translation work for Tyndall was hardly at an end.

With Fritz's birth, Helmholtz now had five children all told: two with Olga (Käthe and Richard) and three with Anna (Robert, Ellen, and Fritz); he had no more. The Helmholtzes certainly found happiness with the birth of their new child. Anna told Tyndall that her baby boy was "so strong and big." Helmholtz told him that he was well and that their "little son" gave them "a lot of pleasure." But it soon proved otherwise: Fritz was in fact physically weak. Between Robert's disability and Fritz's needs, the Helmholtzes (read: Anna) had their hands full. Though Robert's disability was apparently much more serious and at times life-threatening than Fritz's general physical weakness, Fritz also suffered from psychological or emotional problems; he needed his mother's (or a professional's) care throughout his life. The Helmholtzes' daughter Ellen, by contrast, was a physically healthy and emotionally normal child. Helmholtz's daughter Käthe also developed (seemingly) normally: she aspired to be a painter, attended balls, and traveled quite extensively: to Munich, the Tyrol, upper Bavaria, Paris, and England. As for Richard, he took private instruction at home before attending the Humanistisches Gymnasium in Heidelberg. Hermann and Anna at first disapproved of his wanting to pursue a career in railroad technology, wanting him instead to concentrate on pure science, which they considered more valuable and dignified; railroad technology appeared to them a socially inferior profession. Yet they ultimately relented. In 1868, shortly after being confirmed, Richard began mathematical studies at the Polytechnic Stuttgart in preparation for his technological studies.[50]

In health or in illness, Helmholtz's family was always around him, and he and they mutually supported and protected one another. Yet this did not mean that he was constantly involved with their daily problems; that was Anna's role. She assumed the daily care of their two disabled sons, sought to shield Hermann from their (and her) daily problems, and turned to him for advice and support only when the most important decisions concerning the two boys had to be made.[51] In these and other ways, she helped him do his work—science and its popularization.

Learning to See the World

Ophthalmology at Heidelberg

In late 1862, immediately after shepherding the *Tonempfindungen* through the press, Helmholtz turned to working on part three of his *Handbuch der physiologischen Optik*.[1] The research and writing of this part occupied most of his research time during the next four years. He systematically assessed the pertinent secondary literature, conducted his own original research on the horopter (for the ever-changing definition of "horopter," see below) and human eye movements, and wrote at length about theories of perception. He did all this while simultaneously teaching and serving as Heidelberg's prorector (1862–63) and as dean of the Medical Faculty (1865–66) and while delivering popular-science lectures in Heidelberg, elsewhere in Germany, and in London. Though the contents of part three owed nothing to these external circumstances, Helmholtz's ability to find the time to research and write were nonetheless dependent on them; the pace of science depends in good measure on more than science proper.

The immediate context for Helmholtz's work on part three was his position of scientific leadership within the German ophthalmological community, and especially Heidelberg's. Ophthalmology at midcentury was a flourishing enterprise, and Heidelberg was one of its centers. As a member of the Medical Faculty and as already a major contributor to ophthalmology, he stood close to the clinical ophthalmologists at Heidelberg as well as to the entire medical specialty. But, although a medical doctor, he was certainly not a clinical ophthalmologist. Rather, he was a theoretical and empirical investigator of the eye and an innovator and synthesizer of ophthalmological knowledge. In these ways he complemented his clinical colleagues at Heidelberg.[2]

The institutional (and clinical) leader of ophthalmology during the early period of Helmholtz's time at Heidelberg was Chelius. In 1819 he became the university's professor of surgery and, later, director of its surgical clinic. While he was a well-regarded surgeon—he successfully operated on one of Frédéric Chopin's fingers!—he was not a researcher or a medical innovator. Yet he controlled ophthalmology at Heidelberg for nearly a half century, until his retirement in 1864. His replacement, Karl Otto Weber, was a specialist in pathological anatomy and a surgeon; Helmholtz had helped bring him to Heidelberg from Bonn. The fourth and final important figure at Heidelberg was Knapp, who remained subject to annual reappointment by Chelius. When Weber assumed the chair of surgery in 1865, ophthalmology was separated off from it, and Knapp became extraordinary professor of ophthalmology. On Weber's death in 1867, a new ordinary professorship of ophthalmology opened up, but it went to Otto Becker. Knapp felt that he had not received sufficient recognition for his efforts to make ophthalmology an independent discipline at Heidelberg, and so in 1868 he moved to New York, where he developed a flourishing practice and became one of America's foremost ophthalmologists. Before he left, however, his clinical research and aid proved most useful to Helmholtz for his *Handbuch*.[3]

Heidelberg was also a leader in ophthalmology in another way. In 1857, under the leadership of Albrecht von Graefe, it became the site of the first informal meeting of German ophthalmologists. The group met there annually until 1863, when it formalized its activities by becoming the Ophthalmological Society and continued to meet in Heidelberg annually. Its first two presidents were Helmholtz's friends Graefe (1863–70) and Donders (1870–89). Graefe was one of the leading clinical ophthalmologists in the German lands and a founding editor of the *Archiv für Ophthalmologie* (1854), one of the leading journals in the field. Helmholtz's ophthalmoscope (in versions increasingly improved by others) became the prime instrument for investigating the retina and eye movements, and its applications appeared regularly in studies in the *Archiv*'s pages. Helmholtz was even closer to Donders, who was a clinically oriented research ophthalmologist. Donders and Helmholtz studied each other's writings and incorporated some of each other's results into their subsequent works.[4] Yet despite his friendly relations with Graefe in Berlin and Donders in Utrecht, Helmholtz rarely attended the society's meetings, since they were held in late summer, when he was usually away on vacation. He did, however, attend the biweekly meetings of the Heidelberg Natural History and Medical Association. He was elected its chairman and served in that capacity during his entire professorship in Heidelberg.

The Horopter and Eye Movements

The two most controversial (and closely related) points in physiological optics circa 1860 were the horopter problem and eye movements.[5] Helmholtz spent much of his research time between late 1862 and early 1865 working on these two problems. The scientific papers that he published during this period first appeared in preliminary form locally, in the *Verhandlungen des naturhistorisch-medicinischen Vereins zu Heidelberg*, and subsequently in specialized national scientific journals. These publications constituted the first stage of his analysis of these two problems. Here he first tested out his ideas and appealed to his fellow specialists in ophthalmology and physiology. The second stage was to incorporate the work into part three of the *Handbuch*.

In October 1862 Helmholtz presented a short paper to his fellow Verein members on the mathematical form of the horopter. The horopter, he explained, was "the embodiment of those points of external space whose images fall on identical points of the retina, in both eyes, at a given position of the eyes." The paper was largely devoted to definitions or precise descriptions having to do with measuring the eye's structural geometry: for example, it dealt with identical retinal points, lines-of-sight, and various meridians, planes, surfaces, angles, intersections, and fixation points. He noted that Wundt had already taken up the horopter problem and that he (Helmholtz) had adopted certain of Wundt's procedures for making measurements, and he also noted that Johannes Müller had already developed an equation for the horopter conceived as a circle. Helmholtz himself developed several equations for characterizing the horopter or its parts, including the radial horopter, the circular horopter, and the combination of the two into the total horopter.[6] This short paper became the first of several attempts during the next two years to characterize the horopter precisely.

He also studied eye movements. In May 1863 he presented a preliminary communication on this topic to the Verein. "The problem of eye movements," he explained, concerned determining the various rotations (up, down, left, right) that the eyes had to make for the two retinas to achieve a line-of-sight of a given point. In solving this problem, he employed a law developed earlier by Donders (and Georg Meissner), and he also referred to work by Adolph Fick, Friedrich von Recklinghausen, and Wundt. In particular, he claimed to have found "an optical principle for eye movements," one involving probability calculations and using the principle of least squares. The principle—known as the principle of easiest orientation—was that "there must be a position of the eye from which all infinitely small movements of the same can occur without rotation around the line-of-sight." This principle built on earlier laws by Donders and Listing and was in accord with earlier findings by Ruete and Listing, though they had given no proof of it. Helmholtz claimed he had experimentally confirmed Listing's law—that the eye's orientation remains the same when it looks in a specified direction of orientation—with his very own eyes.[7]

In June 1863 he sent Graefe a manuscript entitled "On the Normal Movements of the Human Eye," for Graefe's *Archiv*. Graefe was keen to publish it as soon as possible and had it out that autumn. This sixty-page analysis became Helmholtz's definitive statement on eye movements. The first (of five) sections concerned theory. He argued that the principle of easiest orientation explained eye movements that accorded with Listing's law of eye movements. He cited the work of Donders, Meissner, Fick, and Wundt as having shown the way for his own analyses and findings, though, unlike the latter two, who worked in terms of the eye's muscle mechanisms, Helmholtz invoked an optical principle (that of normal eye movements). He also thought his principle accorded with Darwin's "proven influence of the inheritability of individual characteristics on the fitter formation of organisms."[8]

In the second section, he reported on his "own experiments on eye positions." He used the results of Ruete, Donders, and Wundt on afterimages to give what he labeled "a direct proof" of Listing's law. He made observations that forced him to contort his own eyes into odd positions in order to test the law. (Shades of Newton!) He got Knapp to be an observer and fellow eye contortionist. The third section concerned determining eye positions by means of a binocular double image. He reported on and discussed Meissner's experimental results, some of which he redid and corrected. The fourth section reported experimental results on disturbances to the eye's orientation through rotations (especially involving afterimages). He characteristically indicated that he believed human willpower had a great ability to control eye movements. "Insofar as they are conscious," he wrote, "all of our acts of the will refer not to the contraction of definite muscles, whose existence most people hardly know anything about and which do not always reach a definite body-part position, but rather [refer] much more to the attaining of a perceptual success." He concluded this section, and effectively the paper, with the broad claim that humans *learned* to train the muscles throughout their bodies: "The child who learns to walk and the boy who learns to swim are initially unskilled and uncertain, and they give great effort. Practice finally teaches them to execute these movements. . . . A similar type of forced usage, derived from the needs of orientation, I believe, also rules eye movements. I therefore considered it unnecessary to seek after anatomical arrangements in determining the law of these movements." The paper's fifth and final section consisted of a mathematical appendix.[9]

Having completed his paper on eye movements, Helmholtz presented remarks on the horopter's form—depending on the fixation point, its shape varied—in several physics and ophthalmology journals. His archrival Hering had recently claimed that Helmholtz "may have erroneously ascribed two sections to the general horopter curve, whereas it may have only one." Helmholtz's reply momentarily took the high road. He did not want to claim "that the general result of his [Hering's] calculation of the horopter may be incorrect." He explained, "On the contrary, I can only recommend to my readers his treatment of the problem as very elegant, clear, and complete." Yet he then reversed course and attacked: "Frankly, the asymmetry of the two retinas, which has a very important influence on the formation of the horopter, has been considered [by Hering] only quite incidentally." He conceded that when he had given his preliminary notice on the horopter, he had presented it by using both a second- and a fourth-degree equation; but now he realized that two equations of the second degree would suffice. In his final paper on the horopter, he maintained that he had "given and published this simplified representation even before Herr Hering's corresponding investigations."[10] Hering got under his skin.

Simultaneously, in December 1863, Helmholtz spoke on the horopter before the Verein. He responded to Recklinghausen's recent work on asymmetry in the distribution of identical retinal points of both eyes. He offered a new definition of "identical points in both fields of vision," now concluding, "In general the horopter amounts to a line of two curves which can be represented as the intersections of two surfaces of the second degree."[11]

The following spring he again spoke before the Verein, but this time on a side issue: his experiments on "muscle noise" (*Muskelgeräusch*). The very notion of this phenomenon was arresting and patently related to (and demonstrative of) his ongoing interest in physiological acoustics. He recognized that the idea of such noise had often been doubted, but he reported that, under properly prepared experimental circumstances, and by the use of either the ear or a stethoscope, such noise could in fact be heard "very clearly" and in a variety of muscles. He thought the phenomenon could probably be explained by using du Bois–Reymond's hypothesis of a rapid exchange of electromotoric molecular groupings during muscular contraction. Such proof, he suggested, was largely a matter of using increasingly sensitive tone-detecting instruments. He returned to the topic (now as "muscle sound" [*Muskelton*]), again before the Verein, in July 1866, reporting that he had constructed more delicate listening devices that allowed him to hear (previously undetectable) low muscular tones and to represent their oscillation frequencies visually on strips of paper.[12]

Helmholtz's main research topics in the mid-1860s, however, continued to be the investigation of the horopter and eye movements. In 1864 he again published in Graefe's *Archiv*, where he presented his definitive work on the horopter. He analyzed "the distribution of corresponding points in both fields of vision," confirming some of his own earlier results and some results of Recklinghausen and Volkmann; discussing the horopter's form; analyzing its "meaning for sight"; and, finally (again), giving it a mathematical description. He provided more precise definitions of the "field of vision" and the horopter and emphasized the importance of Lotze's idea of "local signs" (stimulations to and on the retina that create spatial values that aid in transforming sensations into perceptions) as well as his own interpretation of vision as a learned phenomenon.[13]

In June 1864 Helmholtz spoke yet again before the Verein, this time on stereoscopic sight. He gave the first demonstration of his newly constructed stereoscope, which he claimed was twice as powerful as a standard one. He also reported on its use for studying binocular space projection, in particular for looking at a thread or a piece of wire (just as Hering had). Although he obtained different results than Hering had, he spoke positively about some of Hering's findings. He noted, too, that Wundt had shown that there was a limit to exactness when dealing with questions about our eyes converging on a distant point. Finally, he disputed Hering's theory concerning stereoscopic space projection. Here was scientific competition. In November he once again addressed the Verein, in a talk titled "On the Influence of Cyclorotation on the Projection of Retinal Images Outward." He published this address, as well as that on the stereoscope, in the *Verhandlungen*. A summary of the former and much of his other recent work on vision were incorporated into his Royal Society Croonian Lecture from the previous spring. He left the details for the *Handbuch*.[14]

In the last of his appearances before the Verein to report on his latest results on eye movements (January 1865), he declared "Our will's intention can be directed to no other purpose than that of simply and clearly seeing a definite object. We thus learn," he continued, "to execute abnormal eye movements as soon as we permit our eyes to see under abnormal conditions." "The will's intention," he thought, was essential for learning to see objects. Even an empirical law like Listing's was ultimately dependent on it, for he claimed that that law was dependent on training the eye. He reported experimental results issuing, as so often in the past, from observations made using his own eyes. His talk appeared both in the Verein's *Verhandlungen* and in the more popular *Heidelberger Jahrbücher der Literatur*.[15]

Late in December 1864, having completed his research on the horopter and eye movements, and while deeply involved in writing about spatial perception and the theory of perception (that is, the final section of part three), Helmholtz received a letter from the Italian geometrician Luigi Cremona of Bologna. Cremona had made his reputation largely on his work in projective geometry and in understanding the transformation of curves. Cremona wrote because a colleague had sent him Helmholtz's recent paper on the horopter, which he especially liked because Helmholtz's results led to mathematical generalizations concerning various curves. Cremona discussed the mathematics at length.[16] If Helmholtz had not already recognized it independently, he could see from Cremona's letter that observational results on the horopter led to mathematical issues beyond simply characterizing the horopter's form mathematically. Cremona's letter presaged rethinking by Helmholtz and by mathematicians from 1868 onward about the nature of geometry. Helmholtz's work on the horopter and eye movements had primed him to reevaluate his understanding of geometry, space, and motion. A little over four years later, another Italian mathematician, Eugenio Beltrami, also prompted him to realize that there were geometries other than that of Euclid. Herein lay a scientific and mathematical future beyond anyone's dreams.

Hard at Work, Rivalry, and Rest

As Helmholtz's extensive papers on the horopter, eye movements, and other topics between 1862 and 1864 indicate, he was working extremely hard. By the end of 1864, he had eighty-five publications to his name, including several monographs; between 1862 and 1864 alone, he published thirteen items. He was constantly at work. His enormously strong work ethic stemmed from the examples and standards set by his father, his Gymnasium, and his medical school. Those nearest to him knew his work ethic best. In late 1864, Anna characterized him to her brother as "very hardworking" and said that he found it "extremely convenient" to have all his facilities at home, in the Friedrichsbau, "so wonderfully at the ready." His assistant Julius Bernstein recalled that any Heidelberg faculty member or student who went out for the evening to enjoy himself and happened to pass by Helmholtz's laboratory late at night "quite often saw Helmholtz's study lamp still illuminated in the Friedrichsbau." Knapp said that Helmholtz was not only gifted intellectually but extraordinarily industrious. "How often have I looked up at him standing before his desk at 11 at night when I returned from some entertainment. He was a judicious and untiring worker; he did all his work himself. . . . He had no assistant in his scientific work, not even an amanuensis."[17] He was driven.

In late 1864 he informed his friend Soret that he was working "as ardently as possible" to complete part three of the *Handbuch*. "But it is precisely the still incomplete section that requires such great labor and creates so many difficulties." That section almost certainly concerned the large-scale argument that he was creating between nativism and empiricism as alternative theories of perception and the associated role of "unconscious inferences." He also wrote du Bois–Reymond:

> I'm working, insofar as possible, on the third part of my physiological optics [book]. It's a wicked chapter because one must penetrate deeply into the psychological realm, yet one can't, even through the best-considered thoughts, in any way count upon convincing people. At the moment there's a lot of work being done in this area, and, amid lots of weeds, at any given point in time the most unexpected new facts nonetheless shoot through. I myself have gradually also had to take precautions with my eyes [as a result of performing] physiological-optical experiments. I've decided to finish the last part as best as I can for now, even if doing so is not to my own [full] satisfaction. For the moment I've temporarily let all other experimental work remain on the sidelines.

A month later he wrote du Bois–Reymond again that he was still working up the last part, which, "frankly," also had suffered many interruptions during the semester. "Still, I now hope to be finally finished with this thankless work by the Easter break." Then he got personal:

> Herr E. Hering has made me very angry with his shameless way of passing judgment on other people whom he has not, in part, even once taken the trouble to understand properly. I don't, however, want to treat him unfairly since he is, in his own way, an intelligent person. Right now he finds himself at quite cross purposes [with others]; still, at least he works out his standpoint in a consistent way. Moreover, he was, or so I hear, previously mentally ill [*geisteskrank*], and that has restrained me from trumping him until now, as he's indeed deserved it in certain places.

The oddity in this newly emerged hostile relationship—one that soon appeared in part three as a major scientific rivalry—was that, as a young scientist, Hering had held up Helmholtz as the master, indeed as his master, of physiological optics. He was, however, extremely feisty and polemical, and he could (and did) criticize Helmholtz (as well as Wundt, Volkmann, and others) while simultaneously considering him his teacher (though the two had never even met). The gulf between the two men's academic statuses could hardly have been greater: between 1860 and 1865 Hering practiced medicine privately in Leipzig and, on the side, served as a private lecturer at the university there. He managed to publish a series of influential articles on binocular vision and depth perception that he gathered together as his *Beiträge zur Physiologie* (1861–64). As Helmholtz took aim at Hering, he was still more than a year away from completing part three.[18]

Helmholtz especially needed rest and recuperation that summer of 1865. He hiked in the mountains. Anna thought he looked tired and was glad that he would spend several weeks hiking in order to recover his health and strength, but she told him, "Life is so very desolate and dull without you, Hermann!" He replied more existentially and apologetically: "Do be certain," he wrote her from the Piedmont, "that, from a distance, I feel even more clearly how necessary your presence has become to fulfilling my existence and how very much I am grateful to you for it." He apologized to her that, when he suffered from migraines or worried about money or became lost "in scientific speculations," he sometimes got discouraged and thus often made life difficult for her. "I know very well," he confessed, "that on such days I'm a boring or exhausting companion. I'll really try, even more than I've done so far, to become master of the situation." He thought he had lost his sense of spiritual or mental equipoise. He wrote: "I can't be calm without strong intellectual or bodily effort and that is itself a defect for [doing good] scientific work, since the best ideas and plans always emerge only in moments or hours of complete well-being."[19]

After two weeks of hiking in the mountains with his friend and colleague the ophthalmologist Rudolf Schelske, he felt he had recovered his physical strength. He was "rewarded with a brilliant sunset and a magnificent illumination of the entire Mont Blanc mountain chain." Then he went on to Geneva, where he stayed with the Sorets and met, among others, the Genevan physicist Auguste de la Rive, a leading figure in the Genevan scientific community. When he returned from vacation, he began a year-long appointment as dean of the Medical Faculty.[20] The administrative position consumed much of his time and energy.

In mid-January 1866, he sent off the virtually completed manuscript of part three to the publisher. He told du Bois–Reymond that he was "now near the summit of a great mountain, namely, that of Physiological Optics, which has always stood as a barrier in my way."

[I have] only to rework the final paragraph, which, quite frankly, concerns a tricky issue, the competition between fields of vision. I still don't know, when all this is over, how I'll use the [ensuing] freedom to work, and what I want to do. This last part of the *Optik* has pained me extremely because the topic has, so far, not yet been worked through [sufficiently] such that one could approximately sum it up, and also because an individual researcher can't finalize most issues with his own observations because individual differences play a great and essential role here. There are few practiced observers here who are free from theoretical prejudices; naturally, everyone believes his own eyes more than those of others.

I am normally used to doing my work [*Sachen*] completely and clearly until I see essentially nothing more to do. In this case, however, I can't do so, and so I've had no real joy in the work.[21]

That lack of joy was probably further felt by the book's being still nearly a year away from publication. The delay may have lain less with Helmholtz and his publisher than with the historic military-political circumstances emerging between Prussia and Austria.

Wartime

Only once in Helmholtz's lifetime (during the Franco-Prussian War) did external political and social events either stop or slow him down in his work. In the spring of 1866, Bismarck provoked a war to liberate Prussia from Austria's longstanding involvement in Prussia's political affairs, to divide the liberal opposition in Prussia, and, generally, to strengthen Prussia in German-dominated Europe. Prussia's ally in this fight, Italy, itself sought liberation from the Habsburg Empire. The Austrians had France and nearly all of the southern and smaller German states on their side, including (at first) liberal-minded Baden. Since 1859, when the so-called New Era (that is, the new liberal spirit under the new monarch, Prince William) had emerged in Prussia, Baden had called for German political unity. At the start of the Austro-Prussian War, Baden sympathized with and became Austria's ally; but after 3 July 1866, when Prussian troops decisively defeated the Austro-Saxon army at the Battle of Königgrätz, Baden and its people distanced themselves from France, and even those who had previously felt sympathy for Austria now became supporters of Prussia. By late July, Prussian forces occupied Heidelberg; in August, Baden and Prussia signed an armistice followed by a peace agreement. Baden put its troops under Prussia's war command. Among other war prizes, Prussia annexed Hanover, and as a consequence, the University of Göttingen was henceforth governed by the Prussian Ministry of Culture.[22]

Helmholtz, a Prussian living in Baden, had no doubt as to precisely where his loyalties lay. At least in the run-up to the war, the Helmholtzes sought to live as normally as possible. He was still in the final phase of writing part three and spent a good deal of time in his laboratory and study. When Anna inquired, as she often did, as to the practical value of his scientific research, he assured her that its utility was "of a secondary interest . . . [and] that once the facts are established a way will always be found to apply them and to make them useful for humanity."[23]

For several weeks Heidelbergers continued to think that they would have to bivouac Prussian troops in their homes; yet few or none appeared. "The only important thing in this situation," Anna told her brother, "is that my husband doesn't lose his calm and doesn't let himself be driven into a fright; also, that he doesn't let himself get too melancholy owing to the sounds of my crying." Most students abandoned their studies. In Berlin, Georg Quincke estimated that as a result of the war, class sizes had shrunk to a quarter of their normal sizes. Helmholtz wrote du Bois–Reymond:

> For quite some time the possibility of a fight over the Neckar crossing threatened us here. Since this didn't happen, we've gotten through the war period without greater difficulties. It was, frankly, an exciting time, which very much took hold of one's thoughts, and we had here a rather rough opposition of [political] parties with one another. The moneyed Jews of Frankfurt [*die Frankfurter Geldjuden*] threw their black-yellow democracy at us here; our Ultramontanes [those who looked to the Roman Catholic Church for authority] were also excited and, until the Battle of Königgrätz, threatened resistance against the Grand Duke [of Baden], who has here shown more insight than his advisors, but who hasn't had the necessary courage to put his opinion into practice. Baden has thus certainly arrived at a very false position.

In much the same vein, he wrote Thomson:

> Except for some quartering of troops on both sides, we didn't come into too much contact with the war events of this past summer. It was, however, naturally a time of great political excitement because sympathies for both sides were very much divided everywhere, and very intensely collided with one another. I myself was on the side of the land of my birth, Prussia. And even if I too could in no way support everything that happened from that side, I was indeed delighted over the success which, in spite of everything, means progress for the better in our confused German [political] conditions. The quartering of Prussian troops that we ultimately had in the end was easier to bear than the befriended Bavarian [troops] with which we were blessed with in the beginning.[24]

The German states had now moved much closer to a Prussian-led political union. Bismarck replaced the old German Confederation, which had included Austria, with a new North German Confederation (Norddeutscher Bund), led by Prussia and without Austrian participation. Though Baden was not a part of the new confederation, it did become an allied member through its participation in the Customs Union (*Zollverein*) as well as through minor military accords.

In effect, Helmholtz had moved closer to Prussia's political orbit. After the war, he spent three weeks resting and recovering his strength in Engelberg, Switzerland, where he met Wiedemann and Clausius and "somewhat refreshed" his mind and also sought to make his waistline "somewhat slimmer." By October the manuscript of part three had been "long finished." He complained that the publication process was taking too long and that his publisher was looking to save money. He had already received the page proofs of the French translation of the entire *Handbuch* and said, "It will in many regards be better than the original."[25]

In late November 1866, he told both Tyndall and Thomson that he was "nearly finished" with seeing part three through the press, and he looked forward to soon sending each of them a copy. "That was really a long piece of work, and I'm happy to be finally finished with it." For four years he had devoted himself as much as possible to writing this third and final part. Given that part one had originally appeared in 1856 and part two in 1860, all told, the research and writing of the *Handbuch* had taken (or was spread out over) more than a decade. It was all "only a mechanical waste of time," he told du Bois–Reymond. He had little time left for anything else and felt exhausted: "It seems to me as if, with progressive age, the inventive faculty [*Erfindungskraft*] is spent."[26]

The completion of part three marked an intellectual turning point in his life, just as the Austro-Prussian War marked a political turning point in German history. Helmholtz's reputation now deepened and spread wider. It even spread to Australia, where a flower was named after him. Ferdinand von Mueller, a German who had emigrated to Australia in 1847, where he became a leading naturalist and figure in Australian science, had long wanted to honor Helmholtz, whom he revered as an "illuminating *savant*." He christened the flower as "Helmholtzia acorifolia," "as a *lasting* memorial" and, he told Helmholtz, "in order also to immortalize your name in plant geography."[27] The year 1866 ended well for him.

The *Handbuch* (Part Three)

Part three incorporated Helmholtz's papers of 1862–65 on the horopter and eye movements; it initially appeared in two installments in late 1866. In 1867 Leopold Voss, the publisher, reissued all three parts together as a single, consolidated monograph. The entire *Handbuch* (all three parts) ran to 917 pages. It was, however, much more than a handbook.

Helmholtz apologized to his readers for the volume's long delay in appearance. He explained that while working on it he had twice moved residences, had also worked on topics other than physiological optics, and had had to develop a theory of sensory perceptions, which was a difficult and complex subject requiring much time and thought. Although the entire field of physiological optics had made recent and rapid progress, he said, it was still very much under development. Issues of proper observational and experimental technique, protocol, and experience, not to mention the role of psychological factors, abounded and meant that many points remained unresolved. He sought to bring "order and context" to this complex and broad field, and to do so through his empiricist theory of vision. That was the "principle," the "guiding thread" of his massive work, that made it more than a handbook—that was what made it a long sustained argument. He acknowledged that others had preceded him along this line, but he thought their views had in part suffered from what he called the era's "materialist" outlook. He hoped that his work moved beyond theirs through its emphasis on the broader context of the problems analyzed. He stressed that he had done his own physiological observations and experiments and had given full discussion of his associated methods; he was not merely reporting on what others had found. The book contains hundreds of footnotes and bibliographical references to the literature—the vast majority of which cites works by scientists, not philosophers.[28]

At nearly four hundred pages of text, part three was more than twice as large as part one ("The Dioptrics of the Eye," which made up the three-part *Handbuch*'s first sixteen sections) and nearly 70 percent larger than part two ("The Theory of Visual Sensations," constituting sections 17–25). Part three, "The Theory of Visual Perceptions," was divided into eight sections (26–33). Its most controversial were its first (number 26) and last (number 33); they were the volume's most theoretical and rhetorical.

In section 26, Helmholtz addressed human perceptions in general. "We use the sensations that light stimulates in our visual nerve apparatus," he began, "in order to form representations [*Vorstellungen*] from them about the existence, shape, and position of external objects. We call these sorts of representations visual perceptions." The purpose of part three was to discuss and analyze scientifically the conditions that led to visual perceptions of all sorts. Included were a discussion of sensory illusions, the difficulty of making observations on subjective sensations, the influence of experience, the agreement about intuitive images (*Anschauungsbilder*) and objects, and unconscious inductive conclusions.[29]

Helmholtz's fundamental point about perceptions was that they were representations, and that these were "always acts of our psychical activity." To investigate perceptions was for him ultimately to discuss psychological processes. It was the study of the "laws" of the "mind's activities" (*Seelenthätigkeiten*). While there was certainly a physical-physiological substrate underlying the mind that also required investigation, studying the mind's "psychical activities," its "laws and nature," was unavoidable when studying perceptions.[30]

Helmholtz further maintained that some psychical activities (namely, those implied in certain aspects of perception) were "unconscious inferences," a notion that he had gradually developed since the mid-1850s. Human beings experienced only the sensory effects of "external objects," never the objects themselves, sensations being no more "than the means which serve us towards knowledge of the external world." Such sensory effects produced "the idea of a cause," but nothing more. The study of perceptions was thus ultimately the study of "unconscious inferences," which he distinguished from but thought similar to "conscious inferences."[31]

The "facts," he averred, showed the "extended influence . . . which experience, training, and custom have on our perceptions." He admitted that neither he nor anyone else could say precisely how far this "influence" went, but in this regard he thought there was nothing to be learned from the study of newborns and animals. Still, "experience teaches us to recognize a compound aggregate of sensations as a sign [Zeichen] for a simple object." He believed that "the most essential, principled antithesis" among researchers of perception concerned the role of experience. "Some tend to concede the widest possible scope to the influence of experience, namely, to derive all spatial presentation [Raumanschauung] from it. We can designate this view as the empiricist theory." (This was, of course, Helmholtz's view.) "The others, to be sure, have to admit a certain role for the influence of experience. They believe, however, that in all observers and for certain homogeneously entering, elementary presentations [Anschauungen], there is a system of innate [angeborenen] presentations, not based on experience; that is, one must presume spatial relations. We may well designate this latter view, in contrast to the former, the nativist theory of sensory perceptions."[32]

He acknowledged that the difficulty with his psychological interpretation of the perception process was that "so far we generally know next to nothing about the nature of the psychical processes." The task of the empiricist theory was thus to develop knowledge about them. That theory's shortcomings, however, were not so great as those of the nativist theory, since the latter merely assumed some innate mechanism to explain the origins of sensory perception. Whatever drawbacks his own theory may have had, he maintained that one must choose a "standpoint . . . so as to be able to bring at least clear order into the chaos of phenomena." He hoped that whatever interpretation one chose, it would not interfere with "the true observation and description of the facts."[33]

He emphasized that he "designated sensations [Sinnesempfindungen] only as symbols for the relations of the external world." He denied that there was "any type of similarity or equality with what they designate" and the external world. He rejected any presumed "preestablished harmony between nature and the mind [*Geist*]" or "the identity of nature and the mind." He claimed that the nativist theory, with its "innate mechanism," was presuming or was close to presuming "a certain preestablished harmony."[34]

For Helmholtz, representations were the end result of practical human action. Along the route from human action to the representation of things lay symbols, the "naturally given signs for the things which we learn to use for regulating our movements and actions." The only relationship between perceptions and reality was that of "the temporal sequence of events with their different properties. The simultaneity, the sequence, the regular return of simultaneity or sequence occurs in the sensations as well as in the [external] events." But that did not mean, he cautioned, that the "temporal relationships" of perceptions somehow also gave "the true image of the temporal relationships" of the "external events" since variable amounts of time were needed within the body to transmit sensory messages from the body's various organs, its eyes, and its ears to its brain.[35]

Helmholtz did not mean that the representation of things was false. Rather, "every image is similar to its object in one respect but dissimilar in all others," whether the image was that of a painting, a statue, a piece of music, or a theatrical performance. "Thus the representations of the external world are images of lawlike temporal sequences of natural events, and when they are properly formed according to the laws of our thought and when we are again able to properly translate them back through our actions, the representations which we have are then also uniquely true for our reasoning power [*Denkvermögen*]; all others would be false." As important as the bodily organs, the eyes and the ears, and the brain are for the process of perception, in Helmholtz's view it was nonetheless "human understanding [that] masters a great deal in the world and brings it under a rigorous causal law." Yet there was "no guarantee that it can necessarily master everything that exists and can happen in the world."[36]

In attempting to understand "how our representations and perceptions are formed through inductive conclusions," Helmholtz again turned for help to Mill's *System of Logic*, which in his view had given the "best" analysis of such conclusions. In particular, he believed sensory illusions originated from the fact that "induction is formed through an unconscious and ineluctable activity of memory, which precisely for this reason [makes it] seem like a strange, coercive force of nature to our consciousness." The soundness of such induced, unconscious conclusions was tested by individuals through the voluntary actions of their bodies. Just as experiment was the route to "the certainty of our scientific convictions," so was it the route "for the unconscious inductions of our sense perceptions. Only when we bring our sensory organs, by means of our own will, into different relations to objects, do we learn to confidently judge the causes of our sensations." He believed such bodily experimentation directed by the human will began in infancy and continued constantly throughout a person's lifetime.[37]

In the final part of this opening section, Helmholtz offered his purest philosophical beliefs or analysis. He had no doubts about the "law of causality"; it was, he thought, part of "our thinking." "We can, in general, come to no experience of nature's objects without the law of causality already operating in us; it can thus also not be deduced from our experiences of nature's objects." In this sense, at least, he was a Kantian. In contrast to Mill, he believed there was no empirical proof of the law of causality. He assumed "a principle of free will" in humans as well as animals and thought it could not be eliminated.[38] Like Kant, he thought that when considering our own behavior we are free; but when considering events in nature, we automatically invoke the principle of causality.

As in his essay on the conservation of force (1847), Helmholtz again invoked the abstract concepts of matter and force, saying they were necessarily intertwined with one another and could not be observed. They were the ultimate "causes that reveal the facts of experience." For Helmholtz, a scientist could comprehend nature only through finding "generic concepts and natural laws." Only through them could change be explained. Such laws were valid "independently of our observation and thought." The scientist always sought to "reduce natural phenomena to a law." To understand phenomena, they had to be "subjugated . . . to the mastery of our understanding." Moreover, a scientist had to presume that all phenomena could be comprehended. He continued: "The law of sufficient reason is thereby actually nothing other than the drive of our understanding to subjugate all our perceptions to its own mastery, not to a natural law. Our understanding is the ability to form general concepts. It finds nothing to do in our sensory perceptions and experiences if it cannot form concepts and laws, which it then calls objective and causal."[39] That certainly had a Kantian ring to it.

On the other hand, he wrote of Kant as a figure in the philosophical history of the theory of sense perceptions. He referred to him only six times in the entire *Handbuch*—and then only briefly, if quite respectfully. His most positive statement of his views came at the end of section 26, where (in less than two pages) he reviewed the history of theories of perception from Descartes to his own day (including the views of Müller, Johann Georg Steinbuch, Johann Friedrich Herbart, Lotze, Wundt, and others). He wrote of Kant:

> The most essential step concerned analyzing the question [of the theory of sense perceptions] from the right standpoint. This was done by Kant in his *Kritik der reinen Vernunft*. Therein he derived all real content of knowledge from experience, distinguishing this content, however, from what it condi-

tioned through the peculiar abilities of our mind in the form of our intuitions [*Anschauungen*] and representations [*Vorstellungen*]. Pure *a priori* thought can only yield formally correct rules which, to be sure, appear to be absolutely constraining as necessary laws of thought and representation, but have no real meaning for the actual world [*Wirklichkeit*], and thus can also never permit any sort of deduction concerning facts of a possible experience.

In other words, Kant offered (only) a schema that laid out the conditions of any possible experience. It was, however, Helmholtz's teacher Johannes Müller who "transformed, through his theory of specific sense energies," Kant's abstract philosophical structures, so that they became a working, scientific framework that he and his fellow physiologists could pursue to arrive at a natural scientific, empirical understanding of the origins of human perception.[40]

Around mid-1866, at the latest, as he finished writing on the origins of spatial perception, Helmholtz was keenly aware of the general philosophical topic of the origins of space. He commented on Kant's views of space (and time):

> He thus especially considered the geometric axioms as also originally given theorems in spatial intuition [*Raumanschauung*], a view about which one may indeed still dispute. Joh. Müller and a series of physiologists followed his view, which sought to develop the nativist theory of spatial intuition. Joh. Müller himself assumed that the retina, in its spatial expansion, experiences itself by means of an innate ability and that the sensations of both retinas fuse together. E. Hering is the person who, in recent times, has sought to develop this view most consistently and to bring new discoveries in accord with it.

In short, Helmholtz placed Kant and Müller with Hering in the nativist camp.[41] But while he treated Kant and Müller as his greatly respected and honored predecessors, he treated Hering as a hostile opponent.

Already in 1866–67, then, Helmholtz put some distance between himself and Kant (as he had initially done in 1855 and as he was also now doing with Müller, however politely). Some eighteen years later, in 1884, he made his intellectual relationship to Kant clearer still:

> At the beginning of my career I was a more-believing Kantian than I am now. Or rather: I then believed that what I wanted to see changed in Kant were insignificant minor points, as compared to what, even today, I greatly admire as his main achievement and which would not come into question. But then I later found that precisely the strict Kantians of the present era stand solidly there [i.e., by the minor points]; and it is there that they see the philosopher's [i.e.,

Kant's] highest development. In my opinion, [by contrast], that is where Kant had not entirely overcome the insufficient rudiments of the knowledge of his era and had also not entirely overcome its metaphysical prejudices. He had not entirely reached the goal he had aimed for.[42]

That was as close as he came to criticizing Kant directly—or to calling himself a "Kantian."

Helmholtz's theory about human perceptions in general constituted the long opening section of part three. There followed six strictly scientific sections, but since these were largely based on empirical observations and experiments, they implicitly (if silently) provided (or were meant to provide) evidence for Helmholtz's empiricist theory of vision. A good deal of what he published in these sections was a combination of reporting on and analyzing the published work of others, along with a slightly more mature, refined version of what he himself had already published in specialized journals, especially his work between 1862 and 1866 on the horopter, eye movements, cyclorotations, and the stereoscope.

The first of these scientific sections (number 27) reviewed eye movements, including discussion of the eye's torsional rotation, the law of cyclorotation, the influence of convergence, the operational movements of eye muscles, arbitrariness in eye movements, the meaning of the law of movement for orientation, geometric considerations of rotations, the derivation of the law of rotation from the principle of easiest orientation, observational methods for determining the law of rotation, and the measurement of muscle movements via the ophthalmotrope (a model of the eye muscles). In section 28 he addressed the monocular field of vision, in particular the superficial ordering of objects in the field, measuring by sight in direct and indirect seeing, the illusions of measuring by sight through special images, the compensating for the eye's blind spot, the calculation of parallaxes of indirect seeing, and observations on individuals born blind. He then turned (section 29) to the direction of vision, including the feeling of innervation of the eye muscles, the center of visual directions, and the localization of subjective phenomena. In section 30 he analyzed depth perception, including monocular and binocular perception, incomplete judgment of convergence and its consequences, the geometric representation of stereoscopic projection, Recklinghausen's retinal functions (normal surfaces), and different forms of the stereoscope. There followed (section 31) an analysis of binocular double vision: in particular, determining the corresponding points of both fields of vision, the horopter, the exactness of depth perception, the separation and blending of double images, and the geometric representation of corresponding points and of the horopter. The final scientific section (32) concerned the competition of vision, including that of contours, colors, luster, and contrast. To each of these six scientific sections (27–32), Helmholtz appended a short historical sketch of the topic addressed; he did not here engage in any philosophical analysis.[43]

The final section (33, "Critique of Theories") recapitulated the foundations of his own empiricist theory and critiqued those of the nativists' theory (principally as expressed by Peter Panum and, above all, Hering). He claimed that his purpose was to assess the facts that he had presented in the scientific sections against the nativist and empiricist theories of perception. Even as he strongly argued against the former and for the latter, he conceded that, in truth, not enough was known to definitively choose one theory over the other. It was really a matter of "the way certain metaphysical considerations" affected certain researchers, in particular the level of concern about the paramount role of psychological factors. He even conceded that there was currently little of a natural-scientific basis for understanding such factors. Even so, because some denied that such an understanding was possible, he explicitly did not want to deny psychological factors, as the "spiritualists" did, or claim, as the "materialists" did, that there was already such a basis. He saw his own position as that of the factually based scientist seeking after laws. "One must not forget that materialism is as much a metaphysical speculation or hypothesis as spiritualism is," he proclaimed. Neither merited admittance into science, where "factual foundations will decide" claims that were made.[44]

Wherever Helmholtz looked at "the phenomena," he saw the role of "psychical processes." He considered Wheatstone, Volkmann, Heinrich Meyer, Albrecht Nagel, August Classen, and Wundt (and later Donders and Fick) to be on his "empiricist" side of things. Adequate and proper recognition of their work, including Helmholtz's own, suffered "from the antipathy of our era against philosophical and psychological investigations."[45]

Before criticizing nativists, he sought to strengthen his own case by reiterating the empiricist and nativist views. First the empiricist: "The principal theorem of the empiricist view is that sensory perceptions are signs for our consciousness whose meaning is left to our mind [*Verstande*] to learn to understand. As far as concerns signs received through the sense of sight, these vary according to intensity and quality, that is, according to brightness and color. Moreover, there must still persist a difference of the same [i.e., the signs] which is dependent on the position of the stimulated retina, a so-called local sign. The local signs of the sensations of the right eye are thoroughly different from those of the left." Then came the nativist view: "Its central point is that the localization of impressions in the field of vision derives from an innate arrangement: either that the mind [*die Seele*] is supposed to have direct knowledge of the retina's expansions, or that, as a consequence of the stimulation of certain nerve fibers, there exist certain spatial representations by means of an innate, not further definable, mechanism." He noted again that Müller had taken that view and had given empirical expression to Kant's claim that space and time are "originally given forms of our intuitions." After Kant and Müller, other nativists followed, including Recklinghausen, Hering, Panum, and the physicist Kundt.[46]

Helmholtz excused Kant for his nativism by reason of Kant's concentration on the a priori, and he excused Müller because nothing was known in his day about eye movements. He left Recklinghausen and Kundt largely without any direct critique and Panum with but minor criticisms. By contrast, he portrayed Hering as the main nativist culprit and did not let him off the hook so easily. He credited him for his work on the directional sight of objects, but he nonetheless devoted the final ten or so pages of the *Handbuch* to criticizing Hering's results and theory, even declaring that at points Hering had used polemics and had, ironically, resorted to psychical interpretations of the perception process. Helmholtz claimed, rather patronizingly, that he had singled out Hering ("a clear and consistent thinker") only because he represented the nativist theory so well. Yet he could not refrain from adding in a footnote toward the very end of the *Handbuch*:

> I hope that one may not look at this criticism, which in the interest of the subject I was forced to direct against Herr E. Hering's views, as an expression of personal irritation as a result of the attacks which he has directed against my latest works. I believe that the viewpoint of the nativist theory of vision, upon which Herr Hering has taken his stand, must rather necessarily lead any logically thinking mind to the sort of hypotheses which are the basis of his theory. And I have directed the attacks especially against his views because they seem to contain the clearest and most consistent realization of the nativist theory which is presently still possible. I have sought in the course of this last part to answer the objections, insofar as they concern factual matters, which Herr Hering has made against my works. Those objections which are solely of a personal nature, I have preferred to leave unmentioned, except where I had to acknowledge that Herr Hering was right.

This attack on Hering's theory was different from Helmholtz's derogatory if passing remarks on the deleterious influence of Hegel, Schelling, and Schopenhauer on German culture and thought, and even from his sustained critique of Goethe as a physical scientist. This attack was emotionally charged, even if it was seemingly directed only against Hering's theory. Its sources lay perhaps in part in simple fatigue after such a long period of work on his book. ("I've been continually involved in these investigations throughout a good part of my life"). Perhaps in part, too, it may have been a sense of lèse-majesté. But Helmholtz may also have realized that the intellectual differences between himself and Hering over the theory of perception were ultimately not as great as he had portrayed them to be; after all, he had declared at the start of part three that what was then most needed in physiological optics was to bring some sort of intellectual order to the subject. The controversy he created between himself and Hering was in no small part a rhetorical strategy.[47] Be all that as it may, these last pages of the *Handbuch*, with their attack on Hering's theory, did not represent Helmholtz's finest hour.

Reactions and Impact

The publication of the entire *Handbuch* was a landmark event not only for physiology but also for ophthalmology, psychology, and philosophy (epistemology). All physiologists, ophthalmologists, and psychologists studied it, and many philosophers were familiar with its pertinent sections. August Classen, an ophthalmologist and philosopher, wrote Helmholtz about the volume's impact on him: "I'm in a sort of rapture which I must declare to you. The profusion of materials and the pellucid clarity of presentation put the reader in a sort of paradise." He thought such enthusiasm, which came mainly from his "deep honor" for Helmholtz's "scientific greatness," might surprise Helmholtz, since in the past they had had some differences concerning the theory of vision; but he also thought Helmholtz had treated his (Classen's) views fairly.[48]

Wundt, in reviewing part three, declared, "It would be impossible here to give even a hazy overview of the rich content that this outstanding work in its three great Parts . . . presents." Instead, he emphasized the book's "long-recognized meaning" and restricted his review to some critical remarks concerning Helmholtz's empiricist theory. He claimed that already in 1858–59 he had argued for the importance of unconscious inferences in perception. He also disagreed with Helmholtz's interpretation of the process whereby unconscious inference occurs. Yet all in all, he offered a highly complimentary and respectful review. Helmholtz's and Wundt's good working relationship survived Wundt's priority claim in conceiving the idea of "unconscious inference." (Already in 1863 Brücke had warned Helmholtz that it was said that Wundt "may have used [Helmholtz's] ideas with great freedom.")[49]

To take a third and final German example, Friedrich Ueberweg, professor of philosophy at Königsberg, wrote a nearly article-length letter to Helmholtz, saying that he had been studying the *Handbuch* and thought that neither Hering's nativist nor Helmholtz's empiricist view of depth perception was right; indeed, he did not like the black-white distinction between the two. He raised objections against the empiricist theory and sought to build his own theory of space upon Müller's views. He claimed that, like Helmholtz, he was neither a materialist nor a spiritualist (that is, an idealist); instead, he proceeded in the way "which every natural researcher *as such* practices." He closed his letter saying "you know how very much I esteem your great achievements."[50]

The *Handbuch* became de rigueur for anyone concerned with or even hoping to be concerned with physiological optics. The young Freud asked his friend Eduard Silberstein in Berlin to buy him "a secondhand and cheap but still readable copy of *Helmholtz's* physiological optics." Freud was already studying the book, though he found it difficult. However, not every German scientist or philosopher was entirely enamored of it, or, more precisely, of Helmholtz's philosophical views as expressed therein. In the early 1870s, Nietzsche studied it to understand Helmholtz's theory of perception. Though he disagreed with the theory, he nonetheless used it and Helmholtz's epistemology generally to develop parallel ideas about the logic of dreams, the hypothetical nature of truth, and, more generally, his rather un-Helmholtzian views about epistemology.[51]

The *Handbuch* appeared in French (as *Optique physiologique*) in the summer of 1867 and ran to slightly more than one thousand pages. Helmholtz used that edition, moreover, to update the entire volume with the latest facts pertaining to physiological optics as a whole and to supply the latest bibliographical references, in addition to correcting calculation errors in the original German version. When it appeared, he had Javal, one of his translators, make sure that copies were sent to Chevreul, Soret, Regnault, Becquerel, Bernard, and the physicist Léon Foucault, among others.[52] Thus he indicated who his friends were or who he simply considered sufficiently important in the French-speaking world to receive a copy.

Following the *Optique physiologique*'s appearance, Helmholtz dithered about attending the International Congress of Ophthalmology and the International Exhibition being held in Paris that August. He had had an excellent first visit to Paris in 1866, and his works were now being translated into French. His presence in the City of Light would doubtless generate further interest in the *Optique physiologique*. As in the past, Anna urged him to go even though she knew that she would miss him: "Afterwards, you'll come back to me with more peace of mind, and we'll then have a very clear understanding about [how best to use] our free time." She told him that he should make his decision without thinking about her. She knew that a trip like the one he was considering to Paris was stressful and that he had already had plenty of stress to deal with during the past semester.[53]

He decided to go, and he found the city "hellishly" hot and crowded. He regretted that Anna was not with him, in part because he thought the exhibition both stimulating and amusing. It was "something that I've never easily seen before," he told her. "It's really tremendously impressive, tasteful, and at the same time also richly instructive. One lets oneself be amused with a good conscience and one can completely enjoy oneself. One needs a lot of time, however, for one can't see all too much on one day after another, even if one takes breaks in between with Tunisian coffee or Dreher's beer or Neuchâtel chocolate ice cream."

He was relaxing. After visiting the exhibition, he spent the next three days at the congress. When it was announced that he was present, the attendees—a stellar cast of ophthalmologists—applauded and rose to their feet. ("I was very much honored.") Again: "I was honored and received by the Society with acclamation." He was then forced to give an extemporaneous lecture in French, "for there was no time for preparation." On the second day, he presented a paper on binocular vision. At the Ophthalmological Society's banquet, Graefe, its president, gave the first toast: "To the three savants who, during the past ten years, have contributed the most to the progress of scientific ophthalmology: MM. Helmholtz, Donders, and Bowman." Since Donders and Bowman were not there, only Helmholtz could reply, and he did so with diplomatic courtesy (and modesty): "To the savant who . . . has more than anyone contributed to giving ophthalmology the high position that it occupies within the body of the natural sciences: to M. de Graefe." A second toast to Helmholtz sought to evoke Pope's famous encomium to Newton:

> Ophthalmology was in the dark woods
> God spoke, and Helmholtz was born
> And there was light!

Helmholtz felt himself slowly forgetting his sense of shame; he told Anna that he was keen to return to her. Several Parisian scientists hoped to see him before he left. Paul Broca, a medical doctor, brain anatomist, and anthropologist, sent him the program of the upcoming anthropological congress, which opened three days after the ophthalmology congress closed; he hoped Helmholtz would attend it. He probably did not, but by the time he returned to his family, who were vacationing at Tegernsee later that month, Anna "found him very distressed and exhausted, . . . no surprise after such fatiguing exertion." She hoped that "the cool lake, the good mountain air, and sitting around doing nothing" would soon restore him.[54]

Helmholtz's first two visits to Paris (1866 and 1867) changed his attitude toward French science and culture. In addition to the *Handbuch*'s French translation, between 1867 and 1873 four of his popular lectures were translated into French. His name now reached the nonscientific but otherwise cultured part of the French public. In 1868, for example, the Russian writer Ivan Turgenev was in Paris, where he visited Mme. Mohl's exclusive salon. There he met Richard Liebreich, "the famous oculist," who lived in Paris. "He spoke to me," Turgenev continued, "with admiration of Helmholtz, his master." Also that year, Etienne-Jules Marey keenly sought Helmholtz's support in his campaign to be appointed to a chair at the Collège de France. Marey was a medical doctor, a physiologist, and above all a pioneering chronophotographer, who built on the work of

Ludwig, Helmholtz, Vierordt, Bernard, and others, using or innovating, or himself inventing graphical recording instruments and techniques that presented novel and dramatic images of physiological and other phenomena.[55] In 1870 he was appointed to the chair of professor of medicine at the Collège.

The previous year, Auguste Laugel had published *L'optique et les arts*, intended as a sequel to his *La voix, l'oreille, et la musique* (1867). Whereas the latter had used recent discoveries concerning physiological acoustics to understand musical harmony, the former used scientific analyses of vision to understand art. Laugel sought to use optical laws as the basis of aesthetics. "In optics as in acoustics," he wrote, "one cannot have a better guide than M. Helmholtz. His *Optique physiologique* is one of the most beautiful monuments of modern science." Laugel also published an article on Helmholtz's physiological optics for the widely read *Revue des Deux Mondes*, an article that Javal found "detestable" and which he urged Helmholtz to respond to. "I meet some people who speak of the discoveries of messieurs Laugel and Helmholtz. That is really too much!"[56]

Hippolyte Taine read and cited Helmholtz's *Optique physiologique* positively in his own widely read study *De l'intelligence* (1870). He used Helmholtz's work to discuss, inter alia, the issue of visual sensation and retinal excitation. He cited Helmholtz's (and others') experiments showing the passage of a finite amount of time between the physical experience of a sensation and its mental recognition. He even adopted Helmholtz's epistemological outlook: "All this detail leads to the same conclusion: our pure visual sensations are nothing but *signs*. Experience alone teaches us the meaning of them; in other words, experience alone associates with each of them [the signs] the image of the corresponding tactile and muscular sensation." Citing the *Optique physiologique*, he further wrote: "Today, analysis by physiologists and physicians has shown, through a multitude of proofs and counter-proofs, all the routes of this association." Finally, he cited Helmholtz on human vision as well as on energy conservation. He even contrasted him rather favorably to Bernard in the matter of "explicative reason," declaring that whereas Bernard made forced axiomatic pronouncements, Helmholtz adduced facts concerning human vision and their essential role in achieving understanding. The visible and the intellectually conceivable (or explicable) were, for Taine, dependent on one another.[57] Helmholtz now had *adeptes* in France.

In contrast to the immediate translation of the *Handbuch* into French, an English-language translation did not appear until 1924–25. Nonetheless, Anglophones became keen readers of Helmholtz's book in German or in French. Maxwell much admired Helmholtz's work on eye motion and studied the *Handbuch* in this regard (at least). George Henry Lewes enthusiastically read the *Handbuch* through the 1870s; as early as 1857 he had begun reading Helmholtz on color and optics in general and on epistemology; he read him again later in further preparation for his five-volume *Problems of Life and Mind* (1874–79). Darwin, too, read at least parts of the *Handbuch*, for, in the sixth edition of *On the Origin of Species* (1876), he wrote:

> Natural selection will not produce absolute perfection, nor do we always meet, as far as we can judge, with this high standard under nature. . . . Helmholtz, whose judgement no one will dispute, after describing in the strongest terms the wonderful powers of the human eye, adds these remarkable words: "That which we have discovered . . . the external and internal worlds." If our reason leads us to admire with enthusiasm a multitude of inimitable contrivances in nature, this same reason tells us, though we may easily err on both sides, that some other contrivances are less perfect.

Similarly, in the second edition of the *Descent of Man* (1877), Darwin, now citing Helmholtz's *Popular Lectures on Scientific Subjects* (1873), wrote of the unlikelihood of "absolute perfection" in vision, in spite of "that wondrous organ the human eye." He continued, "And we know what Helmholtz, the highest authority in Europe on the subject, has said about the human eye; that if an optician had sold him an instrument so carelessly made, he would have thought himself fully justified in returning it."[58]

Like British scientists, the founders of American pragmatism were also great readers of the *Handbuch*, as well as many of Helmholtz's other writings. Pragmatism, that quintessential American philosophy, was first conceived by members of the Metaphysical Club in the early 1870s. Chauncey Wright, Charles Sanders Peirce, and William James, to recall only the three best-known pragmatists, all read and to one extent or another were influenced by Helmholtz. Wright, a scientist and psychologist and also professor of philosophy at Harvard, read Helmholtz on the physiology of color perception and on the eye's optical structure, and it helped him to better appreciate "habit and discipline; the value of which to life *in general* is obviously in the uses of distinct vision." Peirce, mathematician, scientist, and philosopher, approvingly quoted from the *Handbuch*. Helmholtz's empiricist theory of vision was one source of his pragmatism (though he retained doubts about the empiricist theory). He drew on the *Handbuch* in his own analysis of the importance of signs and sensations as the source of our ideas.[59]

As for James, he read at least some of Helmholtz's *Popular Scientific Lectures* and was much interested in his route to and law of conservation of energy. With regard to Helmholtz's work in physiological optics, he noted as early as 1874, "In the investigation of the senses and their perceptions much has been done by German inquirers, among whom we may mention Wundt . . . and the immortal Helmholtz in his Optics." Two years later he reiterated: "The experiments and conclusions which will make Helmholtz's work on optics immortal, . . . have hardly yet filtered down to the level of the 'reading public.'" But by 1879 he thought "Helmholtz's treatment of perception" contained "rather indefinite and oracular statements about the part played by the intellect therein [that] have momentarily contributed to retard psychological inquiry." In 1881 he seems to have changed his mind again; he commented: "In Helmholtz's great work on physiological optics it is impossible to know what most to admire, the mathematical profundity, the mechanical inventiveness, the experimental originality, the subtle psychological observation, or the erudition." Six years later, however, he expressed publicly his first differences with Helmholtz. While reviewing a recent American textbook on physiological psychology, he commented on Helmholtz on space perception: "Not pretending to be an original experimenter, it is natural that he [the author] should be receptive and respectful of the *facts* of the Wundts and the Helmholtzes. But a dash more of incredulity as to some of their *opinions* would have given his pages a spirit and character which they lack." It was not until 1890, when writing on Helmholtz's work on space perception in his own landmark volume, *The Principles of Psychology* (1890), that James's differences with Helmholtz became fully clear. He declared:

> And now what shall be said of Helmholtz? Can I find fault with a book [the *Handbuch*] which, on the whole, I imagine to be one of the four or five greatest monuments of human genius in the scientific line? If truth impels I must fain try, and take the risks. It seems to me that Helmholtz's genius moves most securely when it keeps close to particular facts. At any rate, it shows least strong in purely speculative passages, which in the *Optics*, in spite of many beauties, seem to me fundamentally vacillating and obscure. The "empiristic" view which Helmholtz defends is that the space-determinations we perceive are in every case products of a process of unconscious inference.

James questioned and criticized Helmholtz's viewpoint. He claimed that Helmholtz had no theory of space perception, though he "makes the world think he has one." He continued:

> And so difficult is the subject, and so magically do catch-words work on the popular-scientist ear, that most likely, had he written "physiological" instead of "nativistic," and "spiritualistic" instead of "empiristic" (which synonyms Hering suggests), numbers of his present empirical evolutionary followers would fail to find in his teaching anything worthy of praise. But since he wrote otherwise, they hurrah for him as a sort of second Locke, dealing another death-blow at the old bugaboo of "innate ideas." His "nativistic" adversary Hering they probably imagine—Heaven save the mark!—to be a scholastic in modern disguise.

James's skepticism toward Helmholtz's theory of vision in no way diminished the *Handbuch*'s influence throughout the first half (at least) of the twentieth century. Edwin G. Boring, leader of experimental psychology at Harvard between 1920 and 1950, declared in the latter year that the *Handbuch* was still "a gospel in this field" and that the *Handbuch* and the *Tonempfindungen* were "still classics for the experimental psychology of sight and hearing."[60]

Popularizing Progress in the Theory of Vision

Editors of new journals and new editors of journals seeking renewal often turn to well-known and stimulating authors in their search to make themselves (re)attractive. Helmholtz was a target of such attempts after he had completed the *Handbuch*. He received three invitations in 1867–68 that gave him a chance to solidify his reputation in physiological optics, reach out more directly to philosophers, or popularize his and others' work in the theory of vision.

In 1868 Eduard Pflüger, the professor of physiology at Bonn, became the founding editor of the *Archiv für die gesamte Physiologie des Menschen und der Tiere*, and he asked Helmholtz to write an essay for the journal's first issue. He told Helmholtz that he intended to publish only truly worthy manuscripts and asked him for his cooperation in this regard. Helmholtz responded immediately and positively to Pflüger's request by offering a piece for the new journal. Indeed, to get the journal started off right, Pflüger was prepared to delay publication of the first issue until Helmholtz was ready with his piece. But there was no delay: Helmholtz's "Mechanics of the Hearing Ossicles and of the Eardrum," which had first appeared in shorter form (without discussion of the eardrum) in the Heidelberg Verein's *Verhandlungen* in August 1867, now appeared as the first item in the first issue of Pflüger's *Archiv* in 1869.[61] Brücke, Donders, and Ludwig also contributed to this first issue; their names, along with Helmholtz's, appeared on the journal's cover page, and the *Archiv* became perhaps the foremost journal of physiology in Germany.

The second invitation came from Julius Bergmann, a philosopher and mathematician, who in 1868 became the new editor of the one-year-old *Philosophische Monatshefte*. Bergmann sought to expand the range of the journal by attracting essays on the import of results of the natural sciences for philosophy. He solicited an essay from Helmholtz, who sent him one entitled "On the Historical Development of Modern Natural Science," which soon appeared in the *Monatshefte*.[62]

And in 1867 Heinrich von Treitschke asked Helmholtz to contribute to the *Preussische Jahrbücher*, of which he had recently become the editor. Treitschke had come to Heidelberg that year as professor of history. He was a well-known chauvinist and propagandist, and the Baden government had brought him to Heidelberg against the faculty's wishes. It chose him to succeed the liberal Ludwig Häusser because it knew Treitschke propagandized for national unity and that he would help build ideological solidarity between the southwestern German lands and Prussia. The Treitschkes and the Helmholtzes became acquainted and began socializing soon after Treitschke arrived in Heidelberg. The Treitschkes were welcome guests in the Helmholtz home; Helmholtz became one of Treitschke's few friends in Heidelberg and the only natural scientist with whom he became friendly at all. Treitschke had enormous respect for Helmholtz and drew on him for advice concerning scientific matters.[63]

When Treitschke took over as editor in 1866—at a moment of heightened political tension within the German lands and above all between Prussia and Austria—he wanted to reinvigorate the *Preussische Jahrbücher* by making it into the equivalent of France's venerable *Revue des Deux Mondes*, which had first appeared in 1829 and served as a general intellectual review, publishing such leading authors as Stendhal, Heine, Balzac, Hugo, Mérimée, Sand, and Renan. "We will never become a *Revue des Deux Mondes*," Treitschke told Wilhelm Wehrenpfenning, soon to become his coeditor, "if we do not also sometimes say something about the natural sciences; but then, only from a classical pen. Here I have been lucky. Helmholtz, one of the first, if not indeed the very highest authority in physiology, and also one with an adroit pen, has promised me . . . three long essays wherein he wants to summarize the results of his investigations on the sense of vision. That would give something really important for the February, March, and April [issues]." Treitschke deemed Helmholtz and (their fellow Heidelberg colleague) Bunsen to be "good patriots." He received Helmholtz's three-part essay in early 1868 and thought it so important that he expected to delay publication of another essay and instead publish Helmholtz's in the February issue. Georg Reimer, the *Jahrbücher*'s publisher, wanted republication rights and asked Helmholtz for such rights. "I'm very grateful to you," Treitschke told Helmholtz, "and look forward with much pleasure to the continuation [of the essay in the next two issues]. It was just what, until now, the *Jahrbücher* lacked."[64]

Helmholtz's essay, "Recent Progress in the Theory of Vision," made his results and views on vision known to a general, highly educated reading public. The essay was a reworked version of a set of lectures that he had recently given in Frankfurt am Main and in Heidelberg and at the same time a condensation of the *Handbuch* for nonscientists. It reappeared, with minor changes, in 1871 in the second collection of his *Populäre wissenschaftliche Vorträge*; with some minor changes, it also appeared in French (in 1869) and in English (in 1873). It presented his views of the "the general, interesting results of [physiological] optics" and also attempted "to overcome some misunderstandings concerning the much-disputed [notion of] unconscious inferences." He thought it unfortunate, as he told du Bois–Reymond, that he could not avoid going into some philosophical issues: "Philosophy is, incidentally, a dreadful wasp's nest, which one should never touch."[65]

The structure of his three-part essay was essentially that of the *Handbuch*. In part one, he described the eye as an optical instrument; in part two, its physiology (that is, the sensation of sight); and in part three, how it functioned psychologically (that is, the perception of sight). The essay as a whole also served as a popular version of his argument for the empiricist theory of perception. He told his readers that the nativist theory was "unnecessary" and that the empiricist theory could explain all pertinent facts simply by reference to "the essential laws through their daily experience according to well-known associations of intuitions and ideas." To be sure, he conceded, "A complete explanation of psychical activities has still not been given and probably will also not be given in the near future." But the nativist theory had simply been inadequate to explain psychical facts, he maintained. So that point should not be used against the empiricist theory, whose advocates were natural researchers seeking to understand "the secrets of mental life" (*Seelenleben*). The empiricist theory alone, he claimed, was without contradictions.[66]

Moreover, he emphasized at the outset that the development of ophthalmology during the past twenty years had played an important role in research about vision; he tied practical, human needs concerning eyesight to purely scientific results. He used the language of progress—the word appeared in the essay's title—and spoke of the unprecedented relief of human suffering that ophthalmology had brought, naming Graefe, Donders, and Bowman as exemplars in advancing ophthalmology. All this progress, he said, had depended intimately on systematic scientific research as its "essential, necessary foundation." Here he cited Schiller's poem "*Archimedes und der Schüler*" (1795), with its message of the importance of science for its own sake and its assertion that practical "fruits" resulted from the pursuit of pure science. In Helmholtz's view, pure scientific research, even research of a theoretical nature or that which seemed otherwise irrelevant for human needs, could ultimately have extraordinarily practical consequences for everyday life.[67]

Soon after the appearance of the third and final installment of his essay, Helmholtz sent the entire set to du Bois–Reymond, who read it "with great pleasure" while traveling by train to and from visiting Ludwig in Leipzig. He commented on the issue of nativism versus empiricism: "Incidentally, already twenty years ago we debated this subject. I maintained that the feeling for beauty was innate to us and you maintained that we called only the useful beautiful—for example, the female breast, only because we look at it as perhaps good for sucking on. I have to concede that on this point my need for causality is susceptible to a greater resignation than yours." Similarly, Donders told Helmholtz that he agreed with virtually everything in his essay "only not with your exclusive empiricist theory." By way of reply, Helmholtz conceded that his empiricist theory was presently "only one of the possible views of the matter"; if the "facts" someday invalidated his theory, then he would give it up, in which case it would at least have been heuristically useful. But for now, at least, he thought that possibility was "not very likely."[68]

In mid-1867, with the publication of the *Handbuch* and its French translation behind him, Helmholtz prepared to leave physiology as his main research field. He told Ludwig that he had "intentionally" paused from further work in physiological optics and psychology and that he again felt stimulated by physiological considerations to recommence his "electrical studies." He had also recently published two papers on the factual foundations of geometry (1868); he now had had enough of theory. He continued:

> I find that too much philosophizing ultimately leads to a certain demoralization and makes one's thinking lax and vague. I first want, for a while, to discipline [my thoughts] again by doing experiments and mathematics; and then, later, go back again to the theory of perception. It's also good in the meantime to listen to what others have to say, to what they object to, to what they misunderstand, etc., and to learn whether they are in general already interested in these questions [the empiricist theory, etc.]. My followers in these matters are, so far, still small in number, but they are good people.

At least as of October 1877, the *Handbuch* had sold out, and Helmholtz began contemplating a second edition. He realized this would take a lot of time; indeed, it appeared, only because of the help of Arthur König, who in 1882 became his trusted assistant, in a series of installments between 1885 and 1896. By 1877, moreover, his empiricist theory was again very much on his mind, for he was preparing the most philosophical address of his career. As for Hering, Helmholtz thought his results were "arbitrary, since he did not care to take the trouble of conducting further experiments towards testing his theory." Helmholtz emphasized the importance of "psychological analysis" in physiological research, especially in the future: "I consider this the single, fruitful way of proceeding. We have to single out the simplest forms of psychological activity and try to find the rules of their development. I myself came to this task only incidentally in the *[Handbuch der] physiologischen Optik*, and I haven't systematically and consistently worked at it. Precisely that is needed." He intended to change the situation. However, because in subsequent years he largely abandoned physiology for physics and for the applications of physics to other disciplines (chemistry and meteorology) and devoted much time to institute building and public service, Hering was free to work unchallenged by Helmholtz. It became evident (by 1880, at the latest) that many physiologists were increasingly inclined to favor Hering's nativist over Helmholtz's empiricist theory. Hering himself led the way in attacking Helmholtz's empiricist theory and observational results and also developed his own theory of color ("opponent theory"), which found wide acceptance in the late nineteenth century and beyond.[69]

Long before the fully revised second edition of the *Handbuch* appeared (1896), the first edition, like the *Tonempfindungen*, was intensely studied by Helmholtz's colleagues, and many of its individual observational and experimental findings were revisited. Even those who disagreed with his theory of perception were forced to come to terms with the *Handbuch*. From their initial appearance onward, both books became fundamental scientific works for the disciplines of physiological optics, physiological acoustics, ophthalmology, psychology, and philosophy (epistemology), which at various points overlapped. Beyond that, the two works became intellectual resources for everyone interested in finding commonalities among the sciences and the arts and for those who looked to science for help in understanding art. In the meantime, however, Helmholtz turned his attention elsewhere.

15

Almost a Professor of Physics

Another Fiasco with Bonn

As the law of conservation of energy gradually became one of the centerpieces of physics, and as Helmholtz's work in other branches of physics—optics, acoustics, hydrodynamics, and electrodynamics—appeared, his reputation as an internationally renowned physicist gradually became further enhanced. Along with Weber, Kirchhoff, and Clausius, he stood among Germany's very best physicists. By 1867 he had but little interest left for physiology; physics became his primary interest.

Yet that standing remained incomplete: he was a professional physiologist, not a professional physicist. At Heidelberg, that meant that he belonged to the medical, not the philosophical, Faculty, and it meant he attracted only an occasional physics or mathematics student to his classes. Those few with whom he did work had come principally to study with Kirchhoff or with Heidelberg's mathematicians. Among these were several Russians, including the physicist Alexsandr Grigorievich Stoletov and the psychophysicist Nikolai Baxt.[1]

Another Russian student who came to Heidelberg principally to study mathematics was Sofia Kovalevskaia, who arrived in 1869. Until late in the century, most universities in Europe and North America refused to admit women for study or were even actively hostile toward admitting them. Even when women were admitted, they were sometimes not allowed to seek or obtain a degree. German universities were among the least progressive: Baden did not regularly admit women until 1900, Prussia not until 1908. Kovalevskaia was very much the exception. She was one of the first women university students in Germany, and she needed special permission from individual university professors to attend their courses as well as the university's general permission to pursue her academic interest in mathematics. Helmholtz and his colleagues helped get her admitted and supported her in other ways. She quickly convinced them of her mathematical aptitude, but she also had to prove that she was, as she claimed to be, a married woman. Helmholtz and several other faculty members doubted this, even as they obligingly accepted her and her partner's assurances. She took courses with Kirchhoff, with the mathematicians Leo Koenigsberger and Paul du Bois–Reymond, and with Helmholtz in physiology. She remained in Heidelberg for only one academic year before moving to Berlin to work with the mathematician Karl Weierstrass. There she again had contact with Helmholtz — she was often invited to his home — after his own transfer there in 1871. But Berlin would not officially admit her as a degree candidate, and so she ultimately took her degree from Göttingen (in absentia in 1874) with a dissertation on differential equations, becoming the first woman since the eighteenth century to receive a PhD degree in Germany. She went on to make noteworthy contributions to mathematics and pursued a university career in Stockholm until 1891, when she died there at age forty-one.[2] Helmholtz, in short, played a role in her early professional success.

In May 1868 Julius Plücker, who simultaneously held ordinary professorships of mathematics and physics at the University of Bonn, died, and the possibility of Helmholtz's becoming his successor in physics suddenly emerged. Pflüger immediately informed Helmholtz of Plücker's death and its potential consequence for him: "Everything here seems splendidly well positioned for you." Friedrich August Kekulé, one of Bonn's two professors of chemistry, also wanted him to come. So, too, did Lipschitz, who had returned to Bonn in 1864 as professor of mathematics. These three led an effort to line up other faculty members to bring Helmholtz there as professor of physics. Pflüger eagerly became Helmholtz's informant about the views of various faculty members about him, and he apprised him of the nature of the courses he might teach and what his salary and income from student fees would amount to. Max Schultze, Bonn's professor of anatomy, told him that he too hoped Helmholtz would succeed Plücker, that the ministry had already decided to build a new physics institute (and other scientific institutes) at Bonn, and that he and Pflüger saw Helmholtz's coming as good for both of them.[3]

Helmholtz responded positively to the Bonn possibility. His initial motivation for wanting to move there was simply to become a professor of physics, which included directing the institute. That was what he had always wanted professionally. Physiology was sufficiently established as a discipline, he thought; most of his students were medical students, but his own strength lay in the mathematical-physical approach to the subject. He thought physics needed younger, mathematically oriented teachers in order to move ahead. He attracted 20 to 25 students per class at Heidelberg; at Bonn, he reckoned, he would have 120 to 150. He believed he could do more to help physics than to help physiology. But he also had another reason for wanting to leave Heidelberg: he judged that the Medical Faculty was not progressive, and he had little faith it would become so. Still, he thought his life was "more pleasant, more comfortable, and more independent" in Heidelberg than it would be in Bonn. At first he did no more than unofficially let the Prussian authorities know that if they met his conditions, he would not necessarily turn down an offer from Bonn. If Prussia wanted him back, it was up to it (and Bonn) to make a sufficiently attractive offer. He apparently did not know that in early August, Wilhelm Beseler (the curator of the University of Bonn, that is, the ministry's local representative) had informed Minister Heinrich von Mühler that Bonn's Philosophical Faculty had warmly recommended Helmholtz's appointment as professor of physics and that the Medical Faculty recommended that, "in the interest of its students," everything should be done to obtain him. Beseler continued: "That his world reputation has been established for years; that, in his researching of nature and its forces, the physical side is in the foreground; that his acquisition by a Prussian university would be a mark of renown for the Prussian administration [as a whole]; that his transfer to Bonn in order to assume the vacant chair for this university would be a great fortune not only for the culture of the natural sciences—all that is certain." Beseler considered Helmholtz to be "of extraordinary importance . . . for science in the [most] eminent sense." Bonn's Medical Faculty added that, in terms of "depth," "comprehensiveness," and "sublime speculations," Helmholtz could "perhaps be compared only with Leibniz."[4] That was tantamount to saying he was perhaps one of the two greatest German philosopher-scientists ever.

In the midst of the feelers from Bonn, the British Medical Association asked Helmholtz to come to its meeting at Oxford that August. Henry Wentworth Acland also wanted to know if (and hoped that) Helmholtz might come to Oxford. Acland championed the cause of science in general at Oxford and sought to advance science-based medicine in particular, though he de-emphasized early specialization and stressed the unity of the sciences. He was also one of the key figures in the planning and design of the University Museum (a modern science center), which opened its doors in 1860. Acland tried to persuade Helmholtz to attend the association's meeting and wanted him to demonstrate the ophthalmometer and an acoustic device there. "Our medical friends would be grateful for a demonstration of either."[5] But this effort failed, perhaps because Helmholtz had his eye on Bonn and so did not want to be distracted by spending several weeks in Britain.

He had another preoccupation that spring. For some twenty years, he had suffered from (springtime) hay fever. In the spring of 1867, he thought he had developed a cure for—or at least he (largely) cured himself of—hay fever. As he explained to Carl Binz, a pharmacologist at the University of Bonn who specialized in studies of quinine, he had suffered from hay fever since 1847. From 20 May to the end of June of each year, he reported, he experienced frequent sneezing, especially when the weather was warm. This sometimes led "to painful inflammation of the mucous membrane and the nose's exterior, and caused a fever with strong, severe headaches and intense exhaustion." He always suffered these symptoms at the same time of year. This led him to think that the illness might be due to an organism. Between 1865 and 1869, therefore, he undertook microscopic examinations of the "vibrion-like little bodies" that appeared in his nasal secretions during hay-fever season. Binz's initial suggestion that quinine might be effective against the infusoria led Helmholtz to concoct a solution of "sulphuric acid quinine," which proved to be sufficiently effective. He developed a protocol: lying flat on his back, he dispensed four cubic centimeters of quinine solution from a pipette into each of his nostrils: "I then turned the head this way and that in order to let the liquid flow around in every direction. Then, after I stand up, the rest [of the liquid] flows over the soft palate into the esophagus." He obtained immediate relief: he could go out in the sun without experiencing any symptoms. He found, however, that he had to apply his solution three times daily to keep himself free of symptoms. Beginning in 1867, he took his self-prescribed dosage on a regular basis during hay fever season, managing to suppress all symptoms before they appeared. He wrote to Binz that he believed that, in all likelihood and under certain conditions, "the living vibrions in the nasal secretion" were responsible for the hay fever. After he explained his cure to the yachting William Thomson, who requested details to pass on to Spottiswoode, the president of the London Mathematical Society and a fellow hay fever sufferer, he described his self-concocted materia medica, along with its proper dosage and method of application. "After having destroyed the malady during several successive summers I am nearly completely free of it."[6]

Over the course of the next seven months, little transpired concerning the Bonn appointment. Helmholtz became increasingly frustrated and disappointed. Minor difficulties emerged at Bonn and in Berlin (Prussia), while Heidelberg and Karlsruhe (Baden) showed unqualified readiness to retain him. As he considered his financial needs, additional motives on his part and that of the other interested parties appeared: the level of his compensation, the political economy of two rival states (Baden and Prussia), his pride, the level of trust between the parties, and perhaps even a certain autonomous logic of negotiation. An academic switch like Helmholtz's—from physiology to physics, from a medical to a philosophical faculty—was extraordinary, if not, perhaps, quite unprecedented; the transfer from one state (Baden) to another (Prussia) only complicated matters further.[7]

Helmholtz was a highly attractive candidate to many at Bonn: to its Medical Faculty and to the mathematicians in the Philosophical Faculty, since he complemented members of both groups and since his "world fame" promised to make Bonn a more attractive institution to potential students. Senior administrators also spoke of Helmholtz's "world fame" and potential drawing power for students, as well as the desirability of having him again associated with Prussia. But several Bonn scientists were less than enthusiastic about him. While perhaps not outright opposed to him, and while mindful of his contributions to physics, these individuals wanted something else in their new physicist. Of all attendees in physics courses at Bonn, 90 percent were medical students, while most of the remaining 10 percent were future secondary-school teachers of mathematics and physics. Several Bonn scientists thus wanted a trained, experienced teacher of experimental physics who could guide students in lectures and, especially, laboratory work. They looked in particular to three of Magnus's former students: Georg Quincke, Gustav Wiedemann, and especially Adolph Wüllner, who was already extraordinary professor of physics at Bonn. To appease Helmholtz's backers and those who wanted a leading physicist, a compromise was reached: Helmholtz was recommended as the top candidate, Clausius was placed second, and the three experimentalists were listed together in the third spot. Helmholtz himself had his own concerns, as he told Lipschitz, namely about abandoning physiology for physics: "I must say, however, that it would be difficult for me to give up sensory physiology because it is the subject with which I've been most involved during my life. Moreover, it would be extremely unpleasant in this regard for me to get into a rivalry with Pflüger." Pflüger then graciously conceded to Helmholtz the right to teach sensory physiology as well as to give his popular lectures on the "results of the natural sciences."[8]

In July, after rumors of Helmholtz's possible departure for Bonn began circulating in Heidelberg, Zeller, the university's prorector and Helmholtz's good friend, issued a statement of concern about the grave loss such a departure would be; the medical students issued a statement calling for Helmholtz's retention; and Julius Jolly, Baden's interior minister, alerted his ministry and the university about Helmholtz's possible departure and the universal desire to retain him in Heidelberg. On 31 July, Beseler informed the ministry in Berlin of the Bonn Philosophical Faculty's recommendations, with Helmholtz listed as its first choice. In early August, Helmholtz and Zeller went to Bonn to represent Heidelberg officially at the celebration of the fiftieth anniversary of Bonn's founding; Helmholtz also went, unofficially, to discuss the position, to meet the Bonn faculty, and to assess firsthand the university's institutional condition. He left feeling that most of the philosophical faculty were against his appointment as Bonn's new physicist. Yet Beseler's recommendation had been made only days before. Max Schultze assured Helmholtz that he (Helmholtz) would be moving to Bonn; but Schultze, it turned out, was not well informed. From Berlin, du Bois–Reymond told Helmholtz that there was confidential talk of his return to Bonn and that next year the Prussian universities could expect an increased budget, "from which, however, a notable position would simultaneously be eliminated due to your salary [demands]. One is determined," he continued, "to have you almost *coûte que coûte*; and if you don't come, that would be your decision." A ministry undersecretary had asked du Bois–Reymond to send him something by Helmholtz that he might read; du Bois–Reymond sent Helmholtz's popular lectures. Du Bois–Reymond also speculated that it would be only a few years before the professorship of physics at Berlin (in Prussia) would become available (i.e., Magnus would retire or die), and so if Helmholtz were to become professor of physics at Bonn (in Prussia), he said, "You would likewise unquestionably be called here, and I would consider this to be a very great piece of luck for everyone, myself included."[9]

As Helmholtz waited for months without hearing a word from Berlin and not much more from Bonn, others honored him: in September, Baden made him a Grand Duke of Baden Privy Counselor, Second Class, and he was also awarded (in 1868) the Commander Cross, Second Class, of the Order of the Zähringer Lions. The American Academy of Arts and Sciences elected him a Foreign Honorary Member of its section for physics and chemistry. The Physical Class of the Royal Society of Göttingen elected him a foreign corresponding member. (In thanking the society, however, he had to inform its secretary that his first name was "Hermann, not Heinrich" and he asked him to make the necessary corrections. So much for "world fame.") In addition, the Italian Society of the Sciences awarded him its first Matteucci prize, which went to "the Italian or foreign scientist who will have made the physics discovery judged to be the most important of recent times."[10] He was, after all, a man of "world fame."

Only at year's end 1868 was Prussia finally ready to offer the position to Helmholtz—and to open negotiations with him. In early January Baden made a counteroffer, raising his salary considerably, while the City of Heidelberg sought to seal the deal by making him an honorary citizen. Then the Bonn negotiations suddenly collapsed; Helmholtz bowed out and the offer went to Clausius on the very day that Helmholtz turned it down. Clausius accepted later that month. Helmholtz told Ludwig that he had not liked the way Prussia had acted during the negotiations, adding: "If I may say so, homesickness for Heidelberg finally prevailed, that is, for its moral atmosphere."[11] That laconic explanation left a great deal unsaid and was misleading.

Du Bois–Reymond had heard two versions of why the negotiations with Helmholtz had broken down. One was that Helmholtz would not agree to name a figure for definitively accepting the Bonn offer, thus leaving open the possibility of future negotiations with Baden. The other was that he had asked for five thousand talers, a ridiculously high amount. In Berlin, du Bois–Reymond reported, "one set of individuals, and others, are ready to rebuke you because they're irritated that Prussia can't get you back, and because it's always fun for people to be able to hold something against an important phenomenon." He asked Helmholtz to speak with the ministry, and he asked for an explanation for himself.[12]

Helmholtz had entered into negotiations, he explained, because it meant he might become a professional physicist, and hence he was prepared "to accept [a position] under relatively modest, external conditions." He believed that he was only required to inform Berlin of his current compensation at Heidelberg, that it was up to Berlin, not him, to make an initial offer. However, Lipschitz, acting as an intermediary for the Bonn and Berlin authorities, twice pressed him in early June to name a salary figure and any conditions he might have; that would then create a basis on which negotiations could occur. Helmholtz initially refused to do so but then relented, presuming that his demands would be kept respectfully confidential. "With this affair I see once again," he wrote Lipschitz, "how difficult it is to separate oneself from the well-nourishing milk cow of the Medical Faculty if one once lies at its breasts and then turns to the chaste muse of the Philosophical Faculty. However, when one has 4½ children [Anna had been pregnant with their fifth child, Friedrich Julius Helmholtz], and many human relationships and duties, and nothing to live from except the products of one's own brain, then it's difficult not to become a vulgar calculator and materialist."[13] He was human after all.

Whether individual members of a Medical Faculty in fact earned more than members of a Philosophical Faculty is, however, far from certain. For any given faculty member's total annual income included not only a definite amount of salary but also an indefinite, variable amount of student-fee income (which depended on class size, number of examinations given, and more) and, in some cases, a housing allowance and perhaps other forms of compensation as well (e.g., a widow's pension). At Heidelberg in 1868, Helmholtz earned a salary of 4,000 guldens: 3,600 guldens (2,057 talers) in direct salary plus a 10 percent housing allowance of 400 guldens (229 talers); and 1,750 guldens (1,000 talers) in student-fee income. That made a grand total of 5,750 guldens (3,286 talers). He also earned retained income toward a widow's annual pension of 990 guldens

(566 talers). But he was not willing to go to Bonn for the equivalent or even a trifling increase. He had thus confidentially (and reluctantly) told Lipschitz in Bonn and Olshausen in Berlin that he would not come for less than a salary of 3,600 talers and that that amount should be the basis for any negotiations. He thought that figure, given his "economic circumstances," was "relatively modest." He told Lipschitz that Prussia's negotiations with him were "an insult."[14]

To Ludwig, Helmholtz summed up and weighed his reasons for staying in Heidelberg and doing physiology versus going to Bonn and doing physics. He thought, first, that physics was more unified than physiology, and so he could teach all of it rather than only parts of the rather intellectually disparate field (and methods) of physiology. However, that made physiology a greater challenge intellectually; and it was also "perhaps more useful to humanity." Second, he noted that Prussia's hard and tight-fisted way of negotiating with him was unpleasant. Baden, by contrast, was willing to make sacrifices to retain him. Third, he preferred the "moral atmosphere" of Heidelberg (Baden) to that of Bonn (Prussia), and much preferred being under Jolly's (Baden's) than Mühler's (Prussia's) ministry. Finally, his friends in Heidelberg wanted him to stay.[15] He made no reference to financial compensation.

Helmholtz did not even bother to ask Heidelberg to make a counteroffer. He thought 3,000 talers would scarcely match what he was already earning, and he reckoned that an additional 600 talers would compensate him for any mistakes he might commit in judging the cost-of-living difference between Heidelberg and Bonn, as well as "for the trouble of exchanging positions." Though he had sought to be cooperative, Berlin responded with "five months of the deepest silence." It was only after he told Schultze that Baden intended to meet all his wishes and to retain him that he learned through third parties that Berlin was awaiting the outcome of a budget approval before formally approaching him. He did not encourage Baden in the least, he claimed, but it "lay in the nature of things" that it would respond. In August, minister Jolly himself came to see Helmholtz and got him to agree not to finalize any agreement with Prussia without first giving Baden the opportunity to make a counteroffer. Helmholtz felt obliged toward Baden, "since it had always dealt with me with the greatest willingness and respect." The only reason he would leave was to obtain for himself "a more appropriate and presumably more fruitful activity."[16]

On 26 December, as he further explained to du Bois–Reymond, the Prussian ministry had asked him to meet with Beseler in Mainz, "on neutral ground," so that they could negotiate and finalize the matter. He agreed to go but felt obliged to inform Baden, which sent a ministerial representative to see him two days later. Before he could even tell the representative what Prussia had offered, Baden immediately increased his salary by 1,750 guldens (1,000 talers) and was prepared to offer still more. Helmholtz told Baden that money was not the issue (though he gladly accepted the salary increase). Baden's way of proceeding had only strengthened his desire to stay in Heidelberg. By contrast, when he asked Berlin to provide him a residence within its new, planned physics institute for Bonn, it treated him "quite coolly."[17]

He immediately telegraphed Beseler about the salary increase from Baden; Beseler insisted they meet the next day (29 December). (Beseler thought Helmholtz was "very charming" and "a noble man," and that he would be the most outstanding acquisition of any Prussian university ever.) But now it was too late, and Beseler had too little to offer. In fact, he never even made an offer. "I explained to him, moreover, that I was not the petitioner; it was not my concern to make an offer. I demanded to know how much one wanted to offer me." Beseler responded only to say that he could offer Helmholtz no more than 3,000 talers; however, after he had telegraphed the minister that the situation with Helmholtz would thus be hopeless, the minister had authorized him to go higher (an additional 500 talers in the form of rental allowance). Helmholtz believed that he had been "willingly and generously cooperative" vis-à-vis Prussia, but that Prussia had been stingy in its salary offer and had distrusted his claim that Baden had increased his salary by 1,000 talers. Instead, Berlin supposedly thought he sought more money; it was, he said, "a process through which my goodwill should be exploited to my disadvantage." His instincts had told him, he claimed, that he should have ended the negotiations right then and there, but out of courtesy to Beseler and to avoid a scene, he did not. Pressed by Beseler to state his salary demands, Helmholtz asked for 7,000 guldens (4,000 talers in salary and a rental allowance), but he also noted that he had promised Jolly the opportunity to make a counteroffer before accepting any Prussian offer. This "was a mistake" on his part, he conceded, because it led him to forget his reasons for wanting the Bonn post in the first place. Beseler also wanted him not to seek any counteroffer from Baden, so as to exclude Prussia from being forced to offer still more money to match a new demand.[18]

This state of affairs left him in a tizzy. He decided, therefore, to end all oral negotiations and to deal with Prussia only in writing. As a result, he spent the following day (30 December) "in bed with the worst migraine, unable to order my thoughts." Though he had recognized the "risky shadow sides" of the situation in Bonn since his recent visit there, it was still not easy for him to decline Prussia's offer. He was convinced that most of Bonn's Philosophical Faculty did not want him: "In fact my name had gotten on the list only through a compromise that was very little flattering for me." His opponents wanted Wüllner, who, Helmholtz pointedly noted, was a Catholic. He thought he could overcome his opponents' concerns if he had the ministry's backing. So he wrote the ministry on 31 December to explain that, "if it had *very* good will," he needed more time than the proposed two weeks to make his decision, which included additional discussions with Baden, as he had promised to do. A few days later, he refused the call to Bonn, and Jolly raised his salary to 5,200 guldens (2,971 talers).[19]

The upshot of the past seven months' worth of waiting and a week's worth of negotiating was that Helmholtz stayed in Heidelberg, where he now had a salary, free residence, and various other benefits that together totaled roughly 3,600 talers. He had gone from a salary (and rental allowance) of 4,000 guldens (2,286 talers) before the negotiations to 5,200 guldens (2,971 talers) at their conclusion. In addition, his assistant, Bernstein, received various benefits that enabled him to assume certain tasks that Helmholtz had previously performed. Helmholtz had never asked for 5,000 talers, he told du Bois–Reymond, and, to help preserve his good name and dignity, he asked him to let their friends know about the negotiations. It soothed his conscience that apparently Clausius, not Wüllner, would be getting the call to Bonn. As in the negotiations with Bonn in 1857–58, Helmholtz felt a measure of distrust and lack of respect on Bonn's and Prussia's side vis-à-vis himself. Beseler, for his part, thought Helmholtz too weak to resist Baden's enticements even though he wanted to be professor of physics.[20] Twice now, his negotiations with Bonn (Prussia) proved to be a fiasco.

Helmholtz told Tyndall that he had "nearly" accepted the Bonn offer, and shortly thereafter he likewise told Knapp that, "after many doubts and struggles," he had finally declined the offer. He was now busy reworking lectures, and in the autumn he wanted to take more care of his health (he had suffered this past winter from migraines again and was less able than ever to withstand them). Anna told her aunt much the same story, though far more briefly. The difficulties had come, she said, from the side of the Prussian government, which responded in a very ungenerous manner. Her husband had "in no way found among them the regard and the confidence" that he had a right to ask for." "He prefers to remain at his position [in Heidelberg], under a government which has always acted very liberally towards him. Here he has only to name the sum that he needs for his instruments and his laboratory, and they agree to it — they have increased his salary by a thousand guldens. In the end, the contrast is very striking." She was glad to be staying in Heidelberg, even as she seemed to suggest that someday they might nonetheless be moving to Berlin.[21]

In late July 1869, Purkyně, who had held the chair in physiology at Prague since 1849, died at age eighty. Though Prague offered Helmholtz "a lot of money" to become Purkyně's successor, he was not interested and turned Prague down immediately. He had no interest in another chair of physiology elsewhere. He was more than comfortable in Heidelberg, and he thought he could work best in a small, sleepy town of its sort rather than in a big urban setting. Nor was he in any mood to move now that he had reached "the age . . . where comfort, health, and energy" were particularly important. His health, he said, was "quite tolerable, if I get my rest, work moderately, and am still more moderate in eating and drinking." To be sure, the previous spring he had suffered heart problems; but that had ended now. When he got upset over something, his sensitive health was affected, but thanks to the treatments he had received, even his migraines were better. Moreover, he had twelve students in his laboratory. Wundt had set up a private laboratory, which also helped Helmholtz, "since otherwise I would now not have enough places." At the very moment when Prague's offer arrived, foreign guests were visiting—Hirst and Tyndall. The Helmholtz home became something of a mecca for foreign, especially British, scientists, who were always warmly received. He had no reason to move. The university in Prague then turned to Helmholtz's nemesis, Hering, as Purkyně's successor; Hering accepted and remained there for the next twenty-five years.[22]

Receiving visitors like Tyndall and Hirst demanded Anna's time and energy. In July 1869, she told her aunt Mary that she was closing her *"boutique"* for a month. While she enjoyed each individual visit, collectively it sometimes became too much for her. Guests often came to the Helmholtz house as early as 11:00 a.m. "They come to us from all parts of Europe, not to mention the Americans. It would be very nice if the natural scientists weren't all more or less taciturn and my husband, being in no way loquacious, . . . [means that] the task of making the conversation go falls on me—and sometimes it's a lot to do." Recently they had received one visitor from Brazil, two from the Netherlands, a family from Nîmes, and a Roman count and his wife: altogether, Anna had to animate conversation for and manage people from five different nationalities. "To have tea at our place is very charming—but as they leave every afternoon, it has the effect of an hotel. None of this concerns the visit of Tyndall, who was the greatest pleasure that I have felt in a long time. He is spending the day here with us—and leaves already this afternoon. What a lovely man, lively and frank—full of interesting things—nearly affectionate. He's won my entire heart, and I hope to see him again often, although it is difficult to keep up with him."[23]

The Rivalries of Thermodynamics

Helmholtz had developed many friendly relationships with British scientists, those with Roscoe, Tyndall, and, above all, Thomson being especially warm.[24] However, his attitude toward Peter Guthrie Tait, the Scottish mathematical physicist at the University of Edinburgh and Thomson's close collaborator and friend, was standoffish at best. Tait made difficulties for Helmholtz. First, Thomson and Tait thought little of Tyndall as a scientist, and they looked down on him for his popularization efforts. Helmholtz, by contrast, valued Tyndall as a scientist and greatly respected and supported his popularizing efforts. He did not, however, want to offend either Thomson or Tyndall by taking sides with one against the other. More importantly, Tait was a terrible chauvinist. His prejudice in favor of his fellow British scientists against certain of their German counterparts at times put Helmholtz in a difficult position vis-à-vis his own countrymen.

In February 1867 Tait sent Helmholtz the first two draft chapters of his "little work" on the history of thermodynamics that he planned to publish as *Sketch of Thermodynamics*. The book owed its origins to Tait's dispute with Tyndall over the recent history of the theory of heat. His purpose was to give Joule and Thomson greater credit than they had received to date for their contributions to the subject. However, he feared that in the process Helmholtz and Kirchhoff might be offended, and so he asked them, "Kindly point out to me anything which appears to you objectionable in the way in which I have spoken of your connexion with the subject, [and] I shall be delighted to correct it before my little work is published." He wrote a similar letter to Clausius, who, thanks to Thomson and Tait's hostile treatment of Tyndall and their priority claims for British scientists, already had a negative view of them. Helmholtz replied critically, advising Tait to avoid polemical remarks and not to enter into priority controversies. Tait rejoined by asking for more criticism. He had already published on priority in the discovery of spectrum analysis, giving more credit to Balfour Stewart (and less to Kirchhoff) than Helmholtz and others thought appropriate. As a result, Tait started to have doubts about his own position vis-à-vis Kirchhoff. He also gave less credit to Clausius than Helmholtz would have liked. Here too Tait became defensive, saying he thought Clausius's work on the mechanical equivalent of electrical discharge and on entropy was perhaps unjustly omitted (though he thought Rankine had priority in conceiving the idea of entropy). Tait also gave less credit to Mayer than Helmholtz would have preferred, to which Tait replied that Mayer had done "profound and most original" research and that he, Tait, had given him due recognition. But above all, Thomson and Tait were big backers of Joule in the priority issue concerning conservation of energy, and Tait thought Tyndall had come out too strongly in favor of Mayer. Still, Tait worried that his little book on the history of thermodynamics would be harmful to him vis-à-vis Clausius. A year later Tait sent Helmholtz the proofs of the parts of his *Sketch* that concerned the priority of Kirchhoff vis-à-vis Stewart. He wanted Helmholtz's approval of his historical analysis and feared Clausius's critique. He also asked Helmholtz for permission to quote him on Mayer (from a letter of Helmholtz's) in his preface, which he was still at work on. "I wish to be perfectly impartial; and, though expressing a decided opinion of my own, to give my readers a view of the other side of the case also." Helmholtz agreed to allow the quotations from his letter but on the condition that he be allowed to review Tait's text in proof. Tait agreed.[25]

In his preface Tait claimed, "It is almost impossible to be strictly impartial, however we may strive to be so," and that "Joule's magnificent, but much neglected, papers of a quarter of a century ago are being rediscovered and attributed to others." He acknowledged that he may have taken "a somewhat too British point of view." He quoted Helmholtz at length on Kirchhoff and, especially, Mayer. Helmholtz did not share Tait's views on Mayer, however. He wrote:

> R. Mayer was not in a position to conduct experiments. He was, furthermore, rejected by the physicists known to him (as likewise happened to me several years later). He could hardly even gain space for the publication of his first compressed presentation of his thoughts. You'll know that as a consequence of these rejections he finally became mentally ill. Today it is difficult to transpose oneself into that time and its range of ideas and to make clear to oneself how absolutely new the matter [of the conservation of force] then appeared. It seems to me that even Joule had to struggle a long time over recognition for his discovery.

Five years later, Helmholtz reprinted the letter himself and commented: "I was to some degree astounded over the resistance that I met within the circles of experts. The acceptance of my work in Poggendorff's *Annalen* was denied to me, and, among the members of the Prussian Academy, it was only C.G.J. Jacobi, the mathematician, who took up my work. Fame and professional advance could at that time still not be gained with these new convictions [about force conservation]; rather the opposite was the case." Tait agreed with "a great part" of Helmholtz's evaluation of Mayer's contribution and said that, were it still possible, he would have changed parts of his published text. He added: "I think it best to retain them, giving Mayer, however, the benefit of the able and weighty advocacy of Helmholtz." Tait also noted that his history of the conservation of energy was largely confined to assessing the contributions of Joule versus those of Mayer, while merely acknowledging that of others, for example, Colding, Séguin, and Helmholtz.[26]

Clausius felt hostile to the book, especially as it concerned Tait's low estimation of Mayer's achievements. He thought its main purpose was to advocate for Joule, Rankine, and especially Thomson. He also found the parts of it concerning his own work to be "extremely insulting." He thought Tait lacked "scientific seriousness" and "sincerity." Though Clausius had a high opinion of Maxwell as a scientist, he also found that Maxwell's *Theory of Heat* (1870) falsified history as it concerned his (Clausius's) creation of the concept and theory of entropy. Two years later, after Maxwell sent him his book on electricity and magnetism and Clausius saw that Maxwell showed understanding for his (Clausius's) complaints, Clausius's opinion of Maxwell softened.[27]

Helmholtz wrote to Thomson (in English) about attempts to write the history of thermodynamics: "I am very glad to see from your letter, that you are again occupied with real scientific work. I wished that you would let alone history of science. You have neither the temper, to be an impartial judge, nor the amount of litterary [sic] knowledge, which is necessary, to distinguish forged documents and real one's [sic]." In the meantime, and within this highly contentious atmosphere, Joule was awarded the Royal Society's Copley Medal for 1870 "for his experimental researches on the dynamical theory of heat." Yet it was precisely for this work that he had already won the Royal Medal in 1852. Mayer then received the Copley in 1871 "for his researches on the Mechanics of Heat." Both descriptions were broad enough to allow claims by either party or their proxies as to who had "discovered" the law of conservation of energy. Helmholtz was a contender—he was nominated for it in 1870, 1871, and 1872—but not then a winner. The behind-the-scenes politics of the award were intense.[28]

In a second edition of his *Sketch* (1877), Tait's dispute with Clausius and his advocacy of Thomson's priority in discovering the second law of thermodynamics came even more to the fore. As to Joule versus everyone else, he concluded: "Thus, in all the scientifically legitimate steps which the early history of the principle records, Joule had the priority. His work has been much extended by others, especially Clausius, Helmholtz, Mayer, Rankine, and Thomson, in the developed applications of the principle in many directions. . . . The experimental foundation of the principle in its generality, and the earliest suggestions of many of its most important applications, belong unquestionably to Joule." As for Helmholtz, Tait declared him to be "one of the most successful of the early promoters of the science of energy on legitimate principles" and his essay of 1847 to be "admirable."[29]

Euclid Dethroned: Geometry as Empirical Science

No later than early 1868, Helmholtz's investigations into the empirical basis of visual perception had led him to a systematic reconsideration of the foundations of geometry. Broadly speaking, because of his work on the "spaces" of sensory perception—for color, for tone, and for tactile feel—from the early 1850s onward, he had sharpened his conceptualization of the problem of measuring sensory contents in their respective spaces. His work in the mid-1850s on the curvature of the eye had the same effect, as did his Kant Memorial Address (1855). Perhaps still more importantly, the completion of part three of the *Handbuch*, and with it his further rethinking of the empirical foundations of visual perception, brought him to intensify his focus on the relationship between geometry and physical reality. He wrote Lipschitz in early 1868: "I've been much occupied with philosophical studies on the theory of sense perceptions; among these, a mathematical, analytical attempt concerning the algebraically possible systems of geometry and the origin of the geometric axioms. . . . All this is, however, still in an embryonic state."[30]

Helmholtz first gave public expression to his rethinking of geometry in May before the Heidelberg Natural History and Medical Association; a brief publication appeared in its *Verhandlungen* soon thereafter.[31] The venue reflected the tentative nature of Helmholtz's contribution as well as its origins in physiology. His article's title revealed perfectly his standpoint: "On the Factual Foundations of Geometry" ("Ueber die thatsächlichen Grundlagen der Geometrie"). The publication, while highly conceptual in nature, contained no mathematical equations or derivations. His opening lines were significant:

> Investigations concerning the way that localization occurs in the field of
> vision have also occasioned the lecturer to reflect in general about the origins
> of the general perception of space. First, there is here a question whose answer
> in any case belongs to the field of the exact sciences, namely, which proposi-
> tions of geometry express truths of factual meaning, and which, on the other
> hand, are only definitions or consequences of definitions and of specially
> chosen ways of expression. This investigation is completely independent of
> the further question as to where our knowledge of the propositions of factual
> meaning originates.

There were more geometric axioms than had been previously imagined, Helmholtz declared. This was not just a matter of completing Euclid's work, but rather of recognizing that there existed many other geometric "facts."[32]

Helmholtz also stated that, after he had largely completed his paper, he discovered a recent publication by the mathematician Bernhard Riemann entitled "On the Hypotheses Which Lie at the Foundation of Geometry" ("Ueber die Hypothesen, welche der Geometrie zu Grunde liegen"), "in which the same investigation is pursued employing an insignificantly varying principal question." Riemann's work, which was originally presented as his habilitation lecture at Göttingen in 1854, but which did not appear in print until 1868, posthumously, had apparently alerted Helmholtz to the fact that Riemann's teacher, Gauss, had earlier written on the subject of the curvature of space. Helmholtz explained that Riemann's findings concerned "the general properties of space, its continuity," as well as the manifold nature of its dimensions. He noted that, like Riemann's geometry, the "system of colors" constituted "a similar, threefold-extended manifold," an insight that owed much to his past work in color perception, and that the eye's field of vision constituted another such system. But there was a fundamental difference between Riemann's and Helmholtz's approaches to the foundations of geometry: whereas Riemann assumed that the expression for the line element *ds* was a hypothesis, Helmholtz assumed that its general form was due to the facts, namely, that space is empirically measurable. While he conceded that his own results concerning geometric systems were largely contained within Riemann's system, he emphasized that Riemann's conclusion was his (Helmholtz's) own initial assumption: viz., "that spatial constructs [*Raumgebilde*] should have that degree of mobility without change of shape which the geometry presupposes." That assumption then limited the number of potential hypotheses for expressing a line element. Congruence and spatial measurement lay at the heart of Helmholtz's understanding of geometry; congruence was needed to make spatial measurement possible. "My starting point," he wrote, "was that all original space measurement rests on the determination of congruence, and that, therefore, the system of space measurement must presuppose those conditions under which alone there can be discussion of the determination of congruence." He then briefly laid out a set of four specific conditions that defined freely mobile rigid bodies. When the number of dimensions is limited to three and space is considered infinite, he explained, Euclidean geometry resulted. What was important for Helmholtz, in contrast to Euclid, Riemann, and mathematicians generally, was that his approach showed especially clearly "how a definite character of rigidity and a special degree of mobility of natural bodies is presumed, so that such a measuring system, like that generally given in geometry, can have a factual meaning. The independence of the congruence of rigid-point systems from their location, translation, and rotation is the fact upon which geometry is based." He concluded: "Like every physical measurement, that of space must be based on an unchanging law of uniformity in natural phenomena."[33]

Helmholtz's route to Riemann's work helps explain why and where he soon published a second paper in 1868 on the foundations of geometry; indeed, how he now, in marked contrast to his first paper that year, published on what was rapidly becoming the field of non-Euclidean geometry. He had first learned of and read Riemann's work in the spring of 1868, while completing his own first work on the factual foundations of geometry. On 21 April, a month before Helmholtz gave his presentation before the Heidelberg Association, he wrote Ernst Christian Julius Schering, a mathematician educated at Göttingen and an editor of Gauss's papers, to inquire about a recent reference by Schering to Riemann's lecture. He said that during the previous two years, while working on part three of his *Handbuch*, he had also been working on "the hypotheses of geometry." He added that his work was "not yet finished and completed," since he still hoped to be able to "generalize isolated points." He asked whether Riemann had published his lecture, or whether it might soon appear. If Riemann's views and results were the same as his own, he said, then his would be "superfluous." "I do not wish, in that case, to spend so much more time on it than they [the spatial issues] have already cost me, [not to mention the] headaches." Schering sent him a reprint of Riemann's lecture and commented that he thought Riemann's results had built on some of Gauss's ideas on the measurement of curvature. He added that Gauss himself had seriously explored the possibility that space was non-Euclidean—Gauss called it *"Astralgeometrie"*—and that he (Schering) was exploring the possibility of expressing "the laws of mathematical physics" in terms of non-Euclidean geometry. On 18 May, four days before his lecture to the Heidelberg Association, Helmholtz thanked Schering for sending Riemann's lecture and included with his letter a new manuscript of his own on facts as the basis of geometry, saying that it concerned matters not discussed by Riemann. He requested that Schering see to its publication in the same venue where Riemann's lecture had appeared. (He noted here that he was a corresponding member of the Göttingen Gesellschaft.) Helmholtz thus decided that his second paper on geometry should appear in the transactions (*Nachrichten*) of the Göttingen Society because that was where Riemann's own work had appeared and therefore was best known. Moreover, its title—"On the Facts Which Lie at the Foundation of Geometry"—was almost identical to Riemann's, except for one decisive difference: where Riemann referred to "hypotheses" (read: analytical or mathematical) as the foundations of geometry, Helmholtz referred to the factual (read: empirical or physiological) foundations. Schering replied on 24 May to thank Helmholtz for his "very important investigation" on geometry, which he would be sending to the society for publication by the end of the week, and he expressed his admiration for Helmholtz's work on this topic. A week later he wrote again to say that the society was most gratified to publish Helmholtz's manuscript and that proofs were on the way. He asked Helmholtz for permission to print a larger number than usual of offprints because the essay was "of such fundamental importance and general interest." He thought that, like Riemann's essay, it would (otherwise) soon be out of print.[34]

Helmholtz's second paper on the foundations of geometry appeared on 3 June 1868. He noted again that it was his work on spatial perception that had led him to think about spatial study per se. In particular, he wondered just how many geometric theorems might have an "objective meaning" (as opposed to being true by definition). As in his first paper, he noted that his approach was like Riemann's, and indeed he here conceded priority to him. He also noted that physiological optics contained two examples—the system of colors and the measurement of the field of vision—that used manifolds in a way that differed from that of geometric measurement.[35] Above all, he stressed the importance of congruence as the heart of his approach to geometry.

In contrast to Riemann, Helmholtz argued that spatial measurement rested on the factual observation of congruence, which required rigid bodies that moved freely. He laid out a set of four mathematical hypotheses and drew various conclusions from them. The most important difference between Helmholtz and Riemann was (as before) that Helmholtz's *conclusions* were the same as Riemann's *assumptions*: viz., in mathematical terms, "that there exists a homogeneous expression of the second degree of the differentials, which remains unchanged with every motion of two points of vanishingly small distance bound rigidly together." That meant "that Riemann's assumption is identical with the assumption that space is monodrome [i.e., the rotation of a rigid body around an axis leads to every point's returning to its initial position] and that infinitely small spatial elements are, in general and apart from their delimitation in space, congruent with one another." Depending on the number of dimensions used, different geometries were possible; Euclid's was only one among several, he pointedly noted. But since physical space has three dimensions, Euclid's geometry does indeed hold there, he emphasized. Moreover, he again emphasized that all spatial measurement was dependent on the use of physical bodies. "The independence of congruence of position, of the direction of the spatial figure superposed on itself, and of the path onto which they [the figures] have been brought to one another," he concluded, "is the fact upon which the measurability of space is based."[36] This second paper, unlike the first, thus also concerned non-Euclidean geometry and was mathematical in its language and appearance.

Though the first attempts to create a non-Euclidean geometry, as this emerging topic was rapidly becoming known, reached back into the late eighteenth and early nineteenth centuries, the field only received serious, sustained attention (by mathematicians as well as by others) through a combination of the translating efforts of the French mathematician Guillaume-Jules Hoüel and Helmholtz's first writings on the subject (1868–70). Hoüel's French translation of Helmholtz's first paper on the foundations of geometry soon appeared (in 1868 or 1869). Hoüel, who held the chair of pure mathematics at the University of Bordeaux, himself published mathematical studies on non-Euclidean geometry, though he was not considered a leading mathematician. In addition to translating Helmholtz on geometry, he translated papers by the mathematicians Nicolai Ivanovich Lobachevsky, János Bolyai, Heinrich Richard Baltzer, Riemann, and Beltrami, as well as pertinent selections from the correspondence of Gauss and Heinrich Christian Schumacher. That correspondence revealed Gauss's own long-standing doubts about Euclidean geometry as the only possible geometry. Between 1866 and 1870, Hoüel thus helped to spread knowledge of non-Euclidean geometry, a subject whose major founders were of several nationalities. Between Hoüel's services as translator and Helmholtz's publications and general reputation, the "new" subject of non-Euclidean geometry was suddenly on the scientific and mathematical map.[37]

Although Hoüel had translated Helmholtz's essay into French, he harbored some doubts about the work of "the celebrated physicist." He sent Beltrami a proof copy of his forthcoming translation. Beltrami told Hoüel that he was dissatisfied with Helmholtz's approach to geometry through spatial measurement by congruence; instead, he preferred the approach of space curvature as developed by Gauss. He also noted one or two other problems with Helmholtz's essay, including his neglect of the works of Lobachevsky and Bolyai. Beltrami, who was strongly interested in mechanics and mathematical physics, and who saw mathematics more or less as Helmholtz did (that is, as a tool for understanding nature), then wrote to Helmholtz directly. He began with compliments, saying he counted Helmholtz as one of "the greats of this world," whom he honored as being among "the princes of the mind who so generously serve the progress of humanity." He had read the French proofs of Helmholtz's paper, he continued, and thought that he had already reached the same conclusions in two previous essays in 1868—"Saggio di interpretazione della geometria non euclidea" and "Teoria fondamentale degli spazii di curvatura costante"—both of which he had already sent to Helmholtz. But he also explained that there was a difference between his own work and Helmholtz's, namely, that regarding pseudospherical or hyperbolic surfaces (that is, those of constant negative curvature). Although his work built on Gauss's representation of a surface in space, he pointed out certain consequences that differed from Gauss's and he brought Gauss's work into agreement with Riemann's. He conceded that non-Euclidean geometry could be approached from different points of view (and referred to the earlier work of Bolyai, Lobachevsky, and Riemann). Despite intellectual differences on these points of geometry, he couched his criticisms ever so graciously: "It's the least of the duties that a pupil can perform vis-à-vis his master."[38]

Helmholtz replied quickly, and Beltrami soon wrote again saying that he was greatly honored by Helmholtz's approval of his point concerning pseudospherical surfaces. Indeed, this had now led Helmholtz to make a change in the introduction to the French translation of his article and to send a note to the Heidelberg *Verhandlungen*, where he had first published on this topic. Beltrami's letter and publications had made Helmholtz realize that his own paper contained a fundamental error: viz., that infinitely expanded space must be flat. Beltrami had also pointed out that there were imaginary constants (as well as the so-called real constants assumed by Helmholtz). Helmholtz thus also noted in an addendum (30 April 1869) to his paper in the *Verhandlungen* that Beltrami had recently published two related essays, one on the interpretation of non-Euclidean geometry and another on the fundamental theory of spaces of constant curvature. Beltrami had studied surfaces and spaces with negative mass curvature; and his results were in precise agreement with the much earlier results of Lobachevsky. He showed Helmholtz that curvatures of space could be not only positive and constant, but also negative and constant. In short, Beltrami's intervention led to major corrections in Helmholtz's work; that intervention enabled Helmholtz to better appreciate non-Euclidean geometry and the subject's historical precedents in the work of others, not least by Beltrami himself. The two now became good friends, sharing an interest not only in non-Euclidean geometry but also in electrodynamics and hydrodynamics.[39] Helmholtz may not have been the "inventor" of non-Euclidean geometry, but he had shown the importance of "congruence" and the empirical conditions of measurement in geometric thinking. He had also shown an intellectual agility and quickness of mind by adjusting to the relevant and recently published ideas of both Riemann and Beltrami, giving their works decisive notice and importance among mathematicians.

The New Geometries of Intelligent Beings

Helmholtz did not simply leave the matter there. While the second paper of 1868 on geometry became Helmholtz's mathematically most important one on the subject, this was not his last word on non-Euclidean geometry. For he now helped make it known to the larger, educated public. In 1870–72, and again in 1876–78, he brought his and others' results concerning non-Euclidean geometry to the public, under the titles "The Axioms of Geometry" and "On the Origin and Meaning of Geometrical Axioms." The first of these was a précis of the subject, the second a philosophical defense.

As to the first, in February 1870 he published a three-page English-language summary of his recent work. "The Axioms of Geometry" appeared in a new English journal entitled *The Academy: A Monthly Record of Literature, Learning, Science and Art* (founded 1869). It aimed to strengthen British intellectual life by spreading knowledge of research in Continental Europe. Its editor was Charles Edward Appleton of Oxford, who was a deep admirer of the Germans—he had studied in Heidelberg and Berlin—in matters of research, in particular their research ethos and their concern to diffuse results to society at large. He advocated for more research in British universities and was a member of the Endowment of Scientific Research movement. *The Academy* was meant as a rival to *Nature* and, soon enough, to *Mind* as well as to the older *Athenaeum*. In September 1869 he cajoled Helmholtz into sending him a paper on the new geometry. He received it two months later and also published abstracts of Helmholtz's recent Innsbruck address (September 1869) and of his paper on Corti's rods. He wanted more. In the end, he got only "The Axioms of Geometry," which was immediately translated into French and appeared in both the *Revue des Cours Scientifiques de la France et de l'Etranger* and *Le Moniteur Scientifique*.[40]

His short essay succinctly summarized not only his own contributions but also those of Gauss, Riemann, and Beltrami, and it alluded to related work by Elwin Bruno Christoffel and Lipschitz. He noted that the development of non-Euclidean geometry was not only a matter of mathematics but was also "immediately connected with the highest problems regarding the nature of the human understanding." To aid the nonmathematician's understanding, he asked his readers to imagine that there existed "intelligent beings" living on two-dimensional surfaces, spheres, and ellipsoids, and he drew some geometric consequences for them. For example, "The sum of the angles of a triangle would be greater than two right angles, and the difference would grow with the area of the triangle." He told them, too, of Beltrami's "elegant exposition" of pseudospherical surfaces and how Beltrami, like Lobachevsky before him, had developed a geometry without employing parallel lines. Euclid's geometry was more or less readily understandable to everyone; but if three dimensions are considered, then "the investigation . . . can be carried on only in the abstract way of mathematical analysis." That meant using abstract notions like "line elements," Riemann's "manifolds," and Gauss's "measure of curvature." He also averred that congruence based on freely mobile rigid bodies was "the original fact, upon which all our notions of space are based." That was the starting point, he said, from which he could derive Riemann's mathematical results. He concluded: "The axioms on which our geometrical system is based, are no necessary truths, depending solely on irrefrangible laws of our thinking. On the contrary, other systems of geometry may be developed analytically with perfect logical consistency. Our axioms are, indeed, the scientific expression of a most general fact of experience, the fact, namely, that in our space bodies can move freely without altering their form. From this fact of experience it follows, that our space is a space of constant curvature, but the value of this curvature can be found only by actual measurements."[41] He and others had dethroned Euclid—the latter was now only one (if immortal) prince among several—and Helmholtz thought he had made geometry an empirical science.

Not quite everyone agreed. For example, William Stanley Jevons, a leading English political economist, pioneer in mathematical economics, logician, and philosopher, took issue with Helmholtz in the pages of rival *Nature*. Jevons averred that Helmholtz's "opinions" "were based upon the latest speculations of German geometers." He scorned Helmholtz's heuristic invention of "intelligent beings" living in the pseudosphere as far too imaginary and irrelevant. He also chided him for appearing to speak in the name of others when, he said, Helmholtz's "speculations" sought merely "to stamp them [other mathematicians] with the authority of his own high name." Jevons himself had no doubt about the truth of "Euclid's elements"; he dismissed the notion of curvature of space. And finally, he found Helmholtz guilty of fallacious reasoning: "With all due deference to so eminent a man as Helmholtz, I must hold that his article includes an *ignoratio elenchi*. He has pointed out the very interesting fact that we can conceive worlds where the Axioms of our Geometry would not apply, and he appears to confuse this conclusion with the falsity of the axioms. Wherever lines are parallel the axiom concerning parallel lines will be true, but if there be no parallel lines in existence, there is nothing of which the truth or falsity of the axiom can come in question." Jevons declared—and this seemed to bother him the most—"that all attempts [like J. S. Mill's] to attribute geometrical truth to experience and induction . . . are transparent failures."[42]

Helmholtz returned fire (in *The Academy*) with his own lessons in geometry, logic, and philosophy: "Where I say that geometrical axioms are true or not true for beings living in a space of a certain description," he pointedly instructed Jevons, "I mean that they are true or not true in relation to those points, or lines, or surfaces, which can be constructed in these spaces, and which can become objects of real perception to those beings." He instanced this general point with details from non-Euclidean geometry and then further instructed "Mr. Jevons" in the epistemological ways of how mathematicians operate: "Our mathematicians . . . speak of imaginary lines and points of intersection . . . and their imaginary co-ordinates, as if such imaginary dimensions of space really existed; and they do this to preserve analogy and homogeneity in the analytical expressions. But for all this no mathematician ever came to the conclusion that a fourth dimension of space exists, even though he find [*sic*] it convenient to write his equations as if it existed. And I cannot see why the mathematical intellects of a spherical or pseudo-spherical world should come to another conclusion." Jevons, he concluded, had failed to distinguish between analytical and empirical truths.[43]

The rejoinder to Jevons's critique shows that, even when placed on the defensive, Helmholtz endeavored to spread knowledge of non-Euclidean geometry. Through the 1870s, many English mathematicians first learned to appreciate non-Euclidean geometry in good part through Helmholtz's writings. (The leading English mathematicians William Kingdon Clifford and Arthur Cayley were also influential in this regard.) Yet most, like Jevons, were generally dubious about or even hostile to the notion of higher-dimensional spaces, empirical or otherwise. Broadly speaking, they considered the empiricist approach to non-Euclidean geometry as a philosophical and cultural challenge to their vision of geometry as "necessary truths." The reinvention and early articulation of non-Euclidean geometry was decidedly a German and an Italian affair, most decidedly not a British or French one.[44]

Helmholtz provided similar instruction to his fellow Germans. In 1869 or 1870 he lectured on non-Euclidean geometry before the Docentenverein in Heidelberg. The resulting published essay underwent a patchy, discontinuous development, emerging in part out of the lecture in Heidelberg and the "essential contents" from the *Academy* article. It first appeared in his *Populäre wissenschaftliche Vorträge* (1876) and differs only slightly from its English translation, "The Origin and Meaning of Geometrical Axioms," which appeared simultaneously in *Mind* (1876).[45] Here was his second philosophical defense of the topic.

In this essay Helmholtz declared that the very fact that geometry could be made into a systematic science had led many to be keenly interested in it as a source for epistemological reflection. He considered geometry to be unmatched in terms of its completeness and without any doubts or contradictions. It concerned the measurement of space, and as long as scientists and others used appropriate data, geometry had never failed them.[46]

Geometry stood at the center of philosophical analysis, Helmholtz believed, because its theorems had real content without being empirically derived. He thought geometric axioms were the best examples for responding to Kant's fundamental question of how synthetic a priori propositions were possible. He claimed that because such geometric propositions obviously exist, they constitute proof "that space is an a priori given form of all external perception [*Anschauung*]." Such formally empty propositions may be filled with any sort of experiential content. He thus thought his listeners and readers would be interested in the epistemological meaning of geometric axioms even if their knowledge of geometry was little more than that learned long ago at school. But he especially sought to inform his listeners and readers about the new non-Euclidean geometry, including its logical relationships and its relationship to experience.[47]

What are the origins of the traditional (i.e., Euclidean) axioms of geometry, he asked. "Are they an inheritance from the divine source of our reason, as the idealist philosophers think, or has the penetrating acumen of generations of mathematicians that has appeared so far simply not been sufficient for finding the proof?" What makes this so difficult to answer, he noted, is that "logical conceptual developments" easily get mixed up with everyday experience. He pointed out that construction problems in geometry lay at the heart of the system of geometry, and while they seemed merely to help children learn the system, in fact "they establish the existence of certain figures," in particular points, lines, and circles. This brought him to the heart of the matter: "The foundation of all proofs in the Euclidean method is establishing the proof of the congruence of lines, angles, plane figures, solid bodies, etc." The key operational word here was "congruence": that the geometric objects in question must be freely movable to one another without changing their shapes or dimensions. That raised the (rhetorical) question "whether this assumption includes any logically unproved presupposition." Helmholtz thought that there was indeed such a presupposition, and an important one at that. Every proof by congruence, he intended to show, "is based upon a fact obtained from experience alone." At the same time he assured his listeners and readers that non-Euclidean geometry was pursued "almost exclusively by means of the pure, abstract method of analytic geometry [i.e., the geometry of manifolds]."[48]

To help his listeners and readers understand this new geometry, Helmholtz asked them to imagine two-dimensional beings living on the surface of a solid body, beings with abilities of perception similar to those of humans. Their geometry, he noted, would necessarily be only two-dimensional. They would be as limited in perceiving only two dimensions as humans are in perceiving only three; they could as little perceive three-dimensional objects as humans can perceive four-dimensional objects.[49]

Such beings could still draw the shortest lines on their surface, known as geodesic lines, which, Helmholtz noted, are not the straightest lines for humans. He used this *Gedankenexperiment* of imaginary intelligent, perceiving, but nonetheless only two-dimensional beings on a sphere to offer a series of examples that highlighted their inability to imagine three-dimensional space. By implication, he said, the same was true of the inability of humans to imagine four (or more) dimensions. At issue, then, was not logical power but instead physical space. The beings in question would establish a type of geometry that fit their type of space—in other words, the origin of geometric axioms was to be found in the type of physical space inhabited by the geometer.[50] Geometry, that seemingly most fundamental of all sciences, is thus a function of human spatial perception, he argued.

He explained that one could imagine a geometry of surfaces in which the axiom of parallel lines (Euclid's fifth postulate) did not hold. He was referring to Beltrami's invention of saddle-like pseudospheres, and he noted that already in 1829 Lobachevsky had conceived of a geometry without the parallel postulate. Beltrami's pseudospherical surfaces were in complete agreement with Lobachevsky's results. Helmholtz drew a conclusion here: "We, as inhabitants of a space of three dimensions and endowed with sensory tools for perceiving all these dimensions, can indeed plainly imagine the different cases in which surface beings may have to form their spatial perception because, toward this end, we need only limit our own views to a narrower field. . . . When we turn to space of three dimensions we are limited in our power of imagination by the structure of our organs and the experiences gained with them, which fit only the space in which we live." He also noted that, as in Riemann's idea of a system of differences in which a thing is determined by n measurements—"an n-fold extended manifold or a manifold of n dimensions"—one can imagine a color system (following Young and Maxwell) as a threefold manifold and, similarly, tones as a twofold manifold (pitch and intensity). He concluded: "Thus it is seen that space, considered as a field of measurable quantities, in no way corresponds to the most general concepts of a manifold of three dimensions; rather, it also involves special determinations which are conditioned by the completely free mobility of rigid bodies with unchanged form toward all places and with all possible changes in direction."[51]

Riemann, Helmholtz explained, had taken a different approach to non-Euclidean geometry than he (Helmholtz) had. Riemann approached it in terms of the geometry of manifolds, whereas Helmholtz came to it "partly through investigations on the spatial representation of systems of colors" and "partly through investigations on the origin of our visual estimates for measurements of the field of visions." Even so, they came "to similar results." Since Helmholtz's work in this area went back to the 1850s and 1860s, he was in effect making something of a priority claim, or at least a claim for independent discovery of the field of non-Euclidean geometry. Yet his and Riemann's differences were not simply the difference between the physiologist and the mathematician. For Riemann derived his theorems about rigid bodies (convergence), whereas Helmholtz proceeded "from the observed fact that in our space the movement of rigid bodies is possible with the degree of freedom which we know, and from this fact derived the necessity of the algebraic expression assumed by Riemann as an axiom." In this way, Helmholtz claimed that his work was more fundamental: he had derived what Riemann had assumed.[52]

Geometric measurements, he again argued, by way of conclusion, depended on congruence, and this in turn led to the problem of measurement itself, that is, to the consideration of measuring instruments. Except for changes of physical form that are due to changes in temperature or the effects of gravity, scientists presume that "our measuring instruments, which we hold to be stable, are really bodies of unchanging form." Geometric axioms are thus not simply about space per se, "but rather simultaneously also about the mechanical behavior of our stablest bodies in motion." His account of the origin and meaning of geometric axioms led him, as it did others, to express doubts about Kant's doctrine of space (geometric axioms) as a priori theorems given through transcendental intuition. Helmholtz was no "strict Kantian." Rather, in his view humankind had arrived at its understanding of geometry, or at least that of Euclid's, only very slowly, through a trial-and-error process of more or less exact, sometimes merely comparative, measurements of objects against empirical reality. It had arrived at this understanding not through logic but through everyday experience, or, more precisely, by its gradual perception and memory of such experiences—read: "unconscious inferences"—as an accumulation of measurements taken by human beings with (or without) their physical instruments. Any geometric system, Euclidean or non-Euclidean in nature, could of course be imagined. But that did not make its propositions real.[53]

Helmholtz's publications of 1868–76, along with his general renown, made non-Euclidean geometry more widely known, both within and beyond mathematics. With all their enormous talent and creativity, Lobachevsky, Bolyai, Gauss, Riemann, Beltrami, and others, who individually proffered ideas that all together helped develop non-Euclidean geometry, had been unable to make the topic sufficiently plausible, convincing, and generally appreciated by others, including their fellow mathematicians. By incorporating parts of the work of Riemann and Beltrami and further developing it, Helmholtz catalyzed investigation of the subject of non-Euclidean geometry and put it on the cultural, mathematical, and scientific map in a way that no one before him had done.

Helmholtz's writings on non-Euclidean geometry influenced the fine arts—painting, music, and even literature—though great attributive caution is in order here. Non-Euclidean geometry was widely sensed as important, and occasionally studied and drawn upon, by various sorts of artists in the late nineteenth and early twentieth centuries. Cubists (for example, Marcel Duchamp), the architect and writer Claude Bragdon, and the composer Edgard Varèse (with his "musical geometry") drew upon Helmholtz. Indeed, everyone who was interested in spatial curvature and the "fourth dimension" in one way or another drew on or felt inspired by non-Euclidean geometry, if not on or by Helmholtz directly.[54]

In this regard, Edwin Abbott Abbott's well-known *Flatland—A Romance of Many Dimensions* (1884) merits special notice. Abbott had studied classics and theology at Cambridge, where he also received a solid education in mathematics. He would have been fully able to understand Helmholtz's essays on geometry, and Helmholtz's essays in *The Academy* (1870) and *Mind* (1876 and 1878) may have been three of the sources that Abbott read before writing *Flatland* or drew on as he wrote it. That work was at once a mathematical fantasy, a piece of science fiction, a social satire, and a spiritual journey. It played off the new notions of space that had emerged with non-Euclidean geometry. Abbott's fantasy world, like Helmholtz's writings, included creatures who sought to comprehend the second, third, and fourth dimensions. Whether he read him or not, as William James declared in 1880: "A discussion of space and the space relations of consciousness which ignores [Shadworth] Hodgson, [Thomas Kingsmill] Abbott, Helmholtz, Wundt, and Lotze necessarily condemns itself to incompleteness."[55]

The Mathematicians Take Over

Neither before nor after his essays on geometry did Helmholtz consider himself a mathematician, not even an applied mathematician. However, his work in geometry, along with the strong mathematical dimension of his analyses in hydrodynamics, electrodynamics, and other topics in theoretical physics, led many mathematicians to count him as one of their own. Indeed, until about the 1870s, to do mathematical physics often effectively meant to be counted as a mathematician. He was, moreover, friends with Lipschitz, Koenigsberger, Weierstrass, and Leopold Kronecker, among other mathematicians. Though he was never close to Plücker, the two had a noteworthy if only indirect relationship through the person of Felix Klein.

Toward the end of his career, Plücker became a geometrician. His last student was Klein, who received his doctorate at Bonn in 1868 with a dissertation on line geometry and mechanics. Following Plücker's death that year, Klein completed the first volume of Plücker's projected two-volume study of spatial geometry based on line elements, and on the basis of Plücker's draft writings for the second volume and their discussions together, Klein also completed the projected second volume. That publication employed algebraic forms to represent line elements.[56] Under Plücker's initial guidance and inspiration, geometry became the focus of Klein's career. Both mentor and student sought to relate geometry to mechanics and optics.

Helmholtz's publications in 1868–70 on non-Euclidean geometry proved important for the general direction of Klein's own career. They inspired Klein, who in 1871 published his first piece on non-Euclidean geometry, showing that if Euclidean geometry was consistent, then so was non-Euclidean; he conceived both types as parts of the more expansive field of projective geometry. His work and that of others during the 1870s helped to mathematize and formalize non-Euclidean geometry in a way that was quite foreign to Helmholtz's conceptually and physiologically oriented papers. Moreover, Klein and other mathematicians, above all Sophus Lie, further extended, and in Lie's case rather sharply criticized, Helmholtz's approach to the subject. Klein showed that different geometries had different groups (mathematically speaking), while Lie deepened the understanding of group transformations along with their relationship to geometry, in particular, what became known as the Helmholtz-Lie space problem. Later in the decade and beyond, still others, like Wilhelm Killing and Henri Poincaré, further mathematized non-Euclidean geometry, rendering the subject increasingly abstract and general. While both men ultimately deviated from Helmholtz's ideas on geometry, both also found much inspiration in them. These and other professional mathematicians wanted and produced far more analytical and developed versions of geometry, and they tied it to other highly analytical and rapidly developing fields, for example, abstract group theory and algebra. Nonetheless, during the remainder of the nineteenth and into the early twentieth century, a series of leading mathematicians, including Poincaré, (the young) Bertrand Russell, David Hilbert, and Hermann Weyl, continued to study Helmholtz's writings as he had developed them between 1868 and 1870, as well as his philosophical writings of 1876 and 1878. They variously favored and developed, or opposed and criticized, Helmholtz's view of the foundations of geometry.[57]

Helmholtz's analysis thus remained an important jumping-off point for all those who wanted to understand geometry in the "modern" sense, including those who sought to understand what, if any, relationship it might have with physical phenomena (motion). Klein himself sought to unite the various types or branches of geometry in terms of group transformations (his so-called Erlangen Program). He also, for example, hired the young mathematical physicist Arnold Sommerfeld as his assistant in 1894, and in time Sommerfeld became the teacher of the theoretical physicists Werner Heisenberg, Wolfgang Pauli, Hans Bethe, and many others. Similarly, Adolf Hurwitz, an expert in Riemannian geometry among other things, was an outstanding Klein student. In 1892 Hurwitz moved to the Zurich Polytechnicum, where four years later he was joined by his close friend Hermann Minkowski. Einstein attended Hurwitz's first-year classes on calculus and later his class on differential equations, and he subsequently enrolled in (but was often absent from) a series of advanced mathematics classes given by Minkowski. Furthermore, Marcel Grossmann, Einstein's classmate, close friend, and later colleague and early collaborator on general relativity, became an expert in non-Euclidean geometry, receiving his first degree at the Polytechnicum and then his doctorate at the University of Zurich. There was, in other words, a broad line of mathematical and pedagogical descent running from Helmholtz's first proclamation on the importance of Riemann's manifold and his articulation of non-Euclidean geometry in 1868 to Einstein's completion of his general theory of relativity in 1915. Helmholtz's rethinking of the foundations of geometry as the empirical measurement of multidimensional physical space and his view of an intimate connection between geometry and physics thus helped lay the basis for Einstein's development of the general theory of relativity (1907–15).[58] In these and other broad ways, Helmholtz's work became foundational for the leaders of twentieth-century mathematics, mathematical physics, and theoretical physics. He had taken up the subject of geometry as a measuring issue concerning sensory physiology; in the hands of others, it evolved after 1870 into a central tool for understanding physical space.

Helmholtz's work in geometry somewhat typified his general approach to problems in science, and it suggests one reason for the extraordinary success of at least some of his work. He managed, for example, to "discover" or "invent" conservation of force/energy by bringing together several subfields of physics, as well as by his physiological concerns regarding bodily heat. Similarly, his invention of the ophthalmoscope arose in large part as a combination of his understanding of the eye's anatomy and physiology, on the one hand, and that of the geometric (nota bene) pathways of light rays, on the other. And his success in physiological optics and acoustics was due above all to his ability to combine results from different areas of science (physics and physiology, especially) and to provide as much mathematical analysis as the subject might allow. Much the same was true for his role in the development of non-Euclidean geometry: motivated initially by physiological concerns and an empiricist outlook in matters of perception, he reconsidered the geometry of the eye's field of vision, hoping to achieve a better understanding of its ability to localize objects. These same synthesizing abilities soon manifested themselves in his work in hydrodynamics and electrodynamics; later in his career they also showed themselves in his path-breaking work in chemical thermodynamics and atmospheric physics.

Alliances and Adversaries in Hydrodynamics and Electrodynamics

Helmholtz's first major contributions to hydrodynamics were made, it may be recalled, in 1858–59; these were immediately well received by mathematical physicists, both in the German lands and in Britain. His work soon constituted the foundational studies in hydrodynamics. Moreover, he and others recognized an analogy between his work in hydrodynamics and ongoing developments in electrodynamics. Thomson and Tait became especially enamored of the possibilities they saw in using his notion of the vortex motion of atoms. Tait also used Helmholtz's hydrodynamics for understanding knot theory, which in turn became an element in the articulation of modern topology.[59]

Except for one piece on friction in fluids in 1860, Helmholtz did not, however, publish on hydrodynamics again until 1868. Instead, he was preoccupied with other tasks: working first on the *Tonempfindungen* and then on part three of the *Handbuch*, rethinking the foundations of geometry, and discharging his heavy administrative obligations as prorector and as dean of the Medical Faculty. Then in 1867–68, with all this work behind him, his interest in hydrodynamics was rekindled by Tait and Thomson. That year Tait translated Helmholtz's memoir of 1858 on hydrodynamic equations into English. He also cleverly built a "smoke box" that produced dramatic vortical smoke rings that concretely illustrated Helmholtz's abstract theorems; these in turn inspired Thomson to imagine such rings as the effective basis of all matter, the "vortex atom," which played a leading role in British theorists' thinking about matter down to the 1890s.[60]

But in mid-1868, a problem arose. Tait alerted Helmholtz that the well-connected French mathematician Joseph Bertrand—professor of analysis at the Ecole Polytechnic, of mathematical physics at the Collège de France, and perpetual secretary for mathematical sciences at the Académie des Sciences—writing in the Académie's prestigious *Comptes Rendus*, had rejected Helmholtz's views on vortices. Bertrand attacked Helmholtz's work and sought "to smash the whole theory." "You will have to finish him off in a complete manner," the ever-quarrelsome Tait advised, adding that he had told Thomson, "We had better leave you to smash him [Bertrand] in your own way." Thomson, who had taken his wife and gone to Kissingen so she could recuperate at its spa, wrote Helmholtz to say that perhaps they might meet in nearby Heidelberg. He was working on vortices, and he doubtless hoped to discuss the subject with Helmholtz in person; Helmholtz in turn offered to visit Thomson in Kissingen.[61]

Helmholtz was thus forced into disputing with Bertrand—"*mon adversaire*," he called him—after Bertrand, Hermite's brother-in-law, claimed that Helmholtz had committed a mathematical error at the outset of his 1858 memoir, an error that supposedly invalidated all of his subsequent results in hydrodynamics. Helmholtz begged to differ, arguing that his approach was a physical one—favoring dynamics over geometry and kinematics—and that it was the same approach that Kirchhoff, Thomson, and Stokes had taken. Bertrand, who was generally hostile to recent mathematics, rejoined that mathematics, not physics, should have the upper hand in this argument about hydrodynamics. Then, after Bertrand's fellow Frenchman Adhémar Barré de Saint-Venant took Helmholtz's side, Bertrand upped the ante, claiming that Helmholtz's memoir contained mathematical falsehoods. Helmholtz now adopted a less diplomatic tone and replied that Bertrand had misrepresented his (Helmholtz's) mathematical statements. It did not help Bertrand's case when the Abbé (François-Napoléon-Marie) Moigno unnecessarily entered the fray in support of Helmholtz in his popular science journal *Les Mondes*, forcing Bertrand into a side dispute. The whole matter then fizzled out, with Helmholtz the undeclared winner: his theory continued to serve as the foundation of hydrodynamics into the twentieth century, until it was eventually transformed by relativity theory. The *touche finale* of this unproductive episode came in early January 1870, when Helmholtz was elected a corresponding member of the Académie des Sciences. He received thirty-seven (of forty-four) votes, far outdistancing even his closest competitors, including Anders Jonas Ångström, Dove, Grove, Henry, Joule, Kirchhoff, Mayer, Stokes, Thomson, and Tyndall. Thomson wrote to congratulate him and asked whether "Bertrand ever had the grace to confess his error?" Whether he did or not is unknown. But George Eliot, the intellectually voracious English novelist, managed to learn of "Helmholtz's splendid hydro-dynamical theorems" that year, indicating that even the most abstract mathematical and scientific ideas could seep into the literary world.[62]

Responding both to his own (corrected but intellectually narrow) theory of organ pipes (1863) and to work by one of his physiology students on viscous fluid (blood flow), in 1868–69 Helmholtz again took up the issue of vortex motion. He now employed energetic and variational principles, à la Thomson and other British natural philosophers, to show that under fixed boundary conditions, real fluid motion is represented by a minimal loss in frictional energy. His analysis of viscosity showed that, under given conditions, there were similarities in the flow of air (e.g., in organ pipes) and water and electric currents. He proposed, independently of Stokes and aided by one of Tyndall's striking empirical findings concerning the behavior of smoky air in relation to sound waves, the notion of discontinuity surfaces, which he showed to be inherently unstable. Furthermore, drawing on and combining diverse results from d'Alembert on complex potentials, from Augustin-Louis Cauchy on complex functions, and from Riemann on the complex plane, Helmholtz employed these and other mathematical tools to advance understanding of discontinuous fluid motion. Kirchhoff, Rayleigh, and Stokes, for example, soon further developed his work on discontinuous fluid motions (though Thomson showed skepticism toward Helmholtz's results). Helmholtz's scientific problems were thus as likely to emerge from a physiological as from a physical context; and his methods came from seemingly distant mathematical or physical concepts, with isolated empirical results suggesting a proof of concept and giving inspiration to pursue a problem further. The final results were of both theoretical and practical interest.[63]

Just as he had previously synthesized studies of heat (in 1847) and physiological optics and acoustics (in the 1850s and 1860s), after 1868 Helmholtz sought to clarify and bring unity to the field of electrodynamics, which he otherwise saw as "a pathless desert." He later wrote: "Observed facts and consequences issuing from highly doubtful theories ran past one another and were without secure limits. In the course of striving to learn how to comprehend this entanglement, I had undertaken to clarify the field of electrodynamics, so far as I could see, and to locate the distinctive consequences of the different theories [of electrodynamics] in order to decide, where possible, between them by appropriately conducted experiments."[64]

He was referring, broadly speaking, to the discovery or invention of a disparate set of electrical and magnetic, and indeed electromagnetic and electrodynamic, phenomena and theories from the 1820s to the late 1860s. These included, to name only the most important names, work by Hans Christian Ørsted, Poisson, Faraday, Ohm, Emil Lenz, Thomson, Wilhelm Weber, Maxwell, Franz Neumann, and the latter's son Carl Neumann. By 1868 the leading German theorists of electrodynamics were Weber and Franz Neumann; theirs was a particle-based theory, and Helmholtz was quite familiar with their writings. By then, too, the leading British theorist was Maxwell, whose theory was based on the idea of a field. After 1853 Helmholtz became the non-British physicist by far most familiar with British physics. He had developed deep personal connections with several British physicists. (Maxwell much admired Helmholtz's work on color, acoustics, and hydrodynamics; he probably read Helmholtz as much as Helmholtz read him.)[65]

Thus as Helmholtz reentered physics and intellectually confronted electromagnetic theory after 1867, he already had a highly appreciative attitude toward Maxwell's electromagnetic theory, at a time when few of his Continental colleagues had much if any acquaintance with it. He wrote Tyndall (in English) in January 1868:

> Maxwell's theory, which unites the theory of magnetism to electricity and
> to light into one hypothesis, is one of the most brilliant mathematical concep-

tions which have ever been made. It is, however, at the same time so deviating from all current ideas about the nature of forces that I really do not yet want to venture publicly into declaring a judgment on it. I do, however, very much believe that one can designate it as a development of Faraday's ideas brought into strongly mathematical form. And therefrom follows the possibility of calculating the speed of light from the electric constants into a condition that terribly impresses one. You can perhaps say that this realization of Faraday's ideas opens up to us presentiments of a deeper connection of all imponderables.[66]

This declaration, written less than five years before the publication of Maxwell's *Treatise on Electricity and Magnetism* (1873) and nearly twenty years before Heinrich Hertz's experimental demonstration of electromagnetic waves, shows that from the start Helmholtz was greatly impressed by Maxwell's theory, even if one could not then (and indeed not before 1888 or so, if ever) call him a "Maxwellian." Maxwell, in turn, admired Helmholtz's work in electromagnetism, even though he did not share his action-at-a-distance approach to electrodynamics.

Helmholtz originally approached the subject of electrodynamics from an unsolved problem in physiology, namely, that of open currents. In his initial study of 1869, he used an induction coil to measure the propagation of impulses in frog nerves (as he had done in the early 1850s). He found a variation in current between the primary and the secondary circuit. The oscillatory currents that he produced and measured suggested to him differences in charge between the frog's skin and its interior mass. This led him to a systematic study of electric oscillations in three-dimensional conductors by exploring Weber's (via Kirchhoff's), Maxwell's, and Neumann's respective theories of electrodynamics in connection with his own experimental results; he recast those three theories as variations of his own, more general, action-at-a-distance potential theory, which, depending on one of three values of a constant, could yield back the original theory of Weber, Maxwell, or Neumann. Whereas Weber's (and Neumann's) theories were particle-based, and Maxwell's field-based, Helmholtz's was based on energy concepts. This allowed him to develop a more general set of equations than the other three, which, depending on the constant's value, could be "reduced" to his own theory once the value of the constant was taken into account. As always, he sought to generalize, synthesize, and unify.

More particularly, Helmholtz generalized Neumann's electrodynamics (the potential law) to deal with open currents. Depending on the motions of electricity, Neumann's potential was arrived at with one constant value, Weber's with another, and Maxwell's with yet a third. But Helmholtz also believed that with Weber's value the system's equilibrium was unstable; and thus Weber's electrodynamics, or so he argued, stood in conflict with energy conservation. When he also considered the polarizability of the medium (including vacuum) between conductors, he was led to support Maxwell's theory, which he saw as a limiting form of his own electrodynamics and which he couched in terms of potential theory. The results initially appeared in three short papers in the *Verhandlungen* of the Heidelberg Association. He sent Borchardt a copy of one of these (on the mathematical theory of electrodynamics). Borchardt declared it to be of the "greatest importance" and asked Helmholtz if he could republish it in his *Journal für die reine und angewandte Mathematik*, a venue that reflected the subject's rather mathematical than physical presentation. Helmholtz agreed to republish it there, but he also decided to write a much longer version. Indeed, Borchardt said that he also wanted to publish other papers by Helmholtz on the mathematical theory of electrodynamics, and so the revised, extended paper became the first in a series of three fundamental papers on the topic.[67]

Helmholtz thought—others begged to differ—that he had clarified electrodynamic theory. He had, at least, raised doubts about Weber's theory and brought Maxwell's electromagnetic theory of light to the attention of Continental physicists; and he had produced a framework in which experiments could be conducted to decide among the three theories. During the 1870s and 1880s, his papers profoundly influenced a few important physicists working on the topic of electrodynamics. Here, too, as in his role of bringing popular British science to Germany, he had become the mediator between British and German physics. However, Weber, Carl Neumann, and Johann Karl Friedrich Zöllner soon came to differ with Helmholtz's theory. Weber, in particular, ably defended himself against the charge that his theory of electrodynamics stood in contradiction to the principle of conservation of energy, and on this Maxwell, who had initially shared Helmholtz's critique, agreed. Part of Helmholtz's critique, it transpired, was mistaken.[68] Like Bertrand, Weber (and his several supporters in Germany) remained Helmholtz's scientific adversary.

The 1860s saw Helmholtz at the height of his intellectual power, creativity, and productivity. Though he had certainly displayed a good deal of all that before he came to Heidelberg and did so again throughout his years at Berlin, it was nonetheless in Heidelberg that he brought all of his talents to bear as he completed the last two parts of his *Handbuch*, wrote the *Tonempfindungen*, rethought the foundations of geometry, created a new fundamental understanding of hydrodynamics, and analyzed competing electrodynamic theories. His creative scientific years in Heidelberg bear comparison to Newton's from the mid-1660s to the late 1680s, to Darwin's from the late 1830s to the late 1850s, and to Einstein's from the years right after 1900 to 1916. During their respective ten to twenty years of work and creativity, each showed (in his own way and for his own subjects and time) what it meant to do creative science at the highest intellectual and creative levels.

16

The Road to Berlin

The Aim and Progress of Natural Science

Though he might otherwise have already been on vacation during September 1869, Helmholtz delivered three public lectures at the Naturforscherversammlung, which was holding its forty-third annual meeting at Innsbruck, in the Austrian Tyrol. Nearly one thousand people attended, making it one of the society's best-attended meetings since its founding in 1822. As was often the case at such meetings, German nationalism and the belief in progress by means of natural science pervaded the Innsbruck gathering. German participants came from every German university except Kiel; foreign participants came from Switzerland, Italy, France, England, Holland, Belgium, Denmark, Russia, Turkey, and the United States. Brücke, who had been invited to give a plenary address, came from Vienna, and he urged du Bois–Reymond to come as well, pointing out that Helmholtz and Mayer, who rarely appeared in public, would also be there. Helmholtz and Mayer had met only once before, at the Naturforscherversammlung meeting in Karlsruhe in 1858. Anna Helmholtz also saw Mayer at Innsbruck, and she was "very much struck—painfully so—with his confused way of speaking." She "had some pains in believing that this was the man" she had heard so much about. In Innsbruck, the Helmholtzes stayed with Leopold Pfaundler, who was the professor of physics at the university there; went to the theater; and visited the Bergisel, just south of the city. Whether they also took the twelve-hour train trip to Bolzano, as more than a thousand other attendees and accompanying persons did, is unknown.[1]

On 18 September 1869, in Innsbruck's National Theater, Helmholtz gave the invited, opening, plenary address before the attendees, what Anna called his "big speech." As he stepped onto the podium, there were "loud and sustained cheers." That he spoke at all was unexpected. That May, Wilhelm Foerster, an astronomer associated with the Berlin Observatory, had invited him on behalf of the Berlin Scientific Union (Wissenschaftlicher Verein) to lecture there in early 1869. Foerster said that "the entire educated world of Berlin" would be "joyous" if Helmholtz, "the man whose unequalled gift for true and illuminating presentations in the field of natural research" was "appreciated by everyone," would be a speaker. But Helmholtz declined, saying that his duties in Heidelberg were so onerous that he lacked the time to take on any new ones; besides, he had "no appropriate theme," and Berlin already had many popular lecturers. Nonetheless he accepted the Innsbruck invitation, even though he was apparently a last-minute substitute speaker. The talk was originally entitled "On the Historical Development of Modern Natural Science," and it appeared in the meeting's proceedings as well as in another, new journal, the *Philosophische Monatshefte*. He subsequently rewrote it substantially, and it appeared, to use its English-language title, as "On the Aim and Progress of Natural Science" in the second collection of his *Populäre wissenschaftliche Vorträge* (1871). The address sought to illuminate "the development of the entire field of the sciences," though Helmholtz conceded that he could only touch on certain areas of science, noting that even Humboldt, who had once overseen all science and sought to bring together the different parts, could himself probably no longer do so. Specialization had now become the norm of scientific life, with each individual scientist concentrating all his efforts on a single topic.[2]

Like scholars whose work is confined to libraries and archives, scientists certainly worked with written materials; but, Helmholtz also argued, unlike them scientists observe or grasp nature itself. Such work took much preparatory effort, including the elimination or measurement of errors in instrumentation and general preparation for doing research; only after completing such tasks could their work proper begin. The acquisition of empirical knowledge also implied that the senses themselves had to be sharpened, and the requisite skills (for example, those of the microscopist's hand, which he tellingly compared to a violinist's hands) had to be developed. At times, the scientist even needed "the courage and cold-bloodedness of the soldier," since he sometimes had to destroy certain parts of men or animals before he could observe or heal certain other parts. Helmholtz probably intended these general if obvious methodological and philosophical remarks to inspire and give recognition to the hard, daily work of the vast array of specialists who constituted the German scientific community, and perhaps too as a defense against the rising antivivisection movement. Yet the more specialization increased, he argued, the greater the felt (psychological) need for understanding the entirety of science and its larger context. Without such an understanding, the specialist lacked confidence in the overall purpose, utility, and lasting value of his own work and findings. He needed to feel "that he too has contributed a building stone to the large entirety of science." He needed, too, to feel that "nature's reasonless powers" were made "serviceable to the moral purposes of humanity."[3]

Helmholtz maintained that the "immediate practical use" of individual scientific results occurred only rarely and that such utilities were virtually impossible to anticipate. While he believed that the practical dimension of the natural sciences had reshaped modern man's life, such results or applications were generally unpredictable, and so it made little sense to predict unless one had very specific indications with which to work. In the more distant past, he claimed, useful practical applications of science had emerged only when highly skilled technologists had worked for a lifetime on an issue or project and when they had lots of luck. More recently, practical inventions were mostly the "fruits of trained scientific knowledge of the objects in question." Nonetheless, such knowledge was "always at first without any direct prospect of possible use." Scientific knowledge should thus be pursued for its own sake.[4]

He welcomed the Naturforscherversammlung meetings as a chance for "participants from the nation's educated circles" to meet with one another as well as with the "influential statesmen" in attendance. This last was a touch misleading, since there were only three "statesmen" present: the head of the regional Tyrolean government, a regional captain, and the mayor of Innsbruck. The statesmen expected to learn from the scientists about "the further progress in civilization, further victories over the natural forces." The state and society provided the material resources for scientists to do their work and were therefore "justified in asking about the results of this work." Hence this was a most appropriate time and place "that an accounting be given over the progress of the entirety of the natural sciences, over the aims that it strove towards and over how close it had approached towards these aims." Here was his theme.[5]

Helmholtz asked how progress in science could be measured. It certainly did not mean the mere accumulation of facts, which were of little relevance to increasing intellectual understanding or "advancing rule of man over nature's powers." Instead, intellectual understanding required order and context, while control over nature required predictable success in areas and under conditions not yet explored. To achieve either of these first required a law of the facts in question, which meant finding a general concept under which such facts became understandable. Only then did such facts acquire value. Such a law was "not merely a logical concept" or a means for furthering speculation; rather, it had to embody a concept that led to discovery of and through facts, and repeatedly did so in a variety of ways and conditions. Nature's laws were "a foreign power," he explained. "They are not chosen arbitrarily and they determine our thinking." A correct natural law admitted no exceptions. This was not a matter of "arbitrariness" or "choice." "The law confronts us as an objective power, and we accordingly call it force."[6]

As examples of objectivity in science, Helmholtz pointed to the laws of light refraction, of chemical affinity, of electrical contact forces between metals, and of the forces of adhesion and capillarity. They were all objective laws, he said, if only at a low level in terms of the range of phenomena they encompassed. They and others served as the basis for further generalizations. Ultimately, scientists sought to determine the mechanical forces between point masses, where it is clearly seen "that the force is only the objectified law of effect." Such laws were necessary and in no way arbitrary. Causality is built into nature, Helmholtz believed. To say that progress has occurred in science "as a whole" is tantamount to saying that the causal context behind a host of phenomena has been understood. He cited Galileo's law of falling bodies and Kepler's laws of planetary motion as examples. These laws not only served as models for others but were also useful practically, for purposes of ship navigation and geodesy. "Through them, many industrial and social interests were advanced."[7]

The development of chemistry during the past century, Helmholtz further noted, had eliminated the four metaphysical elements of antiquity, replacing them with sixty-five others, in part thanks to the recent invention of spectroscopy. This last also led to the discovery of new elements in the heavens, reinforced the various identities of terrestrial and celestial matter, and proved that all mass in the universe is composed of the chemical elements. All change in the world, he continued, is only the spatial rearrangement of the elements or of matter in motion. The most elementary forces are the forces of motion, and so "the final goal" of science is to determine all change in terms of underlining motions and forces, "thus to reduce everything to mechanics." This was the "ideal challenge" that science sought to fulfill, even if it remained a distant one. Astronomy had to date achieved the most in this regard, followed by acoustics, optics, and electrical theory. There had also been important, if lesser, advancements in the theory of heat and in chemistry, while physiology had barely begun. The one law that had "nearly" achieved this goal of reduction to mechanics was the law of conservation of energy. The individual who had first stated this law in its "pure and clear" manner and in "its absolute generality," Helmholtz announced, was seated in the audience before him and, indeed, would be the speaker after him: Julius Robert Mayer of Heilbronn. (The audience showed its goodwill if not agreement by a burst of applause.) Helmholtz noted that while Mayer's achievement issued from his physiological concerns, Joule's approach came from his technical concerns from mechanical engineering. But they and others had developed a new concept, variously referred to as the quantity of force, or of work, or of energy. To be sure, forerunners of this concept had been partially present in both theoretical and practical mechanics, while work in machine technology had also made contributions.[8]

Helmholtz perceived an intimate relationship between the abstractions of theoretical mechanics and the operations of various practical machines, water mills, and windmills, since all three followed the law of conservation of energy. Such mills were, however, also dependent on meteorologic processes that followed that law. There were innumerable ways of obtaining "mechanical motive power" from nature. Many clever people had figured out how to use such motive power in many ways in their industrial machinery. Fully appreciating in precise, measureable terms the relationship between work performed and nature's forces of motion "completed a first and important progress toward the solution of the comprehensive, theoretical task of reducing all natural phenomena to motion." The law indicated, Helmholtz said, that despite any number of changes, the universe was "provided with a supply of energy that can be neither increased nor decreased." However distinct and great a physical change may be, the same quantity of energy always persists in the universe. If energy disappears in one form, it must appear in another. Here, supply and demand were always in equilibrium. As to the second law of thermodynamics, which Carnot had first noted but which Clausius had then "corrected," Helmholtz gave it only the briefest mention. It showed, he said, "that this change, generally speaking, continues constantly in a definite direction, while ever more of the great supply of the universe's energy must turn into the form of heat."[9]

Helmholtz then turned to illustrating the role of the law of conservation of energy in cosmology, meteorology, and physiology. According to the Kant-Laplace hypothesis, loose and widely distributed bits of interstellar matter heated up under the influence of gravitational forces as they moved through space, eventually coagulating into larger masses. Spectrum analysis, itself dependent theoretically on the mechanical theory of heat, revealed the chemical composition of such masses, including meteors, comets, the planets of the solar system, and other bits of interstellar matter. These masses, Helmholtz further explained, contain energy that is continuously radiated as heat and light. As solar radiation strikes the earth's surface, it produces wind in the atmosphere and currents in bodies of water, that is, it sets in motion the earth's entire meteorologic system. Finally, solar radiation in effect provides energy to plants, which become food for animals. Hence organic life and phenomena were ultimately driven by cosmology and physical laws. This portrait of the interrelations of all natural scientific processes indicated, he declared, that an important step forward had occurred in finding nature's laws.[10]

The law of conservation of energy applied not only to plants and animals in external terms (the gaining or transforming of energy) but also internally (to their own physiology). From that understanding he and Mayer had originally, if independently, taken up their respective investigations. There were big issues at stake here: he was uncertain whether all life processes were absolutely lawlike or whether they admitted a certain amount of freedom. Were human organic processes ruled by the soul—variously known as "Lebensseele," "Lebenskraft," "Archäus," "Anima inscia," and so on—that acted through physical and chemical forces? The notion of a directing soul and the law of conservation of energy contradicted one another, Helmholtz declared, since the hypothetical soul could do an arbitrary amount of work without consuming any energy. Yet there was no evidence of this. To the contrary: in terms of work performed, there was complete similarity between steam engines and animals: both needed fuel and oxygen to operate or move, both expelled (consumed) matter, and both produced heat and work. Yet he conceded that, as concerned plants or animals, the law of conservation of energy still lacked any "exact, quantitative investigations of the consumed and produced amounts of force equivalents," and thus it remained mere conjecture.[11]

Even so, he judged that physiological research had progressed more in the past forty years than it had in the past two millennia. To that set of accomplishments he added Darwin's theory, which opened up new vistas in science. It replaced the old explanations of understanding "organic purpose" that employed either a directing soul or "an act of supernatural intelligence." Its novelty lay in showing how organic beings could originate and develop according to nature's laws, as opposed to the intervention of some external intelligence.[12] He understood that Darwin's notion of natural selection implied the end of purposefulness in organic change and of progress as inherent to it and that its explanatory mechanism was materialistic in nature.

Helmholtz's central epistemological point remained his claim that our sensations, depending on their type, provide only signs of external objects, not images. The signs must appear simultaneously with the external object or process, but that is all. In the course of time, individuals learn "through use and experience" to "read" or understand the meanings of these signs. They are like a language. There is no "pre-established harmony" between the inner mental world and the outer physical world; indeed, empirical studies showed that the eye and the ear were imperfect instruments in relating to the world. He thought this (i.e., Müller's) epistemological standpoint was broadly in accord with Darwin's theory.[13]

He now brought his essay full circle. He had argued at the outset that science aimed at establishing laws, ultimately, laws of motion. The temporally ordered signs that we experience put us in a position "to directly portray lawlikeness in the temporal sequence of natural phenomena. If, under the same conditions in nature, the same effect occurs, then the observing person will, under the same conditions, see the same sequence of impressions simultaneously repeated." The sensory organs were a central element not only in man's daily, practical life but also "precisely for the fulfillment of the task of science," to know its laws.[14]

By way of conclusion, Helmholtz left the high philosophical ground for some thoughts about the practical and political implications of science. The fact that "ever more general and comprehensive laws have revealed themselves" meant that science could "clearly prove its great practical consequences." Medicine, for example, was a "practical science" that depended on physiology. Both fields had progressed now that "precise observation" had combined with the discovery of laws. He said he had experienced this transformation personally: as a young medical student and practicing physician, much of medicine had been ruled by either ungrounded theory or "an exaggerated empiricism." Change had come through "the introduction of mechanical concepts": for example, concerning heat phenomena in the body; in nerve physiology; in microscopy; and in ophthalmology, which last had become a model for other medical disciplines.[15] His view of the science-technology relationship, in which the latter depended on the former, drew largely on his own experience.

Finally, he turned to the relations of science and the state. He judged that scientific research about inorganic nature had progressed at roughly equal rates in the various European nations, although Germany led in physiology and medicine. He realized, however, that questions about life were not simply mechanical matters; they were closely connected with psychological and ethical issues, and for this reason science should be pursued "for ideal purposes and without any close concern with practical use." Science required "indefatigable diligence," something that, he averred, had always distinguished the German researcher, "who works for inner satisfaction and not for external success." Indeed, he thought the Germans were also particularly distinguished by "a greater fearlessness before the consequences of the complete and full truth." (Since his audience burst into applause, it appears they agreed with him.) While he conceded that there were, in every regard, "excellent researchers" in England and France, they nonetheless "almost always had to bend before social and church prejudices, and if they wanted to express their convictions openly, they could do so only by injuring their social influence and effectiveness." Germany, he maintained, was "bolder," for it realized that "the fully recognized truth" was the "remedy" against dangers and disadvantages brought by half-truths. The Germans, he thought, liked to work, were moderate, and were closely attentive to morality.

Hence they could "look fully into the face of truth" and not be satisfied merely with hasty and partial theories. These remarks notwithstanding—and the fact, as he himself noted, that this meeting was occurring "on the southern border of the German fatherland"—in the very next breath he maintained that "in science we certainly do not need to ask about political borders, since our fatherland reaches as far as the German tongue is heard and as far as German diligence and German fearlessness find approval in the struggle for truth." Perhaps these chauvinistic remarks were meant to help rebuild cultural and political ties that had been broken through the Austro-Prussian war of 1866. His reference to the German fatherland being coextensive with the German tongue was in any case an allusion to a famous statement in 1813 by Ernst Moritz Arndt, the well-known anti-French, anti-Semitic German nationalist and historian. Arndt had taught at the University of Bonn, where he and Helmholtz had known each other. It was Germany's "cardinal virtues" in science, Helmholtz concluded, that were "the remedy" for its "bodily suffering" and which would make it an intellectual center characterized by "intellectual independence, fidelity to one's conviction, and love of truth." Since the audience again burst into applause, it seemed to approve these points as well.[16]

Though Helmholtz's was the opening address, it was not the only plenary address. He was followed directly by Mayer speaking on the mechanical theory of heat. Mayer's address was short and offered little that was new. He said, however, that an entire series of men—Hirn, Joule, Colding, and Helmholtz among them—"had independently discovered" the mechanical equivalent of heat. From there he moved to the idea that there were three basic "categories of existence"—matter, force (*Kraft*), and the soul (*die Seele*)—and he declared that "a true philosophy may and can be nothing other than a propaedeutic for the Christian religion." The contrast with Helmholtz's address could not have been clearer. While state officials, the Naturforscherversammlung's organizers, and other plenary and opening session speakers spoke of nationalism, progress, reason, liberalism, and freedom, Mayer stressed the unknowable, God, and Christianity. The meeting as a whole was marked by a contrast between the liberal-progressive and the conservative-clerical outlooks. Mayer wrote to his wife from Innsbruck about his and Helmholtz's addresses: "Helmholtz spoke first, before me, in a brilliant, one-and-a-half hour speech 'on the developmental history of modern natural science.' In it, he also naturally thought it befitting to acknowledge (with great applause) my priority. I too was received and discharged with applause."[17]

The Versammlung of course consisted of more than the opening and other plenary addresses; it was also an opportunity for individuals to report on their specialized results within various scientific disciplines. When the Versammlung had held its first meeting in 1822, there were no individual sections devoted to reports on the latest work in a single discipline (or even two or more related ones). In 1828 seven such sections were introduced; by 1869 there were eighteen. This organizational development reflected the rise of specialization in nineteenth-century science. Helmholtz himself spoke in the section devoted to physics and mechanics on the topic of the oscillatory movements of electricity and the theory of electrodynamics. Later, he cochaired (with Brücke, Virchow, and Heidenhain) the section on anatomy and physiology, where he also spoke about a pump for blood discharge and a model of the ear's small bones that are involved in hearing.[18]

Three plenary addresses were also given at the close of the meeting. The Genevan zoologist Karl Vogt, a Darwinist and a materialist, spoke on the latest research results in prehistory; the Viennese psychiatrist Max Leidesdorf spoke on the relations of society to mental illness and how to prevent such illness; and Virchow spoke on the present state of pathology. The plenary addresses by Helmholtz, Vogt, Leidesdorf, and Virchow emphasized the progress of science and were well received; they gave the Naturforscherversammlung the sense of pride and mission that it sought. Against them, Mayer's address stood as the odd man out.[19]

A Friendly Nonnegotiation

Officials within the Prussian Ministry of Culture, as well as various Berlin scientists, had long talked about bringing Helmholtz back to Prussia, to Berlin. Though the ministry had failed to get him for Bonn in 1868, it retained the hope of yet bringing him back. Talk was in the air. Baden's minister of the interior, Jolly, worried openly that Heidelberg would lose him to Berlin. Ludwig told Helmholtz that during a visit to Berlin "everyone, without exception, advocated for your emigration." Magnus, while visiting Helmholtz in Heidelberg in 1869, expressed his hope that Helmholtz would one day return to Prussia.[20]

Helmholtz's ever-growing reputation was making it more difficult for Heidelberg to retain him. He had, it may be recalled, been elected a corresponding member of the Prussian (1857), Bavarian (1858), Göttingen (1859), Vienna (1860), Paris (1870) and Belgian (1870) academies of science. But he was not yet a foreign member of the Prussian Academy of Sciences (since no position became available until 1870). In nominating him to the Prussian academy in March 1870, his backers, led by du Bois–Reymond and Magnus, emphasized that they thought he had already achieved in science what someone with the highest ambitions might perhaps achieve by the end of his career. Making him a foreign member, they said, would only be advantageous for the academy. Their colleagues in the Physical-Mathematical Class voted unanimously for him, and the Plenum voted in favor as well. His election became official on 1 June 1870.[21] That made his connection to Berlin a little closer.

Far more significantly, on 4 April 1870, Magnus died. He had long occupied one of the most important, if ill-defined, scientific positions in Prussia. He originally held a chair in technology associated with chemistry; but in 1843 he had managed to add physics to it, including the directorship of the physics cabinet and laboratory; he was Berlin's de facto professor of experimental physics. Du Bois–Reymond wrote Helmholtz the day Magnus died to say that although Magnus, "our old friend," was in much pain and dying from colon cancer, he had signed, "on his deathbed" the letter supporting Helmholtz as a foreign member of the academy. Du Bois–Reymond began scheming to get Helmholtz's warm body for Berlin as Magnus's was turning cold, or at least before it was buried. He was still irritated with himself for not having approached Heinrich von Mühler, the minister of culture, in 1868 to request that he be allowed to conduct the negotiations with Helmholtz respecting Bonn. His greatest fear now was that old Dove could become the new head of the physics cabinet and that young Quincke, "who has significant forces available within the ministry," could become the second physicist in Berlin. Given the "very muddy" nature of Magnus's position, anything could happen.[22]

Whether or not he himself was offered the position, Helmholtz hoped that Quincke would not be Magnus's successor. He realized, too, that ever since the Bonn negotiations, he had "become indifferent towards physiology" and had a real interest "only for mathematical physics." He hoped somehow to transfer to a physics chair but thought it more likely that Kirchhoff, not he, would get the call. If Kirchhoff were called, Helmholtz would be satisfied if he could get Kirchhoff's position as professor of physics at Heidelberg. He just did not want to see a second-rank person obtain "the top position in physics in Germany." In fact, Berlin was considering only Helmholtz and Kirchhoff, a fact that Werner Siemens confirmed when he came to Heidelberg to discuss the Berlin position.[23]

In addition to being a trained and experienced physicist, Kirchhoff had one other advantage over Helmholtz that appealed to the ministry: he would cost much less. However, he had a disadvantage in the form of a serious foot injury — he had been on crutches for quite some time — which was not unmanageable in a small town like Heidelberg but might become problematic in a big city like Berlin. Helmholtz himself greatly preferred Heidelberg to Berlin, and so he hoped Kirchhoff would get the call. (There was no rivalry; the two remained good friends.) Rumors began to spread: in late April, Jolly learned from the newspapers ("to my not limited alarm") that Helmholtz would be leaving Heidelberg. At that point the rumor was false. But Jolly was prepared to do whatever it would take to retain Helmholtz, and he diplomatically asked him about just that.[24]

The Berlin faculty recommended both Helmholtz and Kirchhoff to the ministry as Magnus's successor. It rightly considered Helmholtz "the more ingenious and comprehensive researcher," Kirchhoff "the better-schooled physicist and proven teacher." It thought that Helmholtz's strength lay in research, Kirchhoff's in teaching (not least in regard to introductory students). It noted that it would probably be easier to attract Kirchhoff than Helmholtz to Berlin. Du Bois–Reymond thought that the ministry did not want to deal with Helmholtz because it was still irritated by its two previous failures with him. But it did not want to say this openly, and so it spread a rumor (not for the first time) that Helmholtz had three shortcomings: he was not a professional physicist; his demands were too great; and he was not an enthusiastic teacher, or at least, not a good teacher for run-of-the-mill students. These points found their way into a seemingly "flattering" newspaper account of Helmholtz, which du Bois–Reymond judged "perfidious." He concluded that the ministry would first approach Kirchhoff. He feared, however, that the entire process could go awry—that Quincke would somehow get the call or at least be appointed head of the instrument collection. He thought that if he, and not some ministry official, were to conduct the negotiations with Helmholtz, he could achieve a successful conclusion. As Helmholtz's close friend and as the current rector of the University of Berlin, du Bois–Reymond had excellent standing for such a task.[25]

Helmholtz considered his election to the academy and the Berlin faculty's nomination of him to succeed Magnus a "true honor." He was moved that Magnus had initiated his nomination to the academy: "I always liked him a lot. As I now see, I had wrongly suspected that he felt a certain opposition vis-à-vis my mathematical direction." He was not, however, in the least surprised about the ministry's alleged maneuvers against his candidacy. He had thought much about the possibility of moving to Berlin, and he concluded that it was "doubtful" that a call would be advantageous for him. He knew that as a teacher he would have far more influence in Berlin than in Heidelberg, but he thought that organizing and running a laboratory would take him away entirely from "productive work." He also thought he would not be nearly as free in Berlin as he was in Heidelberg, since a professor in Berlin was subjected to far more time-consuming administrative and other nonscientific tasks than one in Heidelberg. Nor was he willing to take a loss in pay, not even in exchange for a professorship of physics. Hence he sincerely hoped that Kirchhoff would receive and accept the call, though he had doubts as to whether Kirchhoff would accept.[26]

While Kirchhoff certainly had some professional strengths and experience in physics that Helmholtz lacked, Helmholtz was by far the bigger fish, especially on the international science scene. He was particularly sought after in Britain. In 1869 Acland at Oxford wrote, just as he had written the previous year, that he and others there—including Robert Bellamy Clifton (the professor of physics), William Donkin (Savilian Professor of astronomy), and Henry J. S. Smith (Savilian Professor of geometry)—were "very anxious both to see and to *hear*" Helmholtz in Oxford. They hoped this would occur in the summer of 1870 when Clifton's new physics laboratory, the Clarendon, would be nearly finished. If Helmholtz would in the meantime be coming to England that autumn, Acland hoped to see him then, too. He also wanted Helmholtz to give several lectures in Oxford on physics and medicine, which Oxford University Press would then publish. "We are in a transition state which is at once interesting, anxious & laborious." And he wanted to know if Helmholtz would be going to the BAAS meeting in Exeter that coming summer.[27]

Helmholtz was also invited (through Acland) by Lord Robert Cecil, Third Marquess of Salisbury, to come and accept an honorary doctor of laws degree at Salisbury's installation ceremony as the newly elected chancellor of Oxford. Salisbury, who had taken a (fourth-class) degree in mathematics, was himself an amateur scientist. He supported the natural sciences at Oxford while criticizing an overly great concern with classical education. He had also nominated Darwin, Tyndall, and Huxley for honorary degrees; internal university politics, however, blocked the awards to Darwin and Huxley, while Tyndall's name was not even placed before the university's Hebdomadal Council (though he did receive an honorary doctorate in 1873).[28]

Helmholtz did not go to Oxford that June, presumably because he was in the midst both of teaching and negotiating with Berlin, in the person of du Bois–Reymond. Du Bois–Reymond went to Heidelberg as the ministry's emissary to see both Helmholtz and Kirchhoff. All three, plus Bunsen, the chemist Hermann Kopp, and Koenigsberger first took a walk together in the Heidelberg countryside. (It had long been Helmholtz's custom to take walks with Kirchhoff and Bunsen; as of 1869, Koenigsberger started joining them.) But Kirchhoff was still on crutches, so—leaving Kirchhoff, Bunsen, Kopp, and Koenigsberger to go their own way—du Bois–Reymond and Helmholtz then set off on their own, up the Rhine valley. Du Bois–Reymond found that he liked Helmholtz "as much as ever." "He leaves no trace, at least with me, of playing the great man."[29]

Du Bois–Reymond had first spoken explicitly with Kirchhoff about the Berlin appointment, saying that the ministry wanted to know what it would take to bring him there. He sensed that Kirchhoff thought Berlin's offer of 3,100 talers annually plus free housing was an unexpectedly generous one, and he thought Kirchhoff was prepared to come. Kirchhoff, however, wanted to first discuss the matter with Karlsruhe, which sent its own emissary and immediately offered him an additional 1,100 guldens (about 629 talers) in salary, plus an assistant. Kirchhoff decided to stay put. He said he was staying because of his foot problem, since it would prevent him from performing his duties properly in the big city of Berlin. (Koenigsberger later claimed that he and Bunsen knew all along that Kirchhoff had no intention of leaving.)[30]

Du Bois–Reymond telegraphed the news of Kirchhoff's response to Justus Olshausen, Helmholtz's old friend and colleague from Königsberg and, since late 1858, a senior official in the Prussian ministry. He also let others in Berlin know, "in order to alert them and to exert vigorous pressure in favor of Helmholtz." Du Bois–Reymond thought that dealing with Helmholtz was in a sense easier than dealing with Kirchhoff, since Karlsruhe could not give Helmholtz what he wanted: a professorship of physics. More money would not keep him there.[31]

He and Helmholtz quickly reached preliminary agreement about the terms of his appointment to Berlin as professor of physics: an annual salary of four thousand talers; a promise to build and equip a new physics institute; assurance that Helmholtz alone would be the director and thus would alone determine who might work there—"From my side there would naturally be the very greatest consideration towards Professor Dove"—and who could use the auditorium; a residence for himself within the institute (and, in the meantime, payment of his rent while he lived elsewhere); provisional quarters for the institute near the university, as well as one or more assistants; and moving expenses. If and when the minister agreed to these conditions, Helmholtz was prepared to come to Berlin to discuss details and assess the situation firsthand. If an agreement could be finalized by 1 July, he was prepared to start teaching in Berlin that autumn. He told a correspondent that he would gladly lecture on cosmogony in Cologne next January if he remained in Heidelberg, "which at the moment is rather doubtful." Du Bois–Reymond subsequently reported from Berlin that some faculty members there found Helmholtz's "demands indeed high, but also justified" and that Mühler was seeking funds to meet those demands. In the meantime, du Bois–Reymond single-mindedly devoted himself "to organizing a supporting petition from the Philosophical Faculty" that would bring Helmholtz to Berlin by the start of the winter semester. "Mühler has the notion," he added, "that you're almost as great a *savant* as Humboldt." (Though both Humboldt, who died in 1859, and Helmholtz were important public scientists, the minister might have more appropriately said that Helmholtz was by far the greater scientist.) Mühler certainly considered Helmholtz's appointment to Berlin also to be "politically an act of great importance." Ernst Curtius, the dean of Berlin's Philosophical Faculty, urged the minister to act as speedily as possible in "this highly important matter" and not let too much time elapse, as happened in 1868 with Bonn, and so perhaps lead Helmholtz to lose interest in or appetite for Berlin.[32]

Anna Helmholtz increased the pressure on Berlin by informing du Bois–Reymond that Vienna wanted to bring Helmholtz there, about which du Bois–Reymond immediately informed the ministry; that sped up its decision-making.

Olshausen told du Bois–Reymond that he could tell Helmholtz confidentially that Mühler and the finance minister had agreed to all of Helmholtz's demands. (The personal salary of 4,000 talers included 2,000 from the university, the highest then earned, and the other 2,000 from the academy.) In addition, Helmholtz would receive income from sitting on various examination committees. It was still unofficial and had to be approved by the Prussian legislature, du Bois–Reymond warned, but he told Helmholtz, "You can count on its being as good as cash." The matter could not, however, be finalized before 1 January 1871, at the earliest.[33]

Helmholtz thanked du Bois–Reymond immediately and profusely for his "ceaseless work and for the love" he had shown for Helmholtz, "personally and . . . in this professional opportunity." He was delighted: "This call has now turned out to be as honorable and advantageous as I could only have dreamed of and could never have expected without your active intervention. Leaving aside the expectation that I can now limit my activity to the field of science that has meant the most to me [i.e., physics], I'm also glad that, thanks to this change, I'll again be closer to you. For as valuable as it has been being here together with Bunsen and Kirchhoff, our ways of life accord too little with one another for them to become really warm [toward me]." Du Bois–Reymond thought Helmholtz ascribed too much influence to him in this matter, and he modestly said it was Helmholtz's accomplishments and fame that led to his appointment; all that he did, he said, was shake the tree a bit in the right place so that the fruit, so ripe for the picking, simply fell to the ground. He shrewdly advised Helmholtz not to come to Berlin immediately, because several important people would be away and because, once there, he would seem less of a giant than he appeared to be in Berlin from distant Heidelberg.[34]

To all intents and purposes, then, an agreement was reached in late June to bring Helmholtz to Berlin. Helmholtz informed Mühler on 3 July that he accepted Berlin's offer, conditioned on approval of his salary and the Prussian legislature's voting to fund the construction of a new physics institute. Anna thought they might move to Berlin sometime early in the next year. She was sad to leave little Heidelberg for the big, foreign city of Berlin, where she would have to begin a new life. But she had known all along, she said, that one day they would be moving there.[35]

In the Franco-Prussian War

The normal uncertainty of winning the legislature's approval was complicated by ongoing political tensions between Prussia and France, or more precisely between Bismarck and Napoleon III. Du Bois–Reymond warned Helmholtz of his fears of war with France, and so of further delays that might ensue concerning

Helmholtz's appointment. Berlin was in a rabid anti-French mood, he said. "It's still the same as our parents used to tell us: the officers of the gendarmes of 1806 are sharpening their sabres."[36] Ferdinand Helmholtz's son doubtless appreciated the historical allusion.

Helmholtz himself feared that war would arrive shortly: he and others knew it would involve the Badenese and other southern Germans. He believed the French were looking for an excuse to start a war and that the Prussians, who thought war would come sooner or later, would not seek to avoid it. "That can significantly change all our plans and prospects," he told Anna. His fears were shared by many in Europe. Anna took the Helmholtzes' younger children (Robert, Ellen, and Fritz) with her to Starnberg, outside Munich; if war came, Helmholtz wanted them there with her parents and wanted Anna to return to Heidelberg. In Heidelberg, he reported, there was "great enthusiasm for war." War between France and Germany was soon declared (15 July), and Baden's troops mobilized that day. Soldiers might have to be quartered in the Helmholtz home. The university became a military hospital and its students soldiers.[37]

The majority of the Badenese population, as well as Frederick I, had, like all liberals, long sought German unification and wanted it under Prussia's aegis. Baden had cooperated with Prussia. Prussian troops helped save the grand duke's monarchy in 1848, and they remained in Baden until 1852. There was thus much sympathy for Prussia in Baden. After Königgrätz (1866), Baden willingly became Prussia's ally, and its forces came under Prussian command. The Badenese were keenly aware that they were highly exposed to nearby French forces and that France aimed to eliminate Baden. When war finally began on 29 July, Badenese troops were in the thick of it alongside the Prussians.[38]

Many German medical doctors, Helmholtz included, did military medical duty during the war. (Anna became involved in nursing.) From mid-July to mid-September Helmholtz served in Heidelberg as the director of the Reception and Distribution of the Wounded, as head of a similar office at the railway station, and as the representative of Heidelberg's medical doctors in general. He considered himself fortunate to have such a position and to be able to help. Within ten days of the war's start, the wounded began pouring in to Heidelberg's hospital: Bavarians, Prussians, and French. It was feared that typhus and dysentery would break out. In early August Helmholtz was dispatched to Wörth and Sulz to help with the wounded there, and he accompanied a group of wounded back to Heidelberg. At the bloody battle of Wörth, where the Germans had a major victory, he became "the spectator of a battlefield after the battle." For a while he found this work invigorating, even if he was often called to nighttime service. He felt himself to be a real patriot. But in the end, especially as things slowed down and he had less to do, he felt exhausted. His migraines returned and he suffered fainting spells. His friend and doctor, Nikolaus Friedreich, advised rest and recovery "from the labors and excitements of recent times." In mid-September he joined his family in Starnberg, but it proved too cold for him there, so, as Friedreich had advised, he went to the mountains (Merano) for three weeks to drink whey, to walk, and generally to recover his health. When he returned to Heidelberg in October, he did not resume military medical work: "My own work occupies me sufficiently, and I must still treat my brain a bit cautiously." Since the university had so few students that winter semester, his teaching duties were minimal. By then he could also afford to leave Heidelberg, because the war had gone very well for Prussia, Baden, and their allies. (In early September, Napoleon III was captured at Sedan and his regime collapsed.) Karl Christian Bruhns asked Helmholtz to write an essay on Humboldt's physiology for an edited volume to honor the one hundredth anniversary of Humboldt's birth, but, owing to a lack of spirit or energy, he declined.[39]

Helmholtz made, if unwittingly, another type of contribution to the war effort: thanks to Thomson, he participated in a propaganda campaign. In late July Thomson wrote to thank him for his condolence letter after the death of Thomson's wife and to express his regrets that they could not now meet. Thomson was worried about Helmholtz; he asked him to write, and he hoped that Helmholtz might visit England in the autumn. Helmholtz replied (in English) that he planned to move to Berlin by next April—as long as the war continued successfully for the German states and the Prussian legislature approved the ministry's request for construction funds for a new physics institute. He continued: "At present I am occupied from morning to evening with hospital business. We have here already 500 wounded, and prepare the same number of beds for the sacrifices, slaughtered near Metz. The joy for the German victories is very much dampened by the dreadful number of fallen soldiers." He also had family in the war, he explained: one of Olga's nephews and one of Anna's brothers were "before the enemy," while one of his own sons, Richard, was in training. Richard had volunteered (at Helmholtz's urging) at the start of the war for the Karlsruhe Artillery Regiment. He served in the winter campaign against General Bourbaki and in late January was lightly wounded in the face by a misfired (German) grenade before being discharged from the army in March 1871.[40]

Were Napoleon III to be successful, Helmholtz further wrote, this would mean "political annihilation" for the Badenese and others. That realization drove the Germans on. "Now we are obliged to carry on the war to the end, [so] that a repetition of such a rapacious invasion, as Napoleon had the purpose to do, becomes impossible for a long number of years." The French, he said, had committed atrocities; moreover, the French foreign minister, the Duc de Gramont, "had invented the insidious lie that the Badish [i.e., Badenese] army used exploding musket balls and violated the convention of [Saint] Petersburg." Even though the minister had been told this was not true, he nonetheless threatened the devastation of all Germany. Toward this end, the French even used "the Turcos, these African savages," who "were thrown at our frontier." (The "Turcos," also known as "tirailleurs indigènes," were a tough new infantry regiment bred during France's thirty years of fighting in Africa.) Yet Helmholtz claimed that they could not stand up to the German soldiers; they fought "only against women and children." He thought the war would soon be over, but he could not go to England anytime soon: "Our hospitals will not be so soon evacuated." He closed by thanking Thomson for sending "the specimens of your new recording telegraph."[41]

Then Thomson, on his own initiative and without Helmholtz's prior knowledge or consent, had the letter (with slight emendations) published as "A German View of the War" in the *Glasgow Daily Herald* (9 September 1870). Helmholtz had written it in high dudgeon. Others made similar propaganda efforts—Theodor Mommsen addressed the Italians on the war through open letters to two Milanese newspapers; David Friedrich Strauss addressed the French through an open letter to Ernest Renan; and Max Müller and Thomas Carlyle each addressed the English through open letters to the *Times*, for example. Their names headed their letters, however; Helmholtz's letter appeared anonymously. Thomson himself deplored the French war declaration and said he recalled that, in 1859 in Arran, Helmholtz had told him that France was a constant threat to the unification of the German states. Thomson continued: "I feel that emperor, officials, journalists, and people require a crushing defeat to cure for ever that disease of vanity of which Louis Napoleon as emperor is [sic] was only one symptom. I believe France itself will be better and happier ten years hence for the bitter lesson you are now teaching it. But it is a terrible price you are paying for what is only your right." The only thing Thomson felt neutral about was the wounded on both sides, and he became involved in fund-raising efforts to help them. He also sent Helmholtz fifty British pounds to be used as he thought best to aid either hospital work or bereaved families.[42]

Du Bois–Reymond had heard that Helmholtz was practicing his "old military-medical arts," but all that he himself could give "to the Fatherland" was his speech of 3 August. That speech, "The German War," had already entered its third printing of two thousand copies. It was a brilliant but highly chauvinistic address, in which du Bois–Reymond portrayed the Germans as totally peaceloving innocents who were ready to defend their lands at all costs against the French. The latter, he maintained, were egotistical, militarily aggressive, and bloodthirsty imperialists, uncivilized (appearances to the contrary), and alone responsible for the war. He famously referred to the University of Berlin as "the intellectual regiment of the House of Hohenzollern." The speech constituted a defining moment in the shaping of du Bois–Reymond's public image. It also caused much offense in France but was mostly seen quite positively in Germany and elsewhere. Helmholtz responded to it by thanking his most-trusted friend for sending him a copy. He commented: "It spoke to me from the soul, and the strong demand [in sales] for it shows that that was also the case with many others." He asked rhetorically what should become "of the unfortunate nation" (i.e., France). "How far will it go in consuming itself in its mad vanity and its powerless hate? It is a terrible tragedy and therein a crushing justice!"[43]

By the autumn of 1870, the war was becoming increasingly distant from Helmholtz and Heidelberg, except for the four barracks set up outside Heidelberg's new, semifinished hospital and a few soldiers who hobbled on crutches along the city's streets. German forces had prevailed, and they had done so rapidly and decisively. Had the French not been a "so dreadfully careless people," Helmholtz mused to Knapp in distant New York, they might well have won the war at the outset. On 15 November, Baden dissolved its participation in the North German Confederation and joined the new German Confederation. It also signed an agreement that placed its troops permanently under the command of Prussia, and in mid-December it joined the emerging German Reich. Baden's grand duke (Frederick I) was present on 18 January 1871 in the Hall of Mirrors at Versailles for the declaration of the new Reich and the crowning of King William of Prussia as the German emperor. Baden's monarch and its people, Helmholtz among them, enthusiastically welcomed unification. Paris capitulated on 28 January 1871, and three days later an armistice was announced. For his contributions to the German war effort, the Grand Duke of Baden bestowed upon Helmholtz the state's Remembrance Medal.[44]

Negotiating with Berlin

In October du Bois–Reymond again visited Helmholtz in Heidelberg, and he told him that while his personal compensation was set, the legislature had yet to approve the construction funds for the new institute. Within days, it met and voted the funds, but then a new legislature had to be called and the issue decided anew. Du Bois–Reymond warned that, given the war's uncertainties, funds for the new building necessarily remained uncertain. He suggested that Helmholtz come to Berlin in November to look for a residence, inviting the Helmholtzes to stay with him. Helmholtz was prepared to live with a delay in the funding approval, provided that, once the state's prewar financial condition returned, the government would immediately repropose the issue to the legislature and provisional institute facilities were sufficiently adequate for him to conduct his own experimental work and direct a student laboratory. But he did not intend to go to Berlin and look for a residence until his appointment had been definitively approved. Weierstrass informed him that the ministry had requested the funding in its budget, but he and others at the university became worried that the delay in the funding approval might lead Helmholtz to delay transferring to Berlin—"this terribly important matter for the faculty"—and so he got a ministry official to assure Helmholtz in writing that there was nothing to worry about. The Berlin faculty hoped Helmholtz would be there by Easter of 1871.[45]

"It's very reassuring for me to know this," Helmholtz told du Bois–Reymond, "since from the start I had feared that I would still have to remain here next summer. When one has already decided (once again) to undertake a transplanting, it is of course desirable to do so as soon as possible." He was concerned that the legislature would become so involved with voting procedures and other matters that further delays might follow. Once the legislature voted the funds, he was prepared to sign on for Berlin. In the meantime, he decided to visit Berlin after Christmas to prepare the provisional laboratory, look for a residence, and take care of various minor matters. His experience with Berlin's privy counselors had taught him that once he turned in his resignation to Karlsruhe, "everything would be decided" to his "disadvantage." In mid-December the ministry made a definitive, written offer but also informed him that, owing to war expenses, the legislature's approval of a new building for physics would be delayed by at least one year. Mühler said the legislature would definitely approve the requested funds and that he wanted him to ask for his release from Baden as of Easter. The ever-cautious Helmholtz agreed but said he would not send the letter until after he arrived in Berlin. In the meantime, he asked du Bois–Reymond to find a rental agency that could provide a list of available residences near the Tiergarten and costing up to one thousand talers annually. As always, du Bois–Reymond aided his friend.[46]

Helmholtz told his mother-in-law that they were effectively in the process of moving to Berlin permanently. He thanked her for the recent Christmas present of two books by David Friedrich Strauss (*Voltaire* and *Das Leben Jesus*), which he said he valued "very highly." About the latter book, he told his mother-in-law, "One must indeed occasionally look at it in order to be up-to-date in discussions, otherwise one repeatedly and easily forgets individual facts from one's head. In the end, no one who has worked for understanding human development and represented its interests can avoid religious discussions, since they emerge from time to time, even if one does not seek them out." A month later he wrote her again about the book: "I must also acknowledge that this book shows a much greater love and admiration for the founders of our religion than do all orthodox-ecclesiastical presentations by one or another party. The alleged God remained foreign to me, but I do understand men."[47]

The Helmholtzes left for Berlin on 26 December and stayed with the du Bois–Reymonds. While he visited the ministry, she looked for a place to live "on all the well-laid streets." She found a flat near the Tiergarten for 1,100 talers, though it was far from what she wanted. He got all that he had expected from the ministry, began negotiations for a piece of land for the new building, and surveyed the provisional institute's quarters. Everyone sought to be "accommodating and charming" toward him. The Helmholtzes saw old friends and made new ones. Dinners and entire social evenings were planned for them. The du Bois–Reymonds "were particularly friendly and hospitable." Helmholtz said, "They invited God and the world for us. I saw some very elegant and exciting establishments." They also heard a concert in the home of the violinist Joseph Joachim. All told, they stayed eight days ("our North Pole trip to Berlin," Anna called it), and before they left, they were committed psychologically. In early January Borchardt asked Helmholtz to review a manuscript by Ludwig Boltzmann—Borchardt had reservations about it—for his *Journal*, adding that everyone was looking forward to Helmholtz's being in Berlin as of Easter.[48]

Du Bois–Reymond wrote Ludwig that Helmholtz's visit went very well and that all was set for his move to Berlin next Easter. Indeed, on New Year's Day 1871, Helmholtz asked Baden to release him from its service so that he could begin teaching in Berlin on 1 April. To both Jolly and the grand duke he expressed his great respect and gratitude for all that they and Baden had done for him through the years.[49]

But then a potential new roadblock appeared in the form of a feeler from Cambridge. Britain was in the early stages of its Endowment of Scientific Research movement (1868–1900), which sought to increase financial support for scientific research, especially in the pure sciences. Since its embryonic beginnings in the 1850s, this renewed British emphasis on research had looked especially to the German lands as its standard. Helmholtz, with his extensive and friendly relations with many British men of science, was just what the new movement needed. In late January, Thomson, who knew of Helmholtz's offer from Berlin, wrote him on behalf of Stokes and others at Cambridge to ask if he might consider accepting the new professorship of experimental physics there and the directorship of its equally new Cavendish Laboratory. "It is much desired to create in Cambridge a school of experimental science," Thomson wrote, "not merely by a system of lectures with experimental illustrations, but by a physical laboratory in which students under direction of the professor and his assistant or assistants, would perform experiments, and the professor would have all facilities attainable, for making experimental investigations." The new laboratory building was under construction (the Duke of Devonshire had given £6,000 for it). Helmholtz could anticipate an annual compensation of £500, along with student fees. In addition he would be appointed a fellow of Peterhouse College, which meant additional income (£250–£300). All told, his income would be at least £800 annually, Thomson thought. This was a generous offer, since a British professor could live comfortably on £500 per annum at the time. If he also received a fellowship to Trinity, he would get an additional £600, making a total of about £1,400 per annum. Thomson said that, apart from the financial induce-ment, Britain offered other advantages: "The desire for physical science is grow-ing stronger and stronger in the University, and the force of public opinion is steadily advancing in support of it, and to stimulate it when stimulus is needed." Helmholtz would have to lecture only twenty weeks per year, leaving the remain-ing time for his own research. "I need not say that it would be a great gratifica-tion and advantage to English scientific men to have you among us instead of merely having very rare opportunities of seeing you, and that I myself would consider the difference of distances from Glasgow to Cambridge and Berlin a great gain." Helmholtz declined the offer, and in March 1871 Maxwell accepted the Cavendish professorship. But Helmholtz nonetheless sent a copy of Thom-son's letter to Olshausen at the ministry, who replied that he and the under-secretary were agreed that Helmholtz should encounter no difficulties in regard to his position or the institute. The king would soon be signing the necessary papers to finalize his appointment.[50] The ministry was nervous.

Du Bois–Reymond told Helmholtz that the physical institute's provisional quarters had been arranged. There were six rooms in the east wing of the Uni-versity Building (Prinz-Heinrich-Palais); they were near both the instrument-collection rooms on the ground floor and the auditorium. He had spoken with a senior official in the ministry about purchasing land for the new institute building, but they had yet to settle on a site. The good news about the quarters was that Helmholtz would have a laboratory there in time for the opening of the new semester. He already had three American visitors in Heidelberg who wanted to go with him to Berlin, and they wanted to know when the laboratory would open.[51]

On the Origin of the Planetary System and Life

Helmholtz spent the first three months of 1871 preparing to leave Heidelberg. It was a cold winter there as well as in Berlin, and Heidelbergers felt an earth-quake and its aftershocks for three days. They could use a lift that winter. Before leaving, he delivered several departing lectures. One of these, "On the Origin of the Planetary System," was a fund-raiser for the benefit of disabled soldiers. The auditorium was packed with listeners, the stage covered with laurels and a wreath laid upon them, and "the entire public arose when he appeared."[52] He gave them the lift they needed.

The lecture was in part a retread, however; he had given it, or at least a version of it, as part of the lecture cycle of 1862–63, in which "Conservation of Energy" was the introductory lecture, and he was going to give it again in Cologne a month later. In his lecture, Helmholtz again argued for the validity of the Kant-Laplace hypothesis in explaining the origin of the planetary system and its implications. But he also raised questions about the origin of life and the ultimate purposes of humankind. These touchy topics interested everyone. In presenting a popular account of them, he believed that he had to justify popular science itself, which should normally rest, he said, on "well-secured facts and the finished results of research . . . not on immature guesses, hypotheses, or dreams." Yet the topics he discussed had always been "the playground of the most extravagant speculations." He thus felt somewhat uncomfortable about speculating, since in effect it placed him in the category of the ancient cosmological mythmakers. Yet the fact remained, he said, that everyone wanted to know about human origins and ends as well as the origin and end of the world itself; indeed, he thought the issue of the end had "perhaps still greater practical interest than that of the beginning."[53]

Helmholtz reminded his listeners that until age forty Kant had been a natural scientist and that his philosophical writings had come later. Kant had shown that one could talk about cosmological issues using "the inductive method," rather than resorting "to the lofty speculations of a supposedly 'deductive method.'" Helmholtz thought there was increasing scientific evidence for the plausibility of the Kant-Laplace hypothesis, and so he felt further popularization was warranted.[54]

Helmholtz deemed the law of universal gravitation and spectrum analysis as two especially great discoveries, since they allowed scientists to draw conclusions about the nature and behavior of matter throughout the universe. Spectrum analysis showed the similarities between terrestrial and celestial bodies, indicated that most light radiates through gaseous dust clouds in the sky, and showed the chemical constitution of the sun (and other heavenly bodies) and the widespread presence of nitrogen and hydrogen in the heavens.[55]

Though it had long been believed that the solar system was essentially unchanging, modern scientific research, he explained, had indicated that it is in fact constantly changing, that space is not empty but rather filled everywhere with the ether (a radiation-bearing medium) as well as untold numbers of meteors and that friction (and thus heat) is generated as the planets move. As high-speed meteors pass through space, they heat up, but only on their surfaces, not inside. The earth's cosmological history was no different from that of other celestial bodies in the solar system: as they move(d) through space over millions of years, they attract(ed) loose mass and thus gradually increase(d) in size; that implied that all mass was probably once loosely distributed in small amounts (ultimately as dust) throughout the cosmos.[56]

Helmholtz introduced another argument that favored the Kant-Laplace hypothesis: the orbits and equatorial planes of all planets and their satellites barely deviate from one another, and, a few small moons aside, they all rotate in the same direction. This is a coordinated system, he declared, in which one body depends on another (though he did note that comets and meteor showers show "irregularity"). In the course of time, the various forces and processes in the heavens swept up the smallest bits of matter into larger bits, eventually resulting in the creation of planets and their satellites that revolve around the sun in elliptical paths. According to the Kant-Laplace hypothesis, he further explained, the solar system had begun as "a chaotic ball of mist" in which for "many trillion cubic miles" no mass was to be found. The small balls of mass had slow periods of rotation, but as they rotated they became increasingly larger and, under forces of mutual attraction, their speed of rotation gradually accelerated and they flattened out into disklike objects. In some cases, this process resulted in the formation of a planet, in others of satellites, in the case of Saturn, its rings, and so on. The process was ongoing.[57]

This celestial thermodynamics was also a terrestrial thermodynamics, since it created and drove meteorologic, geological, and organic systems. As Helmholtz had noted in his London lectures of 1864, solar radiation in the form of heat and light maintained all life on earth. Heat was also the driving force of the earth's atmosphere, its bodies of water, its winds, and more. Organic life itself, he said, arose from water in combination with solar energy; plants and animals, in their turn, served man as food and energy. Indeed, he linked carbonaceous matter (fossil fuels) directly to the modern, industrial world: "Even the bituminous coal and the brown coal—the power sources of our steam engines—are remains of primeval plants, old products of sunbeams."[58]

Could it be any wonder, he asked rhetorically, that "our forefathers of the Aryan race in India and Persia looked upon the sun as the most appropriate symbol of the Godhead? They were right to do so if they saw in it the dispenser of all life, the ultimate source of all earthly happenings." However, the sun should be treated as a fully material body, something rationally understandable, not as a mysterious god or mystical entity, and so he asked about the source of the sun's force. The sun's heat, he explained, produces "a great deal of mechanical work" on earth. The sun itself was like an enormously powerful steam engine whose power far exceeded any imaginable here on earth. But where did the sun get its own supply of energy? Helmholtz pointed to the heat generated by meteors and other celestial matter as they moved through space; he noted that far more meteors fell onto the sun than onto the earth. He thus hypothesized that the source of the sun's energy (its radiating heat) was "cosmic forces" arising from the sun's formation via the compression of finite masses under gravitational force.[59]

Enough heat had accumulated in the sun, he further argued, to cover the earth's needs for at least the past 22 million years. He surmised that, given additional compression, the sun would accumulate additional energy to radiate heat toward the earth for at least another 17 million years. He also noted the earth's high internal temperature, "which can hardly be something other than a remainder of an old heat supply from the time of its origin." He judged that all this only confirmed the bold and astoundingly brilliant Kant-Laplace hypothesis. Nevertheless, in the course of consuming its energy supply and radiating heat, the sun would eventually extinguish itself, and so life would cease. Human beings had a difficult time accepting this fact, he observed: "It appears to us like a wound of the salutary creative force, which we otherwise find active in everything, namely, the relations concerning living creatures." But humankind had to come to terms with this finding and cease thinking of itself "as the center point and final purpose of creation." It had to realize that humans "are only bits of dust on Earth, which itself is a bit of dust in a vast spatial realm." The duration of the human species, as well as its various predecessors, is nothing compared to that of the earth and its history, and in comparison to what is yet to come. When the sun exhausts its supply of energy, the earth will grow cold and rigid.[60]

Helmholtz's account of the origins of human life was a non-Christian, pro-evolutionary one. Perhaps human life also came from the sea, he mused. And who knows, he asked, where this may lead during the next 17 million years in terms of new life forms? Perhaps someday human bones would seem as odd to humankind's evolutionary successors as those of the ichthyosaurs seem to humankind today, he opined. "Indeed, if the Earth and the Sun should rigidify into motionlessness, who is to say what new worlds will be ready to take up life?"[61]

The most novel and controversial part of his lecture came, however, when he speculated that life on earth may have initially appeared only after meteors, comets, or some other interstellar matter containing germs of life or microbes (or some other organic material) in their interiors crashed onto earth and, over a long period of time and under the right conditions, began to flourish. Indeed, this process might have occurred not only on earth but also elsewhere in the universe, that is, on other planets. Different beings would, in effect, be related to one another, no matter how different their appearances might be.[62]

He thus promulgated the idea of the extraterrestrial origins of life, an idea also sometimes referred to as "panspermia" (life-endowing seeds floating throughout space) or "the cosmozoic hypothesis." Though his name became closely associated with the idea, he was in fact neither the first nor the last to propose it. Since the mid-1860s, numerous chemists and others had suggested it, though there was disagreement as to whether the organic material within meteorites was itself biological or whether it was due to a prior synthesis of hydrocarbon compounds. Helmholtz may have come to the idea through a recent report by the French organic chemist Marcellin Berthelot. Helmholtz's own hypothesis, however, was highly speculative; the empirical evidence for it was thin. Yet men as professionally different as the zoologist and evolutionary theorist

Ernst Haeckel and the theologian Rudolf Schmid thought it improbable and that it only displaced the problem of the origin of life to another world. More hostile still was the astrophysicist Zöllner, who later ruthlessly attacked it. Thomson, who believed in a Creator of the universe, took up Helmholtz's hypothesis but gave it a very different, nonmaterialistic interpretation.[63]

Helmholtz and Thomson thus drew quite different inferences from his speculative hypothesis, inferences that reflected their broader attitudes toward life. However much this hypothesis may affect "our moral feeling," Helmholtz said, the overriding issue was whether life had any purpose or whether it was "an old, purposeless game" that would eventually end. Darwin's theory of evolution by means of natural selection had shown us, Helmholtz said,

> that not mere lust and joy, but also pain, struggle, and death are the power-
> ful means by which nature shapes its finer and more complete forms of life.
> And we human beings know that in our intelligence, State organizations, and
> morals we consume the inheritance which our forefathers have accumulated
> through work, struggle, and the courage to sacrifice, and that what we achieve
> in the same sense will ennoble the life of our descendants. Thus, the individual
> who works for the ideal purposes of humanity can, if only in a modest way and
> within his own limited domain, bear without fear the thought that the thread
> of his own consciousness will some day be torn. Still, with the thought of the
> final destruction of the living race and thus with all the fruits achieved by
> past generations, even men of so free and great morality as Lessing and
> David Strauss could not reconcile themselves.

That was clearly a statement of his own broad outlook on the purpose of life and work. He maintained that just as a flame appears not to change yet in fact is constantly changing (owing to its consumption and renewal of oxygen), and just as water waves appear not to change but are in fact constantly being reconfigured (owing to new water particles), so living bodies in general are changing through the renewal of their matter over shorter or longer periods of time. "What continues to exist as the special individual is, as with the flame and the wave, only a form of movement which constantly pulls new matter into its vortex and pushes out the old again."[64] For Helmholtz, humankind was an ever-changing set of finite beings.

Helmholtz's Heidelberg audience, Anna reported, was thrilled with his lecture—it "was absolutely wonderful and inspired the entire world"; thirty-five years later, his former assistant Julius Bernstein agreed. Presumably, part of what inspired them was the scientific breadth and clarity of his presentation. His lecture revealed enormous powers of synthesis and imagination: he had invoked and utilized parts of astronomy, cosmology, chemistry, physics, geology, and climatology. He was a sort of scientific dreamer, one who could take known facts and boldly hypothesize explanations for and give meaning to them. But more than this, the lecture fit the age. His focus on the Kant-Laplace hypothesis — the hypothesis of a German and a Frenchman — at this politically tense moment in European history, as the new German Reich emerged and concluded its military-political arrangements with the collapsed French Empire, seemed both intellectually and politically fitting. Though Helmholtz never said so, he may have meant to suggest that science was above and beyond politics, could withstand politics or be independent of it. If so, in this he was typical of many German cultural figures.[65] Certainly he had little interest in political and social events, and to the extent that he did, his views usually focused on their negative aspects.

Finally, his broader account of cosmology, the earth, and human history also fit the naturalistic and materialistic tenor of the age. Darwin had published *On the Origin of Species* in 1859, and the *Descent of Man* appeared in February 1871, the very month in which Helmholtz gave this address; Tyndall lectured on "Scientific Materialism" in 1868 and more famously did so again with his Belfast Address in 1874; and Huxley wrote *On the Physical Basis of Life* (1869). The lecture, like Darwin's, Tyndall's, and Huxley's work, helped shape a new, more secular and scientific outlook. Helmholtz never invoked God and never referred to Christianity. He gave a strictly naturalistic account of the possible origins and fate of the universe and man. Matter and motion were the bases of life, and scientific laws — especially those of thermodynamics, spectrum analysis, gravitation, and evolution by means of natural selection — explained how matter had been shaped into the world as humankind knows it. Just as the German Reich constituted a new, unified political framework for the German states, so, in its own way, Helmholtz's lecture provided a unified interpretation or understanding of inorganic, organic, and human evolution. He himself later said that this lecture topic was a favorite of the day.[66]

Departing Heidelberg

The last three months in Heidelberg were filled with so many dinners, honors, signs of friendship, and warm goodbyes that the Helmholtzes found leaving to be especially hard. In mid-March the packers arrived. In the middle of it all, Helmholtz found time to read du Bois–Reymond's most recent address, "The Empire and Peace." Du Bois–Reymond declared that it was the duty of members of the Academy of Sciences "to uphold quietly the flag of science entrusted to us, although our hearts are also with the Emperor and the army, with sons and brothers outside in the winter field camps." He reiterated the loyalty of the academy's members to the Hohenzollern regime, whose kings had backed the academy from its founding in 1700 onward, and to the new German Empire. Yet in the next breath he declared, "There is only one science, even if the sort of praise given it by different peoples can be different. It is our way not to distinguish between German and foreign discoveries." He now surprisingly declared, We "honor at the same level the heroes of French science—a [Antoine] Lavoisier, Laplace, [Georges] Cuvier, a [Augustin-Jean] Fresnel, [André-Marie] Ampère, [the orientalist Antoine Isaac] Silvestre de Sacy, and [the archaeologist Jean Antoine] Letronne—with our own heroes as well as with those of every other nation." Moreover, he regretted the physical damage done by Germany's armed forces to France's (that is, Paris's) scientific institutions and their holdings and wished only for peace and for France's scientific and artistic renewal in the postwar period. In contrast to the chauvinist, aggressive war speech that he had given the previous August, this speech of 26 January 1871, delivered when the Germans had already won the war and barely a week after Bismarck and the military leadership had declared a new Reich in the Hall of Mirrors at Versailles, was conciliatory. Helmholtz said of it: "I like it even more than the previous one because it doesn't hurt any feelings." On 1 April 1871, he was appointed an ordinary member of the Prussian Academy of Sciences; two days later, he and his family left for Berlin.[67]

Helmholtz considered his years in Heidelberg—apart from those of Olga's final months of illness and then death—the happiest and most successful of his life. He met Anna there, and so renewed his personal life, and he had a stimulating circle of friends. He loved the university, a point that he reiterated as late as 1891. Walks on the beautiful, lush hills and in the woods surrounding Heidelberg helped clear and stimulate his mind and fantasy to conceive new ideas and helped refresh him spiritually and give him peace of mind. They helped bring light and intellectual order to what was previously darkness and chaos.[68]

The 1850s and 1860s were the glory years in the long history of the University of Heidelberg, above all in the natural sciences. Bunsen, Kirchhoff, and Helmholtz had formed a troika that brought the university international fame and students and helped make it a destination for scientists throughout Europe and from America. These three had not only done creative scientific work but had also developed institutional facilities and a favorable intellectual atmosphere for using them. But by the early 1870s, the aging Bunsen, with his emphasis on inorganic rather than organic or physical chemistry, had become scientifically passé. Helmholtz's departure in 1871 was followed by Kirchhoff's (also for Berlin) in 1875. Partly as a result, Heidelberg declined sharply as a leading natural scientific center and underwent an institutional crisis. The death or eventual departure of a series of leading figures in the humanities—Vangerow, Bluntschli,

Zeller, Fischer, Häusser, Treitschke, and Gervinus among them—further undermined Heidelberg's standing. Even Germany's unification in 1871 weakened it, for it now lost its special role as a stronghold for German political unity.[69] Heidelberg's heyday was over, and Berlin now became the rising power center of German academic and political life.

PART III

Scientific Grandee

17

In the Capital of *Geist*

Berlin and the Reich

The Berlin that Helmholtz returned to in 1871 was scarcely the one he had left in 1848. It was now the capital of the German Empire, not just of Prussia, and William I was both Prussia's and the empire's formal leader. The de facto leader, Otto von Bismarck, Germany's chancellor and Prussia's minister president, had forged a national and illiberal state from above, though the Reich remained pervaded by strong and ceaseless domestic political tensions and centrifugal forces. Germany had emerged from the Franco-Prussian War as a great power, one whose considerable material and technological strengths were matched by enormous scientific and cultural achievements and potential. The two realms—industry and culture—were mutually supportive. Berlin itself became a large governmental, military, and diplomatic center. Yet even as the city served as the home of the conservative Prussian monarchy and aristocracy, the upper civil service and army officers, and well-to-do industrialists and businessmen, many Berliners were liberal, and some even socialist-minded. An event early in Helmholtz's new professional life in Berlin illustrates a few of these points.

On 16 June 1871, the Reich's political and military leaders put on

411

a patriotic spectacle in the heart of Berlin to celebrate Germany's victory over France and its founding as an empire. Helmholtz, his family, and his friends had ringside seats at the military parade as it marched past his institute's windows along Unter den Linden—but only after he had stood his ground against a ministry bureaucrat who had sought to control their use. Two weeks before Helmholtz's move to Berlin, the bureaucrat in question had indicated to him who might be invited to sit at the windows to watch the parade as it marched along Unter den Linden and on through the Brandenburg Gate. Helmholtz thought the bureaucrat "naive" to imagine that he would agree to his suggestion; he informed him instead that he was reserving the window seating for his own family and friends, though he did agree to allow certain persons suggested by the ministry to sit at certain windows (but only after he vetted them). This little political power struggle set the tone for his dealings with the ministry; he was concerned that an unfavorable precedent would be set and that the ministry would start limiting his authority even before his arrival. And so, the Helmholtz family sat at the corner window on the ground floor facing the colossal Blücher statue and the Opera House, the du Bois–Reymonds sat at another window, Dove and Magnus's widow at yet another, and, in part out of courtesy to the ministry, two of its officials at still another. To show further cooperation with the ministry, he allowed the university judge and his patron, the wife of Prince Adelbert, to invite several guests to sit at the remaining windows. These were "magnificent seats," Anna said, and she was greatly impressed by the endless squadrons of troops (some forty-two thousand soldiers) that marched past them, led by the emperor and his family, Field Marshal Helmuth von Moltke and other senior officers, and Bismarck.[1] This was Helmholtz's institute—located in the political heart of Prusso-German power.

Berlin grew rapidly as an urban center: in 1849, shortly after Helmholtz left, it had a population of 432,000; in 1871, when he returned, it had nearly doubled to 826,000; and by 1890 it nearly doubled again to 1.6 million. Its elite lived in palaces or large villas in or near the historic center (*die Mitte*) while the working classes lived on an ever-expanding periphery in sprawling tenement blocks. The middle class, as everywhere in Germany, lived around both groups and was squeezed politically between emerging mass movements of workers and an increasingly anachronistic nobility and Prusso-German absolutist monarchy.

Especially after 1871, Berlin rapidly became a modern industrial city, as it had begun to do by the 1860s: it was dominated economically by industrial, trade, financial, and communication enterprises. It boomed between 1871 and 1873 (the *Gründerjahre*): many new firms were founded, and the railroad and residential construction industries flourished. The streets were gradually paved and an underground sewerage system installed, yet poor public health conditions persisted. In 1873 there was a major financial panic and crash, which had significant cultural consequences. A recovery soon followed, thanks in part to burgeoning enterprises like Siemens & Halske and the Allgemeine Elektrizitätsgesellschaft. Berlin became known as the "Electropolis." For Helmholtz, the city, with its ceaseless construction, was "an extremely uncomfortable place to reside in. Incredibly expensive, terribly vast, and still without the organizers and helpful means of an older big city," it caused him to lose a large amount of time for no purpose. He found life in the big city "unnatural."[2] If Berlin in the 1870s was a large construction site, by the early 1880s it had become a *Weltstadt*.

Chez Helmholtz

The Helmholtzes rented a residence at Königin-Augusta-Strasse 45 (modern Reichpietschufer 82). Toward the end of their first year there, they had so little money left over that they could afford only a very modest Christmas party.[3] Still, they remained in this fashionable, well-to-do, upper-crust area, near the city center, just south of the Tiergarten and near the Matthäuskirche, until December 1876. From there, Helmholtz walked to his university institute on Unter den Linden.

As a Badenese native, Anna had spent most of her adult life in a small university town; the big, dark, and raw city of Berlin left her feeling uprooted. Even as she adjusted to the new lifestyle, she felt homesick for family, friends, and *Heimat*. Only the importance of Hermann's being in Berlin made living there bearable for her. She had a passion for creating a beautiful home and maintaining order. (One of her friends thought she had an obsession for order.) She had an excellent sense of color and materials and arranged her furniture harmoniously. Her home was adorned with well-chosen busts, statues, vases of flowers, and perfectly hung paintings, which she realigned whenever the slightest need arose. The latest in literature adorned her tables.[4] Before the year 1871 had ended, the second collection of Helmholtz's *Populäre wissenschaftliche Vorträge* had appeared, and perhaps it too was on display chez Helmholtz.

Four of Helmholtz's five children lived with him and Anna during the first year or so in Berlin. Käthe's, Robert's, and Fritz's mental or physical well-being continued to be of grave concern to their parents. Käthe, who had always been sickly, was an idealist who worshipped her father but could not find her way in life professionally. She briefly tried her hand at translation—helping to translate Tyndall's *Heat*—and she painted. Anna attempted to help her develop her artistic talent, but she remained restless. In 1872 she married Wilhelm Branco (after 1895, von Branco, and after 1907 von Branca), the eldest son of Friedrich Wilhelm Branco, Helmholtz's old friend and one-time superior in the Gardes du Corps in Potsdam. Wilhelm Branco was a military officer and farmer who later became a professor of geology and paleontology. In 1873 they had their first and only child, Edith, making Helmholtz a first-time grandfather (and reminding him, as he told Knapp, that he was getting old). But from the moment of Edith's birth, Käthe's health seriously deteriorated—she suffered from a pulmonary disease, apparently either pleurisy or catarrh (tuberculosis), which was more or less what her mother had died of. Hermann and Anna sent her on extended trips to warmer climates; but these did little or nothing to help, and Helmholtz constantly worried about her. He told Franz Boll, himself gravely ill from tuberculosis, that on several occasions Käthe had stood at death's door. In fact, she died on 25 April 1877, at age twenty-seven, in her aunt's house in Dahlem. Her death sent the Helmholtz household into a depression. As for her step-brother Robert, his frail, malformed body and delicate health continued to plague him and his parents. He underwent a series of surgical procedures and physical therapy, and a special attendant was engaged to help him. He did not attend school until the final year of Gymnasium, when he entered the prima class of the Französisches Gymnasium in order to obtain an *Abitur*. Fritz, the youngest of Helmholtz's children, was also physically frail and often sickly, and he too became a constant worry to his parents. He lagged behind his contemporaries in development and suffered, it appears, from a lack of motivation.[5]

Unlike these three Helmholtz children, his other two brought him no known worries or concerns. Helmholtz sought a practical position to help his son Richard gain some experience in railroad technology. In the autumn of 1871, Richard began an apprenticeship with the Borsig locomotive factory in Berlin. He then spent three years (1873–76) studying mechanical engineering at the new Polytechnische Schule Munich, where Wilhelm von Beetz, one of Helmholtz's oldest friends and a professor of physics, kept an eye on him. He told Helmholtz that he heard only good things about Richard, who was "by far [the Polytechnic's] best pupil." After graduating in 1876, Richard joined the distinguished locomotive factory Krauss & Co. in Munich. As for Ellen, the Helmholtzes' daughter, she was physically healthy and vivacious. Though as a girl/woman she could attend neither a Gymnasium nor a university, the mathematician Leopold Kronecker gave her and his own daughter private mathematical instruction, and he found her to be extraordinarily talented. (Helmholtz found that there were "many examples" of young women who showed that they were "indeed fully capable of [pursuing] mathematical study.")[6] Ellen was a source of much happiness and pleasure to her parents. See figure 17.1.

Helmholtz's children thus brought him love and pleasure, on the one hand, but also, on the other, the considerable burden of looking after and constantly worrying about the emotional or physical well-being of Käthe, Robert, and Fritz. As wife and mother/stepmother, Anna took on the greater share of this burden, but it was nonetheless theirs together. In 1875, while Hermann was in Italy, where he sought to recover his health, she urged him to stay as long as he felt he

Fig. 17.1 Helmholtz in 1876, with his three youngest children, Friedrich Julius, Ellen, and Robert. Siemens Forum, Munich.

needed to: "I find," she wrote him concerning their children's health, "that fate is too brutally against us, and I must at least use the time of being alone for weeping." When her brother, the Orientalist Julius von Mohl, died in January 1876, Anna, then forty-two years old, wrote: "It's as if I'm sixty!" "I've become as old as the hills, and everything is all the same to me." She felt she had lost her youth. A year later she went to Paris to be with her aunt Mary and then reluctantly brought Mary, who was losing her memory, back with her to Berlin for a visit.[7] For better and for worse, Helmholtz's family was always with him. His cease-less social engagements with friends, colleagues, and others and his vacations in southern Germany and abroad helped relieve the stress that family problems and professional life brought him.

So too did music. There was, of course, a piano in the Helmholtz home (both parents played). Shortly after the family moved in, Steinway & Sons delivered its latest, highly experimental baby grand piano. It was on permanent loan to Helmholtz, "in part," Steinway wrote Helmholtz, "because you have done so many useful things for us concerning the phenomena of tone, and still more im-portantly because we hope that this piano may stimulate you to make further valuable discoveries in the empire of sounds." In thanking Steinway, Helmholtz praised the piano's ability to hold its tone, its light, tender touch, its damping ability, and its general precision and clarity. All in all, it was "a perfect instru-ment," especially for playing Bach fugues, which gave him his greatest pleasure. Theodore Steinway himself twice visited Helmholtz in the summer of 1873 to supervise the installation of the latest "high strings" on the piano so as to im-prove it, and to make minor repairs. He showed Helmholtz the piano's inner workings, and Helmholtz planned to send him suggestions for improvements. Steinway asked him to write a letter explaining what he liked about the piano. Helmholtz's letter emphasized the firm's new duplex scale, "based on scientific principles."[8]

Steinway published the letter in its 1873 catalog, prefacing it with a few com-mercial remarks: "A most valuable distinction has been received by Steinway & Sons, in that conferred on their pianos by Professor Helmholtz, who occupied the Chair of Acoustics [sic] in the University of Berlin, and who is unanimously admitted to be the highest authority in the science of acoustics that is known." It further declared that "the various improvements made by Steinway & Sons" "struck the Professor's attention, and, after a careful examination of the Stein-way Pianos, and those of other makers, he arrived at the conclusion that the Steinway piano alone reached the acme of perfection, and he directed the pur-chase of a Steinway grand piano for express use for his experiments and lec-tures on acoustics in the Berlin University," which in fact he did. In 1885 Stein-way & Sons republished two Helmholtz letters, along with an extract from a third. His endorsements had obvious commercial intent, and they appeared in Steinway catalogs for nearly twenty years. In this he was in good company: other celebrity endorsements came from Franz Liszt, Richard Wagner, Anton Rubin-stein, and Hector Berlioz. The firm and Helmholtz cultivated each other: he got a piano, and they got the leading acoustic scientist's endorsement and cultural prestige—and perhaps even increased sales.[9]

Hermann and Anna's love for each other only deepened over the course of their marriage. On the occasion of his fifty-sixth birthday (1877), seventeen years into their marriage, she wrote him, "With each passing year I am more yours, and my love and honor, which at the start was perhaps more instinctual, has turned into joyous conviction." He replied: "I never forget that your love is the greatest jewel of my life." Outsiders also saw the marriage as an extremely happy one. The two were said to complement one another well: she ruled in small matters, he on the big ones. (Boltzmann wrote, pace Koenigsberger, "that no one led him against his will.") He was also on the best of terms with his in-laws, addressing his mother-in-law, Pauline von Mohl, as "Best Mama," and his father-in-law, Robert von Mohl, as "Papa." In 1873 Mohl, who was a strong supporter of Reich unity and was a National Liberal, was elected to the Reichstag from Baden. He spent part of the last two years of his life, until his death in 1875, living in the Helmholtz home. He had the greatest respect and admiration for his son-in-law, not only for his scientific accomplishments but also for his general cultivation, including his appreciation of poetry, music, and languages, and the generally simple, moderate, but disciplined lifestyle that Helmholtz led. Helmholtz reciprocated that respect.[10]

The Rise of the University of Berlin

The Friedrich-Wilhelms-Universität (Frederick William University), to use the University of Berlin's formal name (between 1828 and 1949), lay in the historic heart of the city, its front along Unter den Linden, across from the Opernplatz and the emperor's palace. It was associated by name, location, and the professoriate's generally conservative political outlook with the established authorities of Prussia and the Reich. Founded in 1810 as part of the Reform movement in Prussia and in part to replace Halle, which had been lost in the Napoleonic territorial settlement with Prussia, the university had, before midcentury, only a modestly distinguished faculty and even more modest facilities. It served largely as a place to train professionals for administration, law, the clergy, medicine, and education and was known as a "work" university. To be sure, it preserved an internal sense of academic freedom, and after the 1860s it offered a somewhat more liberal atmosphere, yet it was anything but a place for intellectual or political radicals. Its mentality reflected that of the surrounding military and bureaucrats who ran Prussia and the Reich rather than the far more liberal sentiments held by some of the city's population. Both the student body and the professoriate displayed a strong sense of nationalism, especially after 1870.[11] See figure 17.2.

Already in the 1860s, and especially after 1871, the university dramatically expanded its enrollments, the number of professorial positions, and its physical facilities. In 1860 it enrolled 1,620 students; in 1871, 2,603; in 1884, 5,000;

Fig. 17.2 The University of Berlin, circa 1872, with the old Physics Institute located in the east wing (*on the right*) of the university building. Universitätsarchiv, Universitätsbibliothek, Humboldt-Universität zu Berlin.

and, though its growth slowed in the early 1890s, in 1903, 7,000 students were enrolled. By 1881 it had more students than Leipzig and twice as many as Munich, its two main rivals. The 1870s and 1880s were the university's great decades for new building construction. Moreover, ten new ordinary professorships and associated institutes were established in the natural sciences (chemistry, geography, astronomy, physics, botany, mathematics, meteorology, mathematical astronomy, petrography, and zoology) and an additional eight chairs in the human and social sciences as well.[12] Berlin arguably became the center of science in Germany.

This spectacular institutional growth was accompanied by a marked enhancement of the intellectual reputation of its faculty. Before 1871 its outstanding figures included the mathematicians Ernst Eduard Kummer and Karl Weierstrass; the physiologist Emil du Bois–Reymond; the chemist August Wilhelm von Hofmann; the botanist Alexander Braun; the pathologist and anthropologist Rudolf Virchow; the medical researchers Heinrich Adolf von Bardeleben, Friedrich Theodor Frerichs, and Bernhard von Langenbeck; the historians Johann Gustav Droysen and Theodor Mommsen; the archaeologist Ernst Curtius and the Egyptologist Richard Lepsius; and the classicist Hermann Diels. After 1871 the Prussian Ministry of Culture transformed Berlin into the nation's premier university. Helmholtz was not the only outstanding figure who was now brought there. Others included the physicists Gustav Robert Kirchhoff (1875),

August Kundt (1888), and Max Planck (1889); the mathematicians Leopold Kronecker (1883) and Immanuel Lazarus Fuchs (1884); the historian Heinrich von Treitschke (1874); the philosopher Eduard Zeller (1872); the political economists Adolph Wagner (1870) and Gustav von Schmoller (1882); the anatomist Heinrich Wilhelm von Waldeyer-Hartz (1883); and the internist and neurologist Ernst Viktor von Leyden (1876). Some suggested that many of these new faculty members had their best work behind them or that their heavy teaching and administrative responsibilities consumed energy they might have otherwise devoted to creative research. However true that may be, Berlin, to say the very least, was now Germany's center of mathematics, physics, and physiology. In October 1879, the young Karl Pearson, future mathematician, biostatistician, eugenicist, and more, came there to study (originally) physics, in particular to hear lectures by Helmholtz and Kirchhoff, and to work in Helmholtz's laboratory. He called Berlin "the capital of all *Geist*."[13] Many thought Berlin had become the leader of Germany's dynamic, competitive higher education and research system, one often envied and sometimes emulated.

Helmholtz was determined to contribute to the university's emerging growth, reform, and renown. He did so by opposing the candidacies of those whom he thought less than first-rate and by helping those whom he regarded as such. He was seen as a force for change within the Philosophical Faculty: "My husband has come forward in the faculty as a youthful revolutionary," Anna wrote ten months after their arrival, "and often horrifies the old men." When an opening occurred in 1872 for a second ordinary professor of chemistry, he opposed a suggested nominee because the individual was, he said, "scientifically not important enough," dismissing out of hand the claim that seniority was a reason for considering him.[14]

Apart from guiding Kirchhoff, Kundt, and Planck to Berlin, two of Helmholtz's major recruiting efforts concerned scholars in the humanities: Zeller and Treitschke. In early 1872 Berlin sought to acquire Zeller, the liberal-minded, leading neo-Kantian thinker and specialist in Greek philosophy, who taught at Heidelberg. He and his family had been personally close to the Helmholtzes there. After the ministry's emissary failed to induce Zeller to come, it sent Helmholtz to see him with an improved offer. Zeller then visited Berlin for a week, staying with the Helmholtzes. He chose Berlin and was grateful for the friendship the Helmholtzes had shown him; he joined Helmholtz's circle of close friends and colleagues, which included many of Berlin's leading liberal faculty members.[15]

In looking for a historian to succeed Leopold von Ranke, the ministry first sought to attract Jacob Burckhardt, a well-known Swiss historian; when he declined, the Philosophical Faculty took the matter up directly. In 1873 Helmholtz became the principal (and initially sole) advocate for Treitschke, who was not without his opponents. In time, however, even Treitschke's strongest opponents came to welcome him. When Treitschke received the offer, he wrote Helmholtz to ask him how much income he needed to live there decently and to express his concern that Droysen might feel cramped by his presence in Berlin. He wondered what Helmholtz, who was a friend of Droysen's, thought of this potential conflict. Helmholtz told Treitschke that the university very much needed more "enthusiasm" and that the conservative Prussian bureaucracy effectively robbed it of independent-minded people. This led to a certain amount of "scientific conservatism" among the faculty. The university, he said, "still needs more people with a youthful sense and fresh energy, full of real inspiration for the fields they represent and without fear of attack from people of the right or the left." He thought the recent appointment of Zeller had helped in this regard and that Treitschke could help too. He said that Treitschke "had . . . a mission to fulfill [at Berlin], also in regards to the first *university* of the Prussian state." The faculty committee, including Droysen, voted unanimously for Treitschke, indicating just how much change was desired. As for income, Helmholtz said Treitschke would earn at least six thousand talers. "That's also my income," he revealed. That would have put him in the top 2 percent or so of earners, though Helmholtz claimed not to feel wealthy and that they lived no better than they had in Heidelberg. "One's time and energy are greatly called upon, yet it's a lively activity, much of it not very beautiful but nonetheless against a great background and with a feeling that the effort is not fruitless." Treitschke visited the new minister of culture, Paul Ludwig Adalbert Falk, in Berlin, staying three days with the Helmholtzes. Then he decided for Berlin. The decision was reinforced when he learned that Helmholtz had been selected as the new dean of the Philosophical Faculty.[16]

Helmholtz also played a pivotal role in retaining Mommsen, who was called to Leipzig in 1873. The two had become very good, indeed close friends. Mommsen was not only a philologist, an ancient historian, and a jurist, but also a member of the Prussian legislature. As a professor and as a political figure, he was a major representative of Prussian and German liberalism. After he was called to Leipzig, Helmholtz told him that his leaving Berlin would endanger the plans for completing the *Monumenta Germaniae Historica* (a set of source materials for studying German history, begun under the academy's auspices in 1872) and, furthermore, that he thought the Leipzig faculty or the Saxonian Ministry of Culture may have intrigued against Mommsen, tricking him into a verbal promise but not actually making him a written offer. Mommsen doubted the latter but told Helmholtz how much he valued his friendship. Helmholtz replied that Mommsen's leaving would be a sad day for Berlin.[17] Mommsen stayed.

On Ideals in Culture, History, and Science: The Magnus *Eloge*

Culture had long been at the heart of German national identity, and the notions of *Kulturnation* or *Kulturstaat* had long preceded German political unity. The state was the dominant patron of culture, especially for the arts and sciences. *Kultur* was institutionalized in good measure in the Gymnasia and the universities, thus making the academic elite (especially the Berlin professoriate) of exceptional importance. State schools and universities, through their implementation of *Wissenschaft* and *Bildung*, led the way in fostering *Kultur*, thereby providing a national identity to the otherwise diverse German peoples.[18]

The academies of science, especially the one at Berlin, played their cultural roles largely by promoting research.[19] As an "ordinary" member of the Prussian Academy of Sciences, Helmholtz held full status in Germany's most prestigious scientific institution, one rivaled only by the Royal Society of London and the Académie des Sciences in Paris.

The academy was "protected" by the German emperor and Prussian king, while it operated under the aegis of the Prussian Ministry of Culture. These formalities aside, it was a self-operating institution, which meant that Helmholtz and his fellow ordinary members had to devote considerable time to it. It consisted of two classes—the physical-mathematical and the philosophical-historical—of equal status and administered by four permanent cosecretaries (two per class). Within each class there were twenty-five ordinary memberships, reserved (informally) for the various scientific disciplines. Below these fifty ordinary memberships, which together constituted the full academy (the *Plenum*), there were three other types: foreign, honorary, and corresponding. Foreign members (sixteen per class) were individuals who lived outside the Berlin region and had been invited to join on the basis of their distinguished scientific work (or connections); corresponding members (one hundred per class) were individuals who also lived outside the Berlin region and with whom the academy wanted to maintain a formal relationship. The academy's scientific work consisted of that done by its members and their associates; it maintained no scientific facilities of its own. Its quarters, along with those of the Academy of Arts, were located in a building on Unter den Linden, next to the university.

The full academy met once a week for thirty-nine weeks per year. (Helmholtz regularly attended these meetings.) Each ordinary member was expected to read one finished piece of work per year before the full academy, which then published it in one of its several outlets (the *Monatsberichte*, the *Abhandlungen*, or the *Denkschriften*). Each class held one meeting per month, except during vacation time. Each ordinary member was also expected to read a paper before his class, though these did not have to be publication-ready. Furthermore, the full academy held three annual special sessions to honor its patron saints: a session (on 24 January) to celebrate the birthday of its "renewer," Frederick II (the Great); the Leibniz session (circa 1 July), devoted to honoring the founder, to holding inaugural addresses by the ordinary members elected during the previous year, to reading memorial addresses for deceased members during the past year, and to awarding prizes; and a session to celebrate the reigning Prussian monarch's birthday. (Helmholtz regularly attended these three meetings, too.) Finally, the academy financially supported its members' or others' work, or work that required more than one scientist to conduct, or work that lasted for a more-or-less extended period. Every two years it also announced a prize question; this alternated between the two classes and within each class between its two sections. All in all, the academy represented or symbolized the unity of the sciences, or at least expressed the hope and longing for such unity, an aspiration not shared by its counterparts in Paris and London.[20]

On 6 July 1871, at the academy's annual Leibniz session, Helmholtz gave his memorial address on Gustav Magnus. As Magnus's successor, it was his duty to do so and to speak of him as a physicist (Hofmann spoke of him as a chemist). Though Helmholtz did not say so, Magnus's importance to science as a whole and to physics in particular lay less in his own physics results than in the way he supported and organized others, especially younger scientists. Helmholtz's address was, in fact, more about Helmholtz's ideal man of science, about German history and its affect on science, and about the status and relations of mathematical and experimental physics than about Magnus. While he referred to himself as Magnus's "thankful student" and his friend and successor, in fact he had never studied with him and had spent only three months working in his laboratory as a "postdoctoral" researcher.[21] Rather, Helmholtz apparently had something else to say: he valued a certain attitude toward science and the world that he saw in Magnus.

In discussing Magnus's family background—he noted that Magnus issued from a commercial family but did not mention its Jewish ancestry—Helmholtz evinced great respect toward business and the practical uses of science. This background led Magnus to view science not as an activity restricted to the lecture hall and the laboratory, but as something "that reaches out directly to all relations of life." Helmholtz referred approvingly to Magnus's "lively interest in technology" and his "keen participation in the work of the Landes-Oekonomie-Collegium."[22] He admired men like Magnus, Siemens, and Graefe who could relate theory and practice to each other.

Helmholtz portrayed Magnus as "ruled by prudence," as a man "of artist-like harmony, who avoided the immoderate and the impure," and of "moral and intellectual tact." He achieved most of his goals because he chose those that were achievable. His inner harmony manifested itself in his external person: in his charming behavior, his self-confidence, his clarity, and his human warmth.[23]

Magnus's family had prospered in business, but Magnus had chosen an academic career. He understood "that the comfortable enjoyment of a carefree existence and socializing in the most charming circles of members and friends did not give long-term satisfaction, but rather that only work, and indeed only disinterested work for an ideal goal, did so." He did not work to increase his own riches, "but only for science; not in a dilettantish and ill-humored way, but rather toward a solid goal and indefatigably. He realized that his name could not be made famous quickly through the vain snatching after astonishing discoveries; but rather, and quite the contrary, he became a master of a true, patient, and modest work, constantly testing it and not ceasing until he knew that it could not be improved upon." Helmholtz portrayed Magnus as a methodical, precise, and trustworthy worker.[24]

Magnus's character of "purity and disinterestedness" enhanced his ability to attract young scientists to his private laboratory, Helmholtz said, and to help keep them on their chosen paths. Once he recognized their "keenness for and ability to do science," he opened his laboratory to them and put his instruments and equipment at their disposal. He worried neither about competition nor about the occasional, unwitting damage done by young, unskilled experimenters to his expensive instruments. Nor did Magnus seek to exploit a young scientist's work for himself, as some foreign scientists did. He wanted only the best and most trustworthy results.[25] Helmholtz admired him for all these qualities.

Helmholtz also used this occasion to voice his views on German intellectual and political history. The sciences in Germany, he said, had developed rather "late and hesitantly." As he had claimed previously in Innsbruck, he said again, "[German researchers] are praiseworthy for their passionate, relentless, and disinterested love of truth, one that does not stop before any authority and before any pretext, and that is very unassuming in its demands vis-à-vis external success." Hence the Germans were always, in Helmholtz's view, in search of deep principles and pursued things to their end. They were, in contrast to what he had just said about Magnus, little concerned with the practical ramifications of their work or applying their results so as to yield applications in the commercial world. The long-term development of German intellectual life, Helmholtz maintained, had begun in the political context of the Reformation, where theology had been the principal area of study. Germany had then "liberated" Europe from domination by the Roman Catholic Church, but it did so at a high price. As a result of the Wars of Religion, the German states had been left in physical, political, and commercial shambles, "abandoned without defense," before an "arrogant" neighbor. The struggles against Rome had also injured German intellectual life, and as a result of the dangers faced by the German states, it became impossible for individual Germans to express dissent openly. That had not been a moment to do science, he claimed. But "the German spirit" could not remain satisfied with such a situation. German intellectuals, therefore, had devoted themselves principally to theology, using classical philology and philosophy as tools to clarify "moral, aesthetic, and metaphysical problems." "Criticism of the sources of knowledge had to be undertaken, and this was done with much deeper seriousness than before." The German Enlightenment brought extraordinary results, but this too revealed that "metaphysics has, as is undeniable, a dangerous attraction for the German mind. It could not leave metaphysics again until it had investigated all its hiding-places and convinced itself that, for now, there was nothing more to be found there." After 1750, he further maintained, German artistic life and the German language began to flourish. But as soon as the impoverished German lands and life had begun to recover from the devastation wrought by the Wars of Religion and to hope for a better future, the Napoleonic Wars came upon them. Out of this political and military state of affairs came, in Helmholtz's tragic rendering, profound intellectual consequences. In comparison to "the great conceptions of philosophies and poets," the routine work of German natural scientists (astronomers excepted) seemed small-minded; German intellectuals in the pre-Napoleonic era had shown "moral force": they had helped "break the Napoleonic yoke," and their successors benefited from "the great poetical works, which are the noblest treasure of our nation." Yet they had also bequeathed a sense of unreality, "and individuals, like nations, that want to develop themselves into mature beings, must learn to look reality in the face in order to bend it to the purposes of the mind." This intellectual inheritance had committed "mistakes" concerning the "mind's orientation." To be specific, German romanticism and Hegelianism amounted to "a sentimental hash for nobility and inspiration." Its effects were felt throughout German intellectual life, including the natural sciences, until "one understood that one must first learn the facts before one can erect . . . laws."[26]

This brought Helmholtz back to Magnus, who had taken up the fight on behalf of empiricism and against speculation. He fought a double fight: for physics (that is, the foundation of science) and against speculation, of which the well-attended University of Berlin had been a "self-controlled fortress." Magnus "preached continuously to his students" that to understand nature, one had to observe and to experiment, not merely to reason, however plausibly. Magnus's research in biochemistry (that is, blood gases) was effectively an attack upon vitalism, and his research in "organic physics," in particular that concerning organic metabolism, "laid the scientific foundation for the correct theory of breathing," which others then built upon.[27]

Helmholtz's only criticism of Magnus as a scientist was his skepticism toward and superficial knowledge of mathematical physics, a skepticism born of Magnus's distrust of speculation. Magnus had gone too far in this regard, he thought. Yet Helmholtz conceded that a generation earlier mathematical physics was based on a confusing mixture of empirical facts, semantics, and hypotheses, so Magnus's attitude toward it was perhaps understandable. Helmholtz expressed his own skepticism toward hypotheses about the atomic nature of matter, not because he wanted to argue against them per se, "but only against striving to derive the foundations of theoretical physics from purely hypothetical assumptions about the atomic structure of natural bodies." Even mathematical physics, he said, was now understood as "a science of pure experience," a point that held for Thomson, Maxwell, Kirchhoff, and others, but certainly not for everyone. He continued:

> We find in direct experience before us only extended, manifoldly shaped and composed bodies. We can make our observations and experiments only on such bodies. Their effects are composed from the effects which all their parts contribute to the sum of the whole. And if we thereby recognize the simplest and most general laws of operation of masses and matter upon one another as found in nature, and if we want to free precisely these laws from the accidents of shape, size, and position of the combined bodies, then we must return to the law of operation of the smallest volume parts, or, as the mathematicians characterize it, the volume element. These are not, however, like the atoms, disparate and varied, but rather are continuous and similar.

The discussion revealed that his ideal physicists were Gauss, Neumann (and his pupils, including Kirchhoff), Faraday, Stokes, Thomson, and Maxwell. More than that, it marked a moment when he in effect announced his abandonment of the physics of force for energy physics. In his view, mathematical physics was based as much on empirical phenomena as was experimental physics. In principle, there was little if any difference between the two: "The mathematical continues the business of experimental physics," he asserted, "in order to discover ever simpler and more general laws of phenomena." To do modern experimental physics required an understanding and use of theory, just as to do modern theoretical physics required a similar knowledge of experiment. That was how his own work in physics had proceeded, he said, and even though Magnus had opposed mathematical physics, Helmholtz had "nonetheless always found in him the most willing and friendly recognition." Scientists agree that "the task of science is to find the laws of the facts." Some preferred to theorize about possible new laws on the basis of known facts; others to seek new facts in order to discern new laws. Helmholtz opposed both empty theorizing and an "exaggerated empiricism" that had no interest in discovering new laws.[28]

Magnus, he concluded, had what most men wanted in life (that is, money and material goods). "But he knew to dignify such external goods by placing them in the service of a disinterested purpose. What is dearest to the soul of a noble person — to warm oneself at the center of a charming family, and within a circle of true and important friends — was granted to him. I want, however, to praise the very rare luck that he had of working in pure inspiration for an ideal principle, and [to declare] that he saw the cause he served grow victoriously and to unimagined riches, and into widespread, effective prosperity." That was self-description, or at least a description of the ideal self that Helmholtz hoped to be. His address, in short, was less about Magnus than about himself: his ideal scientist, his heroic view of German history, and his ideals of physics. As the philosopher Wilhelm Dilthey, who heard Helmholtz's address in person, later rightly declared, Helmholtz was "like the embodiment of the natural scientific spirit of the times."[29]

Cruising in the Scottish Isles

Following his first semester at Berlin, Helmholtz needed much rest and relaxation. In March 1871 Thomson, the president-elect of the BAAS, had invited him to attend the society's meeting in Edinburgh and afterward to enjoy a cruise on his new schooner yacht, the *Lalla Rookh*, for several weeks along the West Highlands and the Hebrides. To further enhance the invitation, both the British physiologist and pathologist John Hughes Bennett and the organic chemist and Tait's brother-in-law Alexander Crum Brown invited Helmholtz to stay in their respective Edinburgh homes during the meeting. Helmholtz decided not to attend the meeting, but he did want to go sailing with Thomson, who was delighted and noted that perhaps Helmholtz could "mix a little work" in with the cruise. All of this irritated Tait, who feared he would again fail to meet Helmholtz and who hoped to teach him "the mysteries of golf" at St. Andrews.[30]

In early August, Helmholtz and his family went to the seashore, near Kiel, but at midmonth he left them for Thomson and Scotland. He found the Edinburgh hillsides and St. Andrews's bay most impressive: the area offered a "great life of bathing guests, elegant women and children, and gentlemen who play golf in sport outfits." (He was also astounded by men bathing nude in St. Andrews.) He dined with Tait, with the Belfast chemist Andrews, and with Huxley, "noisy, pleasant, and interesting people." Andrews showed them several noteworthy experiments. Tait was as enamored of golf as Thomson of sailing, and he got Helmholtz to try a round; the results were as might be expected.[31]

On 24 August, Helmholtz joined Thomson in Glasgow. He was quite impressed with the new University of Glasgow but found Thomson's new home, in comparison with the old one where he and the late Mrs. Thomson had lived, rather sad. There was a lovely portrait of her on a dining room wall, and it reminded him of his feelings for Olga: "I had to hold back the tears. It is very sad when men lose their wives and their lives are devastated," he told Anna.

He visited the Duke of Argyle's "splendid park" with its "magnificent water-fall" and saw a fireworks display at the duke's castle. Then they went sailing on the *Lalla Rookh*. On board ship with Thomson and Helmholtz were, at various times, Thomson's brother James, Crum Brown, and others (but not the other original invitees: Maxwell, Huxley, Tyndall, and Tait). Helmholtz had little or no fear of sailing, even in stormy, rainy conditions. He saw his first northern lights while on board ship here. While sailing between Scotland and Ireland, he read Charles Kingsley's *Hypathia* (1853), "an artificial, botched piece of work," and he and Thomson discussed the theory of waves, "which he also most prefers to treat as a sort of 'race' between the two of us." Thomson's competitiveness astounded him. When Thomson went ashore by himself at Inveraray, he apparently warned Helmholtz before departing: "Now, mind, Helmholtz, you're not to work at waves while I'm away."[32]

Thomson and company spent several hours in Belfast Lough before sailing along Scotland's Atlantic coast to the Isle of Skye. Helmholtz felt increasingly comfortable on board the *Lalla Rookh* and was impressed with the speed, maneuverability, and control of all the yachts he saw. In Belfast they visited Andrews and his laboratory, saw a regatta, and dined with James Thomson, a professor of engineering there. They were all, Helmholtz said, "sharp people." Northern Ireland, he noted, was ruled by the Orangemen (a Protestant Irish order), who "if possible are even more Prussian than the Prussians themselves." They hoped that Prussia would establish a new empire, like that of Charlemagne, "in order to keep the Celts and Slavs down and to save civilization." The Scots, he found, spoke even more harshly of the French than the Germans did: they considered them "like savages whose civilization has been completely lost."[33]

They spent a day at the land house of a high-ranking British diplomat and politician, Lord Dufferin (Frederick Temple Hamilton-Temple-Blackwood), at Oban, on the west coast of Scotland. Helmholtz thought Lady Dufferin "unusually beautiful"; her mother, Lady Hamilton, had contacts with the French court, which meant she was not exactly pro-German. Lord Dufferin's house was stocked with Egyptian and Asian pictures and other objets d'art from Egypt and Asia. Dufferin showed it all off to Helmholtz and company, so that he could "get acquainted with everything." He added, "Because I didn't know so well the possibilities of a noble English [*sic*] house, it was all very interesting to me." On Sunday, Helmholtz was forced to attend Anglican services in Dufferin's personal chapel. "It was, however, only a liturgy, without sermon."[34]

From on board the *Lalla Rookh*, he described the bleak mountains and sparse vegetation on the voyage between the island of Skye and mainland Scotland. They sailed to Tobermory and took a walk on the Isle of Mull. They hurried to the countryside estate of Hugh and Jemima Blackburn in Roshven, which was surrounded by "the loneliest mountains." Blackburn was Thomson's closest friend and a professor of mathematics at Glasgow, where he had succeeded to the chair previously held by Thomson's father. Helmholtz found that Jemima, one of the best-known illustrators of nature (especially birds) in Victorian Britain and a first cousin to Maxwell, had a special talent for painting animals: "[She] sympathizes with the animals in all their habits and modes of life." Many of her paintings had been reproduced as lithographs, and some had been exhibited in London. To indicate how much at home Thomson felt there, he said, "He always carries around with him his mathematical notebook, and as soon as something occurs to him, right there, in the middle of a group of people, he begins to calculate, which people here generally regard with a certain awe. How would it be," he asked sarcastically, "if I too got the Berliners used to such a thing!" In the Sound of Mull, while on board the *Lalla Rookh*, Helmholtz and Thomson conducted experiments — they hung a fishing line a few feet into the water — to determine the minimum velocity of wave propagation of ripples on the water's surface, a topic on which Thomson had been working for some time and on which he eventually published. While they experimented, small whales came near their ship, which gave them something (else) to talk about.[35]

Despite all the pleasures of sailing, partying, sightseeing, and even experimenting, Helmholtz eventually felt lonely without the company of a woman. He wrote Anna: "I find that, after a while, a husband who is [especially] no longer young doesn't feel well when he lets himself wander around the world without higher supervision, and that the world would probably not present much beauty if it were peopled only with men who wish only to be very practical and unedifying." He had had enough. When they reached Glasgow in a few days, he left Thomson and headed back to Anna.[36]

The cruise and the trip in general proved most beneficial to Helmholtz. He had met several British men of science, including the chemist John Ferguson (soon to become Regius Professor of Chemistry at the University of Glasgow) and the young chemist John Millar Thomson as well as the Strutt family. The cruise had benefited his health, too: "I began my lectures a week ago," he wrote Thomson in early November, "and I find myself much more comfortable at present, than during the summer. I have time to work for myself."[37]

18

The Burdens of Building Physics

In the Provisional Institute

Helmholtz was no longer a physicist manqué, and he liked that feeling. With him, Berlin now had a theoretical and an experimental physicist in one. His appointment reflected the rising importance of theoretical physics in Germany. He thought the theoretical physicist must understand experiment and the experimentalist must understand theory.[1]

At Berlin as elsewhere, however, the experimental side dominated the daily reality of the physics institute. Helmholtz taught classes in introductory physics to large numbers of first- and second-year students who had little or no interest in theoretical physics. While he propounded the principles of physics in class, their emphasis nonetheless lay decidedly on experiment, both as illustration by him in the lecture hall and as practiced by the students in the laboratory. Even the most advanced students, as well as his research associates at Berlin (as elsewhere), largely pursued experimental, not theoretical, physics. To meet these needs, Helmholtz needed a large and well-equipped physics institute, and it took him seven years to get it.

The Berlin Physics Institute began life in embryonic form in the early 1840s, under Magnus's direction, with his financial support, and in his home near the university. As a man of some means, Magnus bought the instruments himself and formed a cabinet. He later donated or sold some of these instruments to the state, ultimately bequeathing his own collection, apparatus, and library to the university. These items together formed the nucleus of Helmholtz's pre-1878 institute. In the year between Magnus's death and Helmholtz's arrival, Dove, the now elderly experimental physicist and distinguished meteorologist, ran the institute on a temporary basis, while Quincke, who was professor of physics at the Trade Academy (Gewerbeakademie) in Charlottenburg, ran the colloquium. Dove supervised the transfer of Magnus's cabinet and books to the east wing of the University Building, where Helmholtz's provisional institute occupied parts of two floors facing onto Unter den Linden. This provisional institute consisted of an auditorium, instrument-collection rooms, and a library on the ground floor and six laboratory rooms on the first floor. There was also a room for Helmholtz and another for his assistant. The auditorium and the laboratory were far too small to accommodate student demand and the setup of all of Helmholtz's instruments.[2]

Helmholtz, his staff, and his students had to make do with these facilities — so limited that he had to do his own experiments in a corridor — until 1878. He lost much time and sometimes got exhausted from simply moving instruments and apparatus around before and after each lecture, not to mention the long walk between his home and the institute. Anna thought the auditorium was "atrocious." Already in 1875 there were twice as many laboratory applicants as spaces available. Yet at least one student, Arthur Schuster, liked the close quarters. A native of Frankfurt am Main, Schuster was an English "postdoctoral" researcher in Berlin in the autumn of 1874; he went on to a long and distinguished career in British physics. He noted that the research students worked together at tables in small quarters, and he thought this brought them closer and enhanced their experience. Helmholtz, he said, spent an hour a day with his advanced students, discussing their (and his) work with them and generally giving advice.[3]

To run the institute, Helmholtz needed help and money, and the need in both categories kept increasing. Except for Friedrich Neesen, he found his assistants from among his advanced students. During the early to mid 1870s, his assistants included (beside Neesen) Heinrich Friedrich Weber, Paul Glan, Wilhelm Giese, Ernst Hagen, and Heinrich Kayser. Glan was the only one who represented a failure and a unique unpleasantness. In March 1878 Helmholtz found that Glan, who had been his assistant for seven years, was not attending to students' needs and had recently displayed "arbitrariness and negligence," and so he asked him to resign. When Glan refused to do so, Helmholtz fired him. As for the early institute budget, it provided for an attendant, two (soon three) assistants, and laboratory materials, as well as the rent for Helmholtz's residence near the Tiergarten. An assistant earned 300 to 500 talers annually, depending on whether he was the first, second, or third assistant. The total annual budget ranged between about 2,800 and 3,500 talers. By 1876–77, the institute—now consisting of the professor (Helmholtz) and his staff of six (the extraordinary professor, three assistants, an attendant, and a porter)—had a budget, not counting Helmholtz's and the extraordinary professor's salaries, that reached some 33,000 marks per year. There were now about 20 to 24 students who wanted to work in his laboratory but space and a budget enough for only 15. Helmholtz anticipated that once he moved into the new institute, student requests for laboratory places would increase, and therefore the need for assistants and an increased budget would, too. When the extraordinary professor (Poggendorff) died in 1877, Helmholtz sought a replacement. He felt overwhelmed: "Due to lecturing, supervision of the laboratory, directorial and faculty business, along with similar work for the Academy of Sciences, my time," he told the ministry, "is under such demand that I can only do scientific work in the (extremely limited) remaining periods of time."[4]

Politics and Traffic Vibrations

To understand why it took so long (seven years) to establish the new institute building, is to understand something of the intertwined bureaucratic and political problems that all entrepreneurial scientists in nineteenth-century Germany experienced.

The long gestation was due neither to ill will from the higher authorities nor (after 1871) to lack of money. William I showed much greater support for the advancement of the sciences than had his two immediate predecessors (Frederick William III, who had only infrequently supported scientific institutions, and Frederick William IV, who had cared more about art academies than scientific institutions). By 1870, moreover, many political leaders believed that the advancement of science and technology in both teaching and research terms would strengthen their states and societies economically and politically, and many Berlin scientists benefited from the establishment of the new Reich and its effort to turn Berlin into Germany's scientific and cultural capital. Broadly speaking, German victories on the battlefields between 1864 and 1871 (including France's indemnity payment of 5 billion francs, plus interest) helped pave the road that Helmholtz took to Berlin and facilitated the construction of his and other institutes in Berlin. Yet once there, Helmholtz confronted what turned out to be a frustrating seven-year struggle, first in seeking approval of the proposed site and the financing of his institute (1871–73) and then in overseeing its construction (1873–78).

Bureaucratic uncertainty, political turf battles, and concern about the possible effects of vibrations from outside traffic on delicate institute instruments caused the initial delays. In July 1871 the ministry asked Helmholtz to visit a proposed site near the university and assess its fitness for the new institute. By October he was preparing the plans for the institute, including his residence, to be presented to the Prussian legislature. Yet there was disarray in the ministry: in January 1872, Falk had replaced Mühler as the minister, and Falk needed time to master his portfolio. Moreover, the land had yet to be purchased. "Whatever is not military here is most miserable," Anna wrote Tyndall. Furthermore, the plans for the physics and the physiological institutes came into potential conflict with plans for a new central municipal tram system, which found its big backer in the Ministry of Trade. The close proximity of such a system, not to mention the institute's location in the heart of the city, meant that mechanical vibrations and electromagnetic fields originating from outside the institute might well affect its precision scientific work inside, something that Helmholtz and du Bois–Reymond wanted to avoid or minimize at all costs. To assess the potential hazards, du Bois–Reymond conducted underground galvanometric tests for vibrational constancy.[5]

Anna thought she might be able to expedite the land-acquisition process by speaking directly with Crown Princess Victoria about the institute's importance and so push the "dilatory ministries" into action, since Hermann, she said, would never push anyone. The university senate, for its part, appealed directly to the emperor, asking him to get the minister of agriculture to relinquish all land rights to the Ministry of Culture. If that did not work, both Helmholtz and du Bois–Reymond were prepared to tender their resignations, "because Hermann didn't come here to let himself be delayed for years by the Royal Prussian Ministry of Culture." That apparently did the trick, and du Bois–Reymond could proclaim, "Now Helmholtz and I are monarchs of all we survey." The two monarchs, with Paul Spieker, the government architect, then spent the autumn and winter of 1872 drawing up architectural plans. Helmholtz was so deeply involved in the planning that he even asked his friend Beetz to supply him with his plans for the storage cabinets in his Erlangen institute. Du Bois–Reymond promised to send Bence Jones in London a set of the construction plans so that he could show his friends that "after all [Prussians] are not so exclusively military as it may seem at a distance, and that some of the french Millions find their way into a scientific channel." Helmholtz and du Bois–Reymond also had to fight off a new plan by the tram company to install a viaduct that crossed the Spree and thus closely approached their institutes. In this, too, they were successful. Nevertheless, Helmholtz repeatedly told Anna that he would gladly return to Heidelberg if only he would not have to teach physiology. He found daily life in Berlin stressful and bemoaned losing so much time walking two to four times daily between his residence and his institute.[6]

The pressures from work were constant, and the stress affected Helmholtz's health, which remained a never-ending worry to him and Anna. In December 1871 she told her mother that Hermann had been so busy the previous day that he did not come home until midnight. Six months later she wrote in the same vein: "I hardly see Hermann—he has to be at three to four weekly meetings, comes home in the evenings dead tired, eats, and sleeps. It's truly a great hurry that can't be called a life." And so, as in the past, in between semesters he sought out southern Germany, Switzerland, Italy, and occasionally Austria for rest. He routinely spent a few weeks by himself (or with friends) hiking in the Swiss or Italian Alps and a few weeks with his family in Upper Bavaria (at Ambach on Lake Starnberg). In September 1872, for example, he recuperated in Thun. His health improved somewhat, though he still suffered from migraine headaches and heart palpitations: "He's depressed and discouraged," Anna told her aunt. He went to Geneva for several days to visit Auguste de la Rive and made excursions around the lake; he had hoped also to see his good friend Soret in Geneva, but Soret was away.[7] Nonetheless, Switzerland was good for him.

Once back in Berlin, his obligations and burdens resumed. He had neither mornings nor afternoons to himself. Examining students and meetings with colleagues also robbed him of his normal leisure hours. He at least managed to fend off the nonprofessionals, the "importune visitors with all kind of business, for the most part the greatest nonsense," as he told Thomson. "My life consists in making his few hours at home as peaceful as possible," Anna said, "never burdening him with daily matters." Overall, though, he had no regrets about coming to Berlin and taking on enormous responsibilities; he told Lipschitz: "I'm basically satisfied." He was glad to have some influence on students, too, and his health was actually better than it had been in Heidelberg.[8]

The Copley Medal and Other Honors of Science

In Britain, meanwhile, Thomson sought to secure the Copley Medal for Helmholtz.[9] The highest award given by the Royal Society of London, it was arguably Britain's most prestigious scientific award. Only a relatively small number of Germans had received it: Gauss (1838) was the first, followed by Liebig (1840), Ohm (1841), Humboldt (1852), Dove (1853), Müller (1854), Wilhelm Weber (1859), Plücker (1866), von Baer (1867), and Mayer (1871). Thomson's first effort to obtain it for Helmholtz came in October 1871, just four months after Helmholtz had led the way in obtaining a corresponding membership for him at the Academy of Sciences and less than two months after Helmholtz had gone cruising on his yacht. What could be more natural than old friends helping each other get awards? As Thomson wrote Stokes when asking for help in the Helmholtz matter: "The more I see of Helmholtz's work the more I value it, and I am very anxious that what I believe is the right award should be made in this case."[10]

But Thomson was disappointed: instead of Helmholtz, Mayer received the award that year for his work on the mechanical equivalent of heat. That award was, in effect, a response to the award's going to Joule in 1870 for his work on the very same subject. And the following year, another German, the chemist Friedrich Wöhler, received the award, leaving Helmholtz again out in the cold.

While the Copley went to others, Helmholtz received smaller recognitions from elsewhere. In April 1872 he was most pleased to be elected an honorary member of Heidelberg's Natural History and Medical Association. The award brought back fond memories of his years with that association. The same year, he was reelected a foreign honorary member of the Mathematical–Natural Scientific Class of the Imperial Academy of Sciences Vienna. (This was, however, a mere formality: in 1865 he had been elected a member from Baden, but now that he resided in Prussia, it was necessary, according to Vienna's regulations, for him to be reelected.)[11]

In late 1872 Thomson suggested to the Master of Peterhouse and the vice chancellor of the University of Cambridge that Helmholtz be invited to give the annual Rede Lecture ("to a cultivated audience") for 1873. They agreed, and Thomson invited him on the university's behalf. Helmholtz declined: he was too preoccupied with work and doubted that he could successfully speak sufficiently good English in such a lecture setting. The authorities then chose Tait, who told Helmholtz that he would "try to render justice as between Thomson & Clausius in the matter of the Second Law of Thermodynamics."[12]

Thomson also had another request of Helmholtz. He was leading a campaign to have his brother James, a long-serving professor of engineering at Queen's College, Belfast, appointed (as Rankine's successor) to the chair of engineering at Glasgow. He had already obtained letters of recommendation from Andrews, Tait, and Joule, and he was expecting one from Maxwell. He asked Helmholtz for one, too. He planned to take these letters to the Home Secretary, who would make the appointment. This was not the first time that Helmholtz had been asked to help with British engineering. In 1868 James Forrest, the secretary of the Institution of Civil Engineers in London, told him that they were gathering the latest information on the state of engineering education in various countries, and he asked Helmholtz what he knew about such education, especially in Germany.[13]

Helmholtz considered James Thomson, as he said in his letter, "a man of very acute judgment in questions relating to physical, mechanical, and mathematical science, and a very extended amount of knowledge in these same branches." He praised his theoretical work on changes due to pressure in the freezing temperature of water as being "an original idea of first-rate importance." He also highly valued his hydrodynamic and meteorologic studies. James Thomson scrutinized scientific issues deeply, he said; "I should think that such a man would be the very best teacher for young engineers." Two months later he was appointed to the chair of engineering at Glasgow. William Thomson thought Helmholtz's letter "must . . . have had more influence in promoting his appointment, than almost any other document put into the hands of Mr. Bruce, the Home Secretary."[14]

Then in September 1873, Tyndall hinted that the Copley might be forthcoming for Helmholtz. Two days before Helmholtz received the official notice from the society, Tyndall sent a cryptic telegram saying, "If you will come over I will make up Faraday[']s own bed for you." Helmholtz could only understand the message after he received the society's official communication, which said that he was to receive the Copley, for his "researches in Physics and Physiology," and thus not specifically and exclusively for his "discovery" of the law of conservation of energy. A few days later Tyndall wrote again to say that Helmholtz had won the medal and that he hoped Helmholtz could come to London to receive the award in person. Helmholtz told Tyndall that he was greatly honored to receive the Copley but that there could be no question of his traveling to London merely to receive it personally and to express his gratitude. He explained that even if he could manage to free up his work schedule for four or five days to make the trip, the combination of a rapid trip and the festivities in London would be more than his delicate health might allow.[15]

Instead, he wrote an effusive, very flattering letter to the society that, as he knew, would be read aloud at the Fellows' dinner. He wrote (in English), in part:

> I keep firmly written in my memory the pleasant remembrance of former
> times when I was so happy as to assist at your [Royal Society] meetings.
> I have felt deeply both the scientific weight of such a union of eminent men
> and the heartiness of the welcome with which the foreign guest was received.
> On these occasions I have learned to admire the organization of this Society,
> kept together solely by the love of science and the voluntary exertions of its
> fellows, governed solely by itself, the history of which, through a period of two
> hundred years, from Newton to Faraday, to the present moment, has been an
> uninterrupted sequence of glorious victories, which human intellect has won
> over the blind forces of Nature.

And he closed: "It is a prerogative of great men and great natures that they are free to give acknowledgment without a trace of [national] jealousy to whatsoever merits other men have attained. I wish that the Royal Society may continue to enjoy this prerogative, which ensures the highest and best fruits of international scientific intercourse." Given the society's past awards to foreigners, that sentiment was not false. At virtually the same moment, Helmholtz was also elected an Honorary Fellow of the Royal Medical and Chirurgical Society of

London. Six months later, he was again invited to England to receive an award, this time from Cambridge University; he again decided not to go. He explained that neither his current teaching duties nor his duties as dean of the Philosophical Faculty, which made him "responsible for a great amount of business not easily to be transferred upon anybody else," permitted him to leave Berlin for any length of time during the semester.[16]

There were still more honors during these first years in Berlin. He was elected a member of the American Philosophical Society in 1873. In 1875 the venerable but newly reconstituted Reale Accademia dei Lincei in Rome, seeking to boost its prestige by adding distinguished foreign members to its ranks, elected him and a series of other foreign scientists: the astronomer Otto Struve, the mathematician Michel Chasles, Hooker, the geologist James Dwight Dana, Bunsen, and Darwin. Helmholtz garnered more votes in physics than any other foreign physicist, including Kirchhoff, Clausius, Regnault, and Thomson, the other top vote-getters. The academy's president, Quintino Sella, a mineralogist and a leading Italian politician, thought Helmholtz was "among . . . living physicists perhaps the strongest head, the most original" thinker. The anticlerical Sella and other liberal leaders of the Italian Risorgimento sought to use and invoke science and technology to help build the new, post-1870 nation-state of Italy. The academy did, however, have a problem with its newly elected foreign scientists: how to get them to participate in its meetings in Rome. Three months later, while alone on his regular autumn vacation and to Anna's dismay, Helmholtz unexpectedly showed up in Rome. Sella paid a return visit to Berlin and invited Helmholtz to send the Lincei a manuscript for publication. The two men cultivated each other, and Helmholtz became one of the principal German scientists with whom Italian scientists came into contact after 1871. When his friend Beltrami, for example, informed him that a memorial statue for Volta was to be unveiled in Pavia in April 1878, Helmholtz let the Academy of Sciences know about this and asked if it should make some sort of gesture in response. The University of Pavia, in the person of Beltrami, invited him to attend the unveiling, and Beltrami especially noted Helmholtz's importance to advancing physics, a field that Volta had helped further through his invention of the battery, and Helmholtz's "goodwill towards Italy." In 1876 Helmholtz was also elected a foreign member of the Royal Danish Academy of Sciences and Letters.[17]

Finally, there were rewards and honors from his own university. In July 1873 Helmholtz agreed in principle to speak to a friend's local Verein while making a planned journey through western and southern Germany the following spring. (That past March, he had held public lectures at the Verein für wissenschaftliche Vorträge in Barmen and at the Erholungsgesellschaft in Aachen.) But a final decision, he said, would have to wait until he knew whether he would hold a university office the next year. If that was the case, then he would not be able to give the lecture.[18]

The decision came on Friday, 1 August 1873, a particularly busy day for Helmholtz. He oversaw the laying of the foundations of his physics institute; attended an important meeting at the academy, where Zeller lectured in detail on Socrates and Plato; sat in on several student examinations; and attended a faculty meeting where elections for the new university rector and dean of the Philosophical Faculty for the coming academic year (1873–74) were held. These were important positions: the university was formally governed by a board consisting of the rector (elected annually), the university judge, and the four deans (one from each faculty, also elected annually). Helmholtz, Weierstrass, and Mommsen were the finalists for the rectorship; Helmholtz came in third—Weierstrass first—but was then elected dean of the Philosophical Faculty, "which is indeed a great honor, but rather tedious [work]," he explained to Anna. She and the children were already away on vacation while he prepared his final class for the semester. He also read Zeller's *Kirche und Staat* (1873), which argued that the church should be subordinate to the state and indeed saw the church, in particular the Roman Catholic Church, as the state's enemy. "I found the book indeed interesting," Helmholtz wrote Anna, "despite its dealing with a subject that has been discussed a great deal. I've never read anything so well argued on this subject as this is." Along with the visit from Theodore Steinway and his technician to tune his piano, it had been a long day, and Helmholtz, for all the honors he had received that year, had not received the one he coveted most: the rectorship.[19]

As Dean and Construction Supervisor

In August 1873 Anna reported, "Hermann is almost never at home, since, in addition to everything else, this year he also always has some sort of dean's business to attend to." In October, as if to confirm her words, he informed a correspondent from Dortmund, "My time is . . . so taken [by duties as dean] that I have to decline all invitations to visit the Rhine cities next spring."[20]

Part of the "everything else" was overseeing construction of the new physics institute, which was in fact part of a large new university science complex. Along with physics, it included new institutes for physiology, pharmacology, chemistry (the second such), and technology. Physics and physiology, each of which included residences for their respective directors, were equal in size and by far the largest of this five-institute complex. The entire two-building complex occupied 7,763 square meters of terrain. It was within easy reach of Unter den Linden and so the university. The institute's front lay to the north, along the Spree, just across from the newly laid Reichstagufer; the river's dampness and its use by industry, together with the mechanical vibrations and electromagnetic fields generated by street traffic, made this a difficult locale for scientific institutes dedicated to experimental research, especially work that required precision measuring with sophisticated instruments.[21]

Helmholtz, du Bois–Reymond, and Spieker worked from spring 1873 to spring 1874 supervising the laying of the special foundations, which were needed not merely for the building's general support but also to ensure that the physics and the physiology institutes remained as free as possible from vibrations caused by the heavy amount of surrounding street traffic. All this special foundational and associated bridge work near the Spree slowed construction. Helmholtz told Tyndall on more than one occasion how annoyed he was by the slow progress. Moreover, the finance minister raised objections about the size of the residences for Helmholtz and du Bois–Reymond, which meant that their architectural plans had to be revised. Helmholtz complained to Tyndall that he "often regretted" that he had turned down the offer from Cambridge. (He managed to forget that he had signed the formal agreement to go to Berlin before receiving Thomson's informal offer of Cambridge.) He felt quite frustrated: "When one has only ten, or at the most twenty, working years left, and year after year time is wasted on silly and gross obstacles, in the course of time one can become very dejected." Anna shared his frustrations.[22]

Vacations became more than ever a necessity for him. Before leaving town in August 1873, he drafted an article on ballooning, dined with members of the Imperial Cholera Commission, and spent much time with (serious) people who wanted to talk with him. The Helmholtzes first went to Vienna. There they visited the International World Exhibition and the Stephansdom, toured the Ringstrasse, and saw a performance of *Maria Stuart* at the Stadttheater. Then they went to Engelberg, Switzerland, where he relaxed and followed his doctors' recommendation to drink a special, local mineral water, though he doubted its efficacy. Afterward, he traveled alone to Italy, in part to see Rome; even before leaving, he regretted doing so without Anna, and when he reached Florence, whose "charming" art treasures he had barely begun to absorb, he felt even more pained by her absence. By noon of his first day there he was "almost fainting from tiredness; it is more demanding than being outdoors crossing over an Alpine pass." He hoped she would come with him to Florence next year. "May your life in our northern homeland [i.e., Berlin], where you have followed me, be easier for you than it has been so far and may it be somewhat freer from problems," he wrote her. He visited Bologna, where he saw Beltrami. They had much to talk about, since Helmholtz's work in hydrodynamics had an important influence on Beltrami's own work in fluid mechanics.[23]

In the autumn of 1874, Helmholtz visited Venice and, again, Bologna and

Florence, this time with Anna. Brücke met them in Florence and found that Helmholtz looked well, but he reported that Anna was dissatisfied with the state of his health. She had consulted two doctors "on account of Hermann's fainting. Both warn about tiredness of the brain due to work—and also about keeping late hours and too little sleep." To further protect his health, she guarded him against all unnecessary social activities, since she could not stop the late-night committee meetings "and the official mad rush and overburdening with business." She sought to limit "vexations, concerns, and tiredness" in his life; but she was not very successful: five months later, while in Düsseldorf, he became ill— he suffered "a powerful poisoning"—and a local doctor telegraphed her to come at once. She feared for his life. By the time she arrived, however, he had largely recovered; she took him home. That autumn he again went to Engelberg to rest and drink the local mineral water, "which did him good," and again to climb in the Swiss Alps.[24]

During the early winter of 1874, workers completed the laying of the science complex's foundations and began construction of the above-ground structures. The Helmholtzes felt that the construction proceeded at a snail's pace—it lasted until 1878—and they continued to feel frustrated. Helmholtz longed to have "more free time . . . for scientific work." While he met many important and likable people in Berlin, he thought life there was "a mad rush" and that much of his time was simply wasted walking between his home and his provisional institute.[25]

Knapp, his friend and colleague in New York, invited him to lecture in America. While Helmholtz was intrigued by the prospect of visiting America, he could not do so until his institute was completed; besides, he felt exhausted: "The Berlin bustle always makes me very tired." By semester's end, he did not want to speak with anyone and, instead, sought to be alone with his own thoughts. "I still have something left that I want to do for science, and I can't lose all too much more time. So I'm beginning to believe that I will not go to America (in this life)."[26] On this last point, he proved to be mistaken.

Two years later, Knapp's American colleagues tried to bring Helmholtz there, this time for the International Congress of Ophthalmology in New York City. The congress offered to pay his travel expenses. His presence, the organizing committee of doctors said, would honor his many American friends:

We come to you in the *fraternity of science*, and feel that we appeal to the sentiment which has been the motive of your life. We do not ask you to come to witness *our* achievements in science:—it is not for us to speak in this matter; but we solicit your coming that American scientists may personally testify to you how warmly they appreciate your labors, and may exhibit their sense of indebtedness to your achievements. The Committee believe that you probably have little conception of the enthusiasm with which your coming among us would be greeted. There is no voice among workers in science, which would be heard with more pleasure than your own. Not only among the members and friends of the Congress would you be welcomed, but in all the enlightened circles of our society would you be most cordially received.[27]

That was certainly flattering.

William James, independently, simultaneously, and perhaps more effectively, confirmed the ophthalmologists' point: "Helmholtz . . . as well as any living man may claim to give voice to the scientific spirit," he said. But Helmholtz was too busy supervising the construction of his new institute to visit distant America. (Indeed, he even lacked the time and energy to lecture in Aachen.) He sent George Bancroft, the American historian and former envoy to Berlin (1867–74), a copy of volume three of his *Populäre wissenschaftliche Vorträge* — Bancroft had taken the first two volumes back with him to America — which had recently appeared: it contained his lectures on Magnus, on the origin and meaning of geometric axioms, on optics in painting, and on the origin of the planetary system. Bancroft asked if the ministry had finally built him the promised physics institute. "If they don't treat you better in Berlin," he wistfully wrote, "I wish it were within the compass of possibility to win you for my country."[28] American science was growing rapidly and becoming more important; but it was not (yet) important enough for the busy Berlin dean and construction supervisor to visit America.

The *Annalen der Physik und Chemie*

Helmholtz's becoming professor of physics at Berlin led to his close association with Germany's principal journal of physics, the *Annalen der Physik und Chemie*, still edited (since 1824) by Poggendorff. At first, his activities were limited to minor matters. In 1872, for example, he arranged for the publication of a paper on the aurora borealis (northern lights) by Auguste de la Rive. Similarly, after Johann Jakob Müller, the professor of technical physics at the Zurich Polytechnic and a former student of Helmholtz's at Heidelberg, died in 1875, Helmholtz arranged for the posthumous publication of a manuscript by him on electrodynamics. Heinrich Friedrich Weber, Helmholtz's assistant, became Müller's successor in 1875, and in 1896 he became Einstein's principal physics teacher there.[29]

Helmholtz also had occasion to act as the *Annalen*'s diplomat, a role that, thanks to his extensive travels in Italy and wide range of friendships in science there, came to him naturally. In February 1874 there was a celebration of the fiftieth anniversary of Poggendorff's editing of the *Annalen*. Several prominent

Berlin scientists held a party to celebrate the event. Their colleagues in Rome—Corrado Tommasi-Crudeli, Blaserna, Sella, and others—sent a congratulatory telegram to help celebrate the event but became offended when they received no response. Helmholtz admonished du Bois–Reymond, organizer of the celebration event and the telegram's recipient. He argued that the Germans were seen as "arrogant," giving grist "to the French and clerical party." He added, "The friends of German science have a difficult position vis-à-vis their enemies, and it's our duty to support them morally." He urged that Poggendorff and the others in Berlin promptly send a warm response thanking their Italian colleagues. Du Bois–Reymond promptly wrote a thank-you letter.[30]

When Poggendorff died in 1877, Gustav Wiedemann became the new editor of the *Annalen*. Wiedemann was an experimental physicist, a former student of Magnus's, and a good friend of Helmholtz's. He served successively as professor of physical chemistry (1871–87) and of physics (after 1887) at the University of Leipzig. Wiedemann edited the journal—and this was new—in collaboration with the Berlin Physical Society, in particular with Helmholtz. Helmholtz had come a long way: whereas once, in 1847, he had been an outsider whose manuscript had been rejected, he was now an insider who helped oversee the journal's direction and development. In 1878 the society moved into a designated room in the newly built institute, and Helmholtz became its new president, though he often let du Bois–Reymond (now the honorary president) run the biweekly meetings, where students and others presented their work. Moreover, because the society's *Fortschritte der Physik* had fallen so far behind its publishing schedule—at one point up to seven years!—in 1877 the society established the *Beiblätter zu den Annalen der Physik und Chemie* to ensure timely notice of recent work. Wiedemann edited the *Annalen* until his death (in 1899), and Helmholtz collaborated with him until his own death five years earlier. Wiedemann was responsible principally for experimental physics and Helmholtz for theoretical physics, a point that also reflected the gradual emergence of theoretical physics as a subdiscipline within physics as a whole. Under Helmholtz's leadership, the Physical Society became more devoted to physics *tout court* (though most presentations concerned experiment rather than theory) and shed much of its early multidisciplinary nature; physiology and electrotechnology, for example, formed their own societies and journals. Similarly, the *Annalen* became ever more devoted to physics alone.[31] All this involvement with the society and its several publishing venues gave Helmholtz still further institutional prominence.

Intellectual Standard-Bearer: Electrodynamics and Optics

As the institute's leader, as a leading figure in German science, Helmholtz set the intellectual standards and tone for the entire institute—even as he remained an intellectual loner. He neither joined nor formed a "school" of physics; even his earlier relationships with those who pursued physiology as "organic physics" were essentially a matter of informal assent to a general outlook, not of substantive intellectual contact or coordinated, team research. Nor did he tell his students what to do, though he sometimes gave them problems to work on. As Willy Wien, one of his most distinguished and last students, later noted: "Helmholtz had no tendency to stimulate work on any given topic." Rather, he let the students choose for themselves.[32] However, in the 1870s a clear preference emerged both in Helmholtz's own work and that of many of his students, namely electrodynamics.

On the other hand, through his various institutional positions, Helmholtz opened doors for others that led to professional recognition and advancement. As an academy member, he regularly arranged for the publication of manuscripts by other scientists in its *Monatsberichte* or its *Physikalische Abhandlungen*. Between 1871 and 1878 alone, he communicated about twenty-eight papers by others for publication in these journals. Included was work by both German and foreign scientists: among the former were Julius Bernstein, Eugen Goldstein, August Kundt, Emil Warburg, and H. F. Weber; among the latter, the Russians Nikolai Baxt, Sergei Lamanskii, and Wladimir Dobrowolskii; the Americans Henry A. Rowland and Elihu Root; and James Moser of Vienna. Helmholtz himself read about one paper per year before the academy and published thirteen papers in its *Monatsberichte* during this period. All this helped to strengthen the academy and the institute and made them more international.[33]

Between 1871 and 1878, he authored and published a series of substantial and influential articles on electrodynamics (and electromagnetism generally) and on optics (the theory of the microscope and the theory of anomalous dispersion); these articles were closely related to one another, in part because they were exploring the validity of Maxwell's approach. He also published an important (though popular) piece on atmospheric physics.

In the rapidly developing field of electrodynamics, Helmholtz became a leader in critically analyzing and synthesizing work on that topic, doing so in a way that highlighted Maxwell's own recent synthesis. Already at Heidelberg in 1869–70, he had sought to clarify the state of electrodynamics and had suggested that Maxwell's approach, as a limiting case of his own, was probably the most fruitful to pursue. He continued these efforts in Berlin.

His first Berlin publication on the topic—on the propagation speed of electrodynamic effects—appeared less than sixty days after he had moved to the capital. This fact, and the fact that he here exclusively reported experimental results, suggests that he almost certainly did this work in Heidelberg. He hoped that his latest results would help in deciding between Wilhelm Weber's, Carl Neumann's, and Maxwell's versions of electrodynamics. Settling this matter, he said, was "a matter of principal importance for the foundations of natural science."[34]

While he always appeared to definitely favor Maxwell's over Weber's and Neumann's approaches, he also recognized that the matter was far from closed. No single approach or tradition in electrodynamics—whether that of an individual or of a certain metaphysical underpinning—was alone persuasive. What ultimately prevailed was the ability to analyze and draw on different (sometimes antagonistic) approaches (for example, particle-based versus field-based), to communicate hard-won insights among a wide variety of philosophical approaches, and to communicate across national boundaries. In all these respects, Helmholtz distinguished himself, as can be seen especially clearly in three papers (1872–74) that helped establish a framework for electromagnetic theory.[35]

Building on his earlier analysis of 1868–71, Helmholtz sought to provide an exhaustive critical analysis of each of the three main approaches to electrodynamics. He concluded that Weber's work violated the law of conservation of energy and so led to a perpetuum mobile, while Neumann's led to energetic instabilities. Helmholtz's haughty tone, however, cost him dearly. Zöllner at Leipzig vigorously sought to defend his older colleague Weber and even suggested unpatriotic behavior on Helmholtz's part; Weber's colleagues at Göttingen, Ernst Schering and Eduard Riecke, also sought to defend their leader. The controversy ended inconclusively.[36]

Helmholtz did only a modest amount of experimental work, including some on galvanic polarization, whether of platinum or in gas-free liquids. He had hoped to carry out planned experimental work for deciding between Ampère's law of electrodynamics and his own potential law of electrodynamics (that is, his reinterpretation of Maxwell's theory). This last, he said, derived from Gauss's and Neumann's general potential law, "one of the most auspicious and most brilliant accomplishments of mathematical physics." More successfully, he conducted experimental work on polarization currents—and reported on similar work conducted by his former student Baxt, now in Moscow—aimed at deciding between the various theories of electrodynamics. Nonetheless, experimental physics was principally what his students and "postdocs" did. He reported on the experimental work of his American "postdoc" Rowland on the electromagnetic effect of electrical convection and of his American student Root on galvanism, both pieces of work being done in his laboratory. While he was pleased with Rowland's work, he thought its results did not allow for a decision among the various electrodynamic theories. Furthermore, and most importantly—and this was a profound exception to Wien's claim—he set experimental problems for his young student Heinrich Hertz to solve, and he presented him with an electrodynamic system (his version of Maxwell's) that in effect challenged Hertz to confirm it experimentally. Helmholtz's own work, however, became increasingly more that of the theoretical physicist. He could not yet say whether Maxwell's or his own version of electrodynamics could explain the behavior of open-current systems; all he could do was to say that (further) experimental work was needed to decide between them and to encourage others to take this up.[37] If his Maxwell-like views and system convinced no one at the time, they nonetheless set the stage for later Continental developments in electrodynamics, not least by Hendrik Antoon Lorentz and Hertz.

While Helmholtz's views and system were far less noticed in Britain, in 1874 he reviewed and summarized for a broad, English-reading audience the current state of understanding about electromagnetism and electrodynamics. Part of the occasion for his review—it appeared in two general, learned venues, *The Academy* (London) and the Smithsonian Institution's *Annual Report*—was the recent appearance of Maxwell's *Treatise on Electricity and Magnetism* (1873). Helmholtz wrote:

> [These two volumes] contain not only this new theory [i.e., Maxwell's], but a very complete, methodical, and clear exposition of all those parts of electric science which could be brought under precise theoretical conceptions, and be developed mathematically. He has done a great service by this work to every student of physics. . . . I consider his method of forming new theoretical conceptions, which are at the same time perfectly definite in their quantitative determination and yet as general as possible, and not more specified than is needed, or more than our present knowledge of the facts allows, as really a model of cautious scientific progress.[38]

For Helmholtz, Maxwell was a model physicist.

Several times in his career—with Mayer and the conservation of force (energy) in 1847, with Brücke and the ophthalmoscope in 1850 and the theory of accommodation in 1851, and with several mathematicians and non-Euclidean geometry in 1868—Helmholtz independently and more or less simultaneously made an important scientific discovery or invention or innovation that others also made or were about to make. Such a potential clash led him in the spring of 1871 to resolve with Wilhelm Weber that, whatever their differences on scientific matters or of priority, there should be no personal animosity between them.[39] He maintained a cordial, if somewhat distant, relationship with Weber. The issue of priority also emerged in regard to Helmholtz's and Ernst Abbe's more or less simultaneous work on the theory of the microscope in 1873–74. Analysis of their shared topic of investigation is interesting not only for its own sake, but also for the light it throws on the general emergence of applied physics and the dramatically increasing interaction of physics and technology.

Prior to 1873–74, it was almost exclusively physicists and astronomers who pursued optical theory; and in terms of application, they were interested in telescopes, not microscopes.[40] During the half century before 1870, as microscopy gradually emerged as a useful scientific tool, it was a combination of precision instrument makers and life and medical scientists who led the way in analyzing and improving microscopes. They increasingly turned to the microscope for research purposes and became more and more concerned with improving its imaging capabilities. They sought to understand the nature of the relationship between the images produced and the biological preparations from which they issued. In particular, as cell theory emerged in the late 1830s and the organic physicists advocated that organic life could and should be understood solely in terms of the laws of physics and chemistry, the microscope gradually began to be viewed as a crucial instrument both in pursuit of that general program and in studying cellular pathology. A series of botanists and others—Matthias Schleiden, Hugo von Mohl, Virchow, Hermann Schacht, Pieter Harting, Carl Nägeli, Simon Schwendener, and Leopold Dippel—discussed the microscope and its uses and emphasized methodological or epistemological issues concerning the validity and interpretation of the images produced. All these "theories" of the microscope were confined to discussion of its geometric optics. The microscope's physical optics lay beyond their interest and, for most, apparently beyond their understanding. By the late 1860s, the microscope was a well-accepted instrument in scientific and medical research as well as in teaching, and there were more than a few competing guidebooks explaining its nature and use. Then suddenly in 1873, there appeared papers by Abbe and Helmholtz offering a theory of the microscope.

When Abbe's paper appeared, he was a young, virtually unknown physicist and consultant for optical instrumentation in Jena. His dissertation under Weber at Göttingen in 1861 concerned experimental testing of the equivalence of heat and mechanical work; he had no expertise in optics at that point. In 1863 he began teaching at Jena. He was a poor lecturer, and though he succeeded in establishing a physics laboratory, he was quite handicapped by its small size and the poor quality of the few instruments that he had. In 1870 he published two papers—they were his only two to date—in the *Jenaische Zeitschrift für Medicin und Naturwissenschaft*; they were marketed to the Jena locals and designed to get him promoted to extraordinary professor, which occurred later that year. All the while, Abbe was developing a close relationship with Carl Zeiss, who owned a small and financially struggling mechanical and optical instrument-making firm in Jena. Zeiss made top-of-the-line microscopes, but he did so on a trial-and-error basis, without any scientific foundation, and he wanted to change his approach. However, he lacked sufficient scientific training to improve the quality of his microscopes and other optical products. He hired Abbe as a consultant to gain scientific help toward that end.

Abbe's professional situation and reputation in the early 1870s thus stood in stark contrast to Helmholtz's. Everyone knew Helmholtz, while few outside of the provincial city of Jena had ever heard of Abbe. Yet in the early 1870s Helmholtz and Abbe shared three related and similar interests that led them to a common concern with microscope theory and, in time, to advancing the state of scientific instrumentation. First, both were keenly interested in analyzing, creating, and using scientific instruments. Second, Helmholtz had to concern himself with providing instruments for his new Berlin physics institute. Abbe had similar concerns, though on a much smaller scale. Third, after the conclusion of the Franco-Prussian War, Helmholtz was appointed to a small but high-level Prussian commission devoted to improving the state of precision technology in Prussia. In mid-1872, the commission issued a report reviewing the state of precision technology in the kingdom. It called for greater institutional support for increasing the quantity and quality of instruments for both Prussia's scientists and its various industries. The commission's intervention was welcomed by the military, itself much concerned with the state of precision technology. (The military wanted to improve land-triangulation measurements and telegraphy.) In 1873–74 Helmholtz found himself on a new but related committee, this one calling for an institute for the advancement of scientific mechanics, an institute that would foster the creation and improvement of instruments that would not only be useful to physics, astronomy, chemistry, geodesy, and so on, but would also help to advance Prussian industry. For all these reasons, then, in 1873 Helmholtz and Abbe were (independently) primed to think about microscopes, among other instruments. This broad professional and industrial context drove both men to theorize about resolving power and microscope improvement.

In the spring of 1873, Abbe published a long study on the theory of the microscope and the perception of microscopic phenomena. It appeared in the *Archiv für mikroskopische Anatomie*, a journal read not by his fellow physicists but rather by microscopists and life and medical scientists.[41] The paper's general thrust was to show that microscopes could be built on the basis of theory. He argued that, to date, microscope construction had been a completely empirical, trial-and-error process, with only an occasional reflection on using theory. He thought that, under close workshop supervision, the technical difficulties of construction could be overcome and that precisely prescribed lenses and lens systems could be constructed. He claimed that he and Zeiss were doing just that in Zeiss's Jena workshop. Optical theory and practice at the Zeiss workshop, Abbe declared repeatedly, worked hand in hand.

The key to understanding microscope theory and construction, Abbe held, was to appreciate the importance of the size of the lens system's angular aperture. But though he maintained that a proper theory of the microscope required mathematical analysis of its dioptrics, he provided no such analysis. Nonetheless, what was essentially new in Abbe's analysis, and central to it, was the argument that microscopes to date had been built using the same laws of dioptrics as used in telescopes or photographic cameras; their makers had completely neglected to consider the decisive physical process of diffraction. Microscope makers needed not only to eliminate chromatic and spherical aberration by finding the right geometric optics for their instruments; they also needed to take into account the diffraction effects produced. This meant giving serious consideration to the importance of the angular aperture.

Abbe also analyzed the dioptric conditions of microscope performance. The entire theory and construction of the microscope, he argued, should be built around and use a theorem that he then neither named nor represented mathematically. That theorem, however, soon came to be called the sine condition: all rays issuing from an object must produce images with exactly the same magnification, no matter what their angle of incidence was. This was the basis, Abbe said, for a theory of image errors in the microscope. To yield maximum resolving power, he advocated the use of wide apertures and oblique illumination, so that the microscope could maximize the capture of diffracted light, all the while meeting his sine condition. Among his other conclusions was one of particular interest to his intended audience of microscope users: he believed that there was only an arbitrary similarity between an object and its image. The presence of diffraction effects, he argued, meant that an object's image did not stand in any constant relationship to the actual morphological shape of the object under examination. It also meant, he concluded, that there was a limit beyond which microscopic resolution could not go.

In October 1873, six months after Abbe's paper appeared, Helmholtz, unaware of Abbe's paper, published a two-page, preliminary study on the limits of the microscope's resolving power in the Prussian academy's *Monatsberichte*. Like Abbe, he pointed to the central importance of taking diffraction fringes into account. But in contrast to Abbe, he presented an equation for determining microscopically the size of the smallest discernible distance within an object.[42] Six months later, he published a much longer version of this paper in the *Annalen*. The respective publication venues of Abbe's and Helmholtz's papers, and their respective reputations, effectively determined that Helmholtz's analysis, not Abbe's, was the one that became best known to most physicists and others interested in this topic.

In his longer article, Helmholtz, like Abbe, recognized that most advances in microscopy were now achieved "only in very small and delayed individual steps"; furthermore, it was not understood why such advances were made. He also recognized the important work of many "practical opticians," adding that everyone realized the necessity of "a very large aperture." In contrast to Abbe, however, Helmholtz referred to the light-refraction and diffraction processes that occurred in the eye or its various parts. Perhaps more importantly, whereas Abbe constantly appealed to the practicality of his results, Helmholtz was concerned to represent his work's scientific generality. Helmholtz's paper, in contrast to Abbe's, was filled with equations and derivations, culminating in his finding that the smallest resolvable distance within an object was equal to the distance between diffraction fringes. The key to future advancement in microscopy, he argued, lay in understanding and minimizing diffraction. In a postscript, he reported that when he had completed his paper and was about to send it off, he became aware of Abbe's paper and read it. He thought his and Abbe's results were "for the most part" quite similar, although his alone gave a proof of the main theory concerning diffraction. That, and the fact that his paper was part of a volume celebrating the fiftieth anniversary of Poggendorff's editorship of the *Annalen*, justified the publication of his own paper, he thought.[43]

During the next two decades or so, Abbe's paper remained largely unknown, and nearly everyone—except Helmholtz himself—associated Helmholtz's name alone with microscope theory and the limits of the microscope's resolving power. Helmholtz never again published on the topic, while Abbe restricted his publications to discussions of various improvements to the microscopes that were made in his workshop. There is no evidence that his microscopes themselves actually ever embodied his theory or that microscopists were much interested in it (or in Helmholtz's version, either). Microscopists apparently had no interest in the issue of the limits of resolving power until, in the 1880s, Robert Koch's bacterial investigations with Zeiss microscopes drew their attention to it. After Koch's work, no one doubted the functional relationship between aperture and resolution, but microscopists never showed much interest in microscope theory per se; what did interest them was the quality and utility of Zeiss's microscopes.[44]

Though under the circumstances it might have been otherwise, Abbe and Helmholtz developed a cordial and mutually beneficial relationship after 1874. In May 1878, Helmholtz visited Abbe in Jena to offer him a custom-designed position as a special professor of optics at the University of Berlin. Abbe declined, but he continued to enjoy Helmholtz's support in his various projects. In the early 1880s, Abbe and his firm undertook the creation of a new sort of specialized glass for Zeiss microscopes and other optical products. In 1882 he formed a partnership with the glass chemist Otto Schott, their joint aim being to produce commercially viable glass of the highest quality, with the precise optical properties specified by Abbe. When Abbe needed a trained physicist to assist him at the Zeiss Werke, he asked Helmholtz to recommend someone. Helmholtz suggested Siegfried Czapski, who in 1884 had earned his degree under him. Czapski went on to become one of the leading figures at the Zeiss Werke during the coming decades.[45]

The limit of the resolving power of microscopes was not the only issue in optics that then interested Helmholtz. He also sought to understand the phenomenon of anomalous dispersion, where in certain absorbing bodies, the normal behavior of the curve generated by the refractive index versus the wavelength suddenly becomes anormal ("anomalous") near the absorption lines or bands in the medium's absorption spectrum. Between 1840 and 1871, this phenomenon had been repeatedly observed but not understood: in anomalous dispersion, the normal order of colors is somehow changed, and this was somehow dependent on some sort of selective absorption. Then in 1871–72, Wolfgang von Sellmeier proposed a theory to explain such dispersion. He suggested that the ether's vibrations led to vibrations of material oscillators and that this accounted for the observed anomalous dispersion. He also managed to deduce an empirical law describing anomalous dispersion in terms of absorption frequencies.[46]

Sellmeier's theory intrigued Helmholtz. First in shorter form in the *Monatsberichte* (1874), and then in the *Annalen* (1875), he remodeled, generally simplified, and mathematized Sellmeier's theory; he recast it in terms of mechanical equations of motion involving both absorption and dispersion. His theory, which presented a pair of equations capable of accounting for all observed results in matters of dispersion, later came to be of crucial importance for others working in physical optics; indeed, in 1893 it became one aspect of Helmholtz's own more general electromagnetic theory of optical dispersion. Thomson, who initially did not like Helmholtz's theory, calling it "beautiful" but "retrograde," later employed it in his famous Baltimore Lectures of 1884.[47]

There were thus strong structural parallels between Helmholtz's electrodynamic theory, the theory of the resolving power of the microscope, and his explanation of anomalous dispersion. All three displayed his careful and close reading of the contemporary physics literature and his ability to rethink problems first raised by others and then put them in mathematical form. These were the same abilities that he had first displayed in his essay on the conservation of force. They were all works in theoretical, not experimental, physics.

Atmospheric Physics as Public Science

The 1860s and 1870s marked a new phase in both ballooning and understanding the earth's atmosphere. Balloons were first used in warfare for reconnaissance and escape purposes during the American Civil War and the Franco-Prussian War. Far-sighted individuals began to imagine the possibility of flight for civilian and military purposes. In 1874, for example, Heinrich von Stephan, the head of the Imperial German Postal Service, supported the idea of air transportation for mail service, while the German military established an air transport division in Berlin. Both the postal service and the military became early backers of air transportation's possibilities.[48]

In the early 1870s, Helmholtz became involved in a minor way with the issue of guided ballooning and flight. His mountain hiking had long piqued his interest in the physics of the atmosphere and in the climate; his essay "On Ice and Glaciers" (1865) constituted his first published expression of that interest. He employed his hydrodynamic equations to try to understand the relationship between air resistance and the speed needed to initiate and sustain flight. The practical implications of such theoretical considerations were obvious, which was doubtless why the Prussian government asked him to investigate balloon steering and why he published his work on this subject in a journal devoted to commerce in Prussia. Though he spoke only of guided balloons (not airships or airplanes), he envisioned the possibility of motor-driven balloons. He also studied the relationship between animal size and muscle power and concluded that, based on their own muscle power, humans would be unable to fly by wings alone. While his work stimulated interest in the possibility of flight, his conclusions on the limitations of human muscle power led to the widespread and erroneous impression that he had denied the possibility of human flight. Nonetheless, the Prussian government appointed him (in 1875) chairman of a committee to test guided air travel. He reported on flight possibilities in March 1878 and did so again many years later (in 1894) before a commission charged with evaluating Ferdinand Graf von Zeppelin's ideas about flight.[49]

A similar mix of scientific interest and public service expressed itself in Helmholtz's essay on cyclones and thunderstorms, which appeared in the *Deutsche Rundschau* in 1875. The subject (including weather forecasting) was not unrelated to the potential of flight. In 1874 Julius Rodenberg became the founding editor of the *Deutsche Rundschau*, a journal intended to address a well-educated readership interested in high culture; it was to include explanations of specialist scientific results in a way that would stimulate intelligent discussion among such readers and bring enlightenment. Rodenberg, like Treitschke and his *Preussische Jahrbücher*, intended for his journal to serve the German-reading public as the *Revue des Deux Mondes* had long served the French public: as a journal of general culture, to bring the most interesting recent advances and understandings in the sciences and humanities to the generally educated reader and to be a forum for imaginative literature and the arts. Rodenberg persuaded Helmholtz, du Bois–Reymond, Zeller, and the historian Heinrich von Sybel to write essays for it. With these four "greats" on board, he said, he could get others as well. With a circulation of some nine thousand, it became probably the most serious high-brow magazine in Germany. Many of Imperial Germany's leading figures in intellectual and cultural life published essays in the *Deutsche Rundschau*.[50]

Helmholtz wanted to write an essay for Rodenberg, but he was simply too busy to do so. So many people wanted to speak with him, he said, that he lacked blocks of free time to write, so he could not guarantee Rodenberg that he could meet a deadline. A year later, however, he had something presentable to offer. He had originally planned to write on the influence of physical research on the theory of the earth's formation and its interior, but now, instead, he sent him the essay "Cyclones and Thunderstorms" (Wirbelstürme und Gewitter), which he had first delivered orally to an audience in Hamburg.[51]

On one level, Helmholtz's subject was the nature and formation of storms; but on another, deeper level, it concerned accounting for meteorologic phenomena in a lawlike manner and thus enhancing weather forecasting. He began gently and slyly, with a well-known verse from Goethe about the (supposedly) arbitrary nature of rain, one that Helmholtz had long retained because it struck at the intellectual shortcomings of meteorology, and hence of physics. While astronomers and physicists had discovered laws governing celestial phenomena, he noted, the behavior of such meteorologic phenomena as clouds, rain, and wind seemed capricious. The unpredictability and imprecision of knowledge of meteorologic phenomena stood in stark contrast to that of astronomical phenomena. Even though meteorologists understood various physical forces (for example, of hydrodynamics and thermodynamics) and had amassed knowledge of some empirical meteorologic results, he continued, these results remained inapplicable to local times and places. He thus considered meteorology's status as a science to be unsatisfactory. Yet he did not think, as Goethe's verse suggested, that this was a matter of chance overwhelming law. Rather, he sought to show that reason (that is, science) could replace ignorance, superstition, and skepticism in understanding meteorologic phenomena. While he recognized that in certain regions of the earth and in certain seasons of the year, the weather was indeed predictable, he thought the major shortcoming of meteorology was a lack of extensive, long-term observational data (for specific locations) on mean temperature, air pressure, rainfall, wind direction, and the like.[52]

Like others before him—the Thomson brothers, the American geophysicist William Ferrel, the German mathematician Theodor Reye, and the French mining engineering Henri Peslin, for example—Helmholtz became a key figure (and his essay a key text) in introducing thermodynamic thinking (specifically, adiabatic temperature change) into meteorology. He claimed that climatology (i.e., atmospheric physics) was more developed than meteorology; it was easier to explain the causes behind atmospheric data and their relationships to one another. He explained that the earth's radiation was distributed between the equator and the poles, creating a heat exchange with the atmosphere in which the earth lost much of its heat to the cooler atmosphere. The earth's air, he argued, is heated principally at the equator, where it expands the most and is lightest, while it is coldest, densest, and heaviest at the poles. This heat differential produces air circulation around the earth "which, on the ground, is overwhelmingly directed from the pole to the equator, and in the air, by contrast, from the equator to the pole." The earth's diurnal rotation on its axis further shaped the direction of the atmosphere's currents. In short, there were well-understood, general, physical (atmospheric) forces that largely created and, so to speak, stood behind the weather, which could be explained rationally.[53]

Helmholtz also discussed some particular better-known phenomena, including how the trade winds operate. He noted Dove's contribution to understanding the interactive relationship between the polar and the equatorial winds, and he analyzed the thermodynamic processes by which water circulated between the earth and its atmosphere. There were large regions of the earth, he said, where the circulation of the water and the air in the atmosphere occurred in a regular, lawlike manner. But there were also places where disturbances occurred more or less regularly. The tropics and their cyclones, he recognized, frequently brought changes of air circulation into the system. He praised his colleague Dove's leadership in advancing meteorology by having discovered (in the 1820s) the whirling nature of cyclones, which others later confirmed. But he did not mention that by the 1870s the elderly Dove's leadership had been eclipsed: that Dove denied the vortical nature of storms, or that Dove's explanation of the warm, dry foehn wind in the Alps as due largely to moisture brought from the tropics had been superseded by an explanation advanced by the Austrian Julius Hann, who saw the foehn as due to local adiabatic (thermodynamic) changes. (In fact, a decade earlier, Helmholtz himself had developed the essence of Hann's view in his 1865 lecture on ice and glaciers, but it was Hann's version that proved persuasive.) Pointing to the loss of life and property caused by cyclones, Helmholtz also hoped to lessen the anxiety they produced by offering a scientific explanation. And he hoped that development of a telegraphic system that reported regularly on approaching cyclones and other meteorologic events might provide an early-warning system.[54]

Helmholtz gave special attention to the recent and important meteorologic work of Theodor Reye of the University of Strasbourg, which, as a result of the Franco-Prussian War, had become a German university in 1871. Reye was interested in the issue of how variations in temperature that lead to weak changes in atmospheric pressure could nonetheless produce significant storms. He had developed the concept of "unstable equilibrium" (rapid changes in the atmosphere), employing recent developments (especially by Clausius) concerning the mechanical theory of heat to understand the role of change in pressure. Helmholtz used Reye's concept of "unstable equilibrium," his own theoretical results in hydrodynamics, and mechanics to explain the general nature, geometry, and motion of vortices in the atmosphere; he sought to explain both hurricanes and tornadoes in terms of vortices, improving and developing the so-called convective (or thermal) theory of cyclones. His inspiration to analyze the nature and causes of storms was due in part to his many walks in the Alps, but his analysis of the vortical nature of storms applied only to tropical storms, not to those at middle latitudes.[55]

Helmholtz concluded his essay by declaring that the great problem currently facing meteorology had to do less with theory than with acquiring precise and accurate data concerning (often very small) changes in temperature, humidity, wind direction, and so on for specific times and places. Reye's analysis showed that even slight variations in local meteorologic conditions could mean enormous atmospheric changes. Hence data concerning mere averages was inadequate for some of meteorology's most pressing needs, especially for advancing its aim of predicting the weather at a particular time and place. While Helmholtz conceded that chance would continue to play a role in meteorologic analysis, he thought it could be overcome. He concluded his essay in the spirit of Laplace, not of Goethe, by anticipating that someday a fast-working mind, given sufficient and precise data, and working with valid laws, would be able to predict meteorologic events. His convective (or thermal) theory of cyclones, though soon criticized by Hann, remained the dominant theory until it was replaced by the polar-front theory of cyclones in the twentieth century.[56]

The bringing together of Reye's, Clausius's, and his own results into a convective theory of storms was by now a classic instance of Helmholtz's synthesizing ability in science. Here in meteorology, as in ophthalmology, physics, psychology, chemistry, and mathematics, his importance if not insights arose in good measure through synthesizing the work of others, not least by using physical theory and mathematics. He dipped into meteorology again in the late 1880s when he returned to the problems and laws of atmospheric motion. He thought of himself no more as a meteorologist than as an ophthalmologist, a psychologist, a chemist, or a mathematician, even though professionals in these disciplines regarded him as one of their own if not as a leader. His essay helped make atmospheric physics (meteorology and climatology) a public science, and it furthered his position as Humboldt's successor: the authoritative German interpreter of science to the public.

The New Institute

Helmholtz's research and teaching responsibilities, his supervision of the new institute's construction, his leadership of the institute's colloquium (on Wednesdays), his duties and weekly meetings (on Thursdays) at the academy and (every second Friday evening) at the Physical Society—all this and more exhausted his energy and placed him under great stress. By the second half of the 1870s, he increasingly needed recuperation from work, not only at the end of the academic year but also between semesters. "My husband has more to do," Anna wrote an old friend, "than he can manage." She worried that he would become overburdened by all the demands on his time and the consequent effects on his health. Yet in April 1876 he used the semester break not to rest but to prepare a new edition of the *Tonempfindungen*. "And while I am writing," she continued, "he has already sounded out on the piano beside me the most astounding things." His need to work rarely ceased; he cut back only when he felt he absolutely had to. Not surprisingly, when Tyndall invited him to London, he had to refuse: "I overworked during the winter, and now I'm undergoing a medicinal regime, one that requires me to take the greatest amount of rest possible." He continued to hold his regular lectures but temporarily refrained from all other work and all social engagements. He walked in Berlin's forests and took some sun; as long as he rested, he felt well, he said. Brücke had heard "not good [news] about Helmholtz's health. . . . He's supposed to look bad, fainting again, showing intermittent pulse." Helmholtz told Soret that he had worked too much the previous spring and had begun to suffer from irregular heartbeat.[57]

That autumn (1876) he visited Franz Boll, a young German microscopic anatomist and physiologist, in Switzerland. Boll had been appointed to the new professorship in comparative anatomy and physiology at Sapienza University of Rome, but he was gravely ill with tuberculosis. The weather was cold, and Boll's health worsened. Helmholtz visited him twice daily and provided medical advice, which Boll followed. Boll had gotten into a dispute with Helmholtz's successor at Heidelberg, the physiological chemist Willy Kühne, over the discovery of a light-sensitive retinal pigment known as "visual purple" (rhodopsin) used to produce an optogram, or a visual image fixed on the visual purple (also known as a "photogram"). Helmholtz advised him not to be troubled by the matter. (Boll was first to discover the pigment, whereas Kühne invented the closely associated inscription process by means of his optogram.) Helmholtz claimed that "unprejudiced witnesses" were against Kühne. "I had thought him more intelligent." He helped both men publish their independent discovery of rhodopsin.[58]

In 1877 Helmholtz vacationed with his family at Ambach. For six weeks he had been suffering from "pain under the left heel bone," which he treated with warm footbaths. He was healthier than in the previous year, he reported, but his health was still not all that he hoped for. In Belalp, Switzerland, he stayed with the Tyndalls in their summer home and met up with Knapp and Rudolf Schelske, as well as with William Rutherford, the Edinburgh professor of physiology. From there he went to Zermatt for a climb alone before meeting Anna in Lucerne. Then they went to Italy for a month, visiting Stresa, Milan, La Spezia, Rome, and Florence. He had hoped to see Boll, Tommasi-Crudeli, and Blaserna in Rome, but both Anna and Hermann caught a touch of malarial fever, and so they left quickly for Fiesole (above Florence), where they recovered before returning to Berlin.[59]

In December 1876 the new institute residence was ready for him and his

Fig. 18.1 A front view of the Physics Institute at the University of Berlin, including Helmholtz's residence (*front right corner*) and the Technological Institute (*front left corner*). *Illustrierte Zeitung*, 29 December 1877, 532, in Staatsbibliothek Preussischer Kulturbesitz, Berlin.

family, though the classrooms and laboratories remained unfinished. The residence alone had a surface area of 442 square meters and, with furnishings, cost 315,000 marks.[60] The front door opened onto the Neue Wilhelmstrasse, one of Berlin's most exclusive neighborhoods. The Helmholtzes resided here for more than twelve years, until May 1889.

In early 1878, before the institute's construction was fully completed, Helmholtz and his staff began moving into their offices. The institute proper had four stories that covered a surface area of 1,350 square meters. Its central structure, which was in part constructed to allow optical investigations to be done on the top (second) floor, contained a large auditorium (seating 212 persons) and a small lecture hall (seating 60) solely for theoretical physics on the first floor. The second floor also contained two rooms for the library, one of which served as the permanent meeting room of the Berlin Physical Society. All told, the institute building cost 800,000 marks to construct; with equipment and furnishings, it reached a grand total of about 1,264,000 marks. (The construction of the entire science complex cost 4,500,000 marks.) Even before the institute was completed, it had become an object for sightseers.[61] See figure 18.1.

Helmholtz and his staff set up the laboratories and the library. He purchased many new instruments, especially precision instruments. For many of these he turned to Carl Bamberg, a Berlin precision instrument maker who had studied physics and astronomy at the university. In 1871, at the urging of Berlin scientists, Bamberg had established a mechanical instrument-making firm; it was felt that this way Berlin might not be so dependent on Parisian instrument makers. It was a propitious moment to set up shop, when many types of Prussian and Reich scientific and technological institutes, as well as military agencies, had money to spend and would be needing Bamberg's services. In 1873 Bamberg received written guarantees from leaders of the German Navy, the Army General Staff, the army's triangulation agency, and the Astronomical and Astrophysical Observatories, as well as Helmholtz's own institute, that they would give him contracts to supply them with precision instruments.[62]

Helmholtz spent most of 1878 more or less completely setting up the institute. That task forced him, for example, to decline the Chemical Society's invitation to give the Faraday Lecture that year. Instead, he was occupied with such trivial matters as finding out whether cleaning services were included in his rental arrangement with the state. He also arranged for a series of eight sculpted busts of physicists — the only known name being Faraday's — to be displayed around the institute. He intended the busts to serve as symbols of some of physics's greatest heroes and their achievements. An institute, he believed, was more than a place where lectures were heard, equations solved, observations and experiments performed; it was also a place to recall tradition and be inspired. To obtain the bust of Faraday, he asked Tyndall to have a copy made. The Royal Institution owned such a bust, and Tyndall fulfilled the request. The Institution sent an "extraordinarily beautiful bust" of Faraday, which Helmholtz placed in the large auditorium.[63]

The idea of such iconic busts was not unprecedented; in fact, the sculpting and purchase of busts of famous scientists was something of a tradition and was a step beyond the taking and sending of photographs of scientists, which itself had become a common practice among scientists after about 1850. In October 1872, to help celebrate Donders's twentieth jubilee as ordinary professor at Utrecht, the Leiden physiologist Adrian Heynsius prepared an album of pictures "of the greatest physiologists of our century," including Johannes Müller and Helmholtz. Helmholtz readily agreed to send him a photographic portrait of himself. But then, instead, Heynsius asked him to sit for a marble bust portrayal (and also arranged to get one of Müller made from past likenesses of him). They found a sculptor to do the bust for 800 talers (Müller's cost only 400). The busts were finished in time for the Donders celebration in Utrecht. Five years later, when du Bois–Reymond visited Utrecht, he went to Donders's home and saw the impressive (and expensive) bust of Helmholtz, which Donders kept in his study. The locals, he said, thought it "exorbitant."[64]

Helmholtz's disciplinary transition from professor of physiology to professor of physics, and his building of a brand new institute—easily the biggest, best-equipped, and perhaps even the most beautiful of its day—symbolized the increased status of physics by the 1870s. Edward L. Nichols, a young American physicist who studied with Helmholtz between 1876 and 1878, thought the institute was "a structure unsurpassed, probably, among physical laboratories of that time." Karl Pearson thought its laboratories, like the university as a whole, "princely." For many, the new institute was exemplary if not inspirational. Emperor Dom Pedro II of Brazil, a dilettante of scholarship and science and a promoter of education and science in his homeland, asked Helmholtz to send him a copy of the building's plans. When the organic chemist Adolphe Wurtz, a native of Germany and a senator and a member of L'Institut in Paris, visited the University of Berlin in 1882, he called Helmholtz's new institute "a monument." It radiated "an idea of the powerful means available to some of today's German professors." For Helmholtz it was also an enormous administrative responsibility, as reflected in its annual budget (for 1878) of 27,330 marks.[65]

Physics Teacher, Mentor, and Promoter

As the director of a large institute, Helmholtz had enormous responsibilities and demands on his time as teacher, mentor, and promoter of students and their work. For his efforts, in the mid-1870s he earned nearly 19,000 marks annually. This amount included 8,400 in salary from the university, 6,900 from the academy, and about 3,400 for other service (probably student fees). There were built-in raises over the years.[66] He was undoubtedly among the highest-paid university professors of the day.

He worked hard for this compensation. Since youth, he had arisen early in the day; his best working hours (at research) were in the morning. Though his daily routine changed in 1888—when he largely left teaching to become the head of the new Reichsanstalt—he usually devoted the midday hours to teaching. There followed an afternoon break, after which he returned to work. He devoted most evenings to relaxation and family life. Anna often complained that she lost him to his work or that when she did see him he was too tired or rushed to be good company.[67]

Between his first full semester of teaching at Berlin (summer 1871) and his last (winter 1888), he lectured daily (except Thursdays) from 12:00 to 1:00 p.m. and from 1:00 to 2:00 p.m. If he needed to get demonstration apparatus set up, he went downstairs to the lecture hall toward 11:00 a.m. and usually did the setup himself. He also had responsibility for laboratory exercises, which were held daily from 10:00 a.m. to 3:00 p.m. and on Saturdays to 1:00 p.m. At least one of his two (and sometimes three) courses per semester was devoted to experimental or theoretical physics in general; the experimental physics courses (mostly for beginners) drew larger audiences. But sometimes he taught more specialized courses: for example, the mathematical theory of electricity and magnetism, or mathematical optics, or mathematical acoustics, or the mathematical theory of electrodynamics. On a few occasions between 1871 and 1875, he taught courses dealing with some aspect of physiology or, more rarely, the logical principles of the experimental sciences. From the summer semester 1875 onward, however, he taught physics exclusively. As early as the summer semester of 1874, he taught a class on the mathematical theory of electrodynamics, which was also the title of his last announced but never-held class of the winter semester of 1894–95.[68] Unlike a modern professor at a research-oriented university, he received no sabbatical or research semester or teaching reduction through course buyout.

Strictly speaking, Helmholtz had been hired as Magnus's successor, that is, to replace an experimental physicist. As already noted, he was much more a theoretical than an experimental physicist, and he wanted a similarly oriented colleague at Berlin. He had his eye on Kirchhoff. He let Kirchhoff know that the ministry wanted him for Berlin, and he led the way in bringing him to Berlin and in making him an ordinary member of the academy. Kirchhoff was charged explicitly with teaching theoretical physics. Beginning in 1875, he gave a regular cycle of lectures on mechanics, optics, heat, and electricity and magnetism. His appointment, along with Helmholtz's, made Berlin into a genuine center of theoretical physics, perhaps the first such ever.[69]

The majority of the students in Helmholtz's introductory lectures on physics were registered in the Philosophical Faculty, though he also attracted a substantial number of students (roughly one-third of his students) from the Medical Faculty and occasionally even a few from law or theology. For seventeen years running, between 1871 and 1888, about four hundred to six hundred students attended these introductory physics lectures annually, or so he claimed. He not only lectured to them; he also supervised the laboratory (mostly through his assistants) and even examined some of them. About half of all his students were foreigners (especially Americans and Britons, but also a good number of Russians and small numbers from France, Italy, Belgium, the Netherlands, and Japan). Already in the first semesters after the new institute opened, all its available seating and laboratory places were taken. By the mid-1880s, students had to be turned away.[70] The institute thus embodied and symbolized an emerging feature of German academic life: the rise of mass education, in particular for the sciences and medicine.

Helmholtz almost certainly did not enjoy teaching large, introductory classes. On at least one occasion, he allowed Dove to assume his responsibilities for teaching the introductory class so that he could concentrate his own efforts on more advanced topics. (He had about fifty students in his other class that semester.) Anna complained about his teaching those "always very abstruse courses in higher mathematics and physics," since they "brought him strain but little money, though they gave him more pleasure than the lucrative things [that is, the introductory courses]." She added, "He has, in general, a very unfortunate tendency against profitable things, which is very beautiful in itself—but Mammon is also still of value in this expensive vale of tears."[71]

From midcentury onward, student physics laboratories gradually became an established part of physics courses, complementing the professor's lecturing. At Berlin under Helmholtz, the laboratories, especially the introductory general ones, were run by his assistants, though Helmholtz would walk around to see how his advanced students and "postdocs" were doing. By the late 1870s, he had three assistants per year. They sometimes helped him set up the illustrative experiments for his lectures in the main auditorium, but mostly they supervised the student laboratory work and, the easiest task, ran the library. They (not Helmholtz) had the vast majority of contact with the students who took only the introductory physics course. At the provisional institute, where space was extremely limited, there were only a few advanced students working in the laboratory. Heinrich Kayser claimed that in 1877–78 he was the only such and that Helmholtz, who occasionally looked at Kayser's work, only rarely gave him advice, but when he did, Kayser found it most encouraging. Years after he had worked in Helmholtz's old institute, Arthur Schuster visited Berlin again and caught the spirit of the times. He conceded that Helmholtz's new institute was "noble" and "impressive on the outside and perhaps also inside to the casual visitor," yet he also thought "the soul and scientific spirit of the old place had gone." Precisely because everyone had his own workroom, Schuster claimed, the institute lacked a "common bond" and "scientific inspiration," and Helmholtz lacked the time to see everyone.[72] On the other hand, precisely by not supervising his students too closely, he gave them a good measure of freedom and greater confidence (for those who succeeded) in their choice of topics, methods, and work in general. Like Weber, Kayser, and Hertz, quite a few of his assistants went on to distinguished academic careers.

Anecdotal evidence suggests that Helmholtz left much to be desired as a teacher of introductory and perhaps even advanced physics. Hertz declared that as concerned introductory physics instruction, "Helmholtz's reputation was not actually that of a brilliant university teacher." Julius Hirschberg, who later became a leading historian of medicine, claimed that medical students found Helmholtz's introductory physics lectures too difficult. Heinrich Rubens, who became a leading experimental physicist and himself later taught at Berlin, said it was difficult for beginners to follow Helmholtz's lectures and that he had "no understanding" for their difficulties. Friedrich Paulsen, who later became a well-known philosopher and historian of education, attended Helmholtz's lecture course on experimental physics in 1871. He wrote: "His lectures did not help me much although zeal and application were not lacking on my part, either in or outside the lecture room. His delivery left much to be desired, and what I was out for—fundamental concepts and comprehensive ideas—he did not dispense." Carl Runge, who came to Berlin in 1877 to study mathematics and who later became a leading mathematical physicist, did not care for either Helmholtz's or Kirchhoff's lecturing style, and so he went elsewhere.[73]

But Helmholtz had a (much) better effect on the advanced or at least the research students. Hertz said his very person was more important than what he said, and Rubens thought he helped talented, advanced students bring out their creativity. But Max Planck, perhaps Helmholtz's best-known student, did not see it quite that way. After three years of study in Munich, Planck transferred to Berlin in 1877 to hear Helmholtz and Kirchhoff lecture. There he realized how provincial his Munich teachers were: Berlin widened his "scientific horizon." Yet, like his fellow student and friend Runge, he was unimpressed with the lectures he heard (Helmholtz on theoretical physics [mechanics]). "Helmholtz had obviously never properly prepared," Planck later recalled. "He always spoke in a halting manner and had to seek out the necessary data in a small notebook. Moreover, at the blackboard he constantly miscalculated, and we had the feeling that he himself was at least as bored with his lecture as we were. The consequence was that listeners gradually dropped out, until finally there were only three of us left, including myself and my friend, Rudolf Lehmann-Filhés, who later became an astronomer." So Planck left after one semester and took his degree (in 1879) at Munich, with a dissertation on thermodynamics. No one in Munich was interested in it; Kirchhoff disputed its contents; and Helmholtz did not even bother to read it. Even du Bois–Reymond, ever Helmholtz's champion, spoke only obliquely about Helmholtz as a teacher: "He gave in a penetrating way the best of what he had, but, to speak frankly, aimed more at the minority who were in a position to receive and appreciate what he had to offer." Richard Wachsmuth, a physicist who later worked for Helmholtz at the Reichsanstalt and became the *aide de famille* in later years, conceded, "Helmholtz was neither a brilliant speaker nor an easily understandable teacher, and yet no one could avoid the impression of his greatness."[74] In short, even individuals of high intelligence and motivation found his lectures wanting. His lack of experience as a physics teacher before 1871 showed. Neither his one physics course (from Dove, circa 1839–40) as a student nor his twenty-three years of teaching physiology seem to have helped much in this regard.

Nevertheless, he may have been an influence on some of his students in less direct and tangible ways. Among Helmholtz's students was the philosopher August Stadler, who later taught philosophy at the Zurich Polytechnic, where Einstein became Stadler's student. Oskar Bolza studied physics with Helmholtz and Kirchhoff and then later switched to mathematics, eventually moving to the United States, where he had a distinguished career and helped develop the mathematics department at the University of Chicago. For numerous young scientists, Helmholtz's name was a great attraction, even if they later switched to other fields or mentors, or even, as in the case of the young Freud, if they did not actually study with him. In the early 1870s, Freud, while studying with Brücke in Vienna, decided to spend a semester in Berlin to hear du Bois–Reymond, Helmholtz, and Virchow lecture. While still a medical student in Vienna, he told his fiancée that Helmholtz was one of his "idols." Although he never in fact went to Berlin, and although in the mid-1890s he began to take a more psychological approach to studying the mind, the young Freud strongly adhered to the general intellectual outlook of the organic physicists of 1847, and he continued to hope to see growing linkages between the neurophysiological and the psychological aspects of studying the mind. As late as the 1890s, he still called Helmholtz his "scientific idol."[75]

In the early 1870s, too, there were a few individuals, like Ferdinand Karl Braun, who had studied principally with Magnus and then studied briefly with Helmholtz, taking their degrees under Helmholtz. Helmholtz helped Braun get a professorship at Marburg (in 1877); Braun later became well known for his invention of the cathode-ray oscilloscope, for which he won the Nobel Prize in Physics (1909).[76] By contrast, Eugen Goldstein, who came to Berlin in 1870 and worked many years with Helmholtz, did not take his doctorate until 1881. His work concerned electrical discharges in vacuums, and he became well known for his cathode-ray experiments (under Helmholtz) and more generally for his investigations on electric discharges in gases.

Between 1871 and 1878, Helmholtz served on promotion (examination) committees for twenty-seven individuals in a variety of fields. By far the greatest number were in chemistry; physics was next, and there were a few in philosophy, mathematics, astronomy, and the history of astronomy. The best known of these individuals were Braun and Auerbach. Helmholtz also directed or sat on more than a few habilitation examinations. Most were of a routine nature and led to positive results, such as the habilitations of Neesen and Glan in physics, Benno Erdmann in philosophy, Eugen Baumann in chemistry, and Hermann Ebbinghaus in psychology. But on rare occasions things went awry. In 1872 Helmholtz was the second expert, after the chemist Hofmann, on a habilitation review for one Eugen Dreher, a chemist. Helmholtz said Dreher's manuscript was not original, that it was only a revised piece that had already appeared in print and whose true author was probably Dreher's teacher. Helmholtz wrote: "It is obvious that we thus cannot accept this work as a *specimen eruditionis*." It was an "unwieldly and dumb" piece of work, an "attempt to fool the faculty" into appointing someone as a docent. Nearly three years later, Helmholtz again served as the second examiner for Dreher, and again voted against him. He thought Dreher's work was "extremely superficial, probably put together from popular presentations, very incomplete and distorted by the grossest errors." He concluded, "[Dreher] has not the faintest idea of what scientific work is, something which, moreover, I've been able to determine through the many official discussions I've had with him as Dean." He thus again voted against him, and hoped he would stay away from the university.[77] Here, as always, he was a gatekeeper for quality of academic work and academic workers.

Helmholtz also had numerous American students, the period from 1870 to 1900 being the high point of Americans' studies in Germany.[78] They or their teachers were impressed by the German ideals of university research and academic freedom; they admired the German PhD as a sign of high academic distinction, and they liked the rather inexpensive nature of study in Germany. Helmholtz was among the figures who most impressed them: his name attracted them to enroll in his courses, to take a degree with him, to do advanced research in his laboratory, or simply to meet him and look over his institute. With the earnings that Tyndall amassed from his lecture tour in America in 1873, he endowed the Tyndall Fellowship that year for two young American students of physics to study annually at any university in Europe, though he hoped they would study in Germany, either at Heidelberg or at Berlin (that is, at Bunsen's or Helmholtz's universities).[79]

In the early years, Helmholtz's American students included Elihu Root (PhD in 1876) and Edward L. Nichols, who first studied with Helmholtz (1876–78) but took his PhD at Göttingen (1879) and then worked with Edison at Menlo Park, before going on to an academic career first at Kansas and then at Cornell. There, in 1893, he became the founding editor of the *Physical Review*. Francis R. Upton took his degree with Helmholtz (1878) and then, like Nichols, went to work for Edison. He became his principal scientific associate precisely when Edison was trying to perfect the incandescent lamp. Upton's value to Edison lay less in his understanding of the latest in electromagnetic theory than in his experimental skill and in his ability (in contrast to Edison) to make calculations and to understand the physics and related literature. Edison also used him to check for patent infringements and to determine the general state of electric lighting. He proved to be a crucial figure on Edison's team between 1878 and 1881, a team that variously invented, improved, and patented the carbon-filament lamp. Upton also helped Edison in the development of the dynamo and the central station system. In short, Upton's work and understanding helped to lay out the fundamental physical principles and system efficiencies for the incandescent light bulb, the dynamo, and associated electrical systems. Edison himself recognized the general value of being associated with Helmholtz: in 1877, when he got into a priority dispute with David Hughes, a British inventor, over the invention of the microphone, he enlisted Helmholtz's (and Siemens's) name(s) in an effort to strengthen his claim.[80]

Helmholtz had numerous other foreign "postdoctoral" researchers in his institute during the 1870s, some of whom became leaders of the next generation of physicists. Two merit particular mention. Ludwig Boltzmann, then a young professor of physics at the University of Graz, was Helmholtz's first "postdoctoral" researcher. In the spring of 1870, Boltzmann went to Heidelberg to introduce himself to Helmholtz, Kirchhoff, and Koenigsberger. He had studied Helmholtz's writings on electromagnetic and thermodynamic theory and believed that his teacher, Josef Stefan, and Helmholtz were the only Continental physicists then familiar with Maxwell's electromagnetic theory. He thought Maxwell had erred in his determination of the dielectric constants in his theory, and, taking a leave from Graz for the winter semester 1871–72, he went to work in Helmholtz's Berlin laboratory to determine those constants experimentally. As with nearly all of his other "postdoctoral" researchers, Helmholtz only advised Boltzmann on what to do; he did not directly work or collaborate with him. At the institute, Boltzmann found Helmholtz difficult (though not quite impossible) to approach; but when Helmholtz lectured on heat equilibrium among gases at a session of the Physical Society, Boltzmann thought he was the only one who understood Helmholtz, and they had a good discussion afterward. Like many others, Boltzmann was impressed with the instrumentation he found in Helmholtz's laboratory. The stay in Helmholtz's institute helped mark his future, and it put him in contact with part of the German scientific scene and made him feel a part of contemporary physics. After Boltzmann returned to Graz, he and Helmholtz corresponded about Maxwell's theory; Helmholtz helped him get further data for his experimental work, and Boltzmann's experiments on determining the dielectric constants gave support to Maxwell's theory. In 1874 he sent Helmholtz his important paper on the electrostatic action-at-a-distance of dielectric bodies, where he used Helmholtz's idea of the effect of the charged sphere on the dielectric. He also reported that he had confirmed all Maxwell relations between dielectric constants and refraction indexes. Boltzmann's presence in Berlin helped make Helmholtz's institute into something of a center for Maxwell studies.[81]

A second notable "postdoctoral" researcher during these early years in Berlin was Rowland, then a young, unknown physicist just appointed professor of physics at the brand new (1876) Johns Hopkins University in Baltimore. Rowland came to Berlin as part of his tour of European physics laboratories in 1875–76; he was scouting for ideas about physicists, instruments, and laboratory organization in preparation for launching his own department of physics at Hopkins.

Under his leadership, the department soon became the single most important institutional if not intellectual center of American physics during the last quarter of the century.[82]

Rowland arrived unannounced and outlined in detail two experiments on the electromagnetic effect of electric convection currents that he hoped to conduct in Helmholtz's laboratory. Helmholtz welcomed Rowland's proposal, and although the laboratory was more than full, he found space for him in the basement. "Prof. Helmholtz here is the greatest physicist on the continent," Rowland wrote his mother. He found him to be "very quiet and dignified" and said that although he gave "very little attention to those working under him," he did pay attention to Rowland and his experiments. Rowland also wrote to Hopkins's president, Daniel Coit Gilman:

> Prof. Helmholtz' character interests me much. Although the foremost physicist on the continent he is by no means brilliant and in some of his suggestions I can sometimes fancy there is a trifle of stupidity. His wonderful genius lies in his power of concentration. When he thinks about a thing of any complexity his whole mind is on it: I have sometimes thought that one might knock him down without his feeling or knowing it. It is in this way that he brings forth his great results but if he does not think thus deeply his thoughts are almost as other men. As an experimenter, he is quite poor. In these respects he reminds one of Newton who, to all appearances, was not very "smart" but yet had a wonderful power of concentration like Helmholtz and, as I am told, like Tait.

Arthur Schuster, to extend the comparison, thought Helmholtz as quick as Thomson in his ability to shift his mind from one subject to another, but he thought Helmholtz, in contrast to Thomson, always had a good deal of the *grand seigneur* about him. Helmholtz paid the costs of Rowland's experiment, solicited Rowland's paper for publication by the academy, oversaw the manuscript's translation and publication, and had reprints sent to Rowland. Rowland never forgot the service and hospitality, though he later griped, "The experiment is being constantly referred to as Helmholtz's experiment—and . . . if I get any credit for it whatever, it is merely in the way of carrying out Helmholtz's ideas, instead of *all* the credit for ideas, design of apparatus, the carrying out of the experiment, the calculation of results, and *everything* which gives the experiment its value." He had a point. Helmholtz himself had previously thought of Rowland's idea, but he had not been much interested in it. Rowland's experiences and contacts in Berlin proved essential in his build-up and leadership of physics at Hopkins. Rowland helped linked German and American physics. His work in spectroscopy, especially his invention of a much-improved diffraction grating, became a crucial element that helped sustain this linkage. He and his department became indebted to Helmholtz, and the experience in Berlin with him marked the beginning of an open channel between Berlin and Baltimore.[83]

Helmholtz received untold numbers of letters of recommendation or introduction from colleagues and friends, both domestic and foreign, for young students and professionals who hoped to study with him, make his acquaintance, or see his institute. Conversely, he was often asked to write such letters himself. In 1879 Schuster, for example, contemplated applying for the professorship of experimental physics at a new college in Birmingham and asked Helmholtz to write a letter of recommendation for him, one that he could use for both this position and others. He emphasized the importance that such a letter from Helmholtz would have. In 1881 he was appointed to an endowed professorship in applied mathematics at Owens College in Manchester.[84]

There were also letters for those whom he did not know well. For example, I. F. Tsion, a Russian-Jewish physiologist who had studied with Ludwig, received Helmholtz's strong endorsement in 1871 for appointment as Russia's first Jewish professor and as Sechenov's successor at the Military Medical Academy; Tsion got the position (but was forced out in 1875). By no means did all those whom he recommended get the hoped-for position. For example, he wrote in support of Carl von Lemcke, an aesthetician and writer who had habilitated at Heidelberg in 1862 with a work on aesthetics and literary history. The letter was for a position in Weimar that Lemcke did not get. By 1871 Lemcke had, however, become a professor at the Art Academy in Amsterdam. He longed to return to Germany and so again asked Helmholtz to write on his behalf for an available professorship of aesthetics and art history at the Technische Hochschule Aachen. Two months later, he received the professorship there.[85]

Helmholtz also had to deal occasionally with a diverse collection of other individuals who had no direct ties to himself or Berlin. He was often asked to receive foreign visitors who simply wanted to meet him or tour his institute. Tyndall, for example, wrote him in 1874: "A young hero-worshipper, who makes you one of the objects of his idolatry, wishes for two minutes conversation with you." There were also disgruntled or would-be, semi- or nonacademics who wanted to discuss their situation or scientific work; they admired him and wanted his recognition in one way or another, or at least his judgment and help with their own scientific efforts or frustrated careers. John Cleves Symmes, for example, a captain in the US Army, who in 1871 was living in Berlin. He had already called on Helmholtz in Heidelberg in 1867, seeking to relate to him his theory of a perpetual motion machine based on "the rotation of atoms." He recalled to Helmholtz in 1871 that it had drawn "a little whiff of surprised, doubtful, yet genial laughter" from his lips. And he claimed that his idea had now matured to the point where he was building such a machine; if he succeeded, it would "prove many suspected things in Physics—the rotation of atoms and molecules—the actuality of the existence of the ether—the distinction between the electric and the luminiferous ethers—the character of the Gravitic force." He had written a thirty-page manuscript on this topic, and he wanted Helmholtz's opinion of it. He wanted to call upon him and read it to him, not least since he could then "imbibe some 'Physical drops' from the highest source possible." There is no evidence that Helmholtz ever granted a meeting.[86]

Perhaps the most frustrated, and certainly the most self-defeating, of all the nonacademics who sought Helmholtz's recognition was Alexander Wilford Hall, a Methodist minister from New York City, author of the rambling and tendentious *Evolution of Sound: A Part of the Problem of Human Life Here and Hereafter Containing a Review of Tyndall, Helmholtz, and [Alfred] Mayer* (1878) and, in its "revised edition," *The Problem of Human Life: Embracing the 'Evolution of Sound' and 'Evolution Evolved,' with a Review of the Six Great Modern Scientists. Darwin, Huxley, Tyndall, Haeckel, Helmholtz, and Mayer* (1880). As concerned Helmholtz, the book at once acknowledged him as one of the foremost scientists of the era and presented a full-throated attack on his acoustic analysis. Hall represented himself as "an investigator of science and philosophy in the interests of religion, and in opposition to materialism in all its phases." His books exemplified amateur science at its worst. Hall not only enumerated Helmholtz's (and others') "superficiality," "Numerous Self-Contradictions and Inconsistencies," and "Fatal Admissions," but he did so in a tone that threw caution to the winds. He was the great giant slayer, saving the age (and religion) from these scientists and putting them in their place. To ensure that his book came to Helmholtz's attention, Hall, who appears to have been his own publisher, sent him a copy.[87] Helmholtz had to deal with everyone, from the leading members of the younger generation of physicists to the crackpots.

19

Among the Elite

On the Optics of Painting and the Painting of Helmholtz

Slowly in the eighteenth century and then more quickly in the nineteenth, Berlin gradually developed an impressive set of scientific, technological, and medical institutions, while the development of its artistic institutions and its central place in the literary and fine arts came a bit more slowly. To be sure, Berlin had long boasted the Academy of Arts, the Opera House, several impressive theaters, the Singakademie, and several important art museums (for example, the Altes Museum and the Neues Museum) and numerous architectural monuments (for example, the Brandenburg Gate). It also became a vibrant center for newspapers and the publishing industry more generally, and it hosted numerous local literary, artistic, and musical associations. But the combination of Berlin's conservative political leadership, the artistic importance of several smaller cities (for example, Munich and Weimar), and the geographically dispersed nature of German cultural life in various regions meant that it was not until the last third of the century that Berlin rose fully to the fore on the German artistic, musical, and literary scenes. The National Gallery (Nationalgalerie), for example, was built (1866–76) on the Museum Island, where it joined the Altes Museum and the Neues Museum; that helped crystallize Berlin as a general center for the arts. The nearby Opera House on Unter den Linden and the Singakademie also formed part of this center. The city had a good social basis for a strong musical life, since it had many musical associations that encouraged attendance at concerts. But it was the founding of the Advanced School of Music (Hochschule für Musik) in 1869 that noticeably transformed Berlin into a center for the study of music. The king persuaded the violinist Joachim to found and lead the school, and Joachim attracted talented classical musicians to serve as instructors. A new quartet was arranged for concerts at the Singakademie. In addition, the Philharmonisches Orchester was founded in 1882. Berlin became one of the world's great music centers.

Beyond his strictly scientific and academic work, Helmholtz was present in the Berlin cultural scene in other ways: as a specialist in the scientific foundations of painting and music, as the spouse of one of Berlin's leading *salonnières* (and as cohost of that salon), and as a figure who often socialized with patrons of the arts or artists themselves. He was, for example, a leader of the university's Scientific Association (Wissenschaftlicher Verein), an association sponsored by the emperor that held a lecture series in the Singakademie on late Saturday afternoons between January and March. In January 1872 he gave the first lecture there, entitled "On the Optics of Painting."[1]

The lecture was one of three on this topic that he gave (the other two in Düsseldorf and Cologne) between 1871 and 1873. The lectures were later summarized, revised, and published as a single essay, presenting his views of human perception (including physiological optics), especially the problem of binocular vision as it related to painting. It analyzed four main artistic categories as they pertained to painting: form, shade, color, and harmony of color. His overall aim was to understand "the principles of pictorial illumination." Unsurprisingly, he argued that there were substantive connections between sensory physiology and the "theory of the fine arts," and he proposed to illuminate these as he had already done for sensory physiology and music. He believed that to understand painting fully, it was necessary to appreciate the physiological processes by which the senses respond to paintings.[2]

Helmholtz declared at the outset that some of his listeners or readers had greater experience in contemplating works of art or in the study of art history, or had greater "practical experience" in the arts than he had.[3] But that was perhaps false modesty. He brought far more to the subject than his understanding of physiological optics. Already as a child, it may be recalled, his father had emphasized to him the importance of the arts and aesthetics; as a young professional, Helmholtz had lectured on anatomy to artists at the Academy of Arts; and, especially when traveling abroad, he spent much time viewing paintings and other objets d'arts firsthand in museums.

"The painter's first purpose," he explained, "is to call forth in us, through his colored canvas, an active perception of the objects he seeks to present." This meant that the artist had to create "a sort of optical illusion" that awakened in the viewer "an idea of the object," "so lively and sensorily powerful as if in reality we had it in front of us." Sensory physiologists, he believed, had much to learn from the historical traditions of the great artists in terms of their "means and methods of presentation" and the "series of important and more meaningful facts" that they had created. The study of artistic masterpieces could teach as much about physiological optics as the laws of sensory physiology and sensory perception could teach about the theory of art. Moreover, he thought art history could help in understanding how viewers perceptually experienced a painting's different parts, their locations, and their relationships: in short, the means by which art produces its effects.[4]

Helmholtz then turned to the "forms" of painting, the first of his four main artistic categories of painting. By "form" he meant the degree and kind of "similarity" a painter might achieve in representing external objects. A person of more refined taste in art expected not a mere copy of nature; rather, he required "an artistic choice, order, and even idealization of the presented objects." Artistic representations of ordinary human figures, for example, should not be like those in photographs; instead, they should use expressive and beautiful forms "that bring out, in a lively manner, a side of human nature that is in full and undisturbed development."[5]

Helmholtz argued that the very notion of "the really true image of the object of nature" was in fact problematic. For one, an artist painted on a flat surface, whereas the human eye forms its optical view of the external world in response to the light that strikes its curved retina; in contrast to a painting hanging on a wall, if and when the viewing eye (retina) moves, its image changes. Human beings (normally) view the world binocularly, with two eyes, and thus from two slightly different perspectives. This creates the crucial problems of judging distance and depth perception. But when human beings look at a (flat) painting, they combine these two views into one, and as they move in relation to the painting, their eyes adjust their perceptions of objects in it for both distance and depth.[6]

This meant there was an unbridgeable "incongruency," Helmholtz continued, "between the view of a painting and the view of reality." Painters had developed various practical means to lessen this incongruency, and such skills—for example, concerning choice, arrangement, and illumination of objects—could be understood in terms of scientific theory as well. A prime example of a painter's skills lay in the expression of depth dimensions by means of perspective. The painter learned to group objects in relation to one another so that one appeared closer while another appeared more distant; to create objects of different sizes;

and to use shadowing, illumination, and aerial perspective to yield different depth perceptions within the painting. Moreover, though the presentation of objects in a painting was very important, it was ultimately subordinate to the larger purpose of inducing "an undisturbed and lively effect of the painting on the observer's feeling and mood."[7]

Helmholtz then turned to shading, his second artistic category. He emphasized that there is a great difference between what human beings see with their eyes when they stand before particular objects and when they stand before a painting of those objects. An artist's techniques could lessen but not eliminate this difference, and artists knew there were certain things they could not represent at all. In theory, an artist could seek to create a retinal image in a viewer's eyes when he or she viewed a painting that was identical to the retinal image when the person viewed the same field in the real world. Yet, owing to the absorptive power and reflection of various colors of light in the real world as opposed to that on a canvas, this difference could not be fully overcome.[8]

He explained that "the most important factor of the arrangement" of objects was "the different dulling of our eye by light," known also as "tiring" or fatigue. Over time, nerve activity leads to a decline in the level of work capacity—the muscles and the brain tire—and the eye's ability to absorb light gradually weakens. These physiological facts create a problem for the painter, who must somehow use colors to create the same impression on a viewer in a gallery as the person would get when, for example, looking at the moon at night or at a desert scene. The painter has to accommodate the nature and limitations of viewers' physiological-optical systems. Hence the painter's task was not to imitate reality but to translate the sensory impressions from the real world to the canvas, from one "sensitivity scale" to another. Helmholtz here introduced Fechner's "law for the eye's sensitivity scale." Fechner, an "ingenious researcher," had discovered a law that was a key to distinguishing the degrees of brightness of observed bodies, indicating the upper and lower limits of the eye's ability to absorb light. While the eye takes in a wide range of brightness, the painter is concerned only with the appearance to the eye of the constant ratio of brightnesses of surfaces of various colors and therefore tries to imitate this ratio in the painting. It is this ratio that constitutes the basis for judgments as to how dark (or light) a perceived body should be. The painter's goal is not to imitate the color of bodies—as children and members of less-experienced nations do—but rather to recognize the effect of light on the eye. That was what "old masters, above all Rembrandt," had done.[9]

Color, Helmholtz's third artistic category, also had, like shading, its "variations." The two categories are tied to each other physiologically, he maintained, since the sensitivity scale for different colors is different. Color sensitivity depends on the way the nerve apparatus reacts. Thomas Young had sought to demonstrate that all colors are mixtures of three fundamental colors (red, green, and violet), which "are perceived completely independently of one another through three different systems of visual nerve fibers." Helmholtz also explained contrast and discussed various types of contrast phenomena, noting that the retina sustains a certain measure of fatigue for each of the fundamental colors. The artist needed to use her or his colors to take into account contrast phenomena and irradiation phenomena (the spreading of light or color of a very bright object onto a neighboring one) when representing the impressions that strike the eye.[10]

Helmholtz then briefly described the physical understanding of colors and how they are produced, including the wave theory of light, the role of various-sized particles in determining color, the spectrum of colors, and why the heavens (often) appear blue, as well as various other colors or shadings of the atmosphere at various heights and places. Painters of all sorts needed to understand aerial coloring, he noted.[11]

The painter's representation of lights and colors, he repeated, was a matter of "translation," not imitation. The effects of scale alone dictated this. Each painter made this "translation in his own, individualistic way." As prime examples, he cited Rembrandt and Fra Angelico, whose works he had viewed in Florence. "Fra Angelico is overpoweringly charming in his executions," he wrote Anna from Florence. "In every cloister cell there is a fresco image by him or according to his outlines, some of which are of the very highest order. A coronation of Maria by Christ in Heaven, both in white garments, looking under the Sacred, is of a tenderness and purity the likes of which I've never seen." He thought the galleries of the Uffici were the equal of those of the Louvre, not least because of the exquisite Raphaels and the Titians he saw there.[12]

Helmholtz's fourth and final artistic category of painting was the harmony of colors. He believed that the greatest paintings had been "executed with relatively dark tempera and oil colors" and had been "determined for spaces with a modest amount of light," though there was "a natural lust for colors" and light. In contrast to combination tones, he emphasized, there were no known "definite rules" about color combinations. Indeed, it was hard to find empirical data about what combinations produced a harmony of colors, and their significance for painting could therefore be mentioned only in a very general way.[13]

He concluded by suggesting that ultimately a work of art "should latch onto and enliven our attention; it should awaken a rich fullness of slumbering connections of ideas and, thereby, bind feelings in an effortless way and guide them to a common aim." It should unite what were otherwise disparate, isolated points into a sort "of ideal type" and thereby refresh the viewer. To Helmholtz, art thus appeared more powerful than reality, since "reality always mingles annoyances, scatterings, and infringements in its impressions, while art can bring all elements together in an unhindered way so as to achieve the intended impression."

The "intelligibility" of a work of art increases, he argued, "the larger, the more penetrating, the finer, and the richer it is to the truth to nature of the sensory impressions, which last should awaken a series of ideas and the affects bound up with them. It [the work of art] must make its effect surely, quickly, unambiguously, and exactly, if it is to make a lively and powerful impression."[14] Helmholtz had explained his understanding of the optics of painting, and with it he became a part of Berlin's art scene.

Helmholtz's lectures (essay), together with his *Handbuch der physiologischen Optik* and (other) writings on color and depth perception and those on non-Euclidean geometry, apparently influenced a number of painters, especially the impressionists and neoimpressionists. Yet the depth and specificity of that influence is difficult to gauge. Did they read him directly or only learn of his views through the work of others? Artists in the 1870s and 1880s certainly showed an increased interest in the optics and physiology of color, light, the eye, and visual sensation and perception, as well as the chemistry of colors and color mixing. The effect of Helmholtz's writings was probably as much indirect as direct, and doubtless it occurred in conjunction with similar work by other scientists, such as Young, Maxwell, Chevreul, Ogden Rood, and, to a lesser extent, Brücke and Wilhelm von Bezold.[15]

Helmholtz's essay was translated into French, and so it soon had a wider readership. Here the influential writings of Charles Henry, who became noted for his *Cercle chromatique* (1888) and his attempt to develop aesthetics on the basis of mathematics and psychophysiology, undoubtedly played a role. (Henry addressed Helmholtz as "illustre Maître.") The painter Georges Seurat also probably read Helmholtz's essay on painting, in part since it was appended to the French edition of Brücke's book *Principes scientifiques des beaux-arts . . . suivis de l'optique et la peinture par H. Helmholtz* (1878); he may also have read Helmholtz on color in the *Optique physiologique*. The French impressionist Camille Pissarro reportedly studied Helmholtz's work on physiological optics.[16] The rise of sensory physiology, psychophysics, color theory, and new ideas about perception led not a few writers, painters, musicians, and photographers, from the 1870s through the early twentieth century, to explore the possibilities of a "scientific" basis to art and the development of a scientific aesthetics.

Helmholtz's thinking on colors and painting became well known in the English-speaking world through the work of Rood, an American physicist. Rood had studied physics and chemistry in Germany (with Liebig) and had taken drawing and painting lessons. In 1863 he became professor of physics at Columbia, where he conducted research in physiological optics, colors, visual phenomena, thermoelectric currents, electrical discharges, and the photometer. In 1874 he lectured on optics in painting, and after 1879 he became well known among artists and scientists alike for his *Modern Chromatics*, by far the most important such study of the era, which sought to explain color perception on the basis of Young's theory as modified by Helmholtz and Maxwell. Here Rood paid much attention to Helmholtz's work on colors, color mixing, and the perception of color, interpreting these topics for artists.[17]

Similarly, in the 1880s the English photographer Peter Henry Emerson sought to give photography a scientific basis—he called it "Naturalistic Photography." To explain issues of light intensity (including shading) and color in photography, and to understand human vision as it pertained to photography, he turned especially to Helmholtz's "On the Optics of Painting" and to the *Handbuch*. Emerson sought to make photography a rational art.[18]

Whatever influence Helmholtz's scientific analysis of painting may have had on art, he himself became the subject of two well-known paintings: by Franz von Lenbach (1876) and by Ludwig Knaus (1881). He is also featured in Adolph Menzel's (socially revealing) sketch (1874) depicting a soirée. See figures 19.1, 19.2, and 19.3. These portraits reflected a new style in German painting. During the first half of the century, German painting freed itself from its earlier religious and patriotic entanglements; then in the 1870s there came, alongside the emergence of impressionism and the continuation of monumental art, so-called *Gründerzeit* painting, a compromise between intimate and monumental art. *Gründerzeit* emphasized the individual as a public figure, political action, industriousness, and luxuriousness, and it concerned large personalities, as exemplified by Bismarck in politics, Siemens in business, Wagner in music, Nietzsche in philosophy, Mommsen in history and politics, and Helmholtz in science. German painting became distinguished by the cult of personality and heroes.[19] Menzel, Lenbach, and Knaus celebrated Helmholtz as a cult figure, a point that can be further appreciated through understanding Berlin's salon culture and social elite.

Anna's Salon

Although Berlin's rich tradition of salons dated back to the late Enlightenment, it was not until after 1860 that a small set of wealthy families fully took root in the city. As Berlin grew in economic, political, and cultural importance, so did the salons and salon culture. Between 1860 and 1890, there were about twenty-five elite salons, including Anna Helmholtz's. Like their predecessors in Enlightenment Paris, these salons displayed the social and cultural presence of a tiny set of socially and financially prominent couples who pursued high culture and social connections. They brought together politically and artistically talented people and, in the unique case of the Helmholtzes, scientifically talented as well. They allowed for relaxed conversation in elegant settings. Each was led by a woman, the *salonnière*, who sought to stimulate discussion about literary, artistic, po-

Fig. 19.1 Helmholtz, 1876. By Franz von Lenbach.

litical, and other matters, and the salons featured musical performances. Each was, in effect, the lady's court; each met on a *jour fixe*, more or less weekly (except in the summertime); and each had both regular guests with standing invitations and special, one-time guests—for example, a noteworthy out-of-town novelist who had just published a new book. Since guests came from different professions and walks of life, the salons facilitated a certain measure of interaction between different social and political groups, though they were anything but working or middle class, let alone bohemian, in nature.[20] Apart from Anna Helmholtz, the greatest *salonnières*, like Countess Marie ("Mimi") von Schleinitz

Fig. 19.2 Helmholtz, 1881, by Ludwig Knaus, 1881. Several sensory-physiological instruments are included in the painting: a tuning fork on a wooden box resonator, a Helmholtz (brass spherical) resonator, an ophthalmometer (*far right*), and, lying flat near the center of the table, an ophthalmoscope. Helmholtz's left hand touches a simple prism with stand, and a laboratory notebook lies open on the table. This painting hangs in the Alte Nationalgalerie Berlin. The Francis A. Countway Library of Medicine, Harvard Medical Library—Boston Medical Library.

and Anna von Roth, all belonged to the aristocracy or had husbands who were among the wealthiest individuals in Berlin. The salons helped civilize Germany's rapidly growing, rough-and-tumble, capital.

Countess von Schleinitz, whose husband Alexander was the Prussian minister of the Imperial household, ran the most elegant and exclusive salon in Berlin. Her guests were decidedly liberal, and her salon was a center point for Bismarck's opponents. The Helmholtzes were regulars. In May 1873, for example, they spent an evening there hearing the Russian pianist Anton Rubinstein play, just as they had heard him play earlier that week at the home of Gustav Richter, a well-known portrait painter. From the Helmholtzes' point of view, the problem with Rubinstein was that he did not get inspired to play until after midnight, so that "all too often" they didn't get home "before two in the morning." In June 1874 Schleinitz's prominent guests included Crown Prince Frederick William and Crown Princess Victoria, the Helmholtzes, the painters Adolph Menzel and Anton von Werner, and Wilhelm von Bode, a senior official in the state museum system. The crown prince had been particularly keen to discuss art collecting. Schleinitz wanted "to immortalize" the evening, so she asked Menzel to paint a collective portrait of her most prominent guests that evening. (Menzel, it may be recalled, was an old friend of Wilhelm Puhlmann, from Potsdam.) The result was his *Salon of Frau von Schleinitz on 29 June 1874* (fig. 19.3).[21] The portrait became well known and burnished the Helmholtzes' image as members of Berlin's social and cultural elite. Nevertheless, for all of Helmholtz's cultural preeminence, and despite his being one of Germany's best-compensated professors, he could in no financial or social sense be compared to the Schleinitzes or many of their other guests.

The Helmholtzes' social relationships with that elite began from the moment they arrived in Berlin. George Bancroft, an American historian and diplomat, who knew the city's upper crust well, reported shortly after Helmholtz moved to Berlin: "Our circle in Berlin is just enlarged by the arrival of Helmholtz: the great master of natural science." The Helmholtzes soon came to know, among many others, Countess Hildegard von Spitzemberg (and her husband). The count and countess attended Helmholtz's lecture "On the Optics of Painting," and the two couples were soon dining at each other's homes and seeing each other at various salon evenings.[22]

Within six months of their arrival, Anna established her own salon. Every Tuesday evening the Helmholtzes received ten to twelve people as regular guests, along with one or more special guests. In stark contrast to other salons or even special evenings given by some of Helmholtz's academic colleagues, the Helmholtzes kept food and drink to a minimum, generally serving only tea and various types of buttered breads. Anna ran her salon for nearly three decades in

Fig. 19.3 A soirée at the salon of Frau von Schleinitz, 29 June 1874, sketched by Adolph von Menzel. The sketch shows (*left to right*) Helmholtz, Heinrich von Angeli, Countess Marie von Schleinitz (*reclining on the sofa*), Anna Helmholtz, Count Götz von Seckendorff (*in the back*), Countess Hedwig von Brühl, Crown Princess Victoria (*seated in the center*), Count Wilhelm von Pourtalès (*in the back*), Crown Prince Friedrich (*standing, facing left*), Count Alexander von Schleinitz, Anton von Werner, and Prince Victor von Hohenlohe-Langenburg (*seated on the right*). Staatsbibliothek Preussischer Kulturbesitz, Berlin.

her four Berlin residences. Members of Anna's own family, such as her politician-father Robert and her diplomat-brother Ottmar, attended the salon.[23]

Anna's emotional, nervously temperamental, yet warmhearted, sensitive, and energetic nature complemented Hermann's calm, olympian bearing and cerebral manner. Moreover, her great friendship with Mimi von Schleinitz, her father's political reputation and connections, the Helmholtzes' friendship with the crown prince and crown princess, and Anna's charity work attracted many to her home. What distinguished the Helmholtz salon from all others was its mix of participants: leading figures from the scientific and scholarly worlds, the arts, politics, diplomacy, and the military, as well as the court and high society. It was an Enlightenment salon in Imperial Germany. Anna herself viewed it as linked in spirit to the Weimar of the *Goethezeit*. Of course it also reflected Helmholtz's own cosmopolitanism, rationalism, and cultivation of the arts. It was, however, very much a bourgeois salon, one of only two such among Berlin's otherwise most-exclusive set. Though not blind to the social power of royalty and high position in general, Anna and Hermann were not overly impressed by it, either. What they admired, and what she sought in her guests, was a spirited intellect and *Bildung*. Whereas Mimi von Schleinitz's was an aristocratic and wealthy salon, Anna's was a *Gelehrtensalon*, emphasizing the high intellectual quality of its diverse attendees.[24]

Helmholtz's scientific renown, as well as his reputation among musicians and painters, led many to covet an invitation to Anna's salon. Yet if his name brought them in, she kept them there. She ran the evening, animated conversation around one or more themes of general interest, and created a relaxed atmosphere, while Helmholtz himself often stood off in a corner or sat in a chair, responding to questions posed to him and generally enjoying himself. In November 1872, for example, Fanny Lewald, a well-known liberal German writer, had a lively discussion with him about David Friedrich Strauss's *Der alte und der neue Glaube* (1872), a work notorious for replacing the gospel of Christianity with that of nature or science, rejecting both God and teleology. (Strauss himself was a keen reader of Helmholtz's writings.) Lewald, who originated from Königsberg and came to regret her youthful conversion from Judaism to Christianity, was also a freethinker. She pressed Helmholtz about his own views on God and immortality, but he (reportedly) would say only that "no one can prove scientifically that everything ceases after death," though on another occasion he (again reportedly) said that "he could not believe in a continued personal existence after death."[25]

The Helmholtzes invited many kinds of scientists, mathematicians, and technologists, including industrialists and engineering professors, to their salon.[26] Some came as colleagues, others as young but advanced science students. They included the physiologists du Bois–Reymond and Sigmund Exner; the cellular pathologist, anthropologist, and politician Virchow; professors of medicine like Ernst von Bergmann, Ernst von Leyden, and Richard Liebreich; the physicists Kirchhoff, Hertz, Planck, and Kundt; the chemist Hofmann; the botanist Pringsheim; geographers and geologists like Wilhelm Reiss and Ferdinand von Richthofen; mathematicians like Borchardt and Leopold Kronecker; and technologists like Siemens, Friedrich Hefner von Alteneck, and Franz Reuleaux. Planck was first invited in 1877–78, as a nineteen-year-old student. There he met du Bois–Reymond's two eldest daughters, and they talked about music.[27] After 1888, as Helmholtz's new junior colleague, Planck became a regular guest. The Helmholtz salon also gave invitees a chance to meet leading figures in other walks of life. When foreign scientists such as the physicists Blaserna (1872) and

Raoul Pictet were in Berlin, they too were invited, and their presence naturally enhanced the evening.

Professors in the humanities and (occasionally) in the social sciences were also regular guests. The historians Mommsen, Sybel, Treitschke, and Ranke were regularly present; so, too, were the jurist Heinrich Rudolf von Gneist; the political economist Gustav von Schmoller; classical philologists like Ernst Curtius; Egyptologists like Richard Lepsius; sociologists like Wilhelm Dilthey and philosophers like Zeller; theologians like Adolf von Harnack; art historians like Hermann Grimm and Julius Meyer; and Germanists like Erich Schmidt. The English historian Lord John Acton was a one-time guest, as was Max Müller. The Mommsens and the Helmholtzes regularly invited each other to their homes.[28]

Many leading artists were invited as well: painters including Arnold Böcklin, Menzel, Ferdinand von Harrach, Ludwig Passini, Gustav Richter, Werner, and Lenbach; sculptors like Reinhold Begas and Adolf von Hildebrand; composers like Richard Wagner (about whom more presently); musicians like the violinist Joachim and his wife, a highly talented singer; the conductor Karl Eckert; the pianist, composer, and conductor Anton Rubinstein; singers like Marianne Brandt; and writers like Berthold Auerbach, Lewald, and Rodenberg. The Helmholtzes and the Joachims quickly became friends and visited each other at their homes.[29]

In 1872, while the Helmholtzes dined with Lenbach at Mimi von Schleinitz's, Lenbach asked Anna to tell Helmholtz of his desire to paint him. Lenbach had already done portraits of Franz Liszt and of both Wagners (Richard and Cosima). "Lenbach is, by the way, very fashionable among crowned heads," she told her father. "Besides Emperor William he still has his Swan King [Ludwig II], as he calls the Bavarian Majesty, to do, and then to paint the Emperor of Austria and Crown Prince Rudolph in Vienna." However, it was not until September 1876 that Helmholtz agreed to sit for Lenbach in his Munich atelier, and he was annoyed at the time lost. As he painted, Lenbach reportedly queried Helmholtz about the latest discoveries in electricity. Cosima Wagner told Nietzsche that she thought Lenbach's portrait of Helmholtz was "the most beautiful one he has done, perhaps the glorification of simple straightforwardness."[30] See figure 19.1.

In July 1880 Helmholtz again sat for a portrait, this time with Knaus. "I've now sat almost daily for three hours," he told Anna. "(And Mommsen's portrait will be a very effective and interesting counterpiece to mine.) I don't know what I think of my own portrait." (See fig. 19.2.) While the painting lacks warmth, it does portray the "great" scientist with iconic scientific instruments on display, leaving the viewer highly conscious of perception, both as a topic of scientific study and as a relationship between Helmholtz and the viewer, not to mention a sense of precision, exactness, and globality. Knaus's portrait, and a very similar one by him of Mommsen, were sent in 1885 to the World Exposition in Antwerp and in 1893 to the World's Fair in Chicago. They highlighted Germany's cultural prowess.[31]

Music was of course performed regularly in the Helmholtz salon—it was a vital element of Hermann and Anna's life together—and it was performance of a high caliber. "I'll never forget the evening [at the Helmholtzes']," Planck later wrote, "when for the first time I heard Joseph Joachim play his new rendition of Brahm's 'Hungarian Waltzes,' or also, when Marianne Brandt . . . sang 'Wotan's Departure from the Walküre'—of course on another occasion." Helmholtz himself had a range of musical favorites, including such different composers as Brahms and Wagner. In music as in science, Planck astutely remarked, Helmholtz "was averse to everything dogmatic and, when he encountered them, acknowledged the beautiful and the genuine." "The hours are unforgettable," Richard Wachsmuth, one of Helmholtz's later assistants, wrote, "when the rooms were filled with the sweet tones of the Amati and the Stradivari, and when the exquisite piano, that Steinway had brought to the author of *On the Sensations of Tone* as a sign of his admiration, was sounded."[32]

Numerous commanding financial and business figures also attended the salon; Adolf von Hansemann, head of the Disconto-Gesellschaft; Robert von Mendelssohn, head of the Mendelssohn and Company bank; Ernst von Mendelssohn-Bartholdy, grandson of the composer and a senior figure in the bank; the banker Adolph von Rath, and various members of the Siemens family. Anna also invited individuals who were active with her in supporting women's education and nursing or were otherwise engaged in addressing social problems. Many of these guests had ties not only with one another but also, in effect, to the generations that preceded and succeeded them. Hermann von Schelling, the youngest son of the idealist philosopher, and Prince Max von Baden, the last chancellor of the Reich, were also guests at the Helmholtzes' salon.

One person who was never invited was Bismarck. The Helmholtzes never even met him, though Anna was much impressed with a set of his published letters, which she thought Hermann should read. However, Arthur von Brauer, one of the Iron Chancellor's longtime and closest associates, greatly enjoyed evenings at the Helmholtz salon. He and Anna became great friends, and that constituted an indirect if distant link to Bismarck. Brauer estimated that it had taken Anna nearly a decade to build her salon, but that once it had matured, it was the best Berlin had seen since the days of Henriette Herz. He liked it especially because of its wide variety of *gebildete* types, including "the best people of court society." "Everything that has importance in Berlin traffics here," he noted. Included were opposition party figures like the liberal Mommsen and the left-liberal Ludwig Bamberger, a leading Deutsche Freisinnige (German Free Thought) party member of parliament. The Helmholtzes, especially Anna, were also friends with Bamberger, a Jew who had cut his ties to Judaism without becoming a Christian. He was a leading liberal politician who strongly opposed Bismarck. Brauer valued the opportunity that Anna's salon gave him to talk to political opponents like Bamberger. As for Helmholtz himself, Brauer found him a man "of touching modesty and simplicity. Even when one brought up his discoveries and inventions, he spoke of them in such a way as to make his excuses. He was proud only of the fact that he had worked his way up from small, difficult [financial and social] relations." Brauer found him so self-effacing and quiet that if one did not know him beforehand, one might not notice him. Helmholtz told Brauer, however, that he loved the salon evenings: they let him forget about his work and refreshed him.[33] Though the Helmholtz salon and friendships were predominantly of a liberal cast, there were visitors and friendships with people of a conservative outlook as well.

With Wagner

The Helmholtzes and the Wagners first met in January 1873 at the home of the renowned conductor Karl Eckert, while the Wagners were visiting Berlin to raise funds for his planned theater at Bayreuth. At Mimi von Schleinitz's salon, Wagner read from his *Götterdämmerung* before high-ranking princes and princesses, counts and countesses, leading ministers and ambassadors, and high-society and high financial figures, as well as prominent figures in the arts and sciences, including the Helmholtzes. Cosima wrote Nietzsche about the evening and their visit to Berlin: "I've found great enthusiasm in the German Empire for D. Strauss's book [*Der alte und das neue Glaube*], which, on the basis of some quotations from Helmholtz, liberates us from redemption, prayer, and Beethoven's music. I was glad to meet Helmholtz himself; he belongs to the most select audiences that attended the *Götterdämmerung* reading at Frau von Schleinitz's." Two years later, the Wagners again visited Berlin to raise money for his theater. The Helmholtzes dined with them at Mimi von Schleinitz's one evening, and a week later Helmholtz attended a rehearsal for an upcoming performance of a Wagner composition. Cosima wrote in her diary: "Lots of friends; and great emotion (Pr. Helmholtz, in constant tears, heard the godly thing)." On the Wagners' last evening in town, the Helmholtzes gave a party for them following a performance. In March 1876 Wagner again returned to Berlin to seek financial support for his Bayreuth enterprise; he saw the opening of *Tristan* and visited Mimi von Schleinitz, Bismarck, and Helmholtz.[34]

The first Bayreuth festival took place in August 1876, with *Der Ring des Nibelungen* as the first performance. Thousands of people, from crowned heads of state on down, attended. Emperor William I of Germany and Dom Pedro II of Brazil were there, as were King Ludwig II of Bavaria and King Karl of Württemberg, Grand Duke Karl Alexander von Sachsen-Weimar, Grand Duke von Schwerin,

Grand Duke Vladimir of Russia, and numerous German princes and princesses. Socially prominent families like the von Schleinitzes, the von Dönhoffs, the von Keudells, and the von Radowitzes came, as did Austrian noblemen and three German parliamentarians—no socialist and no Catholic Center Party men among them. Leading figures in music and the arts also attended: Franz Liszt, Anton Bruckner, Edvard Grieg, Peter Tschaikovsky, Camille Saint-Saëns, Gottfried Semper, the painters Hans Makart, Lenbach, and Werner, and numerous orchestral and theater directors, actresses, and actors. The managers and conductors of all the German opera houses, as well as some sixty music critics, were present, along with music journalists like Eduard Hanslick, Ernst Dohm, and Ludwig Pietsch. It was extremely hot, crowded, and physically exhausting. But it was probably the most extravagant, talked-about high cultural event of the century and was a major moment in modern cultural history.[35] The Helmholtzes were there, too.

The trains to Bayreuth had been packed full with all sorts of people, including Englishmen, Anna reported. Most attendees had difficulty finding a place to stay or even getting something to eat. The Helmholtzes, by contrast, were received at the station by Eckert, who conducted many of Wagner's works, and stayed in his palace outside of Bayreuth, an hour away from the crowd. Anna wrote her children:

> We ate our mid-day meal under very difficult circumstances, and then hurried to receive the Emperor, found Frau von Schleinitz, Frau von Loe, the Usedoms, the Grand Duke of Weimar and various Altenburg dukes assembled; were greeted; and it was as if we were in Berlin—all the well-known people together, all the artists, Menzel, [the painter and animal lover Paul] Meyerheim, Makart, Lenbach, in short, all of Europe's nice people together. There was also a tremendous mob there and every government official; cordons strung by the fire department; we hurrahed, bowed, and greeted the Emperor as best we could.

In Bayreuth they met friends and acquaintances everywhere. Then they visited the *Meister* himself at his home, where they spent the evening. At the Wagner soirée, they saw Liszt "and many, many dear and magnificent people." The *Meister*, who went to bed early, found time to ask Helmholtz jokingly "whether he was here directing!" The next day the Helmholtzes went to Mimi von Schleinitz's house in Bayreuth, where they had tea with Lenbach and some members of the Rothschild family "and rested in the shadow of culture, in somewhat bohemian conditions." Anna said, "We then got dressed and went to *Das Rheingold*, which was the prelude to the *Nibelungen*. We had very good seats on the Wagner Bank, between [the diplomat Joseph Maria von] Radowitz [Jr.] and the great architect Semper. . . . Lenbach sat in front of us, and then came all the beautiful women,

Liszt, etc." The performance spectacle that evening lasted two hours, and it re-commenced late the next afternoon. All told, *Der Ring* lasted fifteen hours (over three days plus a preliminary evening). From Bayreuth, Helmholtz wrote the mathematician Lipschitz that he had just seen the first two parts of *Der Ring* and that he was greatly impressed. Even so, he had some criticism: "While the many mythological items have been treated in a peculiarly skilled manner, they aren't really musically alive, or at least only partially so. Where Schopenhauer's philosophy or his theoretical whimsicality get mixed in, Wagner is often enough also barren and abstruse."[36]

Helmholtz may have enjoyed his few occasional personal meetings with Wagner more than he prized his music. After 1876 he apparently did not see him again until the spring of 1881, when Wagner came to Berlin to direct *Der Ring*. The Wagners dined one evening with the von Schleinitzes, and the Helmholtzes arrived afterward: "The famous professor gave himself the most obvious trouble to show himself as R's friend." A week later, shortly before the Wagners left town, Helmholtz visited Wagner alone. It was their last meeting. Helmholtz certainly honored Wagner and valued his exchanges with him about musical aesthetics. His appreciation of Wagner's music, as well as Beethoven's, suggests a romantic dimension in him that is otherwise only occasionally visible. He himself noted that as a student he had become enamored of, and constantly practiced play-ing, Gluck's works, which he respected throughout his life; he thought Gluck was Wagner's "true predecessor." Though Wagner's music did bring him to tears once, there is little direct evidence that he considered Wagner to be a historic musical figure; in the *Tonempfindungen*, for example, he made only one refer-ence to him, and that was a neutral one. According to an anecdote, he joked that after listening to *Parzival* or *Tristan*, it became easy to do differential calculus. Wagner felt a sense of friendship toward Helmholtz, whom he honored with a ditty that suffers in translation but presumably refers to Goethe's *Faust*, entitled "To Helmholtz":

> All theory is gray??
> Against that, friend, with pride I say:
> Let harmony sound for us,
> unite to a helmet [*Helm*] a noble piece of wood [*Holz*].[37]

The two also differed on various issues. In 1876 the antivivisection move-ment picked up steam in Germany, and it put du Bois–Reymond, Ludwig, and indeed perhaps all experimental physiologists, on the defensive. Wagner was a committed antivivisectionist, Helmholtz anything but. When Wagner joined an antivivisection committee, Cosima told Mimi von Schleinitz that Helmholtz's actions in favor of vivisection had "paralyzed" Wagner's own. "'Science,' that's the great word with which participation in this agitation is laid ad acta. He [Wagner] asked me whether you might not be in a position to neutralize Helmholtz's influence." Moreover, Helmholtz did not share Wagner's fanatical search for national redemption and a national mission, let alone his racist messianism, anti-Semitism, and *völkisch* outlook. But he did not therefore distance himself from Wagner, who was a notorious anti-Semite and the author of *Das Judenthum in der Musik* (1850), a work Helmholtz perhaps knew. Unlike Wagner, moreover, Helmholtz was neither a political reactionary nor a romantic, nor had he ever been, as Wagner had once been, a revolutionary, a socialist, and an anticapitalist. Whereas Wagner sought to counter modernity with ancient and medieval Germanic myths and legends, Helmholtz sought to advance it by doing science, by relating understanding of science and technology to the larger public, and by treating music and painting as in part susceptible to scientific analysis. Their friendship ended with Wagner's death in February 1883. Anna wrote her sister Ida: "You can imagine that we're completely crushed by the death of the great man."[38]

Helmholtz's and Wagner's contact was friendly, yet intermittent and short-lived, while Anna, like so many others, had come under the spell of Cosima, with whom she remained friends long after Wagner's death.[39] There was some irony in that relationship, since the Wagners—in sharp contrast to the very bourgeois Helmholtzes—were among the most notorious couples of the age. Cosima, the illegitimate daughter of the pianist Franz Liszt, was married off as a teenager to the pianist Hans von Bülow; Wagner, though married, became her lover; they had three illegitimate children together before his wife died and Cosima could get a divorce from von Bülow, and so, in 1870, Richard and Cosima married. That was certainly not the Helmholtzes' lifestyle.

As Lobbyist: Financing the Stazione Zoologica

After 1871 Helmholtz became committed (with others) to helping establish the Marine Zoological Station (Stazione Zoologica) in Naples, Italy. The institution was the brainchild of Anton Dohrn, a German zoologist who had studied with Carl Gegenbaur and Ernst Haeckel in Jena, a man of independent means and of extensive connections among the German scientific elite. He fit in well with Helmholtz's crowd in Berlin. His background in comparative physiology and embryology had led him to see the importance of organizing a laboratory devoted to these topics, and the conception, development, and running of the Station became his life's work and an important contribution to science.[40]

Dohrn's plans for the Station began in 1870, shortly after he first met Helmholtz at the Naturforscherversammlung meeting in Innsbruck. His major problem was how to fund the Station. At first he got his father to put money into it, and he used some of his wife's dowry as well. But he knew that much more was needed and that it would probably have to come from a governmental agency. He began his efforts at a propitious moment: Berlin was then showing great enthusiasm for establishing new institutions; the German lands had just become united, and Berlin sought to expand its cultural presence abroad. The Italian lands, too, had recently become united, and after 1871 Italy looked to Germany as a model in politics and science. An Italo-Germanic scientific project seemed to make sense.[41]

In August 1871 Dohrn went to Berlin and asked Helmholtz for a meeting to explain his plans for and the financing of his fledgling institution. He asked him—as well as du Bois–Reymond, Haeckel, Darwin, Huxley, and others—for a letter of support for his marine zoological station. Helmholtz said that Dohrn's plan seemed to be "of great value and interest for science." He noted various logistical research problems that contemporary individual zoologists and botanists faced. "The necessary observations of the special forms and conditions of marine fauna remain very incomplete," he acknowledged, having often heard such complaints from zoologists and botanists. In an age when Darwin's principal claims were on the minds of many—On the Origins of Species had, to repeat, appeared in 1859 and The Descent of Man in February 1871—Helmholtz, an early and strong supporter of Darwin's views, wrote Dohrn: "Precisely the transformations of the different forms into one another, which have been of such great interest for the entire development of living creation, have occurred, according to the current system of observation, only recently, and have obviously been found to be very incomplete. For this reason, your plan to establish an ongoing observatory seems to me to be very promising, and, especially given the rich fauna of the Mediterranean, Naples seems to be one of the most favorable locations for this purpose." He thought anyone who supported Dohrn's plan would perform "a real and true service to science." Dohrn asked Helmholtz for permission to publish his letter as a sign of support; Helmholtz readily agreed, though he wondered whether it was really advantageous to do so from Dohrn's point of view, since he thought himself "too dilettantish" in zoology to merit the invocation of his name.[42] His numerous visits to Italy and his strong connections with many Italian scientists doubtless also influenced him to support a German scientific institute in Naples.

Dohrn published Helmholtz's as well as other letters of support, successfully using them to raise money. But still his project remained significantly short of cash. And so he conceived a new idea: to rent or sell tables at the Station to universities, national research councils, or individuals. In October 1872 he asked du Bois–Reymond to ask Falk (the minister of culture) and Rudolf von Delbrück (chief of the chancellor's office) to come to the academy to discuss the matter with himself, Helmholtz, and others. The zoological station in Naples, Dohrn explained, would complement the Deutsches Archäologisches Institut (German Archaeological Institute) in Rome. Dohrn felt so grateful for Helmholtz's, du Bois–Reymond's, and Virchow's support that, he said, had the academy consisted only of these three members, he would have named the Station's ship after them. (Instead, he named it after Johannes Müller.) The academy, that is the Reich, provided some money, though not quite enough, for Dohrn's project. The Reich thus now began supporting science financially and politically, something that was constitutionally supposed to be the responsibility of the individual German states alone. Backed by the emperor, Delbrück, Robert von Keudell (the German ambassador in Rome), and Falk, on the political end, and by Helmholtz, du Bois–Reymond, Virchow, and other German scientists on the scientific end, the Stazione Zoologica opened its doors in Naples in 1872.[43]

But Dohrn had to continue seeking more money from Berlin. In June 1875 he again turned to Helmholtz for help, this time asking him to speak with the crown prince on behalf of the project. He hoped the crown prince would make a financial contribution — Darwin, the Royal Society, and other English scientists had already contributed — and that individual German scientists would likewise do so (to date, none had). Helmholtz did not consider himself sufficiently competent in marine zoology to speak with the crown prince about it. However, he, along with du Bois–Reymond and Virchow, was prepared to speak about the Station's "scientific meaning." He thought Dohrn should not expect much in the way of monetary support from the crown prince, since his assets were quite limited. "He [the crown prince] and the Crown Princess do a lot for the arts, which, to be sure, with us [Germans] have to do much more begging than the [natural] sciences do, and for which both [the crown prince and princess] also have more personal interest." Later that year, he again showed himself to be a leading academy figure in seeking financial support for Dohrn's Station. He proved to be the decisive figure, both in his advice on how to deal with the crown prince and in gaining the academy's agreement to finance a boat for the Station.[44] Within a few years of its opening, the Station became the premier marine zoological station, a leading-edge institution (especially for evolutionary research) backed in good part by Germany but also able to operate through its self-generated income that came from selling specimens and renting tables or table space. It was highly international in its composition of researchers and supporters, and it helped advance the understanding of marine life.

Dohrn continued his efforts to find long-term, permanent funding for the Station, and that meant from the Reichstag. He visited Berlin again in early 1879, once again obtaining the support of Helmholtz, du Bois–Reymond, Virchow, Siemens, and others. The first three of these petitioned the Reichstag to support the Station, arguing that since the rise of the descriptive life sciences in the eighteenth century, and especially since 1850, there had emerged an attempt to understand the transformation, development, and metamorphosis of organisms. This effort transcended merely collecting and categorizing objects. Zoologists needed access to the sea and to appropriate facilities and equipment for the analysis of their specimens. Thanks to Dohrn, the petitioners said, an institute devoted to addressing these new problems had been created. To date the Station had received financial support from a combination of sources: private money, the academy, various governments, and various parties in England. It had grown to serve twenty-five researchers per year, a number that soon increased to thirty, two-thirds of whom were German. The Station needed money to continue operating at its current level, and if it did not get it, the petitioners argued, both science and Germany's reputation would be injured. They asked the Reichstag to secure the Station's future.[45]

On two occasions in early 1879, Dohrn saw Helmholtz. He was optimistic about his project. He wrote his wife that he dined one evening at Kronecker's: "I always have a type of honorary place here; yesterday I again sat beside Helmholtz, across from Mommsen, etc. In short, I'm climbing very high in people's estimate." He also attended a reception at Helmholtz's: "Shrewdly, I will not fail to be there, especially on account of the Mrs. [Helmholtz]." He was feted in the city as he sought support from the legislature for his Station. He asked his wife to have the German-language petition signed by Helmholtz and others translated into Italian, and he would have it published in *Rassegna*; he was also arranging for an English translation to be published in *Nature*.[46]

In the spring of 1880, with permanent financing still unachieved, Dohrn made yet another trip to Berlin. "Here too in Berlin, in the end only the great savants remain my sincere friends," he wrote. "The others go into the Envy-Choir. But Helmholtz, especially Mrs. Helmholtz!, Dubois, Virchow, Siemens, and Roth (as a personal friend), they are tied to my heart—the others would like to see, with the greatest delight, that I've at last crashed." He also presented the first volume of the Station's studies on fauna and flora to the academy and to Helmholtz, du Bois–Reymond, and Virchow.[47]

As Dohrn waited for permanent financing to be approved, there was talk that he might get a professorship in Berlin and, if he wanted, unite it with his directorship of the Naples Station. He asked du Bois–Reymond to let this be known publicly. He also hoped the Italians would make a financial contribution to the Station; Virchow might speak in this regard with Guido Baccelli, a leading statesman. Twice more Dohrn visited Helmholtz to obtain his (further) support for his Station. Dohrn again became convinced that final approval was at hand, though Gustav von Gossler, the new minister of culture, felt that Dohrn had not done enough. Gossler wanted him to send around the proclamation that Helmholtz and his coauthors had written and invite a "a large gathering" to the Reichstag, including "all the outstanding men of Berlin, of the government, of the Reichstag, learned and scientific men, medical doctors, and financial leaders." At their meeting, Dohrn and Gossler planned that Helmholtz, Virchow, Siemens, and others would give short speeches "and thereupon set the subscription list in motion." The crown prince sent a letter of support for the Station's financing. While there was general support for it, no one individual was willing to make Dohrn's cause his own. Helmholtz, he wrote, was "cool and, as is his nature, reserved."[48] Not until five years later did the Reichstag at last grant the Station's permanent financing.

Pour le Mérite

In southern Germany, especially Baden, the Helmholtzes had met many royal and aristocratic figures. Such occasions multiplied once they were in Berlin. They were invited to numerous social events by liberal members of the Prussian royalty and nobility. Helmholtz was also a member of Berlin's Monday Club (Montagsclub), a distinguished social club founded in 1749 consisting of elite government officials, scientists, artists, medical doctors, military officers, and the like. It met in the "Englisches Haus" in the Mohrenstrasse on Monday evenings, where the men dined together and afterward heard a lecture on a learned or scientific topic.[49]

In Berlin the Helmholtzes became well acquainted with the monarchy. Emperor William I and Empress Augusta invited them to more or less intimate gatherings. The emperor had been the commanding general of the Gardes du Corps, Helmholtz's old regiment, and that constituted a bond between the two men. The empress, in contrast to her husband, was highly cultured, liberal, an Anglophile, and an opponent of Bismarck's. Anna's younger brother, Ottmar, was a career diplomat and civil servant, who served for six months as Empress Augusta's cabinet secretary. The empress let Anna know that she was very pleased with his work.[50]

Above all, the Helmholtzes were favorites of Prussian Crown Prince Frederick William and Crown Princess Victoria (the eldest child of Britain's Queen Victoria and Prince Consort Albert, and known as Vicky). The crown prince loved study in general as well as the arts but also enjoyed his military duties; the crown princess had studied painting and music, loved the arts in general, and had been instructed in the natural and social sciences. They invited many university professors to their quarters; the Helmholtzes, along with Lepsius, Mommsen, Zeller, and Curtius and their wives were regular guests. So too were the art collector Count Wilhelm von Pourtalès, the von Schleinitzes, the von Radziwills, Prince von Hohenlohe-Langenburg, the British ambassador to Germany, Lord and Lady Odo Russell, and the painters Menzel, Richter, Heinrich von Angeli, and Werner, to name but a few who were also friends or acquaintances of the

Helmholtzes. In general, Vicky, who considered Bismarck her archenemy, supported artists and scientists. The crown prince and crown princess's son, the future Emperor William II, recalled that he often saw Helmholtz (and Virchow) at their house and that Anna and his mother were very good friends and spent much time together.[51]

Helmholtz was also slightly acquainted with another emperor, Dom Pedro II of Brazil. From childhood on, the Brazilian emperor had been greatly interested in science and technology as well as many other aspects of culture. He was a cultivated amateur, a generalist who sought to follow developments in science, technology, and other areas of knowledge. In 1877 he was elected an *associé étranger* of the Académie des Sciences in Paris. He visited Berlin twice during the 1870s and on at least one occasion called upon Helmholtz before going on to see Bismarck.[52]

The scientific reputation that Helmholtz had earned, the public service that he had given, the patriotism he had shown, and no doubt too the social connections he quickly made in Berlin—all of that led to his election on 17 August 1873 to the Orden pour le Mérite für Wissenschaften und Künste (Order of Merit for the Sciences and Arts). The order was a highly select organization that had been established in its civilian form by Frederick William IV in 1842 to honor extraordinarily distinguished figures in the arts, letters, and sciences. It was meant to complement the military order, which had been created by Frederick the Great a century earlier; hence, it was (and is) known as the Civilian or the Peace Class. It was the highest civilian award in Prussia. As originally conceived, it consisted of thirty German and twenty-six non-German members, with Alexander von Humboldt as its first chancellor (1842–59). Members included (at one time or another) the scientists Johannes Müller, Gauss, Bessel, John Herschel, and Faraday; scholars like Jakob Grimm, Ranke, Bancroft, and Mommsen; and artists like Jean-Auguste-Dominque Ingres, Liszt, Giacomo Meyerbeer, Felix Mendelssohn-Bartholdy, and Menzel, to name only a few. The order helped both to solidify Prusso-Germanic nationalism and to enhance the elite cultural and social status of Germany's artists, scientists, and men of letters—there were no women members—in the eyes of their counterparts elsewhere in Europe and in America.[53] The award recognized Helmholtz's scientific accomplishments but also served to bind him ever closer to the state. Nothing could have better shown that he was in and among the cultural elite.

Kulturkampf in Science, I

Kulturkampf

The 1860s and 1870s marked a new high point in European national-ism. Bismarck's victory over the liberals and other constitutionalists, along with his various provocations that led to wars against Denmark, Austria, and France, reinforced a strong national, chauvinist atmo-sphere within Germany. It was equally matched by the nationalist and imperialist forces of its archrivals Britain and France. These political (and cultural) rivalries found their counterparts within the world of sci-ence. As in politics, the German, British, French, and other scientific communities displayed nationalist and chauvinist sentiments and ten-sions. Helmholtz and many other scientists sought to achieve a balance between the ways science was perceived and promoted: as a universal, international phenomenon, on the one hand, and as a matter of na-tional effort and pride, on the other.

Berlin stood at the heart of the *Kulturkampf*, a cultural and politi-cal struggle against the Roman Catholic Church that reached back to the 1850s and 1860s, dominated much of German political and cultural life from 1871 to 1878, and did not fully end until 1887. Its origins lay both in the Church's antimodernist stance—especially as expressed in

Pope Pius IX's Syllabus of Errors (1864) and the doctrine of papal infallibility (1870)—and in Prussia's defeat of Austria in 1866, as south German Catholics came to oppose ever more strongly Prussia's hegemony over them. Bismarck questioned (more or less cynically) German Catholics' political loyalty, claiming that it lay more with Rome and the Church than with Berlin. The emergence of the Catholic Center Party and the Germanizing of Polish Catholics in the Prussian east also reinforced older prejudices against Catholics. Especially in the mid-1870s, Bismarck led his fellow conservatives, as well as his liberal opponents, by helping the Prussian parliament and the German Reichstag enact a series of anti-Catholic laws and bureaucratic regulations. Yet despite the widespread anti-Catholicism, the *Kulturkampf* proved to be a failure for Bismarck and the liberals. As it slowly dissolved after 1878, it even became a threat to the conservatives and liberals themselves.[1]

The *Kulturkampf* forced Helmholtz into the public arena in a way that he had never before experienced or expected. The addresses, writings, and activities of several friends or colleagues (du Bois–Reymond, Virchow, Tait, and Tyndall) as well as two enemies (Zöllner and Eugen Dühring) formed the background and shaped the nature and timing of his own, involuntary involvement in Germany's *Kulturkampf*.

Science and Nationalism: Du Bois–Reymond's and Virchow's Addresses

In October 1869 du Bois–Reymond delivered a rectoral address before the members of the University of Berlin. He spoke on the German university system, claiming that its decisive, unmatched feature was the freedom of professors to teach whatever they considered most appropriate: "the German mind" neither put intellectual limits on research nor feared any intellectual consequences. He praised the position of the private lecturer (*Privatdozent*) as the means for continually renewing the system intellectually and institutionally. He also noted that the system embodied the students' freedom to learn. While he conceded that the German university system was not perfect, he believed it had played a "glorious" role in German history. The German universities, he said, were the centers from which the German "mind" fought against "Roman intellectual slavery" and, after 1800, from which Germany had renewed itself.[2]

Virchow was the most politically active scientist in Bismarckian Germany: he cofounded (in 1861) the Deutsche Freiheitspartei (Progressive Party); served as a longtime member of the Prussian Landtag, the German Reichstag, and the Berlin city council; and was a leader at many of the annual meetings of the Association of German Naturalists and Physicians. He argued repeatedly for keeping

science free from outside interference, for its importance in unifying and educating the German people, for its influence on moral education, for its part in
shaping a liberal society, for its applications in industry and medicine, and, most
generally, for it as a source of progress.[3]

At Rostock in September 1871, Virchow spoke before the Naturforscherversammlung on science and German national life. He recounted the association's past meaning for Germany and science, as he saw it, and explained his
vision of the sciences' new role in the emerging German Reich. German and
Austrian scientists, he claimed, led the way in creating a sense of German-wide
unity in both the scientific and the national-political arenas; by its very nature,
the Naturforscherversammlung brought scientists together and so helped to
form a more unified Reich. His speech reflected his alliance, as a left-liberal, with
Bismarck's anti-Catholic policy. He strongly criticized the Church's attack on the
state through its Syllabus of Errors. Its actions, he claimed, justified the state's
counterattack. He drew a sharp distinction between the outlooks of the Church
and of the natural sciences, with the latter standing for intellectual and moral
progress, individual and societal development, material well-being, the chance
to rid oneself of errors and illusions, and the possibility of national unity. The
meaning of science for Virchow lay not only in the material benefits it could con-
fer but also in its "ideal" side as a basis for national life.[4]

In August 1872 du Bois–Reymond returned to the podium to address, like
Virchow the year before, the Naturforscherversammlung in Leipzig, attended
that year by nearly thirteen hundred natural-science researchers and physicians. He spoke on the limits of natural knowledge. His address—perhaps the
most intellectually distinguished of his outstanding second career as a speaker
about science—concerned the two subjects that, he said, scientists would never
be able to comprehend: the nature of matter and force, and human consciousness. Although scientists would continue to amass empirical results on these
phenomena, they would never be able to explain them. In his judgment, scientists were and always would be ignorant—"Ignoramus" and "Ignorabimus," as
he famously put it, about these phenomena. His epistemological doubts suggest
that the old struggle between the materialist and the vitalist points of view had
entered a new phase in that, like Helmholtz and unlike Haeckel, he wanted to
avoid or minimize pointless philosophical debate. The address immediately became one of his most famous (and his most controversial).[5]

Finally, in January 1873, Virchow made another and equally historic speech,
this one before the Prussian Landtag, where he coined the term *Kulturkampf*,
the "struggle" between science and the Church. This struggle, he said, was shaping modern history. He favored enacting laws against the Church, thought that
no compromise with the Catholic bishops of Germany was possible, and believed

that the interests of the Church (including its Pope and his alleged infallibility) were inimical and opposed to those of the Reich. The Church and the Reich were "already at present in open war."[6]

Zöllner's Attack

In this highly charged political atmosphere, Helmholtz faced the two enemies — "opponents" is far too mild a word — of his career. It was not by chance that they challenged him as he was increasingly glorified as a scientist and as a strong mood of nationalism overtook Germany. One enemy, Johann Karl Friedrich Zöllner, was a professor at Leipzig, an accomplished if minor astrophysicist who resented Helmholtz, his warm relations with British scientists, and what Zöllner saw as Helmholtz's deficient nationalist spirit. The other, Eugen Dühring, was a private lecturer at the University of Berlin, a philosopher and political economist who also fancied himself a scientist but who is better described as a demagogic orator and author, a rabble-rouser, and a future leader of the nascent racial anti-Semitic movement. Dühring was hell-bent on causing as much trouble as possible for the German academic establishment.

In 1872 Helmholtz found himself on the receiving end of a book-length, chauvinistic diatribe by Zöllner. Zöllner had ironically begun his career as a rather obsequious student of Helmholtz's writings. A decade earlier, he had initiated a correspondence with him in which he told him about his new photometer. He had finished a manuscript on eye movements, and Helmholtz's work had given him the explanation he needed for part of the phenomenon. He proudly proclaimed that his psychological theory concerning optical illusions had found Helmholtz's "recognition." In 1864 he habilitated at Leipzig; four years later he became extraordinary professor of "physical astronomy" (that is, astrophysics, a term that he apparently coined in 1865) there; and in 1872 he became ordinary professor.[7]

Just as he was appointed to the ordinary professorship, Zöllner published his *Über die Natur der Cometen: Beiträge zur Geschichte und Theorie der Erkenntniss* (1872), a long, hastily written, quirky book that was far less about comets and epistemology than it was a polemic and a broadside filled with German nationalist rhetoric, along with general hostility toward British scientists (especially Tyndall, Thomson, and Tait) and the treacherous Helmholtz, who supposedly aided and abetted them by speculating on comets and the emergence of life on earth. Zöllner said his book was sparked by a speech given by Thomson and Tait at a meeting of the BAAS in 1871 and by the German-language edition of their book on theoretical physics.[8]

Zöllner also attacked the German academic world in general, including the way it made appointments. He was much exercised by Thomson's claim that

Stokes and Stewart had discovered spectrum analysis before Bunsen and Kirchhoff did, and he defended Kirchhoff's priority in the discovery of the absorption and emission law of radiation, as against Thomson's flimsy claims for British priority. He attacked the German version of Thomson and Tait's *Treatise on Natural Philosophy* (*Handbuch der theoretischen Physik*), which was translated by Gustav Wertheim; Helmholtz supervised and also translated the introduction. Zöllner issued an even stronger critique of the German translation of Tyndall's work. He portrayed this translation work as unpatriotic. Furthermore, he challenged Thomson and Tait's critique of Wilhelm Weber's electrodynamic hypotheses, maintaining that Helmholtz's claim that Weber's theory contradicted the principle of conservation of energy was itself mistaken and that he and (independently) Carl Neumann had shown this to be so.[9]

Zöllner's text was replete with anti-English remarks in regard to science. (He always referred to the English or England, though neither Thomson nor Tait was English; both of them lived and worked in Scotland.) He claimed that the source of all the English shortcomings was the general overemphasis on inductive processes in English science. Moreover, the widespread popularization of science in England had made matters worse, since it (supposedly) spread ignorance. By contrast, he said, "Germany is called to be the carrier and showplace of this age because only the Germanic mind harbors in its depth that fullness of deductive needs and abilities that are needed for successfully overcoming the inductive materials stored up through the exact sciences."[10]

In what later became a typically antimodernist move, Zöllner also criticized non-Euclidean geometry and showed himself to be an anti-Semite and an antivivisectionist. Yet his principal epistemological differences with Helmholtz concerned the comprehensibility of nature and the role of sense perception. He believed that Helmholtz's emphasis on assuming the existence of the law of causality so that unconscious inferences could occur merely copied what Schopenhauer had already said nearly sixty years earlier. He attacked Helmholtz for imitating if not stealing from Schopenhauer, while simultaneously suggesting that Helmholtz had never even read Schopenhauer. He bitterly complained that while Helmholtz devoted eight pages of his *Handbuch* to explaining various optical illusions on the basis of eye movements, he gave only three lines to Zöllner's explanation on the basis of unconscious inferences, and he further noted Helmholtz's "complete misunderstanding" of his own "theory of the unconscious operations of the understanding." He related his earlier experiments to show that Helmholtz's theory about the role of eye movements was dead wrong.[11]

Reactions to Zöllner's book came fast. Tait was delighted by it, precisely because Zöllner had "pitched into" Tyndall, Thomson, Helmholtz, and Hofmann; he looked forward to the "splendid row." He told Helmholtz, "*You* have been even more maltreated than Thomson & myself." He thought it probably useless to respond to Zöllner but insisted that his and Thomson's scientific differences with Weber were in no way "rude or impertinent."[12]

Tyndall dismissed Zöllner's book altogether. Until its appearance, he told Helmholtz, he had thought of him only as "a modest hardworking man who had done some good work in photometry and with the spectroscope." But when he looked at the book, he said, "I could hardly credit my senses; the discharge of animosity against me was so virulent, and the cause of it all so inadequate. From Germany I had small expectation of such an attack as this." Tyndall found Zöllner's objections to his popularization of science irrelevant. "His wrath, I may remark, is a small matter to me when weighted against the fact that *you* [Helmholtz] have thought those popular books worthy of reproduction in Germany." He could not understand how "any sane mind" would object to popularization.[13]

Helmholtz sympathized totally with Tyndall. "Zöllner's book," he replied, "has also set us in Germany into the most extreme state of astonishment." At first it was thought he was mentally ill. "Moreover, for the moment it still hasn't developed so far that one might have to lock him up in a madhouse. Quite the contrary, the Leipziger [faculty] have, following the publication of this book, made him a preliminary professor, and he's supposed to have many friends in Leipzig who are amazed by him." Moreover, there was "a sort of rivalry between the little minds of the University of Leipzig and Berlin, and there one figures that he has perhaps done a service by firing away at the Berliners." Still, Helmholtz thought the book "crazy." Zöllner, he said, suffered from a "poisonous envy" vis-à-vis the success of others while he himself failed to progress in his own work. As for his theory of comets, it was "too nonsensical for a healthy person, one who has so much knowledge and talent, as Zöllner at the start (in any case) had, to have brought it forth." Helmholtz at first decided not to respond to Zöllner at the time. "One makes oneself laughable if one would respond to him." He did, however, respond to him on the topic of Neumann's support of Weber's electrodynamic law. But what he found most objectionable in Zöllner's book was "the hatred of the metaphysicians against natural science." It was "the old natural scientific (the metaphysical) ecstasies that still [lay] hidden in many German temperaments, and that [had] not yet considered to go public." Metaphysically oriented people seemed to believe Zöllner, Helmholtz thought. And in a slightly petulant tone, he added: "I myself have already also had to experience how much envy there slumbers away in some hearts against my own truly very modest external success in life." The philosophers thought Zöllner an "expert authority in natural science," and they gladly used this "to give vent to their oppressed hearts." (Nietzsche liked Zöllner's book and found in him a kindred spirit who attacked German academics; the neo-Kantian philosopher Alois Riehl liked it too, though he deplored its polemical nature and its criticism of Helmholtz.) In addition, Helmholtz continued, there were the German chauvinists who, though they had "done nothing and . . . suffered nothing" in the Franco-Prussian war, "woke up one beautiful morning as the victors of Sedan, and henceforth did not know how to let go of their national pride. For these types, Zöllner designed his book very, very well."[14]

A Low-key Visit to Leipzig

Zöllner's absurd charges against Helmholtz came while the latter was revising his essay on conservation of force for republication. Lipschitz had just pointed out an error in it, and after thanking him for doing so, Helmholtz continued:

> I really don't have to tell you how annoying it was to me to be forced, thanks to *C. Neumann's* arrogant thoughtlessness and *Zöllner's* craziness, into a further polemic against old *W. Weber*, against whom, in my first work, I had intention-ally said as little as possible, thinking that what was said would suffice for every expert in this matter. I have to concede that, in general, Zöllner really achieved his purpose of getting under my skin (for several weeks). If one suddenly sees such a profusion of passionate hate let loose upon oneself, above all from a per-son with whom one believes to have completely friendly relations, one natu-rally becomes mistrustful against oneself, and broods as to whether one may really have given a justified reason for such outbursts.

The differences between Weber and Helmholtz were substantive, scientific ones, not cultural or political, and in fact those differences never affected their lim-ited but cordial and respectful personal relations. In 1873, for example, Weber wrote Helmholtz to say that he looked forward to the opening of Helmholtz's new institute and laboratory, which Weber said would be "a model for everyone." That year, too, Helmholtz hired, in part on Weber's recommendation, Fried-rich Neesen as his first assistant; Neesen had previously been Weber's assistant at Göttingen. Similarly, it was Helmholtz who took the initiative in asking the Prussian Academy of Sciences to congratulate Weber, a foreign member, on his forthcoming doctoral jubilee (1876).[15]

Zöllner's charges against Helmholtz were more than national chauvinist rivalry, personal envy, and (alleged) mental illness. They were also part of a long-standing antagonistic political and cultural relationship between Prussia and Saxony (including, respectively, Berlin and Leipzig), one that dated back to the Seven Years' War and the Napoleonic Wars (when Saxony took France's side and, as a result of the Congress of Vienna, Saxony ended up losing much territory to Prussia). In 1866 Prussian troops occupied Saxony (Austria's ally) in order to control the railroad lines and thereby dispatch its troops to Bohemia. After Saxony lost the war, it became subject to Prussia within the new North German Confederation, thus losing its ability to conduct its own foreign affairs.[16]

The two provinces were also quite different in character. Saxony was much more liberal: after 1831 it was a constitutional monarchy, and it had long been home to many printing presses, book dealers, and a renowned book fair. Saxony sought to counter Prussia by emphasizing its economic, scientific, and cultural strengths: this sometimes led to a general rivalry between the University of Berlin and the University of Leipzig, including tensions over professorial appointments. Zöllner's text and its charges played into this long-standing intra-German political strife. From his stronghold in Leipzig, Zöllner averred: "Today some learned people in Berlin combine the concept of a 'simple professor' with the idea of an elegant man who, in a brilliantly equipped institute, gives great parties and knows how to hold popular lectures before women and men."[17] Few German academics would have failed to see Helmholtz in this passage.

The name Weber was closely associated with Leipzig. The three Weber brothers came from Saxony. Ernst Heinrich Weber (1795–1878) taught at Leipzig for his entire career, serving as professor of physiology and anatomy until 1871; Wilhelm Eduard Weber (1804–91) taught there as professor of physics between 1843 and 1848, while he was in "exile" from Göttingen, where he otherwise spent his career; and Eduard Friedrich Weber (1806–71) taught physiology as a private lecturer at Leipzig. Although Helmholtz had a cordial relationship with all three, he and Wilhelm had rival theories of electrodynamics; he was also perhaps a competitor of Ernst Heinrich and Eduard Friedrich in physiology. By contrast, Helmholtz had an admirer in the new professor of philosophy at Leipzig, Ludwig Strümpell. He also enjoyed cordial relations with Fechner, who had served as professor of physics at Leipzig until an illness had forced him to retire early (in 1839). He was, along with Helmholtz and Wundt, a prominent if not a founding figure in experimental psychology. His sometimes mystical musings and orientations, however, meant that he was also a man cut from a different scientific cloth than Helmholtz. Still, the two were on good, respectful terms, even after Zöllner's outburst. Fechner consulted with Helmholtz about the composition of tones and about the physiology of speech, for instance. Fechner himself thought that Zöllner's photometric work, like his book on comets, was not sufficiently cautious and not grounded enough in facts and that Zöllner's polemic—its tone and its targets—would undermine future scientific discourse. Apparently others in Leipzig shared Fechner's misgivings about Zöllner's book. Zöllner told Fechner that Helmholtz socialized "mainly in diplomatic and rich Jewish homes," not in those of scientists, and that he lived beyond his means.[18]

Above all, at Leipzig Helmholtz had his close friend Carl Ludwig. In June 1872 Ludwig asked him to attend the upcoming Naturforscherversammlung meeting there that August and to stay with him. Helmholtz had been reluctant to do so. Ludwig assured him, however, "If you would come, you'd convince yourself how untrue all the rumors are that have broken out about a rivalry between Leipzig and Berlin. You have as many friends here and people who honor you as there are people here who know you. And if the little people from Berlin to Leipzig (and in reverse) look over their shoulders, then let the better ones show that we know only that envy which improves the heart and the forces that make one competitive." Ludwig and others were delighted when Helmholtz agreed to come. The situation with Zöllner, however, was not good. After Zöllner's most recent publication, which included references and quotations from private letters from du Bois–Reymond, from the Leipzig physicist Berend Wilhelm Feddersen, and from the Leipzig chemist Hermann Kolbe, Ludwig, who at one time had been quite friendly with Zöllner, now came to see him as a polemicist with "ignoble motives." He now distanced himself from him entirely, and Helmholtz decided against going to Leipzig. He wrote Karl Thiersch, professor of surgery at Leipzig and the managing director of the Naturforscherversammlung, that he had to decline Thiersch's request to hold a public lecture in Leipzig before a plenary session. He said that his new work did not fit the requirements of such a lecture. He continued: "I would have gladly put in directly [to give] a lecture at the Naturforscherversammlung despite many of the animosities which have come out of Leipzig, if I would have found a theme that I had thought to be unassailable against the malevolent criticism in certain circles of the University of Leipzig for which I, more than other speakers, must still be prepared." Yet in the end he decided indeed to attend the Leipzig meeting, but to keep a low-key public presence: he stuck to a brief presentation of his recent work on the galvanic polarization of platinum, and he visited the Gewandhaus, where he heard a performance of Felix Mendelssohn's concert overture *A Midsummer Night's Dream*.[19]

Translating the British

Helmholtz had no reason to be defensive about promoting the translation into German of the writings of several British men of science, and indeed he felt proud of it. He not only facilitated and supervised these translations by others; he also served, at least in the cases of Tyndall and Tait and Thomson, as the principal go-between with the German publisher Friedrich Vieweg. The firm was a major scientific publisher, especially of physicists and technologists, including Humboldt, Clausius, Wiedemann, and Alfred Wegener.[20] It published several of Helmholtz's works, including the *Tonempfindungen* (six editions between 1863 and 1913) and his three-volume collection *Populäre wissenschaftliche Vorträge* (1865, 1871, and 1876), which from 1884 on reappeared as the two-volume *Vorträge und Reden* (third edition, with a fourth edition in 1896 and a fifth in 1903).

Later, the firm also published Koenigsberger's three-volume biography *Hermann von Helmholtz* (1902–3), as well as a much abbreviated, popular one-volume version. Vieweg and the Leipzig firms of Leopold Voss, which published his three-part *Handbuch*, and Johann Ambrosius Barth, who published his three-volume collected scientific papers, the *Wissenschaftliche Abhandlungen* (1882, 1883, and 1895), were Helmholtz's principal German-language publishers.

Thanks to Helmholtz's efforts as intermediary, Vieweg published the German translation of Thomson and Tait's textbook on theoretical physics, the *Handbuch der theoretischen Physik* (2 parts, 1871 and 1874) and a series of books by Tyndall: *Faraday und seine Entdeckungen* (1870); *Fragmente aus den Naturwissenschaften* (two volumes and two editions, 1874, with Anna Helmholtz as translator and a foreword by Hermann Helmholtz; and 1898–99, with Anna von Helmholtz and Estelle du Bois–Reymond as translators and a foreword by Hermann Helmholtz); *Die Gletscher der Alpen* (1898, Gustav Wiedemann, translator); *In den Alpen* (two editions, 1872 and 1899, Gustav Wiedemann, translator); *Das Licht* (two editions, 1876 and 1895, Gustav Wiedemann, translator); *Der Schall* (three editions, 1869 to 1897); and *Die Wärme, betrachtet als eine Art der Bewegung* (four editions between 1867 and 1894, with the third enlarged edition, 1875, edited by Helmholtz and Wiedemann, and with a foreword by Helmholtz). Vieweg also published the German translation of Maxwell's *Theorie der Wärme* (1878) and Rayleigh's *Die Theorie des Schalles* (two volumes, 1879–80), with Neesen, Helmholtz's assistant, as the translator of both of these works.[21] In short, Helmholtz did much to promote British physicists in Germany and, through Tyndall, to help popularize science.

Indeed, he promoted science in Britain, too. In 1873 Norman Lockyer, the astronomer and founding editor of *Nature*, began a series of short, hagiographic portraits entitled "Scientific Worthies." The fourth such piece was on Tyndall; its author was Helmholtz. Lockyer had asked Helmholtz to review Tyndall's writings and give a portrait of him for *Nature*, but he declined to do so. "I would not like to meddle purposely and actively into these discussions [that is, the ongoing dispute between Tyndall, on the one hand, and Thomson and Tait, on the other]. I have had much friendly intercourse with Tyndall, Sir W. Thomson and some also with Professor Tait. I must say, I have a strong feeling that my Scotch friends have been misled by a kind of national jealousy, but it is not my calling to stand up publicly and to declare it into [someone's] face, although I have defended very often Tyndall against them in private conversation." Lockyer was on Tait's side, Tyndall thought. What Helmholtz did do, however, was to publish an extract in *Nature* from his preface to the German translation of Tyndall's *Fragments of Science*.[22]

Three years later, Maxwell wrote a portrait of Helmholtz; Lockyer sent Helmholtz an advance, draft copy for his review and requested some "personal information" about him, as he "would like to appear." Maxwell, he said, was "anxious to know what led" him to study science, and Lockyer hoped he would "not object to let me know." Lockyer was having a similar portrait of Thomson prepared, and he scheduled it to appear in September 1876, when Thomson, as president of the Mathematical and Physical Sciences Section at the BAAS, would be an especially prominent figure at that year's meeting in Glasgow. Lockyer wanted to know Helmholtz's "estimate" of Thomson's "place in Science." "I am having a statement drawn up of his work but criticism from neighbours is not what we want in our series." Helmholtz responded:

> [Thomson's] peculiar merit, according to my own opinion, consists in his method of treating problems of mathematical Physics. He has striven with great consistency to purify the mathematical theory from hypothetical assumptions, which were not a pure expression of the facts. In this way he has done very much to destroy the old unnatural separation between experimental and mathematical Physics, and to reduce the latter to a precise and pure expression of the laws of phenomena. He is an eminent mathematician; but the gift to translate real facts into mathematical equations and vice versa is by far more rare, than that to find the solution of a given mathematical problem. And in this direction Sir W. Th. is most eminent and original. His electrical instruments and methods of observation, by which he has rendered, amongst other things, the electrostatical phenomena as precisely measurable as magnetic or galvanic forces give the most striking illustration, how much can be gained for practical purposes by a deep insight into theoretical questions.

This assessment appeared in *Nature* (7 September 1876), as part of a biographical piece compiled by Tait. As for Helmholtz himself, Maxwell flatteringly characterized him in his portrait as "the most illustrious example not merely of extensive acquaintance with science combined with thoroughness, but of a thoroughness which of itself demands the mastery of many sciences, and in so doing makes its mark on each." He was, Maxwell concluded, "the intellectual giant." Here was a mutual-admiration society that helped strengthen the British scientific community, including building a bridgehead to Germany. Macmillan, *Nature*'s publisher, wanted a photograph of Helmholtz. They were "anxious to include your head in a series of portraits of eminent scientific men which we issue from time to time." (They already had the heads of Faraday and Lyell.) Anna sent them the photograph.[23]

In 1876 Helmholtz's closest British friends and colleagues urged him to visit Britain again. Tyndall wanted him, as noted already, to come to London, and so, too, did Richard Liebreich, a German ophthalmologist who lived there. But he declined owing to poor health. As long as he rested, he said, he felt well enough.

He hoped, however, to visit London in September and perhaps also attend the upcoming BAAS meeting in Glasgow. Thomson wrote from New York City, where he had arrived as part of an extended visit to America that included lecturing at Hopkins. He had heard that Helmholtz was going to Glasgow, and he urged him and Anna to come as his guests and meet the new Mrs. Thomson. "We shall have a great deal to talk over when you come." Roscoe extended a similar invitation to visit Manchester, and Helmholtz now indicated that he would indeed travel to England and Glasgow in September. He also intended to spend several days in London, where he wanted to attend the Exhibition of Scientific Apparatus at the South Kensington Museum. He was listed as one of twenty-three Berlin delegates (there were plenty more from other German cities). The crown prince and princess were the German patrons. Two of Helmholtz's own pieces of apparatus (an electromagnetic machine and a pendulum) would be on display there. However, family concerns intervened; Anna was preoccupied with her aunt Mary's health, and Helmholtz was not feeling well. He did not in fact go to Britain that autumn; instead, he did a first sitting for his portrait by Lenbach. By early September he was in Pontresina, not Glasgow. He had wanted to go to Britain, he told Soret. But he expected that the large crowd in Glasgow, like that he had just experienced in Bayreuth, would not be good for his health, and thus he decided to "go to the mountains."[24]

First Counterattack: On Scientific Method and Metaphysics

Helmholtz planned a two-part counterattack to Zöllner's book. The first part appeared as the preface to part two of the German-language translation of Thomson and Tait's *Treatise on Natural Philosophy*. It focused on what Helmholtz considered the philosophical shortcomings of Zöllner's book and the inappropriateness of metaphysics in science. The second part appeared as the preface to the German-language translation of Tyndall's *Fragments*. Here Helmholtz addressed the issue of the popularization of science. It was of course not by chance that he used precisely the German-language translations of works by these British authors to respond to Zöllner.[25]

Thomson and Tait had collaborated on their *Treatise* for six years before the first volume appeared in 1867. Since 1866 Helmholtz and Thomson had corresponded about its progress. When Helmholtz last visited Glasgow, Thomson had shown him the copyedited pages of volume one, and because Helmholtz himself had required a decade to compose both part three of his *Handbuch* and his *Tonempfindungen*, he was especially understanding of the slow pace of Thomson and Tait's work. He encouraged Thomson by saying there was "a pressing need" for such a book on mathematical physics. Under Helmholtz's stewardship, and with a short preface by him, a German translation of part one of volume one appeared in 1871. By the spring of 1873, Helmholtz was sponsoring and supervising the German-language translation of part two of the *Treatise*. That autumn, he prepared his preface to it.[26]

Helmholtz's subject was "Induction and Deduction," and it was intended not only as a defense but also as a lesson to Zöllner and others about a basic issue in the philosophy of science. Helmholtz considered Zöllner's attack on his British colleagues as "more than a lively critique." He himself took the high road, avoiding any personal remarks and focusing instead on scientific issues or principles. That way, "if new facts are brought forward or misunderstandings are explained, in the expectation that if all the data are given, the scientific experts will finally know how to make their judgment, and do so without prolix discussions or sophistic arts of disputatious opponents." Thomson and Tait's *Handbuch* was designed not only for specialists, he said, but also for students, and thus he felt obliged to respond to Zöllner in the book's preface so that students should not be misled about the proper way to express scientific criticism.[27]

Helmholtz thought that a distinguishing feature of Thomson's physics was its attempt "to purify" it "of all metaphysical, surreptitious elements and of all arbitrary hypotheses." Thomson aimed "to make pure and true expression of the laws of facts." That was part of what made him a leading scientist, Helmholtz insisted, and was "a principal service of the book in hand." By contrast, Zöllner saw the book's lack of metaphysical thinking as "its fundamental deficiency." Helmholtz sought to change the terms of debate: he said Thomson wanted to emphasize the inductive method, whereas Zöllner emphasized the deductive method. Zöllner's book embodied the metaphysics of Schopenhauer, where "the stars should love and hate one another, feel pleasure and pain, and strive to move themselves as corresponds to these sensations." Helmholtz's message was obvious: this was not how to do science. Zöllner tried to reduce scientific disagreements "to all possible personal weaknesses of the opponent." Zöllner, he said, argued "in a completely intolerant manner, in which followers of metaphysical articles of faith treat their opponents by hiding, from themselves and the world, the weaknesses of their own standpoint." As Helmholtz wrote to Tyndall: "Among the learned classes in Germany there is little interest in natural scientific principles; rather, there is all the much more a secret tendency towards metaphysics, and Schopenhauer has a lot of followers. While for the most part they are ashamed to recognize [publicly] their metaphysical hopes, it does give them pleasure when they find an advocate who, with a certain appearance of expertise, cleverly pulls about the leading men of natural science who are so very uncomfortable for them." Tyndall was grateful for Helmholtz's support, especially with respect "to this Scotch madness [you] gave me. I shrank from mentioning to you William Thomson's name in connexion with this subject, knowing the intimacy of your friendship with him. With all his brilliancy of intellect he lacks that force of character which Tait possesses, in a coarse and ungentlemanly form it is true, but still strong. Thomson is thus at the mercy of Tait."[28]

Helmholtz also criticized Zöllner for bringing nationalism into a scientific argument. Thomson and Tait had maintained that hypotheses should stick closely to observable facts, Helmholtz noted, and as an example of a hypothesis that strayed too far from the facts, they cited Weber's law of electrical action-at-a-distance, which shared similarities with Newton's emission theory of light. There was nothing here that was anti-German, Helmholtz said. Everyone and every idea should be open to criticism. No matter how widely held a scientific hypothesis may be, if someone thought it false, then he should not only say so but should also show others why that was so. Zöllner's work, Helmholtz argued, was in part a diatribe and did not contribute to clarifying an issue—quite the opposite.[29]

Helmholtz then turned to Zöllner's use of the deductive method. He thought deduction in science was perfectly justified, which meant that it should be used to test a hypothesis and thereby also bring forth new facts: he saw facts and laws in continuous interaction with one another. Using the inductive method should not bring charges of "impiety," "even if the results of the investigation should turn out to be uncomfortable for the Icarus flight of speculation." Zöllner had charged Thomson and Tait with committing "crude thought errors," of speaking "absolute nonsense," and the like. But the latter, Helmholtz said, was precisely what Zöllner himself did.[30]

Helmholtz did, however, raise one possible shortcoming against Thomson: the hypothesis that organic germs were contained in and transmitted by meteors and that they then "became cooled-off world bodies." Thomson had suggested this possibility in an address before the BAAS in Edinburgh in 1871, as Helmholtz himself had done in the spring of that year. Helmholtz realized that this hypothesis was mere speculation, only a possibility, and that he could fault no one for objecting to it. Yet these sorts of speculations, he insisted, were a legitimate part of the scientific process when there was no alternative. Since there was no satisfactory explanation of organic life in terms of inorganic, perhaps life itself had never even originated, perhaps it was simply as old as matter, with "germs transported from one celestial body to another and developing everywhere where they found the most favorable ground." In any case, he thought little of Zöllner's own physical explanation.[31]

Helmholtz sent Tyndall a copy of his response to Zöllner, which Tyndall found "strong and dignified." He also liked the way Helmholtz had defended Thomson (on the transport of germs) and Kirchhoff. Helmholtz's preface was soon translated into English and appeared in *Nature*.[32] It was both a testament to the limits of nationalism within the international scientific community and a sign of the personal networks that the British and Helmholtz had developed with each other.

Second Counterattack: On the Popularization of Science

During the spring of 1873, Helmholtz began supervising the translation of Tyndall's *Fragments*. He told Vieweg that he intended to write a strong foreword that aimed against Zöllner's attack.[33] He recognized privately that at points Thomson and Tait were in part responsible for the controversy that had been engendered. He wrote Tyndall:

> Our Scottish friends, I must say to my regret, are totally filled with national vanity. This history with [the Scottish physicist and glaciologist] Forbes is completely nasty, and I don't see what one will reply against their representation. Because they in Scotland first heard of the glaciers through Forbes, they believe the entire world may have first heard of it through him. Tait is completely prejudiced in scientific in-fighting [*Boxerei*], and unfortunately Sir W. Thomson lets himself be taken around only all too easily. The emphasis [by the British] of Stokes against Kirchhoff in the question of the discovery of spectrum analysis is very unjust, and very badly received by us. Exploiting this point, Zöllner has favorably worked out the most for himself; many who explain the whole matter as [simply] crazy, also say that a detail or two therein may indeed be reasonable, and then they refer back to this story.[34]

Moreover, Tyndall himself could be quite the polemicist. In August 1874, as president of the BAAS, Tyndall gave his Belfast Address. It quickly became (in)famous. It was in part an account of scientific materialism—many thought it promulgated atheism—in part another skirmish in the ongoing battle of science versus religion, and in part a defense of the freedom of science itself, of its independence from religion. This battle for freedom of science in the 1870s was waged in Germany as well as in Britain and America, and Tyndall did not fail to mention Helmholtz, Huxley, and du Bois–Reymond, praising them for the clarity of their writings and "the breadth of literary culture" that they commanded. He also declared that the principle of conservation of energy was "of still wider grasp and more radical significance" than even Darwin's theory of the origin of species. He thought it applicable to organic as well as inorganic phenomena.[35]

Meanwhile, Anna Helmholtz finished translating Tyndall's *Fragments*, requiring a good two years to do this "very difficult" work, with Hermann checking her translation for scientific sense and accuracy. As she was completing the work in the spring of 1874, he wrote his foreword to it. But he soon ran into a problem: Zöllner was "unfortunately" right in one of his points against Tyndall's theory of comets. Two other difficulties arose. Vieweg had not originally included "On Spirits" in his collection of Tyndall's essays, and Helmholtz insisted that he must, for to omit it was to make it appear that Zöllner had been right. Helmholtz also noted a few of the other, shorter essays that Vieweg had omitted, including the one concerning Mayer: "Tyndall has fought [undecipherable] for R. Mayer (Copley Medalist of 1871) in England against the attempt to put him down. Precisely for this reason this essay must not be excluded." He considered this foreword, as opposed to the one in the Thomson and Tait volume, to be his main reply to Zöllner. Despite being scientifically and administratively overworked, Helmholtz had undertaken this project because he considered "the spread of successful popular presentations of the more important and well-developed parts of natural science as a useful work."[36]

In his foreword Helmholtz developed the theme of popularization. As in Britain, he said, there was a growing desire among educated people in Germany "for natural scientific instruction." Popular science should convey more than just bits and pieces of science for entertainment or mere curiosity. It should respond to "a serious intellectual need." The natural sciences had become useful to modern man, he said, "in the formation of the social, industrial, and the political life of civilized nations." It was not merely that the forces of nature were now sufficiently understood so as to harness them for human purposes and to make a difference to material life, a point that many already appreciated. There was also a "much deeper" reason: namely, science's "influence on the direction of the intellectual progress of humanity." While he acknowledged that the natural sciences had created "a split in the intellectual development of modern humanity," he regarded that as simply inherent in the nature of such development.[37]

Language, he claimed, constituted the centerpiece of "the developmental passage of civilized nations to date." Modern European culture was historically tied to classical languages and literature, and the study of language was closely tied to the study of ways of thinking. Yet that study concerned known truths, not how to find new ones. Logic taught how to draw proper conclusions from premises, but it said nothing about the sources of those premises. To achieve the latter, Helmholtz argued, first required "knowledge of individual cases" and from there the formation of a law followed by drawing further logical conclusions from it. Yet none of this went "to the actual source of knowledge"; it did "not put us before reality." Indeed, there was always "an unmistakable danger" that a combination of image and word could lead to mythmaking and metaphysical systems.[38]

Helmholtz thus thought that the study of language, literature, and their associated logic missed something important in shaping and understanding human culture. They lacked methodical training in how to subordinate the disordered, seemingly irrational materials of reality into an ordered conceptual system as expressed in human language. That was what observational and experimental natural science (almost alone) did. As for the "social sciences"—Helmholtz referred specifically to individual and group psychology, education, sociology, and political science—they were at best something that might occur in "a distant future."[39]

The successful pursuit of natural scientific research showed, he maintained, the fecundity of first determining the facts and then proceeding to develop laws out of them. The sciences of inorganic nature had especially shown this to be so; and once fact-based laws of nature were established, the process of deducing logical consequences could follow. They in turn would allow a further testing of those consequences against empirical reality. There was hardly any other "human thought construction" comparable to this process of natural science in terms of "logical consequence, security, exactness, and fruitfulness." Helmholtz thus commented: "The natural sciences are a new and essential element of human development and one of indestructible meaning. . . . A full development of the individual human being, as of nations, will no longer be possible without an association of the prevailing literary-logical with the natural scientific direction."[40]

Most educated people—including statesmen, schoolteachers, society's sundry moral guides, historians, and so on—had not been trained in natural science and its methods, he noted. Yet it was precisely these people who had to introduce changes to modernize the educational system, and they needed guidance. Everyone—and here he pointedly included women—who was interested in the advancement of science and its influence on human education, culture, and general intellectual development should seek to give "the educated classes insight into the art and success of natural scientific research." Already the substantial number of popular science books and lectures showed that those individuals with a strong background in language and literature were indeed interested in acquiring greater knowledge about the natural sciences.[41]

Rigorously trained scientific specialists, with their frequent use of abstract concepts and their analytical tendencies, were not well prepared to meet the broad needs of educated people for the understanding of modern science in general, Helmholtz thought. Rather, someone who had "a certain artistic talent of presentation, a certain type of necessary eloquence" was needed. This was the very opposite of the procedure found in scientific papers and books. He thought there existed "a sort of barrier" between laypeople and natural scientists. It was thus fortunate that a qualified scientist like Tyndall could provide enthusiastic leadership in overcoming this chasm. Tyndall understood how "the newly achieved insights and views of his science" could have an effect on "a wide circle of people," and he explained them "with eloquence and the gift of insightful presentation."[42]

Helmholtz noted that England had an older tradition of popular scientific lecturing than Germany did, one that made science more widespread in English than in German culture. The lectures and associated demonstrations offered by the Royal Institution underscored this point especially well, he said. For these reasons, he had sought to promote Tyndall's popular scientific writings in Germany, and he thought that the successful, broad reception of the German-language edition of Tyndall's *Heat a Mode of Motion* had already justified his decision.[43]

One could investigate nature's laws, Helmholtz continued, either through the use of abstract concepts or by experimenting; but one could only proceed in the first way after the road had been cleared to a certain degree by the second. Some degree of law based on inductive work had to precede the more analytical investigation, that is, the use of mathematical and logical reasoning, of nature. The observational and experimental road to investigating nature, with its "sensory, living insight of the type," was also how the artist proceeded; thereafter, concepts were formed. No matter how talented a physicist may be at inductive or at deductive work, Helmholtz declared, he could never completely neglect one sort of procedure for the other, for to do so meant that the theorist had lost the empirical basis of his conceptual work and the experimentalist had lost sight of science's aim of expressing findings in terms of precise conceptual structures. Tyndall, Helmholtz emphasized, was principally an experimenter, and he used his imagination in his work, providing insight into experimental phenomena. It was precisely these qualities that made him such a successful lecturer.[44]

Helmholtz had been astounded that, of all foreign scientists, Tyndall should be attacked by Zöllner and that this was done on the basis of German national chauvinism. Tyndall had excellent credentials in the world of German science, Helmholtz insisted. He had studied in Germany (principally at Marburg); loved and kept abreast of German literature and science; was quick to recognize publicly the achievements of Mayer in "discovering" the law of the conservation of force and Kirchhoff for his "discovery" of spectrum analysis (and had paid a price in England for doing so); was currently supporting investigative studies on glaciers by the Swiss (Louis Rendu, Louis Agassiz, and Pierre Jean Edouard Desor) at the expense of the Scotsman Forbes; and, finally, had endowed a foundation for American students to study physics in Germany. Zöllner's attack on Tyndall, Helmholtz in turn charged, issued from philosophical sources, from "the opposition against the inductive method of the natural sciences." Natural science had made its most extensive and rapid progress by avoiding a priori, "supposed deductions," he maintained. German natural researchers, especially in comparison to their British counterparts, had been slow to understand this; but now that they had done so, they had decisively moved beyond such ways of doing science. Helmholtz pointed to the success of physiology in Germany as a premier example of the change in style and the associated progress. Zöllner was a throwback to an earlier era in German science, when two sides respectively sought to emphasize either the high work of pure thought or the lowly work of observation and experiment.[45]

Helmholtz thought it dangerous to equate philosophy with metaphysics, comparing the latter to astrology and the former to astronomy. Metaphysics had always attracted "scientific dilettantes" to philosophy, and they had done more injury to philosophy than its greatest opponents ever had. They looked for "relatively quick and painless ways of insight into the deepest relations of things and the nature of the human mind" and for means to understand the past and predict the future. While he agreed that philosophy had something of interest to offer scientists, that something concerned only epistemology: in what manner and to what extent humans could gain knowledge of the world. But he thought it extremely unlikely that dilettantes would turn to epistemology as opposed to metaphysics, just as they were unlikely to turn to celestial mechanics as opposed to "the astrological sophistry of old times." Modern metaphysics, he said, had given up its boldest plans to develop a system that contained all essential knowledge and to do so on the basis of pure thought alone. Some modern metaphysicians were indeed willing to grant that knowledge was advanced through experimental science and to speak in terms of hypotheses. But others—like Zöllner, for example—were determined to retain a priori propositions. Readers with a metaphysical bent, who placed their faith in Zöllner's presentation of scientific method and the history of scientific discovery, merely postponed the day when natural science and the humanities might bridge the chasm that separated them.[46] With that riposte, Helmholtz finished his two-stage counterattack on Zöllner.

In appreciation of Anna's translating *Fragments*, Tyndall sent her a bracelet. She said she had seen the shortcomings of her first-draft efforts, but then Hermann corrected and improved her translation throughout. As for Zöllner, she thought her husband honored him too much by responding to him at all. The first printing of *Fragmente* soon sold out, and Tyndall gave the royalty rights for all new printings to the Helmholtzes as a gift. Already in May 1876, Anna was working on a new edition of *Fragmente*. The book sold well for decades. By 1898–99, it had expanded into two volumes and had a second German translation from the eighth English edition.[47]

Helmholtz's efforts to popularize science had of course neither begun nor ended with his preface to Tyndall's *Fragmente*. Nor were they limited to the written word. In the mid-1870s, he lectured around western Germany to various nonuniversity audiences. He visited Bochum, where his brother Otto was the director of a steel-casting factory. He thought that one of his lectures, in nearby Hamm, did not go well; he found Hamm "a narrow, poor town with a rather barren hotel." After the lecture he "had to dine with the board, but more emphasis was laid on drinking than on eating, and it was only after 1 a.m. that I succeeded in escaping and left the gentlemen seated by themselves. As a consequence, I awoke with a rather heavy head." That may explain why he told the Dortmunders—he served on its Verein board, too—that he (and his brother Otto) would be leaving immediately after his lecture.[48]

The lecture in Dortmund did not go that well either, since his hosts had failed to provide the needed demonstration apparatus/instruments for his experiments. He managed to get some needed items and did the best he could under the circumstances, but the lecture did not flow; "I was dissatisfied with myself." He visited the factory where Otto worked: "We climbed with great labor—quickly sinking into mudpiles, quickly climbing over slush—onto a newly built, not yet ignited high-temperature oven which he [Otto] himself found to be the most interesting of the new ones, and from whose top there was a very good overview of all the connected apparatus and the entire area. It is a tremendous works, very interesting, and it gives one a lot of new views about the workings of natural forces in large-scale dimensions." That afternoon he left for Krefeld to give a lecture titled "Life and Sunlight"; the next day he headed to Barmen, then Essen, and then back to Dortmund. From Elberfeld he wrote that his lecture on the circulation of water had gone as planned. "The worst is over; the rest of the lectures are repetitions."[49] Like various eighteenth-century natural philosophers and some other nineteenth-century scientists, he had become an itinerant lecturer, if only on a temporary basis.

While he was en route, Anna wrote to say that she did not know where he was or where he had taken his "wisdom to market." When he returned, she suggested, they should review his lectures. He might decide to dictate them to Anna to preserve them while they were still fresh in his mind. "Later, they won't be able to be written up. It would be a nice piece of vacation work. Think about it, dear Hermann." He telegraphed her from Düsseldorf that he had become very ill: he had suffered some sort of "terrible poisoning," and there was no one around to help him. She went to him immediately, fearing for his life, but by the time she arrived, he had improved greatly.[50]

Supervising the translations of Tyndall's and others' popular scientific books, writing introductions to Thomson and Tait's volumes, holding popular science lectures in various German cities, and publishing his own such lectures was a lot of work—and was about as far as Helmholtz wanted to go in this regard. In 1871 Tyndall wrote him that Edward L. Youmans, of the Appleton publishing house in New York, wanted to meet him. Youmans's name was already known to Helmholtz, because in 1865 Youmans had edited and introduced a volume entitled

The Correlation and Conservation of Forces: A Series of Expositions, which included essays by Grove, Mayer, Faraday, Liebig, and Carpenter, as well as Tyndall's translation of Helmholtz's "On the Interaction of the Natural Forces" (1854). Youmans wanted to tell Helmholtz personally about a new, high-level but popular science series he was establishing, which would be known as the International Scientific Series. "He has a project on foot," Tyndall said, "which promises very important results." He added that Huxley, Spencer, Alexander Bain, and Darwin all took "a great interest in this project" and hoped for its success. The series eventually produced some 120 titles in four languages. Youmans's representative in Germany for this project was Isidor Rosenthal, a former student of du Bois–Reymond's and now professor of physiology at Erlangen. Rosenthal signed an agreement with Brockhaus for the German-language version of the series and put together a list of authors, starting with Virchow and Aaron Bernstein. He wanted Helmholtz to help him compose the list and generally be involved in getting others to sign on. But Helmholtz had no further energy for this type of popularization work, and he declined to be involved.[51]

By the mid-1870s, Helmholtz's name was becoming even better known among the middle classes. In 1875 his picture appeared on the front page of the Sunday weekly *Über Land und Meer: Allgemeine Illustrirte Zeitung* (Stuttgart), a family magazine. The accompanying article summarized his work as "epoch-making" and referred to the practical importance of his ophthalmoscope for understanding eye disease. Similarly, in 1876 *Daheim*, a conservative German family magazine that often contained articles seeking to relate recent developments in natural science, portrayed Helmholtz in its section on German professors. It lionized him "as one of the main champions in this effort of the human spirit to rule over nature." Nonetheless, Helmholtz's main aim remained to reach an elite readership. In 1875, after he published the third and final collection of his *Populäre wissenschaftliche Vorträge*, he sent a copy to his friend Julius Meyer, director of the Royal Painting Gallery (Königliche Gemäldegalerie) in Berlin and a writer on art. Meyer read the essays in the third collection with great delight. "As far as concerns its contents and its meaning for me, I feel forced to say: not merely in the fields that are foreign to me, but even in those where I believe myself to be rather at home, I learn from no one as much as from you, both through the problem that you pose as . . . the way of treating it and [through] the form of presentation."[52] Meyer was precisely the sort of figure whom Helmholtz had always hoped to reach through his popular scientific writings.

Kulturkampf in Science, II

The Dühring Affair

Eugen Dühring was a jack of several academic trades and master of none. He published books on the philosophy of mathematics, on natural science, on dialectics, and on political economy, yet he was neither mathematician nor scientist nor political economist. The one "trade" he truly mastered was demagoguery and polemics, oral and written. It included smearing Helmholtz and founding racial anti-Semitism. In his day he had few equals.

Dühring earned his living principally as a lecturer in political economy and philosophy at the University of Berlin. His reputation as an academic author came first and primarily with the publication of his prize-winning *Kritische Geschichte der allgemeinen Principien der Mechanik* (1873).[1] With its publication, he hoped that he might at last—he was now forty years old and blind—get promoted to extraordinary professor.

In 1872 Dühring got a second job, lecturing on literature and philosophy at the new Victoria Lyceum, a jerry-built institute devoted to the "higher" education of women. The Lyceum was a secondary or finishing school for young ladies. The institute derived its name from and came under the patronage of Crown Princess Victoria and, with its cofounder and director Miss Georgina Archer, had a decidedly British patina to it. Among other subjects, Lyceum faculty taught the history of art, English literature, Italian poetry, astronomy, and botany, and they also offered moderately priced special lectures in science and literature. But the institute had difficulties in finding qualified lecturers, since many leading Berlin professors reportedly considered lecturing before women to be beneath their professional dignity. The board of directors included the crown princess, Anna and Hermann Helmholtz, Gneist, Virchow, du Bois–Reymond, and Fanny Reichenheim. Anna Helmholtz, who was particularly interested in women's education and social reform, was active in the Lyceum's leadership; she was a moderate promoter of young women's well-being, not a feminist. She also promoted nursing education and helped administer the Victoria House for Nursing.[2]

Dühring's academic problems first emerged in 1874–75, when he attacked Adolph Wagner, a conservative but also a socialist who supported Bismarck. The two had a nasty exchange of views in the columns of a Berlin financial newspaper, which made the affair a public one and led to an administrative review. Dühring's main problems, and the main charges against him, however, arose from his repeated slanders against Helmholtz and his allegations concerning the general conduct of hiring and promoting faculty members at universities.[3]

Dühring claimed that Mayer, not Helmholtz, was the discoverer of the law of conservation of energy, that Helmholtz had done little more than reword Newton's and Leibniz's views, and that he had never even mentioned Mayer at all. Dühring wrote an entire book portraying Mayer as a martyr to science, indeed as the Galileo of the nineteenth century. While it was true that Helmholtz had not mentioned Mayer in his essay of 1847, no one else had mentioned Mayer at the time, either; in fact, it was Helmholtz, in his reviews that appeared in the *Fortschritte der Physik* between 1850 and 1855 and in his popular lecture of 1854 on the conservation of force, who first noted Mayer's work and its priority, and it was he who defended that priority over Joule's in a letter published in Tait's *Sketch of Thermodynamics* (1868). Mayer himself never claimed that Helmholtz had plagiarized from him or otherwise misled readers on the issue of priority; in 1866 he invoked in a positive vein Helmholtz's version of conservation of energy and his popularization of it. Dühring also claimed that at Innsbruck in 1869 there had been a "scene" between Helmholtz and Mayer: that someone yelled out from the audience that Helmholtz (and therefore not Mayer) was the "discoverer" of the law of conservation of energy and that Mayer left the meeting early. None of this was true.[4]

Dühring's rage against Helmholtz and university professors in general was partly that of the scientific amateur against the professional, which explains in part why Dühring became Mayer's unsolicited advocate. Although Dühring had little formal training in natural science, that did not stop him from writing books like *Neue Grundgesetze zur rationellen Physik und Chemie* (1878), wherein he declared Helmholtz to be a "showpiece of physics" who had been "advanced" into his place in Berlin by all-too-friendly colleagues there. He repeatedly attacked Helmholtz, treating him with disdain and misrepresenting his views. In addition to belittling Helmholtz's work on conservation of force, Dühring attacked non-Euclidean geometry and Helmholtz's contributions to it. He wrote: "It's not surprising that the unclear, a little philosophizing, physiological professor of physics, Herr Helmholtz, could not avoid participating in this matter [non-Euclidean geometry], too." His essay on non-Euclidean geometry was "nonsense." Like Zöllner, Dühring praised Weber's electrodynamic ideas. He attacked Berlin mathematicians as a group, as well as the Prussian academy, and made broadside attacks on professors in general. He criticized a textbook of Kirchhoff's as well as Clausius's mathematical presentation of the mechanical theory of heat. He denigrated scientific journals in general, saying that most articles were worthless. Moreover, he made anti-Semitic remarks against Jacobi and others of Jewish ancestry.[5] The attacks on non-Euclidean geometry and on the increasing mathematization and abstract character of physics presaged the anti-relativity and antitheoretical attitudes that some German physicists adopted after 1914. Dühring was, in effect, a forerunner of anti-Semitic and antimathematical physicists like Philipp Lenard, Johannes Stark, and Willy Wien.

In March 1876 Dühring lectured in the Berlin City Hall on higher professional education for women. The lecture, he later claimed, was the principal reason for his subsequent dismissal from the Lyceum. He declared that unnamed "little authorities of science" envied him, that they were ignoramuses, and that they knew nothing of women's hopes. He spoke of "the learned cliques," suggesting that university professors were afraid of free competition, that they constituted a "medieval guild structure," practiced "the science of unfreedom," and constituted a "learned monopoly." At the Lyceum, he declared, his lectures were the best attended of all lectures in 1874–75, and he thought that after three years there he had earned a permanent position. Yet two months after the lecture, he was dismissed. He claimed that the "Jewish writer Hirsch" had worked against him and that there had been a plot against him; it was an "intrigue" led by a combination of university professors and the "Jewish element." Anna Helmholtz later joined them, he said.[6]

Dühring made many enemies in his day, but the one who made *him* most famous was not Helmholtz but Friedrich Engels. Engels and other socialists saw Dühring as a threat to their Marxian socialist ideology and movement. Engels fought him for the leadership of socialist thought and participation, and he did so by publishing a series of articles in the socialist newspaper *Vorwärts*. These soon appeared in book format as *Herrn Eugen Dühring's Umwälzung der Wissen-*

schaft (1878), the famous "Anti-Dühring." Engels himself was something of a student of Helmholtz's writings, though not always a pliant one. After 1870 Engels began studying physics and chemistry, including Helmholtz's essay on the conservation of force, and was especially interested in the topics of causality and the broad purpose of theoretical natural science. Studying the second volume of Helmholtz's *Populäre wissenschaftliche Vorträge*, he focused on the epistemological relationship of the concepts of subject and object and sought to deepen his appreciation of how theory supposedly mirrored nature. He also polemicized against Helmholtz but did so privately, in his *Dialektik der Natur* (first published in 1925).[7] He did not hesitate to pass judgment on Helmholtz's science and also had many philosophical differences with him.

In the spring of 1877, the University of Berlin and the ministry proceeded to investigate Dühring again, and on 7 July they withdrew his teaching privileges and stopped his classes immediately. In late June, before the decision was rendered, the matter somehow appeared in the Berlin newspapers, and on 12 July a large student protest occurred on Dühring's behalf. The Dühring Affair, as it was called, became a cause célèbre, a battle over academic freedom, with the academic establishment on one side and the blind, poorly paid lecturer Dühring and his student followers on the other. In response, the ministry published all its findings concerning Dühring's remotion (removal). This was a highly unusual step, perhaps unprecedented. The basic charges against him concerned his various public statements against Wagner and Helmholtz, as well as a series of statements concerning the university in general. Among much else, the entire affair led to Dühring's lifelong embitterment toward Helmholtz. Georg Brandes, a major Danish literary figure, who knew Berlin well, heard Dühring lecture at the Architect's Association (Architektenverein) after his remotion. He described him as a hater and a fanatic, who, besides attacking Helmholtz, viciously attacked Jews, Lessing, Goethe, and other thinkers and poets.[8] That hate and fanaticism continued until his death in 1921.

Thought (and Empirical Reality) in Medicine

The spring and summer of 1877 proved to be a miserable time for Helmholtz. He had to stand by and watch the Dühring remotion case run its course, and with it the pro-Dühring student protests. He saw his good name dragged through the mud as one of the establishment heavies. Yet he tried to be philosophical about the situation. He told a close female friend that he was "to a certain degree used to bad treatment from the side of metaphysicians of all sects, spiritualists or materialists." He added, "I've tried throughout my life to cut through their sophisms with facts. People are ultimately intelligent enough to believe the facts more than the subtlest of theories." He found it difficult, however, to understand how someone—for example, those who opposed non-Euclidean geometry—could become angry with him when he had done nothing personal to them. Of Dühring he wrote: "For I've never done anything against him, neither in letters nor in lectures. In the Wagner Affair I've mainly spoken for forbearance towards him. . . . The removal from the Victoria Lyceum was brought about by Miss Archer and Prof. Lazarus; my wife first heard about it when it became a matter of protecting her against Dühring's shameless attack." As for the student protests, he was astounded at how easily students succumbed to transparent lies, distortions, and rumors, something, he thought, that the students of his own day would never have done. He had hoped in vain that Dühring's supporters would just disappear. All this brought more suffering to Anna than to himself. While he did keep an eye on the Lyceum, it was Anna, he said, who was truly involved with it.[9]

That spring and summer Helmholtz also suffered on the personal level. His daughter Käthe died that April, at age twenty-seven. A far less dramatic trouble was a foot problem that lasted for weeks; he treated himself with warm footbaths. In early July Anna and the children went to Ambach, leaving him alone in Berlin. Various responsibilities, including a speech before the Friedrich-Wilhelms-Institut, required him to remain at home. He longed to get out of Berlin and join his family and then go on (alone) to Switzerland to meet Tyndall and Knapp. "There's not much joy to be had in Berlin right now," he wrote a friend. "Almost all of my acquaintances are travelling, the Spree stinks, and it is wickedly hot." His mood may have improved before he left town, for on 1 August he was elected rector of the university for the coming academic year (1877–78). His election was not only an honor but a sign of trust in and support for him; he may well have received some sympathy votes in reaction to the Dühring Affair.[10] As rector of the University of Berlin, he had symbolically attained the pinnacle of institutional academic life in Germany. See figure 21.1.

The following day, 2 August 1877, in an address entitled "Thought in Medicine," he spoke to the faculty and students of the Friedrich-Wilhelms-Institut as part of its annual Founding Day celebration. Though the address was ostensibly a comparison of the state of medicine in 1842 and in 1877, it also represented his view of the proper, "scientific" method of doing medicine: namely, to base it on experience and to obtain a correct balance of induction and deduction.

He began with a reference to the student speech he had given exactly thirty-five years earlier on blood vessel tumors. At that time he had necessarily drawn on findings reported from books, he said. "But book-learning then played a more widespread and respected role in medicine than one tends to concede today." By contrast, the modern, scientific spirit relied on experience. Like world history, medicine had changed enormously in the intervening years, and traditional medicine was "written in a forgotten language."[11]

Fig. 21.1 Kronprinz Friedrich auf dem Hofball, 1878/1895, by Anton von Werner. Helmholtz is shown here with Crown Prince Friedrich and others at the Court Ball: the crown prince (*right center, facing left*); Max von Forckenbeck, the president of the Reichstag and mayor of Berlin (*opposite the crown prince*); the National Liberal politician Robert von Benda (*behind Forckenbeck*); the archaeologist Ernst Curtius (*between Benda and Forckenbeck and partly hidden*); Rudolf Virchow (*in an academic gown and slightly in front of Helmholtz*); Helmholtz (*just to the crown prince's right*); the painter Ludwig Knaus (*behind the crown prince*); and the painter Adolph von Menzel (*in the background near the window*). This painting hangs in the Alte Nationalgalerie Berlin. Bildarchiv Preussischer Kulturbesitz, Berlin.

Helmholtz believed that since human suffering and even life itself were at stake in medical diagnosis and treatment, and since the doctor must marshal all knowledge and use any instruments he can, "epistemological questions about scientific methodology can also have a pressing weight and a fruitful practical significance." This was not merely a theoretical issue: the doctor confronted "the hostile powers of reality." "He can only use the glaringly hard light of facts, and must stop himself from indulging in pleasant illusions." The "fundamental error" of medicine (and of other sciences) before the 1840s or so, Helmholtz opined, was its "false ideal of scientific method," in particular its overemphasis on the importance "of deductive method." Hence in medicine, as perhaps in no other discipline, "a correct critique of the sources of knowledge" was also "a practical, extremely important task of true philosophy." He scoffed at followers "of an arbitrary, world-explaining system," at so-called philosophers who disregarded "the facts of reality." Speaking of medical men before 1840, he said they knew little "of the laws of nature." They thought in terms of logic and conceptual systems and generally overestimated the importance of thought in medicine. True enough, he agreed, humankind's superiority over animals lay in its ability to think. So it was only natural that medical men had previously put a high value on deductive systems of thought and had neglected observation and experimentation, "in order to begin the Icarus flight of metaphysical speculation." But the ability to acquire and transmit knowledge over generations meant that truths had to be regularly critiqued to ensure that error, distortion, and mere reliance on authority did not creep in. Truth had continually to be distinguished from falsehood as knowledge was transmitted and its origins forgotten. Human beings, he continued, form mental concepts, which serve "as ordering powers in the objective world of things," but they must constantly test whether these concepts correspond to "the world of things." The doctor, like all other scientists, had to avoid "psychological anthropomorphism." Examples of the latter were Plato's ideas, Hegel's "immanent dialectic of the world process," and Schopenhauer's "unconscious will." Overreliance on the deductive method and on conceptual systems led to ineffective medicine and intolerance of the new. The fundamental shortcoming of all systems was, Helmholtz thought, the belief that there must be one single complete explanation for every form of medical illness and its associated form of treatment. This had led to dogmatic intolerance on the part of traditional medical men, since if their fundamental hypothesis proved false, the entire system would fail. By contrast, those who built their medical understanding on an empirical, nonsystematic basis could afford to make and acknowledge occasional errors.[12]

Helmholtz noted that in his professional lifetime, medicine had undergone major advances in the fields of microscopic and pathological anatomy, physiology, experimental pathology, drug discovery and treatment, ophthalmology, and physiological chemistry, among others. He cited especially the diagnostic importance of new instruments such as the clinical thermometer, the ophthal-moscope, the otoscope, and the laryngoscope. Moreover, the recently developed ability to detect parasitic organisms was replacing "mystical illness-entities," and this was proving of substantial importance to surgeons. Yet the battle against hypothetical systems was hardly over, he thought; it would remain as long as there were individuals who believed they could obtain results "through lightning flashes of genius" rather than "through laborious work," or who failed to criticize what they wished to be true.[13]

Metaphysical systems, he maintained, were characterized by one of two mo-tifs: the spiritualist or the materialist. The physician must avoid such systems and must "seek to know the laws of facts through observations." Once a law has definitely become known through induction by facts, then deduction may proceed, but with the caveat that its deductive results needed to constantly be tested against empirical reality. That in turn meant that a scientist never ob-tained "unconditional truth." Instead, the law in question attained a certain de-gree of probability. It was the inductive method that truly brought results.[14]

The beginnings of the discovery of a law, Helmholtz argued further, consisted in finding some previously hidden or unnoticed similarity in natural processes. It required what was otherwise known as "wit"; it was what artists did at the highest levels when they had "artistic insight" into phenomena. There was no rule or algorithm by which one might proceed, and it often required hard work. Any newly discovered law must then be tested for its agreement with the facts. Every "conscientious worker" must thoroughly test his work before bringing it to market. Helmholtz warned that "the current practice of deciding questions of priority according only to the date of the first publication, without consid-ering the maturity of the work," had "very much favored this abuse." Hundreds of articles were published every year that offered merely unproven hypotheses "on the ether, on the constitution of atoms, on the theory of perception," and so on. It was all speculation and without scientific value; indeed, it was worse than that, since one could not distinguish the valuable parts from the valueless. For anyone seeking new and trustworthy facts, this meant a great waste of time and effort. Without naming names, his remarks sounded like veiled references to Dühring and Zöllner, and metaphysical philosophers of all sorts, perhaps in-cluding Mayer as well.[15]

Helmholtz believed that his own generation had suffered under the spiri-tualist (idealist) version of metaphysics and that the coming generation would probably suffer under the materialist version. Though he thought science had benefited much from materialist hypotheses, he expressly warned that materi-alist metaphysics was itself a dogmatic hypothetical system that could also harm natural science. Invoking the names of Socrates and Kant, he maintained "that all previously stated metaphysical systems" were only "tissues of sophistry" and that Kant's *Kritik der reinen Vernunft* was "a constant preaching against the use of categories of thought beyond the boundaries of possible experience." In his view, natural scientific research meant seeking after "the laws of facts." He continued: "Moreover, we consider the discovered law to be a power which rules over the processes in nature. We objectify it as force."[16]

He concluded by emphasizing that his animosity toward metaphysics should not be seen as an animosity toward philosophy per se, even though metaphysicians always sought to appear as if they were the only, the true philosophers. "Philosophical dilettantes" usually were followers of "the high-flying speculations of the metaphysicians"; they looked for shortcuts to understanding the world. Metaphysics, he said, is to philosophy what astrology is to astronomy: it is a playing to the masses. He believed that if philosophy renounced metaphysics, it would still be left with "a large and important field, the knowledge of the mental and emotional [*geistigen und seelischen*] processes and their laws." Philosophy's true mission, he believed, was to test the scientist's "main instrument," namely, "human thought, which must be studied precisely according to its capacity to work." Individuals who wanted "a truly scientific foundation of their practical activity"—including practitioners of medicine, statesmanship, the law, theology, and teaching—had to build "on the knowledge of the laws of psychological processes." His own work had brought attacks on him—more than to most natural scientists of his generation, he said—from the side of the metaphysicians. He concluded: "I don't need to explain that someone who has such opinions as I've brought before you, who inculcates in his pupils, where he can, the fundamental theorem: 'A metaphysical conclusion is either a sophism or a hidden conclusion concerning experience,' is considered unfavorably by the lovers of metaphysics and of a priori views. Metaphysicians, like everyone who has no decisive reasons to set before their opponents, do not cultivate being courteous in their polemics. One can approximately judge one's own success by the increasing discourtesy of the replies."[17] This was not the last time that he put forth this message.

Haeckel and Virchow on Academic Freedom

Especially in the 1870s, German academics were much exercised about academic freedom. In September 1877, about a month after Helmholtz delivered his address "Thought in Medicine" and as he was about to begin his service as rector, the biologist Ernst Haeckel gave a speech before the Naturforscherversammlung in Munich in which he declared that evolutionary theory was now an accepted part of natural science. He argued that his philosophy, "monism," or "the unity of the world view," was the true philosophy, neither materialist nor spiritualist; that it led to Darwin's theory of evolution; and that it had practical implications for medicine, government, law, theology, and other fields, including education. Haeckel thus urged that Darwin's theory—he was its leading advocate in Germany—be taught in the schools; indeed, he claimed it would be "the most important educational means," because it would teach students to think in developmental terms. The historical natural sciences, he thought, resided between the traditional classical, historical-philosophical sciences and the exact, mathematical-physical sciences. Moreover, the theory of development had practical use for developing moral character. It was a natural religion, he said, and so it stood in opposition to the dogmatic religion of a church. Haeckel believed that in Darwin's theory of development, and not in so-called revelation, lay the resources needed to overcome mankind's social problems.[18]

At the same Naturforscherversammlung meeting four days later, Virchow replied to Haeckel with his own speech. Virchow thought the moment was right for a discussion of issues of science and society. The freedom of science, he said, was (still) threatened by the church, and he wondered aloud whether the freedom of science currently enjoyed in Germany could last. With an eye to Haeckel's speech, he declared that it could—if scientists acted and spoke moderately, and thereby kept the nation's confidence. This meant limiting scientific freedom in the sense of avoiding "arbitrary personal speculation." Scientists needed to remind themselves that there was a great difference between speculation and empirically based, proven ideas or theories in science. The nation should certainly learn about the latter; much of practical life—Virchow referred to steam engines, telegraphy, photography, dye technology, and so on—essentially depended on proven science. That was the "material meaning" of science. By contrast, there was also the "intellectual meaning" of science, which involved an individual's outlook on things and people. The big practical issue here was the content of biology courses in Germany's ever-expanding schools. In contrast to Haeckel, Virchow did not think—nor did du Bois–Reymond, for that matter—that Darwin's theory was so well established empirically as to merit being taught in the schools. Speculations about the origins and descent of life should be restricted to discussions among university researchers; they should not be taught to schoolchildren. There should be a strict line between what was taught and what was being researched: only scientifically well-established results, not scientific problems, should be taught to the public. To do otherwise, Virchow said, was to endanger science's power.[19] The following year Haeckel offered a rejoinder to Virchow's speech. In the meantime, barely a month after the Haeckel-Virchow controversy had arisen, Helmholtz gave his own speech on academic freedom. The *Kulturkampf* was at its height, and it was present in science as well as in politics.

On Academic Freedom (and a Warning to Its Abusers)

As rector, Helmholtz supervised the university's business and represented it to others. He led its governing curatorium; and he met with the deans of each of the four faculties, the previous year's rector, a select group of five ordinary professors, and the university judge. He was responsible for everything that happened both internally and externally. His administrative duties that year (1877–78) increased when he had also to supervise the move into the new physics institute. Among other things, these responsibilities led him to decline the proffered Faraday Lecture at the Chemical Society in 1878.[20]

The day Helmholtz began his rectoral duties, 15 October 1877, he delivered a rectoral speech entitled "On Academic Freedom in the German Universities." Given the recent Haeckel-Virchow exchange on academic freedom and the Dühring Affair, his greatest concern as rector that autumn was to explain the nature of academic freedom and to warn Berlin's students against its abuse. That winter semester, Berlin had about 5,000 students enrolled; the following summer semester it had about 4,300. Trouble might come from any of them. Helmholtz wanted to preserve the freedom of science within the university and within the authoritarian Prusso-German state structure. In so doing, he placed himself in a longer line of thought first announced by J. G. Fichte, Friedrich Schleiermacher, Wilhelm von Humboldt, and others early in the century.[21]

Helmholtz believed there was deep and widespread academic freedom at the German universities, especially in comparison with the relative lack of it at English and French universities. Tracing its history and character, he discussed several matters concerning both German and foreign universities: the freedom to teach (*Lehrfreiheit*) within disciplinary boundaries, the freedom of student inquiry (*Lernfreiheit*), the definition of a good teacher, the best relationships between teaching and research, the role of private lecturers, and competition within the German universities. He presented natural science as "a somewhat foreign element in the circles of university instruction" and argued that it had been responsible for many structural changes in the universities. He emphasized, as he had done in his Heidelberg prorectoral address of 1862, that despite the various differences between the human and the natural sciences, there remained, in terms of method and final purposes, "the closest relations in their innermost natures." Nonetheless, changes were taking place within the universities, not only because of the rise of the natural sciences but also because the outside world of political and social life was making itself felt within the universities. The student population had become more diverse, the state was making new kinds of demands on universities, increased specialization continued apace, and new institutional resources (coming from external sources) were needed. It was difficult to know what sorts of future demands the university might confront. Helmholtz judged that the German universities were honored at home and claimed that "the eyes of the civilized world are directed upon them." Everyone was watching them, so they could not afford to make "a false step." With these thoughts in mind, he asked rhetorically what had made the German universities so successful. What was their institutional core? What was their "untouchable sanctuary"? And what, if necessary, could be jettisoned?[22]

The great transformation of the medieval into the modern German university, Helmholtz thought, was due to the state's becoming its principal provider. The state thereby acquired the right to share in the university's management. By contrast, the English universities—that is, Oxford and Cambridge, since he excluded discussion of either the Scottish or the newer, municipal Red Bricks—had changed the least. Indeed, they basically remained "schools for clerics." Their students and fellows belonged to England's wealthy elite and lived accordingly. The level of instruction was about that of a good German Gymnasium. Their only demanding feature was the examinations, which themselves consisted mainly of regurgitation of selected contents from designated textbooks; only in selected cases did they include advanced knowledge in specialized fields. The college fellows did the bulk of the teaching, and they did so in their capacity as private tutors. There were few professors, some of whom did not even live on or near campus; they gave poorly attended lectures, devoted to specialized topics of interest to only a few advanced students. The colleges themselves were isolated from one another and had in common only the administering of examinations, the awarding of degrees, and the naming of professors. Helmholtz thought the combination of church interests, the fellowship system, and personal connections served to politicize science at the English universities. Nonetheless, he thought there were many outstanding individual scientists at the English universities. The fellowship system, moreover, did give certain talented men the time to do research, even if they had not encountered real research when they themselves were students.[23]

He recognized, however, that the English universities produced educated gentlemen, even though they were politically and religiously narrow in their outlooks. The students acquired an excellent sense of the ancient world, and they learned to speak well. In both these respects, he thought German students were deficient and that the German universities should follow the English example. In addition, English universities emphasized the importance of being physically fit, and he greatly appreciated that (and warned German students about the "misuse of tobacco and of intoxicating drinks"); he also appreciated their beautiful physical environment and the emphasis on game playing and athletic competition. In these areas, too, Helmholtz thought the English system superior to the German.[24]

Helmholtz's harsh judgments on Oxford and Cambridge did not go unnoticed or unopposed; at least a few scholars begged to differ with him. Walter Copland Perry, an Englishman with a solid knowledge of German universities, did so sharply in the pages of the *Deutsche Rundschau*. Helmholtz replied to him, saying that their differences were largely semantic in nature, though, to be sure, he conceded that reform—inspired by the German example, he did not fail to note—had begun at both universities, and so to that extent his viewpoint was indeed a decade or two out of date. He conceded, too, that his knowledge of English universities came from a combination of books and discussions with various people. Perhaps he had painted too bleak a picture, he acknowledged. Max Müller also strongly disputed Helmholtz's account of English universities; he thought that in the matter of governance and independence, they were superior to their state-supported German counterparts. Even one of Helmholtz's English-language translators, Edmund Atkinson, thought some of Helmholtz's claims had gone too far and that changes at the English universities after 1854 spoke against several of his claims.[25] Certainly Helmholtz's claim that Cambridge academics could not discuss Darwinism was false. By 1875 Darwinism had reached the level of inclusion in the student examination in science.

On the other hand, there were more than a few British scholars who shared Helmholtz's uncorrected point of view. In 1868 Huxley had reached more or less the same conclusions about university education and research at Oxford and Cambridge as Helmholtz did in 1877. Roscoe, who knew German science and German universities firsthand, fully shared his outlook on the shortcomings of Oxford and Cambridge, especially in the sciences, and had said so in print some years before Helmholtz did. Similarly, Maxwell had declared in 1873 that although the numbers of British scientists and science students were increasing, and education in and the popularization of science was spreading, in his view "original research, the fountain-head of a nation's wealth," was decreasing in Britain.[26]

British leaders had themselves been formally investigating and criticizing Oxford and Cambridge since 1850—for example, through the Oxford University Commissions of 1850–51 and of 1877 and the Royal Commission on Scientific Instruction and the Advancement of Science (the Devonshire Commission) of 1870 and after. They perceived an insufficient level of research as a primary issue of concern. In 1873 the Devonshire Commission reminded Oxford and Cambridge of how important it was for universities to do scientific research, and it compared them unfavorably to the University of Berlin. Similarly, Matthew Arnold, in his *Higher Schools and Universities in Germany* (first edition 1868, second 1882), saw the English (and French) universities as distinctly inferior to their German counterparts. He thought that, at least in terms of teaching specialized knowledge, they were little more than "*hauts lycées.*"[27]

Helmholtz's friend Henry Wentworth Acland at Oxford, a leading advocate of science reform there, had pushed to have elementary science introduced into the curriculum. By 1877 the liberals at Oxford had even managed to get a majority of the faculty to agree, in principle, to a science degree. In addition, in 1869 Oxford opened its Clarendon Laboratory (under Clifton's direction), and in 1873–74 Cambridge opened its Cavendish Laboratory (under Maxwell's direction). In physics at least, and under a series of outstanding directors (Maxwell, Rayleigh, and J. J. Thomson), Cambridge became one of the world's centers for physics research. Oxford also got a new observatory in 1873 and enhanced its chemistry laboratory in the late 1870s. It appeared, therefore, that in the 1870s Oxford had entered a period of strength. Yet Clifton did no research during his half-century-long career there, and so physics languished. The professor of chemistry, William Odling, was little different from Clifton. All this meant that, while the recognition of science and scientific research had gradually increased in Oxford and Cambridge after midcentury, at Oxford it "failed to keep pace" with other institutions. As far as scientific research at Oxford was concerned, Helmholtz was by no means alone in his judgments.[28]

The ongoing arguments over higher education in Britain represented a new flare-up of an old problem, one that had begun in the late seventeenth century. Known then as the Battle of the Books, and in France as the quarrel of the Ancients and the Moderns, it much later was famously labeled by C. P. Snow in 1959 as "the Two Cultures." It was a conflict between those who, like Arnold, favored more emphasis on teaching the humanities in the schools and universities, and those who, like Huxley, favored more emphasis on the sciences.[29] For Helmholtz, as for German thinkers generally, the greater endeavor was not a matter of "the Two Cultures" and the proper balance between them; rather, it was to find unity among the "sciences" ("die Wissenschaften") as a whole. For the Germans, culture was one.

Helmholtz then turned briefly to French universities, of which he had no firsthand knowledge. He had few French colleagues and no French friends. Still, he opined that the French universities were the opposite of the English. They discarded everything historical and sought instead to make everything rational. They had become "institutes of pure instruction" for specialized professions, with fixed orders and courses of instruction, many rules, and frequent examinations. They existed separately from the grandes écoles, and even the individual faculties at a given university had nothing to do with one another. Instruction at French universities, Helmholtz said, was limited to the definitely known; it was conveyed in a well-ordered and articulate manner, to be sure. But the deeper foundations, and the doubts and uncertainties about well-established knowledge, went unnoticed. The instructor's word was never challenged. It was the sort of instruction well fit for the average student preparing for a well-established career path. The French state gave little monetary support to its universities, and so teachers were largely dependent on student fees for their income. The students had no sense of group solidarity and lived without supervision. What balanced this highly negative picture, Helmholtz condescendingly added, was that France was a "gifted, lively, and ambitious" nation.[30]

His portrait was fair as far as it went, but it lacked balance in that it did not discuss the *grandes écoles*. The nineteenth-century French university, and more precisely the role of science there, indeed contrasted poorly with German university science. French university scientists had great difficulty in reconciling their ideals of science as free and universal with the fact that they were closely controlled state civil servants, largely devoted to teaching, with little opportunity for research. There was a general sense of institutional stagnation. In the 1860s and 1870s, there was some movement for reform in terms of greater scientific independence and the need to establish modern laboratories, journals, societies, and new sources of patronage. But before 1870 French university scientists were by and large savants who, as Helmholtz declared, emphasized lecturing, especially demonstrating oratorical abilities aimed at as broad an audience as possible. Only from the late 1870s onward did French university science begin to place an emphasis on research and play a role independent of the state. Some French university scientists pointed to German university science as their institutional model, though many also saw it as their greatest rival.[31]

Helmholtz then turned to his main subject: the German universities. These, he said, differed from their English and French counterparts in that they lacked their own financial means and political independence from the state; traditionally, therefore, they had been subject to state authority and power, and at times this had been quite hostile. However, during the past century or so, the German states had developed a more positive relationship with their universities, since state cultural and educational officials were mostly trained at the universities and since the fame of a state's universities reflected well on it. During the political upheavals from the end of the Holy Roman Empire (1806) to the formation of the German Empire (1871), Helmholtz claimed, German universities "had retained a much larger nucleus of inner freedom, and indeed the most valuable part of this freedom, in contrast to consciously conservative England and to France," which pursued freedom "impetuously."[32]

More specifically, German students enjoyed greater freedom to learn or inquire as they saw fit (*Lernfreiheit*). That meant that the university treated its students "as themselves responsible young men who out of their own motivation pursue science" and "who were left to freely arrange their own plan of study." Apart from professions that required students to take specific preprofessional courses in preparation for well-defined examination topics, and which were required not by the universities but by the state as part of its oversight of certain professions, students were completely free to take any courses they wanted to or not, in any order that seemed appropriate to them, with any teacher who might be offering a course at any German-speaking university. Furthermore, freedom to learn meant that the university did not supervise the student in his non-university life. This, for Helmholtz, had large implications. It meant that the student must act in an honorable, peaceful, civil, and generally responsible manner. If not, he warned, then the public authorities might have to intervene, or the students could even end up with the type of supervision that the English universities imposed. And so he warned students to do everything in their power to maintain their freedom to learn and to transmit that learning to succeeding generations. Freedom meant responsibility. All this supposedly brought out "the best and the noblest . . . that the human race, to date, had gained in knowledge and insight," as the students stood "in friendly competition" with one another and "in daily intellectual exchange with their teachers." All this fostered intellectual independence, he claimed, and he cited his own student experience with Müller as a case in point.[33]

The other significant characteristic of the German university, Helmholtz continued, was *Lehrfreiheit* (freedom of teaching). Here, too, German universities distinguished themselves from their English and French counterparts. German university teachers generally became qualified to teach as a result of their proven ability to do research. This included the peculiar German university requirement to habilitate. The ability to conduct research was "without question the principal qualification of a teacher," a point that astounded the English and the French, who by contrast emphasized "the ability to analyze the subjects of instruction in a well-ordered, clear form, and, where possible, eloquently, capturing the attention and doing so in a supportive way." The Germans were indifferent to, often neglected, and sometimes were even suspicious of oratorical ability in teaching. Of course, Helmholtz acknowledged, a clear and well-ordered lecture was easier to follow than its inverse, and some German teachers, though scientifically very talented and original individuals, were undeniably poor lecturers. Still, he favored the individual who knew the content of his lectures well and who spoke from firsthand scientific experience over one who merely enveloped such content in the form of a well-delivered lecture. And as a matter of fact, German students seemed to prefer this too, since they mainly attended lectures by scientifically accomplished if pedagogically undistinguished lecturers. The German university system thus trusted its students—and this was another aspect of *Lernfreiheit*—to discriminate between the superficial and the fundamental. It encouraged the students "to go to the sources of knowledge, whether these lay in books and paintings or in observations of and experiments concerning natural objects and processes." That was part of the point of libraries, collections, and laboratories, which last, he added, were in Germany superior to what could be found elsewhere.[34]

"Often enough," Helmholtz reiterated and conceded, the freedom to learn and to teach had not been protected, either in the German lands or in other nations. But times had changed, at least in the new Germany. "Here too the advanced political freedom of the German Empire has brought healing. At this very moment, one can lecture in German universities on the most extreme consequences of materialist metaphysics, or likewise lecture undisturbed on the boldest speculations based on Darwin's theory of evolution or offer the most extreme adoration of papal infallibility." Individual motives were not to be questioned, opponents were not to be personally slandered, libeled, or otherwise defamed, and illegal action was not to be encouraged. Such behavior was not permitted in science. (Helmholtz presumably had Zöllner and Dühring, among others, in mind.) All that aside, any scientific question or matter was open to free, unhindered discussion, something that was not true of English and French universities, even the Collège de France.[35]

Helmholtz also explained to his listeners and readers that habilitated private lecturers had the same legal rights at German universities as their professorial superiors, though the latter did have control over their institutes' apparatus, laboratory, and expense budgets and were responsible for holding student examinations. To foreigners the position of lecturer was the most astounding aspect of the German university, and one that they envied. They were amazed that so many young scholars and scientists were attracted to such positions that offered no fixed salary—instead, they received small lecture fees from student attendees—and no secure future, and that professors readily accepted them (though that potentially meant competition from those who themselves might suddenly become professors). Hence there were "to a certain degree delicate relations" here.[36] That was putting the matter euphemistically. Helmholtz greatly understated the difficult economic and professional circumstances that private lecturers faced.

The method of selection of faculty members and lecturers favored the potentially best lecturers. First, everyone wanted to attract as many students as possible, and the best students, to their classes, not least because a considerable part of any lecturer's income depended on the size of his classes. Second, each wanted to be surrounded by "excellent colleagues." All this meant that the "specter of rivalry between university teachers"—something the public talked about—was not present when teachers and students were "of the right sort." Indeed, such rivalry could only potentially occur at the larger German universities, where there were occasionally two representatives of the same field. But, Helmholtz further claimed, in such instances the two men usually knew how best to divide up the given field and thus together constituted a strong disciplinary center without suffering a loss of students. Only where one man felt professionally insecure did such rivalrous situations have "unhappy effects."[37]

The institutional effectiveness of the universities could be seen, Helmholtz argued, in the fact that in Germany there were few "pathbreaking men" who worked outside the system. Apart from a few individuals—like Strauss, who was excluded for "ecclesiastical or political grounds," the Humboldt brothers, and Leopold von Buch, who were only associated with an academy of sciences, though they had the right to lecture at a university—most scientific figures in Germany worked at the universities. In England the opposite was true. German scientists were attracted to German universities because they thought they would find well-prepared, hardworking, and independent-minded students. "Thus there extends throughout the entire organization of our universities," Helmholtz boasted, respect for "free, independent conviction," which, he believed, was more solidly impressed on the Germans than on "their Aryan relatives of the Romanisch and Celtic tribes. Political and practical motives weigh more heavily with the latter peoples. They manage—it seems in all uprightness—to withhold inquiring ideas from being used in the investigation of such propositions that to them seem undiscussable, since they belong to the necessary foundation of their political, social, and religious organization. They find it completely justified not to allow their young men to look beyond the border that they themselves are unwilling to cross." In contrast to Germany, in France and England "one can talk of free conviction only in a limited way."[38]

In closing, he emphasized the importance of developing and maintaining an independent point of view. Even when the state had intervened in university affairs in the past, it had never completely eradicated freedom of thought at the universities: "In their [the states'] innermost hearts, they have not abandoned the trust that freedom alone can elevate the blunders of freedom, and more mature knowledge the errors over the less immature." It took hard work, he averred, to reach an independent-minded point of view. From the Reformation to the present, Germany had stood "in the forefront of this struggle." He pronounced, "An elevated, world-historical task has fallen to it, and you [the students] are now called to work together on it."[39]

Although Helmholtz's address was aimed, in the first instance, at the students of Berlin, it was also a warning against the sorts of student protests that had occurred as part of the Dühring Affair. The lecture soon became widely available in print (and so became another element in the *Kulturkampf*). It was discussed in the *Nationalzeitung* and also was translated into French, appearing in the *Revue des Cours Littéraires et Scientifiques*. Its all-too-rosy portrayal, and its partisan and chauvinist dimensions, are readily apparent. It was by far the most political of Helmholtz's many addresses. In effect, it belonged to the series of addresses by du Bois–Reymond, Haeckel, Virchow, and others that in one way or another spoke to the issue of academic freedom and thus to worries about the state of the German university system.[40]

Freedom of scientific inquiry, though guaranteed by the Prussian and the Reich constitutions, was clearly being questioned. In the 1860s and 1870s, Germany experienced a crisis of authority, both intellectual and political, to which was added rising religious prejudice and rapid economic change. This inflammatory atmosphere included many elements: claims for the Pope's infallibility and his generally antimodern attitudes; Bismarck's allegations that German Catholics felt a greater loyalty toward Rome than toward Berlin; widespread anti-Semitism; du Bois–Reymond's anti-French speech during the Franco-Prussian war; writings by Renan and Strauss reassessing the historical Jesus; Virchow's speeches on the importance of science to the German nation and his *Kulturkampf* speech to the Prussian parliament; du Bois–Reymond's speeches on the German universities and on the limits of natural science; the issue of teaching Darwinism and its general challenges to the origins of mankind; the recent series of wars against Denmark, Austria, and France, along with the formation of a new German state (the Reich) from above; the attacks on the Social Democratic party; and, last but not least, the financial crisis in Germany after 1873 that resulted from too rapid economic development and rampant financial speculation. All these elements constituted the tense atmosphere in which Helmholtz gave his address on academic freedom and the German universities.

On Philosophy and Philosophers

By and large, Helmholtz had a distant and sometimes edgy relationship with the German philosophical community. He was friendly with Zeller, a historian of ancient philosophy, a neo-Kantian epistemologist, and a theologian, at both Heidelberg (after 1862) and Berlin (after 1872). But apart from Zeller, whose wife was related to Anna Helmholtz, he had no friendships with philosophers, and he certainly did not attend their professional meetings. He had had no relationships, or only minimally collegial ones, with local philosophers when he taught at Köngisberg, Bonn, Heidelberg, and Berlin. He disliked and so dismissed the entire philosophical outlook emanating from Hegel, Schelling, and their epigones, to which group the names Schopenhauer and Eduard von Hartmann should be added. On occasion he said as much. ("Schopenhauer calls himself a 'Mont Blanc beside a molehill,' when he compares himself with a natural researcher. Students are amazed by the great word, and try to ape the master.") To be sure, he received intellectual support from Johann Erdmann, Benno Erdmann, Friedrich Ueberweg (but also criticism), Kuno Fischer, and Zeller. But he had no noteworthy relationship with Immanuel Herrmann Fichte, despite Ferdinand Helmholtz's devotion to Johann Gottlieb Fichte's writings and person, and despite Immanuel Herrmann's positive references to Hermann's views on sense perception and his results in sensory physiology.[41]

Though scholars have much debated, at times vociferously, whether or to what extent Helmholtz could be called a "Kantian," the fact remains that Helmholtz's Kant was a selective one. Helmholtz was unquestionably quite familiar with the *Kritik der reinen Vernunft*, but it is unknown whether he had similar familiarity with the second and third critiques (of practical reason and of judgment, respectively). He probably had decent familiarity with the third critique, since it concerns aesthetics and natural teleology; at some points, his writings on the arts read as if they might be something of a gloss on the third critique. He never commented on or even mentioned in print or publicly anything about Kant's moral or political thought. Kant's importance for Helmholtz thus largely concerned epistemology (and related methodological issues as these pertained to science) as it was to be found in the first critique (of pure reason). Perhaps equally to the point, Helmholtz (also) stressed the importance of Kant's "Pre-Critical Writings," which deal mostly with the exact sciences and natural philosophy. He most definitely read the *Allgemeine Naturgeschichte und Theorie des Himmels* (*Universal Natural History and Theory of the Heavens*, 1755). He portrayed "the youthful Kant" as an outstanding scientist. He thought it was perhaps only a combination of Kant's lack of scientific apparatus, the strong philosophical atmosphere of the times, and life in isolated Königsberg that had led him away from science and into pure philosophy. He bemoaned Kant's transformation, in his fifty-seventh year of life (1781), from natural philosopher to philosopher *tout court*.[42] Moreover, it was hardly "Kantian" to implicitly critique Kant's theory of perception (as he did in his Kant Memorial Address of 1855) and then explicitly (if ever so gently) critique Kant's theory of space and perception (as he did in the *Handbuch* in 1867 and subsequently through his articulation of non-Euclidean geometry). There is all too little of a specific, substantive nature in Helmholtz's science that justifies calling him a "Kantian"—to whatever extent that label may matter—and not much more in his philosophical outlook that justifies, at least unquestioningly and unconditionally, using that label. There were, in fact, specific differences, ambiguities, and ambivalences in Helmholtz's views toward Kant's philosophy. Rather, what Kant represented for Helmholtz was threefold and very broad: a general framework in which, as a young scientist writing the introduction to the *Ueber die Erhaltung der Kraft* (1847), to set his natural scientific work; a stimulus to think about epistemological issues, including an emphasis on the importance for scientists to check their instruments and other methodological tools or sources in their production of knowledge; and an emphasis on the importance of the natural sciences for philosophy, including the aim and hope of ultimately bringing the two closer together. Beyond all that, "Kantianism" for Helmholtz represented a broad cultural stance, one that stood against figures like Hegel and Schopenhauer, for the centrality of reason in life, and for an Enlightenment worldview of tolerance in all human matters.

In 1868 Helmholtz had written du Bois–Reymond: "Philosophy is, incidentally, a dreadful wasp's nest, which one should never touch." Then from 1869 to 1871 or so, Helmholtz seems to have found renewed hope for a better, more productive relationship with philosophers. He wrote Ludwig in 1869 that, after having "intentionally" paused from further work in physiological optics and psychology, he again felt stimulated by physiological considerations to recommence his "electrical studies." He noted that he had just finished publishing two papers on the factual foundations of geometry (1868). He continued: "I find that too much philosophizing ultimately leads to a certain demoralization and makes one's thinking lax and vague." Two years later, in 1871, the philosopher and theologian Johann Erdmann wrote to Helmholtz that he had learned much from his writings. Helmholtz diplomatically replied that he was "very glad that a better understanding between philosophers and natural researchers" was "again gradually developing." He hoped that both would "again become as close to one another as in the old days. I have always felt the need for such an alliance," he continued, "since I've worked at the borders of science, in part on the most general geometric and mechanical axioms, [in part] on the account of the conservation of force, and in part on the theory of perceptions." But he expressed some doubts about Erdmann's view that a natural law should be understood as "a subjective appearance." Then a year later he expressed himself even more pessimistically about philosophy to the historian Wilhelm Oncken: "Here in Germany all philosophy currently gorges itself on the history of philosophy; the few philosophers who try to pursue something else are without influence and isolated (apart from the flattering Schopenhauerian and Hartmannian metaphysics due to the weaknesses of the age)." Nor was that all:

> Due to doubt about metaphysics, one has also given up doing what philosophy could and should do in a completely correct way: namely, the critique of the sources of knowledge and the theory of method of scientific thinking. That is, frankly, a far more unpretentious business in comparison to the speculative construction of God and the world. Yet it [philosophy] is reprimanded from many sides, so that it has already been neglected. In my opinion, a positive content for philosophy is only again to be had through investigation of the factual pathways of our knowledge from its origins in the sensations.[43]

He advocated—and was a recognized leader of—neo-Kantianism (or, more precisely, a version of it), above all with its (and his) rejection of metaphysics and his emphasis on epistemology. His "Kant" was the "epistemological" Kant, not the "metaphysical" one. This had been his general, public position since his address of 1855 on human perception. His own words, both public and private, suggest that—pace numerous philosophers of science who have seen him as one of the great representatives in a long line of philosophically oriented scientists running from Galileo to Einstein and Bohr—by 1868 at the latest, he thought all too little of philosophy and its utility for science. In the mid-1870s, he was even criticized by his fellow neo-Kantians for his conclusions about (the new) non-Euclidean geometry and its import for the truth of Kant's account of space and perception (see below).

This distance from and critical attitude toward philosophers was also reflected in the fact that Helmholtz published only two articles (both solicited) in philosophy journals. In 1874 George Croom Robertson, the professor of mental philosophy and logic at University College London, wrote Helmholtz that he was starting a new journal entitled *Mental Science Quarterly*—it became, instead, *Mind*—and he hoped that Helmholtz would write an article for it on visual perception. Robertson wanted to bring philosophy and psychology closer together, and an article on perception by Helmholtz was precisely what he needed to move toward that goal. Helmholtz declined to write on visual perception and instead recommended that Robertson ask either Donders, Pflüger, Mach, or Fechner ("our honorable veteran of a psychophysics"). He did, however, say that he hoped someday to write something for Robertson on some other topic but that just then he lacked a suitable subject. "Year after year I count on and hope for free time to be able to write down systematically my thoughts on epistemology. But my time is under such demand that I never get to it." In Germany, he complained, all too often his views had met with "so much misunderstanding" and "so much metaphysical prejudice" from both the idealist and the materialist camps that he hoped someday to present his entire epistemological outlook in all its complexity in one single essay. He thought that it would take quite some time before others understood his views, and he wondered if it was even worth the effort.[44]

In late 1875 he finally had something to send Robertson: the page proofs of his forthcoming article on the origin and meaning of the geometric axioms, which was scheduled to appear the following year in a new collection of his popular scientific lectures. The first part of the essay had already appeared in the *Academy*, but the second part was "completely new." He had made substantial changes to it, so that it might be philosophical enough to merit publication in *Mind*. Robertson responded immediately that he wanted it for *Mind*.[45]

Helmholtz's essay in *Mind*, "The Origin and Meaning of Geometrical Axioms," was indeed an English translation of the German version, which also appeared in print in July 1876. The latter was itself a slight extension or reworking of his 1870 address to the Docentenverein at Heidelberg; the German version contained merely an extra four paragraphs in its opening. The essay, Helmholtz later commented to his German readers, was "occasioned by unbelievable misunderstandings and misrepresentations of Riemann's" and his own "work in philosophical polemics."[46]

In the German-language version of the essay, Helmholtz paused (in a footnote) to dismiss Wilhelm Tobias's recent and extended claim (in *Grenzen der Philosophie, constatirt gegen Riemann und Helmholtz, vertheidigt gegen von Hartmann und Lasker* [1875]) that he (Helmholtz) had definitely misunderstood Kant (and other matters). Helmholtz cited three precise and distinct sections from Kant's *Kritik der reinen Vernunft* to argue that he had certainly not misunderstood him. He seemed concerned about criticizing Kant, whose name barely appears in any of the three versions of Helmholtz's essay. Yet the main point of criticism of Kant's account of the origins of geometry, and thus the nature of space, was a decisive one: what was the relationship between Kant's theory of cognition and the origins of geometry? Helmholtz gently questioned Kant's claim that, as concerned geometry, synthetic a priori principles were possible. He conceded that such principles were indeed possible as regarded the *form* of transcendental intuition, that is, without any relationship to empirical content or mechanical propositions; but once real, empirical content was added, all sorts of geometries (Euclidean, spherical, and pseudospherical, for example) were possible. Kant's claim notwithstanding, Euclidean geometry was not the only such geometry.[47] Instead, there were, as Helmholtz and mathematicians like Beltrami, Klein, and Lie had shown, at least several types of geometry. This development was a decisive challenge (and some feared a fatal blow) to neo-Kantian thought.

Kant and the neo-Kantians found another defender in the person of Jan Pieter Nicolaas Land, professor of philosophy at the University of Leiden. In the January 1877 issue of *Mind*, Land declared that Helmholtz's "remarkable modern speculations" about non-Euclidean space were "likely to be hailed as one of the chief difficulties with which the Kantian theory of space" would have to deal. But behind Helmholtz's argument, he averred, lay "a fundamental error" that appeared to hinder "many adepts of positive science from realizing the true nature of problems belonging to the theory of knowledge, or critical metaphysics." That error lay in the impossibility of simultaneous "wanderings on the border between science and philosophy"; "whoever crosses it," the virtually unknown Dutch professor haughtily instructed one of Europe's premier and most experienced scientists, "shifts his problem as well as his method." To attain "truth" in science was a matter of accumulating sufficient observations and using "correct reasoning"; in philosophy, by contrast, "truth" was precisely "the very thing to be settled." Science, however fallible its results may be, was based in the world of sense data. This was also true for geometry, which Land indeed saw as an empirical science. Epistemology had nothing to do with geometry or physics, he maintained. "The notions of 'objectivity' and 'reality,' hitherto equivalent, must be carefully kept asunder, or else it becomes impossible even to understand the questions at issue." In Land's view, the "results of scientific research" could not be "rightly employed as evidence against philosophical tenets" that disclaimed its validity. Coming all too close to an ad hominem, insulting attack on Helmholtz, Land wrote:

> For a scientific man fresh from physiology of the senses, it is hard to keep in mind that the perceiving, imagining, and thinking "subject" of philosophy is not altogether the same as that with which he had to deal in his former pursuits. There he considered it as a unity of body and mind, one of a class of objects in the world we observe. Here it is nothing more than the correlative of every object whatever, the observer and thinker opposed to them all. Unaccustomed to this kind of abstraction, the student of nature [that is, Helmholtz] speedily rounds it off into the full *anthropos* of physiology, not being aware that he has crossed the fatal border; and much of the reasoning current in his own domain is no longer acceptable as lawful tender.

The Kantians, Land declared, could not imagine anything beyond three dimensions. The non-Euclidean objects imagined by Beltrami and Helmholtz, he said, were unimaginable to the Kantians "unless the notion of imaginability be stretched far beyond what Kantians and others understand by the word." The only space imaginable was Euclidean space, that of "sense-experience." "The Kantian theory of space," Land proclaimed, was a form of a priori transcendental intuition.[48]

Helmholtz was no doubt less than delighted to be apprised by the Dutch professor about the science-philosophy border and its forbidden crossings. In April 1877 he told Robertson that he wanted to respond to Land's criticisms in *Mind* but that, "hindered by official business," it was not so easy to do so just then. He hoped to get his response to Robertson by July, however. Later that month, Helmholtz repeated to James Sully the complaint that he lacked the time to write an extended piece on epistemology. Sully, his one-time student, had become a philosopher and a psychologist and later succeeded Robertson at University College. Sully had sought an account of Helmholtz's views on spatial perception. "Time has become for me," Helmholtz replied, "one of the greatest rarities, that is, free time without being tired." He had long sought to write a book on spatial perception and had even made notes toward that end, but he lacked the requisite time. Instead, he recommended to Sully books by August Classen, who was "a strict Kantian." Helmholtz said, "I do not share his views for the most part."[49]

Finally, in April 1878, Helmholtz published in *Mind* a long, stinging rejoinder to Land's critique. He argued against and rejected every one of Land's points, summarizing his own position as essentially consisting of two points: first, physical geometry is a science whose propositions originate in experience; and second, "knowledge of axioms by transcendental intuition" that has no basis in experience is no more than a hypothesis which is neither proven, necessary, nor relevant.[50]

Helmholtz's reply to Land (and his subsequent philosophical address "The Facts in Perception") seem to have done little to satisfy his opponents. There were doubters about the very idea of non-Euclidean geometry, even at first among a few mathematicians (for example, Angelo Genocchi). Both Zöllner and Dühring had criticized and spurned Helmholtz for his creation and popular accounts of non-Euclidean geometry. Albrecht Krause saw Helmholtz's essay of 1876 on the origin and meaning of geometric axioms as an "attack" on the foundations of Kantianism. The Russian philosopher Nikolai Gavrilovich Chernyschevskii—the leader of Russian nihilism, one of the first Russian materialists, and a leading social and literary critic—also criticized Helmholtz's account of geometry and Kant in the same derogatory tone and dismissive rhetoric as was found in Dühring and, moreover, maintained that Helmholtz had failed to understand Kant.[51] Along with Jevons, Tobias, and others, his critics managed to keep him sufficiently annoyed; some even drew a response from him.

By its very nature, non-Euclidean geometry demoted Euclid's geometry into being just one of several (indeed, innumerable) possible geometries. It took someone whose name was as closely associated as any with classical science and, indeed, with classical thought in general, to argue that Euclid's place in geometric argument was only one of several. Moreover, as non-Euclidean geometry gradually became known and developed during the 1870s, the new spiritists latched onto the notion of a "fourth dimension"; in some quarters, this gave a bad odor to non-Euclidean geometry. They pointed to non-Euclidean geometry—with its fourth dimension and, in the case of Helmholtz's writings, the creation of imaginary creatures used to help understand conceiving (or not) different dimensions—as a scientific indicator of the existence of a spiritual world. Given his many critics and his desire to give a fuller statement of his epistemology, Helmholtz had unfinished philosophical business to complete.

"The Facts in Perception"

That completion came a year after Helmholtz had told Robertson and Sully that he lacked the time and the occasion to write an extended piece on epistemology; he now found both. The result was his departing rectoral address, entitled "The Facts in Perception." To appreciate it fully requires an appreciation of its immediate political context.

The year 1878 marked a turning point in Germany's political life. It witnessed the end of free trade, a new rise in anti-Semitism, the death of Pope Pius IX (and with it, reconciliation between Bismarck and the Catholic Center Party), the anti-Socialist law, and two assassination attempts on the life of William I, the second wounding him seriously. That summer, a siege mentality and virtual state of hysteria gripped the country—even off-handed remarks about the emperor were reported to the police, and the authorities brought an unprecedented number of charges of lèse-majesté.[52] Germany, and especially Berlin, was pervaded by a tense political atmosphere. The country took a turn to the right.

Helmholtz himself was affected by the assassination attempts. Anna's brother Ottmar was with the royals, and he told Helmholtz that the emperor, as a result of the attempts, now had his right hand bandaged and walked with a cane. Helmholtz told Anna that he had two things on his mind. He had decided to vote in an upcoming election for the Conservatives (and in particular for their candidate Field Marshal Helmuth von Moltke), who were pro-Bismarck and pro-German, and not, as he usually did, for the National Liberals, who had formed an alliance with the Progressive Party. The other thing was his departing rectoral address, which he had been working on constantly in July. He had various titles in mind; perhaps he would call it "What Is Real?"; or use one of his favorite lines of poetry, "All that is transient is merely an image" (from Goethe's *Faust*); or "A Path to the Mothers," or, more soberly, "Principles of Perception."[53] In any case, its conservative cast paralleled the turn that Germany was taking from liberalism to conservatism that year.

As rector of the university, Helmholtz also became directly involved in the reaction to the assassination attempts. Karl Nobiling, who made the second attempt on the emperor's life (2 June), was a socialist and belonged to a student organization at the university known as the Akademische Lesehalle (Academic Reading Room). Thanks largely to the reading room's socialist custodian, the organization actively collected socialist newspapers and other radical literature. The Lesehalle came under the supervision of the university senate, of which Helmholtz, in his capacity as rector, was the head. Shortly after Nobiling's assassination attempt, Helmholtz asked the two student leaders of the Lesehalle to appear before himself, Mommsen, and Georg Beseler, a liberal jurist. He informed the two students in advance that the Lesehalle would be searched. After socialist newspapers and other radical literature were found there, the students decided to terminate the custodian's employment. They were also advised that they would come under police surveillance and that other measures would be taken against the organization. They then decided to close down the Lesehalle entirely and reported to Helmholtz that they were resigning their office forthwith.[54]

It was in this politically charged atmosphere, as his year-long rectoral appointment drew to a close, that Helmholtz delivered his address. He spoke on 3 August 1878, a university holiday celebrating the birthday of its founder, "the much-tested King Frederick William III." For the ceremony, Helmholtz wore his rectoral robe, with its golden cloak; he spoke for an hour in the fully packed university auditorium, which led, he said, to its being "very hot under my warm cloak."[55]

He entitled his address "The Facts in Perception," a title that harked back to his writings in 1855 and the 1860s on the empirical basis of human perception and the factual basis of geometry. Though his address overwhelmingly concerned philosophical (that is, epistemological) matters, he began with a brief historico-political discussion, portraying the era in which the university was founded and contrasting it with the present era. The university's founding in 1810 came, he said, at the most parlous moment in Prussian history, when much territory was lost and when the nation felt completely exhausted and deeply humiliated as a result of war and foreign occupation. Yet in retrospect, he thought, it was also a spiritually rich moment: there were "inspiration, energy, ideal hopes, and creative thoughts." Today, by contrast, while Germany and its people were in a "relatively brilliant foreign situation," there was a sense of "envy" vis-à-vis the daring of Frederick William's founding of the university at such a moment of national distress. The king had, on Helmholtz's rendering, "risked the throne and his life in order to entrust himself to the nation's resolute inspiration in the struggle against the [French] conqueror," which showed that he was a man who had "won the trust of his people's spiritual forces." Helmholtz also evoked something of the era's culture. He noted that Schiller, Kant, Herder, and Haydn had all only recently died; Goethe and Beethoven were still alive; and there also stood among Germany's leading academic figures men like Wilhelm von Humboldt, Barthold Georg Niebuhr, Friedrich August Wolf, Friedrich Carl von Savigny, Friedrich Schleiermacher, and Johann Gottlieb Fichte, "our University's second rector and the powerful, fearless speaker, [who] swept his listeners away with his stream of moral inspiration and through the bold intellectual flight of his Idealism."[56] Helmholtz did not mention Hegel's or Schelling's names.

Although he pointed to "the weaknesses of Romanticism," he also recognized that it cultivated "beautiful feelings" and advocated the importance of fantasy (though at the expense of understanding), as "compared to dry, calculating egoism." While romanticism had shown "much vanity," it was nonetheless "a vanity that reveled in high ideals." He said he had known men—he did not explicitly name his own father, but he surely had him in mind—who had volunteered for army service and yet "who were always ready to immerse themselves in the discussion of metaphysical problems." Such men read the great German poets and "burned with anger when Napoleon I, inspiration and pride, and the acts of the War of Liberation were discussed." Matters were now completely and astonishingly different, he maintained: it was now an era "in which cynical contempt for the human race's ideals" had spread into the streets and the press, and which had culminated in two "abominable crimes" that chose the emperor as their object "simply because his person united all that humanity to date had considered worthy of honor and gratitude." It was even becoming difficult to remember the sense of unity, sacrifice, and patriotic inspiration that "all classes" of German people felt just a few years before when the German lands had fought France and become a united Reich. The people had done so at "the call" of the monarch, at least according to Helmholtz, who left Bismarck's central, decisive role in this war and unification unmentioned. In the intervening eight years, it was also increasingly difficult to recall that the educated classes had given "the poorer classes of our people a more carefree and humane existence." "It seems to be humanity's nature," Helmholtz dolefully concluded, "that, along with much light, there is also much shadow: political freedom at first gives the baser motives greater license to manifest themselves and to mutually encourage one another—so long as an armed public opinion does not confront them in energetic opposition." Still, he thought there was hope for his own times, even as he warned against getting carried away by such hope. He advised that each person must be "watchful within the circle in which he works and which he knows, as to how the work of humanity's immortal goals is proceeding." The "secondary" and "practically useful tasks" of his own era should not allow its participants to "lose sight of humanity's eternal ideals."[57] These historico-political remarks were distinctly more conservative in nature and tone than those in his speeches of 1862 at Heidelberg and 1869 at Innsbruck, reflecting both the changing times—the collapse of liberalism and the assassination attempts on the emperor—and his own changed professional situation, now as the professor of physics in the Prusso-German capital who stood at the heart of the academic establishment and who had to endure attacks by the likes of Zöllner and Dühring. The speech was part of the *Kulturkampf*.

Then Helmholtz turned from politics and history to his main subject: epistemology. At the turn of the nineteenth century, he declared, epistemology constituted "the starting point of all science." In their different ways, both philosophy and natural science sought to know what was true and which ideas corresponded to reality. Philosophy approached this epistemological matter independently of the physical world; its realm was that of the mind, that is, of definitions, designations, representational forms, and hypotheses. By contrast, natural science was the "pure residue" of "the world of reality, whose laws it seeks." He portrayed the philosophical and natural scientific realms as complementary. While epistemological questions appeared throughout the sciences, they were acutely present, he said, "in the theory of sense perceptions and in investigations on the fundamental principles of geometry, mechanics, and physics." In general, both facts (experience) and theory (judgment) were needed to arrive at "the truth."[58]

He began with an analysis of perceptions, a topic that he had of course treated intensely in part three of the *Handbuch*. He spoke of the different qualities of sensation, which he said were essentially the same as Fichte's "circle of qualities," though he also based these qualities on Müller's ideas, since, he continued, physiology had shown that sensation was due not to "the type of external impression" but rather to "the sensory nerve that has been affected by the impression." Human sensations were "effects produced by external causes in our organs, and the manner in which one such effect expresses itself depends of course essentially on the type of apparatus which is affected." To the extent that the qualities of sensation inform individuals about the external origins that stimulated them, they are signs, not images; there is no similarity of image between the external object and a corresponding sensation in the body, as there would be between, say, a statue and a drawing of it. Even so, he said, "The relations between the two [the sign and 'what it is a sign for'] are so restricted that the same object, taking effect under equal circumstances, produces the same sign, and hence unequal signs always correspond to unequal effects." Yet he insisted that "this residue of similarity" between the two yields an "imaging of the lawlike in the processes of the real world. Each natural law says that, given preconditions which are alike in certain respects, consequences which are alike in certain other respects will always follow. Since likeness in our world of sensation is shown by like signs, then there will also correspond to the natural-law consequence of like effects upon like causes a regular consequence in the field of our sensations." Thus, signs are "certainly not to be dismissed as empty appearance"; rather, they indicate the existence of a law.[59]

Helmholtz then made several key distinctions, the first being the one between spatial and mental relations in perception. Sensations whose sources are external to the body, he argued, are spatially determined, so space appears to be physical and "laden with the qualities of our perceptions of movement." He distinguished between space "as the necessary form of outer intuition, since we understand as the external world precisely what we perceive as spatially determined," and nonperceptible spatial relations, which have to do with "the world of inner intuition," where self-consciousness occurs. He further and finally distinguished between what he called the circle of "current presentable" sensations (*"die zeitweiligen Präsentabilien*," that is, "the entire group of aggregate sensations induced during the said period of time by a certain definite and finite group of the will's impulses") and the circle of the "present" sensations (that is, "the aggregate of sensations from this group which is just coming to perception"). These two "circles" can, under certain conditions, merge into one another. The changes responsible for the former he called, following Fichte, the Ego forces; those responsible for the latter, the Non-Ego forces. Hence, "the most essential characteristics of spatial intuition can be derived." That point put him in opposition to Kant on spatial intuition (and on geometry), and he spelled this out in some detail.[60]

Helmholtz observed, furthermore, that "daily life"—and especially art—was rich with examples of how the thought-formation process worked. The artist's intuition spotted rest or motion, and if this was perceived to repeat itself often enough, it became lawlike. The artist thereby gains "an intuitive image of the typical behavior of the objects which interest him, of which he subsequently knows just as little concerning how it came about as a child knows by which examples he has learnt the meaning of words." The artist and the viewer (or listener) know that they have found "the real" because the artist presents the essential dimensions of the object or action. The mental process involved here, Helmholtz claimed, was similar to "unconscious inferences." (He added, however, that he had otherwise decided to abandon this term after Schopenhauer and his followers had employed it in a confused manner.)[61]

He then briefly explained his use of Lotze's notion of "local sign" in understanding visual perception, as well as his own distinction between nativism and empiricism in regard to the "acquired knowledge of the visual field." The distinction, he said, rested on the effects of "accumulated memory impressions." The nativists believed that knowledge was due largely "to an innate mechanism in the sense that certain sense impressions would cause certain completed spatial representations." Helmholtz believed that he had refuted this view in stating that experience through movement overcame "innate intuition." Nativism explained nothing; it begged the question; it relied too heavily on the belief that ideas are produced "by the organic mechanism" (as opposed to empiricism's assumption that ideas come from sensations issuing from the external world, which are then transformed "according to the laws of thought"); and its assumptions were simply unnecessary. The only possible objection against empiricism, by contrast, lay in the motor abilities "of many newborns or even of animals hatched from eggs." But even here Helmholtz pointed to the learning process in the first days of life. He doubted that much of the visual process was inherited from parents as opposed to being learned by newborns, though he conceded that more research needed to be done in this regard.[62]

He thought that extreme subjective idealism, in the form of imagining life as a dream (for example, as expressed in Pedro Calderón de la Barca's philosophical play *Life Is a Dream* [1635]), was as unbelievable as it was logically irrefutable. The same held for Fichte's distinction between the Ego and the Non-Ego, though Fichte had at least thought that other human beings were real. For Fichte all Egos emanated from "the absolute Ego," which "was the world of ideas, which the World Spirit posits for itself, and could again assume the concept of reality, as occurred with Hegel." The opposite view, that is, Helmholtz's own, he called "the realist hypothesis." It was based on "normal self-observation, according to which the changes of perception following an action are in no way in any mental connection with the previously occurring impulses of the will. It regards the material world external to us as that which seems to endure in daily perception as existing independently of our ideas." Hence "the realist hypothesis" was "the foundation for action." Both idealism and realism were "metaphysical hypotheses which, so long as they are recognized as such, and however injurious they may become when represented as dogma or as supposed necessities of thought, are completely justified scientifically." Science and, even more, "practical action" always involved metaphysical hypotheses, he here conceded, but it was "unworthy of a thinker wanting to be scientific to forget the hypothetical origin of his principles. The arrogance and the passion with which such hidden hypotheses are defended are the usual consequence of the dissatisfied feeling that its defender harbors in the hidden depth of his conscience about the correctness of his cause." By contrast, our perception of "the law-like in the phenomenon" was not hypothetical, since it originated in "the recognition of a law-like connection between our movements and the sensations appearing therefrom." Concepts thus effectively amounted to summaries of "a changing series of phenomena over time, one which remains the same in all its stages." That led him to cite some of his favorite lines of poetry, from Schiller's "Der Spaziergang" ("The Walk," 1795), a poem about freedom, progress, and civilization:

> The wise man
> Seeks the trusted law in chance's horrifying wonders,
> Seeks the resting pole in phenomena's flight.[63]

Similarly, in his reply to Land only four months earlier, he had insisted that human beings have an "intellectual drive to consider everything that happens as law-like, that is, as comprehensible."[64]

For Helmholtz, as he had already outlined in previous works and lectures, law was "the constant relationship between changeable quantities," it was that which "binds them." It came through "reflective understanding of a phenomenon." Well-articulated and well-understood laws were causal in nature. When scientists become convinced that laws are "unambiguously determined" and that they have existed and will exist forever, everywhere, "and in all cases, then we recognize it [the law] as something existing independently of our ideas, and we call it the cause, i.e., that which, behind the change, is the originally abiding and existing." By contrast, he reserved the term "force" for the recognition of "the law as directing our perception and natural processes, as a power equivalent to our will." This brought him back to Fichte, for Helmholtz here too distinguished between acts due to and acts independent of our will. "The emphasis here falls on the facts of observation, that the perceived circle of presentables is not determined by a conscious act of our idea or will. Fichte's Non-Ego," he maintained, "is here exactly the right negative expression. To the dreamer, too, it seems that what he believes that he sees and feels is not produced by his will or by the conscious linking of his ideas, even if, unconsciously, the latter is in reality often enough the case; to him, too, it is a Non-Ego. So, also, with the idealist who looks at it as the world of ideas of the World Spirit." The "real," that is, "the material," is "that which permanently influences us behind the change of phenomena." In the final analysis, Helmholtz thought that all "we can attain is knowledge of the law-like order in the realm of the real; to be sure, this can only be presented in the sign system of our sense impressions." Here too he turned to literary support, this time to one of Goethe's lines from *Faust*: "All that is transient is merely an image." He believed that as a matter of "broad outlook," he and Goethe were "on the same road" toward truth. (That was a rather different assessment than he had given in 1853.) Knowledge, Helmholtz ultimately concluded, remained "fragmentary": humans can never be certain that they have reached "the underlying substances and forces." There will always be an element of trust in our law-like observations, which is "thus simultaneously trust in the conceivability of natural phenomena." Once conceptualization has been completed, he thought, once the final cause of a change has been determined, one can speak of "the causal law" as being that which "expresses trust in the complete conceivability of the world." "Conceiving, in the sense that I have described it, is the method by means of which our thinking subordinates itself to the world, orders the facts, predetermines the future."[65]

For Helmholtz, then, "the causal law is really an *a priori* given, a transcendental law"; yet it was subject to experience. All one could do was "trust and act!" On the whole he found Kant's philosophy the most convincing on the issue of truth, though he differed with him on subsidiary matters, such as the nature of intuition, in particular as regarded the geometric axioms. Whereas Kant thought them transcendental in origin, Helmholtz thought them empirical. But as long as natural science limits itself to finding "the laws of the real," it will remain secure, he maintained, and "will not be confronted by idealistic doubts. In comparison to the high-flying plans of the metaphysicians, such work may seem modest." "The true researcher," he continued, "must always have something of the artist's insight, of the insight which led Goethe, and Leonardo da Vinci, too, to great scientific thoughts. Both artist and researcher strive—even if in different ways—towards the same goal: to discover new lawfulness. One must not, however, want to propagate idle daydreams and crazy fantasies for artistic insight. Both the true artist and the true researcher know how to work properly and how to give their work a stable form and convincing similitude." By contrast, "the monstrous products of Indian daydreaming," scholasticism, and metaphysics in general, say little about reality. Seven years after the appearance of Darwin's *Descent of Man* and of his own essay on the origins of life (both 1871), Helmholtz concluded his rectoral address thus:

> We: bits of dust on the surface of our planet, itself hardly worth calling a grain of sand in the universe's infinite space; we: the most recent race among the living on earth, according to geological chronology barely out of the cradle, still in the learning stage, barely half-educated, declared of age only out of mutual respect, and yet already, through the more powerful force of the causal law, grown beyond all our fellow creatures and vanquishing them in the struggle for existence; we truly have reason enough to be proud that "the inconceivably sublime work" has been given to us to learn to understand slowly by means of constant work. And we need not feel in the least ashamed if this does not at once succeed in the first assault of a flight of Icarus.[66]

Here, as elsewhere and before, he took an evolutionary measure of the universe and mankind.

The rectoral address and its essay form became Helmholtz's most important and fullest statement of his epistemological views, rounding out his earlier efforts of 1855 and 1866–67. His fundamental ideas and viewpoint drew on a mixture of Kant, Fichte, and Müller; empiricism in general; and especially his own results in physiological optics and non-Euclidean geometry. He was opportunistic and pragmatic: he used what he thought worked. He also, in effect, drew on Ferdinand Helmholtz's romantic philosophical outlook concerning the aesthetic education of mankind. As Ferdinand had declared in 1837: "The mind [*Geist*] lives truly only in the act, as a creative force of appearance, which nature forms according to its clarity." Historians of philosophy (and science) have seen Helmholtz, and this essay in particular, as a major figure and statement in the neo-Kantian movement as it developed from the 1840s on; but especially thanks to his anti-Kantian views on space, they have also debated the extent to which he can legitimately be considered a Kantian.[67] "Neo-Kantian" may be too confining a label. His philosophical position did not draw predominantly on any one source (other than his own scientific work).

Helmholtz's intellectually demanding, philosophical address was followed by an hour's worth of singing and awarding of prizes. The speech had proceeded in an orderly enough manner, but since Helmholtz spoke for an hour on such a recondite topic in a warm room on a hot August day in Berlin, it is not surprising to learn that many of his colleagues, also dressed in their heavy academic gowns, had fallen asleep. There was applause from his close colleagues Zeller, du Bois–Reymond, and Kronecker, but most avoided making any comment at all. As Helmholtz told Anna: "Moreover, I knew that it would not accord with the majority's taste. But I had said to myself that, if I have to work, I want to say something whose working out I myself would find interesting. For in the end, it's always better that they find me too learned than too trivial."[68]

Like many such philosophical addresses, it may have read better than it sounded. A good number of people wanted to read it. Within days of the address, the Heidelberg musicologist Ludwig Nohl asked Helmholtz for a copy. Bluntschli at Heidelberg, a fierce nationalist and liberal, had told Nohl that the address would advance the "ideal interests at the German universities and so avoid the bad consequences of a crass materialism." Others, like Borchardt, simply thanked Helmholtz for sending a copy of his speech. Marie von Olfers, a German author and a family friend, wrote Helmholtz that she and her husband were "inspired" by his speech. "I've already had some thoughts of my own. On dream and truth—appearance or reality."[69] Though the speech contained little that was new for those already familiar with his philosophical views, it nonetheless confirmed that by 1878 Helmholtz had become a *maître à penser*.

With the rectoral address behind him, Helmholtz left Berlin, first undertaking a brief stay with his family in Ambach, then going off on his own for five or six weeks to Switzerland and Italy for rest and recuperation. He went to Davos for the first time. He then went on to Pontresina in mid-September, and from there to Italy. He met first with Francesco Brioschi, an Italian mathematician, a leading political figure, and a founder (in 1863) of the Milan Polytechnic; and then with Ruggiero Bonghi, who until 1876 had been the Italian minister of instruction and who hoped to reform the medical faculty there along Germanic lines. From Nervi, he wrote Anna about the local flora; from Siena, about the city's architecture (especially its cathedral) and its collections of paintings. "To do Siena complete justice," he confessed, "one must go deeper into art history than I've so far been able to do."[70]

From there he went to Naples to see Dohrn and the fledgling Zoological Station. He found the natural environment around Naples "really unbelievably beautiful." He visited the local museum and went to see the artworks of Pompei: "Although almost all decorations have been taken off the walls, there remains a peculiar impression, unique in its kind, and it makes a great impression of pre-civilization, of a small provincial town of the time and its artisans' handicrafts and luxury." He spent several days visiting Vesuvius, which he found particularly stimulating. He climbed down into its crater and managed to avoid getting burned by the hot lava. "It's a unique opportunity to see things which one cannot imagine if one hasn't indeed seen them [for oneself]. After having philosophized so much about the Earth's interior heat, I had to have it before my eyes once again." He and Dohrn walked in the Naples countryside, and they (and Dohrn's wife) made "a beautiful sea trip" to Amalfi aboard the Station's steamboat, the *Johannes Müller*. On the return trip to Berlin, he visited Rome and Boll, and, finally, Florence.[71]

The visits to Dohrn and Boll were not merely social calls and opportunities to sightsee with younger, aspiring colleagues. He had tried, since 1871, to help Dohrn get the Station established by supporting the latter's requests for financial aid from Berlin. He had also sought to care for Boll, and when Boll died in 1879, he helped arrange for the publication of some of his unpublished manuscripts on physiological optics.[72] In these practical ways, too, he continued his efforts to advance science.

Neither Helmholtz's defense of non-Euclidean geometry nor his extended philosophical analysis of epistemological issues in perception seem to have made much difference to those already opposed to him. The feeling was more or less mutual. In 1881 Lipschitz sent Helmholtz an epistemological essay that he had recently published, remarking that its views were similar to Helmholtz's own. Helmholtz replied:

> I'm interested to see that you too have landed upon my thoughts in epistemology. That's very nice, and it encourages me, although I've completely given up all hope of ever experiencing a *reformation of philosophy* itself. In my thoughts I complain, like Schopenhauer, about philosophy as a discipline, but I will not put this down on paper. There are at best only important bookworms who have never produced any new knowledge, and so have no idea about it, of how it comes to pass. Each one reads only himself, and is unable to think his way into the thoughts of another. When I see, however, that mathematicians [e.g., Lipschitz] and physicists are gradually turning into my path, then I at least have hope for the future.

His scorn for philosophers, if not philosophy itself, was transparent but private. During the half century following the 1878 essay, his views on epistemology and on space and geometry continued to find supporters, opponents, and a mixture of the two, and he was regarded as a leading Neo-Kantian. Hermann Cohen, Wilhelm Wundt, Henri Poincaré, Ernst Mach, Alois Riehl, Leonhard Nelson, Bertrand Russell, Ernst Cassirer, Moritz Schlick, and others continued to study and to argue in favor of or against individual points expressed in "The Facts in Perception" as well as in his other epistemological writings.[73]

Helmholtz sought to pass on his experiences with philosophy and philosophers to his son Robert. While he was glad to learn that Robert was interested in philosophy, he warned him about studying it before first (and broadly) studying natural science. "Of 100 people who throw themselves into philosophy, 99 are nothing more than intellectual idlers who want to reach high goals without devoting a lot of work to it; they have, however, never reached [their aims]." The prior pursuit of science, he thought, was the basis for subsequently pursuing epistemology, though this brought little agreement "from the mass of philosophy-dilettantes." It was this sort of attitude that led Maxwell to say of Helmholtz that he was "not a philosopher in the exclusive sense, as Kant, Hegel, [and Henry] Mansel are philosophers, but one who prosecutes physics and physiology, and acquires therein not only skill in discovering any desideratum, but wisdom to know what are the desiderata."[74]

22

Anti-Helmholtz, Again

Zöllner's New Attack and Spiritism

Its title notwithstanding, volume one of Zöllner's *Wissenschaftliche Abhandlungen* (1878) had virtually nothing to do with substantive scientific issues; instead, it was another extended attack on Helmholtz and his associates. Using language that was variously sarcastic, scornful, and inflammatory, Zöllner criticized Helmholtz, Thomson, Maxwell, du Bois–Reymond, and Tyndall, individually and jointly. He faulted Helmholtz—the main object of his attack—for supposedly misinterpreting the work of Newton and Faraday on atoms and other topics; for his idea about the origin of life on earth; for his epistemology; for his ideas (of 1847) on action-at-a-distance as central to the principle of conservation of energy, declaring that it was the source of Helmholtz's (allegedly) erroneous critique of Weber's electrodynamics, which, Zöllner noted, Helmholtz later (in 1872) conceded and Maxwell showed (in 1873); and numerous other failings. In Zöllner's eyes, Helmholtz had betrayed his German countrymen in matters of priority or analysis to the benefit of English scientists. Zöllner sympathized with Dühring against Helmholtz. He made much of Menzel's painting of Mimi von Schleinitz's salon (1874, see fig. 19.3), which portrayed the Helmholtzes socializing with various members of the Prussian royalty and aristocracy, and clearly displayed his resentment of Helmholtz's high social status. To make himself appear even-handed and fair-minded, Zöllner praised Newton, Faraday, and Kant, scolding his opponents' misinterpretation of the two British scientific immortals.[1]

There was yet another issue of contention between Helmholtz and Zöllner: that of spiritism. Despite the increasingly widespread positivist and materialist outlook on life from the 1830s onward, between the 1860s and 1880s there was also a notable rise of interest or belief in spiritism (or spiritualism) and psychical research (the two were often indistinguishable). Numerous scientists, especially the new experimental psychologists, became very interested in spiritism and psychical research, some as enthusiasts and believers, others merely as open-minded persons, and some as debunkers. Among them were William Benjamin Carpenter, Simon Newcomb, Alfred Russel Wallace, J. W. Strutt (Lord Rayleigh), William Crookes, Oliver Lodge, William Huggins, Augustus De Morgan, Gustav Theodor Fechner, and William James. The peak in interest in spiritism coincided with psychology's renewed attempt—one that had been gradually developing since the early to mid-nineteenth century—to become an experimental science. Fechner's psychophysical experiments, his law, and his general impact on measurement theory; Ernst Heinrich Weber's analysis of the "just-noticeable difference" in tactile sensations; Helmholtz's determination of the speed of nerve transmission and related work; du Bois–Reymond's electrophysiological studies—these and other developments all together opened up a range of new experiments (for example, by Donders and Wundt) in the study of human sensation, perception, and thought in their temporal dimensions. The "New Psychology," as this experimental physiology applied to the mind was sometimes called, sought full status as a science, equivalent to that of the physical sciences. Wundt's opening of his experimental psychology laboratory at Leipzig in the late 1870s was one important marker in the gradual transformation of psychology as a discipline. While some experimental psychologists sought to distance their field from spiritism, others cultivated it in the hope of gaining financial backing from individuals interested in furthering psychological experimentation. James, for example, had much sympathy for psychical research. "Orthodoxy is almost as much a matter of authority in science as it is in the Church," he wrote. "We believe in all sorts of laws of nature which we cannot ourselves understand, merely because men whom we admire and trust vouch for them. If Messrs. Helmholtz, Huxley, Pasteur, and Edison were simultaneously to announce themselves as converts to clairvoyance, thought-transference, and ghosts, who can doubt that there would be a prompt popular stampede in that direction?"[2] None converted.

One of the most notorious spiritists was Henry Slade, an American practitioner of so-called psychography, or "spirit writing" on paper, slate, or human skin. Slade was a clever conjurer who could supposedly perform a psychograph at will and in a well-lit space. While sitting at a table, he or the medium (or both) placed a slate and a slate pencil under the table's surface and pressed tightly against it. In due course, a writing sound would be heard by the séance participants, and a message would appear on the slate. In New York, he was caught surreptitiously executing the writing under a séance table. He fled to Britain, where in 1876 he was again exposed, found guilty, and sentenced to three months in jail, but the conviction was overturned on a technicality.[3]

He then fled to Germany, and in late 1877 he held séances attended by Zöllner and others in Leipzig. Zöllner reported that Slade used spiritist means to quickly put four knots in a rope that had already been sealed at its two ends. Zöllner managed to tie Slade's "experiments" to his (Zöllner's) theory of space. Slade's experiments supposedly gave Zöllner "factual proof of the correctness of the theory" (discovered by him [Zöllner]) "of 'synthetic judgments a priori.'" He judged Slade a gentleman and certainly not a swindler and maintained that the spiritists were more logical and open-minded and better thinkers than the so-called scientific men. He attacked Helmholtz for his defense of Tyndall against the spiritists and suggested that several of Thomson's and Maxwell's claims about physical phenomena were like those of the spiritists. He suggested that "modern 'men of science'" were creating the rules of what counted as rational, and he sought to challenge their supposedly morally, intellectually, and academically superior position for doing so. He defended Slade and parapsychological investigations in general, calling the latter "transcendental physics."[4]

Zöllner's encounter with Slade got people's attention. G. Stanley Hall, then a young American experimental psychologist studying in Leipzig with Wundt and Ludwig, noted that quite a few Leipzig students had become "ardent disciples of Professor Zöllner." Zöllner did all he could to make it appear that he was defending German, and above all Leipzig, scientists (Fechner and Ernst Heinrich Weber) against foreign and Berlin scientists. He had also been the strongest supporter of bringing Wundt, Helmholtz's former assistant, to Leipzig; Wundt arrived in 1875 as the new professor of philosophy. Zöllner especially liked Wundt because of Wundt's theory of perceptual processes, which opposed both Helmholtz's and Hering's theories. Wundt admired Zöllner's own work in this area, but he was not an admirer of spiritism, and for this reason several years after he came to Leipzig, Zöllner became disappointed in him. Hall later came to see Zöllner as not only deluded for falling prey to Slade and spiritism; he thought him outright insane. While he thought Zöllner's characterization of Helmholtz had violated "collegiate decorum," he too was unsympathetic both to Helmholtz's popular accounts of non-Euclidean geometry (i.e., space) and to non-Euclidean geometry itself, which he thought constituted "an indiscreet appeal to the scientific imagination" that had misled the general public. The Berlin astronomer Wilhelm Foerster judged that Zöllner, in his later years, had lost his mind. As for Helmholtz, he reported to Tyndall: "Again a new book by Zöllner is here, and it already completely belongs in the madhouse. E. du Bois is taken up the most therein; we two [i.e., Helmholtz and Tyndall] also have the honor of being mentioned, at times as nullified sinners." Tyndall replied: "I must procure Zöllner's book so as to inform myself of this new phase of his insanity." There was widespread, if not quite universal, condemnation of Zöllner's belief in spiritism and his linkage of it to physics and non-Euclidean geometry. By the time Zöllner died (in 1882), he had been effectively written off by the scientific community. Both then and before, Helmholtz would not even give the spiritists (of whatever variety) the time of day, though he retained a mild interest in uncovering the participants' source of psychological gullibility.[5]

Jews and Anti-Semitism

The 1860s and 1870s were turbulent decades in German history. Not only did several or all of the German lands fight three wars, but they also united to form the Second Reich. From 1866 to 1873 Germany experienced extraordinary economic growth, to be sure, but also much inflation, in part as a result of the receipt of about 4 billion marks in French indemnity payments to Germany as part of the settlement of the Franco-Prussian War. No city felt the effects of boom and bust more than Berlin.

Although Bismarck had achieved a unified Germany, many considered it a phony unification, a mere centralization of political and military affairs at the highest levels in Berlin rather than an "organic" social and cultural unification among the German peoples themselves. Several provinces (still) felt hostility toward Berlin and Prussia. A considerable number of individual Germans also felt hostility toward the Roman Catholic Church as an alien institution; toward immigrated Poles; toward the growing number of socialists; and toward Jews, who, though the ancestors of many of them had lived there since the Middle Ages, and despite some intermittent concessions in civil matters since the Napoleonic era, had only recently (1869–72) received definitive status as full German citizens. Among university students there was a turn to the right, with increasing emphasis placed on *völkisch* themes: a romantic view of nature, a belief in spiritist forces, a call for "true" spiritual unity among the German people, and anti-Semitism. During the second half of the 1870s, anti-Semitism—the term was coined around that time—in Germany, in particular in Berlin, experienced a new flare-up, with renewed tirades, threats, and ideological changes. As the boom of the *Gründerzeit* turned to bust, a new scapegoat was needed to explain Germany's economic and political problems.[6]

The Helmholtzes had numerous acquaintances and friends of Jewish an-
cestry or background. These included the musician Joseph Joachim, the writer
Fanny Lewald, the Martin Levys, the musicians and brothers Nicolas and
Anton Rubinstein, the writer Berthold Auerbach, the National Liberal politician
Ludwig Bamberger, and various members of the Mendelssohn or Mendelssohn-
Bartholdy family (descendants of the great Jewish philosopher Moses Mendelssohn). All were more or less regular guests in the Helmholtz home. In addition
to being a celebrated novelist, Auerbach was an important promoter of Jewish
culture. He was also the German-language editor of Spinoza's *Sämmtliche Werke*
and the chair, in 1876, of a committee that sought to raise funds to erect a statue
to Spinoza in The Hague marking the bicentenary of the philosopher's death. The
committee established a support list of honorary and corresponding members,
including leading scientists, philosophers, and men of letters from Europe and
elsewhere; Helmholtz joined them. In the spring of 1879, Helmholtz was elected
a foreign member of the Dutch Academy of Sciences in Haarlem.[7]

Helmholtz also had dozens of distinguished scientific friends, colleagues, and
students who were Jewish or of Jewish ancestry. Among them were the mathematicians Carl Gustav Jacob Jacobi, Carl Wilhelm Borchardt, Immanuel Lazarus
Fuchs, Rudolf Lipschitz, Leo Koenigsberger, and Leopold Kronecker; the physicist and chemist Heinrich Gustav Magnus, and the physicists Peter Theophil
Riess and Emil Warburg; the physiologists Hermann Munk, Ludimar Hermann,
Robert Remak, Julius Bernstein, Rudolf Heidenhain, and Richard Liebreich; the
anatomist Jacob Henle; the botanist Nathanael Pringsheim; the chemists Victor
Mayer and Adolf von Baeyer (Nobel Prize in Chemistry, 1905); the immunolo-
gist and chemotherapist Paul Ehrlich (Nobel Prize in Physiology or Medicine,
1908); the physician and medical historian August Hirsch; and the technologist
Leopold Loewenherz. Numerous Helmholtz students (at various levels and for
various periods of study) were of Jewish ancestry, including the physicists Leo
Arons, Felix Auerbach, Emil Wilhelm Cohn, Eugen Goldstein, Heinrich Hertz
(Helmholtz's most prized student), Ernst Pringsheim, and Heinrich Rubens; the
physiologists I. F. Tsion and Hugo Kronecker (Leopold's brother); and the ge-
ographer Franz Boas, who subsequently became an (American) anthropologist.
The chemist Fritz Haber and the electrochemist, industrialist, and statesman
Walther Rathenau—two highly important industrial and political figures in
the early twentieth century—also studied with Helmholtz. Helmholtz's Jewish
"postdoctoral" researchers included the German-born British physicist Arthur
Schuster and the Prussian-born American physicist Albert Abraham Michelson
(Nobel Prize in Physics, 1907). Probably only a few, if any, of these figures were
observant or believing Jews. Some converted to Christianity for professional or
personal reasons, and some were converted as children by their parents. Among
Helmholtz's associates, for example, Magnus, the Kronecker brothers (Hugo as

well as Leopold), Fuchs, and Lipschitz were born Jewish but were later baptized. Still, the knowledge (or even supposed knowledge) that someone was Jewish could be enough to potentially block the person's academic appointment. Du Bois–Reymond wrote to Helmholtz in 1858: "I presume that you [that is, the Heidelberg Medical Faculty] are not considering [Nathanael] Pringsheim, who of course now is, and for a long time has been, the first of plant physiologists. He is, however, a Jew." Nonetheless, in 1864 Pringsheim became professor of botany at the University of Jena.[8]

Helmholtz's many relationships with men (and women) of Jewish ancestry, taken as a whole, might suggest a very positive attitude toward Jewry on his part. Yet for several reasons it is hard to think of him as a friend of Jews per se. First, on a few occasions he made less-than-flattering remarks about Jews. He referred, for example, to the British mathematician James Joseph Sylvester as "a great mathematician . . . in aspect extremely Jewish, but otherwise an important and presentable person." Second, he and Anna did not hesitate to call Richard Wagner, an avowed, public, and indeed notorious anti-Semite, his (their) good friend. Much the same was true for their friend Heinrich von Treitschke, a full-throated conservative and Prusso-Germanic nationalist, a great admirer of Bismarck, and a leading and vocal anti-Semite. As professor of history at the University of Berlin, as well as a journalist and politician, he was one of Germany's leading and most prominent modern historians. In November 1879, as anti-Semitism flared up during the second half of the decade, he wrote a notorious article entitled "Unsere Aussichten" for the prestigious *Preussische Jahrbücher*. Here he (in)famously declared, "The Jews are our misfortune." The article was an attack on Jewry, if "only" along nationalist and religious, not racial, lines. He saw Jews, especially those from eastern Europe, as a foreign element in the new Reich, a danger to its cultural and social integration. He regarded them as inferior and as a threat to Germans and Germany, even as he called for them to better integrate themselves into German life. Treitschke's notorious and influential article helped make anti-Semitism academically and socially acceptable. Like Dühring before him, Treitschke became a hero to many students at the University of Berlin and elsewhere.[9] Helmholtz never criticized Treitschke in this or any other regard; indeed, they remained lifelong friends.

Third and finally, on two occasions Helmholtz avoided speaking out against "Jewish persecution," to use his own words. In November 1880, as part of the countermovement against the flare-up of anti-Semitism in Berlin and Treitschke's provocative article, du Bois–Reymond forwarded to him a petition against the persecution of Jews and asked him to sign it, as du Bois–Reymond himself had just done. Helmholtz refused, and in doing so gave his reasons: "I find it too poorly written. Some individual sentences are not understandable, while others are of extremely doubtful correctness. The entire measure [of public protest] seems to me completely hopeless and, so long as the entire dispute is limited to bigmouthing, also useless. Under these circumstances I find it not necessary, for the sake of the good cause, to overlook individual sentences, which I do not wish to see published with my signature." Seventy-five others — including the historians Droysen and Mommsen; the mayor of Berlin, Max von Forckenbeck; the astronomers Arthur Auwers and Wilhelm Foerster; the chemist Hofmann; the jurist Heinrich Rudolf von Gneist; Georg Siemens (director of the Deutsche Bank); Werner Siemens; and Virchow — were not so particular about the sentences in question. Two days later du Bois–Reymond told Mommsen, who had originally sent it to him, that Helmholtz had returned it unsigned and had given his reasons. Du Bois–Reymond returned the petition with his own name now crossed out. He must have felt guilty about this, for he added that he was the "brother-in-law of [the geologist Julius] Ewald, friend of Riess, Kronecker, Pringsheim [and] so many other Jews that no one can be in doubt about my opinion."[10]

Although anti-Semitism abated between 1882 and 1892, late in 1892 it flared up yet again, stimulated by an anti-Semitic clause inserted in the German Conservative Party's program at its Tivoli Congress in Berlin. In 1893 Bamberger, a leading National Liberal, asked his good friend Helmholtz to join him in support of one of Bamberg's Jewish protégés. Helmholtz responded: "About the so-called Jewish Question, finally, what should one say other than completely vague and indeterminate general sentences? . . . For my part, I know too few individuals, and these only under special conditions of life, for me to feel justified in speaking out about the characteristics of the majority or about the characteristics of the 'soul of a people.'" Hence he refused to support Bamberger's protégé. "Please forgive me," he guiltily added, "that I cannot better fulfill your recommendation for your protégé. You know that I know to value charming, intelligent, and honorable friends without asking about their religion or descent."[11]

In science and in social life, Helmholtz judged people as individuals, that is, in terms of their individual talents and personal attractiveness. But by the same token, he showed no solidarity with or even sympathy for Jews as an ethnic or religious group, though he knew that they had long been subjected to various degrees of discrimination, intolerance, and violence in Germany and Europe. Like many other German academics and intellectuals, he valued his cultured and assimilated acquaintances, colleagues, and students of Jewish ancestry; but that was different from supporting, as, for example, Mommsen did, minority group rights or accepting those who had not adopted German and secular ways.

In Spain and North Africa

In the spring of 1880, in much need of extended rest, Helmholtz traveled to Spain and North Africa in the company of Ernst von Mendelssohn-Bartholdy and the latter's brother-in-law, Dr. Rudolph Schelske. They met up in Lyon, France, and traveled by train through the Rhône Valley. Helmholtz was impressed with both Lyon and Nîmes, as well as with the countryside; in Nîmes he (as usual) visited the local art museum. Two days later, they were in Barcelona, "a big manufacturing and trading city," which he thought similar to various modern Italian cities, though he judged its people to be "smaller, blacker, more energetic than the Italians, also in dress less frenchified." He and his companions went to the theater to see a Passion play; though it was in Catalan, they thought the acting so very well done that they nonetheless understood its gist. After dinner, they followed a Palm Sunday procession through the city. They toured old cathedrals and spent much time viewing a collection of Japanese ivory figures and drawings, which Helmholtz thought were as good as their European counterparts. They visited the laboratory of a local professor of chemistry; the latter and several colleagues gave them a tour of the newly rebuilt university. Helmholtz was pleased to see that his Barcelonan colleagues owned copies of nearly all of his acoustic apparatus. In Madrid, where they went next, he spent four hours in the Prado viewing paintings by Velázquez, Murillo, Titian, and Raphael. He visited the Escorial and found that it gave "a picture of serious greatness and artistic taste, despite the fanatic Philipp II, wherein he was superior to all his successors; one recognizes that he was terribly serious in getting what he wanted." He was impressed with its church and thought it much more tasteful than St. Peter's in Rome. Yet he also noted that Madrid showed great poverty: "In front of the churches one sees terrible examples of people who belong in hospitals." To his own surprise, Catholic worship services made a poor impression on him. All in all, he thought Madrid a rundown version of Munich. "Only the picture collection is impressive. The collection of people who Velásquez portrayed is so tremendously alive and impressive that they seem like living beings."[12]

Toledo, where they went next, contained a gothic church (with dome) that far surpassed any he had ever seen, including Milan's. It also had the impressive Monasterio de San Juan de los Reyes, or as much of it as had remained undestroyed by the French. "The latter have furiously wreaked havoc in Spain." They did not stay long. Their guide in Toledo was intelligent enough, but their only common language was French, which made him "abundantly tired." He took the night train to Cordova. There he saw the Great Mosque, which had become a cathedral, "a phenomenal achievement of architecture, totally strange and fairytale-like—an immense, flat canvas roof." "One can't refrain from asking oneself," he wrote Anna, "how such a well-trained culture [as that of the Muslims] could disappear." He and his companions also made an excursion to the Sierra Morena, north of the city. They visited the "very well cared-for garden" of the Marquis de la Vega as well as a farmer's garden, and on their way back to the city they climbed a cathedral's steeple.[13]

From Granada, he reported on visits to the Alhambra, which was "as completely magical as the descriptions and pictures seek to paint it," and to a bullfight. He thought the torreros were "beautiful fellows, thin, skilled, and daring," and was also much impressed with their "quite magnificent and elegant costumes" and the skill with which they manipulated the bulls. He felt little concern about the killing of the bulls, since they would otherwise have ended up on the butcher's slaughtering rack. However, he found the treatment of horses, both at the bullfight and elsewhere, "truly shocking," "the really terrible side of the spectacle." He and his companions left for Málaga, where for two days they rested before departing for Gibraltar.[14]

From Tangiers he told Anna that they had "now really gone into another part of the world [i.e., North Africa]." They were "suddenly in the middle of the Mohammedan world," everywhere astounded at what they heard and saw. They had had a taste of this a few days earlier in Gibraltar, where they experienced a mixture of ethnic groups — "Scottish regiment men, Spanish priests, and Berber businesspeople along with ship people from all parts of the world." In Tangiers, thanks to "a Berber dragoman," they made the difficult transition from ship to land; there "a large crowd of white, black, and brown naked fellows," who could be understood only through watching their gestures, jumped onto the steamboat and struggled with one another "(uttering the most agitated screams) over the travelers and their luggage." After they settled in their hotel, they went to see the German consul, a former theologian who knew his way around the Arab world and whom Helmholtz knew from Germany. They shopped at a bazaar, toured the city, sat in cafés and drank in the Arab style, watched goods being unloaded from camel caravans, and took donkey rides up a mountain, from where they viewed the sea, the green hills, and the distant, snow-covered Atlas mountains. As for the natives: "The varieties of garbs and nakedness cannot even be described." He liked the look of their turbans (worn only by the Muslims) and the cape-and-cloak outfit (worn by older men). There were not many women on the streets, and those whom he did see were fully covered in what he referred to as their "bathing sheets," "which they do not pull together all too tightly around the face." He thought it understandable that the Muslims considered themselves the elite among the mix of peoples in Tangiers: "They wash themselves, they do not drink [alcohol], and, so long as they care to, are very courteous, and that with a certain naturalness. Even people who ask for gratuities for performed services do it completely gently and modestly." Everyone wore sandals, which was only sensible, he said, since one was supposed to enter another's home barefoot. Tangiers was labyrinthine, consisting of a maze of curved alleyways; foreigners required a guide. He warned Anna that he had become so suntanned that he would return "with the facial color of an Arab." They left Africa ("the dark continent") without incident, and returned to European soil via a steamer and aided by "an outstanding Arab messenger on board ship" who managed to transact all their business needs and keep the unwanted away. Their return journey took them through Cadiz, Seville, Bordeaux, Bayonne, and Biarritz.[15] The six-week visit to Spain and North Africa sharpened, in a historical and comparative way, his sense of Europe and its North African neighbor.

A Father's Advice and Worries

Of his four surviving children—after Käthe, his eldest daughter, died in 1877—two gave him little to worry about and two gave him much. His eldest son, Richard, still worked at the locomotive factory of Krauss & Co. in Munich, where beginning in 1881 he headed its construction bureau.[16] His youngest child and only remaining daughter, Ellen, still lived at home and likewise gave her parents no special concern. But matters were altogether different with the physical and mental health of the other two sons, Robert and Fritz, as had already become clear in their childhoods.

Robert was seriously disabled from birth: he had a malformed backbone and complications therefrom. In 1879 Anna and Hermann considered (as they had done before) how best and by which doctor to have one of Robert's joints realigned. They decided to allow a nonmedical man do the straightening and associated physical therapy. Anna accompanied Robert to a clinic near Augsburg, while Hermann, still in the midst of the semester, remained behind in Berlin with Fritz and Ellen. After the procedure, Robert wore a brace or other apparatus and underwent treatments for several weeks, which seemed to help. Helmholtz wrote Thomson (in English) four years later:

> Our elder son, Robert, is lame and has a curved back, because he suffered
> by suppuration of a vertebra and of the hip joint. The wounds, healed since
> 15 years, began to open again in the beginning of this year under the influence
> of the more rapid evolution of the bones, which his age calls forth. Before he
> left Berlin his wounds were closed again, and he as well as his mother writes
> me, that he feels stronger and better than before this new accident. But you
> can imagine, that we have been very seriously alarmed, and although our care
> [i.e., concern] is much diminished at present, we are obliged to arrange our
> steps principally according to the state of health of my son.[17]

Hermann and Anna suffered as only parents of suffering children can.

Robert's physical disability did not, however, prevent him from otherwise pursuing a more-or-less normal life. He had serious scientific interests, and these showed him, of all the Helmholtz children, to be the most similar to his father in terms of intellect and attitude. He considered spending one semester studying outside of Berlin, either in Heidelberg, Bonn, Munich, or Geneva. Hermann approved of this plan, above all if Robert were to go to Heidelberg, where he had several friends. He told his son that they would receive him lovingly "and, in an emergency, God forbid," would help him. Helmholtz also had no objection to Robert's studying in Bonn, where he said Robert would find equally able teachers but where the environment would not be as attractive as Heidelberg's. In Munich, by contrast, he thought that, except for organic chemistry (that is, Baeyer), Robert would not find as good teachers as in Heidelberg and Bonn. Helmholtz had even more doubts about the qualities of the teachers Robert would find in Geneva, and in general he thought it most advisable to learn "the elements of science" in the country where one expected to eventually work.[18]

Helmholtz also advised Robert about specific subjects of study. He thought a grounding in basic chemistry was absolutely necessary for more advanced work in the natural sciences in general, but that chemistry was interesting only if one could conduct experiments oneself ("in order to have lively insights into the relationships"). He suggested that Robert spend the winter semester in Bunsen's laboratory attending his course in inorganic analysis, and he advised him to keep a carefully worked-out notebook on the mathematical lectures he attended, "thereby forcing oneself to make everything clear to oneself." He said that he himself was interested only in applied mathematics, that is, mathematical physics. "And everything that I have of mathematics and know about it, I've acquired only through occasional study for applied purposes." However, he added that proceeding in this way was also very costly in terms of time and completeness of mathematical knowledge. Robert should first learn differential equations and then later do physics, especially mathematical physics. As for anatomy and physiology, Helmholtz warned him that to obtain sufficiently useful knowledge in these fields, one had to learn them systematically, and not merely try to pick up bits and pieces here and there. He advised Robert to see if Carl Gegenbaur was teaching comparative anatomy, in which he was "a master." Robert could follow such lectures without having first to attend lectures in human anatomy, Helmholtz said. But it was not yet clear to what extent Robert would in fact pursue studying the life sciences; if he were to do so, Helmholtz said, he would also have to take classes in botany and zoology. Like his father, Robert was also interested in philosophy. Helmholtz counseled him to be cautious in this regard, however. For a scientist with broad interests, for one who "has learned to work in a scientifically rigorous way," he can "do something in epistemology." "We can talk about that at vacation time," he added. Robert spent two semesters (1880–81) studying in Heidelberg.[19] Helmholtz was Robert's good adviser.

As for the Helmholtzes' youngest son, Fritz, he suffered from an intangible yet nonetheless real mental or emotional problem, apparently a lack of motivation and an inability to execute his plans, as well as chronic physical frailty. In 1880, when he was twelve, his parents put him in a clinic (the Salzmann Educational Institute) near Gotha so that he could get special care and attention. In late July, Fritz spent time with Hermann in Pontresina. That autumn, Anna took him back to the clinic and wrote Hermann:

> The poor boy was very, very sad; he clung to me and implored me not to leave him. Yet it could do no good; I had to go. As his courage was still in tact, we drove directly out from Gotha by wagon. Then in Schnepfenthal we went into the house. The Director bade me to see the matter as an experiment. He finds Fritz to look and answer beyond his expectations. But he is yet to be convinced that Fritz will fit in at the institute, and he also can't assume the responsibility. Earthly matters worry me no more today, only matters of the soul. [Doctor] Ausfeld promises that, certainly, every caution will be taken. In a few days, it will be clear whether the child will get used to the situation. Leaving the child today seems to be still completely impossible. If the poor thing still needs me, then I must stay.

Exactly how long Fritz stayed there is unclear; later, however, he did resume his education in Berlin.[20] He remained their, above all Anna's, worry for the rest of their lives.

Away from Berlin and on vacation in the summer of 1880, Helmholtz again felt better: a wound had healed, and his usual heart palpitations and other symptoms of stress had ceased. He planned to go again to Pontresina, which drew people from around Europe who were seeking rest and relaxation, for three weeks: "That was always the means that has helped me best," he told du Bois–Reymond. He and Anna stayed (as usual) at the Hotel Saratz. He climbed the nearby Piz Languard with his Italian friends Tommasi-Crudeli and Blaserna. At the hotel, he was "such a beloved regular guest that the proprietor and waiter treated [Hermann and Anna] with special courtesy." They dined with his good friend Blaserna and with the Italian statesman and economist Stefano Jacini, whom they had met last year in Pontresina. Then Helmholtz left Pontresina, alone, for Italy. He visited Monte Generoso, Florence, and Bologna (where he saw Beltrami), and then went to Vienna to meet up with Anna. In Munich that September, he represented Germany at the International Congress for Geodesy and sat for his portrait with Lenbach.[21]

In Italy Helmholtz's name stood near or at the very top of intellectual life. Giacomo Barzellotti, a well-known historian of philosophy, had recently written in the *Nuova Antologia* on the state of contemporary German philosophy, in which, he told Helmholtz, "you stand first in line." He believed that Italian intellectual and social life would be better off when they had placed themselves "in a close and more intrinsic relationship with the higher, more modern culture of Europe." Barzellotti was a neo-Kantian who opposed materialism, empiricism, theological metaphysics, and Hegelianism. He wanted science tied to a neo-Kantian viewpoint, and he sought to promote the latter in Italy. He sent Helmholtz his new article on the origin and development of the contemporary critical movement of German experimental philosophy, in which Helmholtz played "such a great role." His piece popularized Helmholtz's philosophical and related experimental work for a broad Italian audience.[22]

As a foreign member of the Accademia dei Lincei, Galileo's old order, Helmholtz had been asked to join its commission on biology and to help award a prize of ten thousand lire. Tommasi-Crudeli had already spoken to him about this, and Blaserna said that the commission had "difficulties" in making such an award. So the Italians wanted Helmholtz, with his great authority and powers of judgment, as Blaserna put it, to be on the commission. That would be a great service to the Accademia and to Italy, he said; it required that Helmholtz come to Rome. Moreover, Blaserna had a new laboratory—he had established it in 1881—that he wanted to show Helmholtz. He also wanted his advice concerning some laboratory matters.[23] That was certainly a more pleasant activity than being harassed by Dühring again.

Dühring's New Attack

Dühring continued to raise the subject and recount the history of his remotion, pointing to Helmholtz's alleged machinations against him. In *Robert Mayer der Galilei des neunzehnten Jahrhunderts* (1880), he again portrayed Mayer as a martyr to science, in particular as a victim of Helmholtz's alleged untruths in regard to priority of the "discovery" of the law of conservation of energy. As always, Dühring used innuendo, sarcasm, ad hominem disparagements and other rhetoric, biased evidence, and sheer chauvinism to make his case. He sought to belittle Helmholtz—"the resonating Professor Helmholtz with his advertising resonances [read: Steinway & Sons]," who was "a kind of salon physicist." He repeated his suggestion that Helmholtz, like Joule, had sought to make Mayer's work his own. He scolded Helmholtz for having failed to support Mayer in 1849 in an exchange of views over applied physiology between Mayer and Otto Seyffer, a private lecturer in physics at Tübingen. German university professors, Dühring maintained, sought to repress discoveries in Germany, revealing them only after they were exported to foreign markets and then reimported to Germany.[24]

Dühring also reproached Helmholtz for promulgating non-Euclidean geometry, a supposedly confusing subject, and berated him for his sponsorship of the German-language translations of Tyndall's work. Helmholtz, Dühring alleged, unjustifiably sought to make himself appear to be a physicist, indeed "a discovering physicist." Appearances to the contrary, in comparison to Ohm's work in electricity and acoustics (in particular, the theory of tones), Helmholtz did little beyond a couple of experiments of minor importance. Helmholtz's popular essays on the conservation of energy were also useless, Dühring further claimed, and merely had the appearance of being "scientific." Helmholtz himself never went, and could not go, beyond the level of the fashionable in his popular scientific essays. No one had been interested in Helmholtz's "little brochure" on the conservation of force, he also (and for once rightly) claimed. In the 1850s, Helmholtz and his supporters had supposedly discovered the law of conservation of energy and of the mechanical equivalent of heat, and did so without ever mentioning Mayer, who supposedly remained virtually unknown until Dühring first drew attention to him in the 1870s.[25]

Dühring continued to malign Helmholtz for his alleged role in Dühring's remotion. Helmholtz and other professors had been "manipulated by the Press, which lay in Jewish hands." They had sought "to discredit" Dühring by claiming that he was a megalomaniac or suffered from persecution mania. The Jewish press barons (all liberals) were on Helmholtz's side and had smeared Dühring.[26] Especially through the first half of the 1880s, Dühring continued to harass Helmholtz through a trio of repetitive works and charges.

Defending against "A Handful of Lies"

Dühring's claims about the neglect of Mayer's work by Helmholtz and others, and about his own discovery of it, were simply false. As early as 1862, Tyndall had lectured (and soon after published) on Mayer's work. Tyndall wrote both Clausius and Helmholtz to inquire about their views of Mayer's writings and accomplishments.[27]

Helmholtz, by contrast, planned to remain silent on the matter. "Dühring is well known with us here as a man composed of envy and malice, to whom a handful of lies isn't important," he told Tyndall. He thought any answer that he might give would be either weak or written in a tone unfitting for scientific discourse. ("Unfortunately, we already have enough of that through the imitators of Schopenhauer like Zöllner and Dühring.") Besides, were he to respond, he would have to say that he believed his work on energy conservation was better than Mayer's. In the late 1840s, he declared, conservation of energy "was still considered foolishness, and when I became acquainted with Mayer's first writings, he initially seemed to me only like a committed party-comrade." He had avoided a priority dispute and intended to do so in the future. "Here in Germany Dühring's book [on Mayer] has caused only laughter, at least with everybody whose judgment is worth something." Mayer had been (mentally) ill from time to time, Helmholtz said, and he had turned to alcohol for relief, as he confessed to the physicists Poggendorff and Adolph Paalzow.[28]

But in fact this was not Helmholtz's last word on Mayer. Dühring had either struck a raw nerve or, more likely, Helmholtz thought he could not let Dühring's wild charges go completely unanswered, even if he were to forgo a point-by-point rebuttal. Three years later (1883), in an appendix to a reprinting of his popular lecture of 1854, he noted that he was "the first" to have cited Mayer, "who had correctly conceived the law of the conservation of force [*Kraft*] in its generality." He said he had defended Mayer's priority over Joule before his English friends, "who had tended to deny every right of Mayer." He pointed in particular to the publication of his letter to Tait in the introduction to Tait's *Sketch of Thermodynamics* (1868) and in his own recently appeared collection of scientific papers. But he also noted that he thought Mayer's first two publications were of a quite speculative nature. Moreover, except by Mayer himself, Helmholtz and the other "discoverers" of the law of conservation of energy had been treated discourteously: those who had pursued experiments and adduced empirical research to support the law were treated in a denigratory manner, as if they lacked "the gaze of genius."

> I myself have been presented as one of the worst miscreants, and owe this, as I presumed, to the circumstance that, through my investigations on sense perceptions, I, more than other fellow specialists, have been involved with epistemological questions. I have striven to dispel everything that I still came across that had a mist of false scholastic rationalism [about it]. I knew before these disputes over Robert Mayer that I had thereby made myself not loved by both the secret and the open followers of metaphysical speculation, and I already long ago saw that it could not be otherwise.[29]

Helmholtz claimed that in 1847 he had simply not known of Mayer's work but that he later made efforts "to draw the attention of the scientific public to him [Mayer]," through his lectures of 1862 and 1869. But this did him no good as far as his opponents (i.e., Dühring) were concerned. Although Dühring attacked him "vehemently," Helmholtz sought to avoid getting into a rhetorical exchange with him about nonscientific issues. However, on the scientific side, there remained the general methodological issues of speculation versus empiricism and of deduction versus induction that still required some discussion and which had led to different assessments of Mayer's contributions. Helmholtz confessed that he had often praised Mayer because he felt sorry for him on account of his illness; but now that Mayer's name was being exploited to promote "scientific principles" that Helmholtz "held to be radically false," principles that had "unfortunately still not completely lost their seductive force for the educated classes of Germany," he felt determined to speak out on the matter.[30]

He saw his own work, as well as Mayer's and that of all others who had "discovered" the law of conservation of energy, "in no way as a thoroughly new induction, but rather as only the final step in making precise and complete a generalization of an already long, ongoing and rising inductive conviction that had already expressed itself in manifold ways." The law, he said, had emerged gradually through the work of Leibniz, Daniel Bernoulli, Sadi Carnot, Rumford, Davy, and others. This was the situation in the mid-1840s, as he approached the topic of conservation of energy. He added, "I myself had taken this route without knowing anything of Mayer and (at the start) Joule." He thought of his own work on this subject as "purely critical and ordering" and thought its "main purpose could only be to test and to complete an old conviction which developed in an inductive way on the newly won material." He thought his own scientific leitmotif here had not been novel. He had thus also "chosen the designation of my essay as 'On the Conservation of Force' [Ueber die Erhaltung der Kraft], in order to characterize it as an elaboration of the old principle 'of the conservation of living force' [von der Erhaltung der lebendigen Kraft], just as I have joined it in the introduction [in the essay] to the old question of the possibility of a perpetuum mobile."[31]

While he praised Mayer's ability—"an extraordinarily independent and ingenious mind"—and achievement, he also thought Mayer had much in common with his predecessors and warned that, as original science, his results should not be overstated. In particular, Mayer's first essay (1842) "gives no proof, at least nothing that a natural researcher would recognize as proof, but rather puts forth only 'theses.'" Yet Helmholtz did not question Mayer's priority, and he thought Mayer's first essay may have sought only to secure priority rather than prove the law. Particularly in an era that was reacting strongly against speculative, Hegelian approaches to knowledge, Helmholtz said, Mayer's proof was seen as insufficient and led "every enlightened natural researcher" to stop reading before he got to page two. In his first two essays at least, Mayer had shown himself the prisoner of metaphysics and without sufficient empirical results. In retrospect, Helmholtz said, one can see that Mayer's first essay essentially showed an understanding and valid statement of the law of conservation of energy. "Yet [it] also shows, through the way he sought to present his knowledge, a rather strong bias in the false rationalism to be found in the era's medical schools and nature philosophy." He noted, too, that his own work, as well as that of Joule and others, was neglected. "For us [i.e., Helmholtz and Joule], too, overcoming the inertia of the dominant view [concerning conservation of force] was not completely easy and not very quick." Only since 1854, he said, had Mayer's work and priority been recognized. "Mayer's fate" showed how central it was to work through an idea thoroughly, to give "convincing proof" of it. None of Helmholtz's explanations stopped Dühring from repeating and indeed elaborating his charges in his endless attacks on Helmholtz, as he did again some five years later in a second edition of his account of his employment at the Victoria Lyceum and again some twenty years later in the second edition of his autobiography.[32]

In the Scientific Capitals of Europe

Invitations from Britain

In February 1878 the president of the Chemical Society (London) invited Helmholtz to give its Faraday Lecture, which was delivered every three years by "a foreigner of high scientific distinction." The first Faraday Lecture was in 1869, two years after Faraday died, and the previous lecturers were all chemists: the Frenchman Jean-Baptiste André Dumas (1869), the Italian Stanislao Cannizzaro (1872), and the German August Wilhelm von Hofmann (1875). Helmholtz was asked to speak on "chemical philosophy, or chemical physics." He regretfully declined, since as rector of Berlin he had "more than usual to do" and, in addition, "the preparations for the move into the new institute also demand much time."[1]

This was not the end of the matter, however. Nearly three years later, Roscoe, who held the chair of chemistry at Owens College, Manchester, and was now the president of the Chemical Society, again asked him to deliver the Faraday Lecture. He noted that Helmholtz had not been in England for quite some time (since 1871). Helmholtz agreed to give the lecture for 1881, and he told Roscoe, "I've again recently been working partly in electrolysis and partly in electrodynamics, more or less in

Faraday's ways, and will attempt to make as much of that as possible understandable for chemists, without becoming all too abstract."[2]

The trip outlined by Roscoe included dining at William Crookes's place in London; giving the Faraday Lecture (5 April); attending the Faraday Dinner and Royal Society Soirée; dining at the home of Warren De la Rue, a chemist and astronomer and the immediate past president of the Chemical Society; and attending Tyndall's lecture at the Royal Institution. Roscoe also noted, "An invitation from the University of Cambridge is on its way to you to fix a date for conferring the LLD upon you." In the meantime, Roscoe awaited a draft of Helmholtz's lecture for revision and printing. Anna planned to accompany Hermann.[3]

British, Irish, and Italian Anticipations

In the months before Helmholtz's arrival in Britain, there was much anticipation among British and Irish men of science about his visit. Several hoped to see him or have him visit their institutions. For the Faraday Dinner, Roscoe invited William Bowman, Sedley Taylor, Coutts Trotter (a former student of Kirchhoff's and Helmholtz's and now a lecturer in physics at Trinity College, Cambridge), Thomson, Tyndall, "& the chief representatives of Science," along with anyone else whom the Helmholtzes might name. Roscoe became Helmholtz's impresario and detail man for the visit.[4]

Tyndall was keen to see Helmholtz. He hoped to arrange for Helmholtz to dine with him and others as part of their X-Club, a small, private dining group that sought to promote natural selection and scientific naturalism, reform the Royal Society, and generally modernize and develop British science. Though its numbers were small, it exercised considerable pressure to favor its chosen candidates for prestigious awards and positions and to promote its favored policies. Between 1864 and 1893, its nine members met once a month; these included Tyndall, Huxley, Spencer, and Hooker, as well as five lesser figures (Frankland, Lubbock, Hirst, Spottiswoode, and the zoologist and surgeon George Busk). But Helmholtz could not make it; he also told Tyndall that he needed an electrometer and a galvanometer for his Faraday Lecture, "so that it'll be visible to the listeners," and that it would not be a popular-science lecture but rather in the style of a BAAS presidential address. The Helmholtzes looked forward to Tyndall's own Friday evening lecture.[5]

Tyndall often turned to Helmholtz for advice and support. In September 1880, for example, he informed him that he was proposing that the Royal Society award its Royal Medal to Joseph Lister and said, "Your support would materially help me in pleading his cause before the Council of the Royal Society." Helmholtz replied immediately with an extremely positive assessment of Lister's accomplishments and with his support. Two months later Tyndall told Helmholtz that his (their) effort to gain the award for Lister had succeeded. Similarly, Tyndall asked Helmholtz's help concerning the proprietary use of the telephone: "I am always troubling you; but when you can aid the cause of right in this country I am tempted to apply to you." The issue at hand was the post office authorities' attempt "to stop the use of the telephone, contending that it was covered by, and included in, the patents taken out for the electric telegraph." Thomson, Stokes, Rayleigh, "and other good men," had protested this action. Tyndall intended to do so as well, but he first wanted Helmholtz's opinion. Two years earlier, when Helmholtz had first learned of the telephone's invention, he thought a scientific explanation of its physics (the acoustic and electrical relationships) was "obvious," since "for years [he] went to bed and awoke with Fourier series in [his] head." In 1878, without making any claims to originality, he published his results on the telephone's acoustics. He now replied to Tyndall that he was ignorant of English patent law but that the telephone was not novel in terms of known physical processes. But, he said, "The execution is original and noteworthy. Already the telephone by [Johann Philipp] Reis transmitted song within certain limits of the pitch of a note. Bell has made his instrument applicable for a much wider realm of the scale, and thereby the transmission of language has become possible. His mechanical execution is also an original idea. The microphone is rather more only a refinement of Reis's apparatus."[6]

There were other anticipations. Crookes, a distinguished chemist and physicist, the editor of the *Chemical News*, and a not unsympathetic analyst of spiritism, told Helmholtz that he was "anxious to have the honour of entertaining" him at dinner sometime during his stay in London, and he proposed asking some leading British scientific men to meet him. G. Johnstone Stoney, an Irish physicist, wrote to say that the Royal Dublin Society would like to have Helmholtz repeat his Faraday lecture there. In Dublin there was much enthusiasm for his coming. The Dubliners had just set up a "great [27-inch] Vienna refractor," and Stoney thought Helmholtz would be interested in seeing this telescope (then the world's largest). Helmholtz agreed to come. The University of Dublin wanted to bestow upon him an honorary degree, and the Royal College of Surgeons in Ireland wanted to give a banquet in his honor and make him an honorary fellow. Moreover, Lord Rayleigh had heard that Helmholtz would be coming to Britain, and he asked Schuster to write Helmholtz on his behalf to say that he hoped Helmholtz would spend some time in Cambridge. (In 1879 Rayleigh had succeeded Maxwell as professor of experimental physics and director of the Cavendish Laboratory.) Schuster wrote as asked, adding that he did not think Helmholtz would find the Cavendish so interesting; he did not think it could compare to the laboratory in Berlin.[7]

Thomson, too, heard that Helmholtz would be delivering the Faraday Lecture and naturally wanted to see him during his visit to Britain. "I am coming up from

Glasgow to London on purpose to hear you lecture on the 5th of April," he said, "and to meet you at the Chemical Society's dinner on the 6th." He hoped that, after Tyndall's lecture, Helmholtz would return with him to Glasgow. He added that Helmholtz could also join him for a day in his country house at Largs, and they could work in his laboratory in Glasgow for most of the week. Then they might go to Edinburgh to see Tait and attend a Royal Society of Edinburgh meeting. Helmholtz accepted Thomson's invitation, though there was concern that Helmholtz's visit to Cambridge might consume too much of his time.[8]

An invitation now also arrived to visit Italy. Quintino Sella, president of the Reale Accademia dei Lincei, had heard from their mutual friend Tommasi-Crudeli (a pathological anatomist, public hygienist, and bacteriologist, as well as an Italian senator) that Helmholtz would be coming to Rome to participate in the Accademia's decision about awarding a royal prize for work in the biological sciences. But Helmholtz declined to visit Rome, adding that he could not attend the Rome meeting because it was to occur in the middle of the semester. Moreover, he had reservations about participating as a judge in such a decision process, since "for the past 9 years" he had "followed the progress of the biological sciences only in a very fragmented way." On the other hand, if, as he had recently explained in Pontresina to Tommasi-Crudeli, the meeting sessions were held during his vacation time, he would gladly come to the Accademia. Sella and his colleagues decided to delay the meeting to discuss the prize until such time as Helmholtz could come to Rome. Helmholtz was "very surprised" by their decision and explained that he had already agreed to give the Faraday Lecture in April 1881, so that he could not attend a meeting of the Accademia until sometime between mid-September and mid-October. As it turned out, the Accademia's physiological committee met with him in Florence on 10 October. The Italian scientific leadership was determined, at some cost, to get Helmholtz's name associated with Italian science.[9]

Manchester and Cambridge

The *London Times* considered Helmholtz's visit of sufficient note to report on it (twice). In late March the Helmholtzes spent several days with Roscoe in Manchester. Roscoe adored Helmholtz. He had considered him preeminent among Heidelberg's scientists and scholars at midcentury. The two families, he proudly declared, "became intimate." "Helmholtz was a very temperate man; he never smoked, and I remember his saying that he found that the smallest quantity of alcohol dispelled from his mind 'all his good ideas,' as he used to express it, by which he meant that if any great problem had to be thought out, this was only possible when his brain was free from alcoholic taint."[10]

The Helmholtzes then spent two cold, windy, but "wonderful" days in Cambridge. After lunch with the vice chancellor, the Helmholtzes and others went to the Senate House, which Anna compared to one of the doge's palaces. Rayleigh led the ceremony awarding Helmholtz the honorary doctor of laws degree. He had to inscribe his name in the university register, for which he received loud applause from the students. He received his doctorate kneeling before the vice chancellor, and they shook hands. "It was very beautiful," Anna reported home.[11]

The Rayleighs invited the Helmholtzes to their home afterward. Lady Rayleigh was the sister of Arthur James Balfour, who became a leading member of the Conservative Party and a future prime minister of Britain. The Rayleighs and the Balfours belonged to Britain's high nobility. Balfour had visited the Helmholtzes "often" in Berlin. The Helmholtzes and the Rayleighs had tea *à quatre*; later they dined together, along with Trotter, Schuster, and Mrs. Eleanor Mildred Sidgwick (née Balfour; she was Balfour's sister and the wife of Henry Sidgwick, the professor of moral philosophy at Cambridge).[12]

From Rayleigh's point of view, however, Helmholtz's visit seems to have been somewhat disappointing. The two had a great deal in common scientifically; Rayleigh's interests in physics paralleled Helmholtz's in acoustics and wave theory. Yet Rayleigh reported, "There is not very much to be got out of him in conversation, but he has a very fine head." Robert John Strutt, Rayleigh's son, who was then a young boy, was nonetheless greatly impressed when he learned that Helmholtz "was an even cleverer man" than his father. Helmholtz showed him how the attractive force of his horseshoe magnet was concentrated in its ends.[13]

London: Electrochemistry and the Faraday Lecture

On 4 April the Helmholtzes and the Roscoes went down to London together. That afternoon the Helmholtzes went to the Royal Institution, where his talk would be held the following afternoon and where Tyndall had arranged "all Faraday's bits and scraps [to be] ready for him [Helmholtz]." They dined that evening at the Crookes's. Roscoe thought that "Frau von Helmholtz had a highly sensitive and active temperament" and reported, "On the morning of the day on which he [Helmholtz] was to deliver the Faraday Lecture she came down to breakfast and amused us by saying that she had been so nervous about the success of his lecture that she felt in the night as if she should die; upon which he remarked in his calm, equable manner, 'Alas, that doesn't happen so quickly.'" He then went back to sleep.[14]

The Royal Institution's lecture hall was packed to the gills for Helmholtz's lecture; more than eleven hundred tickets were distributed. In the audience were various members of the nobility and many women, as well as many leading scientific figures in Britain, including Thomson and Lubbock. In his introductory remarks, Roscoe portrayed Helmholtz as "eminent as an anatomist, as a physiologist, as a physicist, and as a mathematician" and added, "We chemists are now about to claim him as our own." Tyndall then gave a brief address on Faraday before Helmholtz was finally called to the rostrum, where he was welcomed enthusiastically.[15]

Though Helmholtz never considered himself a chemist, he had done work early in his career in physiological chemistry and animal heat and on the chemical implications of conservation of energy (electrochemistry and thermochemistry). More to the point, during the 1870s and 1880s he conducted research in electrochemistry that sought to adjudicate between the so-called chemical theory of the galvanic cell and the contact theory. In doing so, he sought to understand the nature and sources of electric currents and the relationships among currents, electromotive forces, and resistance within dilute solutions, including work on galvanic polarization and on concentration cells (of electrolytes). He conducted this research with a view, in part, to drawing out implications for electrochemistry (electrolysis) that arose from the discoveries of the first and second laws of thermodynamics, and for electrodynamics. Much of the work, he emphasized, was slow going and delicate in nature; he doubtless roughed up or dirtied his hands while working, as he did, with batteries, electrodes, wires, voltammeters, chemical solutions, and the like. He also carefully followed the literature on electrolysis and on electrochemistry in general, constantly comparing and contrasting his work with that of others.[16] Although his attempted adjudication between the contested contact and the chemical theories of the cell apparently convinced few if any chemists or physicists, it nonetheless set the stage for the influential development (by Helmholtz, in the early 1880s) of chemical thermodynamics and (by others, later in the decade and beyond) of the new and closely related field of physical chemistry.

It also set the stage for his Faraday Lecture of 1881, "On the Modern Development of Faraday's Conception of Electricity." Helmholtz contributed most famously to chemistry by here reintroducing the idea of the atomic nature of electricity—he himself had argued for this idea as early as 1847 and again as recently as 1880. His clever and diplomatic lecture title honored one of Britain's greatest men of science by suggesting that Faraday's (controversial) ideas about electricity were still alive and explaining how they had persisted. He praised Faraday fulsomely.[17] The lecture itself presumed a good deal of chemical knowledge and was the most strictly chemical of all of Helmholtz's work; he could not have given it had he not spent part of the previous decade doing experimental work in and theorizing about electrolysis. It came at a moment when chemists and physicists were looking for a theory or theories that might explain some or all of the enormous amount of chemical data they had been accumulating during the past century or so.

The topic of electrical theory also lay at the heart of Faraday's work. Helmholtz characterized Faraday's approach to scientific theory as being closely informed by observed facts and by avoiding hypotheticals: its aim was to "free natural science from the last remnant of metaphysics." He acknowledged that Faraday was neither the first nor the only scientist to pursue this aim, citing, rather oddly in this context, Goethe and Humboldt as examples.[18]

The essence of Faraday's electrical investigations, Helmholtz maintained, was that they were concerned with "the nature of force," that is, either physical forces, which act at a distance, or chemical forces, which act between molecules close to one another, or with the larger relationship between the physical and the chemical forces. He linked Faraday's electrical investigations to those of Newton, Coulomb, Ørsted, and Ampère, adding that Faraday was "instinctively" led by something like the law of conservation of energy, even before that law was discovered.[19]

Helmholtz also reviewed Faraday's discoveries of induced currents, electromotoric forces, dielectric polarization, and diamagnetism. He noted how Faraday conceptualized a system of tensions among electric, magnetic, and electromagnetic phenomena and forces and how Maxwell then clarified and mathematized Faraday's system. This system agreed both with the relevant facts and with the law of conservation of energy, as well as with Newton's third law (equality of action and reaction). Indeed, Maxwell had shown that light and electromagnetism were only different manifestations of the same thing, the imponderable, space-filling ether.[20]

He explained that the action-at-a-distance theorists (for example, Franz Neumann), were also nonetheless alive and well and that he himself followed this approach but did so using the potential law as a basis. He was thereby able to derive all known electromagnetic laws and explain all electromagnetic phenomena as these concerned closed currents; open currents remained unexplained. However, his results were fully consistent with the laws of mechanics, he claimed. Wilhelm Weber and other "men of outstanding significance" also took an action-at-a-distance approach and offered theories that explained the phenomena of closed currents; but their theories, he thought, contradicted mechanical principles when applied to open circuits.[21]

Helmholtz then turned to electrochemical processes. Like Faraday, he claimed to keep the facts separate from his hypotheses. Yet he introduced a hypothesis soon enough, for he assumed that electricity is a substance and that there are two and opposite kinds of it (positive and negative). He thus linked Faraday's name and his own to the atomic theory of electricity. From here he moved to the relationship between electrical and chemical forces. After reviewing Berzelius's theory of electrolysis, a theory that chemists had rejected because it spoke in terms of positive and negative electricity, Helmholtz pointed out that Faraday had managed to transform Berzelius's theory by asking about the amount of decomposition product that is formed by an electrical current of determined strength and period. This had led Faraday to his law of definite electrolytic action. Helmholtz tested this law and reinterpreted it in terms of the modern theory of chemical valence. All modern understanding of the laws of electrical processes, he explained, had proceeded on the basis of Faraday's well-confirmed law.[22]

Finally, Helmholtz turned to ionic theory, noting that Faraday had first introduced the term "ion" and its related terms. He discussed the theory at length and reinterpreted Faraday's law to say, "Through each cross-section of an electric conductor we always have equivalent electrical and chemical movement." The same amount of positive and negative electricity moves with each monovalent ion. All this concerned observable phenomena and their relationships, but it rested "on the preexistence of indestructible atoms." Atomic theory, Helmholtz insisted, accounts for all these chemical facts and relationships. This meant, he concluded, that electricity comes "in definite, elementary quanta."[23]

The lecture was a great success. Anna thought it went "very brilliantly," though she confessed not understanding it. It was interrupted by a number of "Cheers" as Helmholtz presented his own views and calculations. Roscoe rightly called it "a turning-point in the history of the subject [electricity]" and said it "laid the foundation" of current "ideas of the theory of electrolysis."[24] It greatly furthered the notion of the atomic nature of electricity, but, as Helmholtz realized, that notion did not accord with Maxwell's electromagnetic theory, in which electricity was seen as a strain in the ether.

With the lecture behind him, Helmholtz began his networking, socializing, and touring of London. The Helmholtzes and others were invited to the Tyndalls' for tea. They (and some 150 others) attended the Faraday Dinner and Royal Society Soirée. The evening included speeches, which were interspersed with glee singing. The following evening the Helmholtzes dined at De la Rue's, and a day later the painter and director of the National Gallery, Frederick William Burton, scheduled a visit there for Helmholtz. That evening, the Helmholtzes and the Roscoes dined at the home of William Bowman before they all went off to hear Tyndall's lecture at the Royal Institution. The *London Times* reported that Helmholtz was to meet members of the Society of Telegraph Engineers and Electricians; that society was to hold a soirée at University College. "[It] will be lit up by the electric light, and it is hoped that there will be a full display of all the recent novelties in electrical science." William Grylls Adams, the professor of natural philosophy at King's College and the recent codiscoverer of the photovoltaic effect, also wanted to see Helmholtz and show off the Wheatstone Laboratory at King's. Adams was especially well connected in London scientific circles: he was the brother of the astronomer John Couch Adams, the successor in 1865 to Maxwell at King's, and the recent past president of the Physical Society of London. At the end of their week in London, Anna wrote home, "Our busiest of all weeks lies behind us, and I can't be thankful enough that Papa has survived everything so well and, for a long time, hasn't been as happy as he is now." That June, Tyndall wrote to her: "It must have been in the highest degree gratifying to you to observe the esteem in which your husband is held in this country. May he be long able to prosecute these labours which have secured for him that esteem."[25]

Brief Visits to Ireland, Scotland, and Paris

Helmholtz then went to Ireland and Scotland, alone, while Anna remained behind in England. Stoney scheduled a lecture by him before the Royal Dublin Society one afternoon and planned to have friends over that evening, saying that they "wd very much wish to meet you." But Helmholtz's stay was going to be so short that Stoney did not have enough time to show him Dublin. Later that year, Helmholtz was awarded (in absentia) an honorary fellowship from the King and Queen's College of Physicians in Dublin. The college had wanted to give the award to him while he was visiting Dublin, but its laws had prevented it from doing so.[26]

From Dublin Helmholtz went to Glasgow, where he stayed with Thomson for a few days. Then they went together to Edinburgh, and Helmholtz gave a lecture (perhaps concerning an instrument for determining the relations between electromotive force, resistance, and current intensity) at the Royal Society of Edinburgh. Crum Brown said that Andrews, the Belfast chemist and physicist who became known for his work on phase transitions between gases and liquids, was retiring and that his friends and colleagues were putting together a testimonial to him. They wanted some foreigners to sign on, too, and Crum Brown had been asked to approach Helmholtz in this regard.[27] He presumably did so. In any case, the Helmholtzes then left Britain for a short spring visit to Paris.

Every aspect of Helmholtz's visit to Britain had been a great success. Roscoe wrote him: "Our Society is much indebted to you for what you have done for us. The Editor of the Chemical Society Journal is anxious to have your proof of the Faraday lecture returned to him *as soon as possible* in order that it may appear in the June No." While still in Paris, Helmholtz wrote Roscoe that as he had traveled through Britain and had given talks, "every minute was so completely taken up" that he could not get to writing down "the few notes that are still needed for the completion of the Faraday Lecture." The editor of the *Journal der chemischen Gesellschaft* also wanted it, and there were hopes it would appear in the June issue. Too many people, he explained, wanted to speak with him along the way, and so he was constantly delayed in revising his text for publication.[28]

Berlin

The Helmholtzes were back in Berlin by 11 May. They attended a performance of Wagner's *Der Ring* shortly thereafter and received the Wagners privately in their home. All four cycles — the complete opera — were performed for the first time in Berlin. The opera's performance and the Wagners' presence caused a great to-do in the city, and so the Wagners limited their visits to the Helmholtzes and Mimi von Schleinitz and came "only under the condition that no one else will be there." The Helmholtzes had two "long and extensive" discussions with the Wagners. Anna was extremely taken with both of them, but especially with Cosima. Nearly three weeks later, the Helmholtzes spent a rainy day at the "charming villa" in Charlottenburg belonging to Ernst von Mendelssohn-Bartholdy, who, with his cousin Franz von Mendelssohn, was the cochief of the Bankhaus Mendelssohn and Company. The Simsons, the Schelskes, and Joseph Joachim were there too, "and the gentlemen argued about Wagner, which between Papa and Joachim was quite hopeless," Anna said. In the midst of their friendly discussion, Franz von Mendelssohn telephoned them from the city and performed one of Joachim's compositions on his violin, leading everyone to get as close as possible to the telephone. The sounds, Anna reported, "could be heard clearly, right up to the finest nuances of the violin's sound," so that Joachim said it was "the Amati, not the Stradivari," that Franz von Mendelssohn was playing. "This omnipresence of the distant is uncanny and ghostly."[29]

In July 1881 Thomson, who was president of the Council of the Glasgow Science Lectures, invited Helmholtz on the council's behalf to give a popular-science lecture in its lecture series sometime later that year or early in the next year. Helmholtz declined: it would require too much time to write up in English. He added, "I have every reason to be covetous with my time since, this year, I'll be 60 years old and still have a lot of work before me that I would like to complete." In late August, Hermann went alone to Pontresina and Engadin. Anna and the family thus missed celebrating his sixtieth birthday (31 August 1881). She wrote him: "I have so very much on and in my heart for you, and will try to do my duty towards you [*in Deinem Sinne*], so help me God."[30] They rendez-voused in Klagenfurt, Austria.

Paris: The First and Second International Electrical Congresses

Nationalism rose sharply after 1815, and it experienced further significant inflections after the revolutions of 1848 and, above all, through the unification of the Italian states (1860–70) and the German states (1871). The natural scientific communities of several nations participated in this development. Moreover, from 1880 to 1914 there was a flowering of national scientific enterprises throughout Europe and North America. These were at once highly nationalist and internationalist in nature: scientists often showed themselves to be both proud patriots and nationalists, on the one hand, and avowedly internationalist in their scientific ethos and practices, on the other. The various national science enterprises, as well as individual scientists, were both competitive and cooperative with one another. After 1870 there was a marked rise in international congresses, societies, and projects, especially of an observational and a metrological nature. This increased international activity meant significantly increased international travel by scientists, which could not have been carried on without the establishment of railroad systems and improved mail-delivery systems after 1830. The period also saw a marked increase in the ability to communicate scientific results rapidly and to coordinate scientists and their activities. With these changes came a felt need for new, agreed-upon international norms in matters of scientific nomenclature, units, and standards. The strong sense of internationalism in science continued until 1914; then it was not recaptured until well after World War II.[31]

There was at the same time an especially deep rivalry between the French and German scientific communities, even as their nationalist impulses were tempered by the universalist ethos of science. By midcentury French science had lost much of its vigor and intellectual leadership and had become more insular. Especially after 1860, the German states assumed the leading role once played by their rival beyond the Rhine. Germany now surpassed France in terms of the size of its scientific community and the level of its infrastructure (institutes and periodicals). Throughout the German lands, many new science facilities were constructed, allowing an ever larger number of German (as well as foreign) scientists to pursue their observational and experimental work. By contrast, French scientists after midcentury lacked the higher level of financial support that their German colleagues were receiving from the state. The French both admired and envied German science. A cold war ensued between French and German science in the half century after 1860, varying in intensity until 1914.[32]

France (i.e., Paris), however, became a center for international conferences, congresses, and committees — and so came to be known as a land of talkers — while Germany continued to build laboratories, including the Physikalisch-Technische Reichsanstalt, and so was considered a nation of doers. The recently invented or innovated dynamo, telephone, incandescent electric light bulb, and electrical streetcars, with their associated large-scale technological systems, along with older electrical technologies (such as the overland and undersea telegraph) and recent developments in electrical and magnetic science, made it necessary to establish internationally recognized and agreed-upon sets of electrical nomenclature, units, and standards. It took three decades of intermittent international electrical congresses — in 1881, 1882, 1884, and 1889 in Paris; in 1893 in Chicago; in 1900 in Paris again; in 1904 in St. Louis; in 1908 in London; and in 1911 in Turin—as well as numerous governmental agreements to (more or less) resolve the practical and legal problems associated with electrical metrology.[33] Helmholtz attended the first three of these congresses (in Paris) and the fifth (in Chicago). He was a key figure in international scientific metrology.

The first, and ultimately most important, congress was the one in Paris in 1881. It was here that a set of delicate theoretical and political issues concerning electrical metrology were resolved. The International Electrical Congress, as it was called, was held within the larger framework of the International Exhibition of Electricity, between 15 September and 5 October 1881. It was an official French conference, sponsored by the minister of post and telegraphy. All told, some 250 individuals attended this invitation-only congress. Helmholtz was impressed with many of the invitees.[34]

The congress consisted of three sections: theoretical electricity, which included physicists, chemists, and physiologists; telegraphy and telephony; and all other electrical applications (including lighting). There were three French vice presidents and three foreign vice presidents (the German Helmholtz, the Briton Thomson, and the Italian Gilberto Govi). In addition to France, twenty-seven countries sent delegates. Eleuthère-Elie-Nicolas Mascart, professor of physics at the Collège de France and director of the recently founded (1878) Bureau Central Météorologique, was appointed the congress's secretary and was France's leading representative to the congress; he remained France's leader in electrical metrology matters during the next quarter century.[35] The largest national contingents (apart from France) came from Germany, Belgium, Spain, Britain, and Russia. Other European and Latin American countries, as well as the United States and Japan, were also represented. The leading figures at the congress were physicists. Indeed, this was probably the first time that physicists from many countries had met collectively in any significant number.

The congress's *raison d'être*, Helmholtz explained in a lecture a few weeks later to the Elektrotechnischer Verein back in Berlin—he joined this association in early 1880, within sixty days of its founding, attended its meetings occasionally, and could occasionally read his name in its journal, the *Elektrotechnische Zeitschrift*—issued from the fact that the rapidly growing electrical industry involved enormous sums of money. Serious legal disputes within the industry required that some sort of legal order be established, which in turn required the creation of a system of electrical units and standards. Manufacturers of electrotechnological equipment and apparatus needed to be able to determine precisely, within certain well-defined and trustworthy limits, such properties as the amount of resistance in a wire, the electrostatic capacity of a wire cable, or the electromotive force of a dynamo. Thus there had to be agreement among manufacturers as to the values of different electrical units and standards. Helmholtz argued that scientists would also benefit by the introduction of such standards. For example, they would place greater trust in their mass-produced instruments and would purchase their specialized, custom-made items at lower costs and with a greater sense of control of instrumentational errors. "Once legal measuring units are determined for technology and for practical legal determinations, these are then so arranged that they serve science as well as possible. Science is thus very essentially involved in this matter."[36]

In organizing the congress in Paris, the French performed a diplomatic and political masterstroke worthy of Talleyrand at the Congress of Vienna. For until now they had not really been players in electrical metrology; the British and the Germans had been the dominating powers. It was above all England and the English-speaking world, Helmholtz maintained, that required such legal standards, since these countries were home to many different manufacturers and, hence, standards of resistance. Germany, by contrast, had its own, single standard of resistance, the one developed by Siemens, "whose factory was almost the only one which supplied resistance standards in larger quantities." Therefore, the Germans did not feel quite so pressed as the English did to establish legal resistance or other standards.[37]

Before the congress convened, there were thus several systems of electrical units and standards in use. Among these were two absolute systems (an older one originally developed by Gauss and Weber and a newer one developed by the BAAS) and the centimeter-gram-second (CGS) system used by the BAAS. There were empirical results (for resistance and electromotive force) that did not fit into any of these systems. There was also confusion owing to varying nomenclatures: there were, for example, two Webers (for current), a British and a German one, but they had different bases; and besides, practical electricians considered the Weber a measure of quantity of electricity, not intensity. Furthermore, there was the issue of whether the system of units should be tied to the electrostatic system (as the Germans wanted) or the electromagnetic system (as the British wanted). Scientific, commercial, and national interests, not to mention the force of tradition, all came into play.[38]

The British CGS system, also known as the BA system (short for BAAS), was a "practical system." The French, by contrast, developed a "theoretical system," which, like the recently adopted (1875) metrical system, aimed to be universal and "natural." The Germans, for their part, developed a third system, whose roots lay in the work of Gauss and Weber and which was based on the millimeter, the milligram, and the second. Helmholtz praised Gauss and Weber's efforts, which used Ampère's law concerning current strength to provide an electromagnetic basis for determining resistance. The presence of iron in buildings, as well as the earth's magnetism, could, he noted, easily elicit variable results in electrical metrology. But Helmholtz also argued that the Weberian system of units produced numbers that, in practical terms, were too large to work with. He ultimately favored the BA approach, which was based on the metric system. In the meantime he supported the German system, that is, the Siemens mercury-based unit of resistance, and said that Austria, Russia (in part), and other eastern lands were also behind it. The most important issue, however, was to find the numerical equivalents between these the British, French, and German systems.[39]

This, then, was roughly the unsettled state of affairs among the three major systems for setting electrical metrology when the congress convened in September 1881. For the physicists and other leading theoreticians of electrical metrology who met in Section One of the Congress (Theoretical Electricity), the issue of the international electrical units was "regarded as the most important work of the Congress."[40]

Section One's president was Dumas; Kirchhoff and De la Rue were its vice presidents and Mascart and Eric Gérard its secretaries. By far the most important of the thirty-one delegates were Mascart, Helmholtz, Thomson, and Siemens. During the first two sessions, Thomson spoke about the BA ("practical") system, called for its adoption, and announced that several participants had already agreed to adopt it. Helmholtz said the issue of electrical units and standards was more important for the future than for the immediate present: certainly one needed practical units now, but it would take time to determine them. He praised the English system but also called for the adoption of the Siemens standard of resistance. He spoke, too, about a range of problems and said a commission should study them; the Germans wanted to slow things down and called for further study of issues before definitive decisions were made. This entire discussion took place in French. Du Bois–Reymond, whose mother tongue was French, found it "for a short while amusing, then painful, to hear foreigners like Sir William Thomson and Prof. Helmholtz from Berlin speak terrible French." Helmholtz essentially agreed, telling Anna that it was "fortunate" that she did not have to hear him speaking French. One observer of Thomson's and Helmholtz's heated debate in French thought it an "unforgettable scene of comedy." However inelegantly, the different national positions and interests had been laid out for all to consider.[41]

These daytime disputes in bad French did no harm to the friendship between Helmholtz and Thomson, nor did they keep the various representatives from socializing with one another. Helmholtz was given excellent seats for himself and his guests at the grand opera one evening, and he invited the Thomsons to join him. (They saw Gaetano Donizetti's *La favorite*.) Similarly, Thomson gave Helmholtz, William Siemens, du Bois–Reymond, Kirchhoff, and Clausius his tickets for the Sunday performances — Thomson did not want to desecrate the day, Helmholtz noted — in the Théâtre Français to see plays by Molière (*L'Avare*) and George Sand (*Le Mariage de Victorine*), "both items performed at the highest level of excellence." On another evening, Helmholtz, du Bois–Reymond, Kirch-hoff, Thomson, the Siemens brothers, Clausius, Mach, Johann Wilhelm Hittorf, and Kundt went to the apartment of Rudolph Koenig, a leading acoustic instru-ment maker, "where it was hot and boring." Helmholtz had so far seen only a bit of the electrical exhibition itself, though at some point he did manage to see— as did Thomson, the Siemens brothers, Kirchhoff, Edison, Bell, Tyndall, and the Marquis of Salisbury—the excellent exhibition of the Norwegians Carl Anton Bjerknes and his son, Vilhelm. (Vilhelm later became one of the premier figures in the formation of modern meteorology.) At the exhibition the Bjerkneses pro-duced stunning hydrodynamic analogies by means of electromagnetic instru-ments.[42] These various divertissements were more than simply a matter of relief from the daytime tensions that arose and the tedium of committee meetings in that very hot September in Paris. The friendship that Helmholtz and Thomson had built up over the past three decades proved essential in resolving the con-gress's central dispute: defining and measuring electrical resistance.

The opening session, Mascart thought, had been all about principles, not practicalities; the second session focused on whether the system should be a logical or a conventional one, in particular whether Siemens's unit of resistance should be used. There had been much disagreement, and when Dumas, who presided over the sessions, saw that the second session could not reach any re-sults, he broke it off early. The following Sunday evening, at Thomson's initiative, there was a dinner at the Hôtel Chatham for a select set of figures: Thomson and William Siemens (for the English), Helmholtz, Clausius, Kirchhoff, Wiedemann, and Werner Siemens (for the Germans), and Mascart. It was proposed, and Werner Siemens now agreed, that the British system be adopted but that it be called "for practice." This practical system was based on CGS electromagnetic units. The ohm and the volt were defined, and the standards for them would be set by a commission to be established.

Yet there was still difficulty in agreeing on the unit for current intensity: there were two Webers, one of which had a value ten times that of the other. Helmholtz showed "great wisdom," Mascart said privately to some friends, in helping the group find a compromise, not least thanks to his old friendship with Thomson. Lady Thomson had not eaten lunch that day, and so they all went to a restaurant near the congress hall, and as they sat like a "small committee around an ordinary white marble table, the three . . . units—Ampère (in lieu of Weber), Coulomb and Farad—were agreed upon."[43]

The next morning, Section One issued its report. It recommended that the CGS system form the basic units of electrical metrology; that the "practical units" for the ohm (resistance) and volt (electric potential) be 10^9 and 10^8, re-spectively; that the unit of resistance be determined "by a column of mercury of a square millimetre section" at zero degrees centigrade; that a commission should determine, "for practical purposes," the column's length in order to arrive at the ohm's value; and that "Ampère" be used to designate the name for current, "Coulomb" for the quantity of electricity, and farad for capacity. (As suggested by Helmholtz, "Weber" as the name for a unit was dropped.) The Congress adopted these recommendations. Two days later, a general session was held, and many delegates were surprised to learn the results that had been arrived at privately by the small group over a dinner and a lunch.[44]

The British and the Germans, and in part the French as well, each got what they wanted. The British BA system was adopted, but it was a "practical system"; its key value, for the standard of resistance, remained to be determined, as the Germans had wanted. The French, who were neutral on scientific-technical matters, had hoped to see an international institute for electrical metrology established in France, and while the French did not get such an institute, Paris became the de facto meeting place for several future conferences. The Germans, for their part, sought to delay decision-making as long as possible and hoped ultimately to see such an institute established in Germany. A compromise was reached by establishing a commission to study the practical details of the various propositions agreed upon. As Helmholtz concluded: "That Congress has reached what was reachable under these circumstances. There had to be a compromise." About six weeks after the congress ended, the president of France officially conferred the Ordre National de la Légion d'Honneur on Helmholtz in absentia (he was too ill to attend the ceremony in Paris). Among the several titles and offices held by Helmholtz, the president cited his recent role as a vice president of the International Electrical Congress.[45]

Helmholtz well realized that trying to justify systems of absolute measures for electrical and magnetic quantities, in particular on the basis of Maxwell's theory of electromagnetism, and with it to develop practical measures of the various quantities, was not merely an issue of "pure scientific purposes," though, as he explained to physicists in the *Annalen der Physik und Chemie* (1882), it was that. His article was meant in part as a response to a similar article by Clausius that year and in part as a suggestion for addressing the broadest metrological concerns of the recent International Electrical Congress in Paris.[46] From the outset, Helmholtz had been keenly aware of the commercial and industrial dimensions as well as the scientific importance of electromagnetic theory.

The British were keen to have Helmholtz back in Britain. William Siemens, Werner's brother and the president of the BAAS, issued an official invitation to him to come to the upcoming BAAS meeting in Southampton. Thomson did likewise and said that afterward they should go on to Largs. Instead, Helmholtz went to Pontresina for a month with Anna and their daughter Ellen, and from there to Klagenfurt. He told Thomson he could not visit Southampton, even if he had been "inclined to leave the majestic stillness of woods and mountains for the bustle and the excitation of crowds of learned people." He said, "After ten months of work I want very much some undisturbed time of quietude, for which, I found, Pontresina is one of the best places in the world." Soon enough, however, Helmholtz was going to leave for Paris to participate in the second international electrical standards commission, and he hoped Thomson would be there too. "You and Lord Rayleigh ought to be there, if anything of any value shall be performed there. I am hoping, therefore, that I shall see [y]ou there."[47]

Before leaving for his vacation and the subsequent international electrical standards commission meeting in Paris, Helmholtz wrote Karl Heinrich von Boetticher, the minister of the interior, concerning the international commission for electrical science and practice. He said that he, Friedrich Kohlrausch, and Wiedemann agreed that French physicists were probably seeking to gain an international laboratory for electrical metrology in France; this would almost inevitably have a Frenchman as its director and would eventually lead other nations to relinquish doing such work, instead having it done in France. He thought that only the "Romanisch nations" would support France in this regard; the Germans and the English would oppose it. In reality, he continued, to date the French had done nothing in electrical metrology, while the British and the Germans had done most of the work. The Germans already had a good, if not perfect, standard of resistance (the Siemens mercury column); the French sought to adopt the English measure. Hence the French wanted to speed things up, whereas Helmholtz thought it was in Germany's interest to get a definitive, rather than a quick, result.[48] This was an unsubtle hint that the Germans should establish their own institute, and soon.

The second International Electrical Congress opened in Paris on 16 October 1882, in the offices of the Foreign Ministry. It was a smaller, more modest affair than the previous one. The main issue was the determination of the value of the ohm, that is, to discuss the work of the commission appointed to determine its value. The Germans, who were in the minority, continued to insist on the advantages of the Siemens unit of resistance. They thought it premature to introduce a definitive standard of resistance. Moreover, no government had yet endorsed any of the proposals made at the 1881 congress. The Germans, and Dumas as well, thought the construction of an international institute for determining the resistance standard (and perhaps electrical metrology in general) was inopportune, since it would inhibit individual workers and it would be difficult to attract physicists to it. Moreover, it was noted that while such an institute might be fine for doing precision physics, it might not be good for creative work that demanded freedom of research. In sum, no one now advocated an international institute. The Germans thought that the BA standard would willy-nilly come into practical force, and this would make a general electrical metrology system pointless. Toward the congress's closing, an assembly was held in the Elysée Palace and French president Jules Grévy spoke blithely about "the magnificent strides of scientists and the pride of France in having gathered such a body of men in her capital." No definitive action concerning the ohm's value was taken.[49]

While in Paris, Helmholtz attended a soirée at the home of Etienne-Jules Marey, who demonstrated the new chronophotography to his guests, including Bjerknes, Govi, Crookes, "and several notables of French science." He heard, too, from his friend Gabriel Lippmann, the French physicist whom he had come to know when Lippmann visited Germany in 1873 as part of a French government mission to survey the teaching of science; Lippmann also received his doctorate in Heidelberg (1874) under Kirchhoff. Lippmann and the mathematician Charles Hermite asked Helmholtz to support Lippmann for the vacant chair of mathematical physics at the Sorbonne. Helmholtz spoke very positively about Lippmann to two influential figures, one of whom was Hermite. Helmholtz's recommendation of Lippmann proved "decisive" for Hermite, and Lippmann, who worked in several areas of physics, including optical and electrical instrumentation, was appointed to the chair. He later became renowned for his analysis and methods of the photographic reproduction of color, for which work he won the Nobel Prize in Physics in 1908. Shortly after returning to Berlin, Helmholtz reported on the work of the congress at a session of the Physical Society.[50]

Moving Up: Ennoblement and Wealth

New signs of Helmholtz as a grandee emerged. By 1881 his professional standing was such that two publishers—Barth and Reimer—expressed interest in publishing his collected scientific papers.[51] He chose Barth because it had offered first and because the majority of his papers had appeared in its *Annalen*. Volume one of his *Wissenschaftliche Abhandlungen* appeared in 1882, volume two in 1883, and the third and final volume in 1895.

In late January 1883, apparently at the instigation of the crown prince and crown princess, Helmholtz was ennobled; he henceforth became known as "von Helmholtz." In a society like Prussia and Imperial Germany generally, which placed enormous importance on titles, that put him in a social class above his fellow German scientists (and most everybody else). He claimed and hoped that this would not affect his friendships. "Otherwise this new honor, which I could not deny to my dear children and grandchildren," he told a friend, "would indeed become a punishment for me."[52]

He showed pride in his social elevation, though, and bragged that it had been centuries since anyone in Prussia had been ennobled on the basis of his scientific (as opposed to military or political) achievements. In fact, Bavaria had previously ennobled Liebig, and Prussia the historian Leopold von Ranke, and several learned figures were also ennobled starting in 1883. Even so, the ennoblement of a bourgeois figure was comparatively rare. Helmholtz felt pleased for Anna and the children and "for public reasons." On the day of his ennoblement (28 January 1883), the crown princess gave Anna a medal with portraits of the crown prince and princess as a memento of their silver wedding anniversary. Jules Laforgue, a Frenchman who had lived in Berlin and knew the Prussian-Imperial court well, saw Helmholtz's ennoblement in a rather harsh light. He claimed that "the *savant* Helmholtz had let himself be ennobled in order to satisfy his wife—in getting married, she had lost her particle." Yet for all the social prestige that this "particle" (that is, the "von") carried in Prusso-German life, it meant little politically. Helmholtz never was, nor did he become, political in any formal sense: he never belonged to a political party or held political office, as did Virchow, for example. Nor, except for a few occasional remarks, did he write or speak about political matters. He had always been loyal to the state, however, as his youthful years in Prussian military service indicate, and he had proudly associated himself with Baden's, Bavaria's, and Prussia's kings on occasion, so that doing so fully with the House of Hohenzollern was an easy transition for him. Allowing his name to be associated with the Prussian monarchy and aristocracy, in turn, effectively meant giving them a touch more legitimacy, a point of some importance when he became president of the Physikalisch-Technische Reichsanstalt in 1887. His ennoblement did not affect his working relationships with his scientific colleagues, though it probably did make him seem a bit more exalted to some. It was certainly a new element in the construction of his persona and public identity. He thought this social recognition was useful for science—and no doubt it was. Three months later, in April 1883, he was elected a foreign associate of the National Academy of Sciences of the United States.[53] The Helmholtzes had certainly moved up the social ladder.

Later that year, his financial situation decidedly improved, too. Its source lay in Anna's very close relationship with her aunt Mary von Mohl, especially after Anna's uncle Julius (Mary's husband) died in 1876. Anna visited her in Paris in 1877 and persuaded her to return with her temporarily to Berlin. Mary was lonely and increasingly dependent on her, and this meant that Helmholtz, too, at times had to accommodate his plans to Mary's needs. In Paris, Hermite, France's leading mathematician, knew Madame von Mohl, and he kept an eye on her. In the mid-1870s, Hermite held the professorship of analysis at both the Ecole Polytechnique, where he was one of Henri Poincaré's teachers and emphasized to him to the importance of German mathematics and physics, and at the Sorbonne, where he taught exclusively after 1876. Hermite was well known for his work in the theory of functions and was a great admirer not only of Helmholtz but of Helmholtz's colleagues and friends Weierstrass and Borchardt. He sought to promote German mathematics and physics in France to help modernize these fields there. He wrote Helmholtz at least three times about Mary's health. He visited her from time to time, and in late May 1882 he reported to Helmholtz that he had found her on the couch and barely able to speak, but that she had recovered a bit while he was there. (He noted en passant that he was having one of his students translate one of Helmholtz's works into French.) A year later, at age ninety, Mary von Mohl died. Anna went to Paris to arrange the funeral—Hermite, among others, was present—and to attend to Mary's estate, of which she was a beneficiary.[54]

How much did she inherit? Brücke told du Bois-Reymond that Helmholtz and Siemens had recently visited him in Vienna and that Helmholtz looked well after his Rome visit (see below); and he said, "He has become a rich man, or the husband of a rich woman; for probably this will be the will's heiress." That same month Dilthey told a friend that long-term plans were afoot to create a special position for Helmholtz that would free him from teaching. He added: "He has inherited a large fortune and presently wants a position in which his time belongs to his work." Shortly after they returned from Paris, the Helmholtzes jointly made out their last will and testament. They left the bulk of their estate to their three surviving children in common (Robert, Ellen, and Fritz) and to Hermann's lone surviving child (Richard) from his first wife, Olga. They also left a small annuity to Hermann's spinster sister Julie and something to Anna's brother Erwin. They valued their estate at approximately sixty thousand marks.[55] It seems not unlikely that some, if not most, of these assets came to the Helmholtzes through Anna's inheritance. Together with his high salary, subsidized housing, and tenured teaching position, he was now financially secure.

The Helmholtzes, like the Thomsons, owned equity in the Tharsis Sulphur and Copper Company, a Glasgow-based mining and chemical engineering firm. The expected dividend that year (1886) was half of what it had been in the past, and the stock's value had declined "considerably" in the past year. Helmholtz's equity position in Tharsis may not have been his only one. Kuno Fischer, the Heidelberg philosopher, complained to Bunsen that he (Fischer) had lost his entire investment in South American railroad equities. When Bunsen said nothing, Fischer probed again: "It happened likewise to Helmholtz," he wrote, to which the shocked Bunsen responded: "So, so—I wouldn't have expected that from von Helmholtz."[56] Perhaps he was not so financially secure after all?

That summer (1883), as he had done for the past ten to twelve summers, Helmholtz returned to Pontresina. Doctors had recommended that, for their health, Anna take their three children to the ocean; they went to England. In late August, Helmholtz received a letter from the Ministry of Public Instruction in Berlin, inviting him to be on a commission to help establish a new institute of physics and technology, with him (Helmholtz) as its director. "It is an important thing for science," he told Thomson, "and also for my own personal interest, because I may hope, that I shall be able to concentrate my power of working, as much, as has been left to me, into one purely scientific direction."[57] Money now had less utility for him, time more.

Rome: International Geodesy

That autumn Helmholtz headed for the Internationale Erdmessung (International Geodetic Association) conference in Rome. He put himself into an Italian mood by writing five verses in memory of Casamicciola, a hot-mineral spa on the northern coast of the island of Ischia that had been destroyed in July by a terrible earthquake. He was one of nearly three hundred *"personaggi illustri"* to do so.[58] By mid-October 1883 he was at the conference in Rome.

Modern geodesy effectively began in Newton's day, with attempts to conceive and measure the figure of the earth, in particular to conduct arc measurements. During the first half of the nineteenth century, Gauss and Bessel led such geodetic efforts in central Europe. In 1862 the Prussian government, under the leadership of Major-General Johann Jacob Baeyer, assumed that leadership with the establishment of the Mitteleuropäische Gradmessung (Central European Arc Measurement) association, which in 1867 became the Europäische Gradmessung (European Arc Measurement) association. Baeyer, who had earlier led the Prussian army's topographical work, now developed a second career as the principal leader of these international geodetic measurement efforts. The international group's central office was located at the Prussian Geodetic Institute (Preussisches Geodätisches Institut, established in 1867) in Berlin; Baeyer headed the Prussian institute and was simultaneously president of the Central Bureau of the (central) European organization. During most of his directorship, the bureau was little more than a publication center. The rapid increase in international traffic and communication, however, required more interaction among national bureaus. There was a widely felt need for global standards in matters of time, weights and measures, and electricity. Doubtless in part for this reason, and given his already strong involvement in developing international electrical standards, Helmholtz agreed in 1881 to serve on the board of Baeyer's institute. In 1886 the European Arc Measurement association became the International Geodetic Association.[59]

Already in 1880, Baeyer feared that topographical work in Europe was essentially finished, and he thought that Prussia/Germany had a chance of retaining the Central Bureau after his death only "if," as Siemens, at Baeyer's request, put it to Helmholtz, "an indisputable authority throughout Europe," like him (Helmholtz), "could be fixed upon as the future chief of the Central Bureau." By 1883 the eighty-nine-year-old Baeyer was unable to attend the Rome conference. Helmholtz was sent as his representative. He stayed with his friend Blaserna and resided in Blaserna's new Physics Institute at Sapienza University of Rome.

Blaserna provided for him very nicely: his quarters afforded excellent views of much of southern and western Rome. Blaserna invited all of Helmholtz's Italian friends to dine with them daily. On several evenings, he and Blaserna went to the Teatro Valle, where they saw the "famous actress Signora Eleonora Duse." Helmholtz said, "This theater piece has helped my Italian a great deal, even if I didn't understand much of the dialogue." Blaserna witnessed "what lively interest Helmholtz took in the finest and smallest details of her uncommonly rich acting. (Helmholtz cried without shame at the end.) He also surprised his host with "his broad and fundamental knowledge of modern French theater." To Blaserna's further amazement, Helmholtz also knew well Victorien Sardou's play *Fédora*, which formed the basis of Umberto Giordano's opera *Fedora*. While in Rome, too, he went to see as much artwork and as many antiquities as time would allow. "He was, in general, the richest nature that I have ever come across," Blaserna said, adding that he was simply interested in and knowledgeable "about everything."[60]

Some Italian ministers appeared at the conference's opening, including the minister of instruction, Guido Baccelli, who Helmholtz referred as "my friend" and who addressed the group in Latin! Earlier that year, Baccelli had sought Helmholtz's general advice on higher education as he prepared a piece of legislation for submission to the Italian legislature. Baccelli generally agreed with Helmholtz's recommendations—namely, to make the Italian system more like the German system—and thought very highly of Helmholtz's judgment.[61]

Helmholtz reported that there was, among the official reports read aloud, one on "the introduction of an international time [zone system] and the prime meridian that belongs thereto. This theme," he feared, "will become the playground of all speechifiers at this congress if the matter isn't successfully buried in a commission that was elected yesterday." The Rome conference's principal purpose was to reach agreement on a prime meridian, and it drew observers from beyond Europe, including the United States. Greenwich was selected as the prime meridian, becoming the baseline for determining world time.[62]

There were also social aspects to the conference, which allowed Helmholtz to deepen older friendships with Italian scientists and make new friends. Blaserna invited him to dine one evening with the astronomers Giovanni Schiaparelli and Pietro Tacchini; afterward three young physicists joined them. Helmholtz expected to see Tommasi-Crudeli as well. The congress scheduled a big, official dinner and also arranged a special visit to the Vatican for all congress representatives. A week later the entire congress was scheduled to go to Naples and Vesuvius to sightsee. In Rome, Helmholtz also visited Cannizzaro's chemistry institute as well as the Polytechnic School. He saw the Sistine Chapel at the Vatican. He also attended a session of the Accademia dei Lincei, where "for propriety's sake" he had to appear. Accompanied by a young architect, he went to the

Forum. Baeyer was honored by the congress with a gold medal for his long years of service; Helmholtz accepted it on his behalf. Afterward, the entire congress dined in the Hotel Quirinale. Helmholtz wrote Anna: "I sat beside Contessa [Ersilia Caetani] Lovatelli, a widowed, very stately lady, who as philologist is a member of the Lincei and is also a cousin of the Count Lovatelli, whom we know, and of Prince [Leopoldo] Torlonia, the mayor of Rome. There were many toasts, the last being from me to the municipal officials of Rome; it won lots of applause. That is, however, so far my only real piece of action at the congress. I find myself otherwise rather useless here." The following day he planned to go to the theater again, while on 20 October the city of Rome planned a party for the entire congress. In the meantime, he and Blaserna went to the Colli Albani, traveled to Frascati and over to Castel Gandolfo, where the pope had his summer residence, then on to Albano Laziale, Genzano, and the Lago di Nemi. After the conference he headed home via Florence and Vienna. Late that October, back in Berlin, he received a visit from the Thomsons, their first and only trip there.[63]

Berlin: At the Academy

For twenty-four years, from 1871 to 1894, Helmholtz served as an ordinary member of the Academy of Sciences.[64] In this role, together with his activities as professor of physics, he helped make Berlin, like London, Paris, Edinburgh, and Rome, into one of Europe's scientific capitals. The academy was Prussia's most distinguished scientific organization. Belonging to it, even if an aged academician's creative powers might lie far in the past, meant that one was at the top.

During Helmholtz's first decade as an ordinary member, the academy still operated under the set of statutes that it had adopted in 1838. These remained in force until 1881, when a new, revised set was adopted. The academy functioned under the aegis of the Ministry of Culture. It still consisted of two classes of equal status, the physical-mathematical and the philosophical-historical, that met separately once a month, while the entire academy (the Plenum) met once a week. It also still met in three special sessions: one devoted to the memory of Frederick II (Frederick the Great) as the academy's renewer (24 January); one devoted to Leibniz, the academy's first president (in early July); and one devoted to the reigning Prussian monarch. For these and other services, each ordinary member received, as of 1838, 200 Reichsthalern and, as of 1881, 900 marks.

For the conscientious member, simply attending meetings meant that during the academic year one devoted perhaps several hours per week to academy business. Helmholtz attended his first meeting on 17 April 1871, the month in which he first began teaching at the university. His true *rite de passage* came, however, that July, when he gave the memorial address for his predecessor, Magnus, at the Leibniz session. While he never held office at the academy, like all ordinary members he spent a good deal of time simply attending meetings. For example, in 1872 he attended thirty meetings and in 1882 twenty-four. He also occasionally attended special meetings to help decide matters of broader state interest. For example, in the mid-1870s he and others discussed proposals for advancing industrial precision mechanics and technology, safety from lightning strikes, arc measuring, and the German language and literature. Attending weekly and special meetings, along with teaching, researching, and conducting administrative duties at the university, meant that his day was normally busy with, not to say burdened by, meetings.[65]

Helmholtz advanced the academy's mission and standing in three particular and intimately related ways. First, he lectured there regularly on his current scientific work and published in its *Monatsberichte* or, from 1881, its successor, the *Sitzungsberichte*. This lecturing became his most substantial contribution to the institution's scientific life. Between 1872 and 1892, he gave twenty-one lectures before the physical-mathematical class. Between 1871 and 1894 he also gave sixteen lectures before the Plenum. All told, he published in the *Monatsberichte*, the *Sitzungsberichte*, or the *Denkschriften* fifty-one items, a figure that includes several reports, several articles appearing in parts, and two personal speeches. Except for the years 1885, 1886, and 1894 (the year of his death), he published in an academy journal at least once a year.

Helmholtz also advanced the academy's standing by helping to elect other scientists as academy members, by seeing that their manuscripts got published in the academy's journals, and by helping them gain financial support for their research. In the generation before his, Dove, Poggendorff, and Magnus (and perhaps even Riess) had represented physics at the academy. While he was in the academy, Helmholtz led the way in replacing them with more modern figures. Between 1873 and 1892 he nominated seven individuals to become ordinary members. He wrote the petitions for four of these (Siemens, Kundt, Boltzmann, and Planck) and cosigned the petitions for the remaining three (Hans Landolt, Wilhelm von Bezold, and Hermann Carl Vogel). (Boltzmann, who, along with Kundt, was meant to be Helmholtz's replacement at the university when Helmholtz assumed the presidency of the Reichsanstalt, decided in the end not to come to Berlin.) In 1894, in one of his last acts at the academy, Helmholtz nominated Planck, who in effect took Helmholtz's place at the academy and in many other ways as well. To the extent that any single individual could embody Helmholtz's legacy at the academy, it was Planck, and he did so through much of the first half of the twentieth century. Finally, by playing an important role (along with du Bois–Reymond) in bringing his friend Kirchhoff from Heidelberg to Berlin, Helmholtz in effect helped make him an ordinary member, since (as already noted) foreign members who moved to Berlin automatically became ordinary members of the academy. With these nominations to ordinary membership, Helmholtz helped modernize the Academy's participation in physics.

Helmholtz also joined in recommending the mathematician Georg Frobenius as an ordinary member, and it is extremely likely that he voted for Weierstrass in 1875. At the time he told a friend that he, Kirchhoff, and Mommsen would vote for Weierstrass "as currently Germany's most important mathematician."[66] Equally interestingly, in 1886 Helmholtz took the unusual step of blocking the nomination of a potential ordinary member, expressing doubts about the election of the Berlin anatomist and anthropologist Gustav Fritsch. He argued that others, for example, Robert Koch, were more qualified to represent the biological sciences. A decision was postponed. In 1893 Fritsch's backers again nominated him, and he again failed to be elected. (The world-renowned bacteriologist Koch was not elected an ordinary member until 1904.)

Between 1871 and 1893, Helmholtz was also involved in supporting fourteen scientists to become corresponding members of the academy. In the case of three of these (Thomson, Hittorf, and Hertz), he authored the nominating petition, while for three others (Clausius, August Toepler, and Kohlrausch), he was a cosigner. His support for eight of these figures—the Glasgow-based Irishman (and British Imperialist) Thomson, the Russian P. L. Chebyschev, the Italians Cremona and Felice Casorati, the Frenchman Hermite, the Austrian Hann, and the Dutchmen Donders and C. H. D. Buys-Ballot—show that he strove to foster (or maintain) an international dimension for the academy.

Helmholtz was also the lead nominator of his friend Siemens to become an ordinary member. This was particularly tricky because, while Siemens was one of Germany's leading businessmen and entrepreneurs, his scientific efforts principally, though not exclusively, concerned applied physics and mechanics. The academy stood for pure science, and his nomination and approval—it was unanimous—in 1873 was unprecedented. Equally tricky, but in a different way, was the effort to make Virchow an ordinary member. Here du Bois–Reymond, who had previously tried to get Virchow admitted, led the effort. Virchow had allegedly been rejected because he worked in pathology, that is, a medical science, a subject not represented by the academy. (His left-liberal politics were probably another, if not the real, factor.) Helmholtz was a cosupporter of Virchow, who was elected in 1873 on this second try, but not unanimously. The academy qua institution had long been perceived as indifferent to, if not outright hostile toward, applied science and technology; these were perceived to bear negatively on modern German academic and cultural life. This may be one reason why, in a partial attempt to soften the academy's image in this regard, it admitted Siemens and Virchow.

Helmholtz also played a fundamental role in attracting and communicating manuscripts from others for publication in academy venues. Between 1871 and 1894, he communicated ninety-one papers to the *Monatsberichte* or its successor, the *Sitzungsberichte*, as well as other papers to the *Physikalische Abhandlungen*. Most of these were by German physicists or, in a few instances, physically oriented German physiologists. Some were authored by young American physicists. Many were works of his current or former students. Assuming that a given manuscript had sufficient scientific merit, Helmholtz was unequivocal in his support. During the 1870s and 1880s, for example, Eugen Goldstein worked in Helmholtz's laboratory. Helmholtz saw to the publication of his seminal works, especially his work on electric discharges in gases. Throughout the 1880s and early 1890s, Helmholtz supported the pathbreaking spectroscopic research of Kayser and Runge. In 1887 and 1888 he arranged for the publication of four papers by Hertz that offered experimental proof of Maxwell's electromagnetic theory of light; Hertz's work played a major role in helping to mark a turning point in both physical thought and modern communications. During the late 1880s and early 1890s, he also supported the work of Heinrich Friedrich Weber, Willy Wien, Otto Lummer, and Ferdinand Kurlbaum, all of whom did seminal work toward understanding blackbody radiation and achieving a spectral-energy distribution law; thus they helped shape the emerging quantum physics in the late 1890s.

In sum, either through his own lecturing and publications on physics or through his support and election of new members (physicists), and through communicating their and others' results to the academy, Helmholtz replaced an older generation of academy physicists with a younger one attuned to contemporary problems. Between 1870 and 1883, Magnus, Poggendorff, Dove, and Riess all died. These physicists had all been experimentalists, and the academy's publications in physics under their leadership were nearly all experimental in nature. After 1871 Helmholtz became the new representative and leader of physics at the academy, not only through his own work and communication of manuscripts by colleagues, but also by leading the way in bringing Siemens, Kirchhoff, Kundt, and Planck into the academy as his colleagues in physics and by publishing papers representing theoretical physics, the new subdiscipline of physics, in its journals.

Helmholtz also advanced the academy's mission and standing by serving on sundry committees charged with representing the academy in various ways to the outside world. These services originated either in the academy itself or, more often, with the ministry through requests to the academy. For example, he helped in the academy's supervision or assessment of various other Prussian scientific and technological institutions. In the early 1870s, he was appointed to an academy commission concerned with the establishment of the new Astrophysikalisches Observatorium in Potsdam. In 1876, to cite a second example, he was appointed to an academy committee charged with examining the feasibility of attaching the Prussian Central Office for European Arc Measurement (Preussisches Centralbureau für europäische Gradmessung) to the academy. (This never materialized.) More importantly perhaps, from about 1881 onward he served on an academy committee charged with oversight of the Prussian Geodetic Institute, and during the late 1880s and early 1890s he also served on the academy committee that oversaw the Meteorologic Institute (Meteorologisches Institut).

After 1888 Helmholtz's general service activities for the academy seemed to lessen owing to his other responsibilities and, after his seventieth birthday in 1891, perhaps because of his age. Through the various forms of support that he gave to other scientists, he helped keep the academy abreast of contemporary science and strengthened its reputation in Berlin, the Reich, and beyond Germany.

Britain and Paris Once Again

Early in 1884 the University of Edinburgh invited the Prussian academy to send a representative to help celebrate the university's three hundredth anniversary that April. Helmholtz agreed to go, and he appended a trip to Paris. He had intended to bring Anna, Robert, and Ellen along, but a serious lung problem forced the Helmholtzes to send Robert to a spa for convalescence. So Helmholtz made the month-long trip to Britain with Ellen alone; Anna joined them later on, in Paris. His agenda included visits to London, Manchester, Glasgow, and Edinburgh.[67]

After arriving in London on 3 April, Helmholtz went to visit De la Rue. In the evening he delivered an extemporaneous lecture — perhaps on his construction of a so-called electrodynamic balance for measuring galvanic currents — at the Royal Society. Afterward, Tyndall invited him to his club, where he met and dined with Spencer, Lubbock, Huxley (president of the Royal Society, 1883–85), and Hooker. The next day he drove through the city, visited the British Museum and Sir Henry Verney, and then dined with Tyndall and heard a geology lecture. Roscoe then brought Hermann and Ellen to Manchester, where they stayed with him and his family. The two men talked about chemical thermodynamics. Roscoe was a close friend of Joule, and after Helmholtz said he would like to meet Joule, Roscoe took him to see him. Joule suffered from "extreme shyness and reserve," Roscoe said, and had become "a recluse"; late in life he also suffered from "mental weakness." (Joule was then sixty-four years old and unable to work.) Roscoe reported that the meeting between Helmholtz and Joule was "pathetic" and that Helmholtz was much disappointed.[68]

On the evening of 12 April, Hermann and Ellen reached Netherhall in Largs,

Thomson's country home near Glasgow. They visited Thomson's laboratory at the University of Glasgow. Thomson's scientific instruments, Helmholtz thought, were too sophisticated ("too subtle") and, in effect, wasted on artisans and officials who lacked sufficient training in physics. He thought German instruments, at least those made by Siemens and Friedrich von Hefner-Alteneck, were more appropriate in a university setting. He also thought Thomson put too much of "his eminent penetration" into industrial matters, that he could better use his talents by doing more pure science. He thought, finally, that Thomson's mind was filled with all sorts of theoretical speculations—for, example, on the ultimate nature of matter—but that these never culminated in a satisfactory result. He directed this criticism to himself as well.[69]

Helmholtz found plenty of invitations awaiting him in Scotland; he declined all of the nonacademic ones. In Edinburgh, where innumerable festivities were planned, he and Thomson stayed with Crum Brown. Helmholtz had been asked to give a toast at the big tercentenary banquet and to address the university's student body. These were requests that he could not decline.[70]

There in Edinburgh, he both represented the academy and was awarded an honorary degree. The Edinburgh Harveian Society (named for William Harvey) was to hold its annual meeting, with a dinner to follow, and the society invited Helmholtz to attend. Another dinner was scheduled for 18 April, at which Donders, Bowman, and John Scott Burdon-Sanderson, who held the Waynflete Chair of Physiology at Oxford, would be present. Among others also expected in Edinburgh were Lord Rayleigh, the Thomsons, the Cayleys, the Taits, and Dmitri Mendeleev.[71]

The tercentenary jubilee itself, Helmholtz wrote Anna, "was very magnificent, interesting, and enthusiastic," and he regretted that she had not come along. It brought together "really a large quantity of interesting people from all around the world"—among them were also Virchow and Pasteur. Both Helmholtz and Edinburgh university people were astonished "that so many important people would come together" for this event. He found Edinburgh's students to be quite lively. But he was bored by the many repetitive speeches, which lasted from 8:00 p.m. to midnight; he was one of the last to speak. There were toasts to "Literature, Science, & Art." The university's principal asked the American ambassador to reply briefly to "Literature"; Sir Frederic Leighton, a painter, sculptor, and a leading figure in the arts, to "Art"; and Helmholtz to "Science." He was probably a good choice for such a toast. William James told the Unitarian Ministers' Institute at Princeton, Massachusetts in 1881, "One runs a better chance of being listened to to-day if one can quote Darwin and Helmholtz than if one can only quote Schleiermacher or Coleridge." At a dinner at the Royal Society Club, Helmholtz was asked to give a toast on behalf of all those present. Afterward, he delivered a short lecture, again hastily composed in the morning, at the Royal

Society of Edinburgh. Before leaving Edinburgh, he and Ellen went to a church service in St. Giles' Cathedral. "The Scottish Established Church actually makes a good impression. The clergy are strict in the practical application of Christianity. Preaching and music were essentially a repetition of the Tercentenary Service. The clergyman emphasized," he approvingly reported, "very reasonably that one should not assume that, in the long run, science and religion can contradict one another." They returned to Glasgow and then soon after to London, where the City had invited him and August Wilhelm von Hofmann to a reception and *Conversazione*. They would dine with Bowman, who was also inviting Frederick William Burton in order to get the Helmholtzes a special invitation to the National Gallery. From London, Hermann and Ellen, along with the Thomsons, traveled to Paris, where they were to meet up with Anna.[72]

In Paris, Helmholtz and Thomson participated in the third international conference (1884) devoted to electrical standards. While its results in regard to the ohm were once again not definitive, there was sufficient agreement that a basis for a legal definition could, in principle, be set. Heinrich (von) Wild, the (Swiss) director of the Central Physical (Meteorologic) Observatory in Saint Petersburg, told Helmholtz that the Russian government was asking for a report on the legal value of the ohm as an international standard. Wild said that since in Germany people listened "above all" to Helmholtz, he wanted to know what was happening there in regard to electrical standards. "We here gladly prop ourselves up on what Germany does in such matters, & I would be gladly indebted to you for a couple of lines about the intentions there in this affair." Helmholtz replied that no definite decisions had yet been reached; that the Germans were discussing the establishment of an institute for precision mechanics, perhaps in association with an observatorium; and that he would eventually be involved in making some sort of judgment concerning the German institute. He asked Wild to keep him apprised of any work in Saint Petersburg concerning the value of the ohm. (Wild did so.) It took many years to reach agreement on a legal standard. All the subsequent congresses (1889–1911) dealt with subsidiary, derivative issues and were not, like the 1881 congress, a discussion of broad, theoretical (let alone political) issues. They concerned rather the presentation of finished empirical results. In time, industrialists and electrotechnologists took matters over from the physicists.[73]

On 28 May 1884, two days after the Helmholtzes had completed the legal ordering of their estate, Ellen became engaged to Arnold von Siemens, Werner's oldest son, at a party held in Siemens's home in Charlottenburg. Dohrn was invited—he was again in Berlin seeking support for his Zoological Station—and he visited the Helmholtzes on several occasions that month. That year, too, and perhaps not unrelatedly, Adolf Kohut published in *Westermanns Illustrierte Deutsche Monatshefte: Ein Familienbuch für das gesamte geistige Leben der Gegen-*

wart a double-columned, nine-page article (with photograph) describing Helmholtz's life and work.[74] By now, literate middle-class as well as upper-class persons in Germany would or could have easily heard of Helmholtz.

Late that July, Anna went to Munich, where she saw Lenbach, depressed but "otherwise the dear friend of us all," she wrote Hermann. She persuaded him to do a portrait of Siemens. In his atelier, she saw portraits of Helmholtz, Bismarck, and the pope. At Bismarck's command, Lenbach was going to Berlin to paint Bismarck, and he wanted to visit Helmholtz while there; "He places you, along with Bismarck and Moltke, at the very highest spot. Now he wants to have 'the Siemens' as the fourth of these, as he put it." A week later, in early August, Anna was in Berchtesgaden while Helmholtz went to Bayreuth to hear Parsifal.[75]

Late that autumn, Ellen von Helmholtz and Arnold von Siemens were married. The fathers of the bride and the groom had been friends for some forty years. During this time each had made enormous achievements in his respective field. It presumably gave them both much satisfaction to see their families united through the marriage of their children. Ellen, an "ash-blond, pretty young girl," according to her good friend Marie von Bunsen, and Arnold eventually had five children, and Ellen devoted most of her life to their upbringing and the management of her household on the Wannsee.[76]

For her parents, Ellen's departure meant a great loss of their sense of home. All day long Anna cried at her "loss" of Ellen, who had been Anna's "greatest joy in life" and the center of her activity in the past years. When Ellen left, she took "youth and cheerfulness from the house," leaving Anna with the burden of caring for her (and Hermann's) two ill sons (Robert and Fritz), who needed much attention. "Until then, Ellen's shadow had covered up everything; lots of young girls who came in and out [of the house], the entire life and doings, the cheerfulness, the singing and laughter—now everything has become quiet." As for Hermann, he was "not to be seen, except in the evening," when he was "always very charming," and there were "various, comforting fifteen-minute periods [together]." He saw the situation similarly. He told Thomson that his family was well, even though their house had become "much lonelier through Ellen's marriage." He went on: "We do however see her and her little son often; he's doing well and is very nice. This winter I myself have worked in part on the new edition of my physiological optics [book] and in part mathematically, and so unfortunately have not gotten to experiment."[77]

The close of the year 1884 brought Helmholtz another sort of pleasure, a new edition of his "popular" scientific lectures. In late 1882 he had asked his publisher, Vieweg, to bring out a new edition of the third collection of his *Populäre wissenschaftliche Vorträge* (that is, to include some new items) and gather them all into one volume. A year later Vieweg suggested that, instead, they make the three small collections into two large volumes. Helmholtz explained the new title (*Vorträge und Reden*) to his readers, conceding that some of the lectures were hardly "popular." As he saw it, however, "They all, including the new ones, fit together. . . . They are attempts to communicate the results of mathematical, natural scientific or physical research to a circle of listeners and readers whose studies have not quite lain in this special direction." The new edition contained everything from the older *Populäre wissenschaftliche Vorträge*, some additional popular scientific lectures, some academic speeches, and the prefaces to the German translations of the works by Thomson and Tait and by Tyndall, "because these contain discussions about epistemological questions and the popularization of science which, in terms of content, stand together with other essays in this collection." While he made no changes to the lectures, he conceded that he would now say some things differently.[78]

Institutional Brilliance

Finding New Leaders

The Berlin Physics Institute, like all German university institutes, was hierarchical: at its top stood Helmholtz, its director (and ordinary professor) between 1871 and 1888. Berlin, rather unusually, also had a second ordinary professor (Kirchhoff, 1875–87). Below them came one or more extraordinary professors, private lecturers, assistants, and staff (a housemaster, an attendant, and a mechanic). While Helmholtz was responsible for the entire institute, each one of the professors and lecturers was responsible for one or more well-defined areas of teaching. The vast majority of students attended only an elementary (and experimentally oriented) lecture course in physics; only a small number attended advanced courses.

To make all this happen, Helmholtz needed good logistical support for the institute's personnel as well as adequate financing of the institute as a whole. In the period 1878–88, the institute had an annual budget (not including salaries for its ordinary and extraordinary professors) of about 27,000 marks. Helmholtz's salary (circa 1878) was about 10,703 marks; a decade later, it was about 16,540 marks. Beyond this, he received an additional 8,400 marks (later lowered to 6,900) from the academy, for a total of about 19,103 marks circa 1878 and about 23,440 marks circa 1888. To this must be added other small amounts of income from student fees and from his association with the military academy.[1] His income vastly surpassed not only that of the average working man; it was also considerably higher than that of most of his university colleagues. Kirchhoff, the other ordinary professor, earned (in the mid-1880s) about 16,900 marks (about half of which came from the academy). Rounding out Berlin's set of senior physicists was Wilhelm von Bezold, a physicist turned theoretical meteorologist. Though not, strictly speaking, a member of the institute, he was appointed in 1885 to the first chair in meteorology at Berlin; with Helmholtz, he did much to turn Berlin into a center for studying the thermodynamics of the atmosphere.

From 1884 to 1888, the institute experienced much transition and uncertainty of leadership. Helmholtz was then largely preoccupied with planning the Physikalisch-Technische Reichsanstalt, of which he became founding president (see below). In 1887–88, his principal appointment was switched from the university, where he continued to teach theoretical physics, to the new Reichsanstalt. By 1884 Kirchhoff was suffering badly from palsy and became unable to teach; he died in October 1887. Thus by 1887–88, the institute needed two new ordinary professors, one for theory and one for experiment. In early 1888 Helmholtz initiated offers to Kundt to become Berlin's new professor of physics (representing experimental physics) and Boltzmann to become the second professor of physics (representing mathematical or theoretical physics). Helmholtz also led a group of colleagues in petitioning for these two to become ordinary members of the Academy of Sciences.[2]

Kundt, a former student of Magnus's and one of the leading experimentalists of the day, readily accepted the offer and moved to Berlin. In contrast to Helmholtz and Kirchhoff, Kundt was a dedicated and gifted teacher. He reorganized and expanded the teaching laboratories for introductory students, had the lecture hall enlarged from 240 to 330 seats, and made the colloquium into a "looser," friendlier gathering than it had been under Helmholtz and Kirchhoff.[3] Kundt became the new leader of physics at Berlin.

For Boltzmann, by contrast, the call to Berlin simply proved to be too much. He was principally a theoretician, though he had talent at experiment as well. Under Helmholtz's guidance, the ministry offered the post to Boltzmann, an Austrian, in March 1888, and it looked as if he would come. But then in June, he rejected Berlin's offer, explaining (at first) that he was suffering serious eye problems and would have to limit his teaching; later he said he was simply not up to the job at all, that he was (also) suffering from a nervous condition (depression) and that he could not assume teaching a new, demanding discipline like mathematical physics in Berlin. So Helmholtz and the ministry moved on.[4]

Later that summer and autumn, the ministry, at Helmholtz's urging, nominated first Hertz and then Planck for the position in theoretical physics. The former showed reluctance to come to Berlin and, in becoming Kirchhoff's replacement, to give up experimental physics. Planck, who was then teaching in Kiel, agreed to come as an extraordinary professor. Berlin now had two new physicists: Kundt as the ordinary professor for experiment and Planck as the extraordinary professor for theory. In 1889 Planck became the director of a new Institute for Theoretical Physics at the university. He also came to know Helmholtz personally and professionally. More than a half century later, he wrote of Helmholtz:

> In his personality taken as a whole, in his incorruptible judgment, in his simple nature, he embodied the worth and the truthfulness of his science. He had, moreover, a human goodness that deeply penetrated my heart. When he looked at me in conversation with his penetrating, searching, and yet well-meaning eyes, I was overcome with a feeling of limitless, childlike acceptance. I would have unconditionally granted him everything that was in my heart, in the sure confidence that I would find in him a just and mild judge. A word of recognition, let alone praise, from him could make me more happy than any external success.[5]

Planck adored Helmholtz. Berlin proved to be the right place for him, and, along with Kundt, he brought fresh leadership to physics there.

Students, Assistants, and "Postdocs"

As noted in detail in the section "Physics Teacher, Mentor, and Promoter" in chapter 18, Helmholtz had demanding teaching responsibilities. He claimed — though it seems doubtful — that he "always greatly enjoyed" teaching, not least because it forced him to see the entirety of his science.[6] To help him bear this load, he had three assistants who were paid out of the institute's budget and earned (circa 1885–88) between 1,200 and 1,500 marks (plus housing) annually, depending on their rank and length of service. In addition, Helmholtz often requested special, one-time sums, for example, for equipment. The most extraordinary of these requests came in about 1888 (after he had left and Kundt had arrived), when the institute was remodeled at a cost of about 310,000 marks.[7]

The institute's introductory classes were overwhelmingly attended by students who did not intend to become physicists: there were, typically, many medical and chemistry students in these classes. Most did little or no work in the laboratory. Apart from lecturing before such students and teaching small advanced physics classes, Helmholtz's biggest teaching task was serving on (paid) examination committees for promotions (doctoral dissertations) and habilitations.

The promotions were in most cases in chemistry and, to a much lesser extent, in physics; a few were in astronomy and mathematics, and there were very small numbers in philosophy, music, music history, history, and the history of astronomy. All told, he served on at least 185 such committees during the decade 1878–88. Among the best known of such students were the physicists Weber, Kayser, König, Johannes Pernet, Willy Wien, Theodor Des Coudres, Kurlbaum, Max Wien, Hertz, Goldstein, Ernst Pringsheim, and Lummer; the philosophers Benno Erdmann and Georg Simmel; and the psychologist Hermann Ebbinghaus. Robert von Helmholtz also worked under his father and Kirchhoff; he apparently promoted with Kirchhoff.[8]

On occasion Helmholtz heard from former students, usually when they sought out his help in getting published or when they needed a letter of recommendation. Several of these were trained physicists who went on to do important work in technology, like Paul Nipkow, inventor of an early form of television; Franz Schulze-Berge, whom Helmholtz sent to work with Edison on the development of the phonograph;[9] and Michael Pupin, a distinguished physicist-cum-electrical-engineer, who most notably invented the electrical transmission line that made long-distance telephony possible.

There were also many subsequently noteworthy and well-known students who studied with Helmholtz for one or more semesters but did not take a degree under his aegis. The physiologist Jacques Loeb, who ultimately took a much different approach from Helmholtz's concerning the phenomena of life and who made his career in America, studied with him. Similarly, Julius Stieglitz studied physics with Helmholtz and later became an important physical chemist at the University of Chicago. (Chicago thus benefited from both Stieglitz's and Oskar Bolza's physics education under Helmholtz in Berlin.) Fritz Haber also studied physics with Helmholtz for one semester; he later became one of the most distinguished chemists (and controversial scientists) of the early twentieth century. Walther Rathenau, an electrochemist and future industrialist, essayist, and statesman, also briefly studied physics with Helmholtz. The English statistician (and Germanophile) Karl Pearson spent a few weeks in Helmholtz's and Kirchhoff's physics classes but ultimately decided that his talents lay elsewhere. Franz Boas also came into contact with Helmholtz. Boas began studying physics under Clausius at Bonn, hoping to work with Helmholtz in Berlin. When family circumstances prevented him from doing so, he took his PhD in physics at Kiel. He then studied and habilitated in geography at Berlin, and Helmholtz served on his habilitation committee; Boas's research topic was the nature of arctic ice. (In 1865, it may be recalled, Helmholtz had published on the subject of ice and glaciers.) Boas's intellectual and career arc led him, however, from physics to ethnology and from Germany to America, where he became a leading twentieth-century social scientist at Columbia University.[10]

Numerous foreign physics students came to meet Helmholtz, see his institute, study with him for a semester or so, or even take an advanced degree with him. Pupin has left this evocative portrait of Helmholtz's person in 1885:

> His appearance was most striking; he was then sixty-four years of age, but looked older. The deep furrows in his face and the projecting veins on the sides and across his towering brow gave him the appearance of a deep introspective thinker, whereas his protruding, scrutinizing eyes marked him a man anxious to penetrate the secrets of nature's hidden mysteries. The size of his head was enormous, and the muscular neck and huge thorax seemed to form a suitable foundation for such an intellectual dome. His hands and feet were small and beautifully shaped, and his mouth gave evidence of a sweet and gentle disposition. He spoke in the sweetest of accents, and little, but his questions were direct and to the point.

Like Pupin, many students came from America, and Helmholtz did much to shape the future physics elite there during the Gilded Age. Some, like Michelson, were "postdocs" *avant la lettre*. (It was at the Berlin institute that Michelson, in consultation with the critical but supportive Helmholtz, first designed and developed his interferometer and associated methods for detecting the earth's motion through the ether, for which he later [in 1907] won the Nobel Prize in Physics.) As a new college student, Alfred Stieglitz, an American of German-Jewish origin and a future photographer and art promoter, spent a short semester listening to Helmholtz's lectures on experimental physics, though he found them too abstract for his tastes. Likewise, Nicholas Murray Butler spent the year 1884–85 studying philosophy at the University of Berlin, where he also attended Helmholtz's lectures on wave theory, though these "soon proved to be far too technical . . . to follow." Butler later helped transform Columbia College into Columbia University and served as its president for much of the first half of the twentieth century. Though most of Helmholtz's foreign students came from America, there were also several from Russia, Britain, France, Italy, Japan, and elsewhere. Occasionally the institute was graced by a high-profile nonscientific visitor. The German emperor, empress, and crown prince, for example, visited the institute in early December 1878, shortly after its completion. And Andrew Dickson White, who served as the American envoy to Berlin from 1879 to 1881 and was president of Cornell University, visited in July 1881.[11]

In sum, by 1888 Helmholtz had overseen the construction and use of superb physical facilities; had provided good managerial direction and brought on board highly talented new colleagues; had attracted a sizable, talented student body; and had given outstanding intellectual leadership to the Berlin Physics Institute. Under his direction, the institute succeeded brilliantly.[12]

Shaping a Star Student: Heinrich Hertz

From among Helmholtz's many talented physics students, one stood out in his (and everyone else's) eyes as the star: Heinrich Hertz (1857–94).[13]

In the fall of 1878, Hertz, after first studying engineering in Dresden and Munich, went to study physics with Helmholtz, in the very semester in which the latter's newly completed institute first opened its doors. For someone like Hertz, who preferred experimental to theoretical physics, Helmholtz's institute provided everything he could desire. Hertz told Helmholtz that he wanted to work on an announced prize competition (in the Philosophical Faculty) involving the investigation of the inertial mass of electrical bodies (dielectric currents). Helmholtz had devised the topic as part of his much larger research agenda concerning electrodynamics. Within the compass of his own broad framework of the potential law as the basis of all electrodynamics, Helmholtz hoped to determine, on an empirical basis, the degree to which the ether and matter might be polarizable. He discussed the prize topic with Hertz, suggested pertinent literature for him to read, introduced him to his assistants, and helped him get set up, offering him a small but well-equipped room in which to work. Thereafter, Helmholtz visited him regularly, if briefly, and supervised his work closely. He encouraged Hertz in his experimental work when he (Hertz) had doubts as to its potential success. Hertz also attended Helmholtz's lectures on mathematical acoustics as well as lectures by Kirchhoff on mechanics.[14]

In August 1879 Hertz won the prize competition. His work, however, had been strictly empirical: he had not even included a suggestion that his results might have negative implications for Weber's theory, a point that Helmholtz later made clear to him. Although Hertz himself was apparently not unaware of the larger stakes at issue concerning electrodynamic theory, these were also then beyond what he, as a novice, could address. He went to Helmholtz to discuss whether he should publish his prize work or seek a new thesis topic. Hertz told his parents, "[Helmholtz] was extremely friendly, congratulated me repeatedly, and suggested new work for me to do, which, however, would require fully two to three years. I would not undertake this work if Helmholtz had not asked it of me (and in a way that was so especially filled with honor), and also promised me his support and interest in every way. As I certainly won't have time to undertake a special doctoral work," he reported, "I could use the finished work for that purpose." Helmholtz later declared that he had seen immediately that this was "a student of very unusual talent." Master and disciple found each other. Helmholtz posed the new project for Hertz: to determine the electrodynamics of dielectrics. This, too, was for a prize, but one from the academy. The project was one aiming indeed to decide, in light of Maxwell's theory, about polarization. Hertz said that he found the topic too difficult, and besides, he could not do a three-year "secret work," as Helmholtz wanted it to be. He declined to pursue it and instead took up another dissertation topic, this one to determine induction in rotating spheres. Working on this topic made him thoroughly familiar with Helmholtz's electrodynamics. In early February 1880, he presented himself before his doctoral committee, which consisted of Helmholtz, Kirchhoff, the mathematician Kummer, and the philosopher Zeller, for a preliminary oral examination that lasted two hours and centered on the defense of his dissertation work. He passed easily.[15]

Nonetheless, Helmholtz wanted to shape him further—and Hertz wanted to be shaped by him. After graduation, Hertz continued to work in Helmholtz's laboratory, in search of a new research topic. That summer Helmholtz informed the ministry that Hertz had already worked there for eighteen months, had won the recent physics prize "in a very outstanding manner," and had done his promotion examination "likewise in an excellent manner." He continued: "His Prize Work, like his dissertation, contains experimental sections which testify to an unusual ability in this direction too." While vacationing in Pontresina, he proudly announced that Hertz would be his new, third assistant (after Hagen and Kayser). His main task was to help in the preparation of Helmholtz's lectures and associated experiments; he was also (at first) partly responsible for the library. Hertz told his parents that Helmholtz gave his assistants a lot of responsibility and was also agreeable to purchasing larger and more expensive apparatus for them so long as the purchases could be justified.[16]

Helmholtz knew how to take care of his assistants in other ways, too. Shortly after Hertz's appointment began, he was invited to the Helmholtzes' for tea with the family. The conversation ranged widely, though Hertz found it all rather "painful," constantly threatening to become "scientific" rather than casually conversational. Anna Helmholtz made things go, however. "The residence," Hertz further judged, "is furnished in an extremely sumptuous manner; the rooms are extraordinarily large." He continued: "Helmholtz himself was very charming, but I don't believe he is a good conversationalist. He speaks in such a slow and measured manner that, for me at least, it's impossible to keep one's attention on him, except when it's a matter of things in which every word must be really weighed. Then, however, the conversation stops, since I don't want to put up my opinion against his." Later that month, Hertz had breakfast with the Helmholtzes, only this time things (or Hertz) were much more relaxed.[17]

In 1878, it may be recalled, the Berlin Physical Society had moved its headquarters to the institute's new building, and Helmholtz had been elected its new president. He ran its sessions, which were held every other Friday evening in the institute's small library or in the large lecture hall. Out-of-town guests were occasionally present. The Swedish physical chemist Svante Arrhenius, for example, once attended a meeting, and he heard Helmholtz speak. "He made a real impression of extraordinary originality," Arrhenius told Wilhelm Ostwald, his fellow physical chemist. It was "well known," he said, that Helmholtz "often expressed himself in puzzling ways, which, in any case, is true." Ostwald himself visited Berlin in 1882–83, where he heard Helmholtz lecture and also met him. He described him as "a medium-sized, robustly built gentleman with a bald forehead and graying moustache. With his peculiarly built, gigantic skull, and with his slow, solemn movements and abstract eyes, he had the effect of a statue, almost monumental." Ostwald found Helmholtz's lecture to be linguistically perfect—"a masterpiece of conciseness and exactness; it could have been printed immediately." Yet he also found it to be "colossally boring."[18]

After a session, the society's attendees normally went out for a drink. Du Bois–Reymond and Kirchhoff often took part, but Helmholtz, who was "shy of alcohol," rarely did. "His non-participation in the after-session gatherings was not disadvantageous for their liveliness of tone," as one participant later diplomatically put it. "At least among the Society's younger members, the reverence for Helmholtz was so great that none of them would have dared to speak within his presence and earshot."[19]

Helmholtz also took care of Hertz in another way: he helped him get published. Since Helmholtz was the associate editor (responsible for theoretical physics) of the *Annalen*, any manuscript of Hertz's that Helmholtz thought worthy of publication would be published in the *Annalen*. During his student years (1878–83), Hertz published eight articles in the *Annalen*. Helmholtz was also gatekeeper to the society's own publications. In 1882, under his presidency, the Physical Society began issuing the *Verhandlungen der Physikalischen Gesellschaft zu Berlin*. This new journal contained the published versions of lectures held at its meetings as well as short research reports. Its appearance indicates the society's growth and its desire to be timely in announcing its members' results. Hertz published two articles in the *Verhandlungen* while still a student in Berlin. The society continued to publish its slow-paced, supposedly (but in fact never) annual *Fortschritte der Physik*, for which Hertz also wrote several reviews.[20]

For Hertz, as for others, being at the institute also meant making professional connections in other ways. He met numerous visitors there, including the English physicist Oliver Lodge; later in the 1880s, Lodge, like Hertz, played a central role in recasting "Maxwell's equations." Hertz was also invited to a formal ball while at the institute. Helmholtz, Siemens, and Weierstrass were there, too; the stage had electric lighting, "the first in Germany." Throughout his time as an assistant, Hertz continued to dine occasionally with the Helmholtz family, sometimes as their sole guest, which for him was "the most pleasant type of society," he told his parents.[21] In sum, Helmholtz helped make him feel socially comfortable as he became a professional physicist and au courant with the discipline.

While Hertz spent much of his time as a student learning to create scientific effects in the laboratory (not simply making measurements) and learning to make laboratory setups and manipulate devices without any particular interest in theory,[22] he was certainly aware of Maxwell's theory—even if he could not at the time experimentally verify it (and probably did not even fully understand it). For Helmholtz was the most energetic physicist in bringing Maxwell's theory of electricity and magnetism to the Continent. That effort included urging the translation of Maxwell's masterpiece, *A Treatise on Electricity and Magnetism* (1873), into German. Bernhard Weinstein's translation—Weinstein habilitated under Helmholtz in 1885—of Maxwell's book appeared in 1883.[23] Helmholtz gave Hertz a topic (electrodynamics) to work on, broad scientific guidance, the means and freedom to experiment, and the encouragement and professional connections that he would need to advance his career.

In early 1883 that career began to change decisively. Kirchhoff informed him that Kiel wanted to hire a private lecturer in mathematical physics and that he and Weierstrass had recommended him to the ministry. Hertz went to Helmholtz for advice about the matter. Helmholtz already knew about it but was initially uncertain whether the position would be advantageous for Hertz. Kiel's facilities and equipment were minimal, and Hertz would have no right of access to them, since they were in the domain of the professor of physics there, Gustav Karsten. Helmholtz also thought that Karsten, though an old friend, might not be so well disposed toward Berlin (perhaps because its new, expensive institute had effectively robbed Kiel's institute of new financial support that it might otherwise have received from Prussia for its own new institute). Hertz told his parents that, ultimately, Helmholtz thought it a good position for him and that he had also "promised to recommend" Hertz to other professors. "He moved according to his custom, however, in very reserved limits, and was, I want to say, visibly troubled not to have any influence on my decision, while I wanted precisely such influence!" Helmholtz heard from his old friend Albert Ladenburg, the professor of chemistry at Kiel, who had trained under Bunsen and Helmholtz at Heidelberg, that Karsten and Leo Pochhammer were already so delighted with Hertz's previous work and so strongly favorable toward him that they agreed, were he to come to Kiel, he would not even need to habilitate there. "In the matter of your student Dr. Hertz," Ladenburg wrote, "things look excellent." In April Hertz resigned his assistantship at Berlin and became a lecturer at Kiel. He also completed a seminal piece of work on glow discharge (*Glimmentladung*), that is, on cathode rays: "I've read your work on glow discharge with the greatest of interest," Helmholtz wrote him, "and I can't refrain from sending you my applause in writing. The matter seems to me to be of the greatest importance." It remained so, at least until J. J. Thomson's "discovery" of the electron fourteen years later. In the meantime, Hertz considered Helmholtz's letter of praise to be "the strongest and most pleasing spur to [further scientific] activity" that he could have received.[24]

Hertz's research at Kiel was so distinguished that, about eighteen months after he had arrived, he became a candidate for the recently available professorship of physics at the Polytechnic in Karlsruhe (Baden). He went to Berlin in late 1884 to speak with various people, including Helmholtz, about Karlsruhe. Friedrich Althoff, an autocrat who directed the Prussian higher education system, wanted Helmholtz's opinion about Hertz and Karlsruhe (as well as about Heinrich Friedrich Weber's leaving Breslau, König's move to Breslau, and Auerbach's going to Kiel); all three had been Helmholtz's students, and the first two his assistants.[25] Hertz was offered and accepted the appointment at Karlsruhe, doubtless in part because of Helmholtz's own high standing with Badenese higher-educational authorities.

During his time at Karlsruhe (1885–89) Hertz reemerged (after his strictly theoretical work at Kiel) as an experimentalist, working on spark discharges and induction. In December 1886 he reported to Helmholtz on his latest work: the induction effects of one open, direct current on another. He had, moreover, managed to display standing waves. He told Helmholtz that he had yet to write up his work on rapid electric oscillations: "I don't want to take up your time, honored Herr Privy Counselor, with the details," he told him. "You'll hopefully not consider me immodest that I've reported to you only generally about these experiments before the latter are brought to completion." The results soon appeared in the *Annalen* for 1887. In May of that year, Hertz reported to Helmholtz that he had discovered important experimental results concerning the influence of ultraviolet light on electric discharges. When Helmholtz received Hertz's manuscript soon thereafter, he decided to put it in the next issue of the academy's prestigious *Sitzungsberichte*, to ensure rapid communication of its results to one and all. Hertz's work soon became known as his discovery of the photoelectric effect. It was a major discovery whose importance was not fully appreciated until Einstein's explanation of it in 1905 in terms of quantum phenomena.[26]

In October 1887, Hertz was about to complete his experiments on induction effects and thus answer the prize problem that Helmholtz had first suggested to him in 1879: "At that time, after I had solved the University [prize] problem, Helmholtz challenged me to undertake [the academy prize]," he told his parents, "but I disregarded it since I did not see any practical way [of doing so]. Now I've succeeded, almost easily, in a way that at the time one could frankly not imagine." Days later he recorded in his diary that he had finished writing up his work on the induction effects of insulators (i.e., open currents) and had sent it off to

Helmholtz. He asked Helmholtz to see to the publication of his manuscript on induction effects in the academy's *Sitzungsberichte* (again ensuring prompt and wide announcement) and acknowledged that Helmholtz had first suggested this work to him years ago. Helmholtz replied to him: "Manuscript received. Bravo! I'll hand it in for printing on Thursday H. v. Htz." "That gave us," Hertz's wife Elizabeth wrote, "great pleasure." A month later Hertz told Helmholtz that he had completed new standing wave experiments.[27] This was the demonstration of "effects"—Hertz did not yet speak of "electromagnetic waves"—and so was not (yet) thought of as proof of Maxwell's theory.

Michael Pupin was present at the meeting of the Physical Society when Helmholtz announced and described Hertz's latest work. He wrote: "Many scientific giants of the university were present and Helmholtz presided. There was an atmosphere of expectancy as if something unusual was going to happen. Helmholtz rose and looked more solemn than ever, but I noticed a light of triumph in his eyes; he looked like a Wotan gazing upon the completed form of heavenly Walhalla, and I felt intuitively that he was about to disclose an unusual announcement, and he did." Helmholtz then described Hertz's experimental findings. He had only received "a preliminary report" but nonetheless explained, Pupin continued, the pertinence of the experimental findings for "the Faraday-Maxwell electromagnetic theory, and affirming that these experiments furnished a complete experimental verification of that remarkable theory." Pupin reported, "Everybody present was thrilled, particularly when Helmholtz closed with a eulogy [read: encomium] of his beloved pupil, Hertz, and with a congratulation to German science upon the good fortune of adding another 'beautiful leaf to its laurel wreath.'" In January 1888 Hertz sent Helmholtz another manuscript, this one on the speed of propagation of electrodynamic effects, and again asked him to publish it in the academy's *Sitzungsberichte*. He was concerned about sending Helmholtz yet another manuscript: "I feel embarrassed as to whether I should hereby excuse myself. On the one hand, I think that it very much requires an apology when I so often occasion trouble for you and loss of time; on the other, I think, once again, that a pupil should not have to excuse himself before his teacher due to a fortunate piece of work. So it'll be best if I leave it entirely up to your kindness as to what you want to do for me." As his language indicates, Hertz still viewed himself very much as Helmholtz's "pupil" [*Schüler*], like a schoolboy. Yet the work in question was Hertz's epochal experimental proof of the identity of electromagnetic and light waves, a work that Helmholtz called "absolutely ingenious." Helmholtz asked du Bois–Reymond to rush Hertz's manuscript into print in the academy's *Sitzungsberichte*; it appeared soon thereafter. Hertz gradually (sometime between 1884 and 1892) came to speak of electrodynamic waves and became a "Maxwellian." By 1888 it was clear to him that Maxwell's theory (in comparison to Weber's and Neumann's) was absolutely superior. Yet only gradually between 1887 and 1891 did Hertz stop thinking within a Helmholtzian framework and instead become a "Maxwellian." Like Helmholtz, he still thought in terms of electrodynamic forces as acting at a distance, rather than, as the Maxwellians did, as a field effect; that sort of thinking was, however, coming to an end for him, as it came to an end for other Germans, including Helmholtz himself.[28]

New opportunities now opened up for Hertz. Helmholtz strongly recommended him to the chemist Emil Fischer in the latter's capacity as dean of the Philosophical Faculty at Würzburg, where a position had opened up. Hertz was "a truly ingenious and very promising young man," he said, "one who simultaneously commands the highest abstractions of mathematical physics and who knows how to convey these in a very skilled way and to make them useful in making experimental discoveries." He had nothing but praise for his work. As a sign of Helmholtz's sure judgment about young physicists, it is worth noting that his other recommended candidate ("with complete conviction") for the position there was Wilhelm Conrad Röntgen, who had studied not with him but with Kundt at Zurich and Strasbourg. Helmholtz occasionally communicated Röntgen's manuscripts to the academy (as he also did during these years for his students Braun, Weber, Kayser, König, and Ebbinghaus). Röntgen, Helmholtz said, was "clearly a very resourceful head and skilled experimenter, sharp in his view of problems." He added, "He would probably be a very fitting teacher for an audience whose majority was composed of medical students. Yet he is also fully able to move ahead those individuals who want to study physics further." In the event, Röntgen was awarded the position. There, seven years later (1895–96), he made his spectacular discovery of X-rays, becoming the first winner of the Nobel Prize in Physics (1901). In the meantime, as the position at Würzburg remained under consideration, Hertz reported to Helmholtz that he had reflected electrodynamic waves, and more. He soon thereafter published on the forces of electrical oscillations according to Maxwell's theory, hence showing himself to be increasingly a "Maxwellian."[29]

Though Würzburg preferred Röntgen to Hertz, new opportunities for him arose elsewhere. In September 1888 Helmholtz's friend Lipschitz at Bonn consulted with him about finding a successor for its recently deceased professor of physics, Clausius. (Helmholtz wrote a memorial notice of Clausius for the Berlin Physical Society, one of only two such that he ever wrote.) Assuming that Bonn could not get Kohlrausch, whom Helmholtz thought the best person for the job, he recommended Hertz: "Among the younger physicists, I believe that Professor Hertz in Karlsruhe (currently also a candidate for Giessen) must be considered as the most talented and richest in original ideas. He was previously my assistant for a short time. He is as talented in commanding the most abstract mathematical theories as he is in finding, with great skill and a great gift for discovery, the ways of deriving questions of an experimental type from such theories. His latest investigations on the propagation of electric effects through air show him to a thinker [*Kopf*] of the first order." Less than a month later, Hertz was in Althoff's office in Berlin, where three possible professorships (Berlin, Bonn, and Göttingen) were dangled before him, sometimes in a friendly manner, sometimes not. At noon he was invited into the office of the minister himself, Gustav von Gossler, who at first saw him alone until Althoff later came in. (Althoff, after having been first briefed by Helmholtz on Hertz, in turn had briefed Gossler.) Hertz said he did not want to go to Berlin (that is, to become Kirchhoff's replacement as Berlin's theoretician). Instead, he hoped for Bonn, failing which he preferred Göttingen, both positions being for an experimentalist; and if both of those failed to emerge, he would reluctantly accept Berlin. The meeting lasted three hours; Hertz felt exhausted. To regain his strength, he went out for a meal and then directly to Helmholtz's, where he found the family had itself just begun eating; they insisted that he join them. Afterward, in the few minutes that Helmholtz had available before he and Anna had to rush off to the theater, Helmholtz and Hertz reviewed the latter's situation. Hertz feared that he would have to defend his choices before Helmholtz. To his great relief, Helmholtz said that Hertz had definitely made the right choice, but that if he were to decide to come to Berlin, Helmholtz would arrange space for him at the new Reichsanstalt so that he could also do experimental research.[30] The ministry agreed to Hertz's choice of Bonn and then, as already noted, appointed Planck (but only as an extraordinary professor) as Kirchhoff's successor in Berlin.

In November 1888 Hertz wrote Helmholtz again, this time to report that he had produced (and manipulated) short waves—he had previously been working with long waves—and had shown the connection between light and electricity, "about which I felt pressed to report to you." He closed his brief letter: "Forgive me, greatly honored Privy Counselor, in my rush, as I'm trying to bring these observations to your attention very quickly." Helmholtz arranged to publish Hertz's manuscript on this new work in the academy's *Sitzungsberichte*, before it, like Hertz's other papers, appeared in the *Annalen*. Helmholtz replied to Hertz that he had transmitted the latter's latest manuscript to the academy. The following day he wrote again, this time in a quite different and collegial, rather than professorial, tone:

Dear Friend, I was very pleased to learn about your most recent achievements. They are things about whose possibility I gnawed at for years, trying to find a way as to how I could get at them. I'm therefore quite familiar with the entire range of ideas [here] and am equally clear about its great importance. After officially handing it [Hertz's manuscript] over to the Academy on Thursday, I also lectured on it on Friday at the Physical Society, and did so in connection with a lecture by Dr. [Robert] Ritter, who showed the influence of ultraviolet light on sparks.

He continued:

> I am, personally, sorry that you don't want to come to Berlin. However, as I've previously already said to you, I believe that, in preferring Bonn [to Berlin], you've certainly acted absolutely rightly and in your own best interest. Whoever still sees much scientific work before him that he hopes to accomplish, should keep away from the big cities. At the end of life, when it's more important to make use of an achieved point of view for the building of a new generation and requisitioning for state administration, things are different.[31]

Helmholtz was obviously speaking of himself.

The intellectual path and the problems that Helmholtz had first laid out for Hertz while he was a student at Berlin found their end (or at least one important ending) while Hertz was a professor at Karlsruhe. Under Helmholtz's aegis, during his time at Karlsruhe Hertz had published four papers on his electrical researches in the academy's *Sitzungsberichte* and nine in the *Annalen*. Everyone in the world of physics thus soon knew his experimental results, even if they were still sometimes difficult to reproduce and even if not quite everyone (yet) understood their full implications for the validity of Maxwell's theory. Weber's theory, and soon Helmholtz's framework as well, now disappeared from the scene. In the spring of 1889, Hertz moved to Bonn and began working there. With Helmholtz leading the way, he was appointed a corresponding member of the Prussian Academy of Sciences.[32] That fall, at what became the historic Heidelberg meeting of the Naturforscherversammlung, he explained these results and their theoretical implications to Germany's physicists (and others) in person.

A Second Look at Helmholtz as a Physics Teacher

By 1878 Helmholtz had seven years of experience in teaching physics behind him. It is extremely difficult to assess his (or anyone's) teaching ability and effectiveness, especially when there are no set standards and methodologically defensible bases for doing so. Ultimately, one is left with only the opinions of students and colleagues—and only those opinions that were expressed, and have survived, and have been located.

At the introductory level of physics instruction, Helmholtz as a teacher left much to be desired. There is no known written remark that, at the elementary level, he was a good teacher, despite his many scientific achievements and despite the kindness he showed toward inquiring students. Kayser, who was Helmholtz's assistant for seven and a half years and who soon became one of the world's foremost spectroscopists, knew Helmholtz as a teacher quite well. He reported that Helmholtz at times got lost in the demonstration experiment (accompanying the lecture) before his students. He would, for instance, vary an experiment's parameters just to see what might happen, rather than simply use it as a demonstration of a physical principle. Kayser wrote: "In general, I'm of the opinion that Helmholtz was not a good teacher, or, more precisely stated, that he could only be a good teacher for a very few individuals. He saw everything from such a high plane that the students themselves had to be unusually advanced in order to be able to understand him." But it was precisely that constant desire to vary the experimental conditions that helped distinguish Helmholtz, as it did Hertz, as an experimental scientist.[33]

Various former students had similar assessments. In 1883, for example, Friedrich Martius, who had already earned his MD but returned to the university to do research, attended lectures by du Bois–Reymond and Helmholtz. "I couldn't blame trusted 'pupils' when they began to 'skip' after a few classes," he noted. "They didn't understand him [Helmholtz]. I experienced that, at the start of the hour, he would read [results] off of a complicated apparatus, which he forgot to explain, and then he began calculating on the blackboard until it was entirely covered with the most intricate formulae. Finally, he declared: 'You see, Gentlemen, the result sufficiently agrees with the assumptions.' He then left the auditorium. Only a very few of the remaining students knew what all this was actually about." Similarly, Albert Magnus-Levy, who also became a distinguished medical researcher and studied briefly with Helmholtz, averred that he "was not a good lecturer, at least not for elementary physics." Alexander Tschirch, to cite a final example, studied with both Hofmann and Helmholtz in Berlin; he later became professor of pharmacy and pharmacognosy at the University of Berne. He wrote of Helmholtz as a teacher:

He was a very bad teacher. He spoke somewhat from above, down through his nose, and he always seemed very bored in the main lecture. His experiments, at which Dr. König assisted him, rarely succeeded, but were never repeated. 'There's something wrong with the apparatus. Now, you have indeed heard, what it's all about,' and already he had turned around and was again calculating at the board. By contrast, his colleague Hofmann's class always remained filled, right up to the very last place and until the end of the semester. Helmholtz's class, [on the other hand,] soon emptied out, and, of the remaining students, half were asleep, since it was always very warm in the auditorium in the summer afternoon around 12 o'clock. Helmholtz obviously considered these classes, aimed at beginners, as a burdensome duty, which he himself submitted to only reluctantly.

Even Heinrich Rubens, one of the leading experimental physicists at the turn of the century, conceded that "for beginners it was, frankly, not often easy to follow the master's train of thoughts. He had no understanding for human weaknesses in matters of comprehension and understanding. For this reason, his lectures on experimental physics for large numbers of students were not appreciated according to their merit. In addition, it was also the case that he did not place much value on the external form of the lecture, as many of his colleagues in the Berlin teaching positions did, especially his predecessor Gustav Magnus and his successor August Kundt."[34] Coming from a professor of physics at the University of Berlin (1906–22), who was also the director of the institute Helmholtz built, this was indeed a frank judgment.

The situation does not seem to have been much if any better at the more advanced level. G. Stanley Hall, who spent several years studying in Germany and later became a prominent American experimental psychologist, wrote to William James from Berlin in 1878: "I have heard Helmholtz & Kirchhoff each five times a week in theoretical physics. . . . Helmholtz is a poor lecturer. Makes mistakes, hesitates, etc., while Kirchhoff is the clearest." A few years later, James wrote the psychologist Carl Stumpf that he had recently been in Berlin for a week and had heard various professors lecture there as well as in Leipzig, Prague, and Liège. "In each place I heard all the university lectures I could, and spoke with several of the professors. From some I got very good hints as to how *not* to lecture. Helmholtz, for example, gave me the very worst lecture I ever heard in my life except one (that one was by our most distinguished American mathematician)." The chemist Richard Willstätter, to cite a German student, reported: "Helmholtz spoke softly and only a little in his lectures, and he calculated at the board; his class was too difficult. Hofmann's lecture, [by contrast], was an elegant theater. It united model experiments with descriptive text, and without demanding much reflection or work." Even Planck, who was otherwise a strong supporter and friend of Helmholtz's, found the latter's lectures to be unprepared, almost without profit, and downright boring. The American physicist W. F. Magie, who studied briefly with Helmholtz and was perhaps the most generous judge of him as a teacher, nonetheless wrote:

> In our ordinary sense it can hardly be said that Helmholtz was a great teacher. His lectures were well delivered, in beautiful and intelligible language, but there was a certain halting and uncertainty, a feeling, as it were, for the thought and a strange inaccuracy in stating the details of an argument, that in a popular lecturer or in one of our college professors would be set down to a failure to grasp the subject. His blunders in the use of the simplest mathematical expressions were notorious; and, at least in the later years of his professorship, when the preparation of his experiments was left to his assistants, it was no uncommon

thing for the lecture to come to a standstill, while refractory apparatus was being pushed and pulled, adjusted and readjusted, in the effort to present an illustration which would not come.[35]

Pupin, another American who attended (in 1885–86) Helmholtz's experimental physics lectures, held a minority opinion: "They [the lectures] were most inspiring, not so much on account of the many beautiful experiments which were shown, as on account of the wonderfully suggestive remarks which Helmholtz would drop every now and then under the inspiration of the moment. Helmholtz threw the search-light of his giant intellect upon the meaning of the experiments." Comparing Kirchhoff and Helmholtz as teachers, and continuing the theme of inspiration, Pupin said that Kirchhoff's lectures were letter perfect, but "Helmholtz, on the other hand, rarely, if ever, seemed to have covered the whole ground of the subject which he proposed to discuss during the semester." "Kirchhoff's classes kept constant in number during the whole term. Those of Helmholtz, especially when he began to lecture on Mathematical Physics, grew smaller and smaller as the end of the semester approached. But then, Kirchhoff taught and Helmholtz inspired. Kirchhoff was an apostle, Helmholtz was a prophet." For Helmholtz it was the physical concepts that mattered, not the mathematical structures, analyses, or details. Many students simply could not follow his lectures, Pupin wrote. "Hence the opinion which one hears from time to time that Helmholtz was a poor lecturer. Those who understood him had an unbounded admiration for his lectures and a feeling of deep affection and gratitude for the man to whose inspiring words they owed so much of their intellectual growth. The affection of his pupils for him he reciprocated in a quiet and dignified way." Arrhenius noted much later that Kundt and Kohlrausch (the latter as the second president of the Reichsanstalt, 1895–1905) had suffered from the burdens of office in Berlin; that Emil Warburg (professor of physics at Berlin, 1895–1905, and then the third president of the Reichsanstalt, 1905–22) had somehow withstood the pressure of Berlin; and that Paul Drude, who became professor of physics at Berlin in 1905 and committed suicide the following year, clearly could not. "Helmholtz," for his part, "could only withstand his lecturing obligations by disdain. But no less a figure dared risk doing that."[36]

Founding the Imperial Institute of Physics and Technology

After the Franco-Prussian War, a group of Berlin scientific, technological, military, and industrial leaders, Helmholtz and Siemens among them, hoped to establish an institute for precision technology in Berlin, one devoted above all to improving the state of scientific instrument making in Prussia.[37] Siemens hoped to establish a more comprehensive type of institute, an independent physical-mechanical institute, one that conducted both pure and applied science as well as dealing with more narrowly technological issues. Siemens, Helmholtz, and others were especially attentive to the emerging field of electrical metrology and the intimately related rise of the electrical industry. These fields were linked to national and international political power and prestige. To help establish such an institute and address these national concerns, Siemens originally intended to make a donation to the academy. But the academy, often hostile or indifferent toward technology, not to mention any institutional change that might affect its traditional mission, distanced itself from projects involving applied science and technology.

Meanwhile, a parallel effort to support glass and instrument research in Jena was occurring. Helmholtz strongly supported the combined efforts in Jena of the chemist Otto Schott and the physicist Ernst Abbe. Schott sought to develop, on a more scientific basis, highly improved varieties of glass for microscopes, telescopes, and other instruments. Abbe, at the Zeiss Werke, sought to advance optical theory. In 1882–83 Abbe led efforts to obtain financial support from Berlin to support Schott's research and development work on glass. In part he turned to Berlin because he knew that Helmholtz and Siemens were leading an effort to establish a new sort of national physical and precision-instrument laboratory there. A committee, led by the Berlin astronomer Wilhelm Foerster and including Helmholtz, was formed. Though Abbe at first thought Helmholtz was delaying matters by insisting that the committee approve a new physics laboratory for metrological purposes before he consented to support for Zeiss and Schott, in fact Helmholtz all along supported Abbe's and Schott's request for state financial support to test their optical theory and related ideas about promoting new glass compositions. Abbe reported to Helmholtz on Schott's latest results and asked him to help get state financial aid for Schott to continue his work. He thought such help might come as part of Helmholtz's efforts to establish the new laboratory in Berlin. He explained that Schott's work had advanced to the point where they had either to discontinue it altogether or to expand it into a larger-scale effort. He asked for Helmholtz's help in gaining state aid and visited him in Berlin to discuss the matter with him face to face.

Helmholtz helped Abbe in part by introducing him to Foerster, who headed the Normal-Eichungs-Kommission (Standards and Calibration Commission) and so was responsible for establishing and maintaining standardized weights and measures. Foerster was also interested in the improvement of thermometer glass. He, Abbe, and Schott soon had a memorandum on the desk of the Prussian minister of culture, and the ministry quickly agreed to support the idea of a glass laboratory in Jena. Abbe wrote of his visit to Helmholtz: "I succeeded in getting Helmholtz to become actively interested in the cause and to maintain his support, which in Berlin is very considerable." He also managed to persuade Virchow, himself once a leading medical microscopist and still very much a leading liberal politician in the Prussian Landtag, to become an enthusiastic supporter of his project. The Landtag, too, was persuaded: it awarded Abbe, Schott, et al. sixty thousand marks for research and development of new types of glass. Their Jena partnership now built the Glastechnisches Laboratorium Schott und Genossen. By 1885 their laboratory was in commercial production of numerous types of glass—soon to number in the hundreds—that met the precision optical specifications of the Zeiss Werke and others. Such specifications included chemical composition, refractive index, ratio of refraction to dispersion, specific weight, and hardness. The Prussian government's investment paid off quickly, both in terms of the strengthened state of precision optical instrumentation and scientific results, and in terms of increased sales at Zeiss. Zeiss now became the foremost German producer of microscopes and other assorted optical products, whether measured in terms of everyday goods, high-end scientific and specialized items, or sales.[38]

In 1883–84 Siemens decided to forgo his plan to bequeath a gift to the academy for a physical-mechanical institute. Instead, he decided to found what over the next four years became the Reichsanstalt, an institution completely independent of the academy, universities, or institutes of technology. He offered the Reich—it was to be an Imperial, not a Prussian institute—land that he owned in the Berlin suburb of Charlottenburg, across from the newly formed (1879) Technische Hochschule Charlottenburg and just west of the Tiergarten. In exchange, he asked the Reich to construct, equip, and finance (on an ongoing basis) this Imperial institute of physics and technology. His motives were several. Above all, he wanted to advance both pure physics and physics-based technology. Like Helmholtz, he was especially concerned about the rapidly emerging field of electrical metrology and about preventing the British and the French from gaining the upper hand in setting electrical standards. The ability to set standards would lead to advantages for a nation's electrical industries. In part, too, he wanted to help his old friend Helmholtz, who longed to be relieved of his teaching duties. Anna, then in Paris to dispose of her deceased aunt's personal property, wrote Hermann: "I'm very excited to see what the final result of the Charlottenburg negotiations will be. The more I think about it—that we could lead a life together outside of the city and without you having to lecture—the more obvious it all is to me." Helmholtz replied:

> I was at Siemens's yesterday evening. He showed me the land that he wants to give for the Reichsanstalt project. . . . Early yesterday we had another meeting with [Wilhelm] Wehrenpfennig [in charge of Prussia's Technische Hochschulen] that was about reviewing the form of the budgetary motion. Werner Siemens said that he had originally bequeathed a sum to the Academy in his will. He

now believes, however, to be able to better advance the matter if he gets the state to agree to support the proposal on a permanent basis. In this regard he is very magnificent, and I believe that the gift will carry much weight for the proposed plans.

He also wrote that he had been invited to join a group of thirty leading German figures on a sixty-seven-day trip around the world, culminating in the opening of the Northern Pacific Railway. "If one still wants to see America in this life, this may be the most favorable opportunity. I haven't yet said No, although it's really not necessary to see America, at least not for what I have to do in the world."[39] He subsequently declined to go.

Helmholtz wanted to retire from teaching; that was one important reason he helped found the Reichsanstalt and eventually accepted its presidency. In early September 1883, he wrote (in English) to Thomson that he could not come to Netherhall to see the Thomsons later that month because he had to meet with various ministry officials to discuss the proposed institute: "The aim is to build at Berlin a kind of scientific physical Observatory; this plan has been proposed principally by Dr. Werner Siemens and myself." "I have been urged to accept the direction of it. . . . It is an important thing for science, and also for my own personal interest, because I may hope, that I shall be able to concentrate my power of working, as much as has been left to me, into one purely scientific direction." Eighteen months later, Helmholtz wrote Rayleigh in even stronger words: "I must say that I've now also had more than enough of holding lectures. We here are now possibly getting, in the form of a gift from Dr. Werner Siemens, a scientific, physical observatory, one without teaching purposes and whose direction has been offered to me. This matter is developing only too slowly, in that I'm 63 years old."[40]

As with all institution building, things moved slowly. Two more years passed before the architect's plans and a potential list of board members—Helmholtz called them his "physical parliament"—were ready. At the same time, the Prussian minister of culture worried about "the threatening loss [to the university] of Professor von Helmholtz."[41] Helmholtz spent much time between 1883 and 1887 at meetings with various ministerial figures, with Siemens, and with other scientists and technologists; he even appeared before a Reichstag committee. In all these ways he led in planning and articulating the Reichsanstalt's scientific and technological purposes and problems, its scientific and technological administrative structure, its financing, and its architectural design, and in choosing its personnel. There was some opposition to the proposed Reichsanstalt by Bismarck, various legislators, and various technological groups. Between 1883 and 1887, however, Siemens led the way in overcoming that opposition. Legis-

Fig. 24.1 The proposed Reichsanstalt, drawn sometime between 1884 and 1887.
National Institute of Standards and Technology.

lative approval—and some seven hundred thousand marks in authorization to finance the Reichsanstalt—finally came in March 1887. See figure 24.1.

By June 1887, after about eighteen months of wrangling, the Reichstag finally agreed that the president of the Reichsanstalt would be paid an annual salary of twenty-four thousand marks, which was more than even an undersecretary of state earned, and far more than any other ordinary professor of physics received. Bismarck, however, demanded that Helmholtz—everyone knew that the position was created for him—also teach one course on theoretical physics per year at the university. To this Helmholtz readily agreed. In July he was officially called to head the commission overseeing the new Reichsanstalt. He spent much of 1887–88 working in rooms at the Technische Hochschule Charlottenburg, that is, close to the Reichsanstalt's work site. In June 1888 he was officially named the president of the Reichsanstalt.

The Reichsanstalt formally opened its doors in October 1888, with Helmholtz as its president. It was divided into two sections: the Physical Section (with Helmholtz as its director), devoted to fundamental ("pure") science, and the Technical Section (with Leopold Loewenherz as its director), devoted to applied science (principally electrical, optical, and precision-technological) and testing. Metrology and measurement in general were the Reichsanstalt's *raison d'être*.

On a personal level, however, the Reichsanstalt constituted the culmination of Helmholtz's career as a professional physicist. It also fulfilled his repeated calls for more support for German science by the state. Nevertheless, the transition from the university to the Reichsanstalt—even though his relationship was never entirely severed—was not easy. Helmholtz took Lummer, one of his assistants at the university, with him as his assistant at the Reichsanstalt, and he urged that another one of his assistants, König, be retained and better remunerated for running things at the university. Anna von Helmholtz also played a role in shaping the new institute, in that she helped design their house and garden on the Reichsanstalt's grounds. In late May 1889 the Helmholtzes started moving into their new home at the Reichsanstalt in Charlottenburg. The move depressed Anna. "In order to rid himself of the lectures and examinations that make him so tired," she wrote her sister Ida, "Hermann has taken this leap into the unknown, made this unfinished thing. He felt that his own strength was declining as a result of the wear-and-tear of instructing beginners; he felt that he could do more if he could have his time for himself instead of for those who study."[42]

Toward New Foundations for Theoretical Chemistry and Physics

Between the opening of the new university physics institute in 1878 and his move to the new Reichsanstalt in 1887–88, Helmholtz's own scientific research focused on electromagnetism, chemical thermodynamics, monocyclic systems (mechanics), and the principle of least action.

Helmholtz's published work in electromagnetism, including electrodynamics, during this period was comparatively minor; to judge by his publication record at this time, it appears that he had placed electrodynamic theory on the back burner. Apart from his individual studies on the telephone, on electric boundary layers (double layers), on current movements in polarized platinum, on an electrodynamic scale, on galvanic polarization, and on electrical metrology, he did only one piece of work on electrodynamics proper. He studied forces affecting the interior of magnetic or dielectric polarized bodies, obtaining results that agreed with the contiguous-action viewpoint of Faraday, Thomson, and Maxwell. He also published some observations on how best to use a balance in order to determine magnetic moments.[43]

In addition, he found time to publish an (admiring) review of Thomson's *Mathematical and Physical Papers* in *Nature*. The first two volumes concerned Thomson's publications in electrostatics and magnetism (to 1856) as well as those on the transatlantic telegraph. Helmholtz approvingly declared, "One great merit in the scientific method of Sir William Thomson consists in the fact that, following the example set by Faraday, he avoids as far as possible hypotheses on unknown subjects, and by his mathematical treatment of problems endeavours to express the law simply of observable processes." Especially since the

1860s, Helmholtz had increasingly made that method his own, and he approved of the fact that Thomson, as he put it, avoided hypotheses "respecting the unknown interior mechanism of the phenomena."[44]

After his Faraday Lecture of 1881, Helmholtz turned—as several other physicists and chemists were then also doing—to applying thermodynamics to chemical processes, as he had already done (and did again later) to the earth's atmosphere. His seminal papers on the thermodynamics of chemical processes led the way here. Although several others (e.g., Julius Thomsen and Marcellin Berthelot) had already applied the first law of thermodynamics (then understood to be the equivalence of heat and work) to understanding heat in chemical processes, Helmholtz also applied the second law (in the sense of increase of entropy over time in a closed system), as did both August Horstmann and Josiah Willard Gibbs. But Horstmann's results proved to be too limited and Gibbs's simply went unnoticed.[45] Helmholtz had often, both in his strictly scientific and in his popular addresses, invoked the first law, but it is noteworthy that only twice previously—in his address of 1854, where he discussed the "heat death" of the universe, and even more briefly in his address of 1869 on the origin of the planetary system and life—had he invoked the second. The second law by now showed far greater utility and had attained far greater status in physics than it had when Clausius in 1850 and Thomson in 1851 had first articulated it. Over the intervening decades, and despite some intellectual tussles with Clausius, Helmholtz had clearly come to a deeper appreciation of the second law's importance. He now put it to novel use.

Helmholtz's success and leadership in chemical thermodynamics also derived in good part from his own previous work in electrochemistry. He had sought, during the 1870s, to understand the chemical energy of dilute solutions, in particular the relationship between chemical change and electromotive force. In February 1882 he published the first (and most fundamental) of three papers on the thermodynamics of chemical processes. In this first paper, which was entirely devoted to theoretical issues, he abandoned the old, so-called thermal theory of affinity, which he and everyone else had previously employed in trying to understand chemical energy, and instead employed the concept of entropy to show that chemical energy consisted of two parts: a "bound" part (that which appeared as heat alone) and a "free" part (the "useful" part, or the "work" that drives the direction of the chemical process, noting, incidentally, that such work could be gotten from a galvanic battery). Although others before him—François Massieu, Gibbs, and Thomson—had derived similar thermodynamic results, none of them had applied those to chemistry. By contrast, Helmholtz became the beneficiary of his own earlier work in electrochemistry; it was his idea of "free energy" and its application to chemistry (the "Helmholtz equation") that constituted his great innovation and effectively inaugurated the field of chemical thermodynamics. In a second paper (July 1882) and again in a third (1883), he bolstered his theoretical conclusions by showing their applicability to several observational or experimental issues in electrochemistry (for example, concerning salt solutions) and for understanding galvanic polarization (showing especially that the "free energy" determined the electromotive force of galvanic cells). With these three papers, so-called classical thermochemistry yielded to chemical thermodynamics. Helmholtz published one more piece of research on electrochemistry and continued to teach chemical thermodynamics, but his research in these fields (and in chemistry as a whole) ended after he published this trilogy in 1882–83. He left the field, as he so often did, for others to develop.[46]

Besides theoretical chemistry proper, Helmholtz's innovation in chemical thermodynamics (and its further development by others) greatly stimulated work in a new, intimately related field that soon became known as physical chemistry. From the late 1880s onward, Gibbs, Jacobus Henricus van't Hoff, Pierre Duhem, Ostwald, Arrhenius, Walther Nernst, Planck, and many others built, to one extent or another, upon Helmholtz's results and made major contributions to developing this new field. Yet despite his several contributions to chemistry (and chemists' recognition of these), Helmholtz never showed much interest in the subject; he saw it, like anatomy, as essentially consisting of empirically dispersed facts, not of laws.[47]

Physical chemists began employing the thermodynamically more general "Gibbs-Helmholtz equation" instead of the "Helmholtz equation," and the new theory of solutions became, along with chemical thermodynamics, the second of two pillars of the "new" physical chemistry. Helmholtz, however, harbored doubts about that theory, in particular Arrhenius's theory of ionic dissociation and van't Hoff's of osmotic pressure. When the Swede Arrhenius visited Germany in 1890, he found that German scientists, though broadly acknowledging recent results in the theory of electrolytic dissociation, were nonetheless waiting to learn what Helmholtz thought of that theory before openly committing themselves to it. Arrhenius himself believed that Helmholtz was not thinking about it at all and that, in effect, his authority was holding back the development of physical chemistry. That was not quite true. In 1891 Helmholtz wrote privately that the various theories of Nernst, van't Hoff, and Ostwald had "already proven themselves as extremely fruitful" and had led to "a multitude of factually correct conclusions, although some arbitrary assumptions" lay "hidden within these." Still, he appreciated that physical chemists needed such assumptions to understand chemical processes, "since the tremendously comprehensive system of organic chemistry" had developed "in the most irrational way," with arbitrary facts piling up one upon another. He thought, however, that "a healthy core" lay in this entire direction of "applying thermodynamics to chemistry."

What was needed for that to develop further, he believed, was mathematical analysis. Nernst drew support from and was influenced by Helmholtz: he developed the "Gibbs-Helmholtz equation" and related it to thermodynamic quantities in such a way as to transform it into the third law of thermodynamics (the heat theorem). This became another signal achievement of physical chemistry.[48] For their respective work, van't Hoff (in 1901), Arrhenius (in 1903), Ostwald (in 1909), and Nernst (in 1920) each won the Nobel Prize in Chemistry. To a greater or lesser extent, their work built on or developed out of Helmholtz's. Broadly speaking, in part they earned their prizes for making chemistry more physical and more theoretical, which was precisely what Helmholtz had also done.

While keeping an eye on the emergence of physical chemistry, Helmholtz also witnessed the emergence of another new field, low-temperature physics. He quickly recognized its potential industrial importance. Indeed, as president of the new Reichsanstalt, he had more reason than ever to be sensitive to the potential applications of science to technology and applications to understanding temperature measurement in every aspect, including that of thermometry. It was in these circumstances that, in 1887, he published an analysis of recent low-temperature work (in 1885) by his friend Raoul Pictet and Max Corsepius. Their analysis focused particularly on liquids in freezers. Pictet was an experimental physicist who specialized in liquefying gases at low temperatures and high pressures; he first made his name in 1877, when he became the first to liquefy oxygen (simultaneously with Louis Cailletet but independently). He served then as professor of industrial physics at the University of Geneva and became a leading figure in the liquefaction of gases (low-temperature physics) and refrigeration technology. In 1886, after losing a bitter patent fight in Geneva, he moved to Berlin. He established his own research laboratory there devoted to the industrial-scale production of gas liquefaction and, simultaneously, worked for a German commercial refrigeration firm to which he had sold several patents. Helmholtz, in his analysis of Pictet's and Corsepius's recent work, noted that "for the industrial purposes of ice manufacturing," their work using *liquide Pictet* (a mixture of carbon dioxide and sulfur dioxide) for producing low temperatures "obviously" had great advantages. Yet in studying their paper closely, he found that they had employed calculated (rather than measured) temperature values and that, as a result, their thermodynamic analysis of the chemical and electrochemical processes employing Pictet's liquid in the freezing process actually violated the second law.[49] That critique, as well as Pictet's dismissal of the pertinence of the second law to the refrigeration process, did not help his reputation as a scientist. He left Berlin in 1895 for Paris.

Apart from following developments in physical chemistry and low-temperature physics, from 1883 onward Helmholtz focused most of his intellectual energy on the statics of monocyclic systems, the principle of least action, the epistemology of numbers and measurement, atmospheric physics, and electromagnetic theory.

Rather than extend his results in chemical thermodynamics to physical chemistry, Helmholtz instead used them—specifically, the concepts of "bound" and "free" energy—as starting points in developing a recondite subject that he dubbed the statics of monocyclic systems. He aimed to reconsider the foundations of the second law of thermodynamics in mechanical terms. Between March and November 1884, he published two highly abstract papers—indeed, the most abstract (that is, theoretical) of his career—that made no reference to observational or experimental results. As in his work on chemical thermodynamics, it was mainly the second law that held his attention. In particular, he sought to provide, as Maxwell had done for electromagnetic theory, a mechanical analogy to help understand that law.[50]

By a monocyclic system Helmholtz meant any mechanical system in which there occurred "stationary, reversible movements" whose velocity was dependent upon only one parameter; such systems, assuming that their parameters varied extremely slowly, could be considered hidden and "static" heat systems. As an example of a stationary monocycle, he cited a frictionless spinning top on a fixed axis. He then sought to use such systems to create a mechanical *analogy* for—not to prove, let alone explain—the second law.[51]

Helmholtz had been spurred to find a mechanical analogy for thermodynamics through the study of both Boltzmann's unsuccessful attempt (in 1866) to prove the mechanical foundations of thermodynamics and Boltzmann's quite successful results (in 1871, 1872, and 1877) in overcoming the newly recognized contradiction between classical mechanics (representing systems involving fully reversible behavior) and the second law of thermodynamics (with its irreversible processes). Boltzmann's probabilistic interpretation of the second law laid the foundations of the new field of statistical mechanics. Unlike Clausius, Maxwell, and Boltzmann, Helmholtz avoided entirely the use of probabilistic arguments in his writings. Rather than make a probabilistic interpretation of the second law, he presented his monocyclic analogy (and much mathematical manipulation of Lagrange's equations of motion) to show that the heat of such a system, ultimately consisting of the hidden motions of its molecules, could not be entirely converted into work. In this sense he remained true to classical mechanics as it ran from Newton down to his own time, including his own essay of 1847 on the conservation of force (energy).[52]

Helmholtz's analysis of monocyclic systems drew the attention of Boltzmann and Hertz. Boltzmann reacted very positively to Helmholtz's analogy; like Helmholtz, he admired Maxwell's general usage of analogies in physics. In three papers on monocyclic systems published between 1884 and 1886, Boltzmann found a few shortcomings in Helmholtz's work—his results, he showed, were less general than Helmholtz had thought—but nonetheless extended it by showing that his own probabilistic interpretation of the second law could, using an ensemble of systems, be related to Helmholtz's mechanical analogy. That further clarified the relationship between heat and work. Moreover, Boltzmann showed that Saturn's rings constituted a monocyclic system, thus showing that the idea of a monocyclic system could represent physical reality as well as serve as an imaginary, analogical entity. When Boltzmann submitted the manuscript for one of these papers to Crelle's *Journal*, its coeditor, Helmholtz's friend Leopold Kronecker, asked Helmholtz to review it for possible publication. Helmholtz recommended it, noting that he himself had once (unsuccessfully) tried a similar approach. He added by way of conclusion: "I'm comforted to see, incidentally, that B. hasn't gotten much further [than I have]."[53] At sixty-four years of age, his ambition and competitiveness were (still) alive and well.

After finishing his analysis of the statics of monocylic systems, Helmholtz turned to analyzing another aspect of the foundations of physics, one that emerged out of his invention of monocyclic systems: namely, the physical meaning of the principle of least action (roughly: in a conservative system, the path of a particle having the minimum "action," or the kinetic less the potential energy integrated over time, satisfies Newton's second law of motion). He used Hamilton's form of the principle (1834–35), itself demonstrable by using Newton's laws of motion, to derive both Lagrange's equations of motion and his own set of dynamic equations (i.e., Hamilton's equations). Hamilton had thereby derived most of the physical tools that he and others would need for further analysis in mechanics. Fifty years later, in 1886, inspired by Maxwell's notion of a "kinetic potential" in his electromagnetic equations (analogous to Helmholtz's notion of "free" energy in thermodynamic systems), Helmholtz extended Hamilton's principle. He started his deduction by assuming Hamilton's principle and from there derived all of the equations of motion (Newton's laws as well as Lagrange's and Hamilton's equations), along with the differential equations needed to account for matters of heat and electricity. Although he thought the principle could not account for irreversible processes, he believed that his monocyclic systems could. Unlike Maxwell, Boltzmann, and others, he (once again) found it unnecessary to resort to a probabilistic interpretation of the second law of thermodynamics. Rather, he assumed, as he had in his work on monocycles, that hidden, stationary motions of molecules were ultimately responsible for thermal motion; as with his work in 1884, he claimed that he had provided only an analogical account of the second law, not an explanation of it. His results altogether reinforced (as was doubtless his aim all along) his long-standing belief in explaining all physical phenomena solely in terms of mechanical laws. In sum, he had expanded Hamilton's form of the principle of least action in a way that made it serve as what he hoped would be the new foundation of physics. Poincaré (in 1889) and Paul Ehrenfest (in 1906) later raised objections to Helmholtz's analysis, while his one-time students Hertz and Planck sought other ways to reconstruct a comprehensive mechanical worldview.[54]

Helmholtz sent Boltzmann an offprint of his paper on least action, which the latter graciously acknowledged. Throughout his career Boltzmann admired Helmholtz's work, but he also did not hesitate to say, as he did privately nearly a decade later, that, as concerned the use of the principle of least action, he considered Helmholtz to be his follower, not his leader. In January 1887, less than a year after publishing his analysis, Helmholtz lectured in a public forum at the academy on the history of the principle. He soon discovered that Adolph Mayer, a mathematician at Leipzig, had already published on this very topic; he therefore limited his own published account to a discussion of Leibniz (whom Mayer had not discussed) and Lagrange and Hamilton (about whom Helmholtz differed with Mayer). This unpublished lecture before the academy, Helmholtz's only foray into the history of science, appeared posthumously in the academy's own history in 1900.[55]

The lecture's public forum—it was intended to honor Frederick II (Frederick the Great), the academy's renewer—forced Helmholtz to give a nonmathematical account of the principle's history. He said that his work in recent years on its extension to other parts of physics and on its general meaning had led him to read "the old literature" on it. He explained that though there had been predecessors, the principle was first conceived by Pierre-Louis Moreau de Maupertuis, who, soon after doing so, became a member of the academy. He recounted how the king had brought Maupertuis—then one of Europe's leading mathematicians, natural philosophers, geodeticists, and philosophers—to the academy from Paris in order to renew the academy and increase its international visibility, and he related what transpired there during the Frenchman's contentious tenure as its president (1745–53). Part of Helmholtz's lecture, then, was devoted to explaining in layman's language what the principle was and what it meant in modern terms. He summed up this part: "All that happens [in nature] is represented through the flow, in one direction or another, of the world's eternally indestructible and non-increaseable supply of energy. The laws of this flow are completely summarized in the principle of least action."[56]

With a touch of sarcasm, he characterized Maupertuis himself as a man of brilliance, enormous energy, vaunting ambition, vanity, egotism, arrogance, lack of modesty, jealousy, and pettiness. Maupertuis had indeed, Helmholtz declared, contributed greatly to "the regeneration of the Academy"—and the king had rewarded him munificently for his contributions. He also recalled how Leibniz—who had originally advised Frederick I to establish the academy (in 1700–1701)—had proposed a scientific theorem that came deceptively close to

Maupertuis's version of the principle of least action. Leibniz had also famously declared that this was the best of all possible worlds, and that declaration, Helmholtz said, was seen by some as a sort of general precursor to Maupertuis's more precise and mathematical statement of the principle. Indeed, Leibniz's supposed statement—however primitive it may have been—of the principle *avant la lettre* later led to an unpleasant dispute between Maupertuis and Johann Samuel König, "a man," Helmholtz said, who "appeared to have been of somewhat boorish manners and who, with his republican feeling of equality, loved to show off." In particular, König had revealed a fragment of an unpublished letter that suggested that Leibniz may have "discovered" the principle of least action before Maupertuis. This led Maupertuis to charge König, his one-time client and ally, with being a plagiarist for having introduced a supposedly "falsified" letter. (Helmholtz himself remained neutral about the letter's authenticity; most, but not all, scholars now consider it to be authentic.) After an academy investigation into the matter, König resigned from the academy. Then Voltaire, whom the king had also recently brought to the academy, showed, said Helmholtz, "great courage and great warmth" in defending König, just as he (Voltaire) had done for others previously. Voltaire's satiric pamphlet of the Maupertuis-König affair employed fictitious names, but it was transparent to everyone who and what were involved. It soon led to Maupertuis's undoing and resignation from the academy (not to mention Voltaire's as well).[57]

Helmholtz also reported that in 1744 Euler had come even closer than Leibniz to "discovering" the principle. The king had also managed to attract Euler, the leading mathematician and natural philosopher of the era, to the academy in 1741; three years later he became the director of the academy's physical-mathematical section. Nonetheless, neither Maupertuis nor Euler, according to Helmholtz, had given a mathematically and conceptually satisfactory statement of the principle, which was not achieved, he added, until Lagrange, building on Euler's work, gave a general mathematical proof of the principle in 1760–61. Five years later, in 1766, Lagrange, an Italo-French mathematician and astronomer, succeeded his teacher Euler as the academy's director of the physical-mathematical section. Helmholtz concluded his historical account of the principle's "discovery" by noting that, after Maupertuis's departure, the king "saw that naming a president who himself covets the renown of a writer and philosopher may not be such an innocent matter."[58] His lecture thus described the sometimes rancorous culture and dubious politics of science at the academy. He used it simultaneously to explain the principle of least action to a partially lay audience, to praise the academy and its eighteenth-century royal sponsor, and to warn his contemporaries of the potentially dangerous consequences to the academy when dubious personnel decisions are made or academicians misbehave.

By the late 1880s, as Helmholtz was in the process of leaving the university for the Reichsanstalt, and as he was entering the final phase of his career, he must have realized that he himself had become a part of the history of science. He had come to see his original essay of 1847 on the principle of the conservation of energy (force) as a historical document. Whereas in 1861 he had refused a request to republish it (since he lacked the time to do the necessary updating), in 1882 he had the essay reprinted in volume one of his *Wissenschaftliche Abhandlungen*, but he did so with a set of comments and corrections, and in 1889 he allowed Ostwald to republish it, with comments and corrections, as the first volume in the latter's new series "Klassiker der exakten Wissenschaften."[59] Perhaps he changed his mind because by the mid-1880s he had substituted the principle of least action for that of the conservation of energy as the foundational, all-encompassing principle of physics. In the early 1890s, he returned to further investigating the uses of the principle. In the meantime, he concentrated his attention on another aspect of the foundations of physics, namely, the epistemology of number and measurement.

The Epistemology of Number and Measurement

In 1883–84, Dilthey wrote Helmholtz about his plans for a *Festschrift* to celebrate the upcoming fiftieth anniversary of Zeller's doctorate (1887) and invited him to participate in it. Since Zeller was a good friend of Helmholtz's, he readily agreed to write an essay for the proposed volume. The essay concerned the epistemological analysis of number and measurement.[60] It turned out to be Helmholtz's last major piece of philosophical (epistemological) analysis. That he chose the foundations of measurement as his topic—a topic that had been but rarely treated, and in any case not by physicists—is not surprising, since it fit with his own current scientific program and with the scientific and political-economic times.

During the previous decade, it may be recalled, Helmholtz had represented Germany at international meetings aimed at establishing electrical units and standards. He was currently in the process of establishing the Reichsanstalt, an institute devoted to metrology. The recent setting and legal enactment of international time zones and international standards for weights and measures suggest a further political and institutional need for a broad reconsideration of measurement. At the same time, and quite separately, there were mathematicians who were reconsidering the logical foundations of their field, physicists revisiting the meaning of temperature measurement, and psychologists reevaluating the possibility and means of measuring psychological phenomena. Helmholtz stood at the intersections of scientific thought and political-economic action concerning the foundations of measurement. Measurement had become a hallmark of nineteenth-century physics and, indeed, of science generally, and Helmholtz felt compelled to reassess its foundations, just as he was reassessing those of mechanics and thermodynamics.

In asking, as he did, what it meant to measure something, Helmholtz's chief concerns included ordinal numbers as creating acts of consciousness in time, the derivation of arithmetic axioms, equality (physical as well as mathematical), the meaning of sign numbers, and the nature of quantity and its measure.[61] For Helmholtz, measurement essentially meant dividing quantities into equal units. He sought to impart to these abstract notions an empirical basis or understanding, to reset the foundations of arithmetic in terms of empirically based laws, and yet simultaneously to argue for the role of psychology (intuition) in arithmetic, since he saw psychology as an empirical science. His synthetic notion of counting and measurement meant that he considered number to be definitionally independent of quantity; it helped mark a turn to what became the modern or representational concept, namely, that "quantity and number are defined separately." Finally, Helmholtz's analysis of arithmetic here effectively complemented his analysis of geometry two decades earlier; in both cases he emphasized the centrality of providing an empirical basis to these respective branches of mathematics. This project also gave him a fresh occasion to distance himself—as he had already done in regard to human perception (1855, 1866–67, and 1878) and geometry (since 1868)—from what he here referred to as the "strict adherents of Kant," thus implying that he was not one of them. With regard to the foundations of both arithmetic and geometry, he saw himself as an empiricist; however, his general admiration for Kant forced him to do an intellectual sidestep by declaring that Kant's views on space were not completely invalidated by empirical science.[62]

Helmholtz believed that ultimately both arithmetic and geometry had their roots in the physiology of perception. For initial guidance in his essay, he thus drew on a variety of scientists: the writings of several mathematicians (the brothers Robert and Hermann Grassmann, Ernst Schröder, and Paul du Bois–Reymond); several physicists (Adolf Elsas and perhaps Maxwell and Mach); and several psychologists and physiologists (Fechner, Wundt, and Johannes von Kries) who were interested in the possibility and method of measuring psychological phenomena. However, he by no means accepted all their individual positions and arguments (and they themselves at points disputed one another or were disputed by third parties). Furthermore, Helmholtz had already— apparently in the mid-1840s—addressed some of the mathematical, physical, and philosophical issues that he addressed here; in some sense, then, these concerns had been with him throughout his entire career.[63]

Overall, Helmholtz's essay was either not well received or was simply ignored. It got its best reception from the Russian mathematician Alexander

Wassilieff, who also translated it into Russian. The mathematicians Richard Dedekind, Georg Cantor, Gottlob Frege, and the young Bertrand Russell, who keenly read Helmholtz on geometry but argued for a logicist (or set-theoretical) foundation, were more or less hostile to Helmholtz's empiricized arithmetic. Helmholtz's notion of quantity got a friendlier reception from the little-known Leipzig mathematician Otto Hölder and from Poincaré, who incorporated some of Helmholtz's views on number and quantity into his *La science et l'hypothèse* (1902) and extended them. Ernst Mach, always reflecting on the fundamental concepts of physics, physiology, and psychology, also incorporated Helmholtz's empiricist approach to arithmetic (providing new definitions of mass, temperature, and so on) into his *Erkenntnis und Irrtum* (1905). The French physicist Pierre Duhem and the English physicist Norman Robert Campbell each also found much to agree with in Helmholtz. But more broadly, physicists ignored the essay. In the early 1890s, Helmholtz re-presented his ideas on counting and measurement in his advanced course on theoretical physics, and the essay reappeared, largely unchanged, as volume one of his *Vorlesungen über theoretische Physik*. While it still did not find much of an echo among physicists and others, it remains true that the creators of modern measurement theory (Russell, Campbell, Ernst Nagel, S. S. Stevens, Brian Ellis, and Patrick Suppes, to name only the most prominent) all cited Helmholtz's essay, even as they departed from his approach in their effort to keep quantity and number definitionally separate from (but correlated with) each other. His essay, then, marked an important turning point in the transition from classical to modern measurement theory.[64]

25

Celebrations

Silver Anniversary and Pour le Mérite

Helmholtz spent the month of March 1886 suffering from flulike symptoms, including a fever that laid him very low. In April, with the semester at an end, the Helmholtzes and their son Fritz went to Baden-Baden so that Hermann could recover further. They stayed at the Englischer Hof, where they were joined by Bunsen, Koenigsberger, Fuchs, Kirchhoff, and Virchow. Helmholtz had difficulty shaking off his illness, and in addition he was working on a mathematical problem that he could not solve; Anna thought that both together had made him "somewhat depressed."[1]

Back in Berlin and in the company of family, friends, and colleagues, the Helmholtzes celebrated their silver wedding anniversary on 16 May at the Physics Institute, which was adorned with garlands for the occasion. That morning, the Helmholtz children and sole grandchild Hermann went into the institute's auditorium, which had been transformed for the occasion: it was decorated in silver, with bouquets of flowers everywhere, along with bronze medals, silver baskets, paintings, and the like. With everyone crowded into the auditorium, and Ellen and her baby son, Hermann, standing at the center, the attendees welcomed Anna and Hermann as they entered, while the couple's musical friends performed Handel's *Seht er kommt mit Preis gekrönt*. The Helmholtzes received gifts galore, as well as telegrams, letters, and cards from those who could not come. The festivities lasted all day and well into the evening, including dinner and a party. Music and theatrical arts were performed on a stage erected for the occasion. Finally, a gallery chorus sang from Wagner's *Lohengrin*, and the Helmholtzes returned to their living quarters and drank to everyone's health. It was 2:00 a.m. before they got to bed; seven hours later, Helmholtz was in class lecturing.[2]

One month later, Adolph Menzel, the vice chancellor (since 1882) of Pour le Mérite für Wissenschaften und Künste, became its new chancellor and Helmholtz its new vice chancellor. The vice chancellor's duties consisted in waiting to succeed the chancellor someday. As an influential figure in Prussia's most elite cultural and intellectual association, Helmholtz doubtless had a hand in the awarding of the Pour le Mérite to several of his friends: Kirchhoff and Max Müller (1874), Bancroft (1875), du Bois–Reymond and Zeller (1877), Brücke (1878), Stokes (1879), Hofmann (1883), Thomson (1884), Joseph Lister (1885), Siemens (1886), Treitschke (1887), Clausius (1888), Ludwig (1889), Hildebrand (1891), and Kundt and Pflüger (1893).[3]

Heidelberg's Quincentennial

Helmholtz spent ten days in August 1886 in Heidelberg, where he helped celebrate the university's five hundredth anniversary and where he was himself celebrated as the recipient of the first Graefe medal awarded by the Ophthalmologische Gesellschaft. The quincentennial festivities themselves ran for five days.[4] At the opening ceremony, a church service was held in the Heilige Geistkirche; Helmholtz enjoyed the Bach motets and chorales and even the university minister's sermon. Many of Baden's ruling political figures appeared, including Grand Duke Frederick and his family, nineteen court figures, and Crown Prince Frederick William of Prussia and Germany. So, too, did the entire university teaching staff, some two hundred honorary guests, many city officials, hundreds of students, and some thirty-five hundred other participants. The first day's ceremony, held in the university's recently redesigned Old Hall (Alte Aula), was presided over by the grand duke. Zeller read official greetings on behalf of all German universities, to which Otto Becker, Heidelberg's rector and Helmholtz's old friend, responded. Helmholtz lunched with Becker and his wife, as well as with Theodor Leber of Göttingen, a leading ophthalmologist; Major-General Bernhard Oktav von Beck, of the Baden (later Prussian) medical army corps; and Karl Hoff, a painter and professor at the Kunstschule in Karlsruhe, who served as the quincentennial's ringmaster. That evening, the festivities continued up in the castle and at the villa. The crown prince asked after Anna; Prince Karl of Baden held a long conversation with Helmholtz. The grand duke appeared, and a crowd of people threatened to storm into the villa, where the beer that evening flowed generously and for free. As the crowd entered, the situation became "unsafe"; Helmholtz and the other dignitaries slipped out the back door.[5]

Two days later, Kuno Fischer addressed the entire assembly in the main Protestant church, speaking broadly about the university's "political and scientific adeptness." Helmholtz thought Fischer spoke vigorously and with skill, and he found the speech exciting. He had a good seat and so heard most of it, "without becoming tired." By contrast, Roscoe, who was also there, not only found it very hot inside the church but thought Fischer's two-hour-plus address was "very long and tedious" and that "the effect upon the audience was, as might be expected, soporific."[6]

After lunch, Helmholtz addressed the assembly. Heidelberg lay close to his heart, he said, and his twelve years there had convinced him that its beauty was eternal. He thought it no accident that spectroscopic analysis of the chemical composition of the heavens—an astoundingly bold achievement, he added—had first occurred "from these green hills." The scientific researcher, he continued, for all his "laborious and patient work," also needs the poet's gaze. Ideas do not come through hard work alone; instead, he claimed,

> These jump out, like Minerva from the head of Jupiter, unexpected, without surmise; we don't know when they will come. This alone is certain: they do not come to him who has learned to live life between books and papers, or to him who is tired and annoyed by monotonous work. The feeling of full life and force must be there, as there is above all when wandering in the pure air of the hills. And when the quiet peace of the forest separates the wanderer from the world's commotion, when he grasps in one view the rich, luxurious plain below with its fields and villages at his feet, and the setting sun spins its gold threads over the distant mountains, then the germ of new ideas indeed pours very sympathetically into the dark background of his soul, ideas that are fit to shine light and order on the inner world of the imagination, where before there was only chaos and darkness.

His audience applauded vigorously. Everyone wanted to toast him personally (though in fact he knew few of them). At the banquet that evening, he toasted the city—it was the last such toast of the evening, and the audience had already become "restless." He said that the city, the castle, and the university were effectively one. He and Theodor Mommsen also wrote a statement in Latin—presumably Mommsen's Latin—praising all three.[7]

The next day was devoted to awarding honorary degrees, forty-five in all.

Two months earlier, the Heidelberg Medical Faculty had informed Helmholtz that it intended to award honorary degrees to three physicists—Thomson, Siemens, and August Töpler—and the dean of the faculty had asked him to write a brief description of the work of each, so that he could be factually correct in his own description of their work on their honorary diplomas. Roscoe also received an honorary degree; so, too, did another Helmholtz friend, Francesco Brioschi, who, upon Sella's death in 1884, became the new president of the Accademia dei Lincei. (Others who received degrees but were not present included Helmholtz's colleagues Alexander Graham Bell, Chevreul, Koch, Newcomb, Pflüger, and Rayleigh.) Helmholtz had, or sought to have, private conversations with several of his friends (Trotter, Blaserna, Hermite, and the Swedish physiologist Alarik Frithiof Holmgren), "but the swarm of unknown and indifferent people who were everywhere and wanted to introduce themselves, was too great. They interrupted every reasonable conversation." He did, however, manage to spend part of a day socializing at Koenigsberger's: the latter had invited Helmholtz, Hermite, Fuchs, Bunsen, Roscoe, Brioschi, and others to his place. Late that week, Helmholtz and others were also invited to Karlsruhe to spend an afternoon with the grand duke. The latter hosted "a very pretty garden party.⁸

Pure and Applied Science: The Graefe Medal and Ophthalmology

After the quincentennial week, Helmholtz remained in Heidelberg for several more days to participate in the meeting of the Ophthalmologische Gesellschaft. Only once before (in Paris, in 1867) had he attended this annual meeting. Nearly a year earlier, the society had unanimously chosen him as the first recipient of its new Graefe Medal. While he felt gratified by the recognition, he was also quite surprised, since "so many years had gone by" and his work had been "directed to other fields"; he thought his ophthalmological work had effectively entered "into the realm of the historical." So much time had passed that even his migraines had by now ceased to return.⁹

Perhaps that was why the society's president, Donders, explained the precise reasoning behind its choice of Helmholtz. Among contemporary natural scientists, Donders declared, no one stood above Helmholtz and "only very few stood beside him." Indeed, Helmholtz had "[already] assumed a position among the great men of all times." He was so intellectually gifted, Donders claimed, that he had been able to switch disciplines from physiology to physics. He pointed first and foremost to the *Handbuch der physiologischen Optik* as evidence of Helmholtz's contributions to ophthalmology. Along with giving an understanding of the eye's anatomy, Donders explained, the *Handbuch* constituted "the foundation of our knowledge of the [eye's] pathological processes and the key to the explanation of [medical diagnostic] symptoms."¹⁰

Donders then turned to Helmholtz's invention of the ophthalmoscope. While Helmholtz himself had modestly acknowledged that others had done much of the work in advancing understanding of accommodation, Donders said, it was "Helmholtz alone" who had invented the ophthalmoscope. When Graefe first used the device, he reportedly said: "Helmholtz has opened up for us a new world!" Doctors had previously been ignorant and confused about the disease of "black star" and could offer no effective treatment; now that had changed. Donders quoted Graefe as saying that the ophthalmoscope had "gained for therapy an unimagined field," from which, after only a few years, "beautiful fruits" had come. Graefe had earlier (in 1858) awarded Helmholtz a silver cup with the inscription "To the Creator of a New Science, to the Benefactor of Mankind, in Thankful Remembrance of the Invention of the Ophthalmoscope." Now Helmholtz received the first Graefe medal as a memorial to Graefe. Donders closed by noting that Helmholtz was "in unfaded freshness of body and mind" and that he was surrounded by those who loved him. Helmholtz was moved (but also embarrassed) by Donders's fulsome address.[11]

Helmholtz replied by telling the society that he felt both honored and burdened by its recognition. While his instrument had proven its worth, he recognized that new, subsequent ophthalmological innovations and investigations had overtaken his first efforts and reshaped the discipline. He told the ophthalmologists, "Really one of the great medical doctors should have received the Medal, since I only supplied them with the tool for their art."[12] That was not completely false modesty.

Hence he considered this recognition a piece of luck, since others had worked long and hard in attempting to reach their scientific goals. He continued: "Indeed, it perhaps lies in the nature of human relations that new, original thoughts have all the more difficulty in gaining recognition the more truly original, fruitful, and valuable they are." This was the source of his "embarrassment," he explained. He had made this point once before, in a dedication speech at the memorial ceremony when a statue of Graefe was unveiled in Berlin, at the Charité. He quoted himself: "'The ophthalmoscope was more a discovery than an invention,' that is, when a well-schooled physicist came along, understood the importance of such an instrument, then tested the optical apparatus and developed the knowledge needed to complete the same."[13]

He thought that circa 1850 there were five to ten other young German researchers who might have invented the ophthalmoscope. He ascribed his success not to himself but rather "really to my great teacher, the powerful Johannes Müller." He did not mean any special training that Müller had given him, but rather the confidence Müller had shown in him and his support in helping him obtain an academic position. "Therein was expressed the new insight into the role that physics might then be able to play in physiology. It was, moreover, a bold move. Through the bestowing of your award today you in effect express a recognition of Johannes Müller."[14] That was a more-than-generous interpretation.

Helmholtz characterized his invention of the ophthalmoscope as a pedagogical "necessity" that arose from his having to explain to his physiology students "the theory of the eye's illumination." Once he asked the right questions, he said, he got the right answers, and with them the appropriate instrumental means for illuminating and viewing the dorsal area of the human eye. By contrast, he claimed, ophthalmometry had developed out of disputatious opinions as to the theory of accommodation, and he noted that he himself had made many errors in trying to understand this phenomenon. As for color theory, his third and final contribution to ophthalmology, it was Müller's theory of specific sense energies that had drawn his attention to it. Here, too, pedagogy had played a role, since he did not like lecturing on topics in which he had done no observational or experimental work. So he began mixing colors and got results different from those reported by others. This led to "a radical change of all previously proposed laws of color mixing." In particular, his experimental results when he mixed yellow and blue conflicted with those of Goethe and Brewster, "two masters of the first order." The mixture of those colors, they said, led to green. "This was one of the facts which brought me over to the empirical theory of perception." Neither Goethe nor Brewster, nor for that matter Hering and his followers, understood that a given color did not issue simply out of the mixture of simple components. His color theory work, Helmholtz declared, had in turn led him to rethink the entire field of physiological optics and to write his *Handbuch*. All told, he judged his accomplishments in ophthalmology—the invention of the ophthalmoscope, the theory of accommodation, and color theory—as no more than good and hardly inspired work. He saw himself "as at best, and looked at with the eyes of a very well-meaning judge, an attentive, industrious, and well-schooled worker."[15] That was false modesty.

In his speech, Helmholtz also raised "the issue of the different character which scientific activity shows in different branches of science." He pointed to the different gifts that various scientists have in doing observational as opposed to theoretical work. At the extreme end, theoretical physics provided

the complete rule over matter through exactly defined, exceptionless, ruling laws, whose consequences are to be developed with the finest sharpness of mathematical thought. Insofar as this succeeds, we are freed from the causal context of everything that is dark and mystical. The forces of nature which are so conquered, accommodate themselves not only theoretically to humankind's knowledge but also to the service of his will. The gaining of new insights of this kind therefore demands precisely and often enough the highest development of human understanding, something of which only a few individuals are capable. However, what is achieved in gaining such knowledge and summarized in the exact and clear form of science, can be transmitted securely and completely to others.

Yet he also judged theoretical physics to be a field that was "unfortunately very narrow." For example, it could not, "for the most part," take into account "the organic world." Yet that world included making decisions about war, the state, and medicine; these required action in thousands of ways, whether or not one possessed "clear insight into the context of things." The artist, in contrast to the theoretical scientist, had a different intellectual strength, namely that of representing experience and phenomena. Helmholtz did not mean that the two abilities were mutually exclusive, or that the artist did not have some scientist in him and vice versa. Yet the artist could not explain why and how he did something. And that led Helmholtz to declare that "the great medical doctors" were like artists. One could give "a sort of allegorical presentation" of what great doctors did, but one could not explain it in terms of "physiological theory." To do so was to erect a misleading measure. A physicist could see medicine's shortcomings only in the theoretical realm; even a doctor trained in physics "deals uncertainly and without success, feels unsatisfied and unhappy, and is thus without moral influence on the patients and their caretakers. In short, he recognizes here the limits of his ability." That was, if nothing else, a self-portrait of Helmholtz early in his career as a physician. "However, the geniuses are, as I have always seen, modest precisely in relation to that wherein they are superior to others. Precisely what becomes so easy that they have a hard time understanding why others are not also able to do precisely what becomes so easy for themselves. With great talent," he concluded, "there is however also and always bound the corresponding great sensitivity for the errors of their own works." He compared himself to the simple "blacksmith" who provided the tools or materials for the artists (that is, the geniuses) to work with.[16] More false modesty.

That evening a dinner was held in the Sanatorium Schloss Heidelberg for sixty ophthalmologists and their wives. There was a great deal of drinking and toasting, and from the hotel's terrace, as Helmholtz reported home to Anna in Berlin, one could see a beautiful moon. She bitterly and tearfully regretted that she could not be there with him to experience the honor he had received. But their son Robert was still recuperating from yet another operation on his backbone, "an old bone suffering," as Helmholtz told Lady Thomson. Two days later Anna reported that Carl Ludwig had brought Helmholtz's "really wonderful" Graefe Medal with him to Berlin—Helmholtz himself had gone on to Interlaken—and described the entire event for her. The newspapers reported on his address, which she found much too modest. ("You are certainly not merely a blacksmith.") The young ophthalmologists, she said, well understood that they owed "their scientific specialty" to her husband. For them, the ophthalmoscope was a given, a natural thing, "like how we look at forks and knives."[17]

Near Death at Interlaken

At Interlaken, where he had gone to meet Ellen and her son, Helmholtz became seriously ill. Fortunately, a friend of his, the well-known Swiss ophthalmologist Friedrich Horner, was also there with him. Neither Helmholtz nor his various doctors were ever quite certain of the nature of the illness—variously described as "intestinal cholic" or a "stomach-intestinal disturbance"—that afflicted him. Horner, in any case, greatly aided him.[18]

Helmholtz became so ill that he asked Anna to come, which put her in the dilemma of having to choose between her recuperating son Robert and her husband. She worried greatly about her husband's "gloomy and depressed moods." She felt that Robert had recuperated sufficiently that she could leave him in order to take care of Hermann, who she thought might be in a life-threatening condition. When she arrived in Interlaken, he told her that he had been close to death; but by then he had recovered enough to take at least a step toward her. Twice daily, for six hours at a stretch, he suffered from what he believed to be muscular pains; his nights were nearly unbearable. "He is completely different than he otherwise is," Anna wrote Robert. At Interlaken, he was "morally depressed." After he had recovered sufficiently, Anna and Hermann went to Berne, where he showed "less lassitude." She could barely withstand the responsibility of traveling with and caring for him.[19]

In the Berne Alps (Rigi Kaltbad) Helmholtz recovered a good deal, but still not completely. He was now suffering from a mild "fever excitation" in the mornings and a "heavy belly" in the evenings. He also had heart palpitations and insomnia. He took quinine; that, plus the fresh air, he believed, helped fend off his malaria-like symptoms. The German ambassador to Rome, Robert von Keudell, an intimate of Bismarck's and an excellent concert pianist, was also in the Rigi Kaltbad for a couple of days, and the Helmholtzes discovered him to be "a wonderful musician." During the mornings and evenings, he played Bach and Beethoven—and whatever else the Helmholtzes wanted—for hours on end. The three of them also took a short day trip to nearby Selisberg to meet up with Blaserna and Marco Minghetti, the latter an important Italian financial, political, and cultural figure. When it got too cold in the Rigi, the Helmholtzes went down to Engelberg. Helmholtz still found himself especially sensitive to cold and damp, and he took still greater care with his diet. He still suffered stomach pains, could not sleep, and felt "weak and wretched." They went on to Locarno, but that did not help, either. Anna remained his caretaker and was constantly torn between her concerns for him and for their son Robert, who in the meantime had been sent to a spa in Falkenstein (Taunus).[20]

The Helmholtzes then went to Strasbourg to consult a leading internist and friend, Adolf Kussmaul, who examined and questioned Helmholtz for two hours "and found nothing." His heart beat normally, if "somewhat weakly"; all his organs were basically healthy. As to the illness that he was suffering from, Kussmaul could only suggest that it might have been due to an inflammation that somehow erupted through an internal cavity wall. Following the consultation, Helmholtz's health improved: he slept and ate better, his heart beat more regularly, and Anna thought he now looked "wonderful." Kussmaul judged, however, that Helmholtz's overall health was poor, and he ordered him to spend the next three weeks at a spa in Baden-Baden. He was to drink the local "hot water with *a little bit* of Karlsbader mineral salt (1 teaspoon full for every two glasses)," and at night put "a hydropathic compression around the body." Both treatments were effective, and he regained his appetite. Donders told a colleague a couple of months later that Helmholtz had lost five kilograms and concluded that this showed either that he had not yet fully regained his sense of mental equilibrium or that he was short of blood.[21]

When Baden-Baden got cold and wet, the Helmholtzes returned to Berlin. Being back there helped: "My husband's old nature and face is slowly, day-by-day, coming back." Though he still needed rest, he was able to participate in an international meeting of geodeticists then taking place in Berlin, and he attended academy meetings (even delivering a lecture), university faculty sessions, and doctoral examinations. "And that on the supposed Resting Thursday!" Anna did all she could to eliminate any time-wasting events, "like visits, letters, or smaller business matters." His "mood and sleep" were now "very good" since he was "at home." Anna and Hermann now took the greatest care concerning his diet, sticking to nourishing and frequent-but-light meals aimed at strengthening him.[22]

Hermann, Anna, and Ellen felt great gratitude toward Horner for his help during Helmholtz's near-death illness at Interlaken. Helmholtz wrote Horner early that September to thank him. So, too, did Ellen; she sent him her most-valued portrait of her father, an oil sketch by Lenbach. It was the best one the family had of him, she said. The Helmholtzes had previously displayed it in their family living room; but when Ellen married and moved out, she put it on her writing desk in her own home. "I'll never forget the impression that your [watching] eye made on him, and us all, during his illness, and I'll keep with infinite gratitude this holy memory of the heavy and unique hours where your comforting words spoke so truly to the heart." She reported that in the meantime her father's condition had improved. Anna, too, thanked Horner from the bottom of her heart. Despite all her ongoing concerns, she added, "I have . . . never forgotten how much thanks we owe to the friendly doctors or doctor friends. And I only wanted, if we may, to give you a proof of this thanks." To wit: the Helmholtzes had three copies of his Graefe Medal, and Helmholtz asked if Horner would like to have one of them. Helmholtz was now well enough to look after himself; the previous evening, Anna noted, he had attended a musical soirée with the minister of culture. That Christmas, Helmholtz sent Horner the complete editions of his *Wissenschaftliche Abhandlungen* and his *Vorträge und Reden*, with an inscription in each of the volumes. Anna enclosed a letter saying that Helmholtz wanted to be remembered by Horner at Christmas, and he hoped that Horner would find a place for these volumes on his bookshelves. Horner had little or no time to enjoy them: on 20 December 1886, on about the day that they arrived at his home in Zurich, he died.[23]

Pure and Applied Science: The Fraunhofer *Eloge*

The distinction between pure and applied science is largely a modern one; it emerged in the nineteenth century. Many nineteenth-century scientists either gladly acknowledged that science had a dual nature or made no such distinction at all. To one extent or another, disciplines such as astronomy, mathematics, and botany, for example, had always had relevance to aspects of everyday life. But the economically significant and strikingly practical interaction of science and technology began (and not always successfully) with the application of organic chemistry to agriculture and industry in the 1830s and 1840s, followed by applications of physics to the new electrical and optical industries. This helps explain why, until late in the century, most scientists attached little meaning to the distinction. Like Helmholtz's work in ophthalmology, that of Josef Fraunhofer in optics and optical technology also serves as a case in point.

On Sunday, 6 March 1887, in the auditorium of the Berlin City Hall, the German Society for Mechanics and Optics (Deutsche Gesellschaft für Mechanik und Optik) celebrated the centennial of Fraunhofer's birth. The society's aim was not only to recall Fraunhofer's life and accomplishments in theoretical and practical optics. It also sought to promote the general cause of the mechanical and optical industries by announcing the establishment of a new Fraunhofer Endowment (*Stiftung*) for the support of talented young men who wanted to become optical and mechanical technologists. Wilhelm Foerster, the Berlin astronomer and one of the driving forces behind this celebration, asked Helmholtz, the society's honorary president, to lecture on Fraunhofer. Among the many attendees were leaders of the German mechanical and optical industry; ordinary members of the society, including practicing mechanical and optical instrument makers; ministers of state, including the armed services; municipal leaders; members of the Academy of Sciences; and various professors and lecturers from around Berlin. The celebration in Berlin—there was also one in Munich with Abbe as the lead speaker—was of civic and political importance. At noontime, following the singing of a brief musical hymn by the Gesangvererin Caecilia, Helmholtz welcomed everyone.[24]

He considered this celebration of Fraunhofer as "in fact a memorial day of the German middle class [Bürgerthum]," one that should make them all proud. He called "the art of practical mechanics" one of the outstanding types of "civic work," even though it was not one of the most financially rewarding types. But mechanics had the highest virtue of all, he proclaimed, in that it strove for "the highest exactness, sanitariness, and reliability" of results. He himself had long witnessed "how high these first and highest virtues of civic work" had increased "with the leading mechanics."[25]

Helmholtz maintained that every physicist was at least a bit of a practical mechanic. He noted that when he himself wanted to explore a new research topic or point of view, he found it useful "to produce models of the requisite instruments, admittedly fragile and provisionally thrown together from materials of poor quality, that sufficed at least to perceive the initial traces of an anticipated success and become acquainted with the most important impediments that could frustrate it [such success]." Later, he turned to professional mechanical instrument makers for advice; he learned much from them. It was due to "the intelligent help of practical mechanics," he claimed, that the fields of physics and astronomy had advanced, as had understanding of the "world's structure" [Weltgebäude] and the atmosphere, not to mention the development of such technologies as telescopes, the electrical telegraph and lighting systems, and instruments used for sea travel and land surveys. Precision instrumentation had advanced the sciences both in shaping their overall foundations and through individual measuring results. The sciences as well as technology, he argued, were deeply indebted to instrument makers and the virtues they embodied.[26]

That was why practical mechanics and opticians honored Fraunhofer, an individual who had brilliantly displayed "the best virtues of the German middle class [Bürgerthum]." He portrayed Fraunhofer as a man "who through his own strength and industriousness rose up from the poorest [socioeconomic] conditions, and worked his way up under heavy obstacles to become proprietor of the world's best-known optical workshop of its time and as one who made scientific discoveries." (Though Helmholtz did not say so, the similarities to his own rise in life and successful career were plain enough.) Fraunhofer had advanced the understanding of optical phenomena. He noted especially Fraunhofer's use of Newton's discovery of the color of thin films for understanding the curvature of cut lens surfaces. It was only toward the end of his brief life (1787–1826) that Fraunhofer had turned to pure science, "that is, to investigations that had before them no immediate purpose but strove only towards completeness of our optical knowledge." Fraunhofer's investigation of color phenomena led him to a refined analysis of prismatic spectral colors; this in turn led him to investigate "applied color phenomena which produce the opaque atmosphere around the Sun and Moon." As a result of this and other work, including his founding of the world-class optical workshop in Bavaria, by 1817 Fraunhofer was appointed a member of the Bavarian Academy of Sciences "and was celebrated and became famous throughout scientific Europe." Helmholtz thought, moreover, that there were strong similarities between Fraunhofer's life and career and that of Faraday, who had begun life as an apprentice bookbinder without material resources and, thanks largely to his own efforts, later became one of the "great physicists of Europe." Such people displayed the virtues—hard work, autodidacticism, thrift, intelligence, and dedication to precision and accuracy in their work, in short, the "restless industry and fidelity in work which are the true sources of great achievements"—that Helmholtz so admired.[27]

Yet there was a major difference between Fraunhofer and Faraday, Helmholtz thought. Fraunhofer, unlike Faraday, remained wedded to his artisan roots, and his strictly scientific work issued from there. He never sought to be anything more (or less) than an excellent technologist. David Brewster maintained that Fraunhofer's accomplishments in improving achromatic lenses led to England's loss of its lead in this area, a fact that was painful to the English in both a scientific and an economic sense. Helmholtz used Brewster's warning to Britain as his own warning to Germany: "As soon as a great nation ceases to celebrate triumphs in the arts," he quoted Brewster, "then the fear is not ungrounded that it also wants to cease triumphing through weapons." Helmholtz thought that Britain had eventually heeded Brewster's warning and had "perhaps" recovered its lead. But Germany had not let up, and Helmholtz reminded his audience that the Prussian government had recently provided support to the optical glass works of Abbe and Schott in Jena and had "proposed to the Reichstag a still much more comprehensive plan for the advancement of mechanics" (that is, the Reichsanstalt). The speech was thus also meant as part of Helmholtz's (and others') public efforts to gain support among precision technologists for the proposed Reichsanstalt.[28]

Helmholtz's address was followed by a briefer speech by Foerster containing similar themes: it stressed the importance for German science and technology of continued governmental support for practical mechanics and optics. Then Rudolf Fuess, one of the society's leaders, very briefly addressed the audience. He was especially concerned that Germany find and support talented young men who would look to Fraunhofer as their model. With this goal in mind, he formally announced the society's formation of the Fraunhofer Endowment. It would help support young Germans lacking the financial means to become trained theoretical and practical mechanics and opticians. Two years earlier, he noted, the city of Berlin had already established a "trade school for young mechanics and opticians." The Fraunhofer Endowment would support needy young men to study at this and other schools throughout Germany and would also support them as apprentices in workshops at home and abroad. The celebration, which closed with a performance of a piece by Beethoven, helped set the public stage for the opening of the Reichsanstalt later that year.[29]

26

Doyen

The Year of the Three Emperors

In November 1887, as the Helmholtzes and several aristocrats dined at Corneli and Gustav Richter's place, everyone was much concerned about Crown Prince Frederick William's poor health. He had been suffering from throat maladies that had recently been diagnosed as cancer. The predictions by the crown prince's German doctors that his health would deteriorate were realized more quickly than anyone had anticipated. In February 1888, he underwent a tracheotomy. The Helmholtzes, or at least Anna, occasionally dined with the crown prince and princess. Helmholtz was often in their home. Anna and the crown princess, Victoria, were close friends.[1] As members of the crown prince's circle, the Helmholtzes were in a position to know something of the state of his health.

The crown prince's deteriorating health depressed Anna and many others in her circle. They had rested many of their hopes for change in Germany on him and his wife. The crown prince and princess, Anna thought, stood for the ideal, the beautiful, and the spiritual in life, and against militarism. Bismarck disliked and opposed the crown prince. Yet in truth the crown prince was not a liberal; instead, he deferred to his wife's liberal political views and surrounded himself with liberal-minded people like the Helmholtzes' friend Bamberger. The situation of the crown prince's health was worsened by the crown princess when, following the advice of her English surgeon, Sir Morell Mackenzie, a leading laryngologist who detected no cancer in the crown prince's throat, she decided that the crown prince did not (yet) require surgery.[2]

On 9 March 1888, Emperor William I died. The crown prince, then convalescing in San Remo, arrived two days too late to see his father alive, and that tardiness further undermined his position. He assumed the throne as Frederick III. Bismarck did all he could to isolate him. Anna von Helmholtz's name became closely associated (not least in Bismarck's mind) with the new emperor, not only because of her past friendship with the royal couple, and with Bamberger, who supported and advised them; but also because she became known as the leader of an abortive attempt by some fifty royal, aristocratic, or educated women in Berlin to show their solidarity with the new empress by signing a welcoming letter. Under pressure, probably from Bismarck and army leaders, all the signers withdrew their signatures. When Bismarck went to see Empress Victoria on 21 March, she had three supporting, politically liberal ladies in attendance, one of whom was Anna von Helmholtz. At the empress's request, Anna invited Mackenzie to a dinner party at the Helmholtzes' home; other guests included the empress, Bamberger, Hofmann, Siemens, and du Bois–Reymond. Helmholtz, who along with Ernst von Leyden and du Bois–Reymond had originally welcomed Mackenzie to Berlin, later reportedly said that he had misjudged both Mackenzie's character and ability. This fragile situation lasted ninety-nine days: Frederick III died on 15 June 1888. Out of respect for his death, university lectures were canceled. Helmholtz appeared at a regularly scheduled meeting of the Physical Society on the following evening simply to announce its immediate cancellation. Crown Prince William now became the new monarch and was henceforth known as William II, Prussia's king and Germany's emperor. He decidedly did not share his mother's liberal views. However, he was enamored of science and especially technology: in the late 1870s, he had studied for four semesters at Bonn, where he took courses in physics (with Clausius), who awakened his keen interest in technology, and chemistry (with Kekulé). He was the third of the three emperors in 1888.[3]

More British (and Other) Honors

That year Helmholtz received two new honors from British cultural institutions. In February he was elected an honorary member of the Musical Association of London, and in July the Society of Arts gave him its Albert Medal, to recognize the value of his researches "in various branches of Science, and of their practical results upon Music, Painting, and the useful Arts." The award memorialized Prince Albert, who had long been the society's president. Helmholtz felt much honored by this recognition from "artists and friends of art," but he was so preoccupied with his new duties at the Reichsanstalt that he could not go to London to receive the award. Nor did he travel to Lisbon to become an honorary member (April 1888) of Portugal's Academiae Scientiarum Olisiponensis (Academy of Science, Lisbon). Nor could he travel to Bologna to receive (June 1888) an honorary doctoral degree as part of the celebrations marking the eighth centenary of the University of Bologna (founded 1088). As for German awards that year, he received the Star of the Royal Crown Order, Second Class.[4]

In 1890 two new British honors came Helmholtz's way. In January Thomson informed him that he and Roscoe wanted Helmholtz to become a member of the International Joule Memorial committee. "What we want is your name on the list of the general committee," they said. "The prize will be an international one to encourage the scientific work of younger men of any nation and in any part of the world." Then in February the Academic Senate of the University of Edinburgh recommended that Helmholtz be offered its Gifford Lectureship for the coming two years (ten lectures per year). The Gifford Lectures on Natural Theology became one of Britain's most prestigious lecture series; they were intended to relate the physical and natural sciences and discuss the unity and harmony of nature. Thomson urged him to accept, and he invited Helmholtz to live with him in Glasgow during part of his time in Scotland. Helmholtz declined on the grounds that his administrative and scientific commitments in Berlin were too great.[5]

Helmholtz also declined other British invitations: the Chemical Society's fiftieth birthday (February 1891) and the centenary celebration of Faraday's birth at the Royal Institution (where he was to be awarded a Diploma of Membership in the Royal Institution). He could not attend, he said, because he "was obliged to accept an election as Honorary President of the Commission of jurors [for the Testing Commission] at the Electric exhibition at Frankfurt [am Main]" (i.e., the International Electrotechnical Exhibition), whose first meeting conflicted with those in London. He nonetheless received a diploma making him an honorary member of the Royal Institution.[6]

Pleas for Advice and Help

Especially after 1871, Helmholtz came to embody the German public's image of the scientist. His *Populäre wissenschaftliche Vorträge*, the *Tonempfindungen*, the public announcements of his various honors, and the occasional appearance of his name in general newspapers and magazines—all served to help make him renowned among the German public. One result was that he often received requests for scientific advice from officials or career advice from private individuals. These ranged from mathematics to music, from the socially high to the socially low, and Helmholtz usually responded to them. A few examples may suffice to illustrate his perceived importance for nonscientists.

In May 1888 the Prussian minister of culture, Gustav von Gossler, and a friend visited an art exhibition in Berlin where they became fascinated but puzzled by a hanging Japanese metal mirror that reflected light in unusual ways. They developed their own hypothesis to explain the reflected light's sources and behavior, but they asked Helmholtz for his analysis. Similarly, Rudolf Helmbold, a medical student at Jena, sent Helmholtz his analysis of the trisection of an angle and asked for his opinion. A professor at the Höhere Bürgerschule (Municipal School) on Lake Constance wrote to say he was studying the production of fire, had read a lot of literature on the subject, and wanted to know how mankind had invented fire.[7]

Many individuals turned to Helmholtz for advice on musical issues. Carl Andreas Eitz, a primary school music teacher, planned to build a harmonium, and he wanted Helmholtz to advise him and review his plans. Helmholtz also had an intermittent exchange of letters with a retired Prussian Gymnasium teacher who wanted to discuss his own work on the harmonium. The teacher had nearly completed a manuscript on the harmonium, and he appealed to Helmholtz to help him get it published. From France, Georges Le Gorgeu, a lawyer and archaeologist, asked Helmholtz's help in judging the acoustic performance and maintenance of church vases. From Worms, Friedrich Schön, a former member of the administrative council of Bayreuth's Theater Festival Performances and someone who knew Helmholtz personally, wrote that the local theater needed a gong as part of an upcoming performance and asked if Helmholtz could give him advice about obtaining one. Similarly, Oskar von Chelius, a composer and a military figure who was an acquaintance of Anna's, asked Helmholtz's advice about a new musical instrument that allowed the amplification and transformation of tones. He wanted to know Helmholtz's views on the instrument. Helmholtz also regularly received letters from admiring amateur scientists, minor scientists on the fringe of the scientific world, science teachers who could not find or had lost a job, practicing physicians, and the like—most wanted scientific advice and help, but some wanted merely to thank him for one or more of his writings.[8]

There were also untold requests from nonscientists about physiological or physiological-optical matters who wanted to know Helmholtz's judgment. Dr. E. Jaesche of Dorpat reported (at great length) his own findings and analyses of double images. Henri Stassano of Naples wrote Helmholtz (twice) concerning the eye's accommodation. W. E. Bryan, of Bath, England, wrote Helmholtz as "the highest living authority on the Science of Colour" to ask if he thought it made sense to try to "devise an Art of Colour-melody . . . which by *rhythmical sequences* of suitable colours would produce effects analogous to those of music."[9]

There were also letters from scientific hopefuls. Ferdinand Lorenz, a "simple working man" in Berlin, wrote to inquire about an electrotechnical matter. Helmholtz responded with a handwritten note, and Lorenz wrote again (in semi-literate German) to say "what joy" a note from Helmholtz had given him. Carl Kutschera, a law student from Graz, wanted Helmholtz's opinion of his cosmological ideas concerning space and the cosmological system. Certain unnamed scientists and laypersons had scorned him, he said, and so now he turned to Helmholtz, "an authority of the first order."[10]

There were also letters from those who simply wanted to be associated with Helmholtz's name. Dr. Julius Friedländer, a Berlin physician who had attended (and been greatly influenced by) Helmholtz's lectures in 1874, wrote in 1890 to say that, alongside his independent interest in the philosophy of Spinoza, Helmholtz's lectures inspired him "to uncover the relationships of this great thinker to modern natural science." Friedländer and Martin Berendt had just coauthored *Die rationelle Erkenntniss Spinozas*, and they were working on a manuscript entitled "Spinoza in Light of Modern Natural Science and Philosophy." Their new manuscript incorporated ideas from Helmholtz's popular essays and lectures as well as his writings on sensory physiology. They hoped he might read it sympathetically and grant them permission to dedicate it to him. If Helmholtz were to endorse it, Friedländer said, that would help make their case that Spinoza was an important figure for the development of modern science. Helmholtz granted them permission; and, upon its publication, the authors sent him a copy of their *Spinoza's Erkenntnisslehre in ihrer Beziehung zur modernen Naturwissenschaft und Philosophie* (1891).[11]

A final example of letters asking for advice and help is that of Dr. Phil. Paul Otto Schmidt of Berlin, who was facing a "life question." Schmidt had attended the Realgymnasium in Halle, but at sixteen years of age he joined the navy after his parents said they could no longer afford to support him. He had a hunger to read anything and everything he could. He spotted Helmholtz's *Populäre wissenschaftliche Vorträge* in a Kiel bookstore window, and bought the set. "From that moment on," he wrote, "I've read nothing but these lectures, whose content and manner of presentation charmed me. If at that time, lacking the necessary basic [scientific] knowledge, I did not fully understand many things [in the lectures], I nonetheless surmised what was at issue, and that brought out in me a warm desire to dedicate myself completely to the exact natural sciences, for which, by the way, I had already shown an aptitude and understanding at school." After he finished his tour in the navy, he returned to the Gymnasium and received his leaving certificate. He was determined to study mathematics and physics and was now attending the University of Halle, concentrating on these fields. He did not, however, expect to obtain a teaching position because there were already too many mathematics and physics teachers. He had written a doctoral dissertation on the meaning of space and time in the light of modern physics (1887), followed by *Das aristokratische Prinzip in Natur- und Menschenleben* (1888). Both had been positively reviewed, he claimed, but this broad science writing had led him away from his true interest, specialized studies in the exact natural sciences. He now had come to a point in his life where he had to make a decision. He asked for Helmholtz's advice: "Without you, Highly Honored Sir, I would today be [but] a well-appointed navy paymaster." Might he perhaps work in some capacity for Helmholtz, he asked, for example as an assistant? He requested a personal meeting. Helmholtz, who as a young man had himself been unable to pursue his own first love (physics), almost certainly had sympathy for Schmidt. His handwritten marginal note on Schmidt's letter indicates that Schmidt "appeared personally for a discussion."[12]

Death of the Scion

Such letters, though often burdensome in number, were as nothing in comparison to the responsibility and issues the Helmholtzes felt in regard to their son Robert. Though disabled and often in poor health, Robert Helmholtz had sought to lead as normal a life as possible. After graduating from Gymnasium in Berlin, he first studied science at Heidelberg (with Bunsen) and then returned to Berlin to study mathematics and physics (with Kirchhoff and his father). His close friend Lummer, one of Helmholtz's students and (after 1887) employees, described Robert as intellectually gifted, with great perseverance and a strong work ethic. He had strengths both as an observer and as an experimentalist, and mathematically. He knew early in his career that he wanted to do research. In 1885 he promoted at Berlin with a doctoral dissertation on steam and cloud solutions. In the summer of 1886, however, he (once again) became seriously ill and had to undergo, as already noted, an operation on his deformed spine. During the following three years, he did creative scientific research, publishing a treatise on changes of freezing point, which employed his father's concept of free energy, and a paper on experiments on steam. He also worked on a manuscript on the radiation energy of flames for a competition held by the Verein für Gewerbefleiss zu Berlin (the Berlin Association for Industry). He won (posthumously) the five-thousand-mark prize and a medal. He also wrote a eulogy of Kirchhoff for the *Deutsche Rundschau*. By 1889 Robert had seven publications to his name. According to Lummer, Robert was an excellent speaker, a man of social charm and good company, and a modest, selfless person with a strong sense of duty.[13]

In June 1889 Robert again became seriously ill. He was more "depressed and in low morale," Anna found, than she had "seen him before." He lost all pleasure in life, including that of becoming an assistant at the Reichsanstalt, where he was scheduled to begin work in October: "No wonder that, for the first time in life, he has become bitter," Anna wrote. Late that month, he became exhausted and completely lost his appetite; he stopped working and was confined to bed. Anna and Hermann felt helpless and hopeless. In late July, Robert had a heart attack that led to his final, painful two weeks of life. Anna wrote her mother and sister of the end: "I hold him in my arms, call out, stroke him, have Hermann come over, who then breathes air into him—but no breath comes out, no sound—everything is over." On 5 August 1889, at twenty-eight years of age, Robert Julius von Helmholtz died in the Helmholtz home in Charlottenburg. Four days later, the "physics world of Berlin," Lummer said, attended the funeral to remember him and to comfort his family. Carl Runge, a young mathematical physicist who had briefly studied with Helmholtz and who had married one of du Bois Reymond's daughters, wrote Planck of Robert's funeral: "It's very sad for his parents." Planck and other close friends keenly felt that Robert's health problems and death constituted a "deep tragedy" in his parents' lives.[14] With Robert's death went the possibility that any of Helmholtz's children would carry on the name Helmholtz in science.

The pain of losing Robert brought Anna and Hermann closer than ever. Following the funeral, he went to the mountains, alone, for rest; she planned to join him when both felt ready. While he intended to continue working, he felt himself becoming "indifferent" toward work. From Pontresina, he wrote Treitschke that, although "from youth onwards [Robert] was a child of anxiety and concern for us, and although we were often enough reminded that his suffering was only dormant and not over, nonetheless the last years had gone so well that we had grasped at bold hopes." Thus his death came to his parents as "a relatively unexpected, unforeseen blow." In his final years, Robert showed great entrepreneurial spirit. "In spite of all the bitterness due to pain," Helmholtz continued, "his end has for me something of the reconciliation of a high tragedy. He fought the struggle of the mind against an ailing body, so hard as so rarely happens, without envy toward others. He made a full and rich life with friends, and is fallen in the middle of a victory." Helmholtz remained the proud father, as he conceded to Treitschke. By early September, when Anna joined him, he was suntanned and rested, looking and feeling like a new person. It was, Anna thought, his "ineffable luck that his world of thoughts and his profession" lay "far, far away from life," so he could "flee there when everything else fails." For Anna herself, the recovery, such as it was, took much longer. She wrote Ellen: "How much we talk, Papa and I, about the nature of the body and the spirit [Geist], about the immortality of both, and about being separate and united." Ellen's close friend Marie von Bunsen reported that one summer evening, while she sat at the Helmholtz family table talking with them and Helmuth von Moltke, the subject of the afterlife arose. "Helmholtz denied the possibility and stated: 'I cannot really believe in a personal existence after death.' Then, after a short pause: 'I could rather accept a form of soul migration . . . for individuals . . . who feel in themselves an impulse and a force.'" From Pontresina, Helmholtz traveled alone to the Schwarzwald, where he arranged for Fritz's course of study there, and then to the Naturforscherversammlung in Heidelberg.[15] He was back.

Edison, Hertz, and a Historic Meeting in Heidelberg

The sixty-second meeting of the Association of German Naturalists and Physicians, held in Heidelberg from 18 to 23 September 1889, was a historic one for several reasons. First, nearly two thousand individuals attended, making it one of the largest and most important in the society's history. Hertz noted that "the best physicists were there [at Heidelberg], Helmholtz, Kundt, Kohlrausch, Siemens, etc. almost everyone who has a name." So, too, were the chemists Fischer, Meyer, Baeyer, Hofmann, and Roscoe; medical men like Virchow, Adolf Strümpell, and Ernst von Bergmann; the mathematician Koenigsberger; and the physicists Boltzmann, Wiedemann, and Paalzow. It was a great opportunity to meet others of shared interests or whose papers and books they had studied. For example, Victor August Julius, a professor of mathematical physics from Utrecht and a keen student of both Helmholtz's and Koenig's acoustic work, met Helmholtz at Heidelberg for the first time. Julius cherished the opportunity. Helmholtz and Koenig met again—they had first met in Paris in 1881—when Helmholtz chaired a session in which Koenig presented his latest work on timbre and beats. Helmholtz remained unpersuaded by Koenig's results, and the issue between them over combination tones versus beat tones remained unresolved. Koenig tried to confront Helmholtz over the issue, but Helmholtz was not much interested in discussing it. Even a rainy day could be turned to one's advantage, as Hertz found when he joined in a "cozy" small group with other specialists.[16] Having meals or sitting together over drinks was important to virtually everyone.

Edison's presence also made the meeting historic. After visiting the Universal Exhibition of Paris, he went to Berlin partly to see Helmholtz. He wrote:

> I visited all the things of interest in Berlin; and then on my way home I went with Helmholtz and Siemens in a private compartment to the meeting of the German Association of Science at Heidelberg, and spent two days there. When I started from Berlin on the trip, I began to tell American stories. Siemens was very fond of these stories and would laugh immensely at them, and could see

the points and the humor, by his imagination; but Helmholtz could not see one of them. Siemens would quickly, in German, explain the point, but Helmholtz could not see it, although he understood English, which Siemens could [also] speak. Still the explanations were made in German.

For their first evening in Heidelberg, the three men (and Edison's wife) dined at the Schlosshotel on the hill. Helmholtz described Edison to Anna as "somewhat similar to a beardless Napoleon I, only a more good-natured man with very intelligent eyes." After dinner the group went down to the museum to see a demonstration of the latest version of one of Edison's recent phonographs. A crowd pressed around them as Edison's assistant explained and demonstrated it. It produced an "extraordinarily clear" sound, Helmholtz found. Among other things, they listened to "the Radetzky March, performed by a full military band, so that one could [even] discern the individual instruments." Edison promised to send Helmholtz a phonograph. In fact, he sent him five: one for himself and one for Siemens; Helmholtz distributed the other three to the Urania and the Reichsanstalt for exhibition and demonstration before the public and before scientists, respectively. At the Reichsanstalt, Joachim tested the phonograph using his violin, and various singers did likewise using their voices. Helmholtz reported all this to Edison and also told him of recent acoustic analyses by academic scientists using the phonograph (including confirmations by himself of some of his own earlier theoretical work on tones) as well as various recent technical modifications of it in Germany. "In short," Helmholtz wrote Edison, "the phonograph is absolutely and everywhere the King of the Day in Germany." Edison himself was already at work on an improved version of his phonograph, his private secretary wrote Helmholtz in reply. When it was ready, Edison would send it to him.[17]

The third (and by far intellectually most important) feature that made this meeting historic was Hertz's plenary presentation before the society. His title was "On the Relations between Light and Electricity," and it concerned his recent results on Maxwell's theory and its experimental proof in the form of electric waves. He showed that light was indeed an electromagnetic phenomenon and that light and electricity were identical, just as Maxwell had predicted all along. Hertz's theme, as he told his parents, was the unity of physics. Helmholtz himself gave an extemporaneous, opening address. But it was Hertz's hour-long presentation on his newly discovered electric waves that was easily the intellectual highlight of the meeting. Helmholtz found Hertz's lecture quite impressive. That evening he first put his own talk to paper and then joined the banquet in the museum.[18]

The fourth and final reason why the meeting was historic was that it led to organizational change. The association had always been a rather loosely run operation. A great debate now took place concerning proposed changes to the nature of the annual meeting and its organization. The reformers were led by Virchow (long the association's de facto leader); and he found his chief supporters in Helmholtz, Hofmann, Bergmann, and Wilhelm His. They wanted significant changes in the bylaws: to publish its proceedings on a regular basis; to give the association a permanent home; to make it, in general, more like the BAAS (that is, with a stronger public presence); and to tighten up its organization. The reformers prevailed, and a renewed group emerged.[19]

Montpellier

In December 1889 Anna von Helmholtz went alone to France, mostly to Cannes, where she lived for nearly four months among Europe's social elite and where she sought to lessen the pain from Robert's death. It was the longest that she and Hermann had ever been apart, and she missed him terribly. By Palm Sunday, he had joined her in Cannes. "He too suffers from a great tiredness," she found. In Antibes, Helmholtz rested and regained his equilibrium. "He needs me as a connection between himself and the external world," she wrote Ellen, "and has had to do without me for too long."[20]

The Helmholtzes went back to Berlin in early May but planned to return to France later that month. He had been designated as the University of Berlin's official representative to the forthcoming six-hundredth anniversary celebration of the University of Montpellier. He and Anna traveled together to Paris. They went up the Eiffel Tower—it had opened only a year earlier—stopping at each floor to take long looks around until they reached the top. Afterward they went to a salon in the Champs de Mars to see an exhibition of paintings by Jean-Louis-Ernst Meissonier, Carolus-Duran, and various impressionists. Helmholtz accompanied Anna to Vert Bois and then left her behind for eight days while he traveled alone to Montpellier.[21]

The celebration in Montpellier lasted four days and brought more than two hundred thousand visitors, including politicians—President Sadi Carnot was there—and military figures as well as many leading French and foreign academics. It was a mixture of science and politics. President Marie François Sadi Carnot, as Helmholtz may well have realized, was the nephew of the engineer and physicist Nicolas Léonard Sadi Carnot, one of the founding figures of thermodynamics (1824) and himself the eldest son of Lazare Carnot, a distinguished mathematician, engineer, and politician. At the railroad station in Montpellier, a student reception committee awaited Helmholtz as he got off the train. One of its members first noticed him as "a simple, elderly gentleman" who seemed to be looking for something, and he and the others quickly discussed whether this man could possibly be one of the festival's participants and whether they should approach him; most thought not. But one student nonetheless did, and Helmholtz rather apologetically asked him ("Dear Colleague") whether he could tell him where the festival's meeting place might be. The "Colleague" explained that he was a local medical student, to which Helmholtz reportedly responded in the simplest of manners: "I'm delighted to find a fellow countryman among the students here. My name is Helmholtz." As the news of his arrival quickly became known, a "warm, spontaneous ovation" broke out on the platform. Also awaiting him there were Charles-Paul-Henri Gide, a professor of political economy and member of the university's Law Faculty, and his Swiss wife; they brought him to their home, where he stayed as their guest.[22]

He felt very comfortable there, especially because his room had a balcony that looked out onto the street; he was pleasantly surprised to see not only the French, Belgian, and Italian flags waving, but the German one as well, "for the first time . . . in twenty years." He lunched that first day with several other official delegates before participating in the opening session. In the evening he dined with the zoologist Armand Sabatier of the University of Montpellier, met the university's chancellor, and also met a professor from Marseille who, like Helmholtz's grandmother, was named Sauvage. On the following day he lunched with André Prosper Crova, the professor of physics at Montpellier, and then participated in the "festive session" with President Carnot. It was held in the open atop the highest point in the city, Le Peyrou, from where Helmholtz could see both the sea and Mont Ventoux.[23]

He reported to the Prussian minister of culture about his toast at the festival. He said the festivities in Montpellier showed that in medicine (physiology), "under the influence of the proper scientific methods," the old "theoretical oppositions" between conservative and leftist had disappeared. He also said in his speech: "The unifying force of science, however, may have a much more general and comprehensive meaning. None of us can work scientifically and therefrom reach any kind of a valuable result without the advantages flowing not only to one's own nation but, soon enough, to the whole of civilized humanity as well. All nations that participate in the work of science share a common field of activity and necessarily mutually support one another. Such an understanding will ultimately prevail and must display friendly relations." He told the minister that his words were strongly applauded by all those present, that they resonated with his audience even more than he had anticipated that they would. He ended his report by declaring that medicine had "already, for quite some time, exercised this peacemaking calling of science" and that it did so "before the eyes of the world, even in moments of highly excited passion in war." He concluded his letter by declaring that science was a "unifying force."[24]

The festive procession, the celebration's final event, was held on 25 May, on Le Peyrou and under a gigantic tent. The weather was as beautiful as could be;

it was, Helmholtz said, as if a painting by Titian or Veronese were transformed into real life. The professors' multicolored academic gowns also contributed to making the scene dramatic. He thought the entire event "was enchantingly beautiful." Even the speeches were good. "On the whole, things went very well for me," he told Anna. "I've been overwhelmed with kindness; have had to swallow flatteries as never before; and have thus received the impression that I did good work politically." He left Montpellier the next day. Albert Vigié, the dean of the Law Faculty, sent thanks (in the name of the entire faculty) to Helmholtz for his visit. Louis Lortet, the professor of zoology at Lyon, also wrote afterward to say that he had had to leave Montpellier early and so regretted that he could not see Helmholtz at the Medical Faculty banquet. Had he been there, he would have said to him, in the Lyon faculty's name, "In what great esteem we hold [you] and your person and your remarkable works." He hoped that "the efforts of men of science" would have "more success than those of politicians who direct the destinies of humanity!"[25] He was not alone in his sentiments.

Unreformed Prussian Schools

That autumn, Helmholtz faced the thorny issue of evaluating the Prussian school system. Prussia's leaders had always sought to limit access to the educational system as a whole. Only the relatively few, select individuals who attended a Gymnasium had any chance of taking a state civil service examination for the middle ranks, and only those who attended a university could take it for the upper ranks. Thinkers like Hegel and doers like Bismarck expected education to serve the state. By December 1890, when the first Prussian school conference was convened (and after Bismarck had been dismissed the previous March), there was growing concern about the nature of the Prussian school system and about its producing a rising academic proletariat.[26]

Helmholtz's opinions on education and the role of science in it were valued by several European leaders in education and research. Victor Duruy, the minister of public instruction in France, and Ruggiero Bonghi, the Italian minister of instruction and first director of the Milan Polytechnic, had long ago sought out his opinions. In 1885 Tyndall wrote Helmholtz to say that he knew he had "somewhere" published his "opinion as to the comparative merits of the culture to be derived from science and . . . language," and he asked Helmholtz to remind him where he had written on that issue. Helmholtz referred him to his address of 1862 at Heidelberg on the relations of the sciences. In the intervening twenty-eight years, his opinions remained unchanged. He was unimpressed with the performance of the Realgymnasia (which emphasized modern science and deemphasized classical languages). The classical Gymnasia pupils who later attended his student laboratories were, he thought, better than those from the

Realgymnasia, though he admitted to Tyndall that the latter had better preparation in mathematics and physics than the former. He found it difficult to get the graduates of the Realgymnasia to think independently. He said, expressing special concern, "Members of those classes who have to represent the nation's intellectual and political interests remain almost completely unacquainted with the type of thinking [natural science] that has had the greatest success and developed the surest results, and with the foundations of our natural knowledge, on which the power of civilized peoples essentially rests."[27]

Prussia's leaders were also interested in Helmholtz's opinions on education. The immediate occasion for hearing them emerged from William II's effort to reform the secondary schools. Calls for reform of Prussia's secondary schools reached back to the 1870s and were supported by several groups. For the emperor, however, reform meant strengthening the schools' nationalist and patriotic spirit and fighting the growing socialist challenge. He especially wanted to see more German (and less Latin) taught, and he wanted more modern German (and Prussian) history in the Gymnasia, including the role of the Prussian monarchy. He and like-minded others were worried about the rise of an academic "proletariat," not to mention hostile journalists. Under the guidance of Gustav von Gossler, an advisory conference was convened in December 1890 at the Ministry of Culture. The emperor and the ministry together set a series of specific questions that they asked a carefully selected set of individuals to address. Helmholtz, as one of forty-four members of the School Conference Commission, testified before it on the preparation of secondary-school students for studying natural science at the tertiary level. He and Virchow were the only natural scientists on the commission (Koch declined to participate). A large majority of the commission members favored protecting the status quo (that is, privileges) of the Gymnasium and preventing the Realgymnasia students from gaining the right to enter the universities. The commission was anything but a modernizing, reformist group of advisers.[28]

In his testimony Helmholtz argued that the best students required the best schools. He asserted that "the best means . . . , in order to impart the best intellectual development," was the study of classical language. Greek especially prepared them for the study of natural science and gave them "the fine formation of taste." He noted that he had taught physics for seventeen years; he had examined—this was probably hyperbole—some 400 to 600 students annually and seen many of them in the laboratory as well. His students came from many different nations: roughly half were German and half foreign. Thus he could "make rather good comparisons on the intellectual direction and ability of differently instructed pupils." Roughly one-third of his students were medical students with good mathematical preparation. Another one-third were in need of better preparation in mathematics, physics, and the other natural sciences. Yet classical studies were, he thought, the most valuable studies of all, and it would be better to sacrifice other areas rather than restrict these. While students from the Realgymnasia were better prepared in mathematics and science than those from the classical Gymnasia in those fields, and therefore found the first two semesters of study easier than their counterparts did, by the end of the second year of study those from the classical Gymnasia were not only the equals of their Realgymnasia counterparts; they were also able to work more independently. He conceded that of course there were exceptions, and for those of sufficient talent, it mattered little which type of Gymnasium they attended. He thought it unnecessary to change current mathematical studies at the Gymnasium.[29]

He also believed that the most important thing that the Gymnasium gave was "a humanistic education . . . , that people learn to love the classical sense of meaning [*Sinnesweise*], authors, and works of art." But he also warned that German students needed to improve their mastery of German. When students in his laboratory handed in written reports, he said, in some cases he had to return these—sometimes three or four times—until the student wrote them properly, reported facts accurately, and drew the appropriate conclusions. Indeed, he thought that his own professional colleagues often lacked sufficient writing, reporting, and logical skills. Nor did he hold himself as being above criticism: "I know from my own experience how much effort it has later cost me as I learned to help myself to the necessity of rigorous expression, how often I've rewritten my manuscript until I was fully satisfied. I've let it [a manuscript] lie there again and again, and when it became completely foreign to me, I've read it again until I could no longer find any error in it." He noted, too, that when he had attended Gymnasium, there was a greater freedom to choose which classical authors one preferred. The main problem at the Gymnasia, he thought, was not that students were overburdened with work but that there were too many of them who did not belong in a Gymnasium in the first place. In the study of languages, both classical and modern, he thought that the elite secondary school system in England and France had now equaled and perhaps surpassed the German system. Some of his English, and even American, students were more orderly and precise in their laboratory reports than their German counterparts.[30]

Helmholtz also wanted to see more discussion of mathematical theorems and their explanation in German. "I know from my own experience," he testified, "that when I've worked out a new mathematical proposition I never feel certain [about it] until I've written it down. One obtains real control of the correctness of the mathematical reasoning only when one has put everything down on paper." Though "a feeling for language" certainly required instruction in language, it also required a certain understanding of nature—both factually and methodologically. The shortcoming of language instruction for learning modern natural science, he maintained, was that it was artificial and always contained exceptions; science, by contrast, was built on the results of nature—the laws of natural science allow no exceptions, he here maintained. It was important that students of the human sciences at least understood how the natural sciences worked. "How often has it happened to me," he bemoaned, "that I've tried to explain to some man who is instructed in literature what appeared to me to be a relatively simple natural-scientific proposition, and that I then realized that the man doesn't understand me at all. I have no starting point with him whatsoever, and it's not merely a matter of ignorance of the material but rather that such men have not learned to recognize the facts as such in their unconditioned trustworthiness and to reckon with them." He thought that until sometime shortly after 1800, many educated individuals—and this point also held for the current "more uneducated classes"—suffered "very much indeed from a false rationalism": they failed to understand the difference between mere word games and the real understanding of the causes of natural phenomena. Therefore, education had to "open up the understanding for the application of these intellectual forces and methods that have led to enormously important factual results, whose importance now rules our entire life." This meant emphasizing mathematics education. These were the main points, he thought, that the German Gymnasia needed to be concerned with.[31] His opinions on secondary-school education were, in short, the same as the ones he had stated in his Heidelberg address nearly three decades earlier.

Had it not been for the emperor's personal advocacy and (widely resented) intervention, few might have paid any attention to this advisory conference at all. Among other things, the ministry decided to eliminate the Realgymnasia, but its decision was opposed by both liberals and conservatives in the Reichstag. The University of Berlin, among others, sent in a petition containing sixty-nine signatures (including those of Dilthey, Helmholtz, Mommsen, Schmoller, Treitschke, and Virchow) opposing the decision. They thought it grossly favored the churches as opposed to "enlightened" Germany. In the end several new ministerial regulations (including less Latin and more German) were introduced; Gossler resigned; and William and his new chancellor, Leo von Caprivi, momentarily abandoned "reforming" efforts.[32]

The Endowment, the Medal, and the Bust

In June 1891 Helmholtz and Mommsen unofficially organized a committee to celebrate the twentieth anniversary of the Franco-Prussian War and the founding of the Reich. Helmholtz suggested they ask King Albert of Saxony, General Julius von Verdy du Vernois, the German minister of war, and Bismarck, "who obviously knows more about the most intrinsic nature of war than all the generals taken together," to join them. Helmholtz had a better impression of

Bismarck than the latter had of him: the anti-intellectual Bismarck could not even recall Helmholtz's name, though he proceeded to lump him into a group of three professors—Virchow and du Bois–Reymond being the others—whom he cavalierly described as know-it-alls even as they were unable to speak about the causes of things.[33] Helmholtz's organizing effort went no further than Bismarck's knowledge of science.

That year also marked Helmholtz's seventieth birthday. Several colleagues and friends wanted to celebrate it and memorialize his scientific and cultural achievements by establishing an endowment in his name (the Helmholtz *Stiftung*, or Endowment), by creating and awarding a Helmholtz Medal, and by commissioning a marble bust. To finance the creation of the Endowment, the medal, and the bust, they formed a central committee consisting of 172 members, which in turn was led by a smaller committee consisting of du Bois–Reymond, Leopold Kronecker, Kundt, Ernst von Mendelssohn-Bartholdy, and Zeller. They asked scientists and others for their financial support and goodwill. Though the principal contributors were German, others also made donations; in all, about seven hundred individuals or organizations made contributions.[34]

The Academy of Sciences agreed to formally accept and administer the endowment, which had an initial capitalization of forty-eight thousand marks. Its statutes were drawn up by the two academy secretaries and Helmholtz and Kronecker. Its purpose was to recognize outstanding scientists in the fields covered by the academy's Physical-Mathematical Class as well as in epistemology. The award came in the form of a quite substantial (620-gram) gold medal that had Helmholtz's picture on one side and the name of the awardee and pertinent information concerning him on the other. The plan was to make one award every second year, starting in 1898.[35]

The medal was thus intended to honor outstanding researchers in the scientific fields in which Helmholtz himself had been active. Apart from helping to draw up the statutes, Helmholtz also recommended who should receive the first four medals. These were presented in 1892 to du Bois–Reymond (physiology), Weierstrass (mathematics), Bunsen (chemistry), and Thomson (physics), that is, to his closest living scientific colleagues and friends. It was planned further that these four individuals would constitute the basis of a collegium that would select future awardees.[36] Rewarding Helmholtz's friends and individuals whose scientific accomplishments lay years if not decades in the past was not an auspicious beginning. Although the first medal was awarded nearly nine years before the first Nobel Prize (1901), it never became anywhere nearly as significant as the latter.

The marble bust was the work of Adolf von Hildebrand, who made busts or reliefs of many of the leading economic, political, and social figures (or their wives) of Imperial Germany. Helmholtz had long held a high opinion of Hildebrand's work, and in early 1891, Hildebrand told a friend, "It's worth a lot to me" to do it and that he welcomed this "beautiful opportunity to come closer to this man." Their relationship went well from the start. The mayor of Florence, where Helmholtz did his first sittings, invited Helmholtz, Hildebrand, and other savants and artists to tour the Palazzo vecchio. Hildebrand wrote his friend Conrad Fiedler about Helmholtz: "I could understand him very well, though unfortunately I could not speak with him much more after Frau Helmholtz arrived." She limited Helmholtz's discussions with Hildebrand, perhaps worried that Helmholtz might find the sessions too draining. Still, Hildebrand much valued his discussions with him.[37]

While Hildebrand worked on the bust, the Helmholtzes vacationed in Italy and southern Austria. On 31 August 1891, while in Madonna di Campigno (in the Dolomites), the family celebrated his seventieth birthday, though he confessed to Ludwig, whom he called his "chief authority" in matters of physiology, that the family party "failed completely." Helmholtz received an enormous number of birthday greetings: some 190 telegrams arrived, along with innumerable letters and gifts. Even so, certain family members became irritated when the new Prussian minister of culture, Robert Graf von Zedlitz, did not appear and also forgot to inform the emperor of the event. By contrast, the King of Sweden knew of it, and he awarded Helmholtz the Order of the Large Cross of the Polar Star. Kundt and Hertz each wrote newspaper articles extolling Helmholtz's person and his scientific, medical, and other achievements for the general public. The *Electrician* magazine of London had written Helmholtz previously to say that it wanted to carry a biographical notice (with portrait) in its August issue "in connection with the celebration" of his seventieth birthday. Hugo Kronecker wrote the notice (read: encomium).[38]

Even before these birthday greetings had arrived, Helmholtz felt overwhelmed by his correspondence and the demands on his time. Many of these, he told the historian Alfred Dove (the son of Heinrich Wilhelm Dove and editor of the *Allgemeine Zeitung*) were of no consequence or were outright "silly letters" that could wait for a response. "But the time and strength of a person is finite," he continued, "and the more one has been known to have so-called luck, the more one is forced onto the defensive, whereby the considerate and gladly seen friends must ultimately suffer still more than the intruders and fools."[39]

In September he wrote Ludwig, reporting that for the past two weeks he had been sitting for Hildebrand, his health was good, and he was prepared for his upcoming birthday celebration in Berlin in November. He had "somewhat mixed joy" about this event, which he did not imagine to be festive at all. He wondered whether all the praise he was receiving was justified: "Apart from all the questions of vanity, there is finally, for us, for one who has worked hard throughout his life, the justified question: Is what you did useful and estimable? Only others who have the use and advantage from them [these works] can answer this question."[40]

The Helmholtzes arrived back in Berlin on a rainy late September evening. Within two weeks, Anna's "friend Lenbach" had cleaned up and rearranged her "green room," which for the time being she devoted to displaying Hildebrand's bust of Helmholtz. There would be no party on 2 November, she declared; instead, Helmholtz would merely receive congratulations during the day and participate in a celebratory meal at "Der Kaiserhof" hotel in the evening. Since this would not include any women, she was planning to have a reception and "perhaps also a *dîner*, and that'll be enough of this gruesome game." During November and December the Helmholtzes displayed the bust in a Berlin exhibitor's gallery. One Hildebrand expert judged it monumental in character. Marie von Bunsen thought it Hildebrand's "masterpiece."[41] See figure 26.1.

Making Publicity: The Interview

As part of the publicity surrounding his seventieth birthday, Helmholtz granted an interview to Hanns von Zobeltitz, editor of *Daheim: Ein deutsches Familienblatt mit Illustrationen*. In a previous issue, the magazine had reviewed Helmholtz's scientific work; it now wanted to report on his personal and daily life.[42]

When Zobeltitz appeared for the early-morning interview, he was received by a house attendant and was brought to Anna von Helmholtz in her salon. He learned from her that Helmholtz rose early and that the morning hours were "his best working time." He took his first break around 1:00 p.m.; this lasted until about 4:00 p.m. Then he returned to work until the evening, which was generally devoted to relaxation and family. Zobeltitz also learned that Helmholtz went to the theater regularly, played a good deal of music, and often read the German classics. His favorite authors, Anna further related, were Shakespeare and Goethe. "He is in no way against or unacquainted with our modern literature," however, she pointed out. "Only it must not be exactly Ibsen." He honored Wagner: the two men had discussed music together, and she and Helmholtz had felt close to him. Two Lenbach portraits of Helmholtz hung on her salon's walls. (She said Lenbach liked Helmholtz's head.) Another portrait of him (fig. 26.2) was done in Florence in 1891. She also said her husband was normally against interviews but that he was making an exception for *Daheim*. She then brought Zobeltitz into Helmholtz's study for the interview. He found it to be a moderately large but "simple working room." It contained well-ordered bookshelves on two walls and a tall, large, two-sided oak desk in the middle.[43] See figure 26.3.

Zobeltitz had interviewed Helmholtz a decade earlier for *Daheim*; he was astonished to find that his appearance had changed little: his posture was still quite erect, and his face had a fresh look: "the same sharp-looking, penetrating

Fig. 26.1 Marble bust of Helmholtz, 1891, by Adolf von Hildebrand. Physikalisch-Technische Bundesanstalt, Braunschweig and Berlin.

Fig. 26.2 Helmholtz in 1891 (in Florence). Deutsches Museum, Munich, Nr. 22197.

Das Studierzimmer.

Fig. 26.3 Helmholtz's study in his residence at the Reichsanstalt. *Vom Fels zum Meer* 14 (October 1894–March 1895): 118. Courtesy of the University of Chicago Library.

pair of eyes" and "the same benevolent smile on the lips, which never utter anything but a word of warm recognition." The face was crowned by a head full of white hair.[44]

Their conversation developed naturally, and so Zobeltitz jettisoned the long list of prepared questions he had intended to ask. They spoke first of Helmholtz's youth and early professional years. Helmholtz still vividly remembered how, some forty-four years earlier, he had sent his essay on the conservation of force to Poggendorf's *Annalen* and that it had been rejected: "It happened to me as to other young authors; it was simply rejected." Zobeltitz also reported on Helmholtz's invention of the ophthalmoscope, that he had sent it to old Jüngken, who was, Helmholtz said, "at that time, [Berlin's] best-known ophthalmologist." He could make neither heads nor tails of it "and sent it back . . . rather peevishly." It took the skilled hands and open-mindedness of young Graefe, Helmholtz said, to appreciate the instrument's use and see its possibilities.[45]

Zobeltitz also asked Helmholtz about the relationship between his scientific study of physiological acoustics and his love of music. "'This connection is most definite,' the Herr Geheimrat confirmed." It was as a youth, and despite his piano teacher, that he came to love music. During his student years he came to know the work of Gluck well, "and moreover his *Armide* inspired" Helmholtz so that it "forced" him back to the piano. He saw Gluck as "the genuine forerunner of Richard Wagner." He also tried playing several other instruments, for reasons both scientific and aesthetic. The theater was also important to him as "the surest and most pleasant way" to free himself from his thoughts after working hard, "something . . . otherwise often very hard to do."[46]

After reviewing the Heidelberg years, the conversation turned, finally, to Helmholtz's efforts at popularization. Zobeltitz said he had always greatly admired Helmholtz's essays for bringing scientific understanding, "to make it [the latter] the permanent property of the education of our people, and, what is of perhaps still greater value, to awaken love of the natural sciences and understanding for its study in broader groups of people." Helmholtz responded:

> My [popular] lectures have given me a lot of pleasure! . . . The difficulty of being popular in a good sense essentially lies in the complex composition of the [listeners in the] auditorium. One never knows exactly what foundations one may assume as being known. Hence they [the lectures] require extreme care in preparation, time-robbing work, and this is indeed one of the reasons why so few researchers know how to speak before an auditorium of lay persons. Our time is indeed a costly thing. Moreover, and on the other hand, the success of the spoken word in such a lecture is extremely doubtful—the same lecture has a different, better, and more lasting effect in printed form.[47]

He was sensitive to both listener and reader.

The interview ended, and Zobeltitz re-joined Anna in her salon, where they talked briefly about the forthcoming celebration of Helmholtz's seventieth birthday and the Helmholtz Endowment. Anna mentioned all this "with touching modesty yet at the same time with the visible joy of a woman who is proud of her spouse." She showed him the collection of medals Helmholtz had been awarded during the course of his career, above all the Graefe Medal, and she noted that in connection with the new endowment, a Helmholtz Medal was being created.[48] The precelebration publicity was over, and the public celebration now began.

The Seventieth Birthday's Public Celebration

Since Helmholtz (and most others) would be vacationing at the time of his actual birthday (31 August), the official celebration was postponed until 2 November, the day on which he had received his medical degree forty-nine years earlier and a time when he and his fellow academics would be back at work in Berlin. In the month or so leading up to and following the celebration, Helmholtz (again) received dozens of congratulatory addresses and statements from individuals and institutions as well as honorary memberships in twenty medical, musical, scientific, and technological societies or universities. The honors came not only from institutions around Germany but also from Belgium, Italy, the Netherlands, Congress Poland, Russia, Sweden, Switzerland, and elsewhere beyond the Reich. He was named an honorary citizen of his native Potsdam and received an honorary doctorate from the University of Utrecht. The celebration was an international event. He had so many persons and organizations to thank for their well-wishes that, in some instances, he sent a mere form letter with his accompanying signature. His fellow ophthalmologists produced a *Festschrift* in his honor. There was not only the interview granted to Zobeltitz for *Daheim*; in addition, Emil Schiff, a longtime Berlin correspondent of Vienna's *Neue Freie Presse*, published a lengthy celebratory essay on Helmholtz and his work in the *Deutsche Rundschau*.[49] Virtually everyone in the German cultural world knew of this event. There was no lack of publicity.

For all its pleasures, 2 November 1891 was a long and very public day for Helmholtz. It began with a reception in his home for a series of civil servants, corporate (organizational) representatives, friends, and students. Some seventy people paid their respects and showed their friendship, including the minister of culture and deputations from the university (led by the rector), the Prussian military-medical training institute (the former Friedrich-Wilhelms-Institut), and the Reichsanstalt, as well as various former students, including Hertz and Kronecker. In the evening there was a dinner for 260 guests at the hotel Kaiserhof. There were three dozen toasts, from the Prussian ministers of the interior (Heinrich von Boetticher) and of culture (Zedlitz) on down to representatives of a wide variety of domestic and foreign universities and scientific, medical, and technological organizations. Helmholtz's closest collaborators and colleagues were also in attendance, including du Bois–Reymond, Zeller, Siemens, Kundt, Bezold, Mommsen, Treitschke, Sybel, Virchow, Hofmann, and Pictet, among the Berliners; and Ludwig, Wiedemann, Ostwald, Hertz, Exner, Boltzmann, Kronecker, and Blaserna, among the non-Berliners. His artist friends the violinist Joachim and the painter Menzel also attended. Those present, reported the *New York Times*, constituted "a unique gathering of leaders in German science and philosophy." Berlin and Viennese newspapers also took notice. Helmholtz's picture was made the frontispiece of the current issue of the *Annalen*. A foreigner traveling along the Rhine read about the celebration in German newspapers and said that the press reported every detail and devoted several articles to Helmholtz. One newspaper satirized the celebration, thereby making a political point, but even here Helmholtz was treated respectfully, for the event showed that the German people "had their eyes fixed on the rejoicing in Berlin."[50]

Various European states also honored Helmholtz on his seventieth birthday and Prussia awarded him the title "Real Privy Counselor with the Predicate, Excellency." That title put him in place nineteen (of sixty-two) in the official ranking of Prusso-German court society. He now outranked all other academic officials at court and even most governmental and military figures; it made him (and Anna) officially *hoffähig* (presentable at the royal court). He was touched and pleased by this. He was also awarded the Large Cross of the Order of Zähringer Lions of Baden; the Austro-Hungarian Decoration of Honor for Art and Science; and the Great Cord of the Orders of Saints Maurizio and Lazzaro by the King of Italy, Umberto I. In thanking du Bois–Reymond and the academy for establishing the endowment, Helmholtz said they had permanently linked his name "to the advancement of science," which was "the highest honor that could be shown . . . by men of science." "Science," he claimed, had "become the only uniting tie of modern humanity that preaches peace unconditionally." He further maintained that the scientist did not work for his own well-being but rather "for the well-being of his people; he works for the well-being of all humanity," a claim that in part flew in the face of his own personal career, as Helmholtz himself soon conceded (see below).[51]

Helmholtz's speech before the attendees became, in somewhat changed form, his brief written autobiography. There, too, he claimed the high road for science: "Today science and art are the only remaining ties of peace among civilized nations. Their ever higher, growing cultivation is a common task of all, one that in the common work of all strives after the common advantage of all. A great and holy work!" But that high road was also a shared and a rocky one, he confessed. He pointed out that in his own work he had used methods that others before him had used and that pure luck had sometimes played a role in his getting good results. He especially recognized that his work had developed in stages and that along the way he had taken false paths; the public saw only the finished product. He noted his own falterings as a scientist: "I had to compare myself to a mountain climber who, without knowing the way, slowly and laboriously climbs up, often having to turn around because he can go no further, who then soon discovers, through reflection or through chance, new traces leading him forward a bit, and finally, when he has reached his goal, finds, to his shame, a royal way on which he might have gone upward if he had been clever enough to find the right beginning." His readers, however, saw none of this in the final products of his research efforts. He reflected that, like some other scientists or artists, his achievements doubtless came partly from "a favored nature." Moreover, he preferred to work in fields where one had a reasonable chance of success. But then he mused further on the sources and circumstances of (his) scientific ideas and creativity:

Because I rather often came into the uncomfortable position of having to wait for favorable occurrences to happen, I have, if and when they [ideas] came to me, had some experience that might be useful to others. They often quietly enough sneak into one's thoughts without one at first recognizing their meaning. Later, only an accidental circumstance sometimes helps to recognize when and under which conditions they come. They are otherwise there, without one knowing from where [they come]. In other cases, however, they suddenly appear, without effort, like an inspiration. As far as concerns my own experience, they never came to a tired brain or while at my desk. I always first had to turn my problem around in a lot of different ways and from every possible angle, such that I looked at all its twists and complexities in my head and let them run through freely, without writing. To get things that far usually isn't possible without much preliminary work. Then, after the tiredness from that effort had passed, there had to occur an hour of complete bodily relaxation and quiet well-being before the good occurrences [that is, ideas] came. They were often there, as in the cited verses of Goethe, in the morning at awakening. . . . They also came especially readily, as I already reported in Heidelberg, during easy climbing over woody mountains in sunny weather. They seemed however to be driven off by the smallest quantity of alcoholic drink.

Of course, these were the good times. The bad times, he acknowledged, sometimes lasted weeks or even months.[52]

He then turned to another of his standing intellectual concerns: epistemology. Not only must a scientist investigate and understand the limits of the instruments with which he is working; it was also important to understand the nature of thought processes and their limitations. He considered this to be a *factual*, indeed an objective matter. "My essential result was that the sensations of sense are only signs for the character of the external world, whose meaning must be learned through experience." His interests in epistemological issues, he said, were first awakened in his youth, by his father, "who had been deeply impressed with Fichte's Idealism, and who I often heard argue with colleagues who honored Kant or Hegel." Yet only rarely did he come away satisfied from his epistemological "investigations": for every friend he made in this regard, he also made perhaps ten enemies. "Namely, I've always thereby brought out [against me] all the metaphysicians, even the materialists, and people of hidden metaphysical tendencies." Too often he had seen learned men develop a ruinous "megalomania," and he sought to protect himself against this by not becoming like his enemy. "I knew that sharp self-criticism of my own works and activities is the protective safeguard against this misfortune. One needs, however, only to keep one's eyes open for what others can do and what one cannot do oneself. Then, I find, the danger is not great." He considered all his work as being open to improvement.[53]

But now, in contradiction to what he had said at the outset of his talk, Helmholtz suddenly conceded that it had *not* been "the well-being of humanity" that had "from the beginning" motivated him in his work; instead, it was his scientific curiosity.

> In truth, it was the special form of my curiosity which drove me forward and determined the use of all my remaining time to do scientific work. Both these conditions demanded, moreover, no essential deviation from the goals I strove towards. My position required me to hold university lectures; my family, to solidify my reputation as a researcher. The state, which supported me and provided the auxiliary scientific means and a good deal of free time, thereby had a right to demand my consideration that I communicate, in an appropriate form, freely and completely to my fellow citizens, everything that I discovered with its support.

He always found it laborious to write up his results; he often had to write four to six drafts of a manuscript before it was ready for submission, that is, ready not only in terms of style but also in terms of completeness and logic. In preparing a manuscript, he constantly asked himself what his friendly colleagues would think of his work. "They hovered before me as the embodiment of the scientific mind of an ideal humanity and gave me the standard."[54]

To be sure, he did not mean that, before the midpoint of his career, when he still had to work for his "position in the world, higher ethical reasons might not have contributed along with curiosity" and his "sense of duty as a state official. However, it was harder to become certain of their real existence so long as egotistical motives directed the work." He thought this was true for "most researchers." After one had achieved a secure, tenured position, however, there was a division between those who had no (or nor longer had any) "inward drive to science," and so stopped doing research, and those who put "in the foreground a higher understanding of their relationship to humanity." These latter now began to see, he maintained, how their thoughts circulated within the scientific community, especially among the younger generation, how they took on a life of their own and were developed further by others. This was very rewarding for the individual researcher, and he felt "a sort of fatherly love" toward his thoughts and their younger carriers.[55]

The entire matter transcended the individual researcher and the fate of his thoughts. He maintained: "At the same time, however, there comes before him the entire world of thought of civilized humanity as an on-going and ever-developing whole, one whose lifetime, relative to that brief one of the individual, seems eternal. He sees himself, with his small contributions to the building up of science, in the service of an eternal, holy cause, one with which he is bound by close ties of love. His work itself thereby becomes holy to him." All this, Helmholtz added, was perhaps easy enough to understand in theory; but only those who had experienced it could appreciate it fully.[56] Here his allusive account of the large-scale, long-term, and social structure of science made it sound almost like a religious experience.

In Helmholtz's opinion, most people did not believe in such "ideal motives," but instead thought of such feelings as a desire for fame or glory. The way to decide between these two sentiments, he suggested, was to ask oneself whether or not one thought the research results one obtained were his own or in part due to others. He felt fortunate to receive thanks for doing what interested him. "I have found among you [the scientific community] the standard of intellectual abilities of men, and through their involvement in my work they have awakened in me the living apprehension of the common life of the intellectual world of ideal humanity, which puts before me and into a higher light the value of my efforts. Under these conditions I can only consider the thanks which you want to bring before me as a free gift of love, given without asking for a reciprocal gift and without obligation."[57]

In Vienna, the *Neue Freie Presse* devoted three long columns to Helmholtz's autobiographical remarks at the celebration, briefly noting some of those present and calling Helmholtz "a genius" and one whose remarks gave insight into his world and his creative ways. In Britain, the chemist Edmund Atkinson, one of Helmholtz's principal English-language translators, thought the celebration had "for its universality or importance, . . . , perhaps, never fallen to the lot of a Man of Science during his lifetime." Siemens put the point similarly, if more personally and politically: "We were, perhaps fortunately, finished with the Helmholtz celebration! It's ultimately very gratifying that a savant, and not only a soldier or an official, has been placed on the coat of arms."[58]

The *Kommers* and Mark Twain

If Siemens had finished celebrating, Helmholtz had not. It was arranged to hold a joint student celebration for both Helmholtz's and Virchow's seventieth birthdays. On the evening of 7 November 1891, a great *Kommers* (student beer fest) was held in the large reception hall of a brewery in Friedrichshain, a working-class district. At least two thousand students were there. Both Helmholtz and Virchow spoke, as did (for the faculty) Foerster. Foerster declared that Helmholtz and Virchow had "maintained the intellectual strength of our people, raised the renown and the regard of our Fatherland among the other nations of our Earth, and . . . powerfully helped to raise them in a way perhaps still deeper and more lasting than can happen through the most magnificent political and military successes."[59] The message was clear: science could do it better.

Also present that evening was an unexpected guest, the American humorist Mark Twain (Samuel Langhorne Clemens). Twain was spending the winter of 1891–92 in Berlin, where he was socially well connected; his cousin Alice Bryan Clemens of St. Louis had married a German officer, Maximilian von Versen, who rose to become a full general and belonged to the emperor's court circle. Hence Twain was invited to many diplomatic events—even to a special meeting with the emperor—and became generally well known and popular in the capital. He thought Berlin was "a luminous center of intelligence—a place where the last possibilities of attainment in all the sciences" were to be had "for the seeking. . . . They teach everything here," he recorded in his notebook. "I don't believe there is anything in the whole earth that you can't learn in Berlin except the German language. It is a desperate language."[60]

From Berlin, Twain wrote a series of letters for the *New York Sun* and the McClure newspaper syndicate. In the sixth of these, "The German Chicago," he noted the huge seventieth-birthday celebrations for Virchow (in mid-October) and Helmholtz. Then came the student *Kommers*, meant to honor both men jointly. It was held in a huge hall "beautifully decorated with clustered flags and various ornamental devices, and was brilliantly lighted." Tables (each seating twenty-four) were set up throughout the hall's spacious floor. There were a half dozen tables reserved for some one hundred notable professors; at the center of these sat "the two heroes of the occasion and twenty particularly eminent professors of the Berlin University." Twain himself was honored with a seat at the Helmholtz-Virchow table. He said, "I was not really learned enough to deserve it. Indeed, there was a pleasant strangeness in being in such company; to be thus associated with twenty-three men who forget more every day than I ever knew." The whole place "was a stupendous beehive." The students' "intent and worshiping eyes were centered upon one spot—the place where Virchow and Helmholtz sat. The boys seemed lost to everything, unconscious of their own existence; they devoured these two intellectual giants with their eyes, they feasted upon them, and the worship that was in their hearts shone in their faces." The band played martial music and the students rose. At the call "One—two—three!" all glasses were drained and then brought down with a slam on the tables in unison. The result was as good an imitation of thunder as I have ever heard." The next hour was filled with singing and with the students simultaneously drawing their swords aloft, then slamming them twice on their tables, and then replacing them in their scabbards. The evening drew to a close with speeches by two of the students and replies by Virchow and Helmholtz.[61]

The seventieth-birthday festivities were finally over. In December Helmholtz became ill with the flu. Mimi von Spitzemberg, who dined at the Helmholtz's home during his three-week illness, noted that he and Pictet "held extremely interesting talks, tied up with Pictet's lecture" on the previous Saturday "about the soul as a physical concept and a motor." In the meantime, Twain, who had remained in Berlin, moved into the Hotel Royal, on Unter den Linden. In mid-January he too got the flu and suffered with it for several weeks. By mid-February Twain and Helmholtz had both sufficiently recovered from their respective flus that Helmholtz could at least call upon Twain in his hotel for a private meeting.[62] German science met American humor.

27

Science, Art, and Standards Business

A Free Pass for Goethe in Weimar

In April 1892 the Helmholtzes vacationed in southern Austria and then in Abbazia (modern Opatija), a posh seaside spa area in Croatia favored by Austrian royalty and aristocrats. As always, Helmholtz brought along a research notebook. "Papa is meditating on mathematics," Anna wrote Ellen. Sigmund Exner, who had recently succeeded Brücke in Vienna, joined them for three days. Then they went to Fiume (modern Rijeka) and elsewhere in Istria; they were "astounded" by the mixture of languages and peoples.[1]

In May the chemist Hofmann died, and so the Helmholtzes lost part of their social circle and Helmholtz lost a close colleague. Emil Fischer was called as Hofmann's successor. Against Berlin's advantages—an intellectually stimulating scientific life, excellent material resources, and large numbers of students—stood its disadvantages, as Kundt, Fischer's former teacher, explained to him: "Well Fischer, you'll be amazed at the load of work that's piled on a professor here," a point that Fischer soon enough came to appreciate. After deciding for Berlin, Fischer and his wife visited there to find a place to live; Anna used the occasion to invite them to one of her evening parties. Fischer was im-

675

pressed by the Helmholtzes' "magnificent official residence" and by the company. (He was especially pleased to meet old Siemens.) Helmholtz gave a short speech welcoming the Fischers to Berlin and into their set. In July there was a very large memorial ceremony for Hofmann in the newly opened Philharmonie. Virtually the entire university faculty, including Helmholtz, attended. The empress dowager ordered that her own monument in the Tiergarten be flanked by two others: on one side, Hofmann; on the other, Helmholtz.[2] It did not happen, however.

In the summer of 1891, the Grand Duke of Saxony-Weimar-Eisenach and Carl Ruland, the managing director of the Goethe Society (Goethe-Gesellschaft) in Weimar, visited the International Electrotechnical Exhibition in Frankfurt, where Helmholtz was the honorary chairman of the Electrotechnical Testing Commission. They talked with him about Goethe's attitude toward modern science. The conversation led the grand duke to invite Helmholtz to speak on Goethe before the society. Helmholtz agreed, a lecture was set for 11 June 1892, and he spent the spring drafting a presentation on Goethe and science.[3]

The Goethe Society was relatively new, having been founded in 1885 to maintain Goethe's cultural heritage and promote research on him. It published the Weimar edition of *Goethes Werke* and the *Goethe-Jahrbuch* and was intimately associated with the Goethe Archive, the Goethe Library, and the Goethe Museum, all located in Weimar. It financed research projects on Goethe and promoted related studies. Its "protector" was the grand duke, and its patroness was Grand Duchess Sophie von Weimar, but it also counted a number of leading *salonnières* from Berlin among its other patronesses, and in general the German social elite stood behind it. Eduard von Simson, Helmholtz's old acquaintance from Königsberg and now president of the German Imperial Court, was its president. By 1892 it had nearly three thousand members and was becoming Germany's largest and perhaps most influential literary society. (In 1899 the manuscripts of Schiller and of other German "classical" authors were also housed in the renamed Goethe- und Schiller-Archiv in Weimar.) Favored by the feudal-aristocratic elite, it also promoted, and was promoted by, German cultural nationalism.[4]

Erich Schmidt, a member of the society's board, had suggested to Helmholtz that he publish his society talk—"On Goethe's Presentiments of Coming Natural Scientific Ideas"—in the *Deutsche Rundschau*. Schmidt's suggestion was at the behest of the periodical's editor, Julius Rodenberg. Helmholtz already had other such offers, and so he stated a series of conditions, including the right to republish it in his *Vorträge und Reden*. Rodenberg wanted the manuscript before Helmholtz delivered his talk, but Helmholtz did not finish it until that very day, and he had already granted the society the right to publish it in its yearbook. Nonetheless, Rodenberg accepted Helmholtz's conditions for publication in the

Deutsche Rundschau, which included an honorarium of five hundred marks, and the lecture was published there as well. Its publication in the *Deutsche Rundschau* suggests that interest in Goethe as a scientist and on the part of scientists was still alive.[5]

The Helmholtzes were invited to the traditional *dîner* given by the grand duke and duchess in the Schloss Belvedere on the evening before the lecture. The grand duke, brother of Empress Augusta, was "filled with respect for intellectual greatness, wherever it is to be found," Anna wrote Ellen. The attendees at Helmholtz's talk were a combination of aristocrats and the well-to-do; with a few exceptions, they were not university professors and certainly not intellectuals.[6]

After Helmholtz's first essay on Goethe and the sciences (1853), other prominent scientists—Dove (1853), Virchow (1861), Tyndall (1880), and du Bois–Reymond (1882)—also published essays critical of Goethe's theory of colors or otherwise criticized him as a scientist. For all his appreciation of Goethe as a literary figure and his (over)appreciation of him as a leader in natural history and osteology, Helmholtz's critique of him as a theorist of colors had been devastating. By 1892 there was little left of Goethe's reputation as a scientist. (Haeckel was one of the few to still find scientific inspiration in him.) Already in earlier publications—his *éloge* for Magnus (1871), a postscript to a reprint of his essay on Goethe (1875), and his essay "The Facts in Perception" (1878)—Helmholtz's attitude toward Goethe had shown softening, even as Goethe's theory of colors remained unacceptable to him. What was left of Goethe, as of Humboldt, was as a symbol of science and of *Bildung*, of the common strivings, bonds, and fit of science and art. Helmholtz's lecture emphasized these latter points, which allowed him to find a middle ground for judging him: neither irrelevant to modern science, as du Bois–Reymond and nearly all others would have it, nor still useful in terms of a particular inspiration (transformationism, as Haeckel would have it).[7]

Helmholtz's lecture of 1892 was different in tone as well as viewpoint from that of 1853. He now spoke of "the incomparable man" and declared that Goethe had absorbed all the culture of his era without sacrificing his own sense of independence and natural sensitivity. Goethe set a standard "for the genuine and the original in the intellectual [*geistigen*] nature of human beings."[8]

If, as Helmholtz thought, in the mid-nineteenth century Goethe had been a more influential figure among the *Gebildete* in Europe than natural scientists had been, that situation had changed by late in the century. Science had developed such that its general understanding of nature—here Helmholtz pointed to Darwin's theory of the origin of species and to the principle of the conservation of energy as illustrations—showed "the reliability and fruitfulness of its fundamental principles." Science was now also associated with the "practical circumstances of life." Once it was realized that the "correct perceptions" of human knowledge came by means of the human sensory apparatus, then, Helmholtz maintained, the inductive method became the path of science. However, those senses themselves had to be properly understood and the validity of their effectiveness tested. Research in sensory physiology had led to well-proven, empirically based laws. Science had led not only to understanding nature but also to man's power over it, especially through technology. It had led to the power of predicting the course of nature in the future, just as the "prophets and magicians" had once done.[9]

But, Helmholtz further argued, there was also another way—that of artistic representation—of gaining insight into nature and the human mind that was also convincing and communicable to others. Art, like science, could lead to understanding nature and man. This sort of "truth," he said, was "the inner truth of the soul's represented events, its consistency, its agreement with what it has so far learned of the development of such moods, that is, it is the correctness of the representation of the natural course of these conditions." The artist's truth had to be recognizable by his audiences; it concerned the "emotions [*Seelenbewegungen*], character qualities, and individual decision-making." Such truth derived neither from the empirical sciences nor from philosophical reflection, but from "a special artistic intuition." "Intuition," Helmholtz said, was conceptually the opposite of thinking.[10]

While Helmholtz thought that various principles or techniques of painting—for example, the laws of perspective—contributed much to painting, they did so only as an aid to previous experiences that allowed the artist and viewer to recognize objects in a painting. He conceded that it was often difficult to distinguish between "what belongs to the physiological mechanism of the nerves and what trained experience about the unchanging laws of space and nature has provided." But as an empiricist, he believed that the former was chiefly responsible.[11]

These considerations from sensory physiology were pertinent to understanding artistic intuitions, Helmholtz believed. Such intuitions came without effort or expectation; even the artist himself did not know where they came from. But this did not mean they were unrelated to past experience that might be constituted as a law. Helmholtz thought artistic intuition, which did not emerge through thought and was indescribable in words, was "knowledge of a type that concerned phenomena." It was a much richer type of intuition than that given by mere words. Since art, like science, represented and conveyed truth, it was essential that the artist command "a fine knowledge of the law-like behavior of represented phenomena and also their effect on the listener or viewer." Yet "the artistic representation" could not be "a copy of an individual instance" but rather was "a representation of the type of phenomenon concerned."[12]

What, then, was the nature or secret of beauty in art, Helmholtz asked? He offered only an approximate answer. Since he thought truth in art was not a matter of copying something from nature, the artist had to turn "to the lawlikeness of the type." "Thus the more exact is his intuitive picture of the latter, the freer he will be to meet the demands of beauty and expression." He cited examples from meter and rhyme in poetry and the musical arrangement of a dramatic text or song, where translation from one language to another showed that "the given word-forms of language" were only a secondary issue. In a stage performance, for example, it was mostly the rhythm and rhyme of a poem or song, not the actual words, that led the listener to better appreciate "the emotions of the actors in the much richer, finer, and more expressive movements of sound." "In general, we should not disdain pleasant sensations as elements of beauty, for during the long work of generations, Nature has so trained our bodies that we find a feeling of wellness in an exercise where the perceiving activities of our soul can develop in the freest and most secure activity." To illustrate the point, he considered "the prominent influence of the beautiful on human memory": memory greatly facilitated a sense of the whole of an object and its context.[13]

This analysis brought him to the borderland between the scientist and the artist. The artist's memory for individual things was much more developed than the scientist's, he thought; memory played a far more important role in art than in science. This was especially true in music. Yet Helmholtz doubted that musicians and conductors retained millions of musical notes and pauses in their minds; rather, they retained "only the musical phrases of pieces of music," including their order, linkage, and changing timbres. To achieve their musical aims, conductors had to retranslate the musical score "by giving their musicians the right signs."[14]

In science, by contrast, a good memory was based on concepts and language. "Only the initial inventive thought, which must precede conceptualization in words, must always be formed and appear in both types of activity in the same way. Indeed, that can happen initially, only, and always in a way that is analogous to artistic intuition, as a presentiment of new law-likeness." For example, the scientist discovered new similarities in otherwise different phenomena. The ability to find such similarities was known as "wit." It was "a suddenly appearing insight which one cannot attain through methodical pondering but instead appears like a sudden piece of luck."[15]

In the sciences, Helmholtz maintained, as he had said four decades earlier, that Goethe achieved his greatest successes in animal and plant morphology. Goethe was convinced that bodily animal and plant forms were based on "a common structural plan." His fellow anatomists and zoologists, by contrast, believed that organic forms could change constantly, and so they failed to follow and appreciate Goethe's viewpoint. Helmholtz now portrayed Goethe as a forerunner of Darwin, as someone who liberated anatomists and zoologists from their prejudices. But Goethe was "less fortunate" in his theory of colors. The main reason for his failure, as Helmholtz had argued in 1853, was the inadequate instruments and equipment that he had available to him, so that he "could not observe the decisive facts. He never had fully purified, simple, colored light before his eyes, and so he could not believe in its existence." It now (in 1892) appeared as if the instruments, and not Goethe's "theory" or his attitude toward Newton and Newtonianism (i.e., modern science), were the source of Goethe's theory of and results in colors. (In contrast to his scathing account of 1853 of Goethe's shortcomings as a physicist, Helmholtz here avoided virtually all details.) In physics, as in natural history, Goethe believed that behind the "multiplicity of the phenomena" there was "an original phenomenon [*Urphänomen*]" and that it was the physicist's task to find it. He was hostile to abstractions, which he saw as "intuitionless concepts." Helmholtz, by contrast, saw concepts—like matter and force—as essential to theoretical physics. Goethe had no interest in "such transcendental, unimaginable abstractions." This made sense, Helmholtz claimed, insofar as it appealed to "confused and mystical minds"—for example, those who believed in the notions of animal magnetism and the "theory of the life-force [*Lebenskraft*]."[16]

Helmholtz now suddenly and quite stunningly declared that "contemporary physics has already and entirely followed the roads that Goethe had wanted it to follow." He was referring to its phenomenalistic approach. To be sure, Goethe had made some "incorrect interpretations" in his examples, and so he became involved in an "embittered polemic against the physicists." Had he been familiar with Huyghens's wave theory of light, Helmholtz explained, "this would have put in his hands a much more correct and more intuitive '*Urphänomen*' than the scarcely appropriate and very involved path which he chose toward this end in the colors of opaque media." In short, it was now not so much Goethe's method in physics that was erroneous as it was his poorly chosen example.[17]

Helmholtz linked Goethe's approach to physics with the phenomenalistic approaches to understanding matter of Faraday and Kirchhoff. Kirchhoff's stress on description in physics lay "not far from the Goethean *Urphänomen*." Helmholtz himself had, during the past three decades or so, gradually moved toward a descriptive or phenomenalistic approach to physics, placing less emphasis on concepts per se and more on empirical phenomena; indeed, the very distinction between perception and conception in Helmholtz's thought became increasingly less distinct.[18]

His lecture thus constituted a development of and epistemological shift away from his 1878 address "The Facts in Perception," which itself had evolved from his Kant Address of 1855. Once again he quoted (or rather slightly misquoted) some of his favorite lines of poetry, from Schiller's elegiacal nature poem "The Walk" ("Der Spaziergang"), with its theme of civilization's progress.[19]

A law of nature is "not only a guide for our observing reason; it also rules over the operation of all processes in Nature without our paying attention to, wishing, or wanting it. Indeed, unfortunately, this often enough even happens against our wishes and will." When the right conditions hold, this law is referred to as a force, which in turn means the "cause of changes": "it [the force] is that which remains hidden behind the change of phenomena." These various conceptual terms—"law" and "force"—were justified as long as they corresponded to the facts, "and, when used correctly, the abstract form of designation allows the great advantage of constituting a much shorter linguistic expression than the description of the *Urphänomen* developed in statements concerning conditions." Helmholtz warned, however, against being led astray by mere words, and he repeated his judgment about the shortcomings of Goethe's theory of colors.[20]

Noting Goethe's distaste for Kant, Helmholtz himself here criticized Kant and suggested that a reformed Kant (i.e., Helmholtz's Kant) might have been to Goethe's liking. Helmholtz adored the conclusion of *Faust*, where the *Chorus mysticus* sings of truth: "All that is transient is merely an image," interpreting this to mean that "what happens in time, and what we perceive through the senses, we know only as an image." He added, "I hardly know how to express the upshot of our physiological epistemology in a better way." He now felt himself aligned and allied with Goethe. Drawing on Goethe's reference to the "indescribable," Helmholtz declared that he meant that which could not be expressed in words but only in artistic form. Goethe had ultimately gone beyond epistemological considerations "into a higher realm," that which was "in the service of humanity and of the moral ideal, which is symbolized through the Eternal Feminine." The point was symbolized, he said, by Goethe's famous transformation of "In the Beginning was the Word,"—that is, in different contexts a word, concept, or natural law—into "In the Beginning was the Act." Action was necessary for man, even in matters that concerned knowledge. The epistemological foundations of science required an appreciation, indeed an analysis, of the role of the senses in slowly and inductively forming an individual's knowledge of the facts and laws of daily life and the world.[21]

Helmholtz's ultimate claim was that Goethe had success in science wherever "pictures of intuition" of the sort used by poets came into play, but he was unsuccessful where inductive methods were needed. On the other hand, said Helmholtz, "where it is a matter of the highest questions about the relationship of reason to reality, his healthy adherence to reality protects him from aberrations and leads him securely to insights that reach towards the borders of human reason."[22] In comparison to his previous address on Goethe as a scientist, given nearly forty years earlier in Königsberg, Helmholtz here gave Goethe a free pass as a thinker. A combination of time; the setting in Weimar at the Goethe Society, with its posh and socially prominent patrons; Helmholtz's own evolution toward a more phenomenological understanding of nature; and his own social evolution into becoming one of Europe's leading *Kulturträger*—all this taken together had mellowed him, softened his epistemological critique of Goethe, and increased his estimation of Goethe as scientist and as thinker.

From Weimar, Anna wrote Ellen, "Papa's speech was about the borderland between art and science. It was very beautiful, but perhaps too high in its tone and too abstract for the occasion. The entire court was present; the Grand Duchess was deeply moved and filled with thanks, the Grand Duke and everyone was very satisfied." After the lecture, the society gave another *dîner*, and that evening there was a soirée at Lilly von Richthofen's, which the grand duke also attended. The following morning the Helmholtzes viewed the grand duchess's art collection; the grand duke showed them "hand drawings by the greatest masters of all times." Helmholtz later sent the grand duke and duchess the published version of the lecture, which they appreciated.[23]

After he returned to Berlin, Helmholtz learned from the French chemist Marcellin Berthelot that he had just been nominated to become an *associé étranger* of the Académie des Sciences. He was elected soon thereafter. The nomination could not have occurred until an *associé étranger* had died, thus leaving a position available, since only a small, fixed number were admitted (in this case, the emperor of Brazil, Dom Pedro II, had died). In October the Helmholtzes returned to Weimar to attend the celebration of the fiftieth anniversary of the grand duke and duchess as a governing couple. To mark the occasion, they awarded Helmholtz with a "Remembrance Medal."[24]

Helmholtz's second essay on Goethe found many readers in its subsequent publication in the *Deutsche Rundschau*. One of them, Hildebrand, the sculptor, painter, and theorist of art, wrote his friend Fiedler, an art historian: "I've read the lecture by Helmholtz with interest. What he says about the laws with respect to the plastic arts corresponds entirely to my thinking; it has indeed grown out of his field of work, and proves the correctness of my work. I've always thought that it [my thinking about art] would find a good reader precisely in Helmholtz. If I only had more time to work more on this." Two weeks later, he reread the essay and had some second thoughts and mild criticism of it. He said, even so, "The speech is frankly interesting and full of good ideas."[25]

Hildebrand was a leading German theorist of art in the late nineteenth century. In 1893 he published a little book entitled *Das Problem der Form in der bildenden Kunst* that reveals the considerable influence of Helmholtz's recent essay on Goethe. One of Hildebrand's admirers thought the book showed the importance of intelligence in art, that art was more than just a matter of dexterity with the hands. He thought it important "that people with the stature of Helmholtz and Wundt" took note of it. Hildebrand reported with great pleasure hearing that Helmholtz had read his book: "It has very much interested Helmholtz from the philosophical side—he may have come to the same results in a completely different way."[26]

Dilthey was much cooler toward Helmholtz's essay. He thought it "amusing . . . that Helmholtz means to re-discover his epistemology in Goethe. The fundamental historical ignorance [shown here] can indeed only be condoned in a great natural scientist. I'd fail an historian who said the same things in a doctoral exam." He thought, too, that, historically speaking, Helmholtz's epistemological theory was "essentially" that of J. S. Mill or Thomas Brown. A month later he wrote the same friend that he thought Helmholtz's essay "as a panegyric to Goethe [was] false. That Goethe did not see purified colors neither explains nor 'excuses' his theory of colors. There is no resolving of the contrast Goethe-Helmholtz. Such an attempt suffers from objective falsehood. [Only] Helmholtz's ignorance of history makes [such] complete credulity possible to him." Even du Bois–Reymond, ever the promoter of if not apologist for Helmholtz, thought Helmholtz had been far too generous with Goethe as a scientist. Yet not everyone agreed with Dilthey, either. In 1917 Einstein, then at the height of his own intellectual creativity, reviewed Helmholtz's two essays on Goethe. He said of the second that it would be "read with delight by anyone who can have joy in the scientific contemplation of the world [*Weltbetrachtung*]. . . . Dear Reader! It would be profane to summarize. Read it yourself!"[27]

Standards Business in Edinburgh

In August, two months after Helmholtz delivered his lecture in Weimar, the BAAS held its sixty-second annual meeting in Edinburgh. Helmholtz attended in order to meet with his British counterparts and help further settle the long-standing disagreements between Germany and Britain over electrical standards. As president of the Reichsanstalt, he was Germany's senior representative in such metrological matters, while Thomson was a member (along with Rayleigh, Richard Glazebrook, George Carey Foster, and William Edward Ayrton) of the Board of Trade's advisory committee on electrical standards. That spring, Thomson had written Helmholtz that the board, "in deference to your wish," agreed to delay (until November) announcing its decisions in regard to practical electrical standards of resistance, current, and potential. Thomson told Helmholtz that the committee hoped that he and some of his associates might come to the BAAS meeting to discuss the British group's proposed standards and to reach an agreement between Britain and Germany. He also hoped Anna would accompany Helmholtz and that before or after the meeting the Helmholtzes would give them "as long a visit at Netherhall" as possible. Helmholtz agreed to go (with Anna) to Edinburgh and, afterward, to Netherhall. The Thomsons were delighted, and Thomson wrote Tait to ask him "to arrange about [Helmholtz's] being received in Edinburgh." Siemens, for his part, still remained dissatisfied with the definition of the ohm in terms of the electrical resistance of a partially filled mercury tube, as he once again explained in a long letter to Helmholtz. The Reichsanstalt's new definition, he said, was "a break" with the older one that he had advanced (in terms of a mercury unit), which had been agreed upon internationally in Paris in 1884. He wanted Helmholtz to try to influence the British toward meeting that previously accepted international agreement and to prevent them from acting in a "one-sided" manner. "In any case, such a reckless action would face great political and practical concerns in [terms of] Germany's accession."[28] Helmholtz's visit to Britain was hardly a social call on Thomson.

On their way to Scotland, the Helmholtzes visited London and stayed for a couple of days with Rayleigh at his country estate in the village of Terling, Essex. In London they stayed at Arthur James Balfour's, where there was a lively party. On Sunday they went with a "very civil" group to church. During the otherwise strictly social weekend at Terling, Rayleigh and Helmholtz discussed electrical standards. The Rayleighs also gave a party for the Helmholtzes. Among the guests were Balfour; the Sidgwicks; Michelson; Glazebrook, the Cambridge physicist; and William Robertson Smith, professor of Arabic at Cambridge.[29]

James Sully had hoped that Helmholtz would also attend the International Congress of Experimental Psychology in London. This was the second such congress, the first having been held in August 1889. Some three hundred experimental psychologists would be in attendance, and Henry Sidgwick would serve as the president. Sully wanted Helmholtz to speak to the congress on physiological optics "or any other subject which you might prefer to this." But Helmholtz would not commit to giving a paper, and so Sully wrote a second time to ask if the congress might at least list his name as being among its contributors. Helmholtz agreed only to meet privately (at Balfour's place) with twenty psychologists for lunch. Thereafter, he and Anna immediately went to Edinburgh.[30]

The Edinburgh meeting's distinguishing feature was the presence of Helmholtz and two Reichsanstalt associates who accompanied him there to help settle the electrical standards issue. Michelson, Glazebrook, Stokes, Tait, and Schuster were also present. Thomson thought the meeting about the best he had ever known "in point of instructiveness." A niece of his was there too, and she reported, "Helmholtz and Uncle William were inseparable, and both spoke a good deal in the sections, and Tait sat and smiled serenely at everything." Anna reported home, "The British Association stands socially far above our Naturforscherversammlung, and in form and in almost every way is very English. Women are to be found everywhere; only God knows what they get from the Section sessions, though they are present there as well." There was a big dinner one evening at the home of the Coxes, followed by a reception for two hundred people, including many "Grandies." The Helmholtzes also dined one evening on a yacht.[31]

The Edinburgh meeting also proved to be a good moment for the British and the Americans to advocate for their own Reichsanstalts. That year, Edward Weston had discovered a metal alloy with a temperature coefficient (with respect to resistance) of zero. He later received a patent (in 1893) for its use in electrical resistors. Reichsanstalt scientists soon began doing research on and with electrical resistors employing Weston's new alloy, which they dubbed manganin and duly credited Weston as its discoverer. Thomson, in a speech at the BAAS in which he sought to get the British government to establish its own Reichsanstalt, credited the Germans, thus suggesting that if Britain had had its own Reichsanstalt, a Briton might have discovered the alloy. Helmholtz, who had spoken just before Thomson, "rose quickly. 'The discovery of a metal whose resistance diminishes with temperature was made by an American engineer,' he said succinctly." Ayrton, however, claimed that Weston was English. (He was in fact born in England, but he had emigrated to America in 1870, at age twenty.) The American Henry Carhart, who was also at the meeting, heard discussions there about establishing national laboratories for physics and chemistry. Seven years later, in 1899, the National Physical Laboratory in Teddington, England, was opened, and nine years later, in 1901, the National Bureau of Standards in Washington, DC, was established, both of them imitating (and adapting from) the Reichsanstalt. The American, British, French, and German representatives at Edinburgh agreed to specify the conditions for determining the values of the ohm, the ampere, and the volt, thus setting the basis for values that were to be agreed upon the following year in Chicago.[32]

From Edinburgh, the Helmholtzes went to Glasgow and then on to Thomson's place, Netherhall Largs. Anna was most impressed by the University of Glasgow and its buildings and laboratories, including Thomson's own; he showed them, inter alia, various precision instruments for measuring electricity. They also went on board Sir John Burns's "magnificent Yacht." Helmholtz, Thomson, and three other scientists discussed "mathematical problems" during tea, which Anna disapproved of. "It's time," she told Ellen, "that Papa has a bit more rest in his life; the many demands on his attention tire him out." They returned to Glasgow the next day and from there traveled to London, Cologne, and Bayreuth.[33] So much for rest.

That fall—to be specific, 2 November 1892—marked the fiftieth anniversary of Helmholtz's receipt of his doctorate (that is, his medical degree), and he was again, as he had been a year earlier, honored publicly. A formal celebration was held at the academy, with du Bois–Reymond presiding. He said that the previous year, at Helmholtz's seventieth-birthday celebration, the academy was not among the many who had congratulated Helmholtz, because it was not its custom to take formal notice of birthdays (though it was present in the sense that it had arranged for the formal establishment and public announcement of the new Helmholtz Endowment). So the academy, which now wanted to celebrate the fiftieth anniversary of Helmholtz's doctorate, found itself "in the awkward position" of repeating what others had already said about him and praising him for being the "Meister" of so many students and colleagues. A long list of many of Helmholtz's well-known scientific achievements and other professional accomplishments was recalled, and the ceremony concluded with the hope that Helmholtz would continue to live a long and productive life, one that would continue to bring the academy much fame.[34]

And not only the academy. The emperor sent a congratulatory telegram on 2 November 1892, along with a bust of himself. Further congratulations and signs of respect came from the University of Berlin's medical and philosophical faculties. Gustav von Gossler, the minister of culture, requested an hour of Helmholtz's time so that he could express his "honor and gratitude." Gossler felt fortunate "as a German" and "as a man" to be so near to Helmholtz's "striving and works"; he recognized Helmholtz's importance "to the Fatherland."[35] Once again the state expressed its honor and respect for him.

28

Charismatic Leader

Building and Directing the Reichsanstalt

Helmholtz had hoped to devote the remaining years of his life to doing research in theoretical physics, and to some extent he did just that, concentrating on atmospheric physics, electromagnetic theory, the principle of least action, and mechanics. But in leaving the university for the Reichsanstalt, he had to agree to teach one course per semester on theoretical physics at the university, and, more to the point, he had to spend a very substantial amount of his time planning and supervising the Reichsanstalt's construction and then directing it. He exchanged one set of institutional burdens for another.

Between 1887–88 and 1894, Helmholtz was the defining figure in planning and supervising the construction of the Reichsanstalt, developing its administrative structure, and overseeing its operations.[1] He helped choose the Board of Directors and outlined the institute's formal purposes and annual work agenda. The board members were men of different professional backgrounds (academic physics, the precision-technology industry, and the instrument-making industry generally) as well as from the Reich government. It had about twenty-five members, many of whom—including Foerster, Siemens, Bezold, Clausius,

Kohlrausch, Abbe, Kundt, and Wiedemann—were more or less close colleagues of Helmholtz's.

Helmholtz was also the chief scientific and administrative officer. The Reichsanstalt was divided into two sections, the scientific and the technical. Helmholtz was simultaneously president of the Reichsanstalt, director of the Scientific Section, and supervisor of the director of the Technical Section. To fill all his roles, he created a rule-governed, bureaucratic system of scientific organization with a hierarchical work structure. With the board's approval, he alone specified most of the Reichsanstalt's annual research and testing agenda.

Helmholtz imparted a strong research ethic to the Scientific Section, emphasizing the foundations of metrological physics, especially in the areas of heat, electricity, and optics. He hired the staff for the Scientific Section, as well as the director of the Technical Section (Leopold Loewenherz, until his death in 1892, and then Ernst Hagen). By 1893 the Reichsanstalt employed sixty-five people, several of whom were Helmholtz's former students. The staff operated more as sets of small teams (often two to four) than as individuals, though a few staff members were allowed to operate individually and a few guest researchers were also invited to join. Helmholtz determined what research or other work the staff would pursue, as well as what could be published and where. He secured especially close relationships with the *Annalen der Physik und Chemie* and the *Zeitschrift für Instrumentenkunde*, and he created a new house organ, the *Wissenschaftliche Abhandlungen der Physikalisch-Technischen Reichsanstalt*. To be sure, he delegated some authority to the heads of the heat, electricity, and optics laboratories and to Loewenherz (later, Hagen), who administered both sections. This gave him time to conduct his own physics research. The staffs of both sections were in close contact with their colleagues at the university and at the Technische Hochschule Charlottenburg. Thus Berlin now had three physics research centers.

Initially the two sections were located in provisional quarters at the Technische Hochschule: Helmholtz worked there between 1887 and 1891 and then moved into the Scientific Section's newly completed and very large building, the Observatorium. The Technical Section's buildings were completed two years later. The Helmholtzes started moving into their new residence in May 1889 (see figs. 28.1 and 28.2). When the entire Reichsanstalt was completed in 1897, it occupied 8.45 acres of land; consisted of ten new buildings (five for each section), all located within a well-gardened setting and including a most fitting residence for its president; and cost 3,672,360 marks to establish. The Scientific Section alone provided more floor space and probably better facilities for physics research than any extant academic physics institute, and it had more professional physicists on its staff than any other scientific institution of the day. Yet there was strong concern among some academic physicists that the sec-

Fig. 28.1 Front view of the Reichsanstalt's *Observatorium* (main scientific building). Siemens Forum, Munich.

Fig. 28.2 Helmholtz's residence at the Reichsanstalt. Henry S. Carhart, "The Imperial Physico-Technical Institution in Charlottenburg," *Transactions of the American Institute of Electrical Engineers* 17 (1900): 555–83, on 558.

tion should complement, not compete with, university physics institutes, and so the section was devoted more to helping industry than doing pure science. The Technical Section, for its part, did a great deal of practical testing and measurement work for industry. The entire physical infrastructure and professional staff were thus intended to help secure the metrological foundations of heat, electricity, and optics, as well as to further the practical use of metrological standards for these fields and to conduct practical measurements for the electrical, gas, thermometer, sugar, precision-technology, and instrument-making industries, among others. Along with promoting the metrological needs of German science and technology, the Reichsanstalt also generally promoted the needs and reputation of the German Reich as those concerned physical and industrial standards (both domestic and international). Under Helmholtz's general direction, the Reichsanstalt became a flourishing scientific and technological institution; by the turn of the century, it had imitators in Britain and America. It was an institute for an empire.

(Still a) Leader in Psychology

Helmholtz also remained scientifically active in the field of psychology, and circa 1890 his standing in that discipline was still quite high. When Ebbinghaus and König founded the *Zeitschrift für Psychologie und Physiologie der Sinnesorgane* (1890), for example, they invited Helmholtz to become one of their advisory editors. They had many reasons to turn to him, since his influence on physiological psychology during the previous four decades had been profound. He became more than just an advisory editor, however: between 1890 and 1894, he published a series of five articles (on sensation, colors and color theory, perception, and Fechner's law) in this new journal. He had by no means lost interest in matters of visual perception; psychologists still viewed him as one of their own. With König's help, he revised his *Handbuch der physiologischen Optik*, issuing small installments of it between 1885 and 1894. König supervised the republication of the entire revised volume in 1896. Together with *Die Lehre von den Tonempfindungen*, it remained a work of primary importance for psychologists.[2]

One indicator of Helmholtz's continued high standing in psychology late in the century can be seen in the works of William James, especially in *The Principles of Psychology* (1890) but also in *The Will to Believe* and in his collected *Essays in Psychology*. These works, above all *The Principles*, greatly shaped the discipline of psychology in America. Although James also relied on authorities other than Helmholtz—notably Spencer, Wundt, and Bain—he cited Helmholtz's ideas, empirical results, writings, and name far more often than those of any other author. In *The Principles*, he called special attention to Helmholtz (often with long quotations) on a range of topics: reaction time and nerve conduction, color compounding, unconscious inference, perception, attention, elementary tones, sensations, Weber's law, time perception, afterimages, contrast, inverted vision, illusions, muscle sense, convergence, the cyclopean eye, space, innervation, effort or will, counting, and scientific laws in general. Wherever the issue was a factual or methodological one, James cited Helmholtz positively. However, in matters of theory or interpretation, he differed with Helmholtz, often profoundly. James also carefully studied Helmholtz's individual writings on conservation of energy and on the eye and the ear (including both the *Handbuch* and the *Tonempfindungen*), as well as all of Helmholtz's philosophical and popular scientific essays. Ralph Barton Perry, one of James's early and principal biographers, repeatedly drew attention to James on Helmholtz, judging that Helmholtz was "one of his scientific idols."[3]

Teaching Theoretical Physics

Yet it was physics, not psychology, that mostly preoccupied Helmholtz intellectually during his remaining years.

Helmholtz's duties at the university officially consisted in teaching one lecture course per semester in theoretical physics. As far back as 1874, he had taught a class on the mathematical theory of electrodynamics, which was also the title of his last announced, but never held, class of the winter semester of 1894–95. He also still guided advanced students in their research, participated in (fee-paying) examinations, which he loathed, and did other miscellaneous work.[4]

Between 1871 and 1894, he put his signature to 1,169 promotion and 171 habilitation cases. He personally examined and approved 85 dissertations and 27 habilitations and tested 217 individuals for promotion. The promotions (especially after 1888) were in many cases in chemistry (by far the largest number) and then, to a lesser extent, in physics, with a few in astronomy and mathematics; there was also a small number in philosophy, music, music history, and history, and even one in the history of astronomy. Among Helmholtz's last well-known students were Walther Rathenau (promotion with Kundt as first, Helmholtz as second, reader), Hermann Ebbinghaus (habilitation in psychology), Ernst Pringsheim (habilitation in physics), Leo Arons (habilitation in physics), Heinrich Rubens (habilitation in physics), Michael Pupin (promotion in physics), and Willy Wien (promotion and habilitation in physics). Most of these figures became influential in their fields in the 1890s and the first decade of the 1900s. Rathenau in particular became an electrochemist, the head of the Allgemeine Elektrizitätsgesellschaft, an essayist, the manager of the Agency for Raw Materials in the Imperial German Department of War (during World War I), and foreign minister during the early Weimar years, before being assassinated in 1922. He saw Helmholtz as one of the key influences in his intellectual grounding.[5]

As Helmholtz's teaching career was drawing to a close, the status of women in science stood on the cusp of change in Germany (and elsewhere): whereas previously they had been denied permission to study or take advanced degrees, they were now gradually permitted to do so. About 1889, Sarah Whiting, a young American graduate student in physics, was reportedly denied permission to use Helmholtz's research facilities (though she was permitted to attend Kundt's and Bezold's lectures); by contrast, Christine Ladd-Franklin, an American logician and psychologist, worked briefly in König's laboratory in 1892, where she did research seeking to reconcile the differences between Helmholtz's and Hering's views on color. (She hoped to synthesize the two approaches through an evolutionary tetrachromatic theory of color.) Following the leadership of Helen Lange, in 1893–94 Helmholtz joined other liberal professors in calling for a secondary-school *Realkurs* for women in Berlin. Both Hermann and Anna von Helmholtz joined the board that administered the course. By 1896 Lange's petition to institute such a course had developed into Germany's first state-recognized *Abitur* for women. In principle, women now had the first real opportunity to attend and graduate from a German university.[6]

In the spring of 1892, Helmholtz agreed to work with several of his advanced students to assemble into publishable form his lectures on theoretical physics. Though the set of volumes that ultimately appeared (*Vorlesungen über theoretische Physik*) had Helmholtz's name on the covers and title pages, it is impossible to know (except for part one of volume 1) precisely the extent to which he was their true and full author. Helmholtz himself reviewed no more than one-third of volume 3 and about half of volume 5. The volumes were certainly based on his lectures, held mostly between 1892 and 1894; and they definitely represent and summarize his work in, and views about, theoretical physics since 1847. But he did not write these lectures out (or any of his other lectures). Instead, he lectured freely—he improvised, as his editors put it—and occasionally glanced at a small notebook to remind him of the general topic to be covered. While lecturing, he sometimes dropped a point of view that he had just advocated for and instead took up another if it seemed to be more useful. (This may have been what the American physicist W. F. Magie and others meant when they said that to hear Helmholtz lecture was to see the creative process at work in physics.) Or perhaps it was simply a polite reference to inconsistency. In any case, it was to see someone interested in conceptual presentation and clarification, not in giving absolutely correct mathematical derivations or the orderly presentation that is so necessary for most beginners. Helmholtz's advanced students took notes on what "the master"—they used that expression frequently—said, and they composed stenographic student notes (so-called *Mitschriften*). Carl Runge, who worked with König in bringing out Helmholtz's lectures on acoustics, noted, for example, that only part of these lectures were worked out; the remainder had to be put together from student notes, with Runge doing the mathematical part. This was not easy, especially when the student notes contained remarks like "muttered at the blackboard; used some incomprehensible words." The student notes for each course of lectures were expanded upon and whole sections were labeled or relabeled by the individual editor(s) as the editor(s) saw fit; Helmholtz's notebooks as well as his earlier lectures on experimental physics were consulted or used on some points; that is, these volumes were heavily edited, not to say largely written, by their various individual editors. König, who was not a theoretical physicist but who was Helmholtz's trusted major domo in all editorial matters and who arranged for the volumes' publication, served as his designated general editor for the set. While the good faith of the editors in reproducing Helmholtz's lectures cannot be doubted, the organization and wording of these volumes, and perhaps even some of the content (though not the broad ideas), are the products of several hands.[7]

Like the not-fully-certain authorship, the organization and publication chronology of these volumes is also confusing. The final, published form of the lectures consisted of six volumes (in seven); these appeared between 1897 and 1907, but in no particular order. Volume 1 itself consisted of two parts. Part one, edited by König and Runge, is the only thin volume among the set of seven. Though it is the first volume in the set, it appeared much later in the publication process (in 1903). It was devoted to methodological principles and the foundation of mathematical representations. Its epistemological and other philosophical standpoints and issues, however, repeated those that Helmholtz had been making or developing for nearly a half century, from the introduction, to his essay on the conservation of force, to "The Facts in Perception," and from his essays on the foundations of geometry to his essay on numbers and measurement. Part two, edited by Otto Krigar-Menzel, appeared five years earlier (in 1898); it contained Helmholtz's lectures on the dynamics of discrete mass points. Volume 2, edited by Krigar-Menzel, appeared in 1902 and concentrated on the elasticity of solids. Volume 3, edited by König and Runge, appeared in 1898 and was devoted to the mathematical principles of acoustics. Volume 4, edited by Krigar-Menzel and Max Laue, was the last to appear (1907) and was concerned with electrodynamics and the theory of magnetism. (It thus appeared two years after Einstein's revisionist and revolutionary work in electrodynamics.) In contrast to the previous volumes, however, it was not based on stenographic notes, but on a combination of Helmholtz's notebooks, student notes from an earlier lecture course he had taught in 1888–89, and on some of his own individual papers. Volume 5, edited by König and Runge, appeared in 1897 and was the first of the seven to appear; it was concerned with the electromagnetic theory of light. Finally, volume 6, edited by Franz Richarz, appeared in 1903 and was devoted to the theory of heat. Richarz worked with a combination of Helmholtz's notebooks, stenographic notes of others, and his own notes from Helmholtz's lectures on this topic from the early 1880s. The lectures certainly represent Helmholtz's views on and principal findings in theoretical physics; indeed, they represent, generally speaking, the latest thinking in the subject in the early 1890s. They were thus doubtless of use to physicists as the volumes began appearing after 1897. But by the time the final volume appeared (1907), the field of theoretical physics was in such flux that the lectures had lost their timeliness and gradually became, instead, a memorial to Helmholtz himself.[8]

Leading Physics at the Academy

Helmholtz continued to promote physics at and with the Academy of Sciences.[9] Between 1888 and 1894, he attended sixty-three sessions of its Physical-Mathematical Class and twenty-five of the Plenum, lecturing seven times before the class and twice before the Plenum. His lectures variously concerned electromagnetism, thermodynamics, mechanics, geophysics, and atmospheric physics, as well as one on psychophysics. Six of them were published in the academy's *Sitzungsberichte* or *Abhandlungen*, and they helped give substantive scientific life to the academy, especially to the physical scientists; it seems doubtful that humanists and social scientists at the academy would have understood such advanced, highly scientific lectures. All told, between 1850 and 1893, Helmholtz published forty-four papers in the academy's three journals.

Helmholtz also vitalized the academy by still helping others to become corresponding or ordinary members, by still helping get manuscripts published by nonmembers, and by still trying to find financial support for certain individuals. Between 1888 and 1894, he was the lead nominator or cosigner for Hertz as a corresponding member and for Boltzmann (later nullified), Kundt, Planck, and Vogel as ordinary members. This promotional activity helped support both senior physicists and the upcoming generation. The most dramatic, spectacular example occurred in late January 1888, when Helmholtz asked du Bois–Reymond to rush into print, in the academy's *Sitzungsberichte*, Hertz's manuscript giving experimental proof of the identity of electromagnetic and light waves. Hertz's work proved to be epochal. Helmholtz supported a series of manuscripts by Hertz on this and related topics, and he also furthered the publication of manuscripts by Kundt, Kayser and Runge (a series on spectroscopy), Willy Wien, König, Braun, Lenard, Nernst, Lummer, Kurlbaum, and H. F. Weber, among others. Several of these publications—for example, those by Wien, Weber, and Lummer and Kurlbaum—concerned blackbody radiation and spectral-energy distribution; these soon proved to be important steps on the road to quantum physics. At one point or another, Helmholtz helped publish manuscripts by, or otherwise importantly furthered the careers of, six future Nobel Prize winners in physics—Röntgen (1901), Lenard (1905), Michelson (1907), Braun (1909), Wien (1911), and Planck (1918)—as well as one (Nernst, 1920) in chemistry. In his capacity as vice chancellor of the Orden pour le Mérite für Wissenschaften und Künste, he argued vigorously and successfully to get Kundt appointed to a new opening (1891–92).[10] It helped German science, too, when Helmholtz was appointed (June 1892) a foreign associate (*associé étranger*) of the Paris Academy of Sciences.

Meteorology and Atmospheric Physics: From Krakatoa to Count Zeppelin

At the academy, too, Helmholtz oversaw committees supervising the Meteorologic Institute and the Geodetic Institute (Geodätisches Institut).[11] His interest in meteorology and atmospheric physics was an old, standing one, but from late August 1883 onward he had a more acute reason to pay attention to meteorology. The volcanic island of Krakatoa, located between Java and Sumatra, dramatically erupted at that time, becoming a global geophysical and social event and the cause for much scientific debate about the geophysical status of the earth and the sun. It brought renewed interest in the study of the upper atmosphere. Its enormous series of eruptions not only devastated the island itself; its effects included earthquakes, ash flows, dust and other gas emissions, shock waves, a tsunami, and general darkness, not to mention the large-scale loss of human and other life, all of which was to one degree or another heard, seen, and felt not only in Asia but even in distant North America and Europe. Many came to know or hear of Krakatoa. Like the recent international electrical congresses in Paris and the growing international trade between Eurasia and North America, Krakatoa was further, if here geophysical and catastrophic, proof that the world was becoming ever more tightly bound together, as the increasing emergence of notions of global weather patterns also implied.

News about Krakatoa and its aftermath reached Helmholtz through several sources. His son Robert reported from Berlin on the "cloud glow" that he observed there in late November. Helmholtz also heard about it from Georg von Neumayer, the founder of the Deutsche Seewarte (Marine Observatory) in Hamburg (1875), director of the Deutsche Meteorologische Gesellschaft (Meteorologic Society) there, and a leader of German and international polar expeditions. (The year 1882–83 marked the first International Polar Year.) Neumayer's reputation rested principally on his being the lead organizer of German geophysics initiatives rather than on being a research scientist per se; he was well connected in the world of meteorology and the earth sciences generally and did much to make Germany a global player in these regards. In 1882 he sent Helmholtz, at Helmholtz's request, one of the new magnetic maps of the earth. In 1884 he wrote again to discuss Krakatoa and to send him samples of volcanic ash from the eruption. Similarly, Johann Kiessling, a former student of Franz Neumann's and a researcher in Magnus's laboratory, and now a secondary-school teacher of physics in Hamburg and an associate of the Hamburg-Altona branch of the Deutsche Meteorologische Gesellschaft, wrote Helmholtz in June 1886 concerning his own important study of Krakatoa. Kiessling soon became known for his interpretation of Bishop's Ring, the corona sometimes seen in dust clouds such as those issuing from the Krakatoa eruption. He also sought to simulate this phenomenon in the laboratory, and in doing so his "diffraction chamber" became a forerunner of the cloud chamber.[12] In short, Helmholtz followed (parts of) the literature on meteorology and had good contacts with its practitioners.

For more than a decade after his publication "Cyclones and Thunderstorms" in 1875, however, he did not publish on meteorology or the atmosphere. Then, while hiking in the Rigi mountains in September 1886, he observed cloud and thunderstorm formations in the nearby Jura mountains and felt motivated to present a short, descriptive account of these observations. He imagined the formations as vortex sheets in the atmosphere, effectively transferring his old hydrodynamic concept from the earth's waters to its atmosphere.[13]

This new concept of vortex sheets in the atmosphere manifested itself in three new papers on meteorology and atmospheric physics, papers that embodied Helmholtz's uncanny ability to creatively mix disciplines, in this case using thermodynamics and hydrodynamics (not to mention an analogy with air's motion in an organ pipe, i.e., acoustics) to better understand the atmosphere and ocean surfaces. In the first of these papers, published in 1888, he further developed his insight on atmospheric movements, in particular that discontinuous motion occurred between the upper and lower portions of the trade winds. These discontinuous surfaces in the atmosphere, he argued broadly, were due to a combination of gravitation, solar heat, the earth's rotation, and friction at the earth's surface. In the course of time, an unstable wave surface would be transformed into numerous vortices, producing a damping of the upper trade winds and heat exchange between the upper and the lower winds. These effects prevented more violent winds from circulating in the atmosphere.[14]

A year later, in 1889, Helmholtz published a second paper on atmospheric movements, now clarifying or correcting two or three misstatements or assumptions, further refining individual points, trying to understand how the earth's rotation affected (in mechanical terms) wind damping, and extending his analysis from atmospheric waves to water waves. He showed that billows of water largely originated from a steady wind blowing above a plane surface of water. He found a mathematical relationship between wind velocity and the height and length of sea waves and showed how the energy densities of the wind and the waves influenced each other. The formation of sea waves, he found, was analogous to that of combination tones in acoustics. (Much earlier, in 1871, Thomson had worked on understanding the relationship between wind and sea waves, as Helmholtz knew from his yachting trip with him at the time; Rayleigh reminded him of this in a letter after he read Helmholtz's first paper on atmospheric movements. The phenomenon became known as Kelvin-Helmholtz instability or cloud.)[15]

The third paper appeared in 1890, and it concerned the energy of billows and wind, extending the previous two papers on atmospheric movements. Helmholtz here employed a variational principle similar to that of least action to represent mathematically the formation, shapes, energy, and momentum of the water surface. He also empirically tested some of his theoretical findings about sea wave formation. To do so, he asked the Reich minister of the interior, Heinrich von Boetticher, the government official formally in charge of the Reichsanstalt, for permission to take a five-week vacation on the French Riviera. He noted that his family had been vacationing there that winter, and he conceded that part of his trip would be personal. But he also claimed that the seacoast there presented the most favorable physical and meteorologic conditions for testing his theory. He also noted that the previous year he had read a paper before the academy on the relationship between billows and wind and that he believed his theory could have "notable consequences for meteorology" as well as "useful results" for "nautical theory." He wanted to test that theory empirically. Boetticher agreed to the trip. To conduct his tests, Helmholtz took along a portable anemometer for measuring wind strength. In April, while on the Cap d'Antibes, he took wind measurements and sought to count billows, but both turned out, as he himself conceded, to be highly inaccurate, and the conditions were far too variable to confirm his theory. In the summer after his return, he read another paper before the academy, this one concerning weather under the influence of wind. Despite the future importance of the paper on atmospheric waves for modern meteorology, the two successor papers on wind and waves remained virtually undiscussed in subsequent literature.[16]

The high regard for Helmholtz as one of the leaders of theoretical atmospheric physics and theoretical meteorology in general manifested itself in ways that went beyond the reception of his writings. In June 1892, for example, Samuel Pierpont Langley, secretary of the Smithsonian Institution, wrote him (as he also wrote Rayleigh, Wolcott Gibbs, and Cleveland Abbe) to say that Thomas G. Hodgkins had bequeathed to the Smithsonian an endowment fund devoted "to the increase and diffusion of more exact knowledge of the nature and properties of atmospheric air." To execute the donor's wish, Langley "proposed to offer a number of prizes for scientific investigations of a high merit bearing upon the properties of the atmosphere, to be awarded without regard to the nationality of the author." Such work could concern any branch of natural science as long as it had something to do with the atmosphere: for example, topics in hygiene, anthropology, biology, chemistry, electricity, and geology might qualify. Langley asked Helmholtz to make some suggestions concerning "the nature of the principal relationships existing between Physics and the atmosphere, and indicate one or two subjects arising out of these relations which you consider to be proper for prize essays." He also asked him to serve on the prize committee, simultaneously suggesting that Helmholtz himself might be eligible for such an award if he was doing some sort of atmospheric research. Helmholtz accepted Langley's offer to serve on the prize committee, and he recommended that his former students and current coworkers Lummer and Pringsheim be given an award (initially of $500) for their thermodynamic research (on the specific heat of gases) from the Hodgkins Fund as well as an additional award for meteorologic research. Langley thanked Helmholtz for agreeing to serve and awarded $500 to Lummer and Pringsheim. The award carried the stipulation that their paper first appear in a publication sponsored by the Smithsonian Institution. In June 1894 Lummer and Pringsheim requested, through Helmholtz, a second award of $500 from the fund to continue their research. Helmholtz said their work had already been successful and that this would guarantee that they would complete their planned future work.[17] Their paper appeared in the *Smithsonian Contributions to Knowledge* (1898).

Helmholtz's standing in the field of atmospheric physics was also reflected in the awarding of the new Buys-Ballot Medal. In early 1893 Heike Kamerlingh Onnes, a Dutch experimental physicist who had studied with Bunsen and Kirchhoff at Heidelberg, and who was now the director of the Physics Institute at the University of Leiden, wrote Helmholtz to ask his advice about nominees for this new award, to be given by the Royal Academy of Sciences in Amsterdam. C. H. D. Buys-Ballot (1817–90), as Onnes did not need to explain to Helmholtz, had been (principally) a chemist and a leading meteorologist at Utrecht. Onnes explained that the Dutch academy did not have a meteorologist as a member, so he (Onnes), as a physicist, had been chosen by the academy commission to help name a recipient of the first Buys-Ballot award. (Onnes himself specialized in low-temperature physics and in 1911 discovered superconductivity, for which he won the Nobel Prize in Physics for 1913.) But he did not consider himself sufficiently competent to judge meteorologic work. The new award, Onnes explained, was to go to the individual who, during the past decade, had done the most for meteorology. He mentioned that there was talk about possibly awarding it to Georg von Neumayer. Onnes wanted to know which individuals, in Helmholtz's

opinion, had contributed the most to the physical understanding of meteorology in the past decade.[18]

Helmholtz replied that he could not claim to give any expert opinion as concerned the most outstanding meteorologists; "I have only been involved very
superficially with the normal, practical issues of meteorologists." Instead, he
characterized his relationship with the discipline as restricted to "aerodynamics"
(i.e., the dynamics of the atmosphere). He noted, however, the work of his Berlin
colleague Bezold, whose publications he believed had helped advance "a physical
understanding of atmospheric processes." Bezold himself told Helmholtz that
he thought Julius Hann was the most deserving of the Buys-Ballot Medal. In
thanking Helmholtz for his help, Onnes said he appreciated Bezold's work but
that his commission colleagues had decided to award the first medal to Hann.[19]

Helmholtz's influence in atmospheric physics was reflected in still another
way. In September 1893 Cleveland Abbe sent him a volume of translations that
he had made of papers devoted to the mechanics of the earth's atmosphere. Abbe
was an important meteorologist at the US Weather Bureau (founded in 1870),
its chief signal officer, and editor of its official weather forecasts (and other publications) between 1871 and 1915. Six of the twenty papers in the collection
translated by Abbe were by Helmholtz. Referring to the collection, Abbe told
Helmholtz in a letter, "You yourself contributed the most brilliant memoirs."
(The other essays were by Gotthilf Hagen, Kirchhoff, Anton Oberbeck, Hertz,
Bezold, Rayleigh, Max Margules, and Ferrel.) Abbe included all of Helmholtz's
pertinent or fundamental writings on hydrodynamics and atmospheric physics
between 1858 and 1890.[20]

As human air travel began to seem possible in the late nineteenth century,
understanding the atmosphere found a new or increased importance. Already in
the early 1870s Helmholtz had published on the theoretical possibilities of ballooning and flight, and he had served on a Prussian commission for evaluating
guided air travel. He presumably considered these matters pertinent to his scientific work on vorticity in fluid flows. In the early 1890s, he was again called on
for his expertise in atmospheric physics. Ferdinand Graf von Zeppelin, a retired
senior army officer, together with a young engineering colleague, sought to find
financial backing for a "rigid airship." After Zeppelin thought they had solved
the important technical problems, he sent their plans to the Prussian Ministry
of War for testing and review and for financial support. The military administration, however, was slow to respond, and so Zeppelin urged the emperor to intervene. In November 1893 the military set up a commission to evaluate his design
of a "steerable airship." The commission consisted of members from the Prussian
Airship Division along with various scientists (meteorologists and statics experts); at Zeppelin's request, Helmholtz was appointed chairman, even though

Zeppelin may have known that Helmholtz had shown a rather skeptical attitude toward the possibilities of manned air flight. In March 1894 the commission met in Berlin. Though some members noted various problems—for example, concerning the statics of the airship and its steerability—the commission (Helmholtz included) gave its approval to Zeppelin to try to develop an airship. The war ministry, however, continued to delay matters, and when the commission met again (14 July 1894) to further review Zeppelin's request, it turned down his proposal. Two days before the commission met, Helmholtz had suffered a stroke and so could not participate. Zeppelin did not get the financial support he had hoped for. However, the emperor compensated him with six thousand marks, and two years later the Verein Deutscher Ingenieure supported Zeppelin's dirigible project.[21] Less than two decades later, the Zeppelin was providing commercial air service for passengers and soon also served the German military (not least in World War I).

All in all, during these years from Krakatoa to Zeppelin, Helmholtz had made signal contributions to meteorologic science. Above all through his paper on atmospheric movements, and to a much lesser extent in his studies on cloud and thunderstorm formations, on the further refinement of his understanding of atmospheric motion, and on the energy of billows and wind, he had fruitfully returned to the study of atmospheric physics. More to the point, with his earlier work on hydrodynamics, his address "Cyclones and Thunderstorms," and other related work, his meteorologic studies proved seminal for several major late-nineteenth- and twentieth-century meteorologists. The Norwegian Vilhelm Bjerknes, who studied with Hertz, the Swede Carl-Gustaf Rossby, and the American Harry Wexler—that is, the three leading theoretical atmospheric scientists who, together with their own associates and others (including the Austrian Max Margules, who studied with Helmholtz, and the American Jule Charney)— together created the modern discipline of meteorology. All of these prominent figures expressly cited Helmholtz as an (if not the) inspirational figure for them. These and other atmospheric scientists—for that is what they had become— drew specifically on Helmholtz's writings on hydrodynamics, on cyclones and thunderstorms, and on the mechanics of atmospheric motion as they developed a new theory of storms based on the concept of discontinuous surfaces. This is not, of course, to say that Helmholtz's concept alone was the sole source of modern meteorologic theory or even of the polar-front theory; nor is it to say that they all agreed with every point he made (they did not). But, taken as a whole, Helmholtz's writings formed a central contribution to the development of modern meteorologic theory and atmospheric science.[22]

Last Hurrah in Science: Least Action, Electromagnetic Theory, and the Physics of Principles

Nearly six years passed between Helmholtz's first publication on least action and a second and a third such publication, as well as two intimately related studies of color dispersion and the nature of the ether. During these gap years (1886–93) he was preoccupied with a series of other intellectual issues: the epistemology of number and measurement, physiological psychology and a new edition of his *Handbuch*, a (second) address on Goethe, and atmospheric physics. He was also delayed by his efforts to establish the Reichsanstalt, by his teaching duties in theoretical physics, by concerns issuing from the illness and death of his son Robert, and by many other, much smaller concerns, including research on the history of the principle of least action. But there is yet another, perhaps more important reason that helps explain the gap: it was during this period that Hertz demonstrated electromagnetic waves and that he, along with Oliver Heaviside and Lodge, reformulated "Maxwell's equations" into what became their canonical form. Helmholtz, and virtually all other physicists, now no longer harbored any doubts about Maxwell's theory, even if Helmholtz still saw it as a limiting instance of his own version of electrodynamics. But what precisely did that theory mean? How did it fit into the theoretical foundations of physics as a whole? How did it apply to or how could it be reconciled with other parts of physics, for example, mechanics or electrolytic conduction? From the late 1880s onward, Helmholtz and others thought about such unresolved issues in theoretical physics, and it seems most likely that he devoted part of the gap years to pondering different aspects of them.

He spent at least part of the winter 1891–92 thinking about the implications of the principle of least action for electrodynamics: in March 1892 he read a paper on that subject before the academy, and in May he published a long paper on it in the *Annalen*. Whereas previously, in 1886, he had linked the principle to thermodynamics, he now linked it to electrodynamics, thus extending its scope and further demonstrating its utility in constituting the foundations of physics. In the light of Hertz's recent accomplishments, he derived Hertz's form of Maxwell's electrodynamic equations from the variational form of the principle of least action. He largely avoided hypothetical entities—the ether was still one—and instead emphasized the use of differential equations for linking mechanics not only to thermodynamics but also to electrodynamics. Both Hertz and Planck, for their parts, saw this approach as the best route for future theoretical physics, for it promised to unite several fields of physics together.[23]

At the end of his May 1892 paper, Helmholtz announced that he planned to publish further on this matter—and in October he read a paper on electromagnetic theory before the academy. In the winter of 1892–93, he extended that work into an electromagnetic theory of color dispersion. Here he emphasized the importance of ionic theory, assuming that ions served as oscillating centers of electric force and as light scatterers. The paper's intellectual basis in least action and, more to the point, Maxwell's now well-established electrodynamics (or, more precisely, Helmholtz's version of it), shows that Helmholtz was decidedly au courant with the latest in physics. But it also shows that his earlier work on the theory of anomalous dispersion (1874), on galvanic polarization (in the 1870s and early 1880s), and on ionic theory (1881) was still very much a part of his scientific thoughts. He was now able to give a firmer (if somewhat different) intellectual grounding to these theories. This new dispersion-theory paper proved of much importance to a series of investigators, such as Poincaré, Richard Reiff, and Paul Drude, who sought in the 1890s to understand the general relationship between matter and ether, as well as to account for microoptical phenomena. One result of these investigations into the optical properties of matter and its relationship with the ether was the emergence of a microphysics for electromagnetism, notably J. J. Thomson's "discovery" in 1897 of a new kind of "ion," namely, the electron. Helmholtz himself drew a series of consequences from Maxwell's theory for understanding the properties and motions of the ether, lecturing on this topic before the academy in July 1893.[24]

Broadly speaking, with his development of the generalized principle of least action and the derivations therefrom, Helmholtz simultaneously developed a physics of principles, a physics in which all of the grand physical principles (including those of mechanics, thermodynamics, and electromagnetism) could be derived from a single differential equation and in which hypothetical elements were either unnecessary or at best heuristic aids. (The phrase "physique des principes" apparently came from Poincaré, who drew the idea from Maxwell and Helmholtz.) This approach became the ideal and set the tone for much of theoretical physics (especially among German-speaking physicists) from the early 1890s onward.[25] Among the next generation of physicists, Planck became a leader in similar efforts, seeking in the next two decades to derive all of physics through least action, just as Helmholtz had done. In a similar and even more effective way, Einstein became a leader in developing a physics of principles, except that for him there was of course a different, more fundamental principle than that of least action to set the research agenda for the twentieth century.

29

Atlantic Crossings

The International Electrical Congress in Chicago

The winter and spring of 1892–93 were painful for Helmholtz. The death of Werner von Siemens in December put the entire Helmholtz-Siemens clan into mourning. Fritz von Helmholtz was more ill than his parents had previously realized; they both worried greatly about him. And, seemingly far less important, Helmholtz had to decide whether he should go to Chicago to represent Germany at the International Electrical Congress in August.[1]

The congress was part of the World's Columbian Exposition.[2] Already in the summer of 1892, Knapp had invited Helmholtz to visit the exposition and, along the way, himself in New York City. With his concerns about Fritz and other matters, Helmholtz delayed responding. He wanted to see America, but he had much to do in Berlin. He told Knapp that, apart from the usual aches and pains of a seventy-one-year-old, he felt well. His position at the Reichsanstalt allowed him to dispense with the "more unimportant" duties faced by a university professor, and the Reichsanstalt's development to date pleased him; his daughter Ellen now had four children, and he and Anna spent part of their summers with her and the grandchildren at her villa in Wannsee. He was, moreover, quite busy with his "literary work": gradually bringing out a new edition of the *Handbuch* (with König's help); seeing volume 3 of his *Wissenschaftliche Abhandlungen* through press; and, also with König's help, preparing for publication his notes from his six-semester course of lectures on theoretical physics. "You see," he told Knapp, "I'm cleaning up. When one sees one's friends departing around oneself, it's time for the cleanup." In short, he decided not to go to Chicago.[3]

But in mid-June he changed his mind. After long delays on the part of the Reich government, he had been asked to go to the Congress as its representative. But Anna and his family were worried about his traveling alone. She asked Knapp if he might help Helmholtz upon his arrival in New York, as he had previously offered to do, and accompany him during at least part of his travels in the United States. Helmholtz told Knapp: "I know perfectly well that the country [the United States] represents the real future of civilized humanity and that it includes a large number of interesting people, while we in Europe see chaos or Russian world rule as getting ever closer." Anna added a postscript, suddenly announcing that she intended to accompany her husband, "whether he likes it or not." She thought he felt that if he did not go, it would be a definitive declaration that he was an invalid.[4]

They were scheduled to leave Berlin on 5 August, and the Congress would open on the 20th. Helmholtz's family, as well as his doctors, had opposed his decision to accept the Reich's request on the grounds that it was too dangerous for a man his age to make an ocean voyage at the height of summer. Yet Helmholtz decided to go "in spite of" the family's "protest," Anna wrote her sister. So Kundt asked the minister of the interior for help; and the latter in turn asked Chancellor Leo von Caprivi, who decided to pay for Anna as Helmholtz's officially accompanying person and caretaker. They planned to be in the United States for six weeks. In the meantime, Anna worked daily with Estelle du Bois–Reymond on translating Tyndall's *Fragments of Science*; did the paperwork to get Hermann's very ill sister, Julie, into a nursing home; and saw to the practical preparations for their trip to America. As the departure date approached, she told Ellen that her father looked "so tired" that she feared "every small exertion on his part." She hoped the voyage would help him recover. "He looks more and more like his bust"; it was, she thought, as if Hildebrand had anticipated Helmholtz's future look. She wanted to go out to Wannsee to see Ellen; "But to leave Papa alone now might not be right, and it would rob him of his peace."[5]

They arrived in Hamburg and Bremen with an entourage (Lummer et al.) and were brought on board their ship, the *Lahn*, by a special director's steamer. Among the passengers was the Göttingen mathematician Felix Klein, also headed to Chicago. Klein spent some time with Helmholtz during these voyages. He knew the Norwegian mathematician Sophus Lie, a geometrician and group theorist, rather well; that led to a discussion of why Lie, who had been much stimulated and influenced by Helmholtz's papers on non-Euclidean geometry, had nonetheless rather vehemently attacked Helmholtz's analysis of space. In 1890 Lie had shown the deficiencies of Helmholtz's analysis and criticized it for not employing group-theoretical methods (that topic then became known as the Helmholtz-Lie space problem); and in 1893 Lie published another, similar critique. Klein blamed Lie's vehement attack on the latter's "pathological, touchy temperament" and on his feeling of being "constantly unrecognized" by Berlin mathematicians and others there.[6]

The *Lahn* arrived in the New York harbor early on the morning of 18 August. As the Helmholtzes disembarked, they found Knapp with his carriage and Ernst Pringsheim (with roses) awaiting them. In the quarter century since Knapp had worked under Helmholtz as a student and younger colleague in Heidelberg, he had become one of the leading ophthalmologists in America, had become wealthy, and (since 1888) was a professor at Columbia's College of Physicians and Surgeons. His carriage took the Helmholtzes directly to his grand residence on Fifth Avenue. Tea was served, and then Henry Villard, a German-American journalist, railroad and electrical industry financier, and politico arrived. It was very hot in New York City, and Villard persuaded the Helmholtzes ("completely against our intent," Anna reported) to sleep at his place in Dobbs Ferry. The Helmholtzes spent the following day sightseeing in New York: they took an elevator up to the fifteenth floor of the Lloyd newspaper building to gaze down at the city "from a bird's eye point of view." They visited Knapp's private clinic, the Ophthalmic and Aural Institute, one of the foremost of its kind in America. After two and a half days of sightseeing in New York, they visited Newport, Rhode Island ("terribly elegant," Anna found). Then, four days after their arrival in New York, they departed on an express train bound for Chicago, a nineteen-hour journey away.[7]

Chicago's geographical location and the natural resources in its hinterland led to its growth and prosperity after 1850. By 1890 about 1.1 million people inhabited the city, making it the second largest in the United States. It consisted socially of a mix of a small, wealthy elite and hundreds of thousands of impoverished workers. Led by its elite, it had cultural aspirations; it became a major center of art, literature, publishing, and entertainment. Its elite wanted desperately to show that Chicago's streets could be literally cleaned up and made safe and that Chicago had excellent products to market to the world. After the Great Fire of 1871 that destroyed much of the city, the World's Columbian Exposition became the second epochal event in the history of Chicago. The political and financial elite of Chicago wanted to demonstrate not only that the city had recovered from 1871 but that it was safe and cultured; they wanted everyone to see that Chicago was a city of the future.

The exposition (or "World's Fair") extended from 1 May to 30 October 1893. It marked the four-hundredth anniversary of the "discovery" of America. It was also meant to display the best material resources and industrial goods of each participating nation—more than five thousand German exhibitors showed their goods.

The exposition was (mostly) located in Jackson Park, about seven miles south of downtown and just off Lake Michigan. The largest of all such nineteenth-century expositions, it had numerous buildings, each devoted to a special subject. The Palace of Fine Arts (known after 1933 as the Museum of Science and Industry) housed, among other things, the German Fine Arts exhibit, which displayed Knaus's portraits of Helmholtz and Mommsen. (See fig. 19.2.) The Electricity Building displayed both items of everyday use and the historical course of electrical science and technology. All the leading electrical firms, including Siemens & Halske, sent their best, most advanced pieces of apparatus for display. Electricity dominated the exposition both as a power source and as the Electricity Building's aesthetic look.

As part of the exposition, but located downtown, along the lakefront on Michigan Avenue, was the World's Congress Auxiliary, a set of international congresses. These off-site congresses were held in a newly constructed building, the Memorial Art Palace, which at the close of the exposition was renamed the Art Institute of Chicago. They were intended as the intellectual complement to the material exhibitions at the exposition on the south side. The individual congresses were devoted to such topics as religion, higher education, woman's progress, labor, psychology, mathematics, and electricity. They attracted leading professionals from Europe, the United States, and elsewhere. The more than twelve hundred sessions were attended by an estimated two hundred thousand people.

Helmholtz was news. The *Chicago Sunday Tribune* carried a story on Sunday, 20 August 1893, headlined "Helmholtz Arrives Tomorrow," that showed his picture and offered a potted biography as it sang his praises to Chicagoans. "No name is better known in the scientific world," it declared, "and no man living has done greater work in solving perplexing scientific problems." The International Electrical Congress was opening the next day, and Helmholtz was going to address it.[8]

The Helmholtz entourage—it included Lummer, Pringsheim, and three others from the Reichsanstalt—arrived on the morning of 21 August, as the *Chicago Daily News* reported in detail on its front page. The reporter thought Helmholtz, "one of the leading scientists of the world," "could easily have been taken for Prince Bismarck if he had possessed the height of the iron chancellor." However, he thought Helmholtz's age—he was off by a year—was starting to show: "Seventy-three summers and winters have frosted his mustache, silvered his hair, and wrinkled his strong, intellectual face." He found him to be "slightly below" average height and said he had "lost considerable of the vigor he must have possessed in youth." The electrotechnologist Elisha Gray and Otto William Meysenburg, the founding head of the American branch of Siemens & Halske, met the Helmholtzes' train at the station and at once drove them to Meysenburg's luxurious home on Astor Street, where they resided while in Chicago.[9]

The International Electrical Congress meeting in Chicago was the fourth in a series, which included meetings in 1881, 1882, and 1884 in Paris. None of the previous congresses had been able to definitively resolve the issue of the value of the ohm. The Chicago meeting, like its predecessors, was a government-initiated one, with France, Germany, Great Britain, and the United States leading the way; there were also representatives from Italy, Mexico, Sweden, Switzerland, and elsewhere. Virtually every leader in matters of electrical metrology—with the exception of Thomson—attended. The Americans Rowland and Carhart went to see Helmholtz at Meysenburg's place shortly after his arrival there; they had been delegated by the congress's participants to ask him to serve as its honorary president, and he immediately agreed to do so. When he appeared at the congress's opening session a few hours later, he received a tumultuous welcome. Hundreds listened to him as he formally opened the Congress in his role as honorary president. The *Chicago Daily Tribune* carried a front-page story with a headline reading (in part): "Homage to Genius. Electrical Congress Opens with Enthusiasm. Helmholtz Is Here. His Presence Creates a Deep Stir of Admiration." He had to take repeated bows. Elisha Gray proudly linked the American electrical luminaries Franklin, Morse, and Henry to their European counterparts Galvani, Volta, Ampère, Faraday, Thomson, de la Rive, and Helmholtz. When Helmholtz was at last able to speak, he declared, "I feel nearly more ashamed about these excessive honors which you give me so that I am not quite assured that I have [the] necessary merits." He acknowledged that he had long "been occupied by electricity," but he thought perhaps he was being honored rather for his age than his merits. He spoke about the development of electrical instrumentation and technology during the course of his career, claiming, "The present generation, if I include myself into the present generation [laughter], has seen a greater development of science than any generation before us." During the remainder of the week, the official delegates and a limited number of others met in small groups to discuss proposed definitions and values for various electrical standards. Helmholtz actively participated in these discussions, bringing to bear his decades of experience as a physicist and as a participant in the previous congresses, as well as his leadership of the Reichsanstalt.[10]

On the evening of 24 August, the American electricians hosted a banquet for their foreign colleagues at the Grand Pacific Hotel. Edison was among them, "and at the mention of his name the hall rang with enthusiastic applause." After the meal, Gray offered the first toast, with Helmholtz responding to it. "We Europeans have come over here," he said, "with the feeling of a good father rejoicing in the success of his children, to which he himself could not attain. Europe is too narrow for the splendid march of electrical progress and America has grandly performed the task set before it. We see in you the result of better conditions and prospects than we have enjoyed, and we rejoice with you in your remarkable advancement. Gentlemen, I drink my glass to the great American Nation." That was good diplomacy. His colleagues rose and cheered.[11]

The congress's last day brought its second (and final) plenary session. As the congress's report of its accomplishments was read aloud, Helmholtz sat center stage, with the other leading figures grouped around him. The congress had built on the work of the earlier congresses as well as on Helmholtz's meeting with leading British figures in Edinburgh at the BAAS the previous year. In particular, it agreed to definitions and definitive values for the units of resistance (ohm), current (ampere), electromotive force (volt), quantity (coulomb), capacity (farad), work (joule), and power (watt). This fourth congress was therefore a historic one from the viewpoint of the electrical industry and electrical metrology. It brought to a conclusion discussions that reached back not only to the first congress in 1881 but to initial discussions in the 1860s and even earlier. Although increased precision measurements later brought some changes to the standards agreed upon in Chicago, what followed in succeeding decades was far more a matter of routinely updating or reforming electrical metrology. (It did, however, take fifteen years for the various individual countries to make these definitions and values the legal basis of all electrical metrological matters in their respective countries, and not until 1908 did they receive full international recognition.) Helmholtz now brought the Congress to a formal close: "We have performed a really important work and one which I hope will bear good fruits for the future in correcting the incongruities of electric science and electric motions so that all scientific and industrial men can understand each other in the simplest and best way." That evening, Meysenburg held a soirée in his home for select individuals to spend the evening with the Helmholtzes.[12] (See fig. 29.1.)

After the congress closed, Helmholtz spent several more days in Chicago visiting the exposition and socializing with friends. Carhart saw him there, sitting in a wheelchair. Anna reported that Helmholtz was impressed by the exposition and saw much that was new for him. The physicist Thomas Corwin Mendenhall tried to get him to do sounding experiments on board a steamboat on Lake Michigan, but Helmholtz was so busy with "other friends" that he was "obliged to decline."[13]

On the evening of 30 August, to honor and celebrate Helmholtz's seventy-second birthday (31 August) and his presence in Chicago, the Imperial German World's Fair commissioner gave a private banquet at the swanky Hotel Richelieu

Fig. 29.1 Helmholtz (*front row, just left of center*) in Chicago, with other official delegates to the International Electrical Congress, 1893. George W. Vinal, "Transition from International to Absolute Electrical Units as It Affects the Physical Chemist," *Journal of the Washington Academy of Sciences* 38:8 (1948): 265–69, on 266.

on the Michigan Avenue lakefront. He invited "thirty-five distinguished Germans," including senior German civil servants and provincial government representatives, Knapp, and Helmholtz's five Reichsanstalt associates. The commissioner declared that the work of Helmholtz and the other scientists in Chicago had strengthened "the fraternity of nations." The following day, the Helmholtzes left by train for Denver and other points west.[14]

Colorado and the West

The Helmholtzes, Knapp, and Knapp's daughter traveled by train together out to Colorado and the Rockies.[15] The railway travel conditions, their stays in various luxury spas, and the towns and sites they visited in Colorado were typical of those experienced by well-to-do Europeans who toured the American West in the late nineteenth century. The trip across the Midwest and the Great Plains, from Chicago to Denver, took thirty hours; Helmholtz withstood the travel well. In Denver they rested for a day, relaxing in a luxury eight-story hotel with a spectacular view of the mountains, then toured the city with a group of local ophthalmologists. From there they went to Manitou, and then made a six-hour round-trip journey up to see Pike's Peak. Like many other Europeans, they compared the biogeography and travel in the Rockies to that in the Alps. They spent an afternoon in Colorado Springs and then traveled twelve hours by railway to scenic Glenwood Springs, on the edge of the Colorado River.[16]

From Colorado Springs they went to Grand Junction, where there was "a hot saltwater bath with a large swimming basin; huge, barren mountains on all sides; coal pits; and every element of ugliness. Plus a gorgeous hotel!" The pianist at their hotel introduced himself to the Helmholtzes: he had been a school friend of their son Richard in Heidelberg. They visited an Indian school. Then they traveled further through the Rockies by railway, through the Marshall Pass (eleven thousand feet). Shortly after they began their descent, their train's locomotive and first car became accidentally decoupled from the rest of the train, leaving them abandoned on the track. Help soon arrived, but the delay meant that they had to spend a day and a half waiting in a nearby town before the train could again proceed. The Helmholtzes seemed to enjoy it all, happy to sleep in a coach at night if need be and waking refreshed the next morning. But Helmholtz also tired of drinking the local water and decided he did not want to travel any further west; he apparently had no great desire to see California. So on 9 September they headed back east with Knapp and his daughter. It took thirty hours to reach Kansas City and an additional ten to reach St. Louis. Anna reported that she and Hermann looked "like vagabonds." "Sun and dust have unbelievably affected our outward appearance," she asserted. Everyone, Europeans as well as Americans, but above all Coloradans, was concerned about the fall in the price of silver. So much had already been extracted from Colorado mines that some were being shut down. This in turn led to a further weakening of the economy. Arriving in Kansas City, they were astonished to find local ophthalmologists and their wives awaiting them with roses. On board the train across Missouri bound for St. Louis, they met a farmer of German descent who, as they were departing, said to Anna, "And now Lady, tell me which is Professor Helmholtz?" And Anna reported to Ellen, "To my question as to how he might know that he might be on the train, he laughed and pulled out his newspaper, wherein it was reported that Papa is expected in Kansas and St. Louis."[17]

Helmholtz had found the West "more interesting than beautiful and pleasant." To his German eyes, the landscape was filled with "endless, bleak, and desolate places far from one another." He found it boring, hot, and dirty. Yet he was impressed by the sheer size of America. He thought the country, as he told Ellen, was still at its cultural beginnings. "To date, everything is still very unfinished and appears partly very unreasonable and paradoxical." But he also thought that was only to be expected, since America's culture, he said, had begun with the steam engine and electric lighting, leaving matters like the culinary arts and household management in rather primitive states. In St. Louis, a reporter questioned him about currency and national economic issues, even though Helmholtz assured him, "I have never studied national-economic questions."[18]

Boston and Harvard

The Helmholtzes spent three days traveling (now alone) from St. Louis to Boston, passing through Canada for a few hours and stopping in Niagara Falls, which greatly impressed them both. "Niagara Falls," Helmholtz wrote, "is the first thing in America that really makes a simultaneously powerful, great, beautiful, and refreshing impression." Moreover, he reported that throughout their trip people had treated them extremely well. On the evening of 15 September, they arrived in Boston.[19]

They spent four days there, extending their visit in part so that Henry Pickering Bowditch, the physiologist and the dean of the Harvard Medical School, could return in time to see Helmholtz. Bowditch had studied with Ludwig in Leipzig, it may be recalled, and had married a German woman. Ludwig had greatly inspired him, and Bowditch's laboratory in Boston, modeled on Ludwig's, became the first physiological laboratory in the United States. Bowditch asked several Boston-area friends to show Helmholtz around until he returned, when he would seek out Helmholtz at the Hotel Vendome.[20]

The Helmholtzes' visit to Boston quite naturally included a visit to Cambridge, that is, to Harvard. Twenty-two years earlier, Helmholtz had been so little aware of Harvard that he had mistakenly referred to it as the University of Cambridge, in Cambridge, Massachusetts. It is a measure of Helmholtz's improved awareness of science (and scholarship) as a non-European phenomenon, or of how much Harvard had risen in the world of international science, that he now decided to devote a day to visiting there and to spend still more time with some of its faculty. Several of them gave the Helmholtzes an extensive tour of the campus, including visits to various laboratories.[21]

Among these was the psychological laboratory of Hugo Münsterberg, who was a physiological psychologist but one with strong epistemological interests. He had studied physiology (also with Ludwig), medicine, and psychology at Leipzig and Heidelberg and was especially influenced by Wundt, in whose laboratory he had worked. His early experimental investigations concerned the eye. At Heidelberg he met Josiah Royce, and in Paris he met William James. James arranged for him to teach at Harvard and to lead its Psychological Laboratory for three years and perhaps permanently. That allowed James to devote himself to philosophy and to spend a year in Europe. At Harvard, Münsterberg became close to Royce and to George Santayana; their set also included the philosopher George Herbert Palmer, the art historian Charles Eliot Norton, and Bowditch. Like Helmholtz, Münsterberg was especially interested in acoustics and knowledgeable about music and the arts. He had apparently seen Helmholtz in Chicago, and Helmholtz now visited him at his home in Cambridge.[22]

Helmholtz also met James again. By mid-September, having spent fifteen months in Europe, James had arrived back in Cambridge. He told Carl Stumpf, who had just been appointed professor of psychology at the University of Berlin: "We had Helmholtz here, by the bye, in the autumn. A fine looking old fellow, but with formidable powers of holding his tongue, and answering you [with?] a friendly inclination of the head. His wife was a *femme du monde*, however, and fully made up for his lack of conversation." Helmholtz was, in any case, "a *herrlicher Mensch*." James wrote similarly to his brother Henry from his country place in Chocorua, New Hampshire:

> We had the great Helmholtz and his wife with us one afternoon, gave them tea and invited some people to meet them; she, a charming woman of the world, brought up by her aunt, Madame Mohl, in Paris; he the most monumental example of benign calm and speechlessness that I ever saw. He is growing old, and somewhat weary, I think, and makes no effort beyond that of smiling and inclining his head to remarks that are made. At least he made no response to remarks of mine; but Royce, Charles Norton, John Fiske, and Dr. [Henry Pickering] Walcott, who surrounded him at a little table where he sat with tea and beer, said that he spoke. Such power of calm is a great possession.[23]

James was impressed.

Washington and Johns Hopkins

From Boston, the Helmholtzes traveled to Washington and Baltimore (possibly stopping over in Philadelphia).[24] Before they arrived in Washington, Mendenhall, a leading science administrator and head of the US Coast and Geodetic Survey, left a solicitous letter at their hotel asking Helmholtz precisely when he would be arriving and saying that he had organized some of Washington's leading scientists and science administrators into a group that awaited Helmholtz's visit. These included Langley, the Smithsonian's secretary; Newcomb, an astronomer, the director of the National Almanac Office, and a professor of mathematics at Johns Hopkins; and John Shaw Billings, a surgeon, library director, and bibliographer who headed the library of the surgeon general's office (later it became the National Library of Medicine). Billings also had designed the recently opened (1889) Johns Hopkins Hospital in Baltimore and later directed and reorganized the New York Public Library. In addition there were others who would be "much pleased" to see Helmholtz. The group would await Helmholtz at the Cosmos Club—"composed mostly of gentlemen engaged in Scientific work"—located across the street from the Arlington Hotel, where Helmholtz would be staying. Mendenhall and his wife also offered to guide the Helmholtzes around Washington, "to ride about the City and suburbs for a few hours to see some of the beauties upon which we pride ourselves." He also wanted Helmholtz to visit the US Coast Survey.[25]

Villard was also in Washington, and he too wanted to introduce Helmholtz to the city. His connection to Helmholtz came not only through Knapp, but also through his extensive involvement in financing electrical power and lighting systems. Along with also financing railroads, Villard was an early stockholder and director of the original Edison Light Company; he knew Siemens and Emil Rathenau of the Allgemeine Elektrizitätsgesellschaft in Berlin; and with them and others he reorganized and recapitalized (in 1889) a new American firm, the Edison General Electric Company, which in turn absorbed all the prior American Edison lighting companies. He had served as its president until 1892, when it combined with the Thomson-Houston Electric Company to become the General Electric Company, which later collapsed. His own financial backers, including Georg Siemens of the Deutsche Bank, were in Germany. Villard was also a major figure in the Democratic party. In the 1892 presidential election, he worked on the Democratic campaign, raising money for Grover Cleveland and helping him get the nomination and win the election. This in turn gave him entrée to Cleveland when he took office in March 1893. He advised the president about cabinet and diplomatic appointments and warned him about the coming financial crisis. Villard knew the Washington political scene well.[26] He showed the Helmholtzes around Washington, introduced them to various politicians, and arranged for them to meet Cleveland at the White House on the afternoon of 26 September. Anna wrote Ellen: "Washington was very amusing and interesting under Villard's guidance. We saw the entire political machine, including President Cleveland, the silver-haired Senators, a session of Congress, the [leading] scientific men and institutions, the city's magnificent neighborhoods, and left feeling very satisfied. The President received us and gave us his portrait."[27] In short, they met Washington's scientific and political elite.

The Helmholtzes also spent a day visiting Baltimore, that is, at Johns Hopkins. President Gilman himself gave Helmholtz a tour of the campus, while Billings showed Anna the Hopkins hospital, "this wonderwork," in east Baltimore. She was also impressed with its nursing school, and Billings gave her a great deal of literature about the medical institutions as well as a copy of his annual catalog and index for medical literature.[28]

New York City: Steinway, Columbia, and AT&T

By Sunday, 1 October, the Helmholtzes had returned to New York City, where they again stayed with Knapp.[29] They visited both Steinway Hall and the Steinway piano factory, meeting with William Steinway, the firm's president. At the factory, Helmholtz was especially interested in viewing the mechanical insides of a piano under construction, to assess how much the latter "agreed with the results of my acoustic studies." In particular, he carefully observed the lengths and pressures of the strings and their resulting influence on sound tone. Steinway insisted that Helmholtz choose a new grand piano for himself, even though the Helmholtzes did not need one. Anna reported: "The entire firm gathered around and it was very beautiful. Our current grand piano is sufficient for us, but Mr. Steinway assured us that he has learned so much from Papa's acoustics that he can only show his thanks through the success of his instrument." That evening a "large dinner" was given for them: Carl Schurz, a former German revolutionary and now an American journalist and statesman; the Villards; "and important medical doctors in the city," among others, attended. Steinway recorded in his diary: "Beautiful day." Late that October, Steinway sent Helmholtz a new grand piano, in exchange for the one he already had. The piano arrived at the Helmholtz home in early December, and Helmholtz thanked him for it. He wrote in English:

> We have already had occasion to hear it played under the hands of several of our musical friends. We cannot but admire the fullness and strength of its tone as well as the mildness and softness of the same, qualities which one rarely finds united. The instrument found the same degree of applause and admiration also among my friends, who tried it. One of them, my colleague Professor Planck[,] is a very skilful [sic] musician, and at the same time a first rate mathematician and physicist. He is my successor at Berlin University in teaching the acoustical theory of Music. In his judgment I have the greatest confidence and he was quite of the same opinion as my other friends, my wife and myself. We feel ourselves, therefore, certain to have become the owners of one of the most accomplished musical instruments by your kindness.[30]

Steinway and Helmholtz were indebted to each other, as they had been since 1871.

While in New York, Helmholtz delivered two lectures and participated in other events at Columbia College. His appearance there came as Columbia was in the midst of profound institutional changes. It was an opportune moment for Columbia's leaders as they sought to advance their plans for radically reconfiguring the institution.

Columbia College began in 1754 as King's College, in lower Manhattan; in 1857 it moved to new quarters in Midtown. Well into the 1880s Columbia remained little more than a local institution that catered to the well-to-do sons of New York's social and financial elite. Unlike the new Johns Hopkins or the venerable but reforming Harvard under Charles Eliot, Columbia had remained a local college and had no national or research mission. Its quarters had become cramped, and it had financial problems that required attention.

To be sure, in the 1880s, Frederick A. P. Barnard, the college's longtime president, sought to transform Columbia by adding to its collegiate mission the goal of becoming a research, German-style university that stressed graduate education. He and others sought to make Columbia into a first-rate national if not international institution, while at the same time emphasizing its need to participate in New York City's cultural life and address its social needs. Three of Barnard's key allies—John William Burgess, Ogden Rood, and Nicholas Murray Butler—were great admirers of Helmholtz. Burgess had studied philosophy, economics, law, and history (with Mommsen) in Berlin, Leipzig, and Göttingen from 1871 to 1873, and he had taken Helmholtz's course on logic. At Columbia, he became a leading political scientist and constitutional lawyer. Rood was Columbia's professor of physics. Butler, an educator who had taken a class with Helmholtz at Berlin, was Barnard's student and protégé, as well as Burgess's student. He later became Columbia's long-serving president (1901–45). All three had studied at and were greatly impressed by German universities, as had a number of their colleagues within the so-called university (as opposed to the college) faction.

Barnard did not get far in his attempt to change Columbia; but after his death in 1889, the struggle between the two factions reached a climax when the trustees narrowly elected Seth Low, a Columbia graduate, well-to-do merchant, former mayor of Brooklyn, and representative of the university faction, as the new president. Low aimed to realize what Barnard had only begun: to emphasize the importance of academic research and graduate education and to make Columbia at once a leader in New York's civic culture and in solving the nation's social problems, all the while turning Columbia into a first-rank national and international institution. Toward these ends, Low reorganized, consolidated, and expanded Columbia administratively during the 1890s. Among the many administrative restructurings, in 1891 the College of Physicians and Surgeons was fully absorbed into Columbia, which thus now had its own, fully integrated medical school. Low also commissioned an entirely new campus: in 1892 Morningside Heights, on the Upper West Side, was selected and purchased as the new site; it was dedicated in 1896. The institution was renamed Columbia University in the City of New York, and in 1897 the Midtown campus was moved to its new Morningside Heights location.

Hence, when Helmholtz appeared in New York in October 1893, Columbia as a research institution was still in its infancy. Its trustees, faculty, and students still had all too few examples among Columbia's own scientists and scholars who embodied the research ethos and had made major research contributions to their fields. And Columbia's plan to construct an entirely new campus meant that Low and the other trustees needed to raise funds on an ongoing basis. A visit by Helmholtz, however brief, fit their needs perfectly.

Late in the afternoon of 3 October, Helmholtz spoke at Columbia's College of Physicians and Surgeons. He addressed not only its students but also its medical professionals and various scientists. Knapp had invited him to lecture there, and he asked that Helmholtz speak on his invention of the ophthalmoscope. Knapp chaired the event and, in his introductory remarks, told the audience that Helmholtz "had for years been the acknowledged leader of German science." He briefly outlined the history of ophthalmology and discussed Helmholtz's contributions to physiological optics. Helmholtz was then greeted "with most enthusiastic applause." In his prefatory remarks, Helmholtz apologized for speaking about himself and for his English. He said he had often been asked to speak on this subject and that he had virtually nothing new to say on it. He noted that others before him had investigated the eye's luminosity; he thought his own novelty in making the invention lay essentially in his understanding of the mathematics and physics of such luminosity, as well as in constructing a crude version of the instrument to capture the reflected light from the anterior part of the eye. He reportedly said:

> All that was original with me in the matter was that I went on to ask how the optic images could be produced by the light coming back from the illuminated eye. All my predecessors had failed to put this question to themselves. They had stopped in the middle of their way instead of going on to the end. As soon as I answered that question I saw also how an ophthalmoscope could be constructed, and it took me two days to do it and successfully experiment with the new instrument. I say this to impress upon you how necessary and how useful it is to go on to the end when investigating natural phenomena. You must not go half-way and then stand still or go back; you must finish your meditations, go to the end, so that you may see clearly the full relation of the several phenomena to one another.

The large audience, whose numbers included Alexander Graham Bell and Seth Low, received his talk enthusiastically. *Scientific American* reported on it in its news column, declaring that Helmholtz's invention of the ophthalmoscope was "one of the crowning achievements of medical science in the nineteenth century." It had "saved the eyesight of thousands." Helmholtz's work in physiological optics was judged to be "of inestimable value." That evening the Helmholtzes dined at the Lows' on the Upper East Side.[31]

After dinner the two couples went to the Law Library of Columbia College, where Low had arranged a reception for the Helmholtzes. The *New York Times* reported that Low had invited more than five hundred of New York's most socially, politically, financially, and intellectually prominent individuals, not a few of whom were from Old Knickerbocker families, to meet Helmholtz. Among those present were lawyers like Joseph H. Choate and the Franco-American Frédéric René Coudert, who had played a decisive role in France's recent (1886) donation of the Statue of Liberty to the United States; the statesman Elihu Root; E. L. Godkin, the editor of *The Nation*, and the publisher Henry Holt; financial tycoons and Wall Street types such as J. Pierpont Morgan; Edison; other industrialists, like Abram S. Hewitt, who was also a mayor and a congressman; a few (other) politicians, like Mayor Thomas F. Gilroy; supporters of New York's major cultural institutions; philanthropists of all sorts; many of Columbia's trustees and senior administrators; and, finally, many of the college's leading faculty members, including Burgess, Rood, and Butler. "Prof. Helmholtz," the *Times* told its readers, "is admired wherever science has a friend." It characterized him as "the son of a modest teacher at the Potsdam Gymnasium" and said he had "risen by merit alone to an elevated station." That was true enough, though it also happened to fit the American Horatio Alger stereotype of success. The newspaper emphasized Helmholtz's intellectual honesty and his great scientific productivity.[32] What the article did not say was that the evening's purpose was not simply to meet and greet Helmholtz; it was also an opportunity for Low and the trustees to illustrate the kind of place Columbia was (or, better, hoped to be) and the kind of company it kept. In other words, Low and his fellow leaders at Columbia utilized Helmholtz's presence in the city to advance Low's drive to solidify financial and political support as he led the transformation of Columbia College into Columbia University.

Helmholtz's second and final lecture was to Columbia's student body, given on 5 October in the college's main library, with a standing-room-only audience. Among the listeners was Robert A. Millikan, then a graduate student studying physics. Bell and Steinway were also present, as was Josiah Willard Gibbs, professor of mathematical physics at Yale and America's foremost talent in that subject. Although Gibbs had studied with Kirchhoff in Heidelberg and knew Helmholtz's work in electromagnetic theory and thermodynamics well, he and Helmholtz had apparently never met in Heidelberg and had met only briefly in New York. Helmholtz regretted that there was no time for a sustained discussion with him that day.[33]

As Low and Helmholtz entered the lecture hall, the audience broke out into a thunderous applause. Low introduced Helmholtz as someone who had given "high service" not only to "science" but also to "humanity." He noted that Helmholtz was an expert in matters of sound and that Bell, inventor of the telephone, had come all the way from Halifax, Canada, just to see Helmholtz. Low then turned the podium over to Helmholtz, who, "in spite of his great age," showed "great vigorousness." He spoke for an hour, "with a clear, audible voice, . . . no trace of tiredness," in "a rather good, if also naturally not very flowing English."[34]

Helmholtz believed his lecture topic—the acquisition of knowledge—was of interest to all students. He said it applied not only to the sciences but to the arts as well. In his handwritten lecture notes, he posed the broad questions: "What is science? and why [has] the influence of science on all kinds of human activity . . . been so increased?"[35]

His themes and major points were (unsurprisingly) the same ones he had been advocating for the past half century: science is multifold in scope and nature; daily experience is a rich source of knowledge; human memory must be supplemented by written sources, which give a sort of permanency to what becomes known; mankind's ability to use knowledge acquired over generations is one reason for its "superiority . . . over all other living creatures," hence the importance of the "conservation of all knowledge" in a variety of written formats; the laws of nature must be sought on the basis of facts; science has brought technological, industrial, and civilizational progress; owing to the inherent involvement of psychological factors, the humanities cannot produce laws; intuition in science is essential but also must be checked; and it is inherent to science, he added, that it is always changing, developing. He ended his lecture by advising Columbia's students to "go and follow the example given by the scientific and industrial development of modern times. Endeavour to find everywhere the effects of eternal laws." Low led the audience in giving Helmholtz three loud "Hurrahs" and "C-o-1-u-m-b-i-a! Helmholtz!"[36]

Bell had long admired Helmholtz. He reportedly "explained that, without Helmholtz's researches in the realm of sound, he could never have invented the telephone." (Steinway also reportedly said that "without Helmholtz's epoch-making discoveries the manufacturing of pianos would never have reached such development and completeness.") After the lecture, Bell immediately took Helmholtz to the downtown office of the American Telephone and Telegraph Company (AT&T). He wanted to demonstrate the latest in AT&T's long-distance telephony operations. Telephone company officials and others, including at least one reporter, awaited them. Calls were placed to Boston, where Helmholtz spoke with the president of the company; to Washington, where he spoke with Langley (they talked about the Smithsonian and its researchers); and to Chicago. Helmholtz reportedly characterized the telephone as "a wonderful tool, for which a great future" lay ahead and declared his amazement at "how far telephony" had come in America. After seeing the rest of the telephone center, he went to Low's home for dinner.[37]

Before leaving New York, Helmholtz visited Pupin's laboratory at Columbia and also visited him at his home on the Jersey shore. He had been Pupin's principal teacher—"my revered teacher," Pupin called him—and had been instrumental in Pupin's receiving his position at Columbia. Pupin was then in the process of inventing a method (apparently the first) of electrical tuning resonators (that is, like the once "modern" form of adjusting a knob on a radio to transmit or receive the right wavelength). "When Helmholtz visited this country in 1893," he later wrote, "I showed him my electrical resonators and the research which I was conducting with their assistance. He was quite impressed by the striking similarity between his acoustical resonance analysis and my electrical resonance analysis, and urged me to push on the work and repeat his early experiments in acoustical resonance, because my electrical method was much more convenient than his acoustical method." The electrical resonator subsequently became used for selectively detecting alternating currents of definite frequency and also for harmonic telegraphy. Pupin and Rood were the only scientists in New York City who had a chance to spend any amount of time with Helmholtz. Most had only a brief introduction, and many were disappointed.[38] There simply was not enough time for him to spend with his colleagues.

Dark Passage Home

Late on Friday evening, 6 October 1893, the Helmholtzes boarded their ship, the *Saale*, which departed early the next morning for Bremen. Steinway sent the Helmholtzes "a fine souvenir of flowers." From the start of the voyage, they traveled "on the edge of a typhoon [i.e., a hurricane]." But Helmholtz felt well and refreshed, even as Anna suffered, as always, from seasickness. Six days out, on Thursday evening, 12 October, while Anna lay ill in her berth, Helmholtz went upstairs to the men's smoking lounge, carrying a copy of Kuno Fischer's new book, *Arthur Schopenhauer*. He sat for a while with Felix Klein, the ship's captain, and Dr. William J. Morton, an American neurologist who was a friend of Knapp's. When it became too warm for him, however, he decided to leave, said goodnight, and headed for his cabin. Morton, who thought Helmholtz had looked a little red in the face as he left, then "heard a heavy fall, hurried outside, and found Papa lying below, at the [foot of] the steps," as Anna wrote Ellen. He had fainted and fallen down a steep set of stairs, with perhaps as many as fourteen steps. Morton had Helmholtz brought to the ship doctor's quarters. In the meantime, Klein told Anna of the incident and said that blood was streaming from Helmholtz's forehead and that two doctors were attending him. Together they went to Helmholtz, where they found him conscious—he could answer questions—and his forehead already bandaged. Anna thought the ship's doctor (one Frobenius) but above all Morton, who attended to the wound itself, had saved Helmholtz's life. The ship's captain then had six of his stewards carry Helmholtz to his (the captain's) own berth, where Helmholtz more or less slept that night. Still, blood continued to seep out of his wound—it was a deep cut—and Anna and three others took turns watching over and attending him. Frobenius kept a finger on the wound to try and staunch the bleeding, which in fact only finally ended after Morton closed the wound using a compress. Through it all, Helmholtz's mind and speech remained intact. Everyone was extremely solicitous. Everyone but Anna at first feared that Helmholtz had had a stroke; by contrast, she suspected that his fall was due to a fainting fit, something that he had experienced once before, long ago. She surmised that he was already unconscious before the fall, "because he had not stretched out his hands before him for protection."[39]

Helmholtz spent the following day resting in bed, with two stewards alternately watching over him. The seas were still very stormy; Anna's seasickness worsened until Morton gave her "an American remedy against seasickness, bromo soda," that nearly cured her. She wrote, "Today I could stand up straight." By the following afternoon, Helmholtz's condition had improved markedly. In the meantime, however, they were on the high seas, without sunlight or starlight—the captain navigated by compass—with the ship rolling amid tossing waves, "in the thickest clouds," at least four days out from Southampton.[40]

The Helmholtzes had originally intended to visit the Thomsons in London and then go to Paris, where Hermann was to meet with the Forty Immortals of the Académie Française. Instead, they now intended—they were four days into his recovery—to go directly to Bremen. "And indeed, we can only be thankful if we can bring him back home alive," Anna told her sister. Morton thought that the coming five days would tell Helmholtz's fate. Anna felt guilty about the accident, that she had not paid attention to him precisely when he needed her most. It was a grim ending to what had otherwise been an excellent trip.[41]

By 17 October they had reached Bremen. Helmholtz was in a "somewhat subdued" state. A carriage brought him from the ship to their hotel in Bremen ("a sad procession!"). He was still under a doctor's care and needed "absolute rest," Anna explained to Fritz. She was glad that he had survived and that his faculties were intact, but he was "very weak." She sent a telegram to one of the secretaries of the academy announcing and detailing Helmholtz's accident. Richard Wachsmuth, the physicist *aide de famille*, and Ellen rushed to Bremen to help out. Many telegrams arrived—from the emperor, from the mayor and the senate of Bremen, and from many others whom the Helmholtzes knew or even did not know—all wishing them the best. Flowers and fruits arrived, too. The chief medical doctor of the Bremen city hospital personally cared for Helmholtz. "And we were in other ways overwhelmed with kindnesses and refreshments in the most charming ways," Helmholtz later told Knapp. They remained in Bremen for eight days, as Helmholtz gradually regained his strength. They finally left for Berlin on or about 23 October. His doctor arranged for them to make the seven-hour trip "in a director's carriage, with a bed and living comforts like princes

[have]." However, Helmholtz was barely able to read and could only occasionally even be read to.[42]

By 5 November, Helmholtz was considerably though not yet completely recovered. During the month after his accident, he told Klein, he had "suffered very much from dizziness, even if the head wound had closed very quickly." He told Knapp that during the first weeks of recuperation he could barely walk without feeling tired and dizzy; that he had difficulty reading, and so had to limit it; "and had a very disfigured face with blood beneath it." Nonetheless, he was able to receive friends, and he had a visit from the empress.[43]

Mutual Honor: A Dedication and Two Forewords

During the winter of 1892, Hertz was preparing, at the urging of the publisher J. A. Barth, to republish as a single volume his numerous individual articles on electric waves that had appeared in the *Annalen*. After finishing his introduction, he wrote Helmholtz:

> I believe that from this Introduction, rather than from the individual articles, it will emerge better and more clearly that my work has itself emerged not so much directly from the study of Maxwell's works, as I'm always hearing, as much more and essentially from the study of your Excellency's works. Indeed, it was your personal suggestion that provided the first initiative hereto. It would be a great joy to me, both in regard to my general gratitude to you and my honor of you, if you would permit me on this special occasion to dedicate the present work in its collected form [of articles] to you. The individual articles themselves lack the expression of thanks that I owe you in respect to them, as indeed in every respect. I would be very sorry if you should want to refuse me this request because it is a way to atone for this purely accidental omission [of an earlier thanks]. Any sign of agreement on your part—for example, your calling card—would satisfy me.

Hertz's sense of debt to and admiration for Helmholtz could hardly have been clearer. Perhaps even "love" is not too strong a word here. Certainly he saw Helmholtz as his intellectual guide in matters scientific and as his mentor in professional matters generally. Helmholtz was, it seems not unreasonable to suggest, something of a father figure to Hertz, and perhaps Hertz was something of a son to Helmholtz. Be that as it may, Helmholtz, who responded immediately to Hertz's letter, now addressed him as "Honored Friend and Colleague," that is, as his professional equal. He wrote: "I accept with great thanks your offer to dedicate to me your collection of articles on electric waves. This is, for once, a dedication for which I can have undiluted joy. For those in need of instruction, whatever their country, it is a great advantage to have all the articles collected together." Helmholtz himself, as he incidentally noted in this letter to Hertz, was in the process of writing a piece on the reformulation of Maxwell's equations in terms of the principle of least action. Hertz's collection of papers, *Untersuchungen über die Ausbreitung der elektrischen Kraft* (1892) with its accompanying dedication, appeared shortly thereafter and was "dedicated by the author to his Excellency, the Real Privy Counselor Herr Hermann von Helmholtz, in deepest respect and gratitude."[44]

A few months later, in July 1892, Hertz began suffering from a chronic cold and a painful ear-and-nose infection, which then spread to his teeth and jaw (and, in time, elsewhere). By November he was so incapacitated that he could neither meet his teaching responsibilities nor conduct scientific work. Helmholtz heard about Hertz's illness and wrote to their mutual colleague Pflüger at Bonn, asking about Hertz's health. When Hertz learned of Helmholtz's inquiry, he wrote him to say that Helmholtz's concern had helped him "feel much better." His recovery, however, was going slowly, and he told Helmholtz that he expected he would not regain his full physical strength and his ability to work until sometime the next year. He also told him that, despite his illness, he was engaged in theoretical work inspired by Helmholtz's work on the principle of least action. He assured Helmholtz that he was working as hard as his sickly body would allow.[45]

In early January 1893, Hertz underwent a series of medical procedures or surgical operations. He nonetheless managed to continue working on theoretical mechanics, which at least gave him some relief from his medical problems, he said. By March he was again able to see students. His afflictions persisted and indeed spread to other parts of his body, and although he underwent at least two further medical procedures, these were of little or no help.

By December he had managed to complete a manuscript on mechanics. He arranged for its publication under the title *Die Prinzipien der Mechanik in neuem Zusammenhange dargestellt* (1894). Shortly thereafter, on 1 January 1894, at age thirty-six, emaciated and bedridden, Heinrich Hertz died of blood poisoning, probably the end result of the multiple infections he had suffered during the previous eighteen months. Hertz's death left Helmholtz feeling depressed during what were the remaining eight months of his own life. That April, he recovered to a certain degree with a trip to Venice and Klagenfurt. It was his last trip ever.

Hertz had thought that only the first half of *Die Prinzipien* was fully ready for publication; he had hoped to have more time to rewrite the second half, but that was not to be. Instead, he had to entrust all further editorial work to his assistant, Philipp Lenard, who thought that Hertz would have changed only two (unnamed, and so apparently small) points, but Lenard did not want to make any changes without first consulting Helmholtz. Helmholtz already had a copy of the manuscript in his hands, so that simplified things. But it took him some time to grasp Hertz's aims, and he felt uncertain enough with Hertz's text that he refused to edit it in any way or even to agree to Lenard's two proposed editorial changes, instead advising him to simply mark the doubtful passages for the reader's attention. Hertz's book appeared late that summer, with two forewords. One, by Hertz, noted that, with a couple of important exceptions, the entire book was "essentially influenced and dependent upon" Helmholtz's earlier studies on the principle of least action as the foundation of physics and on Helmholtz's related work on monocyclic systems.[46] The other foreword was by Helmholtz.

Even before the book's appearance, however, and indeed immediately after Hertz's death, Bonn began looking for a successor. Lipschitz, Helmholtz's old friend at Bonn, consulted with him in this regard. In mid-January, Helmholtz told Lipschitz that he thought Hertz ("this unique man") was irreplaceable. Lipschitz had asked about Eilhard Wiedemann as a possible replacement; Helmholtz replied that he thought Wiedemann was right for Bonn—that is, he was a skilled experimentalist, a good teacher, and well versed in the literature—but that Wiedemann was young. He said that if Bonn wanted someone older, then he would recommend Kohlrausch and, if that did not work out, Quincke. However, he also thought that neither of the latter two might want to leave their present positions, and so he also recommended Kayser, who had done work in spectroscopy, on the adhesion of gases, and on the velocity of sound in tubes—and who happened to be Kundt's and Helmholtz's former student and (for many years) Helmholtz's former assistant.[47] Kayser got the appointment.

In early July 1894, with the academic year behind him, Helmholtz composed his foreword (read: memorial notice) to Hertz's book. He declared, "The news of the death of this favored darling of genius [i.e., Hertz] was deeply shocking for everyone who has grown accustomed to witnessing the progress of humanity as consisting in the broadest development of its intellectual capabilities and in the rule of mind [*Herrschaft des Geistes*] over both the natural passions and the antagonistic forces of nature." He spoke, too, of Hertz's "timid modesty," despite his "outstanding natural talents," and of his ambition "to achieve high aims." He thought that people like Hertz and other young scientists who have "well-endowed natures" were "all the more dissatisfied with their own work" precisely because of their "greater . . . capabilities and their [higher] ideals." Nor was that quite all: "The most gifted [individuals]," he continued, "can only therefore clearly reach the highest [levels of achievement] because they are the most sensitive about any sort of shortcomings, and so they work indefatigably" to overcome them. That statement about Hertz might as well have been an assessment of his own ambitions and achievements. Helmholtz felt greatly pained by

Hertz's death: "Among all the students whom I have had, I always considered Hertz as the one who had penetrated furthest into my own circle of scientific thoughts and upon whom I ventured to place the most certain hopes for their further development and enrichment."[48]

Helmholtz devoted much of his foreword to recounting the history of nineteenth-century electrodynamics, both as it concerned his own clarifying efforts and as it set the stage for Hertz's work on electric waves and related matters. He stood in awe, he said, before Hertz's ability to analyze problems and to complement that analysis with the most appropriate type of experimental procedures. "Through his discoveries," he said, "Heinrich Hertz . . . secured permanent fame for himself in science."[49] He was a proud *Doktorvater*.

Then, in less than one page of remarks, he turned his attention to Hertz's book on the principles of mechanics. He called it "the last memorial of his [Hertz's] earthly activity." He spoke of Hertz's "attempt" to develop a new, complete system of mechanics deriving "from a single fundamental law, which, from a logical point of view," could "only be considered as a plausible assumption." He praised Hertz's "great acuity" in "striving" to derive all mechanics from a single law and noted his "very admirable development" of mechanical concepts. But when he referred to Hertz's use of "imperceptible masses and invisible [hidden] motions" as a substitute for his rejection of the concept of "force" in mechanics—this rejection and a geometrization of mechanics in terms of mass points were the book's principal innovations—he noted that this was merely a "hypothesis" and that, "unfortunately," Hertz had not given any "individual [concrete] examples" to instantiate it. Though this approach had similarities to Helmholtz's own invention of monocyclic systems, even Helmholtz had doubts about Hertz's approach to mechanics. He placed Hertz's proposed hidden motions sans the concept of force alongside the approaches to microphenomena put forth by Thomson (atomic vortices) and Maxwell (a system of rotating cells). Helmholtz himself favored another approach, one in which facts and laws were represented generally by differential equations. He said that while he did not mean to close discussion on the possibilities advocated by Hertz, Thomson, and Maxwell for understanding the foundations of mechanics, he thought Hertz's approach entailed "great difficulties" that had to be overcome. "In the future, this book will still possibly be of great heuristic value as a guide for the discovery of new, general characteristics of the natural forces," he opined. He welcomed Hertz's imaginative, alternative approach to mechanics, but he did not adopt it as his own or even endorse it.[50]

In fact, he even found this final, book-length work by his stellar former student and colleague difficult to understand. Though it naturally attracted the attention of physicists, it found little if any resonance or application in their actual work; some criticized it as being far too formal and abstract in its approach, not to mention impractical in its abandonment of the concept of force and its replacement by the hidden motions of invisible masses. It was, instead, largely philosophers like Ernst Cassirer and (the young) Ludwig Wittgenstein, and especially philosophers of science (the Vienna Circle or, more generally, the logical positivists), who found something of heuristic value in Hertz's book—namely, the notion that a scientific theory merely represented a "picture" (*Bild-Theorie*) or was merely a symbol of reality and that such pictures or symbols would naturally vary among persons, places, and times. That conception of theory as constituted by symbols (or pictures or images) of reality was different from Helmholtz's own. Rather, he viewed theory as constituted by signs of reality, even though he later came to speak in terms of images. On this philosophical point, Hertz's final piece of work—with its lack of empirical referents and its sharply deductivist structure, with no reference to the role of perception in scientific investigation and with its openness to the possibility of more than one interpretation of the underlying empirical reality—departed significantly from Helmholtz's philosophy of science.[51]

Helmholtz's foreword to Hertz's *Die Prinzipien der Mechanik* was thus a memorial notice for his most favored if not beloved student. It became his final publication (during his lifetime). Along with his posthumously published piece on the principle of least action in electrodynamics, it brought to a close more than a half century of creative scientific work and thinking.[52]

Dying, Death, and a Funeral

During the second half of November 1893, Helmholtz had more or less resumed his administrative duties at the Reichsanstalt and had taken up scientific work again. Klein sent him several of his articles on geometry and a report on the recent meeting of the International Mathematical Congress in Chicago. He wanted to alert him that Lie had published a new book containing many "subjective opinions." As much as he deplored Lie's renewed attack on Helmholtz, Klein thought that Lie did important work and that it had not been recognized sufficiently in Germany. Helmholtz ignored Lie's attack and instead replied, "On the whole, my accident turned out relatively favorably, but in this life I definitely won't be traveling for a second time to America."[53]

After he had recovered sufficiently, he found a great deal of other correspondence awaiting him. For example, George J. Romanes, a physiologist and evolutionary biologist, had written to ask unofficially, but on behalf of the vice-chancellor of Oxford, whether Helmholtz would give the Romanes Lecture (initiated in 1892) there in 1894. Romanes noted that Gladstone and Huxley had each previously delivered a Romanes lecture. "The aim of this foundation," he explained, "is to secure written discourses from the most eminent men of each successive generation, irrespective of nationality; so that some centuries hence these be formed [*sic*] a numerous series of lectures, presenting great historical, as well as intrinsic, value." The entire leadership of Oxford wanted Helmholtz to come. Helmholtz declined, and the German evolutionary biologist August Weismann gave the lecture instead. Helmholtz also declined D. Argyll Robertson's flattering invitation to the International Ophthalmological Congress in Edinburgh for August 1894. He did, however, accept honorary membership in Britain's Institution of Civil Engineers. In January both Helmholtzes received awards from the state, Hermann the Crown Order with large blue band, Anna the Order of Louise, presumably for her work in supporting the Victoria-Haus. Then in April the Peter Wilhelm Müller-Stiftung (Endowment) of Frankfurt am Main awarded Helmholtz its eponymous medal for his work in mathematics, leaving it up to him to decide whether it should be for pure or for applied work.[54]

In March the Helmholtzes went to Illenau in Baden to see his sister Julie, who was being treated for mental illness at the Heil- und Pflegeanstalt. Then they spent Easter taking the sun in Abbazia (modern Opatija), a resort town near Trieste, on the Adriatic coast, where the emperor and empress also sojourned. They relaxed and convalesced for two weeks, and Helmholtz hiked in the nearby mountains.[55]

Helmholtz's mood was much colored by the recent deaths of several close friends and colleagues. Already Leopold Kronecker (1891), Siemens, Hofmann, and Brücke (all in 1892) had died; in December 1893 Tyndall passed. Louisa Tyndall, John's widow, thanked the Helmholtzes for their condolences and replied: "There was no appreciation my husband valued so much as that coming from Germany and above all from *him*."[56] Then, less than a month later, Hertz died.

In May yet another friend and colleague, Kundt, died. Helmholtz gave a eulogy before Kundt's casket in the auditorium of the Berlin Physics Institute. Kundt died young, at fifty-five, and his death, Helmholtz said, left them all — family, friends, and students — "filled with pain, in fear and anxiety before the dark door through which no human eye can see into the unknown beyond." His eulogy bears further quotation:

> [Kundt's death] only fills us, the living, with sadness, doubt, and misery, and our heart is torn asunder with contradictory feelings. We honor the departed as one of the rare, gifted spirits of whom we hope that, as a leader of the human race, he would influence wide circles, and indeed would do so through his intelligence in the struggle against the blind forces of nature, through his ideal way of thinking, and through his warm-hearted participation in the struggle against the egotistical passions of men. Taken as a whole, our hope is linked to a gradual increase in the morality of our race and to elevating into the realm of

reason and humanity a few outstanding individuals who bring the others along through their intelligence and example.

These eulogistic words say at least as much about Helmholtz as about Kundt. Kundt, Helmholtz continued, "came from pure science and had a pure, ideal character. He was not, however, [a person] without interest for the practical endeavours of technology, for valuable scientific findings for the use of the human race." Helmholtz ended by noting that Kundt took the greatest of pride and pleasure in seeing his students advance. One of Helmholtz's former students, C. Riborg Mann, an American, who attended the eulogy, later remarked of Helmholtz: "Many thought him cold. . . . Those who heard him as he spoke at the funeral of his friend and colleague, Professor Kundt, will never for an instant admit that he was cold or unsympathetic."[57]

In the final months of his life and career, Helmholtz committed what was for him an unprecedented act of defiance with respect to the authorities. In 1889 Leo Arons, a nonpracticing Jew and a political activist, became a private lecturer in physics at the University of Berlin. In 1891 Arons joined the Social Democratic Party, which he strongly supported financially. He was also a member of the Berlin City Council and was especially involved with school issues. Then in 1892 the Berlin Philosophical Faculty recommended him for an extraordinary professorship, but the ministry refused to make an offer. Instead, Althoff, the head of higher-education affairs within the ministry, formally questioned Arons about his political activities and raised the possibility of withdrawing his *venia legendi*. The faculty opposed Althoff's cross-examination and the threat of remotion, and in May 1894 several figures—Helmholtz, Dilthey, and Mommsen among them—signed a petition protesting Arons's treatment. To be sure, the faculty warned Arons not to be too enthusiastic in his political activities, but it did not question his right to engage in such activity, and it would not agree to remoting him. The issue straggled on for several years and gradually became an issue more about faculty rights than individual political rights. Still, Helmholtz and others made their point. A stalemate between the ministry and the university emerged, with the emperor himself seeking Arons's dismissal in 1897. In 1898 Prussia passed a law (the "Lex Arons") aimed at punishing or dismissing any academic teacher it deemed to have "wounded" or acted "unworthy" of his position. In early 1900 Arons had his teaching privileges withdrawn.[58]

In Bremen, the day before he sailed for America, Helmholtz had responded to a request from Sigmund Exner, one of his former students and (since 1891) Brücke's successor as professor of physiology at the University of Vienna. Exner was the co-organizer of the sixty-sixth Naturforscherversammlung, and he invited Helmholtz to give a lecture before a general session, to be held in Vienna in late September 1894. Helmholtz said he wanted to participate but that at his age he could not now give a definitive response. The matter remained pending. In June 1894 Exner wrote again to ask what might be the subject of Helmholtz's lecture and on what day of the meeting might he care to speak. Helmholtz had been preparing a paper for the meeting entitled "On Continuous Forms of Motion and Apparent Substances." Less than a year previously, in July 1893, in his paper on the consequences of Maxwell's theory for the ether, he had ended by informing readers that he would again be publishing, "in later essays," on this topic.[59] He continued to produce papers until virtually the end of his life; his desire to be scientifically creative virtually never stopped. On 14 June 1894, Helmholtz delivered his last lecture at the academy. It was on the principle of least action in electrodynamics and was published posthumously. On 7 July he delivered his last university lecture. Mann was among the listeners, and afterward he asked Helmholtz to pose for a photograph; it was apparently the last one taken of him. (See figs. 29.2 and 29.3.) On 11 July, Helmholtz wrote Koenigsberger that he wanted to recommend (the late) Hertz for the Peter Wilhelm Müller-Stiftung prize of fifteen thousand marks. (Helmholtz, Koenigsberger, and Lipschitz constituted the endowment's prize jury.) He believed that Hertz's overall set of scientific achievements had received far more recognition abroad than at home, or at least less than Hertz had deserved, and he sought to correct that balance. Furthermore, he proposed that Koenigsberger meet him in Bonn to discuss the matter there (with Lipschitz), since he thought Lipschitz the least able of the three to travel. Before that could happen, however, the endowment's directors decided that a posthumous award did not accord with its statutes. Helmholtz also told Koenigsberger that he planned to leave for Bad Gastein (Austria) in early August and that he intended to lecture in Vienna in mid-September at the Naturforscherversammlung. He was still filled with plans.[60]

But on 12 July, while at home, he suffered a stroke that paralyzed him on the left side of his body. His doctors hoped that, to one degree or another, he might slowly recover. (He managed, at least, to dictate several letters to Wachsmuth.) Anna wrote her sister: "Poor Hermann is so patient, and unfortunately so clear about his condition. His consciousness has in no way atrophied; his speaking is somewhat changed, but in any case the symptoms are not advanced, and no part [of the body] necessary for life has been struck. The helplessness and immobility are excruciating." Though he was tired, he was not in pain. After learning the news of her father's condition, Ellen, then away from home, returned to Berlin to help care for him. Wachsmuth and a household employee were already doing so. Everyone asked or wrote to Anna about Hermann's condition. She was not surprised by the stroke, she said, because before its appearance she had noticed small but dangerous signs of just such a possibility. She and Ellen alternated

Fig. 29.2 Helmholtz's last university lecture (7 July 1894), at the University of Berlin Physics Institute. American Institute of Physics, Emilio Segrè Archives, College Park, Maryland.

Fig. 29.3 The Helmholtzes (*far left*) on their veranda at the Reichsanstalt, 1894, with an unknown figure (*on Helmholtz's left*) and three of Helmholtz's Reichsanstalt colleagues: Ferdinand Kurlbaum (*standing*), Ernst Hagen (*center right*), and Otto Lummer (*far right*). Archiv zur Geschichte der Max-Planck-Gesellschaft zur Förderung der Wissenschaften, Berlin.

taking night watch. Helmholtz now lived and slept upstairs alone, except for his caregivers' visits and the night watch person.[61]

Two days later Anna wrote her sister Ida: "Hermann's thoughts wander around restlessly; reality and dreams, wishes and past events, place or time—all are in cloudy, unsteady movement before his soul. Mostly he doesn't know where he is; he believes he's on a trip, in America, onboard ship. During the day, everything is quiet and peaceful, almost festive here above [in his quarters], his fantasies are friendly; at night, he is more restless and demands all sorts of impossible things." His speech remained "always rigorously logical," Anna wrote. "He can't think in any other way. It's as if his soul was always far, far away, in a very beautiful, noble sphere, where only science and eternal laws rule—then, suddenly, none of that fits with what's around him and he becomes restless." He had no shortage of medical men and other caregivers around him: in addition to several doctors and at least one of their assistants, himself also a doctor, who slept in the Helmholtz house, there were two indefatigable nurses who came from the Victoria-Haus and joined Wachsmuth, Ellen, Anna, and the household employee in the caregiving. ("Only Ellen is a visible joy to him; if his eye falls on her person, he is happy.") Eight days after the stroke, Hermann was doing better; one of his doctors, Ernst von Leyden, even assured Anna that Hermann's life was no longer in danger, though he did predict Helmholtz would require a long recuperation, "until the overtired brain can again come to order. Whether that will ever completely happen, one naturally cannot know." Anna was simply glad that he was still alive; beyond that, she and the others tried to make him as comfortable as possible. His sons Fritz and Richard came to see him, but they had to leave because "every change in the room or speech" made him "agitated." Nor could there be any question of telling him that his sister Julie, whom he had recently visited, had died three days earlier. "I honor and love my husband too much," Anna told her friend Rosalie Braun-Artaria. "I'm just happy about the days of grace that are still left to us" and want to "show him what he is to us." In the meantime, he lay "helpless on his bed. His mind, half-awake, struggles with dreams, which fog the reality around him and pain him, until, after a few hours, he shakes them off and then he is completely his old, superior self." As in his prestroke days, though he tired easily, "he was scientifically the same as before: clear, involved, decisive, and deciding. Everything that belonged to the personal, to general humanity (except for his family), was for him immensely laborious. He enjoyed music, painting too, but he became tired quickly and we did everything for only a half-hour long."[62] His family loved him, and he them.

On 31 August, Helmholtz and his family celebrated his seventy-third birthday together; he was sufficiently alert mentally to send greetings and thanks to his sister-in-law Ida and other family relations for the flowers they had sent. He received visits from several of his Reichsanstalt colleagues and kissed his five grandchildren who were present (a sixth lived away from Berlin). It was a good day for him. The emperor dispatched an undersecretary of state that day to assure Anna and Hermann that the latter's pension was guaranteed. She repeated this Imperial declaration to Helmholtz numerous times "in order to calm him." He knew the end was near: "He talks a lot about it, that he will soon be no more, and, alas, he makes one feel frightfully sorry." Yet his thirst for science never abated: he had Anna read to him — twice — Lord Rayleigh's and William Ramsay's preliminary report (13 August) from the Oxford meeting of the BAAS of their recent discovery of argon. "I always thought there must be something more in the atmosphere," he reportedly commented. Even as the end approached, his passion for reason was still alive. Then, during the night of 4 September, he took a turn for the worse. He uttered his last words: "It's hard — it's hard"; "I wish that you [Anna] will still find something beautiful." He suffered a second stroke the following morning (at 9:00 a.m. on 5 September), became further paralyzed, and lost consciousness. He died three days later, on Saturday, 8 September, at 1:00 p.m., without having regained consciousness. From the moment of the second stroke until his death three days later, Anna did not leave the room, while his children (Richard, Fritz, and Ellen), his son-in-law (Arnold von Siemens), his sister-in-law (Betty Johannes, née von Velten), and his three doctors, as well as

Fig. 29.4 Sketch of Helmholtz by Franz von Lenbach, 29 April 1894. Siemens Forum, Munich.

Lummer and Wachsmuth, were all there at the end.[63] It was a hard death, but, as nearly always in his life, he was surrounded by family.

The family was inundated with condolences in the form of personal visits, telegrams, and letters, not to mention articles and announcements in newspapers, magazines, and journals. Mommsen, Helmholtz's friend and next-door neighbor, was reportedly the first to arrive. Empress Frederick and the Grand Duchess of Baden both telegrammed seconds before Helmholtz died; shortly thereafter the emperor himself telegraphed Anna to say, "The scientific world, the Fatherland, and your King mourn" for their loss. Much the same was true for many in the educated and cultured worlds of Germany and beyond. Those closest to him, and not only his family, were in shock. Planck told Runge that Hertz's and Kundt's deaths earlier that year, and then Helmholtz's second stroke, affected his own ability to think and to do scientific and administrative work. "The year [18]94 is a very mournful one for my science," he wrote. He thought that "German physics" had taken strong "hits." "Especially in Berlin, things look dark and sad."[64]

Burial services were scheduled for 12 September, 2:00 p.m., with the funeral procession proceeding from the Helmholtz home at the Reichsanstalt. Joining the family in their mourning, and acknowledging their respect for him, were all the state ministers of Germany; municipal representatives from Potsdam, Heidelberg, and Charlottenburg; and official delegates from learned organizations. Lenbach sent his last portrait of Helmholtz (see fig. 29.4), which complemented the one he had done for the Helmholtzes the year before. Senior members of the nobility, including the grand dukes of Baden and Weimar, sent wreaths, telegrams, or letters. Flags were flown at half-staff. In Potsdam all clocks rang (for an hour) while the burial took place, and the Potsdam Gymnasium canceled all classes and declared the day a memorial holiday. The funeral was not well attended, since most Berlin academics and the Helmholtzes' friends were then away on vacation. Planck was there, however. He had more a feeling of "horror" than "mourning," he said, because he did not know how physics in Berlin could live "without this man." The mourning came later. Anna, too, was in shock. She had "lost everything," she said, "and cannot grasp that I should continue my life in darkness and alone." "Everyone says," she wrote a friend, "he is irreplaceable, but except for myself, only a few know that."[65] Planck was one of those few, and he now took Helmholtz's place as the leader of Berlin physics, and then as one of the leaders of early-twentieth-century theoretical physics.

Epilogue: Helmholtz in Modern German Memory

In the days and weeks following Helmholtz's death, large numbers of friends, colleagues, and others visited Anna privately or sent their condolences—some days she received fifty letters. More than one hundred obituary notices and memorial addresses appeared in newspapers, magazines, and scientific and cultural journals throughout Europe and America.[1] The sheer unprecedented number of these private and public items and their emotional content, heightened by a sense of mourning and depression, indicates that Helmholtz's death marked a highly emotional moment within the scientific community, indeed the recognition of an ending to a distinguished and unprecedented leadership in science and an increased sense of concern about the future of science.

This epilogue explores Helmholtz in modern German memory: how various Germans chose to remember him, how they used (and occasionally abused) his name, his image, and his private and public self to create what effectively became a myth or idol or icon of the man, the scientist, and sometimes even his work. From his funeral onward, the public persona that he himself had first (and gradually) created was amplified and developed into an often ossified, idolized remembrance of him as an unqualifiedly flawless human being and scientist. Largely separated from his historical context with his family, his teachers, his colleagues, and

German culture and society at large, and largely without a critical word about him as a person or a scientist, the Helmholtz who appears here as remembered by Germans is an idol or icon whose creators seemed (and seem) interested only in inventing an inspiring ideal for nonhistorical purposes.

Imperial Germany

After Helmholtz's death, the general and learned scientific, medical, engineering, and cultural journals, societies, and organizations of Imperial Germany (and well beyond) all memorialized him. These public articles were overwhelmingly encomia, filled with fulsome praise and even heroizing. Helmholtz's picture appeared widely, momentarily becoming the image of the scientist. He was, one journal said, the "illustrious Prince of Science." The organizers of the Vienna Naturforscherversammlung in September 1894, where he had been scheduled to speak, requested a picture of him for its opening session; Anna sent Wachsmuth there as her representative, with a Lenbach portrait of Helmholtz to be placed on display. Such displays occurred not only in public: Lord Kelvin, as William Thomson had been known since 1892, placed a portrait (also sent by Anna) of Helmholtz on his study wall; and Roscoe had one hanging over his desk. In lieu of Helmholtz's planned lecture in Vienna, the scientists and medical doctors assembled there held a "mourning session for the greatest departed of our time." In Berlin there was a grand memorial service at the Singakademie, while smaller services were held at the Physical Society and the Academy of Sciences (see below). Memorial addresses (lasting through 1896) were given at or by the Physical-Economic Society at Königsberg, the German Chemical Society, the German Society for Mechanics and Optics, the Silesian Society for Fatherland Culture, the University of Utrecht, the Royal Society of London, the Chemical Society (London), the Natural Research Society (Zurich), the Museum in Berne, the Budapest Royal Society of Medical Doctors, and at other venues.[2] Mourning for Helmholtz focused on his intellectual contributions to science and his public role as spokesman for science. His death induced not only a heightened sense of community among scientists, but also apprehension if not depression about science's future. In recalling his life, the obituarists and memorialists helped give meaning and purpose to science; Helmholtz had embodied the life of science and the image of the scientist.

Helmholtz's death and Kundt's death earlier that year meant that successors were needed at both the university and the Reichsanstalt. The Prussian Ministry of Culture moved quickly. Kohlrausch was the initial choice to replace Kundt at the university, but Kohlrausch raised issues that entailed negotiations and delays. Then, when Helmholtz died in September (creating the need for a successor at the Reichsanstalt), that further complicated the situation at the university. The authorities finally managed to attract Kohlrausch, an experimental physicist, not to the university but as the Reichsanstalt's new president, and Emil Warburg, another experimentalist, as Kundt's successor at the university. Warburg declared that he worked in the tradition of Helmholtz (as well as that of Magnus and Kundt), adding: "I owe a large part of my scientific views to Helmholtz's published writings."[3] What exactly that meant, he did not say.

Shortly after Helmholtz's death, the leadership of Berlin's physical, medical, physiological, chemical, and electrotechnical societies—fifteen in all—planned a joint memorial celebration of his life. It was no easy task finding someone who could give a keynote address authoritatively reviewing Helmholtz's multifaceted accomplishments. They ultimately settled on the theoretical meteorologist and physicist Bezold. (Anna concurred, since Bezold had known Helmholtz's work and personality well.) The celebration took place on the afternoon of 14 December 1894, at the Singakademie. The house was packed; there was not an empty seat available. Along with Helmholtz's family, the emperor and empress (indeed the entire court), everyone of importance in the world of science in Berlin was there, not to mention many of Helmholtz's friends, colleagues, and general admirers. A large bust of him (presumably the one by Hildebrand) was displayed on stage and decorated with palm branches and laurel wreaths. The program opened with choral music; Bezold then gave his address devoted to Helmholtz's scientific work; the violinist Joachim (accompanied by Planck on the organ) followed with a performance of Schumann's *Abendlied*; and the program closed with the choir singing a piece by Brahms. Virtually everyone thought the celebration appropriate and touching. Planck thought Bezold's address was worthy of the occasion. The classical philologist Hermann Diels thought it all came off perfectly and that Bezold's speech was beautiful, pitched at just the right level for the diverse audience. Dilthey agreed that all was brilliant—except for the central event, Bezold's speech, which he judged "a conglomerate [of points] that gave no idea of the context of [Helmholtz's] lifework and times." Anna thought it had been a "beautiful, sad day" and that Joachim's (and Planck's) performance "was simply divine."[4]

As if the Singakademie event had not been enough, six months later the academy held a similar event, as part of its annual Leibniz Session (4 July 1895). Du Bois–Reymond gave the obligatory address commemorating the deceased member (Helmholtz). His speech, though much longer than Bezold's, was, on one level, the same as Bezold's, consisting mainly of a narration of Helmholtz's scientific accomplishments. But on another level, it reflected the importance of Helmholtz to the academy and so helped extend the Helmholtz of myth. Helmholtz's death, du Bois–Reymond declared, was "not merely . . . an immense loss for science, but has also been felt as a national misfortune." Nobody felt this more than the academy, he claimed. Helmholtz had had no interest in influencing the world, du Bois–Reymond said, and when that did occasionally happen, as with his invention of the ophthalmoscope, it was a mere accident. Rather, Helmholtz's aim had been to understand nature theoretically, and to do so in all nature's complexity, including the physical, the biological, and the psychological dimensions. Helmholtz possessed "the unmatched ability to seek out those questions, and to answer them victoriously, which at every point were precisely the most important and whose treatment promised the best success."[5] No unfinished scientific business, let alone failures, here.

In du Bois–Reymond's view, "the secret" of Helmholtz's productivity lay not merely in the "calm and uniform nature of his professional work" but much more "in his indefatigable diligence and in his ability to always keep in mind, to lay hold of, and to keep ready for assessment an enormous diversity of facts and thoughts." He thought that only Descartes and Leibniz could be compared to Helmholtz as a scientist. "It should be noted, however, that since those [early modern] days the content of science has become incomparably richer and more variegated, and so more difficult to conquer." He closed his idolizing address by declaring: "We will never again see his equal. Indeed, it is questionable as to whether a figure like him can ever again appear."[6] How could he possibly know that Helmholtz was (supposedly) the last generalist of science?

Five years later, Adolf von Harnack, a theologian and church historian, similarly lionized Helmholtz in his history of the academy. He declared that, at least in the recent past, no other academy could claim such a name as Helmholtz's. Perhaps no other living German scientist or scholar stood as high in the opinions of educated Germans and non-Germans alike. That was why the academy had placed his statue next to those of the Humboldt brothers. Harnack, who was also head of the Royal Library, thought Helmholtz was "indisputably the greatest natural scientist" the academy had ever had.[7] He portrayed Helmholtz in superlatives, as simply the academy's greatest member ever.

These two memorial events in Berlin were certainly not the only ones. At the University of Heidelberg (November 1895), for example, Koenigsberger gave an address on Helmholtz's "investigations on the foundations of mathematics and mechanics." Arthur Rücker, too, gave a memorial lecture on Helmholtz (at the Royal Institution in March 1895), calling him "the most widely cultivated of all students of nature, the acknowledged leader of German science, and one of the first scientific men in the world." Roscoe had urged Rücker to write a review of Helmholtz's work, and he assured Anna that Rücker was "well able to do it, & is an enthusiastic admirer of Helmholtz and his Works." Roscoe acted as a friendly gatekeeper: he promised Anna that he would forward the proofs of Rücker's manuscript before publication.[8]

Addresses were only one form of remembrance and idolization. Another was to "analyze" Helmholtz's brain. In late-Imperial Germany, an age when brain anatomy and physiology, as well as psychiatry, began to thrive, it became fashionable to examine the brains of "geniuses" (for example, Siemens, Bismarck, Mommsen, and Menzel). The emphasis then was not on racial or eugenic ideology, as it came to be after 1914; rather, it was on finding the sources of genius, whether healthy or pathological. A postmortem examination of Helmholtz's brain showed great destruction on the left side owing to the two strokes he had suffered, while the right side had been destroyed by blood hemorrhaging. A wax cast of the left side was made so that "this once so powerful thinking organ could be passed on to posterity." David Hansemann, who was an assistant in Virchow's institute but was not a brain anatomist, along with four others (two brain anatomists, a psychiatrist, and a paleoanthropologist), did independent analyses. They each measured and weighed the skull and brain (circumference, height, width, weight, number and shape of convolutions, and so on) and drew conclusions—read: speculated—about the relationship between Helmholtz's brain and his scientific creativity and musicality. These "conclusions" were almost all certainly based on or biased by knowing Helmholtz's writings or reputation rather than his brain's anatomy, let alone its functioning. Hansemann and his colleagues also hinted at comparisons between Helmholtz's brain and those of other "geniuses." The details were reported not only in a scientific journal but also to a wider audience in the general periodical *Die Umschau*, which of course only furthered the belief that these scientists had indeed found a correlation between Helmholtz's brain and his intellectual accomplishments. Later, in 1907, Edward Spitzka used a simple illustration to compare the many folds of Helmholtz's brain with the lesser number of folds found in the brain of a Papuan man from New Guinea and the still fewer folds of a gorilla's brain. Brain fetishism was in, and in Helmholtz's case it was used to good hagiographic effect.[9]

Anna von Helmholtz, as already suggested, played perhaps the single largest, most important role in creating the posthumous image of Helmholtz as the flawless man and scientist, as idol and icon. From her marriage to Hermann in 1861 until his death in 1894, she had lived in his shadow, for and around him but never fully with him, as she herself well recognized. Already six months before his death, she felt like she was near the end of her own life and that she had done all that she could do in this one. Two weeks after he died, she wrote her sister: "It's now my duty to make my life, if not also among friends, as he had wished, then peaceful and worthy of his dear memory." She took comfort in being surrounded by Lenbach's portraits of Helmholtz, to which she now added one of herself as the widow in mourning. She felt physically weak, suffered from insomnia, and was depressed. Hermann had been her *raison d'être*, and now the center of her existence was gone. Especially during the final eighteen months of his

Fig. E.1 Anna von Helmholtz in 1895. Portrait by Franz von Lenbach. Siemens Forum, Munich.

life, she had given herself over to his needs entirely. In January 1895 she wrote Henry Pickering Bowditch: "I am weary and wretched and wish my life was over too."[10] She was grief-stricken, in mourning, and in depression. See figure E.1.

Helmholtz's death, and the appointment of Kohlrausch as his successor at the Reichsanstalt, meant that Anna had to move out of the presidential villa. In early 1895, as she prepared to do so, she had Helmholtz's library evaluated for sale; the Reich accepted the valuation and bought it (for seventy-five hundred marks). When the books were taken away, she commented, "To me it was as if his soul was taken out." Cleaning up his desk was also hard. She gave his beloved pump organ (harmonium) to the Physics Institute. In February 1895 she moved into a new residence, on the southern edge of the Tiergarten. She brought along his desk and the Steinway piano, and naturally the bust of him as well. She placed pictures of him everywhere.[11]

After the move, Anna traveled a great deal: to the Riviera, the Antibes, Italy, western and southern Germany, and Switzerland. In early 1898 she undertook a two-month trip to Egypt. That summer she vacationed in the Low Countries. In February 1899 she made her final trip to Paris.[12] Notwithstanding the friends and travel, she continued to mourn and to be lonely, yet she also found a new purpose in life, or rather four specific tasks, all intimately tied up with Helmholtz.

First, she sought to assure the future well-being of their son Fritz. He had obtained a leaving-certificate from the Königliches Luisen-Gymnasium in Berlin; then studied at the Königliche Landwirtschaftliche Hochschule Berlin; then did a practicum in Ludwigsruhe-Langenburg; and, finally, did further studies at the Landwirtschaftliche Hochschule Hohenheim near Stuttgart (though he never received a diploma). Anna worried about him constantly and often had to take care of him. He had, she said, serious psychological problems (for example, he could not "relate" to or understand others). One of his doctors reportedly described his "illness" as one of nerves, or of his soul. Anna thought that he simply could not be on his own. She visited him regularly and was afraid to leave him alone. She and his doctor thought his plans to farm were unrealistic. Fritz became depressed. Finally, Anna discovered a small country house ("Porta Maria") near Baden-Baden that she and Fritz thought would be perfect for him. She purchased it and he moved in by the summer of 1899; she thought his future looked settled if not good.[13]

Anna bequeathed her personal assets to her two children, Fritz and Ellen, and directed that the assets she held in common with Hermann—about one hundred thousand marks, which was, she noted, about the value of Helmholtz's estate at the time of his death—be split into four equal shares among Fritz, Ellen, Richard Helmholtz, and Hermann's granddaughter Edith von Branco (Käthe Helmholtz's only child). Arnold von Siemens was asked to supervise Fritz's share, and Richard to support him in this. She left it up to Ellen as to how to distribute her pictures and other artwork among all of Helmholtz's children, Anna's siblings, and Hermann's friends. She did, however, ask that the Hildebrand bust of Helmholtz be given to the University of Heidelberg for exhibition in its Aula, "as the place where he was happiest," and that the Lenbach portrait go to the Nationalgalerie in Berlin. Among other minor provisions, she also bequeathed five thousand marks to the Victoria-Haus für Krankenpflege.[14]

Second, Anna became Helmholtz's literary executor, as he had requested. She dealt with his publishers. She edited (very lightly) the new (fourth) edition of his *Vorträge und Reden* (1896) and generally oversaw the publication of the third

Fig. E.2 Helmholtz family gravesite. Waldfriedhof, Berlin-Wannsee. Siemens Forum, Munich.

volume of his *Wissenschaftliche Abhandlungen* (1895) and (through 1899) of his *Vorlesungen über theoretische Physik*, naturally leaving the substantive matters or editorial decisions to qualified others. She also maintained and ordered Helmholtz's unpublished writings, including his correspondence with his father.[15]

Third, Anna had a new gravesite built for Hermann and their family in Wannsee. She commissioned Hildebrand to design it and oversee its construction. It was completed in the fall of 1897. Helmholtz and his son Robert were re-buried there, in the Landeseigener Friedhof. Helmholtz's gravestone was placed in the center, with a wreath surrounding his name. He was flanked by Robert on one side and a place that was left vacant for Anna on the other, while extra burial room was also left for Fritz. See figure E.2.[16]

Fourth and finally, and most obvious and important for the idolizing and icon-making, Anna sought to help establish a public monument to her husband. After Bezold's address, the emperor decided to establish a public memorial to Helmholtz in the form of a statue. He was prepared to spend ten thousand marks and to provide a public space for it. The "Central Committee for the Establishment of a Memorial for Hermann von Helmholtz" was formed, consisting of 179 leading academic and business figures, more than one-quarter of whom were foreigners. The committee publicly sought financial support from domestic and foreign sources. Münsterberg, Wolcott Gibbs (president of the National Academy of Sciences), and Knapp constituted the American section of the committee, for example. *Science* explained that the memorial was for "one of the greatest scientific geniuses of all time, whose name will not be forgotten as long as men care for the knowledge of Nature," and readers were called upon to help fund a monument to Helmholtz in Berlin. It was meant as "a universal expression of devotion to the spirit of natural science." A similar announcement appeared in *Nature*. In Holland, the physicist Lorentz also helped collect money for the memorial.[17]

But the process moved slowly. It took time to collect the necessary funds—the final cost was about ninety-three thousand marks—and to decide on a fitting design. The emperor selected Ernst Herter as the sculptor. He created a larger-than-life marble figure of Helmholtz, in a standing position wearing an academic gown and, as Anna saw it, "somewhat theatrical in his hand-movement"; for her the statue had "something of the look of a false Luther." (See fig. E.3.) The statue was finally unveiled at a well-attended public ceremony on 6 June 1899, in the forecourt of the University of Berlin's main building and its entrance-way on Unter den Linden. Diels (and apparently others) thought it blocked the way into the building and was also "dissatisfying" as a work of art. Perhaps what irked Diels was that Helmholtz's statue in front of the university's main building symbolized that the natural sciences, and not classical philology, now stood at the entranceway to the university. And with the enhanced position of the sciences came a different set of practical economic and cultural consequences for the new German state and for German society at large, ones that the humanities could never know or realize. The Helmholtz statue, like several others (for example, Virchow's) in the city's center, perhaps also symbolized a new sense of German nationalism.[18] Helmholtz had, so to speak, become official art. Even so, the statue was also a symbol of the hopes, dreams, and ideals of all those who honored natural science.

Aesthetic and political judgments aside, Anna's life's work, as she told her sister Ida, was now complete. She took the ever-ailing Fritz to Heidelberg for continued medical care and to his country place nearby and moved, herself, to Baden-Baden. She died unexpectedly of heart failure on 1 December 1899, in her sixty-sixth year. She was buried next to Hermann in Wannsee. Fritz died less than two years later and was also buried there.[19]

The gravesite and the statue were not the only physical representations of Helmholtz's legacy, which was further preserved in various monuments and building inscriptions. In 1896 the Urania, a science museum, constructed a new building in the heart of Berlin. Its upper facade exhibited busts of five heroic figures in German culture, with Humboldt in the center, Helmholtz on his im-

Fig. E.3 Helmholtz Memorial Statue by Ernst Herter in the courtyard of the University of Berlin. Unveiled and dedicated on 6 June 1899. Medizinhistorisches Institut und Museum der Universität Zürich, Switzerland.

mediate left, Siemens on his immediate right, Kepler on his far left, and Copernicus on his far right. In the Schöneberg district of Berlin, furthermore, the new "Helmholtz-Realgymnasium" opened its doors in 1902.[20]

Helmholtz's name also lived on elsewhere as public inscription, and so as an inspirational legend, icon, and idol. In 1897, when the Library of Congress in Washington, DC, completed its new building (now known as the Thomas Jefferson [or Main] Building), visitors could see, inscribed on tablets above the windows in the southwest gallery on the second floor, a series of names of immortal figures in the arts and sciences; Helmholtz's was there alongside Homer, Aristotle, Michelangelo, Galileo, Mozart, and others. In 1912 two large statues representing Art and Science were placed on the platform in front of the Boston Public Library (Copley Square). The science statue was engraved with the names "Newton, Darwin, Franklin, Morse, Pasteur, Cuvier, Helmholtz, and Humboldt," indicating the high esteem in which Helmholtz (and German culture) was held. In 1916, when the Massachusetts Institute of Technology moved from its old campus in Boston to its new one in Cambridge, the designers inscribed Helmholtz's name, as well as those of a select set of other scientists and technologists, near the top of its new main building. Somewhat later, in 1924, the National Academy of Sciences dedicated its new building in Washington, DC. A set of thirty-seven low-relief bronze panels there depict the "greats" of science from the ancient Greeks to Helmholtz. Germans were not the only idolaters.

Helmholtz's *Vorlesungen über theoretische Physik* constituted, at least to a certain degree, another element in the iconic realm. This work appeared, it may be recalled, as a set of six (mostly thick) volumes in seven, between the years 1897 and 1907 and, as concerned their individual subject matters, in no logical order. They proved to be of value to any number of physicists at the turn of the twentieth century. Arnold Sommerfeld, on the threshold of becoming one of the twentieth century's leading mathematical physicists, wrote to Runge, one of the editors, to thank him for sending two of the volumes. "You can imagine how valuable their possession is for me," he said.[21] Did he mean intellectually valuable or valuable as an icon for remembering Helmholtz and his way of doing physics? For these volumes were quickly becoming a memorial and a monument to Helmholtz as much as (if not more than) a source of intellectual sustenance for contemporary theoretical physics. In retrospect, the set effectively represented a capstone to what soon enough became known as "classical physics." By the time of their completion, both quantum physics and special relativity had appeared on the scene, as well as the discovery and study of radioactivity, which soon helped lead to new understandings of atomic structure. Indeed, these and other new developments in physics were themselves rapidly undergoing their own further development. Although Helmholtz's work had helped make both "classical" and "modern" physics possible, these thick, imposing volumes most definitely represented "classical" physics; they were (also) iconic volumes, representing the memory of Helmholtz, not (at least by 1907) the state of contemporary theoretical physics.

Biographies (read: hagiographies) of Helmholtz also began to appear. The first full-length one, by John Gray McKendrick, professor of physiology at Glasgow University, was published in 1899. McKendrick revered Helmholtz. While six additional biographies of varying length and documentation subsequently appeared, the one by Koenigsberger became by far the most important and influential.[22] What they all have in common is a hagiographic, Whiggish presentation of their subject. For Koenigsberger, it could hardly have been otherwise.

Anna von Helmholtz's concern about Helmholtz's legacy meant in part a concern about who would write about her husband's life and work. She had already played a gatekeeper role with respect to the speaker at the memorial service at the Singakademie. She had supported Roscoe's choice of Rücker to write a substantial English-language address (and probably a biography); she had been ready to put any items she could at Rücker's disposal and generally help him. Along these lines, too, she expressed "disappointment" with du Bois Reymond's speech, and while she liked the obituaries done by Hugo Kronecker, Theodor W. Engelmann, and Hans Landolt, she thought each one wrote only from the perspective of his own discipline and noted that none of those men were physicists. She had been anxious to hear from Koenigsberger about any efforts he might make toward writing a biography.[23]

In 1895, after Koengisberger had published an article on Helmholtz's geometry and mechanics, Anna repeatedly conveyed to him her worry that someone might write an unauthorized biography of Helmholtz. After reading Koenigsberger's article—she readily admitted she could not understand much of it—she concluded that he was one of the few individuals who could write "a memorial" and do so in Helmholtz's "spirit," in regard to both his science and his life more broadly. She thought Koenigsberger's article was "the most beautiful monument that could be done for the dear man [Helmholtz]." Koengisberger had shown the love and understanding for her late husband and his work, she thought, that was needed to write a full-length biography. Koenigsberger fully appreciated the many-sided nature of Helmholtz's scientific work, and so she felt that he could bring everything together into "a unified whole."[24]

Koenigsberger, however, was preoccupied with his own scientific work, and he gave Anna no definitive, positive answer. After her death, however, he remained in regular contact with Fritz von Helmholtz. Then when Fritz died (17 November 1901), in a hospital in Heidelberg, at his funeral the next day Koenigsberger became acutely conscious that the Helmholtz family was dying out, and he resolved then and there to write a biography of Helmholtz and to do so within the span of one year.[25]

One year was not quite enough, but in fact, it took Koenigsberger merely sixteen months (November 1901—March 1903) to assemble the necessary primary material, write the manuscript, and publish a three-volume biography of Helmholtz. He was able to do so for three primary reasons, including his own mathematical expertise and, to a lesser extent, his scientific understanding of Helmholtz's writings (especially concerning physics) and his simple determination to complete the project. But he also had the full cooperation of Richard von Helmholtz and Ellen von Siemens-Helmholtz, including provision of all Helmholtz correspondence (in their possession) and papers; similar cooperation came from various Helmholtz colleagues and state authorities (Prussia, Baden, and the Reich). Koenigsberger, who was not a historian (nor a physicist, nor a physiologist, for that matter), was also able to complete his biography so quickly in part because he utilized many and long (sometimes pages-long) direct quotations, which he simply copied from Helmholtz's correspondence or from his scientific, philosophical, and popular writings. At some points he even failed (probably unwittingly) to provide quotation marks. Moreover, he wrote in an utterly uncritical manner and provided all too little or even no historical or scientific context or analysis of the various people, scientific issues, and historical events in the story. Using this Victorian life-and-letters approach, Koenigsberger rapidly completed his three-volume biography of "one of the God-blessed princes in the empire of intellectual and moral power." That was the hagiographical spirit in which he wrote his biography. Yet, despite its shortcomings, it has long served (and still serves) as a useful work in the history of science. Virtually everyone who has been interested in Helmholtz has used it as a work of reference or has cited Helmholtz quotations from it.[26] The biography itself became part of the Helmholtz legend and an essential part in the making of Helmholtz into an idol or icon, a mythical figure.

When Koenigsberger's biography appeared in 1902–3, it received nothing but public praise. One reviewer typically called it "masterly" in its portrayal "of the great researcher and thinker," revealing much of the intimate life of one of the "princes in the empire of the mind." The reviewer aptly compared it to the Helmholtz memorial statue in the University of Berlin's forecourt. Like the statue, it created a presence. Roscoe called it "excellent." One critical if private note came from Boltzmann, who chastised Koenigsberger for claiming that "Helmholtz was accustomed to yielding unconditioned authority to his wife in all matters." Instead, Boltzmann declared that Helmholtz did nothing against his own will and that he was hardly ruled by his wife. Two other critics, moreover, noted that there was too little of the man, of the personality, in Koenigsberger's volumes: they called for Helmholtz's correspondents and acquaintances to help build a Helmholtz archive by depositing any letters they might still have from him and by writing down their reminiscences of him as a person.[27] But these were by far the exceptions.

For those who could not read German or who did not care to wade through Koenigsberger's three volumes, there were soon alternatives. Oxford University Press issued an abridged, one-volume, English-language translation in 1906, with a short but fulsomely praising preface by Kelvin. In 1911 Vieweg published a one-volume, abridged German version. It is likely that the abridged English-language version did more than any other single item to spread knowledge of Helmholtz's life and work around the world, enhancing him in posterity as a legend and a myth. One reviewer maintained that the nineteenth century was "characterized" by "the advance of science and the spread of democracy"; that its "greatest men . . . were its scientific leaders," not its statesmen, soldiers, artists, or writers; and that Helmholtz, along with Darwin, was one of the two great leaders of nineteenth-century science. Writing in a similar vein in 1912, William Osler, as he was establishing a medical library at McGill University in Montreal, wrote one of his former pupils there in regard to the one-volume English-language version: "I think all students and young doctors should read the lives of Pasteur and Helmholtz."[28]

The unveiling of Helmholtz's statue and the publication of Koenigsberger's biography were not the only means by which Helmholtz's legacy, not to say his presentation as idol and icon, were furthered in Imperial Germany. There was also, for example, the academy's awarding of the Helmholtz Medal, its highest scientific recognition. After the initial four medals were awarded in 1892, there was no awardee again until 1898. Between then and 1918, ten individuals became Helmholtz Medalists: Virchow, Stokes, Santiago Ramon y Cajal, Antoine-Henri Becquerel, Emil Fischer, van't Hoff, Simon Schwendener, Planck, Richard Hertwig, and Röntgen.[29] Several of these individuals had been more or less closely associated with Helmholtz, and six of them (Röntgen, Becquerel, van't Hoff, Fischer, Ramon y Cajal, and Planck) later won the new Nobel Prize. No further awards were made for the next forty years (until 1959), reflecting Germany's status as a political outcast from the international scientific community for much of the period between 1918 and 1945 and its political disarray during the early Cold War. In theory, the Helmholtz Medal might have become the scientific world's foremost award; instead, the Nobel Prize assumed that distinction.

Taken as a whole, moreover, the publication or republication of virtually all of Helmholtz's oeuvre—the *Wissenschaftliche Abhandlungen* (three volumes), the three-part *Handbuch der physiologischen Optik*, the *Lehre von den Tonempfindungen*, the posthumous *Vorlesungen über theoretische Physik* (six volumes in seven), and the *Vorträge und Reden* (two volumes)—also helped shape the image and memory of Helmholtz. Little wonder that Thomas Mann could declare in 1910 that Helmholtz, along with Theodor Fontane, Bismarck, Moltke, William I. Wagner, Menzel, Zola, Ibsen, and Tolstoy, was "a member of the European race of heroes."[30] But although the substantive impact of Helmholtz's work within the scientific community continued for a decade or so after his death, and indeed in some sense has never fully ended, the 1890s marked the beginning of significant changes in physics, sensory physiology, and psychology, and so of his standing. Though he was anything but forgotten, his dominance in these fields was effectively over.

Nonetheless, during the three decades following his death, Helmholtz's epistemological views, his explanations of energy conservation, and other scientific ideas continued as a source of interest and disputes among philosophers and other academics. Studies of his understanding of and relationship to Kant appeared; sometimes, as in the work of Ludwig Goldschmidt, it was argued that in his epistemological views, Helmholtz was not a Kantian at all. Ernst Haeckel claimed that Helmholtz's law of conservation of energy constituted part of the foundation of his own system of monism and that Helmholtz was a freethinker; he cited him at numerous points in his international bestseller *Die Welträthsel* (1899) and elsewhere. Even Mach, whose philosophy of science differed so radically from Helmholtz's, praised parts of Helmholtz's work and in general spoke of Helmholtz's "consummate critical lucidity."[31]

To those who aspired to become scientists or medical doctors, and to the great majority of nonscientists who simply sought to understand something of science, Helmholtz and his work continued to represent an ideal. He retained an undiminished stellar reputation among ophthalmologists, and the fiftieth anniversary of the invention of the ophthalmoscope in 1901 became another moment for celebration. At the Zurich Polytechnic, the young Einstein came to appreciate Helmholtz, both through his writings and through former Helmholtz students—the physicists Heinrich Friedrich Weber and Johannes Pernet, as well as the philosopher of science August Stadler, all of whom were among Einstein's teachers. In distant India, C. V. Raman, as a twelve-year-old high-school boy, read Helmholtz's *Popular Lectures on Scientific Subjects*; he judged Helmholtz to be "the supremest figure" in the modern world. Raman himself became a preeminent authority in modern optics (he won the Nobel Prize in Physics in 1930) and was also a keen student of musical instrumentation. He studied *On the Sensations of Tone* carefully: "It profoundly influenced my intellectual outlook," he declared. Likewise, the seventeen-year-old Max Laue also read Helmholtz's *Vorträge und Reden* with great enthusiasm. That book gave him many of the first elements of his physics education, and for decades he continued to study Helmholtz's philosophical and other essays. Like Einstein and Raman, he too later won the Nobel Prize in Physics (1914). Indeed, in a survey of physics by Germany's leading physicists in 1915, Helmholtz's was the most-cited name. Even a scientific amateur (to put it kindly) like the historian and public intellectual Henry Adams, when he sought to explain the dynamics of history in terms of thermodynamics, read Helmholtz and spoke of him in the same breath as Galileo, Newton, Maxwell, Hertz, and Kelvin.[32]

At the turn of the century, Helmholtz's writings continued to inspire artists as well. The aesthetic notions of the French poet Paul Valéry, who was also a keen student of the sciences, were greatly stimulated by Helmholtz's empiricist view of perception. Patrick Henry Bruce, an American painter of the modernist school who was much influenced by the impressionists and the neo-impressionists, drew directly on the color theories and findings of Chevreul and Rood, and so indirectly on Helmholtz. Gustav Mahler, to cite a final example, read Helmholtz not only on music, but on physics as well; his library included all of Helmholtz's works.[33]

Yet in some quarters, the intellectual and cultural climate of the fin de siècle opposed or simply departed from much that Helmholtz had presumed and stood for, such as the idea that mechanisms constituted nature and that art naturally offered representations of objects. The rise of the avant-garde and of modernism in general—expressionism, vorticism, cubism, futurism, Fauvism, the German secessionist movement, and freely associating consciousness, to name only a few—did not fit well with Helmholtz's outlook on the world or his style. As figures like Nietzsche, Mach, Bergson, Arthur Rimbaud, and the postimpressionists came into vogue, Helmholtz had little or nothing to offer them or their admirers. Modernism in the arts tended to collapse the distinction between self and world, as did Mach and the monists and, later, Niels Bohr and quantum physics; "objectivity," if it existed at all, was said to be something created by man, not something present in material reality (nature). The increasing attacks on "traditional culture" and on conceptual thinking and reason itself, along with the questioning of "bourgeois" values, made Helmholtz irrelevant as a model. Modernist thought, with its opposition to traditional logic and narration, its challenge to representation and, with it, a world made up of stable things governed by causality and causal laws, had little or no accommodation for Helmholtz's epistemological outlook.[34] The new emphasis on notions of decay and decadence—some even invoked Helmholtz's own "heat-death" theory of the universe as an example—would have been anathema to him. His unquestioning belief in science, technology, and industry, and the social progress that he believed flowed from these, could not be reconciled with the new antiscience attitudes that emerged in the 1890s.

Then there were those increasingly central if diverse figures in the social sciences, like Pavlov, Dilthey, James, and Max Scheler, who, though not modern-ists, simply did not share Helmholtz's epistemological views, and especially his view on perception. From an entirely different quarter, moreover, Lenin's epistemological essay *Materialism and Empirio-Criticism: Critical Comments on a Reactionary Philosophy* (1909) argued for a thoroughgoing materialist doctrine and criticized Helmholtz's (alleged) philosophical idealism. Finally, Helmholtz's old enemy Eugen Dühring continued to polemicize against and mock him until his own end (1921).[35] However, as intellectually important and politically powerful as some of these and other figures became late in the nineteenth and early in the twentieth century, they did little overall harm to Helmholtz's reputation.

The coming of World War I also presented an opportunity to use the name and memory of Helmholtz, and this in a decidedly partisan way. Shortly after the assassination of Archduke Francis Ferdinand of Austria and his wife, and with nationalist feelings running at a fevered pitch across Europe, the Prussian minister of culture informed the minister of finance that it was vital to establish a Kaiser Wilhelm Institute for Theoretical Physics; it would fend off challenges to Germany's high standing in theoretical physics, a field "which has passed down to German science through the efforts of Helmholtz." The French and the English, the minister somehow believed, were now cultivating theoretical physics more fully than the Germans.[36] The institute, with Einstein as director, was established in 1917 (at least on paper, since not until 1937 were a building constructed and staff hired).

The Weimar Republic and Nazi Germany

World War I was a disaster for German scientists. The participation by some of their leaders and others in war propaganda, in chemical warfare, and in other ways helped make Germany an outcast from the international scientific community. During the early Weimar years, hyperinflation left many individual German scientists and their institutes without adequate means to conduct science and, since some could no longer even afford the subscription price of a foreign journal, unable to keep abreast of the latest developments in their field. In the immediate postwar years, with the economy depressed, a psychologically depressive atmosphere reigned. To try to find new trust and support, some leaders of German science turned to exploiting Helmholtz's name in both the domestic and the international realms. As a result, the Helmholtz Society for the Advancement of Physical-Technological Research (Helmholtz-Gesellschaft zur Förderung der physikalisch-technischen Forschung E. V.) was founded in 1920 by leaders of German industry and science. This society raised about 1 million dollars in capital, though some 80 percent of it was wiped out by the hyperinflation of the early 1920s. The society became an important supporter of German physics, however, especially that of senior physicists pursuing pure physics.[37]

With the centennial of Helmholtz's birth in 1921, a wide variety of medical doctors and scientists were further reminded of his name and the inspiration and trust it evoked. Scientists, medical doctors, philosophers, and others in Germany invoked Helmholtz's name to an extent that had not been seen since his death nearly three decades earlier. Ophthalmologists and medical doctors were perhaps the most active in this regard. Julius Hirschberg, one of Helmholtz's former students and a well-known historian of ophthalmology, noted that there were three memorial statues of Helmholtz in the center of Berlin: in the forecourt of the university, in front of the Brandenburger Gate, and on the Potsdamer and Victoria-Brücke (near the Philharmonie). These statues, as well as a portrait-relief of him in the Berlin Ophthalmic Hospital (Augenheilanstalt), helped keep memory of him alive. Hirschberg declared, "Helmholtz was a glory of Germany; he was an ornament of humanity." The *Vossische Zeitung* published an entire supplement marking the centennial of Helmholtz's birth, with articles by Ellen von Siemens-Helmholtz and others.[38]

The University of Berlin, too, held a special memorial service and laid a wreath at its Helmholtz statue. The physical, physiological, and philosophical societies sponsored their own celebrations of the centennial, at which they recalled his accomplishments in their respective fields. Emil Warburg, Helmholtz's successor once removed at the Reichsanstalt, recalled him as a physicist and the leader of the Reichsanstalt; Max Rubner spoke of him as a physiologist; and Moritz Schlick spoke of him as a philosopher. Benno Erdmann provided a major analysis of Helmholtz's theory of perception for the academy that year, too. In Bonn the physicist Heinrich Konen and the physiologist August Pütter spoke on Helmholtz's accomplishments in their respective disciplines, recalling his years at Bonn. Konen hoped that Helmholtz's legacy would inspire scientists and help end the bitterness and nationalist chauvinism left among them by the war. Major general scientific and cultural journals—like *Die Naturwissenschaften*, the *Deutsche Revue*, and *Die Umschau*—as well as numerous daily newspapers reminded their still war-weary and economically anxious readers of Helmholtz and of what he had stood for in a far more peaceful era. Even the socialists made a point of remembering him.[39]

Perhaps no single academic did more to keep Helmholtz's name alive in the interwar period than Schlick, a physicist—he had studied with Planck—as well as a philosopher. Schlick closely followed and wrote about developments in geometry and relativity by Einstein and others. In 1921 he and Paul Hertz edited a set of Helmholtz's epistemological writings. In 1922 Schlick was appointed professor of philosophy at the University of Vienna. That year, too, he joined and helped formalize a group of highly talented scientists, mathematicians, and philosophers that became known as the Vienna Circle. Members included Rudolf Carnap, Kurt Gödel, Hans Hahn, Otto Neurath, and Friedrich Waismann, among others. The multifaceted members of the Circle opposed metaphysics and all rhetoric in philosophy; but conversely, they emphasized the importance of epistemology and philosophy of science; indeed they considered science in general essential for analyzing and advancing philosophy; and they stressed empirical knowledge, the use of logic and mathematics, and the unity of the sciences. These emphases lay at the heart of Helmholtz's intellectual outlook, of course, and the members of the Circle, and more generally the "Logical Positivists" or "Logical Empiricists," as they and others beyond Vienna were also known, looked to Helmholtz as a seminal and inspirational figure for their intellectual movement. Certainly that was the case for Schlick, even though his views on geometry and epistemology in part disagreed with Helmholtz's.[40]

But it was not only the intellectual elite who recalled Helmholtz. So too did various lesser-known philosophers, medical figures, and applied meteorologists, to name but a few. Helmholtz's name was also remembered in *Die Grossen Deutschen*, as one of only 160 of the most prominent Germans, from Arminius to Hindenburg, of all time. Helmholtz also served as a model figure for the volume *Deutsche Männer*.[41] And he was recalled in another way as well, as the character "Helmholtz Watson" in Aldous Huxley's dystopian novel of 1932, *Brave New World*. In all these ways, the example—and myth—of Helmholtz as a model scientist and a model German man was projected and strengthened.

During the Weimar Republic and in Nazi Germany, no science was more seriously attacked than modern physics—that is, relativity theory, quantum mechanics, and the increasingly abstract, mathematical nature of physics in general. The attackers themselves were led by physicists. Two of the most prominent were leading experimentalists from the prewar era: Philipp Lenard, who had studied (for one semester) with Helmholtz, had served as Hertz's assistant, and had won the Nobel Prize in Physics (1905); and Johannes Stark, who in 1907 had promoted Einstein's work on relativity, who also won the Nobel Prize in Physics (1919), who served on the Helmholtz Society's board, and who became president of both the Reichsanstalt (1933–39) and the German Research Community (Deutsche Forschungsgemeinschaft). Lenard, Stark, and other supporters of National Socialism called for their colleagues to do "*Deutsche Physik*" ("German Physics"), that is, experimental physics, exclusively, as opposed to the abstract, mathematical physics that they denounced as "*Jüdische Physik*" ("Jewish Physics"). They succeeded in poisoning the atmosphere for doing physics—no one wanted to cite (at least by name) "Einstein's theory of relativity." Indeed, Einstein was the chief personal object of their attack, and there was much concern among his many supporters that, as a result of the antirelativistic and anti-Semitic attacks on him, he might emigrate from Germany. As early as 1920, the chemist Fritz Haber feared that such attacks might drive Einstein to leave. He wrote Einstein of his and their colleagues' support of him: "All judicious people have attributed a leadership capacity to you and do so in such great measure as has not been true since Helmholtz." Similarly, the theoretical physicist Paul Volkmann said that, under the circumstances, a popular account of relativity theory was needed to justify and explain it to a broad audience, in the way that Helmholtz had once done with other areas of science.[42] At least in the intellectual dimension of physics, Einstein had inherited Helmholtz's mantle. The proponents of "*Deutsche Physik*" attacked Einstein in an unscrupulous way that (the anti-Semitism aside) is reminiscent of the way Dühring had once attacked Helmholtz.

Lenard and other promoters of "*Deutsche Physik*" themselves invoked Helmholtz's name as they saw fit. Lenard considered Helmholtz one of the "Great Men of Science." He claimed that the fact that Helmholtz had never formally studied mathematics at the university but had nonetheless done outstanding mathematical work in physics showed "the complete uselessness of the extensive mathematical and other courses of training at present-day universities." There was even a petty nonevent in metrology involving Helmholtz and the Nazis. After the Nazi government had signed (1933) an international agreement officially designating the "Hertz" as the unit of frequency, several Nazi leaders in electrotechnology and their bureaucratic governmental allies sought to withdraw from or evade that agreement. In their (mistaken) view, Hertz was a "half Jew" and had nothing to do with radio technology and other matters concerning frequency. Moreover, they believed that to give his name international prominence was to do harm to Nazi Germany's racial image and "worldview." Nazi administrators thus proposed to expunge his name (while avoiding a critical international review of their action) by employing the elegant fiction that the designated abbreviation for frequency ("Hz") should stand for "Helmholtz" rather than "Hertz." The *Führer*, however, ruled in 1941 against the proposed change.[43]

Notwithstanding Lenard's use (abuse) of Helmholtz's name and the pettiness of the issue of metrological nomenclature, Helmholtz and all he stood for—reason, tolerance, and a certain measure of liberalism—offered nothing to the Nazis and their supporters. In January–February 1935, a Nazi student rally was held in the forecourt of the University of Berlin, with the Helmholtz statue unwittingly in its midst (see fig. E.4). Yet someone—presumably a Nazi official—was unhappy with the statue's prominent place there, for later that year it was removed to the garden on the west wing of the university, near where (the liberal) Mommsen's statute was already located and where it remained for nearly six decades. As for Helmholtz's Gymnasium in Potsdam, it openly "coordinated" itself with the Third Reich, and its director aimed to make its pupils "into proper Germans."[44]

Fig. E.4 Helmholtz Memorial Statue in the Main Court of the University of Berlin during
a Nazi Student Rally, 1935. Andreas Feickert, the Reichsführer of the German Students,
is speaking from the balcony of the University of Berlin. Bildarchiv Preussischer Kulturbesitz,
Berlin/Art Resource, NY.

Four years later, in 1939, Ludwig Glaser—a former student of Stark's, a Nazi, and a full-throated supporter of "German Physics"—celebrated "Nordic man in Germany." "Nordic man" had done so much for physics, Glaser claimed, though he deplored that it had also resulted in Einstein, "a Jewish politician," for which he blamed Helmholtz! In a tirade against "Jews in Physics: Jewish Physics," he declared: "From the patriotic [*völkisch*] point of view, the life of this unquestionably great scholar [Helmholtz] is marred by certain facts which have pernicious consequences that are only now becoming apparent." These "facts" were Helmholtz's "refusal to acknowledge the merits of Robert Mayer"; his "weak stance as the representative of German science at international congresses, resulting in Gauss's and Weber's names being eliminated in the naming of the conventional international electrical units; and the complete lack of appreciation for one of his most important German contemporaries, the physicist and mathematician Hermann Grassmann. That the blame must be laid on Helmholtz for having granted the Jews admission to physics," Glaser continued, "is proven by the case of the physicist and anti-Jewry pioneer Eugen Carl Dühring. His removal from the academic faculty at Berlin University in 1877 is indicative and marks the beginning of the Jews' unscrupulous invasion of German professorships in physics." Glaser, whose outburst placed him on the extreme end of even "German Physics," also listed the names of several of Helmholtz's Jewish (or partly Jewish) students. "How welcome Helmholtz's influence was to Jewry is demonstrated by the fact that the Eastern Jewish mathematician Leo Koenigsberger became his biographer." Moreover, Glaser blamed Helmholtz for the appointment—after Helmholtz's death!—of Warburg (who was of Jewish ancestry) to positions in Berlin. It was Helmholtz, Glaser said, who trained, supported, or influenced Jewish physicists like Hertz, Eduard Wertheimer, Goldstein, Leo Grunmach, Rubens, James Franck, and Max Born, to name only some. Indeed, and still worse in Glaser's view, Helmholtz had "prepared long ago" the ground for "a Jew like Einstein," that is, for the rise of "Jewish Physics." Glaser was not alone in his absurd diatribe, no matter its disregard for the facts and its non sequiturs. One year later, Wilhelm Müller, an aerodynamicist and fluid dynamicist who, despite much objection from the faculty, had succeeded Sommerfeld as professor of theoretical physics at the University of Munich, similarly sought to defame Helmholtz for his supposedly pernicious influence on theoretical physics and his support of Jewish science students.[45]

To be sure, militarists and others in Nazi Germany could also find some "good" use for Helmholtz's name. The Versailles Treaty had forced the German government to close the Prussian Friedrich-Wilhelms-Institut, but Hitler reopened it in 1935 under the name "Militärärztliche Akademie." In October 1939, shortly after the Nazi Blitzkrieg on Poland and the start of World War II, Erich Hoffmann, a dermatologist and a graduate of the reopened academy, recalled the centennial of Helmholtz's and Virchow's entrance into the institute (1838). He believed that both men owed their "universal meaning," "certainly in the first instance, to their excellent inheritance [*Erbgut*] from the German race," but then he also noted the environment in which they were raised and trained. Helmholtz's name (and potted biography) even appeared in a reference work on artillery and ballistics; it seems that his work on hydrodynamics was relevant in the context of ordnance.[46]

The Helmholtz family name was kept alive in other ways. Richard von Helmholtz helped do so through his distinguished career as a locomotive construction engineer in Munich; he eventually became chief engineer of the highly successful Krauss & Company (later KraussMaffei) and received an honorary doctorate of engineering degree from the Technische Hochschule Danzig in 1913. After his retirement in 1917, Richard wrote several historical articles and two books on the history of railroad engineering. During his adult lifetime, he had little contact with his father or his stepmother; he never married and instead had a longtime partner. With his death in Munich in 1934, the line of direct male descent from Hermann von Helmholtz ended.[47]

After her mother's death in 1899, Ellen von Siemens-Helmholtz became the family's literary executor. During the 1920s, as she was reviewing her father's papers, she discovered her mother's own correspondence, which she published in a two-volume edition. These letters constituted a rich source not only on Anna's life but also on her husband's. In reviewing them, Planck thought they showed not only something of Helmholtz's intellectual abilities but also his sensitivity to human issues, to nature, to art, and more. "Above all, time and time again the impression intrudes upon one of the judgment of his entire personality, that, among all the heroes of science to whom the distinction of character and the depth of soul [*Gemüt*] stand so completely at the height of the intellect, it [his personality] may yield way to few." Before Ellen's own death in 1941, she placed most of her father's papers permanently in the archive of the Academy of Sciences.[48]

In 1941, two years into World War II and at a moment when Nazi Germany had conquered much of Europe, a dozen or so articles appeared on Helmholtz's life and work, marking both the 120th anniversary of his birth and the 90th of his invention of the ophthalmoscope. Hans Schimank recounted Helmholtz's life and achievements in a journal tellingly entitled *Deutschlands Erneuerung* (Germany's renewal). Schimank, a young historian of science, analyzed the "heroic fate" of this "German researcher in the age of the founding of the [Second] Reich." Friedrich Schomerus of the Zeiss Werke used the occasion to recall the (thin but important) professional relationship between Ernst Abbe and Helmholtz. Hermann Pistor, a leading German ophthalmologist, called Helmholtz "an intellectual *Führer* of universal meaning in the empire of the sciences . . . a German man to whom the entire world of culture owes thanks for new knowledge and progress of a powerful magnitude." Pistor concluded: "Germany will always remember [Helmholtz] with pride and reverence, and thanks its great son, one of the greatest of all time in the empire of the mind." Friedrich A. Zschau expressed a similar thought: "Hermann von Helmholtz, whose name can never be extinguished from history, will always remain a model for every new generation."[49] These quotations suggest the use of the memory of Helmholtz for both adapting to the regime and recalling imagined days of past glory for German science under his leadership.

By September 1944, bombs were raining down upon German cities, and Allied ground forces had begun invading Germany from both east and west. It was becoming clear to all but the most fanatical Nazis that the war would end sooner rather than later and that Germany would be defeated. Jonathan Zenneck and Laue, both physicists, found time nonetheless to recall the fiftieth anniversary of Helmholtz's death. Laue was, among physicists who had remained in Germany, the most outspoken opponent of the Nazi regime. He was a former student of Planck's and Lummer's, a leading exponent of relativity theory and other aspects of modern physics, and winner of the Nobel Prize in Physics. As a friend of Einstein's and a professor of theoretical physics at Berlin, Laue carried forth into Nazi and postwar Germany the spirit of science that Helmholtz, along with Planck and Einstein, had embodied.[50]

Even before the war's end, in 1944, the German Physical Society (Deutsche Physikalische Gesellschaft, as the Berlin society had been renamed in 1899), under the editorial leadership of Ernst Brüche, sought to help rehabilitate German physics by publishing a new journal, the *Physikalische Blätter*; it became a forum for news and discussion of the state of the profession. Its very appearance, as well as its contents, marked a sign of change in Germany. It sought a broad audience, it was especially concerned with training the next generation of physicists, and it called for a renewed emphasis on doing fundamental research. Toward that end, in its midyear issue of 1944, it displayed a full-page portrait of Helmholtz, noting that he had died fifty years earlier and, as one of the "great" physicists of the previous century, had led Germany to the forefront of international science. Helmholtz and a handful of other German scientists were held up as role models for young German scientists, to show what German science once was and could again be. Indeed, Brüche also intended the journal to show how, with a few exceptions like Lenard and Stark, German physicists had not cooperated with the Nazis.[51] His point is more than debatable.

Notwithstanding all this heroizing and mythicizing of Helmholtz, World War II damaged his legacy and memory on an institutional level and in a concrete way. In late August 1944, the British Royal Air Force thoroughly bombed and largely destroyed the city of Königsberg, including the university and other parts of the city's historic center. Then in April 1945, the Soviets bombed, invaded, and destroyed whatever was left, killing or expelling most everyone who had remained. Königsberg became a Russian city, known as of 1946 as Kaliningrad; the physical manifestations of its cultural history and much else, including any trace of Helmholtz's world there between 1849 and 1855, was extinguished. In Berlin, Helmholtz's old Physics Institute (along the Reichstagsufer) was almost completely destroyed in 1945; only the external walls remained. What had once been Germany's premier physics institute now lay in ruins. Also in early 1945, the buildings of the Reichsanstalt, which under both Stark and his successor Abraham Esau had become a fully "coordinated" institution of the Nazi regime, were also either completely destroyed or burned out as a consequence of the war. None remained in a usable state at war's end. In anticipation of the expected bombing of Berlin, however, in 1943–44 nearly all of the Reichsanstalt's laboratories had been removed to several small cities in Germany. Finally, during the night of 14–15 April 1945, as the war in Germany was drawing to a close, Potsdam was heavily bombed by the British Royal Air Force's Bomber Command, and a week later it sustained still further destruction when the Red Army invaded. The city's inner core was largely destroyed; nearby Hoditzstrasse, where Helmholtz had grown up, was destroyed. In all likelihood, both the house in which he had been born and the one in which he grew up were destroyed.[52] Helmholtz's former Gymnasium, as well as his former residences and institutes in Bonn and Heidelberg, survived the war. Nonetheless, by the end of World War II, German science, if not Helmholtz's legacy and memory, had reached a historic low point.

The Two Germanies and Today

The end of World War II led, of course, to a divided Germany: the Federal Republic of Germany (West Germany) and the German Democratic Republic (East Germany) were established in 1949. Each built up its own social, political, and economic systems as well as systems of higher education and research. Each exploited Helmholtz's name and memory as the Germans sought to (re)build their separate Germanies, their respective disciplines, and their individual careers.

The first signs of Helmholtz's use as a historical figure in rebuilding science, mainly modern physics and medicine, appeared even before the founding of the two Germanies. As the postwar era emerged, the *Physikalische Blätter* pointed to the founding of the Helmholtz Society as "a warning for our times," namely, that money was needed in Germany for research in physics and technology. In 1947, slightly more than a century after the Berlin (German) Physical Society had been founded, Brüche recalled how Helmholtz had spoken before it on numerous occasions, and he used the society's centennial to call for renewal. A year later he returned to Helmholtz, now portraying the similarities between him and the British physicist Ernest Rutherford and emphasizing how an individual personality could leave his mark on an era and how science belonged to no nation. Brüche called the first fifty years of the society "the Helmholtzian Era": "It was," he said, "the time of classical and measuring physics, which was continued on especially by Helmholtz's numerous [former] students at the Reichsanstalt." It was a strained if perhaps useful attempt in a Germany occupied in part by the British and completely preoccupied by a search for moral and scientific renewal. The use of Helmholtz as a model physicist in the *Physikalische Blätter* was also an attempt by the society to distance itself and its science from the Nazi regime, to maintain that it had not cooperated with the regime, and to rebuild the society and the socioeconomic foundations of its science. Helmholtz, as Planck noted in a posthumous article in the *Physikalische Blätter*, had shown "incorruptible judgment. . . . [His] simple nature embodied the worth and truthfulness of his science." He was, Planck said, a peaceful man. Other West German physicists also soon recalled his memory.[53]

The years 1950–51 marked the centennial of Helmholtz's invention of the ophthalmoscope, and scholars in both West and East Germany used the occasion to recall that seminal instrument for the history of medicine—and did so in their respective national domains. In 1950, moreover, the German Academy of Sciences (Deutsche Akademie der Wissenschaften zu Berlin), located in East Berlin, the successor institution to the Prussian Academy of Sciences, celebrated its 250th anniversary in part by publishing a book of portraits of its most famous members, with Helmholtz naturally among them. It was an explicit claim that his legacy also belonged to East Germany, and it helped legitimate that state.[54] Indeed, the East German regime controlled four important institutions or items that were potentially useful for linking itself to Helmholtz's legacy and thus raising the regime's status: the University of Berlin, the Helmholtz statue, the Helmholtz Medal, and the bulk of Helmholtz's papers.

In 1949 the East Germans renamed the Friedrich-Wilhelms-Universität zu Berlin, that is, Helmholtz's old university, as the Humboldt University of Berlin. When they celebrated its 150th anniversary in 1960, Helmholtz was given a prominent place through a historical paper by Laue heroizing Helmholtz.[55] The East Germans did not, however, relocate Helmholtz's statue to its previously prominent spot in the forecourt; to have done so would have meant associating his name too closely with the university, and thus at the expense of Marx's, Engels's, and Lenin's names.

The Helmholtz Medal and the papers were under the control of the academy. Not until 1954 did the academy decide to renew the awarding of the medal, and, as noted earlier, it did not make the first new award until 1959, after a forty-year hiatus (the last award having been in 1918). It gave the medal to three of the academy's ordinary members (Otto Hahn, Gustav Hertz, and Laue), all very senior statesmen of German science; Hahn was a leader in opposing West Germany's use of nuclear energy and its acquisition of nuclear weapons, Hertz was a professor of physics in East Germany, and Laue (again) was an outspoken opponent of the Nazis. Of the eighteen individuals who received the medal between 1959 and 1990, nine were Germans and nine were foreigners. Of the nine Germans, seven were East Germans and two were West Germans (Hahn and Laue). Of the nine foreigners, one was Danish (Niels Bohr), one was British (Paul Dirac), one was French (Louis de Broglie), and six were Soviet (Nikolai Bogoliubov, Viktor Ambartsumian, Vladimir Fock, Andrei Kolmogorov, Peter Kapitza, and Aleksandr Prokhorov).[56] The facts that the award was not renewed until the post-Stalinist era; that the Cold War was raging and East Germany, in particular, was hemorrhaging skilled workers and professionals and thus greatly in need of whatever international status and prestige it could muster; that one-third of those who received the medal were Soviets (all but one were Russians); and that only three other foreigners received the award (with no Americans among them), suggests that more than ability and merit were among the criteria used for making the award. The academy, it would appear, also had a political agenda in awarding the Helmholtz Medal: it used Helmholtz's name to raise the status of East German and Soviet science during the Cold War.

The academy restricted use of the Helmholtz papers (with a few exceptions) until the late 1980s, allowing access to only a handful of authorized East German scholars. This was a decidedly political, and self-interested, use of the papers on the part of several East German Marxist ideologues; they sought to interpret Helmholtz's epistemological and scientific writings as stepping stones on the way to Marxism-Leninism, even as they recognized that Helmholtz himself remained innocent (and perhaps even effectively unaware) of the thought and writings of Marx and Engels. In 1971, on the occasion of the 150th anniversary of Helmholtz's birth, the East German Marxist philosophers Herbert Hörz and Siegfried Wollgast brought out an edition of Helmholtz's philosophical lectures and essays, including an introduction that portrayed him as a central figure in epistemology and in the formation of a "progressive" (that is, "materialist") worldview; through his physiological studies and philosophical essays, Helmholtz had supposedly pointed toward the epistemological direction that Marx, Engels, and Lenin later developed into its full dialectical-materialist mode. The academy, likewise, celebrated the 150th anniversary with a set of papers devoted to analyzing Helmholtz's creative powers and his efforts to effectively unite different branches of science together. In 1973, moreover, Humboldt University published a collection of fourteen articles devoted to Helmholtz's work "from the viewpoint of Dialectical Materialism and of Modern Science." Many of the East German Marxist ideologues who regularly used and abused Helmholtz's name made contributions. The deputy head of the section devoted to Marxist-

Leninist philosophy at the university hoped the volume's contributions would link Helmholtz's scientific work and philosophical outlook to dialectical materialism, that is, with "the highest accomplishment of theoretical thought." Finally, in 1986 a group of East German scholars brought out an (excellent) edition of Helmholtz's correspondence with du Bois–Reymond by combining their own holdings from the Helmholtz papers in the academy with those located in West Berlin. Hörz and Wollgast again included a tendentious introduction to the volume, placing these letters within their supposedly "materialist" and Marxist-Leninist contexts.[57]

West German journalists, scientists, and institutional leaders, while less heavy-handed in terms of ideology than their East German counterparts, were generally no less keen to call Helmholtz their own. They recalled the sixtieth anniversary of his death. An entry on him appeared in *Die Grossen Deutschen: Deutsche Biographie*. West German scholars treated him as one of the great scientists and medical doctors, while the University of Heidelberg and the city of Berlin sought to remind the world of their historical ties to him.[58]

Helmholtz's name also became a part of daily life in postwar Germany. The city of Potsdam remembered him with a Helmholtzstrasse and renamed its Viktoria Gymnasium as the Hermann-von-Helmholtz Gymnasium Potsdam. In 1945 the city of Heidelberg also renamed one of its Gymnasia after him and has, broadly speaking, sought to associate Helmholtz's name with the city. It too has a Helmholtzstrasse. The city of Berlin has doubled that: it has two streets named "Helmholtzstrasse," not to mention a Helmholtzpark. The Harnack-Haus in Dahlem contains a large Helmholtz Lecture Auditorium (Helmholtz-Saal) as well as rooms and apartments named in honor "of great Germans," all scientists, including one for Helmholtz. All told, today there are at least nine German cities that have a Gymnasium named after Helmholtz and at least forty cities (including all major ones) that have a street named after him; many of these streets are located at or near universities or other scientific institutions.[59]

Since the unification of the two Germanies in 1990, the Germans have maintained Helmholtz's role as an important cultural icon. The centennial of his death (1994) again brought recognition of him. The Federal Republic issued two postage stamps bearing his portrait. The statue of him was moved back into the forecourt of Humboldt University. The Berlin-Brandenburg Academy of Sciences (Berlin-Brandenburgische Akademie der Wissenschaften), to use the society's latest and full name, began once again—after the academy was reorganized in 1991 following political reunification—to award the Helmholtz Medal, its highest award. It now does so every two years, and of the twelve medals awarded between 1994 and 2016, seven have gone to Germans and five to foreigners (three Americans, one Briton, and one Canadian). Both Knaus's original painting of Helmholtz (1881) and Werner's original painting *Kronprinz Friedrich auf dem*

Hofball, 1878/1895, which prominently features Helmholtz, now hang in the Alte Nationalgalerie, in the traditional cultural heart of Berlin. See figures 19.2 and 21.1. The city of Heidelberg again celebrated its connection with Helmholtz through a symposium devoted to his life and work there, and it supported the efforts of a local scholar to document that life and times.[60]

Starting in the late 1960s, various American, British, and Israeli historians and philosophers of science, soon joined by their colleagues in France, Germany, Italy, and other European countries, began publishing much more scholarly, more contextualized, and less heroic studies of Helmholtz. These aimed not at merely listing and briefly describing (for the nth time) his vast array of scientific achievements, or restating his philosophical views, or retelling his life story in undocumented heroic format, or discussing his life and work minus their historical contexts. Aided in part by the discovery or republication of his letters, and by access to and (largely unprecedented) use of archival materials, in particular those held by the academy, modern critical scholarship about Helmholtz reached the level that historians and philosophers of science have managed to achieve for other notable scientists, such as Galileo, Newton, Darwin, and Einstein. By 1994 two substantial edited volumes of studies on Helmholtz appeared, pushing forward the effort to treat his life and work in a modern historiographical manner, one that aimed to be scholarly, critical, and contextual—not merely celebratory—in its approach.[61]

Most recently, since 1995 the Helmholtz Association of German Research Centers (Helmholtz Gemeinschaft der deutschen Forschungszentren) has gradually been established. It aims to organize and increase the profile of large-scale research centers around Germany (there are currently eighteen). It has become the largest scientific organization (in terms of employment and budget) in Germany today.[62] By attaching Helmholtz's name to it, the founders have sought to emphasize the importance of long-term, fundamental, interdisciplinary scientific and technological research within Germany and to link the association to various European and non-European partners. Moreover, the Reichsanstalt's direct successor, the Bundesanstalt, has endeavored to maintain its own historical ties with Helmholtz. Since 1973 the Bundesanstalt has awarded an annual Helmholtz Prize for outstanding work in the field of precision measurement in physics, chemistry, or medicine. Institutionally speaking, the Helmholtz Association has helped maintain (and exploit) Helmholtz's name alongside those of several of Germany's greatest cultural figures: Goethe, with the Goethe Institute for the promotion of the German language and culture; Humboldt, with the Alexander von Humboldt Stiftung (endowment) for the support of individual domestic and foreign researchers; and Planck, with the many institutes that constitute the Max Planck Gesellschaft (society) for advanced scientific and scholarly research. This public and institutional prominence (and exploitation) of the name "Helmholtz" has, ipso facto, extended and amplified his memory into that of a mythical figure, or idol, or icon; he has become a patron saint. Indeed, by extracting Helmholtz's name and achievements out of their personal and historical context, and thus by neglecting those who supported and opposed him, or those whom he (legitimately) borrowed from or competed against, these institutions have helped to further the earlier efforts to make him a demigod, idol, or icon, as others have done to Goethe, Humboldt, and Planck.

In sum, Helmholtz's family; members of the German scientific and technological community; German institutions of all sorts; historians, philosophers, and journalists; and a series of German political regimes from Imperial Germany to the Federal Republic have sought to shape Helmholtz's legacy for their own purposes. They have (collectively speaking) renewed and extended a persona that has (at some points and for all too many people) made Helmholtz into a mythical figure, or idol, or icon in German public culture; and they have used (and occasionally abused) his name to advance a variety of familial, scientific, cultural, and political causes in Germany. In no small measure and all too often, they have produced hagiography rather than critical historical scholarship. They are not alone. As biographies of numerous other scientists who lived in centuries since the sixteenth show, Helmholtz is neither the first nor the last to be portrayed in such a manner. Contemporary scholarship in the history of science has, however, sought another, more contextual and critical way of understanding the actors who have produced scientific knowledge since the early modern era; it has attempted to show that more than memory or nonscholarly purposes are required for understanding how scientific lives are made and shaped and how those lives made and shaped the scientific and cultural worlds in which they lived and worked.

Acknowledgments

It gives me great pleasure to thank publicly the many institutions and individuals who supported me in my work and so helped make it possible.

I received generous support from the National Endowment for the Humanities (RH-21168–94), the National Science Foundation (SES-0450718), and the (alas, now defunct) Dibner Institute for the History of Science and Technology at the Massachusetts Institute of Technology. In addition, I received extensive cooperation and further financial support from the University of Nebraska–Lincoln (UNL).

I thank, too, the many archivists, librarians, and other staff members at the archives and libraries listed in this volume's references. I would like to mention especially the staff at the archive of the Berlin-Brandenburgische Akademie der Wissenschaften, which houses the principal collection of Helmholtz's *Nachlass* (literary remains). I am most grateful, in particular, to three of its successive directors — Klaus Klauss, Wolfgang Knobloch, and Vera Enke — as well as staff members for the help they gave me during many working visits there over many years. The Interlibrary Loan services at UNL were equally indispensable.

Portions of this book draw on some of my previous work on Helmholtz, and I acknowledge and thank the Franz Steiner Verlag for permit-

765

ting me to adapt parts of the introduction to my edition of *Letters of Hermann von Helmholtz to His Parents: The Medical Education of a German Scientist, 1837–1846* (Cahan 1993b) for use here in parts of chapters 1–4; Cambridge University Press for permitting me to adapt my discussion of Helmholtz and the Reichsanstalt from my book on that subject (Cahan 1989; copyright Cambridge University Press, reprinted with permission) for use here in parts of chapters 24 and 28; deGruyter Verlag for permitting me to adapt part of an article (Cahan 1999) in various portions of the text; Taylor and Francis for permitting me to adapt parts of an article (Cahan 2010); and the Royal Society of London for permitting me to adapt parts of two other articles (Cahan 2012, 2012a). I also gratefully acknowledge access to materials held in the archives, libraries, and other institutions listed in this volume's references and permission from the museums, archives, libraries, publishers, and others that have allowed me to reprint illustrations and other items from their works or holdings.

I thank, furthermore, the many colleagues, friends, and acquaintances who helped me to locate letters from, to, or about Helmholtz, as well as various other items, including secondary sources pertinent to this biography. Most of this research happened long ago, when my investigations were at an early stage, and sometimes our exchanges were quite brief. Thus I suspect that a few of those mentioned here may not even recall assisting me. Nevertheless, their readiness to help and the scholarly leads they provided deserve acknowledgment: Simon Bailey, Marco Beretta, Rainer Bloch, Rainer Brömer, Joe Burchfield, Kenneth Caneva, Enrique M. De La Cruz, Peter Dohrn, Ariane Dröscher, Michael Eckert, John T. Flynn, D. W. Fostle, Christiane Groeben, Ralf Hahn, Jane Harrison, Werner Heegewaldt, Dieter Hoffmann, Giorgio Israel, Paul Israel, Frank A. J. L. James, Erika Krausse, Richard L. Kremer, Peter T. Landsberg, Kathryn M. Olesko, Theodore Porter, David Robinson, Henning Schmidgen, Winfried Schultze, Henry Z. Steinway, William Y. Strong, Paul Theerman, R. Steven Turner, Sir Ralph Verney, Scott Walter, and David B. Wilson. I owe special thanks to Ruprecht von Siemens for supplying me with copies of Helmholtz's original essays for the Hebrew and German parts of his Gymnasium final examinations and, more generally, for discussions concerning his maternal grandfather (Hermann von Helmholtz), the Helmholtz family, and the disposition of Helmholtz's literary estate; to the late Hermann Ehret for his invaluable transcription of letters (especially those from Ferdinand Helmholtz to Immanuel Herrmann Fichte); to Gabriel Finkelstein for alerting me to numerous relevant items concerning Helmholtz (among other topics) in Emil du Bois–Reymond's correspondence with third parties; and to Guy Strutt, Lord Rayleigh's grandson and the administrator of the Rayleigh Archive at Terling Place, for providing information about Helmholtz's visit there. In addition, I thank Henry Z. Steinway and Guy Strutt for permission to cite materials in the Steinway and Rayleigh archival collections, respectively. Unless otherwise noted, all translations from German (as well as from French and Italian) into English are my own. Where I thought it advisable, I have not hesitated to make slight changes to previous English-language translations (including my own) to make these, I hope, more understandable to readers.

My very able undergraduate assistants Robert Nichols, Andrew Hansen, and Aaron Pattee, each of whom was supported by the UCARE program at UNL, provided valuable research assistance, as did Jenna Schmaljohn, a history graduate student at UNL. Ranelle Maltas and Brad Severa of the New Media Center at UNL also deserve thanks for their help in formatting this book's manuscript.

On a more substantive intellectual level, I thank Paolo Brenni and David Pantalony for their expertise about various nineteenth-century acoustic and optical instruments and apparatus, not least Helmholtz's. Kathryn M. Olesko graciously corrected me or commented on a number of factual points. I thank Gabriel Finkelstein for his careful reading of an initial draft of chapters 1–6 and for our many thoughtful discussions about Helmholtz and science in nineteenth-century Germany; Alexandra Hui for her likewise careful reading of chapter 12; Bernard Lightman for his close reading of and useful comments on chapter 13 (as well as several other sections of this book concerning Victorian science and John Tyndall); and Klaus Hentschel for his critical reading of the epilogue. In response, I made factual corrections and, as best I could, sought to incorporate everyone's thoughtful criticisms. Similarly, I thank Kenneth Caneva for many challenging and enlightening discussions about Helmholtz in general and, more particularly, the history of the principle or law of conservation of energy. The work of many scholars past and living, as my endnotes indicate, has enabled me to deepen my understanding of Helmholtz's intellectually wide-ranging, complex, and often difficult scientific work and philosophical ideas; the notes are meant not only as evidence for claims made in the text but also as grateful acknowledgment for the guidance I found in the works cited and for the facts and learned analyses that, in many cases, others before me or independently of me have reached. In this regard, I wish to single out the very helpful guidance I received from Olivier Darrigol's authoritative analyses of nineteenth-century physical science and measurement theory, both as they concern Helmholtz's own writings (that is, internally) and as they concern their wider scientific context.

Kenneth Caneva, Olivier Darrigol, and Robert J. Richards read large portions of the penultimate draft of this book. Each brought his deep understanding of various intellectual dimensions of nineteenth-century science and philosophy to their readings of the draft. I have sought throughout to meet their challenging comments and criticisms, while keeping to my own views. I thank them for taking valuable time from their own scholarly work to review mine. Finally, I am very grateful to the University of Chicago Press's two anonymous and insightful reviewers of the draft. They reminded and encouraged me to stress the big issues concerning Helmholtz, and I have sought to do that.

For their general moral support and encouragement during various phases of this book's long gestation, I wish to thank my friends and colleagues Jed Buchwald, Robert Kargon, Richard L. Kremer, Laura Otis, and Robert J. Richards, as well as several members of the Department of History at UNL (Thomas "Tim" Borstelmann, James D. Le Sueur, Timothy R. Mahoney, William Thomas III, and Kenneth J. Winkle). I also thank Jeannette Jones, another department colleague and friend, for her excellent advice at a crucial moment in this volume's realization. I would like, too, to record my deep thanks to my good friends in Berlin, Friedrich and Margarethe Hagemeyer and Jens and Eva Reich, for their warm hospitality during many of my numerous trips to Berlin over the years; for their general encouragement of my work; and for imparting to me something of their invaluable knowledge of their cherished city's past and present.

Karen Merikangas Darling, executive editor, Books Division, at the University of Chicago Press, has encouraged this project throughout much of its long history. Her faith in it and her general support have proved absolutely essential. I am most grateful to her. I thank Lois Crum for her superb copyediting work. Relatedly, I thank UNL (both its College of Arts and Sciences and its Research Council) for generous subvention awards that have helped make the publication of this book possible.

I come, finally and far above all, to thanking my family for the support and love they have given me, and that not only during the research and writing of this book. My wife, Jean Axelrad Cahan, provided me with enduring support and her excellent taste (and not only in matters scholarly) and bore with me through the long years during which I sought to realize this project. She read the entire text as commentator, critic, and literary stylist. In short, I thank her for her love. Mine for her cannot, in truth, be expressed in words. Our daughter Lara has likewise been my staunch and loving supporter, as has my sister Elizabeth Grace Aron, who, in this regard, came to replace the role once played by our parents, Haskell and Sylvia Cahan. Likewise I thank my brother-in-law Stan Aron for his unwavering support.

Although the research and writing of this book could not have occurred without the generous support of the institutions listed above and in the references, nor without the collegiality, friendship, support, and good advice of many individuals, it is no mere cliché to say that I take full responsibility for the facts presented and the interpretations proffered in this study.

Abbreviations

AD	Anton Dohrn
AdK	Akademie der Künste, Berlin
AH	Anna Helmholtz (née von Mohl)
AHA	Akten der Historischen Abteilung des Akademiearchivs, Berlin-Brandenburgische Akademie der Wissenschaften, Berlin
AJE	Alexander John Ellis
AS	*Annals of Science*
BGLA	Badische Generallandesarchiv, Karlsruhe
BSBMH	Abteilung für Handschriften und Seltene Drucke, Bayerische Staatsbibliothek, Munich
CH	Caroline Auguste Helmholtz
CL	Carl Ludwig
CR	*Comptes Rendus Hebdomadaires des Séances de l'Académie des Sciences*
CUL	Department of Manuscripts and University Archives, Cambridge University Library
Dep. 5	Handschriftenabteilung, Depositorium Runge-du Bois-Reymond, Haus Potsdamer Strasse, Staatsbibliothek Preussischer Kulturbesitz, Berlin (Nachlass Runge-du Bois-Reymond)

DM	Handschriften-Bestand (Handschriftenabteilung), Archiv, Deutsches Museum, Munich
EB	Ernst Brücke
EdBR	Emil du Bois–Reymond
ESH	Ellen von Siemens-Helmholtz (née Helmholtz)
EUL	Helmholtz Letters, Gen. 2169, Special Collections, Edinburgh University Library, Edinburgh, UK
FCD	Franciscus Cornelis Donders
FH	August Ferdinand Julius Helmholtz
GEGHL	MS Vault Eliot, George Henry Lewes, George Eliot and George Henry Lewes Collection, Beinecke Rare Book and Manuscript Library, Yale University, New Haven, CT
GM	Gustav Magnus
GRK	Gustav Robert Kirchhoff
GSPK	Geheimes Staatsarchiv Preussischer Kulturbesitz, Berlin (unless otherwise indicated, references are to I. HA Rep. 75Va)
GUL	William Thomson Papers, University of Glasgow Library, Glasgow, UK
HBJ	Henry Bence Jones
HeHz	Heinrich Hertz
HER	Henry Enfield Roscoe
HH	Hermann Ludwig Ferdinand Helmholtz (In January 1883 Helmholtz was ennobled. He and his family members thereupon added "von" to their name.)
HK	Hermann Knapp
HKL	Hermann Knapp Letters, J. M. Wheeler Library, Edward S. Harkness Eye Institute, College of Physicians and Surgeons, Columbia University, Department of Ophthalmology, New York Presbyterian Hospital, New York
HN	Hermann von Helmholtz Nachlass, Akademiearchiv, Berlin-Brandenburgische Akademie der Wissenschaften, Berlin
HT	Heinrich von Treitschke
HUB	Universitätsarchiv, Humboldt Universität, Berlin
IHF	Immanuel Herrmann Fichte
ISZ	Ida Freifrau von Schmidt-Zabiérow (née von Mohl)
JT	John Tyndall
KM	Kultusministerium (Königliches Ministerium der geistlichen-Unterrichts- und Medizinalangelegenheiten). All the KM papers are kept at the GSPK.
KP	Kelvin Papers, University of Glasgow Library
LK	Leo Koenigsberger, *Hermann von Helmholtz*, 3 vols. (Braunschweig: Friedrich Vieweg, 1902–3)
MM	Mary von Mohl
OH	Olga Helmholtz (née von Velten)

PB	*Physikalische Blätter*
PGT	Peter Guthrie Tait
PM	Pauline von Mohl
RC	Rudolph Clausius
RH	Robert Helmholtz
RKUHA	Archiv, Ruprecht-Karls-Universität Heidelberg, Heidelberg
RL	Rudolf Lipschitz
RM	Robert von Mohl
RSC	Henry Roscoe Collection, Library, Royal Society of Chemistry, London
SBPK	Handschriftenabteilung, Haus Potsdamer Strasse, Staatsbibliothek Preussischer Kulturbesitz, Berlin
SF	Archiv, Siemens Forum, Munich (formerly Siemens Archiv)
SHPS	*Studies in History and Philosophy of Science*
SP	Steinway Papers, Fiorello H. LaGuardia Archives, LaGuardia Community College, City University of New York, Long Island City, NY
SZADNA	Archives, Stazione Zoologica "Anton Dohrn," Naples
TAED	Thomas A. Edison Papers, Digital Edition, http://edison.rutgers.edu /singldoc.htm
TAEM	Thomas A. Edison Papers, Rutgers University, Piscataway, NJ, Microfilm Edition
TM	Theodor Mommsen
TMN	Theodor Mommsen Nachlass, Handschriftenabteilung, Haus Potsdamer Strasse, Staatsbibliothek Preussischer Kulturbesitz, Berlin
TPRI	Tyndall Papers, Royal Institution of Great Britain, London
UCL	Manuscripts Room, Library, University College London
VG	Bibliothèque de Genève, Ville de Genève
VR	Hermann von Helmholtz, *Vorträge und Reden*, 5th ed., 2 vols. (Braunschweig: Friedrich Vieweg, 1903)
VV	Archiv, Vieweg Verlag, Wiesbaden
WA	Hermann von Helmholtz, *Wissenschaftliche Abhandlungen*, 3 vols. (Leipzig: Johann Ambrosius Barth, 1882, 1883, 1895)
WGA	Archiv, Walter de Gruyter & Co. (formerly Georg Reimer), Berlin
WLBS	Cod. Hist. 4 593, Handschriftenabteilung, Württembergische Landesbibliothek, Stuttgart
WT	William Thomson
WTP	William Thomson Papers, CUL

Notes

Introduction

1. Other biographies include McKendrick 1899; Reiner 1905 (in German); Ebert 1949 (in German); Lazarev 1959 (in Russian); Lebedinskii, Frankfurt, and Frenk 1966 (in Russian); Rechenberg 1994 (in German); and the semibiographical Meulders 2001 (in French), 2010 (in English).

2. The Schiller lines are from "Der Spaziergang" ("The Walk," 1795). These were among Helmholtz's favorite lines of poetry, and he quoted them again and again throughout his life. See chapter 21, note 63.

3. HH 1891, 9.

4. Cahan 1993.

Chapter One

1. [Loewy] 1894, 1044; EdBR 1912, 2:518; ESH 1929, 1:83–84; Cahan 1993b, plates 3–4; FH to IHF, 10 September 1819, in WLBS, 1e, no. 180 (first quote); FH to IHF, 3 November 1821, WLBS., no. 185 (second quote); FH to IHF, 10 October 1833, ibid., no. 196; IHF to HH, 7 June 1859, in HN 145; HH 1886, 2:314; 1891, 1:17; Epstein 1896, 39.

2. Lenz 1910–18, 1:490–92; Fuchs 1990, esp. 182; Clark 2007, 362–63, 374; FH to IHF, 16 June 1813 (all quotes except last), in WLBS, 1e, no. 179; HH 1995, 343 (last quote).

3. Johann Gottlieb Fichte to FH, 2 January 1814, in Fichte 1967, 2:609–10; Schulprogramm 1857, 22; Kusch 1896, 37–38; Otto Helmholtz in Epstein 1896, 38.

4. Herrmann-Schneider 1928, 1; Johanne Fichte to FH, 21 July 1815, in WLBS5, 1; FH to Johanne Fichte, 25 August 1815, in WLBS5, 2 (quotes).

5. FH to IHF, 10 September 1819 (quotes), in WLBS, 1e, no. 180; Otto Helmholtz in Epstein 1896, 40; Herrmann-Schneider 1928, 3.

6. HH, "Curriculum vitae," in Kusch 1896, 26, mentions his maternal grandfather but makes no reference to the Anglo-American Penn family; HH to Norman Lockyer, 28 August 1876, in Lockyer Collection, University Library, University of Exeter, Exeter, UK (quote).

7. CH to IHF, n.d. (ca. 1822–23), in WLBS, 1e, no. 200; Otto Helmholtz in Epstein 1896, 40 (first quote); IHF to Eduard Schuderoff, 17 May 1820 (second, third, and fourth quotes), in WLBS, 2, no. 5; IHF to Schuderoff, 28 January 1820, ibid., no. 3 (fifth quote); IHF to Schuderoff, 10 March 1820, ibid., no. 4.

8. FH to IHF, 12 October 1820, in WLBS, 1e, no. 181.

9. Howitt 1842, 434–39; Springer 1878, 242–47; Holmsten 1971.

10. Haeckel 1912, 119–40; Müller 1968, 25–30; Holmsten 1971, 100–105; Barclay 1995, 42–44, for this as well as the previous paragraph.

11. Paepke 1969, 112, 119–21; Lammel 1993, 47–49, 128–29; Adress-Kalender 1837, 34; Cahan 1993b, 104n5; Wirth 1965, 81; HH 1891, 6–7; Kusch 1896, 35–39.

12. FH to IHF, 12 October 1820, in WLBS, 1e, no. 181; Cahan 1993b, plate 4; Kohut 1884, 722; Schulprogramm 1839, 8; Gesamtkirchenbuch der Heilig-Geist-Kirche zu Potsdam, Taufen Jahrgang 1821, no. 53, S. 98; FH to IHF, 3 April 1823, in WLBS, 1e, no. 191.

13. FH to IHF, 30 September 1821, in WLBS, 1e, no. 184 (quotes); HH 1995, 358.

14. Mielke 1972, 1:5, 57, 59; 2:xxv, T73; FH to IHF, 12 October 1822 (quotes), in WLBS, 1e, no. 187; ibid., no. 181; FH to IHF, 6 August 1821, ibid., no. 183.

15. FH to IHF, 12 October 1822; FH to IHF, 19 February 1823, in WLBS, 1e, no. 189; Epstein 1896, 38, 40, 192.

16. FH to IHF, n.d. (probably early 1822), in WLBS, 1e, no. 182 (first quote); FH to IHF, 28 June 1822, ibid., no. 186 (second quote); FH to IHF, 12 October 1822 (third, fourth, and fifth quotes).

17. FH to IHF, 25 December 1822, ibid., no. 188.

18. FH to IHF, 19 February 1823, ibid., no. 189 (first five quotes); FH to IHF, 26 February 1823, ibid., no. 190 (sixth quote); FH to IHF, 3 April 1823, ibid., no. 191 (seventh quote).

19. CH to IHF, n.d. (probably late winter or spring of 1823), ibid., no. 200.

20. Epstein 1896, 38.

21. FH to IHF, 23 August 1823, in WLBS, 1e, no. 192; FH to IHF, 28 September 1823, ibid., no. 193 (quote); Cahan 1993b, 108n6, plate 4.

22. FH to IHF, 30 December 1823, in WLBS, 1e, no. 194; Cahan 1993b, plate 4; ESH 1929, 1:83–84; Otto Helmholtz in Epstein 1896, 38, 40; Schulprogramm 1837, 56; Helmholtz 1837, 24 (quote); FH to IHF, 10 October 1833, in WLBS, 1e, no. 196.

23. Hansemann 1899, 5; "Die Enthüllung" 1899, 557; HH, "Curriculum vitae," in Kusch 1896, 26 (second, third, and fourth quotes); HH 1891, 6 (first, fifth, sixth, and seventh quotes); Helmholtz quoted in Zobeltitz 1891, 770 (eighth, ninth, and tenth quotes).

24. Kusch 1896, 35; HH 1891, 7 (first quote); 1868a, 354–55; 1855b, 114 (second quote).

25. Craig 1978, 187–88; HH 1891, 6–7 (quotes).

Chapter Two

1. Schulprogramm 1831, 43–44, 47, 49; 1832, 3, 18–22, 25, 27; 1833, 20–23, 27–29; 1839, 1–8.

2. Blume 1834, 25, 27, 29–30; Schulprogramm 1835, 1–20, 23 (quotes), 27.

3. Schulprogramm 1827, 24; Kusch 1896, 37; Schulprogramm 1828, 22–23.

4. Schulprogramm 1825, 12–13; 1827, 20; 1828, 22–23; 1829, 22–25, 27; FH to IHF, 26

February 1823 (quote), in WLBS, 1e, no. 190; Otto Semler in Epstein 1896, 192; Otto Helmholtz in Epstein 1896, 39.

5. Kusch 1896, 17, 37–38; Otto Helmholtz in Epstein 1896, 38–39; Clark 2007, 350–53, 362–65, 374, 378–80, 383–87; HH 1891, 7 (first quote); Schulprogramm 1857, 22 (second quote).

6. Otto Helmholtz in Epstein 1896, 39; Otto Semler ibid., 192; HH in Wiedemann 1989, 134–35 (quote).

7. Helmholtz 1837, 1–2 (first three quotes), 3, 5 (last three quotes), 17.

8. Ibid., 6–8.

9. Ibid., 8–12 (first quote on 8, second on 11), 28 (third quote).

10. Ibid., 13–16, 37.

11. Ibid., 21–24.

12. Ibid., 29–32.

13. Ibid., 32–36.

14. HH quoted in Epstein 1896, 192–93; Otto Helmholtz ibid., 39; cf. ibid., 192; AH to MM, sometime between 2 January and 21 February 1862, in ESH 1929, 1:102–3, on 102 (quote).

15. Kusch 1896, 23 (first two quotes); FH to IHF, 10 October 1833 (third quote), in WLBS, 1e, no. 196; on self-control and science, cf. Daston and Galison 2007.

16. HH, "Curriculum Vitae," in Kusch 1896, 25–26.

17. Ibid., 25–26 (quotes), 27; HH 1891a, 205–6.

18. Kusch 1896, 24–25, 27; Schulprogramm 1837, 45–53; HH, "Curriculum vitae," in Kusch 1896, 27 (quotes).

19. HH 1891, 6–7 (first four quotes); HH, "Curriculum vitae," in Kusch 1896, 25, 27, 29; HH 1891a, 202 (fifth and sixth quotes), 205–6.

20. HH 1891, 7 (quotes), 13; HH to RH, 4 July 1880, in ESH 1929, 1:248–50, on 249.

21. HH 1886, 314 (first quote); 1891, 7–9 (second quote on 8); 1871a, 41; Epstein 1896, 38.

22. See Ansprachen 1892, 21; HH to CH, 26 July 1837, in Cahan 1993b, 38–40; HH 1891, 7–9 (first quote on 8, second on 9), 17.

23. HH 1891, 17 (quote); cf. Ansprachen 1892, 21; HH to CH, 26 July 1837, in Cahan 1993b, 38–40; HH 1891, 7–9.

24. HH, "Curriculum vitae," in Kusch 1896, 27, 36; Blume 1834, 23; Olesko 1989, 104–12, 116–17; 1991; Dirichlet 1881, 3–4; Koenigsberger 1904, 2–5; Meyer 1838; HH 1883, 406; 1847a, 73–74; 1891, 8 (quote).

25. HH to FH, 16, 20 July 1838, in Cahan 1993b, 41–43; ESH 1921.

26. Kusch 1896, 6, 29–34 (quotes on 31); Meyer in Hirschberg 1921, 1116; HH 1891a, 207; HH, leaving-certificate examination essays for Hebrew and German, in Siemens Family Papers.

27. "Deutsche Prüfungsarbeiten der Abiturienten Helmholtz und Klotz: Potsdam: Michaelis 1838," in Siemens Family Papers (all quotes except last); Kusch 1896, 28–29 (last quote).

28. Rassow 1912, 33–35.

29. Ibid., 3, 33–35 (quotes); Browne 1995–2002, 1:97; Schulprogramm 1839, 8.

30. HH 1886, 314 (quote); Zobeltitz 1891, 769; HH to RM, 2 June 1868, in ESH 1929, 1:147; HH to Norman Lockyer, 28 August 1876, in Lockyer Collection, University Library, University of Exeter, Exeter, UK; HH 1877, 169, 179; 1891, 9.

31. Cahan 1993b, 10; HH to FH, 30 March 1837, ibid., 35–37 (quote on 36–37); Hansemann 1899, 1; EdBR 1912, 2:569.

32. HH 1891, 9 (quote); HH to FH, 30 March 1837, 35, 37.

Chapter Three

1. Cahan 1993b, 11–14.

2. HH to his parents, 5 November 1838; 7, 16 October 1842, in Cahan 1993b, 48–52, 92–93, 93–94, resp.; HH to FH, 31 October, 1 December 1838; 17 March, 11 December 1839, ibid., 43–48, 53–54, 55–56, 74–75, resp.; AH to MM, sometime between 2 January and 21 February 1862, in ESH 1929, 1:102–3.

3. Cahan 1993b, 14–15; HH to FH, 31 October 1838, ibid., 48; HH to his parents, 5 November 1838, ibid., 48; Virchow 1906, 22, 23 (quotes), 27–28, 30, 45–46; Goschler 2002, 36–48.

4. HH to his parents, 31 October 1838, 45.

5. HH to his parents, 5 May 1839, in Cahan 1993b, 56–60, on 56–57; HH to FH, 31 October 1838; 24 June 1842, ibid., 45–48 (quotes on 46–48), 90, resp.; HH's parents to HH, 2 November 1838, in LK, 1:23–25; Howitt 1842, 429–34; Springer 1878, 84, 142–51.

6. HH to his parents, 5 November 1838; 5 May 1839, in Cahan 1993b, 48–51 (all quotes except last), 57, resp.; HH to FH, 1 December 1838, ibid., 53; CH to HH, [1838], in Elbogen [1956], 156 (last quote).

7. Cahan 1993b, 11–16.

8. Ibid., 15, 17.

9. HH to FH, 1 December 1838 (first quote); 15 May 1839, in Cahan 1993b, 60–63 (second through ninth quotes); HH quoted in Neumann 1925, 14 (tenth quote); HH to his parents, 5 May 1839, 59 (eleventh quote).

10. HH's parents to HH, 2 November 1838, 24; HH to his parents, 5 November 1838; HH to FH, 1 December 1838, 53; HH to FH, 15 May 1839 (quotes); Zobeltitz 1891, 770.

11. HH to FH, 1 December 1838, 53; HH to FH, 17 March 1839, in Cahan 1993b, 55–56 (quote).

12. HH to FH, 1 December 1838, 53–54 (quotes); HH to FH, 17 March 1839; HH to his parents, 5 May 1839, 56; HH to FH, 15 May 1839, 60.

13. HH to FH, 15 May 1839, 60, 62; EdBR 1912, 2:569 (first quote); HH to his parents, 5 May 1839, 58 (second and third quotes); HH 1891, 1:10 (fourth and fifth quotes).

14. HH to his parents, 11 July 1839, in Cahan 1993b, 65–66 (first three quotes); HH to his parents, 6 September 1839, ibid., 70–72 (fourth and fifth quotes).

15. Ibid., plate 11, table 1; HH to his parents, 5 November 1838, 49–51; HH to his parents, 5 May 1839, 58–9.

16. HH to FH, 8 December 1839, ibid., 72–74 (first quote on 73); HH to FH, 31 October 1838, ibid., 47; HH to his parents, 5 November 1838, ibid., 52; HH to FH, 30 July 1840, ibid., 77–78, on 78; HH to FH, 11 December 1839, ibid., 74–75 (all other quotes); ibid., plate 9.

17. HH 1877, 180; du Bois–Reymond 1918, 4–7; Cahan 1993b, 15–16.

18. Cahan 1993b, 16; HH 1877, 167 (first quote), 170 (second quote), 176–79 (third quote on 178); 1869, 395–96.

19. HH to his parents, 1 September 1840?, 79–81, in Cahan 1993b.

20. HH to his parents, 11 January 1841, ibid., 82–84 (quotes); Barclay 1995.

21. FH to IHF (including HH's addendum), 2 June 1841 (all quotes except last), in WLBS, 1e, no. 197; HH to IHF, 29 July 1841 (last quote), ibid., no. 198.

22. HH to FH, 7 August 1841, in Cahan 1993b, 85–87 (first two quotes on 85, 87); HH to his parents, 12 August, 17 September 1841, ibid., 87–88 (third quote), 88–89, resp.; C. Besser to Ew. Wohlgebohren [FH], 16 August 1841, ibid., 113; Dr. Knapp to [FH], 16 August 1841, ibid., 113–14 (fourth and fifth quotes); Knapp to [FH], 24 August 1841, ibid., 114–15; FH to IHF, 27 August 1841, in WLBS, 1e, no. 199 (sixth, seventh, and eighth quotes).

23. HH 1877, 180 (quote); Otis 2007, 58, 68; AH to MM, sometime between 2 January and 21 February 1862.

24. Cahan 1993b, plate 10; HH to FH, 24 June 1842, ibid., 90; ibid., 19–20; HH 1869, 395–96; 1877, 167 (first two quotes), 169 (third quote), 178, 180; 1886, 314–15 (fourth quote); 1887, 324; Virchow 1906, 20–21.

25. HH to FH, 1 June 1842, in Cahan 1993b, 89 (quotes); HH to FH, 1 August 1842, ibid., 91–92.

26. Alexander von Humboldt to Eichhorn, 10 December 1840, in Biermann 1985, 90–92 (quote on 92); Humboldt to Altenstein, 29 April 1833, ibid., 61–62; Haberling 1924, 156–57; Lohff 1992; Holmes 1994; Otis 2007; Finkelstein 2013, 45–54, 61.

27. HH, "Collegienhefte aus den Vorlesungen von Johannes Müller: Vergleichende Anatomie, Pathologische Anatomie. Berlin 1840," in HN, 538; Holmes 1994; Virchow 1906, 24.

28. Haberling 1924, 203, 330; the essays in Hagner and Wahrig-Schmidt 1992; HH 1877, 181–82 (quote).

29. HH to FH, 1 August 1842, 91 (quote); Holmes 1994, 16–17; Finger and Wade 2002, 144–46; Otis 2007, 116, 169–70.

30. Cahan 1993b, plate 11 (quote); cf. Daston and Galison 2007, esp. 229–30.

31. HH 1842; Clarke and Jacyna 1987, 84–86, 98; HH to his parents, 7 October 1842; HH to FH, 1 August 1842, both in Cahan 1993b, plates 12, 13; Helmholtz's diploma, 2 November 1842, in "N. N. Original des Doctor Diploms. 1831," Diploma: Doctoris Medicinae et Chirurgiae, SF, Signatur Helmholtz SLL 494; Königliche Universitätsbibliothek 1899, 245.

32. HH 1877, 181 (first quote); 1886, 315 (second quote); 1891, 9; HH to EdBR, 29 May 1858, in Kirsten et al. 1986, 186–87, on 186; HH 1877a, 202–3.

Chapter Four

1. HH to his parents, 7 October 1842, in Cahan 1993b, 92–93.

2. Ibid., 21–22; Guttstadt 1886, 343, 345–47, 349, 353, 359.

3. HH to his parents, 7 October 1842; HH to his parents, 16 October, 8 December 1842, in Cahan 1993b, 93–94 (quote), 95–96.

4. CH to HH, first half of January 1843, in LK, 1:52; HH to his parents, 16 January, 8 February 1843, in Cahan 1993b, 98–99, 99–100.

5. HH to FH, 10 March 1843, in Cahan 1993b, 101–2; HH to FH, 17 March 1843, ibid., 102–3 (quotes).

6. HH to CH, 21 May 1843, ibid., 103–4 (first two quotes); HH to FH, 25 July 1843, ibid., 105–6 (third quote).

7. HH to CH, 21 May 1843; HH to FH, 25 July 1843; HH 1877, 179.

8. Helmholtz in Schiff 1893; HH to his parents, 7 October 1842, 92; Helmholtz's Personalakte, in RKUHA, PA 1700, Bl. 7–8; Treitschke 1927, 5:430 (quote).

9. Helmholtz in Schiff 1893; Lenoir 1982, 197–99; Kremer 1990a, 239–40; Caneva 1993, 49–159; McDonald 2001a; HH 1843, 727 (quotes); Fruton 1972, 45, 48–63; Olesko and Holmes 1993, 50–54. (Four decades later, in 1882, Helmholtz backed away from some of his conclusions [see HH 1843, 734]). Cf. Justus von Liebig to HH, 8 [?] 1870, in HN 275; HH to Liebig, 28 April 1870, copy in BSBMH, Liebigiana IIB, Helmholtz, Hermann (1); original at SBPK.

10. HH 1845 (quotes on 735, 736, 744, resp.); EdBR 1912, 2:563; Holmes 1992, 25–26, 29, 31–36, 41, 46–47; Olesko and Holmes 1993, esp. 50–59.

11. HH 1846, 700 (quote); Olesko and Holmes 1993, 59–64.

12. HH to his parents, 30 October 1845, in Cahan 1993b, 106–8, on 108.

13. Ibid., 26–29.

14. Ibid., 108.

15. HH to FH, 19 December 1845, ibid., 109–10 (first quote); HH to FH, 25 January 1846, ibid., 111–12 (second and third quotes).

16. HH to FH, 19 December 1845, 110n6; Hofmann 1870, esp. 995; HH 1871a, esp. 38, 41; Hoffmann 1995.

17. Hofmann 1870, 1009; Haberling 1924, 153–54, 165; HH 1871a, 38 (first quote); 1893, v; HH to FH, 19 December 1845; Helmholtz's three-page manuscript "Versuche über Gährung bei Magnus" (1845–46), in HN 666; HH to FH, 25 January 1846, 111 (second quote).

18. Karsten 1847, iii–x; 1850, ix; 1853, vii; HH 1847; 1848a; 1850e; 1893, v.; Bezold 1896, 20–21, 23; Warburg 1925; Goldstein 1925; Scheel 1935, 5–6; Jungnickel and McCormmach 1986, 1:110–11, 254, 258–59; Schreier and Franke 1995; Fiedler 1998; Finkelstein 2013, 79–83.

19. HH to FH, 19 December 1845, 110; HH to FH, 25 January 1846, 111; Hofmann 1870, 1096; Schiff 1883 (first quote); Helmholtz in Schiff 1893; Brücke 1928, 20–21, 25; HH 1891, 13; Epstein 1896, 37; EdBR to Hallmann, 25–26 December 1845, in du Bois–Reymond 1918, 122 (second quote).

20. EdBR to Hallmann, May 1842, in du Bois–Reymond 1918, 107–8; Cranefield 1957, 1959, 1966; Mendelsohn 1965, 1974; Galaty 1971, 1974; Lenoir 1988; Holmes and Olesko 1995, esp. 199; Olesko 1995; Finkelstein 2013, 45–54, 64.

21. HH to EdBR, 1 August, 5 October, 21 December 1846; 12 February 1847, in Kirsten et al. 1986, 74; 74–75, on 74; 75–78, 78–79 (quotes on 78); HH 1847a, 74.

22. Betty Johannes (née von Velten), in Kremer 1990, 193–201, on 193–95 (first quote on 194–95); ibid., 3, 193, 194; HH to OH, 20–21, 22 May; 6–8 April 1847, ibid., 5–9, on 5; 9–10, on 9; 1–5 (second quote on 2).

23. HH to OH, 6–8 April 1847.

24. Kremer 1990, 1, 5; HH to OH, 1 December, 20–21 May 1847, ibid., 31–33, on 32; 6–8, 9 (first two quotes); HH 1982, vols. 1 and 2, title pages; HH to OH, 20–21 May 1847, 8; HH to OH, 22 May 1847, 9 (fourth, fifth, and sixth quotes).

25. Kremer 1990, 17, 19; Helmholtz's Personalakte, in RKUHA, PA 1700, Bl. 7–8; HH to OH, 12–13 June 1847, in Kremer 1990, 10–13 (quotes on 10, 13).

26. HH to EdBR, 21 July 1847, in Kirsten et al. 1986, 81–82, on 81; HH 1847a, 12; Karsten 1850, ix; HH to OH, 22 July 1847, in Kremer 1990, 18–20 (quote on 18).

27. Darrigol 2001, esp. 285–87, 341–43; HH 1891, 10 (quotes); 1877, 177; Lenoir 1982, 197–215; Kremer 1990a, 237–55, 293–307; Caneva 1993, 49–159; Bevilacqua 1993, 297–304.

28. HH 1884, 1:viii–ix. There is much disagreement as to the sense(s) in which or the degree to which, or even whether at all, Helmholtz was a Kantian; about Kant's general influence on him; and about Helmholtz's changed attitude toward Kant later in his career. See, e.g., Heimann 1974a; Fullinwider 1990; Hatfield 1990, 165–234; Heidelberger 1993; Darrigol 1994; Krüger 1994; Schiemann 1997; Hyder 2009; Jurkowitz 2010; De Kock 2016.

29. Breger 1982, esp. 129–58, 214–23; Bevilacqua 1993, 301, 309–12; Brain and Wise 1994; Wise 1999; Darrigol 2001, 342–43.

30. Kuhn 1959; Elkana 1970, 1974; Heimann 1974, 1974a; Cantor 1976; Harman 1982; Kremer 1990a; Bevilacqua 1993; Caneva 1993; Darrigol 2001; Jurkowitz 2002, 2010; HH to Berend Wilhelm Feddersen, 16 November 1859, in DM. Already in 1850, and again in 1852, 1853, and 1855, Helmholtz acknowledged Mayer's work in *Die Fortschritte der Physik*. See Weyrauch 1893, 316–21; Gross 1891, 19–32; 1898, 141–74.

31. HH 1847a, 67–68, 74–75 (this last added in 1881); 1883, 1:407.

32. HH 1847a, 12–13 (quote). In an addendum of 1881, Helmholtz distanced himself from this Kantian outlook on causality and phenomena. (See ibid., 68–75, on 68.) He wrote an undated, untitled sixteen-page manuscript on natural science generally and on general concepts of nature (objects and their perception, time, space, geometry, and so on). Though he does not refer to Kant by name, the manuscript has a broad if loose Kantian flavor to it. (See HN 705 [4.17].) Koenigsberger printed it (*LK*, 2:126–38) and claimed that Helmholtz wrote it several years before 1847. (See also Darrigol 1994, esp. 217–18; 2000, 215; Krüger 1994.)

33. HH 1847a, 14.

34. Ibid., 14–17.

35. Ibid., 17–18, 24–25 (quote), 66; Bevilacqua 1993, 319–32; Buchwald 1994, 397–400; Smith 1998, 126–49, 176–77, 185–86.

36. HH 1847a, 68.

37. J. C. Poggendorff to GM, 1 August 1847, draft, in HN 537.

38. GM to EdBR, 2 August 1847, in HN 536. At least as late as 1858, Magnus thought Helmholtz's essay contained "lots of *hypothetical . . . arbitrary*" notions. (See GM to Alexander von Humboldt, 1 March 1858, in SPKB, Nachlass Alexander von Humboldt, Karten 8, no. 15.)

39. HH to Feddersen, 16 November 1859 (first quote); HH 1847a, 72–74 (Helmholtz quoted his letter of 1868 to Tait in ibid., 71–73 [second and third quotes]); 1891, 11 (fourth and fifth quotes); EdBR 1912, 2:524; Jacobi 1996, 104–5, including Helmut Pulte's introduction on xxvii–xxix.

40. HH to OH, 29–30 July 1847, in Kremer 1990, 20–24 (first three quotes); HH 1848; HH to OH, 4–5 August 1847, in Kremer 1990, 24–26 (fourth, fifth, and sixth quotes).

41. EdBR to HH, 4 August 1847, in Kirsten et al. 1986, 82–83, on 82 (first quote); HH to EdBR, 6 August 1847, ibid., 84 (second quote); Helmholtz in Zobeltitz 1891, 769; EdBR 1912, 2:524; Finkelstein 2013, 85.

42. HH to G. A. Reimer, 14, 20 August, 6 November 1847, in WGA.

43. HH to OH, 7–8, 21–22 September 1847, in Kremer 1990, 26–28 (first five quotes on 26–27), 29–31 (last quote on 30).

44. HH 1891, 11; Darrigol 2000, 216; EdBR to CL, 4 January 1848, in du Bois–Reymond 1927, 4–6 (first quote on 6); CL to EdBR, 4 February 1848, ibid., 7–10, on 10; Finkelstein 2013, 72–74, 89–96; Julius Robert Mayer to Carl Gustav Reuschle, 12 January 1848, in Weyrauch 1893, 288 (second quote); EdBR 1912, 2:524–25; Gariel 1894, 430; Binz 1869, with a letter by Helmholtz on 100–102.

45. HH 1848; Olesko and Holmes 1993, 64–74; Finkelstein 2013, 57–75, 89–96.

46. HH to OH, 15–16 December 1847; ? January 1848, in Kremer 1990, 33–35, 39–40 (quote on 40); EdBR to CL, 4 January 1848, ibid., 6.

47. HH to OH, 1 December 1847, ibid., 31–32; Johannes quoted ibid., 194 (first quote); HH to OH, n.d., quoted ibid., 195–96 (second quote).

48. Clark 2007, 468–85; Prittwitz 1985, 59, 60, 74, 113, 122, 150, 344, 393–94, 405–6, 426, 436, 439, 481, 486–87; Haeckel 1912, 126; Müller 1968, 34–47; Holmsten 1971, 108–10.

49. Schreier and Franke 1995, 61–64; Prittwitz 1985, 425; Otis 2007, 161, 231; HH to his parents, 11 January 1841, in Cahan 1993b, 82–84, on 83; CL to Jacob Henle, 3 November 1851, in Dreher 1980, 105–9 (first quote on 106); Goschler 2002, 58–92; HH to OH, 18–19 July 1848, in Kremer 1990, 42–44 (second quote on 44); HH to OH, 20 June 1848, ibid., 41–42 (third quote on 41); Epstein 1896, 36 (fourth quote); HH to OH, 18–19 July 1848, 43.

50. Johannes Müller to the Ministry of Culture, [June 1848], in GSPK, Rep. 76 Va, Sekt. 2, Tit. X, No. 11, Bd. VI, Bl. 144; Müller to [the Ministry of Culture], 6 May 1848, printed in *LK*, 1:94–95; Ordentliche Sitzung des Senats der Königliche Akademie der Künste, 22 July 1848, PR AdK 45, Bl. 127–127r; KM to the AdK, 12 July 1848, PR AdK 192, Bl. 60 (quotes).

51. HH, "Vortrag gehalten in der Akademie der Künste am 19.8.48," in HN 539; printed in *LK*, 1:95–105 (quotes on 95–96).

52. Ibid. (quotes on 99–100).

53. Ibid. (quotes on 101–102).

54. Ibid. (quotes on 102–103).

55. Ibid. (quote on 104).

56. Ordentliche Sitzung des Senats der Königlichen Akademie der Künste, 19 August 1848, PR AdK 45, Bl. 134 (first quote); Königliche Akademie der Künste, 22 August 1848, PR AdK 96, Bl. 165; HH 1886, 2:315–16 (second quote).

57. See, e.g., Biermann 1985, 1990; Müller 1928; Barclay 1995, 42–43, 52, 65–66, 110, 186, 230–31; Werner 2004.

58. Humboldt to Karl Sigismund Freiherr vom Stein zum Altenstein, 29 April 1833, in Biermann 1985, 61–62; Humboldt to [Minister of Culture] Johann Albrecht Friedrich Eichhorn, 10 December 1840, ibid., 90–92, on 90; Humboldt to Eichhorn, 10 December 1840, ibid., 93–94; Helmholtz in Schiff 1893; Stevens 1863, 295; Werner 2004, 185, 266; Treitschke 1927, 5:28–29; KM to the AdK, 2 September 1848, PR AdK 192, Bl. 62; EdBR 1912, 2:564.

59. HH 1889a, 2 (first quote); EdBR to Eduard Hallmann, 6 January 1849, in du Bois–Reymond 1918, 128–32, on 130 (second quote); Finkelstein 2013, 102; Helmholtz in Schiff 1893; Silliman 1853, 2:336 (third and fourth quotes).

60. CL to Jacob Henle, 22 November 1848, in Dreher 1980, 56–60, on 57–58; EB to EdBR, 9 December 1848, in Brücke et al. 1978–81, 1:17 (first quote); EB to EdBR, 25, 30 March 1849, ibid., 21–23; EdBR to CL, 17 May 1849, in du Bois–Reymond 1927, 49–53, on 50 (second quote).

61. Medicinische Fakultät to KM, 1 April 1849, in KM, GSPK, I. HA, Rep. 76 Va, Sekt. 11, Tit. IV, no. 13, Bd. 1, Bl. 104–5; EdBR to Hochgeehrter Herr Geheimrath, copy 22 April 1849 (quotes), ibid., Bl. 106–7; original in Darmstaedter Sammlung, SPKB.

62. Johannes Müller to Kultusminister, 7 May 1849, in Haberling 1924, 329–30.

63. CL to Henle, 9 July 1849, in Dreher 1980, 77–80 (quote on 78–79); Schröer 1967, 44–45; Friedrich Wilhelm to Staats-Minister Adalbert von Ladenberg, 19 May 1849, in GSPK, Rep. 76 Va, Sekt. 11, Tit. IV, no. 13, Bl. 115; KM to HH, 8 June 1849, ibid., Bl. 119–20; Kremer 1990, 3n12; Werner 1997, 60; EdBR to CL, 7 August 1849, in du Bois–Reymond 1927, 66–68, on 67.

64. HH to EdBR, 19 August 1849, in Kirsten et al. 1986, 85 (first quote); Johannes, in Kremer 1990, 196 (second and third quotes); Melms 1957, 91, 94.

Chapter Five

1. HH 1857a, 134–35.

2. Friedlaender 1896, 41–42; Kossert 2005, 112–16, 138–44; HH 1871, 56 (quote).

3. Prutz 1894, 183, 193; Titze 1995, 384, 388; Lexis 1893, 1:118–19; Eulenburg 1904, 303, 305; Friedlaender 1896, 42.

4. Prutz 1894, 124–26, 194, 197–98, 201–2; Olesko 1991; 1994, 26–27.

5. Stieda 1890, 38, 74–75, 83; HH to EdBR, 14 October 1849, in Kirsten et al. 1986, 86–88, on 87; Clark 2007, 248; "Der Verein" 1859, 1:1–3, 5.

6. Kremer 1990, 3n13, 80n9; Johannes, ibid., 193–201, on 196–97; DM, Nachlass Helmholtz, no. 39: Copir-Buch, NR, P1, Bl. 1; HH to EdBR, 14 October 1849, 86 (first quote), 88; HH to EdBR, 22 April 1850, in Kirsten et al. 1986, 96–97, on 97 (second quote).

7. Eulner 1970, 54, 61, 528; HH to EdBR, 14 October 1849, 86; HH to EdBR, 24 March 1852, in Kirsten et al. 1986, 127–28 (quotes).

8. HH to EdBR, 14 October 1849, 87 (quotes); Olesko 1991, 202–3; Olesko and Holmes 1993, 83; HH to GRK, mid-November 1852, in *LK*, 1:178–79.

9. Kremer 1990, 97n2; HH to EdBR, 15 January 1850, in Kirsten et al. 1986, 90–92, on 91–92 (quote); HH 1850e.

10. Kremer 1990, 46n4; HH to FH, 29 March 1850, in *LK*, 1:120–21 (quote on 121); HH to Verehrter Onkel, 29 March 1850, in Ebstein 1920, 153; Friedlaender 1896, 57–58; Weisfert 1975, 108, 206–7; Stieda 1890, 83; HH to OH, 20 August 1853, in Kremer 1990, 112n9; Johannes, ibid., 196; Barclay 1995, 46–47, 53–54, 58, 70; Manthey 2005, 424–31, 442–60, 487–92; Kossert 2005, 128–35.

11. HH to FH, [December 1849], in *LK*, 1:114 (quotes).

12. CL to HH, 10 November 1850, in HN 293 (all quotes except first); CL to Jacob Henle, 3 November 1851, in Dreher 1980, 105–9 (first quote on 106).

13. HH 1891, 11–12.

14. Titze 1995, 392; Lexis 1893, 1:120; HH to FH, [December 1849] (first quote); FH's reply, in *LK*, 1:114–16; HH to EdBR, 14 October 1849, 86; HH to EdBR, 22 April 1850, in Kirsten et al. 1986, 96–97; HH 1877, 168 (second quote). Uhthoff 1902, 7, lists Helmholtz's courses for 1850–51.

15. EB to EdBR, 9 July 1849, in Brücke et al. 1978–81, 1:25–26, on 25 (first quote); CL to Henle, 3 November 1851, 107 (second quote); Seeig Rathke, W. Cruse, and Hirsch, "Gehorsamtes Gesuch der medizinischen Fakultät, betreffend den Professor Dr. Helmholtz," 8 November 1851, in GSPK, I. HA Rep. 76 Va, Sekt. 11 Tit. IV, no. 13, Bd. 1, Bl. 126–27, in Werner 1997, 170 (third quote).

16. HH to EdBR, 14 October 1849, 87–88 (quote on 88); Johannes, in Kremer 1990, 197; HH 1850, 1850a, 1850b, 1850c, 1850d. The present discussion largely follows the analysis of Olesko and Holmes 1993, 74–108. See also Brain and Wise 1994; Holmes and Olesko 1995; Finger and Wade 2002, 149–52; Wise 2007; Schmidgen 2009, 2015. On the physical side, Helmholtz's experiments here also provided unprecedented quantitative results concerning electromagnetic self-induction. (See Darrigol 2000, 218–20.)

17. HH to FH, [December 1849]; HH to EdBR, 15 January 1850, in Kirsten et al. 1986, 90–92 (quote on 90–91); HH 1850a, 1850b, 1850d.

18. Johannes Müller to HH, 7 February 1850, in HN 320 (first quote); CL to HH, 10 [February?] 1850, in HN 293; EdBR to HH, 19 March 1850, in Kirsten et al. 1986, 92–94, on 92 (other quotes); HH to EdBR, 5 April 1850, ibid., 94–95, on 94; Alexander von Humboldt to EdBR, 18 January, 4 March 1850, in Schwarz and Wenig 1997, 101–2, 104 (quote), resp.; Humboldt to HH, 12 February 1850, in *LK*, 1:118; S[chiff] 1891. Humboldt's library included HH 1850 and 1850b (Stevens 1863, 295.)

19. EdBR to HH, 25 August 1850, in Kirsten et al. 1986, 98–100, on 100 (first, second, and fourth quotes); IIII to EdBR, 28 August 1850, ibid., 100–102, on 101 (fifth quote); EdBR to CL, 9 April 1850, in du Bois-Reymond 1927, 87–99, on 88–99 (third quote); Finkelstein 2013, 97–114; Fox 1973, esp. 445, 452–64, 469. See also Schmidgen 2009; Schestag 2003; Otis 2007, 123–24.

20. HH to Verehrter Onkel, 29 March 1850 (first two quotes); HH to FH, 29 March 1850 (third and fourth quotes); HH to FH, n.d. [spring 1850], in *LK*, 1:123–25; FH to HH, 3 April 1850, ibid., 121–23; Olesko and Holmes 1993, 93–108.

21. HH 1850; HH to EdBR, 28 August 1850, 102 (first three quotes); HH to EdBR, 17 September 1850, in Kirsten et al. 1986, 104–6, on 105 (all other quotes).

22. CL to HH, 10 November 1850 (first quote), in HN 293; HH 1850c; Klauss 1981; Karsten 1855, vii; Tyndall 1898a, 2:351 (second and third quotes); HH to EdBR, 17 September 1850, 106.

23. HH to FH, 17 December 1850, in *LK*, 1:133–34; HH 1886, 315–16 (second quote); see HH 1891, 12 (first quote).

24. HH 1891, 12; Brücke 1928, 25; Helmholtz in Snyder 1964, 576; HH 1887, 324 (quote).

25. HH to FH, 17 December 1850 (first quote); HH, "Mitteilung für die physikalische Gesellschaft zu Berlin über eine Methode, die Netzhaut und das auf ihr entworfenen Bild einer Flamme am lebende Auge sichtbar zu machen," 25 November 1850, in HN 557, read by EdBR as "Über das Leuchten der Augen"; Klauss 1981; FH to HH, n.d. [mid-December 1850 to autumn 1851], in *LK*, 1:134–35 (second and third quotes).

26. HH 1851; AH to ISZ, 2 November 1890, in ESH 1929, 2:32–34, on 33; Helmholtz in Snyder 1964, 575–76; Friedenwald 1902, 549–51; Hirschberg 1918; Tuchman 1993, 29–41; Cahan 1998.

27. HH 1891, 12 (quotes), 19; 1886, 314.

28. HH 1869, 395–96; 1886, 314, 317–18 (quotes); 1891, 12–13, 18.

29. Cf. HH 1877, 185.

30. HH to EdBR, 12 June 1851, in Kirsten et al. 1986, 114–16 (quotes on 115–16).

31. HH to OH, 6 August 1851, in Kremer 1990, 46–53 (quotes on 46–47); Heintz's letters to HH, in HN 193.

32. HH to OH, 6 August 1851, 48.

33. Ibid., 49, 49n13.

34. Ibid., 49–52 (all quotes but Ruete's); Ruete 1852, 2 (quote).

35. HH to OH, 6 August 1851, 52–53; HH to OH, 10 August 1851, in Kremer 1990, 53–60 (quotes on 54–56).

36. HH to OH, 10 August 1851, 56, 56n12; cf. Silliman 1853, 2:291–95.

37. HH to OH, 10 August 1851, 57–58.

38. Ibid., 58–59.

39. Ibid., 59–60; cf. Silliman 1853, 2:298–301.

40. HH to OH, 10 August 1851, 60.

41. HH to OH, 16 August 1851, in Kremer 1990, 61–66 (quotes on 61–63); Tuchman 1993a, 118, 123.

42. CL to Henle, 14 July 1851, in Dreher 1980, 101–5 (all quotes but last on 101–3); CL to Henle, 3 November 1851; EdBR to CL, 6 February 1852, in du Bois-Reymond 1927, 107 (last quote).

43. HH to OH, 16 August 1851, 63. For Helmholtz's use of "institute" for equipment or laboratories, see chapter 7, note 16.

44. Ibid., 63–64.

45. Ibid., 65–66.

46. Ibid., 66 (first, fourth, fifth, and sixth quotes); CL to HH, 20 June 1851 (second quote), in HN 293; EdBR to CL, 5 August 1851, in du Bois-Reymond 1927, 100–101 (third quote).

47. OH to HH, 18–20 [August 1851], in Kremer 1990, 67–70, on 68–70.

48. HH to OH, 22 August 1851, ibid., 70–74, on 70–71.

49. Ibid., 71–72.

50. CL to Henle, 3 November 1851, on 105–8 (first two quotes); CL to EdBR, 6 February 1852, in du Bois-Reymond 1927, 105–7, on 105 (last two quotes).

51. HH to OH, 22 August 1851, 72–74.

52. HH to OH, 26–31 August 1851, in Kremer 1990, 75–81 (quotes on 75–77).

53. Ibid.

54. Ibid., 80–81.

55. HH to OH, 3–8 September 1851, ibid., 81–88 (quotes on 81–83).

56. Ibid., 83–84.

57. Ibid., 84–85.

58. Ibid., 85–86.

59. Ibid., 87.

60. Ibid., 87–88 (quotes on 87).

61. HH to OH, 14–15 September 1851, ibid., 88–91, on 88–89.

62. Ibid., 89.

63. Ibid., 89–90.

64. Ibid., 90.

65. Ibid., 91.

66. OH to HH, 17 [September 1851], ibid., 92–93.

67. HH to OH, 21 September 1851, ibid., 93–97, on 94–95.

68. Ibid., 95.

69. EB to HH, 24 June 1851, in HN 74; HH to OH, 21 September 1851, 95 (quotes); Brücke 1928, 27–29, 35, 137–40.

70. HH 1851b; HH to OH, 21 September 1851, 96 (quote); Brücke 1928, 42.

71. HH to OH, 21 September 1851, 96 (first four quotes); Brücke 1928, 139–40; HH to CL, shortly after 21 September 1851, in *LK*, 1:158 (fifth quote); Rudolph Wagner to EdBR, 19 February 1852 (sixth quote), in SPKB, 3k 1854(5), cited in Kremer 1990, 96n15; EB to EdBR, 14 October 1851, in Brücke et al. 1978–81, 1:47.

72. HH to OH, 21 September 1851, 96–97, n16.

Chapter Six

1. Seeig Rathke, W. Cruse, and Hirsch, "Gehorsamtes Gesuch der medizinischen Fakultät, betreffend den Professor Dr. Helmholtz," 8 November 1851, in GSPK, I. HA Rep. 76 Va, Sekt. 11 Tit. IV, no. 13, Bd. 1, Bl. 126–27, in Werner 1997, 170; HH 1895, 607–9; EdBR to HH, 16 May 1851; 30 May 1853, in Kirsten et al. 1986, 113–14, on 113 (first two quotes), 142–44, on 144, resp.; EB to EdBR, 8 April 1853, in Brücke et al. 1978–81, 1:55–57, on 55; Schulze quoted in Kremer 1990, 103n26 (third quote).

2. CH to HH, [January 1852], in Elbogen [1956], 156; FH to HH, 5 April 1852, [July–September 1852], in *LK*, 1:165–66, 168.

3. HH to EdBR, 3 February, 20 June 1852, in Kirsten et al. 1986, 116–19, on 119, 131–32, resp.; Johannes Müller to HH, 1 June 1852 (first quote), in HN 320; CL to HH, 15 May 1852 (second quote), in HN 293; EdBR to CL, 2 August 1852 (third and fourth quotes) and CL to EdBR, 1852, both in du Bois-Reymond 1927, 111–13, on 113; 113–17, on 115, resp.

4. HH to EdBR, 3 February 1852, 119; EdBR to HH, 9 February 1852, in Kirsten et al. 1986, 120–23, on 123; HH to EdBR, 16 July 1852, ibid., 134–35, on 135; Jacob Henle to Rudolph Wagner, 1, 7 July 1852, in Eulner and Hoepke 1979, 47–50, on 48; 50–52, on 51, resp.; HH to EdBR, 20 June 1852; Tuchman 1993a, 113–28.

5. HH to EdBR, 23 January 1853, in Kirsten et al. 1986, 139–41, on 139–40 (quotes); Franz Eichmann to Raumer, 16 February 1853; KM to Eichmann, 5 March 1853, both in KM, GSPK, I. HA, Rep. 76 Va, Sekt. 11, Tit. IV, no. 13, Bd. 1, Bl. 139–41, Bl. 142, resp.; HH to OH, 31 August [1853], in Kremer 1990, 121–25, on 121n6 and 103n26.

6. Weve and ten Doesschate 1935, 90; Albrecht von Graefe to HH, 7 November 1850, in HN 172; EB to HH, 24 June 1851, in HN 74; P. J. Kipp to HH, 3 March 1852, in HN 230; HH to EdBR, 20 June 1852; Jaeger 1997, 11, 16–17; "Der Verein" 1859, 5; Ruete 1852; Eulner 1970, 337; Ravin and Kenyon 1992.

7. Graefe to FCD, 11 February 1852, in Weve and ten Doesschate 1935, 8–12, on 10 (first quote); Graefe to FCD, 4 October 1852, ibid., 12–14, on 13; HH to FCD, 9 October 1853, in *LK*, 1:191–92; Theunissen 2000; EB to HH, 18, 21 April 1852, in HN 74; CL to EdBR, 9 December 1852; EdBR to CL, 9 January 1853, both in du Bois-Reymond 1927, 117–19, on 118 (second quote), 119–21, on 119; Ruete 1852; HH 1852; Friedenwald 1902, 551–52, 566–69; James Clerk Maxwell to WT, 15 May 1855, in Maxwell 1990, 1:305–13, on 308–9; HH 1877, 179; Bader 1933, 58.

8. Wade and Finger 2001; Tscherning 1910; HH 1855, esp. 283, 296, 345; HH to EdBR, 22 March 1855, in Kirsten et al. 1986, 155–56, on 156; HH 1855b; Tuchman 1993, 37–38; Lenoir 1993.

9. HH to EdBR, 11 April 1851, in Kirsten et al. 1986, 110–12, on 111; EdBR to his future wife (Jeannette du Bois-Reymond), 14 January 1853, in Dep. 5, K. 11, no. 5; HH to EdBR, 23 January 1853, 140; HH 1853; Krönig 1856, v.

10. FCD to HH, 26 June 1853, in HN 116 (quote); HH to CL, 3 July 1853, in *LK*, 1:191; EB

to HH, 23 October 1853, in HN 74; Brücke 1928, 24–25; HH to EdBR, 11 April 1851, 111; EdBR to CL, 31 March 1851, in du Bois–Reymond 1927, 97–100, on 99; Olesko and Holmes 1993; Lenoir 1993a; 1994, 185–88; Otis 2001, 2002; Hoffmann 2003.

11. Münchow 1978; Graefe 1854, vi–vii; Bader 1933, 55–56, 59–62, 82; Eulner 1970, 324–46; Fahrenbach 1987; Lenoir 1993; Theunissen 2000, esp. 563–73; AH to ISZ, 2 November 1890, in ESH 1929, 2:32–34 (quote on 33); FCD to HH, 1 December 1858, in HN 116.

12. Kipp to HH, 9 May 1852, in HN 230; CL to HH, 26 May 1853, in HN 293; FCD to HH, 20 September 1855, in HN 116; Bonner 1963, 30; HBJ to EdBR, 19 March 1852, in SPKB, Sammlung Darmstaedter 3k 1852 (4), Bl. 116–17 (quote); EdBR to HH, 15 June 1852, in Kirsten et al. 1986, 130–31, on 130; HH to EdBR, 20 June 1852; Weve and ten Doesschate 1935, 96.

13. EdBR to his future wife (Jeannette du Bois–Reymond), 4 May 1853 (first quote), in Dep. 5, K. 11, no. 5; Ernst Haeckel to his parents, 13 October 1853, in Haeckel 1921, 70–71 (second quote on 71); HH 1891, 12–13 (third quote).

14. HH to OH, 18–19 July 1848, in Kremer 1990, 42–44, on 43; HH 1852e; Holmes and Olesko 1995; Olesko and Holmes 1993, 103–4; Chadarevian 1993, 279–84; Schmidgen 2015. See also HH 1850c, 1851c, 1854a.

15. HH 1851a and 1851b (on induced currents); HH 1852c and 1853c (on equivalent circuits). Cf. J. C. Poggendorff to HH, 4 April 1854, in HN 355; HH 1852b; Finkelstein 2013, 121–22; Holmes 2003, 761–62; Boring 1950, 42, 45; Turner 1982; Heidelberger 2004; Hui 2013.

16. EdBR to HH, 16 May 1851, 113; Johannes, in Kremer 1990, 193–201, on 197; Gregory 1977; Daum 1998.

17. HH 1853a, 25; HH to EdBR, 11 April 1851, 112. On Goethe's extensive scientific work and Goethe as a scientist, see, e.g., Amrine, Zucker, and Wheeler 1987; Richards 2002, esp. 327–29, 367–76, 407–502.

18. HH 1853a, 26–27.

19. Ibid., 27–30 (quotes); Mandelkow 1989, 184–88; Richards 2002.

20. HH 1852g; esp. Kremer 1993, 206–20, 227, which the present analysis closely follows; Turner 1996.

21. HH to EdBR, 24 March 1852, in Kirsten et al. 1986, 127–28; EdBR to HH, mid-April 1852, ibid., 128–30, on 129–30.

22. HH 1852d; Kremer 1993, 221–33; CL to HH, 29 January 1854, in HN 293; FCD to HH, 26 June 1853, in HN 116; HH 1852h; William Barton Rogers to Henry Rogers, 7 January 1853, in Rogers 1896, 1:330–31, on 330.

23. HH to EdBR, 20 June 1852, 132 (first quote); HH 1852f, 608 (second quote).

24. HH 1853a, 30–32, 38–40.

25. HH 1867e, 268.

26. HH 1853a, 31, 33–34.

27. Ibid., 34–38.

28. Ibid., 40.

29. Ibid., 40–44.

30. Ibid., 44–45.

31. Ibid., 46–47; *Kieler Allgemeine Monatsschrift für Wissenschaft und Literatur* (1853); Mandelkow 1980, 194.

32. HH to Verehrter Herr Onkel, [spring] 1853, in Germanisches Nationalmuseum, Archiv, Nürnberg; HH to EdBR, 22 May 1853, in Kirsten et al. 1986, 141–43 (quote on 141); Johannes, in Kremer 1990, 196.

33. HH to EdBR, 24 June 1852, in Kirsten et al. 1986, 133 (first quote); EdBR to HH, 16 January 1853, ibid., 138–39, on 138 (second quote); HH to EdBR, 23 January 1853, ibid., 139–41, on 141 (third quote).

34. HH to EdBR, 22 May 1853, 141; Tyndall 1904, xiii–xiv (quote); Jackson 2015; Frank Turner 1993 ("public scientist").

35. HH 1852i; August Krönig to HH, 9 July 1853 (first two quotes), in HN 247; HH 1847, 1848a, 1850e, 1852a, 1852i, 1854, 1855a, 1856, 1857, 1858, 1859; Krönig to HH, 6 October 1854; 3, 17 January, 11 April 1855; 17 January; 11, 15 February; 2 March, 20 November, 21 December 1856; 2 January 1857; 12, 16 September 1858, all in HN 247; HH to OH, 6 [August 1853], in Kremer 1990, 97–104, on 97–100 (third quote on 98). Weyrauch 1893, 316–21, noted Helmholtz's articles, as did Theodor Gross (see Gross 1891, 19–32, esp. 22–25; 1898, 141–74, esp. 146–52), Mayer's partisan, who argued that Helmholtz learned of and read Mayer's work between 1847 and 1851, and who sought to minimize or devalue its contributions to conservation of energy.

36. HH to OH, 6 [August 1853], 100–101.

37. Ibid., 101–2.

38. Ibid., 102–4.

39. HH to OH, 18 August [1853], ibid., 104–9 (quotes on 104–5).

40. EdBR to CL, 17 February 1852, in du Bois-Reymond 1927, 107–11, on 110; HH to OH, 18 August [1853], 105–6 (first two quotes); Holmes and Olesko 1995, 213–16; HH to EdBR, 22 May 1853 (third quote).

41. HH to OH, 18 August [1853], 106.

42. Ibid., 106–8.

43. Ibid., 105, 108–9.

44. HH to OH, 20 August 1853, in Kremer 1990, 110–15, on 110–11 (quotes).

45. Ibid., 112.

46. Ibid., 113.

47. Ibid., 113–14.

48. Ibid., 114.

49. Ibid., 115–16 (quotes); HH 1854c.

50. Brock and Meadows 1998, 110–45; HH to OH, 25–29 August [1853], in Kremer 1990, 115–20 (quote on 116–17).

51. HH to OH, 25–29 August [1853], 117–18.

52. Smith 1998, 141; Heinrich Wilhelm Dove to Edward Sabine, 9 August 1853, in Public Record Office, BJ3.2, fol. 53, Kew, quoted in Kremer 1990, 118n12 (first quote); HH to OH, 25–29 August [1853], 118–19 (second quote).

53. HH to OH, 25–29 August [1853], 119–20 (first, third, and fourth quotes); EdBR quoted in Finkelstein 2013, 152 (second quote); HH to OH, 31 August [1853], 121 (fifth and sixth quotes).

54. HH to OH, 31 August [1853], 122.

55. Ibid., 122–23.

56. Ibid., 124.

57. Ibid., 125.

58. HH to OH, 8 September 1853, in Kremer 1990, 125–31, on 126–27.

59. Ibid., 127.

60. Ibid., 128.

61. Ibid.

62. Ibid., 129.

63. Ibid., 130.

64. Ibid., 130–31.

65. Thomson 1882, 1:182–83n; Thompson 1910, 1:288, 308; Smith 1998, 2, 13, 126–49, 178–79, 249, and passim; Smith and Wise 1989, esp. 282–347; HH 1853b. Thomson's article originally appeared in the *Transactions of the Royal Society of Edinburgh* (March 1851) and then, in slightly but crucially different form, in the *Philosophical Magazine* 4 (1852).

66. HH to OH, 8 September 1853, 131.

67. HH to OH, 14–16 September 1853, in Kremer 1990, 132–37 (quotes on 132, 134–35).

68. Ibid., 132–33.

69. Ibid., 133.

70. Ibid., 133–34 (quotes); HH 1854c.

71. HH to OH, 14–16 September 1853, 130, 133, 135, 136.

72. Ibid., 136 (first quote); HH to CL, 2 June 1854, in McKendrick 1899, 90–92 (second quote on 92).

73. HH to CL, 2 June 1854, 92 (first quote); HH to OH, 22–25 September 1853, in Kremer 1990, 137–43, on 137–38 (other quotes).

74. HH to OH, 22–25 September 1853, 139–40.

75. Ibid., 140–42.

76. Ibid., 137, 142–43.

77. Ibid., 138–39.

78. HH to OH, 27 September 1853, in Kremer 1990, 143–44; HH to CL, 2 June 1854 (quotes).

Chapter Seven

1. Bevilacqua 1993, 317–18; 1994. HH 1854b was his principal reply to Clausius.

2. GM to HH, 20 May 1855, in HN 295; HH 1855; GRK to HH, 4 March 1854, in HN 231; Wilhelm Weber to HH, 2 February 1855 (quote), in HN 500; HH to EdBR, 13 June 1854, in Kirsten et al. 1986, 144–45.

3. HH to EdBR, 13 June 1854, 145 (quotes); HH 1854a; EdBR to HH, 21 June, 16 August 1854, in Kirsten et al. 1986, 145–47 (on 145–46), 149, resp.; Krönig 1857, v.

4. HH 1854d, 51.

5. Ibid., 52–53, 57–58.

6. Ibid., 54, 57–58.

7. Ibid., 54, 57–58, 58–59 (quote), 61–63.

8. Ibid., 62–63, 65.

9. Ibid., 65–67; Brush 1966–67, esp. 494.

10. HH 1854d, 67–74, 79.

11. Ibid., 73–80 (quotes on 76, 80). In an "Afterword" (1875) to his republished essay on Goethe, Helmholtz crowned Goethe a forerunner of Darwin. HH 1903, 1:46–47.

12. HH 1854d, 80–83.

13. EdBR to HH, 21 June 1854, 147; EdBR to HH, 26 October 1856, in Kirsten et al. 1986, 162–65, on 162.

14. Büchner 1856, 105–6; Karl von Vierordt to HH, 24 July 1854, in HN 482; Lange 1866, 381–82, 388; Gregory 1977, 160–62; Ernst Haeckel, *Notizbuch*, in Ernst-Haeckel-Haus, Friedrich-Schiller-Universität Jena, Best. B/Abt. 1, no. 162; FH to HH, n.d. [shortly after 7 February 1854], in *LK*, 1:211–12 (first two quotes on 212); Brock and MacLeod 1980, folio 1206, entry for 1 July 1856 (third quote); HH 1856a; Smith and Wise 1989, 500–501, 520–33, 542–49, 551, 559–61, 593; Smith 1998, 148–49; James 1982, esp. 173–76, 178–81.

15. Schnädelbach 1991, 15–21, 32–33, 49–50, 88–89, 95, 100–105, 131–35; Köhnke 1993.

16. Schnädelbach 1991, 70–71, 88–89, 95, 100–105, 110–11, 118–19; Gregory 1977, 145–48. The new German institute — more precisely, the natural science institute — stood in contrast to the old "cabinets," or collections of instruments and associated equipment maintained in a glass-enclosed case (the cabinet). These latter were (mostly) privately owned collections, and nearly always used only for a scientist's own lecture demonstration and personal research purposes; they were not for student use or use by others (colleagues). Between the

1830s and the 1870s, these "cabinets" gradually evolved into "institutes," even if at first it was often seemingly little more than a name change.

The modern institute that arose around midcentury in the German lands was in the first instance constituted by a change from a private to a public or state entity, not only in a legal sense but also in terms of acquiring a new name and location: the new institutes, however undeveloped and unorganized at first, were located in university buildings rather than in a scientist's home. The instruments and equipment—often bequeathed by or purchased from a recently deceased scientist at a given local institution—became public (university) property. The new university-based institute was headed by one individual, the ordinary professor. He had full control over the institute, which included permission to enter; use its laboratories, physical instruments, other equipment and apparatus, library, and other facilities; determine the teaching topics and schedules of its members; and so on. All other institute members—of which there were an increasing number, including extraordinary professors and private lecturers, if any, and custodial staff—were subordinate to the institute director (i.e., the ordinary professor). For those who had the director's permission, the institute's facilities were open to other institute members for research and to students for pedagogical purposes. Unlike an American department, the institute was the director's absolute fiefdom. Though institutes were initially located within an extant university building, in time most were housed in new, purpose-built university buildings, some as stand-alone entities and some as part of a larger science complex. Indeed, some institutes, such as Helmholtz's at Heidelberg for physiology and at Berlin for physics, had residential quarters for the director and his family within the institute, and so in a sense brought this entire set of developments full circle. From the late 1860s onward, the legal, organizational, material, and pedagogical dimensions of the new institutes were a far cry from the cabinets of previous generations. For further discussion, see Cahan 1985.

17. Gregory 1977.

18. Sudhoff 1922; Daum 1998, 119–37; Steif 2003, esp. 102–12; Degen 1954; Gregory 1977, 73–5.

19. Wagner 1997, 8, 11, 12, 147, 148 (quote); HH to EdBR, 20 June 1852, in Kirsten et al. 1986, 131–32; Degen 1954, 272–74; Gregory 1977, 72–73.

20. Degen 1954, 275–77; Gregory 1977, 73–75; Steif 2003, 106–9.

21. CL to HH, 5 November 1854 (first quote), in HN 293; HH to CL, n.d. [circa September 1854], in *LK*, 1:216 (second quote); cf. ibid., 158; HH to EdBR, 23 December 1854, in Kirsten et al. 1986, 151–52 (third quote on 152).

22. Vorländer 1977, 332–34, 336–37, 348.

23. HH 1855b, 87.

24. Ibid., 87–8 (quotes); on Helmholtz and Fichte, see HH to FH, n.d. [ca. 1 March 1855], in *LK*, 1:242; Heidelberger 1993, 482–95; Köhnke 1993, 152–53, 193; De Kock 2014, 2014a.

25. Richards 2002, 114–16, 125–26, 130–35, 140–45, 157–58.

26. HH 1855b, 89, 99.

27. Ibid., 90–94.

28. Ibid., 93–94, 96–99 (quote on 98); on Helmholtz and Müller, see Heidelberger 1977; Lenoir 1992; 1993, esp. 112–21.

29. HH 1855b, 99; Lenoir 1992; 1993, esp. 112–21; 2006.

30. HH 1855b, 99–103 (quote on 100); cf. HH 1877, 186.

31. HH 1855b, 103–5.

32. Ibid., 105–7; Heidelberger 1977, esp. 43–47; Lenoir 1992.

33. HH 1855b, 110–15 (all quotes except penultimate); HH to EdBR, 3 May 1856, in Kirsten et al. 1986, 159–61, on 159–60 (penultimate quote).

34. HH 1855b, 115–17 (quotes); on the history of neo-Kantianism, see Köhnke 1993, esp. (for Helmholtz) 131, 146–49, 151–63.

35. James Clerk Maxwell to George Wilson, 4 January 1855, in Maxwell 1990, 267–74 (quote on 267); cf. Maxwell 1890, 1:141–45, 152, 243, 414–15; see also Maxwell 1995, 775; HH 1855c, 1855d, 1856f; Kremer 1993, 233–37; Turner 1996; Lenoir 2006, esp. 164–72.

36. EB to HH, 23 October 1853, in HN 74; *Totenbuch der St. Nikolai-Kirche zu Potsdam* (1854), no. 203, 137; HH to FH, ca. 1 October; FH to HH, 3 October 1854, in *LK*, 1:220, 220–21, resp., both partly in ESH 1929, 1:94; cf. Epstein 1896, 41.

37. Sobotta 1933, 56–58, 60; Lützeler et al., 1968, 48.

38. HH to EdBR, 5 November 1854, in Kirsten et al. 1986, 150–51 (quotes on 150); Finkelstein 2013, 156–57.

39. HH to Ew. Hochwohlgeboren [Schulze], 3 December 1854, in GSPK, Rep. 76 Va, Sek. 11, Tit. 4, no. 20, Bd. 2, Bl. 1–2.

40. HH to Ew. Hochwohlgeboren [Schulze], 19 December 1854, ibid., Bl. 3–4.

41. Ibid.

42. CL to EdBR, 12 October 1854, in du Bois–Reymond 1927, 132; CL to HH, 24 December 1854, in HN 293; CL to Jacob Henle, 16 January 1855, in Dreher 1980, 139–42, on 140–41 (quotes); Schröer 1967, 63.

43. Alexander von Humboldt to EdBR, 13 March 1855, in Schwarz and Wenig 1997, 139–40 (quote); EdBR to Humboldt, 16 March 1855, ibid., 141–42; EdBR to HH, 16 March 1855, in Kirsten et al. 1986, 154–55; Finkelstein 2013, 156–58; HH to EdBR, 22 March 1855.

44. Humboldt to the minister, 24 March 1855, quoted in *LK*, 1:249–50.

45. Humboldt to HH, 24 March 1855, quoted in ibid.:250.

46. HH to Ew. Excellenz [Raumer], 3 April 1855, in GSPK, Rep. 76 Va, Sek. 11, Tit. 4, no. 20, Bd. 2, Bl. 19–20.

47. Draft letter KM [Raumer] to HH, [5 May] 1855, ibid., Bl. 32; Draft letter KM to Curator, 9 October 1855, ibid., Sek. 3, Tit. 4, no. 39, Bd. 2, Bl. 223; HH to FH, 25 April 1855, quoted in *LK*, 1:251–52, on 252 (first quote); Moritz Naumann to HH, 20 May 1855 (second, third, and fourth quotes), in HN 324; Moritz Ignaz Weber to HH, 24 May 1855, in HN 498.

48. CL to HH, 9 May 1855 (quotes), in HN 293; EB to HH, 22 May 1855, in HN 74.

49. "Der Verein" 1859, 6 (first quote); HH to OH, 17–19 July 1855, in Kremer 1990, 145–50 (second quote on 145–47).

50. HH to OH, 17–19 July 1855, 148–49 (first three and seventh quotes), and n14; HH, "Toast am 18ten Juli 1855," in SF, Sign.: Helmholtz 6.LL496 (fourth, fifth, and sixth quotes). On Helmholtz and Olshausen's friendship, see the letters of Justus Olshausen to HH, in HN 339.

51. HH to OH, 26 July 1855, in Kremer 1990, 150–51, on 151.

52. "Lokales und Provinzielles," *Ostpreussische Zeitung* (Königsberg), no. 181, 4 August 1855, n.p. (quotes); HH to OH, 26 July 1855, 150.

Chapter Eight

1. HH to OH, 4–5 August 1855, in Kremer 1990, 152–56, on 152 (first quote), 154; HH to OH, 6 [7]–8 August 1855, ibid., 156–59; ibid., 146n6, plate 8; HH to EdBR, 14 October 1855, 5 March 1858, in Kirsten et al. 1986, 157, 176–78, resp., on 178 (second quote); GM to HH, 7 November 1855, in HN 295; Wilhelm Heintz to HH, 16 July 1856, in HN 193; Häfner 1934, 384; [1962?], 111.

2. Julius Budge to HH, 8 July 1855, in HN 77; HH to OH, 4–5 August 1855, 153–54 (quotes); HH to EdBR, 14 October 1855; HH to FH, 6 March 1856, in *LK*, 1:261–62; HH to FH, December 1855, ibid., 257–58; HH to Karl Otto von Raumer, in GSPK, Bl. 220–23, copy, cited in Hörz 1994, 11–12.

3. WT to HH, 24 July 1855, in HN 464; Kremer 1990, 154–55.

4. HH to WT, 3 August 1855, in KP, H13.

5. HH to OH, 6[7]–8 August 1855, 158–59; cf. HH to OH, 4–5 August 1855, 154.

6. HH to OH, 6[7]–8 August 1855, 158–59 (quote); HH to WT, 3, 11 August 1855, in KP, H13, H14; WT to AH, 11, 12 October 1894, in HN 464.

7. HH to FH, 6 March 1856, in *LK*, 1:261–62, on 262 (quotes); HH to FCD, n.d. [summer 1855], October 1855, ibid., 252; HH to FH, 6 March 1856, ibid., 261–62, on 262; Otto Jahn to HH, 21 February 1856, 19 November 1859, in HN 216; HH to EdBR, 26 May 1857, in Kirsten et al. 1986, 171–73, on 171–72; Johannes, in Kremer 1990, 193–201, on 197–98; Bowman 1890–91, xix–xx.

8. Scharlau 1986, ix–xviii; RL to HH, 22 November 1856, in HN 281; Olesko 1991, 361.

9. HH to RL, 2 December 1856, 113–14 (first four quotes); RL to HH, 8 February 1857, in HN 281; HH to RL, 3 March 1857, in Scharlau 1986, 115–16 (fifth and sixth quotes); RL to HH, 16 March, 16 April 1857; 18 January 1858 (seventh quote), all in HN 281; Wenig 1969, 179.

10. Bezold 1920, 515–19; Sobotta 1933, 56–58, 60; Lützeler et al. 1968, 48; Georg Meissner to Jacob Henle, 18 May 1855, in Eulner and Hoepke 1975, 8–13, on 11; *Vorlesungsverzeichnis*, in Universitätsarchiv Bonn, Bestand: Rektorat, Sign.: U 70, cited in Hörz 1994, 13–14.

11. HH to EdBR, 14 October 1855, in Kirsten et al. 1986, 157; Kremer 1990, 107n12; HH to EdBR, 5 March 1858, 177 (quote).

12. EdBR to CL, 1 October 1855, in du Bois-Reymond 1927, 137–39, on 139; CL to HH, 14 October 1855, in HN 293; EdBR to HH, 27 April 1856, in Kirsten et al. 1986, 158–59 (first quote on 158); HH to EdBR, 3 May 1856, ibid., 159–61 (third, fourth, and fifth quotes on 160–61); EB to HH, 15 June 1856, in HN 74 (second quote). Cf. HH to FH, 6 March 1856, in *LK*, 1:261–62, where he defends the quality of his teaching in anatomy; EB to EdBR, 7 May 1856, in Brücke et al. 1978–81, 1:82–83, on 82.

13. HH to EdBR, 15 October 1856, in Kirsten et al. 1986, 161–62 (quotes on 162). See also Kremer 1990, 153–54n4.

14. HH to EdBR, 3 May 1856, 160–61 (first, second, and third quotes); HH to Hofrath Wöhler (Secretair der Königlichen Gesellschaft der Wissenschaften zu Göttingen), shortly before 22 February 1856, in Niedersächsische Staats- und Universitätsbibliothek Göttingen, Handschriftenabteilung, 4° Cod.Ms.Hist.lit.116:III, Bl. 210r. (hr. 106); HH to EdBR, 14 October 1855; HH to EdBR, 15 October 1856, 162 (fourth quote); HH to EdBR, 15 April 1858, in Kirsten et al. 1986, 182–84, on 183.

15. HH to EdBR, 15 October 1856, 162.

16. HH to EdBR, 18 May 1857, in Kirsten et al. 1986, 167–69 (first two quotes on 168); HH to EdBR, 5 March 1858, 177 (third quote).

17. HH to Prof. Julius Plücker, 7 October 1856, in "Bericht über die Verwaltung des anatomischen Instituts während des Wintersemesters 1855/56 und Sommersemester 1856"; HH to Prof. Deiters, 4 October 1857, "Bericht über das Anatomische Institut während des Universitätsjahrs 1856–1857," both in Rektorat A7, 1 Bd. XVI; A7,1 Bd. XVII, resp., Universitätsbibliothek Bonn, Handschriften; HH to EdBR, 19 March 1858, in Kirsten et al. 1986, 179–81 (first quote on 180); HH to EdBR, 5 March 1858, 177 (second quote); Titze 1995, 105; HH to EdBR, 15 April 1858, 183 (third, fourth, and fifth quotes).

18. CL to EdBR, 26 May 1854, in du Bois-Reymond 1927, 128 (first two quotes); HH to EdBR, 22 March 1855, in Kirsten et al. 1986, 155–56, on 156 (third quote); HH 1856e, 1860b, 1867d.

19. Gustav Karsten to HH, 20 June 1852; 6 February, 3 March 1853, all in HN 227; HH to Adolf Fick, 4 September 1854, in *LK*, 1:265–67; HH 1886, 317 (quotes); Turner 1987.

20. HH 1856e; 1867e, vii–viii (on Helmholtz's personal observations).

21. HH to WT, 18 June 1856, in CUL, WTP, Add 7342 H63.

22. WT to HH, 30 July (quote); 6, 11 August 1856, all in HN 464; HH to WT, 1 August 1856, in CUL, WTP, Add 7342 H64.

23. HH to OH, 15 August 1856, in Kremer 1990, 159–61, on 160–61 (first three quotes);

HH to FH, [15? August 1856], in *LK*, 1:273–74 (fourth quote); HH to WT, 16 January 1857, in GUL, H-15.

24. HH to OH, 15 August 1856, 161.

25. HH to OH, 17–18 August 1856, in Kremer 1990, 162–65 (quotes on 162–63).

26. Ibid., 164.

27. Ibid., 164–65.

28. HH to OH, 21 August [1856], ibid., 166–69, on 166–67.

29. Ibid., 167–68.

30. HH to OH, 31 August 1856, ibid., 169–71 (quotes); HH to EdBR, 15 October 1856, 162.

31. EdBR to HH, 26 October 1856, 162 (first quote); HH to EdBR, 8 November 1856, in Kirsten et al. 1986, 166 (second quote); EdBR to HH, 15 January 1857, ibid., 167; EdBR to HH, 24 May 1857, ibid., 169–71, on 170–71; HH to EdBR, 26 May 1857 (third, fourth, and fifth quotes), ibid., 171–72; Harnack 1970, 3:123.

32. Bernard 1967, 74, 84–85 (quotes).

33. HH to EdBR, 15 October 1856, 162; EdBR to HH, 26 October 1856, 165; James Clerk Maxwell to WT, 18 December 1856, in Maxwell 1990, 487–91 (first quote on 491); cf. Maxwell to George Gabriel Stokes, 27 January 1857, ibid., 492; WT to HH, 30 December 1856 (second quote), 16 January 1857, in HN 464; HH to WT, 16 January 1857, in KP, H15; HH 1852i; Vogel 1993, 259–66; Darrigol 2005, 146–47; Ferdinand Sauerwald to HH, 22 March 1856, in HN 397; HH to EdBR, 3 May 1856, 160; cf. HH to Wilhelm Heinrich von Wittich, 21 May 1856, in *LK*, 1:267–68; HH 1856b, 1856c, 1856d; HH to EdBR, 8, 11 November 1856, in Kirsten et al. 1986, 166–67.

34. Darrigol 2007.

35. The account in this and the following paragraph draws on the analysis in Turner 1977. See also Vogel 1993; McDonald 2001; Steege 2012, 43–79.

36. Turner 1977.

37. Borscheid 1976, esp. 50–70; Riese 1977; Wolgast 1985a; Doerr 1985; Tuchman 1987, 1988, 1993a; James 1995, 5–7; Werner 1996, 1997.

38. Clark 2007, 530–31; Tuchman 1993a; Werner 1996; 1997, 26–56; Werner 1997 also contains (on 170–203) transcripts of many pertinent archival items; Robert Bunsen to HH, 9 May 1857, in HN 79; Bunsen to the Badisches Ministerium des Innerns, 28 May 1857, in BGLA: 235/29872, Bl. 22–30s, in Werner 1997, 182–84 (quote on 182).

39. HH to EdBR, 18 May 1857, in Kirsten et al. 1986, 167–69, on 167–68.

40. Ibid., 168–69.

41. EdBR to HH, 24 May 1857, 170 (quote); Finkelstein 2013, 162–64; HH to EdBR, 26 May 1857, in Kirsten et al. 1986, 171–73, on 172.

42. HH to EdBR, 26 May 1857, 172; EdBR to HH, 26 July 1857, in Kirsten et al. 1986, 174–76, on 174–75.

43. HH to Johannes Schulze, 29 May 1857, in GSPK, Sekt. 3 Tit. IV no. 39, Bd. III, Bl. 18–19, printed in Pieper 1997a, 5–7 (first quote); Willdenow to Schulze, 30 May 1857, in GSPK, Sekt. 3 Tit. IV no. 39, Bd. III, Bl. 20–21, printed in Pieper 1997a, 8–10 (second and third quotes); Schultze to HH, 10 June 1857, in HN 441; Erlass, 19 June 1857, in GSPK, Sekt. 3 Tit. IV no. 39, Bd. III, Bl. 24; draft letter KM to HH, [3 July 1857?], ibid., Bl. 27; Kultusminister to Bonner Universitätskuratorium, Kultusminister to HH, and HH to the Minister, 13 July 1857, ibid., Bl. 26–28.

44. GRK to HH, 20 June 1857 (first quote), in HN 231; HH to Ew. Excellenz [Karl Otto von Raumer], 13 July 1857, in GSPK, Sek. 3, Tit. 4, no. 39, Bd. III, Bl. 28; HH to EdBR, 14 July 1857, in Kirsten et al. 1986, 173–74 (second quote on 173).

45. CL to HH, 2 March 1857 (first quote), in HN 293; Justus Olshausen to HH, 18 August 1857 (second quote), in HN 339.

46. *LK*, 1:295; IHF, *Tagebüchern* (23 July 1857; 13 August 1857), in WLBS, Cod. Hist. 4 593, VIII; HH to OH, 21 August 1857, in Kremer 1990, 171–74; HH to OH, 25 August 1857, ibid., 175–77 (quote on 175).

47. IHF, *Tagebüchern* (1 July 1855 [first quote]; 22 May 1857 [second quote]); Schul-programm 1857, 22–23; Herrmann-Schneider 1928, 109–12, 116–17; Fichte 1970; IHF to FH, 23, 26 May 1857 (third, fourth, and fifth quotes), in HN 145, also in WLBS, 1e, no. 200; Fichte 1856.

48. Schnädelbach 1991, 83, 118; Köhnke 1993, 109, 112–18, 120–21, 382, 403, 610; Frauen-städt in Schopenhauer 1873, i–cxxxiii, esp. xii–xv; Hörz 1995a.

49. Arthur Schopenhauer to Ernst Otto Lindner, 9 June 1853, in Hübscher 1978, 312–14, on 314; Schopenhauer to Julius Frauenstädt, 22 May 1854, ibid., 342–43; Schopenhauer to Frauenstädt, 15 July 1855, ibid., 368–69, on 368 (first quote); Schopenhauer to Johann August Becker, 20 January 1856, ibid., 380–81, on 380 (second quote); Schopenhauer to Frauenstädt, 7 April 1856, ibid., 390; FH to HH, 27 September 1856, in *LK*, 1:278.

50. FH to HH, 27 September 1856, 277–78; HH to FH, 17, 31 December [1856], in *LK*, 1:283–85 (quotes); Schopenhauer to Frauenstädt, 28 June 1856, in Hübscher 1978, 395–96.

51. FH to HH, 8 February 1857, in *LK*, 1:285–91 (quotes on 285–89); FH to HH, 8 May 1858, in *LK*, 1:333–42; FH to HH, n.d. [1859], ibid., 1:319.

52. HH to FH, 4 March 1857, in *LK*, 1:291–93 (quotes on 291–92); HH 1877, 173, 184; 1878, 233, 403; 1892, 358; Köhnke 1993, 156.

53. HH 1857b, 12 (first quote); HH to EdBR, 14 July 1857, 173 (second quote); J. C. Poggendorff to HH, 16 July 1857, in HN 355; HH 1857c.

54. HH to FH, 3 October 1857, in *LK*, 1:296–97, on 296 (first three quotes); HH to Ew. Excellenz [Karl Otto von Raumer], 4 March 1858, in GSPK, Sekt. 3, Tit. 4, no. 39, Bd. III, Bl. 32–33v, reprinted in Werner 1997, 188–89; HH to EdBR, 18 May 1857, 169; Noeggerath and Kilian 1859, 70; Lampe and Querner 1972, 47–48; EB to HH, 26 July 1857 (fourth quote), in HN 74; Anton Danga to HH, 4 June 1861 (fifth quote), in HN 103; HH to FH, 3 October 1857 (sixth quote); Bavarian Minister von Zweel to HH, 7 February 1858, in *LK*, 1:298.

55. HH 1857a, 121–22.

56. Ibid., 128–35.

57. Ibid., 121–22 (quote), 135.

58. Ibid., 135–46, 153–54.

59. Ibid., 143–47 (quotes on 146–47).

60. Ibid., 147–48, 154 (quote).

61. Ibid., 149–54 (quote on 151).

62. Ibid., 154.

63. Ibid., 154–55 (quotes on 154).

64. HH to the Redaktion der illustrirten deutschen Monatshefte, 22 December 1857 (first quote), Herzog August Bibliothek, Wolfenbüttel: Mittlere Briefsammlung no. 698 + 699; Rudolf Haym to HH, 9 December 1857 (second and third quotes), in HN 192; Haym to HH, n.d. (fourth and fifth quotes), in HN 192.

65. HH 1858a, 1859a, 1859b; Darrigol 2005, passim, esp. v–viii.

66. HH 1954, 251 (first quote); 1875, 26 (second quote).

67. HH to EdBR, 5 March 1858, 177 (quote); HH to Carl Wilhelm Borchardt, 8 January 1858, in *LK*, 2:123; Borchardt to HH, 14 January, 6 February 1858, in HN 58.

68. The following (and subsequent) discussion(s) of Helmholtz's hydrodynamics draw(s) on the analysis of Darrigol 2005, esp. 144–54.

69. Ibid.; Bokulich 2015; Epple 1999, 94, 98–101, 116, 120, 149; Maxwell 1890, 1:451–88, on 488.

70. CL to HH, 10 October 1857, in HN 293; Bunsen to HH, 15 December 1857 (first quote);

28 February 1858, in HN 79; GRK to HH, 4 February 1858 (second quote), in HN 231; HH to Ew. Excellenz [Karl Otto von Raumer], 4 March 1858, in GSPK, Sekt. 3, Tit. 4, no. 39, Bd. III, Bl. 32–33, printed in Werner 1997, 188–89 (third and fourth quotes).

71. Haelscher and Willdenow (for the Kuratorium), 6 March 1858, in GSPK, Sekt. 3 Tit. IV no. 39, Bd. III, Bl. 29–31, cited in Pieper 1997a, 11n28; Kuratorium to the Bonn Medizinische Fakultät, 6 March 1858, UA Bonn, MF 3102, printed in Pieper 1997a, 10–11 (first and second quotes); *Allgemeine Medicinische Central-Zeitung*, 10 April 1858, Sp. 231, cited in Pieper 1997, 13–14 (third and fourth quotes).

72. HH to EdBR, 5 March 1858, 176–77 (quotes).

73. Ibid., 177–78 (quote on 178).

74. EdBR to HH, n.d., in Kirsten et al. 1986, 181–82; Weizel to Baden Ministry of the Interior, 22 June 1858, UA HD: A-553/1, in Werner 1997, 199; Finkelstein 2013, 164; HH to EdBR, 15 April 1858, in Kirsten et al. 1986, 182–84, on 184; Borscheid 1976, 78; Goschler 2002, 94; HH to RH, 4 July 1880, in ESH 1929, 1:248–50, on 248; Zobeltitz 1891, 770; CL to HH, 10 April 1858 (quote), in HN 293; GRK to HH, 2 May 1858, in HN 231.

75. EdBR to HH, 15 March, 28 April 1858, in Kirsten et al. 1986, 178–79, 185, resp.; Finkelstein 2013, 164–68, 172.

76. EdBR to HH, 26 April 1858, in Kirsten et al. 1986, 184 (first quote); Prinzregent [Wilhelm] to Staatsminister von Raumer, 17 April 1858, in *LK*, 2:112–13; Raumer to Prinz-regent, 28 May 1858, in *LK*, 2:113; HH to FCD, 21 June 1858, in *LK*, 1:302; Hugo Philipp Haelschner to HH, 20 May 1858 (second quote), in HN 182; HH to KM, 20 May 1858, in Rep. 76 Va, Sekt. 3, Tit. 4, no. 39, Bd. III, Bl. 60, in GSPK; HH to EdBR, 29 May 1858, in Kirsten et al. 1986, 186–87.

77. Draft letter HH to Ew. Excellenz [Franz Freiherr von Stengel, the Minister of the Interior], 1 June 1858, in GSPK, Sekt. 3, Tit. 4, no. 39, Bd. III, Bl. 67 (quote); HH to Ew. Excel-lenz [Karl Otto von Raumer], 4 July 1858, ibid., Bl. 65–66; HH to Hochverehrter Herr Geheim-rath, 16 July 1858, ibid., Bl. 131–32.

78. EdBR to his father, July 1858, Dep. 5 K. 11 no. 5, Bl. 27–28 (first quote); EdBR to HH, 14 July 1858, in Kirsten et al. 1986, 187–89, on 188 (second and third quotes).

79. HH to Geheimrath, 16 July 1858 (first and second quotes); HH to EdBR, 21 July 1858, in Kirsten et al. 1986, 189–92 (third and fourth quotes, on 189, 191, resp.).

80. GRK to HH, 3 July 1858, in HN 231; HH to FH, 23 July 1858, in *LK*, 1:302–6 (quote on 303).

81. Wahlvorschlag des Professors Karl Theodor von Siebold 1858, VI. Wahlsitzung, Beil. 7, 10 July 1858 (first two quotes), in Archiv, Bayerische Akademie der Wissenschaften, Munich; Draft letter KM to HH, 27 July 1858, in GSPK, Sekt. 3, Tit. 10, no. 14, Bd. V, Bl. 389; Karl Otto von Raumer to the Medizinische Fakultät, 27 July 1858, UA Bonn, MF 3102, in Pieper 1997a, 34; HH 1859e (third quote); 1859f (fourth quote).

82. HH to Moritz Ignaz Weber, 23 September 1858, in GSPK, Sekt. 3 Tit. IV no. 39, Bd. III, Bl. 131–32, in Pieper 1997a, 38–39 (first quote); Protokoll der Sitzung der medizinischen Fakultät, 25 September 1858, UA Bonn, MF 3102, ibid., 39–40; Medizinische Fakultät to Karl Otto von Raumer, 25 September 1858, in GSPK, Sekt. 3 Tit. IV no. 39, Bd. III, Bl. 123–27, ibid., 40–43; August von Bethmann-Hollweg to Prinz Wilhelm von Preussen [Entwurf], 24 January 1859, in GSPK, Sekt. 3 Tit. IV no. 39, Bd. III, Bl. 154–56, ibid., 46–48; Eduard Pflüger to HH, 16 December 1858; 21 March, 10 August 1859, all three in HN 351; Max Schultze to HH, 10 March 1859, in HN 440; EdBR to CL, 7 November 1858, in du Bois–Reymond 1927, 149 (sec-ond quote); CL to EdBR, 10 November 1858, ibid., 150–53, on 152.

Chapter Nine

1. Silliman 1853, 2:285–89 (quote on 285); EdBR 1912, 2:569; Bernstein 1906, 289; Helmholtz in Kussmaul 1902, 37–38; in Pfaff 1995, 196–97.

2. Weech 1877, 8, 82–84, 93–95; 1890, 584–85, 620; Gall 1968; Wolgast 1985; 1985a, 87–107; Tuchman 1993a.

3. Hintzelmann 1886, 61–62; Lexis 1893, 1:118; Eulenburg 1904, 304; Riese 1977, 24–25, 342; Wolgast 1985, 8, 24–25; 1985a, 87–107.

4. Silliman 1853, 2:287–88; Weech 1877, 82–83; Weber 1886, 228; Riese 1977, 98–101, 112–33; Wolgast 1985, 11, 15–16.

5. Wolgast 1985, 11 (quote), 12, 13; Moleschott 1894, 253–58; Riese 1977, 99; Gregory 1977, 80–99; Weber 1886, 194–95, 198, 200, 206–7, 227, 247, 263–67; Pfaff 1995, 183; Stiefel 1977, 2:1859–60; Tuchman 1993a.

6. Stübler 1926, 277–330; Lexis 1893, 1:121; Riese 1977, 346.

7. Borscheid 1976, 14, 77–80, and passim; Tuchman 1993a, esp. 91–112; James 1995.

8. Heunisch 1857, 591; Lockemann 1949, 165, 190–92, 194–95; Borscheid 1976, 62–71; Riese 1977, 218–19; Wolgast 1985, 14.

9. Robert Helmholtz 1890, 529–30; Roscoe 1900, 530–32; Borscheid 1976, 76–77; Jungnickel and McCormmach 1986, 1:297–99; James 1995.

10. HH 1868, 310 (quote); WT to HH, 2 January 1859, in HN 464; HH to WT, 15 November 1859, in KP, GUL, H16; Thompson 1910, 1:300–303; Smith and Wise 1989, 397–98, 505; Helmholtz in Kussmaul 1902, 37–38; James 1995 ("cultural ornament").

11. HH to Alexander Pagenstecher, 29 December 1881, in Universitätsbibliothek Heidelberg, Handschriftenabteilung, Heid. Hs. 840,3; HH to Dr. von Kirchenheim, 7 March 1886, ibid., Heid. Hs. 3865; HH to Prof. Dr. Pfitzer, 13 October 1891, ibid., Heid. Hs. 840,2; Hintzelmann 1886, 243–44; Kussmaul 1902, 187; 1903, 63–64.

12. Rudolph Wagner, in *Allgemeine Zeitung*, 8 January 1859, in Eulner and Hoepke 1979, 102–3 (n229); Rudolph Wagner to Jacob Henle, 23 January 1860, ibid., 102–5, on 103; Henle to Karl Pfeufer, 18 January 1860, ibid., 60–61 (n173), on 60; HH to RL, 29 December 1858, in Scharlau 1986, 116–18, on 118; Berthold et al. to the Königliche Sozietät der Wissenschaften [Göttingen], 1 October 1859, Bl. 179 (quote); die physikalischen Classe der Königlichen Societät der Wissenschaften, Bl. 180, Königlichen Societät der Wissenschaften, and accompanying document (Bl. 181), both in *Pers* 12, Archiv der Akademie der Wissenschaften, Göttingen; Max Schultze to HH, 10 December 1858, in HN 440.

13. HH to EdBR, 29 May, 29 October 1858, in Kirsten et al. 1986, 186–87, 193–94 on 194 (quote), resp.; HH to FH, 23 July 1858, in *LK*, 1:302–6, on 305–6; RL to HH, 4 October 1858, in HN 281; Werner 1997, 64, 89–93.

14. Kurbjuweit 1985, 327, 330; Tuchman 1993a, 50, 74–75, 103–4, 168; Kronecker 1891, 437; Albrecht 1985, 197; HH to OH, 16 August 1851, in Kremer 1990, 63; GRK to HH, 4 February 1858, in Werner 1997, 194–95; Werner 1997, 92–95; Weizel to Baden Ministry of the Interior, 22 June 1858, in UA HD: A–553/1, in Werner 1997, 199; Karl Steinheil to HH, 17 May, 12 September 1859, in HN 447. For the distinction between a "cabinet" and an "institute," see chapter 7, note 16.

15. HH to the Bau- u. Ökonomie-Kommission, University of Heidelberg, 13 January 1859, in UA HD: A–451/9, printed in Werner 1997, 195–97; Werner 1997, 95–96; Tuchman 1993a, 168–69.

16. HH to Wilhelm Wundt, 5 August 1858, in Wundt-Archiv, no. 1209, Karl-Marx-Universität, Archiv, Leipzig, and printed in Schlotte 1955–56, 335–36, on 335; Wundt 1921, 149–61; Schlotte 1955–56; Bringmann, Bringmann, and Cottrell 1976; Bringmann, Bringmann, and Balance 1980.

17. HH to EdBR, 29 May 1858, 187 (first quote); EdBR to HH, 14 July 1858, in Kirsten

et al. 1986, 187–89 (second and third quotes on 187–88); HH to Wundt, 5 August 1858 (fourth quote); Wundt 1921, 153.

18. Aranjo 2014; HH to EdBR, 29 October 1858, 193; Wundt 1921, 150–58, 160–61; Sechenov 1965, 89–90.

19. HH to EdBR, 29 October 1858, 193 (quotes); HH to RL, 29 December 1858, 117.

20. Hugo Kronecker to HH, 9 August 1872, in HN 249; Kronecker 1891, 437; EB to HH, 13 October 1858, in HN 74; CL to HH, 13 October 1858, in HN 293; Kronecker to HH, 13 August 1863, in HN 249; EdBR to HH, 11 November 1858, in Kirsten et al. 1986, 194.

21. Wundt 1921, 159; Vogt 1994; Birkenmaier 1995, 25; Gordin 2008, esp. 27–32; Albert von Bezold to HH, [? July] 1862, in HN 46; Nikolai Egorovich (Eduard) Junge to HH, 11 October 1858; 29 August, 28 October, 30 November 1859, all in HN 224; Sechenov 1965, 65, 67–70, 81–83, 88–90 (quotes); 93–95, 106–7; CL to Ivan Mikhailovich Sechenov, 14 May 1859, in Schröer 1967, 248–51.

22. Carl Wilhelm Borchardt to HH, 16 March 1859, in HN 58; HH 1859a, 1859b.

23. "Säcularfeier" 1859, cols. 473–74; HH 1859c; Richard Lepsius to Elisabeth Lepsius, 29 March 1859, in Lepsius 1933, 218–19, on 218; HH to OH, 27 March 1859, in Kremer 1990, 177–78.

24. HH to OH, 30 March 1859, in Kremer 1990, 178–81 (quotes); Richard Lepsius to Elisabeth Lepsius, 29 March 1859, 219; Lepsius 1933, 220; *LK*, 1:318–19.

25. EB to EdBR, March–April 1857, in Brücke et al. 1978–81, 1:89–90, on 90; Bowman 1890–91, xvii; HH to OH, 27, 30 March 1859; HH 1859c, 399 (quotes); 1860a; Pantalony 2009, 28–34, 86–88, 113, 214–19; GM to HH, 11 September 1859, in HN 295.

26. WT to HH, 12 May 1859, in HN 464.

27. HH to OH, [4 June 1859], in Kremer 1990, 181–82; HH to OH, 6 [June] 1859, ibid., 183–84.

28. IHF, *Tagebüchern*, 4, 6 June 1859; 15, 16 April 1860 (fourth quote from 16 April); I. H. Fichte über Julius Ferdinand und Hermann Helmholtz, in WLBS, Cod. Hist. 4 593, VIII; IHF to HH, 7 June 1859 (first three quotes), in HN 145.

29. Weech 1877, 10; WT to HH, 11 July, 18 August 1859, both in HN 464.

30. HH to WT, 30 August 1859 (quotes), in CUL, WTP, Add 7342 H65; WT to HH, 6 October 1859, in HN 464; HH to Gustav Michaelis, 30 August 1859, in Staatsbibliothek Preussischer Kulturbesitz, Haus Unter den Linden, Musikabteilung, Mendelssohn Archiv, Mus. ep. H. Helmholtz.

31. HH to "Mein geliebtes Mamachen," 9 [September] 1859, in Kremer 1990, 184–87 (first quote on 185); HH to OH, [12 August 1859], ibid., 187–89 (second quote on 188); HH to OH, [13 September 1859], ibid., 189–91 (third, fourth, fifth, and sixth quotes on 189–90); HH to OH, [20 September 1859], ibid., 191–92 (seventh quote on 191).

32. HH to WT, 15 November 1859.

33. Johannes, in Kremer 1990, 193–201, on 198–99 (first two quotes on 198); HH to RL, 29 December 1858, in Scharlau 1986, 116–18, on 117–18; RL to HH, 16 (and 22), 19 March, 19 April 1859, all in HN 281; HH to Mrs. WT [Margaret Thomson], 16 January 1861, in CUL, WTP, Add 7342 H66; Evangelisches Kirchengemeindeamt Heidelberg, Kirchenbuch entry in "Todtenbuch der ev.-prot. Gemeinde zu St. Peter und Providenz in Heidelberg," Bd. 46, S. 123, no. 186, cited in Werner 1997, 64; HH to Carl Binz, n.d. [circa early 1860] (third quote), in *LK*, 1:346–47, on 347.

34. Wilhelm Heintz to HH, 2 January 1860, in HN 193; EdBR to HH, 4 January 1860 [misdated 1859], in Kirsten et al. 1986, 195; WT to HH, 22 January 1860, in HN 464; GM to HH, 2 January 1860, in HN 295; RL to HH, 6 January 1860 (first quote), in HN 281; CL to HH, 3 January 1860 (second and third quotes), in HN 293.

35. HH to FCD, 9 April 1860 (first two quotes quoted in ESH 1929, 1:97, and in *LK*, 1:347–

48); HH to HK, 2 April 1872 (third quote), in HKL; HH to HK, 5 January 1873 (fourth quote), in HKL.

Chapter Ten

1. E. Brücke, Ca.[?] Littrow, C. Ludwig, Schrätter, 10 May 1859, 448/1859, no. 4, Bibliothek, Österreichische Akademie der Wissenschaften, Vienna; EB to EdBR, 21 June 1859, in Brücke et al. 1978-81, 1:109-10, on 109; EB to HH, 20 June 1859 (first quote); 6 January 1860, both in HN 74; Kühlenthal (Ministry of Foreign Affairs) to Stengel (Minister of the Interior), 4 August 1859, in BGLA: 76/9939, and Stengel's response, 6 August 1859, printed in Werner 1997, 202-3; HH to H. W. Miller, 10 June 1860 (second quote), in Royal Society, London, Archives, MC.6.91; GM to HH, 2 August 1860, in HN 295; honorary diploma (no. 36), in "Acta der Königl. Friedrich-Wilhelms-Universität zu Berlin betreffend: Promotionen honoris causa: Vom 1860 bis 1886," Philosophische Fakultät, vol. 2, 1380, in HUB; Königliche Universitätsbibliothek 1899, 753, 768.

2. Thompson 1910, 1:411; HH to Mrs. WT [Margaret Thomson], 16 January 1861 (quote), in CUL, WTP, Add 7342 H66; William Barton Rogers to Henry Rogers, 19 October, 27 November 1860; 13 October 1862, all in Rogers 1896, 2:43, 52-54 (on 53), 133-34 (on 134), resp.

3. For this and the following two paragraphs, see Schulze 1886, 3-6, 31; Weber 1886, 252-55; Dilthey 1900, 226-27; Wachsmuth 1900, 15; Mohl 1902, 1:63-64, 221, 225; ESH 1929, 1:11-44, 17, 61-63, 71; AH to MM, 26 August 1856, ibid., 1:63-64; Roscoe 1906, 45-47, 59-62, 75.

4. AH to ISZ, 11, 18, 25, 30 October; November 1852, in ESH 1929, 1:39-40 (on 39), 40-41, 42-43, 43-44, 46-47, resp.; AH to PM, 24 October 1850; 9 January 1853; 3, 5, 8 February 1853, ibid., 25, 49, 51-52, 52-53, resp.; ESH 1929, 1:15, 18, 19, 27, 32-35, 57; Simpson 1887, 1, 46-49, 81; Weech 1906, 295-96; Braun-Artaria 1899; Dilthey, 1900, 227-28; Braun-Artaria 1918, 130-31; Lesser 1984, 3, 11-12, 25-27, 102, 103, 116, 119, 128, 135, 164.

5. HH 1867e.

6. This is the theme and analysis of Kremer 1993, which the present account follows closely in this and the following two paragraphs.

7. Ibid., 237-47; HH 1858b, 1859e.

8. Kremer 1993, 247-55; HH 1859d.

9. Rentsch to HH, 28 August 1860 (first quote), in HN 369; Dr. Med. Ripps to HH, 7 April 1861, in HN 376; EB to EdBR, 3 November 1864, in Brücke et al. 1978-81, 1:140; EdBR to HH, 25 March 1862, in Kirsten et al. 1986, 200-203 (second quote on 200); Kremer 1993, 206-7, 256-57; Turner 1987.

10. Wachsmuth 1900, 16; Braun-Artaria 1899; 1918, 133; AH to MM, July 1860, in ESH 1929, 1:72 (quotes); Dilthey 1900, 228; Weech 1906, 296.

11. HH to WT, 13 February 1861, in CUL, WTP, Add 7342 H67.

12. EdBR to HH, 26 February 1861, in Kirsten et al. 1986, 196; HH to EdBR, 2 March 1861, ibid., 196-97 (quotes).

13. AH to MM, 8 February 1861, in ESH 1929, 1:72-73.

14. AH to MM, 21 February 1861, 74-75, on 74 (quotes); AH to HH, 18, 22 March 1861, 77-78, on 77; 78-79, on 79, all ibid.

15. AH to MM, 1 March 1861, ibid., 75-76.

16. Devrient 1964, 2:371.

17. Wilhelm Heintz to HH, 24 February 1861, in HN 193; EB to HH, 9 March, 21 June (first quote) 1861, in HN 74; Justus Olshausen to HH, 10 April 1861 (second and third quotes), in HN 339; GM to HH, 7 April (fourth quote), 9 May 1861, both in HN 295; William Carpenter to HH, 19 February 1861, in HN 85; HBJ to HH, 4 March 1861, in HN 222.

18. Johannes, in Kremer 1990, 193–201, on 198–99.

19. AH to MM, [sometime between 2 and 17] March 1861, in ESH 1929, 1:76–77 (quotes on 76).

20. James 2002, 136–40; 2004.

21. HBJ to HH, 23 November 1860 (quotes); 5, 21 December 1860; 4 March 1861, all in HN 222; HH to EdBR, 2 March 1861; AH to MM, [sometime between 2 and 17] March 1861; HBJ to EdBR, 15 February [1861], in SBPK, SD 3 k 1852 (3), Bl. 311–13; HH 1860. Cf. WT to George Gabriel Stokes, 20 April 1864, in Wilson 1990, 1:319–23, for Thomson's use of Helmholtz's method for studying a vibrating violin string.

22. Brock and MacLeod 1980, entry for 24 March 1861, folio 1572.

23. Mrs. WT to HH, 4 April 1861, in Thompson 1910, 1:416; HH to Verehrte Freundin [Mrs. WT], 22 March 1861 (quote), in CUL, WTP, Add 7342 H68.

24. HH to Verehrte Freundin [Mrs. WT], 22 March 1861; HER to HH, 1 April 1861, in HN 385; William Sharpey to HH, 6 March 1861 (quotes), in HN 402.

25. AH to HH, 18, 21 March 1861, in ESH 1929, 1:77–78; AH to HH, 22 March 1861 (quotes).

26. HH 1861; Mrs. WT to HH, 4 April 1861; Brock and MacLeod 1980, entry for 14 April 1861 (first quote), folio 1575; HBJ to EdBR, 7 July 1861 (second quote), in SBPK, SD 3 k 1852 (3), Bl. 314–16.

27. HH 1861, 565–69 (all but fourth quote on 565–66), 572 (fourth quote).

28. Ibid., 570–79 (quotes on 571, 573, 575, 578); on the contrast between Helmholtz (and Tyndall) and others, see Smith and Wise 1989, 617–19, 626; Smith 1998, 171–72, 253; Lightman and Reidy 2014.

29. EdBR to Jeannette du Bois–Reymond, 23 April (first and second quotes), 26 April (third quote) 1861, both in Dep. 5, K. 11, no. 5; Epstein 1896, 41; Roscoe 1906, 92; GM to HH, 9 May 1861, in HN 295.

30. Helmholtz's Personalakte, in RKUHA, PA 1700, Bl. 10, cited in Werner 1997, 74–75.

31. Zobeltitz 1891, 768 (first quote); HH to HER, 6 August 1861 (second quote), in RSC; AH to MM, 15 October 1861, in ESH 1929, 1:101–2 (third quote).

32. AH to MM, 15 October 1861, in ESH 1929, 1:101–2 (quote); AH to MM, between 2 January and 21 February 1862, ibid., 102–3; AH to Julius von Mohl, 24 June 1861, ibid., 100–101; Braun-Artaria 1899; 1918, 134; Weech 1877, 21; 1890, 595–601; Zeller 1908, 181, 192–93; Dilthey 1900, 229.

33. Braun-Artaria 1899; 1918, 132–33, 139; Dilthey, 1900, 231; Wachsmuth 1900, 15, 19; AH to ISZ, 11 October 1852, in ESH 1929, 1:39–40; AH to HER, 18 January 1863, in RSC; Bunsen 1899; ESH 1929, 1:67.

34. HH to EdBR, 15 March 1862, in Kirsten et al. 1986, 198–99 (quote); HH to WT, 27 May 1862, in CUL, WTP, Add 7342 H69; Alexander Crum Brown to HH, 31 March, 4 August 1862, in HN 72.

35. HH to EdBR, 15 March 1862; HH to EdBR, 13 April 1862, in Kirsten et al. 1986, 203–4, on 203 (quote); HH to WT, 27 May, 14 December 1862, in CUL, WTP, Add 7342 H69, H70, resp.; CL to HH, 28 August 1862, in HN 293; AH to HH, 13 August 1862, in ESH 1929, 1:104–5; EB to HH, 31 August 1862, in HN 74; Vleminx to HH, telegram, October 1862, in HN 487.

36. AH to HH, 28 March 1864, in ESH 1929, 1:112–13, on 113; AH to HH, [Holy Thursday] 1864, ibid., 112; AH to Julius von Mohl, 29 May, 1 December 1864, ibid., 123–24, 124–25 (second quote), resp.; AH to ISZ, 2 November 1890, ibid., 2:32–34, on 32; HH to EdBR, 15 May 1864, in Kirsten et al. 1986, 208–9 (first quote); HH to HER, 18 March 1863, in RSC; EB to HH, 22 July 1864; 5 February 1865, in HN 74; Carl David Wilhelm Busch to HH, 22, 26 May; 20, 27 July; 6 August, 17 November 1864; 5 January, 24 February, 10 August 1865; 15 August 1867, all in HN 82; Otto Jahn to HH, 13 September 1864; 4 January 1865, both in HN 216; CL

to HH, 6 February 1865, n.d. [1865], both in HN 293; Braun-Artaria 1899; 1918, 135; Weech 1906, 298.

37. HH to EdBR, 3 January, 13 February 1865, in Kirsten et al. 1986, 213–14, on 214 (first quote), 215, resp.; HH to Jacques-Louis Soret, 23 December 1864; 12 August 1865, in VG, Ms. fr. 4175, f. 338–39 (second and third quotes), f. 340 (fourth quote); Soret to HH, 19 September 1865, in HN 412; AH to HH, 7, 9, 13 September 1865, in ESH 1929, 1:127, 128 (fifth quote), 128–29, resp.

38. HH to EdBR, 11 January 1866; 22 April, 2 June 1867; 20 April 1868, all in Kirsten et al. 1986, 219–21, 223–24 (on 223), 227, 228, resp.; HH to WT, 24 November 1866, in KP, GUL, H17 (first two quotes); HH to WT, 23 August 1883, in CUL, WTP, Add 7342, H75; AH to PM, 1 April 1867, in ESH 1929, 1:140–41; AH to HH, 6 August 1867, ibid., 142–43 (third quote); AH to MM, 20 January 1868, ibid., 146 (fourth quote).

Chapter Eleven

1. Wielandt to the Ministerium des Innern, 17 January 1862, in BGLA: 235/397, in Werner 1997, 200–201; Toepke 1907, 712; Riese 1977, 72–93.

2. Wolgast 1985a, 89; HH to EdBR, 13 April 1862, in Kirsten et al. 1986, 203–4; Doerr 1985, 4:365; Hintzelmann 1886, 62; "Heidelberg" 1862.

3. AH to Julius von Mohl, 3 May 1862, in ESH 1929, 1:103–4 (quote on 103); HH to WT, 27 May 1862, in CUL, WTP, Add 7342 H69; Eduard Pflüger to HH, 27 April 1862, in HN 351.

4. HH to EdBR, 13 April 1862; Helmholtz's notebook "Allgemeine Resultate der Naturwissenschaften," in HN 720; HH to WT, 14 December 1862, in CUL, WTP, Add 7342 H70; HH to HER, 18 March 1863, in RSC (first two quotes); Bernstein 1906, 288; Stern 1932, 5–6 (third quote); Gregory 2000, esp. 25–28; Cahan 2000.

5. HH 1862, 159.

6. Ibid., 159–62 (quote on 160).

7. Ibid., 162–63.

8. Ibid., 164.

9. Ibid., 164–66; Phillips 2012, 228–53.

10. HH 1862, 166–67; Daston 1999; Jurkowitz 2002.

11. HH 1862, 166–67.

12. Ibid., 167–68.

13. Ibid., 169–71.

14. Ibid., 172–74, 178; HH 1867e, 447, 453; Köhnke 1993, 136, 468; Hatfield 1993, 543–45.

15. HH 1862, 174–79.

16. Ibid., 180 (quotes); Helmholtz 1837, 17.

17. HH 1862, 180–81.

18. Ibid., 181.

19. Ibid., 181–83 (quotes); EB to HH, 21 December 1862, in HN 74.

20. HH 1862, 183–85.

21. HH 1884, 1:x; HH to Verehrter Herr, 5 May 1863 (quote), in VV; EB to HH, n.d. [late 1862], in HN 84.

22. HH to the Bau- und Ökonomiekommission, 13 January 1859, UA Heidelberg: G II 83/14, in Albrecht 1985, 201–2; Tuchman 1993a, 168–71; HH to WT, 14 December 1862 (quote).

23. Bernstein 1906, 283; Albrecht 1985, 127, 336–37, 345–49; Werner 1997, 95–97; Wurtz 1870 (quotes on 60–61).

24. Wundt 1921, 154–61; Bringmann, Bringmann, and Cottrell 1976; Bringmann, Bringmann, and Balance 1980.

25. HH to J. Friedrich August von Esmarch, 21 October 1863, in Schleswig-Holsteinische Landesbibliothek, Kiel, Cb 18 F2/97.

26. HH to Medicinische Facultät (University of Heidelberg), 25 October 1863, in RKUHA, PA Wilhelm Wundt.

27. Rudolf Dohrn to HH, 14 January, 24 February 1872, both in HN 113; HH to Wilhelm Oncken, 1 December 1872, in Justus-Liebig-Universität Giessen, Universitätsbibliothek, Handschriften-Abteilung, 139/100–107; HH to Adolf Fick, 16 December 1872; HH to ?, 1873, both cited in Schlotte 1955–56, 336–37.

28. HH to EdBR, 20 April 1868, in Kirsten et al. 1986, 228.

29. Tschermak 1919; Lenoir 1986; de Palma and Pareti 2011.

30. Engelmann quoted in ESH 1929, 1:105–6 (first quote); Bernstein 1906, 288 (second quote); Knapp 1902, 557; Albrecht von Graefe to Johann Friedrich Horner, 24 July 1868, in Bader 1933, 110–11 (third quote on 111).

31. Wundt 1921, 157; Sechenov 1965, 88–90 (quotes on 89).

32. Knapp 1894, 515–16.

33. Bernstein 1906, 288; cf. Knapp 1902, 557. For other accounts of Helmholtz as a physiology teacher, see also Kussmaul 1903, 43–45, 55, 58–60; Emil Berthold to HH, 30 October 1866; 8 December 1871; 15 May 1889, all in HN 43.

34. Bernstein 1906, 289; Merz 1922, 106–7 (first quote), 118–24 (second quote on 119).

35. Kussmaul 1902, 186; 1903, 61–62.

36. Jaeger 1985.

37. Knapp 1902 (quote); HH to HK, 26 April 1868, in HKL; Riese 1977, 121–24, 132, 231–34.

38. Bonner 1963, 16–19, 33–34, 82–83; Frank 1987, esp. 14, 18–22, 25–28, 38–39; Jeffries Wyman to Morrill Wyman, 26 June 1870, in Francis A. Countway Library of Medicine, H MS c12.1, fd.8 (quote).

39. Frank 1987, 18; William James to Thomas W. Ward, autumn 1867, in Perry 1935, 1:118–19 (first quote); James to Henry Pickering Bowditch, 5 April (May?) (second quote), 15 June (third quote) 1868, both in Francis A. Countway Library of Medicine, H MS c5.2, fd. 1 (copy) and fd. 2 (copy); James to his parents, 3, 9 July 1868, both in Perry 1935, 1:282–83, 283, resp. (fourth quote on 283).

40. Vogt 1994; Kussmaul 1902, 186. For a Polish student, see HH 1869i.

41. AH to Julius von Mohl, 1 December 1864, in ESH 1929, 1:124–25 (first quote); Bernstein 1906, 292; Wundt 1921, 158–59 (second quote).

Chapter Twelve

1. HH 1863; 1954, vi, 225, 247, 297, 306–7, 308, 326, 327; HH to CL, 30 March 1865, in LK, 2:60–61 (first four quotes); HH jotting on a piece of paper, 3 March 1891 (fifth quote), in Smithsonian Institution Libraries, Washington, DC, Dibner Manuscript Collection, Helmholtz Papers, MSS 683A; Epstein 1896, 41; Zobeltitz 1891, 770; cf. Hui 2013, 55–87.

2. HH to Mrs. WT [Margaret Thomson], 16 January 1861, in CUL, WTP, Add 7342 H66; HH to FCD, 1860; 29 April 1862, in LK, 1:360 (quote), 2:11–12, resp.; HH to EdBR, 15 March 1862, in Kirsten et al. 1986, 198–99.

3. Eduard Vieweg to HH, 3 July 1861 (quote), in HN 483; Dreyer n.d. [1936], 45–46, 88–89, 95; Rocke 1993, 68–71.

4. HH to Vieweg, 15 July 1861 (all quotes except last), in VV; Vieweg to HH, 23 July 1861 (last quote), in HN 483.

5. HH to the Vieweg Firm, 21 November; 3, 15, 27 December 1861; 20, 25 February; 5, 7, 13 April; 18 June; 6, 31 July; 3, 29 October; 18 November 1862; 5 May 1863; 14 November, 5 December 1864, all in VV; Vieweg to HH, 29 November; 9, 24 December 1861; 19 February

1862, all in HN 483; HH to Hochgeehrter Herr (probably Georg Ernst Reimer), 28 February 1862 (quote), in WGA; Dreyer n.d. [1936], 60, 94, 107, 109, 130.

6. HH to Mrs. WT, 16 January 1861 (first quote), in CUL, WTP, Add 7342 H66; for a slightly different version, see HH to Margaret Thomson, late May 1862, in ESH 1929, 1:104; HH 1954, vi; AH to RM, 6 December 1862, in ESH 1929, 1:106–7 (second quote on 106); HH to WT, 14 December 1862 (third quote), in CUL, WTP, Add 7342 H70.

7. For example, Trayser and Company to HH, 25 February 1861, in HN 471; J & P Schiedmayer to HH, 18 March, 31 December 1860; 9 January 1862, all in HN 426; HH 1954, vi, 121–26, 316–20; 1859c, 400–402; HH to EdBR, 15 April 1858, in Kirsten et al. 1986, 182–84; Jackson 2006; Pantalony 2009.

8. Rudolph Koenig to HH, 2 December 1859; 29 February 1860; 18 May 1861, all in HN 238; HH 1859c, 404–5; Ferdinand Sauerwald to HH, 3 August 1860, in HN 397; HH 1954, 11–15, 39–49, 51–54, 74–102, 120–27, 161–65, 174–75, 372–74, 380–96, 413–14, 418–21; Pantalony 2009, 1–4, 30–34, 53–55, and passim.

9. HH 1860, 1861a; 1954, 45–49, 51–54, 74–102, 374–77, 380–87, 388–96; Lawergren 1980; Jackson 2006, 272–77.

10. AH to MM, 1 January 1862, in ESH 1929, 1:102; American Philosophical Society 1891, 149–51, 154 (quote), 160–62; HH 1954, 104, 116; Seiler 1868, 7–9, 13–14, 31–35, 41–104, 148–49; 1875, 7–8, 11–34, 62, 64, 83, 86–87.

11. HH 1954, 1–3 (quotes on 2–3).

12. Ibid., 3–4.

13. Ibid., vi, 4–6 (quotes on 5–6).

14. Cf., e.g., Dahlhaus 1970, 49–52.

15. HH 1954, 7, 8 (first quote), 10, 22 (second and third quotes), 23–24.

16. HH 1867c, 1869a, 1869b; 1954, 25, 33–34, 44, 49–65 (quote on 58), 63, 128–51, 158, 166, 172, 227, 406–11; McDonald 2001, 2002; Zimmermann 1993, 27–32; Georg von Békésy, "Concerning the Pleasures of Observing, and the Mechanics of the Inner Ear," Nobel Lecture, 11 December 1961, http://nobelprize.org/nobel_prizes/medicine/laureates/1961/bekesy -lecture.pdf (accessed 6 August 2009). For the post-1863 history of modifications of or challenges to Helmholtz's theory, see Zurmühl 1930.

17. HH 1857d, 1859c, 1860a; 1954, 103–28 (quotes on 103), 398–400.

18. HH 1954, 152–53 (first quote), 159–60 (second quote on 160), 194, 204 (third and fourth quotes), 205–6. See Maley 1990, esp. 120–36.

19. HH 1954, 226 (first quote), 227 (second and third quotes), 228 (fourth quote).

20. Ibid., 229 (quotes), 231–33.

21. Ibid., 234.

22. Ibid., 234–49 (first two quotes on 234, third through sixth on 235, eighth and ninth on 249), 256 (seventh quote), 280–85, 309; HH 1862a.

23. HH 1954, 251–52; Hatfield 1993.

24. HH 1954, 362–63 (first quote on 362), 365–66 (all other quotes).

25. Ibid., 366–67.

26. Ibid., 367.

27. Ibid., 368 (first quote), 370–71 (second and third quotes on 371).

28. Vieweg to HH, 1 July, 29 November 1864, both in HN 483; HH to Vieweg Firma, 11, 31 March; 19 October, 31 December 1876; 18 May 1877; 22 July 1891; 28 August 1892, all in VV.

29. EB to HH, 21 December 1862 (first quote), in HN 74; Brücke 1871; EB to HH, 16 April 1863 (second quote), in HN 74; CL to HH, 3 January 1864 (third, fourth, fifth, and sixth quotes), in HN 293; HH to CL, 27 February 1864 (seventh quote), in *LK*, 2:31; CL to HH, 6 February 1865, in HN 293; Schramm 1998, 217–18, 257n112, 265–66; Klaus Groth to HH, 19 September 1870, in HN 176.

30. Lotze 1868, 277–82 (first quote on 279, second on 281), 463–76, esp. 463.

31. CL to HH, 3 January 1864, in HN 293; Mach 1866; HH to Verehrter Herr College [Ernst Mach], 31 December 1866, in Ernst-Mach-Institut, Fraunhofer-Institut für Kurzzeit-dynamik, Freiburg, printed in Thiele 1978, 35; Mach in Blackmore 1978, 406, 409–10, 415; Mach 1943, 19, 28–29, 35, 99, 305, 307, 375, 382–84; Swoboda 1988, 369–70, 392, 399; Hui 2013, 96–98, 103–5, 192n24.

32. "Zu Helmholtz" 1867; Paul 1868, 4–5 (quote on 4), 11–13, 32–33, 36–41, 82–85, 224–25.

33. Hauptmann 1863; Mendel and Reissmann 1870–79, 5:191–92 (first two quotes on 191); Schubring 1872, 5 (third quote), 6, 70; Manz 1903–4, 22 (fourth quote).

34. Gustav Theodor Fechner to HH, 6 June (first quote), 12 July 1869, in HN 142; HH to Fechner, 3 July 1869, in LK, 2:62–64; Zimmermann 1865, 2:42–43; William Preyer to HH, 19 March 1879 (second and third quotes), in HN 358.

35. Oettingen 1866; 1913, iii–v, 9, 19–21, 24, 26, 27, 40; Schubring 1872; Bagge 1867, 467; Auerbach 1881, 232.

36. Krüger 1863; Rummenhöller 1963, v–xii, 13–17, 43–45, 110–12; 1971; Bagge 1867, 465, 467–79.

37. Mach 1886; AJE to HH, 1 December 1873, in HN 131; Pantalony 2009; Ries [Riemann] 1873; Hugo Riemann to HH, 2 August 1876, in HN 374; Riemann 1900, 25, 31–33, 38–39, 52–57, 66–67, 83–85, 96–97, 107–8, 119–20; Wuensch 1977; Rehding 2003, 19–24, 27, 31–33, 48, 70–71, 79–83, 108, 167, 181; 2005; Partch 1974, 144–45; Ash 1995, 25–41; Reinecke 1999; Green and Butler 2002, 246, 262–66; Kursell 2008, 2013 (also for Helmholtz's influence on Arnold Schoenberg).

38. Mittasch 1952, 366; "H.L. Helmholtz" 1875; "Deutsche Professoren" 1876 (quote on 680).

39. Radau 1865 (quotes on 193–94); Rodolphe Radau to HH, 24 May 1869, in HN 361.

40. MM to M. C. M. Simpson, 23 February 1863, in Simpson 1887, 200–201 (first quote on 200); MM to Emma Weston, 8 April 1866, ibid., 222–24 (second quote on 223); HH to AH, 11 April 1866, in ESH 1929, 1:130–31.

41. HH to AH, 11 April 1866, in ESH 1929, 1:130–31 (quotes); Hannabuss 2000, 443, 445, 450–55; Louis Grandeau to HH, 5 August 1868, in HN 173.

42. HH to AH, 17 April 1866, in ESH 1929, 1:131–32 (first quote); HH to Verehrter Herr College [Ernst Mach], 31 December 1866 (second quote).

43. HH to AH, 17 April 1866.

44. HH to AH, 20 April 1866, in ESH 1929, 1:133–34 (quotes); Fox 1973; 1990, esp. 14–15.

45. HH to AH, 20 April 1866.

46. HH 1866, 1867, 1867a, 1867b; Louis-Emile Javal to HH, 1 January 1867, in HN 217; Georges Guéroult to HH, 10, 29 April; 3 July 1867, in HN 180; HH 1868; 1874, preface.

47. Emile Alglave to HH, 14 November 1867 (first quote); 23 April 1877, both in HN 10; Blaserna 1877, 1 (second quote); Pietro Blaserna to Leo Koenigsberger, May 1902, in LK, 1:364–68; Louis Pérard to HH, 3 November 1868, in HN 347; RC to HH, 17 October 1868, in HN 91; HH 1903, 1:i–iii, on ii; 1869c.

48. Laugel 1867, v, 3, 4 (first quote), 5, 6, 7 (second and third quotes), 28 (fourth quote), 67, 79, 80–81.

49. Gustave Bertrand 1868, 85–99 (quote on 94), 106, 108–9.

50. HH to the Secrétaire de l'Académie Royale de Médecine de Belgique, 19 January 1864, plus accompanying "Notice biographique: Hermann Ludwig Ferdinand Helmholtz," in Acadé-mie Royale de Médecine de Belgique, Palais des Académies, Brussels; GM to HH, 2 August 1860, in HN 295; Académie des Sciences 1869, 1870, 1870a.

51. JT to HH, fragment, n.d., in RI MS JT/1/T/503, TPRI (typescript); JT to HH, 16 January 1864, in HN 477; Tyndall 1904, xiv.

52. Vieweg to HH, 23 March 1863, 17 December 1864, both in HN 483; HH to Vieweg Firma, 22 March 1862; 5 December 1864, both in VV.

53. Friedrich Max Müller to HH, 19 March [1863] (first three quotes), in HN 319; Müller to JT, 14 March 1863, in Müller 1902, 1:287, 288, 290 (fourth quote); HH 1875, title page.

54. AJE to HH, 22 August 1863; 10 April 1864, both in HN 131; JT to AJE, 30 March 1863 (first four quotes), in HN 131 (copy); copies of letters from various publishers (Longmans, 8 April 1863; Baillière, 18 April 1863; W. Churchill, 22 May 1863; Walton & Maberby, 29 May 1863; Chapman & Hill, 20 June 1863; Robert Cock, 8 July 1863; Henry Bohn, 14 July 1863) to AJE, all in HN 131; HH to H. W. Acland, 15 March 1865 (fifth quote), in Bodleian Library, University of Oxford, MS Acland, d. 67, fol. 121R–122R; JT to Madame Mohl [AH], 22 March 1871, in HN 477.

55. Taylor 1873; AJE to HH, 5 July 1875 (first quote), in HN 131; Sedley Taylor to HH, 19 March, 30 April (second and third quotes) 1870, in HN 460; HH to Taylor, 3 May 1870 (fourth, fifth, sixth, and seventh quotes), in CUL, Add 6259/66; Taylor to HH, copy, 7 May 1870 (eighth quote), in CUL, Add 6255/45; Chappell 1876; HH to Taylor, 24 May 1875, in CUL, Add 6259/80; HH to Geehrtes Fräulein, 7 August 1878 (ninth and tenth quotes), in Uppsala Universitet, Wallers samling, Okat 651 F:1.

56. AJE to HH, 13, 18, 20 February; 1 December 1873; 12 August, 1 September, 10 October 1874; 10 March, 5 July 1875, all in HN 131; copy of James Nixon to AJE, 19 February 1873, ibid.; HH 1954, v, 430–556.

57. Ellis 1875, v–vi (first four quotes), vii; 1895, n.p. (fifth and sixth quotes).

58. William Pole to HH, 10 January 1879 (quotes), in HN 357; Pole 1879.

59. Beer 1992; Dale 1989, 68, 78, 88, 102–7, 113, 118, 131, 134, 302n8; Noble 1976, 47P, 49P, citing Haight 1954–78, vol. 4 (1955), 415–16, which in turn cites Lewes's journal (10 January 1868), in Beinecke Library, Yale University; George Henry Lewes, Diary 1869 (11 February, 18 August; 27, 29 September), Section VI, box 40; Diary 1871 (12–15 July; 29–30 November; 2, 4, 11, 21, 27 December), Section VI, box 42; Diary 1872 (3–5 June), Section VI, box 43; Diary 1874 (6–8 April; 17, 24–26, 28–31 August; 10–13, 15, 22 September; 31 December), Section VI, box 45, all in GEGHL; Eliot 1998, 135 (diary entry for 24 February 1869 [quote]); Shuttleworth 1984; Beer 1985, esp. 149–235; Otis 2001, 8, 81–120. The George Eliot-George Henry Lewes Library contains Helmholtz's *Handbuch*, *Tonempfindungen* (in French as well as in German), *Populäre wissenschaftliche Vorträge*, and the essays "Das Denken in der Medicin" and "Wirbelstürme und Gewitter mit einer Illustration." See Baker 1977, 91–92.

60. Darwin 1988a, 67; Charles Darwin to ?, 13 August 1878 (first quote), in Houghton Library, Harvard University, Autograph file, cited in Burkhardt et al. 1985, 493–94; Darwin 1988, 592 (second quote).

61. Rayleigh 1945, 2:viii (third and fourth quotes), 220 (fifth quote), 432 (sixth quote), and passim; Rayleigh 1968, esp. 50, 55, 80–81, 84–85, and passim; Ku, 2006; James Clerk Maxwell to Rayleigh, 26 May 1873, in Maxwell 1995, 856–57 (first quote on 856); 2002, 3:651, 655, 662–63 (second quote); HH 1878a, 1878d, 1878e; HH to EdBR, [1877], in Kirsten et al. 1986, 260; Eduard Schaer to HH, 18 April 1884, in HN 421.

62. Maxwell quoted in Campbell and Garnett 1969, 363–64 (first and second quotes); Ellis 1895, n.p. (third quote); Broadhouse, n.d. [1892?], v (fourth quote), 427–35 (fifth quote, for the questions); Steege 2012, 193–206.

63. Bruce 1973, 47–48, 50–51, 64, 73–74, 76, 93–95, 100, 103–4, 110, 123, 131, 135, 209, 251, 369; Israel 1998, 124, 130–31.

64. Robert Spice Testimony, Telephone Interference, Evidence for Thomas A. Edison, 298–308, in TAEM, reel 11:126–131 (quote on 129 [TAED, TI1018]); D. van Nostrand bill dated 15 November 1875, Thomas A. Edison—Bills & Receipts (D-75-02), in TAEM, reel 13:240 (TAED, D7502AAA, image 18); Edison 1991, 524–26, Doc. 599; Israel 1998, 109–10, 525n22; Baldwin 1995, 318–21; Kursell 2012, 183–85.

65. Loesser 1954, 420, 494–96, 511–14, 564–66; Dolge 1911–13, 1:302–7.

66. Steinway & Sons to HH, telegram, 11 March 1871, in HN 448; Mendel and Reissmann 1870–79, 9:422 (quotes); Dolge 1911–13, 1:303–6; Lieberman 1995, 60–62; and/but cf. Fostle 1995, 288–89.

67. Hiebert 2003, 2013; Hiebert and Hiebert 1994; Dolge 1911–13, 1:425–26, 2:207–13 (quote on 210).

68. Steege 2012, 223; Weber 1958; Christoph Braun 1992.

69. Dr. Schlemmer to HH, 10 November 1871, in HN 429; Surmann to HH, 19 June 1881, in HN 418; John Worthington to HH, 29 March 1889, in HN 515; C. Evans to HH, 3 March 1890, in HN 139; Karl Antolik to HH, 1 March 1892; 13 March 1894, in HN 13; Adolf Menk to HH, 20 September, 14 October 1892, in HN 309; J. A. Zahm to HH, 21 December 1892, in HN 518; Bell and Truesdell 1980, 673–74; Bell 1980; Beyer 1999, 108–23, 154–55, 177–91.

Chapter Thirteen

1. HBJ to HH, 12 April (first quote), 1 June 1863, both in HN 222; HH to EdBR, 26 February 1864, in Kirsten et al. 1986, 207–8 (other quotes on 207); HH 1862–63. The lectures that constitute HH 1862–63 appeared in *Medical Times and Gazette*, April 1864 (HH 1864; 1903a, xi–xii).

2. HH 1862–63, 189.

3. Ibid., 190–91 (first four quotes), 225–29 (fifth quote on 228, sixth and seventh quotes on 225–27).

4. HBJ to HH, 10 December 1863; 25 January 1864, both in HN 222.

5. HH to AH, 3, 9 September 1863, in ESH 1929, 1:107, 108–10, resp.

6. JT to HH, 16 January 1864, in HN 477; William Benjamin Carpenter to HH, 29 February 1864, in HN 85; HH to Carpenter, 6 March 1864, in Bodleian Library, University of Oxford, MS d.36, f.152; WT to HH, 16 March 1864 (quote), in HN 464; Margaret Thomson to HH, 25 March [1864], in Thompson 1910, 1:429 (misdated as 1863); HH to George Gabriel Stokes, 27 March 1864, in CUL, Add 7656 H88.

7. AH to HH, 11 March 1864, in ESH 1929, 1:111–12 (first quote on 111); HH to AH, 14, 19 March 1864, in *LK*, 2:49, 113–15, resp. (other quotes on 113–14); Michael Faraday to HH, 7 April [1864?], in HN 141; Bellmer 1999.

8. HH to AH, 17, 19 March 1864, in ESH 1929, 1:113, 114–15 (quotes).

9. HH to AH, 19 March 1864, 115 (first, third, and fourth quotes); Sheehan 1993, 890–96; HH to EdBR, 26 February 1864 (second quote).

10. HH to AH, 22, 25 March 1864, both in ESH 1929, 1:115–16 (first two quotes), 116–17 (third and fourth quotes), resp.

11. HH to AH, 25 March 1864 (quotes). See also HH to H. W. Acland, 15 March 1865, in Bodleian Library, University of Oxford, MS Acland, d. 67, fol. 121R–122R; on Oxford and scientific research, see Fox 1997, 664, 675–77; 2005, 37, 39, 41–42, 73; Gooday 2005; MacLeod 1972, 114–26.

12. HH to AH, 31 March, 2 April 1864, in ESH 1929, 1:117–18 (first quote), 118 (second quote); HH to AH, n.d., in Thompson 1910, 1:429–30.

13. HER to HH, 28 February, 23 March 1864, in HN 385; HH to HER, 6 March 1864, in RSC; HH to AH, 5 April 1864; 11 April 1866, in ESH 1929, 1:118–19 (quotes), 130–31, resp.; Fox 1997, 676–82, 686, 690; 2005, 42–79; Gooday 2005.

14. HH to AH, 5 April 1864; HH 1864, 385–88 (quotes on 385).

15. HH 1864, 415–18 (quotes on 415–17).

16. HH to AH, [circa 8–10], [15] April 1864, in ESH 1929, 1:119–20 (first quote), 120–22 (second quote on 120–21), resp.

17. HH 1864, 443–46 (quotes on 443, 444).

18. Ibid. (quotes on 446).

19. HH to AH, [15] April 1864, 120–21 (quotes); Mackenzie 1979, 339.

20. HH to AH, [15] April 1864, 121.

21. HH 1864, 471–74 (first quote on 471; second, third, and fourth quotes on 472).

22. Ibid. (quote on 473).

23. HH 1864a, 25.

24. Ibid., 25–42.

25. HH to AH, [15] April 1864, 121; Brock and MacLeod 1980, entry for 17 April 1864, folio 1668; Edward William Brayley to HH, 21 April 1864, in HN 65; AJE to HH, 10, 18 April 1864, in HN 131.

26. James Clerk Maxwell to HH, 12 April 1864 (second quote), in HN 305; HH to AH, 19 April 1864, in ESH 1929, 1:122–23 (other quotes).

27. HH 1864, 499–501 (quotes on 499).

28. Ibid., 527–30 (quotes on 529).

29. HH to EdBR, 15 May 1864, 209 (quotes); HH to EdBR, 5 June 1864, both in Kirsten et al. 1986, 212; Carlo Matteucci to HH, 17 May, 9 June 1864, in HN 302; HH to Matteucci, n.d. [circa late May or early June 1864], in LK, 2:55–56; on EdBR's rivalry and priority battle with Matteucci, see Finkelstein 2013, 57, 58, 66–69, 72–73, 78–82, 90, 104–8, 144, 145–46, 195.

30. HH to EdBR, 15 May 1864, 209 (first five quotes); HH to EdBR, 11 January 1866, in Kirsten et al. 1986, 219–21, on 220; EdBR to HH, 24 May 1864, ibid., 210–11 (sixth quote).

31. HH to EdBR, 26 February, 15 May 1864 (first four quotes); HH to Alexander William Williamson, [circa late 1873], printed in Tilden 1930, 238–40 (fifth quote on 239).

32. HH 1865. See also HH 1865b. For the invitations to Frankfurt am Main, see Gustav Adolf Spies to HH, 22 October, 9 November 1864; 25 January 1865; 13 October 1866; 13 July 1867, all in HN 415.

33. HH 1865, 233.

34. Ibid., 233 (first two quotes), 234 (third and fourth quotes), 235; Kutzbach 1979, 58–62; Darrigol 2005, 167–68.

35. HH 1865, 240–46 (quotes on 242–43), 248–51, 253–54, 260–61.

36. Ibid., 234–37, 242–43.

37. Ibid., 251, 253 (first quote), 253, 254, 261–63 (other quotes).

38. Hlaviwetz [?], Rochleder [?], Redtenbacher, E. Brücke, and Littrow, 7 May 1865, 479/1865, in Bibliothek, Österreichische Akademie der Wissenschaften, Vienna; EB to HH, 22 July 1864; 5 February, 12 November 1865; 1 January 1866, all in HN 74; Joseph Hyrthl to HH, 14 May 1865, in HN 210; Justus von Liebig to HH, 27 December 1866, in HN 275; CL to HH, 24 March, 2 April, 2 May, 25 December 1865, in HN 293; HH to CL, 30 April 1865, in LK, 2:59; HH to the Baden Interior Ministry (August Lamey), 17 April 1865, printed in Werner 1997, 197–98; Baumann 2002, 37–38.

39. HH to Eduard Vieweg, 5 May 1863; 10 October, 5 December 1864; 12 February, 29 April, 19 August, 13 November 1865, all in VV; Vieweg to HH, 29 November, 17 December 1864, both in HN 483; HH to Ew. Wohlgeboren (probably Georg Ernst Reimer), 4 October 1865, in WGA.

40. Gregory 1977; Bayertz 1985; Tiemann 1991; Daum 1998; HH 1865a, v–vi; Frank Turner 1993 (for "public scientist"); Cahan 1993.

41. Vieweg to HH, 13 December 1865, in HN 483.

42. EB to HH, 12 November 1865, in HN 74; CL to HH, 25 December 1865 (first quote), in HN 293; EdBR to HH, 8 January 1866, in Kirsten et al. 1986, 218–19, on 218; JT to HH, 29 November 1865 (second quote), in HN 477; Tyndall 1874, 84 (third quote); Joseph Henry to Felix Flügel, 20 March 1866, in Smithsonian Institution, Washington, DC, Archives, Record Unit 33, Office of the Secretary, Outgoing Correspondence 1865–1891, 3:259–62; see also

Flügel to Henry, 12 April, 24 June 1866, ibid., Record Unit 26, Office of the Secretary (Joseph Henry, Spencer F. Baird), Incoming Correspondence 1863–1879, box 5; C. F. Kroeh to Henry, 1 May 1871, ibid., box 39; Di Gregorio 1990, 219, 227, 234, 269, 366, 390.

43. Moore 2004, esp. 8–9; Brobjer 2004, esp. 21, 28; Mittasch 1952, 366; Schlechta and Anders 1962, 67–68, 108, 125; Nietzsche 1974a, 76; 1974, 77.

44. Sigmund Freud to Eduard Silberstein, 13 August 1874, in Boehlich 1990, 47–52, on 49; Sulloway 1979, 13–15, 65–66, 138, 139, 170, 235, 490.

45. Tyndall 1873, 377–404; HH to JT, 29 May 1866, in RI MS JT/1/H/43; HH to JT, 24 November 1866 (quote), RI MS JT/1/H/44, both in TPRI. Tyndall's and Helmholtz's lectures on ice and glaciers were published both in the *Philosophical Magazine* and elsewhere (together, in French). See Tyndall 1873a.

46. AH to HH, April 1866, in ESH 1929, 1:132–33; AH to Julius von Mohl, 1 January 1867, ibid., 139; AH to PM, 29 October 1867, ibid., 144–45, on 145; JT to HH, 20 November 1867, in HN 477; HH to JT, 29 May, 24 November 1866; HH to JT, 24 November 1867, all three in RI MS JT/1/H/45, TPRI (copy, in AH's hand); HH to the Vieweg Firma, 1 December 1865; 23 October, 12 November 1866; 28 October 1867; 3 February 1871; 21 October 1875, all in VV; HH 1874c; Tyndall 1867; see also Lightman 2015, esp. 407–9.

47. AH to JT, 18 November 1867, in RI MS JT/1/H/31, TPRI; JT to HH, 20 November 1867, in HN 477 (first quote); HH to JT, 24 November 1867 (second quote); JT to Gentlemen, 27 November 1867, both in VV, Archiv, 311 T; AH to JT, 17 March 1869 (third quote), in RI MS JT/1/H/33, TPRI; HH to Vieweg Firma, 26 December 1867; 1 January 1868; 29 January 1869; 19 October 1874, all in VV; Tyndall 1869.

48. HH to JT, 24 November 1867 (quote); JT to HH, 24 December [1867]; 13 January 1868, both in HN 477; AH to JT, 17 March 1869; HH to Vieweg, 26 December 1867; 11 April 1870, both in VV; JT to Gentlemen, 27 November 1867; JT to Dear Sir, 24 December 1867, both in VV, Archiv, 311 T; Tyndall 1870.

49. AH to JT, 17 November 1868, in RI MS JT/1/H/32, TPRI; HH to JT, 9 February 1869, in RI MS JT/1/H47, TPRI (in AH's hand); Tyndall 1870, with a preface by Helmholtz (HH 1870a); JT to HH, 10 October 1870 (quote), in HN 477.

50. EdBR to HH, 22 October 1868, in Kirsten et al. 1986, 231; AH to JT, 17 November 1868 (first quote); HH to JT, 9 February 1869 (second quote); Braun-Artaria 1899; 1918, 135; Weech 1906, 298; AH to Eduard Zeller, 26 November 1895, in ESH 1929, 2:116–17, on 117; Johannes, in Kremer 1990, 193–201, on 199–200; Häfner 1934, 384; [1962?], 112; HH to EdBR, 20 April 1868, in Kirsten et al. 1986, 228; Werner 1997, 65–66.

51. ISZ to ?, n.d., in *LK*, 2:119–20. See also AH to MM, 20 January 1868, in ESH 1929, 1:146.

Chapter Fourteen

1. HH to WT, 14 December 1862, in CUL, WT, Add 7342 H70; AH to RM, 6 December 1862, in ESH 1929, 1:106–7, on 106.

2. Engelking 1950; Lenoir 1993, 126–33.

3. Hörz 1997, 353n141; Stübler 1926, 293, 322; HH 1867e, 526, 715, 831, 834, 836.

4. Weve and ten Doesschate 1935, 96; Donders 1864, viii, 13, 17, 38, 369, 473, 556; HH 1867e, 13, 65, 108, 110, 119, 122, 153, 156, 162, 163, 163, 185–86, 188, 198, 210, 213, 740, 763, 776, 802, 830–31.

5. See esp. Lenoir 1993; R. Steven Turner 1993; Turner 1994; Lenoir 2006, 190–200.

6. HH 1862b, 420 (quote); R. Steven Turner 1993, 156–73.

7. HH 1863b, 352 (first quote), 354–56 (second quote on 354, third on 356), 358.

8. HH 1863c, 360–76 (quote on 366); Albrecht von Graefe to HH, 4 July 1863, in HN 172;

HH, "Allgemeine Resultate der Naturwissenschaften: Biologischer Theil," n.d. [1860s], in MS, Helmholtz-Gemeinschaft, Berlin, Archiv.

9. HH 1863c, 376–419 (first quote on 376, the others on 394–96). Hering successfully challenged Helmholtz's derivation of least orientation, forcing him later (in part three) to provide a new derivation. In 1919 Horace Lamb provided a new, definitive proof. (Lenoir 1993, 146n89.)

10. HH 1864b, 478 (first quote), 480 (the other quotes); 1864c, 1864e; Turner 1994.

11. HH 1864c, 22–23.

12. HH 1864d (quote on 924); 1866a.

13. HH 1864e, 427 (first quote), 448 (second quote), and passim; Hatfield 1990, 158–62, 172, 174–77, 204–5; Lenoir 1993, 110, 122–23; R. Steven Turner 1993, 178–79, 187.

14. HH 1865d, 1865c.

15. HH 1865e, 45–46 (quotes).

16. Luigi Cremona to HH, 20 December 1864, in HN 95.

17. "Titelverzeichniss" 1895, 607–16; AH to Julius von Mohl, 1 December 1864, in ESH 1929, 1:124–25 (first quote on 124); Bernstein 1906, 289 (second quote); Knapp 1902, 557.

18. HH to Jacques-Louis Soret, 23 December 1864, in VG, Ms. fr. 4175, f. 338–39 (first two quotes); HH to EdBR, 3 January, 13 February 1865, in Kirsten et al. 1986, 213–14, on 214 (third quote), 215 (fourth, fifth, and sixth quotes), resp.; R. Steven Turner 1993; esp. Turner 1994, 54–58 and passim; Baumann 2002, esp. 21–22, 33–34, 133, and passim.

19. AH to HH, 4 August 1865, in ESH 1929, 1:126–27 (first quote); HH to AH, 14 September 1865 [misdated as 1864], ibid., 129 (other quotes).

20. HH to AH, 17 September 1865, ibid., 129–30 (quote); Soret to HH, 19 September 1865, in HN 412; Rudolf Schelske to HH, 17 August [1865?], in HN 423; Doerr 1985, 4:365.

21. HH to EdBR, 11 January 1866, in Kirsten et al. 1986, 219–21 (quote on 219–20).

22. Weech 1877, 10–12; 1890, 587, 591–92, 595–96, 603–7; Craig 1978, 1–12; AH to PM, 12 July 1866, in ESH 1929, 1:136–37.

23. AH to MM, 1 June 1866, in ESH 1929, 1:135.

24. Ibid., 136 (first quote); AH to RM, 20 July; 1, 2 August 1866, ibid., 137–38, 138; Georg Quincke to HH, 12 July 1866, in HN 360; HH to EdBR, 2 October 1866, in Kirsten et al. 1986, 222–23 (second quote on 222); and see 302nn1–4; HH to WT, 24 November 1866 (third quote), in KP, H17.

25. HH to EdBR, 2 October 1866, 222 (quote); HH to Verehrter Freund, 14 August 1866, in Karl-Marx-Universität, Universitätsbibliothek, Briefsammlung, Autographensammlung Taut, Gelehrte.

26. HH to JT, 24 November 1866, in RI MS JT/1/H/44, TPRI 8 (first two quotes); HH to WT, 24 November 1866; HH to EdBR, 22 April 1867, in Kirsten et al. 1986, 223–24 (last two quotes).

27. Mueller 1998, 7–10; 2002, 28, 393–94; Ferdinand von Mueller to HH, New Year's Eve 1866 (quotes), in HN 316.

28. HH 1867e, v–vii (quotes on vi).

29. Ibid., 427. Among the many analyses of Helmholtz's theory of perception (and the nativism-empiricism controversy), see, for example and especially, Hatfield 1990; Lenoir 1993; R. Steven Turner 1993; Turner 1994; Lenoir 2006.

30. HH 1867e, 427.

31. Ibid., 430, 432 (quotes); Hatfield 1993, 547–51.

32. HH 1867e, 431 (first three quotes), 433 (fourth quote), 435 (fifth, sixth, and seventh quotes); Friedman 1997; McDonald 2002.

33. HH 1867e, 441–42.

34. Ibid., 442.

35. Ibid., 445.

36. Ibid., 446–47.

37. Ibid., 447 (first two quotes), 450 (third quote), 452 (fourth and fifth quotes).

38. Ibid., 453–54.

39. Ibid., 454–55.

40. Ibid., 207, 208, 428, 456 (quotes), 594, 805; Hatfield 1990.

41. HH 1867e, 441, 456 (quote).

42. HH 1903a, viii–ix.

43. HH 1867e.

44. Ibid., 796.

45. Ibid., 797 (quotes), 820.

46. Ibid., 797 (first quote), 804–5 (second and third quotes), 809.

47. Ibid., 805–19 (first quote on 812, second on 818, third on 819); for "rhetorical strategy," see R. Steven Turner 1993, esp. 154–56, 183–97; Turner 1994.

48. A. Classen to HH, 10 May 1867, in HN 90.

49. Wundt 1867, 326–27 (first two quotes); 1921, 161–77; EB to HH, 16 April 1863 (third quote), in HN 74.

50. Friedrich Ueberweg to HH, 14 July 1868, in HN 478.

51. Sigmund Freud to Eduard Silberstein, 13 June 1875, in Boehlich 1990, 117–18 (quote); Freud to Silberstein, 15, 28 June 1875, ibid., 119, 119–21, on 120; Schlechta and Anders 1962, 122–27, on 125; Krummel 1983, 81–82, 283, 615–16, 647–48; Treiber 1994; Schiemann 2014; Reuter 2014.

52. HH to EdBR, 22 April 1867, in Kirsten et al. 1986, 223–24, on 223; HH 1867, ii–iv; Louis-Emile Javal to HH, 13 August 1867, in HN 217.

53. AH to HH, 31 July 1867, in ESH 1929, 1:142.

54. HH to AH, 14 August 1867, ibid., 143–44 (first six and ninth quotes); Giraud-Teulon and Wecker 1868, 11–12, 20–21, 72, 196–97 (seventh and eighth quotes), 199; HH 1868j; Bernstein 1906, 291; Paul Broca to HH, 16 August 1867, in HN 69; AH to HH, 6 August 1867, in ESH 1929, 1:142–43; AH to PM, 27 August 1867, ibid., 144 (tenth quote).

55. HH 1867, 1867a, 1867b, 1870, 1873; Ivan Turgenev to Pauline Viardot, 27, 28 March 1868, in Turgenev 1972, 141–42 (quote on 141); Richard Liebreich to HH, 20 September 1862; 23 May 1863, both in HN 277; Etienne-Jules Marey to HH, 1 March 1868, in HN 298.

56. Laugel 1869, v, vi, 7 (first quote); Javal to HH, 15(?) August 1868 (second and third quotes), in HN 217.

57. Taine 1870, 1:223–35; 2:74, 91–92, 120 (first two quotes), 126–27, 220, 461, 481–83 (third quote).

58. HH 1962; Maxwell 1995, 341, 837, 935–36; 2002a, 140–41; George Henry Lewes, Journal XII (1866–70), Section VI, box 39, entry for 10 January 1868, 89; Notebook (February 1857), n.p., Section VI, box 49; Notes Physiological, Begun February 1873, Section VI, box 54; Diary 1877 (18–22, 24–25 March; 10 September; 14–15, 20 November), Section VI, box 48; Section VI, box 55, Section 23, Fragmentary Manuscripts; Section VI, box 56, Section 31, "Problems of Life and Mind"; Part of autograph MS of Vol. III, Part I: etc., in folder labeled "Colour Sensation"; George Henry Lewes Diary 1873 (15, 18 January; 2, 6–10, 15–17 April), MS Vault Eliot, Section VI, box 44, all in GEGHL; Darwin 1988b, 171 (first quote); 1988, 456 (second quote).

59. Chauncey Wright to Grace Norton, 29 July 1874, in Thayer 1878, 272–83 (quote on 281); Fisch 1964, 452, 464, 465.

60. James 1987, 273 (first quote), 331 (second quote); 1978, 44; 1983, 47–48 (third and fourth quotes); 1988, 154; 1987, 377 (fifth quote), 399 (sixth quote); 1890, 2:278 (seventh quote), 280 (eighth and ninth quotes); Boring 1950, 301–3 (tenth quote on 302, eleventh on 301).

61. Eduard Pflüger to HH, 8, 16 April; 18, 22 May; 21 July 1868, all in HN 351; HH 1867c, 1869a; HH to HK, 31 December 1869, in HKL; in English, HH 1873; HH 1873h.

62. Julius Bergmann to HH, 4 May 1869, and the accompanying editorial notice in *Philosophische Monatshefte*, Bd. 3 (1869), n.p., both in HN 38; HH 1869–70.

63. AH to PM, 1 April 1867, in ESH 1929, 1:140–41; Treitschke 1914–20, 1:188, 3:181–82; Dorpalen 1957, 131–32; Wolgast 1985a, 104–5.

64. HT to Wilhelm Wehrenpfenning, 20 November 1867, in Treitschke 1914–20, 3:192 (first quote); HT to Salomon Hirzel, 28 October 1867, ibid., 188 (second quote); HT to HH, 2 February 1868 (third quote), in HN 473.

65. HH 1868a; 1871b, 1–98, v; 1884, xii; Pastore 1973, 190–93; Schickore 2001; HH to EdBR, 20 April 1868, in Kirsten et al. 1986, 228 (quote).

66. HH, 1868a, 353–55 (quotes on 353).

67. Ibid., 267–70 (quote on 269), 275–76, 280–93.

68. HH to EdBR, 20 April 1868; EdBR to HH, 25 April 1868, in Kirsten et al. 1986, 229–30 (first two quotes on 229); FCD to HH, 18 May 1868 (third quote), in HN 116; HH to FCD, 26 May 1868, in *LK*, 2:87–88, on 87 (fourth, fifth, and sixth quotes).

69. HH to CL, 28 March 1869, quoted in *LK*, 2:162 (first three quotes); HH to Hochgeehrter Herr, 16 October 1877 (fourth, fifth, and sixth quotes), in SF, Sign.: Helmholtz 6.LL496; Turner 1982, 1994.

Chapter Fifteen

1. HH 1867f, 1870h, 1871e; Debru 2001.

2. Koenigsberger 1919, 100–101, 111–16; Mittag-Leffler 1923, 133–36, 148; Boehm 1958, esp. 306; Koblitz 1993, 89, 100.

3. Eduard Pflüger to HH, 22 May (quote), 11, 21 June; 21 July 1868, all in HN 351; RL to HH, 29 May; 1, 6 (two), 8, 14, 23, 27 June 1868, all in HN 281; Max Schultze to HH, 11, 22? June 1868, both in HN 440.

4. HH to RM, 2 June 1868, in ESH 1929, 1:147 (first quote); Helmholtz's similar remarks quoted ibid., 147–48; Pieper 1998a, 43–46; HH to Wilhelm Beseler, shortly after 28 May 1868, in *LK*, 2:115–16; Beseler to Minister von Mühler, 4 August 1868, ibid., 116–17 (other quotes); HH to EdBR, 7 April 1870, in Kirsten et al. 1986, 237–38; AH to ISZ, 30 December 1896, in ESH 1929, 2:130.

5. H. M. Tuckwell to HH, 22 June 1868, in HN 523; Henry W. Acland to HH, 22 July 1868 (quote), in HN 4; Fox 1997, 643, 646–47, 650–53, 659–64.

6. Binz 1869, with Helmholtz's letter on 100–102 (all quotes except last on 100–101 [HH 1869e]); Carl Binz to HH, 27 August 1867; 2 March 1869; 11 October 1871, all in HN 49; Binz 1874; EdBR 1912, 2:542; WT to HH, n.d. [sometime between October 1871 and June 1872], in HN 464; HH to WT, 17 June 1872 (last quote), in EUL.

7. Pieper 1998a, 68–78.

8. Ibid., 9–42 (first quote on 33–34); HH to RL, 7 June 1868 (second and third quotes), in Mathematisches Institut der Universität Bonn, Bibliothek; also printed in Scharlau 1986, 124–25.

9. Statement from Heidelberg's medical students (19 July 1868), in HN 457; statement from Zeller in re Helmholtz (28 July 1868); Jolly to the Interior Ministry (29 July 1868), all in RKUAH, Personalakte Helmholtz, in BGLA, and printed in Werner 1997, 204–5; Pflüger to HH, 21 July, 9 August 1868, in HN 351; Pieper 1998a, 56; Zeller 1908, 185; Schultze to HH, 7 October; 24, 27 November 1868, all in HN 440; EdBR to HH, 22 October 1868, in Kirsten et al. 1986, 231 (quote).

10. *LK*, 2:117; Pieper 1998a, 82–83; "November 11, 1868" 1868, 103; Henle and Meissner to the Secretär der Königl. Sozietät Herrn Geh. Obermedicinalrath Dr. Wöhler, 25 October

1868, Bl. 255; "Wahlen am 7. November 1868," Bl. 259; Königliches Universitäts-Curatorium an die Königliche Societät der Wissenschaften, Göttingen, 10 November 1868, Bl. 260, in *Pers* 12, all in Archiv der Akademie der Wissenschaften, Göttingen; HH to Hochgeehrter Herr, 20 December 1868, in Niedersächsische Staats- und Universitätsbibliothek Göttingen, Handschriftenabteilung, 4° Cod. Ms. Hist.lit. 116:IV, Bl. 187r +v (first quote); Francesco Brioschi to HH, 29 November 1869 (second quote), in HN 68.

11. BGLA to Ministry of the Interior, 28 December 1868, in Werner 1997, 205; ibid., 135; Mayor of Heidelberg to HH, 21 January 1869, in Stadtarchiv Heidelberg, no. 20, Fasc. 6, printed ibid., 201; Pieper 1998a, 57–68, 79–82; RC to HH, 12 January 1869, in HN 91; HH to CL, quoted in ESH 1929, 1:148.

12. EdBR to HH, 9 January 1869, in Kirsten et al. 1986, 231–32 (quote on 231).

13. HH to EdBR, 14 January 1869, ibid., 232–35 (first quote on 232); Pieper 1998a; HH to RL, 2, 5, 7 June 1868, in Mathematisches Institut der Universität Bonn, Bibliothek, and printed in Scharlau 1986, 122–23, 123–24, 124–25, resp. (second quote on 125).

14. HH to EdBR, 14 January 1869, 232–33 (first two quotes); HH to RL, 2 June 1868; 4 January 1869, in Mathematisches Institut der Universität Bonn, Bibliothek, and printed in Scharlau 1986, 122–23, 126–28, resp.; HH to RL, 4 January 1869, 126; HH to RL, 14 January 1869, in Universitäts- und Landesbibliothek Bonn, Abteilung Handschriften und Rara, Nachlass Rudolf Lipschitz, Kapsel 12, in Hörz 1995, 33–34 (third quote).

15. HH to CL, 27 January 1869, in *LK*, 2:118–19.

16. HH to EdBR, 14 January 1869, 233.

17. Ibid. (second quote); Pieper 1998a, 60–61 (first quote on 60).

18. HH to EdBR, 14 January 1869, 233–35; Pieper 1998a, 61–66; for Beseler on Helmholtz, Pieper 1998, 94–97 (quotes on 96–97).

19. HH to EdBR, 14 January 1869, 234–35 (quotes); Pieper 1998a, 64–68. See also *LK*, 2:118, citing a letter/telegram from Jolly to HH, 2 January 1869.

20. Pieper 1998a, 70–75; HH to EdBR, 14 January 1869, 235.

21. HH to JT, 9 February 1869 (first quote), in RI MS JT/1/H/47, TPRI (copy in AH's hand); HH to HK, 5 March 1869 (second quote), in HKL; AH to MM, 8 January 1869, in ESH 1929, 1:148–49 (third quote).

22. Sablik 1989, 83–86; HH to HK, 31 December 1869, in HKL (first three quotes); AH to MM, 17 December 1869, in ESH 1929, 1:152; Brock and MacLeod 1980, entry for 26 July 1869 (fourth quote), folio 1851; Koenigsberger 1919, 111; Baumann 2002, 53–55.

23. AH to MM, 25 July 1869, in ESH 1929, 1:149–50. See also Koenigsberger 1919, 100.

24. Cahan 2012.

25. PGT to HH, 2 February (first and second quotes), 1 March (third quote), 27 March 1867; 4 March (fourth quote); 13, 25 July 1868, all in HN 459; Knott 1911, 208–17; RC to JT, 11 May, 13 June 1864, in Eidgenössische Technische Hochschule, Zurich, Hs 227:5–163 (copies at the Royal Institution of Great Britain, London).

26. HH 1882, 71–4 (first five quotes); Gross 1891, 26; 1898, 155–58; Gross thought Helmholtz's defense of Mayer was weak; Tait 1868, iii–viii, on v–vii (sixth and seventh quotes).

27. RC to JT, 6 January 1869; 18 June, 11 December 1868; 22 September 1872 (quotes); 30 December 1871; 25 June 1873, in Eidgenössische Technische Hochschule, Zurich, Hs 227:5–163 (copies at the Royal Institution of Great Britain, London).

28. HH to WT, 4 November 1871; 17 June 1872; 3 August [1872?] (first quote), all in EUL; for source quotes, Cahan 2012a, 128–29 (second and third quotes).

29. Tait 1877, xiii–xviii, 68 (quotes).

30. HH to RL, 18 February 1868, in Mathematisches Institut der Universität Bonn, Bibliothek, partially printed in Scharlau 1986, 121–22.

31. HH 1868 f. The date of publication given in Helmholtz's *WA* (1866) is a misprint and should read 1868; the Verein so corrected it in 1871 (Volkert 1993). There is an extant draft

manuscript (dated 1868) of Helmholtz's concerning the foundations of geometry ("Prinzipien der Naturforschung," in HN 700/1). Koenigsberger printed one long manuscript (HN 705 [4.17]), which he dated as sometime before 1847, and two fragments, which he dated prior to 1868; he argued that these showed Helmholtz's long-standing interests in the related topics of geometry, arithmetic, and mechanics. See *LK*, 2:125–62.

32. HH 1868f, 610–11 (quote on 610); cf. HH 1870e, 130. See Richards 1977, esp. 235–41; DiSalle 1993; Hyder 2001; 2009, 105–61.

33. HH 1868f, 611–16 (quotes on 611–14).

34. HH to Ernst Christian Julius Schering, 21 April (first three quotes), 18 May 1868, in *LK*, 2:138–39; Schering to HH, 29 April (fourth quote), 24 May (fifth quote), 31 May (sixth quote) 1868; n.d., all in HN 425; cf. HH 1868g.

35. HH 1868g, 618–19.

36. Ibid., 620–21, 637–39 (quotes).

37. HH 1868h [or 1869?], 1868i [or 1869?]; Boi, Giacardi, and Tazzioli 1998; Voelke 2005, 59–72, 223–50; Richards 1977; Torretti 1978, esp. 155–62, 206; Scholz 1980, 113–23; Gray 1989; DiSalle 1993; Gray 2008, esp. 44–58.

38. Jules Hoüel to Luigi Cremona, 3 February 1869, in Cremona 1992, 81–84 (first quote on 82); Eugenio Beltrami to Hoüel, 8 January, 14 February 1869, both in Boi, Giacardi, and Tazzioli 1998, 71–74, on 72, 74–78, resp.; also ibid., 27–28; Beltrami to HH, 24 April 1869 (second, third, and fourth quotes), in HN 36.

39. HH 1868f, 617, for the addendum, which is there marked 1868 but which first appeared in the Heidelberg Verein's transactions on 30 April 1869 (ibid., 610); Beltrami to HH, 16 May 1869; 18 October 1874, both in HN 36; Boi, Giacardi, and Tazzioli 1998, 27–28, and printed on 205–7; Hoüel to Cremona, 17 July 1870, in Cremona 1992, 85–86, on 85.

40. MacLeod 1972, 132, 137–38, 147–48; Beer 2004, esp. 185–86; Charles Edward Appleton to HH, 11 September, 27 November 18[??], both in HN 14; HH 1870e, 1870g.

41. HH 1870e.

42. Jevons 1871; Richards 1988, 85–88.

43. HH 1872c; Richards 1988, 88–90.

44. Richards 1988, 55–57, 78, 84–86, 91–95, 113, 117, 152, 157; Voelke 2005, 6–7, 277–320; Gray 2013, 38–40.

45. HH 1876, v–vii, on vi; 1903, 2:v–ix, on v; 1870f, 1; 1876, 20.

46. HH 1870f, 3.

47. Ibid., 3–5.

48. Ibid., 6–7.

49. Ibid., 8.

50. Ibid., 9–10.

51. Ibid., 11–19 (quotes on 15–17).

52. Ibid., 19; cf. 21.

53. Ibid., 23 (first quote), 30–31 (second and third quotes on 30). Cf. Friedman 2000.

54. Henderson 1983, passim, esp. 10, 12–17, 26, 51–54, 132, 143, 154n, 200, 225.

55. Abbott 2002, esp. xix–xxii; Abbott 2010, esp. 3–4, 99, 121, 157, 189, 203, 241–42, 257–61; James 1987, 370 (quote).

56. Plücker 1868–69, 1:iii–iv, 2:226 ff.

57. See, for example, Torretti 1978, 154–55, 161, 169, 171–85, 393–94, 404; Richards 1988, 204, 208–14, 221; Richards 1988a; Carrier 1994; Hawkins 2000, 103, 104, 111–19, 124–29, 186, 326, 333, 345–47, 434–40; Heinzmann 2001; Stubhaug 2002, esp. (on Lie) 340–41, 355, 372, 380–81, 384, 385, 396, 423–24; Friedman 2002, 197–99; Voelke 2005, 239–382; Darrigol 2007a; Gray 2013, 13–14, 39–47, 56, 76–82, 85–100, 217–18, 223, 235, 238, 245–46.

58. Friedman 2002, 2009.

59. Hermann Hankel to HH, 23 September 1861, in HN 187; WT to George Gabriel Stokes,

6 October 1859; [February? 1867], both in Wilson 1990, 1:248–49, 331–32, resp.; Epple 1999, 94–95, 98–109, 115–18, 120–23, 126, 130, 149; Gray 2008, 237–39.

60. HH and Piotrowski 1860; Stokes to WT, 22 February 1862, in Wilson 1990, 1:283–85; James Clerk Maxwell to PGT, 1 December 1873, in Maxwell 1995, 944–48, on 944; Maxwell 2002a, 621, 806; PGT to HH, 22 April, 9 May 1867, in HN 459; Thompson 1910, 1:510–12; Knott 1911, 68, 105, 177; Silliman 1963; Smith and Wise 1989, 379–80, 412–25, 427, 431; Kragh 2002, esp. 33–43; Darrigol 2005, 155–56, 191–92.

61. PGT to HH, 13 July, 3 September (quotes) 1868, in HN 459; WT to HH, 24 July 1868, in HN 464; HH to WT, 10 September 1868, in KP, H18.

62. HH to ?, 3 October 1868 (first quote), in L'Institut de France, Bibliothèque, Paris, MS 4225, No. 3; Joseph Bertrand 1868; HH 1868c; Bertrand 1868a; HH 1868c; Bertrand 1868b; 1868c; Adhémar Barré de Saint-Venant to HH, 27 September 1868, in HN 480; Maxwell to PGT, 18 July 1868, in Maxwell 1995, 391–93, on 391; Darrigol 2005, 156–58; Académie des Sciences 1869, 1870, 1870a; JT to HH, 15 March 1870, in HN 477; WT to HH, 23 January 1870 (second quote), in HN 464; George Eliot Holograph Notebook, MS 707, ff. 89–90, cited in Baker 1976, 1:15–16 (third quote).

63. HH 1863a; 1869f, 224; 1868e; Darrigol 2005, 158–66.

64. HH 1894, 368.

65. Maxwell 1995, 22, 124, 241, 321, 391, 398–404, 426, 432, 434, 439–40, 445, 446–48, 530, 545, 593, 596, 778, 944.

66. HH to JT, 18 January 1868, in RI MS JT/1/H/46, TPRI (copy, in AH's hand).

67. HH 1869g, 1869h, 1870c, 1870d; Carl Wilhelm Borchardt to HH, 14, 19 March 1870, in HN 58.

68. The present account largely follows Darrigol 2000, 221–30, 262–63, 412–19; Buchwald 1985, 177–86; 1994, 7–16, 356–57, 375–88. Maxwell called Helmholtz 1870d "a masterly paper." See Maxwell 1995, 355, 596, 686–87, 773 (quote); 2002a, 883.

Chapter Sixteen

1. Querner 1970, 14–15, 17, and passim; EB to EdBR, 20 June 1869, in Brücke et al. 1978–81, 1:155–56, on 156; Mayer in Weyrauch 1893, 477; AH to JT, 27 February 1872 (quote), in RI MS JT/1/H/35, TPRI; AH to PM, 19 September 1869, in ESH 1929, 1:151.

2. AH to PM, 19 September 1869 (first quote); Querner 1970, 18; Daum, Ebner, and Enzenberg 1869, 36 (second quote); Wilhelm Foerster to HH, 13 May 1869 (third, fourth, and fifth quotes), in HN 151; HH to Foerster, 18 May 1869 (sixth quote), in Uppsala Universitet, Wallers samling, Okat 651 F:1; HH 1869, 369 (seventh quote), 374; 1871b, 181–211; Lampe and Querner 1972, 47–48; HH 1903a, xii–xiii; 1869d; Leopold Pfaundler in LK, 1:91.

3. HH 1869, 370–72 (quotes on 372).

4. Ibid., 372.

5. Ibid., 373.

6. Ibid., 374–76 (quotes). Daston and Galison 2007, 27–28, 31, 34–35, 42–44, 206, 214, 230, 242, 253–54, argue that Helmholtz's address (but also writings by Bernard and Huxley) marks a key moment in the language and history of objectivity.

7. HH 1869, 376–78.

8. Ibid., 378–80 (quotes); Daum, Ebner, and Enzenberg 1869, 40; Julius Robert Mayer to his wife, 18 September 1869, in Weyrauch 1893, 445–46.

9. HH 1869, 381–84.

10. Ibid., 384.

11. Ibid., 385–87.

12. Ibid., 387–88.

13. Ibid., 388–94 (quotes on 392–93).

14. Ibid., 394–95.

15. Ibid., 395–96.

16. Ibid., 373, 396–98 (quotes); Daum, Ebner, and Enzenberg 1869, 40; Jurkowitz 2002.

17. Querner 1970, 15, 20–27; Sudhoff 1922, 63; Julius Robert Mayer's speech, in Daum, Ebner, and Enzenberg 1869, 40–44 (first three quotes on 43–44); Weyrauch 1893, 441–46 (final quote on 445–46); Caneva 1993, 331.

18. Querner 1970, 27–28, 30; Steif 2003, 91–93; Daum, Ebner, and Enzenberg 1869, 6.

19. Querner 1970, 22–27; Sudhoff 1922, 63.

20. Julius August Isaac Jolly to HH, 2 January 1869, in HN 221; CL to HH, 31 January 1869 (quote), in HN 293; GM to HH, 19 October 1869, in HN 295.

21. Cahan 1999, 286 (for source documentation).

22. EdBR to HH, 4 April 1870, in Kirsten et al. 1986, 236–37 (quotes on 236); EdBR to HH, 15 May 1870, ibid., 238–39; Jungnickel and McCormmach 1986, 2:19–21.

23. HH to EdBR, 7 April 1870, in Kirsten et al. 1986, 237–38 (quote on 237); HH to Carl Wilhelm Borchardt, 7 May 1870, in *LK*, 2:179; Werner Siemens to Karl Siemens, 28 April 1870, in Siemens 1916, 2:320–21.

24. AH to PM, 18 April 1870, in ESH 1929, 1:153; Jolly to HH, 1 May 1870 (quote), in HN 221.

25. Philosophical Faculty of the University of Berlin to the Prussian Minister of Culture, Heinrich von Mühler, n.d. [May 1870], in *LK*, 2:179–80 (first two quotes); EdBR to HH, 15 May 1870 (last two quotes).

26. HH to EdBR, 17 May 1870, in Kirsten et al. 1986, 239–40.

27. Henry W. Acland to HH, 20 April 1869 (quotes), in HN 4; Fox 2005; Gooday 2005; Hannabuss 2000, 447–48.

28. Fox 1997, 659; Browne 1995–2002, 2:337–39; Simon Bailey (Keeper of the University Archives, Bodleian Library, Oxford) to author, 11 April 2016.

29. EdBR to Jeannette du Bois-Reymond, 10 June 1870, in Dep. 5, K. 11, no. 5.

30. Ibid.; Koenigsberger 1919, 117–18.

31. EdBR to Jeannette du Bois-Reymond, 10 June 1870.

32. HH to EdBR, 12 June 1870, in GSPK, Rep. 76 Va, Sekt. 2, Tit. IV, no. 47, Bd. 11, Bl. 78/78v, printed in Werner 1997, 206 (first quote); HH to Hochgeehrter Herr, 21 June 1870 (second quote), in Stadt Karlsruhe, Stadtarchiv; EdBR to HH, 14 June 1870, in Kirsten et al. 1986, 240–41 (third, fourth, fifth, and sixth quotes); Minister of Culture Heinrich von Mühler to the Minister of Finance, Otto von Camphausen, 14 June 1870, in *LK*, 2:182 (seventh quote); Ernst Curtius to Heinrich von Mühler, 17 June 1870, ibid., 182–83 (eighth quote).

33. EdBR to Jeannette du Bois-Reymond, 22 June 1870, in Dep. 5, K. 11, no. 5; EdBR to HH, 23, 27 June 1870, both in Kirsten et al. 1986, 241–42 (quote), 243–44; Justus Olshausen to HH, 8 January 1871, in HN 339.

34. HH to EdBR, 25 June 1870, in Kirsten et al. 1986, 242–43 (quote); Karl Weierstrass to HH, 8 December 1870, in HN 504; Biermann 1973, 113n149.

35. AH to PM, 26 June 1870, in ESH 1929, 1:153; Mühler to HH, 28 June 1870, in *LK*, 2:183; HH to EdBR, 3 July 1870, in Kirsten et al. 1986, 244; Kirsten et al. 1986, 307n2.

36. EdBR to HH, 15 July 1870, in Kirsten et al. 1986, 245.

37. HH to AH, 11 July (first quote), 12, 13 July (second quote), 18 July 1870; AH to PM, 20 July 1870, all in ESH 1929, 1:154–55; EdBR 1912, 1:395; Koenigsberger 1919, 118.

38. Weech 1877, 47–49; 1890, 622–23; Stiefel 1977, 1:290–97; Craig 1978, 6–7, 14, 19, 21, 27–28, 33; Clark 2007, 393, 395–96, 498, 545–46.

39. Weech 1906, 298; HH, "Die Aussichts-Commission für die Reserve-Lazarethe," 10 August 1870, in Stadtarchiv Heidelberg; HH to EdBR, 17 October 1870, in Kirsten et al. 1986, 247 (first quote); AH to PM, 10 August, 18 September 1870, in ESH 1929, 1:155–56; Nikolaus Friedreich to HH, 20 September 1870 (second quote), in HN 160; Zeller 1908, 185–87; HH to

HK, 17 November 1870 (third quote), in HKL; AH to PM, 3 October 1870, in ESH 1929, 1:156–57, on 156; Hintzelmann 1886, 62; Karl Christian Bruhns to HH, 26 August 1870, in HN 75.

40. WT to HH, 29 July 1870, in HN 464; HH to WT, copy, 21 August 1870 (quotes), in CUL, WTP, Add 7342 H72; the letter (in English) contains two annotations: one that it was sent to Thomson in Glasgow and postmarked "Heidelberg 21 August. 10–12V 3N Glasgow Au. 24 70," and another, in Thomson's hand, reading, "Extract published in Glasgow Herald Frid. Sep. 9/70," which in fact it was (see [HH] 1870b); HH to EdBR, 14 February 1871, in Kirsten et al. 1986, 251–52, on 252; HH to Otto Helmholtz, 22 February 1871, in Smithsonian Institution Libraries, Washington, Dibner Manuscript Collection, Helmholtz Papers, MSS 683A; Häfner 1934, 384; 1962?, 112.

41. HH to WT, 21 August 1870.

42. [HH], 1870b; Mommsen et al. 1871; WT to HH, 8 September 1870 (quotes), in HN 464.

43. EdBR to HH, 13 October 1870, in Kirsten et al. 1986, 245–46 (first two quotes); EdBR 1912, 1:418 (third quote); HH to EdBR, 17 October 1870 (fourth, fifth, and sixth quotes); Finkelstein 2013, 211–19.

44. HH to HK, 17 November 1870 (quote), in HKL; Weech 1890, 625–27; Stiefel 1977, 1:296–97; CV, Personalakt Helmholtz, in SF, Signatur: Helmholtz SLL 494.

45. AH to PM, 3 October 1870, in ESH 1929, 1:156–57; EdBR to Jeannette du Bois–Reymond, 23 October 1870, in Dep. 5, K. 11, no. 5; EdBR to HH, 13 October 1870; HH to EdBR, 17 October 1870; Weierstrass to HH, 8 December 1870 (quote), in HN 504.

46. HH to EdBR, 11 December 1870, in Kirsten et al. 1986, 249 (quotes); Minister of Culture Mühler to HH, 16 December 1870, in *LK*, 2:186; HH to EdBR, 20 December 1870, in Kirsten et al. 1986, 250; EdBR 1912, 2:565.

47. HH to PM, 25 December 1870, in ESH 1929, 1:157 (first two quotes); HH to PM, 21 January 1879, in DM: 1953/13 (third quote). Helmholtz's original German reads: "Die vermeintliche Gott blieb mir fremd, aber den Menschen verstehe ich."

48. AH to PM, 6 January 1871, in ESH 1929, 1:158 (quotes); Borchardt to HH, 4 January 1871, in HN 58.

49. EdBR to CL, 18 January 1871, in du Bois–Reymond 1927, 164; HH to Jolly, 1 January 1871, in BGLA: 76/9939, Bl. 15/15v; HH to Hochgeehrter Herr Legationsrat, in BGLA: 60/221, Bl. 3, printed in Werner 1997, 206–7, 207, resp.

50. MacLeod 1972; WT to HH, 28 January 1871 (quotes), in HN 464; Olshausen to HH, 5 February 1871, in HN 339; AH to PM, 9 February 1871, in ESH 1929, 1:159.

51. EdBR to HH, 11 February 1871, in Kirsten et al. 1986, 250–51, on 250; ibid., 308–9n1; HH to EdBR, 14 February 1871.

52. HH to EdBR, 14 February 1871; AH to PM, 9, 15 February 1871, in ESH 1929, 1:159, 160 (quote); HH 1871.

53. HH 1862–63, 229; 1903b, vi; 1871, 55 (quotes).

54. HH 1871, 56–57.

55. Ibid., 62–64, 66.

56. Ibid., 70–76.

57. Ibid., 77–78 (quotes on 77).

58. Ibid., 78.

59. Ibid., 79–82 (quotes on 78–80).

60. Ibid., 82–89 (first quote on 82, second and third on 83, fourth on 88).

61. Ibid., 89.

62. Ibid.

63. Crowe 1986, 400–405; Thomson 1889–94, esp. 2:200–202; Smith and Wise 1989, 638–41; Pulte 2008, esp. 124–29.

64. HH 1871, 89–91.

65. AH to PM, 15 February 1871 (quote); Bernstein 1906, 291; Craig 1978, 214–15.

66. Burrow 2000, 40–41 and passim; HH 1903b, vi.

67. AH to PM, 15 February 1871; EdBR 1912, 1:421 (first three quotes), 429; HH to EdBR, 17 March 1871, in Kirsten et al. 1986, 252–53 (fourth quote); Harnack 1970, 3:123; Charles Pickering Putnam to James Jackson Putnam, 3 April [1871], in Francis A. Countway Library of Medicine, H MS c4.2, fd. 1.; ESH 1929, 1:160.

68. Zobeltitz 1891, 770; HH to Fr. Schöll (Exprorector, University of Heidelberg), 12 October 1891, in RKUHA, PA Hermann von Helmholtz; [Anna von Helmholtz], "16. Febr. 1897. Mein Testament," typescript, in SF, Sign.: Helmholtz 6.LL496 (original held, as of 29 December 1936, by Dr. Hermann von Siemens); Helmholtz in Kussmaul 1902, 38.

69. Kussmaul 1903, 57–58; Riese 1977, 24, 134, 218–19; Wolgast 1985, 17; 1985a, 87–107.

Chapter Seventeen

1. HH to EdBR, 17 March 1871, in Kirsten et al. 1986, 252–53; AH to ISZ, 24 June 1871, in ESH 1929, 1:164–65 (quote); Mohl 1922, 1:20.

2. HH to HK, 5 January 1873 (first quote); 7 October 1877 (second quote), both in HKL.

3. AH to [PM?], 5 April 1871, in ESH 1929, 1:163; AH to PM, 23 April, 7 December 1871, ibid., 163, 175–76; Brandes 1989, 431–32.

4. AH to PM, 6 March 1872; 10 September 1873, in ESH 1929, 1:177–78, on 177, 172; Wachsmuth 1900, 18–19; Planck 1929–1930.

5. WT to HH, 29 October 1871, in HN 464; HH to WT, 4 November 1871, in EUL; Carl Wilhelm Borchardt to HH, 23 April 1873, in HN 58; HH to Georg von Liebig, 6 July 1873, in BSBMH, Sign.: Ana 377, II, B, Helmholtz, Hermann von (1B); EB to HH, 9 January 1874, in HN 74; Nikolaus Friedrich to HH, 29 May, 31 July 1874, both in HN 160; HH to Jacques-Louis Soret, 7 May 1875; 7 September 1876, both in VG, Ms. fr. 4175, f. 347, f. 349–50, resp.; HH to Franz Boll, 1 November 1876, in Belloni 1982, 131–32; AH to HH, 7 March 1877; AH to PM, 21 March 1877, both in ESH 1929, 1:212–13; AH to Heidelberger friends, probably shortly after 23 May 1877, in ESH 1929, 1:213–14; ESH 1929, 1:176; HH to HK, 5 January 1873; 21 July 1877, both in HKL; Kremer 1990, 53n27; Johannes, in Kremer 1990, 193–201, on 199–201; Melms 1957, 12, 24–25; AH to PM, 9 September 1875; HH to RH, n.d. [ca. 1878–79], both in ESH 1929, 1:199, 233, resp.; L[umme]r 1889, 567; Mohl 1902, 200; Roscoe 1906, 95; AH to HH, 8 October 1875, in ESH 1929, 1:199–200.

6. Johann Heinrich Meidinger to HH, 15 March 1870, in HN 306; HH to HK, 5 January 1873, in HKL; AH to PM, 10 September 1873, in ESH 1929, 1:172; Wilhelm von Beetz to HH, 8 July 1873; 12 March 1874; 30 May 1878 (first quote), all in HN 34; HH to Soret, 7 May 1875; HH to Soret, 7 September 1876; Häfner 1934, 384; [1962?], 112; HH to Beetz, 24 June 1878 (second and third quotes), in DM, 1932–17/24; Bunsen 1932, 41.

7. AH to HH, 8 October 1875 (first quote); AH to Franz, 8 January 1876, in ESH 1929, 1:202–3 (second and third quotes on 202); AH to a friend in Heidelberg, 4 April 1876, ibid., 203; AH to HH, 22 February, 7 March 1877, ibid., 212, 212–13; AH to PM, 21 March 1877, ibid., 213; HH to TM, 7 March 1877, in Hörz 1997, 386.

8. Steinway & Sons to HH, telegram, 11 March 1871 (first quote), in HN 448; HH to Hochgeehrter Herr [Steinway], 9 June 1871 (second quote), in SP; EdBR 1912, 2:541; HH to AH, 1 August 1873, in ESH 1929, 1:186 (third quote); HH to Gentlemen [the Steinway firm], 13 August 1873 (fourth quote), copy, in SP (transcribed from the 1888 catalog, 15; the original text was presumably in German).

9. Steinway catalog (1874) (quotes), in Henry Z. Steinway Private Collection; Lieberman 1995, 60–62; extract from HH to the Steinway firm, 16 March 1885, copy, in SP; Steinway Promotional Materials, Hermann von Helmholtz, no date, in SP; Fostle 1995, 288–89.

10. AH to HH, 28 August 1877 (first quote); HH to AH, 1 September 1877 (second quote),

both in ESH 1929, 1:221, 222, resp.; Wachsmuth 1900, 16–17; Weech 1906, 296; Ludwig Boltz-mann to Leo Koenigsberger, cited in Blackmore 1995, 98 (third quote); HH to RM, 16 April 1872, in RKUHA, Heid. Hs. 3715,7; HH to PM, 1 September 1873 (fourth and fifth quotes), in DM, 1951/25C1; Weber 1886, 256; Schulze 1886, 69, 73, 75, 83; Mohl 1902, 1:240, 2:130, 200; ESH 1929, 1:81.

11. McClelland 2017.

12. Lexis 1893, 1:118; Eulenburg 1904, 260, 261, 263, 304, 306; Lenz 1910–18, 2:358; Titze 1995, 80–81, 72–77.

13. Karl Pearson to William Herrick Macaulay, 29 September [1879], in Karl Pearson Papers 920, Manuscripts Room, University College London; Pearson to Robert Parker, 19 October [1879] (quote), in Pearson Papers 922, ibid.; Porter 2004, 59–60, 64, 67, 219.

14. AH to PM, 30 January 1872, in ESH 1929, 1:176 (first two quotes); HH to TM, 3 March 1872 (third quote), in HN 315.

15. HH to TM, 4, 19 March 1872, in TMN; AH to PM, 6 March 1872, in ESH 1929, 1:177–78, on 177; Eduard Zeller to HH, 10 March, 18 April, 11 June 1872, in HN 520; Zeller 1908, 188–89, 192–93.

16. HT to HH, 22 February, 24 March, 21 October 1873, all in HN 473; HH to HT, 23 February 1873 (quotes), typescript copy in SF, Sign.: Helmholtz 6.LL496; HT to Salomon Hirzel, 3 August 1872; HT to Franz Overbeck, 28 October 1873, both in Treitschke 1914–20, 3:352–53, on 353; 375–79, on 378–79, resp.; AH to Frau Prof. Otto Becker, 2 January 1873, in ESH 1929, 1:183–84; Treitschke 1914–20, 3:181–82; Dorpalen 1957, 192.

17. HH to TM, 12, 15 December 1873, both in TMN; TM to HH, 14 December 1873, in HN 315.

18. McClelland 2017.

19. Unless otherwise noted, the following three paragraphs draw on Cahan 1999, esp. 288–304, which provides primary source documentation.

20. On this last point, see Daston 1999, 74–75, 80, 83–84.

21. HH 1871a, 35 (quote), 40; Hofmann 1870; August Wilhelm von Hofmann to HH, 8 November 1870, in HN 206.

22. HH 1871a, 36–37, 39 (quotes).

23. Ibid., 37–38 (first two quotes on 37), 50 (third quote).

24. Ibid., 37–38.

25. Ibid., 38–39.

26. Ibid., 41–44.

27. Ibid., 44.

28. Ibid., 45–47; on Helmholtz's embrace of energy physics, Buchwald 1994, 400–402.

29. HH 1871a, 50; D[ilthey] 1900, 230.

30. WT to HH, 30 March, 14 June (first quote) 1871, both in HN 464; Alexander Crum Brown to HH, 19 April 1871, in HN 72; John Hughes Bennett to HH, 6 July 1871, in HN 37; PGT to HH, 20 July 1871 (second quote), in HN 459; JT to WT, 9 April 1871, in RI MS JT/1/T/18, TPRI.

31. AH to ISZ, 13 August 1871, in ESH 1929, 1:166; Thompson 1910, 2:612; WT to George Gabriel Stokes, 6 October 1871, in WTP, K178; HH to AH, 20 August 1871, in ESH 1929, 1:166–67 (quotes on 166); Knott 1911, 56, 196–97.

32. Thompson 1910, 2:612, 614–15; HH to AH, 24 August 1871, in ESH 1929, 1:167 (first quote); HH to AH, 27 August, 1 September 1871, ibid., 168 (second and third quotes), 168–69 (fourth and fifth quotes), resp.; Smith and Wise 1989, 736–40; PGT to HH, 20 May, 17 August 1871, both in HN 459; WT to Stokes, 6 October 1871; Thomson 1912, lxiii; James Clerk Maxwell to PGT, 19 October 1871, in Maxwell 1995, 681.

33. HH to AH, 1, 3 September 1871, in ESH 1929, 1:169 (quotes); WT to his sister [Mrs. King], 31 August 1871, in Thompson 1910, 2:615.

34. HH to AH, 6 September 1871, in ESH 1929, 1:169–70 (quotes); WT to his sister [Mrs. King], 31 August 1871.

35. HH to AH, 10 September 1871, in ESH 1929, 1:170–71 (quotes); Fairley 1988, 18, 20, 28–29, 68, 133–34; James Thomson to his wife, 8, 14 September 1871, both in Thomson 1912, lxiii–lxiv, on lxiv; lxiv–lxv, on lxv, resp.; HH to AH, 15 September 1871, in ESH 1929, 1:171–72; John Tatlock (Thomson's assistant) to Stokes, 6 October 1871 (the first part of a letter from Thomson to Stokes, 7 October 1871), in Wilson 1990, 2:362–65, on 363–64; Thompson 1910, 2:747; Smith and Wise 1989, 739.

36. HH to AH, 15 September 1871.

37. WT to HH, 29 October 1871, in HN 464; James Thomson to his wife, 8 September 1871, lxiv; HH to WT, 4 November 1871 (quote).

Chapter Eighteen

1. HH to HK, 5 January 1873, HKL; Jungnickel and McCormmach 1986, 2:21–22, 42–43, 51–52.

2. Documents in Auth and Kossack 1983; HH 1871a, 36; Guttstadt 1886, 139–41, 143; Pistor 1890, 63; Rubens 1910, 283; Neumann 1925, 13, 21, 23, 31; Kayser 1936, 93–44; Hars 1999, 25. See also Georg Quincke to HH, 19 April 1860; 18 November 1863, in HN 360.

3. AH to JT, 27 February 1872, in RI MS JT/1/H/35, TPRI; AH to PM, 6 March 1872; 30 April 1875, in ESH 1929, 1:177–78 (quote), 198–99; Rubens 1910, 284, 286; Hirschberg 1921, 1117; Kayser 1936, 94; Schuster 1932, 244; 1975, 16–77.

4. RC to HH, 22 May 1873, in HN 91; Friedrich Neesen to HH, 13, 22 May 1873, in HN 325; HH to Wilhelm Weber, 20 May 1873, in Deutsches Postmuseum, Archiv, Frankfurt am Main; Weber to HH, 24 May 1873, in HN 500; GRK to HH, 13 May 1873, in HN 231; HH, in I. HA Rep. 76, KM, Va Sekt. 2 Tit. X no. 79, Bd. 1, Bl. 82–82r, 94–95, 98–98r, 106–106r, 109–109r, 143–143r, 145, 153–56 (second quote), 164–67r, 188–89, 190–92 (first quote); HH to Geehrter Herr Doctor, 26 February 1871, in Karl-Marx-Universität, Universitätsbibliothek, Briefsammlung, Autographensammlung Nebauer; Guttstadt 1886, 141; Rubens 1910, 287.

5. EdBR to Jeannette du Bois-Reymond, 27 July 1871, in Dep. 5, K. 11, no. 5; Finkelstein 2013, 187–90; AH to PM, 6 October 1871; 18 June 1872, in ESH 1929, 1:173–74, 180, resp.; AH to JT, 27 February 1872 (quote); EdBR to HH, 26 February 1872, in Kirsten et al. 1986, 257, 310n1; EdBR to HBJ, 5 February 1872, in SBPK, SD 3k 1852 (4) Bl. 91–93, on 92; Cahan 1985, 1989; Dierig 2006.

6. AH to PM, 6 March, 18 June 1872, both in ESH 1929, 1:177–78 (first quote), 180 (second quote), resp.; AH to Frau Prof. Otto Becker, 2 January 1873, ibid., 183–84; HH to WT, 17 June 1872, in EUL; EdBR to HBJ, 13 August 1872, in SBPK, SD 3k 1852 (4) Bl. 93–94, on 94; EdBR to HH, 5, 19 September 1872, both in Kirsten et al. 1986, 254–55; 256–57, on 256, resp; Wilhelm von Beetz to HH, 17 October 1872, in HN 34; EdBR to HBJ, in SBPK, SD 3k 1852 (4) Bl. 99a—100 (third and fourth quotes); Dierig 2006, 233–37.

7. AH to PM, 7 December 1871; 18 June 1872, in ESH 1929, 1:175–76, 180 (first quote); AH to MM, 16 September 1872 (second quote), ibid., 181; HH to August Kundt, 18 April 1894, in *LK*, 3:101–2; HH to Auguste de la Rive, 4, 25 September 1872, in VG, Ms. fr. 2317, f. 69, f. 71, resp.; HH to Jacques-Louis Soret, 18 September 1872; 16 April 1873, ibid., Ms. fr. 4175, f. 345, f. 346, resp.

8. AH to MM, 12 December 1872, in ESH 1929, 1:181–82, on 182; AH to RM, 13 December 1872, ibid., 182; HH to WT, 17 June 1872 (first quote); AH to Frau Prof. Otto Becker, 2 January 1873 (second quote); HH to RL, 16 October 1872 (third quote), in Scharlau 1986, 129–30.

9. The following discussion draws on Cahan 2012a.

10. WT to George Gabriel Stokes, 31 October 1871, in Wilson 1990, 2:365–66 (quote on 365).

11. HH to Naturhistorisch-Medicinischer Verein zu Heidelberg, 26 April 1872, in Universitätsbibliothek Heidelberg, Handschriftenabteilung, Heid. Hs. 840,1; Stefan, Loschmidt, Lang, Brücke, Schrätter, 6 June 1872, 435/1872, in Bibliothek, Österreichische Akademie der Wissenschaften, Vienna.

12. WT to HH, 11 December 1872 (first quote); 8 January 1873, both in HN 464; PGT to HH, 4 April 1873 (second quote), in HN 459.

13. WT to HH, 8 January 1873, in HN 464; Thomson 1912, lxvi–lxvii; James Forrest to HH, 7 July 1868, in HN 152.

14. HH to WT, 14 January 1873 (first three quotes), copy, in KP, T131; and in Thomson 1912, lxvii–lxviii; WT to HH, 16 March 1873 (fourth quote), in HN 464.

15. JT to HH, 25 September (first quote), 13 November 1873, both in HN 477; W. H. Miller (of the Royal Society) to HH, 8 November 1873 (second quote), in Awards, Honorary Memberships, etc. in SF, Sign.: Helmholtz. 6.LL 494; AH to RM, 16 November 1873 (misdated as 1871), in ESH 1929, 1:174–75; JT to HH, 18 November 1873, typescript only, in RI MS JT/1/T/492, TPRI; HH to JT, 22 November 1873, in RI MS JT/1/H/51, TPRI; Cahan 2012, 2012a.

16. HH to Alexander William Williamson, circa late 1873, printed in Tilden 1930, 238–40 (first and second quotes); HH to E. Thompson (honorary secretary to the society), 15 December 1873 (third quote); 27 March 1874, both in Royal Society of Medicine, Library, London; HH to Sir, 5 May 1874, in SF, Sign.: Helmholtz 6.LL496.

17. HH to secretary of the American Philosophical Society, 8 June 1873, in American Philosophical Society, Philadelphia, Library, Archives, Letters Acknowledging Election to the Society; American Philosophical Society 1891, 6; HH to Werner von Siemens, 17 November 1873, in SF, Sign.: Helmholtz 6.LL496; Carte Quintino Sella, serie Accademia dei Lincei, mazzo h, fascicolo 12, Carte varie inerenti le elezioni di soci italiani e stranieri nelle varie classe dell'Accademi 1875–1877, Fondazione Sella, Biella (Italy); HH to Hochgeehrter Herr Präsident [Quintino Sella], ibid., mazzo 6, fascicolo 20, Carteggio 1875, sottofascicolo Helmholtz; Quazza 1992, 530–31 (first quote on 531), 533; AH to PM, 7 October 1875, in ESH 1929, 1:199; AH to HH, 8 October 1875, ibid., 199–200; Quintino Sella to HH, 7, 18 February 1876, both in HN 399; Sitzung der physik.-mathem. Klasse, 18 March 1878, in AHA, Protocoll-Buch der physik.-math. Klasse 1877–1880, Sign. II–V, 120, Bl. 18; Eugenio Beltrami to HH, 12 March 1878 (second quote), in HN 36; Thor A. Bak, Secretary of the Royal Danish Academy of Sciences and Letters, to the author, 6 July 1993.

18. HH to Hochverehrter Freund, 5 July 1873, in HN 532/3. Helmholtz himself was on the board of the Berlin Naturwissenschaftlicher Verein. See HH to [Soret?], n.d., in VG, Ms. fr. 4175, f. 351.

19. HH to AH, 1 August 1873, in ESH 1929, 1:186 (quote); Biermann 1973, 78; Guttstadt 1886, 62.

20. AH to RM, 16 November 1873 (first quote); HH to Hochgeehrter Herr, 31 October 1873 (second quote), in Stadt- und Landesbibliothek Dortmund, Handschriften-Abteilung, 6017.

21. Guttstadt 1886, 135, 143; Pistor 1890, 63–64; for EdBR's physiology institute, Dierig 2006, 67–88 and passim.

22. Kleinwächter 1881; Guttstadt 1886, 135–39, 143; Junk 1888, 143; Architekten-Verein 1896, 2:266–67; Pistor 1890, 63–64; EdBR to HBJ, in SBPK, SD 3k 1852 (4) Bl. 99a–100; JT to HH, 25 September 1873, in HN 477; HH to JT, 22 November 1873 (quotes); AH to Julius von Mohl, 28 January 1874, in ESH 1929, 1:192–93, on 193.

23. HH to AH, 1 August 1873; AH to PM, 8, 11 September 1873, both in ESH 1929, 1:187, 187–89, resp.; HH to AH, 15, 16 September 1873, ibid., 189 (first quote), 190 (second and third quotes); HH to HK, 10 November 1873, in HKL; Beltrami to HH, 16 September 1873, in HN 36.

24. Beltrami to HH, 18 October 1874, in HN 36; EB to EdBR, 23 October 1874, in Brücke

et al. 1978–81, 1:194–95, on 195; AH to PM, 1 November 1874 (first three quotes); 20, 22 March (fourth quote); 9 September (fifth quote) 1875, all in ESH 1929, 1:193–94, 198, 199, resp.

25. HH to Soret, 7 May 1875 (quotes), in VG, Ms. fr. 4175, f. 347; AH to PM, 30 April 1875.

26. HH to HK, 5 January 1873, in HKL.

27. Henry Dewey Noyes to HH, 26 June 1875, in HN 334.

28. James 1987, 324–25 (first quote on 324); HH to HK, 10 November 1875, in HKL; HH to Otto Brans, 25 October 1875, in Senckenbergische Bibliothek, Frankfurt am Main; George Bancroft to HH, 15 April 1876 (second quote), in HN 26; HH 1876, v–vii.

29. HH to de la Rive, 2 June 1872, in VG, Ms. fr. 2317, f. 67–68; Wilhelm Fiedler to HH, 30 January, 2 March, 27 April 1875, all in HN 146; HH to Fiedler, 8 February 1875, in Eidgenössische Technische Hochschule, Zurich, Bibliothek, Wissenschaftshistorische Sammlungen, Hs 87: 402; Cahan 2000, 47–48.

30. TM to HH, 14 March 1874, in HN 315; HH to EdBR, 25 March 1874, in Kirsten et al. 1986, 258 (quotes); see also ibid., 310nn1, 2; HH to TM, 6 May 1874, in TMN 315.

31. HH 1893; Bezold 1896, 23–24; Warburg 1925, 13 November 1871, 35, 37; Goldstein 1925, 39; Scheel 1925; 1935, 6–8; Jungnickel and McCormmach 1986, 2:3–6, 21–22; Fiedler 1998, 61–74, 93–95, 113–15, ccxxiii–ccxxiv.

32. Wien 1930, 15.

33. Cahan 1999, 294–97; "Titelverzeichniss" 1895; Harnack 1970, 3:123–24. Except for coauthoring one paper with Gustav von Piotrowski on fluids (HH and Piotrowski 1860) and coauthoring or reporting on three papers with Nikolai Baxt in physiology (HH 1867f, 1870h, 1871e), Helmholtz published exclusively as an individual scientist (HH 1878b).

34. HH 1871d (quote on 629).

35. Darrigol 2000, passim, esp. viii–ix.

36. HH 1872a, 1873g, 1874 f. See also HH 1873e. HH 1873g, 652, shows that Zöllner's polemical work on comets (Zöllner 1872) even penetrated into an important mathematical (and technical) journal, Borchardt's *Journal für reine und angewandte Mathematik*. Helmholtz further criticized the electrodynamic work of Zöllner and others in HH 1874g. HH 1873g, 687, explicitly declares Helmholtz's intellectual respect for Weber but his disdain for some of Weber's students and physicist friends. Similarly, as HH 1873e and HH 1874e indicate, his critique of Carl Neumann did not extend to the latter's father, Franz, whom Helmholtz continued to honor. For a detailed analysis, see especially Buchwald 1994, 7–24, 340–47, and passim; Darrigol 2000, 231–33, 412–19.

37. HH 1872b, 1873f; 1873e, 700–701; 1874f, 712 (quote), 759–60; 1875b; HH to Soret, 7 May 1875; AH to PM, 30 April 1875; HH 1876b, 1876c; Jungnickel and McCormmach 1986, 1:25–28; Buchwald 1994, 75–77, 351–55.

38. HH 1874e.

39. A. Clebsch to ?, 3 January 1872, in Wiedemann 1989, 165–67, on 166–67.

40. The following account draws on Feffer 1994, 1996, especially for its analysis of Abbe's relationships with Zeiss, with microscopists, and with technical optics; Cahan 1996; and the literature cited in those sources.

41. Abbe 1873.

42. HH 1873d; HH 1874.

43. HH 1874, 185–86 (first three quotes), 211–12 (fourth quote).

44. Feffer 1994, 185–91; 1996.

45. Schomerus 1941, 102; Cahan 1996; Feffer 1994, 255–57, 265; 1996, 59.

46. Buchwald 1985, 233–34; Darrigol 2000, 320; 2012, 250–52.

47. HH 1874d; Buchwald 1985, 198, 234–36; Smith and Wise 1989, 443n112, 469 (quotes); Darrigol 2000, 320–21; 2012, 252.

48. Bélafi 1990, 32–34; Hallion 2003, 66–72.

49. HH 1864, 527–30; 1872, 1873b; HH to AH, 1 August 1873; Johann Fischer to HH, 25 December 1874, in HN 147; Ed. Vidal to HH, 20 May 1875, in HN 481; Peter Klünder to HH, 6 September 1875 (based on the Russian calendar; by the western European calendar, it was 18 September), in HN 235; Kehler 1926; Klein 1926, 230; Darrigol 2005, 257.

50. Fontane 1969, 165–67 (quote on 165); Spiero 1921, 82–100; Brandes 1989, 437–38.

51. HH to Julius Rodenberg, 4 November 1874; 4 December 1875, both in Nationale Forschungs- und Gedenkstätten der klassischen deutschen Literatur in Weimar, Goethe- und Schiller-Archiv; HH 1875a; 1903b, vi.

52. HH 1875a, 139–40, 142.

53. Garber 1976, 51–57; Kutzbach 1979, 45–58; HH 1875a, 142–44 (quote on 143–44); Darrigol 2005, 168–69.

54. HH 1875a, 145–51; Kutzbach 1979, 11–16, 58–62, 80–82, 88–90; Darrigol 2005, 167–70.

55. HH 1875a, 151–54, 158–62; Kutzbach 1979, 63–64, 84, 96–99; Darrigol 2005, 168–72.

56. HH 1875a, 162–63; Kutzbach 1979, 1, 9, 119–45; Friedman 1989.

57. AH to a friend, 4 April 1876, in ESH 1929, 1:203 (first two quotes); HH to JT, 17 May 1876 (third quote), in RI MS JT/1/H/53, TPRI; EB to EdBR, 25 June 1876 (fourth quote), in Brücke et al. 1978–81, 1:210–11, on 210; HH to Soret, 7 September 1876, in VG, Ms. fr. 4175, f. 349–50.

58. Franz Boll to EdBR, 12 November 1876, in Belloni 1980, 394–96; Boll to Felix Lewald, 9 November 1876, in Boll Nachlass, private possession, printed in Belloni 1982, 129; HH to Boll, 28 October 1877, in Belloni 1982, 133 (quote); HH to Willy Kühne, 13 March 1887, in *LK*, 2:233; Kremer 1997.

59. HH to HK, 21 July (quote), 13 August 1877, both in HKL; AH to HH, 23, 28 August 1877, both in ESH 1929, 1:220, 221, resp.; HH to AH, 30 August; 1, 4 September 1877, all in ESH 1929, 1:221–22; HH to Boll, [23 September 1877], in Belloni 1982, 132; Corrado Tommasi-Crudeli to HH, 14 November 1876, in HN 469; AH to PM, 8 October 1877, in ESH 1929, 1:223.

60. Endell 1883, 148–49; Architekten-Verein 1896, 2:268.

61. Endell 1883, 148–49; Guttstadt 1886, 143–48; Pistor 1890, 63–68; Architekten-Verein 1896, 2:266–68; Junk 1888, 143, 146; Rubens 1910, 285; I. HA Rep. 151 Finanzministerium, IV no. 1980: Bauten für das physiologische und physikalische Institut, Dorotheenstrasse 35 und Neue Wilhelmstrasse 16a, 1873–1936, Bd. 2, Bl. 20, 58–9r, in GSPK; Springer 1878, 176; EdBR to Jeannette du Bois-Reymond, 11 July 1877, in Dep. 5, K. 11, no. 5.

62. Feldhaus 1929, 36–37, 39, 41–45, 64.

63. HH to JT, 18 February 1878, in RI MS JT/1/H/55, TPRI; HH to Geehrter Herr?, 22 February 1877, in SF, Sign.: Helmholtz 6.LL496; HH to JT, 18 February 1878; HH to JT, 14 March 1878, in RI MS JT/1/H/56, TPRI; JT to HH, 22 February 1878, in HN 477; HH to Hoch-verehrte Herren, 5 August 1878 (quote), in Royal Institution, London, Archives, CG2/i/1.

64. Thomas Place to HH, 19 May 1873, in HN 354; on Darwin's photograph, see Browne 1995–2002, 2:272–73, 300–303, 362–63, 423–24, 451; Adrian Heynsius to HH, 6 June (first quote), 20 June; 3, 9 October; 14 November 1871; 29 March, 9 May 1872, all in HN 201; EdBR to Jeannette du Bois-Reymond, 1 April 1877 (second quote), in Dep. 5, K. 11, no. 5; Bowman 1890–91, xxii.

65. Cahan 1985; N[ichols] 1894, 225 (first quote); Karl Pearson to Robert Parker, 19 October [1879] in Karl Pearson Papers 922, Manuscripts Room, University College London (second quote); Baron Jauru (Imperial Legation of Brazil, Berlin) to Baron Nogueira da Gama, 1 August 1880, in Maço 183, Doc. 8345, Arquivo da Casa Imperial do Brasil (POB), Arquivo Histórico, Museu Imperial, Petropolis, Brazil; Pedro to HH, 23 April 1880, in HN 523; Wurtz 1882, 14–17 (third and fourth quotes on 15); Rubens 1910, 286.

66. I. HA Rep. 76 KM, Va Sekt. 2 Tit. X no. 79, Bd. 3, "Etat der königlichen Friedrich-Wilhelms-Universität zu Berlin für 1875/8," Bl. 175r–77, in GSPK.

67. Zobeltitz 1891, 768; AH to RM, 13 November 1871, in ESH 1929, 1:174.

68. Schultze 1993; Kayser 1936, 94. One of Helmholtz's notebooks (HN 721), dated Winter 1872–73, is entitled "Logische Principien der Naturwissenschaft" and appears to be an outline of his lectures on that topic.

69. GRK to HH, 17 November 1874, in HN 231; Sitzung der physik.-mathem. Klasse, 21 July 1873, in AHA, Protocoll-Buch der physik.-math. Klasse 1874–1877, Sign. II–V, 119, Bl. 28–29; HH to EdBR, 25 March 1874; Planck 1910, 276; Jungnickel and McCormmach 1986, 2:30–32, 49, 125–29.

70. HH 1891a, 203; Guttstadt 1886, 141.

71. AH to RM, 13 November 1871; AH to RM, December 1872, in ESH 1929, 1:183 (quotes).

72. Kayser 1936, 94; Schuster 1975, 17–18 (quotes).

73. Hertz 1894–95a, 368 (first quote); Hirschberg 1921, 1117; Rubens 1910, 287 (second quote); Paulsen 1938, 213–14 (third quote on 214); Runge 1949, 27–29.

74. Hertz 1894–95a, 368; Rubens 1910, 287; Planck 1948, 8 (first three quotes); 1986; EdBR 1912, 2:568 (fourth quote); Wachsmuth 1900, 17 (fifth quote).

75. Einstein 1987, 46, 49, 318, 364, 365–66; Sigmund Freud to Edward Silberstein, 24 January 1875, in Boehlich 1990, 84–85, on 84; Freud to Martha Bernays, 28 October 1883 (first quote), cited in Jones 1953–57, 1:41; Sulloway 1979, 13–15, 65–66, 94, 138–39 (second quote on 139), 170, 235.

76. Hars 1999, 55–56; Kurylo and Susskind 1981, 14–21; HH to T. Zincke, 31 October 1876, in Hessisches Staatsarchiv Marburg: 307 d no. 113, Bd. II; HH to A. Falk, 10 May 1877, in SBPK, Slg. Darmst. F1a 1847 (2), both cited in Hars 1999, 15, 26–27, 38–40.

77. Helmholtz in Schultze 1994, 3–6 (quotes on 3–4); for Ebbinghaus, Unterlagen der Philosophischen Fakultät, in HUB, Promotionsverfahren der Phil. Fak. 291 ff. Helmholtz, Gutachten, Prüfungen, und Promotion.

78. Unless otherwise noted, this paragraph draws on Cahan 2004.

79. Tyndall 1873b; HH to JT, 11 November 1873, in RI MS JT/1/H/50, TPRI; JT to HH, 13 November 1873; HH to JT, 22 November 1873.

80. Upton's student notebook from Helmholtz's course on light and electricity during June–July 1878 ("E Helmholtz Upton 26/6 78," in TAEM, reel 95:290–313 [TAED, MUN000]); Bryan 1926, 113; Jehl 1938, 2:619, 733; Friedel and Israel 1986, 27, 29, 36–37, 56, 104–5, 122–28, 134–35, 137, 141–42, 180, 194–95, 229; Israel 1998, 92, 157–60, 176, 179, 188, 191, 195, 197–98, 201–4, 211–12, 254, 448, 491n9; Friedel and Israel 2010, 19, 21, 24, 27, 35, 40, 42–44, 53–54, 58, 62, 64–66, 71, 73, 74, 76, 78–81, 82, 84, 85, 87, 95–104, 110–12, 116–17, 119, 120, 125–26, 139, 142, 149, 161–62, 197.

81. Ludwig Boltzmann to HH, 1, 20 November 1872; 26 February, 21 April 1874; 13 February 1875, all in HN 57; Boltzmann to Leo Koenigsberger, 1902, in Boltzmann 1994, I:21–24, II:8–12, 347; Boltzmann 1905, 96–97, 102; Jungnickel and McCormmach 1986, 1:212–13; Hörz and Laass 1989, esp. 15–16, 43–47; Buchwald 1994, 208–14; Flamm 1995, 24, 33–35.

82. For documentation (and further details) on Rowland and his relationship with Helmholtz, see Cahan 2004, 19–24.

83. Schuster 1932, 56–65, 243–44; HH 1876c; HH to Henry A. Rowland, 28 April 1877 (quote), in Henry A. Rowland Papers, Special Collections and Archives, Manuscripts, Milton S. Eisenhower Library, Johns Hopkins University, Baltimore, MD; Buchwald 1994, 354–55.

84. See, for example, Johann Nepomuk Czermak to HH, 2 May 1872, in HN 101; WT to HH, 26 February 1874, in HN 464; Paul du Bois-Reymond to HH, 8 August 1876, in HN 123; James Clerk Maxwell to HH, n.d. (sometime between 1871 and 1879), in HN 305; G. Pirie to

HH, 7 August 1878; 20 September 1880, in HN 353; William Sharpey to HH, 3 April 1873, in HN 402; PGT to HH, 21 September 1871; 23 July 1879, in HN 459; Karl Friedrich Otto Westphal to HH, 5 December 1873, in HN 509; HH to Charles Augustus Young, 20 March 1877, in Dartmouth College Library, Hanover, New Hampshire, Special Collections, Charles Augustus Young Papers; Arthur Schuster to HH, 29 November 1879, in HN 442.

85. A. Kosloff to HH, 16 April 1872, in HN 244; Carl von Lemcke to HH, 19 December 1869; 7 March 1870; 28 February, 30 April 1876, all in HN 271.

86. JT to HH, 10 February 1874 (quote), in HN 477; John Cleves Symmes to HH, 15 April 1871 (other quotes), in HN 420. For other examples, see W. Clemm to HH, 11 February 1872, in HN 92; Marcel Croullebois to HH, 8 November 1871; 6 May 1872, both in HN 98; Friedrich Eugen Weber-Liel to HH, 22 April 1877, in HN 502; G. H. Schneider to HH, 17 October 1877, in HN 434.

87. Hall 1880, vii, 47 (first quote); 1878, xi–xii (other quotes), 96–100, 104, 113, 125, 139, 160, 174–76, 183–84, 186, 188, 191, 199, 200, 201, 203, 211–13, 220, 231, 237, 240–41, 243–45, 247, 254, 286–87, 293–96, 318, 321, 325, and passim; Hall & Co., Publishers to HH, 4 October [ca. 1877–80], in HN 184.

Chapter Nineteen

1. See, for example, HH and Rudolf von Gneist to Hochgeehrter Herr College, 21 October 1876, in Smithsonian Institution Libraries, Washington, Dibner Manuscript Collection, Helmholtz Papers, MSS 683A; Spitzemberg 1963, 126, 129, 135, 139, 140.

2. HH 1873a, 93, 95–96, 116 (quote); HH to Friedrich Max Müller, 15 November 1872; 18 March 1873, both in Senckenbergische Bibliothek, Frankfurt am Main; HH to Hochgeehrter Herr, 7 July 1874, in Stadt- und Landesbibliothek Dortmund, Handschriften-Abteilung, 6342.

3. HH 1873a, 95.

4. Ibid., 96–97.

5. Ibid., 97–98.

6. Ibid., 98–100 (quote on 98).

7. Ibid., 100–103 (quotes on 100–101).

8. Ibid., 107–11.

9. Ibid., 111–16 (first six quotes on 111–12, seventh on 116).

10. Ibid., 118–24 (quote on 118).

11. Ibid., 104–7.

12. Ibid., 125–26 (first two quotes); HH to AH, 15, 16 September, in ESH 1929, 1:189 (third and fourth quotes), 190, resp.

13. HH 1873a, 126–33 (first quote on 126–27, second on 128, third on 131).

14. Ibid., 134–35.

15. Wilhelm von Bezold to HH, 3 April, 8 November 1874, both in HN 47; Vitz and Glimcher 1984, 46–48, 70, 72–77, 83, 86–87, 97, 100–101, 163–64; Kemp 1990, 234, 241–42, 250, 253, 262–63, 312–20, 338; Gage 1999, 48, 185, 212–14, 218, 219–23, 300n17.

16. Brücke 1878; Charles Henry to HH, 4 September 1885 (first quote), in HN 196; Homer 1964, 114, 130–31, 214–16, 246, 248, 289–90, 301–2; Nochlin 1966, 112–15; Pool 1967, 15; Argüelles 1972, 20, 70–71, 72, 87–89, 98–99, 100, 117, 119, 125; Hamann and Hermand 1973, 222–23; Rewald 1978, 133, 137 (second quote); Vitz and Glimcher 1984, 46–48, 70, 72–77, 83, 86–87, 97, 100–101, 163–64; Kemp 1990, 262–63, 312–20, 338; Lebensztejn 1992; Gage 1999, 48, 185, 212–14, 218, 219–23, 300n11, 301n17.

17. Rood 1973, 11–18 (esp.), 38, 113–15, 126–27, 175–76; Fleming 1968, 482, 483, 498, 499; Gage 1999, 78, 209–14, 218, 222–23, 255–56.

18. Emerson 1980; 1993, 103–11, 134–46.

19. Pietsch 1889–90, 666; Hamann 1914, 1, 5–7, 171; Craig 1978, 59.

20. Wilhelmy 1989, passim, esp. 24–27, 268–69, 271. See also Brandes 1989, 124–40; Siebel 1999.

21. AH to PM, 16 May 1873, in ESH 1929, 1:185 (quote); Vasili 1884, 163–65, 229; Bode 1930, 1:107–8; Wilhelmy 1989, 274–81, 820–21.

22. George Bancroft to Frederica King Davis (Mrs. John Chandler Bancroft Davis), 2 May 1871 (quote), in American Antiquarian Society, Worcester, MA, George Bancroft Papers; Nye 1944, 241, 276, 278; Spitzemberg 1963, 126, 129, 135, 139, 140.

23. AH to RM, 13 November 1871, in ESH 1929, 1:174; AH to PM, 6 March 1872, ibid., 177–78; Mohl 1902, 2:200; Wilhelmy 1989, 659.

24. EdBR to Jeannette du Bois–Reymond, 5 June 1873, in Dep. 5, K. 11, no. 5; Bunsen 1899; Dilthey 1900, 231–32; Wachsmuth 1900, 16–17, 19; Weech 1906, 297; Planck 1929–30, 38; Wilhelmy 1989, 283–88, 342, 415, 451, 453, 659–60.

25. Strauss 1872, 210–11, 213–15, 218–21; AH to ISZ, 18 January 1892, in ESH 1929, 2:40; Braun-Artaria 1899; Bunsen 1932, 40–41; Wilhelmy 1989, 285–86 (quotes).

26. Unless otherwise noted, all guest names for Berlin (1871–94) are culled from ESH 1929 and from Wilhelmy 1989, 660, 662–69.

27. Max Planck to Carl Runge, 6 October 1885, in Hentschel and Tobies 1999, 97–98.

28. Bunsen 1929, 43; HH to TM, 7 March 1877; AH to TM, 3 June 1877; 22 June 1889; HH to TM, 13 May 1890, all in TMN.

29. AH to ISZ, 24 June 1871, in ESH 1929, 1:164–65; AH to PM, 23 June 1872, ibid., 180–81; Bernard Cracroft to HH, 20 May 1872, in HN 94; Joseph Joachim to HH, n.d., in HN 220.

30. AH to HH, 19 July 1867; 8 September 1869, both in ESH 1929, 1:141–42, 151, resp.; AH to RM, December 1872 (first two quotes), ibid., 183; HH to JT, 29 September 1876, in RI MS JT/1/H/54, TPRI; Cosima Wagner to Friedrich Nietzsche, 1 January 1877, in Wagner 1940, 75–78 (third quote on 77); 1976–77, 1:1019.

31. HH to AH, 12 July 1880, in ESH 1929, 1:250–51 (quote on 251); Hamann 1914, 166–71, 177, 179–80, 183–85; Hamann and Hermand 1971, 34; Herbst 2006, 91, 95.

32. Bunsen 1899; Planck 1929–30, 38 (first two quotes); Wachsmuth 1900, 17 (third quote).

33. AH to HH, 8 July 1877, in ESH 1929, 1:215; Brauer 1936, v–ix, 208–9 (quotes); Ludwig Bamberger to HH, n.d., 19 December 1886, in HN 25.

34. AH to Frau Prof. Otto Becker, 2, 16 January 1873, in ESH 1929, 1:183–84, 184, resp.; AH to TM, n.d. (Sonntag), in TMN; Cosima Wagner to Nietzsche, 12 February 1873, in Wagner 1940, 43–46 (first quote on 45–46), 141n618; Newmann 1969, 4:386; Schüler 1971, 127; Wagner 1976–77, 1:629, 910, 912 (second quote), 913, 917, 977; Gregor-Dellin 1983, 400–401; 1987, 603–4, 622, 626, 629, 664; Werner and Irmscher 1993.

35. Gregor-Dellin 1987, 651–53, 658, 826; Spotts 1994, 66–68, 70, 71, 76, 77.

36. AH to her children, 14 August 1876, in ESH 1929, 1:204–6 (first, second, and fourth quotes); Wagner 1976–77, 1:999 (third quote); HH to RL, 15 August 1876, in Mathematisches Institut, Rheinische Friedrich-Wilhelms-Universität Bonn, Bibliothek, partly printed in Scharlau 1986, 130 (fifth quote).

37. HH 1954, 339; Wagner 1976–77, 2:732, 733 (first quote), 737; Zobeltitz 1891, 768, 770 (second quote); Epstein 1896, 198; Wagner 1912 (third quote); Gregor-Dellin 1983, 487; 1987, 722–23; cf. Steege 2012, 224–34. In German:

Grau wäre alle Theorie?
Dagegen sag' ich, Freund, mit Stolz:
uns wird zum Klang die Harmonie,
fügt sich zum Helm ein edles Holz.

38. Alarik Frithiof Holmgren to HH, 3 May 1881, in HN 207; Cosima Wagner to Marie von Schleinitz, n.d., in du Moulin Eckart 1929–31, 1:882–85 (first two quotes on 884); AH to ISZ, 17 February 1883, in ESH 1929, 1:263–64 (third quote on 263).

39. Werner and Imscher 1993.

40. Heuss 1948; Bauer 1991, 232–84; Nyhart 1995, 258–59, 263–65, 269, 272, 273n87, 275.

41. AD to Fanny Lewald, 18 September 1869, in SZADNA, Ba 1258 E, Bd. 60; Heuss 1948, 111; Groeben 1985, xxv.

42. Groeben 1985, xxiii–xxvii, 1–2; AD to Lewald, 18 September 1869; Heuss 1948, 111; AD to HH, 2 August 1871, in HN 112; AD to EdBR, [2 August 1871], in Groeben 1985, 1–2, on 2; Simon 1980, 15–20; HH to [AD], 5 August 1871 (first four quotes), in VG, no signature; AD Nachlass, in SZADNA, Ba 764; AD to HH, 18, 28 September 1871, both in HN 112; HH to AD, 26 September 1871 (fifth quote), in BSBMH.

43. AD to EdBR, [18 September 1871]; [19 September 1872], 8 October 1872; 27 January 1878, all in Groeben 1985, 6–7; 12–14, on 12; 16; 125–26, on 125, resp.; ibid., xxvi, 10–11.

44. Ibid., 13; AD to HH, 5 June 1875, in HN 112; HH to AD, 19 June 1875 (quotes), in Dohrn Familienarchiv, BSBMH; Sitzung der physik.-mathem. Klasse, 6 December 1875, in AHA, Protocoll-Buch der physik.-math. Klasse 1874–1877, Sign. II–V, 119, Bl. 71–73; Heuss 1948, 142, 205–6; Browne 1995–2002, 2:383, 480.

45. AD to Marie Dohrn, [February] 1879, fragment, in SZADNA, Bd. 132; H. Helmholtz, Rud. Virchow, and E. du Bois–Reymond, "[Eingabe an den Reichstag. 6.3.1879.] Die Zoologische Station in Neapel," in Groeben and Wenig 1992, 114–17; Heuss 1948, 237–39.

46. AD to Marie Dohrn, 2, 12, 16, 21 March 1879, in SZADNA, Bd. 135.a, Bd. 143, Bd. 145 (first quote), Bd. 148 (second quote), resp.

47. AD to Marie Dohrn, 13 March 1880, ibid., Bd. 155; AD to Marie Dohrn, 22 April 1880 (quote), fragment, ibid., Bd. 173; AD to EdBR, 5 August 1880, in Groeben 1985, 194–95, on 194.

48. AD to EdBR, 17 May 1883, in Groeben 1985, 241–48, on 247; Heuss 1948, 272, 425; AD to Marie Dohrn, 13, 18 May 1884, in SZADNA, Bd. 259, Bd. 262; AH to AD, ibid., in SZADNA: A.1884.H.; AD to Marie Dohrn, 30 May, 12 June, early July, 6 July, 12 November 1884, in SZADNA, Bd. 266 (first three quotes), Bd. 271 (fourth quote), Bd. 284, Bd. 285, Bd. 296, resp.

49. Waldeyer-Hartz 1920, 273, 275.

50. AH to PM, 7 January, 4 February 1874, in ESH 1929, 1:191–92, 193; Mohl 1922, 1:12, 48.

51. HH to TM, 3 March 1872, in HN 315; AH to PM, 4 February 1874, in ESH 1929, 1:193; Vasili 1884, 229; Mohl 1922, 1:61–62; William II 1927, 19.

52. Barman 1999, 51, 92, 109, 117–18, 137, 237, 246, 248, 275–76, 279–82, 303; Mohl 1922, 1:88.

53. *Orden pour le mérite* 1978, 2:316; Barclay 1995, 73; Herbst 2006, esp. 23.

Chapter Twenty

1. Craig 1978, 70–78, 92–93; Clark 2007, 568–76.

2. EdBR 1912, 2:356–60, 367–68.

3. See, e.g., Virchow 1866; Sudhoff 1922a; Ackerknecht 1953; Goschler 2002.

4. Virchow 1871, 73–79, 81; Ackerknecht 1953, 184–86.

5. EdBR 1912, 1:441, 460–61, 464 (quotes); Steif 2003, 309, 115–27; Finkelstein 2013, 265–72.

6. Virchow 1873, 631, 632, 634 (quote).

7. Johann C. F. Zöllner to HH, 21 October, 20 December 1862, both in HN 522; Personal-

akt Johann Karl Friedrich Zöllner, Universitätsarchiv Leipzig, Sign.: UAL, PA 1093, Bl. 15 (quote); Herrmann 1982; Meinel 1991.

8. Zöllner 1872, xcix–c; HH 1995a, 277–78.

9. Zöllner 1872, xii–xiii, xxx–xxxvi, xlvii, xlvii–liv, xlix–1, 329–35; Thomson and Tait 1871, 1874; WT to HH, 23 January, 20 February 1870, both in HN 464; PGT to HH, 15 November 1870; 25 March 1871, both in HN 459; HH 1871c, 1874b.

10. Zöllner 1872, liv, lv–lvi, lvii–lviii, lix–lx, lviii–lxi, lxii, lxii–lxiii (esp. n2), lxvii–lxviii, lxx (quote). For Weber's and Helmholtz's approaches to electrodynamics and the polemical role of Zöllner therein, see Buchwald 1993; 1994, 7–20, 356–57, 395–97, 402–4; Darrigol 2000, 63–64, 223–30, 232.

11. Zöllner 1872, 317–21, 344–53, 378–425 (quote on 405).

12. Tait quoted in Eve and Creasey 1945, 162; PGT to HH, 23 May 1872, in HN 459; PGT to Gentlemen (Vieweg & Sohn), 25 September 1872, in VV; Knott 1911, 256.

13. JT to HH, 13 April 1872, in HN 477.

14. HH to JT, 23 June 1872, in RI MS JT/1/H48, TPRI (typescript copy [quotes]); Schlechta and Anders 1962, 122–27; Hörz 1997, 231.

15. HH to RL, 16 October 1872 (first quote), in Mathematisches Institut, Rheinische Friedrich-Wilhelms-Universität Bonn, Bibliothek; Ludwig Boltzmann to HH, 1 November 1872, in HN 57; Buchwald 1993, 337–45, 363–73, and the literature cited therein; Wilhelm Weber to HH, 24 May 1873 (second quote), in HN 500; Sitzung des Plenums, 15 June 1876, in AHA, GesammtsitzungsProtocolle 1876, Sign. II–V, 53, Bl. 52; 13 July 1876, Bl. 63.

16. Clark 2007, 186, 192–93, 196, 199–200, 305–6, 309–10, 365–67, 369, 388–89, 427, 496, 498, 535, 537, 542.

17. Zöllner 1872, lviii.

18. HH to Geehrter Herr, 12 March 1872, in Karl-Marx-Universität, Universitätsbibliothek, Briefsammlung, Nachlass Fechner; Fechner diary entry for 23 April 1873, in Fechner 2004, 2:993–96 (quote on 995); Ludwig Strümpell to HH, 27 November 1871, in HN 454.

19. CL to HH, 10 June (first quote), 23 June, 14 July 1872 (second quote), all in HN 293; HH to Karl Thiersch (Geschäftsführer der Deutschen Naturforscherversammlung), n.d. (third quote), in Universitätsbibliothek Freiburg im Breisgau, Handschriften- und Inkunabelabteilung, Autogr. 404; HH 1872b; McKendrick 1899, 284.

20. JT to HH, 25 September 1873, in HN 477; HH to Friedrich Vieweg, 14 [May?] 1873, VV; Dreyer n.d. [1936], 47, 60, 107–9.

21. Dreyer n.d. [1936], 108.

22. HH 1874c, 299; J. Norman Lockyer to HH, 28 May 1874, in HN 284; HH to Lockyer, 7 June 1874 (quote), in Norman Lockyer Papers, Norman Lockyer Observatory, Sidmouth, Devon, cited in Meadows 1972, 34; JT to HH, 25 September 1873, in HN 477; HH 1874.

23. Lockyer to HH, 25 June, 26 July 1874; 13 August (first six quotes), 20 August [1876], all in HN 284; HH to Lockyer, 28 August 1876 (seventh quote), in Lockyer Collection, University Library, University of Exeter; Maxwell 1877, 389 (eighth quote), 391 (ninth quote); S. Grove to HH, 2 October 1875 (tenth quote), in HN 177; AH to HH, 8 October 1875, in ESH 1929, 1:199–200.

24. HH to JT, 17 May, 29 September 1876, in RI MS JT/1/H/53, H/54, resp., TPRI; Richard Liebreich to HH, 15 May 1876, in HN 277; WT to HH, 30 May, 9 August (first quote), 18 September 1876, all in HN 464; Biedermann 1877, xii, 312, 333; HH to HER, 25 July 1876, in RSC; JT to HH, 18 September 1876, in HN 477; HH to Jacques-Louis Soret, 7 September 1876 (second quote), in VG, Ms. fr. 4175, f. 349–350.

25. HH to JT, 22 November 1873, in RI MS JT/1/H/51, TPRI; HH to Vieweg, 22 October 1873, in VV; HH 1873c, 421.

26. Smith and Wise 1989, 348–95; Smith 1998, 192–210; HH to WT, 24 November 1866 (quote), in KP, H17; HH to the Vieweg Firma, 13 March, 11 June, 12 August 1869; 24 February

1870; 1 October 1871, all in VV; HH to Vieweg, 17 November 1873; 22 February, 2 March 1874, all in VV; HH to [PGT or WT?], 10 April 1873, in EUL; PGT to Gentlemen (Vieweg & Sohn), 7 September 1873, in VV.

27. HH 1873c, 413.

28. Ibid., 414 (first eight quotes); HH to JT, 17 October 1873 (ninth quote), in RI MS JT/1/H/49, TPRI (typescript copy); JT to HH, 26 October 1873 (tenth quote), in HN 477.

29. HH 1873c, 415–16.

30. Ibid., 416–18.

31. Ibid., 418–21 (first quote on 418, second on 419).

32. JT to HH, 16 February [1874?] (quote), in HN 477; HH 1874–75.

33. HH to Vieweg, n.d., 14 [May?], 21 October 1873; 8 January, 22 February; 2, 12 May; 1, 26 June 1874, all in VV; Tyndall 1876.

34. HH to JT, 17 October 1873.

35. Tyndall 1898, 198–99 (first quote); 1874, 65–66 (second quote).

36. AH to JT, 17 February 1872, in RI MS JT/1/H34, TPRI; AH to PM, 30 January 1872, in ESH 1929, 1:176 (first quote); HH 1874a, 422 (fourth quote); HH to Vieweg, 21 October 1873 (third quote), 22 October 1873; 12 May 1874 (second quote), all in VV; HH to JT, 22 November 1873.

37. HH 1874a, 422–23.

38. Ibid., 423–24.

39. Ibid., 424–25.

40. Ibid., 425.

41. Ibid., 425–26.

42. Ibid., 426–28.

43. Ibid., 428–30.

44. Ibid., 430–31.

45. Ibid., 431–32.

46. Ibid., 432–34.

47. AH to JT, 7 January 1875, in RI MS JT/1/H/36, TPRI; HH to JT, 17 May 1876; Tyndall 1898–99; HH 1874h.

48. HH to Hochgeehrter Herr, 7 July 1874, in Stadt- und Landesbibliothek Dortmund, Handschriften-Abteilung, 6342; HH to O. Overbeck, 21 July 1874, ibid., 6343; HH to Hochgeehrter Herr Doctor, 3 March 1875, ibid., 6344; HH to AH, March 1875, in ESH 1929, 1:196–97 (quotes).

49. HH to AH, 10 March 1875, in ESH 1929, 1:197 (first and second quotes); HH to Verehrter Herr Regierungsrath, 15 March 1875, in Stadt- und Landesbibliothek Dortmund, Handschriften-Abteilung, 8986; HH to AH, 16 March 1875, in ESH 1929, 1:198 (third quote).

50. AH to HH, 19 March 1875, in ESH 1929, 1:198 (first two quotes); AH to PM, 20, 22 March 1875 (third quote), ibid.

51. JT to HH, 2 November 1871 (quotes), in HN 477; Youmans 1865; Isidor Rosenthal to HH, 3 June 1872, in HN 387.

52. "H.L. Helmholtz" 1875, 451 (first quote); "Deutsche Professoren" 1876, 679 (second quote); Julius Meyer to HH, 1 June 1876 (third quote), in HN 312.

Chapter Twenty-One

1. Dühring 1873; 1877, 36; Biermann 1973, 90.

2. Dühring 1877, 60–62, 70; AH to HH, 8 October 1875, in ESH 1929, 1:199–200; ESH 1929, 1:209, 274, 282–83; AH to Florence Nightingale, 24 May 1872, in Nightingale Collection, H1/ST/NC1/72/10, Greater London Record Office, London; HH to Wilhelm von Beetz,

24 June 1878, in DM, 1932–17/24; Bunsen 1899, n.p.; Dilthey 1900, 233–34; Weech 1906, 298; Albisetti 1988, 117–21; Wilhelmy 1989, 428; Siebel 1999, 83.

3. Dühring 1875a, 552–62; 1879, 562–64; Philosophische Fakultät 1877, 10–36.

4. Dühring 1877a, 444–45; 1880; HH 1883, 402; 1847a, 71–74; Weyrauch 1893, 253, 442–55; Leopold Pfaundler's cited remarks in *LK*, 1:91–92.

5. Dühring 1877a, 438–47, 459–60 (third and fourth quotes on 460), 474, 476–77, 504–5, 526–30, 535, 546–48, 550; 1878, 102–23 (first two quotes on 103), 126–27; Dühring 1881.

6. Dühring 1877, 3 (first quote), 36 (second and third quotes), 37 (fourth and fifth quotes), 67 (sixth quote), 68 (seventh quote), 69 (eighth quote); Philosophische Fakultät 1877, 10–36.

7. Engels 1978; 1978a, 307, 308, 355, 356, 358–71, 375, 378, 381, 382, 397, 430, 506–7, 541, 544, 556, 559–60; Reiprich 1969, 20, 37, 40–41, 62, 83, 123, 127, 130; Liedman 1986, 105, 108–9, 155–56, 159–60, 163–64, 191.

8. Philosophische Fakultät 1877, 10–36; Lewenstein 1877; "Die Facultät" 1877; Brandes 1989, 23–30.

9. HH to Verehrte Frau, 24 July 1877 (quotes), in BSBMH, Sign.: Autogr. VA; HH to Beetz, 24 June 1878.

10. AH to HH, 8, 10, 31 July; 3 August 1877, in ESH 1929, 1:215, 215–16, 216–17, 219–20, resp.; FCD to HH, 1 May 1877, in HN 116; HH to Verehrte Frau, 24 July 1877 (quotes); HH to HK, 21 July 1877, in HKL; KM to Seine Majestät den Kaiser und König, 24 August 1877, in KM, GSPK, I. HA, Rep. 89, no. 21487, Bl. 78.

11. HH 1877, 167–68.

12. Ibid., 169–76 (quotes on 169–73).

13. Ibid., 182–83.

14. Ibid., 183–84; for this theme, Schiemann 1997.

15. HH 1877, 185–86 (quotes); HH 1883, "Robert Mayer's Priorität," 401–14; Gross 1891, 26–30; 1898, 160–68.

16. HH 1877, 186–87, 189.

17. Ibid., 188–89.

18. Haeckel 1902b, esp. 2:134–35 (quotes); Richards 2008, 312–31; Steif 2003, 127–38; Daum 1998, 66–83.

19. Virchow 1877, 5–15, 18–22, 26, 28–29, 31–32 (quotes on 7, 8, 9, resp.); Daum 1998, 65–83; Steif 2003, 127–38; Richards 2008, 312–31; Finkelstein 2013, 240, 245–64, 271–74, 280–81, 285.

20. Guttstadt 1886, 62, 64; HH to JT, 18 February 1878, in RI MS JT/1/H/55, TPRI.

21. Guttstadt 1886, iv; Schnädelbach 1991, 36–48.

22. HH 1877a, 193–94.

23. Ibid., 196–99 (first quote on 196, second on 197).

24. Ibid., 198–99 (quote on 199).

25. Perry 1878; HH to J. Rodenberg, 6 January 1878, in Nationale Forschungs- und Gedenkstätten der klassischen deutschen Literatur in Weimar, Goethe- und Schiller-Archiv; Perry 1877; HH 1877a, 196; Müller 1881, esp. 5:7–10, 39–40; Atkinson 1880, v.

26. Huxley 1968, esp. 103–7; Roscoe 1869; Maxwell 1873, 397, quoted in Frank Turner 1993, 174–75.

27. Fox and Guagnini 1998, 58–68, 116–17; MacLeod 1972, 126–39; Haines 1958, 222–26; Kim 2002, xv; Arnold 1892, 133–52, 208–13 (quote on 209), 219.

28. Ward 1965, 148, 289–90, 303, 307; Fox 1997, 680–91 (quote on 690); Fox 2005; Gooday 2005.

29. Huxley 1968a; Arnold 1986, 456–71.

30. HH 1877a, 199–200.

31. Fox 1973; Shinn 1979, esp. 291–304; Weisz 1983; Fox 1984, esp. 70–72, 78–84, 87–88, 91–95, 97–103, 112; Fox 1990, esp. 14–16; Fox and Guagnini 1998, 107–115.

32. HH 1877a, 200–201.

33. Ibid., 201–3 (quotes on 202–3).

34. Ibid., 203–6.

35. Ibid., 205–6.

36. Ibid., 206–7 (quote on 207).

37. Ibid., 207–8 (quotes), 211.

38. Ibid., 209–10.

39. Ibid., 210–12.

40. Königliche Universitätsbibliothek 1899, 742; Emile Alglave to HH, 31 January 1878, in HN 10; HH 1884; Albert Ladenburg to HH, 2 December 18[??], in HN 259. In 1878 Haeckel spoke on and rejoined Virchow's "Die Freiheit der Wissenschaft im modernen Staat." See Haeckel 1902.

41. HH 1877, 184 (quote); Fichte 1970, 313–14; 1873, ix–x, 210–11; Fichte 1876, 426–27.

42. HH 1871, 55–56.

43. HH to EdBR, 20 April 1868, in Kirsten et al. 1986, 228 (first quote); HH to CL, 28 March 1869, in *LK*, 2:162 (second, third, and fourth quotes); Johann Eduard Erdmann to HH, 20 April 1871, in HN 137; HH to Erdmann, 6 May 1871 (fifth and sixth quotes), in Martin-Luther-Universität Halle-Wittenberg, Universitäts- und Landesbibliothek Sachsen-Anhalt, Sondersammlungen, Yi 4 I 147; HH to Wilhelm Oncken, 1 December 1872 (seventh and eighth quotes), in Justus-Liebig-Universität Giessen, Universitätsbibliothek, Handschriften-Abteilung, 139/100–107.

44. HH to George Croom Robertson, 7 December 1874, in UCL, George Croom Robertson Papers, MS. Add 88/11.

45. HH to Robertson, 26 December 1875 (quote); 8 January 1876, both in ibid.; Robertson to HH, 5 January, 14 February, 29 June 1876, all in HN 380; HH 1876a.

46. HH 1876a; 1903b, v (quote).

47. HH 1870f, 3–4, 30–31; 1876a, 319–21.

48. Land 1877 (quotes on 38–42, 45).

49. HH to Robertson, 6 April 1877 (first quote), in UCL, George Croom Robertson Papers, MS. Add 88/11; HH 1878, Appendix III, 394–406; 1878c; 212–24; 1883a, 640–60; 1903b, vii; HH to James Sully, 29 April 1877 (other quotes), in UCL, James Sully Papers, MS. Add 158; Sully to HH, 3 May 1877, in HN 417. In 1881 Helmholtz publicly downgraded his assessment of Kant's epistemological thinking, now distancing himself from philosophical remarks in the introduction to his essay on the conservation of force. (See HH 1882, 68.) In 1887 Helmholtz again distanced himself explicitly from "the strict adherents of Kant." (HH 1887a, 356.)

50. HH 1878c, esp. 659–60 (quote).

51. Boi, Giacardi, and Tazzioli 1998, 58, 78; Voelke 2005, 31; Krause 1878, unpaginated "Vorwort" (quote); Torretti 1978, 286; Hörz 1997, 232, 260–61.

52. Craig 1978, 93, 95–96, 145–50; Niemeyer 1963, 72.

53. HH to AH, 23 July 1878, in ESH 1929, 1:223–24, on 224.

54. Niemeyer 1963, 70–73.

55. HH 1995, 342 (first quote); AH to HH, 1, 4 August 1878, in ESH 1929, 1:224–25, 225 (second quote).

56. HH 1995, 342.

57. Ibid., 344.

58. Ibid., 344–45.

59. Ibid., 345 (first two quotes), 347–48 (other quotes).

60. Ibid., 349–53. Heidelberger 1993, 1994; De Kock 2014, 2014a emphasize Fichte's importance for Helmholtz; by contrast, Schiemann 1997 downplays it. In HH 1995, Appen-

dix I, 367–69, Helmholtz provided a discussion of "the localization of sensations of the inner organs." As noted above, Krause, like Land, had attacked him for criticizing Kant in regard to the geometric axioms and space (see Krause 1878). Helmholtz responded to Krause in Appendix II of his essay (HH 1995, 369–71).

61. HH 1995, 355–56.

62. Ibid., 357–58.

63. Ibid., 359–60. Cf. Helmholtz's citation of these lines in HH to Aureole (a family member), 12 November 1879, in DM, 1951/25C3; in Albumblatt, but in Anna Helmholtz's handwriting, 27 February 1891, in Stadtarchiv Bonn, Ii 98/511; in his speech of 1892 on Goethe (HH 1892) and in an undated autograph card, in Landesarchiv Berlin, Rep. 200 Acc. 980 no. 74. (See also AH to Cosima Wagner, 25 July, 14 December 1893, in Werner and Irmscher 1993, 43–44, 44–45, resp.) Cf., too, the ditty he composed (9 March 1872), quoted in Wiedemann 1989, 134–35:

> Magnetes Geheimniss, erkläre mir das!
> Kein grösser Geheimniss, als Lieb und Hass.
> Warum tanzen Bübchen mit Mädchen so gern?
> Ungleich dem Gleichen bleibet nicht fern.
> Dagegen die Bauern in der Schenke
> Prügeln sich gleich mit den Beinen der Bänke.
> Die endliche Ruhe wird nur verspürt,
> Sobald der Pol den Pol berührt.
> Drum danket Gott, ihr Söhne der Zeit,
> Dass er die Pole für ewig entzweit.

64. HH 1878c, 642.

65. HH 1995, 361–63.

66. Ibid., 363–66 (quotes). In ibid., Appendix III, 371–80, devoted to "the applicability of axioms to the physical world," Helmholtz responded to Land's criticisms. These thoughts had originally appeared a few months earlier in *Mind* (see HH 1878c) and were later republished in German in HH 1883a, 640–60. (Cf. HH 1903b, vii.) For Helmholtz's different views (from Kant's) of free will and intuition, see De Kock 2016.

67. Helmholtz 1837, 17 (quote); see, for example, Schwertschlager 1883; Erdmann 1921; Heimann 1974a; Hatfield 1990; Heidelberger 1993; Köhnke 1993; Krüger 1994; Boi 1996; Schiemann 1997; D'Agostino 2001, 2004; Friedman 2002, 2009; Darrigol 2003a, 549–50; DiSalle 2006; Lenoir 2006; Giovanelli 2008; Hyder 2009; Hatfield 2011, among others.

68. HH to AH, 4 August 1878.

69. Ludwig Nohl to HH, 7 August 1878 (first quote), in HN 332; Carl Wilhelm Borchardt to HH, 6 November 1878, in HN 58. See also Ludwig Christian Wiener to HH, 13 April 1879, in HN 510; Marie von Olfers to HH, 10 February 1879 (second and third quotes), in HN 338.

70. AH to HH, 28 August 1878; HH to AH, 21, 24 September 1878, all in ESH 1929, 1:225–26, 227, 227–28 (quotes), resp.; HH to Franz Boll, 7 August, [late August]; 13, 23 September; 1 October 1878, all in Belloni 1982, 132–35.

71. HH to AH, 27 September (first quote on 229), 29 September (third quote on 230, second and fourth quotes on 231), 2 October 1878, all in ESH 1929, 1:229–30, 230–31, 231, resp.; HH to Franz Boll, 23 September, 1 October 1878, in Belloni 1982, 135–36.

72. HH to AH, 12 July 1880, in ESH 1929, 1:250–51; Margarethe Boll to HH, 5 December 1880, in HN 56; HH 1881d.

73. HH to RL, 2 March 1881 (quote), in Mathematisches Institut, Rheinische Friedrich-Wilhelms-Universität Bonn, Bibliothek; and in Scharlau 1986, 131–32, on 131. In addition to the references cited in chapter 15 and elsewhere in this chapter, see also Laas 1884, 572–97;

Riehl 1904, 1904a, 1921; Reinecke 1903; Study 1914; Cassirer 1950, 41; Torretti 1978, esp. 155, 163, 255, 264–71, 281, 314–16, 393n33; Russell 1983; Richards 1988, 208–14 (re Russell); Friedman 1997; Gray 2008, 25–26, 97–101, 210–12, 297–98, 389, 392–93, 395–97; Neuber 2012; Gray 2013, 13–14, 39–47, 56, 76–82, 85–100, 217–18, 223, 235, 238, 245–46; Biagioli 2014, 2014a.

74. HH to RH, 4 July 1880, in ESH 1929, 1:248–50, on 249 (first quote); James Clerk Maxwell to Lewis Campbell, 21 April 1862 (second quote), in Campbell and Garnett 1969, 335–36, cited in Garber, Brush, and Everitt 1986, 337–38, on 337.

Chapter Twenty-Two

1. Zöllner 1878–81, 1:91, 106, 114–16, 119–21, 127–33, 137–39, 141–44, 149–50, 156–60, 162–65, 171, 176, 186, 193, 290, 297–98, 308–14, 316–21, 323–27, 399, 711.

2. See, for example, Gauld 1968; Oppenheim 1985; Coon 1992, esp. 144–45; Hatfield 1997; Canales 2001; Heidelberger 2004; Schmidgen 2002, 2003, 2005; James 1986, 99–100 (quotes).

3. Gauld 1968, 124–26; Oppenheim 1985, 22–23, 32, 61, 126, 232, 241, 297, 331.

4. Zöllner 1878–81, 1:1, 172–76, 180–81, 183, 186–87, 203 (fourth quote), 260, 276, 726–29 (first three quotes on 729), 3:xxiv–xxv (fifth quote); Marshall and Wendt 1980, 160–62; Meinel 1991, 38–43; Epple 1999, 166–73; Staubermann 2001; Heidelberger 2004, 67–69, 323; Wolffram 2009, esp. 23, 37–41.

5. Hall 1912, 265–8 (first three quotes on 266); Robinson 1988, 57–62; Marshall and Wendt 1980, 160–73; Stromberg 1989; Foerster 1911, 97–98; HH to JT, 18 February 1878 (fourth quote), in RI MS JT/1/H/55, TPRI; JT to HH, 22 February 1878 (fifth quote), in HN 477; EB to EdBR, 17 February 1878, in Brücke et al. 1978–81, 1:216–17, on 216; Leopold von Pfaundler to HH, 28 March 1878, in HN 350; Friedrich Kohlrausch to HH, 22 November 1881, in HN 241; Karl Franzos to HH, 17 January 1891, in HN 155; HH 1892a.

6. See, for example, Craig 1978, 83–85, 153–55; Pulzer 1988, 3–57.

7. Pollock 1880, 451–55; HH to AH, 12 July 1880, in ESH 1929, 1:250–51; Berthold Auerbach to HH, 16 January 1872; 17 May 1875, both in HN 19; HH to Auerbach, 17 May 1875; 21 December 1878; HH to Auerbach, "am 2ten Pfingstag" [i.e., ca. mid-May] 1875, in Deutsche Schillergesellschaft Marbach, Berthold Auerbach Nachlass, Deutsches Literaturarchiv Marbach am Neckar, Handschriften-Abteilung, A: Auerbach Z 3270/1–2, printed in Hörz 1997, 294–96; Scheuffeln 1985, esp. 92; HH to Baumhauer (perpetual secretary of the Hollandsche Maatschappij der Wetenschappen), 18 June 1879, in Hollandsche Maatschappij der Wetenschappen, Haarlem.

8. Kohut [1900–1901]; Wenkel 2007; Ebert 2008, 127, 244, 252, 264–67, 300–302; for anti-Semitism in German physics, see Jungnickel and McCormmach 1986, 2:40, 50, 279, 286–87; in German chemistry, see Rocke 1993, 350–63; EdBR to HH, 20 October 1858, in Kirsten et al. 1986, 192 (quote).

9. HH to AH, 24 August 1871, in Thompson 1910, 2:613–14, on 613 (quote), cited in Parshall 2006, 217; Boehlich 1965, esp. 5–12; Craig 1978, 42, 48–49, 153–55, 204–5; Pulzer 1988, 83–97, 240–45.

10. HH to EdBR, 12 November 1880, in Kirsten et al. 1986, 262 (first quote); Boehlich 1965; EdBR to TM, 14 November [18]80, in TMN, Bl. 14 (second quote); Finkelstein 2013, 219–20.

11. HH to Verehrter Freund [Ludwig Bamberger], 28 March 1893, in Bundesarchiv, Abteilung Potsdam, Nachlass Bamberger.

12. HH to AH, 19 March 1880, in ESH 1929, 1:235–36; AH to HH, 18 March 1880, ibid., 236; ibid., 236–37; HH to AH, 21, 24, 26 March 1880, ibid., 238 (first two quotes), 238–39, 239–40 (all other quotes).

13. HH to AH, 30 March, 2 April 1880, ibid., 240–41 (first four quotes), 241 (fifth quote).

14. HH to AH, 2, 4, 6 April 1880, ibid., 242 (first quote), 242–43 (other quotes), 243–44.

15. HH to AH, 13, 16, 20, 21 April 1880, ibid., 244–46 (all quotes except last two), 246–47 (last two quotes), 247, 248.

16. Häfner 1934, 384; [1962?], 112.

17. AH to HH, 19 April; 4, 9, 14 May 1879, all in ESH 1929, 1:232–33, 233, 234, 234–35, resp.; RH to AH, 30 June 1879, ibid., 235; HH to WT, 23 August 1883 (quote), in WTP, Add 7342, H75.

18. HH to RH, 4 July 1880, in ESH 1929, 1:248–50 (quotes on 248–49).

19. HH to RH, 4 July 1880, 249–50 (quotes); AH to HER, 15 November [1880], in RSC; AH to HH, 11 May 1881, in ESH 1929, 1:258.

20. HH to Julie Helmholtz, 31 July 1880, in Karl-Marx-Universität, Universitätsbibliothek, Briefsammlung, Autographensammlung Taut, Gelehrte; AH to HH, 16 October 1880, in ESH 1929, 1:253 (quote); ESH 1929, 1:252; Weech 1906, 298; Bunsen 1899; Braun-Artaria 1899; 1918, 135.

21. HH to EdBR, 18 August 1880, in Kirsten et al. 1986, 261–62 (first quote on 262); AH to ESH, August 1880, in ESH 1929, 1:251–52 (second quote); HH to AH, n.d., in DM, 1951/25C2; A. Jenenz? to HH, 2 October 1880, in HN 219.

22. Giacomo Barzellotti to HH, n.d. [ca. 1880] (quotes), in HN 28; Barzellotti 1880.

23. Pietro Blaserna to HH, 13 November 1880, in HN 50.

24. Dühring 1878a, 447–53; 1880, 38–39, 47–48, 50, 57–58, 98, 101 (first quote), 102–4, 106–7 (second quote), 109–13, 117–23, 126–30, 147–51, 153–54, 162, 205; 1903, 469–76.

25. Dühring 1880, 95–96, 99–100 (quotes), 103, 105, 118–21, 214–16; 1875, 67–68.

26. Dühring 1880, 114 (quotes), 115–16, 207.

27. JT to HH, 5 May 1862; 25 January 1880, both in HN 477; Weyrauch 1893, for a refutation; RC to JT, 10, 17 November; 11 December 1879, all in Eidgenössische Technische Hochschule, Zurich, Hs 227:5–163 (copies at the Royal Institution).

28. HH to JT, 28 January 1880 (quotes), in RI MS JT/1/H/58, TPRI. Charles Hermite, speaking of Zöllner's and Dühring's attacks against Emil du Bois–Reymond and Helmholtz, wrote his fellow mathematician Paul du Bois–Reymond, Emil's brother, that Emil du Bois–Reymond and Helmholtz had nothing to fear; quite the contrary, they represented "the honor of German science." Charles Hermite to Paul du Bois–Reymond, 2 August 1882, in Lampe [1916], 213–14, on 213.

29. HH 1883, 401–2 (quotes); 1882, 71–73; 1869, 385; 1891, 11; Cahan 2012a.

30. HH 1883, 402–3.

31. Ibid., 403–7 (quotes on 403, 406–7).

32. Ibid., 407–14 (first and fifth quotes on 413, second quote on 407, third quote on 408–9, fourth quote on 410, sixth and seventh quotes on 414); Dühring 1885, iii–iv, 74, 81; Dühring 1903, 96, 162–77, 179–211, 263, 278.

Chapter Twenty-Three

1. John Hall Gladstone to HH, 1 (first two quotes), 5 February 1878, both in HN 169; HH to JT, 18 February 1878, in RI MS JT/1/H/55, TPRI.

2. HER to HH, 5, 20, 21 (quote) November; 28 December 1880; 27 January 1881, all in HN 385; HH to HER, 27 November 1880; 3 January 1881, in RSC; Roscoe 1906, 90–91; AH to HER, 15 November [1880], in RSC.

3. HER to HH, 4, 10 March (quote) 1881, both in HN 385.

4. HER to HH, 20 March (quote), 27 January 1881, in HN 385.

5. JT to HH, 7 December 1880; Thursday, [31 March 1881?], both in HN 477; HH to JT, 11 March 1881 (quote), in RI MS JT/1/H/61, TPRI; MacLeod 1970; Barton 1990.

6. JT to HH, 26 September (first quote); 11 November 1880 (second, third, and fourth quotes), both in HN 477; HH to JT, 29 September 1880, in RI MS JT/1/H/59, TPRI; HH to JT, 22 November 1880 (seventh quote), in RI MS JT/1/H/60, TPRI; HH to EdBR, n.d. [1878], in Kirsten et al. 1986, 260 (fifth and sixth quotes); HH 1878a. S. P. Thompson also asked Helmholtz about priorities concerning the telephone's invention. See Silvanus P. Thompson to HH, 7 April 1882, in HN 463.

7. William Crookes to HH, 18 February 1881 (first quote), in HN 97; G. Johnstone Stoney to HH, 16 February (second quote), 19 March 1881, both in HN 452; HH to Robert McDonnall, 9 March 1881, in Wellcome Institute for the History of Medicine, London, 57468; Arthur Schuster to HH, 3 March 1881, in HN 442.

8. WT to HH, 13 (quotes), 29 March 1881, both in HN 464; WT to HH, 4 April 1881, in Thompson 1910, 2:764–65.

9. Quintino Sella to HH, 27 October 1880; 19 January, 7 February 1881, all in HN 399; HH to Hochgeehrter Herr [Sella], 14 November 1880 (first quote); 31 January 1881 (second quote), in Carte Quintino Sella, serie Accademia dei Lincei, mazzo 8, fascicolo 35, Carteggio 1880, sottofascicolo Helmholtz, and mazzo 9, fascicolo 39, Carteggio 1881, resp., Fondazione Sella, Biella (Italy); Corrado Tommasi-Crudeli to HH, 26 May, 7 June; 11, 25 August; 26 September 1881, all in HN 469. Augustus Lowell sought in vain to attract Helmholtz to Boston during the winter of 1882–83 in order to hold the Lowell Lectures there. See Augustus Lowell to HH, 21 November 1881, in HN 291.

10. FCD to HH, 15, 25 March 1881, in HN 116; HH to HER, 22 March 1881, in RSC; HER to HH, 26 March 1881, in HN 385; HH to JT, 11 March 1881; "Professor Helmholtz" 1881; "Professor Helmholtz in London" 1881; HH 1881h, 535; Roscoe 1906, 89–92 (quotes on 91–92).

11. Rayleigh to HH, 26 March 1881, in HN 364; V.C. [i.e., Vice-Chancellor, Cambridge University] to HH, 28 March 1881, in HN 523; AH to "Lieben Leute!," 29 March 1881, in ESH 1929, 1:254–55; AH to [her children], 2 April 1881, ibid., 255–56 (quotes); Rayleigh to ?, 12 April 1881, in Rayleigh 1968, 130–31, on 130.

12. AH to [her children], 2 April 1881.

13. Rayleigh to ?, 12 April 1881.

14. AH to [her children], 5 April 1881, in ESH 1929, 1:256–57, on 256; JT to HH, 7 December 1880, in HN 477; JT to HER, 4 April 1881 (first quote), ibid. (typescript at RI MS, JT/1/T/1265, TPRI); Roscoe 1906, 92 (other quotes); Schuster 1932, 244, who reports a slight variation on this story.

15. AH to [her children], 5 April 1881; AH to [her children], 12 April 1881, in ESH 1929, 1:257; "Professor Helmholtz in London" 1881; HH 1881h, 536 (quote). Roscoe 1906, 89, for the same quote but with the addition of "and as a philosopher."

16. HH 1872b, 1873f, 1876c, 1877b, 1879, 1879a, 1880, 1881f; Kragh 1993, 406–14; Darrigol 2000, 266–73; 2003, 163–68; Nernst 1921. Three times after 1881, Helmholtz reported on new instrumental apparatus, procedures, or experimental results concerning electrolysis: see HH 1884a, 1887d, 1887e.

17. HH 1880; 1881, esp. 290–91. The latter appeared in shortened form in *Nature* (7 April 1881) and in the *Chemical News* (8 April 1881); in June 1881 it appeared in much fuller form in the *Journal of the Chemical Society: Transactions*.

18. HH 1881, 251–52.

19. Ibid., 253–54 (first quote), 255 (second quote).

20. Ibid., 256–58.

21. Ibid., 258–61 (quote on 259).

22. Ibid., 262–69.

23. Ibid., 270–72 (quotes on 272).

24. AH to [her children], 12 April 1881 (first two quotes); Roscoe 1906, 89–90 (third and fourth quotes).

25. AH to [her children], 12 April 1881 (second quote); "Professor Helmholtz in London" 1881; William Bowman to HH, Thursday, [no date], 18 March 1881, both in HN 61; "Professor Helmholtz" 1881; HH 1881h, 536 (first quote); William Grylls Adams to HH, 5, 6 April 1881, both in HN 5; JT to AH, 5 June 1881 (third quote), in HN 477.

26. Stoney to HH, 6 April 1881 (quote), in HN 452; Laurence Parsons Rosse to HH, 15 March 1881, in HN 388J; Magee Finny (Fellow and Registrar, King and Queen's College of Physicians in Ireland) to HH, 5 November 1881, in HN 523.

27. HH 1884a, 88; Alexander Crum Brown to HH, 28 February, 5 April 1881, in HN 72; HH to HER, n.d. but shortly after April 1881, in Roscoe 1906, 91; PGT to HH, 9 June 1881, in HN 459.

28. HER to HH, 3 May 1881 (first quote), in HN 385; HH 1881; 1881h; HH to HER, n.d. [sometime between 3 and 10 May 1881] (second quote), in RSC; Roscoe 1906, 91.

29. Williams 1994, 136; AH to RH, 11 May, 20 June 1881, in ESH 1929, 1:258 (first two quotes), 259 (last three quotes).

30. WT to HH, 9 July 1881, in HN 464; HH to WT, 15 July 1881 (first quote), in WTP, Add 7342 H73; ESH 1929, 1:259; AH to HH, 27 August 1881 (second quote), in ESH 1929, 1:260.

31. Crawford 1992, esp. 27–32, 36–43.

32. Fox 1990, esp. 14–16, 18, 21–23; Cahan 1985.

33. Blondel 1990; Fox 1990, 18.

34. Original accounts of the 1881 congress include "International Exhibition" 1881; "Congrès scientifiques" 1881; "Die Eröffnung" 1881; HH 1881a, 1881b. See also Janet 1909, 513–17; Blondel 1990; Fox 1990, 1996; HH to AH, 16 September 1881, in ESH 1929, 1:260–61.

35. Janet 1909, 487, 511–12, 518.

36. HH 1881b, 295–96 (quote on 296), first published in slightly different form as HH 1881a (see HH 1903, viii).

37. HH 1881b, 296 (quote), 302–5. As early as 1863, the firm of Siemens & Halske informed Helmholtz and other physicists, technologists, and telegraph services that it was keenly interested in reaching agreement on a resistance standard for electrical currents. Siemens & Halske to HH, 15 November 1863, in HN 407.

38. Janet 1909, 513–14.

39. HH 1881b, 296–306; Smith and Wise 1989, 684–98; Blondel 1990; esp. Olesko 1996.

40. "International Exhibition" 1881, 564.

41. Ibid.; "Congrès scientifiques" 1881, 407–8, 412–14; EdBR to Jeannette du Bois-Reymond, 18 September 1881 (first quote), in SBPK, Handschriftenabteilung, Dep. 5, K. 11, no. 5; HH to AH, 19 September 1881, in ESH 1929, 1:261–62 (second quote); Thompson 1910, 2:775 (third quote); "Congrès scientifiques" 1881, 409–12.

42. HH to AH, 16, 19 September 1881, in ESH 1929, 1:260–61 (quotes); Pantalony 2009, 127, 145–46; Friedman 1989, 12; Fleming 2016, 15–16.

43. "Congrès scientifiques" 1881, 409–14; HH 1881b, 307–8; Janet 1909, 515–16 (quotes).

44. "International Exhibition" 1881, 512; "Congrès scientifiques" 1881, 414; Janet 1909, 517.

45. Fox 1990, 18; Blondel 1990; HH 1881b, 309 (quote); envelope marked "Légion d'honneur: Commandeur 1881"; HH to Monsieur le Ministre, n.d., draft, both in SF, Sign.: Helmholtz 6.LL496; HH to A. Pagenstecher, 29 December 1881, in Universitätsbibliothek Heidelberg, Handschriftenabteilung, Heid. Hs. 840,3.

46. HH 1882a (quote on 994).

47. William Siemens to HH, 8 June 1882, in HN 406; WT to HH, 11 August 1882, in WTP, Add 7342 H73; HH to WT, 18 September 1882 (quotes), in WTP, H74.

48. HH to K. H. Boetticher, 24 August 1882, in ZstA, RmdI, no. 13157, Bl. 68–69, printed in Buchheim 1977, 29–30 (quote on 29).

49. Ludewig 1882, esp. 404–8; HH 1881b, appendix, 411–12; 1882b; Thompson 1910,

2:788–90; John Trowbridge to Daniel Coit Gilman, 12 November 1882 (quote), in Henry A. Rowland Papers, Special Collections and Archives, Manuscripts, Milton S. Eisenhower Library, Johns Hopkins University, Baltimore, MD.

50. Marey 1899, 23–24 (first quote); Marta Braun 1992, esp. 12, 52; Gabriel Jonas Lippmann to HH, 25 October 1882, in HN 279; Charles Hermite to Gösta Mittag-Leffler, 12 November 1882, in Dugac 1984, 182–84 (second quote on 182); Mitchell 2012; HH 1882b.

51. HH to Georg Ernst Reimer, 3 May 1881, WGA.

52. Vermerk, 27 January 1883, in KM, I. HA, Rep. 89, no. 21488, Bl. 52; HH to EdBR, 30 January 1883, in Kirsten et al. 1986, 264; Siebmacher 1981, 100, table 84; Werner 1997, inside cover; HH to Verehrter Freundinn, 2 February 1883, in Märkisches Museum, Berlin, IV 61/2936 Q (quote).

53. HH to Arthur Freiherr Schmidt-Zabiérow, 28 January 1883, in ESH 1929, 1:262–63, on 262 (first quote); AH to Fritz Helmholtz, 28 January 1883, ibid., 263; Laforgue 1922, 56 (second quote); Bunsen 1932, 40–41; Brocke 1996, esp. 268, 277, 279, 281, 289, 301, 312; Cahan 2006; National Academy of Sciences 1884, 6–7; 1913, 345; HH to A[lexander]. Agassiz, 15 May 1884, in Archives, National Academy of Sciences, National Research Council, Washington, DC.

54. AH to Franz, 8 January 1876, in ESH 1929, 1:202–3; HH to HER, 25 July 1876, in RSC; Simpson 1887, 355; Charles Hermite to HH, 29 January, 17 April, 30 May 1882, all in HN 198; Archibald 2002; Gray 2013, 7, 13–14, 382–85, 425–26, 468–81; AH to HH, 18, 26, 31 May; 14 June 1883, in ESH 1929, 1:264, 264–65, 265–66, 267, resp.

55. EB to EdBR, 15 [November] 1883, in Brücke et al. 1978–81, 1:251–52 (first quote); Wilhelm Dilthey to Paul Graf Yorck von Wartenburg, 10 November 1883, in Dilthey 1923, 36 (second quote); Wechselseitiges Testament, 26 May 1884, in Akten betreffend die letztwillige Verfügung des Wirklichen Geheimen Raths und Präsidenten Professor Dr. Hermann von Helmholtz und dessen Ehefrau Anna geb. von Mohl zu Charlottenburg, 63/52 IV 1357/1894; Königliches Amtsgericht. I. Testam.-Repert. No. 12687/84, No. 1, 26 May 1884, both in Amtsgericht Charlottenburg, Abt. 1 Verwaltung.

56. HH to WT, 11 May 1886, in WTP, Add 7342 H77; WT to HH, 23 May 1886 (quote), in HN 464; Fischer quoted in Lockemann 1949, 218–19.

57. WT to HH, 16 August, 17 June 1883, both in HN 464; HH to WT, 2 September 1883 (quote), in WTP, Add. 8812/4.

58. Sorbelli 1936, 62:2; "Casamicciola" 1931.

59. Hirsch and Oppolzer 1884; Helmert 1913; Hörz 1996; Torge 2005; Johann Jacob Baeyer to HH, 23 January 1881, in HN 24.

60. Werner von Siemens to HH, 21 June 1880 (first two quotes), in HN 405; Helmert 1913, 403; Torge 2005, 563; HH to AH, 16 October 1883, in ESH 1929, 1:271 (third quote); Pietro Blaserna to Leo Koenigsberger, n.d. [ca. 1900?], in LK, 2:310–12, on 311–12 (fourth, fifth, and sixth quotes).

61. HH to AH, 16 October 1883 (quote); Guido Baccelli to HH, 12 March 1883, in HN 22.

62. HH to AH, 16 October 1883 (quotes); A. Jenenz? to HH, 21 March 1886, in HN 219; Helmert 1913, 405–6; Torge 2005, 564.

63. HH to AH, 16 October 1883; HH to AH, 19, 21 October 1883, both in ESH 1929, 1:271–72 (quotes on 272), 273; Helmert 1913, 403; Torge 2005, 562; Thompson 1910, 2:798.

64. Unless otherwise indicated, this section is drawn from Cahan 1999, 290–305, where full documentation is to be found.

65. Cf. EdBR to AH, 29 July 1893, in Kirsten et al. 1986, 273.

66. HH to Verehrter Freund, 27 April 1875, in Karl-Marx-Universität, Universitätsbibliothek, Briefsammlung, Rep. IX, 3 (Bendemann), 47 B 33.

67. Sitzung des Plenums, 21 February 1884, in AHA, GesammtsitzungsProtocolle 1884,

Sign. II–V, 61, Bl. 6; HH to HER, 16, 27 March 1884, both in RSC; AH to ISZ, 4, 10 March; 8 April 1884, all in ESH 1929, 1:275, 276.

68. HH 1881e, 1881g; HER to HH, 2 April 1884, in HN 385; HH to AH, 6 April 1884, in ESH 1929, 1:276–77; Roscoe 1906, 120 (quotes).

69. HH to AH, 12 April 1884, in ESH 1929, 1:277–78.

70. Ibid.

71. AH to ISZ, 10 March 1884; Crum Brown to HH, 22 March, 5 April 1884, both in HN 72; William Rutherford to HH, 24 March 1884, in HN 395.

72. HH to AH, 19 (first three and fifth quotes), 22 April 1884, both in ESH 1929, 1:278–79, 279, resp.; A. Fraut to HH, 4 April 1884, in HN 156; James 1979, 91 (fourth quote).

73. HH 1881b, esp. 411–12; Jaeger 1932, 29–30; Heinrich Wild to HH, 4, 16 October; 22 November, 4 December 1884 (quotes), in Academy of Sciences, Saint Petersburg, Archive, F. 210, Op. 1, No. 376, 2–5; HH to Wild, 20 October 1884, ibid., 1, 1 f.; Blondel 1990.

74. ESH 1929, 1:280; AD to Marie Dohrn, 13, 18 May 1884, in SZADNA, Bd. 259, Bd. 262; AD to Marie Dohrn, 30 May 1884, in SZADNA, Bd. 266; Kohut 1884.

75. AH to HH, 31 July, 6 August 1884, in ESH 1929, 1:280–81 (quotes); 281–82, on 281.

76. Bunsen 1932, 40–1 (quote on 40).

77. AH to ISZ, 1 January 1885, in ESH 1929, 1:283–84 (first five quotes); Braun-Artaria 1918, 135; HH to WT, 11 May 1886 (sixth quote), in WTP, Add 7342 H77.

78. HH to Vieweg, 2, 13 November; 14 December 1882; 13 February, 16 July; 5, 20 August; 7, 9, 28 November 1883; 8 September 1884; 27 November 1886; 12 December 1888, all in VV; HH 1884, 1:vii (first quote), viii (second quote).

Chapter Twenty-Four

1. Rep. 76 KM, Va Sekt. 2, Tit. X, no. 79, Bd. 1, Bl. 205–6; ibid., Tit. XV, no. 27, Bd. 3, Bl. 175r–77, in GSPK.

2. Robert Helmholtz 1890; HH 1975a, 107–8; 1975b, 109–10.

3. Cahan 1990; Rubens 1910, 288–90; Boltzmann 1994, I:97n444.

4. Hörz and Laass 1989, 54–59; Ludwig Boltzmann to HH, 6 June, 10 December 1888, ibid., 94–95; Boltzmann 1994, I:97–103n, 109–14, 116, II:120–23; Flamm 1995, 48, 51–53.

5. HeHz to his parents, 5 October 1888, in Hertz [1927], 195–98; HH to HeHz, 15 December 1888, ibid., 261; Planck 1948, 15 (quote).

6. Schultze 1993; HH quoted in Epstein 1896, 194–95.

7. Rep. 76 KM, Va Sekt. 2, Tit. XV, no. 27, Bl. 175r–77, 216r–18; ibid., Bd. 4, Bl. 51, 310r; ibid., Bd. 5, Bl. 16r, 51, 87r–89, 107r–9, 122r, in GSPK.

8. "Unterlagen der Philosophischen Fakultät," in HUB, Promotionsverfahren der Phil. Fak. 291 ff. Helmholtz, Gutachten, Prüfungen, und Promotion; Guttstadt 1886, 141; Lummer 1889, 567–68; Rubens 1910, 287; Schultze 1994.

9. HH to [Thomas A. Edison], 15 January 1886, copy, in TAEM, reel 79:240 (TAED, D8613A); Israel 1998, 283–84, 505n5.

10. HH to Franz Boas, 15 September 1879, in American Philosophical Society, Philadelphia, Library, Franz Boas Papers, Professional Series; Stocking 1982; Liss 1996, 163, 166–67, 171–77; Cole 1999, 51–55, 61, 65–80, 89–93.

11. Pupin 1951, 231 (first quote); Cahan 2004; Lowe 1983, 73; Butler 1939–40, 1:123–24 (second quote); Cahan 2010, 21–32; HeHz to his parents, 10 December 1878, in Hertz [1927], 77–78; A. D. White, letter headed "American Legation. Berlin. 19 July 1881," in SF, Sign. Helmholtz 6.LL496.

12. Rubens declared the period 1878–88 as "the most brilliant epoch" in the institute's history. See Rubens 1910, 286.

13. Among numerous accounts of Hertz's life and work, see especially Jungnickel and McCormmach 1986, vol. 2, esp. 29–30, 35–36, 44–48, 85–89, 92–97, 101, 141–43, 211; Buchwald 1994; Fölsing 1997; Darrigol 2000, 234–64. The present account generally follows these writings.

14. EdBR 1912, 2:546; HeHz to his parents, 6, 17 November 1878; 12 January; 5, 22 May 1879, all in Hertz [1927], 72–75, 75–77, 78–79, 83–85, 85–86, resp.

15. HeHz to his parents, 11 August, 4 November 1879; 6 February 1880, all in Hertz [1927], 87–88 (first quote), 89–90 (third quote), 91–94, resp.; HH 1894, 367 (second quote); "Gutachten Helmholtz (Teil) über die Doktorarbeit von Heinrich Hertz," in HUB.

16. HH to Kultusminister, 22 August 1880 (quotes), in GSPK, Rep. 76 KM, Va Sekt. 2, Tit. X, no. 79, Bd. 2, Bl. 35–35r (typed original in Autographen Sammlung Darmstaedter); HH to Ernst Hagen, 18 August 1880, in Staats- und Universitätsbibliothek Hamburg-Carl von Ossietzky, Literatur-Archiv: Hermann von Helmholtz; HH to HeHz, 8 August 1880, in Hertz [1927], 259; HeHz to his parents, 30 September, 14 October 1880, both in Hertz [1927], 101, 102–4, resp.

17. HeHz to his parents, 9, 24 October 1880; 17 March 1881, in Hertz [1927], 101–2 (quotes), 104–5, 112, resp.

18. Svante Arrhenius to Wilhelm Ostwald, 29 October 1886, in Ostwald 1969, 24–26 (first three quotes on 24); Ostwald 1926–27, 1:187 (last three quotes).

19. EdBR to HH, 26 May 1880, in Kirsten et al. 1986, 261; ibid., 311n1; Goldstein 1925, 40 (quote); Planck 1935, 12, 14.

20. Warburg 1925, 36; Jungnickel and McCormmach 1986, 2:4–5, 21–22; Fiedler 1998, 73–74; Fölsing 1997, 183–85, 324.

21. Lodge 1931, 154; WT to HH, 11 December 1882, in HN 464; HeHz to his parents, 2 February, 1 May 1882, in Hertz [1927], 118–21 (first quote on 119), 124 (second quote), resp.

22. For this theme and analysis, see Buchwald 1994.

23. Maxwell 1883. As early as 1876, Helmholtz had (in vain) urged Vieweg to publish a German translation of Maxwell's book. See HH to Vieweg, 12 May 1876, in VV. In 1881 Isidore Fröhlich, one of Helmholtz's former students, wanted to do a translation of Maxwell's *Treatise* and hoped to gain Helmholtz's support in this regard. See Isidor Fröhlich to HH, 29 May 1881, in HN 162.

24. HeHz to his parents, 1 March 1883, in Hertz [1927], 135–37 (first quote); Fölsing 1997, 195; Albert Ladenburg to HH, 20 March 18[83] (second quote), in HN 259; HH to Kultusminister, 7 April 1883, in GSPK, Rep. 76 KM, Va, Sekt. 2, Tit. X, no. 79, Bd. 2, Bl., 72–72r; HH to HeHz, 29 July 1883 (third quote), typescript copy, in SF, Sign.: Helmholtz 6.LL496; HeHz to HH, ca. August 1883, in *LK*, 2:306–9 (fourth quote on 306).

25. HeHz diary, entries of 24–27 December 1884, in Hertz [1927], 154; Friedrich Althoff to HH, 4 January 1889, in HN 12.

26. HeHz to HH, 5 December 1886, in Hertz [1927], 164–66 (quotes); Hertz 1894–95, 2:32–58, 68–86; HH to HeHz, 8 June 1887, typescript copy, in SF, Sign.: Helmholtz 6.LL496.

27. HeHz to his parents, 30 October 1887, in Hertz [1927], 178–79 (first quote); HeHz to HH, 5 November 1887, ibid., 179–80; Elisabeth Hertz to Heinrich Hertz's parents, 9 November 1887, ibid., 180 (third quote); Helmholtz's postcard, 7 November 1887, ibid., 181 (second quote); HeHz to HH, 8 December 1887, ibid., 182–83; Hertz 1894–95, 2:102–14.

28. Pupin 1951, 263 (first quote), 264 (second and third quotes); Goldstein 1925, 44; HeHz to HH, 21 January 1888, in Hertz [1927], 187–88 (fourth quote); HeHz to his parents, 29 January 1888, ibid., 188–89; HH to HeHz, 23 January 1888, typescript copy, in SF, Sign.: Helmholtz 6.LL496; HH to EdBR, 24 January 1888, in Kirsten et al. 1986, 266–67 (fifth quote); Hertz's "Über die Ausbreitungsgeschwindigkeit der elektrodynamischen Wirkungen," *Sitzungsberichte* (2 February 1888): 197–209, republished in the *Annalen* and later in Hertz

1894–95, 2:115–32; Jungnickel and McCormmach 1986, 2:86–89, 95–97; Darrigol 2000, 247–57, 263–64.

29. HH to Emil Fischer, 23 February 1888 (quotes), in Emil Fischer Papers, Bancroft Library, University of California, Berkeley; W. C. Röntgen to HH, 19 February 1885, in HN 382; Ferdinand Braun to HH, 16 July 1888, in HN 64; H. F. Weber to HH, 26 April, 25 May 1880; 10 July 1887, all in HN 496; Cahan 1999, 294–95; HeHz to HH, 19 March 1888, in Hertz [1927], 191–92.

30. HH 1889a; HH to RL, 9 September 1888 (quote), typescript in SF, Sign.: Helmholtz 6.LL496, reprinted in Hertz [1927], 262–63; HeHz to his parents, 5 October 1888, ibid., 195–98.

31. HeHz to HH, 30 November 1888, in Hertz [1927], 201–2 (first two quotes); HeHz to his parents, 16 December 1888, ibid., 204–5; HH to HeHz, 15 December 1888 (last two quotes), ibid., 261; HH to HeHz, 14 December 1888, typescript copy, in SF, Sign.: Helmholtz 6.LL496; Hertz 1894–95, 2:184–98.

32. Cahan 1999, 292.

33. Kayser 1936, 138–39, 142, 169–73 (quote on 139); Buchwald 1994; Fölsing 1997, 181.

34. Martius 1923, 112–13 (first two quotes on 113); Hoffmann 1948, 72; Magnus-Levy 1944, 335 (third quote); Tschirch 1921, 140 (fourth quote); Rubens 1910, 287 (fifth quote); Hertz 1894–95, 1:368.

35. G. Stanley Hall to William James 1878 (first quote), in William James Papers, Houghton Library, Harvard University, bMS Am 1092.9; Hall to Charles Eliot Norton, 3 February 1879, in Charles Eliot Norton Papers, Houghton Library, Harvard University, bMS Am 1088; Hall to Henry Pickering Bowditch, 23 December 1878, in Francis A. Countway Library of Medicine, Boston, H MS c5.2; Hall 1912, 250–51, 302–3; James to Carl Stumpf, 26 November 1882, in Perry 1935, 2:60–61 (second quote, on 60); Willstätter 1958, 241–42 (third quote, on 242); Planck 1949, 3–4; Magie 1894, 329–30 (fourth quote).

36. Pupin 1951, 232 (first quote); Hertz 1894–95, 1:368; Pupin 1894, 541–42 (second, third, and fourth quotes, on 541); Carhart 1894, 543; Arrhenius to Jacobus Henricus van't Hoff, 11 July 1906 (fifth quote), in Museum Boerhaave, Leiden, Arch. 208f, Correspondence Svante Arrhenius.

37. The following account of the Reichsanstalt's origins draws on Cahan 1989, 24–58.

38. Ernst Abbe to HH, 15 May 1883, in HN3; Kühnert 1961, esp. 323 (quote); Cahan 1989, 14–16; Feffer 1994, 203–34; Cahan 1996.

39. AH to HH, 14 June 1883, in ESH 1929, 1:267 (first quote); HH to AH, 17 June 1883, ibid., 267–68 (second and third quotes on 268).

40. HH to WT, 2 September 1883 (first quote), in WTP, Add. 8812/4; HH to J. R. Strutt (Lord Rayleigh), 2 March 1885 (second quote), in Rayleigh Papers, The Old Rectory, Terling, Chelmsford, England.

41. AH to ISZ, 8 April 1887, in ESH 1929, 1:306; HH to AH, 7 August 1887, ibid., 307–8 (first quote on 307); HH to Verehrter Freund [Werner von Siemens], 31 March 1887, in SF, Sign.: Helmholtz 6.LL496; Gossler to Seine Majestät den Kaiser und König, 19 November 1887, in KM, GSPK, Rep. 89, no. 21489, Bl. 123–25 (second quote on 124v).

42. HH to Kultusminister, 25 September 1887; HH to the KM, 14 September 1887, both in GSPK, Rep. 76, KM, Va, Sekt. 2, Tit. X, no. 79, Bd. 2, Bl. 165–66; AH to ISZ, 8 April 1887; 18 October 1888; 3, 16 June 1889, all in ESH 1929, 1:318–19, 2:9, 2:9–10 (quote), 10–11, resp.

43. HH 1878a, 1879, 1879a, 1880, 1881e, 1881f, 1881a, 1881b, 1882a, 1882b.

44. HH 1881c, 1883b; 1885, 588 (quotes).

45. Kragh 1993, 417–23. See also Dolby 1984, esp. 375–86; Bordoni 2013.

46. HH 1882c, 1882d, 1883c, 1887e; Kragh 1993, 422–23; Dolby 1984; on Pierre Duhem's great admiration of Helmholtz and as a follower of his chemical thermodynamics, see Pierre Duhem to HH, 26 April 18??, in HN 126.

47. Kragh 1993, 424–31.

48. Arrhenius to van't Hoff, 28 October 1890, in Museum Boerhaave, Leiden, Arch 208 f., Correspondence Svante Arrhenius; HH to ?, n.d. [1891], in *LK*, 2:297–98 (quotes); Barkan 1999, 23, 43, 50, 54–56, 72, 74, 85, 88–90; see also Nernst 1921.

49. Sloane 1920, 152–71; HH 1887f (quote on 282).

50. HH 1884b, 1884c. In re the latter's publication, see HH to Leopold Kronecker, 2, 11 July 1884, in Houghton Library, Harvard University, bMS Ger 198(3), folder 1. Helmholtz also published a third paper on the statics of monocyclic systems that year (HH 1884d), but he forbade its republication in his *Wissenschaftliche Abhandlungen* since, as he reportedly stated, it contained an erroneous conclusion. See HH 1895, 3:628; cf. also HH 1886c, 226n1.

51. HH 1884b (quote on 119).

52. Klein 1972, esp. 59–67; 1974, esp. 156–62; Jungnickel and McCormmach 1986, 2:130–31; Bierhalter 1981, 1983 1987, 1992; 1993, esp. 432–47; Bordoni 2013.

53. Klein 1972, 59–60, 63, 67–71; 1974, 162–66; Bierhalter 1993, esp. 447–50; HH to Kronecker, 8 November 1885 (quote), in Houghton Library, Harvard University, bMS Ger 198(3), folder 2.

54. HH 1886c; Koenigsberger 1898, 115–16, 119–23; Jungnickel and McCormmach 1986, 2:131–33; Bierhalter 1992, 59–67; 1993, 450–56; Hecht 1994; Gray 2013, 515–16.

55. Boltzmann to HH, 5 February 1887, in Boltzmann 1994, II:106–7; Boltzmann to Leo Koenigsberger, late 1896, ibid., II:267–68, on 267; HH 1887c, 1970; Harnack 1970, 1:334n2, 3:282–96 (HH 1970); Jungnickel and McCormmach 1986, 2:133–34.

56. HH 1970, 282–87 (quotes on 282, 287).

57. Ibid., 287–96 (quotes on 288, 294–95); for a modern study of Maupertuis, Terrall 2002, esp. 176–89, 231–309.

58. HH 1970, 287–96 (quote on 296).

59. HH to Hochgeehrter Herr (probably Georg Ernst Reimer), 28 February 1862; 3 May 1881 (quote), both in WGA; HH 1882, 12–75; 1889b.

60. Wilhelm Dilthey to HH, n.d. [late 1883, 1884], in HN 110; HH 1887a.

61. Unless otherwise noted, the following account is drawn from Darrigol 2003a. See also the literature cited therein, esp. Michell 1993; see Gray 2008, 93–97, 328–29, 344–45.

62. Darrigol 2003a, 516; HH 1887a, 356–57.

63. Darrigol 2003a, 520–49.

64. Alexander Wassilieff to HH, 1–13 December 1892, in HN 494; HH 1893b; HH to Duhem, 8 February 1893, in Pierre Duhem Papers, Archives, Institut de France, Académie des Sciences; Darrigol 2003a, 518–20, 555–70; Michell 1993; Biagioli 2014.

Chapter Twenty-Five

1. HH to HK, 5 May 1886, in HKL; AH to ISZ, Easter Monday 1886, in ESH 1929, 1:289–90 (quote on 289); HH to Leopold Kronecker, 26 April 1886, in Houghton Library, Harvard University, bMS Ger 198(3), folder 2.

2. AH to ISZ, 18 May 1886, in ESH 1929, 1:290–91; Spitzemberg 1963, 225.

3. HH to AH, 18 June 1886, in ESH 1929, 1:291–92; *Orden* 1978, 316; Wirth 1965, 146–48; RC to HH, 28 January, 20 April 1888, both in HN 91.

4. This and the subsequent discussion of the week's festivities draw on Bartsch 1886.

5. HH to AH, 5 August 1886, in ESH 1929, 1:294–95.

6. Ibid. (first two quotes); Roscoe 1906, 76 (third quote).

7. HH to AH, 5 August 1886 ("restless" quote); Pfaff 1995, 209; Bartsch 1886, 150–51 (for the speech quotes); Kussmaul 1902, 37–39.

8. Theodor Georg Freiherr von Dusch to HH, 12 June 1886, in HN 127; Roscoe 1906, 75;

HH to AH, 5 August 1886; HH to AH, 7, 9 August 1886, both in ESH 1929, 1:296 (second quote), 296–97 (first quote on 296), resp.; Bartsch 1886, 155; Koenigsberger 1919, 181–82.

9. HH to Geehrter Freund, 19 August 1885, in HN 532/3; HH to Kronecker, 4 September 1885, in Houghton Library, Harvard University, bMS Ger 198(3), folder 1; AH to ESH, 17 September 1885, in ESH 1929, 1:286–87, on 286; Der Ausschuss der Ophthalmologischen Gesellschaft to HH, telegram, 15 September 1885, in HN 340; Ophthalmological Society 1886, 8; HH to Verehrter College [FCD?], n.d., in Francis A. Countway Library of Medicine, Boston Archives, Helmholtz; HH to FCD, 31 January 1886, in *LK*, 2:337 (quotes).

10. Ophthalmological Society 1886, 5–42 (first and second quotes on 34, third on 38); FCD to HH, 13 January, 12 July, 28 November 1886, all in HN 116.

11. Ophthalmological Society 1886, 36 (first quote), 38–42 (other quotes), 43–44; HH to AH, 9 August 1886; Bowman 1890–91, xix–xx.

12. HH 1886, 314; HH to AH, 9 August 1886 (quote).

13. HH 1886, 313.

14. Ibid., 315–16.

15. Ibid., 316–18.

16. Ibid., 318–20.

17. HH to AH, 9 August 1886; AH to HH, 11, 13 August 1886, in ESH 1929, 1:297–98, 298–99 (all quotes except the first); Lummer 1889, 567; HH to My Lady [Mrs. WT], 23 September 1886 (first quote), in WTP, Add 7342 H78.

18. HH to My Lady [Mrs. WT], 23 September 1886 (first quote); Bader 1933, 214 (second quote); FCD to Friedrich Horner, 13 August 1886; Otto Becker to Horner, 3 October 1886; HH to Horner, 4 September 1886, all in Bader 1933, 152–53, 175–77 (on 177), 215–18 (on 215), resp.

19. AH to ISZ, 19 August 1886, in ESH 1929, 1:300 (first quote); AH to RH, 22 August, 2 September 1886, ibid., 300–301 (second quote), 301 (third and fourth quotes).

20. HH to Horner, 4 September 1886, 215–17 (first two quotes); AH to Horner, 10 October 1886, in Bader 1933, 219–21, on 219 (fourth quote); AH to ISZ, 14 September 1886, in ESH 1929, 1:301–3 (third quote); HH to My Lady [Mrs. WT], 23 September 1886.

21. AH to RH, 29 September 1886, in ESH 1929, 1:303–4 (first three quotes); AH to Horner, 10 October 1886, 219 (fourth and fifth quotes); Adolf Kussmaul to HH, 7 September [?] 1886, in HN 257; FCD to HH, 28 November 1886, in HN 116.

22. AH to Horner, 10 October 1886.

23. Becker to Horner, 3 October 1886, in Bader 1933, 177; HH to Horner, 4 September 1886; ESH to Horner, 15 September 1886 (first quote), in Bader 1933, 218–19; AH to Horner, 10 October 1886 (second quote); AH to Horner, 16 December 1886, in Bader 1933, 222.

24. Wilhelm Foerster to HH, 12 January, 27 February; 1, 5 March 1887, all in HN 151; "Festbericht" 1887, 114; AH to ISZ, 10 March 1887, in ESH 1929, 1:306; HH 1887 (first printed in "Festbericht" 1887); HH 1903, 2:viii–ix; Jackson 2000, 181–94.

25. HH 1887, 323.

26. Ibid., 324–26, 330 (first quote on 324, others on 326).

27. Ibid., 326 (first two quotes), 330–31 (other quotes on 331); Jackson 2000.

28. HH 1887, 331–33 (quotes on 333); Cahan 1989, 51; Jackson 2000, 186–88.

29. "Festbericht" 1887, 122–28 (quote on 127–28).

Chapter Twenty-Six

1. Spitzemberg 1963, 234–35; AH to ISZ, 12 February 1887, in ESH 1929, 1:305–6, on 305; William II 1926, 13–14.

2. AH to ISZ, 19 November 1887, in ESH 1929, 1:310–11, on 310; Craig 1978, 164–65, 169–70; AH to ISZ, 11 November 1887, in ESH 1929, 1:309–10.

3. AH to ISZ, 12 March; 6, 16 April 1888, in ESH 1929, 1:313, 314, 315, resp.; Craig 1978, 119–20, 162, 164–66; diary entries for 21, 22, 25 March 1888, in Spitzemberg 1963, 245–47; Nichols 1958, 197; AH to Lieber Freund [Ludwig Bamberger], 31 March 1888, in Bundesarchiv, Abteilung Potsdam, Nachlass Ludwig Bamberger; Gerhardt n.d., 106; EdBR to Jeannette du Bois–Reymond, 17 June 1888, in Dep. 5, K. 11, no. 5; William II 1927, 13–14, 158, 160–61.

4. F. Davenport to HH, 8 February 1888, in HN 105; Marciatori 2009, 314–16; H. J. Wood to HH, 10 July 1888 (first quote); HH to Hochgeehrter Herr [H. J. Wood], draft, n.d. (second quote); Awards, Honorary Memberships, etc.; Helmholtz's Personalakte, all in SF, Sign.: Helmholtz. 6.LL494.

5. WT to HH, 11 January (quotes), 24 February 1890, both in HN 464; Minutes of the Academic Senate, University of Edinburgh, 22 February 1890, in EUL; Alexander Crum Brown to HH, 22 February 1890, in HN 72; PGT to HH, 22 February 1890, in HN 459; Friedrich Max Müller to HH, 24 February 1890, in HN 318; John Kirkpatrick to HH, 22, 25 February; 23 April 1890, all in HN 232.

6. Henry E. Armstrong to HH, 24 January 1891, in HN 16; HH to Frederik Bramwel, 4 June 1891 (quote), in Archives, Royal Institution of Great Britain, London, C929/2; HH to Henry Young, 9 August 1891, ibid., C929/3.

7. Gustav von Gossler to HH, 23 May 1888, in HN 171; Rudolf Helmbold to HH, 24 June 1889, in HN 194; Schellenberg to HH, 19 July 1890, in HN 422.

8. Carl Andreas Eitz to HH, 2 September 1889, in HN 130; Hörz 1997, 131, 311; I. Mueller to HH, 12 January 1890; 8 December 1891; 20 March 1892; 28 April 1894, all in HN 320; Georges Le Gorgeu to HH, 8 August 1889, in HN 270; Friedrich Schön to HH, 9 March 1891, in HN 435; Oskar von Chelius to HH, 29 March 1891, in HN 88; ESH 1929, 2:63; Percy C. Gilchrist to HH, 25 May 1894, in HN 168; Alfred Stelzner to HH, 26 February 1892, in HN 449; A. Forster to HH, 13 March 1892, in HN 153; Hermann Schmidt to HH, 3 June 1892, in HN 432; Dr. Landgraf to HH, 9 February 1893, in HN 261; Bernhard Krüger to HH, 24 February 1893 (based on the Russian calendar; by the western European calendar, it was 8 March), in HN 250; Arthur Joachim von Oettingen to HH, 10 July 1893 (based on the Russian calendar; by the western European calendar, it was 22 July), in HN 337; Collyns Simon to HH, n.d., in HN 408. See also Franz Reuleaux to HH, 9 June 1879, in HN 370; HH to Carl Stoeckel, 14 March 1880, in Yale University Library, Beinecke Rare Book and Manuscript Library, General Manuscript Miscellaneous Collection, MS Vault, Section 5, Drawer 1.

9. Dr. E. Jaesche to HH, 3 March 1890 (based on the Russian calendar; by the western European calendar, it was 15 March), in HN 215; Dr. L. Kugel to HH, 27 February 1879, in HN 254; Henri Stassano to HH, 30 May, 26 July 1882, both in HN 446; W. E. Bryan to HH, 18 May 1889 (quote), in HN 76; also concerning color, cf. J. W. Reynolds to HH, 23 May 1888, in HN 371.

10. Ferdinand Lorenz to HH, 5 November 1891 (first two quotes), in HN 288; Carl Kutschera to HH, 14 January 1882 (third quote), in HN 258.

11. Julius Friedländer to HH, 13 December 1890 (quotes); 31 July 1891, both in HN 159.

12. Paul Otto Schmidt to HH, 29 October 1890, in HN 433.

13. Lummer 1889; AH to ISZ, 16 April, 30 December 1888, both in ESH 1929, 1:315, 320, resp.

14. AH to RH, 4 May 1889, in ESH 1929, 1:321–22, on 321; AH to ISZ, 21 July 1889, ibid., 2:11–2 (first two quotes); HH to [HT], 23 August 1889, ibid., 16; AH to Lieber und verehrter Freund, 1 August 1889, ibid., 12; AH to ISZ, 4 August 1889, ibid., 12; AH to PM and ISZ, 7 August 1889 (third quote), ibid., 13; Lummer 1889, 567–68 (fourth quote on 567); Helmholtz Family gravestone, Landeseigener Friedhof, Wannsee, Lindenstr. 1 und 2, Grab: A.T.52. Ehrengrab; Helmholtz family death notice in re Robert von Helmholtz, 5 August 1889, in RI MS JT/1/H/68b, TPRI; William Bowman to HH, 8 September 1889, in HN 61; Emma von

Treitschke to AH, 9 August 1889, in SF, Sign.: Helmholtz 6.LL496; Treitschke 1914–20, 3:603 (with a letter by HH); Carl Runge to Max Planck, 8 October 1889, in Hentschel and Tobies 1999, 113–14 (fifth quote on 114); Planck 1929–30, 38–39 (sixth quote on 38); Braun-Artaria 1918, 135.

15. AH to PM and ISZ, 13 August 1889, in ESH 1929, 2:13–15, on 14–15; AH to HH, 19 August 1889, ibid., 15; AH to ESH, 9, 13 September 1889, ibid., 17 (fifth quote), 18 (sixth quote), resp.; HH to AH, mid-August 1889, in *LK*, 3:24 (first quote); HH to [HT], 23 August 1889 (second, third, and fourth quotes); Bunsen 1932, 42–43 (seventh quote).

16. Steif 2003, 309; Victor August Julius to HH, 14 October 1889, in HN 225; Pantalony 2009, xxxiii–xxxiv, 34, 96–98, 129, 133, 138–40, 143–52, 155–57, 168–70; HeHz to his parents, 26 September 1889, in Hertz [1927], 221–23 (quotes on 222).

17. Edison quoted in Dyer and Martin 1929, 2:744–45 (first quote); HH to AH, 20 September 1889, in ESH 1929, 2:18–19, on 19 (second, third, and fourth quotes); HH to Thomas Alva Edison, 2 January [18 March] 1890 (fifth quote), in TAEM, reel 130:133 (TAED, D9055AAA); Private Secretary of Edison to HH, 27 August 1890, in HN 129.

18. Hertz 1894–95, 1:339–59; Jungnickel and McCormmach 1986, 2:91–92; HeHz to his parents, 26 September 1889; HH to AH, 20 September 1889, 19; Koenigsberger 1919, 184.

19. HH to AH, 20 September 1889; Sudhoff 1922, 275–76; 1922a, 41–42; Steif 2003, 212–24.

20. AH to ESH, 22 November 1889; Good Friday, April 1890; 9 April 1890, in ESH 1929, 2:20, 26–27 (second quote), 27, resp.; AH to HH, 20, 30 December 1889, ibid., 20–21, 22, resp.; AH to ?, 10 January 1890, ibid., 22; AH to Rosalie Braun-Artaria, 4 February 1890, ibid., 22–23; AH to Frau Otto Becker, 10 February 1890, ibid., 23–24; AH to Wanda von Mohl, 25 March 1890, ibid., 24–25; AH to ISZ, Palm Sunday 1890, ibid., 25–26 (first quote).

21. AH to ISZ, 11, 21 May 1890, ibid., 28, 28–29, resp.

22. Rouzaud 1891, 91–105, 208, 244; Charles Gide to HH, 5 June 1890, in HN 166; Epstein 1896, 194 (quotes); HH to AH, 23 May 1890, in ESH 1929, 2:29–31, on 30.

23. HH to AH, 23 May 1890.

24. HH to Staatsminister von Bötticher, n.d., mid-to-late 1890, in *LK*, 3:30–31.

25. HH to AH, 25 May 1890, ibid., 31 (first three quotes); Albert Vigié to HH, 1 August 1890, in HN 484; Louis Lortet to HH, 4 May 1890 (fourth and fifth quotes), in HN 290.

26. Riese 1977, 57; Craig 1978, 191–92.

27. JT to HH, 30 May 1885 (first two quotes), in HN 477; HH 1862, 179–80; HH to JT, 7 June 1885 (third quote), in RI MS JT/1/H/63, TPRI.

28. Albisetti 1983, 171–243; Friedrich Otto Rudolf Sturm to HH, 4 December 1890, in HN 458.

29. HH 1891a, 202 (first two quotes), 203 (third quote), 204.

30. Ibid., 205 (quotes), 206–7.

31. Ibid., 207 (first quote), 208 (second quote), 209 (all other quotes).

32. Riese 1977, 57; Albisetti 1983, 226–43; Nichols 1958, 175–77.

33. HH to TM, 21 June 1891 (quote), in TMN; Bismarck 1926, 554.

34. Ansprachen 1892, 11–12; EdBR in ESH 1929, 2:38; EdBR to HH, 2 November 1891, in *LK*, 3:44–46; "Prof. Helmholtz Honored" 1891; 26 February 1891, Minutes of the Board of Trustees of Steinway and Sons, 166, in SP; Mezhdunarodnyi komitet dlia chestvovaniia Germana fon-Gelmholttsa 1892.

35. Harnack 1970, 2:559–63; Hartkopf and Wangermann 1991, 342–46.

36. "Errichtung" 1970; Harnack 1970, 2:559–63; HH to WT, 4 July 1892, in WTP, Add 7342 H79; WT to HH, 12 July 1892, in HN 464; EdBR to HH, 31 May 1892, in Kirsten et al. 1986, 271.

37. HH to Leopold Kronecker, 26 April 1886, in Houghton Library, Harvard University,

bMS Ger 198(3), folder 2; Adolf von Hildebrand to N. Kleinenberg, 11 February 1891, in Sattler 1962, 358–59 (first two quotes on 359); Hugo Schiff to HH, 1 April 1891, in HN 427; Hildebrand to Conrad Fiedler, 9 April 1891, in Sattler 1962, 362 (third quote).

38. HH to CL, 21 September 1891, in *LK*, 3:43 (first two quotes); Mohl 1922, 1:257; Kundt 1891; Hertz 1894–95a; HeHz in his diary, 8, 9 August 1891, in Hertz [1927]; Alexander P. Trotter to HH, 21 January 1891 (third quote), in HN 474; Kronecker 1891.

39. HH to Alfred Dove, 9 August 1891, in BSBMH, Sign.: E. Petzetiana V.

40. HH to CL, September 1891, in ESH 1929, 2:36.

41. AH to ISZ, 21 September 1891, in ESH 1929, 2:35–36; AH to ISZ, 15 October 1891, in ESH 1929, 2:37 (first three quotes); HH to Hildebrand, 26 December 1891, in Sattler 1962, 373–74; Heilmeyer 1902, 83, 92–93; Bunsen 1899 (fourth quote).

42. Zobeltitz 1891, 768–69; HH to Verehrter Freund [Ludwig Bamberger], 28 March 1893, Bundesarchiv, Abteilung Potsdam, Nachlass Bamberger.

43. Zobeltitz 1891, 768.

44. Ibid., 770.

45. Ibid., 769–70.

46. Ibid., 770.

47. Ibid.

48. Ibid.

49. Kundt 1891; Ansprachen 1892, 60–63, which provides a detailed list of twenty-three formal honors as well as sixty-three formal addresses and congratulatory statements; HH to the Royal Society of London, 2 November 1891, in Royal Society, London, Archives, MC.15.219; HH to the Königliche Gesellschaft der Wissenschaften zu Göttingen, 9 November 1891, in Akademie der Wissenschaften zu Göttingen, Archiv, Pers 52,3, Helmholtz 7; [Ophthalmological Society], 1891; Schiff 1891.

50. Ansprachen 1892, 11; Lungo 1903, 79; "Helmholtz Celebration" 1891; HeHz in his diary, 1, 2 November 1891, in Hertz [1927], 238; Hugo Kronecker to HH, 31 October 1892, in HN 249; "Prof. Helmholtz Honored" 1891 (first quote); G. 1890–91; Schiff 1891a (second quote), whose headline is a misnomer; Raveau 1894, 801.

51. KM to Königliche Akademie der Wissenschaften, 26 October 1891, in AHA, Personalia, Mitglieder 1890–1893, Sign. II–III, 30, Bl. 37; 12 December 1891, Rep. 89 (2.2.1.), Geheimes Zivilkabinett, in GSPK, no. 21490, Bl. 183; Röhl 1994, 87–90; AH to Fritz von Helmholtz, 18 October 1891, in ESH 1929, 2:37; AH to ISZ, 15 December 1891, in ESH 1929, 2:39–40, on 39; Der Staatssekretär des Innern (Nieberding) to HH, in SF, Sign.: Helmholtz. 6.LL494; ? to AH, 29 August 1891, in SF, Sign. Helmholtz 6.LL494, "Grosskreuz dell'Ordini dei Santi Maurizio e Lazzaro. 1891. Rom. Universität Padua"; HH 1891, 3; Ansprachen 1892, 12–13 (quotes).

52. Ansprachen 1892, 47–59; HH 1891, 3–5 (first quote on 4), 14–16 (other quotes); HH 1904, vii.

53. HH 1891, 16–18.

54. Ibid., 18–19.

55. Ibid., 19 (quotes). Cf., too, Helmholtz's remarks in Schiff 1891; HH to Frankfurter Journalisten- & Schriftsteller-Verein, January 1893, in Uppsala Universitet, Wallers samling, Okat 651 F:1.

56. HH 1891, 19–20.

57. Ibid., 20–21.

58. Schiff 1891a (first quote); "Helmholtz Celebration" 1891; HH 1904, vii (second quote); Werner von Siemens to Karl Siemens, 7 November 1891, in Siemens 1916, 2:960–61 (third quote on 960).

59. Adolf Tobler to HH, 23 July 1891, in HN 466; Foerster 1911, 214, 216 (quote).

60. Paine 1912, 933–39; Twain 1935, 219–24 (quotes on 219).

61. Twain 1896 (quotes on 502–4, 512–15); Paine 1912, 936–39.

62. HH to Verehrter Freund [Ludwig Bamberger], 16 December 1891, in Bundesarchiv, Abteilung Potsdam, Nachlass Ludwig Bamberger; HH to Hildebrand, 26 December 1891; Spitzemberg 1963, 297 (quote); Twain 1935, 219–22.

Chapter Twenty-Seven

1. AH to ISZ, 5 March 1892, in ESH 1929, 2:40–1, on 40; AH to ESH, 4, 14, 18 April 1892, ibid., 41–42, 42–43, 43–44 (quotes on 43), resp.

2. AH to ISZ, 8 May 1892, ibid., 44–45; Fischer 1987, 141–42 (quotes on 142), 150–51; Vilhard and Fischer 1902, 192–94.

3. Carl Ruland to HH, 23 March [1892], in HN 391.

4. Geiger 1886, 1892, 1893; Leppmann 1961, 119–24, 128–29; Mandelkow 1980, 224–32.

5. HH to die Redaktion der *Deutschen Rundschau* [Julius Rodenberg], 27 May 1892, in Nationale Forschungs- und Gedenkstätten der klassischen deutschen Literatur in Weimar, Goethe- und Schiller-Archiv; HH to Julius Rodenberg, 30 May, 29 June 1892, ibid.; HH 1892; 1903b, ix; Theodore Stanton to HH, 30 July 1892, in HN 445.

6. AH to ESH, 11 June 1892, in ESH 1929, 2:45–47 (quote on 46); Ruland to HH, 6 June 1892, in HN 391; Mandelkow 1980, 192–93.

7. Mandelkow 1980, 189–90; Richards 2008; D'Agostino 1986, 278–81; Finkelstein 2013, 238–42; Bauer 1991, 282–83.

8. HH 1892, 337.

9. Ibid., 339–40.

10. Ibid., 337–9 (quotes on 338–39).

11. Ibid., 342–43 (quote on 343).

12. Ibid., 344–45.

13. Ibid., 345–46.

14. Ibid., 347–48.

15. Ibid., 348.

16. Ibid., 349–51.

17. Ibid., 351–52.

18. D'Agostino 1986; Barnouw 1987; Schiemann 1997, 1998; Heidelberger 1998; Giacomoni 2002; D'Agostino 2005.

19. See chapter 21, note 63.

20. HH 1892, 353–55.

21. Ibid., 356–60.

22. Ibid., 361.

23. AH to ESH, 11 June 1892, 46 (quotes); Wedel to HH, 15 July 1892, in HN 503.

24. Undated letter to Marcellin Berthelot in AH's hand, in SF, Sign.: Helmholtz 6LL494; HH (also in AH's hand) to Messieurs les Secrétaires perpétuels de l'Académie des Sciences, 17 June 1892, ibid.; HH to Eure Königlichen Hoheiten, 7 November 1892, in Staatsarchiv Weimar, Hausarchiv A XXVI, no. 368[a].

25. Adolf von Hildebrand to Conrad Fiedler, 24 July, 6 August 1892, both in Sattler 1962, 384 (first quote), 384–85 (second quote), resp.

26. Carl von Pidoll to Hildebrand, 9 May 1893, ibid., 401–3 (quote on 402); Hildebrand to Fiedler, 1 August 1893, ibid., 413.

27. Wilhelm Dilthey to Paul Graf Yorck von Wartenburg, 8 June, 18 July 1892, both in Dilthey 1923, 142–45 (first two quotes on 142–43), 147–49 (third quote on 149), resp.; EdBR 1912 2:562; Einstein 1917 (fourth quote).

28. WT to HH, 20 June 1892 (first two quotes), in HN 464; HH to WT, 4 July 1892, in

WTP, Add 7342 H79; WT to HH, 12 July 1892 (third quote), in HN 464; Werner von Siemens to HH, 28 July 1892 (fourth, fifth, and sixth quotes), copy in SF, Firmenarchiv, Z371. Physikalisch-Technische Reichsanstalt III. 1889–1933.

29. HH to WT, 4 July 1892; HER to HH, 15 July 1892, in HN 385; AH to ESH, 1 August 1892, in ESH 1929, 2:47–48 (quote); Visitor's Book, 29 July 1892, in The Old Rectory, Terling, Chelmsford, England; Henry Sidgwick to HH, 22 July 1892, in HN 403; Rayleigh 1968, 126.

30. James Sully to HH, 30 May 1891 (quote), in HN 417; International Congress of Experimental Psychology 1892 [1974]; Sully to HH, 25 November 1891, in HN 417; AH to ESH, 1 August 1892.

31. WT to Lord Rayleigh, 23 August 1892, in Thompson 1910, 2:925–26 (first quote on 925); Margaret E. Gladstone quoted in Thompson 1910, 2:926–28 (second quote on 926); AH to ESH, 5, 8 August 1892, in ESH 1929, 2:48–49 (third quote), 49–50 (fourth quote), resp.

32. Woodbury 1949, 178–79 (quote); Carhart 1894, 542; Cahan 1989, 2–3; Rayleigh 1968, 126–27.

33. AH to ESH, 14 August 1892, in ESH 1929, 2:50–51.

34. EdBR 1912, 2:643–48 (quotes on 643), 648; [Akademie der Wissenschaften, Berlin] to Hochgeehrter Herr College [HH], n.d. [ca. 2 November 1892], in LK, 3:56–62; HH to EdBR, 4 November 1892, in Kirsten et al. 1986, 271–72; EdBR to HH, 7 November 1892, ibid., 272–73.

35. Friedrich Althoff to Euerer Excellenz, 30 October 1892, in KM, I. HA, Rep. 89, no. 21491, Bl. 94–95, on 94; Vermerk, 2 November 1892, ibid., Bl. 60; Kaiser Wilhelm II to HH, n.d.; Minister des Innern von Boetticher, n.d. [both ca. 2 November 1892], both in LK, 3:53, 53–54, resp.; Jolly (dean of the medical faculty) to HH, n.d. [ca. 2 November 1892], in LK, 3:54–55; HH to Jolly, 3 November 1892, in LK, 3:55; HH to the philosophical faculty, in LK, 3:56; Gustav von Gossler to HH, 2 November 1892 (quotes), in HN 171; AH to ISZ, 6 November 1892, in ESH 1929, 2:53.

Chapter Twenty-Eight

1. The following account of Helmholtz and the Reichsanstalt draws on Cahan 1989, 70–125.

2. Turner 1994; Ash 1995, 8–9, 25, 52–59, 61–62, 70–72, 115, 132, 175–76; HH 1890a, 1891b, 1891c, 1891d, 1894a, 1894b; see also HH 1909–11; HH to Vieweg Verlag, ca. 21 November 1890; 22 July 1891, both in VV; HH 1885a, 1886b, 1887b, 1889c, 1892b, 1896.

3. James 1981, 1:92, 159n, 171, 217n, 226, 274, 398–400, 414, 416–18, 431, 476–77, 487–94, 512, 579n, 590–91, 607n, 610n; 2:665–67, 669, 671–72, 680n, 727n, 736, 737n, 743, 749, 757n, 809n, 814n, 836n, 848n, 850, 852–53, 857, 860, 867, 869, 873–76, 878n, 879–82, 888n, 896n, 897n–898n, 899, 908–10, 912, 1105, 1117, 1120n, 1248n, 1260–61; 3:1439, 1440, 1462, 1493–94; 1979, 72, 76; 1983, 1–37, on 17; 204–15, on 210; 38–61, on 47–48; 62–82, on 67; 80, 82; 83–124, on 95; 127–41, on 133–34; William James to Henry Bowditch, 30 May 1880, in Perry 1935, 2:54–55, 82 (quote on 55).

4. Planck 1929–30, 38; Jungnickel and McCormmach 1986, 2:51–52, 254; Schultze 1993.

5. Schultze 1994, 1; "Unterlagen der Philosophischen Fakultät," in HUB, Promotionsverfahren der Phil. Fak. 291 ff. Helmholtz, Gutachten, Prüfungen, und Promotion; HH, expert opinion in the Leo Arons habilitation review, between 28 May and 17 June 1890, in HUB, Phil. Fak., no. 1217, Bl. 102v, ibid., 17–18; HH, expert opinion in the Willy Wien habilitation review, between 21 November 1890 and 27 July 1891, in HUB, Phil. Fak., no. 1218, Bl. 252v–253, ibid., 19; Wien 1930, 11, 12, 15, 20–21; HH, expert opinion in the Heinrich Rubens habilitation review, between 11 and 21 December 1891, in HUB, Phil. Fak., no. 1218, Bl. 317v, in Schultze 1994, 19; Promotion Walther Rathenau, in HUB, Phil. Fak. no. 291, Bl. 272–291, esp. 275–76; Kessler 1928, 26–27.

6. Singer 2003, 98–99, 151–52; Albisetti 1988, 206–7, 225–26; Kargon 2014; Greven-Aschoff 1981, 54.

7. HH 1897–1907; König and Runge 1897–1907a, v; 1897–1907, v; Krigar-Menzel 1897–1907, v–vi; 1897–1907a, v; Richarz 1897–1907, v–vi; HH to Geehrter Herr Doctor [probably Otto Krigar-Menzel], 22 June 1892, in Stadtarchiv Bonn Ii 98/512; Otto Krigar-Menzel to HH, 24 June 1892, in HN 246; Runge 1949, 95–96 (quote).

8. Krigar-Menzel and Laue, 1897–1907, v; Richarz 1897–1907, v–vi; esp., Jungnickel and McCormmach 1986, 2:134–41.

9. Unless otherwise noted, this discussion of Helmholtz at the academy draws on Cahan 1999, 290–305.

10. HH to EdBR, 24 January 1888; 29 December 1891; [before 28 April 1892], all in Kirsten et al. 1986, 266–67, 267–68, 270–71, resp.; Hertz 1894–95, 2:115–32, which appeared first in the *Sitzungsberichte* (2 February 1888) and then in the *Annalen*; H. F. Weber to HH, 17 June, 21 October 1888, both in HN 496; Cahan 2000; EB to EdBR, 29 December 1891, in Brücke et al. 1978–81, 1:288; HH to Adolf von Hildebrand, 26 December 1891, in Sattler 1962, 373–74; Adolph von Menzel to HH, 2 May 1891; 12 February, 29 March 1892, all in HN 310.

11. Sitzung der physik.-mathem. Klasse, 31 July 1890, in AHA, Protocoll-Buch der physik.-math. Klasse 1886–1891, Sign. II–V, 123, Bl. 194–95; ibid., 30 October 1890, Bl. 197; Sitzung des Plenums, 31 May 1894, ibid., GesammtsitzungsProtocolle 1894, Sign. II–V, 69, Bl. 33; Sitzung des Plenums, 5 July 1894, ibid., Bl. 39.

12. Robert Helmholtz 1883, 130 (quote); Georg von Neumayer to HH, 20 April 1882; 17 January 1884, both in HN 327; Johann Kiessling to HH, 6 June 1886; [?] 1888, both in HN 229; Schröder and Wiederkehr 2000.

13. HH 1886a; Darrigol 2005, 172.

14. HH 1888. The present account follows the analysis in Darrigol 2005, 172–78; see also Garber 1976, 61–62; Kutzbach 1979, 197–98.

15. HH 1889; Rayleigh to HH, 29 October 1889, in HN 364; HH 1890; Darrigol 2005, 88, 175–76.

16. HH 1890; HH to Heinrich von Boetticher, 9 March 1890, in *LK*, 3:26–27 (quotes); Sitzung der physik.-mathem. Klasse, 17 July 1890, in AHA, Protocoll-Buch der physik.-math. Klasse 1886–1891, Sign. II–V, 123, Bl. 193; Darrigol 2005, 180–81.

17. G. Brown Goode (on behalf of Samuel Pierpont Langley) to HH, 30 June 1892 (quotes); Langley to HH, 7 October 1892; 19 January, 20 March 1893, all in Smithsonian Institution, Washington, DC, Archives, Record Unit 34, Office of the Secretary 1887–1907, Outgoing Correspondence, Book 25.1, 39–41, 49, 104–7, 134–35, resp.; HH to Langley, 23 November 1892; 25 February 1893; 22 June 1894, ibid., Record Unit 31, Office of the Secretary 1891–1906, Incoming Correspondence, box 82.

18. Heike Kamerlingh Onnes to HH, 4 January 1893, in HN 226.

19. HH to Onnes, 17 February 1893 (quote), in Museum Boerhaave, Leiden, Arch. 8, Correspondence Heike Kamerlingh Onnes; Wilhelm von Bezold to HH, 7 February 1893, in HN 47; Onnes to HH, 6 March 1893, cited in Hörz 1997a, 14–15.

20. Cleveland Abbe to HH, 27 September 1893 (quote); 13 July 1888, both in HN 2; Abbe 1891, which includes translations of HH 1858a, 1868e, 1873b, 1888, 1889, 1890.

21. Bélafi 1990, 70–74; Hallion 2003, 95–96.

22. Friedman 1989, esp. 19–21, 131–33, 174, 221; Darrigol 2005, 177–78, 283–86; Kutzbach 1979, 159–71, 194–99; esp., Fleming 2016, 2–3, 7–10, 18–20, 47–48, 54, 61–63, 81, 102, 113, 115, 135.

23. Sitzung der physik.-mathem. Klasse, 10 March 1892, in AHA, Protocoll-Buch der physik.-math. Klasse 1891–1894, Sign. II–V, 124, Bl. 34; HH 1892c, 1894c (an addendum to HH 1892c); Jungnickel and McCormmach 1986, 2:133–34, 251–53.

24. HH 1892c, 504; Sitzung der physik.-mathem. Klasse, 27 October 1892, in AHA,

Protocoll-Buch der physik.-math. Klasse 1891–1894, Sign. II–V, 124, Bl. 54; HH 1892d, 1892e, which last appears in *WA* as an attachment (additions and corrections) to 1892d (523–25); HH 1892c and HH 1892d were also both published separately in the *Annalen* in 1893; HH 1893a; Sitzung des Plenums, 6 July 1893, in AHA, GesammtsitzungsProtocolle 1893, Sign. II–V, 69, Bl. 33. For analysis see Buchwald 1985, 237–41, 250–58; Darrigol 2003, 171–72. Poincaré read many of Helmholtz's writings in physics. See Hermann Ebert to Henri Poincaré, 29 March 1892; Lucien de la Rive to Poincaré, 10 April 1892; Camille Raveau to Poincaré, 5 November 1899, all at the Poincaré Estate, Paris; Poincaré 2001, 95, 98, 123, 168, 172–74, 177–78; Gray 2013, 15–16, 172–73, 199, 318, 327–29, 335, 343, 347, 511–13.

25. Jungnickel and McCormmach 1986, 2:133–34; Bierhalter 1993, 456–58; Darrigol 2000, 216–17, 262–64; Principe 2012.

Chapter Twenty-Nine

1. AH to ISZ, 21 December 1892; 20 January 1893, both in ESH 1929, 2:52–53, on 52, 53–54, resp.; AH to HH, 29 March 1893, ibid., 54; AH to Adolf von Hildebrand, 10 December 1892, typescript copy in SF, Sign.: 6.LL496; HH to Verehrter Freund [Ludwig Bamberger], 28 March 1893, in Bundesarchiv, Abteilung Potsdam, Nachlass Bamberger; AH to Cosima Wagner, 8 April, n.d., 25 July 1893, all in Werner and Irmscher 1993, 41–42, 42, 43–44, resp.

2. The following discussion draws on Cahan 2010, 6–15.

3. HH to HK, [spring 1893], in HKL.

4. HH to HK, 20 June 1893 (quote), in HKL; AH to Cosima Wagner, 25 July 1893.

5. AH to ISZ, 22 June 1893, in ESH 1929, 2:54–55 (first quote on 54); AH to ISZ, 26 June 1893, ibid., 55; HH to Minister of the Interior Karl Heinrich von Boetticher, 22 July 1893, in *LK*, 3:73; Kussmaul 1903, 58; AH to ESH, 2 August 1893, in ESH 1929, 2:56–57 (second, third, and fourth quotes).

6. AH to ESH, 9 August 1893, in ESH 1929, 2:57; Klein 1926, 226; Felix Klein to Leo Koenigsberger, n.d., probably circa 1902–3, in *LK*, 3:80–81 (quotes); Lie 1888–93, 3:xii–xiii, 437–523; Fritzsche 1999; Stubhaug 2002, 372, 379–81, 384, 385, 396, 423–24; Gray 2008, 126.

7. AH to ESH, 19, 22 August 1893, both in *LK*, 3:81–82, 82–83, resp., on 82 (quotes); AH to PM, 23 August 1893, in ESH 1929, 2:59–60; "Helmholtz Arrives Tomorrow" 1893; Villard 1904, 366; Shastid 1928.

8. "Helmholtz Arrives Tomorrow" 1893.

9. "H. von Helmholtz Arrives" 1893 (quotes); "Devoted to Science" 1893; AH to PM, 23, 24 August 1893, both in ESH 1929, 2:59–60, 60–63, resp.

10. Jaeger 1932, 29–30; Carhart 1894, 542; "Homage to Genius" 1893 (quotes); N[ichols] 1894, 227.

11. "Electricians Meet" 1893 (quotes); Carhart 1894, 542; "Banquet" 1893.

12. "One Congress" 1893 (quote); Jaeger 1932, 30, 42–46; AH to PM, 23 August 1893, in ESH 1929, 2:59–60; invitation card of Mr. Meysenburg to Henry A. Rowland, 25 August 1893, in Henry A. Rowland Papers, Special Collections and Archives, Manuscripts, Milton S. Eisenhower Library, Johns Hopkins University, Baltimore, MD.

13. Carhart 1894, 543; AH to ESH, 31 August 1893, in ESH 1929, 2:63; HH to Thomas Corwin Mendenhall, 29 August [1893] (quotes), in Thomas Corwin Mendenhall Papers, box 14, folder 8, Archives & Special Collections, George C. Gordon Library, Worcester Polytechnic Institute, Worcester, MA.

14. *New-Yorker Staats-Zeitung*, 6 October 1893, 12 (quotes); AH to ESH, 31 August 1893, in ESH 1929, 2:63.

15. The following discussion draws on Cahan 2010, 16–17.

16. AH to ESH, 2, 6 September 1893, in ESH 1929, 2:63–66, 66–68, resp.; HH to ESH, 12 September 1893, in *LK*, 3:89–91.

17. AH to ISZ, 7, 9 September 1893, both in ESH 1929, 2:68–69 (first quote on 68), 69–72 (second quote on 71), resp.; AH to ESH, 6, 13 September 1893, ibid., 68, 72–73 (third quote), resp.; HH to ESH, 12 September 1893, ibid., 90.

18. HH to ESH, 12 September 1893, 89–90.

19. Unless otherwise noted, the following discussion draws on Cahan 2010, 17–19; HH to ESH, 12 September 1893, 90–91 (quote on 90); AH to ESH, 6, 13 September 1893, 68; AH to ESH, 17 September 1893, in ESH 1929, 2:73–74.

20. Henry Pickering Bowditch to HH, 15 September 1893, in HN 60; HH to Dear Sir [Bowditch?], September 1893, Department of Molecular Biophysics and Biochemistry, Yale University, New Haven, CT; Schröer 1967, 237–38 (n203).

21. HH 1871, 62; AH to ESH, 17 September 1893.

22. Hugo Münsterberg to HH, 17 January 1894, in HN 323; Münsterberg 1922, 22–27, 29, 32–36, 39–40, 42–44, 46–47; Hale 1980, 45–55; Spillmann and Spillmann 1993.

23. William James to Carl Stumpf, 12 September 1893; 24 January 1894, both in Perry 1935, 2:186–87, 187–89 (first two quotes on 188–89), resp.; William James to Henry James, 22 September 1893, in James 1920, 1:346–48 (third quote on 347–48).

24. The following discussion draws on Cahan 2010, 19–21.

25. HH to Mendenhall, 14 September 1893, in Thomas Corwin Mendenhall Papers, box 14, folder 8, Archives & Special Collections, George C. Gordon Library, Worcester Polytechnic Institute, Worcester, MA; Mendenhall to HH, 22 September 1893, in HN 308 (quote).

26. Villard 1904, 325–26, 328, 362–65, 367–68.

27. Henry Hunter?, Private Secretary to President Grover Cleveland, to Henry Villard, 26 September 1893, copy in SF, Sign.: Helmholtz 6.LL496; AH to ESH, 3 October 1893, in ESH 1929, 2:74–75 (quote on 74).

28. AH to ESH, 3 October 1893 (quote on 74).

29. The following discussion draws on Cahan 2010, 21–32.

30. HH to Hochgeehrter Herr [William Steinway], 6 October 1893 (first quote), in Fiorello H. LaGuardia Archives, LaGuardia Community College, City University of New York, Long Island City, New York, Steinway Papers; entry for 2 October 1893 (fifth quote), in William Steinway Diary, ibid.; Minutes of the Board of Trustees of Steinway and Sons, 25 October 1893, 296, ibid.; HH to Mr. [William] Steinway, 4 December 1893 (sixth quote, slightly edited to improve its readability), in private collection of Henry Z. Steinway, ibid.; AH to ESH, 3 October 1893 (second, third, and fourth quotes); Fostle 1995, 288–89.

31. Snyder 1964, 573 (first two quotes), 575–76 (third quote on 576); "Professor Helmholtz" 1893 (fourth, fifth, and sixth quotes); AH to ESH, 3 October 1893.

32. "Welcome to Helmholtz" 1893.

33. *New-Yorker Staats-Zeitung*, 6 October 1893, 12; AH to ESH, 3 October 1893; Millikan 1951, 33–41, 292–93; Ogden N. Rood to J. Willard Gibbs, 8 October 1893, in Ms Vault Gibbs, box 14, #53, Beinecke Rare Book and Manuscript Library, Yale University Library, New Haven, CT; Ogden N. Rood to Mathilde Rood, 8 October 1893, copy, in Charles Sanders Peirce Edition Project, Indiana University–Purdue University, Indianapolis.

34. *New-Yorker Staats-Zeitung*, 6 October 1893, 12.

35. HH, ["Notes of an address by Prof. Dr. von Helmholtz"], 1893 (quote), in Columbia University, New York, Rare Book and Manuscript Library, X375.5 H57; *New-Yorker Staats-Zeitung*, 6 October 1893, 12. Steinway judged that the newspaper's account of the lecture was "excellent." (See entry of 6 October 1893, William Steinway Diary.)

36. HH ["Notes of an address by Prof. Dr. von Helmholtz"] 1893 (first three quotes); *New-Yorker Staats-Zeitung*, 6 October 1893, 12 (fourth and fifth quotes).

37. Alexander Graham Bell to HH, 4 October 1893, in HN 35; *New-Yorker Staats-Zeitung*, 6 October 1893, 12 (quotes).

38. Michael I. Pupin to HH, 15 April 1889 (first quote), in HN 359; Pupin 1951, 294, 298–300 (second quote on 299); Ogden N. Rood to Mathilde Rood, 8 October 1893.

39. Entry of 7 October 1893 (first quote), in William Steinway Diary; AH to ESH, 3, 14 October 1893, in ESH 1929, 2:74–75, 75–77 (second, third, and fourth quotes), resp.; *New-Yorker Staats-Zeitung*, 6 October 1893, 12; "May Prescribe" 1893; AH to ISZ, 15 October 1893, in ESH 1929, 2:77; AH to Arthur von Auwers, telegram, 18 October 1893, 9:48 a.m., Bremen, in AHA, Personalia, Mitglieder 1890–1893, Sign. II–III, 30, Bl. 171; Klein to Koenigsberger, n.d., circa 1902–3, in *LK*, 3:93–94; Mohl 1922, 272.

40. AH to ESH, 14 October 1893.

41. AH to ISZ, 15 October 1893.

42. AH to ISZ, 17, 26 October 1893, in ESH 1929, 2:78 (first quote), 79 (sixth quote), resp.; AH to Fritz von Helmholtz, 21 October 1893, ibid., 78–79 (second, third, and fourth quotes); AH to Auwers, telegram, 18 October 1893; HH to HK, 4 December 1893 (fifth quote), in HKL.

43. HH to Klein, 17 December 1893 (first quote), in Niedersächsische Staats- und Universitätsbibliothek Göttingen, Handschriftenabteilung, Cod. Ms. Klein IX, 678; HH to HK, 4 December 1893 (second quote); AH to ISZ, 5, 19 November 1893, in ESH 1929, 2:79–80, 80–81, resp.

44. HeHz to HH, 24 February 1892, in Hertz [1927], 240–41 (first quote); HH to HeHz, 26 February 1892, in Hertz [1927], 261–62 (second and third quotes); Hertz 1894–95, 2:n.p. (fourth quote); (probably) HH 1892c.

45. For this and the following two paragraphs: HeHz to HH, 15 December 1892, in Hertz [1927], 248–50 (quote); Fölsing 1997, 485–89, 494–99, 506, 513–17; ESH 1929, 2:81; AH to ISZ, 24 January 1894, in ESH 1929, 2:81–82, on 82; HH to August Kundt, 18 April 1894, in *LK*, 3:101–2.

46. Philipp Lenard to HH, 28 April 1894, in HN 272; HH to Lenard, 21 May 1894, in *LK*, 3:104–5; Hertz 1894–95, 3:xxv–xxvi (quote on xxv); Fölsing 1997, 497, 507.

47. HH to RL, 16 January 1894, in Hertz [1927], 263.

48. HH 1894, 363–66 (quotes on 365–66); 1903b, ix.

49. HH 1894, 366–75 (quote on 374).

50. Ibid., 375–78 (quotes); Klein 1972, 73–75; 1974, 156, 167–69; Lützen 2005.

51. Klein 1972, 73–75; 1974, 156, 167–69; Majer 1985; Jungnickel and McCormmach 1986, 2:142–43; Fölsing 1997, 503–11; Heidelberger 1998; Mulligan 1998; Schiemann 1998 (who also notes the epistemological similarities between Helmholtz and Hertz); D'Agostino 2001, 2004; Leroux 2001 (who, like Schiemann, sees Hertz's "symbol" as rather close to Helmholtz's "sign" theory); Lützen 2005, 278–89; Patton 2009. See also many of the (other) essays in Baird, Hughes, and Nordmann 1998.

52. HH 1894c.

53. Klein to HH, 1 December 1893 (first quote), in HN 233; HH to Klein, 17 December 1893 (second quote). For similar postrecovery letters, see also HH to HK, 4 December 1893; HH to Mr. [William] Steinway, 4 December 1893.

54. George J. Romanes to HH, 11 October 1893 (quote), in HN 383; D. Argyll Robertson to HH, 9 December 1893, in HN 379; James Forrest, secretary of the Institution of Civil Engineers, to HH, 20 December 1893; 10 January 1894, in SF, Sign.: Helmholtz. 6.LL494; AH to ISZ, 19 November 1893; AH to Fritz von Helmholtz, 30 January 1894, in ESH 1929, 2:82–83; L. Aug. Müller and Hans Müller to HH, 29 April 1894, in HN 322; (draft of) HH to the Peter Wilhelm Müller-Stiftung, 12 May 1894, in HN 322.

55. Schüle 1933; AH to Wanda von Dallwitz, 23 March, 3 April 1894, both in ESH 1929, 2:84–85, 85–86, resp.

56. Louisa T. Tyndall to AH, 29 January 1894, in SF, Sign.: Helmholtz 6.LL496.

57. AH to ISZ, May 1894, in ESH 1929, 2:86; "Hochgeehrte Trauerversammlung" (first three quotes); Mann 1895, 570 (fourth quote). The eulogy document, which is not in Helmholtz's hand, contains a note at the top ("Versuch der Rekonstruktion am Sarge von Kundt 1894") and is in an envelope marked "Fragment . . . N.N. am Sarge von Kundt. 1894," in SF, Sign.: Helmholtz 6.LL496.

58. Fricke 1960; Wolff 1999, 192–99.

59. HH cited in a letter to Sigmund Exner and cited in Epstein 1896, 193; Sigmund Exner to HH, 13 June 1894, in SF, Sign.: Helmholtz 6.LL496; HN 715; *LK*, 3:125–34; Koenigsberger 1898, 124; 1910; HH 1893a, 535 (quote).

60. HH 1894c; Mann 1895, 568–69 (for 7 July); Krigar-Menzel 1897–1907a, v (suggests 11 July); HH to Koenigsberger, 11 July 1894, in *LK*, 3:121–22.

61. AH to ISZ, 16 July 1894, in ESH 1929, 2:87 (quote); AH to Cosima Wagner, 15 July 1894, in Werner and Irmscher 1993, 47; Richard Wachsmuth to Koenigsberger, n.d., in *LK*, 3:122–23; Mohl 1922, 1:277; Pernet 1895, 35; AH to ESH, 5 August 1894, in ESH 1929, 2:90.

62. AH to ISZ, 18, 20 July 1894, in ESH 1929, 2:87–88 (first and second quotes), 88 (fourth and fifth quotes), resp.; AH to Fritz von Helmholtz, 24 July 1894, ibid., 88; AH to Rosalie Braun-Artaria, 30 July 1894, ibid., 89–90 (third, sixth, seventh, and eighth quotes); AH to Hildebrand, 21 September 1894, in Sign.: Ana 550, Nachlass Adolf von Hildebrand, BSMH; Mohl 1922, 1:277; Dr. Heinrich Schüle to HH, 9 March 1894, in HN 439; Schüle 1933, 618.

63. AH to ISZ, 3 September 1894, in ESH 1929, 2:90–91 (first and second quotes); Rayleigh 1895, 538 (third quote); ESH 1929, 2:91 (fourth and fifth quotes); Helmholtz Family Death Announcement, in SPKB, Sig.: Slg. Darmstaedter. Fia 1847: Helmholtz (8 September 1894); "Death of Professor" 1894; Otto Lummer to Koenigsberger?, 11 January 1902, in SPKB, Slg. Darmstaedter. F2c 1890: Lummer; *LK*, 3:123–24.

64. "Death of Professor" 1894; Emperor William's telegram, 8 September 1894 (first quote), in Stadtgemeinde Heidelberg, Stadtrats-Acten: Hermann von Helmholtz, Archiv, no. 20, Fasc. 6, 1869; "Hermann von Helmholtz" 1894, 479; Mohl 1922, 1:278; Max Planck to Carl Runge, 29 August 1894, in Hentschel and Tobies 1999, 131–32, on 132 (other quotes), and cf. 57–58.

65. Death announcement for HH, copy in Stadtgemeinde Heidelberg, Stadtrats-Acten; Mohl 1922, 1:278; AH to ISZ, 4 November 1894, in ESH 1929, 2:96; AH to Fritz von Helmholtz, 1 October 1894, in ESH 1929, 2:93; Hermann Diels to Eduard Zeller, 15 September 1894, in Ehlers 1992, 2:66–67, on 66; Planck to Runge?, 31 December 1894, in Hentschel and Tobies 1999, 134 (first three quotes); Runge to Bernhard Karsten, 8 October 1894, in Hentschel and Tobies 1999, 133; AH to Dallwitz, 17 September 1894, in ESH 1929, 2:91–92 (last two quotes).

Epilogue

1. AH to ISZ, 31 October 1894, in ESH 1929, 2:95–96, on 95. There is a large collection of these notices in SF, Sig. Lc 589.

2. "Late Professor" 1894 (first quote); AH to ISZ, 22 September 1894, in ESH 1929, 2:92–93; WT to AH, 30 November 1894, in HN 464; Thompson 1910, 2:939; HER to AH, 2 September 1894, in HN 385; "Hermann von Helmholtz (im memoriam)" 1894, 531 (second quote); Bezold 1895, 1896; EdBR 1912, 2:516–70; Hermann and Volkmann 1894; Fischer 1894; Krüss 1894; Heidenhain 1894; Engelmann 1894; Thomson 1894, on which see also WT to AH, 30 November 1894, in HN 464; Pernet 1895; Kronecker 1894; FitzGerald 1896; Goldzieher 1896.

3. Ernst Hagen to Heinrich Kayser, 25 November 1894, in SPKB, Slg. Darmstaedter, Fic 1902: Bessel-Hagen; Cahan 1989, 131–32; Warburg 1975, 182–84 (quote on 184).

4. AH to Fritz von Helmholtz, 22 October 1894, in ESH 1929, 2:94; AH to ISZ, 16 December 1894, ibid., 97–8 (second and third quotes); Spitzemberg 1963, 329; AH to EdBR, 22 Octo-

ber 1894, in Kirsten et al. 1986, 274; "Gedächtnissfeier für Hermann von Helmholtz veranstaltet von wissenschaftlichen Vereinen Berlins am Freitag, den 14. Dezember 1894, Mittags 12 Uhr, im Saale der Singakademie," in GSPK, I. HA Rep 89 (2.2.1), no. 21324; Bezold 1895; "Acta der Königl. Friedrich-Wilhelms-Universität zu Berlin betreffend: Das im Vorgarten der Universität aufzustellenen von Helmholtz Denkmal," HUB, R/S no. 349, Bl. 93; Max Planck to Carl Runge, 31 December 1894, in Hentschel and Tobies 1999, 134; Hermann Diels to Eduard Zeller, 14 December 1894, in Ehlers 1992, 1:77–78; Wilhelm Dilthey to Paul Graf Yorck von Wartenburg, 14 December 1894, in Dilthey 1923, 175–76, on 176 (first quote).

5. EdBR 1912, 2:517 (second quote), 534, 570 (first quote).

6. Ibid., 534, 561 (fourth quote), 567–68 (first three quotes), 570 (fifth quote).

7. Harnack 1970, vol. 1:979, 984 (quote); Cahan 1999.

8. Koenigsberger 1898 (first quote); Rücker 1895, 472 (second quote); 1896; HER to AH, 14 November 1895 (third quote), in HN 385. See also University of Heidelberg, "Gedächtnisfeier für Hermann von Helmholtz: Berlin, 14. Dezember 1894," http://ub-fachinfo.uni-hd.de /math/edd/helmholtz/singakademie.pdf (accessed 23 May 2014).

9. Hagner 2004, 212–15, 218–19, 229–30, 234, 245; "Das Gehirn" 1895; Hansemann 1899 (quote on 116); "Die Enthüllung" 1899; Marcuse 1899; Spitzka 1908.

10. Dilthey 1900, 235 (quoting Anna von Helmholtz to this effect); Bunsen 1932, 41; AH to HT, 17 October 1894, typescript copy, in SF, Sign.: Helmholtz 6.LL496 (partially printed in ESH 1929, 2:95); AH to Cosima Wagner, *Sonntagfrüh* 1894, in Werner and Irmscher 1993, 47–48, on 47; AH to ISZ, 27 March 1894, in ESH 1929, 2:86; AH to ISZ, 22 September 1894, 92–93 (first quote); AH to Fritz von Helmholtz, 5 October 1893, in ESH 1929, 2:94; AH to ISZ, 31 October 1894, in ESH 1929, 2:95–96; AH to ESH, 21 August 1896, in ESH 1929, 2:126–27, on 126; AH to Henry Pickering Bowditch, 6 January 1895 (second quote), in Francis A. Countway Library of Medicine, Boston, H MS c5.2.

11. AH to Rosalie Braun-Artaria, 11 November 1894, in ESH 1929, 2:96–97, on 97; AH to ISZ, 13, 24 January; 22 February, 14 May 1895, all ibid., 101–2, 103 (quote), 102–3, 108–9, resp.; AH to Fritz von Helmholtz, 29 February 1895, ibid., 103–4; Williams, General-Verwaltung der Kgl.-Bibliothek to R[eichs]. S[chatz].-A[mt]. I 6324, 26 November 1894, in Bundesarchiv: R2 12376, Abteilung Potsdam; Boetticher to Posadowsky-Wehner, 12 December 1894, ibid.; Posadowsky-Wehner to Hohenlohe, 8 January 1895, ibid.; AH to Cosima Wagner, 23 January 1895, in Werner and Irmscher 1993, 49–50, on 50; Bunsen 1899.

12. AH to ISZ, 22 February, 28 April, 16 September 1895; 8, 19 May 1897; 17 January 1898; 30 January; 8, 25 February; 1 March 1899, in ESH 1929, 2:103, 107, 114, 136, 140–41, 174, 174–75, 180–81, 181–82, resp.; AH to Fritz von Helmholtz, 29 March 1895, ibid., 105; AH to Wanda [von Dallwitz?], 7 April 1895; 8 April 1898, ibid., 105–6, 157–58, resp.; AH to ESH, 6 May; 15, 16, 17, 21 July 1895; 3 April 1897; 20, 26, 27, 31 January; 9 February 1898, [9 or 10 February], 10, 12, 18, 20, 21, 27 February; 4, 9 March; 13, 15 July 1898; 17, 18 February 1899, ibid., 108, 110–11, 112, 112–13, 133, 141–43, 143–44, 145–46, 146, 147, 147–48, 148–49, 149, 150, 151–52, 152, 152–53, 154–55, 156, 157, 161, 161–62, 178–79, 179–80, resp.; AH to Ottmar [von Mohl], 4 August 1898, ibid., 163.

13. AH to Wagner, *Sonntagfrüh* 1894; 12 August, 19 November 1897, all in Werner and Irmscher 1993, 47–48, 53, 53–54, resp.; Wachsmuth 1900, 20; Planck 1929–30, 39; Fritz von Helmholtz to Charles Bally, 17 July [1892]; 29 September 1896; 4 January 1897; 1 January 1900, in Bibliothèque de Genève, Ms. fr. 5002, f. 272–273v, f. 268–269v; f. 270–270v, 271–71v, resp.; Werner 1997, 84; AH to Frau Eduard [Emilie] Zeller, 26 November 1895; 7 July 1899, both in ESH 1929, 2:116–17, on 117, 187, resp.; AH to ISZ, 1 July 1896; 20 April, 31 August, 1 December 1898; 7 April, 23 August 1899, all in ESH 1929, 2:124–25, 159–60, 164–65, 171–72, 183–84, 189, resp.; AH to ESH, 8 September 1897; 19 August, 5 September 1898, all in ESH 1929, 2:138, 164, 165–66; Braun-Artaria 1899; 1918, 135.

14. [AH], "16. Febr. 1897. Mein Testament," typescript, in SF, Sign.: Helmholtz 6.LL496.

15. AH to ISZ, 31 October 1894; 14 May 1895, both in ESH 1929, 2:95–96, 108–9, resp.; AH to ESH, 21, 30 July 1895, ibid., 112–13, 113; AH to Emilie Zeller, 26 November 1895, ibid., 116–17, on 117; Anna Helmholtz 1896, xiv; Wachsmuth 1900, 18.

16. AH to ISZ, 16 September 1895, in ESH 1929, 2:114; AH to ESH?, 2, 3 August 1896, ibid., 125, 125–26; AH to ESH, 8 September 1897, ibid., 138; Adolf von Hildebrand to his wife, 28 October 1896, in Sattler 1962, 457; Esche-Braunfels 1993, 384–91; Ernst and Stümbke 1986, 211.

17. EdBR to AH, 15 December 1894, in Kirsten et al. 1986, 274–75; AH to ISZ, 16 December 1894, 98; Das Central-Comité zur Errichtung eines Denkmals für Hermann von Helmholtz, "Aufruf zur Errichtung eines Denkmals für Hermann von Helmholtz," in HUB, Phil. Fak., no. R.IS349, Bl. 107; "Acta der Königl. Friedrich-Wilhelms-Universität zu Berlin betreffend: Das im Vorgarten der Universität aufzustellenen von Helmholtz Denkmal," in HUB, R/S no. 349, Bl. 111, 115, 117, 120, 153, 158; Münsterberg 1895, 547 (quotes); Hugo Münsterberg to Dear Sir [Samuel Pierpont Langley], 24 April, 12 May 1895, in Smithsonian Institution, Washington, DC, Archives, Record Unit 31, Office of the Secretary 1891–1906, Incoming Correspondence, box 48; "Notes" 1895, 613; Hendrik Antoon Lorentz to Ludwig Boltzmann, draft, [January 1896?], in Boltzmann 1994, II:251–52, on 251.

18. AH to ISZ, All Souls' Day, 1896; 6 May 1899, both in ESH 1929, 2:129–30 (first two quotes), 184, resp.; Fd. 1899; "Die Enthüllung" 1899; AH to ISZ, 7 June 1899, in ESH 1929, 2:184–86; *LK*, 3:138–39; Hermann Diels to Eduard Zeller, 14 October 1898, in Ehlers 1992, 1:220–22, on 221 (third quote); Goschler 1998, 85–89.

19. AH to ISZ, 7, 8 June 1899, both in ESH 1929, 2:186–87; AH to Emilie Zeller, 7 July 1899, ibid., 187; Bunsen 1899; Death notice re AH, 1 December 1899, in TMN, Bl. 27; Wachsmuth 1900, 17; Fritz von Helmholtz to Bally, 1 January 1900; Zobeltitz 1922, 1:300–301; Spitzemberg 1963, 391; Helmholtz-Family gravestone, in Landeseigener Friedhof, Wannsee, Lindenstr. 1 und 2; Grab: A.T.52. Ehrengrab; Weech 1906, 298.

20. Molvig 2010, 333–35; Thouret 1903, 3–13.

21. Arnold Sommerfeld to Carl Runge, 3 November 1898, in Arnold Sommerfeld Korrespondenz, SBPK, Nachlass 141.

22. McKendrick 1899; *LK*; Reiner 1905 (in German); Ebert 1949 (in German); Lazarev 1959 (in Russian); Rechenberg 1994 (in German); the semibiographical Meulders 2001 (in French), 2010 (in English).

23. AH to HER, 26 November [probably 1895], in RSC.

24. Koenigsberger 1919, 194–96 (quotes). See also the letters from AH to Leo Koenigsberger in SPKB, Sammlung Darmstaedter, Helmholtz Correspondence.

25. Koenigsberger 1919, 196–97; *LK*, 3:v ("Vorwort").

26. Koenigsberger 1896; Leo Koenigsberger to HER, 25 December 1901, in University of Manchester, John Rylands Library, Manchester, Henry Roscoe Papers; Koenigsberger to Felix Klein, 14, 28 January 1902, in Niedersächsische Staats- und Universitätsbibliothek Göttingen, Handschriftenabteilung, Cod. Ms. Klein 10, no. 519. 520, resp.; *LK*, 2:v ("Vorwort"); 3:v ("Vorwort") (quote); Koenigsberger 1919, 196–200; cf. Kremer 1994, 382–85.

27. "Intimes" 1903, 487 (first three quotes); Roscoe 1906, 94 (fourth quote); Boltzmann to Koenigsberger, 7 December 1902 (fifth quote), in Boltzmann 1994, II:362; *LK*, 1:375; "Ein Helmholtz-Archiv" 1905.

28. Koenigsberger 1965, iii–v; 1911; "Progress of Science" 1907, 283 (first three quotes); William Osler to Casey Wood, June 1912, in Cushing 1940, 1004 (fourth quote).

29. Akademie der Wissenschaften 1986; Cahan 1999, 309–11.

30. Mann 1974, 9:12–13.

31. Goldschmidt 1898; Riehl 1904; Mach 1898, 19, 35, 83–84, 99, 138, 164–65, 184, 247 (quote), 305, 307.

32. Katscher 1901; Uhthoff 1902; Friedenwald 1902; Cahan 2000; Raman 1947, 21–29,

quoted in Ramaseshan 1990, 519; Herneck 1966, 278; Warburg 1915; Adams 1974, 460; Henry Adams to My dear Sir, 1 January 1909, in Adams 1988, 6:205–8, on 207.

33. Virtanen 1973, passim and, in re Helmholtz, 374; 1974, 88; Agee 1977, esp. 20–22; La Grange 1973, 100–101; 1995, 601, 647–48; 1999, 460.

34. In general on modernism, see Burrow 2000, 238–40, 243–48.

35. Dühring 1895, 106–7, 109–10.

36. August von Trott zu Solz (Minister of Culture and Education) to August Lentze (Minister of Finance), June 1914, in "Das Kaiser Wilhelm Institut für theoretische Physik in Dahlem," Bl. 21–24 (quote on 23r), Rep. 76 Vc, Sekt. 2, Tit.23, Litt. A, GSPK, cited in Seth 2003, 33.

37. Forman 1974.

38. Dimmer 1921; Erggelet 1921; Höber 1921; L 1921; Krückmann 1922; Mamlock 1921; Schröder 1921; Hirschberg 1921, 1117–18 (quote); ESH et al. 1921.

39. "Akten der Friedrich-Wilhelms-Universität zu Berlin betreffend: o. Prof. Dr. Hermann von Helmholtz," in Phil. Fak., HUB, UK Personalia H 211; Warburg, Rubner, and Schlick 1922; Erdmann 1921; Konen and Pütter [1922]; Kries 1921; Wien 1921; Nernst 1921; Riehl 1921; Goldstein 1921 appeared as separate articles in *Die Naturwissenschaften* under the general title "Dem Andenken an Helmholtz. Zur Jahrhundertfeier seines Geburtstages"; Wiener 1921 and Kaufmann 1922 both appeared in the *Deutsche Revue*; Rk 1921 appeared in *Die Umschau*; for the newspaper articles, see Ehrenhaft 1921; Feldhaus 1921; ESH 1921; Stumpf 1921; Lummer 1921; Rothe 1921; Graeff 1921. See also Günther, Dannemann, and Sudhoff 1922; Krannhals 1921; Lau 1922; Möller 1923; Schoen 1921; Sz 1921; Wien 1921a.

40. HH 1921, 1977; Stadler 2001; Friedman 1997; 1999, esp. xi, 6, 19, 42, 60, 64–65, 72; 2002, esp. 203–6; Coffa 1991, esp. 47–61, 140, 171–72, 181–82, 185, 189–90, 198, 200, 204, 207, 262, 309, 350; Boi 1996; Neuber 2012; Oberdan 2015.

41. Riehl 1922; Gumprecht 1927; Badermann 1939; Karlson 1935–36; Ballin 1938.

42. Fritz Haber to Albert Einstein, 30 August 1920, in Einstein Archives, Institute for Advanced Study, Princeton, NJ, quoted in Stern 1999, 131–32 (first quote); Paul Volkmann to Sommerfeld, 16 January 1925 (second quote), in Arnold Sommerfeld Korrespondenz, DM, 89,014.

43. Lenard 1933, 292–95, on 294–95 (quote); Fölsing 1997, 12–13; esp. Wolff 2012.

44. Kania 1939, 160–63, 166–67 (quote).

45. Glaser 1996 (quotes on 223–25); Müller 1996, 253; Hentschel 1996, 6, 51, 251, 261, 265–66, 276–77, 286, 290–92, 301, 340–43, 350, 404, xxviii, xl.

46. Pistor 1941, 276; Hoffmann 1939, 1562 (quotes); "Helmholtz" 1939.

47. Häfner 1934; [1962?], 117; "R. v. Helmholtz" 1934; Johannes, in Kremer 1990, 193–201, on 199; Helmholtz and Staby 1930–37.

48. ESH 1929, 1:"Vorwort"; Planck 1929–30, 37 (quote).

49. "Hermann v. Helmholtz" 1941; "Hermann von Helmholtz" 1941–42; Hansen 1941, 1941a; Lejeune 1941; Lohmann 1941; Pistor 1941, 275 (third quote), 277 (fourth quote); Schering 1941; Schimank 1941, 505 (first two quotes); Schomerus 1941; Steudel 1941; Zschau 1941, 114 (fifth quote).

50. Zenneck 1944; Laue 1944.

51. *PB* 1(1944): 1–2; "Unsere Physikbilder" 1944; Simonsohn 2007, 280–89; Albrecht 1993, 60–63.

52. Kossert 2005, 303–4, 321–22, 326, 338, 340–42, 348; Manthey 2005, 667, 669–71; Clark 2007, 676–78; Ostertag 1993, esp. 489–90.

53. Ck 1946–47; Creutz and Steudel 1948; "Die Helmholtz-Gesellschaft" 1946 (first quote); Br[üche] 1947, 3; 1948 (second and third quotes); Planck 1950, 433 (fourth quote); Gerlach 1950; Pupke 1951–52.

54. East German scholarship: Karsch 1950; Comberg 1951; Deutsche Akademie 1950, 52;

West German scholarship: Diepgen 1950; Engelking 1950; Bochalli 1950; Esser 1950; Hofe 1950; Gerlach 1951; Rohrschneider 1951.

55. Laue 1960.

56. Deutsche Akademie 1957; Dunken 1960, 76; Deutsche Akademie 1960, 28; Akademie der Wissenschaften 1986; Hartkopf 1992, 423–24.

57. Dieter Hoffmann, personal communication to the author, 25 August 1994; Kremer 1994, 382; Hörz 1957; Wittich 1964; Wagner 1965; Hörz and Wollgast 1971, 1971a; Scheel 1972; Meinel 1972; Wirzberger 1973 (quotes facing 277); Kuchling 1973; Hörz and Wollgast 1986; Kirsten et al. 1986.

58. Schoffa 1954; Schmitz 1954; Gerlach 1956, 1962, 1969; Leutner 1955–57, 1:275–77; Hermann 1959; "Helmholtz, Hermann" 1962; Brossmer 1958; Reicke 1971.

59. Schobess 1962; Weber 1997; Blum 1997; Werner 1997; Doerr 1985; Tzschaschel 1994; Breger 1985, 27; Henning 2000, 45, 49 (quote); Albrecht and Hermann 1990, 398; "Helmholtz-Gymnasium," http://de.wikipedia.org/wiki/Helmholtz-Gymnasium (accessed 21 March 2016).

60. "Träger der Helmholtz-Medaille," Berlin-Brandenburgische Akademie der Wisssenschaften, www.bbaw.de/die-akademie/auszeichnungen/medaillen/helmholtz-medaille /traeger-h-m (accessed 2 January 2017); Eckart and Volkert 1996; Werner 1997, 1998; Hoffmann and Ebeling 1994.

61. Cahan 1993a; Krüger 1994a.

62. Helmholtz, www.helmholtz.de (accessed 6 August 2017); Helmholtz-Fond 2013; Hoffmann and Trischler 2015.

References

Archival Holdings

By far the largest collection of unpublished Helmholtz archival holdings is his *Nachlass* (literary remains) at the Archive of the Berlin-Brandenburgische Akademie der Wissenschaften. Its contents help give sense to the extensive but much-scattered holdings elsewhere—and vice versa. It therefore deserves special notice.

In 1931 Ellen von Siemens-Helmholtz, Helmholtz's fourth child and his (and Anna von Helmholtz's) literary heir, gave his *Nachlass* to the Akademie; it has remained there ever since. It occupies about five running meters of shelf space and consists of about forty boxes of various sizes containing letters, scientific manuscripts, and small notebooks (for research, laboratory, or teaching purposes). There is a finding aid. The vast majority of the letters are *to* Helmholtz (almost 1,700 letters from approximately 530 senders); the small remainder (some in draft form) are *from* him (about 35 letters to approximately 30 recipients). There are also 19 third-party letters from and to individuals other than Helmholtz that concern him. All these letters are ordered according to sender or, in the case of those written by Helmholtz himself, recipient. There is also a chronological listing of the letters.

Helmholtz's scientific manuscripts and related materials, including unpublished lectures and notebooks, are of essentially four types. First, there are two sets of lecture notes (*Kollegienhefte*) that he took while attending Johannes Müller's courses (HN 538). Second, there is a very large number of manuscripts that pertain to (and are arranged by) individual scientific disciplines (e.g., "Physiological Optics" and "Electrodynamics"). Many of these (HN 539–701) are largely the final or near-final drafts of published articles or books by him; some are lecture manuscripts or notebooks. Among these, too, are many laboratory or research notebooks or manuscripts as well as individual sheets of calculations. The third type of manuscript consists of lectures or notes on epistemology, the history of science, individual scientists, and the general nature and results of science (HN 702–721). Here, too, many if not most of these materials are advanced drafts of subsequently published items. Fourth, there is a small miscellany of manuscript fragments, drawings, notes, notebooks, and calculations (HN 722–728), as well as several manuscripts or extracts by individuals other than Helmholtz (HN 729–733); in addition, there are photocopies of several ministerial or university items concerning Helmholtz (HN 734–736).

By contrast, the other archival holdings referred to below contain, for the most part, letters *from* Helmholtz. These range in number from a single letter to, in several cases, as many as two dozen or so. A few of these holdings also include letters or other materials about Helmholtz, Helmholtz-related issues, or documents concerning him. Though most of these holdings are located in Germany, some are elsewhere in Europe or in the United States; one is in Brazil.

In addition to the holdings at the Akademie, two other sets of archival holdings merit special notice: the one at the Staatsbibliothek Preussischer Kulturbesitz in Berlin, Handschriftenabteilung, which constitutes the second-largest holding of Helmholtz archival materials and contains a significant amount of Helmholtz correspondence and other materials; and the one at the Geheimes Staatsarchiv Preussischer Kulturbesitz in Berlin (Dahlem, and including items concerning Prussian matters formerly held at Merseburg), which is the third-largest holding and contains official correspondence, reports, budgets, notes on facilities, construction plans, personnel matters (including Helmholtz himself), the Dühring Affair, and the like, by Helmholtz and others, pertaining to his time at the Königsberg Anatomical Institute (1849–55), the Bonn Physiological Institute (1855–58), and the Berlin Physics Institute (1871–94).

Finally, as indicated in this book's endnotes and in the "Printed Works" section below, some of Helmholtz's letters and other correspondence have been published, either as scholarly articles or edited books devoted explicitly to making such letters available, or as parts of such works. All together, the archival holdings listed here (including the specifics about them provided throughout this volume), along with the Helmholtz letters in book format (as collections of letters) or in article format that are listed in "Printed Works," not only take in a large portion of the original letters cited by Leo Koenigsberger in his three-volume biography of Helmholtz (often enough in snippets and without any documentation) but also reveal newly found letters. It is noteworthy, however, that the search for Helmholtz's many letters to Carl Ludwig has yielded no results, while some correspondence between Helmholtz and his father appears to be lost; to that extent, at least, scholars remain dependent on Koenigsberger's quotations or citations.

AUSTRIA

Bibliothek, Österreichische Akademie der Wissenschaften, Vienna

BELGIUM

Académie Royale de Médecine de Belgique, Brussels

BRAZIL

Arquivo Histórico, Musen Imperial, Petropolis

DENMARK

Biblioteket, Danmarks Tekniske Museum, Helsingør

FRANCE

Archives, Académie des Sciences, Institut de France, Paris
Bibliothèque, Académie des Sciences, Institut de France, Paris

GERMANY

Berlin
Akademiearchiv, Berlin-Brandenburgische Akademie der Wissenschaften, Berlin
Amtsgericht Charlottenburg, Berlin
Archiv, Akademie der Künste (Stiftung Archiv der Akademie der Künste and Preussische
  Akademie der Künste), Berlin (AdK)
Archiv, Helmholtz-Gemeinschaft, Berlin
Archiv, Humboldt-Universität, Berlin
Archiv, Walter de Gruyter & Co. (formerly Georg Reimer), Berlin (WGA)
Archiv zur Geschichte der Max-Planck-Gesellschaft, Berlin
Berlin Document Center, Berlin
Deutsche Physikalische Gesellschaft, Magnus-Haus, Berlin
Evangelisches Zentralarchiv, Berlin
Geheimes Staatsarchiv Preussischer Kulturbesitz, Berlin (GSPK)
Handschriftenabteilung, Haus Potsdamer Strasse, Staatsbibliothek Preussischer Kulturbesitz,
  Berlin (SBPK)
Landesarchiv, Berlin
Märkisches Museum, Berlin
Musikabteilung, Haus Unter den Linden, Staatsbibliothek Preussischer Kulturbesitz, Berlin
Zentral- und Landesbibliothek, Berlin

Bonn
Archiv, Rheinische Friedrich-Wilhelms-Universität, Bonn
Bibliothek, Generaldirektion, Deutsche Telekom AG, Bonn
Bibliothek, Mathematisches Institut, Rheinische Friedrich-Wilhelms-Universität, Bonn
Stadtarchiv und Wissenschaftliche Stadtbibliothek, Bonn
Universitätsbibliothek, Rheinische Friedrich-Wilhelms-Universität, Bonn

Bremen
Staats- und Universitätsbibliothek, Bremen

Cologne
Stadt Köln, Historisches Archiv, Cologne

Dortmund
Handschriften-Abteilung, Stadt- und Landesbibliothek, Dortmund

Dresden
Sächsische Landesbibliothek, Staats- und Universitätsbibliothek, Dresden

Frankfurt am Main
Deutsches Postmuseum, Frankfurt am Main
Freies Deutsches Hochstift, Frankfurt am Main
Senckenbergische Bibliothek, Universitätsbibliothek Johann Christian Senckenberg, Goethe
    Universität Frankfurt am Main

Freiburg im Breisgau
Archiv, Ernst-Mach-Institut, Fraunhofer-Institut für Kurzzeitdynamik, Freiburg im Breisgau
Handschriften- und Inkunabelabteilung, Universitätsbibliothek, Freiburg im Breisgau

Giessen
Handschriftenabteilung, Universitätsbibliothek, Justus-Liebig-Universität, Giessen

Göttingen
Archiv, Akademie der Wissenschaften, Göttingen
Handschriftenabteilung, Niedersächsische Staats- und Universitätsbibliothek, Göttingen

Halle-Wittenberg
Sondersammlungen, Universitäts- und Landesbibliothek Sachsen-Anhalt, Martin-Luther-
    Universität, Halle-Wittenberg

Hamburg
Literatur-Archiv, Staats- und Universitätsbibliothek Hamburg Carl von Ossietzky, Hamburg

Heidelberg
Archiv, Ruprecht-Karls-Universität Heidelberg, Heidelberg (RKUHA)
Handschriftenabteilung, Universitätsbibliothek, Ruprecht-Karls-Universität Heidelberg,
    Heidelberg
Stadtarchiv, Stadtgemeinde Heidelberg, Heidelberg

Karlsruhe
Archiv, Sammlungen, Stadtbibliothek, Karlsruhe
Badisches Generallandesarchiv, Karlsruhe (BGLA)

Kiel
Schleswig-Holsteinische Landesbibliothek in Kiel, Kiel

Koblenz
Bundesarchiv, Koblenz

Leipzig
Archiv, Karl-Marx-Universität, Leipzig
Handschriften und Inkunabeln, Universitätsbibliothek, Karl-Marx-Universität, Leipzig
Wundt-Archiv, Karl-Marx-Universität, Leipzig

Marbach am Neckar

Handschriften-Abteilung, Deutsches Literaturarchiv Marbach am Neckar (Deutsche
    Schillergesellschaft Marbach), Marbach am Neckar

Marburg

Hessisches Staatsarchiv, Marburg

Munich

Abteilung für Handschriften und Seltene Drucke, Bayerische Staatsbibliothek, Munich
    (BSBMH)
Archiv, Bayerische Akademie der Wissenschaften, Munich
Archiv, Siemens Forum (Siemens Museum), Munich (SF)
Handschriften-Bestand (Handschriftenabteilung), Archiv, Deutsches Museum, Munich (DM)
Siemens Family Papers, Private Collection, Munich

Murnau

Private Collection, Prof. Dr.-Ing. Walter Henn, Murnau

Nürnberg

Germanisches Nationalmuseum, Archiv, Nürnberg

Potsdam

Abteilung Potsdam, Bundesarchiv, Potsdam
Heilig-Geist-Kirche zu Potsdam, Potsdam
Sankt Nikolai-Kirche zu Potsdam, Potsdam

Stuttgart

Württembergische Landesbibliothek, Stuttgart

Tübingen

Universitätsbibliothek, Tübingen

Weimar

Goethe- und Schiller-Archiv, Nationale Forschungs- und Gedenkstätten der klassischen
    deutschen Literatur in Weimar, Weimar

Wiesbaden

Archiv, Vieweg Verlag (formerly Friedrich Vieweg & Sohn), Wiesbaden (VV)
Hessische Landesbibliothek, Wiesbaden

Wolfenbüttel

Herzog August Bibliothek, Wolfenbüttel

ITALY

Archives, Stazione Zoologica "Anton Dohrn," Naples (SZADNA)
Dipartimento di Matematica, Università degli Studi "La Sapienza," Rome
Serie Accademia dei Lincei, Carte Quintino Sella, Fondazione Sella, Biella

NETHERLANDS

Hollandsche Maatschappij der Wetenschappen, Haarlem
Rijksarchief, Haarlem
Rijksmuseum voor de Geschiedenis van de Natuurwetenschappen en van de Geneeskunde,
    Museum Boerhaave, Leiden

POLAND

Archiwum Państwowe w Olsztynie, Olsztyn
Biblioteka Jagiellońska, Uniwersytet Jagielloński, Cracow

RUSSIA

Archive, Russian Academy of Sciences, Saint Petersburg

SWEDEN

Center for History of Science, The Royal Swedish Academy of Sciences, Stockholm
Uppsala University Library, Uppsala

SWITZERLAND

Bibliothek, Eidgenössische Technische Hochschule, Zurich
Département des manuscrits, Bibliothèque de Genève (formerly Bibliothèque Publique
    et Universitaire), Geneva
Öffentliche Bibliothek der Universität Basel, Basel
Wissenschaftshistorische Sammlungen, Eidgenössische Technische Hochschule, Zurich

UNITED KINGDOM

Archive, Institution of Electrical Engineers, London
Archive, The Royal Society, London
Archives, Royal Institution of Great Britain, London
Department of Manuscripts, National Library of Scotland, Edinburgh
Department of Manuscripts and University Archives, Cambridge University Library (CUL)
Department of Western Manuscripts, Bodleian Library, University of Oxford, Oxford
John Rylands University Library, University of Manchester, Manchester
Library, Royal Society of Chemistry, London
Library, Royal Society of Medicine, London
Manuscripts Room, Library, University College London, London (UCL)
Rayleigh Papers, The Old Rectory, Chelmsford, Terling, Essex
Special Collections, Edinburgh University Library, Edinburgh
University of Exeter Library, Exeter

University of Glasgow Library, Glasgow
Wellcome Institute for the History of Medicine, London

UNITED STATES

Air Force Geophysics Laboratory (AFSC), Department of the Air Force, Hanscom Air Force
    Base, Massachusetts
American Antiquarian Society, Worcester, MA
American Institute of Physics, College Park, MD
Archives, National Academy of Sciences, Washington, DC
Archives & Special Collections, George C. Gordon Library, Worcester Polytechnic Institute,
    Worcester, MA
Bakken Museum, Minneapolis
Bancroft Library, University of California, Berkeley
Beinecke Rare Book and Manuscript Library, Yale University, New Haven, CT
Boston Medical Library, Boston
Department of Molecular Biophysics and Biochemistry, Yale University, New Haven, CT
Fiorello H. LaGuardia Archives, LaGuardia Community College, City University of New York,
    Long Island City, New York
Francis A, Countway Library of Medicine (Boston Medical Library and Harvard Medical
    School), Boston
Houghton Library, Harvard University, Cambridge, MA
J. M. Wheeler Library, Edward S, Harkness Eye Institute, Columbia University, College of
    Physicians and Surgeons, Department of Ophthalmology, New York Presbyterian Hospi-
    tal, New York
Library, American Philosophical Society, Philadelphia
Manuscript Collections, Rare Book and Manuscript Library, Butler Library, Columbia Univer-
    sity in the City of New York, New York
Manuscript Division, Library of Congress, Washington, DC
Manuscripts of the Dibner Collection, Dibner Library of the History of Science and Tech-
    nology, Washington, DC
Private Collection, Henry Z. Steinway, New York
Smithsonian Institution Archives, Washington, DC
Special Collections, Dartmouth College Library, Hanover, NH
Special Collections and Archives, Manuscripts, Milton S. Eisenhower Library, Johns Hopkins
    University, Baltimore
Thomas A. Edison Papers, Rutgers University, Piscataway, NJ

Printed Works

Helmholtz's scientific papers are listed below as they appear in his *Wissenschaftliche Abhandlungen*, published in three volumes in 1882, 1883, and 1895 and abbreviated *WA*; that is, the
original place of publication is not given here (except for items not published in *WA*), though the
original date of publication is given. The same points pertain to his popular scientific lectures,
most of which were (re)published in his *Vorträge und Reden* (1903), abbreviated *VR*. All entries
for Hermann Helmholtz are listed under "HH." Readers interested in pertinent items beyond
those listed here may refer to the bibliography in Cahan 1993a.

Abbe, Cleveland. 1891. *The Mechanics of the Earth's Atmosphere: A Collection of Translations*.
    Washington, DC: Smithsonian Institution.

Abbe, Ernst. 1873. "Beiträge zur Theorie des Mikroskops und der mikroskopischen Wahrnehmung." *Archiv für Mikroskopische Anatomie* 9:413–68.

Abbott, Edwin A. 2002. *The Annotated Flatland: A Romance of Many Dimensions*. Introduction and notes by Ian Stewart. Cambridge, MA: Perseus.

———. 2010. *Flatland*. With notes and commentary by William F. Lindgren and Thomas F. Banchoff. Cambridge: Cambridge University Press; Washington, DC: Mathematical Association of America.

Académie des Sciences. 1869. "Comité Secret." *CR* 69:1385.

———. 1870. "Correspondance," 17 January. *CR* 70:123.

———. 1870a. "Nominations." *CR* 70:27.

Ackerknecht, Erwin H. 1953. *Rudolf Virchow: Doctor, Statesman, Anthropologist*. Madison, WI: University of Wisconsin Press.

Adams, Henry. 1974. *The Education of Henry Adams*. Edited by Ernest Samuels. Boston: Houghton Mifflin.

———. 1988. *The Letters of Henry Adams*. Edited by J. C. Levenson, Ernest Samuels, Charles Vandersee, and Viola Hopkins Winner. 6 vols. Cambridge, MA: Belknap Press of Harvard University Press.

Adress-Kalender, 1837[?]. *Adress-Kalender 1837 für die Königl. Haupt- und Residenz-Städte Berlin und Potsdam auf das Jahr 1837*. Berlin: Rücker and Pückler.

Agee, William C. 1977. "Patrick Henry Bruce: A Major American Artist of Early Modernism." *Arts in Virginia* 17:12–23.

Akademie der Wissenschaften. 1986. "Träger von Akademie-Auszeichnungen: Die Träger der Helmholtz-Medaille." In *Akademie der Wissenschaften der DDR Jahrbuch 1985*, 180–81. Berlin: Akademie-Verlag.

Albisetti, James C. 1983. *Secondary School Reform in Imperial Germany*. Princeton, NJ: Princeton University Press.

———. 1988. *Schooling German Girls and Women: Secondary and Higher Education in the Nineteenth Century*. Princeton, NJ: Princeton University Press.

Albrecht, Bettina. 1985. "Die ehemaligen Naturwissenschaftlichen und Medizinischen Institutsgebäude der Universität Heidelberg im Bereich Brunnengasse, Hauptstrasse, Akademiestrasse und Plöck." PhD diss., Ruprecht-Karls-Universität in Heidelberg.

Albrecht, Helmuth. 1993. "'Max Planck: Mein Besuch bei Adolf Hitler'—Anmerkungen zum Wert einer historischen Quelle." In *Naturwissenschaft und Technik in der Geschichte: 25 Jahre Lehrstuhl für Geschichte der Naturwissenschaft und Technik am Historischen Institut der Universität Stuttgart*, edited by Helmuth Albrecht, 41–63. Stuttgart: Verlag für Geschichte der Naturwissenschaften und der Technik.

Albrecht, Hermann, and Armin Hermann. 1990. "Die Kaiser-Wilhelm-Gesellschaft im Dritten Reich (1933–1945)." In *Forschung im Spannungsfeld von Politik und Gesellschaft: Geschichte und Struktur der Kaiser-Wilhelm-/Max-Planck-Gesellschaft: Aus Anlass ihres 75 jähriigen Bestehens*, edited by Rudolf Vierhaus and Bernhard vom Brocke, 356–406. Stuttgart: Deutsche Verlags-Anstalt.

American Philosophical Society. 1891. *Proceedings of the American Philosophical Society Held at Philadelphia for Promoting Useful Knowledge*. Vol. 29. Philadelphia: MacCalla.

Amrine, Frederick, Francis J. Zucker, and Harvey Wheeler, eds. 1987. *Goethe and the Sciences: A Reappraisal*. Dordrecht: D. Reidel.

Ansprachen. 1892. *Ansprachen und Reden gehalten bei der am 2. November 1891 zu Ehren von Hermann von Helmholtz veranstalteten Feier nebst einem Verzeichnisse der überreichten Diplome und Ernennungen, sowie der Adressen und Glückwunschschreiben*. Berlin: Hirschwald.

Aranjo, Saulo de Freitas. 2014. "Bringing New Archival Sources to Wundt Scholarship: The Case of Wundt's Assistantship with Helmholtz." *History of Psychology* 17:50–59.

Archibald, Thomas. 2002. "Charles Hermite and German Mathematics in France." In *Mathematics Unbound: The Evolution of an International Mathematical Research Community, 1800–1945*, edited by Karen Hunger Parshall and Adrian C. Rice, 123–37. Providence, RI: American Mathematical Society and London Mathematical Society.

Architekten-Verein zu Berlin und Vereinigung berliner Architekten, comp. and eds. 1896. *Berlin und seine Bauten*. 3 vols. in 2. Berlin: Wilhelm Ernst.

Argüelles, José A. 1972. *Charles Henry and the Formation of a Psychophysical Aesthetic*. Chicago: University of Chicago Press.

Arnold, Matthew. 1892. *Higher Schools and Universities in Germany*. 2nd ed. Reprint, London: Macmillan.

———. 1986. *Matthew Arnold*. Edited by Miriam Allott and Robert H. Super. Oxford: Oxford University Press, 1986.

Ash, Mitchell G. 1995. *Gestalt Psychology in German Culture, 1890–1967: Holism and the Quest for Objectivity*. Cambridge: Cambridge University Press.

Atkinson, Edmund. 1880. Preface to *Popular Lectures on Scientific Subjects*, by Hermann von Helmholtz, v–vi. Translated by E. Atkinson. 1st ser. London: Longmans, Green.

Auerbach, Felix. 1881. "Hermann Helmholtz und die wissenschaftlichen Grundlagen der Musik." *Nord und Sud* 19:217–44.

Auth, Joachim, and Heinz Kossack. 1983. "Zur Lage der Physik an der Berliner Universität vor der Errichtung des Instituts am Reichstagsufer." *Wissenschaftliche Zeitschrift der Humboldt-Universität zu Berlin: Mathematisch-Naturwissenschaftliche Reihe* 32:555–67.

Bader, Alfred. 1933. *Entwicklung der Augenheilkunde im 18. und 19. Jahrhundert mit besonderer Berücksichtigung der Schweiz (Nachlass von Prof. Horner, Zürich)*. Basel: Benno Schwabe.

Badermann, G. 1939. "Zur Erinnerung an einen grossen Forscher." *Zeitschrift für Angewandte Meteorologie* 56:33–35.

Bagge, Selmar. 1867. "Zur Theorie der Musik: Die Physiker und die Musiker." *Leipziger Allgemeine Musikalische Zeitung* 2:465–69.

Baird, Davis, R. I. G. Hughes, and Alfred Nordmann, eds. 1998. *Heinrich Hertz: Classical Physicist, Modern Philosopher*. Dordrecht: Kluwer Academic.

Baker, William. 1976. *Some George Eliot Notebooks: An Edition of the Carl H. Pforzheimer Library's George Eliot Holograph Notebooks, MSS 707, 708, 709, 710, 711*. 4 vols. Salzburg: Institut für Englische Sprache und Literatur, Universität Salzburg.

———. 1977. *The George Eliot–George Henry Lewes Library: An Annotated Catalogue of Their Books at Dr. Williams's Library, London*. New York: Garland.

Baldwin, Neil. 1995. *Edison: Inventing the Century*. New York: Hyperion.

Ballin, Herbert. 1938. "Hermann Helmholtz." In *Deutsche Männer: 200 Bildnisse und Lebensbeschreibungen*, edited by Wilhelm Schüssler, 342–43. Berlin: Ernst Steiniger.

"Banquet in Honor of von Helmholtz." 1893. *Chicago Tribune*, 31 August, 3.

Barclay, David E. 1995. *Frederick William IV and the Prussian Monarchy, 1840–1861*. Oxford: Clarendon Press.

Barkan, Diana Kormos. 1999. *Walther Nernst and the Transition to Modern Physical Science*. Cambridge: Cambridge University Press.

Barman, Roderick J. 1999. *Citizen Emperor: Pedro II and the Making of Brazi1, 1825–91*. Stanford, CA: Stanford University Press.

Barnouw, Jeffrey. 1987. "Goethe and Helmholtz: Science and Sensation." In Amrine, Zucker, and Wheeler 1987, 45–82.

Barton, Ruth. 1990. "'An Influential Set of Chaps': The X-Club and Royal Society Politics, 1864–85." *British Journal for the History of Science* 23:53–81.

Bartsch, Karl, ed. 1886. *Ruperto-Carola 1386–1886: Illustrirte Fest-Chronik der V. Säcular Feier der Universität Heidelberg*. Heidelberg: Otto Petters.

Barzellotti, Giacomo. 1880. *La nuova scuola del Kant e la filosofia scientifica contemporanea in*

Germania. Rome: Tipographi Barbèra. Also published in *Nuova Antologia*, 15 February 1880.

Bauer, Franz J. 1991. *Bürgerwege und Bürgerwelten: Familienbiographische Untersuchungen zum deutschen Bürgertum im 19. Jahrhundert*. Göttingen: Vandenhoeck & Ruprecht.

Baumann, Christian. 2002. *Der Physiologe Ewald Hering (1834–1918): Curriculum Vitae*. Frankfurt am Main: Dr. Hänsel-Hohenhausen.

Bayertz, Kurt. 1985. "Spreading the Spirit of Science: Social Determinants of the Popularization of Science in Nineteenth-Century Germany." In *Expository Science: Forms and Functions of Popularisation*, edited by Terry Shinn and Richard Whitley, 209–27. Dordrecht: D. Reidel.

Beer, Gillian. 1985. *Darwin's Plots: Evolutionary Narrative in Darwin, George Eliot, and Nineteenth-Century Fiction*. London: ARK.

———. 1992. "Helmholtz, Tyndall, Gerard Manley Hopkins: Leaps of the Prepared Imagination." *Comparative Criticism* 13:117–45.

———. 2004. "The Academy: Europe in England." In *Science Serialized: Representations of the Sciences in Nineteenth-Century Periodicals*, edited by Geoffrey Cantor and Sally Shuttleworth, 181–98. Cambridge, MA: MIT Press.

Bélafi, Michael. 1990. *Ferdinand Graf von Zeppelin*. Leipzig: BSB B. G. Teubner.

Bell, James F. 1980. "Helmholtz, Hermann (Ludwig Ferdinand) von." In Sadie 1980, 8:466–67.

Bell, James F., and C. Truesdell. 1980. "Physics of Music: The Age of Helmholtz." In Sadie 1980, 14:673–74.

Bellmer, Elizabeth Henry. 1999. "The Statesman and the Ophthalmologist: Gladstone and Magnus on the Evolution of Human Colour Vision, One Small Episode of the Nineteenth-Century Darwinian Debate." *AS* 56:25–45.

Belloni, Luigi. 1980. *Franz Boll, scopritore della porpora retinica: Sue lettere a Emil du Bois-Reymond, Camillo Golgi e Ernst Haeckel*. Milan: Istituto Lombardo di Scienze e Lettere.

———. 1982. "Hermann Helmholtz und Franz Boll." *Medizinhistorisches Journal* 17:129–37.

Bernard, Claude. 1967. *The Cahier Rouge of Claude Bernard*. Cambridge, MA: Schenkman.

Bernstein, Julius. 1906. "Hermann von Helmholtz." In *Badische Biographien*: Part 5, *1891–1901*, edited by Fr. von Weech and U. Krieger, 1:281–94. Heidelberg: Carl Winter's Universitätsbuchhandlung.

Bertrand, Gustave. 1868. "Un nouveau système d'acoustique musicale." *Revue moderne* 4 (1 January): 85–109.

Bertrand, Joseph. 1868. "Théorème relatif au mouvement le plus général d'un fluide." *CR* 66: 1227–30.

———. 1868a. "Note relative à la théorie des fluides: Réponse à la communication de M. Helmholtz." *CR* 67:267–69.

———. 1868b. "Observations nouvelles sur un mémoire de M. Helmholtz." *CR* 67:469–72.

———. 1868c. "Réponse à la note de M. Helmholtz." *CR* 67:773–75.

Bevilacqua, Fabio. 1993. "Helmholtz's *Ueber die Erhaltung der Kraft*: The Emergence of a Theoretical Physicist." In Cahan 1993a, 291–333.

———. 1994. "Theoretical and Mathematical Interpretations of Energy Conservation: The Helmholtz-Clausius Debate on Central Forces, 1852–54." In Krüger 1994a, 89–106.

Beyer, Robert T. 1999. *Sounds of Our Times: Two Hundred Years of Acoustics*. New York: Springer.

Bezold, Friedrich von. 1920. *Geschichte der Rheinischen Friedrich-Wilhelms-Universität von der Gründung bis zum Jahr 1870*. Bonn: A. Marcus and E. Weber.

Bezold, Wilhelm von. 1895 *Hermann von Helmholtz: Gedächtnissrede gehalten in der Singakademie zu Berlin am 14. Dezember 1894 von Wilhelm von Bezold mit einem Porträt von Franz von Lenbach*. Leipzig: Johann Ambrosius Barth.

———. 1896. A toast/speech published in "Feier des fünfzigjährigen Stiftungsfestes der

Physikalischen Gesellschaft zu Berlin am Januar 1896." *Verhandlungen der Physikalischen Gesellschaft zu Berlin* 15:19–25.

Biagioli, Francesca. 2014. "Hermann Cohen and Alois Riehl on Geometrical Empiricism." *HOPOS: The Journal of the International Society for the History of Science* 4:83–105.

———. 2014a. "What Does It Mean That 'Space Can Be Transcendental without the Axioms Being So'? Helmholtz's Claim in Context." *Journal for General Philosophy of Science* 45:1–21.

Biedermann, Rudolf, comp. 1877. *Bericht über die Ausstellung wissenschaftlicher Apparate im South Kensington Museum, zu London, 1876.* . . . London: John Strangeways.

Bierhalter, Günter. 1981. "Zu Hermann von Helmholtzens mechanischer Grundlegung der Wärmelehre aus dem Jahre 1884." *Archive for History of Exact Sciences* 25:71–84.

———. 1983. "Die v. Helmholtzschen Monozykel-Analogien zur Thermodynamik und das Clausiussche Disgregationskonzept." *Archive for History of Exact Sciences* 29:95–100.

———. 1987. "Wie erfolgreich waren die im 19. Jahrhundert betriebenen Versuche einer mechanischen Grundlegung des zweiten Hauptsatzes der Thermodynamik?" *Archive for History of Exact Sciences* 37:77–99.

———. 1992. "Von L. Boltzmann bis J.J. Thomson: Die Versuche einer mechanischen Grundlegung der Thermodynamik (1866–1890)." *Archive for History of Exact Sciences* 44:25–75.

———. 1993. "Helmholtz's Mechanical Foundations of Thermodynamics." In Cahan 1993a, 432–58.

Biermann, Kurt-R. 1973. *Die Mathematik und ihre Dozenten an der Berliner Universität 1810–1920: Stationen auf dem Wege eines mathematischen Zentrums von Weltgeltung.* Berlin: Akademie-Verlag.

———, ed. 1985. *Vier Jahrzehnte Wissenschaftsförderung: Briefe an das preussische Kultusministerium 1818–1859/Alexander von Humboldt.* Berlin: Akademie-Verlag.

———. 1990. *Miscellanea Humboldtiana.* Berlin: Akademie-Verlag.

Binz, Carl. 1869. "Pharmakologische Studien über Chinin." *Archiv für Pathologische Anatomie und Physiologie und für Klinische Medicin* 46:67–105, 129–68.

———. 1874. "An Experimental Observation on Hay Fever." *Nature* 10:26–27.

Birkenmaier, Willy. 1995. *Das russische Heidelberg: Zur Geschichte der deutsch-russischen Beziehungen im 19. Jahrhundert.* Heidelberg: Wunderhorn.

Bismarck, Otto von. 1926. "413: Gespräch mit dem Schriftsteller Dr. Moritz Busch am 27. Januar 1887 in Berlin." In *Bismarck: Die gesammelten Werke*: Vol. 8, *Gespräche*, edited by Willy Andreas, 550–54. 2nd ed. Berlin: Otto Stollberg & Verlag für Politik und Wirtschaft.

Blackmore, John. 1978. "Three Autobiographical Manuscripts by Ernst Mach." *AS* 35:401–18.

———, ed. 1995. *Ludwig Boltzmann: His Later Life and Philosophy, 1900–1906*: Book 1, *A Documentary History.* Dordrecht: Kluwer.

Blaserna, P[ietro]. 1877. *Le son et la musique . . . suivis des causes physiologiques de l'harmonie musicale par H. Helmholtz.* Paris: G. Baillière.

Blondel, Christine. 1990. "Négociations entre savants, industriels et administrateurs: Les premiers congrès internationaux d'électricité." *Relations Internationaux* 62:171–82.

Blum, Peter. 1997. Foreword to Werner 1997, ix–x.

Blume, Guilielmus Arminius. 1834. *Narratio de Lycurgo Oratore: Qua Examina Publica in Gymnasio Regio Postdamiensi.* . . . Potsdam: Deckerscher Geh. Oberhofbuchdruckerei.

Bochalli, Richard. 1950. "Vor 100 Jahren—Hermann von Helmholtz, der Erfinder des Augenspiegels." *Hippokrates* 21:724–26.

Bode, Wilhelm von. 1930. *Mein Leben.* 2 vols. Berlin: Hermann Reckendorf.

Boehlich, Walter, ed. 1965. *Der Berliner Antisemitismusstreit.* Frankfurt am Main.: Insel.

———, ed. 1990. *The Letters of Sigmund Freud to Eduard Silberstein, 1871–1881.* Cambridge, MA: Belknap Press of Harvard University Press.

Boehm, Laetitia. 1958. "Von den Anfängen des akademischen Frauenstudiums in Deutschland: Zugleich ein Kapitel aus der Geschichte der Ludwig-Maximilians-Universität München." *Historisches Jahrbuch* 77:298–327.

Boi, Luciano. 1996. "Les géométries non euclidiennes, le problème philosophique de l'espace et la conception transcendantale: Helmholtz et Kant, les néo-kantiens, Einstein, Poincaré et Mach." *Kant-Studien* 87:257–89.

Boi, Luciano, Livia Giacardi, and Rossana Tazzioli, eds. 1998. *La découverte de la géometrie non euclidenne sur la pseudosphère: Les lettres d'Eugenio Beltrami à Jules Hoüel (1868–1881)*. With an introduction, notes, and critical commentaries by Luciano Boi, Livia Giacardi, and Rossana Tazzioli. Paris: Librairie Scientifique et Technique.

Bokulich, Alisa. 2015. "Maxwell, Helmholtz, and the Unreasonable Effectiveness of the Method of Physical Analogy." *SHPS* 50:28–37.

Boltzmann, Ludwig. 1905. *Populäre Schriften*. Leipzig: Johann Ambrosius Barth.

———. 1994. *Leben und Briefen*, edited by Walter Höflechner. Graz: Akademische Druck- und Verlagsanstalt.

Bonner, Thomas Neville. 1963. *American Doctors and German Universities: A Chapter in International Intellectual Relations, 1870–1914*. Lincoln: University of Nebraska Press.

Bordoni, Stefano. 2013. "Routes Towards an Abstract Thermodynamics in the Late Nineteenth Century." *European Physics Journal H* 38:617–60.

Boring, Edwin G. 1950. *A History of Experimental Psychology*. 2nd ed. New York: Appleton-Century-Crofts.

Borscheid, Peter. 1976. *Naturwissenschaft, Staat und Industrie in Baden (1848–1914)*. Stuttgart: Ernst Klett.

Bowman, William. 1890–91. Obituary of Frans Cornelis Donders. *Proceedings of the Royal Society of London* 49:vii–xxiv.

Brain, Robert M., and M. Norton Wise. 1994. "Muscles and Engines: Indicator Diagrams and Helmholtz's Graphical Methods." In Krüger 1994a, 124–45.

Brandes, Georg. 1989. *Berlin als deutsche Reichshauptstadt: Erinnerungen aus den Jahren 1877–1883*. Edited by Erik M. Christensen and Hans-Dietrich Loock. Berlin: Colloquium Verlag.

Brauer, Arthur von. 1936. *Im Dienste Bismarcks: Persönliche Erinnerungen*. Edited by Helmuth Rogge. Berlin: E. S. Mittler.

Braun, Christoph. 1992. *Max Webers "Musiksoziologie."* Laaber: Laaber-Verlag.

Braun, Marta. 1992. *Picturing Time: The Work of Etienne-Jules Marey (1830–1904)*. Chicago: University of Chicago Press.

Braun-Artaria, Rosalie. 1899. *Anna von Helmholtz: Ein Erinnerungsblatt*. N.p.: N.p., 14 December.

———. 1918. *Von berühmten Zeitgenossen: Lebenserinnerungen einer Siebzigerin*. 2nd ed. Munich: C. H. Beck'sche Verlagsbuchhandlung.

Breger, Herbert. 1982. *Die Natur als arbeitende Maschine: Zur Entstehung des Energiebegriffs in der Physik 1840–1850*. Frankfurt: Campus Verlag.

———. 1985. "Streifzug durch die Geschichte der Mathematik und Physik an der Universität Heidelberg." In *Auch eine Geschichte der Universität Heidelberg*, edited by Karin Buselmeier, Dietrich Harth, and Christian Jansen, 27–50. Mannheim: Edition Quadrat.

Bringmann, Wolfgang G., Gottfried Bringmann, and David Cottrell. 1976. "Helmholtz und Wundt an der Heidelberger Universität 1858–1871." *Heidelberger Jahrbücher* 20:79–88.

Bringmann, Wolfgang G., Norma J. Bringmann, and William D. G. Balance. 1980. "Wilhelm Maximilan Wundt 1832–1874: The Formative Years." In Bringmann and Tweney 1980, 13–32.

Bringmann, Wolfgang G., and Ryan D. Tweney, eds. 1980. *Wundt Studies: A Centennial Collection*. Toronto: C. J. Hogrefe.

Broadhouse, John. N.d. [1892?]. *Musical Acoustics; or, The Phenomena of Sound as Connected with Music: The Student's Helmholtz*. Fifth impression. [3rd ed.?] London: William Reeves.

Brobjer, Thomas H. 2004. "Nietzsche's Reading and Knowledge of Natural Science: An Overview." In *Nietzsche and Science*, edited by Gregory Moore and Thomas H. Brobjer, 21–50. Aldershot: Ashgate.

Brock, William H., and Roy M. MacLeod, eds. 1980. *Natural Knowledge in Social Context: The Journals of Thomas Archer Hirst FRS*. N.p: Mansell.

Brock, William H., N. D. McMillan, and R. C. Mollan, eds. 1981. *John Tyndall: Essays on a Natural Philosopher*. Dublin: Royal Dublin Society.

Brock, William H., and A. J. Meadows. 1998. *The Lamp of Learning: Two Centuries of Publishing at Taylor and Francis*. 2nd ed. London: Taylor & Francis.

Brocke, Bernhard vom. 1996. "Hermann von Helmholtz und die Politik." In Eckart and Volkert 1996, 267–326.

Brossmer, Karl. 1958. "Hermann von Helmholtz (1821–1894)." *Ruperto-Carola* 24:147–48.

Browne, Janet. 1995–2002. *Charles Darwin*. 2 vols. Princeton, NJ: Princeton University Press.

Bruce, Robert V. 1973. *Bell: Alexander Graham Bell and the Conquest of Solitude*. Boston: Little, Brown.

Br[üche], [Ernst]. 1947. "Hundert-Jahrfeier der Physikalischen Gesellschaft." *PB* 3:2–4.

———. 1948. "Helmholtz und Rutherford." *PB* 4:21–22.

Brücke, Ernst. 1871. *Die Physiologischen Grundlagen der neuhochdeutschen Verskunst*. Vienna: Carl Gerold's Sohn.

———. 1878. *Principes scientifiques des beaux-arts, essais et fragments de théorie par E. Brücke … suivis de l'optique et la peinture par H. Helmholtz*. Paris: Librairie Germer Baillière.

Brücke, E. Th. 1928. *Ernst Brücke*. Vienna: Julius Springer.

Brücke, Hans, Wolfgang Hilger, Walter Höflechner, and Wolfram W. Swoboda, eds. 1978–81. *Ernst Wilhelm von Brücke. Briefe an Emil du Bois-Reymond*. 2 vols. Graz: Akademische Druck- u. Verlagsanstalt.

Brush, Stephen G. 1966–67. "Science and Culture in the Nineteenth Century: Thermodynamics and History." *Graduate Journal* 7:477–565.

Bryan, George S. 1926. *Edison: The Man and His Work*. Garden City, NY: Garden.

Buchheim, Gisela. 1977. "Die Entwicklung des elektrischen Messwesens und die Gründung der Physikalisch-Technischen Reichsanstalt." *NTM: Zeitschrift für Naturwissenschaften, Technik und Medizin* 14:16–32.

Büchner, Ludwig. 1856. *Kraft und Stoff: Empirisch-naturphilosophische Studien*. 4th ed. Frankfurt am Main: Meidinger Sohn & Ch.

Buchwald, Jed Z. 1985. *From Maxwell to Microphysics: Aspects of Electromagnetic Theory in the Last Quarter of the Nineteenth Century*. Chicago: University of Chicago Press.

———. 1993. "Electrodynamics in Context: Object States, Laboratory Practice, and Anti-Romanticism." In Cahan 1993a, 334–73.

———. 1994. *The Creation of Scientific Effects: Heinrich Hertz and Electric Waves*. Chicago: University of Chicago Press.

———, ed. 1996. *Scientific Credibility and Technical Standards*. Dordrecht: Kluwer Academic.

Bunsen, Marie von. 1899. *Zur Erinnerung an Frau Anna von Helmholtz*. 10 December. N.p.: n.p.

———. 1929. *Die Welt in der ich lebte: Erinnerungen aus glücklichen Jahren 1860–1912*. Leipzig: Koehler & Amelang.

———. 1932. *Zeitgenossen die ich erlebte 1900–1930*. Leipzig: Koehler & Amelang.

Burkhardt, Frederick, Sydney Smith, David Kohn, and William Montgomery, eds. 1985. *A Calendar of the Correspondence of Charles Darwin, 1821–1882*. New York: Garland.

Burrow, J. W. 2000. *The Crisis of Reason: European Thought, 1848–1914*. New Haven, CT: Yale University Press.

Butler, Nicholas Murray. 1939–40. *Across the Busy Years: Recollections and Reflections*. 2 vols. New York: Charles Scribner's.

Cahan, David. 1985. "The Institutional Revolution in German Physics, 1865–1914." *Historical Studies in the Physical Sciences* 15 (2): 1–65.

———. 1989. *An Institute for an Empire: The Physikalisch-Technische Reichsanstalt, 1871–1918.* Cambridge: Cambridge University Press.

———. 1990. "From Dust Figures to the Kinetic Theory of Gases: August Kundt and the Changing Nature of Experimental Physics in the 1860s and 1870s." *AS* 47:151–72.

———. 1993. "Helmholtz and the Civilizing Power of Science." In Cahan 1993a, 559–601.

———, ed. 1993a. *Hermann von Helmholtz and the Foundations of Nineteenth-Century Science.* Berkeley: University of California Press.

———, ed. 1993b. *Letters of Hermann von Helmholtz to His Parents: The Medical Education of a German Scientist, 1837–1846.* Stuttgart: Franz Steiner Verlag.

———. 1996. "The Zeiss Werke and the Ultramicroscope: The Creation of a Scientific Instrument in Context." In Buchwald 1996, 67–115.

———. 1998. "Ophthalmoscope." In *Instruments of Science: An Historical Encyclopedia*, edited by Robert Bud and Deborah Jean Warner, 425–27. New York: Science Museum.

———. 1999. "Helmholtz als führender Wissenschaftler an der Preussischen Akademie der Wissenschaften." In Kocka 1999, 277–314.

———. 2000. "The Young Einstein's Physics Education: H.F. Weber, Hermann von Helmholtz, and the Zurich Polytechnic Physics Institute." In Howard and Stachel 2000, 43–82.

———. 2004. "Helmholtz and the Shaping of the American Physics Elite in the Gilded Age." *Historical Studies in the Physical and Biological Sciences* 35 (1): 1–34.

———. 2006. "The 'Imperial Chancellor of the Sciences': Helmholtz between Science and Politics." *Social Research* 73 (4): 1093–1128.

———. 2010. "Helmholtz in Gilded-Age America: The International Electrical Congress of 1893 and the Relations of Science and Technology." *AS* 67 (1): 1–38.

———. 2012. "Helmholtz and the British Scientific Elite: From Force Conservation to Energy Conservation." *Notes and Records of the Royal Society of London* 66:55–68.

———. 2012a. "The Awarding of the Copley Medal and the 'Discovery' of the Law of Conservation of Energy: Joule, Mayer, and Helmholtz Revisited." *Notes and Records of the Royal Society of London* 66:125–39.

Campbell, Lewis, and William Garnett. 1969. *The Life of James Clerk Maxwell.* Reprint, New York: Johnson Reprint.

Canales, Jimena. 2001. "Exit the Frog, Enter the Human: Physiology and Experimental Psychology in Nineteenth-Century Astronomy." *British Journal for the History of Science* 34: 173–97.

Caneva, Kenneth L. 1993. *Robert Mayer and the Conservation of Energy.* Princeton, NJ: Princeton University Press.

Cantor, Geoffrey. 1976. "William Robert Grove, the Correlation of Forces, and the Conservation of Energy." *Centaurus* 19:273–90.

Carhart, Henry S. 1894. "Professor von Helmholtz." *Electrical World* 24 (21): 542–43.

Carrier, Martin. 1994. "Geometric Facts and Geometric Theory: Helmholtz and 20th-Century Philosophy of Physical Geometry." In Krüger 1994a, 276–91.

"Casamicciola." 1931. In *Enciclopedia Italiana*, 9:282.

Cassirer, Ernst. 1950. *The Problem of Knowledge: Philosophy, Science, and History since Hegel.* New Haven, CT: Yale University Press.

Chadarevian, Soraya de. 1993. "Graphical Method and Discipline: Self-Recording Instruments in Nineteenth-Century Physiology." *SHPS* 24:267–91.

Chappell, William. 1876. *Helmholtz's New Musical Theories: A Review of Professor Helmholtz's Treatise on the Sensations of Tone as a Physiological Basis for the Theory of Music.* London: Novello, Ewer.

Ck. 1946–47. "Hermann v. Helmholtz und die moderne Physik." *Kosmos* 42–43:129–30.

Clark, Christopher. 2007. *Iron Kingdom: The Rise and Downfall of Prussia, 1600–1947*. New York: Penguin.

Clarke, Edwin, and L. S. Jacyna. 1987. *Nineteenth-Century Origins of Neuroscientific Concepts*. Berkeley: University of California Press.

Coffa, J. Alberto. 1991. *The Semantic Tradition from Kant to Carnap: To the Vienna Station*. Edited by Linda Wessels. Cambridge: Cambridge University Press.

Cole, Douglas. 1999. *Franz Boas: The Early Years, 1858–1906*. Vancouver: Douglas & McIntyre; Seattle: University of Washington Press.

Comberg, Wilhelm. 1951. "Hermann von Helmholtz als Erfinder des Augenspiegels." In *Hundert Jahre Augenspiegel: Fünf Vorträge gehalten von Augenärzten der Deutschen Demokratischen Republik gelegentlich einer Gedenkfeier und Festsitzung zu Berlin am 19. Dezember 1950*, edited by Wilhelm Comberg, 7–17. Leipzig: Georg Thieme.

"Congrès scientifiques: Le congrès international des électriciens." 1881. *Revue Scientifique*, 3rd ser., 1:382–84, 407–15, 525–29.

Coon, Deborah J. 1992. "Testing the Limits of Sense and Science: American Experimental Psychologists Combat Spiritualism, 1880–1920." *American Psychologist* 47:143–51.

Craig, Gordon A. 1978. *Germany, 1866–1945*. New York: Oxford University Press.

Cranefield, Paul F. 1957. "The Organic Physics of 1847 and the Biophysics of Today." *Journal of the History of Medicine and Allied Sciences* 12:407–23.

———. 1959. "The Nineteenth-Century Prelude to Modern Biophysics." In *Proceedings of the First National Biophysics Conference, Columbus, Ohio, March 4–6, 1957*, edited by Henry Quastler and Harold J. Morowitz, 19–26. New Haven, CT: Yale University Press.

———. 1966. "The Philosophical and Cultural Interests of the Biophysics Movement of 1847." *Journal of the History of Medicine and Allied Sciences* 21:1–7.

Crawford, Elisabeth. 1992. *Nationalism and Internationalism in Science, 1880–1939: Four Studies of the Nobel Population*. Cambridge: Cambridge University Press.

Cremona, Luigi. 1992. *La corrispondenza di Luigi Cremona (1830–1903)*. Edited by Ana Millán Gasca. Rome: Università degli Studi di Roma "La Sapienza."

Creutz, Rudolf, and Johannes Steudel. 1948. "Hermann von Helmholtz." In *Einführung in die Geschichte der Medizin in Einzeldarstellungen*, 297–302. Iserlohn, Germany: Silva-Verlag.

Crowe, Michael J. 1986. *The Extraterrestrial Life Debate, 1750–1900: The Idea of a Pluarlity of Worlds from Kant to Lowell*. Cambridge: Cambridge University Press.

Cushing, Harvey. 1940. *The Life of Sir William Osler*. Oxford: Oxford University Press.

D'Agostino, Salvo. 1986. "Scienza e cultura nella Germania dell'800: Le basi filosofiche della fisica-matematica nell'opera di Helmholtz." *Cultura e Scuola* 25 (99): 265–81.

———. 2001. "The Bild Conception of Physical Theories from Helmholtz to Hertz." In *The Dawn of Cognitive Science: Early European Contributors*, edited by Liliana Albertazzi, 151–66. Dordrecht: Kluwer Academic.

———. 2004. "The Bild Conception of Physical Theory: Helmholtz, Hertz, and Schrödinger." *Physics in Perspective* 6:372–89.

———. 2005. "Il difficile ricupero dell'Anschaulichkeit di Goethe nell'opera di Helmholtz." *Nuncius* 20:401–14.

Dahlhaus, Carl. 1970. "Hermann von Helmholtz und der Wissenschaftscharakter der Musiktheorie." In *Über Musiktheorie: Referate der Arbeitstagung 1970 in Berlin*, edited by Frieder Zaminer, 49–58. Cologne: Arno Volk Verlag Hans Gerig K.G.

Dale, Peter Allan. 1989. *In Pursuit of a Scientific Culture: Science, Art, and Society in the Victorian Age*. Madison: University of Wisconsin Press.

Darrigol, Olivier. 1994. "Helmholtz's Electrodynamics and the Comprehensibility of Nature." In Krüger 1994a, 216–42.

———. 2000. *Electrodynamics from Ampère to Einstein*. Oxford: Oxford University Press.

———. 2001. "God, Waterwheels, and Molecules: Saint-Venant's Anticipation of Energy Conservation." *Historical Studies in the Physical and Biological Sciences* 31 (2): 285–353.

———. 2003. "The Voltaic Origins of Helmholtz's Physics of Ions." *Nuova Voltiana* 5:163–76.

———. 2003a. "Number and Measure: Hermann von Helmholtz at the Crossroads of Mathematics, Physics, and Psychology." *SHPS* 34:515–73.

———. 2005. *Worlds of Flow: A History of Hydrodynamics from the Bernoullis to Prandtl*. Oxford: Oxford University Press.

———. 2007. "The Acoustic Origins of Harmonic Analysis." *Archive for History of Exact Sciences* 61:343–424.

———. 2007a. "A Helmholtzian Approach to Space and Time." *SHPS* 38:528–42.

———. 2012. *A History of Optics: From Greek Antiquity to the Nineteenth Century*. Oxford: Oxford University Press.

Darwin, Charles. 1987–89. *The Works of Charles Darwin*. Edited by Paul H. Barrett and R. B. Freeman. 29 vols. London: William Pickering.

———. 1988. *The Descent of Man: Selection in Relation to Sex*. 2nd ed. In Darwin 1987–89, vol. 21. London: John Murray.

———. 1988a. *The Expression of the Emotions in Man and Animals*. 2nd ed. In Darwin 1987–89, vol. 21. London: John Murray.

———. 1988b. *The Origin of Species by Means of Natural Selection*. . . . 6th ed. In Darwin 1987–89, vol. 16. London: John Murray.

"Das Gehirn von Helmholtz." 1895. *Philosophisches Jahrbuch* 8:115–16.

Daston, Lorraine. 1999. "Die Akademien und die Einheit der Wissenschaften: Die Disziplinierung der Disziplinen." In Kocka 1999, 61–84.

Daston, Lorraine, and Peter Galison. 2007. *Objectivity*. New York: Zone Books.

Daum, Andreas. 1998. *Wissenschaftspopularisierung im 19. Jahrhundert: Bürgerliche Kultur, naturwissenschaftliche Bildung und die deutsche Öffentlichkeit 1848–1914*. Munich: R. Oldenbourg.

Daum, J., V. von Ebner, and Hugo Graf Enzenberg, eds. 1869. *Tageblatt der 43. Versammlung Deutscher Naturforscher und Aerzte in Innsbruck vom 18. bis 24. September 1869*. Innsbruck: Wagner'sche Universitaets-Buchhandlung.

"Death of Professor von Helmholtz." 1894. *Standard* (London), 10 September.

Debru, Claude. 2001. "Helmholtz and the Psychophysiology of Time." *Science in Context* 14: 471–92.

Degen, Heinz. 1954. "Vor hundert Jahren: Die Naturforscherversammlung zu Göttingen und der Materialismusstreit." *Naturwissenschaftliche Rundschau* 7:271–77.

De Kock, Liesbet. 2014. "Hermann von Helmholtz's Empirico-Transcendentalism Reconsidered: Construction and Constitution in Helmholtz's Psychology of the Object." *Science in Context* 27 (4): 709–44.

———. 2014a. "Voluntarism in Early Psychology: The Case of Hermann von Helmholtz." *History of Psychology* 17 (2): 105–28.

———. 2016. "Helmholtz's Kant Revisited (Once More): The All-Pervasive Nature of Helmholtz's Struggle with Kant's *Anschauung*." *SHPS* 56:20–32.

De Palma, Armando, and Germana Pareti. 2011. "Bernstein's Long Path to Membrane Theory: Radical Change and Conservation in Nineteenth-Century German Electrophysiology." *Journal of the History of the Neurosciences* 20:306–37.

"Der Verein für wissenschaftliche Heilkunde in Königsberg während der ersten sechs Jahre seines Bestehens." 1859. In *Königsberger medicinische Jahrbücher*, edited by Verein für wissenschaftliche Heilkunde zu Königsberg, 1:1–14. Königsberg: Gräfe und Unzer.

Deutsche Akademie der Wissenschaften zu Berlin, ed. 1950. *Bildnisse berühmter Mitglieder der Deutschen Akademie der Wissenschaften zu Berlin*. Berlin: Akademie-Verlag.

———. 1957. "Statut der Helmholtz-Medaille." In *Jahrbuch der Deutschen Akademie der Wissenschaften zu Berlin 1955*, 299–300. Berlin: Akademie-Verlag.

———. 1960. "Verleihung der Helmholtz-Medaille und der Leibniz-Medaille." In *Jahrbuch der Deutschen Akademie der Wissenschaften zu Berlin 1959*, 28. Berlin: Akademie-Verlag.

"Deutsche Professoren: Hermann Ludwig Helmholtz." 1876. *Daheim* 12:679–81.

"Devoted to Science." 1893. *Chicago Daily Inter Ocean*, 22 August, 12.

Devrient, Eduard. 1964. *Eduard Devrient aus seinen Tagebüchern: Karlsruhe 1852–1870*. Edited by Rolf Kabel. 2 vols. Weimar: Hermann Böhlaus Nachfolger.

"Die Enthüllung des Helmholtz-Denkmals." 1899. *Vossische Zeitung* (Berlin), 6 June.

"Die Eröffnung des Kongresses" and "Entwurf des Programmes für den Kongress." 1881. *Elektrotechnische Zeitschrift* 2:326–31.

"Die Facultät und Dr. Dühring." 1877. *Vossische Zeitung* (Berlin), 29 July.

"Die Helmholtz-Gesellschaft, eine Mahnung für unsere Zeit." 1946. *PB* 2:148.

Diepgen, Paul. 1950. "100 Jahre Augenspiegel: Prof. Hermann von Helmholtz zum Gedenken." *Du und die Welt* 1:7.

Dierig, Sven. 2006. *Wissenschaft in der Maschinenstadt: Emil Du Bois–Reymond und seine Laboratorien in Berlin*. Göttingen: Wallstein.

Di Gregorio, Mario A. 1990. *Charles Darwin's Marginalia*. Vol. 1. New York: Garland.

Dilthey, Wilhelm. 1900. "Anna von Helmholtz." *Deutsche Rundschau* 102:226–35.

———. 1923. *Briefwechsel zwischen Wilhelm Dilthey und dem Grafen Paul Yorck v. Wartenburg 1877–1897*. Halle: Max Niemeyer.

Dimmer, Friedrich. 1921. "Theodor [*sic*] Helmholtz." *Wiener Medizinische Wochenschrift* 71: 1486–87.

Dirichlet, Peter Gustav Lejeune. 1881. "Gedächtnisrede auf Carl Gustav Jacob Jacobi." In *C. G. J. Jacobi's Gesammelte Werke*, edited by C. W. Borchardt, 1:3–28. Berlin: G. Reimer.

DiSalle, Robert. 1993. "Helmholtz's Empiricist Philosophy of Mathematics: Between Laws of Perception and Laws of Nature." In Cahan 1993a, 498–521.

———. 2006. "Kant, Helmholtz, and the Meaning of Empiricism." In Friedman and Nordmann 2006, 123–39.

Doerr, Wilhelm, ed. 1985. *Semper Apertus: Sechshundert Jahre Ruprecht-Karls-Universität Heidelberg 1386–1986: Festschrift in sechs Bänden*. 6 vols. Berlin: Springer-Verlag.

Dolby, R. G. A. 1984. "Thermochemistry versus Thermodynamics: The Nineteenth Century Controversy." *History of Science* 22 (4): 375–400.

Dolge, Alfred. 1911–13. *Pianos and their Makers*. 2 vols. Covina, CA: Covina.

Donders, Franciscus Cornelis. 1864. *On the Anomalies of Accommodation and the Refraction of the Eye: With a Preliminary Essay on Physiological Dioptrics*. London: New Sydenham Society.

Dorpalen, Andreas. 1957. *Heinrich von Treitschke*. New Haven, CT: Yale University Press.

Dreher, Astrid. 1980. "Briefe von Carl Ludwig an Jacob Henle aus den Jahren 1846–1872." PhD diss., Heidelberg University.

Dreyer, Ernst Adolf. N.d. [1936]. *Friedr. Vieweg & Sohn in 150 Jahren deutscher Geistesgeschichte 1786–1936*. Braunschweig: Friedr. Vieweg.

Du Bois–Reymond, Emil (EdBR). 1912. *Reden von Emil du Bois-Reymond*. Edited by Estelle du Bois–Reymond. 2nd ed. 2 vols. Leipzig: Veit.

Du Bois–Reymond, Estelle, ed. 1918. *Jugendbriefe von Emil du Bois-Reymond an Eduard Hallmann zu seinem hundertsten Geburtstag dem 7. November 1918*. Berlin: Dietrich Reimer [Ernst Vohsen].

———, ed. 1927. *Zwei grosse Naturforscher des 19. Jahrhunderts: Ein Briefwechsel zwischen Emil du Bois-Reymond und Karl Ludwig*. Leipzig: Johann Ambrosius Barth.

Dugac, Pierre. 1984. "Lettres de Charles Hermite à Gösta Mittag-Leffler (1874–1883)." *Cahiers du Séminaire d'Histoire des Mathématiques* 5:49–285.

Dühring, Eugen. 1873. *Kritische Geschichte der allgemeinen Principien der Mechanik: Von der philosophischen Facultät der Universität Göttingen mit dem ersten Preise der Beneke-Stiftung gekrönte Schrift.* 2nd ed. Berlin: L. Heimann's Verlag.

——. 1875. *Cursus der Philosophie als streng wissenschaftlicher Weltanschauung und Lebensgestaltung.* Leipzig: Erich Koschny.

——. 1875a. *Kritische Geschichte der Nationalökonomie und des Socialismus.* 2nd rev. ed. Berlin: Theobald Grieben.

——. 1877. *Der Weg zur höheren Berufsbildung der Frauen und die Lehrweise der Universitäten.* Leipzig: Fues's Verlag.[formerly part of 1877 and 1885]

——. 1877a. *Kritische Geschichte der allgemeinen Principien der Mechanik: Von der philosophischen Facultät der Universität Göttingen mit dem ersten Preise der Beneke-Stiftung gekrönte Schrift.* 2nd, partially rev. ed. Leipzig: Fues's Verlag.

——. 1878. *Neue Grundgesetze zur rationellen Physik und Chemie.* Leipzig: Fues's Verlag.

——. 1878a. *Logik und Wissenschaftstheorie.* Leipzig: Fues's Verlag.

——. 1879. *Kritische Geschichte der Nationalökonomie und des Socialismus.* 3rd ed. Leipzig: Fues's Verlag.[formerly part of 1875a]

——. 1880. *Robert Mayer der Galilei des neunzehnten Jahrhunderts: Eine Einführung in seine Leistungen und Schicksale.* Chemnitz: Ernst Schmeitzner.

——. 1881. *Die Judenfrage als Racen-, Sitten- und Culturfrage: Mit einer weltgeschichtlichen Antwort.* Karlsruhe: H. Reuther.

——. 1885. *Der Weg zur höheren Berufsbildung der Frauen und die Lehrweise der Universitäten.* 2nd enl. ed. Leipzig: Fues's Verlag.

——. 1895. *Robert Mayer der Galilei des neunzehnten Jahrhunderts und die Gelehrtenunthaten gegen bahnbrechende Wissenschaftsgrössen*: Vol. 2, *Neues Licht über Schicksal und Leistungen.* Leipzig: C. G. Naumann.

——. 1903. *Sache, Leben und Feinde: Als Hauptwerk und Schlüssel zu seinen sämmtlichen Schriften.* 2nd enl. ed. Leipzig: C. G. Naumann. 1903.

Du Moulin Eckart, Richard Graf. 1929–31. *Cosima Wagner: Ein Lebens- und Charakterbild.* 2 vols. Berlin: Drei Masken.

Dunken, Gerhard. 1960. *Die Deutsche Akademie der Wissenschaften zu Berlin in Vergangenheit und Gegenwart.* 2nd ed. Berlin: Akademie-Verlag.

Dyer, Frank Lewis, and Thomas Commerford Martin. 1929. *Edison: His Life and Inventions.* 2 vols. New York: Harper.

Ebert, Andreas D. 2008. *Jüdische Hochschullehrer an preussischen Universitäten (1870–1924): Eine quantitative Untersuchung mit biografischen Skizzen.* Frankfurt am Main: Mabuse-Verlag.

Ebert, Hermann. 1949. *Hermann von Helmholtz.* Stuttgart: Wissenschaftliche Verlagsgesellschaft.

Ebstein, Erich, ed. 1920. *Ärzte-Briefe aus vier Jahrhunderten.* Berlin: Julius Springer.

Eckart, Wolfgang, and Klaus Volkert, eds. 1996. *Hermann von Helmholtz: Vorträge eines Heidelberger Symposiums anlässlich des einhundertsten Todestages.* Pfaffenweiler: Centaurus-Verlagsgesellschaft.

Edison, Thomas Alva. 1991. *The Papers of Thomas A. Edison*: Vol. 2, *From Workshop to Laboratory: June 1873–March 1876.* Edited by Robert A. Rosenberg, Paul B. Israel, Keith A. Nier, and Melodie Andrews. Baltimore: Johns Hopkins University Press.

Ehlers, Dietrich, ed. 1992. *Hermann Diels, Hermann Usener, Eduard Zeller: Briefwechsel.* 2 vols. Berlin: Akademie Verlag.

Ehrenhaft, Felix. 1921. "Hermann v. Helmholtz." *Neue Freie Presse* (Vienna), 31 August, 1–3.

"Ein Helmholtz-Archiv." 1905. *Gaea* 41:1–3.

Einstein, Albert. 1917. Review of *Vorträge über Goethe*, by H. v. Helmholtz, edited by W. König. *Die Naturwissenschaften* 5:675.

———. 1987. *The Collected Papers of Albert Einstein*: Vol. 1, *The Early Years, 1879–1902*. Edited by John Stachel. Princeton, NJ: Princeton University Press.

Elbogen, Paul, ed. [1956]. *Liebster Sohn . . . liebe Eltern: Briefe berühmter Deutscher*. Hamburg: Rowohlt.

"Electricians Meet at a Banquet." 1893. *Chicago Daily Tribune*. 25 August, 6.

Eliot, George. 1998. *The Journals of George Eliot*. Edited by Margaret Harris and Judith Johnston. Cambridge: Cambridge University Press.

Elkana, Yehuda. 1970. "Helmholtz' 'Kraft': An Illustration of Concepts in Flux." *Historical Studies in the Physical Sciences* 2:263–99.

———. 1974. *The Discovery of the Conservation of Energy*. Cambridge, MA: Harvard University Press.

Ellis, Alexander J. 1875. "Notice by the Translator." In HH 1875, v–xi.

———. 1895. "Translator's Notice to the Second English Edition." In *On the Sensations of Tone as a Physiological Basis for the Theory of Music*, by Hermann L. F. Helmholtz, v. 3rd ed. Translated from the 4th German edition by Alexander J. Ellis. London: Longmans, Green.

Emerson, Peter Henry. 1980. "Photography, A Pictorial Art." *Amateur Photographer* 3:138–39.

———. 1993. *Naturalistic Photography for Students of the Art and The Death of Naturalistic Photography*. Vol. 1. London: S. Low, Marston, Searle & Rivington.

Endell, K. F. 1883. "Statistische Nachweisungen, betreffend die in den Jahren 1871 bis einschl. 1880 vollendeten und abgerechneten Preussischen Staatsbauten." *Zeitschrift für Bauwesen* 33:1–182.

Engelking, E[rnst]. 1950. "Hermann von Helmholtz in seiner Bedeutung für die Augenheilkunde." *Berichte der Deutschen Ophthalmologischen Gesellschaft* 56:12–30.

Engelmann, Th. W. 1894. *Gedächtnisrede auf Hermann von Helmholtz: Gehalten am 28. September 1894 in der Aula der Universität Utrecht*. Leipzig: Wilhelm Engelmann.

Engels, Friedrich. 1978. *Dialektik der Natur*. In *Werke*, by Karl Marx and Friedrich Engels, 20: 305–620. 43 vols. Berlin: Dietz Verlag.

———. 1978a. *Herrn Eugen Dühring's Umwälzung der Wissenschaft ("Anti-Dühring")*. In *Werke*, by Karl Marx and Friedrich Engels, 20:1–303. 43 vols. Berlin: Dietz Verlag.

Epple, Moritz. 1999. *Die Entstehung der Knotentheorie: Kontexte und Konstruktionen einer modernen mathematischen Theorie*. Braunschweig: Friedr. Vieweg.

Epstein, S. S. 1896. "Hermann von Helmholtz als Mensch und Gelehrter." *Deutsche Revue* 21 (2): 31–41, 192–202, 328–39.

Erdmann, Benno. 1921. *Die philosophischen Grundlagen von Helmholtz' Wahrnehmungstheorie: Kritisch erläutert*. Berlin: Verlag der Akademie der Wissenschaften.

Erggelet, H. 1921. "Hermann v. Helmholtz und die Augenheilkunde." *Die Naturwissenschaften* 9:967–72.

Ernst, Helmut, and Heinrich Stümbke. 1986. *Wo sie ruhen . . . Kleiner Führer zu den Grabstätten bekannter Berliner in West und Ost*. Berlin: Stapp Verlag.

"Errichtung der Helmholtz-Stiftung und Verleihung ihrer ersten vier Medaillen." 1970. In Harnack 1970, 2:559–63.

Esche-Braunfels, Sigrid. 1993. *Adolf von Hildebrand (1847–1921)*. Berlin: Deutscher Verlag für Kunstwissenschaft.

Esser, A. 1950. "Zur Geschichte der Erfindung des Augenspiegels." *Klinische Monatsblätter für Augenheilkunde* 11:1–14.

Eulenburg, Franz. 1904. *Die Frequenz der deutschen Universitäten von ihrer Gründung bis zur Gegenwart*. Leipzig: B. G. Teubner.

Eulner, Hans-Heinz. 1970. *Die Entwicklung der medizinischen Spezialfächer an den Universitäten des deutschen Sprachgebietes*. Stuttgart: Ferdinand Enke.

Eulner, Hans-Heinz, and Hermann Hoepke, eds. 1975. *Georg Meissners Briefe an Jacob Henle 1855–1878*. Göttingen: Vandenhoeck & Ruprecht.

————, eds. 1979. *Der Briefwechsel zwischen Rudolph Wagner und Jacob Henle 1838–1862*. Göttingen: Vandenhoeck & Ruprecht.

Eve, A. S., and C. H. Creasey. 1945. *Life and Work of John Tyndall*. London: Macmillan.

Fahrenbach, Sabine. 1987. "Die Herausbildung der Ophthalmologie in Preussen und die wissenschaftliche Schule Albrecht von Graefes (1828–1870)." In *Der Ursprung der modernen Wissenschaften: Studien zur Entstehung wissenschaftlicher Disziplinen*, edited by Martin Guntau and Hubert Laitko, 315–28. Berlin: Akademie-Verlag.

Fairley, Robert, ed. 1988. *Jemima: The Paintings and Memoirs of a Victorian Lady*. Edited and with an introduction by Robert Fairley. Edinburgh: Canongate.

Fd. 1899. "Das Helmholtz-Denkmal in Berlin." *Illustrirte Zeitung* 112 (15 June): 812.

Fechner, Gustav Theodor. 2004. *Tagebücher 1828 bis 1879/Gustav Theodor Fechner*. 2 vols. Edited by Anneros Meischner-Metge and compiled by Irene Altmann. Verlag der Sächsischen Akademie der Wissenschaften zu Leipzig. Stuttgart: Franz Steiner Verlag.

Feffer, Stuart Michael. 1994. "Microscopes to Munitions: Ernst Abbe, Carl Zeiss, and the Transformation of Technical Optics, 1850–1914." PhD diss., University of California–Berkeley.

————. 1996. "Ernst Abbe, Carl Zeiss, and the Transformation of Microscopical Optics." In Buchwald 1996, 23–66.

Feldhaus, F[ranz]. M. 1921. "Hermann Helmholtz." *Illustrirte Zeitung* 157:167.

————. 1929. *Carl Bamberg: Ein Rückblick auf sein Wirken und auf die Feinmechanik*. Berlin-Friedenau: Askania-Werke A.-G./Bambergwerk.

"Festbericht über die Gedenkfeier zur hundertjährigen Wiederkehr des Geburtstages Josef Fraunhofer's am 6. März 1887 im Berliner Rathause: Veranstaltet von der Deutschen Gesellschaft für Mechanik und Optik." 1887. *Zeitschrift für Instrumentenkunde* 7:114–28.

Fichte, Immanuel Hermann. 1856. *Anthropologie: Die Lehre von der menschlichen Seele: Begründet auf naturwissenschaftlichem Wege für Naturforscher, Seelenärzte und wissenschaftlich Gebildete überhaupt*. Leipzig: F. A. Brockhaus.

————. 1876. *Anthropologie: Die Lehre von der menschlichen Seele: Begründet auf naturwissenschaftlichem Wege für Naturforscher, Seelenärzte und wissenschaftlich Gebildete überhaupt*. 3rd ed. Leipzig: F. A. Brockhaus.

————. 1970. *Psychologie: Die Lehre vom Bewussten Geiste des Menschen, oder Entwicklungsgeschichte des Bewusstseins, begründet auf Anthropologie und Innerer Erfahrung*. Two parts. Leipzig: N.p. Reprint, Scientia Verlag Aalen, 1970.

Fichte, J[ohann]. G[ottlieb]. 1967. *Briefwechsel*. Edited by Hans Schultz. 2 vols. Hildesheim: Georg Olms.

Fiedler, Annett. 1998. *Die Physikalische Gesellschaft zu Berlin: Vom lokalen naturwissenschaftlichen Verein zur nationalen Deutschen Physikalischen Gesellschaft (1845–1900)*. Aachen: Shaker Verlag.

Finger, Stanley, and Nicholas J. Wade. 2002. "The Neuroscience of Helmholtz and the Theories of Johannes Müller: Part 1, Nerve Cell Structure, Vitalism, and the Nerve Impulse." *Journal of the History of the Neurosciences* 11:136–55.

Finkelstein, Gabriel. 2013. *Emil du Bois-Reymond: Neuroscience, Self, and Society in Nineteenth-Century Germany*. Cambridge, MA: MIT Press.

Fisch, Max H. 1964. "Supplement: A Chronicle of Pragmaticism, 1865–1879." *Monist* 48: 441–66.

Fischer, Emil. 1894. ["Hermann von Helmholtz"]. *Berichte der Deutschen Chemischen Gesellschaft* 27:2643–52.

————. 1987. *Aus meinem Leben*. Reprinted from *Gesammelte Werke*, by Emil Fischer, edited by M. Bergmann. Berlin: Springer-Verlag.

FitzGerald, George Francis. 1896. "Helmholtz Memorial Lecture." *Journal of the Chemical Society* 69 (2): 885–912.

Flamm, Dieter, ed. 1995. *Hochgeehrter Herr Professor! Innig geliebter Louis! Ludwig Boltzmann Henriette von Aigentler Briefwechsel*. Vienna: Böhlau.

Fleming, James Rodger. 2016. *Inventing Atmospheric Science: Bjerknes, Rossby, Wexler, and the Foundations of Modern Meteorology*. Cambridge, MA: MIT Press.

Fleming, William. 1968. *Arts and Ideas*. New York: Holt, Rinehart and Winston.

Foerster, Wilhelm. 1911. *Lebenserinnerungen und Lebenshoffnungen*. Berlin: Georg Reimer.

Fölsing, Albrecht. 1997. *Heinrich Hertz: Eine Biographie*. Hamburg: Hoffmann und Campe.

Fontane, Theodor. 1969. *Theodor Fontane: Briefe an Julius Rodenberg: Eine Dokumentation*. Berlin: Aufbau-Verlag.

Forman, Paul. 1974. "The Financial Support and Political Alignment of Physicists in Weimar Germany." *Minerva* 12:39–66.

Fostle, D. W. 1995. *The Steinway Saga: An American Dynasty*. New York: Scribner.

Fox, Robert. 1973. "Scientific Enterprise and the Patronage of Research in France, 1800–1870." *Minerva* 11:442–73.

———. 1984. "Science, the University, and the State in Nineteenth-Century France." In *Professions and the French State, 1700–1900*, edited by Gerald L. Geison, 66–145. Philadelphia: University of Pennsylvania Press.

———. 1990. "The View over the Rhine: Perceptions of German Science and Technology in France, 1860–1914." In *Frankreich und Deutschland: Forschung, Technologie und industrielle Entwicklung im 19. und 20. Jahrhundert*, edited by Yves Cohen and Klaus Manfrass, 14–24, Munich: C. H. Beck.

———. 1996. "Thomas Edison's Parisian Campaign: Incandescent Lighting and the Hidden Face of Technology Transfer." *AS* 53:157–93.

———. 1997. "The University Museum and Oxford Science, 1850–1880." In *The History of the University of Oxford*: Vol. 6, *Nineteenth-Century Oxford*, part 1, edited by M. G. Brock and M. C. Curthoys, 641–91. Oxford: Clarendon Press.

———. 2005. "The Context and Practices of Oxford Physics, 1839–77." In Fox and Gooday 2005, 24–79.

Fox, Robert, and Graeme Gooday, eds. 2005. *Physics in Oxford, 1839–1939: Laboratories, Learning and College Life*. Oxford: Oxford University Press.

Fox, Robert, and Anna Guagnini. 1998. "Laboratories, Workshops, and Sites: Concepts and Practices of Research in Industrial Europe, 1800–1914, Part 1." *Historical Studies in the Physical and Biological Sciences* 29:55–140.

Frank, Robert G., Jr. 1987. "American Physiologists in German Laboratories, 1865–1914." In *Physiology in the American Context, 1850–1940*, edited by Gerald L. Geison, 11–46. Bethesda, MD: American Physiological Society.

Fricke, Dieter. 1960. "Zur Militarisierung des deutschen Geisteslebens im wilhelminischen Kaiserreich: Der Fall Leo Arons." *Zeitschrift für Geschichtswissenschaft* 8:1069–1107.

Friedel, Robert, and Paul Israel. 1986. *Edison's Electric Light: Biography of an Invention*. New Brunswick, NJ: Rutgers University Press.

———. 2010. *Edison's Electric Light: The Art of Invention*. Baltimore: Johns Hopkins University Press.

Friedenwald, Harry. 1902. "The History of the Invention and of the Development of the Ophthalmoscope." *Journal of the American Medical Association* 38 (9): 549–52, with an appendix, 566–69.

Friedlaender, L. 1896. "Aus Königsberger Gelehrtenkreisen." *Deutsche Rundschau* 88:41–62, 224–39.

Friedman, Michael. 1997. "Helmholtz's Zeichentheorie and Schlick's Allgemeine Erkenntnislehre: Early Logical Empiricism and Its Nineteenth-Century Background." *Philosophical Topics* 25:19–50.

————. 1999. *Reconsidering Logical Positivsm*. Cambridge: Cambridge University Press.

————. 2000. "Geometry, Construction, and Intuition in Kant and His Successors." In *Between Logic and Intuition: Essays in Honor of Charles Parsons*, edited by Gila Sher and Richard Tieszen, 198–217. Cambridge: Cambridge University Press.

————. 2002. "Geometry as a Branch of Physics: Background and Context for Einstein's 'Geometry and Experience.'" In *Reading Natural Philosophy: Essays in the History and Philosophy of Science and Mathematics*, edited by David B. Malament, 193–229. Chicago: Open Court.

————. 2009. "Einstein, Kant, and the Relativized a Priori." In *Constituting Objectivity: Transcendental Perspectives on Modern Physics*, edited by Michel Bitbol, Pierre Kerszberg, and Jean Petitot, 253–67. Dordrecht: Springer.

Friedman, Michael, and Alfred Nordmann, eds. 2006. *The Kantian Legacy in Nineteenth-Century Science*. Cambridge, MA: MIT Press.

Friedman, Robert Marc. 1989. *Appropriating the Weather: Vilhelm Bjerknes and the Construction of a Modern Meteorology*. Ithaca, NY: Cornell University Press.

Fritzsche, Bernd. 1999. "Sophus Lie: A Sketch of His Life and Work." *Journal of Lie Theory* 9:1–38.

Fruton, Joseph S. 1972. *Molecules and Life: Historical Essays on the Interplay of Chemistry and Biology*. New York: Wiley-Interscience.

Fuchs, Erich. 1990. "Fichtes Einfluss auf seine Studenten in Berlin zum Beginn der Befreiungskriege." *Fichte-Studien* 2:178–92.

Fullinwider, S. P. 1990. "Hermann von Helmholtz: The Problem of Kantian Influence." *SHPS* 21:41–55.

G., R. 1890–91. "Zum 70. Geburtstage von Professor Helmholtz." *Der Bär* 17:645–46.

Gage, John. 1999. *Color and Meaning: Art, Science, and Symbolism*. Berkeley: University of California Press.

Galaty, David H. 1971. "The Emergence of Biological Reductionism." PhD diss., Johns Hopkins University.

————. 1974. "The Philosophical Basis of Mid-Nineteenth Century German Reductionism." *Journal of the History of Medicine and Allied Sciences* 29:295–316.

Gall, Lothar. 1968. *Der Liberalismus als regierende Partei: Das Grossherzogtum Baden zwischen Restauration und Reichsgründung*. Wiesbaden: Franz Steiner.

Garber, Elizabeth. 1976. "Thermodynamics and Meteorology (1850–1900)." *AS* 33:51–65.

Garber, Elizabeth, Stephen G. Brush, and C. W. F. Everitt, eds. 1986. *Maxwell on Molecules and Gases*. Cambridge, MA: MIT Press.

Gariel, C. M. 1894. "Biographies scientifiques: Les travaux de H.-L.-F. Helmholtz." *Revue Scientifique* 54:429–32.

Gauld, Alan. 1968. *The Founders of Psychical Research*. New York: Schocken.

Geiger, Ludwig, ed. 1886. *Goethe-Jahrbuch*. Vol. 7. Frankfurt am Main: Literarische Anstalt.

————, ed. 1892. *Goethe-Jahrbuch*. Vol. 13. Frankfurt am Main: Literarische Anstalt.

————, ed. 1893. *Goethe-Jahrbuch*. Vol. 14. Frankfurt am Main: Literarische Anstalt.

Gerhardt, Carl. N.d. *Erinnerungsblätter für die Seinen*. Privately printed; available at Yale Medical Library.

Gerlach, Walther. 1950. "Hermann v. Helmholtz als Naturforscher." *Berichte der Deutschen Ophthalmologischen Gesellschaft* 56:3–12.

————. 1951. "Hermann von Helmholtz als Naturforscher." In "Bericht über die 56. Zusammenkunft der Deutschen Ophthalmologischen Gesellschaft in München am 18., 19. und 20. September 1950," by H. J. Voss. *Ophthalmologica* 121:228–30.

————. 1956. "Hermann von Helmholtz 1821–1894." In *Die Grossen Deutschen: Deutsche Biographie*, edited by Hermann Heimpel, Theodor Heuss, and Benno Reifenberg, 3:456–65. 4 vols. Berlin: Propyläen-Verlag bei Ullstein.

———. 1962. "Hermann Helmholtz als Naturforscher." In *Humanität und naturwissenschaftliche Forschung*, by Walther Gerlach, 123–32. Braunschweig: Friedrich Vieweg.

———. 1969. "Helmholtz v." In *Neue Deutsche Biographie*, edited by Historische Kommission bei der Bayerischen Akademie der Wissenschaften, 8:498–501. Berlin: Duncker & Humblot.

Giacomoni, Paola. 2002. "Goethe e Helmholtz sulla Percezione." In *Le leggi del pensiero tra logica, ontologia e psicologica: Il dibattito Austro-Tedesco (1830–1930)*, edited by Stefano Poggi, 13–33. Milan: Edizioni Unicopli.

Giovanelli, Marco. 2008. "Kant, Helmholtz, Riemann und der Ursprung der geometrischen Axiome." *Philosophia Naturalis* 45:236–69.

Giraud-Teulon, Félix, and Louis de Wecker. 1868. *Congrès périodique international d'ophtalmologie compte-rendu. . . .* Paris: J.-B. Baillière.

Glaser, Ludwig. 1996. "Jews in Physics: Jewish Physics." In Hentschel 1996, 223–34. First published 1939.

Goldschmidt, Ludwig. 1898. *Kant und Helmholtz: Populärwissenschaftliche Studie*. Hamburg: Leopold Voss.

Goldstein, Eugen. 1921. "Helmholtz. Erinnerungen eines Laboratoriumspraktiken." *Die Naturwissenschaften* 9:708–11.

———. 1925. "Aus vergangenen Tagen der Berliner Physikalischen Gesellschaft." *Die Naturwissenschaften* 13:39–45.

Goldzieher, W. 1896. "Hermann Helmholtz." *Wiener Medizinische Wochenschrift* 46:1–8, 44–48, 98–103.

Gooday, Graeme, 2005. "Robert Bellamy Clifton and the 'Depressing Inheritance' of the Clarendon Laboratory, 1877–1919." In Fox and Gooday 2005, 80–118.

Gordin, Michael. 2008. "The Heidelberg Circle: German Inflections on the Professionalization of Russian Chemistry in the 1860s." *Osiris*, 2nd ser., 23:23–49.

Goschler, Constantin. 1998. "'Die Verwandlung': Rudolf Virchow und die Berliner Denkmalskultur." *Jahrbuch für Universitätsgeschichte* 1:69–111.

———. 2002. *Rudolf Virchow: Mediziner—Anthropologe—Politiker*. Cologne: Böhlau Verlag.

Graefe, A. von. 1854. "Vorwort." *Archiv für Ophthalmologie* 1:v–x.

Gray, Jeremy. 1989. *Ideas of Space: Euclidean, Non-Euclidean, and Relativistic*. 2nd ed. Oxford: Oxford University Press.

———. 2008. *Plato's Ghost: The Modernist Transformation of Mathematics*. Princeton, NJ: Princeton University Press.

———. 2013. *Henri Poincaré: A Scientific Biography*. Princeton, NJ: Princeton University Press.

Greeff, Richard. 1921. "Augenheilkunde und Optik: Helmholtz' bahnbrechende Arbeiten." In ESH et al. 1921, n.p.

Green, Burdette, and David Butler. 2002. "From Acoustics to Tonpsychologie." In *The Cambridge History of Western Music Theory*, edited by Thomas Christensen, 246–71. Cambridge: Cambridge University Press.

Gregor-Dellin, Martin. 1983. *Richard Wagner: His Life, His Work, His Century*. San Diego: Harcourt Brace Jovanovich.

———. 1987. *Richard Wagner: Sein Leben, Sein Werk, Sein Jahrhundert*. Berlin: Henschelverg, 1987.

Gregory, Frederick. 1977. *Scientific Materialism in Nineteenth Century Germany*. Dordrecht: D. Reidel.

———. 2000. "The Mysteries and Wonders of Natural Science: Aaron Bernstein's Naturwissenschaftliche Volksbücher and the Adolescent Einstein." In Howard and Stachel 2000, 23–41.

Greven-Aschoff, Barbara. 1981. *Die bürgerliche Frauenbewegung in Deutschland 1894–1933*. Göttingen: Vandenhoeck & Ruprecht.

Groeben, Christiane, ed. 1985. *Anton Dohrn (1840–1909): Briefwechsel*. Berlin: Springer-Verlag.

Groeben, Christiane, and Klaus Wenig, eds. 1992. *Anton Dohrn und Rudolf Virchow: Briefwechsel 1864–1902*. Berlin: Akademie Verlag.

Gross, Theodor. 1891. *Über den Beweis des Prinzips von der Erhaltung der Energie*. Berlin: Mayer & Müller.

———. 1898. *Robert Mayer und Hermann v. Helmholtz: Eine kritische Studie*. Berlin: M. Krayn.

Gumprecht, Ferdinand. 1927. "Hermann Helmholtz, 1821–1894." In *Leben und Gedankenwelt grosser Naturforscher*, by Ferdinand Gumprecht, 31–64. Leipzig: Quelle & Mayer.

Günther, Siegmund, Friedrich Dannemann, and Karl Sudhoff, eds. 1922. ["109. Sitzung. 9 Dezember 1921."]. *Mitteilungen zur Geschichte der Medizin und der Naturwissenschaften* 21:154–55.

Guttstadt, Albert, comp. 1886. *Die naturwissenschaftlichen und medicinischen Staatsanstalten Berlins. Festschrift für die 59. Versammlung deutscher Naturforscher und Aerzte*. Berlin: August Hirschwald.

Haberling, Wilhelm. 1924. *Johannes Müller: Das Leben des Rheinischen Naturforschers*. Leipzig: Akademische Verlagsgesellschaft.

Haeckel, Ernst. 1902. "Frei Wissenschaft und frei Lehre: Eine Entgegnung auf Rudolf Virchow's Münchener Rede über 'Die Freiheit der Wissenschaft im modernen Staat': 1878." In Haeckel 1902a, 2:199–323.

———. 1902a. *Gemeinverständliche Vorträge und Abhandlungen aus dem Gebiete der Entwickelungslehre*. 2nd ed. 2 vols. Bonn: Emil Strauss.

———. 1902b. "Ueber die heutige Entwickelungslehre im Verhältnisse zur Gesammtwissenschaft." In Haeckel 1902a, 2:121–46.

———. 1921. *Entwicklungsgeschichte einer Jugend: Briefe an die Eltern 1852/1856*. Leipzig: K. F. Koehler.

Haeckel, Julius, ed. 1912. *Geschichte der Stadt Potsdam*. Potsdam: Gropiusche' Hofbuchhandlung.

Häfner, Walter. 1934. "Der Altmeister des deutschen Lokomotivbaues: Dr. Ing. Richard von Helmholtz." *Organ für die Fortschritte des Eisenbahnwesens* 71:384–85.

———. [1962?] "Richard von Helmholtz 1852–1934." In *Pioniere des Eisenbahnwesens*, edited by Erhard Born, 111–17. Darmstadt: Carl Röhrig.

Hagner, Michael. 2004. *Geniale Gehirne: Zur Geschichte der Elitegehirnforschung*. Göttingen: Wallstein.

Hagner, Michael, and Bettina Wahrig-Schmidt, eds. 1992. *Johannes Müller und die Philosophie*. Berlin: Akademie Verlag.

Haight, Gordon Sherman. 1954–78. *The George Eliot Letters*. 9 vols. New Haven, CT: Yale University Press.

Haines, George, IV. 1958. "German Influence upon Scientific Instruction in England, 1867–1887." *Victorian Studies* 1:215–44.

Hale, Matthew, Jr. 1980. *Human Science and Social Order: Hugo Münsterberg and the Origins of Applied Psychology*. Philadelphia: Temple University Press.

Hall, A. Wilford. 1878. *Evolution of Sound: A Part of the Problem of Human Life Here and Hereafter Containing a Review of Tyndall, Helmholtz, and Mayer*. New York: Hall.

———. 1880. *The Problem of Human Life: Embracing the "Evolution of Sound" and "Evolution Evolved," with a Review of the Six Great Modern Scientists, Darwin, Huxley, Tyndall, Haeckel, Helmholtz, and Mayer*. New York: Hall.

Hall, G. Stanley. 1912. "Hermann L.F. von Helmholtz 1821–1894." In *Founders of Modern Psychology*, by G. Stanley Hall, 247–308. New York: D. Appleton.

Hallion, Richard P. 2003. *Taking Flight: Inventing the Aerial Age from Antiquity through the First World War*. Oxford: Oxford University Press.

Hamann, Richard. 1914. *Die deutsche Malerei im 19. Jahrhundert*. Leipzig: B. G. Teubner.

Hamann, Richard, and Jost Hermand. 1971. *Gründerzeit*. Munich: Nymphenburger Verlagshandlung.

———. 1973. *Stillkunst um 1900*. Munich: Nymphenburger Verlagshandlung.

Hannabuss, K. C. 2000. "Mathematics." In *The History of the University of Oxford*: Vol. 7, *Nineteenth-Century Oxford, Part II*, edited by M. G. Brock and M. C. Curthoys, 443–55. Oxford: Clarendon Press.

Hansemann, David. 1899. "Ueber das Gehirn von Hermann v. Helmholtz." *Zeitschrift für Psychologie und Physiologie der Sinnesorgane* 20:1–12.

Hansen, Fritz. 1941. "Hermann von Helmholtz." *Deutsche Optiker-Zeitung* 17:133.

———. 1941a. "Hermann von Helmholtz. Zum 120. Geburtstag." *Wissen und Fortschritt* 15: 442.

Harman, P. M. 1982. "Helmholtz: The Principle of the Conservation of Energy." In *Metaphysics and Natural Philosophy: The Problem of Substance in Classical Physics*, by P. M. Harman, 105–26. Brighton: Harvester Press.

Harnack, Adolf von. 1970. *Geschichte der Königlich Preussischen Akademie der Wissenschaften zu Berlin*. 3 vols. in 4. Hildesheim: Georg Olms.

Hars, Florian. 1999. *Ferdinand Braun (1850–1918): Ein wilhelminischer Physiker*. Berlin: Diepholz.

Hartkopf, Werner. 1983. *Die Akademie der Wissenschaften der DDR: Ein Beitrag zu ihrer Geschichte: Biographischer Index*. Berlin: Akademie-Verlag.

———. 1992. *Die Berliner Akademie der Wissenschaften: Ihre Mitglieder und Preisträger 1700– 1990*. Berlin: Akademie Verlag.

Hartkopf, Werner, and Gert Wangermann. 1991. *Dokumente zur Geschichte der Berliner Akademie der Wissenschaften von 1700 bis 1900*. Berlin: Spektrum Akademischer Verlag.

Hatfield, Gary. 1990. *The Natural and the Normative: Theories of Spatial Perception from Kant to Helmholtz*. Cambridge, MA: MIT Press.

———. 1993. "Helmholtz and Classicism: The Science of Aesthetics and the Aesthetics of Science." In Cahan 1993a, 522–58.

———. 1997. "Wundt and Psychology as Science: Disciplinary Transformations." *Perspectives on Science* 5:349–82.

———. 2011. "Kant and Helmholtz on Primary and Secondary Qualities." In *Primary and Secondary Qualities: The Historical and Ongoing Debate*, edited by Lawrence Nolan, 304–38. Oxford: Oxford University Press.

Hauptmann, Moritz. 1863. "Ein Brief M. Hauptmann über Helmholtz's 'Tonempfindungen.'" *Allgemeine Musikalische Zeitung* 1:669–73.

Hawkins, Thomas. 2000. *Emergence of the Theory of Lie Groups: An Essay in the History of Mathematics, 1869–1926*. New York: Springer.

Hecht, Hartmut. 1994. "Actio, Quantité d'action und Wirkung: Helmholtz' Rezeption dynamischer Grundbegriffe." In Krüger 1994a, 107–23.

"Heidelberg, 12. März." 1862. *Heidelberger Zeitung*, no. 62: 14 March.

Heidelberger, Michael. 1977. "Beziehungen zwischen Sinnesphysiologie und Philosophie im 19. Jahrhundert." In *Philosophie und Wissenschaften: Formen und Prozesse ihrer Interaktion*, edited by Hans-Jörg Sandkühler, 37–58. Frankfurt am Main: Peter Lang.

———. 1993. "Force, Law, and Experiment: The Evolution of Helmholtz's Philosophy of Science." In Cahan 1993a, 461–97.

———. 1994. "Helmholtz' Erkenntnis- und Wissenschaftstheorie im Kontext der Philosophie und Naturwissenschaft des 19. Jahrhunderts." In Krüger 1994a, 168–85.

———. 1998. "From Helmholtz's Philosophy of Science to Hertz's Picture-Theory." In Baird, Hughes, and Nordmann 1998, 9–24.

———. 2004. *Nature from Within: Gustav Theodor Fechner and His Psychophysical Worldview*. Pittsburgh: University of Pittsburgh Press.

Heidenhain, R. 1894. "Gedächtnisrede auf Hermann von Helmholtz." *Jahresbericht der schlesischen Gesellschaft für vaterländische Cultur* 72:32–51.

Heilmeyer, Alexander. 1902. *Adolf Hildebrand*. Bielefeld: Velhagen & Klasing.

Heimann, Peter. 1974. "Conversion of Forces and the Conservation of Energy." *Centaurus* 18: 147–61.

———. 1974a. "Helmholtz and Kant: The Metaphysical Foundations of Über die Erhaltung der Kraft." *SHPS* 5:205–38.

Heinzmann, Gerhard. 2001. "The Foundations of Geometry and the Concept of Motion: Helmholtz and Poincaré." *Science in Context* 14:457–70.

Helmert, F. R. 1913. "Die Internationale Erdmessung in den ersten fünfzig Jahren ihres Bestehens." *Internationale Monatsschrift für Kunst und Technik* 7:397–424.

Helmholtz, Anna von. 1896. "Vorrede zur vierten Auflage." In HH 1903, 1:xiv–xv.

Helmholtz, Ferdinand. 1837. "Die Wichtigkeit der allgemeinen Erziehung für das Schöne." In Schulprogramm 1837, 1–44.

"Helmholtz, Hermann Ludwig Ferdinand von H." 1962. In *Biographisches Lexikon der hervorragenden Ärzte aller Zeiten und Völker*, edited by August Hirsch, 3:151–52. 5 vols. Munich: Urban & Schwarzenberg.

Helmholtz, Hermann von (HH). 1842. "De Fabrica Systematis nervosi Evertebratorum." In *WA*, 2:663–79.

———. 1843. "Ueber das Wesen der Fäulniss und Gährung." In *WA*, 2:726–34.

———. 1845. "Ueber den Stoffverbrauch bei der Muskelaction." In *WA*, 2:735–44.

———. 1846. "Wärme, physiologisch." In *WA*, 2:680–725.

———. 1847. "Bericht über 'die Theorie der physiologischen Wärmeerscheinungen' betreffende Arbeiten aus dem Jahre 1845." In *WA*, 1:3–11.

———. 1847a. "Ueber die Erhaltung der Kraft. Eine physikalische Abhandlung." In *WA*, 1:12–75 (including, on 68–75, addenda composed in 1881).

———. 1848. "Ueber die Wärmeentwicklung bei der Muskelaction." In *WA*, 2:745–63.

———. 1848a. "Bericht über 'die Theorie der physiologischen Wärmeerscheinungen' betreffende Arbeiten aus dem Jahre 1846." *Fortschritte der Physik* 2:259–60.

———. 1850. "Messungen über den zeitlichen Verlauf der Zuckung animalischer Muskeln und die Fortpflanzungsgeschwindigkeit der Reizung in den Nerven." In *WA*, 2:764–843.

———. 1850a. "Ueber die Fortpflanzungsgeschwindigkeit der Nervenreizung." In *WA*, 3:1–3.

———. 1850b. "Vorläufiger Bericht über die Fortpflanzungsgeschwindigkeit der Nervenreizung." *Archiv für Anatomie, Physiologie, und wissenschaftliche Medizin*, 71–73.

———. 1850c. "Über die Methoden, kleinste Zeittheile zu messen, und ihre Anwendung für physiologische Zwecke." In *WA*, 2:862–80.

———. 1850d. "Note sur la vitesse de propagation de l'agent nerveux dans les nerfs rachidiens." *CR* 30:204–6.

———. 1850e. "Bericht über 'die Theorie der physiologischen Wärmeerscheinungen' betreffende Arbeiten aus dem Jahre 1847." *Fortschritte der Physik* 3:232–45.

———. 1851. "Beschreibung eines Augen-Spiegels zur Untersuchung der Netzhaut im lebenden Auge." In *WA*, 2:229–60.

———. 1851a. "Ueber den Verlauf und die Dauer der durch Stromschwankungen inducirten elektrischen Ströme." In *WA*, 3:554–57.

———. 1851b. "Ueber die Dauer und den Verlauf der durch Stromschwankungen inducirten elektrischen Ströme." In *WA*, 1:429–62.

———. 1851c. "Deuxième note sur la vitesse de propagation de l'agent nerveux." *CR* 33: 262–65.

———. 1852. "Ueber eine neue einfachste Form des Augenspiegels." In *WA*, 2:261–79.

———. 1852a. "Bericht über 'die Theorie der physiologischen Wärmeerscheinungen' betreffende Arbeiten aus dem Jahre 1848." *Fortschritte der Physik* 4:222–23.

———. 1852b. "Die Resultate der neueren Forschungen über thierische Elektricität." In *WA*, 2:886–923.

———. 1852c. "Ein Theorem über die Vertheilung elektrischer Ströme in körperlichen Leitern." In *WA*, 3:562–64.

———. 1852d. "Ueber die Theorie der zusammengesetzten Farben: Physiologisch-optische Abhandlung, wodurch zu seiner am 28. Juni d.J. 11 Uhr Vorm. stattfindenden Habilitation als ordentlicher Professor der medicinischen Facultät an der Universität zu Königsberg ergebenst einladet Dr. H. Helmholtz." In *WA*, 2:3–23.

———. 1852e. "Messungen über Fortpflanzungsgeschwindigkeit der Reizung in den Nerven: Zweite Reihe." In *WA*, 2:844–61.

———. 1852 f. "Ueber die Natur der menschlichen Sinnesempfindungen." In *WA*, 2:591–609.

———. 1852g. "Ueber Herrn D. Brewster's neue Analyse des Sonnenlichtes." In *WA*, 2:24–44.

———. 1852h. "On the Theory of Compound Colours." *Philosophical Magazine*, ser. 4, 4:519–34.

———. 1852i. "Bericht über 'die Theorie der Akustik' und 'akustisch Phänomene' betreffende Arbeiten vom Jahre 1848." In *WA*, 1:233–55.

———. 1853. "Ueber eine bisher unbekannte Veränderung am menschlichen Auge bei veränderter Accommodation." In *WA*, 2:280–82.

———. 1853a. "Ueber Goethe's naturwissenschaftliche Arbeiten." In *VR*, 1:23–47.

———. 1853b. "On the Conservation of Force: A Physical Memoir." In *Scientific Memoirs, Selected from the Transactions of Foreign Academies of Science, and from Foreign Journals*, edited by John Tyndall and William Francis, 114–62. London: Taylor and Francis.

———. 1853c. "Ueber einige Gesetze der Vertheilung elektrischer Ströme in körperlichen Leitern mit Anwendung auf die thierisch-elektrischen Versuche." In *WA*, 1:475–519.

———. 1854. "Bericht über 'die Theorie der Akustik' betreffende Arbeiten aus dem Jahre 1849." In *WA*, 1:251–55.

———. 1854a. "Ueber die Geschwindigkeit einiger Vorgänge in Muskeln und Nerven." In *WA*, 2:881–85.

———. 1854b. "Erwiderung auf die Bermerkungen von Clausius." In *WA*, 1:76–93.

———. 1854c. "On the Mixture of Homogeneous Colours." In *Report of the Twenty-Third Meeting of the British Association for the Advancement of Science, held in Hull, September 1853*, 23 (2): 5. London: John Murray.

———. 1854d. "Ueber die Wechselwirkung der Naturkräfte und die darauf bezüglichen neuesten Ermittelungen der Physik." In *VR*, 1:49–83.

———. 1855. "Ueber die Accommodation des Auges." In *WA*, 2:283–345.

———. 1855a. "Bericht über 'die Theorie der Wärme' betreffende Arbeiten aus dem Jahre 1852." *Fortschritte der Physik* 8:369–87.

———. 1855b. "Ueber das Sehen des Menschen: Ein populär-wissenschaftlicher Vortrag, gehalten zu Königsberg in Preussen am 27. Febr. 1855." In *VR*, 1:85–117.

———. 1855c. "Ueber die Zusammensetzung von Spectralfarben." In *WA*, 2:45–70.

———. 1855d. "Über die Empfindlichkeit der menschlichen Netzhaut für die brechbarsten Strahlen des Sonnenlichts." In *WA*, 2:71–77.

———. 1856. "Bericht über 'die Theorie der Wärme' betreffende Arbeiten aus dem Jahre 1853." *Fortschritte der Physik* 9:404–32.

———. 1856a. "On the Interaction of Natural Forces." *Philosophical Magazine* 11:489–518.

———. 1856b. "Ueber Combinationstöne." In *WA*, 1:256–62.

———. 1856c. "Ueber die Combinationstöne oder Tartinischen Töne." In *WA*, 3:7–9.

———. 1856d. "Ueber Combinationstöne." In *WA*, 1:263–302.

———. 1856e. *Handbuch der physiologischen Optik*. Part one. Hamburg: Leopold Voss.

———. 1856 f. "Ueber die Erklärung des Glanzes." In *WA*, 3:4–5.

———. 1857. "Bericht über 'die Theorie der Wärme' betreffende Arbeiten aus dem Jahre 1854." *Fortschritte der Physik* 10:361–98.

———. 1857a. "Ueber die physiologischen Ursachen der musikalischen Harmonie." In *VR*, 1:118–55.

———. 1857b. "Ein Telestereoskop." In *WA*, 3:10–12.

———. 1857c. "Das Telestereoskop." In *WA*, 2:484–91.

———. 1857d. "Ueber die Vocale." In *WA*, 1:395–96.

———. 1858. "Bericht über 'die Theorie der Wärme' betreffende Arbeiten aus dem Jahre 1855." *Fortschritte der Physik* 11:361–73.

———. 1858a. "Ueber Integrale der hydrodynamischen Gleichungen, welche den Wirbelbewegungen entsprechen." In *WA*, 1:101–34.

———. 1858b. "Ueber die subjectiven Nachbilder im Auge." In *WA*, 3:13–15.

———. 1859. "Bericht über 'die Theorie der Wärme' betreffende Arbeiten aus dem Jahre 1856." *Fortschritte der Physik* 12:343–59.

———. 1859a. "Theorie der Luftschwingungen in Röhren mit offenen Enden." In *WA*, 1:303–82.

———. 1859b. "Ueber Luftschwingungen in Röhren mit offenen Enden." In *WA*, 3:16–20.

———. 1859c. "Ueber die Klangfarbe der Vocale." In *WA*, 1:397–407.

———. 1859d. "Ueber Farbenblindheit." In *WA*, 2:346–49.

———. 1859e. "Ueber Nachbilder." In *Amtlicher Bericht über die 34. Versammlung deutscher Naturforscher und Aerzte zu Carlsruhe im September 1858*, 225–26. Carlsruhe: Chr. Fr. Müllersche Hofbuchhandlung.

———. 1859 f. "Ueber die physikalische Ursache der Harmonie und Disharmonie." In *Amtlicher Bericht über die 34. Versammlung deutscher Naturforscher und Aerzte zu Carlsruhe im September 1858*, 157–59. Carlsruhe: Chr. Fr. Müllersche Hofbuchhandlung.

———. 1860. "On the Motion of the Strings of a Violin." In *WA*, 1:410–19.

———. 1860a. "Ueber Klangfarben." In *WA*, 1:408–9.

———. 1860b. *Handbuch der physiologischen Optik*. Part two. Hamburg: Leopold Voss.

———. 1861. "On the Application of the Law of the Conservation of Force to Organic Nature." In *WA*, 3:565–80.

———. 1861a. "Zur Theorie der Zungenpfeifen." In *WA*, 1:388–94.

———. 1862. "Ueber das Verhältniss der Naturwissenschaften zur Gesammtheit der Wissenschaft." In *VR*, 1:157–85.

———. 1862a. "Ueber die arabisch-persische Tonleiter." In *WA*, 1:424–26.

———. 1862b. "Ueber die Form des Horopters, mathematisch bestimmt." In *WA*, 2:420–26.

———. 1862–63. "Über die Erhaltung der Kraft: Einleitung zu einem Cyclus von Vorlesungen, gehalten zu Karlsruhe im Winter 1862–63." In *VR*, 1:186–229.

———. 1863. *Die Lehre von den Tonempfindungen als physiologische Grundlage für die Theorie der Musik*. Braunschweig: Friedrich Vieweg. Subsequent publication included the 2nd ed., 1865; the 3rd rev. ed., 1870; and the 4th rev. ed., 1877.

———. 1863a. "Ueber den Einfluss der Reibung in der Luft auf die Schallbewegung." In *WA*, 1:383–87.

———. 1863b. "Ueber die Bewegungen des menschlichen Auges." In *WA*, 2:352–59.

———. 1863c. "Ueber die normalen Bewegungen des menschlichen Auges." In *WA*, 2:360–419.

———. 1864. "Lectures on the Conservation of Energy, by Professor Helmholtz, Delivered at the Royal Institution on April 5, 7, 12, 14, 19 and 21, 1864." *Medical Times and Gazette* 1:385–88, 415–18, 443–46, 471–74, 499–501, 527–30.

———. 1864a. "On the Normal Motions of the Human Eye in Relation to Binocular Vision." In *WA*, 3:25–43.

———. 1864b. "Bemerkungen über die Form des Horopters." In *WA*, 2:478–81.

———. 1864c. "Ueber den Horopter." In *WA*, 3:21–24.

———. 1864d. "Versuche über das Muskelgeräusch." In *WA*, 2:924–27.

———. 1864e. "Ueber den Horopter." In *WA*, 2:427–77.

———. 1865. "Eis und Gletscher." In *VR*, 1:231–63. With an addendum, "Zusätze zu dem Vortrag (S. 231) 'Eis und Gletscher,'" 1:418–22.

———. 1865a. *Populäre wissenschaftliche Vorträge*. First series. Braunschweig: Friedrich Vieweg.

———. 1865b. "Ueber Eigenschaften des Eises." In *WA*, 1:94–98.

———. 1865c. "Ueber den Einfluss der Raddrehung der Augen auf die Projection der Retinalbilder nach Aussen." In *WA*, 2:482–83.

———. 1865d. "Ueber stereoskopisches Sehen." In *WA*, 2:492–96.

———. 1865e. "Ueber die Augenbewegungen." In *WA*, 3:44–48.

———. 1866. "La glace et les glaciers." *Revue des Cours Scientifiques de la France et de l'Etranger* 3:433–47.

———. 1866a. "Ueber den Muskelton." In *WA*, 2:928–31.

———. 1867. *Optique physiologique*. Translated by Emile Javal and N. Th. Klein. Paris: Victor Masson.

———. 1867a. "De la relation des sciences naturelles avec la science en général." *Revue des Cours Scientifiques de la France et de l'Etranger* 4:693–701.

———. 1867b. "Sur l'origine physiologique de l'harmonie musicale." *Revue des Cours Scientifiques de la France et de l'Etranger* 4:177–89.

———. 1867c. "Ueber die Mechanik der Gehörknöchelchen." In *WA*, 2:503–14.

———. 1867d. *Handbuch der physiologischen Optik*. Part three. Hamburg: Leopold Voss.

———. 1867e. *Handbuch der physiologischen Optik*. Hamburg: Leopold Voss.

———. 1867 f. "Mitteilung, betr. Versuche über die Fortpflanzungs-Geschwindigkeit der Reizung in den motorischen Nerven des Menschen, welche Herr N. Baxt aus Petersburg geführt hat." In *WA*, 2:932–38.

———. 1868. *Théorie physiologique de la musique fondée sur l'étude des sensations auditives*. Translated by M. G. Guéroult. Paris: Victor Masson.

———. 1868a. "Die neueren Fortschritte in der Theorie des Sehen: Vorlesungen gehalten zu Frankfurt am Main und Heidelberg, Ausgearbeitet für die Preussischen Jahrbücher." In *VR*, 1:265–365.

———. 1868b. "Sur le movement le plus général d'un fluide: Réponse à une communication précédente de M. J. Bertrand." In *WA*, 1:135–39.

———. 1868c. "Sur le movement des fluides: Deuxième réponse à M. Bertrand." In *WA*, 1:140–44.

———. 1868d. "Réponse à la note de Monsieur Bertrand du 19 octobre." In *WA*, 1:145.

———. 1868e. "Ueber discontinuirliche Flüssigkeitsbewegungen." In *WA*, 1:146–57.

———. 1868 f. "Ueber die thatsächlichen Grundlagen der Geometrie." In *WA*, 2:610–17.

———. 1868g. "Ueber die Thatsachen, die der Geometrie zum Grunde liegen." In *WA*, 2:618–39.

———. 1868h [or 1869?]. *Sur les faits qui servent de base à la géométrie*. Translated by J. Hoüel. Bordeaux: G. Gounouilhou.

———. 1868i [or 1869?]. "Sur les faits qui servent de base à la géométrie." *Mémoires de la Société des Sciences Physiques et Naturelles de Bordeaux* 5 [1867–69]: 372–78.

———. 1868j. "De la production de la sensation du relief dans l'acte de la vision binoculaire." In *WA*, 3:581–86.

———. 1869. "Ueber das Ziel und die Forschritte der Naturwissenschaft: Eröffnungsrede für die Naturforscherversammlung zu Innsbruck 1869." In *VR*, 1:367–98.

———. 1869a. "Die Mechanik der Gehörknöchelchen und des Trommelfelles." In *WA*, 2:515–81.

———. 1869b. "Ueber die Schallschwingungen in der Schnecke des Ohres." In *WA*, 2:582–88.

———. 1869c. *Mémoire sur la conservation de la force, précédé d'un exposé élémentaire de la transformation des forces naturelles*. Translated by Louis Pérard. Paris: V. Masson.

———. 1869d. "Über die Entwicklungsgeschichte der neueren Naturwissenschaften." In *Tageblatt der 43. Versammlung Deutscher Naturforscher und Aerzte in Innsbruck vom 18. bis 24. September 1869*, edited by J. Daum, V. von Ebner, and Hugo Enzenberg, 36–40. Innsbruck: Wagner'sche Universitäts-Buchhandlung.

———. 1869e. "Ueber das Heufieber." *Virchow's Archiv für Pathologische Anatomie* 46:100–102.

———. 1869 f. "Zur Theorie der stationären Ströme in reibenden Flüssigkeiten." In *WA*, 1:223–30.

———. 1869g. "Ueber die physiologische Wirkung kurz dauernder elektrischer Schläge im Innern von ausgedehnten leitenden Massen." In *WA*, 1:526–30.

———. 1869h. "Ueber elektrische Oscillationen." In *WA*, 1:531–36.

———. 1869i. "Lettre de M. H. Helmholtz, professeur de physiologie à l'université de Heidelberg à M. A. Wurtz, doyen de la faculté de médecine de Paris." In *De la transfusion du sang défibriné, nouveau procédé pratique*, by L. de Belina, 67. Paris: Adrien Delahaye.

———. 1869–70. "Ueber die Entwicklungsgeschichte der neueren Naturwissenschaft." *Philosophische Monatshefte* 4:160–67.

———. 1870. "Goethe Naturaliste et Physicien." *Revue des Cours Scientifiques de la France et de l'Etranger* 7:18–25.

———. 1870a. "Vorrede zur Übersetzung." In Tyndall 1870, v–xi.

[———]. 1870b. "A German View of the War." *Glasgow Daily Herald*, 9 September, 5.

———. 1870c. "Ueber die Gesetze der inconstanten elektrischen Ströme in körperlich ausgedehnten Leitern." In *WA*, 1:537–44.

———. 1870d. "Ueber die Theorie der Elektrodynamik: Erste Abhandlung: Ueber die Bewegungsgleichungen der Elektricität für ruhende leitende Körper." In *WA*, 1:545–628.

———. 1870e. "The Axioms of Geometry." *Academy* 1:128–31.

———. 1870 f. "Ueber den Ursprung und die Bedeutung der geometrischen Axiome." In *VR*, 2:1–31. With an appendix, 381–83.

———. 1870g. "Les axiomes de la géométrie." *Revue des Cours Scientifiques de la France et de l'Etranger* 7:489–501.

———. 1870h. "Neue Versuche über die Fortpflanzungs-Geschwindigkeit der Reizung in den motorischen Nerven der Menschen, ausgeführt von N. Baxt aus Petersburg." In *WA*, 2:939–46.

———. 1871. "Ueber die Entstehung des Planetensystems." In *VR*, 2:53–91.

———. 1871a. "Zum Gedächtniss an Gustav Magnus." In *VR*, 2:33–51.

———. 1871b. *Populäre wissenschaftliche Vorträge*. Second series. Braunschweig: Friedrich Vieweg.

———. 1871c. "Vorrede zur deutschen Übersetzung." In Thomson and Tait 1871, x–xii.

———. 1871d. "Ueber die Fortpflanzungsgeschwindigkeit der elektrodynamischen Wirkungen." In *WA*, 1:629–35.

———. 1871e. "Ueber die Zeit, welche nötig ist, damit ein Gesichtseindruck zum Bewusstsein kommt, Resultate einer von Herrn N. Baxt in Heidelberger Laboratorium ausgeführten Untersuchung." In *WA*, 2:947–52.

———. 1872. "Ueber steuerbare Luftballons." In *Verhandlungen des Vereins zur Beförderung des Gewerbfleisses in Preussen, November–December*, 72:289–92. Later reprinted under the title "Theoretische Betrachtungen über lenkbare Luftballons" in *Polytechnisches Journal* 207 (1873): 465–69.

———. 1872a. "Ueber die Theorie der Elektrodynamik." In *WA*, 1:636–46.

———. 1872b. "Ueber die galvanische Polarisation des Platins." In *Tageblatt der 45. Versammlung deutscher Naturforscher und Aerzte zu Leipzig im August 1872*, edited by A. Winter, 110–11. Leipzig: G. Reusche.

———. 1872c. "The Axioms of Geometry: [Letter] To the Editor of the Academy." *Academy* 3 (41): 52–53.

———. 1873. *Popular Lectures on Scientific Subjects*. Translated by E. Atkinson. 2nd ser. London: Longmans, Green.

———. 1873a. "Optisches über Malerei." In *VR*, 2:93–135.

———. 1873b. "Ueber ein Theorem, geometrisch ähnliche Bewegungen flüssiger Körper betreffend, nebst Anwendung auf das Problem, Luftballons zu lenken." In *WA*, 1:158–71.

———. 1873c. "Induction und Deduction: Vorrede zum zweiten Theile des ersten Bandes der Uebersetzung von William Thomson's and Tait's Treatise on Natural Philosophy." In *VR*, 2:413–21.

———. 1873d. "Ueber die Grenzen der Leistungsfähigkeit der Mikroskope." In *WA*, 2:183–84.

———. 1873e. "Vergleich des Ampère'schen und Neumann'schen Gesetzes für die elektrodynamischen Kräfte." In *WA*, 1:688–701.

———. 1873 f. "Ueber galvanische Polarisation in gasfreien Flüssigkeiten." In *WA*, 1:823–34.

———. 1873g. "Ueber die Theorie der Elektrodynamik. Zweite Abtheilung: Kritisches." In *WA*, 1:647–87.

———. 1873h. *The Mechanism of the Ossicles of the Ear and Membrana Tympani*. New York: William Wood.

———. 1874. "Die theoretische Grenze für die Leistungsfähigkeit der Mikroskope." In *WA*, 2:185–212.

———. 1874a. "Ueber das Streben nach Popularisirung der Wissenschaft: Vorrede zu der Uebersetzung von Tyndall's Fragments of Science." In *VR*, 2:422–34.

———. 1874b. "Vorrede zum zweiten Theile des ersten Bandes: Kritisches." In Thomson and Tait 1874, v–xiv.

———. 1874c. "Scientific Worthies: IV.—John Tyndall." *Nature* 10:299–302.

———. 1874d. "Zur Theorie der anomalen Dispersion." In *WA*, 2:213–26.

———. 1874e. "Recent Problems in the Theory of Electrodynamics." *Academy* 5:290–92.

———. 1874 f. "Ueber die Theorie der Elektrodynamik: Dritte Abhandlung: Die elektrodynamischen Kräfte in bewegten Leitern." In *WA*, 1:702–62.

———. 1874g. "Kritisches zur Elektrodynamik." In *WA*, 1:763–73.

———. 1874h. Preface to *Advancement of Science: The Inaugural Address of Prof. John Tyndall, D.C.L., LL.D., F.R.S., Delivered before the British Association for the Advancement of Science, at Belfast, August, 19, 1874. with Portrait and Biographical Sketch: Opinions of the Eminent Scientist, Prof. H. Helmholtz, and Articles of Prof. Tyndall and Henry Thompson on Prayer*, 11–18. New York: Asa K. Butts.

———. 1874–75. "On the Use and Abuse of the Deductive Method in Physical Science." *Nature* 11 (24 December): 149–51; 11 (14 January): 211–12.

———. 1875. *On the Sensations of Tone as a Physiological Basis for the Theory of Music*. Translated from the 3rd German edition by Alexander J. Ellis. London: Longmans, Green.

———. 1875a. "Wirbelstürme und Gewitter." In *VR*, 2:137–63.

———. 1875b. "Versuche über die im ungeschlossenen Kreise durch Bewegung inducirten elektromotorischen Kräfte." In *WA*, 1:774–90.

———. 1876. *Populäre wissenschaftliche Vorträge*. Third series. Braunschweig: Friedrich Vieweg.

———. 1876a. "The Origin and Meaning of Geometrical Axioms." *Mind* 1 (3): 301–21.

———. 1876b. "Bericht über Versuche des Hrn. Dr. E. Root aus Boston, die Durchdringung des Platins mit elektrolytischen Gasen betreffend." In *WA*, 1:835–39.

———. 1876c. "Bericht betreffend Versuche über die elektromagnetische Wirkung elektrischer Convection, ausgeführt von Hrn. Henry A. Rowland." In *WA*, 1:791–97.

———. 1877. "Das Denken in der Medicin." In *VR*, 2:165–90; and "Anhang," 384–86.

———. 1877a. "Ueber die akademische Freiheit der deutschen Universitäten." In *VR*, 2:191–212.

———. 1877b. "Ueber galvanische Ströme, verursacht durch Concentrationsunterschiede: Folgerungen aus der mechanischen Wärmetheorie." In *WA*, 1:840–54.

———. 1878. "Die Thatsachen in der Wahrnehmung." In *VR*, 2:213–47; and "Beilagen," 387–406.

———. 1878a. "Telephon und Klangfarbe." In *WA*, 1:463–74.

———. 1878b. "Ueber die Bedeutung der Convergenzstellung der Augen für die Beurtheilung des Abstandes binocular gesehener Objecte." In *WA*, 2:497–500.

———. 1878c. "The Origin and Meaning of Geometrical Axioms: (II)." *Mind* 3 (10): 212–25. Reprint (in German) in *WA*, 2:640–60.

———. 1878d. "Rayleigh's 'Theory of Sound.'" *Nature* 17 (24 January): 237–39.

———. 1878e. "Lord Rayleigh's 'Theory of Sound.'" *Nature* 19 (12 December): 117–18.

———. 1879. "Ueber elektrische Grenzschichten." In *WA*, 3:49–51.

———. 1879a. "Studien über elektrische Grenzschichten." In *WA*, 1:855–98.

———. 1880. "Ueber die Bewegungsströme am polarisirten Platina." In *WA*, 1:899–921.

———. 1881. "Die neuere Entwickelung von Faraday's Ideen über Elektricität." In *VR*, 2:249–91.

———. 1881a. "Ueber die Berathungen des Pariser Kongresses, betreffend die elektrischen Masseinheiten." *Elektrotechnische Zeitschrift* 2:482–89.

———. 1881b. "Ueber die elektrischen Maasseinheiten nach den Berathungen des elektrischen Congresses, versammelt zu Paris 1881." In *VR*, 2:293–309, and appendix (added in 1884), 411–12.

———. 1881c. "Ueber die auf das Innere magnetisch oder diëlektrisch polarisirter Körper wirkenden Kräfte." In *WA*, 1:798–820.

———. 1881d. "Vorbemerkung zu einer nachgelassenen Abhandlung von Franz Boll: These und Hypothesen zur Licht- und Farbenempfindung." *Archiv für Anatomie und Physiologie*, 1–3.

———. 1881e. "Eine elektrodynamische Waage." In *WA*, 1:922–24.

———. 1881 f. "Ueber galvanische Polarisation des Quecksilbers und darauf bezügliche neue Versuche des Herrn Arthur König." In *WA*, 1:925–38.

———. 1881g. "On an Electrodynamic Balance." *Proceedings of the Royal Society of London* 32: 39–40.

———. 1881h. "Professor Helmholtz's Faraday Lecture." *Nature* 23 (7 April 1881): 535–40.

———. 1882. *Wissenschaftliche Abhandlungen* (WA). Vol. 1. Leipzig: Johann Ambrosius Barth.

———. 1882a. "Ueber absolute Maassysteme für electrische und magnetische Grössen." In *WA*, 3:993–1005.

———. 1882b. "Bericht über die Thätigkeit der internationalen elektrischen Commission." *Verhandlungen der Physikalischen Gesellschaft zu Berlin* (Sitzung vom 17. November), 101–102.

———. 1882c. "Die Thermodynamik chemischer Vorgänge." In *WA*, 2:958–78.

———. 1882d. "Zur Thermodynamik chemischer Vorgänge (zweiter Beitrag)." In *WA*, 2:979–92.

———. 1883. "Anhang zu dem Vortrag 'Ueber die Wechselwirkung der Naturkräfte und die darauf bezüglichen neuesten Ermittelungen der Physik.'" In *VR*, 2:401–17.

———. 1883a. *Wissenschaftliche Abhandlungen (WA)*. Vol. 2. Leipzig: Johann Ambrosius Barth.

———. 1883b. "Bestimmung magnetischer Momente durch die Waage." In *WA*, 3:115–18.

———. 1883c. "Zur Thermodynamik chemischer Vorgänge: Folgerungen, die galvanische Polarisation betreffend." In *WA*, 3:92–114.

———. 1884. *Vorträge und Reden*. 3rd ed. 2 vols. Braunschweig: Friedrich Vieweg.

———. 1884a. "On Galvanic Currents Passing through a Very Thin Stratum of an Electrolyte." In *WA*, 3:88–91.

———. 1884b. "Studien zur Statik monocyclischer Systeme." In *WA*, 3:119–41, 163–72, 173–78.

———. 1884c. "Principien der Statik monocyclischer Systeme." In *WA*, 3:142–62, 179–202.

———. 1884d. "Verallgemeinerung der Sätze über die Statik monocyclischer Systeme." *Sitzungsberichte der Königlich Preussischen Akademie der Wissenschaften zu Berlin*, 18 December, 1197–1201.

———. 1885. "Report on Sir William Thomson's Mathematical and Physical Papers: Vol. I and II." In *WA*, 3:587–96.

———. 1885a. *Handbuch der physiologischen Optik*. 2nd rev. ed. First installment. Hamburg: Leopold Voss.

———. 1886. "Antwortrede [gehalten beim Empfang der Graefe-Medaille zu Heidelberg 1886]." In *VR*, 2:311–20.

———. 1886a. "Ueber Wolken- und Gewitterbildung." In *WA*, 3:287–88.

———. 1886b. *Handbuch der physiologischen Optik*. 2nd rev. ed. Second and Third Installments. Hamburg: Leopold Voss.

———. 1886c. "Ueber die physikalische Bedeutung des Princips der kleinsten Wirkung." In *WA*, 3:203–48.

———. 1887. "Josef Fraunhofer: Ansprache gehalten bei der Gedenkfeier zur hundertjährigen Wiederkehr seines Geburtstages, Berlin den 6. März 1887." In *VR*, 2:321–33.

———. 1887a. "Zählen und Messen, erkenntnisstheoretisch betrachtet." In *WA*, 3:356–91.

———. 1887b. *Handbuch der physiologischen Optik*. 2nd rev. ed. Fourth Installment. Hamburg: Leopold Voss.

———. 1887c. "Zur Geschichte des Princips der kleinsten Action." In *WA*, 3:249–63.

———. 1887d. "Versuch, um die Cohäsion von Flüssigkeiten zu zeigen." In *WA*, 3:264–66.

———. 1887e. "Weitere Untersuchungen, die Elektrolyse des Wassers betreffend." In *WA*, 3:267–81.

———. 1887 f. "Zu dem 'Bericht über die Untersuchung einer mit der Flüssigkeit Pictet arbeitenden Eismaschine, erstattet von Hrn. Dr. Max Corsepius.'" In *WA*, 3:282–86.

———. 1888. "Ueber atmosphärische Bewegungen." In *WA*, 3:289–308.

———. 1889. "Ueber atmosphärische Bewegungen (Zweite Mittheilung): Zur Theorie von Wind und Wellen." In *WA*, 3:309–32.

———. 1889a. In memory of Rudolph Clausius. *Verhandlungen der Physikalischen Gesellschaft zu Berlin* 8:1–6.

———. 1889b. *Ueber die Erhaltung der Kraft*. Ostwald's Klassiker der exacten Wissenschaften, Band 1. Leipzig: Wilhelm Engelmann.

———. 1889c. *Handbuch der physiologischen Optik*. 2nd rev. ed. Fifth Installment. Hamburg: Leopold Voss.

———. 1890. "Die Energie der Wogen und des Windes." In *WA*, 3:333–55.

———. 1890a. "Die Störung der Wahrnehmung kleinster Helligkeitsunterschiede durch das Eigenlicht der Netzhaut." In *WA*, 3:392–406.

———. 1891. "Erinnerungen." In *VR*, 1:1–21.

———. 1891a. "Bemerkungen über die Vorbildung zum akademischen Studium." In *Verhandlungen über Fragen des höheren Unterrichts: Berlin, 6. & 17. December 1890*, 202–9, 763–64. Berlin: Wilhelm Hertz.

———. 1891b. "Versuch einer erweiterten Anwendung des Fechner'schen Gesetzes im Farbensystem." In *WA*, 3:407–37.

———. 1891c. "Versuch, das psychophysische Gesetz auf die Farbenunterschiede trichromatischer Augen anzuwenden." In *WA*, 3:438–59.

———. 1891d. "Kürzeste Linien im Farbensystem." In *WA*, 3:460–75.

———. 1892. "Goethe's Vorahnungen kommender naturwissenschaftlicher Ideen." In *VR*, 2:335–61.

———. 1892a. Letter to the editor. In *Die Suggestion und die Dichtung: Gutachten über Hypnose und Suggestion*, edited by Karl Emil Franzos, 69–71. Berlin: F. Fontane.

———. 1892b. *Handbuch der physiologischen Optik*. 2nd rev. ed. Sixth and Seventh Installments. Hamburg: Leopold Voss.

———. 1892c. "Das Princip der kleinsten Wirkung in der Elektrodynamik." In *WA*, 3:476–504.

———. 1892d. "Elektromagnetische Theorie der Farbenzerstreuung." In *WA*, 3:505–23.

———. 1892e. "Zusätze und Berichtigungen zu dem Aufsatze: Elektromagnetische Theorie der Farbenzerstreuung." In *WA*, 3:523–25.

———. 1893. "Gustav Wiedemann beim Beginn des 50. Bandes seiner Annalen der Physik und Chemie, gewidmet von H. v. Helmholtz." *Wiedemann's Annalen der Physik und Chemie* 50:iii–xi.

———. 1893a. "Folgerungen aus Maxwell's Theorie über die Bewegungen des reinen Aethers." In *WA*, 3:526–35.

———. 1893b. *H. von Helmholtz, Schet i izmerenie* [counting and measuring] *and L. Kronecker, Poniatie o chislie* [the idea of numbers]. Kasan, Russia: Tip. Imperatovskago universiteta.

———. 1894. "Heinrich Hertz: Vorwort zu dessen Prinzipien der Mechanik, Berlin, Juli 1894." In *VR*, 2:363–78.

———. 1894a. "Ueber den Ursprung der richtigen Deutung unserer Sinneseindrücke." In *WA*, 3:536–53.

———. 1894b. *Handbuch der physiologischen Optik*. 2nd rev. ed. Eighth Installment. Hamburg: Leopold Voss.

———. 1894c. "Nachtrag zu dem Aufsatze: Ueber das Princip der kleinsten Wirkung in der Elektrodynamik." In *WA*, 3:597–603.

———. 1895. *Wissenschaftliche Abhandlungen* (WA). Vol. 3. Leipzig: Johann Ambrosius Barth.

———. 1896. *Handbuch der physiologischen Optik*. 2nd rev. ed. Hamburg: Leopold Voss. First published in three parts, 1856, 1860, 1866, respectively; supplemented 2nd edition published in eight installments, 1885–94; further supplemented for publication in 1896.

———. 1897–1907. *Vorlesungen über theoretische Physik von H. von Helmholtz*. 6 vols. in 7. Leipzig: Johann Ambrosius Barth.

———. 1903. *Vorträge und Reden* (VR). 5th ed. 2 vols. Braunschweig: Friedrich Vieweg.

———. 1903a. "Vorrede zum ersten Bande der dritten Auflage 1884." In *VR*, 1:vii–xiii.

———. 1903b. "Vorrede zum zweiten Bande." In *VR*, 2:v–ix.

———. 1904. "Author's Preface." In *Popular Lectures on Scientific Subjects*, translated by E. Atkinson, vii–x. 1st ser. London: Longmans, Green.

———. 1909–11. *Handbuch der physiologischen Optik*. 3rd ed. Edited by A. Gullstrand, J. von Kries, and W. Nagel. 3 vols. Hamburg: Leopold Voss.

———. 1921. *Schriften zur Erkenntnistheorie*. Edited by Paul Hertz and Moritz Schlick. Berlin: Springer.

———. 1954. *On the Sensations of Tone as a Physiological Basis for the Theory of Music*. 2nd rev. ed. Translated from the 4th German edition by Alexander J. Ellis. With an introduction by Henry Margenau. New York: Dover.

———. 1962. *Helmholtz's Treatise on Physiological Optics*. Translated from the 3rd German edition by James P. C. Southall. 3 vols. Reprint, New York: Dover, 1962.

———. 1970. "Rede über die Entdeckungsgeschichte des Princips der kleinsten Action." In Harnack 1970, 2:282–96.

———. 1975a. "Wahlvorschlag für August Kundt (1839–1894) zum OM." 8 February. In Kirsten and Körber 1975, 107–8.

———. 1975b. "Wahlvorschlag für Ludwig Boltzmann (1844–1906) zum OM." 8 February. In Kirsten and Körber 1975, 109–10.

———. 1977. *Hermann von Helmholtz: Epistemological Writings*. Edited and with an introduction and a bibliography by Robert S. Cohen and Yehuda Elkana. Dordrecht: D. Reidel.

———. 1982. *Über die Erhaltung der Kraft*. 2 vols. With an introduction by Hans-Jürgen Treder. Transcribed by Christa Kirsten. Berlin: Akademie-Verlag.

———. 1995. "The Facts in Perception." In HH 1995a, 342–80. First published 1878.

———. 1995a. *Science and Culture: Popular and Philosophical Essays*. Edited and with an introduction by David Cahan. Chicago: University of Chicago Press.

HH, and Gustav von Piotrowski. 1860. "Ueber Reibung tropfbarer Flüssigkeiten." In *WA*, 1:172–222.

Helmholtz, R[ichard] von, and W. Staby. 1930–37. *Die Entwicklung der Lokomotive im Gebiete des Vereins Deutscher Eisenbahnverwaltungen*. 2 vols. Munich: R. Oldenbourg.

Helmholtz, Robert von. 1883. "The Remarkable Sunsets." *Nature* 29 (6 December): 130–33.

———. 1890. "A Memoir of Gustav Robert Kirchhoff." In *Annual Report of the Board of Regents of the Smithsonian Institution . . . July 1889*, 527–40. Washington, DC: Government Printing Office.

"Helmholtz." 1939. In *Artillerie und Ballistik in Stichworten*, edited by Hans-Hermann Kritzinger and Friedrich Stuhlmann, 147. Berlin: Julius Springer.

"Helmholtz Arrives Tomorrow." 1893. *Chicago Sunday Tribune*, 20 August, 2.

"Helmholtz Celebration and Medal." 1891. *American Journal of Science* 141:521.

Helmholtz-Fond. 2013. "100 Jahre Helmholtz-Fonds e.V. und 40 Jahre Helmholtz-Preis." *PTB Mitteilungen* 4, n.p.

Henderson, Linda Dalrymple. 1983. *The Fourth Dimension and Non-Euclidean Geometry in Modern Art*. Princeton, NJ: Princeton University Press.

Henning, Eckart. 2000. *Beiträge zur Wissenschaftsgeschichte Dahlems*. Berlin: Druckhaus am Treptower Park.

Hentschel, Klaus, ed. 1996. *Physics and National Socialism: An Anthology of Primary Sources*. Edited and translated by Ann M. Hentschel. Basel: Birkhäuser Verlag.

Hentschel, Klaus, and Renate Tobies, eds. 1999. *Brieftagebuch zwischen Max Planck, Carl Runge, Bernhard Karsten und Adolf Leopold . . . mit den Promotions- und Habilitationsakten Max Plancks und Carl Runges im Anhang*. Berlin: ERS Verlag.

Herbst, Katrin, ed. 2006. *Pour le mérite: Vom königlichen Gelehrtenkabinett zur nationalen Bildnissammlung*. Berlin: Staatliche Museen zu Berlin—Stiftung Preussischer Kulturbesitz and G + H Verlag.

Hermann, Armin. 1959. *Grosse Physiker: Vom Werden des neuen Weltbildes*. Stuttgart: Ernst Battenberg Verlag.

Hermann, L., and P. Volkmann. 1894. "Hermann von Helmholtz. Rede gehalten bei der von der physikalisch-ökonomischen Gesellschaft zu Königsberg in Pr. veranstalteten Gedächtnissfeier am 7. December 1894." *Schriften der Physikalisch-ökonomischen Gesellschaft* 35: 63–83.

"Hermann v. Helmholtz, der Erfinder des Augenspiegels: Zur 120. Wiederkehr seines Geburtstages am 31. August 1941." 1941. *Die Blindenwelt* 29:206–7.

"Hermann von Helmholtz." 1894. *Nature* 50:479–80.

"Hermann von Helmholtz (im memoriam)." 1894. *Wiener Medizinische Blätter: Zeitschrift für die gesammte Heilkunde* 17:531–32.

"Hermann von Helmholtz: Zu seinem 120. Geburtstages am 31. August 1941." 1941–42. *Technik für Alle* 32:207.

Herneck, Friedrich. 1966. *Bahnbrecher des Atomzeitalters: Grosse Naturforscher von Maxwell bis Heisenberg*. Berlin: Buchverlag der Morgen.

———. 1982. *Karl Friedrich Zöllner*. Leipzig: BSB B. G. Teubner Verlagsgesellschaft.

Herrmann-Schneider, Hildegard. 1928. *Die Philosophie Immanuel Hermann Fichtes: Ein Beitrag zur Geschichte der nachhegelschen Spekulation*. Berlin: Reuther & Reichard.

Hertz, Heinrich. 1894–95. *Gesammelte Werke*. 3 vols. Edited by Ph. Lenard. Leipzig: Johann Ambrosius Barth.

———. 1894–95a. "Zum 31. August 1891." In Hertz 1894–95, 1:360–68.

———. [1927]. *Heinrich Hertz: Erinnerungen, Briefe, Tagebücher*. Compiled by Johanna Hertz. Leipzig: Akademische Verlagsgesellschaft.

Heunisch, A. J. V. 1857. *Das Grossherzogthum Baden, historisch-geographisch-statistisch-topographisch beschrieben*. Heidelberg: Julius Groos'scher Universitätsbuchhandlung.

Heuss, Theodor. 1948. *Anton Dohrn*. 2nd ed. Stuttgart: Rainer Wunderlich Verlag.

Hiebert, Elfrieda F. 2003. "Helmholtz's Musical Acoustics: Incentive for Practical Techniques in Pedaling and Touch at the Piano." In *The Past in the Present: Papers Read at the IMS Intercongressional Symposium and the 10th Meeting of the CANTUS PLANUS, Budapest & Visegrád, 2000*, 1:425–43. Budapest: Liszt Ferenc Academy of Music.

———. 2013. "Listening to the Piano Pedal: Acoustics and Pedagogy in Late Nineteenth-Century Contexts." *Osiris*, 2nd ser., 28:232–53.

Hiebert, Erwin, and Elfrieda Hiebert. 1994. "Musical Thought and Practice: Links to Helmholtz's Tonempfindungen." In Krüger 1994a, 295–311.

Hintzelmann, Paul, ed. 1886. *Almanach der Universität Heidelberg für das Jubiläumsjahr 1886*. Heidelberg: Carl Winter's Universitätsbuchhandlung.

Hirsch, A., and Th. V. Oppolzer, eds. 1884. *Verhandlungen der vom 15. bis zum 24. Oktober 1883 in Rom abgehaltenen Siebenten Allgemeinen Conferenz der europäischen Gradmessung*. Berlin: Georg Reimer.

Hirschberg, Julius. 1918. *Handbuch der gesamten Augenheilkunde*, vol. 15. Edited by A. Graefe, Th. Saemisch, C. Hess, et al. 2nd rev. ed. 15 vols. Berlin: Springer.

———. 1921. "Hermann von Helmholtz. Ein Gedenkwort." *Berliner Klinische Wochenschrift* 58:1116–18.

"H.L. Helmholtz." 1875. *Über Land und Meer* 33:451.

Höber, R. 1921. "Zu Helmholtz' 100. Geburtstage." *Deutsche Medizinische Wochenschrift* 47: 1001–2.

Hofe, Karl vom. 1950. "100 Jahre Helmholtzscher Augenspiegel." *Forschungen und Fortschritte* 26:102–3.

Hoffmann, Christoph. 2003. "Helmholtz' Apparatuses: Telegraphy as Working Model of Nerve Physiology." *Philosophia Scientia* 7:129–49.

Hoffmann, Dieter, ed. 1995. *Gustav Magnus und sein Haus*. Stuttgart: Verlag für Geschichte der Naturwissenschaften und der Technik.

Hoffmann, Dieter, and Werner Ebeling. 1994. "'Reichskanzler der Wissenschaften': Zum 100. Todestag des Physikers, Physiologen und Philosophen Hermann von Helmholtz." *PB* 50: 827–32.

Hoffmann, Dieter, and Helmut Trischler. 2015. "Die Helmholtz-Gemeinschaft in historischer Perspective." In *20 Jahre Helmholtz-Gemeinschaft*, edited by Jürgen Mlynek and Angela Bittner, 9–47. Berlin: HGF.

Hoffmann, Erich. 1939. "Hundertjährige Wiederkehr des Eintritts von H. v. Helmholtz und

R. Virchow in die militärärztliche Akademie." *Münchener Medizinische Wochenschrift* 86: 1560–62.

———. 1948. *Wollen und Schaffen: Lebenserinnerungen aus einer Wendezeit der Heilkunde 1868–1932*. Hannover: Schmorl & Von Seefeld.

Hofmann, August Wilhelm. 1870. "Zur Erinnerung an Gustav Magnus." *Berichte der Deutschen Chemischen Gesellschaft* 3:993–1101.

Holmes, Frederic Lawrence. 1992. *Between Biology and Medicine: The Formation of Intermediary Metabolism*. Berkeley Papers in History of Science, 14. Berkeley: University of California–Berkeley.

———. 1994. "The Role of Johannes Müller." In Krüger 1994a, 3–21.

———. 2003. "Fisiologia e medicina sperimentale." In *Storia della scienza*, edited by Sandro Petruccioli, 7:748–66. 10 vols. Rome: Istituto della Enciclopedia Italiana.

Holmes, Frederic L., and Kathryn M. Olesko. 1995. "The Images of Precision: Helmholtz and the Graphical Method of Physiology." In *The Values of Precision*, edited by M. Norton Wise, 198–221. Princeton, NJ: Princeton University Press.

Holmsten, Georg. 1971. *Potsdam: Die Geschichte der Stadt, der Bürger und Regenten*. Berlin: Haude & Spenersche Verlagsbuchhandlung.

"Homage to Genius." 1893. *Chicago Daily Tribune*, 22 August, 1–2.

Homer, William Innes. 1964. *Seurat and the Science of Painting*. Cambridge, MA: MIT Press.

Hörz, Herbert. 1957. "Über die Erkenntnistheorie von Helmholtz." *Aufbau* 13:423–32.

———. 1994. *Hermann von Helmholtz und die Bonner Universität*: Part 1, *Helmholtz als Professor der Anatomie und Physiologie in Bonn (1855–1858)*. Berlin: Akademienvorhaben Wissenschaftshistorische Studien Helmholtz-Editionen.

———. 1995. *Hermann von Helmholtz und die Bonner Universität*: Part 3, *Briefe von Lipschitz, Helmholtz und Pflüger zur Berufung von Helmholtz als Physikprofessor nach Bonn*. Berlin: Akademienvorhaben Wissenschaftshistorische Studien Helmholtz-Editionen.

———. 1995a. *Schopenhauer und Helmholtz: Bemerkungen zu einer alten Kontroverse zwischen Philosophie und Naturwissenschaften*. Berlin: Akademienvorhaben Wissenschaftshistorische Studien Helmholtz-Editionen.

———. 1996. "Helmholtz und die Geodätische Assoziation." In *Global Change and History of Geophysics*, edited by Wilfried Schröder and Michele Colacino, 137–54. Bremen-Roennebeck: Interdivisional Commission on History of the IAGA and History Commission of the German Geophysical Society.

———. 1997. *Brückenschlag zwischen zwei Kulturen: Helmholtz in der Korrespondenz mit Geisteswissenschaftlern und Künstlern*. Marburg/Lahn: Basilisken-Presse.

———. 1997a. *Helmholtz und die Meteorologie. Bemerkungen zu Briefe von Meteorologen*. Berlin: Akademienvorhaben Wissenschaftshistorische Studien Helmholtz-Editionen.

Hörz, Herbert, and Andreas Laass. 1989. *Ludwig Boltzmanns Wege nach Berlin: Ein Kapitel österreichisch-deutscher Wissenschaftsbeziehungen*. Berlin: Akademie-Verlag.

Hörz, Herbert, and Siegfried Wollgast. 1971. "Einleitung." In *Hermann von Helmholtz: Philosophische Vorträge und Aufsätze*, edited by Herbert Hörz and Siegfried Wollgast, v–lxxix. Berlin: Akademie-Verlag.

———. 1971a. "Zu den philosophischen Auffassungen von Hermann von Helmholtz." *Deutsche Zeitschrift für Philosophie* 19:40–65.

———. 1986. "Hermann von Helmholtz und Emil du Bois-Reymond: Wissenschaftsgeschichtliche Einordnung in die naturwissenschaftlichen und philosophischen Bewegungen ihrer Zeit." In Kirsten et al. 1986, 11–64.

Howard, Don, and John Stachel, eds. 2000. *Einstein: The Formative Years, 1879–1909*. Boston: Birkhäuser.

Howitt, William. 1842. *The Rural and Domestic Life of Germany: With Characteristic Sketches of Its Cities and Scenery*. London: Longman, Brown, Green, and Longmans.

Hübscher, Arthur, ed. 1978. *Arthur Schopenhauer: Gesammelte Briefe*. Bonn: Bouvier Verlag Herbert Grundmann.

Hui, Alexandra. 2013. *The Psychophysical Ear: Musical Experiments, Experimental Sounds, 1840–1910*. Cambridge, MA: MIT Press.

Huxley, Thomas Henry. 1968. "A Liberal Education; And Where To Find It." In *Collected Essays*, by Thomas Henry Huxley, 3:76–110. 9 vols. New York: Greenwood. First published 1868.

———. 1968a. "Science and Culture." In *Collected Essays*, by Thomas Henry Huxley, 3:134–59. 9 vols. New York: Greenwood. First published 1880.

"H. von Helmholtz Arrives." 1893. *Chicago Daily News*, 21 August, 1.

Hyder, David Jalal. 2001. "Physiological Optics and Physical Geometry." *Science in Context* 14: 419–56.

———. 2009. *The Determinate World: Kant and Helmholtz on the Physical Meaning of Geometry*. Berlin: Walter de Gruyter.

International Congress of Experimental Psychology. 1974. International Congress of Experimental Psychology, Second Session, London, 1892. Reprint, Nendeln/Liechtenstein: Kraus Reprint.

"The International Exhibition and Congress of Electricity at Paris." 1881. *Nature* 24:511–12, 533, 563–64, 585–89, 607–8.

"Intimes aus dem Leben von Hermann von Helmholtz." 1903. *Die Gartenlaube*, 487–89.

Israel, Paul. 1998. *Edison: A Life of Invention*. New York: John Wiley.

Jackson, Myles. 2000. *Spectrum of Belief: Joseph von Fraunhofer and the Craft of Precision Optics*. Cambridge, MA: MIT Press.

———. 2006. *Harmonius Triads: Physicists, Musicians, and Instrument Makers in Nineteenth-Century Germany*. Cambridge, MA: MIT Press.

Jackson, Roland. 2015. "John Tyndall and the Early History of Diamagnetism." *AS* 72:435–89.

Jacobi, Carl Gustav J. 1996. *Vorlesungen über analytische Mechanik (Berlin 1847/48): Nach einer Mitschrift von Wilhelm Scheibner*. Edited by Helmut Pulte. Braunschweig: Deutsche Mathematiker-Vereinigung (Vieweg).

Jaeger, Wilhelm. 1932. *Die Entstehung der internationalen Masse der Elektrotechnik*. Berlin: Julius Springer.

Jaeger, Wolfgang. 1985. "Theodor Leber und die Begründung der Experimentellen Ophthalmologie." In Doerr 1985, 2:321–31.

———, ed. 1997. *Die Erfindung der Ophthalmoskopie: Dargestellt in den Originalbeschreibungen der Augenspiegel von Helmholtz, Ruete und Giraud-Teulon*. Heidelberg: Universitäts-Augenklinik.

James, Frank A. J. L. 1982. "Thermodynamics and Sources of Solar Heat, 1846–1862." *British Journal for the History of Science* 15:155–81.

———. 1995. "Science as a Cultural Ornament: Bunsen, Kirchhoff, and Helmholtz in Mid-Nineteenth-Century Baden." *Ambix* 42:1–9.

———. 2002. "Running the Royal Institution: Faraday as an Administrator." In *"The Common Purposes of Life": Science and Society at the Royal Institution of Great Britain*, edited by Frank A. J. L. James, 119–46. Aldershot: Ashgate.

———. 2004. "Reporting Royal Society Lectures, 1826–1867." In *Science Serialized: Representations of the Sciences in Nineteenth-Century Periodicals*, edited by Geoffrey Cantor and Sally Shuttleworth, 67–79. Cambridge, MA: MIT Press.

James, William. 1890. *The Principles of Psychology*. 2 vols. New York: Henry Holt.

———. 1920. *The Letters of William James*. Edited by Henry James. 2 vols. Boston: Atlantic Monthly Press.

———. 1978. *Essays in Philosophy*. Cambridge, MA: Harvard University Press.

———. 1979. *The Will to Believe and Other Essays in Popular Philosophy*. Cambridge, MA: Harvard University Press.

———. 1981. *The Principles of Psychology*. 3 vols. Cambridge, MA: Harvard University Press.

———. 1983. *Essays in Psychology*. Cambridge, MA: Harvard University Press.

———. 1986. *Essays in Psychical Research*. Cambridge, MA: Harvard University Press.

———. 1987. *Essays, Comments, and Reviews*. Cambridge, MA: Harvard University Press.

———. 1988. *Manuscript Lectures*. Cambridge, MA: Harvard University Press.

Janet, P. 1909. "La vie et les œuvres de E. Mascart." *Bulletin de la Société Internationale des Electriciens*, 2nd ser., 9 (88): 481–527.

Jehl, Francis. 1938. *Menlo Park: Reminiscences*. 2 vols. Dearborn, MI: Edison Institute.

Jevons, W. Stanley. 1871. "Helmholtz on the Axioms of Geometry." *Nature* 4:481–82.

Jones, Ernest. 1953–57. *The Life and Work of Sigmund Freud*. 3 vols. New York: Basic Books.

Jungnickel, Christa, and Russell McCormmach. 1986. *Intellectual Mastery of Nature: Theoretical Physics from Ohm to Einstein*. 2 vols. Chicago: University of Chicago Press.

Junk, Carl. 1888. "Physikalische Institute." In *Handbuch der Architektur*, Teil 4, Halbband 6, Heft 2, 101–58. Darmstadt: A. Bergsträsser.

Jurkowitz, Edward. 2002. "Helmholtz and the Liberal Unification of Science." *Historical Studies in the Physical and Biological Sciences* 32:2:291–317.

———. 2010. "Helmholtz's Early Empiricism and the Erhaltung der Kraft." *AS* 67:39–78.

Kania, Hans. 1939. *Geschichte des Viktoria-Gymnasiums zu Potsdam*. Potsdam: Bonness & Hachfeld.

Kargon, Jeremy. 2014. "The Logic of Color: Theory and Graphics in Christine Ladd-Franklin's Explanation of Color Vision." *Leonardo* 47 (2): 151–57.

Karlson, Paul. 1935–36. "Hermann von Helmholtz. 1821–1894." In *Die Grossen Deutschen: Neue Deutsche Biographie*, edited by Willy Andreas and Wilhelm von Scholz, 3:524–41. 4 vols. Berlin: Propyläen-Verlag.

Karsch, Johannes. 1950. "Zum hundertsten Jahrestag der Erfindung des Augenspiegels durch Hermann von Helmholtz." *Deutsches Gesundheitswesen* 5:1267–69.

Karsten, Gustav, ed. 1847. *Die Fortschritte der Physik im Jahre 1845*. Berlin: G. Reimer.

———, ed. 1850. *Die Fortschritte der Physik im Jahre 1847*. Berlin: G. Reimer.

———, ed. 1853. *Die Fortschritte der Physik im Jahre 1849*. Berlin: G. Reimer.

Katscher, Leopold. 1901. "Helmholtz. Zum 50jährigen Jubiläum der Erfindung des Augenspiegels." *Wiener Medizinische Presse* 42:1465–70, 1501–7.

Kaufmann, W. 1922. "Hermann v. Helmholtz als Physiker." *Deutsche Revue* 47:263–67.

Kayser, Heinrich. 1936. "Erinnerungen aus meinem Leben." American Philosophical Society, Philadelphia.

Kehler, R. v. 1926. "Helmholtz in seinem Urteil über Flugzeuge und Luftschiffe." *Zeitschrift für Flugtechnik und Motorluftschiffahrt* 17:407–9.

Kemp, Martin. 1990. *The Science of Art: Optical Themes in Western Art from Brunelleschi to Seurat*. New Haven, CT: Yale University Press.

Kessler, Harry Graf. 1928. *Walther Rathenau: Sein Leben und sein Werk*. Berlin-Grunewald: Verlagsanstalt Hermann Klemm.

Kim, Dong-Won. 2002. *Leadership and Creativity: A History of the Cavendish Laboratory, 1871–1919*. Dordrecht: Kluwer Academic.

Kirsten, Christa, and Hans-Günther Körber, eds. 1975. *Physiker über Physiker: Wahlvorschläge zur Aufnahme von Physikern in die Berliner Akademie 1870 bis 1929 von Hermann v. Helmholtz bis Erwin Schrödinger*. Berlin: Akademie-Verlag.

Kirsten, Christa, Herbert Hörz, Klaus Klauss, Wolfgang Knobloch, Marie-Luise Körner, Andreas Laass, and Siegfried Wollgast eds. 1986. *Dokumente einer Freundschaft: Briefwechsel zwischen Hermann von Helmholtz und Emil du Bois-Reymond 1846–1894*. Berlin: Akademie-Verlag.

Klauss, Klaus. 1981. "Ein neuentdecktes frühes Dokument zur Geschichte der Erfindung des

Augenspiegels durch Hermann v. Helmholtz." *NTM: Zeitschrift für Naturwissenschaften, Technik und Medizin* 18:58–61.

Klein, Felix. 1926. *Vorlesungen über die Entwicklung der Mathematik im 19. Jahrhundert.* Part 1. Edited by Richard Courant and Otto Neugebauer. Berlin: Julius Springer.

Klein, Martin J. 1972. "Mechanical Explanation at the End of the Nineteenth Century." *Centaurus* 17:58–82.

———. 1974. "Boltzmann, Monocycles, and Mechanical Explanation." In *Philosophical Foundations of Science*, edited by Raymond Seeger and Robert S. Cohen, 155–75. Dordrecht: D. Reidel.

Kleinwächter, Friedrich. 1881. "Die Fundirung der Universitäts-Institute in Berlin." *Centralblatt der Bauverwaltung* 1:359–61.

Knapp, Hermann. 1894. Obituary of Hermann von Helmholtz. *Archives of Ophthalmology* 23: 514–16.

———. 1902. "A Few Personal Recollections of Helmholtz." *Journal of the American Medical Association* 38:557–58.

Knott, Cargill Gilston. 1911. *Life and Scientific Work of Peter Guthrie Tait.* . . . Cambridge: Cambridge University Press.

Koblitz, Ann Hibner. 1993. *A Convergence of Lives: Sofia Kovalevskaia: Scientist, Writer, Revolutionary.* New Brunswick, NJ: Rutgers University Press.

Kocka, Jürgen, ed. 1999. *Die Königlich Preussische Akademie der Wissenschaften zu Berlin im Kaiserreich.* Berlin: Akademie Verlag.

Koenigsberger, Leo. 1896. *Hermann von Helmholtz's Untersuchungen über die Grundlagen der Mathematik und Mechanik.* Leipzig: B. G. Teubner.

———. 1898. "The Investigations of Hermann von Helmholtz on the Fundamental Principles of Mathematics and Mechanics." In *Annual Report of the Board of Regents of the Smithsonian Institution . . . to July 1896*, 93–124. Washington, DC: Government Printing Office.

———. 1902–3. *Hermann von Helmholtz.* 3 vols. Braunschweig: Friedrich Vieweg. Abbreviated *LK* in endnotes.

———. 1904. *Carl Gustav Jacob Jacobi: Festschrift zur Feier der hundertsten Wiederkehr seines Geburtstages.* Leipzig: B. G. Teubner.

———. 1910. "Über Helmholtz's Bruchstück eines Entwurfes betitelt 'Naturforscher-Rede.'" *Sitzungsberichte der Heidelberger Akademie der Wissenschaften: Mathematisch-Naturwissenschaftliche Klasse* 14:3–8.

———. 1911. *Hermann von Helmholtz: Gekürzte Volksausgabe.* Braunschweig: Friedrich Vieweg.

———. 1919. *Mein Leben.* Heidelberg: Carl Winters Universitätsbuchhandlung.

———. 1965. *Hermann von Helmholtz.* Translated by Frances A. Welby, with a preface by Lord Kelvin. Reprint, New York: Dover.

Köhnke, Klaus Christian. 1993. *Entstehung und Aufstieg des Neukantianismus: Die deutsche Universitätsphilosophie zwischen Idealismus und Positivismus.* Frankfurt am Main: Suhrkamp.

Kohut, Adolf. 1884. "Hermann Ludwig Ferdinand v. Helmholtz." *Westermanns Illustrierte Deutsche Monatshefte* 55:720–28.

———. [1900–1901]. *Berühmte israelitische Männer und Frauen in der Kulturgeschichte der Menschheit: Lebens- und Charakterbilder aus Vergangenheit und Gegenwart: Ein Handbuch für Haus und Familie: Mit zahlreichen Porträts und sonstigen Illustrationen.* 2 vols. Leipzig-Reuditz: A. H. Payne.

Konen, Heinrich, and August Pütter. [1922]. *Gedenkfeiern der Universität Bonn für einstige Mitglieder: Hermann von Helmholtz, 1855–1858, Professor der Physiologie und Anatomie an der Universität Bonn.* Bonn: Universitäts-Buchdruckerei Gebr. Scheur.

König, Arthur, and Carl Runge. 1897–1907. "Vorrede." In Helmholtz 1897–1907, 3:v–vi.

————. 1897–1907a. "Vorrede." In Helmholtz 1897–1907, 5:v–vi.

Königliche Universitätsbibliothek zu Berlin, ed. 1899. *Verzeichnis der Berliner Universitätsschriften 1810–1885*. Berlin: Commissions-Verlag von W. Weber.

Kossert, Andreas. 2005. *Ostpreussen: Geschichte und Mythos*. Munich: Siedler.

Kragh, Helge. 1993. "Between Physics and Chemistry: Helmholtz's Route to a Theory of Chemical Thermodynamics." In Cahan 1993a, 403–31.

————. 2002. "The Vortex Atom: A Victorian Theory of Everything." *Centaurus* 44:32–114.

Krannhals, Paul. 1921. "Vom Genie in der Wissenschaft: Zum 100. Geburtstage von Hermann v. Helmholtz." *Hellweg; Westdeutsche Wochenschrift für Deutsche Kunst* 1:165–67.

Krause, Albrecht. 1878. *Kant und Helmholtz über den Ursprung und die Bedeutung der Raumanschauung und der geometrischen Axiome*. Lahr: n.p.

Kremer, Richard L., ed. 1990. *Letters of Hermann von Helmholtz to His Wife, 1847–1859*. Stuttgart: Franz Steiner Verlag.

————. 1990a. *The Thermodynamics of Life and Experimental Physiology, 1770–1880*. New York: Garland.

————. 1993. "Innovation through Synthesis: Helmholtz and Color Research." In Cahan 1993a, 205–58.

————. 1994. "Gleaning from the Archives? The 'Helmholtz Industry' and Manuscript Sources." In Krüger 1994a, 379–400.

————. 1997. "The Eye as Inscription Device in the 1870s: Optograms, Cameras, and the Photochemistry of Vision." In *Biology Integrating Scientific Fundamentals: Contributions to the History of Interrelations between Biology, Chemistry, and Physics from the 18th to the 20th Centuries*, edited by Brigitte Hoppe, 360–81. Munich: Institut für Geschichte der Naturwissenschaften.

Kries, Johannes von. 1921. "Helmholtz als Physiologe." *Die Naturwissenschaften* 9:673–93.

Krigar-Menzel, Otto. 1897–1907. "Vorwort." In Helmholtz 1897–1907, vol. 1, part 2, v–vi.

————. 1897–1907a. "Vorwort." In Helmholtz 1897–1907, 2:v–vi.

Krigar-Menzel, Otto, and Max Laue. 1897–1907. "Vorwort." In Helmholtz 1897–1907, 4:v.

Kronecker, Hugo. 1891. "Hermann von Helmholtz." *Electrician* 27:437–39, 465–68.

————. 1894. "Hermann von Helmholtz: Akademischer Vortrag gehalten im Saale des Museums zu Bern." *Schweizerische Rundschau* 4:510–38.

Krönig, A., ed. 1856. *Die Fortschritte der Physik im Jahre 1853*. Vol. 9. Berlin: Georg Reimer.

————, ed. 1857. *Die Fortschritte der Physik im Jahre 1854*. Vol. 10. Berlin: Georg Reimer.

Krückmann, E. 1922. "Gedenkworte zum hundertsten Geburtstag von Helmholtz." *Medizinische Klinik* 18:30–32.

Krüger, E. 1863. Review of *Die Lehre von den Tonempfindungen*, by Hermann Helmholtz. *Allgemeine Musikalische Zeitung*, n.s., 1:467–71, 483–89, 495–501.

Krüger, Lorenz. 1994. "Helmholtz über die Begreiflichkeit der Natur." In Krüger 1994a, 201–15.

————, ed. 1994a. *Universalgenie Helmholtz: Rückblick nach 100 Jahren*. Berlin: Akademie Verlag.

Krummel, Richard Frank. 1983. *Nietzsche und der deutsche Geist*: Vol. 2, *Ausbreitung und Wirkung des Nietzscheschen Werkes im deutschen Sprachraum vom Todesjahr bis zum Ende des Weltkrieges: Ein Schrifttumsverzeichnis der Jahre 1901–1918*. Berlin: Walter de Gruyter.

Krüss, H. 1894. "Hermann von Helmholtz." *Zeitschrift für Instrumentenkunde* 14:342–45.

Ku, Ja Hyon. 2006. "British Acoustics and Its Transformation from the 1860s to the 1910s." *AS* 63:395–423.

Kuchling, Heinz. 1973. "Vorwort." In Wirzberger 1973, 277.

Kuhn, Thomas S. 1959. "Energy Conservation as an Example of Simultaneous Discovery." In *Critical Problems in the History of Science*, edited by Marshall Clagett, 321–56. Madison: University of Wisconsin Press.

Kühnert, Herbert. 1961. "Ein unbekannter Brief von Ernst Abbe an Hermann v. Helmholtz: Ein Beitrag zur Vorgeschichte des JENAer Glaswerks Schott und Genossen." *Sprechsaal für Keramik, Glas, Email* 94:213–16, 322–23.

Kundt, August. 1891. "Zum 70. Geburtstag von Hermann von Helmholtz." *National-Zeitung* 44 (30 August), n.p.

Kurbjuweit, Lutz. 1985. "Das Haus 'Zum Riesen' Hauptstrasse 52." In Doerr 1985, 5:323–35.

Kursell, Julia. 2008. "Hermann von Helmholtz und Carl Stumpf über Konsonanz und Dissonanz." *Berichte zur Wissenschaftsgeschichte* 31:130–43.

———. 2012. "A Gray Box: The Phonograph in Laboratory Experiments and Fieldwork, 1900–1920." In *The Oxford Handbook of Sound Studies*, edited by Trevor Pinch and Karin Bijsterveld, 176–97. Oxford: Oxford University Press.

———. 2013. "Experiments on Tone Color in Music and Acoustics: Helmholtz, Schoenberg, and Klangfarbenmelodie." *Osiris*, 2nd ser., 28:191–211.

Kurylo, Friedrich, and Charles Susskind. 1981. *Ferdinand Braun: A Life of the Nobel Prizewinner and Inventor of the Cathode-Ray Oscilloscope.* Cambridge, MA: MIT Press.

Kusch, Ernst. 1896. *C.G.J. Jacobi und Helmholtz auf dem Gymnasium: Beitrag zur Geschichte des Victoria-Gymnasiums zu Potsdam.* Potsdam: Kramer.

Kussmaul, Adolf. 1902. "Ein Dreigestirn grosser Naturforscher an der Heidelberger Universität im 19. Jahrhundert." *Deutsche Revue* 27:35–45, 173–87.

———. 1903. *Aus meiner Dozentenzeit in Heidelberg.* Edited by Vinzenz Czerny. Stuttgart: Adolf Bonz.

Kutzbach, Gisela. 1979. *The Thermal Theory of Cyclones: A History of Meteorological Thought in the Nineteenth Century.* Boston: American Meteorological Society.

L. 1921. "Zur Erinnerung an Helmholtz und Virchow." *Allgemeine Medizinische Central-Zeitung* 90:251–52.

Laas, Ernst. 1884. *Idealismus und Positivismus: Eine kritische Auseinandersetzung: Dritter Theil: Idealistische und positivistische Erkenntnisstheorie.* Berlin: Weidmannsche Buchhandlung.

Laforgue, Jules. 1922. *Berlin: La cour et la ville* Paris: Editions de la Sirène.

La Grange, Henry-Louis de. 1973. *Mahler.* Vol. 1. Garden City, NY: Doubleday.

———. 1995. *Gustav Mahler: Vienna: The Years of Challenge (1897–1904).* Vol. 2. Oxford: Oxford University Press.

———. 1999. *Gustav Mahler: Vienna: Triumph and Disillusion (1904–1907).* Vol. 3. Oxford: Oxford University Press.

Lammel, Gisold. 1993. *Adolph Menzel und seine Kreise.* Dresden: Verlag der Kunst.

Lampe, Emil. [1916]. "Briefe von Ch. Hermite zu P. du Bois-Reymond aus den Jahren 1875–1888." *Archiv der Mathematik und Physik* 24:193–220, 289–310.

Lampe, Hermann, and Hans Querner, comps. 1972. *Die Vorträge der allgemeinen Sitzungen auf der 1.-85. Versammlung 1822–1913.* Hildesheim: Dr. H. A. Gerstenberg.

Land, J. P. N. 1877. "Kant's Space and Modern Mathematics." *Mind* 2 (5): 38–46.

Lange, Friedrich Albert. 1866. *Geschichte des Materialismus und Kritik seiner Bedeutung in der Gegenwart.* Iserlohn: J. Baedeker.

"The Late Professor von Helmholtz." 1894. *Christian Age* (London), 3 October.

Lau, Ernst. 1922. "Helmholtz." *Sozialistische Monatshefte* 58:48–49.

Laue, Max von. 1944. "Zum 50. Todestage von Hermann v. Helmholtz. (8. September 1944.)." *Die Naturwissenschaften* 32:206–7.

———. 1960. "Über Hermann von Helmholtz." In *Forschen und Wirken: Festschrift zur 150-Jahr-Feier der Humboldt-Universität zu Berlin 1810–1960*, edited by Willi Göber and Friedrich Herneck, 1:359–66. 3 vols. Berlin: VEB Deutscher Verlag der Wissenschaften.

Laugel, Auguste. 1867. *La voix, l'oreille, et la musique.* Paris: G. Baillière.

———. 1869. *L'optique et les arts.* Paris: G. Baillière.

Lawergren, B. 1980. "On the Motion of Bowed Violin Strings." *Acustica* 44:194–206.

Lazarev, Petr Petrovich. 1959. *Helmholtz* [in Russian]. Moscow: Akademia Nauk USSR.

Lebedinskii, A. V., U. I. Frankfurt, and A. M. Frenk. 1966. *Helmholtz (1821–1894)* [in Russian]. Moscow: Nauka.

Lebensztejn, Jean-Claude. 1992. "L'optique du peintre (Seurat avec Helmholtz)." *Critique* 48 (540): 404–19.

Lejeune, Fritz. 1941. "Helmholtz, der Erfinder des Augenspiegels." *Praktische Gesundheitspflege* 10:37.

Lenard, Philipp. 1933. *Great Men of Science: A History of Scientific Progress*. New York: Macmillan.

Lenoir, Timothy. 1982. *The Strategy of Life: Teleology and Mechanics in Nineteenth-Century German Biology*. Chicago: University of Chicago Press.

———. 1986. "Models and Instruments in the Development of Electrophysiology, 1845–1912." *Historical Studies in the Physical and Biological Sciences* 17:1–54.

———. 1988. "Social Interests and the Organic Physics of 1847." In *Science in Reflection: The Israel Colloquium: Studies in History, Philosophy, and Sociology of Science*, edited by Edna Ullmann-Margalit, 169–91. Dordrecht: Kluwer Academic.

———. 1992. "Helmholtz, Müller und die Erziehung der Sinne." In Hagner and Wahrig-Schmidt 1992, 207–22.

———. 1993. "The Eye as Mathematician: Clinical Practice, Instrumentation, and Helmholtz's Construction of an Empiricist Theory of Vision." In Cahan 1993a, 109–53.

———. 1993a. "Farbensehen, Tonempfindung und der Telegraph: Helmholtz und die Materialität der Kommunikation." In *Die Experimentalisierung des Lebens: Experimentalsysteme in den biologischen Wissenschaften 1850/1950*, edited by Hans-Jörg Rheinberger and Michael Hagner, 50–73. Berlin: Akademie Verlag.

———. 1994. "Helmholtz and the Materialities of Communication." *Osiris*, 2nd ser., 9:183–207.

———. 2006. "Operationalizing Kant: Manifolds, Models, and Mathematics in Helmholtz's Theory of Perception." In Friedman and Nordmann 2006, 141–210.

Lenz, Max. 1910–18. *Geschichte der königlichen Friedrich-Wilhelms-Universität zu Berlin*. 4 vols. in 5. Halle: Waisenhaus.

Leppmann, Wolfgang. 1961. *The German Image of Goethe*. Oxford: Clarendon Press.

Lepsius, Bernhard. 1933. *Das Haus Lepsius: Vom geistigen Aufstieg Berlins zur Reichshauptstadt: Nach Tagebüchern und Briefen*. Berlin: Klinkhardt & Biermann.

Leroux, Jean. 2001. "'Picture Theories' as Forerunners of the Semantic Approach to Scientific Theories." *International Studies in the Philosophy of Science* 15:189–97.

Lesser, Margaret. 1984. *Clarkey: A Portrait in Letters of Mary Clarke Mohl (1793–1883)*. Oxford: Oxford University Press.

Leutner, Karl. 1955–57. *Deutsche auf die wir stolz sind*. 2 vols. Berlin: Verlag der Nation.

Lewenstein, Gustav. 1877. "Dühring contra Fakultät." *Deutsches Montagsblatt*, 18 June, n.p.

Lexis, Wilhelm, ed. 1893. *Die deutschen Universitäten: Für die Universitätsausstellung in Chicago 1893*. 2 vols. Berlin: Asher.

Lie, Sophus. 1889–93. *Theorie der Transformationsgruppen*. 3 vols. Leipzig: Teubner.

Lieberman, Richard K. 1995. *Steinway and Sons*. New Haven, CT: Yale University Press.

Liedman, Sven-Eric. 1986. *Das Spiel der Gegensätze: Friedrich Engels' Philosophie und die Wissenschaften des 19. Jahrhunderts*. Frankfurt: Campus Verlag.

Lightman, Bernard. 2015. "Lost in Translation: Evolutionary Naturalists and Their Language Games." *History of Science* 53:395–416.

Lightman, Bernard, and Michael S. Reidy, eds. 2014. *The Age of Scientific Naturalism: Tyndall and His Contemporaries*. London: Pickering and Chatto.

Liss, Julia E. 1996. "German Culture and German Science in the Bildung of Franz Boas." In *Volksgeist as Method and Ethic: Essays on Boasian Ethnography and the German Anthropologi-*

cal Tradition, edited by George W. Stocking Jr., 155–84. Madison: University of Wisconsin Press.

Lockemann, Georg. 1949. *Robert Wilhelm Bunsen: Lebensbild eines deutschen Naturforschers*. Stuttgart: Wissenschaftliche Verlagsgesellschaft.

Lodge, Oliver. 1931. *Past Years: An Autobiography*. London: Hodder and Stoughton.

Loesser, Arthur. 1954. *Men, Women and Pianos: A Social History*. New York: Simon and Schuster.

[Loewy, Maurice]. 1894. Remarks on the death of Helmholtz. In *Comptes Rendus Hebdomadaires des Séances de l'Académie des Sciences* . . . 119:1044–46.

Lohff, Brigitte. 1992. "Johannes Müller und das physiologische Experiment." In Hagner and Wahrig-Schmidt 1992, 105–23.

Lohmann, W. 1941. "Hermann von Helmholtz, das letzte Universalgenie." *Deutsche Mineralwasser-Zeitung* no. 35 (29 August): 276–77.

Lotze, Hermann. 1868. *Geschichte der Aesthetik in Deutschland*. Munich: J. G. Cotta Buchhandlung.

Lowe, Sue Davidson. 1983. *Stieglitz: A Memoir/Biography*. New York: Farrar Straus Giroux.

Ludewig, J. 1882. "Notizen aus der elektrischen Konferenz in Paris." *Elektrotechnische Zeitschrift* 3:404–8, 459–72.

Lummer, Otto. 1889. "Robert von Helmholtz." *Naturwissenschaftliche Rundschau* 4:567–68.

———. 1921. "Der Physiker und die Technik." In ESH et al. 1921, n.p.

Lungo, Carlo del. 1903. "Ermanno Helmholtz." In *Goethe ed Helmholtz*, 79–108. Turin: Fratelli Bocca.

Lützeler, Heinrich. 1968. *Die Bonner Universität: Bauten und Bildwerke*. Bonn: H. Bouvier and Ludwig Röhrscheid.

Lützen, Jesper. 2005. *Mechanistic Images in Geometric Form: Heinrich Hertz's Principles of Mechanics*. Oxford: Oxford University Press.

Mach, Ernst. 1866. *Einleitung in die Helmholtz'sche Musiktheorie: Populär für Musiker dargestellt*. Graz: Leuschner und Lubensky.

———. 1886. *Beiträge zur Analyse der Empfindungen*. Jena: G. Fischer.

———. 1898. *Popular Scientific Lectures*. 3rd rev. ed. Chicago: Open Court.

———. 1943. *Popular Scientific Lectures*. 5th ed. La Salle, IL: Open Court.

Mackenzie, Norman, and Jeanne Mackenzie. 1979. *Dickens: A Life*. Oxford: Oxford University Press.

MacLeod, Roy M. 1970. "The X-Club: A Social Network of Science in Late-Victorian England." *Notes and Records of the Royal Society of London* 24:305–22.

———. 1972. "Resources of Science in Victorian England: The Endowment of Science Movement, 1868–1900." In *Science and Society, 1600–1900*, edited by Peter Mathias, 111–66. Cambridge: Cambridge University Press.

Magie, W. F. 1894. "Hermann von Helmholtz—The Man and the Teacher." *Electrical World* 24 (14) (6 October): 329–30.

Magnus-Levy, Adolf. 1944. "The Heroic Age of German Medicine." *Bulletin of the History of Medicine* 16:331–42.

Majer, Ulrich. 1985. "Hertz, Wittgenstein und der Wiener Kreis." In *Philosophie, Wissenschaft, Aufklärung: Beiträge zur Geschichte und Wirkung des Wiener Kreises*, edited by Hans-Joachim Dahms, 40–66. Berlin: Walter de Gruyter.

Maley, V. Carlton, Jr. 1990. *The Theory of Beats and Combination Tones, 1700–1863*. New York: Garland.

Mamlock, G. 1921. "Helmholtz als Klassiker der Naturwissenschaften: Zu seinem 100. Geburtstag am 31. VIII. 1921." *Deutsche Medizinische Wochenschrift* 47:1002–3.

Mandelkow, Karl Robert. 1980. *Goethe in Deutschland: Rezeptionsgeschichte eines Klassikers*. Vol. 1 of 2. Munich: C. H. Beck.

————. 1989. *Goethe in Deutschland: Rezeptionsgeschichte eines Klassikers*. Vol. 2 of 2. Munich: C. H. Beck.

Mann, C. Riborg. 1895. "Professor von Helmholtz." *Scribner's Magazine* 18:568–70.

Mann, Thomas. 1974. *Gesammelte Werke*. 2nd ed. 13 vols. Frankfurt am Main: S. Fischer.

Manthey, Jürgen. 2005. *Königsberg: Geschichte einer Weltbürgerrepublik*. Munich: Carl Hanser Verlag.

Manz. Gustav. 1903–4. "H. von Helmholtz und die Musik." *Die Musik* 3:17–25.

Marciatori, Sandra. ed. 2009. *Laureati honoris causa (1888–2008)*. Bologna: CLUEB: Alma mater studiorum Università di Bologna, Archivio storico.

Marcuse, Julian. 1899. "Das Gehirn von Helmholtz." *Die Umschau* 3:404–5.

Marey, Etienne-Jules. 1899. *La chronophotographie*. Paris: Gauthier-Villars.

Marshall, Marilyn E., and Russell A. Wendt. 1980. "Wilhelm Wundt, Spiritism, and the Assumptions of Science." In Bringmann and Tweney 1980, 158–75.

Martius, Friedrich. 1923. "Friedrich Martius." In *Die Medizin der Gegenwart in Selbstdarstellungen*, edited by L. R. Grote, 1:105–40. 4 vols. Leipzig: Felix Meiner.

Maxwell, James Clerk. 1873. "Faraday." *Nature* 8:397–99.

————. 1877. "Scientific Worthies. X.—Hermann Ludwig Ferdinand Helmholtz." *Nature* 15:389–91.

————. 1883. *Lehrbuch der Electricität und des Magnetismus*. Translated by Bernhard Weinstein. 2 vols. Berlin: Julius Springer.

————. 1890. *The Scientific Papers of James Clerk Maxwell*. Edited by W. D. Niven. 2 vols. Cambridge: Cambridge University Press.

————. 1990. *The Scientific Letters and Papers of James Clerk Maxwell*. Edited by P. M. Harman. Vol. 1 of 3. Cambridge: Cambridge University Press.

————. 1995. *The Scientific Letters and Papers of James Clerk Maxwell*. Edited by P. M. Harman. Vol. 2 of 3. Cambridge: Cambridge University Press.

————. 2002. "Rede Lecture on 'The Telephone' May 1878." In Maxwell 2002a, 3:651–64.

————. 2002a. *The Scientific Letters and Papers of James Clerk Maxwell*. Edited by P. M. Harman. Vol. 3 of 3. Cambridge: Cambridge University Press.

"May Prescribe for Bismarck." 1893. *New York Times*, 8 October, 10.

McClelland, Charles E. 2017. *Berlin, the Mother of All Research Universities*. Lanham, MD: Lexington Books.

McDonald, Patrick. 2001. "Epistemology of Experiment and the Physiological Acoustics of Hermann von Helmholtz." PhD diss., University of Notre Dame.

————. 2001a. "Remarks on the Context of Helmholtz's 'Ueber das Wesen der Fäulniss und Gährung.'" *Science in Context* 14:493–98.

————. 2002. "Helmholtz's Methodology of Sensory Science, the Zeichentheorie, and Physical Models of Hearing Mechanisms." In *History of Philosophy of Science: New Trends and Perspectives*, edited by Michael Heidelberger and Friedrich Stadler, 159–83. Dordrecht: Kluwer.

McKendrick, John Gray. 1899. *Hermann Ludwig Ferdinand von Helmholtz*. New York: Longmans, Green.

Meadows, A. J. 1972. *Science and Controversy: A Biography of Sir Norman Lockyer*. Cambridge, MA: MIT Press.

Meinel, Christoph. 1991. *Karl Friedrich Zöllner und die Wissenschaftskultur der Gründerzeit: Eine Fallstudie zur Genese konservativer Zivilisationskritik*. Berlin: SIGMA.

Meinel, U. 1972. "Einige Bemerkungen zur visuellen Empfindung im Blickpunkt der marxistisch-leninistischen Abbildtheorie unter Berücksichtigung der Helmholtzschen Zeichentheorie." *Zeitschrift für Ärztliche Fortbildung* 66:587–89.

Melms, C. Ph. 1957. *Chronik von Dahlem*. Berlin-Grunewald: Arani.

Mendel, Hermann, and August Reissmann, eds. 1870–79. *Musikalisches Conversation-Lexikon.* 11 vols. Berlin: L. Heimann; New York: J. Schuberth.

Mendelsohn, Everett. 1965. "Physical Models and Physiological Concepts: Explanation in Nineteenth-Century Biology." *British Journal for the History of Science* 2:201–19.

———. 1974. "Revolution and Reduction: The Sociology of Methodological and Philosophical Concerns in Nineteenth Century Biology." In *The Interaction between Science and Philosophy,* edited by Y. Elkana, 407–26. Atlantic Highlands, NJ: Humanities Press.

Merz, John Theodore. 1922. *The Reminiscences of John Theodore Merz.* Edinburgh: William Blackwood.

Meulders, Michel. 2001. *Helmholtz: Des lumières aux neurosciences.* Paris: Odile Jacob.

———. 2010. *Helmholtz: From Enlightenment to Neuroscience.* Cambridge, MA: MIT Press.

Meyer, Karl Ferdinand. 1838. "Über die Brennlinien, welche durch die Zurückwerfung des Lichtes von Curven der zweiten Ordnung entstehen." In Schulprogramm, *Zu der öffentlichen Prüfung der Zöglinge des hiesigen Königlichen Gymnasiums. . .* , 1–26. Potsdam: Decker.

Mezhdunarodnyi komitet dlia chestvovaniia Germana fon-Gelmholttsa. 1892. *German fon-Gelmgoltts, 1821–1891.* Moscow: Izd. Imp. Moskovskago universiteta.

Michell, Joel. 1993. "The Origins of the Representational Theory of Measurement: Helmholtz, Hölder, and Russell." *SHPS* 24:185–206.

Mielke, Friedrich. 1972. *Das Bürgerhaus in Potsdam.* 2 vols. Tübingen: Ernst Wasmuth.

Millikan, Robert A. 1951. *The Autobiography of Robert A. Millikan.* London: Macdonald.

Mitchell, Daniel Jon. 2012. "Measurement in French Experimental Physics from Regnault to Lippmann: Rhetoric and Theoretical Practice." *AS* 69:453–82.

Mittag-Leffler, G. 1923. "Weierstrass et Sonja Kowalewsky." *Acta Mathematica* 39:133–98.

Mittasch, Alwin. 1952. *Friedrich Nietzsche als Naturphilosoph.* Stuttgart: A. Kröner.

Mohl, Ottmar von. 1922. *Fünfzig Jahre Reichsdienst.* 2 vols. Leipzig: Paul List.

Mohl, Robert von. 1902. *Lebens-Erinnerungen von Robert von Mohl 1799–1875.* 2 vols. Stuttgart: Deutsche Verlags-Anstalt.

Moleschott, Jacob. 1894. *Für meine Freunde: Lebens-Erinnerungen.* Giessen: Emil Roth.

Möller, W. 1923. "Hermann von Helmholtz." *Nature* 11:46–54.

Molvig, Ole. 2010. "The Berlin Urania, Humboldtian Cosmology, and the Public." In *The Heavens on Earth: Observatories and Astronomy in Nineteenth-Century Science and Culture,* edited by David Aubin, Charlotte Bigg, and H. Otto Sibum, 325–43. Durham, NC: Duke University Press.

Mommsen, T., D. F. Strauss, F. Max Müller, and T. Carlyle. 1871. *Letters on the War between Germany and France.* London: Trübner.

Moore, Gregory. 2004. Introduction to *Nietzsche and Science,* edited by Gregory Moore and Thomas H. Brobjer, 1–20. Aldershot: Ashgate.

Mueller, Ferdinand von. 1998. *Regardfully Yours: Selected Correspondence of Ferdinand von Mueller,* edited by R. W. Home, A. M. Lucas, Sara Maroske, D. M. Sinkora, J. H. Voigt, and Monika Wells. Vol. 1. Bern: Peter Lang.

———. 2002. *Regardfully Yours: Selected Correspondence of Ferdinand von Mueller,* edited by R. W. Home, A. M. Lucas, Sara Maroske, D. M. Sinkora, J. H. Voigt, and Monika Wells. Vol. 2. Bern: Peter Lang.

Müller, Conrad, ed. 1928. *Alexander von Humboldt und das preussische Königshaus: Briefe aus den Jahren 1835–1857.* Leipzig: Koehler.

Müller, F. Max. 1881. *Chips from a German Workshop.* 5 vols. New York: Charles Scribner's, 1881.

Müller, Georgina, ed. 1902. *The Life and Letters of the Right Honourable Friedrich Max Müller.* 2 vols. New York: Longmans, Green.

Müller, Harald. 1968. *Zur Geschichte der Stadt Potsdam von 1789 bis 1871.* Potsdam: Bezirksheimatmuseum.

Müller, Wilhelm. 1996. "Die Lage der theoretischen Physik an den Universitäten." In Hentschel 1996, 246–59. First published 1940.

Mulligan, Joseph F. 1998. "The Reception of Heinrich Hertz's *Principles of Mechanics* by His Contemporaries." In Baird, Hughes, and Nordmann 1998, 173–81.

Münchow, Wolfgang. 1978. *Albrecht von Graefe*. Leipzig: BSB B. G. Teubnerverlagsgesellschaft.

Münsterberg, Hugo. 1895. "Correspondence: The Helmholtz Memorial." *Science*, n.s., 1:547–48.

Münsterberg, Margaret. 1922. *Hugo Münsterberg: His Life and Work*. New York: D. Appleton.

National Academy of Sciences. 1884. *Annual Report of the National Academy of Sciences for 1883*. Washington, DC: Government Printing Office.

———. 1913. *A History of the First Half-Century of the National Academy of Sciences, 1863–1913*. Washington, DC: Lord Baltimore Press.

Nernst, W. 1921. "Die elektrochemischen Arbeiten von Helmholtz." *Die Naturwissenschaften* 9:699–702.

Neuber, Matthias. 2012. "Helmholtz's Theory of Space and Its Significance for Schlick." *British Journal for the History of Philosophy* 20:163–80.

Neumann, Hans. 1925. *Heinrich Wilhelm Dove: Eine Naturforscher-Biographie*. Liegnitz: H. Krumbhaar.

Newmann, Ernest. 1969. *The Life of Richard Wagner*. 4 vols. New York: Knopf.

N[ichols]., E[dward]. L. 1894. "Hermann von Helmholtz." *Physical Review* 2:222–27.

Nichols, J. Alden. 1958. *Germany after Bismarck: The Caprivi Era 1890–1894*. Cambridge, MA: Harvard University Press.

Niemeyer, Theodor. 1963. *Erinnerungen und Betrachtungen aus drei Menschenaltern*. Edited by Annemarie Niemeyer. Kiel: Walter G. Mühlau.

Nietzsche, Friedrich. 1974. "On Eternal Recurrence." In *Existentialism*, edited by Robert C. Solomon, 77–78. New York: Modern Library.

———. 1974a. "On the Will to Power." In *Existentialism*, edited by Robert C. Solomon, 73–76. New York: Modern Library.

Noble, D. 1976. "George Henry Lewes, George Eliot and the Physiological Society." *Journal of Physiology* 263 (Suppl.): 45P–54P.

Nochlin, Linda. 1966. *Impressionism and Post-Impressionism 1874–1904: Sources and Documents*. Englewood Cliffs, NJ: Prentice-Hall.

Noeggerath, J., and H. F. Kilian, eds. 1859. *Amtlicher Bericht über die drei und dreissigste Versammlung Deutscher Naturforscher und Ärzte zu Bonn im September 1857*. Bonn: Carl Georgi.

"Notes." 1895. *Nature* 51:612–16.

"November 11, 1868.—Statute Meeting." 1868. *Proceedings of the American Academy of Arts and Sciences* 8:103.

Nye, Russel B. 1944. *George Bancroft: Brahmin Rebel*. New York: Knopf.

Nyhart, Lynn K. 1995. *Biology Takes Form: Animal Morphology and the German Universities, 1800–1900*. Chicago: University of Chicago Press.

Oberdan, Thomas. 2015. "From Helmholtz to Schlick: The Evolution of the Sign-Theory of Perception." *SHPS* 52:35–43.

Oettingen, Arthur von. 1866. *Harmoniesystem in dualer Entwickelung: Studien zur Theorie der Musik*. Tartu, Estonia: W. Gläser.

———. 1913. *Das duale Harmoniesystem*. Leipzig: C. F. W. Siegel's Musikalienhandlung (R. Linnemann).

Olesko, Kathryn M. 1989. "Physics Instruction in Prussian Secondary Schools before 1859." *Osiris*, 2nd ser., 5:94–120.

———. 1991. *Physics as a Calling: Discipline and Practice in the Königsberg Seminar for Physics*. Ithaca, NY: Cornell University Press.

———. 1994. "Civic Culture and Calling in the Königsberg Period." In Krüger 1994a, 22–42.

———. 1995. "The Meaning of Precision: The Exact Sensibility in Early Nineteenth-Century Germany." In *The Values of Precision*, edited by M. Norton Wise, 103–34. Princeton, NJ: Princeton University Press.

———. 1996. "Precision, Tolerance, and Consensus: Local Cultures in German and British Resistance Standards." In Buchwald 1996, 117–56.

Olesko, Kathryn M., and Frederic L. Holmes. 1993. "Experiment, Quantification, and Discovery: Helmholtz's Early Physiological Researches, 1843–50." In Cahan 1993a, 50–108.

"One Congress at End." 1893. *Chicago Tribune*, 26 August, 8.

Ophthalmological Society. 1886. *Festsitzung der Ophthalmologischen Gesellschaft in der Aula der Heidelberger Universität am 9. August 1886: Ueberreichung der Graefe-Medaille an Hermann von Helmholtz*. Rostock: Universitäts-Buchdruckerei.

[———]. 1891. *Festschrift zur Feier des siebzigsten Geburtstages von Hermann von Helmholtz*. Hamburg: Leopold Voss.

Oppenheim, Janet. 1985. *The Other World: Spiritualism and Psychical Research in England, 1850–1914*. Cambridge: Cambridge University Press.

Orden pour le mérite für Wissenschaften und Künste: Die Mitglieder des Ordens. 1978. Vol. 2. Berlin: Gebr. Mann.

Ostertag, Heiger. 1993. "Vom strategischen Bombenkrieg zum sozialistischen Bildersturm: Die Zerstörung Potsdams 1945 und das Schicksal seiner historischen Gebäude nach dem Kriege." In *Potsdam: Staat, Armee, Residenz in der preussisch-deutschen Militärgeschichte*, edited by Bernhard R. Kroener, 487–99. Berlin: Propyläen.

Ostwald, Wilhelm. 1926–27. *Lebenslinien: Eine Selbstbiographie*. 3 vols. Berlin: Klasing.

———. 1969. *Aus dem wissenschaftlichen Briefwechsel Wilhelm Ostwalds*: Part II, *Briefwechsel mit Svante Arrhenius und Jacobus Henricus van't Hoff*. Edited by Hans-Günther Körber. Berlin: Akademie-Verlag.

Otis, Laura. 2001. *Networking: Communicating with Bodies and Machines in the Nineteenth Century*. Ann Arbor: University of Michigan Press.

———. 2002. "The Metaphoric Circuit: Organic and Technological Communication in the Nineteenth Century." *Journal of the History of Ideas* 63:105–28.

———. 2007. *Müller's Lab*. Oxford: Oxford University Press.

Paepke, Karola. 1969. "Künstlervereinigungen und Kunstvereine in Potsdam: Ein Beitrag zu ihrer Geschichte zwischen 1800 und 1950." In *Beiträge zur Potsdamer Geschichte*, edited by Bezirksheimatmuseum, 109–50. Potsdam: Bezirksheimatmuseum.

Paine, Albert Bigelow. 1912. *Mark Twain: A Biography: The Personal and Literary Life of Samuel Langhorne Clemens*. New York: Harper.

Pantalony, David. 2009. *Altered Sensations: Rudolph Koenig's Acoustical Workshops in Nineteenth-Century Paris*. Dordrecht: Springer.

Parshall, Karen Hunger. 2006. *James Joseph Sylvester: Jewish Mathematician in a Victorian World*. Baltimore: Johns Hopkins University Press.

Partch, Harry. 1974. *Genesis of a Music: An Account of a Creative Work, Its Roots and Its Fulfillments*. 2nd ed. New York: Da Capo.

Pastore, Nicholas. 1973. "Helmholtz's 'Popular Lectures on Vision.'" *Journal of the History of the Behavioral Sciences* 9:190–202.

Patton, Lydia. 2009. "Signs, Toy Models, and the a priori: From Helmholtz to Wittgenstein." *SHPS* 40:281–89.

Paul, Oscar. 1868. *Geschichte des Claviers vom Ursprunge bis zu den modernsten Formen dieses Instruments nebst einer Uebersicht über die musikalische Abtheilung der Pariser Weltausstellung im Jahre 1867*. Leipzig: A. H. Payne.

Paulsen, Friedrich. 1938. *An Autobiography*. New York: Columbia University Press.

Pernet, Johannes. 1895. "Hermann von Helmholtz. 31. August 1821 bis 8. September 1894: Ein Nachruf." *Neujahrsblatt der Naturforschenden Gesellschaft in Zürich* 97:1–36.

Perry, Ralph Barton. 1935. *The Thought and Character of William James*. 2 vols. Boston: Little, Brown.

Perry, Walter Copland. 1877. "German Universities." *Macmillans Magazine* 37:148–60.

———. 1878. "Professor Helmholtz' Rectoratsrede und die Englischen Universitäten." *Deutsche Rundschau* 14:332–36.

Pfaff, Karl. 1995. *Heidelberg und Umgebung*. 3rd rev. ed. Reprint, Heidelberg: Brigitte Guderjahn.

Philipps, Denise. 2012. *Acolytes of Nature: Defining Natural Science in Germany, 1770–1850*. Chicago: University of Chicago Press.

Philosophische Fakultät der Königlichen Universität zu Berlin. 1877. *Aktenstücke in der Angelegenheit des Privatdocenten Dr. Dühring*. Berlin: G. Reimer.

Pieper, Herbert K. O., ed., 1997. *Die Berufung von Eduard Pflüger an die Universität in Bonn*. Berlin: Berlin-Brandenburgische Akademie der Wissenschaften.

———, ed. 1997a. *Dokumente zur Berufung von Eduard Pflüger an die Universität in Bonn*. Berlin: Berlin-Brandenburgische Akademie der Wissenschaften.

———, ed. 1998. *Nach Plückers Tod: Eine Sammlung von neuen Dokumenten zur Wiederbesetzung der vakanten Plückerschen Lehrstühle insbesondere zu den Bemühungen der preussischen Regierung, Helmholtz als Nachfolger Plückers für die physikalische Professur an der Universität Bonn zu gewinnen (1868/1869)*. Berlin: Akademievorhaben, Wissenschaftshistorische Studien, Helmholtz-Editionen.

———, ed. 1998a. *Die schwere Trennung "von der gut nährenden Milchkuh der medicinischen Facultät": Helmholtz: Von der Physiologie zur Physik*: Part 1, *Der gescheiterte Wechsel von der medizinischen Fakultät in Heidelberg zur philosophischen Fakultät in Bonn: Ein Beitrag zur Helmholtz-Biographie*. Berlin: Akademievorhaben, Wissenschaftshistorische Studien, Helmholtz-Editionen.

Pietsch, Ludwig. 1889–90. "Ludwig Knaus." *Velhagen & Klasings Neue Monatsheft* 4:641–68.

Pistor, [Hermann]. 1941. "Hermann von Helmholtz." *Optische Rundschau und Photo-Optiker* 32:275–77.

Pistor, M., ed. 1890. "Das physikalische Institut." In *Anstalten und Einrichtungen des öffentlichen Gesundheitswesens in Preussen: Festschrift zum X. internationalen medizinischen Kongress Berlin 1890 . . .*, 62–68. Berlin: Julius Springer.

Planck, Max. 1910. "Das Institut für theoretische Physik." In *Geschichte der königlichen Friedrich-Wilhelms-Universität zu Berlin*: vol. 3, *Wissenschaftliche Anstalten. Spruchkollegium. Statistik*, edited by Max Lenz, 3:276–78. Halle: Waisenhaus.

———. 1929–30. "Erinnerungen an Anna von Helmholtz." *Velhagen & Klasings Monatshefte* 44:37–39.

———. 1935. "Persönliche Erinnerungen." *Verhandlungen der Deutschen Physikalischen Gesellschaft* 16:11–16.

———. 1948. *Wissenschaftliche Selbstbiographie*. 2nd ed. Leipzig: Johann Ambrosius Barth.

———. 1949. "Persönliche Erinnerungen aus alten Zeiten." In *Vorträge und Erinnerungen*, by Max Planck, 1–14. 5th ed. Stuttgart: S. Hirzel.

———. 1950. "Die Physikalische Gesellschaft um 1890." *PB* 6:433–34.

———. 1986. *Max Planck: Hörer bei Hermann Helmholtz in Berlin: Wintersemester 1877/78*. With an introduction by Hans-Jürgen Treder. Berlin: Akademie Verlag.

Plücker, Julius. 1868–69. *Neue Geometrie des Raumes gegründet auf die Betrachtung der geraden Linie als Raumelement*. 2 vols. in 1. With a foreword by A. Clebsch to vol. 1. Vol. 2 edited by Felix Klein. Leipzig: B. G. Teubner.

Poincaré, Henri. 2001. *The Value of Science: Essential Writings of Henri Poincaré*. Edited by Stephen Jay Gould. New York: Modern Library.

Pole, William. 1879. *The Philosophy of Music: Being the Substance of a Course of Lectures Delivered at the Royal Institution of Great Britain, in February and March 1877*. London: Trübner.

Pollock, Frederick. 1880. *Spinoza: His Life and Philosophy*. London: Kegan Paul.

Pool, Phoebe. 1967. *Impressionism*. New York: Frederick A. Praeger.

Porter, Theodore M. 2004. *Karl Pearson: The Scientific Life in a Statistical Age*. Princeton, NJ: Princeton University Press.

Príncipe, João. 2012. "Sources et nature de la philosophie de la physique d'Henri Poincaré." *Philosophia Scientiae* 16 (2): 197–222.

Prittwitz, Karl Ludwig von. 1985. *Berlin 1848: Das Erinnerungswerk des Generalleutnants Karl Ludwig von Prittwitz und andere Quellen zur Berliner Märzrevolution und zur Geschichte Preussens um die Mitte des 19. Jahrhunderts*. Berlin: Walter de Gruyter.

"Professor Helmholtz." 1881. *London Times*, 5 April, 10.

"Professor Helmholtz." 1893. *Scientific American* 69:247.

"Professor Helmholtz in London." 1881. *London Times*, 11 April, 4.

"Prof. Helmholtz Honored." 1891. *New York Times*, 3 November, 1.

"The Progress of Science: Hermann von Helmholtz." 1907. *Popular Science Monthly* 71:283–84.

Prutz, Hans. 1894. *Die Königliche Albertus-Universität zu Königsberg i. Pr. im neunzehnten Jahrhundert*. Königsberg: Hartungsche Verlagsdruckerei.

Pulte, Helmut. 2008. "Darwin's Relevance for Nineteenth-Century Physics and Physicists: A Comparative Study." In *The Reception of Charles Darwin in Europe*, edited by Eve-Marie Engels and Thomas F. Glick, 1:116–34. 2 vols. London: Continuum.

Pulzer, Peter. 1988. *The Rise of Political Anti-Semitism in Germany and Austria*. Rev. ed. Cambridge, MA: Harvard University Press.

Pupin, Michael I. 1894. "Hermann von Helmholtz." *Electrical World* 24 (21): 541–42.

———. 1951. *From Immigrant to Inventor*. New York: Charles Scribner's.

Pupke, H. 1951–52. "Hermann von Helmholtz als Physiker: Zur 130. Wiederkehr seines Geburtstages." *Mathematische und Naturwissenschaftliche Unterricht* 4:211–12.

Quazza, Guido. 1992. *L'utopia di Quintino Sella: La politica della scienza*. Turin: Comitato di Torino dell'Istituto per la Storia del Risorgimento Italiano.

Querner, Hans. 1970. "Die Versammlung der Gesellschaft deutscher Naturforscher und Ärzte 1869 in Innsbruck." *Berichte des Naturforscher-Medizinischen Vereins Innsbruck* 58:13–34.

Radau, Rodolphe. 1865. "Sur la base scientifique de la musique: Analyse des recherches de M. Helmholtz." *Moniteur Scientifique-Quesneville* 197:193–214.

Raman, C. V. 1947. *Books That Have Influenced Me—A Symposium*. Madras: G. A. Natesan.

Ramaseshan, S. 1990. "Hermann von Helmholtz and His Student in Absentia—C.V. Raman." *American Journal of Physics* 58:519.

Rassow, H., ed. 1912. *Festschrift zur Feier der 100jährigen Anerkennung als Gymnasium am 10. und 11. November 1912*. Potsdam: Edmund Stein.

Raveau, M. 1894. "Hermann von Helmholtz." *Le Moniteur Scientifique-Quesneville* 44:801–6.

Ravin, James G., and Christie Kenyon. 1992. "From Von Graefe's Clinic to the Ecole des Beaux-Arts: The Meteoric Career of Richard Liebreich." *Survey of Ophthalmology* 37:221–28.

Rayleigh, Lord [John William Strutt]. 1895. "Argon." *Notices of the Proceedings at the Meetings of the Members of the Royal Institution, with Abstracts of the Discourses* 14:524–38.

———. 1945. *The Theory of Sound*. 2nd rev. ed. 2 vols. in 1. New York: Dover.

Rayleigh, Lord [Robert John Strutt]. 1968. *Life of John William Strutt: Third Baron Rayleigh*. Augmented ed. Madison: University of Wisconsin Press.

Rechenberg, Helmut. 1994. *Hermann von Helmholtz: Bilder seines Lebens und Wirkens*. Weinheim: VCH.

Rehding, Alexander. 2003. *Hugo Riemann and the Birth of Modern Musical Thought*. Cambridge: Cambridge University Press.

———. 2005. "Wax Cylinder Revolutions." *Musical Quarterly* 88:123–60.

Reicke, Ilse. 1971. "Hermann von Helmholtz und Berlin: Zum 150. Geburtstag am 31. August 1971." *Mitteilungen des Vereins für die Geschichte Berlins* 67:53–56.

Reinecke, Hans-Peter. 1999. "Hermann von Helmholtz, Carl Stumpf und die Folgen von der musikalischen Akustik zur Tonpsychologie. Ganz persönliche Anmerkungen zu einem Kapitel Berliner Wissenschaftsgeschichte." *International Review of the Aesthetics and Sociology of Music* 30:95–109.

Reinecke, Wilhelm. 1903. "Die Grundlagen der Geometrie nach Kant." *Kantstudien* 8:349–95.

Reiner, Julius. 1905. *Hermann von Helmholtz*. Leipzig: Theod. Thomas.

Reiprich, Kurt. 1969. *Die philosophisch-naturwissenschaftlichen Arbeiten von Karl Marx und Friedrich Engels*. Berlin: Dietz Verlag.

Reuter, Sören. 2014. "Nietzsche und die Sinnesphysiologie und Erkenntniskritik." In *Handbuch Nietzsche und die Wissenschaften: Natur- geistes- und sozialwissenschaftliche Kontexte*, edited by Helmut Heit and Lisa Heller, 79–106. Berlin: Walther de Gruyter.

Rewald, John. 1978. *Post-Impressionism: From van Gogh to Gauguin*. 3rd ed. New York: Museum of Modern Art.

Richards, Joan L. 1977. "The Evolution of Empiricism: Hermann von Helmholtz and the Foundations of Geometry." *British Journal for the Philosophy of Science* 28:235–53.

———. 1988. *Mathematical Visions: The Pursuit of Geometry in Victorian England*. Boston: Academic Press.

———. 1988a. "Bertrand Russell's Essay on the Foundations of Geometry and the Cambridge Mathematical Tradition." *Russell* 8:59–80.

Richards, Robert J. 2002. *The Romantic Conception of Life: Science and Philosophy in the Age of Goethe*. Chicago: University of Chicago Press.

———. 2008. *The Tragic Sense of Life: Ernst Haeckel and the Struggle over Evolutionary Thought*. Chicago: University of Chicago Press.

Richarz, Franz. 1897–1907. "Vorwort." In Helmholtz 1897–1907, 6:v–vi.

Riehl, Alois. 1904. "Helmholtz et Kant." *Revue de Métaphysique et de Morale* 12:579–603.

———. 1904a. "Helmholtz in seinem Verhältnis zu Kant." *Kant-Studien* 9:261–85.

———. 1921. "Helmholtz als Erkenntnistheoretiker." *Die Naturwissenschaften* 9:702–8.

———. 1922. "Helmholtz." In *Führende Denker und Forscher*, by Alois Riehl, 223–40. Leipzig: Quelle & Mayer.

Riemann, Hugo. 1900. *Die Elemente der musikalischen Aesthetik*. Berlin: W. Spemann.

Ries, Hugibert [Hugo Riemann]. 1873. "Tonverwandtschaft." *Neue Zeitschrift für Musik* 69: 29–31, 42–43, 54–56.

Riese, Reinhard. 1977. *Die Hochschule auf dem Wege zum wissenschaftlichen Grossbetrieb: Die Universität Heidelberg und das badische Hochschulwesen 1860–1914*. Stuttgart: Ernst Klett.

Rk. 1921. "Hermann von Helmholtz: Zu seinem 100. Geburtstag am 31. August 1921." *Die Umschau* 25:512–13.

Robinson, David K. 1988. "Wilhelm Wundt and the Establishment of Experimental Psychology, 1875–1914: The Context of a New Field of Scientific Research." PhD diss., University of California–Berkeley.

Rocke, Alan. 1993. *The Quiet Revolution: Hermann Kolbe and the Science of Organic Chemistry*. Berkeley, CA: University of California Press.

Rogers, William Barton. 1896. *Life and Letters of William Barton Rogers*. Edited by his wife. 2 vols. Boston: Houghton, Mifflin.

Röhl, John C. G. 1994. *The Kaiser and His Court: Wilhelm II and the Government of Germany*. Cambridge: Cambridge University Press.

Rohrschneider, W. 1951. "Die Erfindung des Augenspiegels durch Hermann Helmholtz vor 100 Jahren." *Deutsche Medizinische Wochenschrift* 76:1409–10.

Rood, Ogden N. 1973. *Modern Chromatics: Students' Text-Book of Color with Applications to Art and Industry*. Facsimile of the first American edition of 1879, with a preface by Faber Birren, "Ogden Nicholas Rood (1831–1902): His Life." New York: Van Nostrand Reinhold.

Roscoe, Henry E. 1869. "Science Education in Germany: I, The German University System." *Nature* 1:157–59.

———. 1900. "Bunsen Memorial Lecture." *Journal of the Chemical Society* 77:513–54.

———. 1906. *The Life & Experiences of Sir Henry Enfield Roscoe, D.C.L., LL.D., F.R.S. Written by Himself*. London: Macmillan.

Rothe, Rudolf. 1921. "Helmholtz als Mathematiker." In ESH et al. 1921, n.p.

Rouzaud, Henri. 1891. *Les fêtes du VIe Centenaire de L'Université de Montpellier*. Montpellier: Camille Coulet; Paris: G. Masson.

Rubens, Heinrich. 1910. "Das physikalische Institut." In *Geschichte der königlichen Friedrich-Wilhelms-Universität zu Berlin*: vol. 3, *Wissenschaftliche Anstalten: Spruchkollegium: Statistik*, edited by Max Lenz, 3:278–96. Halle: Waisenhaus.

Rücker, Arthur W. 1895. "Physical Work of Hermann von Helmholtz." *Nature* 51:472–75, 493–95.

———. 1896. "Obituary Notices of Fellows Deceased [Hermann von Helmholtz]." *Proceedings of the Royal Society of London* 59:xvii–xxx.

Ruete, Christian Georg Theodor. 1852. *Der Augenspiegel und das Optometer für praktische Aerzte*. Göttingen: Dieterische Buchhandlung.

Rummenhöller, Peter. 1963. *Moritz Hauptmann als Theoretiker: Eine Studie zum erkenntniskritischen Theoriebegriff in der Musik*. Wiesbaden: Breitkopf & Härtel.

———. 1971. "Die philosophischen Grundlagen in der Musiktheorie des 19. Jahrhunderts." *Beiträge zur Theorie der Künste im 19. Jahrhundert* 1:44–57.

Runge, Iris. 1949. *Carl Runge und sein wissenschaftliches Werk*. Göttingen: Vandenhoeck & Ruprecht.

Russell, Bertrand. 1983. *Cambridge Essays, 1888–99*. Vol. 1 of *The Collected Papers of Bertrand Russell*, edited by Kenneth Blackwell. London: G. Allen & Unwin.

"R. v. Helmholtz." 1934. *Die Lokomotive Vereinigt mit Eisenbahn und Industrie* 31:157–58.

Sablik, Karl. 1989. "Hering, Vintschgau und das Problem der Nachfolge Purkinjes." *Sudhoffs Archiv* 73:78–87.

"Säcularfeier der f. Akademie der Wissenschaften 28. und 29. März 1859." 1859. *Gelehrte Anzeigen der f. Bayerischen Akademie der Wissenschaften* 48, cols. 473–506.

Sadie, Stanley, ed. 1980. *The New Grove Dictionary of Music and Musicians*. 20 vols. London: Macmillan.

Sattler, Bernhard, ed. 1962. *Adolf von Hildebrand und seine Welt: Briefe und Erinnerungen*. Munich: Verlag Georg D. W. Callwey.

Scharlau, Winfried, ed. 1986. *Rudolf Lipschitz: Briefwechsel mit Cantor, Dedekind, Helmholtz, Kronecker, Weierstrass und anderen*. Wiesbaden: Friedr. Vieweg; Braunschweig: Deutsche Mathematiker-Vereinigung.

Scheel, Heinrich, ed. 1972. *Gedanken von Helmholtz über schöpferische Impulse und über das Zusammenwirken verschiedener Wissenschaftszweige*. Berlin: Akademie-Verlag.

Scheel, Karl. 1925. "Die literarischen Hilfsmittel der Physik." *Die Naturwissenschaften* 13: 45–48.

———. 1935. ["Aus der Geschichte der Gesellschaft."] *Verhandlungen der Deutschen Physikalischen Gesellschaft* 16:2–11.

Schering, W. M. 1941. "Zur 120. Wiederkehr des Geburtstages von Hermann von Helmholtz." *Lebenserfolg* 37[?]:149–51.

Schestag, Thomas. 2003. "Retrouvé: Du temps perdu: Note sur l'origine du temps perdu dans la Recherche de Marcel Proust." *Philosophia Scientiae* 7:115–27.

Scheuffelen, Thomas, comp. 1985. "Berthold Auerbach 1812–1882." *Marbacher Magazin* 36:1–113. (Sonderheft).

Schickore, Jutta. 2001. "The Task of Explaining Sight—Helmholtz's Writings on Vision as a Test Case for Models of Science Popularization." *Science in Context* 14 (3): 397–417.

Schiemann, Gregor. 1997. *Wahrheitsgewissheitsverlust: Hermann von Helmholtz' Mechanismus im Anbruch der Moderne: Eine Studie zum Übergang von klassischer zu moderner Naturphilosophie*. Darmstadt: Wissenschaftliche Buchgesellschaft.

———. 1998. "The Loss of World in the Image: Origin and Development of the Concept of Image in the Thought of Hermann von Helmholtz and Heinrich Hertz." In Baird, Hughes, and Nordmann 1998, 25–38.

———. 2014. "Nietzsche und die Wahrheitsgewissheitsverluste im Anbruch der Moderne." In *Handbuch Nietzsche und die Wissenschaften: Natur- geistes- und sozialwissenschaftliche Kontexte*, edited by Helmut Heit and Lisa Heller, 46–75. Berlin: Walther de Gruyter.

Schiff, Emil. 1883. "Das du Bois-Reymond-Bankett." *Neue Freie Presse* (Vienna), no. 6833, 25 October, 4.

———. 1891. "Hermann von Helmholtz." *Deutsche Rundschau* 69:42–64.

———. 1891a. "H.v. Helmholtz' Doctor-Jubiläum." *Neue Freie Presse* (Vienna), 4 November, 2–3.

———. 1893. "Emil Du Bois–Reymond's fünfzigjähriges Doctor-Jubiläum." *Neue Freie Presse* (Vienna), 23 February, n.p.

Schimank, Hans. 1941. "Hermann von Helmholtz: Leben und Leistung eines deutschen Forschers im Zeitalter der Reichsgründung." *Deutschlands Erneuerung* 15:505–15.

Schlechta, Karl, and Anni Anders. 1962. *Friedrich Nietzsche: Von den verborgenen Anfängen seines Philosophierens*. Stuttgart-Bad Cannstatt: Friedrich Frommann Verlag.

Schlotte, Felix. 1955–56. "Beiträge zum Lebensbild Wilhelm Wundts aus seinem Briefwechsel." *Wissenschaftliche Zeitschrift der Karl-Marx-Universität Leipzig: Gesellschafts- und Sprachwissenschaftliche Reihe* 5:333–49.

Schmidgen, Henning. 2002. "Of Frogs and Men: The Origins of Psychophysiological Time Experiments, 1850–1865." *Encounter* 26:142–48.

———. 2003. "Time and Noise: The Stable Surroundings of Reaction Experiments, 1860–1890." *Studies in History and Philosophy of Biological and Biomedical Sciences* 34:237–75.

———. 2005. "Repetitions and Differences: Psychophysiological Time Machines, 1850–1865." In *Variantology 1: On Deep Time Relations of Arts, Sciences and Technologies*, edited by Siegfried Zielinski and Silvia M. Wagnermaier, 145–57. Cologne: Walther König.

———. 2009. *Die Helmholtz-Kurven: Auf der Spur der verlorenen Zeit*. Berlin: Merve.

———. 2015. "Leviathan and the Myograph: Hermann Helmholtz's 'Second Note' on the Propagation Speed of Nervous Stimulations." *Science in Context* 28 (3): 357–96.

Schmitz, Emil-Heinz. 1954. "Vor 60 Jahren Starb Hermann von Helmholtz." *Der Augenoptiker* 9:11–13.

Schnädelbach, Herbert. 1991. *Philosophie in Deutschland 1831–1933*. Frankfurt am Main: Suhrkamp.

Schobess, Joachim. 1962. "Den Vater rügte die Reaktion . . . der Sohn aber setzte sich durch." *Brandenburgische Neueste Nachrichten*, no. 100, 1 May, 7.

Schoen, Max. 1921. "Helmholtz: Zum hundertjährigen Geburtstag des grossen Naturforschers." *Die Neue Zeit* 2:505–10.

Schoffa, G. 1954. "Vor 60 Jahren Starb Hermann von Helmholtz." *Tägliche Rundschau*, 9 September, n.p.

Scholz, Erhard. 1980. *Geschichte des Mannigfaltigkeitsbegriffs von Riemann bis Poincaré*. Boston: Birkhäuser.

Schomerus, Friedrich. 1941. "Abbe und Helmholtz." *Zeiss-Werk-Zeitung* 16:101–3.

Schopenhauer, Arthur. 1873. *Arthur Schopenhauer's Sämmtliche Werke*: vol. 1, *Schriften zur Erkenntnisslehre*. Edited by Julius Frauenstädt. Leipzig: F. A. Brockhaus.

Schramm, Michael. 1998. *Otto Jahns Musikästhetik und Musikkritik*. Essen: Die Blaue Eule.

Schreier, Wolfgang, and Martin Franke. 1995. "Geschichte der Physikalischen Gesellschaft zu Berlin 1845–1900." *PB* 51 (1): F9–F59.

Schröder, Hermann. 1921. "Helmholtz. (Zu seinem hundersten Geburtstage am 31. August.)." *Münchener Medizinische Wochenschrift* 68:1086–87.

Schröder, Wilfried, and Karl-Heinrich Wiederkehr. 2000. "Johann Kiessling, the Krakatoa Event, and the Development of Atmospheric Optics after 1883." *Notes and Records of the Royal Society of London* 54:2:249–58.

Schröer, Heinz. 1967. *Carl Ludwig: Begründer der messenden Experimentalphysiologie 1816–1895*. Stuttgart: Wissenschaftliche Verlagsgesellschaft.

Schubring, Gustav. 1872. "Biographisches: Hermann Helmholtz." *Musikalisches Wochenblatt* 3:5–7, 51–2, 67–70.

Schüle, Prof. 1933. "Zwei Briefe aus dem Leben von Hermann v. Helmholtz." *Psychiatrisch-Neurologische Wochenschrift* 35:618–19.

Schüler, Winfried. 1971. *Der Bayreuther Kreis von seiner Entstehung bis zum Ausgang der Wilhelminischen Ära: Wagnerkult und Kulturreform im Geiste völkischer Weltanschauung*. Münster: Verlag Aschendorff.

Schulprogramm. 1825. *Zu der öffentlichen Prüfung, welche in dem hiesigen Gymnasium . . . ladet . . . ein J.S. Büttner, Rector*. Potsdam: N.p.

———. 1827. *Zu der öffentlichen Prüfung, welche in dem hiesigen Gymnasium . . . ladet . . . ein J.S. Büttner, Rector*. Potsdam: N.p.

———. 1828. *Guilielmi Arminii Blumii Oratio Ad Munus Directoris Rite Auspicandum D. XV. Octob. MDCCCXXVII Habita. Einladungsschrift zu den diesjährigen öffentlichen Prüfungen in dem Königlichen Gymnasium zu Potsdam. . . .* Potsdam: N.p.

———. 1829. *Zur Theilnahme an der diesjährigen öffentlichen Schulprüfung in dem hiesigen königlichen Gymnasium. . . .* Potsdam: Deckerscher Geh. Oberhofbuchdruckerei.

———. 1831. *Zu der öffentlichen Prüfung der Zöglinge des hiesigen Königlichen Gymnasiums. . . .* Potsdam: Deckerscher Geh. Oberhofbuchdruckerei.

———. 1832. *Zu der öffentlichen Prüfung der Zöglinge des hiesigen Königlichen Gymnasiums. . . .* Potsdam: Deckerscher Geh. Oberhofbuchdruckerei.

———. 1833. *Zu der öffentlichen Prüfung der Zöglinge des hiesigen Königlichen Gymnasiums. . . .* Potsdam: Deckerscher Geh. Oberhofbuchdruckerei.

———. 1835. *Zu der öffentlichen Prüfung der Zöglinge des hiesigen Königlichen Gymnasiums. . . .* Potsdam: Deckerscher Geh. Oberhofbuchdruckerei.

———. 1837. *Zu der öffentlichen Prüfung der Zöglinge des hiesigen Königlichen Gymnasiums. . . .* Potsdam: Deckerscher Geheimen Oberhofbuchdruckerei.

———. 1839. *Zur Feier des Säcularfestes des hiesigen Königlichen Gymnasiums. . . .* Potsdam: Deckerscher Geh. Oberhofbuchdruckerei.

———. 1857. *Zu der öffentlichen Prüfung der Zöglinge des diesigen Gymnasiums, am 7ten April 1857. . . .* Potsdam: N.p.

Schultze, Winfried, comp. 1993. "Verzeichnis der Vorlesungen und Übungen von Hermann Helmholtz an der Berliner Universität vom Sommersemester 1871 bis zum Sommersemester 1894." HUB.

———, comp. 1994. "Dokumentation anlässlich des wissenschaftlichen Kolloquiums zum 100. Todestag von Hermann von Helmholtz am 8. September 1994 an der Humboldt-Universität zu Berlin." HUB.

Schulze, Hermann. 1886. *Robert von Mohl: Ein Erinnerungsblatt, dargebracht zur fünfhundertjährigen Jubelfeier der Ruperto-Carola*. Heidelberg: Carl Winter.

Schuster, Arthur. 1932. *Biographical Fragments*. London: Macmillan.

———. 1975. *The Progress of Physics during 33 Years (1875–1908)*. New York: Arno Press.

Schwarz, Ingo, and Klaus Wenig, eds. 1997. *Briefwechsel zwischen Alexander von Humboldt und Emil du Bois-Reymond*. Berlin: Akademie Verlag.

Schwertschlager, Joseph. 1883. *Kant und Helmholtz erkenntniss-theoretisch verglichen*. Freiburg im Breisgau: Herder.

Sechenov, I. M. 1965. *Autobiographical Notes*. Edited by Donald B. Lindsley. Washington, DC: American Institute of Biological Sciences.

Seiler, Emma. 1868. *The Voice in Singing*. Philadelphia: J. B. Lippincott.

———. 1875. *The Voice in Speaking*. Philadelphia: J. B. Lippincott.

Seth, Suman. 2003. "Principles and Problems: Constructions of Theoretical Physics in Germany, 1890–1918." PhD diss., Princeton University.

Shastid, Thomas Hall. 1928. "Knapp, Jacob Hermann." In *Dictionary of American Medical Biography*, edited by Howard A. Kelly and Walter L. Burrage, 706–8. New York: D. Appleton.

Sheehan, James J. 1993. *German History, 1770–1866*. Oxford: Clarendon Press.

Shinn, Terry. 1979. "The French Science Faculty System, 1808–1914: Institutional Change and Research Potential in Mathematics and the Physical Sciences." *Historical Studies in the Physical Sciences* 10:271–332.

Shuttleworth, Sally. 1984. *George Eliot and Nineteenth-Century Science: The Make-Believe of a Beginning*. Cambridge: Cambridge University Press.

Siebel, Ernst. 1999. *Der grossbürgerliche Salon 1850–1918: Geselligkeit und Wohnkultur*. Berlin: Dietrich Reimer.

Siebmacher, Johann. 1981. *Die Wappen des preussischen Adels*. Part 2 of 2. Reprint, Neustadt an der Aisch: Bauer & Raspe.

Siemens, Werner von. 1916. *Ein kurzgefasstes Lebensbild nebst einer Auswahl seiner Briefe: Aus Anlass der 100. Wiederkehr seines Geburtstages*. Edited by Conrad Matschoss. 2 vols. Berlin: Julius Springer.

Siemens-Helmholtz, Ellen von (ESH). 1921. "Mein Elternhaus." In ESH et al. 1921, n.p.

——— (ESH), ed. 1929. *Anna von Helmholtz: Ein Lebensbild in Briefen*. 2 vols. Berlin: Verlag für Kulturpolitik.

Siemens-Helmholtz, Ellen von (ESH), Richard Greef, Otto Lummer, Rudolf Rothe, and Carl Stumpf. 1921. "Helmholtz zum 100: Geburtstag." *Erinnerungsblatt der Vossischen Zeitung*, 28 August, no. 404, suppl. 4.

Silliman, Benjamin. 1853. *A Visit to Europe in 1851*. 2 vols. New York: G. P. Putnam.

Silliman, Robert H. 1963. "William Thomson: Smoke Rings and Nineteenth-Century Atomism." *Isis* 54:461–74.

Simon, Hans-Reiner, ed. 1980. *Anton Dohrn und die Zoologische Station Neapel*. Frankfurt am Main: Edition Erbrich.

Simonsohn, Gerhard. 2007. "Die Deutsche Physikalische Gesellschaft und die Forschung." In *Physiker zwischen Autonomie und Anpassung*, edited by Dieter Hoffmann and Mark Walker, 237–301. Weinheim: Wiley-VCH.

Simpson, M. C. M. 1887. *Letters and Recollections of Julius and Mary Mohl*. London: Kegan Paul, Trench.

Singer, Sandra L. 2003. *Adventures Abroad: North American Women at German-Speaking Universities, 1868–1915*. Westport, CT: Praeger.

Sloane, T. O'Conor. 1920. *Liquid Air and the Liquefaction of Gases*. New York: Norman W. Henley.

Smith, Crosbie. 1998. *The Science of Energy: A Cultural History of Energy Physics in Victorian Britain*. Chicago: University of Chicago Press.

Smith, Crosbie, and M. Norton Wise. 1989. *Energy and Empire: A Biographical Study of Lord Kelvin*. Cambridge: Cambridge University Press.

Snyder, Charles. 1964. "Helmholtz at Columbia University." *Archives of Ophthalmology* 72: 573–76.

Sobotta, Johannes. 1933. "Das Anatomische Institut." In *Geschichte der Rheinischen Friedrich-Wilhelms-Universität zu Bonn am Rhein*: vol. 2, *Institute und Seminare, 1818–1933*, edited by Friedrich von Bezold, 56–71. Bonn: Marcus, Cohen.

Sorbelli, Albano. 1936. *Inventari dei manoscritti delle biblioteche d'Italia*. Florence: Olschki.

Spiero, Heinrich. 1921. *Julius Rodenberg: Sein Leben und seine Werke*. Berlin: Gebrüder Paetel.

Spillmann, Jutta, and Lothar Spillmann. 1993. "The Rise and Fall of Hugo Münsterberg." *Journal of the History of the Behavioral Sciences* 29:322–38.

Spitzemberg, Hildegard von Varnbüler. 1963. *Das Tagebuch der Baronin Spitzemberg, geb. Freiin v. Varnbüler: Aufzeichnungen aus der Hofgesellschaft des Hohenzollernreiches*. Edited by Rudolf Vierhaus. 3rd ed. Göttingen: Vandenhoeck & Ruprecht.

Spitzka, Edw. Anthony. 1908. "A Study of the Brains of Six Eminent Scientists and Scholars Belonging to the American Anthropometric Society, Together with a Description of the Skull of Professor E.D. Cope." *Transactions of the American Philosophical Society* 21: 175–308.

Spotts, Frederic. 1994. *Bayreuth: A History of the Wagner Festival*. New Haven, CT: Yale University Press.

Springer, Robert. 1878. *Berlin die deutsche Kaiserstadt nebst Potsdam und Charlottenburg mit ihren schönsten Bauwerken und hervorragendsten Monumenten: Eine malerische Wanderung in Buch und Bild für Einheimische und Fremde*. Darmstadt: Friedrich Lange.

Stadler, Friedrich. 2001. *The Vienna Circle: Studies in the Origins, Development, and Influence of Logical Empiricism*. Vienna: Springer.

Staubermann, Klaus. 2001. "Tying the Knot: Skill, Judgement, and Authority in the 1870s Leipzig Spiritistic Experiments." *British Journal for the History of Science* 34:67–79.

Steege, Benjamin. 2012. *Helmholtz and the Modern Listener*. Cambridge: Cambridge University Press.

Steif, Yvonne. 2003. *Wenn Wissenschaftler Feiern: Die Versammlungen deutscher Naturforscher und Ärzte 1822 bis 1913*. Stuttgart: Wissenschaftliche Verlagsgesellschaft.

Stern, Alfred. 1932. *Wissenschaftliche Selbstbiographie*. Zurich: A.-G. Gebr. Leemann.

Stern, Fritz. 1999. *Einstein's German World*. Princeton, NJ: Princeton University Press.

Steudel, Johannes. 1941. "Hermann Helmholtz. Zur 120. Wiederkehr des Geburtstages des grossen Physikers." *Wiener Pharmazeutische Wochenschrift* 74:371–72.

Stevens, Henry. 1863. *The Humboldt Library: A Catalogue of the Library of Alexander von Humboldt*. London: Henry Stevens.

Stieda, L. 1890. "Zur Geschichte der physikalisch-ökonomischen Gesellschaft: Festrede, gehalten am 22. Februar 1890." *Schriften der Physikalisch-Öknomischen Gesellschaft zu Königsberg in Pr.* 31:38–84.

Stiefel, Karl. 1977. *Baden 1648–1952*. 2 vols. Karlsruhe: Verein für oberrheinische Rechts- und Verwaltungsgeschichte.

Stocking, George W., Jr. 1982. "From Physics to Ethnology." In *Race, Culture and Evolution: Essays in the History of Anthropology*, 133–60. Chicago: University of Chicago Press.

Strauss, David Friedrich. 1872. *Der alte und der neue Glaube: Ein Bekenntnis*. 2nd ed. Leipzig: S. Hirzel.

Stromberg, Wayne H. 1989. "Helmholtz and Zoellner: Nineteenth-Century Empiricism, Spiritism, and the Theory of Space Perception." *Journal of the History of the Behavioral Sciences* 25:371–83.

Stubhaug, Arild. 2002. *The Mathematician Sophus Lie: It Was the Audacity of My Thinking*. Berlin: Springer.

Stübler, Eberhard. 1926. *Geschichte der medizinischen Fakultät der Universität Heidelberg 1386–1925*. Heidelberg: Carl Winters Universitätsbuchhandlung.

Study, Eduard. 1914. *Die realistische Weltansicht und die Lehre vom Raume: Geometrie, Anschauung, und Erfahrung*. Braunschweig: Friedr. Vieweg.

Stumpf, Carl. 1921. "Sinnespsychologie und Musikwissenschaft." In ESH et al. 1921, n.p.

Sudhoff, Karl. 1922. *Hundert Jahre Deutscher Naturforscherversammlungen: Gedächtnisschrift zur Jahrhundert-Tagung der Gesellschaft Deutscher Naturforscher und Ärzte Leipzig, im September 1922*. Leipzig: F. C. W. Vogel.

———. 1922a. *Rudolf Virchow und die Deutschen Naturforscherversammlungen*. Leipzig: Akademische Verlagsgesellschaft.

Sulloway, Frank J. 1979. *Freud: Biologist of the Mind: Beyond the Psychoanalytic Legend*. New York: Basic Books.

Swoboda, Wolfram W. 1988. "Physik, Physiologie und Psychophysik—Die Wurzeln von Ernst Machs Empiriokritizismus." In *Ernst Mach—Werk und Wirkung*, edited by Rudolf Haller and Friedrich Stadler, 356–403. Vienna: Verlag Hölder-Pichler-Tempsky.

Sz., 1921. "Zum 100sten Geburtstage H. von Helmholtz." *Zeitschrift des Vereines Deutscher Ingenieure* 65:952.

Taine, Hippolyte. 1870. *De l'intelligence*. 2 vols. Paris: Librairie Hachette.

Tait, Peter Guthrie. 1868. *Sketch of Thermodynamics*. Edinburgh: Edmonston and Douglas.

———. 1877. *Sketch of Thermodynamics*. 2nd rev. ed. Edinburgh: David Douglas.

Taylor, Sedley. 1873. *Sound and Music: A Non-Mathematical Treatise on the Physical Constitution of Musical Sounds and Harmony, Including the Chief Acoustical Discoveries of Professor Helmholtz*. London: Macmillan.

Terrall, Mary. 2002. *The Man Who Flattened the Earth: Maupertuis and the Sciences in the Enlightenment*. Chicago: University of Chicago Press.

Thayer, James Bradley. 1878. *Letters of Chauncey Wright with Some Account of His Life*. Cambridge, MA: Press of John Wilson.

Thiele, Joachim. 1978. *Wissenschaftliche Kommunikation: Die Korrespondenz Ernst Machs*. Kastellaun: A. Henn.

Theunissen, Bert. 2000. "Turning Refracting into a Science: F.C. Donders' 'Scientific Reform' of Lens Prescription." *Studies in History and Philosophy of Biological and Biomedical Sciences* 31 (4): 557–78.

Thompson, Silvanus P. 1910. *The Life of William Thomson, Baron Kelvin of Largs*. 2 vols. London: Macmillan.

Thomson, James. 1912. *Collected Papers in Physics and Engineering: Selected and Arranged with Unpublished Material and Brief Annotations by Sir Joseph Larmor . . . and James Thomson*. Cambridge: Cambridge University Press.

Thomson, William. 1882. *Mathematical and Physical Papers*. 6 vols. Cambridge: Cambridge University Press.

———. 1889–94. *Popular Lectures and Addresses*. 3 vols. London: Macmillan.

———. 1894. "The Anniversary Meeting of the Royal Society." *Nature* 51:132–34.

Thomson, W., and P. G. Tait. 1871. *Handbuch der theoretischen Physik*. Translated by H. Helmholtz and G. Wertheim. Vol. 1, part 1. Braunschweig: Friedrich Vieweg.

———. 1874. *Handbuch der theoretischen Physik*. Translated by H. Helmholtz and G. Wertheim. Vol. 1, part 2. Braunschweig: Friedrich Vieweg.

Thouret, Georg. 1903. *Helmholtz-Realgymnasium in Schöneberg: Jahresbericht I. Ostern 1903*. Schöneberg: Emil Hartmann.

Tiemann, Klaus-Harro. 1991. "Institutionen und Medien zur Popularisierung wissenschaftlicher Kenntnisse in Deutschland zwischen 1800 und 1933 (ein skizzenhafter Überblick)." In *Probleme der Kommunikation in den Wissenschaften*, edited by Annette Vogt, 165–84. Berlin: N.p.

Tilden, William A. 1930. *Famous Chemists: The Men and Their Work*. London: G. Routledge.

"Titelverzeichniss sämmtlicher Veröffentlichungen von Hermann von Helmholtz." 1895. In HH, *WA*, 3:605–36.

Titze, Hartmut. 1995. *Wachstum und Differenzierung der deutschen Universitäten 1830–1945*. Göttingen: Vandenhoeck & Ruprecht.

Toepke, Gustav, comp. 1907. *Die Matrikel der Universität Heidelberg . . . , Sechster Teil von 1846 bis 1870 fortgesetzt und herausgegeben . . . von Paul Hintzelmann. . . .* Heidelberg: Carl Winter's Universitätsbuchhandlung.

Torge, W. 2005. "The International Association of Geodesy 1862 to 1922: From a Regional Project to an International Organization." *Journal of Geodesy* 78:558–68.

Torretti, Roberto. 1978. *Philosophy of Geometry from Riemann to Poincaré*. Dordrecht: D. Reidel.

Treiber, Hubert. 1994. "Zur 'Logik des Traumes' bei Nietzsche: Anmerkungen zu den Traum-Aphorismen aus Menschliches, Allzumenschliches." In *Nietzsche-Studien: Internationales Jahrbuch für die Nietzsche-Forschung*, edited by Ernst Behler, Eckhard Heftrich, and Wolfgang Müller-Lauter, 23:1–41. Berlin: Walter de Gruyter.

Treitschke, Heinrich von. 1914–20. *Heinrich von Treitschkes Briefe*. Edited by Max Cornicelius. 4 vols. in 3. 2nd ed. Leipzig: S. Hirzel.

———. 1927. *Deutsche Geschichte im Neunzehnten Jahrhundert*. 5 vols. Leipzig: G. Hirzel.

Tschermak, A. von. 1919. "Julius Bernstein's Lebensarbeit: Zugleich ein Beitrag zur Geschichte der neueren Biophysik." *Pflüger's Archiv für Physiologie* 174:1–89.

Tscherning, M. 1910. *Hermann von Helmholtz und die Akkommodationstheorie*. Leipzig: Johann Ambrosius Barth.

Tschirch, A[lexander]. 1921. *Erlebtes und Erstrebtes: Lebenserinnerungen*. Bonn: Friedrich Cohen.

Tuchman, Arleen. 1987. "Experimental Physiology, Medical Reform, and the Politics of Education at the University of Heidelberg: A Case Study." *Bulletin of the History of Medicine* 61: 203–15.

———. 1988. "From the Lecture to the Laboratory: The Institutionalization of Scientific Medicine at the University of Heidelberg." In *The Investigative Enterprise: Experimental Physiology in Nineteenth-Century Medicine*, edited by William Coleman and Frederic L. Holmes, 65–99. Berkeley: University of California Press.

———. 1993. "Helmholtz and the German Medical Community." In Cahan 1993a, 17–49.

———. 1993a. *Science, Medicine, and the State in Germany: The Case of Baden, 1815–1871*. New York: Oxford University Press.

Turgenev, Ivan. 1972. *Lettres inédites de Tourgueney à Pauline Viardot et à sa famille*. Edited by Henri Granjard and Alexandre Zviguilsky. Lausanne: Editions L'Age d'Homme.

Turner, Frank M. 1993. *Contesting Cultural Authority: Essays in Victorian Intellectual Life*. Cambridge: Cambridge University Press.

Turner, R. Steven. 1977. "The Ohm-Seebeck Dispute, Hermann von Helmholtz, and the Origins of Physiological Acoustics." *British Journal for the History of Science* 10:1–24.

———. 1982. "Helmholtz, Sensory Physiology, and the Disciplinary Development of German Psychology." In *The Problematic Science: Psychology in Nineteenth-Century Thought*, edited by William R. Woodward and Mitchell G. Ash, 147–66. New York: Praeger.

———. 1987. "Paradigms and Productivity: The Case of Physiological Optics, 1840–94." *Social Studies of Science* 17:35–68.

———. 1993. "Consensus and Controversy: Helmholtz on the Visual Perception of Space." In Cahan 1993a, 154–204.

———. 1994. *In the Eye's Mind: Vision and the Helmholtz-Hering Controversy*. Princeton, NJ: Princeton University Press.

———. 1996. "The Origins of Colorimetry: What *Did* Helmholtz and Maxwell Learn from

Grassmann?" In *Hermann Günther Grassmann (1809–1877): Visionary Mathematician, Scientist and Neohumanist Scholar*, edited by Gert Schubring, 71–85. Dordrecht: Kluwer Academic.

Twain, Mark. 1896. *The American Claimant, and Other Stories and Sketches*. New York: Harper.

———. 1935. *Mark Twain's Notebook*. 2nd ed. Edited by Albert Bigelow Paine. New York: Harper.

Tyndall, John. 1867. *Die Wärme betrachtet als eine Art der Bewegung*. Edited by H. Helmholtz and G. Wiedemann. Braunschweig: Friedrich Vieweg.

———. 1869. *Der Schall*. Edited by H. Helmholtz and G. Wiedemann. Braunschweig: Friedrich Vieweg.

———. 1870. *Faraday und seine Entdeckungen: Eine Gedenkschrift*. Edited by H. Helmholtz. Braunschweig: Friedrich Vieweg.

———. 1873. *Hours of Exercise in the Alps*. New York: D. Appleton.

———. 1873a. *Les glaciers et les transformations de l'eau par J. Tyndall . . . suivis d'une conférence sur le même sujet par M. Helmholtz, avec la réponse de M. Tyndall [et d'une lettre de Helmholtz à Em. Alglave]. . . .* Bibliothèque scientifique internationale, vol. 1. Paris: Librairie Germer Baillière.

[———]. 1873b. "Professor Tyndall's Deed of Trust." *Popular Science Monthly* 3:100–101.

———. 1874. *Advancement of Science: The Inaugural Address of Prof. John Tyndall, D.C.L., LL.D., F.R.S., Delivered before the British Association for the Advancement of Science, at Belfast, August, 19, 1874: With Portrait and Biographical Sketch: Opinions of the Eminent Scientist, Prof. H. Helmholtz, and Articles of Prof. Tyndall and Henry Thompson on Prayer*. New York: Asa K. Butts.

———. 1876. *Das Licht: Sechs Vorlesungen gehalten in Amerika im Winter 1872–1873*. Edited by Gustav Wiedemann. Translated by Clara Wiedemann. Braunschweig: Friedrich Vieweg.

———. 1898. "The Belfast Address." In Tyndall 1898a, 2:135–201.

———. 1898a. *Fragments of Science: A Series of Detached Essays, Addresses, and Reviews*. 2 vols. New York: D. Appleton.

———. 1898–99. *Fragments of Science: Fragmente aus den Naturwissenschaften: Vorlesungen und Aufsätze: Zweite autorisirte Deutsche Ausgabe*. From the 8th English edition. Translated by A. von Helmholtz and E. Du Bois–Reymond. 2 vols. Braunschweig: Friedrich Vieweg.

———. 1904. Introduction to *Hermann von Helmholtz, Popular Lectures on Scientific Subjects*. Translated by E. Atkinson. 1st ser. London: Longmans, Green.

Tzschaschel, Ingeborg. 1994. "Reichskanzler der Wissenschaften: Hermann von Helmholtz und seine Heidelberger Zeit von 1858 bis 1871." *Rhein-Neckar Zeitung*, 8 March, 16.

Uhthoff, Wilhelm. 1902. "Bemerkungen zur Erfindung des Augenspiegels vor 50 Jahren." In *Bericht über die neunundzwanzigste Versammlung der ophthalmologischen Gesellschaft: Heidelberg 1901*, edited by A. Wagenmann, 3–8. Wiesbaden: J. F. Bergmann.

"Unsere Physikbilder." 1944. *PB* 1: following 52, 69.

Vasili, Comte Paul. 1884. *La société de Berlin*. Paris: Nouvelle revue.

Vilhard, Jacob, and Emil Fischer. 1902. *August Wilhelm von Hofmann: Ein Lebensbild, im Auftrage der Deutschen chemischen Gesellschaft*. Berlin: R. Friedländer.

Villard, Henry. 1904. *Memoirs of Henry Villard: Journalist and Financier, 1835–1900*. Boston: Houghton, Mifflin.

Virchow, Rudolf. 1866. "Ueber die nationale Entwickelung und Bedeutung der Naturwissenschaften." In *Amtlicher Bericht über die vierzigste Versammlung deutscher Naturforscher und Ärzte zu Hannover im September 1865*, edited by C. Krause and K. Karmarsch, 56–65. Hannover: Hahn'sche Hofbuchhandlung.

———. 1871. ["Ueber die Aufgaben der Naturwissenschaften in dem neuen nationalen Leben Deutschlands"]. In *Tageblatt der 44. Versammlung Deutscher Naturforscher und Ärzte in Ros-*

tock vom 18. bis 24. September 1871, edited by H. Aubert and W. Flemming, 73–81. Rostock: Universitäts-Buchdruckerei von Adler's Erben.

———. 1873. Untitled speech in the Prussian *Landtag* on 17 January 1873. In *Stenographische Berichte über die Verhandlungen der . . . beiden Häuser des Landtages: Haus der Abgeordneten: 1872–73*, 629–35. Vol. 1. Berlin: W. Moeser.

———. 1877. *Die Freiheit der Wissenschaft im modernen Staat*. Berlin: Wiegandt, Hempel, & Parey.

———. 1906. *Rudolf Virchow: Briefe an seine Eltern 1839 bis 1864*. Edited by Marie Rabl. Leipzig: Wilhelm Engelmann.

Virtanen, Reino. 1973. "Paul Valéry's Scientific Education." *Symposium* 27:362–78.

———. 1974. *The Scientific Analogies of Paul Valéry*. Lincoln: University of Nebraska.

Vitz, Paul C., and Arnold B. Glimcher. 1984. *Modern Art and Modern Science: The Parallel Analysis of Vision*. New York: Praeger.

Voelke, Jean-Daniel. 2005. *Renaissance de la géometrie non euclidienne entre 1860 et 1900*. Bern: Peter Lang.

Vogel, Stephan. 1993. "Sensation of Tone, Perception of Sound, and Empiricism: Helmholtz's Physiological Acoustics." In Cahan 1993a, 259–87.

Vogt, Annette. 1994. "Hermann von Helmholtz' Beziehungen zu russischen Gelehrten." In Krüger 1994a, 66–86.

Volkert, Klaus. 1993. "On Helmholtz' Paper 'Ueber die thatsächlichen Grundlagen der Geometrie.'" *Historia Mathematica* 20:307–9.

Vorländer, Karl. 1977. *Immanuel Kant: Der Mann und das Werk*. 2nd ed. Hamburg: Felix Meiner.

Wachsmuth, R. 1900. "Von Helmholtz, Anna, geb. von Mohl." In *Biographisches Jahrbuch und Deutscher Katalog*, edited by Anton Bettelheim, 4:14–20. 18 vols. Berlin: Georg Reimer.

Wade, Nicholas J., and Stanley Finger. 2001. "The Eye as an Optical Instrument: From Camera Obscura to Helmholtz's Perspective." *Perception* 30:1157–77.

Wagner, Cosima. 1940. *Die Briefe Cosima Wagners an Friedrich Nietzsche*. 2 vols. Edited by Erhart Thierbach. Weimar: Nietzsche-Archiv.

———. 1976–77. *Die Tagebücher*. Edited by Martin Gregor-Dellin and Dietrich Mack. 2 vols. Munich: R. Piper.

Wagner, Kurt. 1965. "Zur richtigen Deutung der Helmholtzschen Zeichentheorie." *Deutsche Zeitschrift für Philosophie* 13:162–72.

Wagner, Richard. 1912. "An Helmholtz." In *Richard Wagner, Sämtliche Schriften und Dichtungen*. Edited by Richard Sternfeld. 12:387. 6th ed. 12 vols. Leipzig: Breitkopf & Härtel.

Wagner, Rudolf. 1997. *Physiologische Briefe (1851–1852)*. Edited, annotated, and with an introduction by Norbert Klatt. Reprint, Göttingen: Norbert Klatt.

Waldeyer-Hartz, Wilhelm von. 1920. *Lebenserinnerungen*. Bonn: Friedrich Cohen.

Warburg, Emil, ed. 1915. *Physik*. Vol. 1. Leipzig: B. G. Teubner. Also published as *Die Kultur der Gegenwart: Ihre Entwicklung und Ihre Ziele*: Third part, *Mathematik, Naturwissenschaften, Medizin*, Third division, *Anorganische Naturwissenschaften*, edited by Paul Hinneberg and E. Warburg. Leipzig: B. G. Teubner.

———. 1925. "Zur Geschichte der Physikalischen Gesellschaft." *Die Naturwissenschaften* 13: 35–39.

———. 1975. "Antrittsrede." In Kirsten and Körber 1975, 182–84.

Warburg, Emil, Max Rubner, and Moritz Schlick. 1922. *Helmholtz als Physiker, Physiologe und Philosoph: Drei Vorträge gehalten zur Feier seines 100. Geburtstags im Auftrage der Physikalischen, der Physiologischen und der Philosophischen Gesellschaft zu Berlin*. Karlsruhe: C. F. Müllersche Hofbuchhandlung.

Ward, W. R. 1965. *Victorian Oxford*. London: Frank Cass.

Weber, Beate. 1997. "Geleitwort." In Werner 1997, vii.

Weber, Georg. 1886. *Heidelberger Erinnerungen: Am Vorabend der fünften Säkularfeier der Universität*. Stuttgart: J. G. Cotta'schen Buchhandlung.

Weber, Max. 1958. *The Rational and Social Foundations of Music*. Carbondale: Southern Illinois University Press.

Weech, Friedrich von. 1877. *Baden in den Jahren 1852 bis 1877: Festschrift zum fünfundzwanzigjährigen Regierungs-Jubiläum Seiner Königlichen Hoheit des Grossherzogs Friedrich*. Karlsruhe: A. Bielefeld's Hofbuchhandlung.

———. 1890. *Badische Geschichte*. Karlsruhe: A. Bielefeld's Hofbuchhandlung.

———. 1906. "Anna von Helmholtz." In *Badische Biographien*: Part 5, *1891–1901*. Edited by Fr. von Weech and U. Krieger, 294–301. Heidelberg: Carl Winter's Universitätsbuchhandlung.

Weisfert, Julius Nicolaus. 1975. *Biographisch-litterarisches Lexikon für die Haupt- und Residenzstadt Königsberg und Ostpreussen*. Königsberg: Gustav Schadlofsky; Hildesheim: Georg Olms.

Weisz, George. 1983. *The Emergence of Modern Universities in France, 1863–1914*. Princeton, NJ: Princeton University Press.

"Welcome to Prof. Helmholtz: Reception to the German Scientist at Columbia College." 1893. *New York Times*, 4 October, 5.

Wenig, Hans Günter. 1969. "Medizinische Ausbildung im 19. Jahrhundert." PhD diss., University of Bonn.

Wenkel, Simone. 2007. "Jewish Scientists in German-Speaking Academia: An Overview." In *Jews and Scientists in German Contexts: Case Studies from the 19th and 20th Centuries*, edited by Ulrich Charpa and Ute Deichmann, 265–95. Tübingen: Mohr Siebeck.

Werner, Franz. 1996. "Die Berufung von Hermann Helmholtz an die Universität in Heidelberg." In Eckart and Volkert 1996, 63–96.

———. 1997. *Hermann Helmholtz' Heidelberger Jahre (1858–1871)*. Berlin: Springer.

———. 1998. "Zum Tod des Physiologen und Physikers Hermann von Helmholtz." *Zeitschrift für die Geschichte des Oberrheins* 146:544–51.

Werner, Petra. 2004. *Himmel und Erde: Alexander von Humboldt und sein Kosmos*. Berlin: Akademie Verlag.

Werner, Petra, and Angelika Irmscher, eds. 1993. *Kunst und Liebe müssen sein: Briefe von Anna von Helmholtz an Cosima Wagner 1889 bis 1899*. Bayreuth: Druckhaus Bayreuth.

Weve, H. J. M., and G. ten Doesschate, eds. 1935. *Die Briefe von Albrecht von Graefe's an F.C. Donders (1852–1870)*. Stuttgart: Ferdinand Enke.

Weyrauch, Jacob J., ed. 1893. *Kleinere Schriften und Briefe von Robert Mayer: Nebst Mittheilungen aus seinem Leben*. Stuttgart: J. G. Cotta.

Wiedemann, Hans-Rudolf. 1989. *Briefe grosser Naturforscher und Ärzte in Handschriften*. Lübeck: Verlag Graphische Werkstätten.

Wien, Wilhelm. 1921. "Helmholtz als Physiker." *Die Naturwissenschaften* 9:694–99.

———. 1921a. "Hermann von Helmholtz." *Zeitschrift des Vereines Deutscher Ingenieure* 65: 1239–40.

———. 1930. *Aus dem Leben und Wirken eines Physikers: Mit persönlichen Erinnerungen von. . . .* Leipzig: Johann Ambrosius Barth.

Wiener, Otto. 1921. "Hermann v. Helmholtz." *Deutsche Revue* 46:97–114.

Wilhelmy, Petra. 1989. *Der Berliner Salon im 19. Jahrhundert (1780–1914)*. Berlin: Walter de Gruyter.

William II. 1926. *My Early Life*. London: Methuen.

———. 1927. *Aus meinem Leben, 1859–1888*. Berlin: K. F. Koehler.

Williams, Simon. 1994. *Richard Wagner and Festival Theatre*. Westport, CT: Greenwood Press.

Willstätter, Richard. 1958. *Aus meinem Leben: Von Arbeit, Musse und Freunden*. Edited by Arthur Stoll. 2nd ed. Weinheim: Verlag Chemie.

Wilson, David, ed. 1990. *The Correspondence between Sir George Gabriel Stokes and Sir William Thomson, Baron Kelvin of Largs*. 2 vols. Cambridge: Cambridge University Press.

Wirth, Irmgard. 1965. *Mit Adolph Menzel in Berlin*. Munich: Prestel-Verlag.

Wirzberger, Karl-Heinz, ed. 1973. "Hermann von Helmholtz' philosophische und naturwissenschaftliche Leistungen aus der Sicht der dialektischen Materialismus und der modernen Naturwissenschaften." *Wissenschaftliche Zeitschrift der Humboldt-Universität zu Berlin: Mathematisch-Naturwissenschaftliche Reihe* 22:279–361.

Wise, M. Norton. 1999. "Architectures for Steam." In *The Architecture of Science*, edited by Peter Galison and Emily Thompson, 107–40. Cambridge, MA: MIT Press.

———. 2007. "Neo-Classical Aesthetics of Art and Science: Hermann Helmholtz and the Frog-Drawing Machine." The Hans Rausing Lecture 2007. Uppsala University.

Wittich, Dieter. 1964. "Das Gesetz der sogenannten spezifischen Sinnesenergie und die beiden philosophischen Grundrichtungen." *Deutsche Zeitschrift für Philosophie* 12:682–91.

Wolff, Stefan L. 1999. "Leo Arons—Physiker und Sozialist." *Centaurus* 41:183–212.

———. 2012. "Jüdische oder nichtjüdische Deutsche?" In *Heinrich Hertz: Vom Funkensprung zur Radiowelle*, edited by Ralph Burmester and Andrea Niehaus, 43–57, 133–35. Bonn: Deutsches Museum Bonn.

Wolffram, Heather. 2009. *The Stepchildren of Science: Psychical Research and Parapsychology in Germany, c. 1870–1939*. Amsterdam: Rodopi.

Wolgast, Eike. 1985. "Das bürgerliche Zeitalter (1803–1918)." In Doerr 1985, 2:1–31.

———. 1985a. *Die Universität Heidelberg 1386–1986*. Berlin: Springer-Verlag.

Woodbury, David O. 1949. *A Measure for Greatness: A Short Biography of Edward Weston*. New York: McGraw-Hill.

Wuensch, Gerhard. 1977. "Hugo Riemann's Musical Theory." *Studies in Music* 2:108–24.

Wundt, Wilhelm. 1867. Review of *Handbuch der physiologischen Optik* (1867), by Hermann Helmholtz. *Deutsche Klinik* 19:326–28.

———. 1921. *Erlebtes und Erkanntes*. 2nd ed. Stuttgart: Alfred Kröner.

Wurtz, Adolphe. 1870. *Les hautes études pratiques dans les universités Allemandes: Rapport présenté à Son Exc. M. Le Ministre de L'Instruction Publique*. Paris: Imprimerie Impériale.

———. 1882. *Les hautes études pratiques dans les universités d'Allemagne et d'Autriche-Hongrie: Deuxième rapport présenté à M. Le Ministre de L'Instruction Publique*. Paris: G. Masson.

Youmans, Edward L., ed. 1865. *The Correlation and Conservation of Forces: A Series of Expositions by Prof. Grove, Prof. Helmholtz, Dr. Mayer, Dr. Faraday, Prof. Liebig and Dr. Carpenter*. New York: D. Appleton.

Zeller, Eduard. 1908. *Erinnerungen eines Neunzigjährigen*. Stuttgart: Uhland'schen Buchdruckerei.

Zenneck, Jonathan. 1944. "Zum 50. Todestag von Hermann von Helmholtz. 31. August 1821–8. September 1894." *Stahl und Eisen* 64:581–84.

Zimmermann, Peter. 1993. "Entwicklungslinien der Hörtheorie." *NTM: Zeitschrift für Naturwissenschaften, Technik und Medizin* 1:19–36.

Zimmermann, Robert. 1865. *Allgemeine Aesthetik als Formwissenschaft*. 2 vols. Vienna: Wilhelm Braumüller.

Zobeltitz, Fedor von. 1922. *Chronik der Gesellschaft unter dem letzten Kaiserreich*. 2 vols. Hamburg: Alster-Verlag.

Zobeltitz, Hanns von. 1891. "Eine Stunde bei Prof. v. Helmholtz." *Daheim* 27:768–70.

Zöllner, Johann Karl Friedrich. 1872. *Über die Natur der Cometen: Beiträge zur Geschichte und Theorie der Erkenntniss*. Leipzig: Wilhelm Engelmann.

———. 1878–81. *Wissenschaftliche Abhandlungen*. 4 vols. in 6 parts. Leipzig: L. Staackmann.

Zschau, Friedrich A. 1941. "Vor neunzig Jahren schenkte ein Deutscher der Welt den Augenspiegel: Zum 120. Geburtstag von Hermann von Helmholtz." *Leipziger Populäre Zeitschrift für Homöopathie* 72:112–14.

"Zu Helmholtz' Lehre von den Tonempfindungen." 1867. *Niederrheinische Musik* 15:25–28.

Zurmühl, Georg. 1930. "Abhängigkeit der Tonhöhenempfindung von der Lautstärke und ihre Beziehung zur Helmholtzschen Resonanztheorie des Hörens." *Zeitschrift für Sinnesphysiologie* 60:40–86.

Index

Note: References to figures are denoted by an "f" in italics following the page number.

academic freedom, 462; and the Dühring Affair, 516, 523, 530; Haeckel and Virchow on, 521–23, 530; Helmholtz on, 523–31; *Lehrfreiheit*, 42, 523, 528; *Lernfreiheit*, 42, 523; and the rise of the University of Berlin, 417; in the University of Heidelberg, 211–12. *See also* education; *Kulturkampf*; universities

Académie des Sciences, Paris, 281, 375–76, 389, 421, 489, 682

Academy of Arts, Berlin (Akademie der Künste), 62, 74, 77, 166–67, 421, 467–68

Academy of Sciences, Bavarian (Bayerische Akademie der Wissenschaften), 206, 218, 642

Academy of Sciences, Prussian (Preussische Akademie der Wissenschaften), 71–79, 92, 121, 150, 153, 185, 218, 232, 359, 389, 406–7, 421–36, 442, 447, 458, 484–88, 497, 515, 589–93, 600–617, 626–27, 639–40, 660, 668, 760, 762, 765. *See also* Prussia

Accademia dei Lincei, 280, 436, 561, 570, 588, 634

accommodation, theory of, 119–22, 150, 159, 167, 181–82, 444, 635–36, 648. *See also* eye movements; physiological optics; vision

acoustics, 2, 183, 186–88, 281, 293, 384, 696; architectural, 289; electricity and, 562; experimental, 144; and hearing, 186, 198–99, 265, 268, 274, 341, 389; Helmholtz on, 274, 280–88, 347, 377, 571, 693; mathematical, 458, 604; and medicine, 7; and music, 244, 262–64, 285–88, 416, 711, 714; and optics, 186, 338, 374, 377; physiological, 4, 186, 190–92, 197, 200–201, 222, 244, 262, 266, 319, 338, 345, 665; sea waves and, 697; of the telephone, 569; theoretical, 186; and tones, 183, 186–88, 197–98, 218–19, 264–76, 287–88, 310, 319, 370, 471, 480, 498, 562, 648–53, 697. *See also* combination tones; music; optics; physics; science; tones; vowels

Bonn, University of, 47–48, 211, 276, 388–90; a fiasco with, 347–57; jockeying for a position at, 164–68; laboratory facilities in, 215; unhappy intermezzo in, 171–207. *See also* Bonn; universities

botany, 17, 33, 125, 485, 640; in Berlin, 418; of Goethe, 124, 130. *See also* science; zoology

Bowman, William, 119, 123, 337, 343, 568, 574, 594–95. *See also* anatomy; medicine; ophthalmology

Brewster, David, 123, 126–28, 147, 240, 636, 642. *See also* optics; science

Britain, 567–68, 593–97, 647; the materialism dispute in, 157; the ophthalmoscope in, 122; physics in, 73; popularizing science in, 291–313; response to the *Tonempfindungen* in, 281–86; the Royal Institution in, 234; trips to, 131–32, 141, 146

British Association for the Advancement of Science (BAAS), 131–44, 156, 163, 174–75, 220, 309, 392, 494, 502–5, 568, 579, 582, 654, 683–85, 708, 731. *See also* Britain; science

Budge, Julius, 131, 134, 164–65, 173–74, 178–80, 189. *See also* physiology

Bülow, Hans von, 484. *See also* piano

Bunsen, Robert, 114–15, 132, 164, 175, 183, 188–90, 237–39, 253, 262, 265, 279, 297–98, 307, 342, 407, 436, 495, 559, 586, 650, 660; in Heidelberg, 203–17, 226–29, 392–94; and spectroscopy, 229, 237, 297–98. *See also* astronomy; chemistry; science

Burdach, Karl Friedrich, 86. *See also* anatomy; physiology

Butler, Nicholas Murray, 603, 715, 717. *See also* Columbia University; universities

Buys-Ballot, C. H. D., 591, 698. *See also* chemistry; meteorology

Cambridge, University of, 36, 141–42, 282, 285–86, 293, 372, 400–401, 434–38, 524–26, 568–71. *See also* universities

Campbell, Norman Robert, 630. *See also* mathematics; philosophy of science; physics

Cantor, Georg, 630. *See also* mathematics

Carnot, Lazare, 654. *See also* mathematics; physics

Carnot, Sadi, 68, 151, 385, 564, 654. *See also* engineering; physics; thermodynamics

Cauchy, Augustin-Louis, 376. *See also* mathematics; physics

causality: the concept of, 67; and free will, 248; and laws, 5, 69, 249, 329–30, 384, 495, 543–45, 750; and phenomena, 3, 33, 67, 384, 778n32; the principle of, 330; and science, 516; the ultimate account of, 69. *See also* epistemology; laws; philosophy of science; science

Charité (hospital), Berlin, 37, 39, 42, 48, 53, 56*f*, 60, 635; intern at the, 55–57

Chemical Society (London), 456, 523, 567–70, 647, 736. *See also* chemistry

chemistry: and electrochemistry, 143, 254, 553, 571–75, 602, 621–23, 691; Faraday Lecture in, 571–75; foundations of theoretical, 620–28; physical, 623; physiological, 63, 155, 212, 454, 520, 572; and thermochemistry, 143, 572, 622; thermodynamics of, 3, 254, 375, 572, 593, 620–24, 835n46. *See also* heat; physics; physiology; science

Chevreul, Michel Eugène, 278, 336, 472, 634, 750. *See also* chemistry

Chicago, 480, 578, 685, 711, 718; Helmholtz in, 709*f*; International Electrical Congress in, 703–9; International Mathematical Congress in, 725. *See also* United States

Chicago, University of, 461, 602. *See also* Chicago; universities

Christianity, 15, 20–28, 31–36, 153, 192, 388, 478; conversions to, 553; and evolution, 404, 406; myths of, 154–56. *See also* beauty; Helmholtz, Ferdinand; theology

civilization, 3, 27–30, 147, 383, 427; and culture, 32; modern, 80; poetry about, 543, 680; and science, 718; tonal systems in, 269. *See also* culture; Germany

Classen, August, 333, 335, 536. *See also* ophthalmology; vision

Clausius, Rudolph, 62, 75, 79, 149, 307, 347, 352–60, 436, 453, 499, 591; hostility of, 73, 359; memorial notice of, 610; in Paris, 580–82; thermodynamics of, 149, 152, 307, 385, 452, 515, 621. *See also* physics; thermodynamics

Clifton, Robert Bellamy, 296, 391–92, 526. *See also* science

du Bois-Reymond, Emil (*continued*)
electrophysiology, 302–3, 319, 550; as
a friend of Helmholtz, 58, 62–65, 72,
79–81, 88, 92–93, 105, 121, 138–39, 206,
223, 238, 391–92, 399–400; on Goethe,
677, 683; introductions of, 79; as a lecturer, 91, 93, 95, 461, 613; and the ophthalmoscope, 120–22; as an orator, 397,
406, 492–94, 531, 737–38, 746; in physiology, 67, 73–74, 80–81, 92–94, 119, 121,
124, 181, 207, 254, 259; work ethic of,
117–18. *See also* physics; physiology
Duhem, Pierre, 622, 630, 835n46. *See also*
chemistry; physics
Dühring, Eugen, 492, 494, 513–17, 520, 523,
529, 537, 540, 549, 554, 756; the attacks
of, 561–65, 751, 754. *See also* academic
freedom; anti-Semitism; philosophy;
physics

earth: the atmosphere of the, 152, 237, 305,
403, 449, 451–52, 621, 696, 699; dimensions of the, 135; energy and the, 297–
300, 306, 385, 403–4; the geodesy of
the, 587; the geology of the, 143, 153,
237, 244, 298–300, 451, 546; the geophysics of the, 695–96; the history of
the, 152–53, 237, 305, 402–4; magnetic
maps of the, 696; magnetism of the, 579;
meteorology and the, 451; multiplicity of
objects on the, 78; origin of life on, 404,
494, 549; sciences of the, 143, 291–92,
406, 695–96. *See also* geography; magnetism; meteorology; physics
Ebbinghaus, Hermann, 461, 602, 610, 690–
91. *See also* psychology
Eckert, Karl, 479, 481. *See also* music
Edinburgh, University of, 357, 593, 647. *See
also* universities
Edison, Thomas, 287–88, 309, 462, 550, 581,
602, 652–54, 707, 717. *See also* electricity; telegraph; telephone
education: aesthetic, 27–30, 272, 545; classical, 392; and Darwin's theory of evolution, 522, 525; engineering; 434; and
the German *Bildungsbürgertum*, 8, 25,
277; mathematics in, 659; medical, 42,
50, 55, 91; music, 277, 286; in Prussia,
26, 55, 74, 79, 81, 417, 608, 656–59; science in, 155, 278, 457–58, 493, 507, 525,
656, 659; and social improvement, 245,

514, 666; as a social science, 507; unity of
all sciences in, 526; of women, 480, 507,
513–15. *See also Bildung*; culture; science;
universities
Ehrenfest, Paul, 626. *See also* physics
Einstein, Albert, 4–5, 67, 244, 374, 379, 440,
460, 608, 683, 702, 749, 751–58, 763;
and electrodynamics, 693; general theory
of relativity of, 374, 753; philosophical
orientation of, 534. *See also* physics
electricity, 115, 384, 389, 576–84, 607, 688–
90, 698, 705–8; and acoustics, 562; and
action-at-a-distance, 504; animal, 74,
123–24, 133, 250; atomic nature of, 572–
74; and chemistry, 572–73; discoveries
in, 479; forces of, 151; and heat, 625; and
industry, 8, 616–17, 640, 690, 705–8,
713; laws of, 187, 378, 384, 574; and light,
160, 310, 378, 462, 606, 653; and magnetism, 359, 377–78, 458, 607, 620; mathematical theory of, 458; the measurement
of, 685; and nerve activity, 254; processes
of, 70, 300, 389, 574; and resistance, 175,
295–96, 683–85; standards of, 576–84,
587, 595, 617, 628, 683–84, 707; studies
in, 344, 378, 444, 533; and the telegraph,
121, 136, 250, 287, 569, 641; theory of,
572–74, 604–12; waves of, 653, 721, 724.
See also electrochemistry; electrodynamics; electromagnetic theory; magnetism;
physics; science
electrochemistry: and the work of Helmholtz
in chemical thermodynamics, 254, 621–
23; and the Faraday Lecture, 571–75. *See
also* chemistry; electricity; electrodynamics; science; thermodynamics
electrodynamics: of dielectrics, 604, 620;
Einstein's work in, 693; and electromagnetism, 620; Helmholtz's critique
of Weber's, 549; Helmholtz's critique
of Zöllner's, 817n36; Helmholtz's work
in, 347, 567–68, 604, 620; history of
nineteenth-century, 724; and hydrodynamics, 366, 372, 374–79; Maxwell's
work in, 701–2; Müller's work in, 440;
and optics, 441–49; principle of least
action in, 701, 725, 728; theory of, 389,
449, 458, 498, 572, 604, 620, 691, 701;
Weber's work in, 101, 175, 379, 495–96,
515. *See also* electricity; electrochemistry;
electromagnetic theory; physics; science

electromagnetic theory: of Helmholtz, 377, 443, 449, 462–63, 573, 624, 687, 701–2; of Hertz, 609, 653, 694; of Maxwell, 377–79, 463, 573–74, 592, 609, 624; of Neumann, 89; scientific importance of, 582; of Weber, 101. *See also* electricity; electrodynamics; magnetism; physics; science

Elektrotechnischer Verein, 578. *See also* electricity

Eliot, George, 7, 284, 305, 309, 376. *See also* literature

Ellis, Brian, 630. *See also* mathematics; philosophy of science

energy: of animals, 152, 302; biochemical, 153; chemical, 621; concept of, 378, 385, 425; conservation of, 141–51, 244, 291–304, 310, 338–39, 347, 358–60, 374, 378–79, 384–86, 401, 443, 495, 505, 514, 549, 561–64, 572–73, 624, 628, 677, 691, 749, 785n35; free, 621–22, 624–25, 650; frictional, 376; kinetic, 70, 297, 625; nuclear, 761; potential, 70, 297, 625; solar, 154, 237, 297, 300, 302, 310, 385, 403–4; of the universe, 237, 297, 385; of waves, 697, 700; of wind, 697, 700. *See also* conservation of force; heat; physics; thermodynamics

Engels, Friedrich, 7, 515–16, 760–61. *See also* philosophy; politics

engineering, 61, 68, 736; British, 141–42, 200, 237, 434; chemical, 586; French, 200; fundamental questions of, 238; mechanical, 385, 414. *See also* physics

Enlightenment: German, 226, 424; patriotic and economic societies of the, 87; popularization of science and the, 308; salon of the, 473, 477

entropy, 358; defined, 621; theory of, 359, 621

epistemology: of Bacon, 194; criticisms of Goethe by Helmholtz in, 129, 682; and evolutionary theory, 386; and Kant, 158, 532–33, 826n49; of Lenin, 751; in the life of Helmholtz, 2–3, 21, 23, 683; of Mach, 274; and Marxism, 761; and mathematics, 368–69, 534–37; and the microscope, 445; of Nietzsche, 310, 336; of numbers and measurement, 624, 628–30, 701; and philosophy of science, 124–25, 194–95, 335, 509, 516, 519, 532, 537–48, 660, 669, 681, 749–53; physiological, 681,

711; and the popularization of science, 597; of Schopenhauer, 194; sign theory of, 198; and the theory of perception, 19, 127, 160, 338, 345, 386, 495, 537–48, 751. *See also* philosophy; philosophy of science; science

Euler, Leonhard, 202, 270, 280, 627. *See also* mathematics

evolution, theory of, 522, 529; and natural selection, 339, 386, 405; and the origin of species, 505, 677. *See also* Darwin, Charles

eye movements: and coordination, 122; the horopter and, 315–21, 326, 332; lecture of Helmholtz on, 300; the retina and, 316; Zöllner on, 494–95. *See also* accommodation, theory of; horopter; ophthalmology; physiological optics

Faraday, Michael, 68, 135, 143, 235, 293–312, 377–78, 425, 435, 456, 489, 549–50, 567–75, 609, 620, 642, 707; and the Royal Institution, 234–35, 296; understanding of matter of, 680. *See also* electricity; electromagnetic theory; magnetism; science

Fechner, Gustav Theodor, 124, 228, 275, 470, 498, 534, 550–51, 629; theory of afterimages of, 228. *See also* philosophy; physics; psychophysics; science

Fichte, Immanuel Herrmann, 13–16, 19, 47, 194, 220, 531, 766. *See also* philosophy

Fichte, Johann Gottlieb, 13, 15, 20, 194, 531, 539; theory of scientific knowledge of, 16. *See also* philosophy

Fick, Adolf, 181, 317–18, 333. *See also* medicine; physiology

Fischer, Emil, 610, 652, 675–76, 748. *See also* chemistry

Fischer, Ernst Gottfried, 34. *See also* physics

Foerster, Wilhelm, 382, 552, 555, 616, 640, 642, 671, 687. *See also* astronomy

Fortschritte der Physik, 62, 72, 89, 132–33, 186, 441, 514, 606, 778n30. *See also* Berlin Physical Society; physics

France: the Franco-Prussian War (1870–71) and, 324, 394–98; declining importance in science in mid-century, 131, 577; education in, 658; and the Electrical Congresses, 576–84, 707; Helmholtz's reputation in, 73, 277–81, 338; mathematics

holtz of, 44; theory of colors of, 159, 193, 636, 677, 680–81; unifying principle in nature and, 67. *See also Bildung*; literature; *Naturphilosophie*; science

Goethe Society (Goethe-Gesellschaft), 676, 681. *See also* Goethe, Johann Wolfgang von

Gossler, Gustav von, 487–88, 611, 648, 657, 686. *See also* science

Graefe, Albrecht von, 119–20, 122–23, 132, 255, 257, 316, 318, 337, 343, 422, 635, 665. *See also* ophthalmology

Graham, Thomas, 236, 298. *See also* chemistry

Grassmann, Hermann Günther, 127, 163–64, 756. *See also* mathematics; physics

Grévy, Jules, 584. *See also* politics

Grimm, Jakob, 489. *See also* literature

Hall, Alexander Wilford, 466. *See also* philosophy; philosophy of science

Halske, Johann Georg, 63, 65, 81, 92. *See also* industry; instruments, scientific; Siemens & Halske

Hamilton, William Rowan, 147, 625–26. *See also* mathematics; physics

Handbook of Physiological Optics (*Handbuch der physiologischen Optik*), 180–82, 186–87, 194, 218–29, 262, 266, 279, 284, 300, 308, 315–45, 360, 362, 375, 379, 495, 500, 502, 690, 701, 704, 748; analysis of perception in the, 326–335, 343, 541; criticism of Kant's theory of space in the, 532; and ophthalmology, 634, 636; painting and the, 472–73. *See also* ophthalmology; optics; perception; physiological optics

Hann, Julius, 306, 452, 699. *See also* meteorology; physics

Hansemann, David, 739. *See also* politics

Hanslick, Eduard, 266, 273, 482. *See also* music

Hartmann, Eduard von, 531, 533. *See also* philosophy

heat: animal, 59, 70, 74, 101, 152, 238, 302, 386, 572; electrochemical, 572, 621–23; formation in muscles of, 59, 71, 73–74, 302; mechanical equivalent of, 237, 388, 434, 562; mechanical theory of, 279, 307, 385, 388, 452, 515; physical nature of, 59; and work, 74, 302. *See also* chemistry;

conservation of force; energy; thermochemistry; thermodynamics

Hegel, Georg Wilhelm Friedrich, 15, 32–33, 67, 110; hostility of Helmholtz towards, 158, 195, 216, 246, 334, 519, 531–32; idealism of, 194; nature philosophy of, 71, 128, 194; opposition to, 154–55, 669; philosophy of identity of, 245; and science, 128, 158–59, 194–95. *See also* idealism; philosophy

Heidelberg, 104–5, 114, 183, 210f; congress of ophthalmology in, 123; the rise of, 209–14. *See also* Germany

Heidelberg, University of, 118–19, 210–14, 252f, 738; anatomy institute in the, 188; chemical institute in the, 188; Helmholtz's departure from the, 351–57, 381–408; Helmholtz's move to the, 188–92, 202–7; Helmholtz's research in the, 218–19, 227–29, 305–24, 326–45, 357–79; Helmholtz's teaching in the, 214–17, 255–60; ophthalmology in the, 315–17; physiological institute in the, 5, 214–17, 252–54; reform of the, 243–52

Heidenhain, Rudolf, 207, 254, 389, 553. *See also* physiology

Heintz, Wilhelm, 61, 75, 100, 103, 223. *See also* chemistry

Helmholtz, Anna (née von Mohl, second wife), 30, 225–42, 357, 646, 651. *See also* Helmholtz, Hermann von; women

Helmholtz, Caroline Auguste (mother), 16

Helmholtz, Ellen (daughter), 241, 303, 312, 395, 414, 558, 582, 593, 595–96, 638–39, 730

Helmholtz, Ferdinand (father), 13–29, 154, 192–95, 220–21, 531; on aesthetic education, 30, 32, 545; as pedagogue and philosopher, 26–30, 118

Helmholtz, Friedrich Julius ("Fritz") (son), 312, 395, 413–14, 558, 560, 631, 703

Helmholtz, Hermann Ludwig Ferdinand (from 1883, von): appointment to Königsberg of, 80–82; attitude towards women of, 103, 108, 132, 164, 240, 293, 348, 414, 507, 514, 692; and the *Bildungsbürgertum*, 8, 20, 25, 41, 64, 277; the brain of, 570, 731, 739; and Britain, 131–32, 141, 146, 291–313; and British science, 131–44, 156–75, 220, 281–86, 309, 387, 392, 494, 502–5, 567–75, 579,

literature (*continued*)
and science, 121–22, 124, 130, 309, 450,
505–8, 514, 659; and science for Helmholtz, 2–3; study by Helmholtz of, 26,
29, 32. *See also* art; culture; philosophy;
poetry; science
Lodge, Oliver, 550, 606. *See also* physics
Lorentz, Hendrik Antoon, 444, 743. *See also*
physics
Lotze, Rudolf Hermann, 102, 195, 273, 320,
330, 372, 542. *See also* philosophy
Low, Seth, 715–18. *See also* Columbia
University
Ludwig, Carl, 63, 73, 80–81, 90–95, 120,
123, 127, 181, 258–59, 498, 637; and
the arts, 273; on Helmholtz, 91, 95, 105,
118, 180, 223, 273, 309; materialism of,
156–57, 166, 168; and politics, 75, 81, 90,
157, 168; in Zurich, 106–8, 166. *See also*
medicine; physiology

Mach, Ernst, 268, 274, 276, 547, 630. *See also*
acoustics; philosophy of science; physics
Mackenzie, Morell, 646. *See also* medical
doctors
magnetism, 70, 607; of the earth, 579; electricity and, 359, 377–78, 458, 607, 620,
693. *See also* earth; electricity; electromagnetic theory; science
Magnus, Heinrich Gustav, 15, 60–61, 70, 131,
223, 234, 239, 390. *See also* chemistry;
conservation of force, essay on the;
physics
Magnus-Levy, Albert, 613. *See also* medicine
Mahler, Gustav, 750. *See also* music
Marey, Etienne-Jules, 337, 584. *See also* photography; science
Marx, Karl, 7, 42, 155, 163, 515, 760. *See also*
philosophy
mathematics, 26, 32, 34, 36, 248, 362, 365;
and acoustics, 187–88, 197–98, 274,
282; in Berlin, 176; and electrodynamics,
379; and Helmholtz, 372–75; and music,
197; and physics, 62, 73, 94, 183, 202,
349, 377; and physiology, 277, 321. *See
also* arithmetic; electromagnetic theory;
geometry; optics; physics
Maupertuis, Pierre-Louis Moreau de, 626–27.
See also Academy of Sciences, Prussian;
mathematics; physics

Maxwell, James Clerk, 120, 235, 286, 305,
359, 425, 463, 607, 609, 625, 653; electrodynamics of, 202, 359, 377–79, 442–
44, 573, 610, 701–2; electromagnetic
theory of, 377–79, 442–44, 463, 573–74,
582, 592, 607–9, 624, 653; equations of,
606, 701, 722; on Helmholtz, 501, 548;
and Helmholtz's acoustics, 285, 377;
and Helmholtz's law of conservation of
force, 141; and Helmholtz's physiological optics, 338; and Helmholtz's theory
of color, 163–64, 228, 301, 377. *See also*
electricity; electrodynamics; mathematics; Maxwellians; physics
Maxwellians, 378, 609–10. *See also* Maxwell,
James Clerk; physics
Mayer, Julius Robert, 68, 237, 301, 310,
358–60, 381, 385, 388–89, 444; and
the attacks of Zöllner and Dühring,
506–20, 561–65, 756. *See also* conservation of force; medicine; physics; thermodynamics
medical doctors: and acoustics, 262; and
anatomy, 42–46, 49–51, 60; and Helmholtz, 39–82, 181, 316, 395, 749, 762;
and materialism, 155; and military
medical duty, 395; and the Naturforscherversammlung, 156; studies in Germany for, 258. *See also* anatomy; medicine; physiology
medicine: and acoustics, 268; clinical, 258;
and evolutionary theory, 522; history of,
459; and industry, 493; the inventions of
Helmholtz in, 90–116, 120, 122, 170; and
pathology, 178; and physics, 392, 637,
759; and physiology, 188, 258, 387; and
science, 120, 122, 212–13, 258–59, 279,
349, 458, 655; thought in, 516–21. *See
also* acoustics; anatomy; medical doctors;
ophthalmology; ophthalmoscope; physiology; science
Mendelssohn-Bartholdy, Felix, 489, 576. *See
also* music
Menzel, Adolph von, 18, 477*f*, 489, 667. *See
also* art; painting
metaphysics: and epistemology, 158, 509,
533; in Germany, 304, 424, 520; Helmholtz's rejection of, 502–5, 533, 545, 753;
hypothetical entities in, 701; materialist,
520, 529; and science, 502–5, 521, 573.